5th International Scientific Conference on Integration, Partnership and Innovation in Construction Science and Education 2016

MATEC Web of Conferences Volume 86 (2016)

Moscow, Russia
16 - 17 October 2016

Part 1 of 2

Editor:

V. Andreev

ISBN: 978-1-5108-3362-3

Printed from e-media with permission by:

Curran Associates, Inc.
57 Morehouse Lane
Red Hook, NY 12571

Some format issues inherent in the e-media version may also appear in this print version.

Printed by Curran Associates, Inc. (2017)

For additional information, please contact EDP Sciences – Web of Conferences at the address below.

EDP Sciences – Web of Conferences
17, Avenue du Hoggar
Parc d'Activité de Courtabœuf
BP 112
F-91944 Les Ulis Cedex A
France

Phone: +33 (0) 1 69 18 75 75
Fax: +33 (0) 1 69 28 84 91

contact-edps@webofconferences.org

Additional copies of this publication are available from:

Curran Associates, Inc.
57 Morehouse Lane
Red Hook, NY 12571 USA
Phone: 845-758-0400
Fax: 845-758-2633
Email: curran@proceedings.com
Web: www.proceedings.com

TABLE OF CONTENTS

PART 1

PART 2

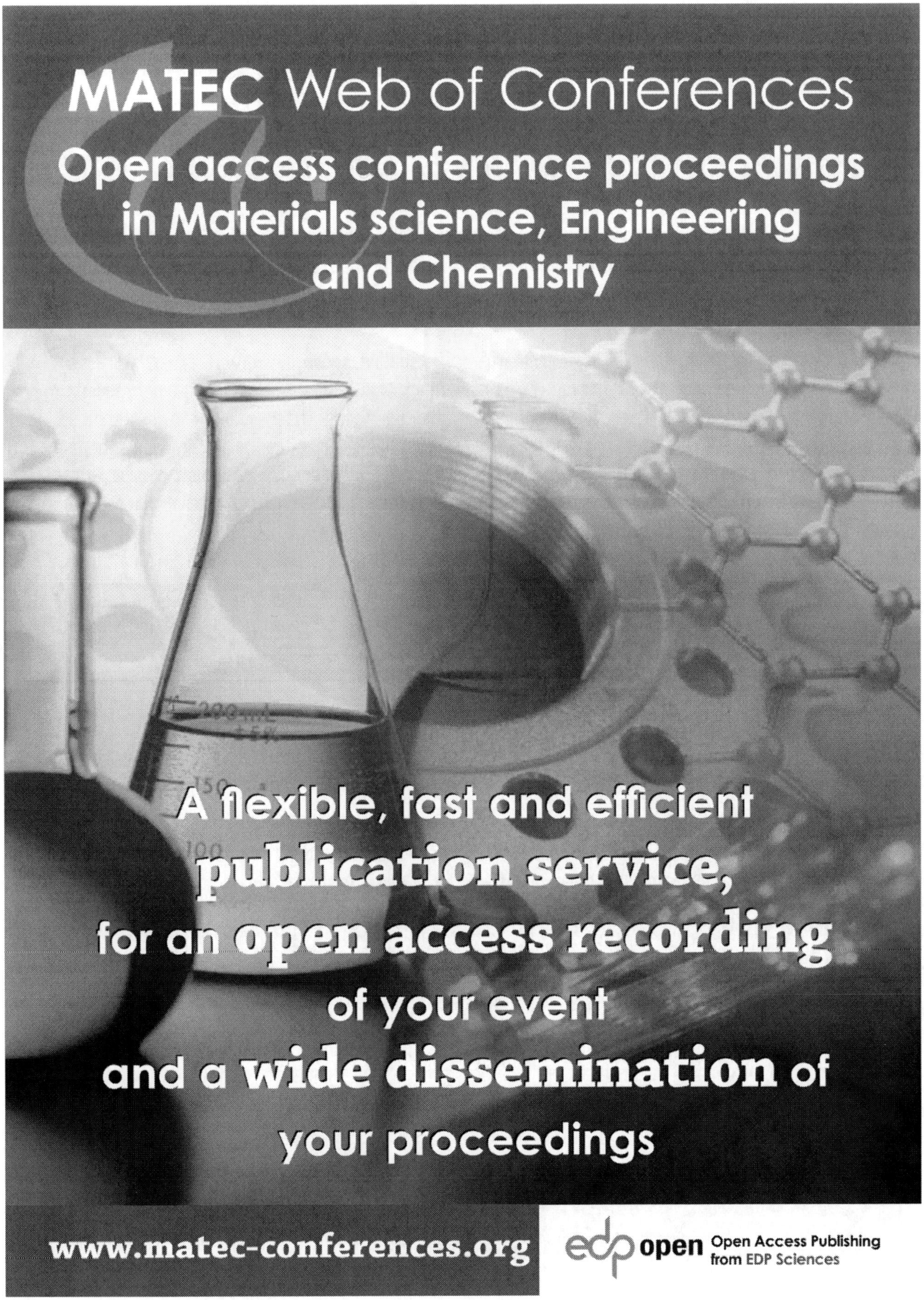
MATEC Web of Conferences
Open access conference proceedings
in Materials science, Engineering
and Chemistry
A flexible, fast and efficient
publication service,
for an open access recording
of your event
and a wide dissemination of
your proceedings
www.matec-conferences.org
edp open Open Access Publishing
from EDP Sciences

edp sciences

MATEC Web of Conferences

Statement of Peer review

In submitting conference proceedings to MATEC Web of Conferences, I certify to the Publisher that I adhere to the **Policy on Publishing Integrity** of the journal in order to safeguard good scientific practice in publishing.

1. All articles have been subjected to peer review administered by the proceedings editors.
2. Reviews have been conducted by expert referees, who have been requested to provide unbiased and constructive comments aimed, whenever possible, at improving the work.
3. Proceedings editors have taken all reasonable steps to ensure the quality of the materials they publish and their decision to accept or reject a paper for publication has been based only on the merits of the work and the relevance to the journal.

Title, date and place of the conference

5th International Scientific Conference
"Integration, Partnership and Innovation in Construction Science and Education"
Short name : IPICSE-2016
16 to 17 of October 2016
Moscow State University of Civil Engineering, 29337, 26, Yaroslavskoye Shosse, Moscow, Russia

Proceedings editor(s):

DSc, Prof. Vladimir Andreev

Date and editor's signature

31.10.2016 Vladimir Andreev

Digitally signed by Vladimir Andreev
DN: cn=Vladimir Andreev, o=Moscow State University of Civil Engineering, ou, email=prof.vladimir_andreev@mail.ru, c=RU
Date: 2016.11.01 00:27:49 +01'00'

V International Conference

"Integration, Partnership and Innovation in Construction Science and Education" (IPICSE–2016)

Preface

The International conference "Integration, Partnership and Innovations in Construction Sciences and Education" (IPICSE) is held for the 5th time by now. It is one of the biggest industry-specific science events in Russian Federation, which annually attracts over 400 specialists participating as speakers or listeners.

Moscow State University of Civil Engineering, the leading university of Russia in the civil engineering field, which celebrates its' 95th anniversary this year, was chosen as a venue for IPICSE-2016. Representatives of several distinguished civil engineering universities of Poland, Italy, Slovakia and Russia assisted in the organization of this event. Among the universities are Warsaw University of Technology, Moscow State University of Civil Engineering, Wrocław University of Technology, University of Zilina, University of Padova, Samara State University of Architecture and Civil Engineering.

Within IPICSE-2016 the Scientific Seminar on modern construction technologies and materials will take place for the first time, in the course of which professional academic community representatives from over than ten countries will discuss significant problems of such disciplines as

- Structural mechanics
- Building structures
- Geodesy and Geotechnics
- Building materials and Technologies,
- Organization and Management in construction
- Recycle of inorganic waste in construction

We would like to express our gratitude to Editorial Board of MATEC Web of Conferences and personally to Isaline Augusto for his generous efforts.

Wish warm companionship, valuable discussions and success in scientific activities to all participants of IPICSE-2016!

Dr. Aleksey Adamtsevich
Coordinator IPICSE-2016

Geometrical parameters influence on behavior of the sandwich plates with corrugated core

Jelena M. Djoković[1], *Ružica R.* Nikolić[2,3,*], *Jan* Bujnak[4], *Katarina* Živković[2]

[1]Technical Faculty in Bor, University of Belgrade, V. Jugoslavije 12, 19210 Bor, Serbia
[2]Faculty of Engineering, University of Kragujevac, Sestre Janjić 6, 34000 Kragujevac, Serbia
[3]Research Center, University of Žilina, Univerzitna 8215/1, 010 26 Žilina, Slovakia
[4]Faculty of Civil Engineering, University of Žilina, Univerzitna 8215/1, 010 26 Žilina, Slovakia

Abstract. The influence of geometric parameters on behavior and stiffness of sandwich plates with corrugated core is considered in this paper. The following parameters were analyzed: ratio of core sheet and facing sheet thickness and the ratio of the core's pitch to the core's depth, as well as the corrugation angle. It is shown that changes of these parameters can contribute to increase or decrease of the corrugated sandwich plate stiffness.

1 Introduction

The sandwich plates are the three-layer structures, which consist of the thick core and the two flat plates – faces. In general, case, the faces are made of the highly resistant materials, while the core is made of materials of the lower resistance and density. Due to high resistance of the faces and small mass of the core, one of the most important characteristics of these structures is a good ratio of the structural resistance and mass.

Advantages of the sandwich plates are the high stiffness and resistance, with respect to their mass, good heat and acoustic insulating properties, good aerodynamics, energy absorption for different types of cores, increase of the inner space and simple mounting. Due to those properties, the sandwich structures have wide application in construction industry, aircraft industry, ship building, in the field of the acoustic and heat insulation, the fire protection, etc.

The sandwich plates' core keeps the two separated faces together and stabilizes them by resisting vertical deformations. It enables the whole structure to behave like a thick plate, when the shear resistance is considered. The core also carries the part of the load in the case when the structure is subjected to bending loading. There are several types of core – the foam or the solid core, the honeycomb, prismatic, truss and corrugated core, Vinson [1].

The light and soft materials are usually used for manufacturing the core. For sandwich plates with the core made of foam or in the honeycomb form, all the planar and flexural loads are carried by the faces, while for the sandwich plates with the prismatic and corrugated core, the core also carries the part of the load.

[*]Corresponding author: ruzicarnikolic@yahoo.com

Large number of researchers was dealing with analysis of the sandwich structures, subjected to various loadings: Allen [2]; Evans et al. [3]; Brittain et al. [4]; Wicks and Hutchinson [5-6]; Valdevit et al. [7]; Tumino et al. [8] and Cheon and Kim [9]. The most used shape of the core is the honeycomb one, since it possesses the highest stiffness and the shear strength with respect to its mass; however, it requires special attention during the manufacturing. The most studied type of the sandwich plate's failure is wrinkling, which comes to the core's resistance to deformation perpendicular to faces. It is especially prominent with plates whose cores are made of expanded materials. The cores that are made of the highly resistant materials, on the other hand, possess very good resistance to wrinkling, where especially good are the corrugated cores. They are sufficiently stiff to contribute to increase of the flexural stiffness of the whole structure. Unlike the honeycomb cores, the corrugated cores sustain well not only the vertical shear, but the bending and torsion, as well.

Influence of geometrical parameters on behavior of the sandwich plate with the corrugated core, shown in Fig. 1, is analyzed in this paper. The plate's behavior depends on its stiffness. Some geometrical parameters, like ratio of core sheet and facing sheet thickness (t_c/t_f), ratio of the core's pitch and the depth (p/h_c), corrugation angle α, can contribute to increase or decrease of the stiffness (and thus the resistance) of the plate with the corrugated core.

2 Corrugated plate geometrical characteristics and stiffness

The sandwich plate with the corrugated core, Fig. 1, consists of two flat plates – faces of thickness t_f and the corrugated core sheet thickness t_c. The dimensions of the plate are as shown in Fig. 1: α – the corrugation angle, h_c – the core's depth measured from the center line at crest to center line at trough, h – distance between the middle planes of the face sheets, $2p$ – the face sheets and f – the length of corrugation flat segment. Subscripts c and f refer to core's and the faces' properties, respectively.

In analysis of the sandwich plates with the corrugated core, for the case of flexural loading, the following assumptions were made: the analyzed corrugated core is symmetric, the faces are made of the same material and with the same thickness, the core material is isotropic-elastic, the sandwich plate's deformations are small, the plate's thickness is significantly smaller than the core's depth, the core influences the flexural stiffness in the x-direction but not in the y-direction, the plate is infinitely long in the z-direction, the core can sustain the lateral shear stress and it also affects the flexural and axial stiffnesses, while the lateral shortening of the core is not taken into account.

The flexural stiffnesses in the x and y directions are determined as, Libove and Hubka [10]:

$$B_x = 2E_f I_f + E_c I_c, \quad B_y = \frac{E_f I_f}{1 - \nu_f^2 \left(\frac{E_f I_f}{B_x} \right)}, \tag{1}$$

respectively, where E_c and ν_c are the Young modulus of elasticity and the Poisson's ratio of the core's material, respectively, while E_f and ν_f are the same characteristics for the faces' material.

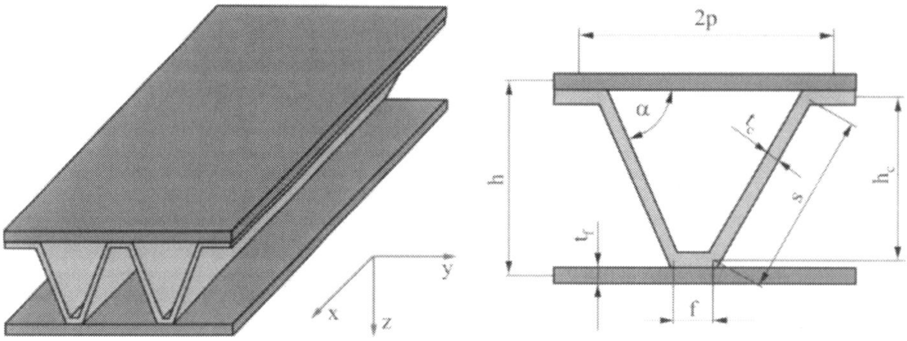

Fig. 1. The geometry of the sandwich plate with the corrugated core.

The cross-section's moment of inertia of the core, parallel to the *yz* plane, per unit width, taken for the symmetry axis of the cross section, I_c, and moment of inertia of the face, considered as the membrane, with respect to the middle plane of the sandwich plate, I_f, respectively are given by:

$$I_f = 2p \cdot t_f \cdot \left(\frac{h_c}{2}\right)^2, \quad I_c = \frac{f \cdot t_c \cdot h_c^2}{2} + \frac{2 \cdot t_c \cdot h_c^3}{12 \cdot \sin\alpha}. \tag{2}$$

The torsional stiffness is determined as:

$$B_{xy} = 2G_f I_f, \tag{3}$$

where G_f is the shear modulus of the face.

3 Results and discussion

In Figs. 2, 3 and 4 are presented variations of the sandwich plate's stiffnesses B_x, B_y and B_{xy} in terms of the (t_c/t_f) ratio, the core angle α and the (p/h_c) ratio, obtained according to equations (1) to (3), by the programming package *Mathematica*®. The case considered is that of the rectangular sandwich plate, loaded by the distributed load q = 10 kN/m^2. The plate is simply supported along the edges, which means that the boundary conditions are: for x = *const.* the deflection is w = 0, the bending moment is M_x = 0, as well as the transversal force Q_y = 0. The face and the core materials are identical with the elastic characteristics: E = 208 *GPa* and ν = 0.3.

In Fig. 2 is shown influence of variation of the of core sheet and facing sheet thickness ratio on flexural stiffnesses B_x and B_y, as well as on the torsional stiffness B_{xy}, for the case of the corrugation angle α = 60°, the core's pitch to depth ratio (p/h_c) =1 and for the core's depth to core thickness ratio (h_c/t_c) =10.

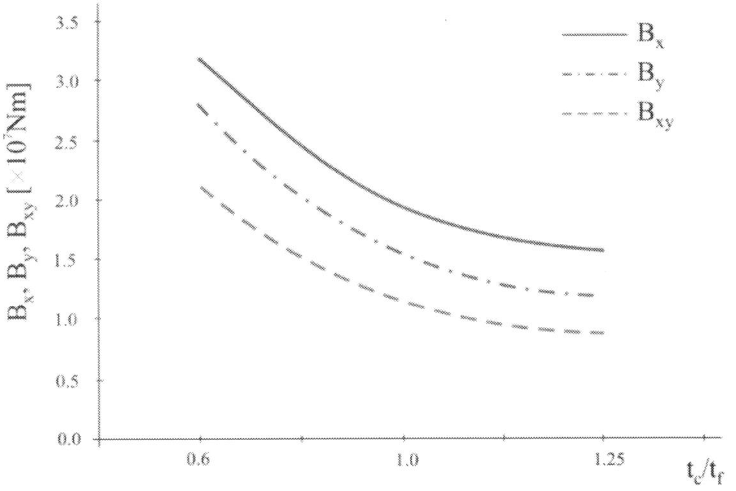

Fig. 2. Variation of the flexural and torsional stiffnesses of the sandwich plate in terms of the (t_c/t_f) ratio.

From Fig. 2 one can see that all the three stiffnesses are decreasing with the face thickness to the core depth ratio increase, for the same values of the corrugation angle α and the ratios (p/h_c) and (h_c/t_c). As the (t_c/t_f) ratio increases, for the certain value of the core's sheet thickness t_c, the facing sheet thickness t_f decreases, what causes the lower stiffness of the corrugated sandwich plate, since the faces carry larger portion of the bending loading. If the tendency of the stiffnesses decrease is compared to tendency of the (t_c/t_f) ratio increase, one can notice that for the (t_c/t_f) increase from 0.6 to 1.0, the flexural stiffnesses B_x and B_y decrease for about 45 %, while the torsional stiffness B_{xy} decrease for about 40 %. The same trend is continued for the (t_c/t_f) ratio increase from 1.0 to 1.25, where B_x and B_y decrease for about 22 %, while the torsional stiffness B_{xy} decrease for about 17 %. This leads to conclusion that the (t_c/t_f) ratio has stronger influence on the flexural stiffnesses than on the torsional one.

In Fig. 3 is shown the influence of the corrugation angle α variation on the flexural and the torsional stiffnesses, for the following geometrical parameters of the sandwich plate: the core sheet and facing sheet thickness ratio (t_c/t_f) =1, the core pitch to depth's ratio (p/h_c) =1 and the core's depth to core thickness ratio (h_c/t_c) =10.

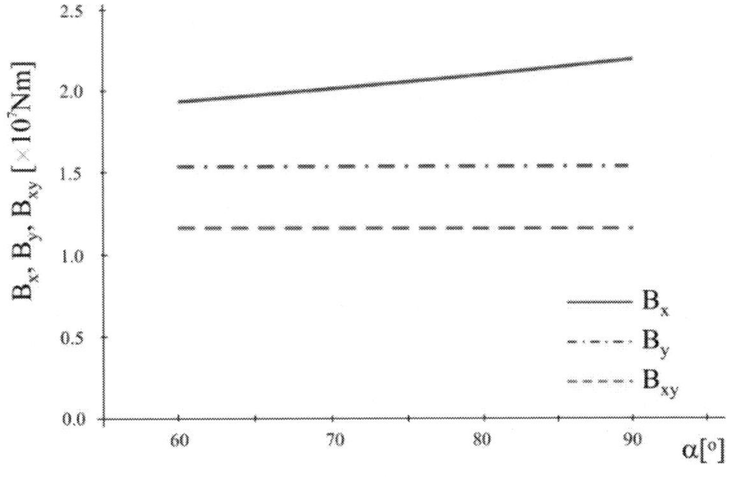

Fig. 3. Variation of the sandwich plate stiffnesses in terms of thecorrugation angle α.

From Fig. 3 could be noticed that flexural stiffness B_x increases with increase of the corrugation angle α, while the stiffnesses B_y and B_{xy} remain constant, for the same values of ratios: (t_c/t_f), (p/h_c) and (h_c/t_c).

In Fig. 4 is presented the influence of the core's pitch to depth ratio (p/h_c) on flexural stiffnesses B_x and B_y and on the torsional stiffness B_{xy}, for the following values of the geometrical parameters of the sandwich plate: $\alpha = 60°$, $(t_c/t_f) = 1$ and $(h_c/t_c) = 10$.

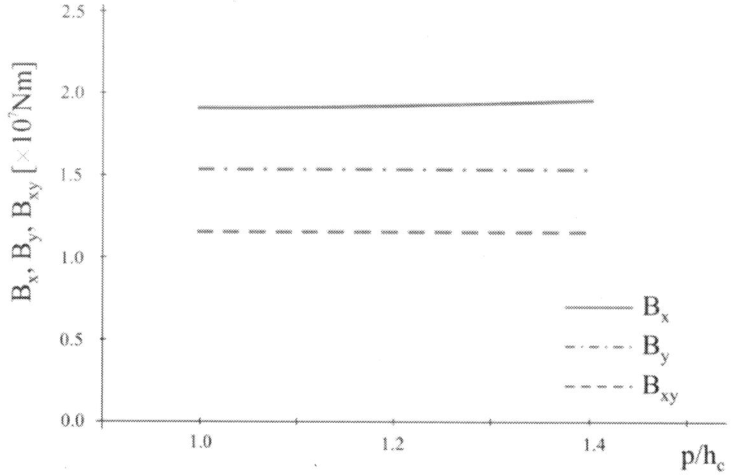

Fig. 4. Variation of the sandwich plate stiffnesses in terms of the (p/h_c) ratio.

From Fig. 4 one can notice that the flexural stiffness B_x slightly increases with increase of the core's pitch to depth ratio, while the stiffnesses B_y and B_{xy} remain constant, for the same values of the corrugation angle α and ratios (t_c/t_f), (p/h_c) and (h_c/t_c).

4 Conclusion

The influence of the geometrical parameters of a sandwich plate with the corrugated core on its behavior was considered in this paper. The following parameters' influences were analyzed: ratio of the core sheet thickness to the sandwich plate face's thickness, the corrugation angle and the ratio of the core's pitch to depth.

Based on results obtained from that analysis, one can notice that all the stiffnesses of the sandwich plate (two flexural and a torsional) are decreasing with increase of the core sheet thickness to the face's thickness ratio, though the influence is more prominent for the flexural stiffnesses than for the torsional one. As that ratio increases, for the constant value of the core sheet thickness, the face's thickness decreases, what causes the decrease of the sandwich plate's flexural stiffness, since for the corrugated plates the faces are carrying the larger portion of the bending loading.

For the case of the corrugation angle increase and increase of the core's pitch to depth ratio, the main flexural stiffness increases slightly, while the one in the perpendicular direction and the torsional stiffness remain constant.

Based on analysis that was conducted within this research, certain recommendation can be given for the selection of the geometrical parameters of the sandwich plates with the corrugated core. It is shown that the lower values of the investigated parameters make the sandwich plate more resistant. It would be the best to choose the plate's parameters in such a way that face and core sheet thicknesses would be the same, $(t_f = t_c)$, that the corrugation angle value would be about 60° and that the value of the core's pitch to depth ratio (p/h_c) would be between 1 and 1.2.

This research was partially supported by the Ministry of Education, Science and Technological Development of Republic of Serbia through Grants ON174001, ON174004 and TR32036 and by European regional development fund and Slovak state budget by the project "Research Center of the University of Žilina" - ITMS 26220220183. The authors are very grateful for this funding.

References

1. J.R. Vinson, Appl. Mech. Rev. l. **54**, 201–214 (2001)
2. H.G. Allen, *Analysis and Design of Structural Sandwich Panels* (Pergamon Press, Oxford, 1969).
3. A.G. Evans, J.W. Hutchinson, M.F. Ashby, Prog. Mater. Sci **43(3)**, 171–221 (1998)
4. S.T. Brittain, Y. Sugimura, J.A. Schueller, A.G. Evans, G.M. Whitesides, J. Microelectromech. Syst. **10**, 113–120 (2001)
5. N. Wicks and J.W. Hutchinson, Int. J. Sol. Struct. **38**, 5165–5183 (2001)
6. N. Wicks and J.W. Hutchinson, Mech. Mater. **36**, 739–751 (2004)
7. L. Valdevit, J.W. Hutchinson, A.G. Evans, Int. J. Sol. Struct. **41**, 5105–5124 (2004)
8. D. Tumino, T. Ingrassia, V. Nigrelli, G. Pitarresi, V. Urso-Miano, Frat. Integr. Strut. **30**, 317-326 (2014)
9. Y.-J.Cheon, H.-G. Kim, J. Mech. Sci. Techn. **29(3)**, 1217-1223 (2015)
10. C. Libove, R.E. Hubka, *Elastic constants for corrugated core sandwich plates, Techn. Note 2289* (National Advisory Committee for Aeronautics, Washington, 1951)

Behavior of the sandwich beam subjected to bending as a function of the core density

Ružica R. Nikolić[1,2,*], *Jelena M.* Djoković[3], *Jan* Bujnak[4], *Katarina* Živković[1]

[1]Faculty of Engineering, University of Kragujevac, Sestre Janjić 6, 34000 Kragujevac, Serbia
[2]Research Center, University of Žilina, Univerzitna 8215/1, 010 26 Žilina, Slovakia
[3]Technical Faculty in Bor, University of Belgrade, V. Jugoslavije 12, 19210 Bor, Serbia
[4]Faculty of Civil Engineering, University of Žilina, Univerzitna 8215/1, 010 26 Žilina, Slovakia

Abstract. The sandwich beam can be considered as the multi-layered structure with a symmetrical cross-section. In this paper is assumed that the structure is created by periodical repetition of a unit cell. The influence of its size on the beam's static behavior in bending was analyzed. The variation of the unit cells number affects the size of the cell, so the static analysis was performed – the flexural stiffness and the beam's deflection were determined as functions of the unit cells number. The two configurations of the sandwich beams were considered: the beam with the constant cross-section along its length and the beam with the periodically variable cross-section. The graphs of the beam's flexural stiffness and deflection variations in terms of the unit cells number were obtained. It was concluded that after a certain number of the cells, the core's density does not further influence the behavior of the sandwich beam, under the given loading conditions. The conclusion from comparison of the two configurations is that the sandwich beam with the variable cross-section behaves somewhat better than the beam with the constant cross-section. The FEM analysis has verified all the conclusions from the analytical solution about the sandwich beams behavior when subjected to bending.

1 Introduction

Sandwich structures are the laminar structures that usually consist of the two thin, rigid plates – faces, which are separated by the softer layer – the core. The core material can be homogeneous isotropic foam, which contributes to the exceptionally small weight of the sandwich structure. Sometimes the core material is reinforced by the lateral fibers or the core consists of the thin-walled layers in the form of honeycomb or a truss.

Lightweight sandwich structures were studied by numerous investigators: Allen (1969) has set basics of the sandwich structures' design and analysis. Evans et al. (1998) have performed a survey of research on mechanical and thermal characteristics of the metal cellular systems. Brittain et al. (2001) have analyzed mechanical properties of the micro

*Corresponding author: ruzicarnikolic@yahoo.com

plates with the truss core and proposed a new method for their manufacturing. Wicks and Hutchinson (2001) concluded that characteristics of the optimized sandwich plates with the truss core are superior to characteristics of plates with the honeycomb core or the stringer-stiffened plates. Wicks and Hutchinson (2004) have further extended their analysis to sandwich plates with truss core subjected to fracture straining and combination of the bending moment and transversal force. Valdevit et al. (2004) were conducting the structural optimization of the sandwich plates with the truss core.

Sandwich structures have a wide application in different branches of industry, like the optical, biomechanical, microelectronics and highly resistant components, due to their high stiffness, superior strength, heat conductivity and small mass. To analyze the functionality and advantages of the sandwich structures, one should first comprehend the geometrical and mechanical factors, which influence the behavior of these structures.

The sandwich beam can be considered as the multi-layered structure with symmetrical cross-section. In this paper is assumed that the sandwich structure is created in such a way that a unit cell is periodically repeated. The attention is devoted to influence of the unit cell size, namely the number of the unit cells in the cross-section, on the static behavior of a beam in the bending loading conditions. The variation of the unit cells number affects the size of the cell, so the static analysis was performed – the flexural stiffness and the beam's deflection were determined as functions of the unit cells number, i.e. the core's density.

The two configurations of the sandwich beams were considered for the sake of comparison of their behavior under the same loading conditions: (i) the sandwich beam with the cross-section that is constant along the beam's length, which consists of the two faces while the core is made of the unit cell with the two square openings and (ii) the sandwich beams with the cross-section that changes periodically along the beam's length – one cross section is full, while the other has square openings.

2 Analytical solution of the problem

In Figures 1 and 2 are shown the two considered configurations of the sandwich beam: in Figure 1 the sandwich beam with a cross section which is constant along the beam's length L and in Figure 2 the sandwich beam with variable cross-section. The cross section of the former beam consists of the two flat plates, the beam's faces, upper and lower, of thickness t_f and width b. The beam's core, of height H_C, is made of a unit cell with two square holes of size a. This is a sandwich structure with $n = 1$, where n represents a number of unit cells, i.e. the core's density. As can be seen in Figure 1 the structure with $n = 2$ has eight square holes. If the structure could be built with sufficient (arbitrary) number of holes, n could be practically limitless ($n = \infty$). The cross section of the latter beam is periodically changing along its length. In the A-A section, the beam has the solid cross section, while in the B-B section the cross section has the square holes. The upper and the lower beam faces have the same thickness, t_f, while the beam's core height is H_C.

The faces and the core are made of different materials, so the elastic characteristics of the materials, the Young modulus and the Poisson's ratio, have indices 1 for the faces' material and 2 for the core's material, i.e. E_1 and ν_1, and E_2 and ν_2, respectively.

The differential equation of the beam's elastic (deflection) line reads:

$$B^{(n)} \frac{d^2u(x)}{dx^2} = M(x)\,, \tag{1}$$

where $B^{(n)}$ is the bending stiffness, $u(x)$ is the deflection and $M(x)$ is the bending moment for the cross section defined by coordinate x.

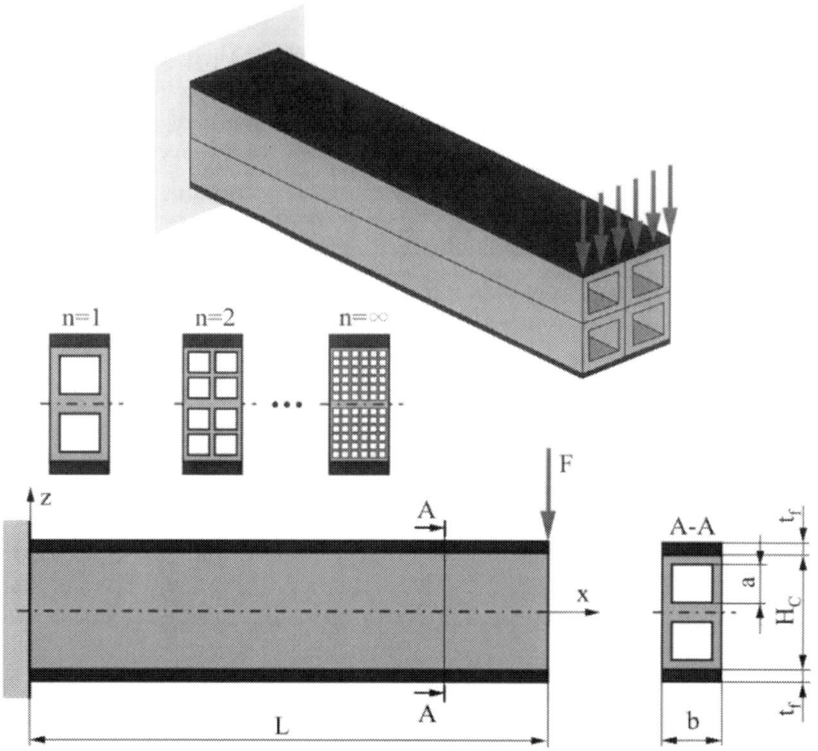

Fig. 1. Schematic presentation of the sandwich beam with the constant cross section.

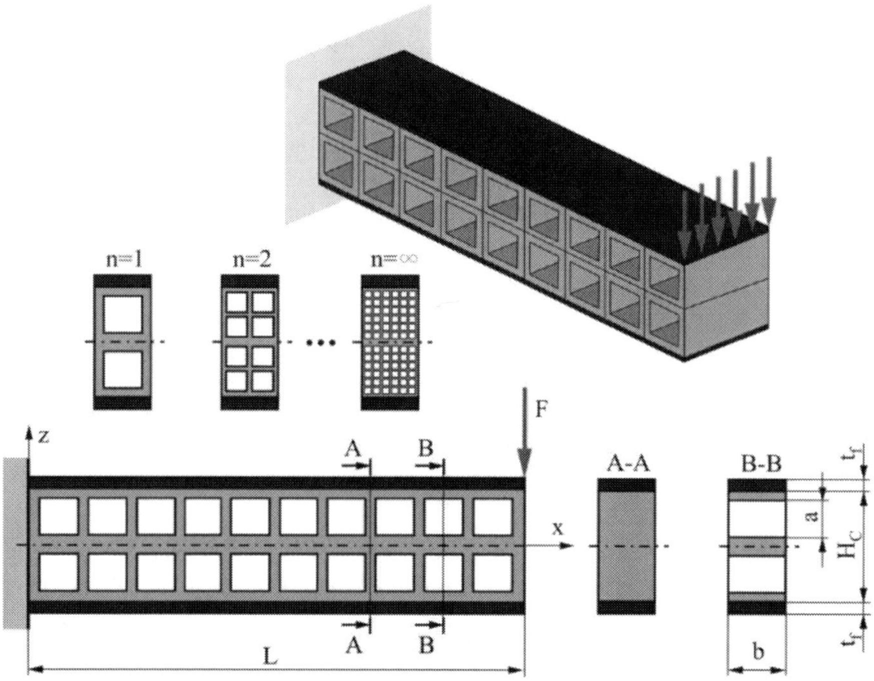

Fig. 2. Schematic presentation of the sandwich beam with the variable cross-section.

For the beam whose cross section is constant along its length, the bending stiffness is calculated as:

$$B_c^{(n)} = \sum E_i I_i \, , \tag{2}$$

while for the beam whose cross section is variable along its length the bending stiffness is calculated as:

$$B_v^{(n)} = \left(\frac{1}{l} \int_0^l \frac{1}{B(x)} dx \right)^{-1} = l \left(\sum_i \frac{l_i}{B_i} \right)^{-1} , \tag{3}$$

where l is the period's length.

For the loading conditions of beams presented in Figures 1 and 2, the largest deflection is obtained at the beams' free ends, i.e. for $x = L$, and can be calculated as:

$$u_{\max} = \frac{FL^3}{3B^{(n)}} \, . \tag{4}$$

For the beam configuration presented in Figure 1 – the beam with the square cross section with the square holes, which does not vary along the beam's length, the bending stiffness in terms of the core's density is obtained as:

$$\begin{gathered} B_c^{(n)} = \sum E_i I_i = 2E_1 I_1 + 2E_2 I_2 \, , \quad I_1 = \frac{1}{3} b \left[\left(t_f + \frac{H_c}{2} \right)^3 - \left(\frac{H_c}{2} \right)^3 \right] , \\ I_2 = \frac{1}{3} b \left(\frac{H_c}{2} \right)^3 - \frac{1}{3} \frac{a}{n} \sum_{m=1}^{n} n \left[\left(\frac{(2m-1)H_c}{4n} + \frac{a}{2n} \right)^3 - \left(\frac{(2m-1)H_c}{4n} - \frac{a}{2n} \right)^3 \right] . \end{gathered} \tag{5}$$

For the beam configuration presented in Figure 2 – the beam with the square cross section with the square holes, which is varying along the beam's length, the bending stiffness in terms of the core's density is obtained as:

$$\begin{gathered} B_v^{(n)} = 2E_1 I_1 + B_{periodic}^{(n)} \, , \quad B_{periodic}^{(n)} = 2E_2 \frac{I_2 I_3 H_c}{(H_c - a) I_3 + 2a I_2} \, , \\ I_1 = \frac{1}{3} b \left[\left(t_f + \frac{H_c}{2} \right)^3 - \left(\frac{H_c}{2} \right)^2 \right] , \quad I_2 = \frac{1}{3} b \left(\frac{H_c}{2} \right)^3 , \\ I_3 = \frac{1}{3} b \left(\frac{H_c}{2} \right)^3 - \frac{1}{3} \frac{a}{n} \sum_{m=1}^{n} n \left[\left(\frac{(2m-1)H_c}{4n} + \frac{a}{2n} \right)^3 - \left(\frac{(2m-1)H_c}{4n} - \frac{a}{2n} \right)^3 \right] . \end{gathered} \tag{6}$$

In Figure 3 is presented variation of the flexural stiffness in terms of the core density for both beam configurations and for the same loading conditions. Results were obtained according to equations (5) and (6) by application of the *Mathematica*® programming routine.

In Figure 4 is presented variation of the beam's free-end deflection in terms of the core density for both beam configurations and for the same loading conditions. Results were obtained according to equations (4), (5) and (6) by application of the *Mathematica*® programming routine.

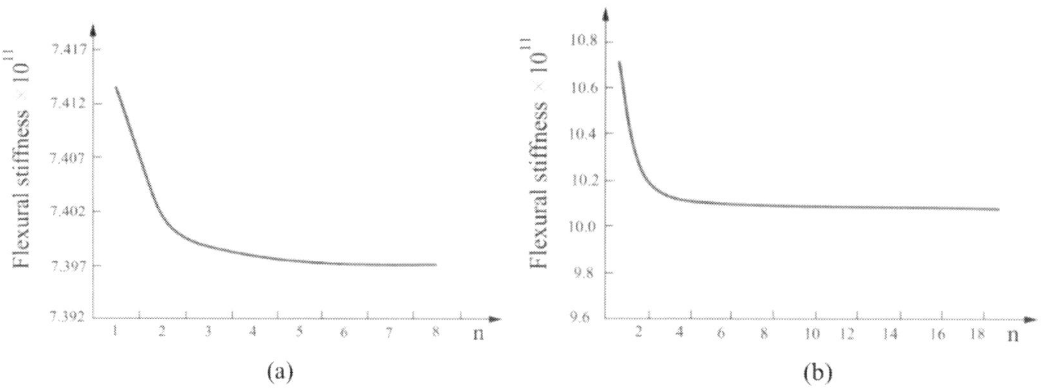

Fig. 3. Beam flexural stiffness in terms of the core density n for the beam: (a) with the cross-section that is constant along the length; (b) with the cross-section that is variable along the length.

From Figure 3 one can notice that the flexural stiffness decreases with the core's density increase. It can also be noticed that after reaching the value of $n > 5$, the stiffness remains almost constant. From Figure 4 one can notice that the value of the beam's free-end increases with increase of the beam's core density. Same as for the flexural stiffness, after reaching the value of $n > 5$, the deflection also remains constant. From that one can conclude that further increase of the core density does not influence the behavior of the sandwich beam in the given loading conditions.

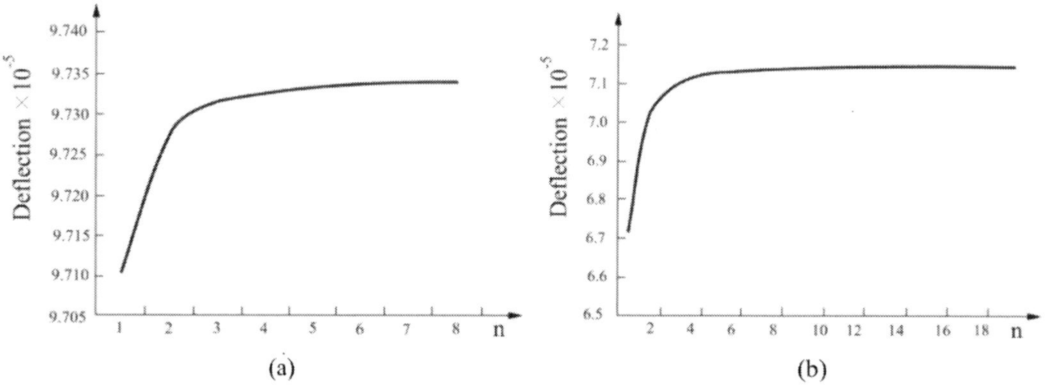

Fig. 4. The free-end deflection in terms of the core density n for the beam: (a) with the cross-section that is constant along the length; (b) with the cross-section that is variable along the length.

3 Finite Element Analysis of the problem

For the purpose of verification of the analytical results, the FME analysis of the two beams' behavior was also preformed. In Figure 5 is presented the beam's free-end deflection, for both beam configurations, for three different values of the core's density. The ANSYS program was used for analysis within the package AutoDesk Invertor.

If one compares the flexural stiffness and the beam's free-end deflection for the two beam configurations, i.e. considers Figures 3, 4 and 5, one can conclude that the sandwich beam configuration with the variable cross-section is better than the configuration with the constant cross-section, for the same loading conditions.

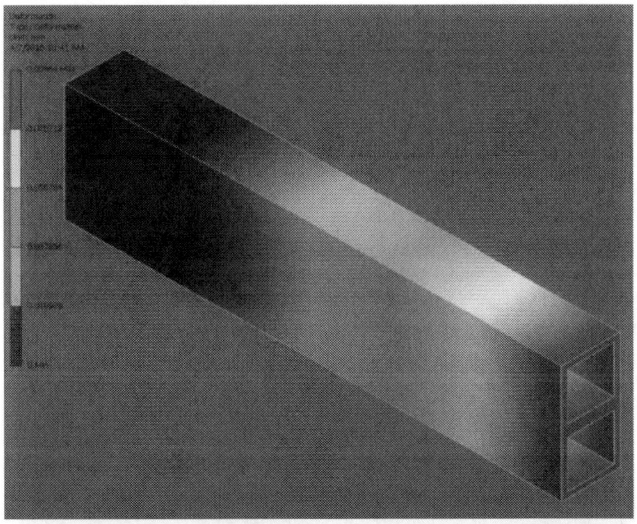
(a)

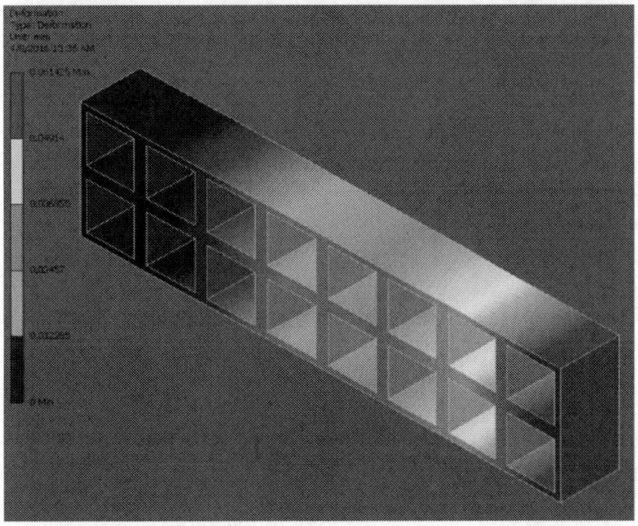
(a)

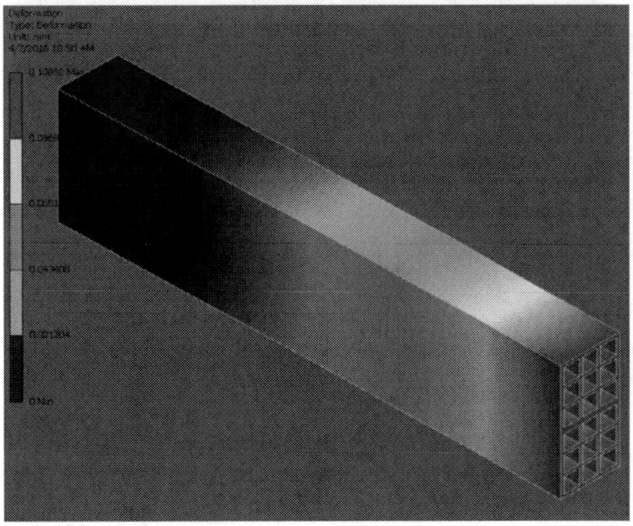
(b)

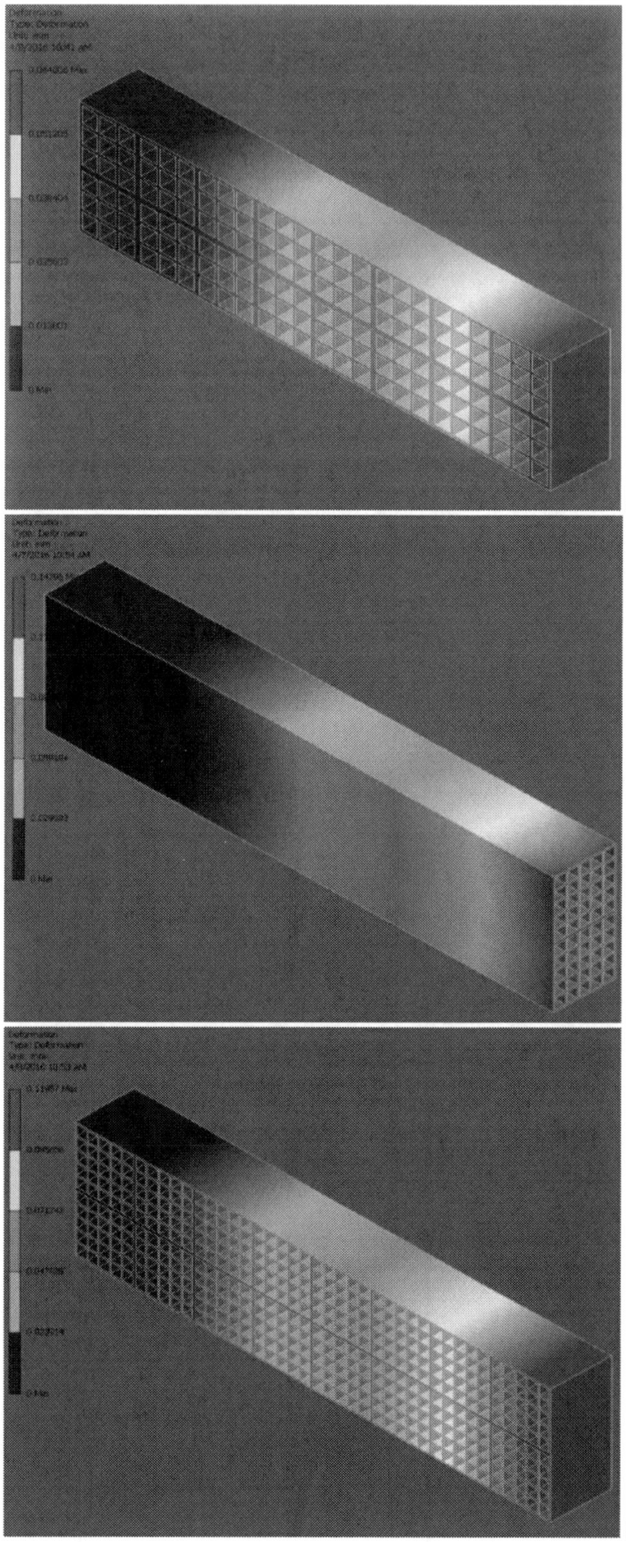

(b)

(c)

(c)

Fig. 5. The free-end deflection in terms of the core density *n* for various core densities, for the beam with the constant cross-section (up) and for the beam with the variable cross-section (below): (a) n = 1, (b) n = 3 and (c) n = 5.

4 Conclusion

In the paper is analyzed the influence of the core density on behavior of the two sandwich beam configurations subjected to bending. The two sandwich beams were considered as the laminar structures with the symmetrical cross-sections. The static analysis was performed to establish the influence of the core's unit cell size on the sandwich beams flexural stiffness and the beam's free-end deflection.

From results obtained both by analytical calculations and by the FME numerical analysis it was concluded that the core density does influence the sandwich beam behavior, but only until the number of the unit cells reaches the value of 5. The further increase of the core's density (above $n = 5$) seems to have no effect on behavior of the sandwich beams subjected to bending by the lateral force.

From comparison of the two beams' configurations behavior, it was also concluded that the sandwich beam with the variable cross-section exhibits better behavior than the sandwich beam with the constant cross-section, under the same loading conditions.

This research was partially supported by the Ministry of Education, Science and Technological Development of Republic of Serbia through Grants ON174001, ON174004 and TR32036 and by European regional development fund and Slovak state budget by the project "Research Center of the University of Žilina" - ITMS 26220220183. The authors are very grateful for this funding.

References

1. H.G. Allen, *Analysis and Design of Structural Sandwich Panels* (Pergamon Press, Oxford, 1969).
2. A.G. Evans, J.W. Hutchinson, M.F. Ashby, Prog. Mater. Sci **43(3)**, 171–221 (1998)
3. S.T. Brittain, Y. Sugimura, J.A. Schueller, A.G. Evans, G.M. Whitesides, J. Microelectromech. Syst. **10**, 113–120 (2001)
4. N. Wicks and J.W. Hutchinson, Int. J. Sol. Struct. **38**, 5165–5183 (2001)
5. N. Wicks and J.W. Hutchinson, Mech. Mater. **36**, 739–751 (2004)
6. L. Valdevit, J.W. Hutchinson, A.G. Evans, Int. J. Sol. Struct. **41**, 5105–5124 (2004)

MATEC Web of Conferences 86, 01003 (2016) DOI: 10.1051/matecconf/20168601003
IPICSE-2016

The measurement of thermal changes on concrete box girder bridge

Lukáš Krkoška[1,*], *Martin* Moravčík[1]

[1]University of Zilina, Faculty of Civil Engineering, Department of Structures and Bridges, Univerzitna 2, 01026 Žilina, Slovakia

Abstract. Thermal load, especially vertical temperature gradient is an important factor that should be considered during the bridge design procedure. In the world several researches focusing on thermal influences was performed, mostly in USA and China, but only a few analysed real measured cross-section temperature gradient at European bridges. In this paper, the temperature field along two decisive cross-section of concrete box girder bridge caused by heat of hydration or exposure to sunlight has been investigated by field monitoring. The measured temperature gradients has been compared to the Slovak bridge design eurocode.

1 Introduction

In practice, the variations of temperature affects bridges significantly. Uniform temperature changes cause large total length changes, expansion or contraction. These changes of length influence internal forces, structural dynamic characteristics and the continuous expansion and contraction may damage critical members of the bridge, such as expansion joint, bearing and anchor head. If these deformations are restricted, temperature variations result in thermal stresses, what may leads to unexpected tensile cracks and consequent possible reinforcement corrosion. Vertical and horizontal temperature difference result in additional bending moments in vertical plane and horizontal plane, respectively.

2 Bridge and Monitoring System Description

In this paper it is analyzed the prestressed concrete box girder highway bridge structure SO201-10, highway part from Hricovske Podhradie to Lietavska Lucka. Bridge has nine spans (48.0 m + 3x58.0 m + 60.0 m + 2x68.0 m + 54.0 m + 37.1 m), with total length of 835.1 m, built by incremental launching method. The main spans (68.0 m) bridging the highway and railway. The length of steel launching nose is 39.0 m, see Figure 2. The deck of the bridge consists of a continuous box girder seated on simple bearings on the piers and abutments. Width of the top slab is 13.75 m and the bottom slab is 8.0 m. The height of the single cell box cross-section is variable, from 3.85 m to 4.5 m. Thickness of the webs varies

* Corresponding author: lukas.krkoska@fstav.uniza.sk

from 450 mm (middle of span) to 1000 mm (above the supports). The top and bottom slabs have constant thickness 300 mm with 500 mm haunches 1.5 m from the webs.

During the construction of the bridge, total 19 embedment vibrating wire strain gauges for concrete, 8 thermocouples and 2 air thermometers were arranged at two characteristic cross sections of the first span, as shown in Figs. 1~3. The first measured cross-section (MP1) is approximately at 3/5 of the first span and second measured cross-section (MP2) is very next to the sleeper. Vibrating wire strain gauges type TES/5.5/T are able to measure strain (as the function of measured value) and temperature. In cross-section MP1 we have measured strains and temperature of the concrete (S1~S7) and in cross-section MP2 we have measured the strains and temperature of the concrete (S1~S11), the temperature of the concrete surface (TP1~TP7), temperature of the interior and exterior air (TE1, TE2) and the temperature of the web in depth 400 mm (TC400). All of these values was measured continually.

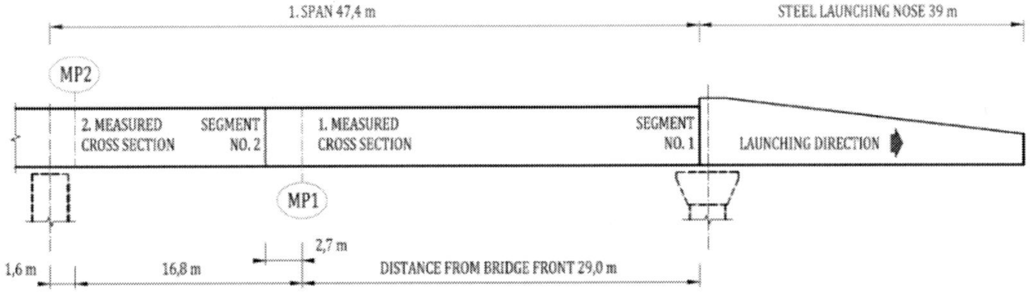

Fig. 1. Position of measured cross sections in longitudinal section.

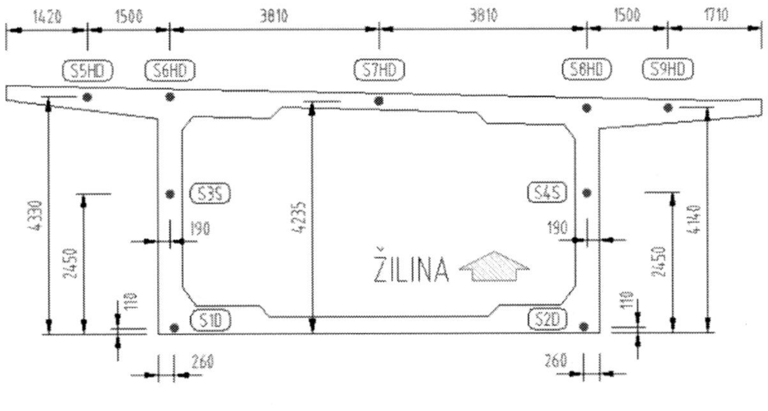

Fig. 2. Strain gauges arrangement at measured cross section MP1.

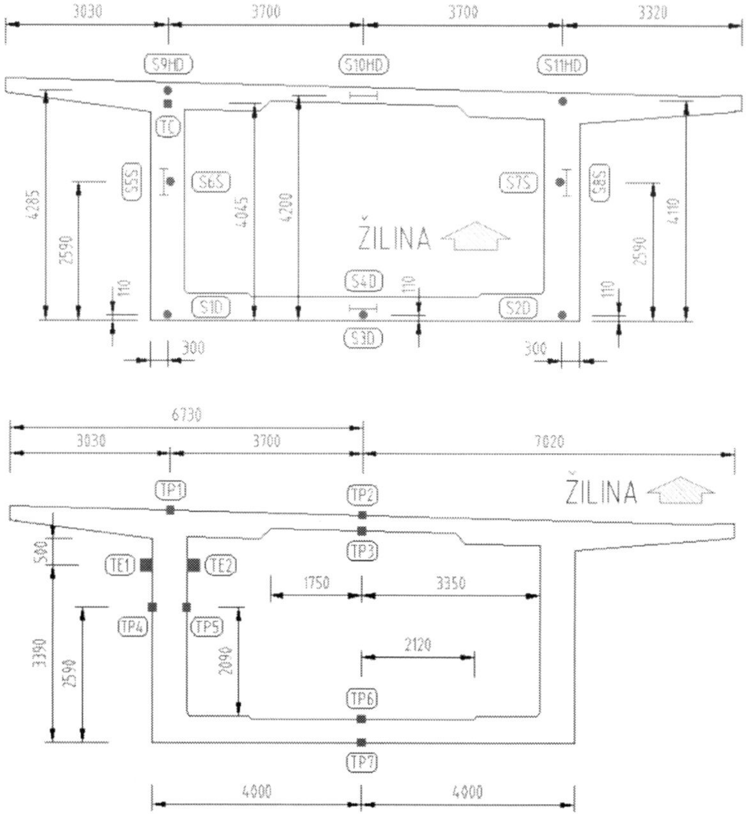

Fig. 3. Strain gauges and thermocouples arrangement at measured cross section MP2.

3 Measurement results

3.1 Variation of temperature caused by hydration process of the concrete

The concrete during the hydration produce a significant amount of hydration heat, what is clearly visible in Fig. 4. These figures shows the variation of temperature caused by concrete hydration in cross-section MP1. The temperature in concrete of bottom slabs and webs rises immediately after casting of the concrete and after hydration process decreases and regresses to the normal temperature of the air, seed Fig. 8. After 14 days was casted the top slab, which temperature rapidly rises due to produced hydration heat and after hydration process again decreases and regresses to the normal temperature of the air, see Fig. 4.

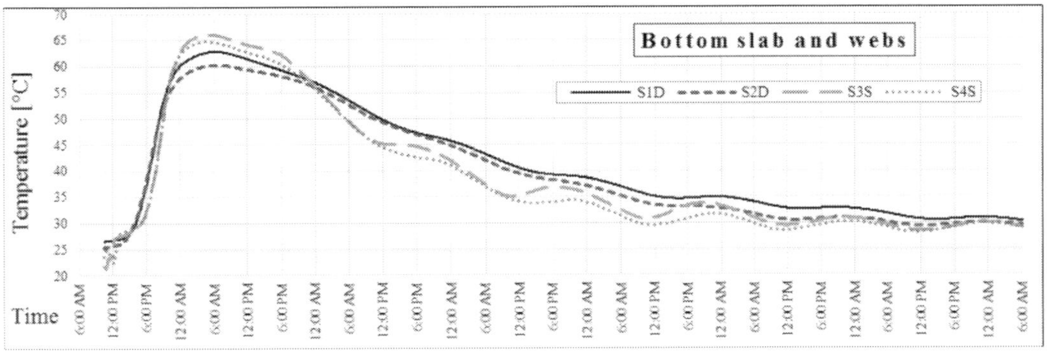

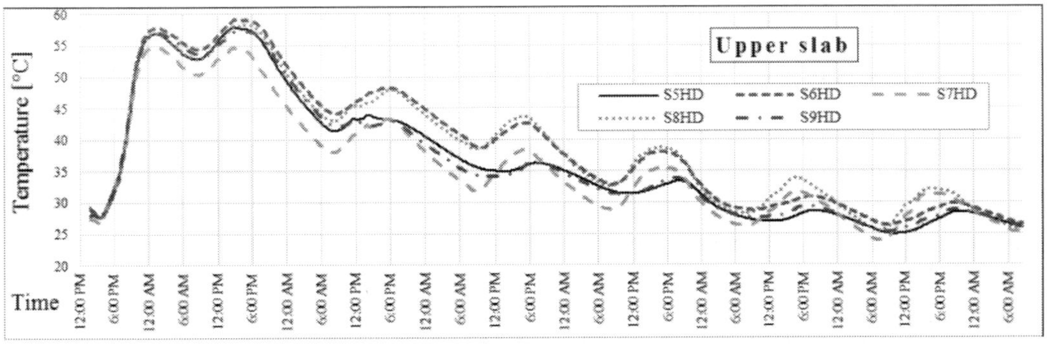

Fig. 4. Temperature in concrete of bottom slab, upper slab and web during the hydration process.

Field test results (see Fig. 4.) show that the variation of hydration heat temperature consist of four temperature periods during the hydration of concrete:

1. First 0-16 h after concrete casting, when temperature of the interior concrete reaches the peak value rapidly. The temperature increment is about 30°C ~ 35°C.
2. 16-20 h after concrete casting, when temperature of the interior concrete is approximately stable about the peak value.
3. 20-48 h after concrete casting, when temperature decreases rapidly.
4. 48-96 h after concrete casting, when temperature slow falls and gradually regresses to the normal temperature of the air.

3.2 Temperature variations caused by air temperature

The variations in temperature of the concrete at the top and bottom slabs (gauges are about 110 mm beneaththe surface) as well as in the webs are shown in Fig. 5. This figure shows temperature of the concrete during three summer days from 10.8.2015 until 12.8.2015 with maximum day air temperature 36 °C. The temperature of the interior concrete increases with air temperature, the rate of the rise in temperature of the top slab is the highest becauseof the perpendicular incidence of sunshine. Maximum measured temperature of the top and bottom slab was 38.04°C (at 16:30 p.m.) and 28.94°C (at 17:30 p.m.) respectively and the maximum temperature of the webs was 30.0°C (at 18:00 p.m.). The peak temperature induced by air temperature in the slabs and webs of the concrete box-girder lags by about 2~3 h behind the maximum air temperature, mainly because of the low thermal conductivity of concrete. It is clearly visible that measured temperature at the same level varies across the width (the difference of temperature of the top slab is 1.5°C ~ 5.0°C and difference in bottom slab and webs is about 1.0°C), mainly because of different shading of the cross section and different angle of sunshine.

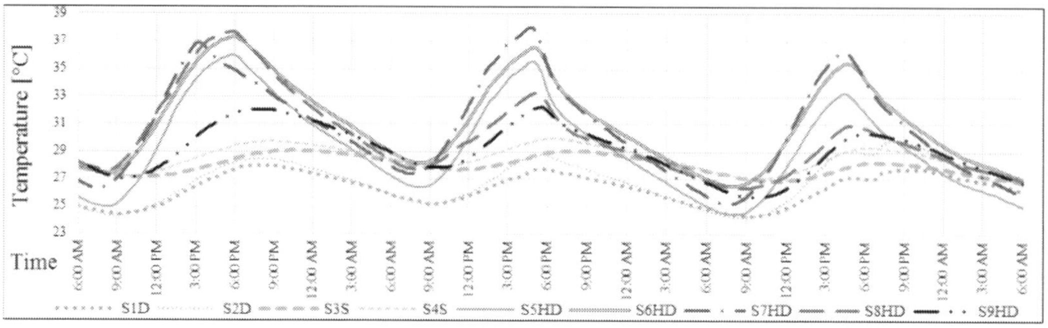

Fig. 5. Peak temperatures in concrete during summer days.

In Fig. 6 are shown the variations of temperatures of different cross-section parts during the period from 17.9.2015 to 19.9.2015. During this period the air temperature reached the maximum value from the date on which the surface thermocouples (TP1-TP7) were installed.

As the result of most intense solar radiation, the concrete temperature of the surface of upper slab of the box is higher than the air temperature outside of the box at the same time and the variations of these temperatures is consistent with each other, with a sinusoidal trend. The range of outer surface temperature was 12°C. Because low thermal conductivity of concrete, the range of temperature changes of measuring points inside the concrete and at the inner surface is smaller (6°C and 4°C respectively) and the peak temperature lags by about 2~3 h behind the maximum air temperature.

Due to shading of the flanges, the webs are not directly affected by the sun, when the sun is more intense. This caused that the temperature changes of the outer surface of the web correspond to changes of the air temperature outside of the box, with a deviation of less than 1°C. Because the thickness of the web, the temperature inside the concrete and at the inner surface change slightly, with a daily fluctuation of less than 2°C.

The bottom slab is protected against solar radiation by upper slab and webs of box girder, so the temperature of the bottom slab is affected only by exterior air temperature. Temperature of outer surface does not reach the temperature of the exterior air and varies with a daily fluctuation of 7°C. The temperature inside the concrete and at the inner surface change slightly, with a daily fluctuation of less than 3°C.

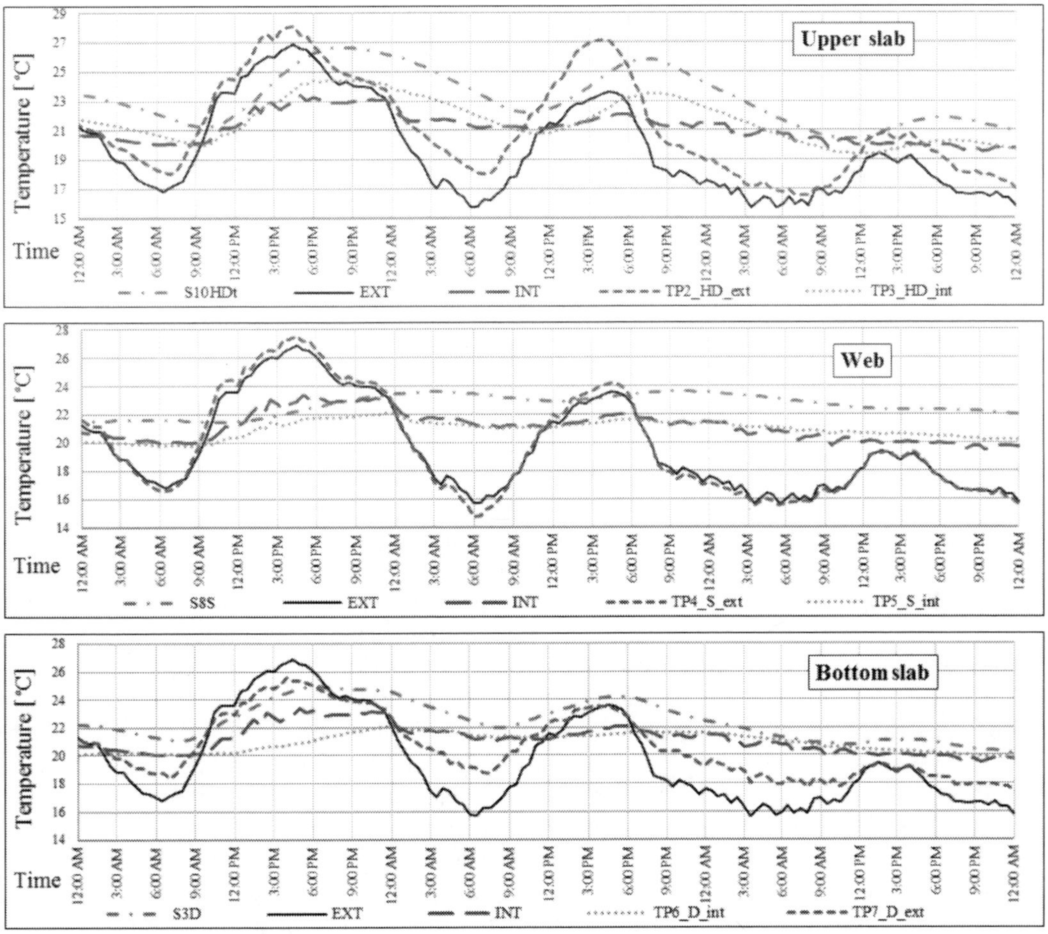

Fig. 6. Temperature changes in concrete at different parts of cross section.

3.3 Distribution of temperature gradient

The variation of air temperature may cause great temperature gradient, which can induce high thermal stress in the concrete box-girder. The vertical nonlinear temperature gradient obtained by subtracting a minimum temperature from the vertical temperatures along the depth of the cross section and comparison of these obtained gradients with temperature gradient for the same concrete box girder cross section recommended in Slovak bridge eurocode STN EN 1991-1-5 [6] are presented in Fig. 7. Maximum measured value of the positive temperature gradient at the top surface of the box-girder was 17.3°C, while the value recommended in Slovak bridge eurocode for this shape of cross-section is only 15.4°C. The distribution of the gradient is different from the recommended one and it is changing during the day with the temperature variations (see Fig. 6) The negative temperature gradient did not reach the recommended one, because the winter temperatures were very weak.

DOI: 10.1051/matecconf/20168601003

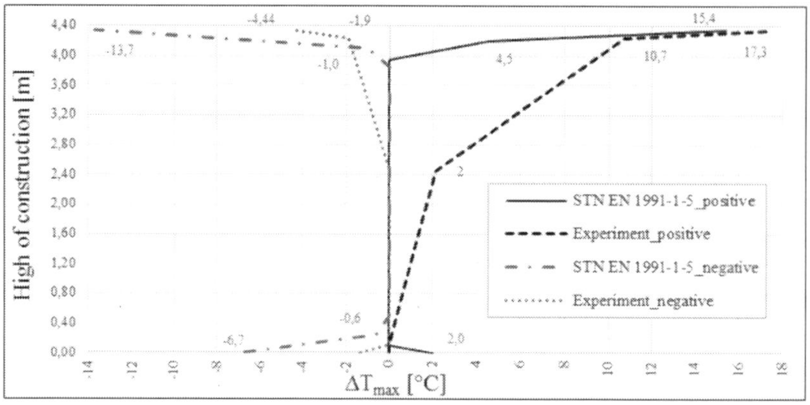

Fig. 7. Comparison of obtained temperature gradients with Slovak bridge eurocode.

4 Conclusion

The concrete during the hydration produce a significant amount of hydration heat what significantly impact on overall temperature and stress distribution of the concrete box-girder cross-section in early age of the concrete. The variation of hydration heat temperature consist of four periods: temperature rising period, constant temperature period, rapid temperature fall period and slow temperature fall period.

Caused by air temperature variations there are large temperature gradient over the height of the box-girder. The value of the temperature gradient can exceed the temperature gradient recommended in bridge standards specifications. Furthermore the temperature at the same level across the width is not constant, but varies, mainly because of different shading of the cross section and different angle of sunshine. This difficult temperature distribution may impact significantly the stress condition of the bridge and the tensile cracks can occur. Therefore is necessary to perform continuous temperature measurements and analysis of temperature gradient on bridge structures of various types, at least in periods with predicted maximum and minimum air temperatures.

The research is supported by the Slovak Research and Development Agency under contract No. APVV-0736-12 and by Research Project No. 1/0336/15 and 2/0033/15 of Slovak Grant Agency.

References

1. M. J. N. Priestley, Amer. Concrete Inst. Jour. **5**, 209-217 (1978)
2. J. H. Emanuel, J. L. Hulsey, Jour. of the Struc. Div. **104(ST1)**, 65-78 (1987)
3. M. M. Elbadry, A. Ghali, Jour. of the Struc. Eng. **109(10)**, 2355-2374 (1983)
4. M. Emerson, Transp. Road Res. Lab. Sup. Rep. **747**, 5-10 (1982)
5. *STN EN 1991-1-5: Actions on Structures. Part 1-5: General Actions – Thermalactions* (1991)
6. *STN EN 1990: Eurocode. Basis of structural design* (1990)

MATEC Web of Conferences **86**, 01004 (2016) DOI: 10.1051/matecconf/20168601004
IPICSE-2016

Synthesis of the force and displacement methods of structural mechanics. The planar truss case

Krzysztof Hetmański[1,*]

[1]Faculty of Civil Engineering, Warsaw University of Technology, 16 Armii Ludowej St., Warsaw 00-637, Poland

Abstract. The papers delivers a new computational method of statical analysis of frames, well suited for both the 2D and 3D settings and easily generalizable to the case of the presence of big axial forces. The method is based on the compatibility equations which have their roots in the force method, while the unknowns refer to the displacement method. These equations involve also the unknown cross sectional forces which are the force method unknowns. The method proposed can be viewed as the displacement method with an implicit procedure of deriving the stiffness equations.

1 Introduction

In contrast to the elasticity theory in which the stress-based and displacement-based approaches are complementary, the force method (FM) and the displacement method (DM) are usually treated independently, even in separate chapters, cf. the literature cited [1-6]. The aim of the present paper is to put forward a synthesis of the FM and DM methods by focusing attention on the statics of planar trusses. Yet the ideas behind refer to all skeletal structures, including 3D case. The main assumption is the linear elasticity setting. The load is assumed in the form such that the truss bars are not subject to bending. The transverse load is allowed only in specific cases.

2 The approach free from the matrix algebra formalism

Independently of the approach applied the first step is determination of the unknown kinematic data (or the degrees of freedom) as independent and non-zero nodal displacements, cf. Fig.1.

* Corresponding author: k.hetmanski@plusnet.pl

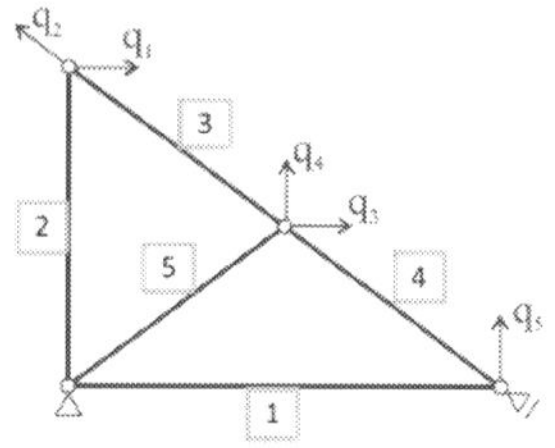

Fig. 1. An exemplary choice of the degrees of freedom of a given planar truss.

The degrees of freedom are collected in a column vector as follows

$$\mathbf{q} = [q_1, \cdots, q_s]^T, \tag{1}$$

where s represents the number of degrees of freedom. At each bar we determine one node as a first node (the index of this starting node is : "i") and the other node (of index "k") as the end node (or second node). At the node "i" we release the constraint associated with the axial force and replace this constraint by the pair of mutual reactions, see Fig.2. The left axial force is denoted by N_i^e. Let us note that the axial force can vary along the bar axis if the axial external load is applied along the bar. Then the axial force is not equal to N_i^e. The resultant of the external axial load is denoted by Q_w^e, see Fig.2. This resultant force is attached to the node "k".

The number of truss members equals p.

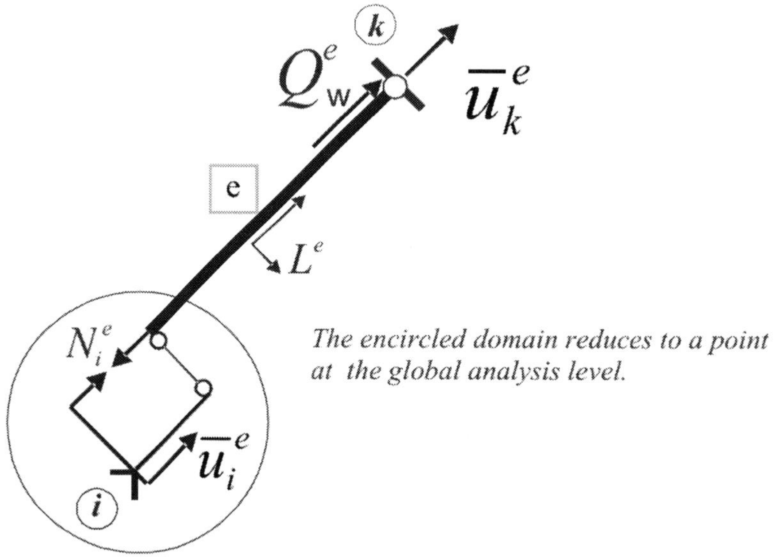

Fig. 2. Releasing the constraint associated with the axial force at point i.

The process of releasing the constraints transforms the truss into a mechanism for which the virtual displacements can be subsequently imposed, see Eq.(2) below. Let the virtual displacements be taken in the form of the following linearly independent column vectors.

$$\bar{\mathbf{q}}_j = [0, \cdots, \bar{q}_j = 1, \cdots, 0]^T, j = 1, \cdots, s. \tag{2}$$

Thus, the number of virtual states is equal to the number of degrees of freedom. For each of these states we associate one equilibrium equation which follows from the virtual work equation. The latter reads

$$-\sum_{e} N_i^e \underbrace{\left(\overline{u}_{jk}^e - \overline{u}_{ji}^e\right)}_{\overline{\Delta}_j^e} + \overline{L}_j = 0, j = 1, \cdots, s. \tag{3}$$

where $\overline{u}_{ji}^e, \overline{u}_{jk}^e$ are nodal virtual displacements projected onto the bar direction e while $\overline{L}_j$ is the virtual work of forces applied to the nodes and of the forces Q_w^e reduced to the nodes. If, exceptionally, there are applied transverse loads, then the load should be replaced by the statically equivalent point loads applied to the nodes. Then these forces will contribute to the expression for the virtual load $\overline{L}_j$. Let us introduce the quantity $\overline{\Delta}_j^e$ which expresses the change of the distance between the nodes i and k measured along the bar axis, caused by the virtual displacements $\overline{\mathbf{q}}_j$. By the method of elimination of the pivotal element from the equilibrium equations (3) one can eliminate s forces N_i^e, provided that the subsequent equations are solvable. If undetermined or mutually contradictory equations appear, this would mean that the truss is geometrically instable. This case is excluded from our discussion. The member forces which remain upon the mentioned elimination process may be treated as redundants of the force method and their appearance indicates that the basic structure is correctly constructed. On the other hand, if all the member forces can be computed, the initial structure is statically determinate. Hence we conclude that the degree of statical indeterminacy equals the difference between the number of cross sectional forces of the truss bars and the number of degrees of freedom of its nodes. One can note that this statement holds for all other types of skeletal structures. The next key step is to construct the compatibility equations referring to the released constraints, where the gap displacements δ_i^e appear, cf Fig.3. δ_i^e measures the distance between the sections being cut.

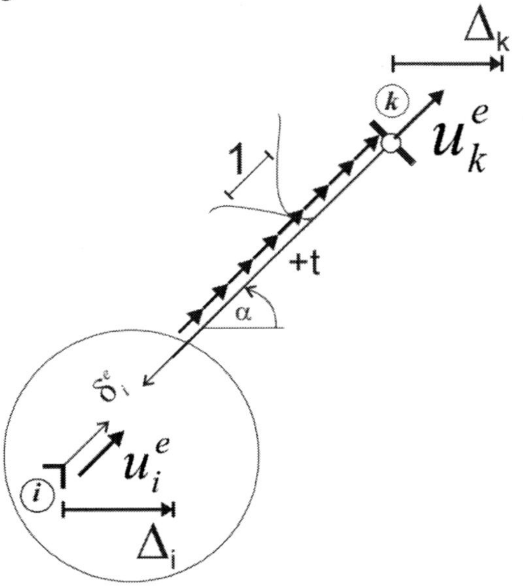

Fig. 3. Relative displacement at point i.

The same figure shows selected span loads and the kinematic loads referring to the nodes of the given bar. This set of loads affects the unknown displacement in the term δ_{i0}^{e}. The exemplary formula is given below

$$\delta_{i0}^{e} = \begin{cases} \alpha_t t L^e \\ \Delta_i \cos\alpha - \Delta_k \cos\alpha \\ 1 \end{cases} \tag{4}$$

The unit error of assembling of the members makes it possible to determine the influence line of the axial force of the truss. Thus the compatibility equations read

$$\delta_{i0}^{e} + N_i^e \frac{L^e}{(EA)^e} + \underbrace{u_i^e - u_k^e}_{-\Delta^e} = 0,\ e = 1, \cdots, p, \tag{5}$$

Let us note that Eqs.(5) still hold if the bars are inextensible. The expression $\Delta^e = u_k^e - u_i^e$ should be expressed by the components of vector , by projecting the displacements of nodes on the directions of the axes of the connecting bars. This can be concisely written in the matrix form in the next section. Now let us mention only that these relations are linear with respect to the components of $\mathbf{q}$. Thus, in equations (5) there appear the bar forces N_i^e not eliminated from (3) as well as all components of $\mathbf{q}$. Then, by a simple elimination one can compute these unknowns, which ends the analysis. Let us check the balance of unknowns and equations. The unknowns are the bar forces for which we have the same number of compatibility equations. On the other hand we have one equilibrium equation for each displacement, which proves that the balance holds. The presented algorithm is not to support the done- by -hand computations, but it was efficiently used in the program code, extended also to the bending frames in 2D.

Let us discuss now the matrix form of (3), (5). The matrix equations are attractive, yet a direct implementation is non effective. A lot of zero components appear which makes the computations ineffective. The problem can be solved by using some refined algorithms, but then the conciseness of the matrix equations is lost.

3 The matrix formulation

The geometric equations linking the bars elongations (6) with the nodal displacements are crucial here. They read

$$\mathbf{\Delta} = \begin{bmatrix} \Delta^1 \\ \vdots \\ \Delta^p \end{bmatrix}_{px1} = \begin{bmatrix} u_k^1 - u_i^1 \\ \vdots \\ u_k^p - u_i^p \end{bmatrix} = \mathbf{B} \cdot \mathbf{q}, \tag{6}$$

where $\mathbf{B}$ is the rectangular matrix of dimensions p by s. Let us introduce the column vectors as follows

$$\mathbf{N} = \begin{bmatrix} N_i^1 \\ \vdots \\ N_i^p \end{bmatrix}, \ \overline{\mathbf{L}}_z = \begin{bmatrix} \overline{L}_1 \\ \vdots \\ \overline{L}_s \end{bmatrix}, \tag{7}$$

The equilibrium equations assume the concise form below

$$\mathbf{B}^T\mathbf{N} = \overline{\mathbf{L}}_z \tag{8}$$

Upon multiplying (5) $\frac{(EA)^e}{L^e}$ by one finds

$$N_i^e = \frac{(EA)^e}{L^e}\Delta^e - \frac{(EA)^e}{L^e}\delta_{i0}^e,\ e = 1, \cdots, p \tag{9}$$

Let us introduce the matrices

$$\mathbf{E} = \begin{bmatrix} \frac{(EA)^1}{L^1} & \cdots & 0 \\ \vdots & \ddots & \vdots \\ 0 & \cdots & \frac{(EA)^p}{L^p} \end{bmatrix}_{p x p}, \boldsymbol{\partial}_{\mathbf{o}} = \begin{bmatrix} \delta_{i0}^1 \\ \vdots \\ \delta_{i0}^p \end{bmatrix}_{p x 1} \tag{10}$$

By using (6) we rearrange the above formulae to the form

$$\mathbf{N} = \mathbf{EBq} - \mathbf{E}\boldsymbol{\partial}_{\mathbf{o}} \tag{11}$$

Substitution of (11) into (8) leads to the governing equation in which the displacements are the main unknowns

$$\underbrace{\mathbf{B}^T\mathbf{EB}}_{\mathbf{K}}\mathbf{q} = \underbrace{\overline{\mathbf{L}}_z + \mathbf{B}^T\mathbf{E}\boldsymbol{\partial}_{\mathbf{o}},}_{\mathbf{Q}}$$

$$\mathbf{Kq} = \mathbf{Q} \Rightarrow \mathbf{q} = \mathbf{K}^{-1}\mathbf{Q}. \tag{12}$$

Coming back to (11) we compute the forces in bars.

4 Extension to other types of skeletal structures

Let us consider planar frames. The displacements adjacent to a node are now introduced as in Fig.4. We note that the independent angles of rotation appear.

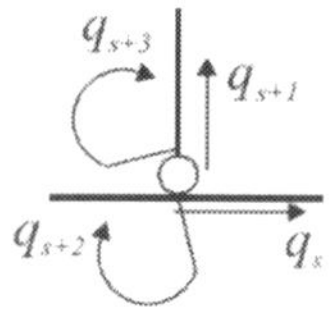

Fig. 4. Degrees of freedom in frame node.

We assume that the bars of frames have arbitrary shape and are clamped at two different nodes of the structure. Thus releasing the constraints around the node i requires introduction of pairs of cross sectional forces, see Fig.5. Here L stands for the length of the chord of the bar.

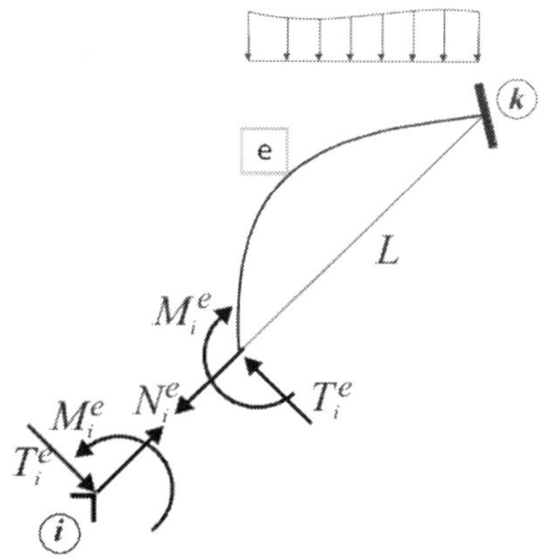

Fig. 5. Releasing the constraints at a node in the direction of one bar.

By imposing virtual displacements according to (2) and applying the virtual work equation one finds the equations of equilibrium and its number is equal to the number of unknown degrees of freedom of the frame. On the other hand the compatibility equations are cleared up in Fig.6.

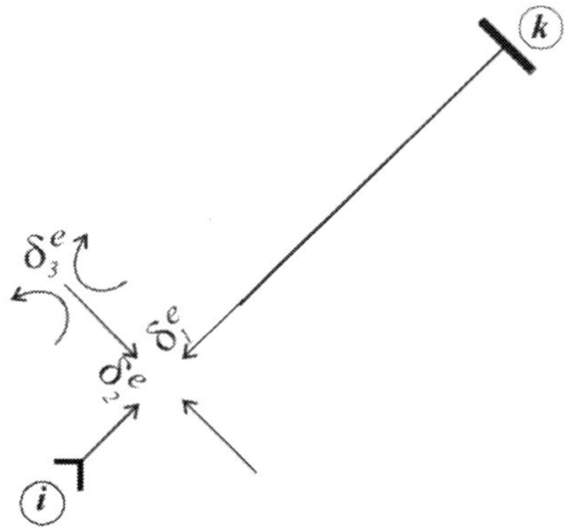

Fig. 6. Compatibility equations.

The compatibility equations are expressed by

$$
\begin{aligned}
\delta_1^e &= \overbrace{N_i^e\delta_{11}^e + T_i^e\delta_{12}^e + M_i^e\delta_{13}^e + \delta_{10}^e}^{\text{the terms corresponding to the loads applied to the element:}} \quad \overbrace{+u_i\cos\alpha + v_i\sin\alpha - u_k\cos\alpha - v_k\sin\alpha}^{\text{the terms determined by nodal displacements:}} = 0 \\
\delta_2^e &= N_i^e\delta_{21}^e + T_i^e\delta_{22}^e + M_i^e\delta_{23}^e + \delta_{20}^e \quad +u_i\sin\alpha - v_i\cos\alpha - u_k\sin\alpha + v_k\cos\alpha + \varphi_k L = 0 \\
\delta_3^e &= N_i^e\delta_{31}^e + T_i^e\delta_{32}^e + M_i^e\delta_{33}^e + \delta_{30}^e \quad -\varphi_i + \varphi_k = 0
\end{aligned}
\tag{13}
$$

The physical meaning of the selected coefficients is cleared up in Fig.7.

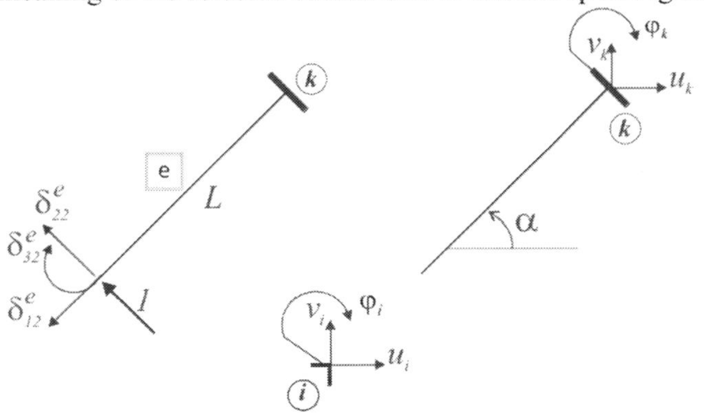

Fig. 7. Selected coefficients in Eqs.(13)

As already mentioned the computer program has not been constructed upon the matrix equations.

The method can be extended to the case of frames subject to big axial forces and to 3D frames in general.

5 Concluding remarks

The compatibility equations (5), (13) deliver a synthesis of the FM and DM methods. The compatibility equations have their roots in the FM while the unknowns involved refer to the DM method. These equations involve also the unknown cross sectional forces which usually are FM unknowns. One can say that the method shown is the DM method with an implicit procedure of deriving the stiffness equations. The two approaches discussed reveal a certain paradox. The basis of the computer program is an algorithm which does not refer to the matrix equations formalism.

References

1. W. Nowacki, *Mechanika Budowli, t. I, II* (PWN, Warszawa, 1965)
2. C. Branicki, R. Ciesielski, Z. Kacprzyk, J. Kawecki, Z. Kaczkowski, G. Rakowski, *Mechanika Budowli. Ujęcie Komputerowe* (Arkady, Warszawa, 1991)
3. A. Gawęcki, *Mechanika materiałów i konstrukcji prętowych Tom I i II* (Wydawnictwo Politechniki Poznańskiej,1998)
4. T. Lewiński, J.Theoret.Appl.Mech. **2(39)**, 307-322 (2001)
5. G. Dzierżanowski, W. Gilewski, K. Hetmański, T. Lewiński, *Zbiór zadań z mechaniki konstrukcji prętowych. Zagadnienia Statyczne* (Oficyna Wydawnicza PW, 2014)

Calculation of filtration of polydisperse suspension in a porous medium

Ludmila Kuzmina[1], and *Yuri* Osipov[2,*]

[1]National Research University Higher School of Economics, 101000 Moscow, Russia
[2]Moscow State University of Civil Engineering, 129337 Moscow, Russia

Abstract. The problem of filtering the suspension in a porous medium is considered. The proposed model for the transport of particles of different sizes is a generalization of deep bed filtration model for a monodisperse suspension with size-exclusion particle capture mechanism. Exact and asymptotic solutions are constructed at the filter inlet and on the concentration front of the suspended and retained particles. Numerical calculation for a suspension with 2-size particles shows that the distribution of deposit in the filter depends on the particle size.

1 Introduction

Filtration problem is an essential part of the underground hydrodynamics, its study is necessary for the design and construction of dams and hydraulic structures [1]. During filtering of the suspension in the porous medium some particles are retained in the pores and form a deposit. Depending on the physical properties of the porous medium and the suspension particle retention may be caused by diffusion, electrostatic and gravitational forces, etc [2]. If particle and pore size distributions overlap, the predominant particle retention mechanism is size-exclusion [3]. It is assumed that the particles pass freely through the larger pores and get stuck at the inlet of pores which are smaller than the particle diameter.

The classical model of deep bed filtration considers the flow of fluid with 1-size particles in a porous media with constant values of porosity and permeability [4]. In a modified filtration model the porosity and permeability of the filter decrease with increasing deposit. Exact and asymptotic solutions of this problem are obtained in [5-9]. Examples of numerical calculation see in [10]. The filtration models for monodisperse suspension are confirmed by many experimental data [11, 12]. However, there are laboratory studies, the results of which can not be explained within the framework of this model [13].

In this paper one-dimensional filtration of polydisperse suspension in a porous medium is considered. Suspended solid particles of the suspension differ, for example, by their sizes. All types of particles transported by fluid flow move with the same velocity, porosity and permeability of the porous medium depend on the particles deposits. It is assumed that filter coefficients depend on the type of particles. Filtering of each type of particles is

* Corresponding author: yuri-osipov@mail.ru

determined by mass balance equation and kinetic equation of particle capture, similar to the equations for a monodisperse suspension. These equations form a system of $2n$ quasi-linear differential equations of the first order. It is assumed that the suspension with constant concentrations of suspended particles is supplied at the filter inlet, and at the initial time the porous medium does not contain any particles.

2 Governing equations

In the domain $\Omega = \{0 < x < 1,\ t > 0\}$ the suspended and retained particles concentrations $C_i(x,t)$, $S_i(x,t)$, $i = 1,...,n$ satisfy the equations

$$\frac{\partial}{\partial t}\left(g(\vec{S})C_i\right) + \frac{\partial}{\partial x}\left(f(\vec{S})C_i\right) + \frac{\partial S_i}{\partial t} = 0 \tag{1}$$

$$\frac{\partial S_i}{\partial t} = \Lambda_i(\vec{S})C_i;\ i = 1,...,n \tag{2}$$

with border and initial conditions

$$x = 0 : C_i = p_i \tag{3}$$

$$t = 0 : C_i = 0,\ S_i = 0 \tag{4}$$

Here $\vec{S} = (S_1,...,S_n)$; porosity $g(\vec{S})$, permeability $f(\vec{S})$, and the filter coefficients $\Lambda_i(\vec{S})$ are continuous positive functions; $p_i > 0$ are constant.

Concentration front of suspended and retained particles moves with constant velocity

$$v = f(\vec{0})/g(\vec{0}) \tag{5}$$

and divides the domain Ω into two subdomains $\Omega_0 = \{0 < x < 1,\ 0 < t < x/v\}$ and $\Omega_1 = \{0 < x < 1,\ t > x/v\}$. In Ω_0 the system (1) - (4) has a zero solution; in Ω_1 the concentrations C_i, S_i are positive. Since the conditions (3) and (4) disagree at the origin, the solution $C_i(x,t)$ has a strong discontinuity on the characteristic line $t = x/v$ - the concentration front. The solution $S_i(x,t)$ is continuous in Ω and has a weak break on the concentration front.

Consider the condition on the concentration front

$$S_i\big|_{t=x/v} = 0. \tag{6}$$

In the domain Ω_1 the problem (1) - (4) is equivalent to the Goursat problem (1) - (3), (6).

3 Exact and asymptotic solution on the concentration front

Differentiating the terms of equations (1)

$$g(\vec{S})\frac{\partial C_i}{\partial t} + C_i\sum_{k=1}^{n}\frac{\partial g}{\partial S_k}\cdot\frac{\partial S_k}{\partial t} + f(\vec{S})\frac{\partial C_i}{\partial x} + C_i\sum_{k=1}^{n}\frac{\partial f}{\partial S_k}\cdot\frac{\partial S_k}{\partial x} + \frac{\partial S_i}{\partial t} = 0;\ i = 1,...,n. \tag{7}$$

Denote

$$f_0 = f(\vec{0}),\ g_0 = g(\vec{0}),\ \lambda_i = \Lambda_i(\vec{0}),\ g_i' = \frac{\partial g}{\partial S_i}(\vec{0}),\ f_i' = \frac{\partial f}{\partial S_i}(\vec{0})\,.$$

According to (2), (6) on the concentration front the equations (7) take the form

$$g_0 \frac{\partial C_i}{\partial t} + C_i \sum_{k=1}^{n} g_k' \lambda_k C_k + f_0 \frac{\partial C_i}{\partial x} + C_i \sum_{k=1}^{n} f_k' \frac{\partial S_k}{\partial x} + \lambda_i C_i = 0;\ \ i = 1,\ldots,n\,. \tag{8}$$

The transition to the characteristic variables

$$\tau = t - \frac{g_0}{f_0} x,\ \ y = x \tag{9}$$

gives

$$\frac{\partial C_i}{\partial t} = \frac{\partial C_i}{\partial \tau};\ \frac{\partial C_i}{\partial x} = \frac{\partial C_i}{\partial y} - \frac{g_0}{f_0}\frac{\partial C_i}{\partial \tau}\,. \tag{10}$$

Differentiating (6) along the concentration front, we obtain

$$\left.\frac{\partial S_k}{\partial x}\right|_{t=x/v} = -\frac{g_0}{f_0}\lambda_k C_k,\ \ k = 1,\ldots,n\,. \tag{11}$$

Substitution of (10), (11) in (8) gives a system of equations

$$f_0 \frac{\partial C_i}{\partial y} + \lambda_i C_i + C_i \sum_{k=1}^{n} A_k \lambda_k C_k = 0;\ \ A_k = g_k' - \frac{f_k'}{v},\ \ i = 1,\ldots,n\,; \tag{12}$$

with conditions

$$y = 0: C_i = p_i\,. \tag{13}$$

The solution of Bernoulli's equations (12), (13)

$$C_i(y) = \frac{p_i e^{-\frac{\lambda_i}{f_0} y}}{1 + \sum_{k=1}^{n} A_k p_k (1 - e^{-\frac{\lambda_k}{f_0} y})},\ \ i = 1,\ldots,n\,. \tag{14}$$

(14) presents formulas for the suspended particles concentrations on the concentration front.

From (2)

$$\left.\frac{\partial S_i}{\partial \tau}\right|_{\tau=0} = \lambda_i C_i(y);\ i = 1,\ldots,n\,. \tag{15}$$

Therefore, the main terms of the asymptotics near the concentration front have the form

$$C_i(y,\tau) = C_i(y) + O(\tau);\ \ S_i(y,\tau) = \lambda_i C_i(y)\tau + O(\tau^2)\,; \tag{16}$$

4 Example. Model of 2-size particle filtration

Consider a filtration model of the suspension with 2-size particles in the porous medium [14]. The coefficients of the equations (1), (2)

$$g(\vec{S}) = a - S_1 - S_2, \quad a > 0; \quad f(\vec{S}) = b(a - S_1 - S_2)^3, \quad b > 0; \tag{17}$$

$$\Lambda_i(\vec{S}) = \lambda_i - S_1 - S_2, \quad \lambda_i > 0; \; i = 1,2\,. \tag{18}$$

Then the velocity of concentration front propagation is $v = a^2 b$; coefficients $A_k = 2,\, k = 1,2$.

Summing equations (2) and substituting (3) and (18) we obtain an equation for the total deposit $S = S_1 + S_2$ at the filter inlet $x = 0$

$$\frac{\partial S}{\partial t} = \lambda_1 p_1 + \lambda_2 p_2 - (p_1 + p_2)S; \; S\big|_{t=0} = 0\,. \tag{19}$$

The solution of (19)

$$S(0,t) = \frac{\lambda_1 p_1 + \lambda_2 p_2}{p_1 + p_2}\left(1 - e^{-(p_1+p_2)t}\right). \tag{20}$$

According to (16) the asymptotic solution near the concentration front in the variables (*x*, *t*)

$$C_i(x,t) = \frac{p_i e^{-\frac{\lambda_i}{a^3 b}x}}{1 + 2\sum_{k=1}^{2} p_k\left(1 - e^{-\frac{\lambda_k}{a^3 b}x}\right)} + O\left(t - \frac{x}{v}\right); \tag{21}$$

$$S_i(x,t) = \frac{\lambda_i p_i e^{-\frac{\lambda_i}{a^3 b}x}}{1 + 2\sum_{k=1}^{2} p_k\left(1 - e^{-\frac{\lambda_k}{a^3 b}x}\right)}\left(t - \frac{x}{v}\right) + O\left(t - \frac{x}{v}\right)^2, \; i = 1,2\,. \tag{22}$$

Calculations of retained particles concentrations were made for

$$\lambda_1 = 1,\ \lambda_2 = 300,\ p_1 = 0.25,\ p_2 = 0.002,\ a = 4,\ b = 1,\ \tau = t - \frac{x}{v} = 1\,.$$

The graph of total deposit $S(0,t)$ at the filter inlet is shown in figure 1.

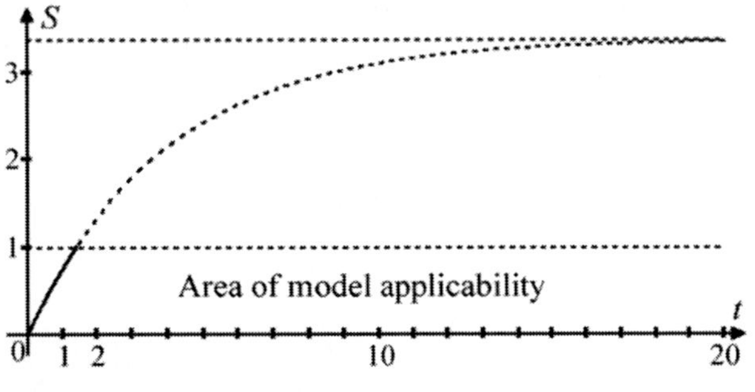

Fig. 1. Total deposit at the filter inlet.

Figure 2 shows the graphs of retained particles concentrations S_1, S_2 and total retained concentration $S = S_1 + S_2$.

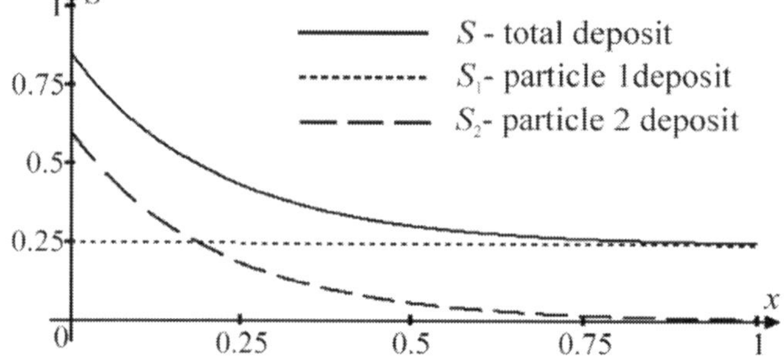

Fig. 2. Retained concentrations near the concentration front.

5 Conclusion

The exact and asymptotic solutions of the filtration problem for polydisperse suspension in a porous medium are constructed on the concentration front. The multiparticle model of size-exclusion particle retention allows differing particle deposition rate for the particles of different sizes. For suspension with 2-size solid particles computation of retained particles concentration near the concentration front was carried out. Physical conditions $\Lambda_i(\vec{S}) > 0;\ \ i = 1,2$ limit the applicability of the example in section 4 (see Figure 1). Figure 2 shows that the penetration depth of retained particles in a porous medium decreases with increasing particle size. Large particles are stuck close to the filter inlet and the smaller particles are uniformly deposited throughout the porous medium.

It is of interest to construct next terms of asymptotic solution and to compare it with laboratory data. This will be considered elsewhere.

References

1 K.S. Basniev, N.M. Dmitriev, *Mechanics of fluid flow* (Wiley, 2012)
2 C. Tien, B.V. Ramarao, *Granular filtration of aerosols and hydrosols* (Elsevier, Amsterdam, 2007)
3 A. Badalyan, Z. You, K. Aji, P. Bedrikovetsky, T. Carageorgos, A. Zeinijahromi, Rev. Sci. Instrum **85**, ID 015111 (2014).
4 J. P. Herzig, D. M. Leclerc, and P. Legoff, Ind. Eng. Chem. **62** (1970)
5 Z. You, P. Bedrikovetsky, L. Kuzmina, Abstr. Appl. Anal. **2013**, ID 680693 (2013)
6 Z. You, Y. Osipov, P. Bedrikovetsky, L. Kuzmina, Chem. Eng. J. **258** (2014)
7 L.I. Kuzmina, Yu.V. Osipov, Procedia Eng. **111** (2015)
8 L.I. Kuzmina, Yu.V. Osipov, Procedia Eng. **153** (2016)
9 L.I. Kuzmina, Yu.V. Osipov, Int. J. Comp. Civil Str. Eng. **12**, 1 (2016)
10 Yu.P. Galaguz, G.L. Safina, Procedia Eng. **153** (2016)
11 P.G. Bedrikovetsky, D. Marchesin, F. Checaira, A.L. Serra, E. Resende, Petrol. Sci. Eng. **32**, 3 (2001)
12 D. Mays, J.R. Hunt, J. Environ. Sci. Technol. **39** (2005)
13 L. Bailey, E. Boek, S. Jacques, SPE 54762, (1999)
14 V.I. Golubev, D.N. Michailov, Trudi MFTI **3**, 2 (2011)

Numerical algorithm for solving of nonlinear problems of structural mechanics based on the continuation method in combination with the dynamic relaxation method

Sergey Trushin[1,*]

[1]NRU MGSU, Department of Structural and Theoretical Mechanics, Yaroslavskoe sh. 26, 129337 Moscow, Russia

Abstract. A numerical algorithm of strength and stability analysis of nonlinear deformable bar systems and thin-walled spatial structures is proposed. A numerical technique is based on the continuation method and the dynamic relaxation method. When using the method of dynamic relaxation the state of static equilibrium of structures is defined after the damped oscillations by integrating over leading parameter. The solution of the initial equation system describing the motion of the mechanical system is reduced to the solution of Cauchy problem for systems of ordinary differential equations. At an each step in the leading parameter the vector of nodal displacements and the time parameter are defined. Several examples of numerical analysis for bar, shell and plate are given.

1 Introduction

In the study of stability of bar systems and thin-walled structures in a nonlinear formulation there is a need to determine the maximum and bifurcation points on equilibrium curves and study the behavior of structures in an overcritical stage. There is a very effective class of methods for creating the equilibrium curves and studying of stability forms of equilibrium of structures. The main idea of these methods is to find a sequence of solutions $\mathbf{u}_m=\mathbf{u}(p_m)$ (m=0,1, 2, ... , M) of the system of nonlinear equations $\mathbf{F}(\mathbf{u},p)=0$ on the basis of available initial solution $\mathbf{u}_0=\mathbf{u}(p_0)$ when the leading parameter p is changed from p_0 to p_M. Various variants of this approach were developed by the researchers [1-5] and several other authors. The continuation method in stability problems of isotropic shells were used in the works [6, 7].

Analysis of the solutions of nonlinear problems of the theory of plates and shells has shown that different schemes based on the Newton-Raphson or Runge-Kutta procedures are given the best results. They use an arc length of the equilibrium curve as the leading parameter and include an auxiliary equation for the iteration on the sphere. Using an arc length as a parameter of the continuation provides an unified procedure of passing through regular, limit and bifurcation points. There is no need to change the leading parameter when

* Corresponding author: trushin2006@yandex.ru

passing a limit point and the very notion of a limit point in this formulation is meaningless.

However, the approach based on the implicit procedures of the Newton-Raphson method has, like any other, certain disadvantages. In particular, when such numerical procedures are used the stiffness matrix becomes singular near critical points.

2 Solution method

When solving static problems with strongly nonlinearity an efficient numerical algorithm based on a combination of the continuation method and the method of dynamic relaxation can be built. The main idea is that the static equilibrium of structures is defined after the damped oscillations with help of integration by time [8-10]. Thus, for solving problems of static stability the equations of motion can be used.

The equations describing the motion of the mechanical system have the form:

$$\mathbf{M}\ddot{\mathbf{x}} + \mathbf{C}\dot{\mathbf{x}} + \mathbf{K}\mathbf{x} = \mathbf{P}(t), \tag{1}$$

where $\mathbf{x}$ is the displacement vector; $\mathbf{M}$ is the mass matrix; $\mathbf{C}$ is the damping matrix; $\mathbf{K}$ is the stiffness matrix of the system; $\mathbf{P}(t)$ is the vector of nodal loads.

We introduce the notation $\mathbf{z}_1 = \mathbf{x}$ and $\mathbf{z}_2 = \dot{\mathbf{z}}_1 = \mathbf{v}$ where $\mathbf{v}$ is the velocity vector. Transforming the equation (1) as shown below:

$$\begin{cases} \dot{\mathbf{z}}_1 = \mathbf{z}_2 \\ \mathbf{z}_2 = -\mathbf{M}^{-1}(\mathbf{K}\mathbf{z}_1 + \mathbf{C}\mathbf{z}_2 - \mathbf{P}(t)) \end{cases} \tag{2}$$

or

$$\dot{\mathbf{z}} = \mathbf{f}(\mathbf{z}, t), \tag{3}$$

where

$$\mathbf{z} = (\mathbf{z}_1\ \mathbf{z}_2)^T,\ \mathbf{z} = (\mathbf{z}_1\ \mathbf{z}_2)^T,\ \mathbf{f} = \begin{pmatrix} \mathbf{z}_2 \\ -\mathbf{M}^{-1}(\mathbf{K}\mathbf{z}_1 + \mathbf{C}\mathbf{z}_2 - \mathbf{P}(t)) \end{pmatrix}. \tag{4}$$

We transform the vector $\mathbf{f}$ to the form:

$$\mathbf{f} = \begin{pmatrix} \mathbf{z}_2 \\ -\mathbf{M}^{-1}(\mathbf{C}\mathbf{z}_2 - \mathbf{R}) \end{pmatrix}, \tag{5}$$

where $\mathbf{R} = \mathbf{K}\mathbf{z}_1 - \mathbf{P}(t)$ represents the residual of the solution of static problem.

Assuming that the nodal displacements $\mathbf{x}$ and time t depend on some parameter s, we introduce an additional equation for the iterations on the cylindrical surface in the system (3):

$$\begin{cases} \dfrac{d\mathbf{z}}{dt} = \mathbf{f}(\mathbf{z}, t) \\ \left(\dfrac{d\mathbf{z}}{ds}, \dfrac{d\mathbf{z}}{ds}\right) = 1 \end{cases} \tag{6}$$

Multiplying the first equation of system (6) by $\frac{dt}{ds}$, we get:

$$\frac{d\mathbf{z}}{ds}=\mathbf{f}(\mathbf{z},t)\frac{dt}{ds}. \tag{7}$$

Substituting (7) into the second equation of system (6) we have:

$$\left(\frac{dt}{ds}\right)^2(\mathbf{f},\mathbf{f})=1 \tag{8}$$

or

$$\frac{dt}{ds}=\frac{1}{\sqrt{(\mathbf{f},\mathbf{f})}} \tag{9}$$

Substituting (9) into (7) we get finally an ordinary differentiation system with the initial conditions:

$$\begin{cases}\frac{d\mathbf{x}}{ds}=\frac{\mathbf{v}}{\sqrt{(\mathbf{f},\mathbf{f})}}\\ \frac{d\mathbf{v}}{ds}=-\frac{1}{\sqrt{(\mathbf{f},\mathbf{f})}}\mathbf{M}^{-1}(\mathbf{Kx}+\mathbf{Cv}-\mathbf{P}(t))\\ \frac{dt}{ds}=\frac{1}{\sqrt{(\mathbf{f},\mathbf{f})}}\\ \mathbf{x}(0)=\mathbf{x}_0\\ \mathbf{v}(0)=\mathbf{v}_0\\ t(0)=t_0\end{cases} \tag{10}$$

For the solution of the Cauchy problem (10) Runge-Kutta or Euler methods can be used. Scheme of Euler method has the form:

$$\begin{cases}\mathbf{v}_{k+1}=\mathbf{v}_k-\frac{\Delta s}{\sqrt{(\mathbf{f}_k,\mathbf{f}_k)}}\mathbf{M}^{-1}(\mathbf{R}_k+\mathbf{C}\mathbf{v}_k)\\ \mathbf{x}_{k+1}=\mathbf{x}_k+\frac{\Delta s}{\sqrt{(\mathbf{f}_k,\mathbf{f}_k)}}\mathbf{v}_{k+1}\\ t_{k+1}=t_k+\frac{\Delta s}{\sqrt{(\mathbf{f}_k,\mathbf{f}_k)}}\\ \mathbf{v}(0)=\mathbf{v}_0\\ \mathbf{x}(0)=\mathbf{x}_0\\ t(0)=t_0\end{cases} \tag{11}$$

The solution of scheme (11) gives the opportunity to find the components of displacement vector **x** and time parameter t in each step of leading parameter s.

If **C**=0, then

$$\mathbf{f} = \begin{pmatrix} \mathbf{z}_2 \\ -\mathbf{M}^{-1}\mathbf{R} \end{pmatrix} \tag{12}$$

and the scheme of Euler method will be presented in the form:

$$\begin{cases} \mathbf{v}_{k+1} = \mathbf{v}_k - \dfrac{\Delta s}{\sqrt{(\mathbf{f}_k, \mathbf{f}_k)}} \mathbf{M}^{-1}\mathbf{R}_k; \\ \mathbf{x}_{k+1} = \mathbf{x}_k + \dfrac{\Delta s}{\sqrt{(\mathbf{f}_k, \mathbf{f}_k)}} \mathbf{v}_{k+1}; \\ t_{k+1} = t_k + \dfrac{\Delta s}{\sqrt{(\mathbf{f}_k, \mathbf{f}_k)}}; \\ \mathbf{v}(0) = \mathbf{v}_0; \\ \mathbf{x}(0) = \mathbf{x}_0; \\ t(0) = t_0. \end{cases} \tag{13}$$

3 Numerical analysis

3.1 Straight bar

The straight bar with a constant cross section under the action of longitudinal forces is considered. Initial data: M=1, C=0, K=1, P=1, Δs=0,01. The analysis is performed with help of the procedure (13).

The displacement x, velocity v, kinetic energy $T = Mv^2/2$ and residual $R = Kx - P$ are shown in Fig. 1. The solution is obtained at 157 steps of leading parameter s.

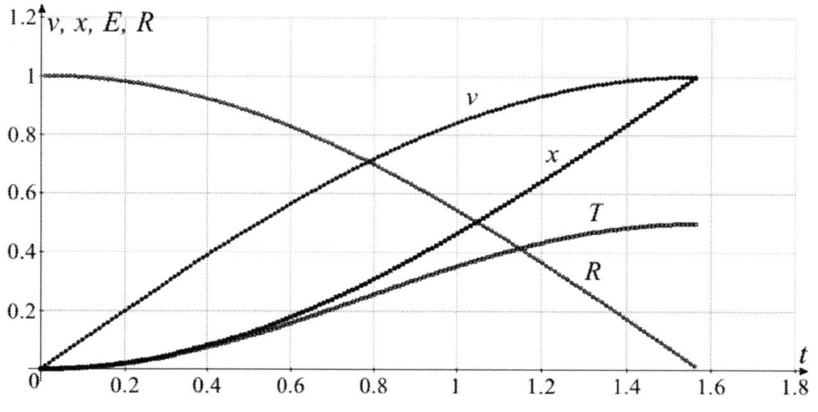

Fig. 1. The results of solution problem of straight bar deformation.

3.1 Shell and plate

Deformation of a flexible spherical shell on a rectangular plan under an action of uniformly distributed transverse load intensive q is considered (Fig. 2).

Boundary conditions are hinged support along the contour. Shallow shell sizes on the plane l_1 and l_2, the radius of curvature R, thickness h, elastic modulus of the material E are given.

The solution of nonlinear differential equations in the mixed form made by the Bubnov-Galerkin method is used [11]. In accordance with a given approximating function the maximum deflection f will be in the center of the shell.

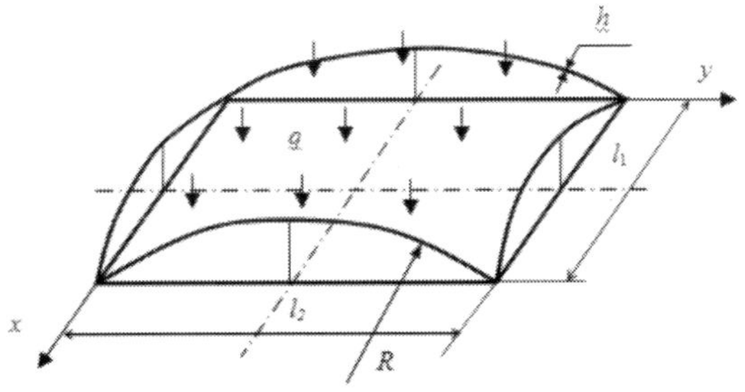

Fig. 2. Shallow shell.

For a shell on a square plan if l_1=l_2 the dependence between the nondimensional load parameter P and the nondimensional deflection at the centre x when Poisson's ratio ν=0,3 is:

$$\left(8.77x^2 - 2.46kx + 0.154k^2 + 22\right)x = P, \tag{14}$$

where

$$k = \frac{2l_1^2}{Rh}, \; P = \frac{q}{E}\left(\frac{l_1}{h}\right)^4, \; x = \frac{f}{h}. \tag{15}$$

The solution is performed using the procedure (11) with increment ΔP. The data for the analysis of shell is follows: M=10^4, C=10, ΔP=50, k=36, Δs=0,01.

The equilibrium curve of the flexible shell is shown in Fig. 3. On the fourth step "snap-through" of shell takes place and there is a transition into a postbuckling equilibrium state.

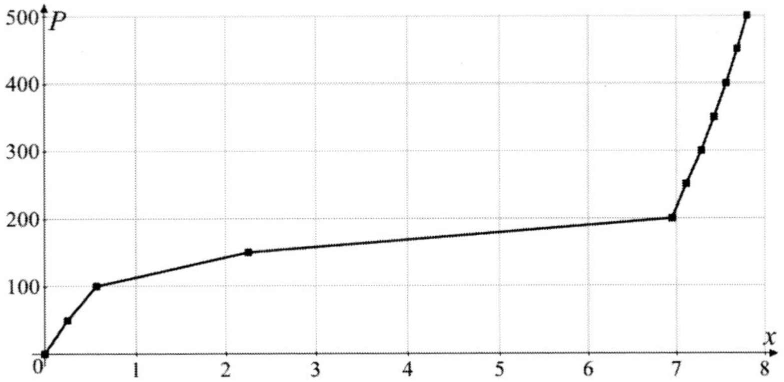

Fig. 3. The equilibrium curve of shallow shell

From the expression (14) as a particular case the solution for flexible plates should be. The data for the analysis of plate is follows: M=10^4, C=10, ΔP=10, k=0, Δs=0,01. The equilibrium curve of the flexible plate is shown in Fig. 4.

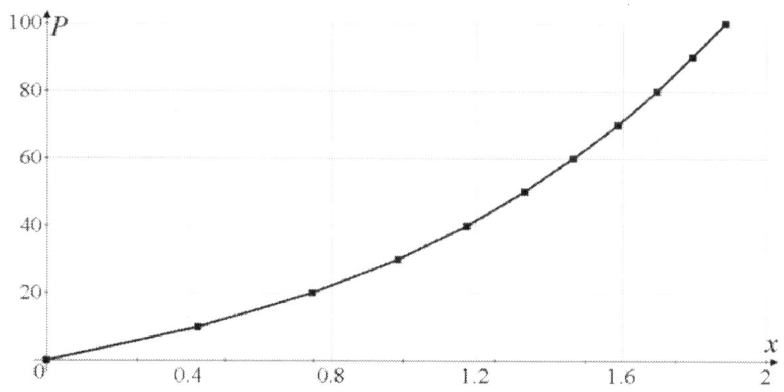

Fig. 4. The equilibrium curve of plate.

4 Conclusion

The continuation method is designed for building of equilibrium curves, determination of limit and bifurcation loads and evaluation of stability of equilibrium in nonlinear analysis of bar systems, plates and shells. One of the most effective numerical methods for solving these types of problems with help of the continuation method is a procedure in which at each step the solution is corrected by Newton-Raphson method. However, using such implicit algorithms may cause some difficulties in numerical implementation of the solution as the stiffness matrix becomes singular near critical points. So the explicit algorithm of dynamic relaxation method can be successfully applied in the analysis of nonlinear behavior of structures near critical points. It does not use the procedure of a matrix inversion or the solution of simultaneous linear algebraic equations with the matrix **K** in the formulation (10).

The fundamental feature of the dynamic relaxation method is that the condition of static equilibrium of the structure is determined after its damped oscillations with help of quasistatic integration by time or by another leading parameter. But with the rejection of the use of matrix operations connected the main disadvantage of the dynamic relaxation method, consisting in a relatively low speed of convergence of the solution.

References

1. M.A. Crisfield, Comput. Struct. **13**, 55 (1981)
2. E. Ricks, J. Appl. Mech. **39**, 1060 (1972)
3. G.A. Wempner, Int. J. Sol. Struct. **7**, 1581 (1971)
4. J.L. Batoz, G. Dhatt, Int. J. Numer. Meth. Eng. **14**, 1262 (1979)
5. V.V. Petrov, *Nonlinear incremental structural mechanics* (Infra-Inzheneriya, Moscow, 2014)
6. I.E. Mileikowskii, S.I. Trushin, *Analysis of Thin-Walled Structures* (A.A. Balkema Publishers, Rotterdam, 1994)
7. S.I. Trushin, *Structural mechanics: finite element method* (INFRA-M, Moscow, 2016)
8. A.S. Day, The Engineer **219**, 218 (1965)
9. G. Ramesh, C.S. Krishnamoorthy, Int. J. Numer. Meth. Eng. **36**, 1339 (1993)
10. K.S. Lee, S.E. Han, T. Park, Comput. Struct. **89**, 216 (2011)
11. A.S. Vol'mir, *Stability of deformable systems* (Scince, Moscow, 1967)

Experimental Investigation of the Deformability of Reinforced Concrete Slabs under Punching

Valery Filatov[1,*], and *Eugene* Bubnov[1]

[1]Samara State Technical University, Institute of Architecture and Civil Engineering, Molodogvardeyskaya str. 194, 443001 Samara, Russia

Abstract. The results of investigations on the finite element models and test samples of flat reinforced concrete slabs of beamless frame under punching by columns of square and rectangular cross-section were shown in this paper. The research technique of reinforced concrete slabs deflections on finite element models at various loading schemes is given. Taking into consideration these results the optimum variant of load application has been chosen. Procedure and the tests results of experimental samples received, when studying of rigidity of fragments of flat reinforced concrete slabs by the action of punching force, are considered. A comparative analysis of the experimental values of the deflections of reinforced concrete slabs under punching by column of rectangular section, depending on the ratio of its sides, is completed. The research results allow to expand experimental data base as well as promote perfecting of modern positions in the theory of deformation of reinforced concrete.

1 Research Deformability of Slabs on Finite Elements Models

Numerical research on finite element models was carried out to define loading scheme when samples tests. It is known that deformations of slabs under uniformly distributed load and under concentrated forces is different. In the design of reinforced concrete structures is generally considered the action of uniformly distributed loads, while the test samples during the test load is applied in the form of concentrated forces. When testing linear construction (beams and columns) this distinction isn't so essential or can be compensated by a design procedure, however, for slab constructions, working in two directions, influence of the loading scheme can be considerable [1]. Numerical methods are the effective tool for the research of stress-strain condition of reinforced concrete constructions and are successfully applied by the authors [2–5].

Investigation was carried out on the finite element models developed for three test samples. The test samples made of heavy concrete represent a fragment of monolithic joint of a column of rectangular section with flat plate. The varied parameter was the ratio of the sides of the column cross-section. Testing of samples was supposed to be carried out in the test frame by the application of loading when reinforced concrete slab will be punching by

* Corresponding author: vb_filatov@mail.ru

the column. The aim of the numerical research was to define such scheme of application of concentrated forces on the plate, at which displacements of slab points would best correspond to displacements of the same slab points under the action of uniformly distributed load.

Finite element models were developed in the LIRA-SAPR 2014 environment and represented a reinforced concrete flat plate (finite element - "shell") in the middle of which the column fragment (finite element - "rod") is located. Plate and column connection was modeled by the options "absolutely rigid body" which was set in plate nodes and "absolutely rigid element" which was set in rods (column). The sizes of "absolutely rigid body" varied and corresponded to column cross-section – 200×200 mm, 200×500 mm and 200×800 mm.

At the first investigation stage the plate of finite element models was loaded with the uniformly distributed load to reveal vertical slab displacements. At the second stage of slab loading was made by the concentrated forces on the sites provided by a construction of the test frame. In this case the position of loading points along external side of a slab varied (on two points of loading along each side). Position of loading points on a plate was chosen in such a way that vertical displacements of finite element nodes of a plate corresponded to vertical displacements of these nodes under uniformly distributed load. The fields of equal vertical slab displacements of finite element models received by numerical research results are presented in Fig. 1, 2.

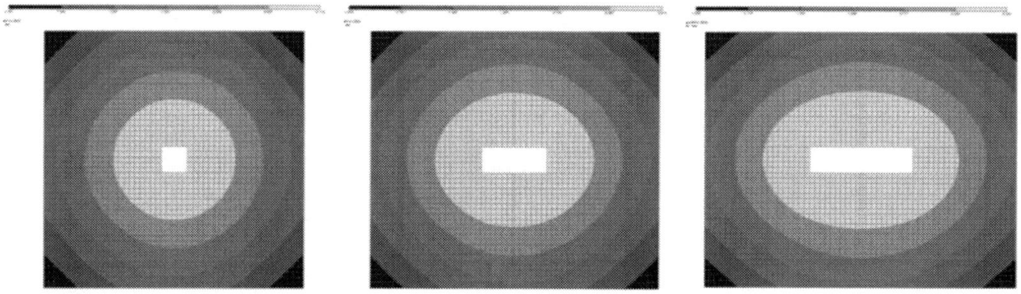

Fig. 1. Fields of equal vertical slab displacements under uniformly distributed loading.

Analysis of the results of finite element modeling shows that when a uniformly distributed load fields of equal vertical slab displacement for a square column are circumferentially and for a rectangular columns - an ellipse, the ratio of the major and minor axes of the ellipse varies with the ratio of the column cross-section side (Fig. 1). These features of plate deformation under load correspond to the theoretical assumptions about the nature of the plate deformation, presented in [6–8]. Presented in fig. 2 fields of equal vertical slab displacements under action of concentrated forces show that the chosen position of loading points on the plate provides satisfactory compliance of deformation slab nature at both schemes of loading.

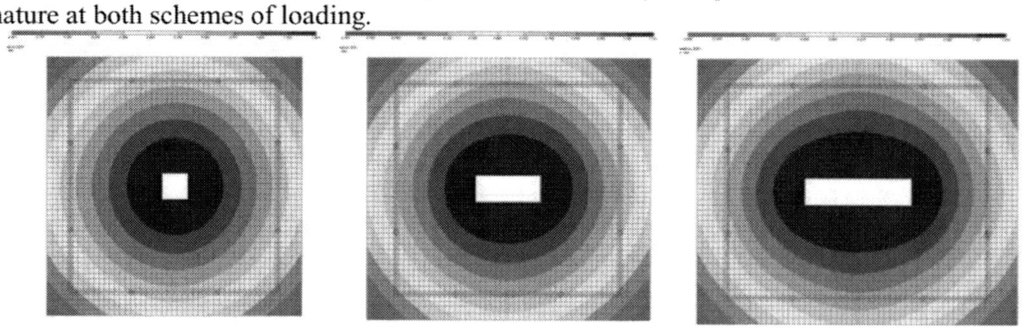

Fig. 2. Fields of equal vertical slab displacements under the action of concentrated forces.

2 Research Deformability of Slabs on Test Specimens

Force resistance research of reinforced concrete structures under complex stress states (combined action of normal and shear forces, bending moments) has an empirical basis due to the work of concrete as a composite elastic-plastic material with discontinuities (cracks formation) [9, 10]. Experimental investigation is performed to study character of the work construction to test the theoretical assumptions and further development of computational models and calculation methods [11–14]. The results of experimental and theoretical investigations of the strength and deformability of flat reinforced concrete slabs under punching, including rectangular columns, presented by the authors [1, 15–20].

Test samples investigated by the authors represented a fragment of reinforced concrete monolithic joint of rectangular cross-section column connection to a flat plate. The varied parameter was the ratio of the sides of the column cross-section. Three samples were tested:

- CMR-1 – column with square section size of 200 mm faces.
- CMR-2,5 – a column of rectangular section of the following size - 200×500mm.
- CMR-4 – a column of rectangular section of the following size - 200×800mm.

Slabs thickness of all test samples was 140 mm. Reinforcing at the top (tensioned) side of a slab was carried out by reinforcing bars grade A500 with a diameter of 14 mm at 100 mm spacing, at lower (compressed) side reinforcing bars grade B500 with a diameter of 5 mm at 100 mm spacing is accepted. Transverse reinforcement in the slab was absent. Cube strength of the concrete was: for CMR-1 sample – 40 MPa, for samples of CMR-2.5 and CMR-4 – 30 MPa.

Test samples loading was carried out in a test frame. Point load was created by a hydraulic jack with the loading capacity of 1000 kN and applied to the bottom edge of the column. Load on test samples slab was transmitted through four metal beams having two hinged points of support on the upper face of the plate. Beams were held by tie rods passing through the sleeves in the sample plate and fixed in the power floor with the help of the traverses. To transfer uniform pressure upon a plate surface to eight points all beams, tie rods and traverses were of one standard size.

Deflections of slab were measured by means of dial gauges as the difference in slab displacements at a given point and near the surface of the column. Arrangement schemes of dial gauges and load application points on the test samples slabs are presented in Fig. 3. Width of cracks opening on a slab surface was measured by means of microscope.

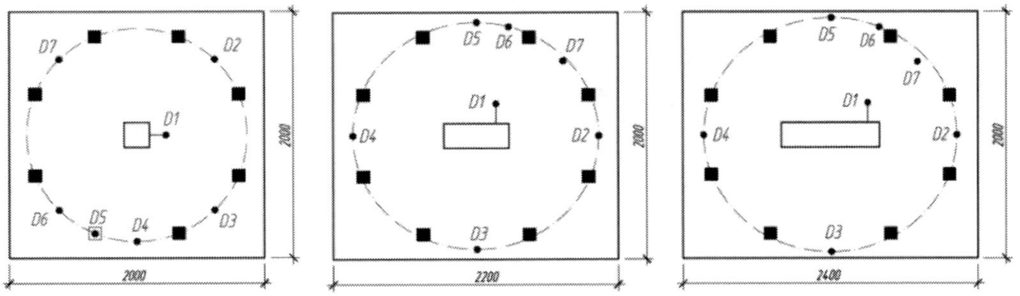

Fig. 3. Arrangement schemes of dial gauges and the points of application of concentrated forces.

The loading of the sample was carried out by steps of 5% of the theoretical failure load. The exposure time of the sample under load at each step was 15 minutes. At each loading stage was made fixing values of strain gauges, indications of dial gauges, schemes of cracks formation and development. CMR-4 sample in the process of testing is shown in Fig. 4.

Fig. 4. Testing of sample CMR-4.

Failure of the samples was brittle due to punching plate by column. The values of failure loads reached 272 kN, 281.3 kN, 328.4 kN for CMR-1, CMR-2.5 and CMR-4 samples, respectively.

The values of the plate deflections of sample CMR-1 before failure varied in the range 72.5 – 78 mm, and the smallest deflection being noted on axis perpendicular to external edge of a slab (D4), and the greatest – on a diagonal axis of a slab (D7). Maximum divergence of values of deflections onto the opposite sides of a slab (D3 – D7 and D6 – D2) doesn't exceed 4% that allows to draw a conclusion about almost symmetric vertical slab displacements of CMR-1 sample in relation to a column in fixing points (Fig. 3). The values of deflections plates samples CMR-2.5 and CMR-4 before the failure exceeded deflections plate of the sample CMR-1 for 4 - 8% on the short side of the plate, and 8 - 11% on the long side. In general, the greatest deflections are noted on the long sides and diagonal axes of test samples slabs. Average values of deflections of test samples slabs depending on the level of loading are presented in Fig. 5.

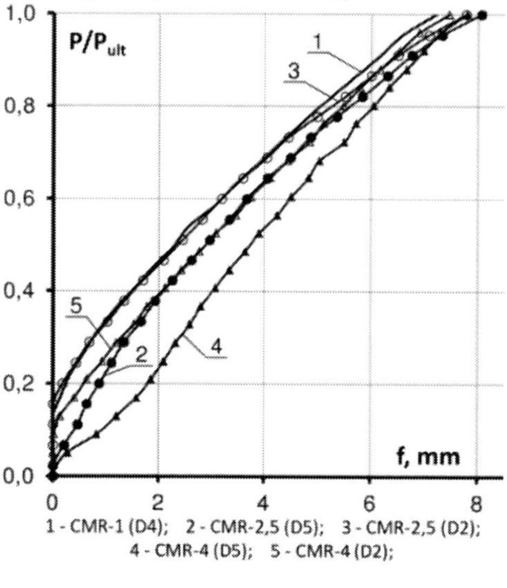

Fig. 5. Deflections of slabs CMR-1÷ CMR-4 samples depending on loading level.

The offered testing technique reinforced concrete slabs on punching allows to model slabs work under uniformly distributed loading by the application of concentrated forces. The results of these studies indicate satisfactory convergence of the qualitative picture of the power resistance of reinforced concrete slabs to punching under the action of a uniformly distributed load and the concentrated forces. New experimental data on the strength and deformability of flat reinforced concrete slabs under punching by column of rectangular section, depending on the ratio of its sides, were obtained.

References

1. J. Sagaseta, L. Tassinari, M. Ruiz, A. Muttoni, Eng. Struct. **77**, 17–33 (2014)
2. E.N. Kodysh, N.N. Trekin, D.N. Trekin, Concrete and reinforced concrete – glance at future: III All-Russian (II International) Conference on concrete and reinforced concrete, **I**, 69–75 (2014)
3. A.G. Tamrazyan, Industrial and civil engineering **7**, 26–27 (2012)
4. N.I. Karpenko, B.S. Sokolov, O.V. Radaykin, Industrial and civil engineering **1**, 28–30 (2013)
5. V.B. Filatov, A.A. Suvorov, Procedia Engineering **153**, 144–150 (2016)
6. V.B. Filatov, Proceedings of the Samara scientific center of the Russian Academy of Sciences **14(4-5)**, 1322–1324 (2012)
7. V.B. Filatov, Bulletin of civil engineers **5(40)**, 80–84 (2013)
8. V.B. Filatov, E.P. Bubnov, A.K. Alekseev, M.A. Bruskov, Z.Sh. Galyautdinov, A.G. Proydin, Proceedings of the Samara scientific center of the Russian Academy of Sciences **16(4-3)**, 646–649 (2014)
9. V.I. Travush, V.I. Kolchunov, N.V. Klyueva, Industrial and civil engineering **3**, 4–11 (2015)
10. V.I. Travush, S.G. Emel'yanov, V.I. Kolchunov, Industrial and civil engineering **7**, 20–27 (2015)
11. A.G. Tamrazyan,_Yu.N. Zvonov, Industrial and civil engineering **7**, 24–28 (2016)
12. V.B. Filatov, Yu.V. Zhiltsov, Proceedings of the Samara scientific center of the Russian Academy of Sciences **14(4-5)**, 1325–1328 (2012)
13. V.B. Filatov, E.V. Blinkova, Industrial and civil engineering **3**, 39–42 (2014)
14. V.B. Filatov, Concrete and reinforced concrete – glance at future: III All-Russian (II International) Conference on concrete and reinforced concrete **I**, 389–396 (2014)
15. N.N. Trekin, D.A. Pekin, Industrial and civil engineering **7**, 17–20 (2014)
16. A.N. Bolgov, A.Z. Sokurov, Concrete and reinforced concrete – glance at future: III All-Russian (II International) Conference on concrete and reinforced concrete **IV**,_139–149 (2014)
17. A.N. Bolgov, D.M. Yakimovich, Concrete and reinforced concrete – glance at future: III All-Russian (II International) Conference on concrete and reinforced concrete **I**, 224–233 (2014)
18. T. Susanto, H.K. Cheong, K.L. Kuang, J.Z. Geng, ACI Struct. J. **101(5)**, 678–687 (2004)
19. A. Muttoni, ACI Struct. J. **105(4)**, 440–450 (2008)
20. C.E. Broms, ACI Struct. J. **87(3)**, 292–304 (1990)

The analysis of deflections and cracking resistance in pre-stressed slabs of off-shuttering formation

Denis Panfilov[1*], *Irina* Piskareva[1] and *Julia* Fomina[1]

[1]Samara State Technical University, Institute of Architecture and Civil Engineering, Molodogvardeyskaya str. 194, 443001 Samara, Russia

Abstract. The paper presents an example of calculations of an assembled reinforced concrete hollow core slab of off-shuttering formation which has been reinforced with pre-stressed wire in its top and bottom areas. The numerical modelling of statically determinate pre-stressed assembled reinforced concrete hollow core slab exposed to short duration uniform loading is shown as a finite element model in PC "LIRA SAPR 2014 R3". This article contains information about the finite elements used, fixing conditions of the slab and of the order in which links between concrete and steel reinforcement were modelled. The calculation results are presented in the form of isofields of finite elements motion in this calculation model. Table 1 presents research results as a finite element model calculations according to the methodology given in specification documents.

1 Introduction

Assembled pre-stressed hollow-core slabs of off-shuttering formation (SSF) are widely used in building construction. The reasons for that lie in their advantage in manufacturing technology; in the possibility of covering comparatively big spans (up to 12 m); in economic use of materials and in the possibility of architectural re-planning of apartments. Their main peculiarity is the absence of crosswise reinforcement in vertical ribs, as well as the increased hardness due to use of pre-stressed steel reinforcement both in the low and in the top zones.

At the moment there are no specification documents in Russia regulating issues of design and calculation of slabs of off-shuttering formation. EN 1168-2012 [1] gives some recommendations of such slabs design. It is somehow not possible to use formula given in these recommendations as we didn't manage to find coherent coefficients there. Besides, the suggested methodology differs greatly from methodologies used by Russian standards. It is recommended to calculate these structures using SP 63.13330.2012 [2]. But since this regulation does not include requirements for calculation of such specific constructions as hollow core slabs, it also raises a number of questions concerning both modeling these products and peculiarities of their work at all stages of their life cycle, especially their service limit state.

* Corresponding author: panda-w800i@yandex.ru

2 Main Part

This paper continues previous research [5-16] and aims to analyze the existing methodology of calculating deflections and cracking resistance in SSF [2, 18[, to introduce an algorithm for calculating SSF in their service limit state using Mathcad 15 nonlinear deformation model [20, 21], to model SSF in PC "LIRA SAPR 2013» and to compare the obtained results.

For theoretical calculations we took a slab with 220 mm section height, 3580 mm length, 1500 mm width (see Figure 1).

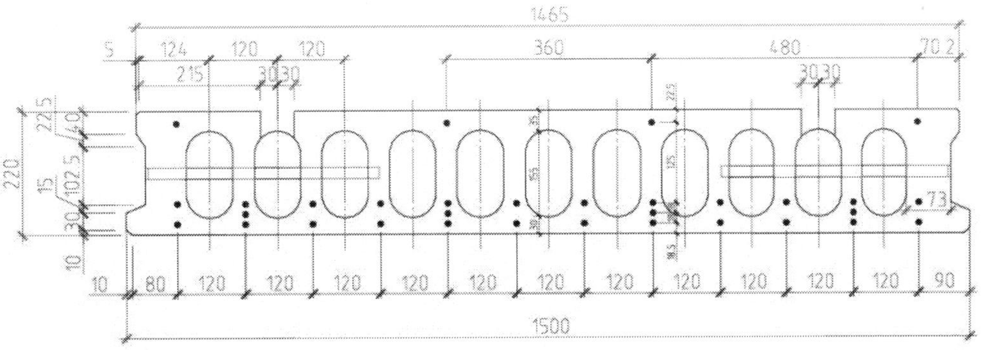

Fig. 1. Cross section of the slab and its reinforcement.

Samples are made of B25heavy concrete with the following mechanical strength characteristics: R_{bn}= 18.5 MPa; R_{btn} = 1.55 MPa.

The floor slab panel here is reinforced at the top (4Ø5) and bottom (28Ø5) zones with high tensile steel wire (GOST 7348 Vr1400 [3]), with the following strength and deformation characteristics: R_{sn}=1400 MPa; E_s= 2×105 MPa.

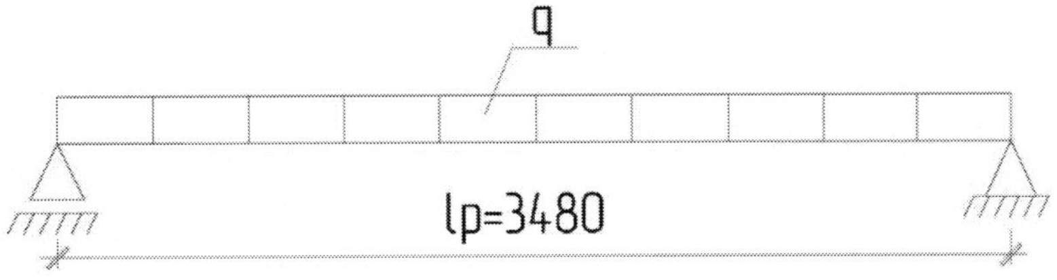

Fig. 2. Design model.

This floor slab panel is design to be used under uniformly distributed loads (in excess of its own mass) - 15, 7kN/m^2. The maximum possible load for the construction equals 44.0 kN/m^2.

In our calculations prestressing first and second losses were taken into account. Thus, final values of reinforcement tension were as follows: 720.0 MPa in the lower zone, 720.0 MPa in the upper zone.

The calculation was made on the stages of manufacture, transportation and use according to SP 63.13330.2012 [2] and a guidebook to SP-102 52-2004 [4].

Deflections of prestressed steel elements which have no cracks in concrete tensile zones are calculated as for an elastic body with rigidity, equal to the product of the modulus on the

 DOI: 10.1051/matecconf/20168601008

moment of the transformed section inertia. Calculation for areas with no cracks is made according to Formula 9.42 [2].

Loading of the structure was done in a stepwise fashion. The choice of control points is based on the terms of logging. Each stage does not exceed 10% of the reference load when checking strength and deformation.

Calculation results are given in Table 1.

Simulation of pre-stress condition was performed in PC "LIRA SAPR 2013» by means of temperature exposure.

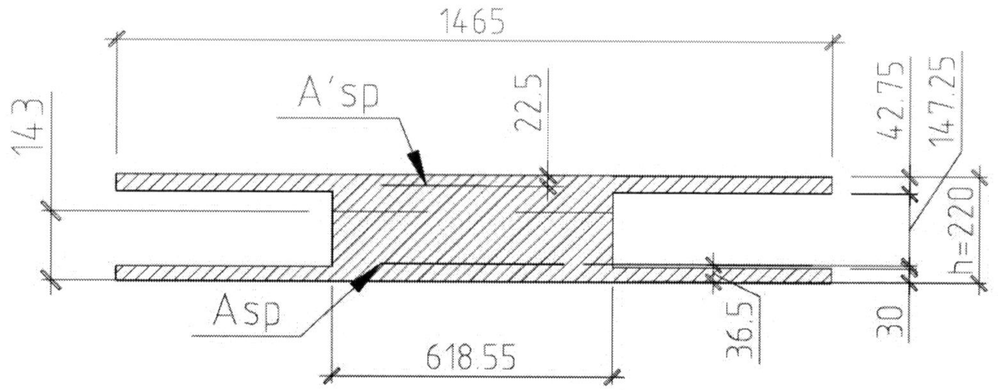

Fig. 3. The transformed section of a slab for calculations in its service limit state.

The slab is modeled with the use of core finite elements as a statically determinate bar in which two rows of pre-stressed high-strength wire were replaced by core finite elements equivalent in terms of area. The joint work of concrete and reinforced finite elements was modeled by the introduction of rigid inserts.

The construction was loaded as follows:

- uploading 1 – its own weight + uniformly distributed load;
- uploading 2 - with allowance for pre-stressed compressed (top) steel reinforcement;
- uploading 3 - with allowance for pre-stressed stretched (bottom) steel reinforcement;

It is not possible to choose proper value of temperature exposure for modelling pre-stressed condition of the whole structure as we do not yet know flexibility of all the elements in the reference direction. Therefore, at first we set arbitrary initial temperature and calculated the construction to obtain the value of the longitudinal effort in the stressed element. As efforts in the stressed cores under thermal effects are opposite to the efforts of prestressing, the calculation for the specified thermal effects was performed as for a separate uploading.

Automatic calculation results are given in Table 1.

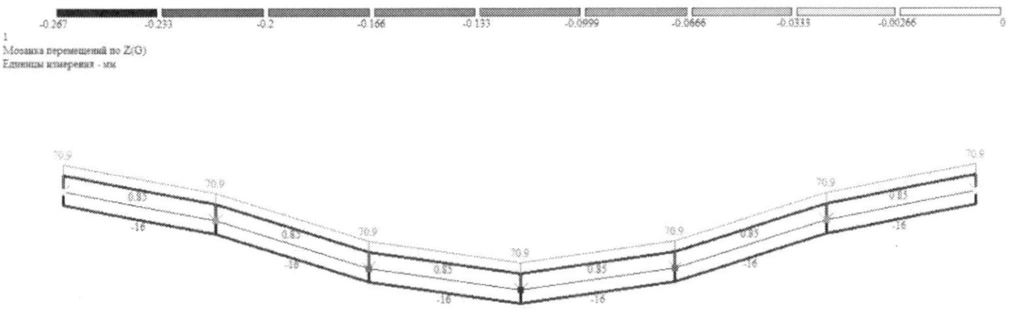

Fig. 4. Mosaic of movement along Z at maximum loading.

Table 1. The results of the study.

Loading №	Load applied, kN/m2	Amount of deflection, mm		Crack width, mm
		according to SP 63.13330.2012	PC "LIRA SAPR 2013»	
0	Own weight	-0.977	-1.17	-
1	4	-0.518	-0.952	-
2	8	-0.059	-0.83	-
3	12	0.400	-0.708	-
4	16	0.859	-0.586	-
5	20	1.318	-0.464	-
6	24	1.776	-0.343	-
7	28	2.235	-0.221	-
8	32	2.694	-0.108	-
9	36	3.153	0.023	-
10	40	3.612	0.145	-
11	44	6.854	0.267	0.041

3 Conclusion

Thus, according to the obtained data, we can conclude that the methodologies we analyzed show dramatic difference in results. It seems incorrect to carry out comparative analysis here.

Numerical modeling of pre-stressed slabs in a linear supply in PC "LIRA SAPR 2013» does not make possible to obtain pre-stressed state realization typical for such a structure. This multiparametric task can only be solved when using more complex software systems. It should also be noted that according to the results of calculations the slab has shown itself as a design that works without cracks in the extended zone, which is characterized by high rigidity of the element. In our further research we plan to perform non-linear numeric modeling and to test prototype samples to compare practical and empirical results. As a result we also plan to to introduce a refined methodology for calculating pre-stressed slabs of off-shuttering formation under uniformly distributed load with account of crack formation and nonlinear work of concrete.

References

1. *STB EN 1168-2012. Standards of Belarus, "Reinforced assembled units. Hollow core slabs"* (2012)
2. *SP 63.13330.2012. Concrete and reinforced concrete structures. Fundamental principles. Revised edition. SP 52-101-2003* (2012)
3. *GOST 7348-81. Carbon steel wire used for reinforcement of pre-stressed reinforced concrete structures. Specification* (1981)
4. *Handbook on design of pre-stressed concrete structures of heavy concrete (for SP-102 52-2004)* (2004)
5. D.A. Panfilov, N.A. Borodachev, Trad. and inn. in arch. and civ. eng., 70th All-Russia conf. proc., Samara State Un. of Arch. and Civ. Eng. **2**, 314-316 (2013)
6. D.A. Panfilov, E.N. Khudyaev, Act. Prob. of arch. and civ. eng. Int. sc. pract. conf.proc., 9-14 (2014)
7. D.A. Panfilov, A.A. Pischulev, Proc. Eng. **111**, 626-631 (2015)
8. D.A. Panfilov, A.A. Pischulev, V.V. Romanchikov, Proc. Eng. **153**, 531-536 (2016)
9. D.A. Panfilov, G.V. Murashkin, N.A. Borodachev, Trad. and inn. in arch. and civ. Eng., 69th All-Russia conf. **2,** 320-322 (2012)
10. D.A. Panfilov, A.A. Pischulev, Proc. Eng. **111**, 619-625 (2015)
11. G.V. Murashkin, V.G. Murashkin, D.A. Panfilov, Herald of Volga-region Dep. of Russian Academy of Arch. and Civ. Eng. **14**, 144-150 (2011)
12. V.G. Murashkin, D.A. Panfilov, A.I. Snegireva, Trad.s and inn. in arch.and civ. eng., 69th All-Russia conf. proc., Samara State Univ. of Arch. and Civ. Eng., 311-315 (2012)
13. D.A. Panfilov, V.G. Murashkin, Traditions and innovations in architecture and civil engineering, 70th All-Russia conference proceedings **2**, 311-314 (2013)
14. V.G. Murashkin, D.A. Panfilov, A.A. Suvorov, Traditions and innovations in architecture and civil engineering, 70th All-Russia conference proceedings **2**, 307-308 (2013)
15. G.V. Murashkin, D.A. Panfilov, V.G. Murashkin, Vestnik VolgGASU, **31(50)**, 388-391 (2013)
16. G.V. Murashkin, V.G. Murashkin, D.A. Panfilov, Int. Jour. for Comp. Civ. and Str. Eng. **8(4)** 89-95 (2012)
17. Yu.V. Veruzhsliy, V.I. Kolchunov, M.S. Barabash, Yu.V. Genzerskiy, *Computer technologies in designing reinforced concrete structures. Coursework engineering design* (2006)
18. S.V. Bosakov, Civil engineering: science and practice, scientific and technical journal pub. by the Ministry of architecture and civil engineering of Belarus Republic **6**, 41-46 (2010)
19. A.S. Gorodetskiy, V.S. Shmukler, A.V. Bondarev, Information technologies in calculations and design of building constructions, (NTU HPI, 2003)
20. D.S. Toshin, Town-planning, reconstruction and engineering support of Povolzhje cities and towns sustainable development, All-Russia conference proceedings, 312-313 (2004)
21. D.S. Toshin, Town-planning, reconstruction and engineering support of Povolzhje cities and towns sustainable development, II All-Russia conference proceedings, 224-229 (2009)

A flexible beam with corrugated web and its performance under bending: an experimental study

Dmitry Shlyakhin[1], and *Alexey* Lukin [1,*]

[1]Samara State Technical University, Institute of Architecture and Civil Engineering, 443001, Samara, Molodogvardeyskaya Str., 194, Russia

Abstract. The paper studies the state of stress of a corrugated web which is a part of a hinged beam taking bending moment. Numerical results in this study were obtained by using the finite elements method. The main parameters of the research are the magnitude of the current workload, geometrical parameters of the corrugation and the type of corrugating. While collecting data on geometry calculation models of beams with corrugated webs the authors used an authorial software «Gofro». Triangulation of a finite element grids was performed by GMSH preprocesser. After through analysis of each parameter, the researchers determined the degree of participation of the corrugated web of a beam which is taking bending moment, and proved the advantage of the sinusoidal shape of the corrugation.

1 Introduction

The basic element of all buildings and structures is usually a frame structure consisting of beams. In order to to reduce the consumption of steel in manufacturing such a structure perforated [1-5] and corrugated [6-10] constructions have been invented. Steel beams with corrugated web are now actively used in buildings and structures of different application [3, 11-14]. When a beam with corrugated web takes a bending moment, it is typical that the web is not included in the work, i.e. there is no normal tension σ_x [15, 16]. Some areas which are close to the girders are an expection here.

Researchers of Kazakh branch of CNIImetallokonstrukcii [17] empirically determined the influence of corrugation parameters (that is corrugation length and height) on the degree of participation of the corrugated web in taking bending moment in comparison with the flat web of the same thickness. All research carried out in Kazakhstan was aimed at corrugated webs of triangular shape. If the shape of corrugation is different (wavy or trapezoid) the corrugated web degree of participation in taking the bending moment may differ from the results for triangular corrugation. Another parameter that can affect the corrugated web is its thickness. Thickness for wavy corrugation is about 2-3 mm [18] and for triangular corrugation it is 3-10 mm [19]. Therefore, a study of the participation of corrugated walls when taking a bending moment is quite relevant.

*Corresponding author: a.o.lukin@yandex.ru

The purpose of this research is to study the state of stress of corrugated webs while taking a bending moment. For this purpose the authors worked out a finite-elemental (FE) model of a beam with corrugated web. The model makes it possible to analyze the impact of the magnitude of the external load, thickness and parameters of the corrugation on the degree of the web participation in bending, as well as to explore the work of a web along the length of one wave of corrugation.

2 Development of a Numerical Model of Beams with Corrugated Web

To achieve this aim the authors carried out a numerical experiment to analyze the state of stress in corrugated webs based on finite elements method (FEM). Numerical results were received by studying hinged beams with L =6m span which were loaded uniformly by distributed load of q intensity (See Figure 1). The cross-section parameters (Figure 1): h_w = 500 mm, b_f = 200 mm, t_f = 10 mm.

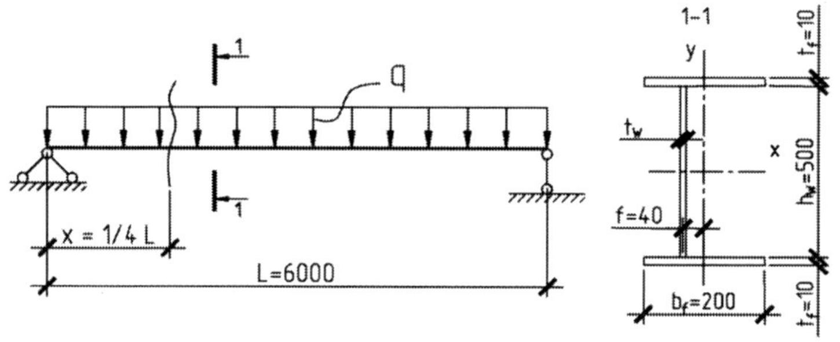

Fig. 1. Calculation model of a beam and cross-section area.

To analyze the stress state of the web two types of corrugation shape were taken: triangular and wavy. Finite elements number in the calculation model, as well as the geometry of corrugations (half wave length and height) are considered identical (see Figure 2): a = 150 mm and f = 40 mm, a/f = 150/40.

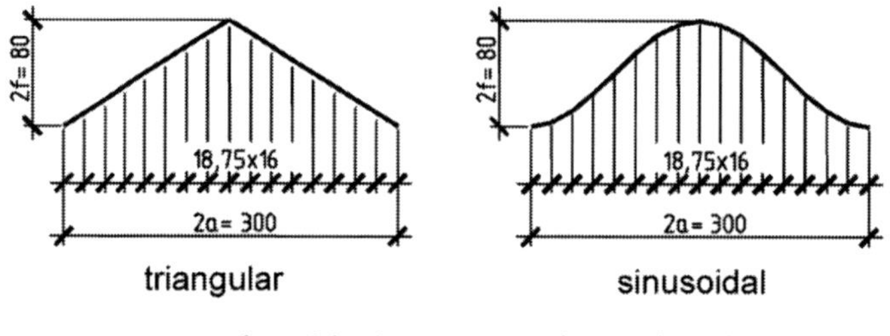

Fig. 2. Corrugations parameters and partitioning scheme of wave-length elements.

The calculated section to determine the height of the web is at a distance of ¼ of a span (half-wave here is 1.35 m to 1.5 m). The values of web participation in the work of the construction were determined graphically, as given below. To do this, the orthographic representation of direct stresses was placed across the marked grid of the web height.

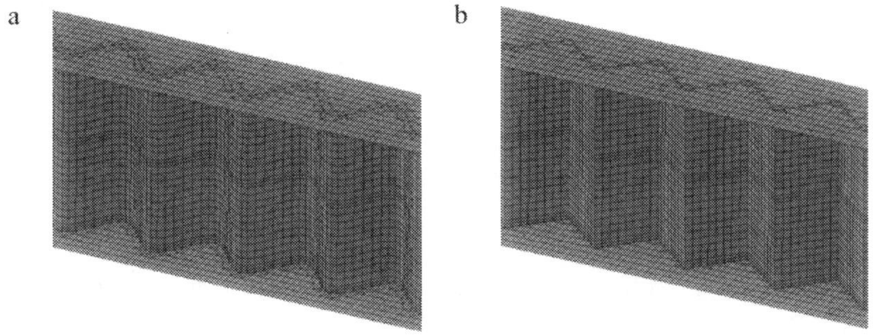

Fig. 3. EC model of beams: (a) with a wavy web; (b) with a triangular-shaped web.

Then the authors introduced calculation models for a FEM study and conducted a series of numerical experiments while using "LIRA" computational system. While collecting data on geometry calculation models of beams with corrugated webs the authors used an authorial software «Gofro» [20]. Triangulation of a finite element grids was performed by GMSH preprocessor [21]. The calculation model based on triangular and rectangular finite elements of universal enveloped type has six degrees of freedom. FE length is approximately 20 mm. Figure 2 shows a scheme of partitioning of wave-length elements FE network tends to concentrate in the places where the web is adjoined to flanges. Physical characteristics of finite elements correspond to those of S255 steel: R_y = 240 mPas, E = 206000 mPas. Calculations were performed in the elastic stage.

3 Analysis of the magnitude of the web taking a bending moment

Initially, the researches made an analysis of the web work and checked the dependency between the height of the part and the applied load (bending moment) with 3 mm web thickness. For this we took a beam with wavy outlines of corrugations (see Fig. 3 a) and different values of distributed load. A section was taken at the top of the waves (as shown in Table 1). The results showed that the load does not affect the degree of participation of web in its work. Figure 4 illustrates normal tensions along the web depth.

Table 1. Magnitude of the web work under different loading.

№	q, kN/m	$M_{1/4}$, kNm	a_w, mm
1	40	135	38
2	50	168.5	38
3	60	202.5	38
4	70	236.5	38
5	80	270	38

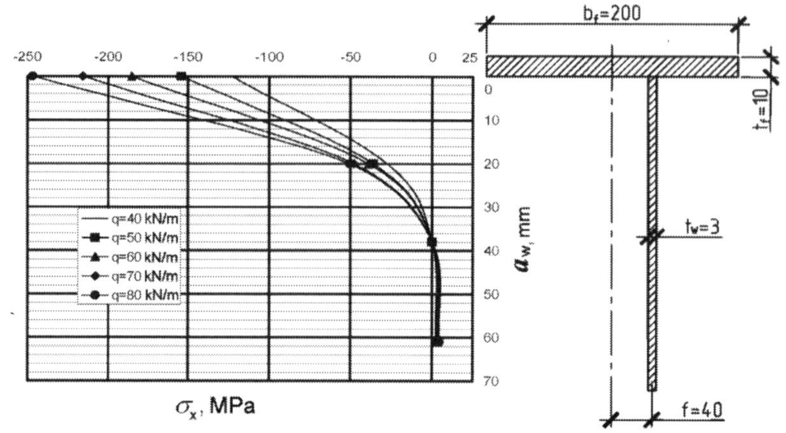

Fig. 4. Normal tensions along the web depth under different loading.

In the next phase of the research the work of the web along the length of one wave corrugation was studied. The height of the bending moment of the web was studied in 5 sections (see Figure 5). These following section are:

- two section at the tops of corrugation (1-1, 5-5);
- one section on the axis of the beam (3-3);
- wo sections at a distance $a/2$ form the top of corrugation (2-2, 4-4).

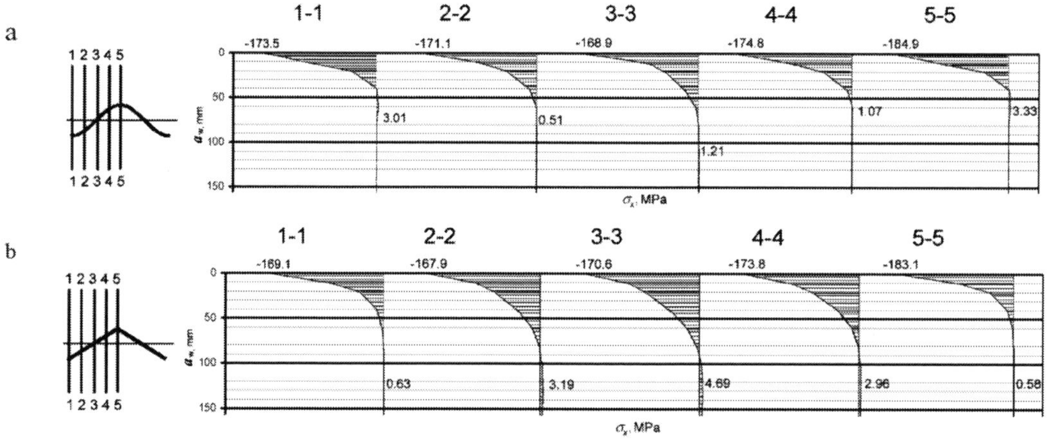

Fig. 5. The place of the section and its influence on the height of web work along the corrugation length: (a) sinusoidal shape of the corrugation; (b) triangular hape of the corrugation.

Calculation results revealed some peculiarities of the web work along the wave-length corrugation. For triangular-shaped corrugations the web bending moment within one wave changes in 1.5 times. For sinusoidal-shaped corrugations it changes from 1.5 times in t_w = web thickness is 1 ... 3 mm to 2 times when web thickness is t_w = 4 ... 10 mm. And for both shapes of corrugation the lowest value occurs on the tops of corrugation (fig. 5, section 1-1, 5-5), and the value is highest at the location of the axis of symmetry of the beam (fig. 5, section 3-3). It should be noted that for wavy webs (described by sinusoidal law) the change of the value takes place on a similar law describing the geometry of the web, but with the period 2 times less as in the previous case (fig. 5 a).

The authors also made an analysis of the influence of web thickness on the degree of participation of the web in the work. To calculate the influence of web thickness on the degree of participation of the corrugated web the researchers took constant parameters of length and height of the corrugation while changing web thickness within the range of 1 mm to 10 mm. Only the section on the top of corrugation was analyzed (section 1-1, fig. 5), because in this place the value of web participation is the lowest. In the calculations this circumstance gives some excess strength to the remaining sections of the corrugated web.

Table 2. Magnitude of the web work with different web thickness.

№	t_w, mm	$\bar{\lambda}_w$	a_{wsin}, mm	a_{wtrian}, mm	δ, %
1	1	17.07	30	60	100
2	2	8.53	34	61	79.4
3	3	5.67	38	61	60.5
4	4	4.26	41	61	48.8
5	5	3.41	46	62	34.8
6	6	2.85	51	63	23.5
7	7	2.44	53	64	20.8
8	8	2.13	55	65	18,2
9	9	1.89	57	70	22.8
10	10	1.71	59	76	28.8

The results given in Table 2 show that in triangular-shaped corrugations the part of the web taking the bending moment remains unchanged and begins to increase abruptly only with the web thicknesses of more than 8 mm (as Figure 6 shows). In wavy corrugations the height of the web depends on its thickness.

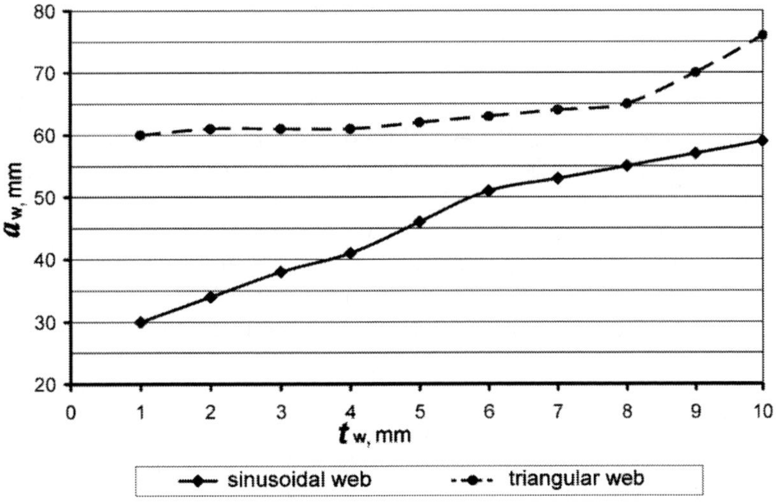

Fig. 6. The part of the web taking the bending moment.

Analyzing the results of the work of corrugated web of wavy and triangular shapes (Table 2), it should be noted that a triangular-shaped web work is more intense than a wavy web work (on condition they are of the same flexibility) $\bar{\lambda}_w$. Defining the difference between the values of the bending moment of a triangular-shaped web to a wavy web, it can be concluded that:

- in thin webs $\bar{\lambda}_w > 5.5$ ($t_w = 1, 2, 3$ mm) the difference is 60-100%;
- in medium webs it is $3.2 \leq \bar{\lambda}_w \leq 5.5$ ($t_w = 4, 5$ mm)-35-49%;
- in thick webs it is $\bar{\lambda}_w < 3.2$ ($t_w = 6, 7, 8, 9, 10$ mm) - 18-29%.

The last stage of the study was held to analyze the impact of the corrugation parameters on the degree of web participation in the work. To the corrugation parameters we refer half wave length a and half wave height f. The geometry of the corrugation is characterized by a/f ratio. Figure 2 shows that if corrugation height is $f = 0$, $a/f \rightarrow \infty$. In this case the web is flat and, consequently, its work equals $a_w = h_w/2$. Papers [23, 24] present calculations of beams with corrugated web, where corrugation is $a/f = 3.875$ and $a/f = 30$. These calculations show that the lower a/f is the smaller part of the web works at a bending moment.

4 Conclusion

The research yielded the following conclusions:

1. The height of the web under normal tension does not depend on the magnitude of the current load. It is determined by the geometrical parameters of the corrugation, web thickness and the corrugation type.
2. The degree of participation of the web differs within one half wave. In the tops of the corrugation the height of the web is 1.5-2 times lower than in the section located on the axis of symmetry.
3. Triangular-shaped web work is more intense than a wavy web work (on condition they are of the same flexibility): in thin webs the difference is 60-100%, in medium webs it is 35-49%, in thick webs it is 18-29%.

References

1. L.V. Katin, *Tradition and innovation in construction and architecture. Proceedings of the 71th of All-Russian scientific and technical conference on the results of R & D in 2013*, 845-848 (2014)
2. G.M. Ostrikov, Building mechanics and structures design **5**, 10-14 (1988)
3. I.S. Kholopov, A.O. Lukin, V.Yu. Alpatov, A.V. Solovjev, K.N. Gudkov, Building materials, equipment, XXI centurry technologies **12(155)**, 40-45 (2011)
4. A.V. Solovjev , I.S. Kholopov, A.O. Lukin, Indistrial and civil engineering **8**, 27-30 (2010)
5. I.S. Kholopov, A.A. Lukin, R.R. Valkaev, *Tradition and innovation in construction and architecture. Proceedings of the 72th of All-Russian scientific and technical conference on the results of R & D in 2014*, 15-17 (2015)
6. I.I. Krylov, A.N. Kretinin, News of Higher Educational Institutions. Construction **6**. 11–14 (2005)
7. I.L. Kuznetsov, A.A. Aktuganov, A.P. Trofimov, Kazan State University of Architecture and Engineering news **1(13)**. 117–121 (2010)
8. A.O. Lukin, V.Yu. Alpatov, D.D. Chernyshev, Vestnik SGASU, Town Planning and Architecture **2**, 4-9 (2016)
9. I.S. Kholopov, A.A. Lukin, P.N. Kozyrev, *Tradition and innovation in construction and architecture. Proceedings of the 72th of All-Russian scientific and technical conference on the results of R & D in 2014*, 68-71 (2015)
10. S.F. Pichugin, V.P. Chichulin, K.V. Chichulina, Resource-intensive materials, constructions, buildings and structures **23**, 390-395 (2012)
11. H. Pasternak, G. Kubieniec, Journal of Civil Engineering and Management **16(2)**, 166–171 (2010)
12. R. Aschinger, J. Lindner, Stahlbau **66(3)**, 136-142 (1997)
13. D.D. Zaborova, J.P. Dunaevskay, Construction of Unique Buildings and Structures **7(22)**, 36-53 (2014)

14. A.O. Lukin, A.A. Suvorov, Construction of Unique Buildings and Structures **2**, 45-67 (2016)
15. G.A. Azhermachev, Industrial and civil engineering **4**, 54–56 (1963)
16. A.V. Solovjev, A.O. Lukin, V.Yu. Alpatov, Indistrial and civil engineering **6**, 27-30 (2010)
17. G.M. Ostrikov, Yu.S. Maksimov, V.V. Dolinskiy, Building mechanics and structures design **1**, 68-70 (1983)
18. *TU U V.2.6-28.1-30653953-007.2007. I-beam corrugated lightweight. Design Guidelines* (2008)
19. *RDS RK 5.04-24-2006. Mix of fabricated I-Beam section of regular type and with corrugated walls meeting strenght properties of rolling section* (2007)
20. A.O. Lukin, Gofro, Certificate of computer programs of state registration, RU 2012660270 (2012)
21. C. Geuzaine, J.-F. Remacle,International Journal for Numerical Methods in Engineering **79(11)**, 1309-1331 (2009)
22. A.O. Lukin, Building mechanics and structures design **5**, 10-17 (2013)
23. M.I. Balzannikov, I.S. Kholopov, V.Yu. Alpatov, A.A. Lukin, Proc. Engineering **111**, 72-81 (2015)

Semianalytical analysis of shear walls with the use of discrete-continual finite element method. Part 1: Mathematical foundations

Pavel Akimov[1,2*], *Mojtaba* Aslami[3], *Marina* Mozgaleva[1], and *Zhiuli* Mskhalaya[1]

[1]National Research Moscow State University of Civil Engineering, Department of Applied Mathematics, 26, Yaroslavskoe Shosse, 129337 Moscow, Russia
[2]Russia Academy of Architecture and Construction Sciences, Presidium, 24, ul. Bolshaya Dmitrovka, 107031 Moscow, Russia
[3]Fasa university, Daneshjou blvd, Fasa, Fars province, Iran

Abstract. The distinctive paper is devoted to the two-dimensional semi-analytical solution of boundary problems of analysis of shear walls with the use of discrete-continual finite element method (DCFEM). This approach allows obtaining the exact analytical solution in one direction (so-called "basic" direction), also decrease the size of the problem to one-dimensional common finite element analysis. The resulting multipoint boundary problem for the first-order system of ordinary differential equations with piecewise constant coefficients is solved analytically. The proposed method is rather efficient for evaluation of the boundary effect (such as the stress field near the concentrated force). DCFEM also has a completely computer-oriented algorithm, computational stability, optimal conditionality of resultant system and it is applicable for the various loads at an arbitrary point or a region of the wall.

1 Introduction

Problems of structural analysis normally lead to complex two-dimensional and three-dimensional systems of equations, and solutions of these systems can be obtained only by numerical approaches [1-3]. Generally it is impossible to find the correct analytical (exact) solution [4, 5] while corresponding experimental researches are more expensive and non-complete. Recent developments in computer industry and mathematics resulted in emerging new branches in computational mechanics under the name of semi-analytical or discrete-continual methods [6-11]. Semi-analytical approaches allow obtaining the solution in correct analytical form, which is more efficient, especially in the zones of boundary effects. Results have normally fast variational behaviour there. Therefor accuracy and convergence of solutions obtained by numerical methods are often heavily dependent on the type of selected basis functions (shape functions) for approximating of unknowns, and the number of discretization nodes (elements) [12, 13]. Convergence in the areas of boundary effects, located near the concentered forces and stress concentrations (i.e. in the most critical areas)

* Corresponding author: pavel.akimov@gmail.com

are very slow for numerical approaches and weakly depends on the number of elements. Even, for example, if the convergence for displacements is relatively high, it is much smaller for internal strains and stresses. Besides, preliminary simplified analytical research of complicated problem can be very useful for understanding of solution properties or behaviour of structure. Applications of discrete-continual methods provides substantial reduction of size of problems, especially for multidimensional analysis. Fields of the applications of discrete-continual method comprises structures with regular (constant or piecewise constant) physical and geometrical parameters (characteristics) in one (so-called "basic") direction (for examples, tall buildings, tunnels, railroads, dams, foundations, etc.). Exact analytical evaluation of the behaviour (stress-strain state) of such structure in the basic direction is of paramount importance. Discrete-continual methods provide correct analytical solution of the considering problem along basic direction of structure and numerical (for instance, finite element) approximation along other directions. As a result, more accurate solutions can be obtained, especially in zones of boundary effect.

The distinctive paper is devoted to two-dimensional (plane stress condition) semi-analytical analysis of shear wall with the use of discrete-continual finite element method (DCFEM) [14, 15, 16]. It allows obtaining the exact analytical solution in one direction and decreasing the size of the problem to conventional one-dimensional finite element analysis.

2 Formulation of the problem

Shear walls, are vertical cantilevered lateral-force-resisting elements, which are normally used as supports for the roof and floor diaphragms, or for the transfer of loads into diaphragms (where the wall cannot continue to the foundation [17, 18]). Shear walls can be full-length walls or short wall sections that can occur at any location along a line of lateral force resistance. When a diaphragm is loaded by wind or seismic forces, the loads are transferred from the diaphragm into the top of the supporting shear walls or other boundary elements at each end of the diaphragm. Shear walls resist in-plane loads that are applied along its height. In many cases, shear walls can be simplified into several two-dimensional problem. Considering this point, a plane stress condition is used for the simulation within DCFEM. The common shear walls have a rectangular or a piecewise rectangular shape with long regular length and therefor can be solved efficiently by this method. Due to substantial height of the shear walls, the vertical direction is used for the analytical solution.

2.1 Continual operational formulation of the problem

Let x_2 be coordinate corresponding to basic direction; $x^b_{2,k}$, $k=1,\ldots,n_k$ are coordinates of boundary points i.e. cross-sections (including points of changes in physical and geometrical parameters). Operational formulation of corresponding resultant multipoint boundary problem of two-dimensional theory of elasticity [19] at extended domain [20], embordering considering structure, within DCFEM has the form:

$$\begin{cases} \mathbf{U}'_k = \widetilde{L}_k \mathbf{U}_k + \mathbf{S}_k, \quad x_3 \in (x^b_{3,k}, x^b_{3,k+1}), \quad k=1,\ldots,n_k-1 \\ \widetilde{B}^-_k \mathbf{U}_{k-1}(x^b_{3,k}-0) + \widetilde{B}^+_k \mathbf{U}_k(x^b_{3,k}+0) = \widetilde{\mathbf{g}}^-_k + \widetilde{\mathbf{g}}^+_k, \quad k=2,\ldots,n_k-1 \\ \widetilde{B}^+_1 \mathbf{U}_1(x^b_{3,1}+0) + \widetilde{B}^-_{n_k} \mathbf{U}_{n_k-1}(x^b_{3,n_k}-0) = \widetilde{\mathbf{g}}^+_1 + \widetilde{\mathbf{g}}^-_{n_k}, \end{cases} \tag{1}$$

$$\widetilde{L}_k = \begin{bmatrix} 0 & E \\ L^{-1}_{k,vv}(L_{k,uu}+C_k) & L^{-1}_{k,vv}\widetilde{L}_{k,uv} \end{bmatrix}; \ \mathbf{S}_k = -\begin{bmatrix} 0 \\ L^{-1}_{k,vv}\widetilde{\mathbf{F}}_k \end{bmatrix}; \ \mathbf{U}_k = \begin{bmatrix} \mathbf{u}_k \\ \mathbf{v}_k \end{bmatrix}; \tag{2}$$

$$L_{k,vv}=\begin{bmatrix}\bar{\mu}_k & 0\\ 0 & \bar{\lambda}_k+2\bar{\mu}_k\end{bmatrix};\quad L_{k,uv}=\begin{bmatrix}0 & \partial_1^*\bar{\lambda}_k\\ \partial_1^*\bar{\mu}_k & 0\end{bmatrix};\quad L_{k,uv}=\partial_1^*\begin{bmatrix}\bar{\lambda}_k+2\bar{\mu}_k & 0\\ 0 & \bar{\mu}_k\end{bmatrix}\partial_1; \tag{3}$$

$$C_k=(\theta_k+\delta_{\Gamma,k})\begin{bmatrix}c_{k,1} & 0\\ 0 & c_{k,2}\end{bmatrix};\quad \theta_k(x)=\begin{cases}1, & x\in\Omega_k\\ 0, & x\notin\Omega_k;\end{cases}\quad \delta_{\Gamma,k}(x)=\frac{\partial\theta_k}{\partial\mathbf{n}_k};\quad \widetilde{\mathbf{F}}_k=\theta_k\mathbf{F}_k+\delta_{\Gamma,k}\mathbf{f}_k; \tag{4}$$

$$\widetilde{L}_{k,uv}=L_{k,uv}-L_{k,vu};\quad L_{k,vu}=L_{k,uv}^*;\quad \mathbf{v}_k=\partial_2\mathbf{u}_k;\quad \mathbf{U}_k'=\partial_2\mathbf{U}_k; \tag{5}$$

Ω is the domain, occupied by structure; $x=(x_1,x_2)$; x_1,x_2 are coordinates (x_2 corresponds to basic dimension); l_2 is the length of structure along basic dimension, $x_2\in[0,l_2]$; Ω_k, $k=1,\ldots,n_k-1$ are corresponding fragments of domain Ω with boundaries Γ_k, obtained by separation from domain Ω by cross-sections $x_2=x_{2,k}^b$ and $x_2=x_{2,k+1}^b$; ω_k, $k=1,\ldots,n_k-1$ are extended domains, embordering fragments Ω_k, $k=1,\ldots,n_k-1$; $\theta_k=\theta_k(x_1,x_2)$ is the characteristic function of domain Ω_k; $\delta_{\Gamma,k}=\delta_{\Gamma,k}(x_1,x_2)$ is the delta-function of border $\Gamma_k=\partial\Omega_k$ [21]; $\mathbf{n}_k=[n_{k,1}\ n_{k,2}]^T$ is unit normal vector of domain boundary $\Gamma_k=\partial\Omega_k$; $\mathbf{u}_k$, $k=1,\ldots,n_k-1$ is the unknown vector of displacements in domain Ω_k; $\widetilde{B}_k^-,\widetilde{B}_k^+$, $k=2,\ldots,n_k-1$, $\widetilde{B}_1^+$, $\widetilde{B}_{n_k}^-$ are matrices (operators) of boundary conditions of the sixth order (x_3-independent); $\widetilde{\mathbf{g}}_k^-,\widetilde{\mathbf{g}}_k^+$, $k=2,\ldots,n_k-1$, $\widetilde{\mathbf{g}}_1^+$, $\widetilde{\mathbf{g}}_{n_k}^-$ are right-side vectors of boundary conditions of the sixth order (x_3-independent); $\overline{\overline{F}}_k$ is the right-side vector in domain Ω_k; $\mathbf{F}_k$ is the vector of body forces in domain Ω_k; $\mathbf{f}_k$ is the boundary traction vector in domain Ω_k; $\bar{\lambda}_k$, $\bar{\mu}_k$ are Lame coefficients of material in domain Ω_k; C_k is the matrix of elastic parameters of the supports (if any); $c_{k,i}$ is the coefficient of resistance in the direction of the axis Ox_i; $\partial_k=\partial/\partial x_k$, $\partial_k^*=-\partial/\partial x_k$, $k=1,2$; $\mathbf{v}_k=\partial_2\mathbf{u}_k=\mathbf{u}_k'$; $\mathbf{v}_k'=\partial_3\mathbf{v}_k$.

2.2 Discrete-Continual operational formulation of the problem

DCFEM presupposes finite element approximation of extended domain along directions of structure perpendicular to the basic direction, while along basic direction problem remain continual (thus extended domain is divided into discrete-continual finite elements). Resultant multipoint boundary problem for the first-order system of ordinary differential equations with piecewise-constant coefficients within DCFEM [14] has the form:

$$\begin{cases}\mathbf{y}_k^{(1)}-A_k\mathbf{y}_k=\mathbf{f}_k,\ x_3\in(x_{3,k}^b,x_{3,k+1}^b),\ k=1,2,\ldots,n_k-1\\ B_k^-\mathbf{y}_{k-1}(x_{3,k}^b-0)+B_k^+\mathbf{y}_k(x_{3,k}^b+0)=\mathbf{g}_k^-+\mathbf{g}_k^+,\ k=2,\ldots,n_k-1\\ B_1^+\mathbf{y}_1(x_{3,1}^b+0)+B_{n_k}^-\mathbf{y}_{n_k-1}(x_{3,n_k}^b-0)=\mathbf{g}_1^++\mathbf{g}_{n_k}^-,\end{cases} \tag{6}$$

where $N-1$ is the number of discrete-continual finite elements; A_k, $k=1,2,\ldots,n_k-1$ are matrices of constant coefficients of order $n=4N$ (discrete analogs of operators $\widetilde{L}_k$, $k=1,2,\ldots,n_k-1$); $\mathbf{f}_k$, $k=1,2,\ldots,n_k-1$ are vectors of size $n=4N$ (discrete analogs of vector functions $\mathbf{S}_k$, $k=1,2,\ldots,n_k-1$); B_k^-,B_k^+, $k=2,\ldots,n_k-1$, B_1^+, $B_{n_k}^-$ are matrices of boundary conditions of order $n=4N$ (discrete analogs of $\widetilde{B}_k^-,\widetilde{B}_k^+$, $k=2,\ldots,n_k-1$, $\widetilde{B}_1^+$,

$\widetilde{B}^-_{n_k}$); $\mathbf{g}^-_k, \mathbf{g}^+_k$, $k = 2,\ldots,n_k - 1$, $\mathbf{g}^+_1$, $\mathbf{g}^-_{n_k}$ are right-side vectors of boundary conditions of size $n = 4N$ (discrete analogs of $\widetilde{\mathbf{g}}^-_k, \widetilde{\mathbf{g}}^+_k$, $k = 2,\ldots,n_k - 1$, $\widetilde{\mathbf{g}}^+_1$, $\widetilde{\mathbf{g}}^-_{n_k}$);

$$\mathbf{y}_k = \mathbf{y}_k(x_2) = [\,\mathbf{u}^T_k(x_2)\;\;\mathbf{v}^T_k(x_2)\,]^T\,; \tag{7}$$

$$\mathbf{u}_k = \mathbf{u}_k(x_2) = [\,(\mathbf{u}^{(k,1)}_n)^T\;(\mathbf{u}^{(k,2)}_n)^T\;\ldots\;(\mathbf{u}^{(k,N)}_n)^T\,]^T\,; \tag{8}$$

$$\mathbf{v}_k = \mathbf{v}_k(x_2) = [\,(\mathbf{v}^{(k,1)}_n)^T\;(\mathbf{v}^{(k,2)}_n)^T\;\ldots\;(\mathbf{v}^{(k,N)}_n)^T\,]^T\,; \tag{9}$$

$$\mathbf{u}^{(k,p)}_n = \mathbf{u}^{(k,p)}_n(x_2) = [\,u^{(k,p)}_1\;u^{(k,p)}_2\,]^T,\;\; p = 1,2,\ldots,N\,; \tag{10}$$

$$\mathbf{v}^{(k,p)}_n = \mathbf{v}^{(k,p)}_n(x_2) = [\,v^{(k,p)}_1\;v^{(k,p)}_2\,]^T,\;\; p = 1,2,\ldots,N\,; \tag{11}$$

$u^{(k,p)}_i = u^{(k,p)}_i(x_2)$, $p = 1,2,\ldots,N$ are components of displacement u_i in the node with coordinate (x^p_1, x_2) on the interval $x_2 \in (x^b_{2,k}, x^b_{2,k+1})$.

3 Analytical solution of resultant multipoint boundary problem

Solution of problem (6) is accentuated by numerous factors. They include boundary effects (stiff systems) and considerable number of differential equations (several thousands). Moreover, matrices of coefficients of a system normally have eigenvalues of opposite signs and corresponding Jordan matrices are not diagonal. Special method of solution of multipoint boundary problems for systems of ordinary differential equations with piecewise constant coefficients in structural analysis has been developed. Not only does it overcome all difficulties mentioned above but its major peculiarities also include universality, computer-oriented algorithm [22, 23], computational stability, optimal conditionality of resultant systems and partial Jordan decomposition of matrix of coefficient, eliminating necessity of calculation of root (principal) vectors [24].

3.1 Construction of partial Jordan decomposition

Jordan decomposition of matrix A_k has the form [25-30]

$$A_k = T_k J_k T_k^{-1},\;\; \text{where}\;\; J_k = \{J_{k,1}, J_{k,2}, \ldots, J_{k,u_k}\}\,; \tag{12}$$

T_k is the matrix of order n , which columns are eigenvectors and root vectors of matrix A_k ; J_k is Jordan matrix of order n ; $J_{k,p}$ is Jordan cell corresponding to eigenvalue $\lambda_{k,p}$; $\dim J_{k,p} = m_{k,p}$.

As we have already mentioned above, specificity of problems of structural analysis comprises in presence of multiple eigenvalues of matrix A_k and consequently in necessity of calculation of root vectors. However at the present time there are no effective numerical method of calculation of Jordan decomposition in the general case [23, 24]. Meanwhile the number of multiple eigenvalues in the considering type of problems is normally limited. Besides these multiple eigenvalues are generally zeros. In this connection special alternative approach to solution has been developed.

Partial Jordan decomposition is based on computation of right and left eigenvectors of matrix A_k.

$$A_k = A_{k,1} + A_{k,2}, \quad \text{where} \quad A_{k,1} = T_{k,1} J_{k,1} \widetilde{T}_{k,1}; \quad A_{k,2} = A_k - A_{k,2}; \tag{13}$$

$T_{k,1}$ is the matrix containing right eigenvectors corresponding to non-zero eigenvalues of matrix A_k; $\widetilde{T}_{k,1}$ is the matrix containing left eigenvectors corresponding to non-zero eigenvalues of matrix A_k; $J_{k,1}$ is diagonal Jordan matrix corresponding to non-zero eigenvalues of matrix A_k; $A_{k,2}$ is the part of matrix A_k corresponding to prime and multiple zero eigenvalues. It is necessary to note here that matrices $T_{k,1}$ and $\widetilde{T}_{k,1}$ in general case are rectangular.

3.2 Construction of fundamental matrix-function of system of equations

Eigenvalues $\lambda_{k,p}$, $p = 1,...,u_k$ are renumbered according to the condition

$$\left.\begin{array}{l} \forall \lambda_{k,p}, \;\; p = 1,..., l_k \;\; \exists m_{k,p} = 1 \\ \forall \lambda_{k,p}, \;\; p = l_k + 1,...,u_k \;\; \exists m_{k,p} > 1 \end{array}\right\}; \quad l_k = \sum_{p=1}^{u_k} \delta_{1,m_{k,p}}, \tag{14}$$

where $\delta_{i,j}$ is Kronecker delta.

Due to distinctive procedure, we should properly modify matrices $T_{k,1}$, $\widetilde{T}_{k,1}$ and $J_{k,1}$.

Let $P_{k,1}$ and $P_{k,2}$ be projectors to subspaces of left and right eigenvectors and root vectors of matrix A_k corresponding to non-zero and zero eigenvalues. They may be denoted as (E is identity matrix)

$$P_{k,1} = T_{k,1}(\widetilde{T}_{k,1} T_{k,1})^{-1} \widetilde{T}_{k,1}; \quad P_{k,2} = E - P_{k,1}. \tag{15}$$

After sorting and biorthogonalization of eigenvectors and eigenvalues fundamental matrix-function $\varepsilon_k(x)$ of system from (6) for arbitrary k is constructed in the special form convenient for problems of structural mechanics [16]

$$\varepsilon_k(x) = T_{k,1} \widetilde{\varepsilon}_{k,0}(x) \widetilde{T}_{k,1} + \chi(x,0)[P_{k,2} + \sum_{j=1}^{m_{k,\max}-1} \frac{x^j}{j!} A_{k,2}^j], \tag{16}$$

where $\quad m_{k,\max} = \max\limits_{l \le i \le u_k} m_{k,i}; \quad \chi(x, \lambda_{k,p}) = \begin{cases} \text{sign}(x)\theta(-\text{Re}(\lambda_{k,p})x), & \lambda_{k,p} \neq 0 \\ 0.5 \cdot \text{sign}(x), & \lambda_{k,p} = 0; \end{cases} \qquad (17)$

$$\text{sign}(x) = \begin{cases} 1, & x > 0 \\ -1, & x < 0; \end{cases} \qquad \theta(x) = \begin{cases} 1, & x > 0 \\ 0, & x < 0; \end{cases} \tag{18}$$

$$\widetilde{\varepsilon}_{k,0}(x) = diag\{\chi(x, \lambda_{k,1}) \exp(\lambda_{k,1} x), \; ..., \; \chi(x, \lambda_{k,l_k}) \exp(\lambda_{k,l_k} x)\}. \tag{19}$$

It should be stated that the sum in the right side of (16) contains four or lower components and corresponds to so-called “beam” part of solution of system.

3.3 General solution of the multipoint boundary problem

Solution of considering problem (6) on the interval (x_k^b, x_{k+1}^b) is defined by formula

$$\mathbf{y}_k(x) = (\varepsilon_k(x - x_k^b) - \varepsilon_k(x - x_{k+1}^b))\mathbf{C}_k + \varepsilon_k * \mathbf{f}_k, \quad x \in (x_k^b, x_{k+1}^b), \tag{20}$$

where $\mathbf{C}_k$ is the vector of constants of order n; $*$ is convolution notation [21];

$$\bar{f}_k(x) \equiv f(x)\theta(x, x_k^b, x_{k+1}^b); \quad \theta(x, x_k^b, x_{k+1}^b) = \begin{cases} 1, & x \in (x_k^b, x_{k+1}^b) \\ 0, & x \notin (x_k^b, x_{k+1}^b). \end{cases} \tag{21}$$

We can rewrite (20) in the form

$$\mathbf{y}_k(x) = E_k(x)\mathbf{C}_k + \mathbf{S}_k, \quad x \in (x_k^b, x_{k+1}^b); \tag{22}$$

$$E_k(x) = \varepsilon_k(x - x_k^b) - \varepsilon_k(x - x_{k+1}^b); \quad \mathbf{S}_k(x) = \varepsilon_k * \mathbf{f}_k. \tag{23}$$

Substituting (22) in boundary conditions (6) and taking into account that

$$E_{k-1}(x_k^b - 0) = \varepsilon_{k-1}(h_{k-1}^b) - \varepsilon_{k-1}(-0), \ k = 2,...,n_k; \quad h_k^b = x_{k+1}^b - x_k^b, \ k = 1,...,n_k - 1; \tag{24}$$

$$E_k(x_k^b + 0) = \varepsilon_k(+0) - \varepsilon_k(-h_k^b), \ k = 1,...,n_k - 1; \tag{25}$$

we have the following system of linear algebraic equations for $\mathbf{C}_k$, $k = 1,...,n_k - 1$

$$K\mathbf{C} = \mathbf{G}, \tag{26}$$

where matrix K can be divided into so-called main K^0 and additional K^1 members,

$$K^0 = \begin{bmatrix} K_{1,1}^0 & 0 & 0 & .. & 0 & K_{1,n_k-1}^0 \\ K_{2,1}^0 & K_{2,2}^0 & 0 & .. & 0 & 0 \\ 0 & K_{3,2}^0 & K_{3,3}^0 & \ldots & 0 & 0 \\ \ldots & \ldots & \ldots & \ldots & \ldots & \ldots \\ 0 & 0 & 0 & .. & K_{n_k-1,n_k-2}^0 & K_{n_k-1,n_k-1}^0 \end{bmatrix}; \tag{27}$$

$$K^1 = \begin{bmatrix} K_{1,1}^1 & 0 & 0 & .. & 0 & K_{1,n_k-1}^1 \\ K_{2,1}^1 & K_{2,2}^1 & 0 & .. & 0 & 0 \\ 0 & K_{3,2}^1 & K_{3,3}^1 & \ldots & 0 & 0 \\ \ldots & \ldots & \ldots & \ldots & \ldots & \ldots \\ 0 & 0 & 0 & .. & K_{n_k-1,n_k-2}^1 & K_{n_k-1,n_k-1}^1 \end{bmatrix}; \tag{28}$$

$$K_{k,k-1}^0 = -B_k^- \varepsilon_{k-1}(-0); \quad K_{k,k}^0 = B_k^+ \varepsilon_k(+0); \quad K_{1,1}^0 = B_1^+ \varepsilon_1(+0); \quad K_{1,n_k-1}^0 = -B_{n_k}^- \varepsilon_{n_k-1}(-0); \tag{29}$$

$$K_{k,k-1}^1 = B_k^- \varepsilon_{k-1}(h_{k-1}^b); \quad K_{k,k}^1 = -B_k^+ \varepsilon_k(-h_k^b); \quad K_{1,1}^1 = -B_1^+ \varepsilon_1(-h_1^b); \quad K_{1,n_k-1}^1 = B_{n_k}^- \varepsilon_{n_k-1}(h_{n_k-1}^b); \tag{30}$$

$$\mathbf{G} = [\mathbf{G}_1^T \ \mathbf{G}_2^T \ \ldots \ \mathbf{G}_{n_k-1}^T]^T; \quad \mathbf{C} = [\mathbf{C}_1^T \ \mathbf{C}_2^T \ \ldots \ \mathbf{C}_{n_k-1}^T]^T; \tag{31}$$

$$\mathbf{G}_1 = \mathbf{g}_1^+ + \mathbf{g}_{n_k}^- - B_1^+ \mathbf{S}_1(x_1^b + 0) - B_{n_k}^- \mathbf{S}_{n_k - 1}(x_{n_k}^b - 0); \tag{32}$$

$$\mathbf{G}_k = \mathbf{g}_k^- + \mathbf{g}_k^+ - B_k^- \mathbf{S}_{k-1}(x_k^b - 0) - B_k^+ \mathbf{S}_k(x_k^b + 0), \ \ k = 2, \ldots, n_k - 1 . \tag{33}$$

Symbol $\otimes$ imply direct product of matrices. It is necessary to note that matrices $\varepsilon_k(+0)$ and $\varepsilon_k(-0)$ are independent of x.

We find it vital to note that diagonal blocks of matrix K are practically singular. This fact leads to several problems. Iterative methods of solution can't be applied in particular. Gaussian elimination method with pivoting is required. It is useful to specify ways of disposal of this disadvantage.

Let us transform considering system of equation as follows: each equation of system, since the first (and finishing next to last), we will replace with the sum of this equation with the subsequent (instead of the initial first equation we take the sum initial the first with initial the second, instead of initial the second – the sum initial the second with initial the third and so on). Instead of initial last equation we take the sum of the initial last with initial the first. Finally we removed the singularity mentioned above. Finally we have:

$$K = \begin{bmatrix} \widetilde{K}_{1,1} & \widetilde{K}_{1,2} & 0 & 0 & .. & 0 & 0 & \widetilde{K}_{1,n_k-1} \\ \widetilde{K}_{2,1} & \widetilde{K}_{2,2} & \widetilde{K}_{2,3} & 0 & .. & 0 & 0 & 0 \\ 0 & \widetilde{K}_{3,2} & \widetilde{K}_{3,3} & \widetilde{K}_{3,4} & \ldots & 0 & 0 & 0 \\ \ldots & \ldots & \ldots & \ldots & \ldots & \ldots & \ldots & \ldots \\ \widetilde{K}_{n_k-1,1} & 0 & 0 & 0 & .. & 0 & \widetilde{K}_{n_k-1,n_k-2} & \widetilde{K}_{n_k-1,n_k-1} \end{bmatrix}. \tag{34}$$

The Reported study was Funded by Government Program of the Russian Federation "Development of science and technology" (2013-2020) with in Program of Fundamental Researches of Ministry of Construction, Housing and Utilities of the Russian Federation and Russian Academy of Architecture and Construction Sciences, the Research Project 7.1.1.

3 Conclusions

Development, research and verification of correct mathematical models and methods of structural mechanics are the most important aspects of ensuring safety of buildings and complexes. The field of application of discrete-continual finite element method (DCFEM) comprises structures with regular (in particular, constant or piecewise constant) physical and geometrical parameters in some dimension (so-called "basic" direction (dimension)). Considering problems remain continual along "basic" direction while along other directions DCFEM presupposes mesh approximation. DCFEM has completely computer-oriented algorithm, computational stability, optimal conditionality of resultant system and applicable for the various loads at an arbitrary point or a region of the shear wall. This method allows increasing the accuracy of the solution (especially in solution of high-rise shear walls) and provides reduction of the computational efforts.

References

1. O.C. Zienkiewicz, R.L. Taylor, *The finite element method: Solid mechanics. Vol. 2* (Butterworth-heinemann, 2014)

2. R. LeVeque, *Finite Difference Methods for Ordinary and Partial Differential Equations: Steady-State and Time-Dependent Problems* (Classics in Applied Mathematics, SIAM, 2007)
3. S.A. Sauter, C. Schwab, *Boundary element methods* (Springer Berlin Heidelberg, 2011)
4. J. Kevorkian, *Partial differential equations: Analytical solution techniques* (Springer, 2000)
5. A.B. Zolotov, P.A. Akimov, V.N. Sidorov, M.L. Mozgaleva, *Numerical and analytical methods for design of structures* (ASV Publishing House, 2009)
6. W.-X. Zhong, X.-X. Zhong, Computers & Structures **37(6)**, 993-1004 (1990)
7. A.B. Zolotov, P.A. Akimov, M.L. Mozgaleva, *Multilevel discrete and discrete-continual variational-difference methods* (ASV Publishing House, 2013)
8. P.A. Akimov, M.L. Mozgaleva, Applied Mechanics and Materials **351-352**, 13-16 (2013)
9. P.A. Akimov, M.L. Mozgaleva, Applied Mechanics and Materials **405-408**, 3165-3168 (2013)
10. P.A. Akimov, M.L. Mozgaleva, M. Aslami, O.A. Negrozov, Applied Mechanics and Materials **670-671**, 720-723, (2014)
11. P.A. Akimov, M.L. Mozgaleva, M. Aslami, O.A. Negrozov, Applied Mechanics and Materials **670-671**, 724-727, (2014)
12. U.M. Ascher, R.M. Mattheij, R.D. Russell, *Numerical Solution of Boundary Value Problems for Ordinary Differential Equations* (Classics in Applied Mathematics, SIAM, 1995).
13. I.M. Smith, *Programming the Finite Element Method* (John Wiley & Sons Ltd, 2004)
14. P.A. Akimov, A.M. Belostosky, M.L. Mozgaleva, M. Aslami, O.A. Negrozov, Advanced Materials Research **1040**, 664-669 (2014)
15. M.H. Sadd, *Elasticity: theory, applications, and numerics* (Academic Press, 2014)
16. A.B. Zolotov, P.A. Akimov, V.N. Sidorov, M.L. Mozgaleva, *Mathematical Methods in Structural Mechanics* (ASV Publishing House, 2008)
17. R.T. Malone, R. Rice, *The Analysis of Irregular Shaped Structures Diaphragms and Shear Walls* (McGraw Hill Professional, 2011)
18. M.Y.H. Bangash, T. Bangash, *Elements of spatial structures: analysis and design* (Thomas Telford, 2003)
19. D.S. Chandrasekharaiah, L. Debnath, *Continuum mechanics* (Elsevier, 2014).
20. P.A. Akimov, M.L. Mozgaleva, Applied Mechanics and Materials **580-583**, 2898-2902 (2014)
21. P. Teodorescu, W.W. Kecs, A. Toma, *Distribution Theory: With Applications in Engineering and Physics* (John Wiley & Sons, 2013)
22. S. Blair-Chappell, A. Stokes, *Parallel Programming with Intel Parallel Studio XE* (John Wiley & Sons, 2012)
23. M. Metcalf, J. Reid, M. Cohen, *Modern FORTRAN Explained (Numerical Mathematics and Scientific Computation)* (Oxford University Press, 2011)
24. P.A. Akimov, V.N. Sidorov, Advanced Materials Research **250**, 3652-3655 (2011)
25. R. Piziak, P.L. Odell, *Matrix theory: From generalized inverses to Jordan form* (Chapman and Hall/CRC, 2007)
26. S.H. Weintraub, Synthesis Lectures on Mathematics and Statistics **2(1)**, 1-108 (2009)
27. J.N. Franklin, *Matrix theory* (Courier Corporation, 2012).
28. F. Zhang, *Matrix theory: basic results and techniques* (Springer Science & Business Media, 2011).
29. A. Bjorck, *Numerical methods in matrix computations* (Springer, 2015).
30. D. Poole, *Linear algebra: A modern introduction* (Cengage Learning, 2014).

Semianalytical analysis of shear walls with the use of discrete-continual finite element method. Part 2: Numerical examples, future development

Pavel Akimov[1,2*], *Mojtaba* Aslami[3], and *Marina* Mozgaleva[1]

[1]National Research Moscow State University of Civil Engineering, Department of Applied Mathematics, 26, Yaroslavskoe Shosse, 129337 Moscow, Russia
[2]Russia Academy of Architecture and Construction Sciences, Presidium, 24, ul. Bolshaya Dmitrovka, 107031 Moscow, Russia
[3]Fasa university, Daneshjou blvd, Fasa, Fars province, Iran

Abstract. The distinctive paper is devoted to the two-dimensional semi-analytical solution of boundary problems of analysis of shear walls with the use of discrete-continual finite element method (DCFEM). This approach allows obtaining the exact analytical solution in one direction (so-called "basic" direction), also decrease the size of the problem to one-dimensional common finite element analysis. Two numerical examples of structural analysis with the use of DCFEM are considered, conventional finite element method (FEM) is used for verification purposes. The presented examples show some of the advantages of the suggested approach to semianalytical analysis of the shear wall. Future development of DCFEM, particularly associated with multigrid approach, is under consideration as well.

1 Numerical examples

Numerous examples have been considered for illustrating the efficiency of discrete-continual finite element method (DCFEM) [1-8] and two of them are presented below. Corresponding software has been developed with the use of Intel Parallel Studio XE [9] (FORTRAN programming language [10]). Conventional finite element method (FEM) has been used for verification purposes.

1.1 Semianalytical solution of the shear wall subjected to horizontal forces

The end-column rectangular $3.6\times7.0\,\mathrm{m}$ concrete shear wall, subjected to two horizontal forces, is considered (Fig. 1). Module of elasticity and Poisson ratio of the wall are equal to $E_{wall} = 26500$ Mpa and $\nu = 0.3$ respectively [11]. For the simulation of module of elasticity of the end column in two-dimensional model, the module of elasticity of the wall is multiplied to the ratio of the thickness of the column with respect to the wall.

* Corresponding author: pavel.akimov@gmail.com

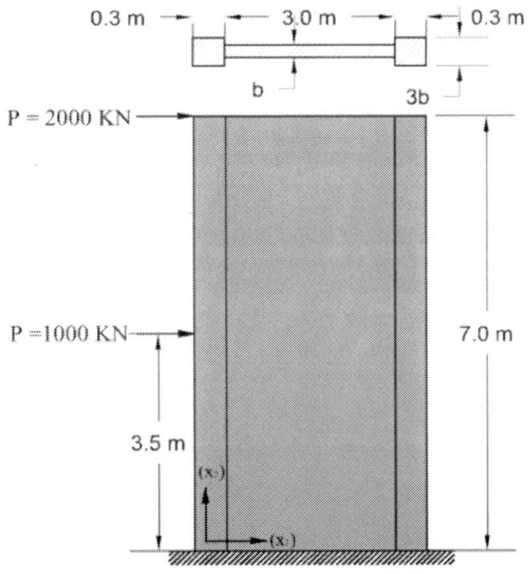

Fig. 1. End-column rectangular concrete shear wall subjected to two horizontal forces.

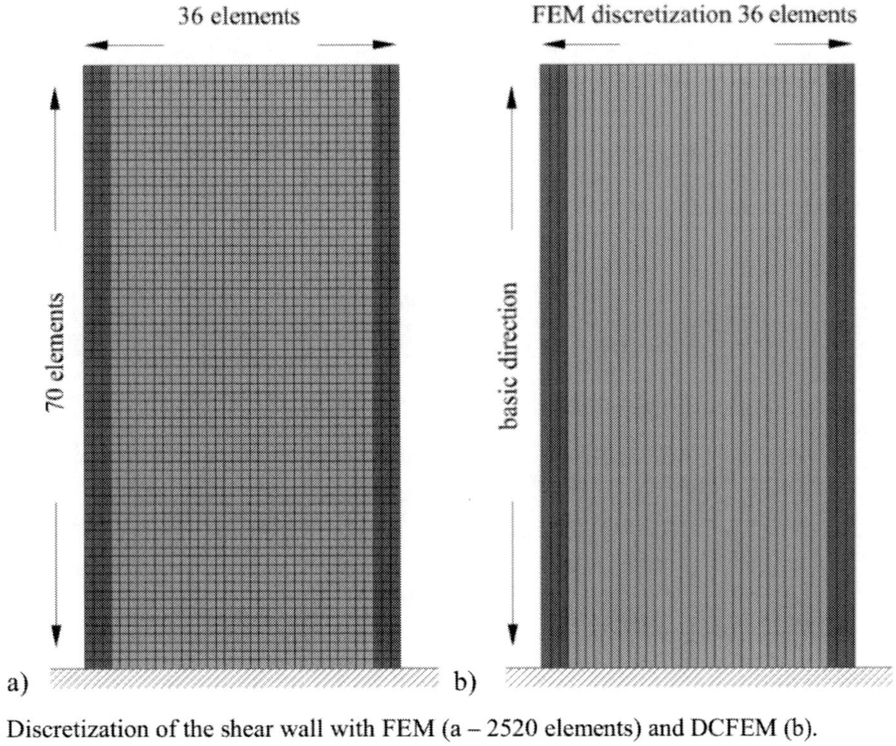

Fig. 2. Discretization of the shear wall with FEM (a – 2520 elements) and DCFEM (b).

Considering the Fig. 1, we have $E_{columnl} = 3E_{wall}$ for the column module. The approximation models of the shear wall within FEM and DCFEM are shown in Fig. 2. With respect to the length of the wall, the analytical solution is constructed along the vertical direction. The two-dimensional stress distribution in the shear wall is shown in Fig. 3. The results of DCFEM solution and comparison with the FEM solution (ANSYS Mechanical [12-14]) along the vertical length of the wall for $x_1 = 0.1$ m and along the horizontal length of the wall for $x_2 = 0.3$ m are represented in the Fig. 4 and Fig. 5, respectively.

DOI: 10.1051/matecconf/20168601011

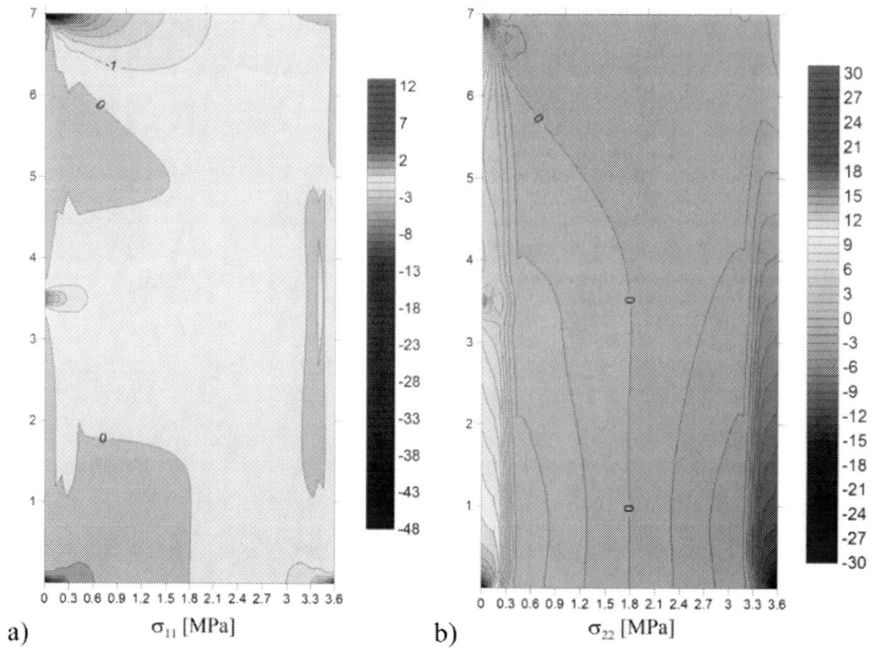

Fig. 3. Two-dimensional stress distribution in the shear wall (Mpa): a) σ_{11}; b) σ_{22}.

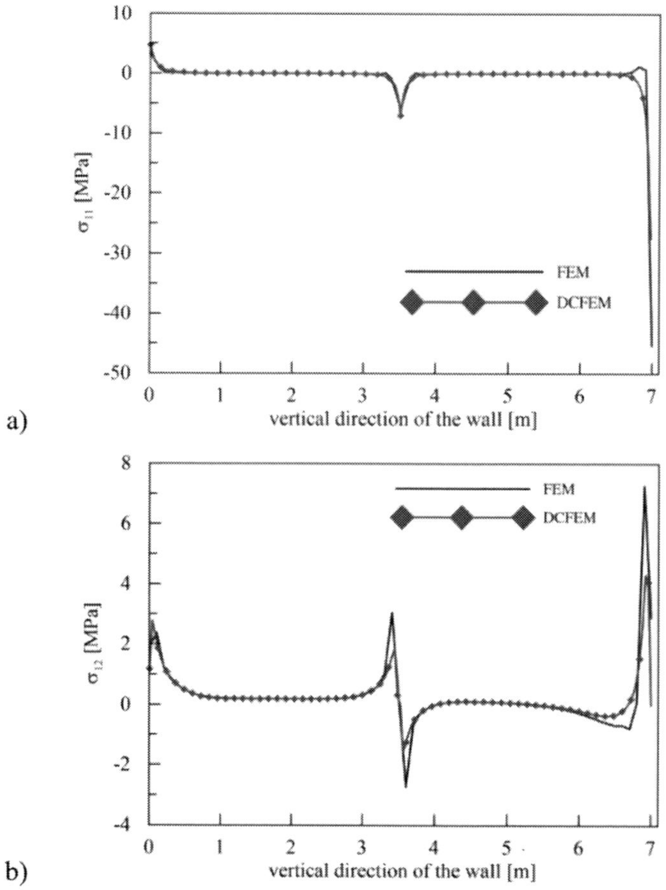

Fig. 4. DCFEM and FEM solutions along vertical length, $x_1 = 0.1$ m (Mpa): a) σ_{11}; b) σ_{12}.

 DOI: 10.1051/matecconf/20168601011

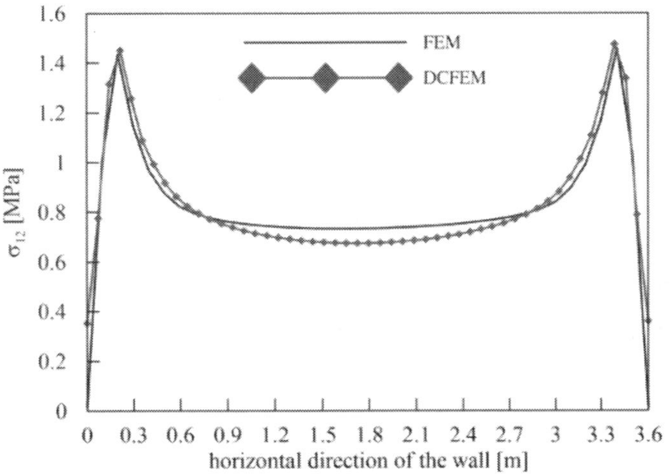

a)

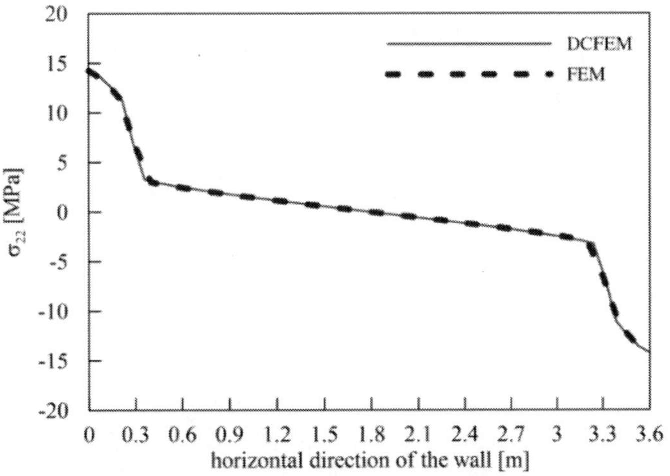

b)

Fig. 5. DCFEM and FEM solutions along vertical length, $x_2 = 0.3$ m (Mpa): a) σ_{12}; b) σ_{22}.

1.2 Semianalytical solution of the shear wall subjected to vertical distributed load and horizontal force

The simple rectangular 3.0×3.5 m concrete shear wall subjected to a horizontal force and distributed load on the top, is considered (Fig. 6).

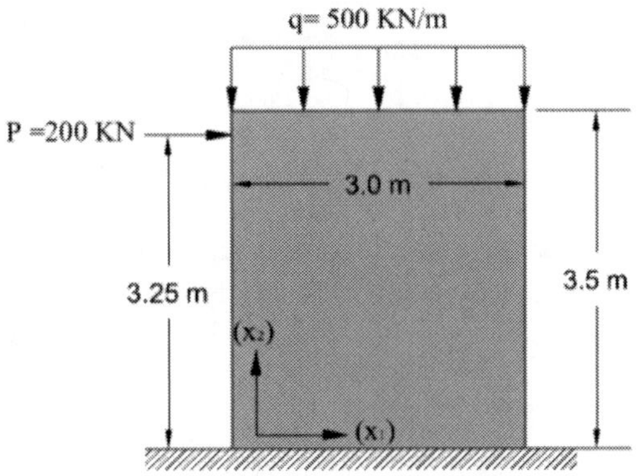

Fig. 6. Simple concrete shear wall, subjected to a horizontal force and distributed load.

DOI: 10.1051/matecconf/20168601011

The shear wall has constant thickness $0.15\,\mathrm{m}$, its module of elasticity is equal to $E_{wall} = 26500\,\mathrm{Mpa}$, Poisson ratio is equal to $\nu = 0.3$. 4200 eight nodded quadrilateral elements is used for the FEM solution, while the higher-precision DCFEM solution requires only 60 discrete-continual finite elements. The two-dimensional stress distribution in the shear wall is shown in Fig. 7.

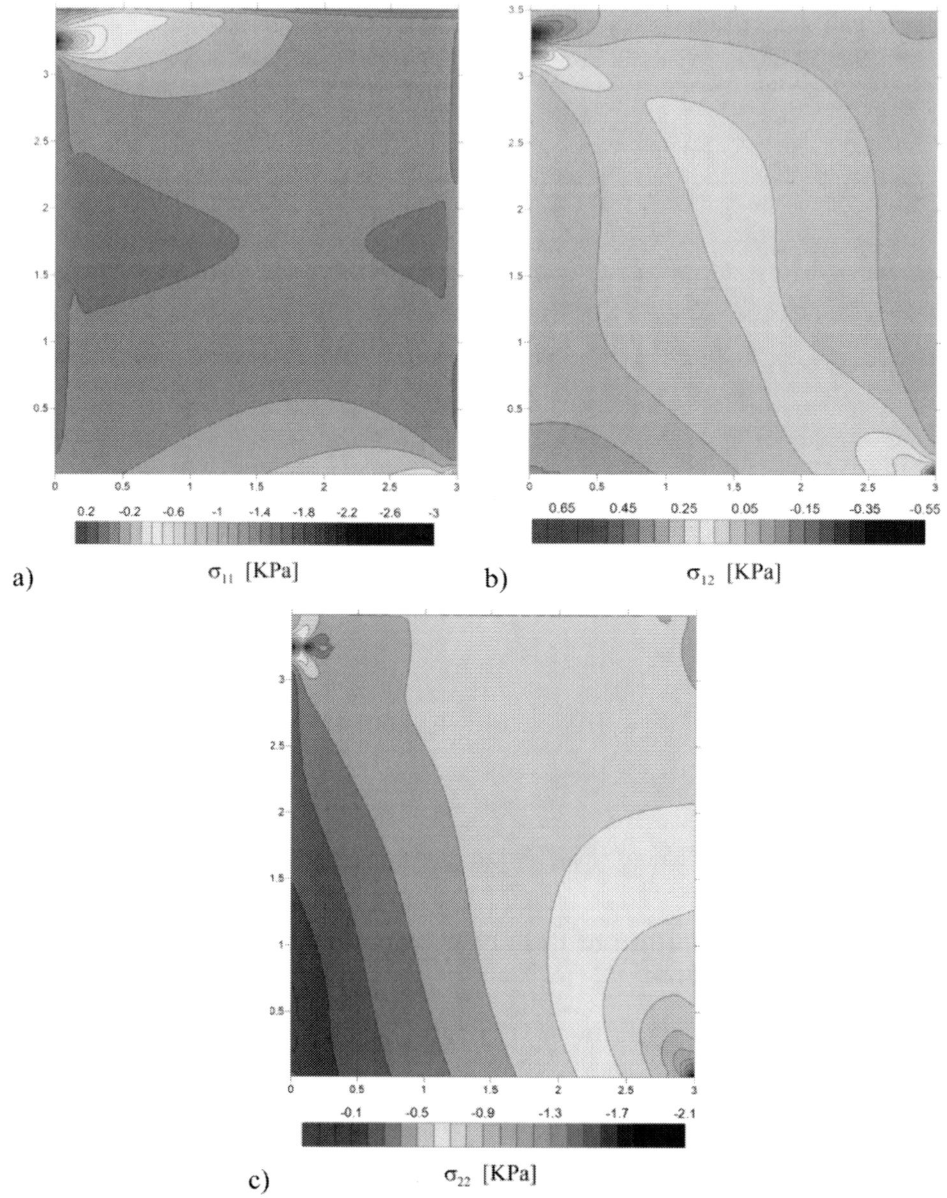

Fig. 7. Two-dimensional stress distribution in the shear wall (Kpa): a) σ_{11}; b) σ_{12}; c) σ_{22}.

The results of DCFEM solution and corresponding comparison with the FEM solution (ANSYS Mechanical [12-14]) along the vertical length of the wall for $x_1 = 0.05\,\mathrm{m}$ and along the horizontal length of the wall for $x_2 = 0.2\,\mathrm{m}$ are represented in the Fig. 9 and Fig. 10 respectively.

2 Future development of discrete-continual finite element method

Vital tendency of future development of DCFEM is associated with the use of multigrid approach for semianalytical structural analysis. Since the pioneering work of R.P. Fedorenko [15], multigrid literature has grown at an astonishing rate.

Multigrid methods are motivated by the observation that standard iterative methods like Gauss-Siedel, damped Jacobi and block-Jacobi are effective in reducing the high frequency error, but are not effective in reducing the low frequency content of the error. These standard solvers are called smoothers as they render error smooth by reducing high frequency content of the error. The in-effectiveness of standard iterative methods can be overcome by projecting the solution onto a coarse space, where the low frequency error becomes a high frequency error. Thus, the reason for using multigrid methods is to accelerate convergence.

Multigrid methods are among the most efficient iterative algorithms for solving the linear algebraic systems associated with elliptic partial differential equations [16-23]. As is known, the algorithm requires a series of problems be "solved" on a hierarchy of grids with different mesh sizes. For many problems, it is possible to prove that its execution time is asymptotically optimal. The niche of multigrid algorithms is large-scale problems where this asymptotic performance is critical. The need for high-resolution partial differential equations (PDE) simulations has motivated the parallelization of multigrid algorithms. However, these algorithms require to handle the operations on several coarser levels where the communication costs are higher than the computation costs.

Multigrid algorithms can be classified into two categories: geometric and algebraic. The primary difference is that algorithms of the former type use an underlying mesh for constructing coarser multigrid levels (coarsening), and algorithms of the latter type use the entries of the fine-grid matrix for coarsening in a black-box fashion. Algebraic multigrid (AMG) methods [24-27] are gaining prominence due to their generality and the ability to deal with unstructured meshes. Geometric multigrid methods are less general but have low overhead, are quite fast, and are easy to parallelize (at least for structured grids) [28-30].

The main steps in developing multigrid methods include: coarsening the fine grid (corresponding fine grid operator can be generated by finite element application), choosing grid transfer operators to move between meshes (i.e., the restriction and prolongation operators), determining the coarse mesh discretization matrices, and finally developing appropriate smoothers. Developing effective multigrid methods often boils down to striking a good balance between setup times, convergence rates, and cost per iteration. These features in turn depend on operator complexity, coarsening rates, and smoother effectiveness.

3 Discussion of the results. Conclusions

The presented examples show some of the advantages of the suggested approach to semi-analytical analysis of the shear wall. Discrete-continual finite element method is rather efficient for evaluation of so-called boundary effects (such as, for instance, the stress field near the concentrated force). It and also has completely computer-oriented algorithm, computational stability, optimal conditionality of resultant system and applicable for the various loads at an arbitrary point or a region of the wall. The structural discontinuities in the analytical direction can be taken into account, by addition of new appropriate boundary condition at the relevant section. Discrete-continual finite element method allows increasing the accuracy of the solution (especially in solution of high-rise shear walls) and provides reduction of the computational efforts [1-8, 31].

The Reported study was Funded by Government Program of the Russian Federation "Development of science and technology" (2013-2020) with in Program of Fundamental Researches of Ministry of Construction, Housing and Utilities of the Russian Federation and Russian Academy of Architecture and Construction Sciences, the Research Project 7.1.1.

References

1. P.A. Akimov, M.L. Mozgaleva, O.A. Negrozov, Advanced Materials Research **1025-1026**, 89-94 (2014)
2. P.A. Akimov, M.L. Mozgaleva, M. Aslami, O.A. Negrozov, Advanced Materials Research **1025-1026**, 95-103 (2014)
3. P.A. Akimov, M.L. Mozgaleva, M. Aslami, O.A. Negrozov, Procedia Engineering **91**, 2-7 (2014)
4. P.A. Akimov, M.L. Mozgaleva, V.N. Sidorov, Procedia Engineering **91**, 14-19 (2014)
5. P.A. Akimov, O.A. Negrozov, Procedia Engineering **111**, 8-13 (2015)
6. P.A. Akimov, O.A. Negrozov, Procedia Engineering **111**, 14-19 (2015)
7. P.A. Akimov, O.A. Negrozov, International Journal of Applied Engineering Research **10(20)**, 40824-40829 (2015)
8. P.A. Akimov, O.A. Negrozov, Materials Physics and Mechanics **26**, 5-8 (2016)
9. S. Blair-Chappell, A. Stokes, *Parallel Programming with Intel Parallel Studio XE* (John Wiley & Sons, 2012)
10. M. Metcalf, J. Reid, M. Cohen, *Modern FORTRAN Explained (Numerical Mathematics and Scientific Computation)* (Oxford University Press, 2011)
11. O.C. Zienkiewicz, R.L. Taylor, *The finite element method: Solid mechanics. Vol. 2* (Butterworth-heinemann, 2014)
12. E.M. Alawadhi, *Finite Element Simulations Using ANSYS* (CRC Press, 2015)
13. R.B. Choudary, *Introduction to ANSYS 16.0* (I K International Publishing House, 2016)
14. E. Madenci, I. Guven, *The Finite Element Method and Applications in Engineering Using ANSYS* (Springer, 2015)
15. R.P. Fedorenko, USSR Computational Math. and Math. Phys. **1(5)**, 1092-1096 (1962)
16. W. Hackbusch, *Multigrid methods and applications* (Computational Mathematics, Springer – Verlag, 1985)
17. U. Trottenberg, C. Oosterlee, A. Schuller, *Multigrid* (Academic Press, 2001)
18. W.L. Briggs, V.E. Henson, S. McCormick, *A multigrid tutorial* (SIAM, 2000).
19. H. Gua, J. Rethorea, M.-C. Baiettoa, P. Sainsota, P. Lecomte-Grosbrasb, C.H. Vennerc, A.A. Lubrechta, Computational Materials Science **112**, Part A, 230-237 (2016)
20. K.S. Kang, Computers & Mathematics with Applications **70(11)**, 2701-2708 (2015)
21. H. Boffy, C.H. Venner, Tribology International **74**, 121-129 (2014)
22. A.K. Garg, I. Kumar, S. Bansal, I. Goyal, International Journal of Science and Research **3(4)**, 473-475 (2014)
23. C. Rodrigo, F.J. Gaspar, F.J. Lisbona, Archives of Computational Methods in Engineering **19(4)**, 499-538 (2012)
24. S. Gratton, P. Henon, P. Jiranek, X. Vasseur, Numerical Linear Algebra with Applications **23(3)**, 501-518 (2016)
25. K. Stuben, Journal of Computational and Applied Mathematics **128(1-2)**, 281-309 (2001)
26. H. Sundar, G. Stadler, G. Biros, Numerical Linear Algebra with Applications **22(4)**, 664-680 (2015)
27. Y. Xiao, Z. Zhou, S. Shu, Numerical Linear Algebra with Applications **22(3)**, 465-482 (2015)

28. Y. Mesri, H. Guillard, T. Coupez, Applied Mathematics and Computation **218(21)**, 10500-10519 (2012)
29. R.S. Sampath, G. Biros, SIAM J. Sci. Comput. **32(3)**, 1361-1392 (2010)
30. Y. Xiao, S. Shu, T. Zhao, Numerical Linear Algebra with Applications **16(7)**, 535-559 (2009)
31. P.A. Akimov, A.M. Belostosky, M.L. Mozgaleva, M. Aslami, O.A. Negrozov, Advanced Materials Research **1040**, 664-669 (2014)

Modeling of weighing and filtration forces in the "dam-foundation" system

Elephan Agakhanov[1,*], *Murad* Agakhanov [2]

[1] Dagestan State Technical University, Imam Shamil Ave. 70, 367015 Makhachkala, Russia
[2] Moscow State University of Civil Engineering, Yaroslavl sh. 26, 129337 Moscow, Russia

Abstract. The authors carried out the modeling of weighing and filtration forces in the "dam-foundation" system through the methodology of modeling of piecewise-potential volume forces based on the equivalence of effects. They considered a uniform "dam-foundation" system with a drainage structure. The depression curve is taken as a broken line, therefore the filtration flow causes the influence of weighing and filtration forces on the part of this system. The modeling of the effect of filtration forces (which are potential ones) is performed by the method of the equivalence of effects. The experimental studies showed that the effects of the filtration flow at the level of the drainage structure were considerably weaker than those at the level of the foot of the dam. The effects by both the weighing forces and the filtration ones with the given domain configuration considerably reduce the compressive stresses in the foot of the dam at its upstream face caused by the effect of the dam mass, which makes worse the static work of the dam.

1 Introduction

Among the effects determining the form and the maintenance characteristics of full-scale hydro-engineering gravity structures, the weighing and filtration forces are quite significant, and their importance may be compared with that of the main effects – the hydrostatic pressure and the own mass of the structure.

The existing examples of analysis of the stressed state caused by the effects of weighing and filtration forces through the method of modeling include the recommendations of replacement of aforesaid effects by the surface loadings applied in the centre of gravity of their loading profiles [1]. This method leads to the reduction of precision of the results obtained and requires the tests of a number of models (e. g. in the process of construction of the influence lines of backpressure).

The methodology of modeling of piecewise-potential volume forces developed in [2] is based on the equivalence of effects; it allows us to obtain an experimental solution of the problem without using single effects.

2 State of the problem

* Corresponding author: muradak@mail.ru

Let us consider a uniform "dam-foundation" system with a drainage structure at the point A (Fig. 1). The depression curve is taken in accordance with [1] in the form of a broken line, therefore the filtration flow causes the effects of weighing W and filtration F forces on the system part shaded in Fig. 1. The aforesaid forces are connected with the voids ratio of the material and the loss of head at the filtration path by the following formulas [3]:

$$W = \gamma_b (1-n)\vec{j}\ , \quad F = -\gamma_b gradh \tag{1}$$

where γ_b – the volume mass of the water; h – the piezometric head; n – the voids ratio of the material; $\vec{j}$ – the unit vector of the y-axis.

2 Method of research

According to the elastic theory [5], the determination of stress-and-strain state with consideration of the creep caused by the weighing W and filtration F forces requires the solution of corresponding elastic problem.

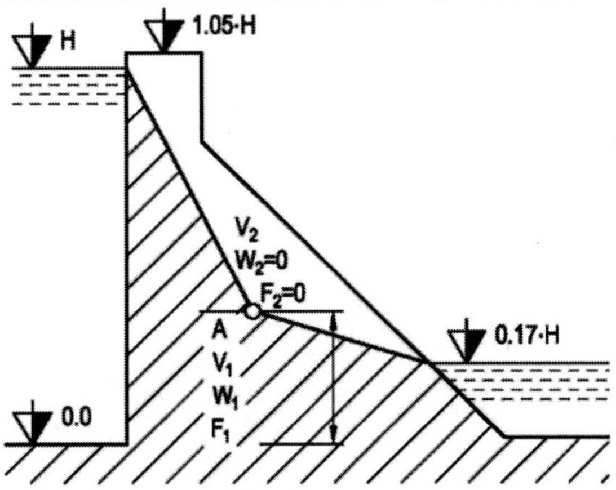

Fig. 1. The "dam-foundation" system under the influence of the filtration flow.

For the simulation of a weighing effect, the models of the domains V1 and V2 are made of a standard optically sensitive material with the following optical and mechanical properties in a high elasticity state:

$$T_{fr} = 130^0\ C,\ \ E_{fr} = 18.3\ MPa,\ \ \nu_{fr} = 0.5\ ,\ \ \sigma_{0,fr}^{(1,0)} = 0.0386\,MPa\ .$$

We freeze the model of domain V1 placed into the field of centrifugal forces, the direction of which is opposite to the weighing action. After reading the necessary experimental data, the model of domain V2 in its natural non-deformed state is glued to the V1-model, then the "burning" of the combined model is performed. The methods of normal translucence and the numerical integration of the equivalence equation gave the components of normal stresses at the outline and cross-sections located at the levels of foots and drainage structures of the models.

In accordance with the modeling criteria, the stress values in the full-scale structure caused by the weighing forces are determined by the formula:

$$\sigma_{fsc}^{(W)} = \frac{K_l W}{K_c \gamma_{mod}} \left(\overset{=(W)}{\sigma}_{mod} - \overset{-(W)}{\sigma}_{mod} \right) \tag{2}$$

where

$K_l = \frac{l_{fsc}}{l_{mod}}$; $K_c = \frac{R\omega^2}{g}$; ω – the angular velocity of the centrifuge rotation;
R – the centrifuge radius;
g – the gravity acceleration;

γ_{mod} – the volume mass of the model material;

$\overset{-(W)}{\sigma}_{mod}$ – the stress values in the model of domain V_1;

$\overset{=(W)}{\sigma}_{mod}$ – the stress values in the model combined of the both V_1 and V_2 models after burning.

According to the elasticity theory, the values of the stress with consideration of the creep coincide with the elastic stress values $\sigma_{fsc}^{(W)}$. Fig. 2 represents a dam with the head value of H=110 m and shows the loading diagrams for normal stresses caused by the action of the weighing pressure.

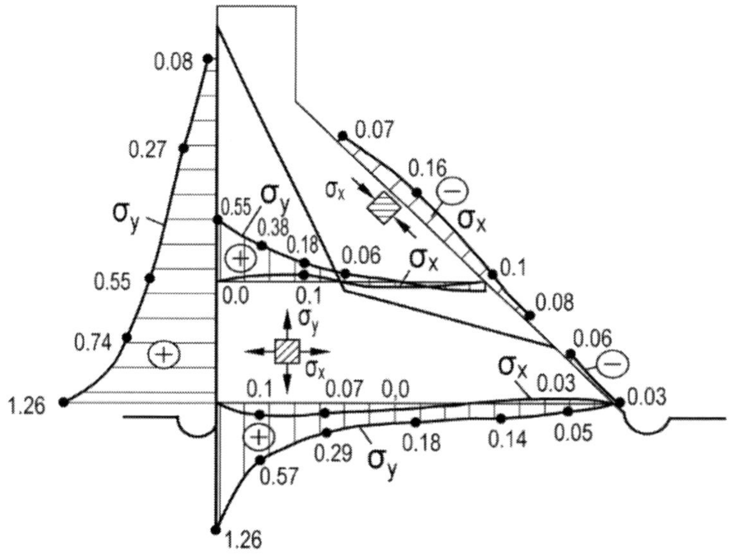

Fig. 2. The loading diagrams for normal stresses σ_x and σ_y at the cross-sections and at the outline of the dam caused by the action of weighing forces (MPa).

The simulation of the action of filtration forces F, which are potential ones, is performed by the method of the equivalence of effects. Here the stresses from the filtration forces F in the domain V_1 are presented as the sum of three components:

$$\sigma_{ij}^{(F)} = \sigma_{ij}^{(P)} + \sigma_{ij}^{(\xi)} + \sigma_{ij} \tag{3}$$

where $P = -\gamma_b h$ – the loading normal to the surface of the domain V_1;

$\xi = \frac{1-\nu}{E}\gamma_b h$ – the forced deformations; $\sigma_{ij} = \delta_{ij}\gamma_b h$.

The simulation of action of the forces normal to the surface of the domain V_1 is performed with the help of a special loading device. The effect of uniformly distributed and piecewise-constant linear distributed loading was obtained through the transmission of the concentrated loading through the stepwise systems including a number of dies and tie-rods. In order to avoid some local stress concentrators which could distort the results, we put a thin rubber layer under the dies closely adjacent to the model surface. For the production of the models of domains V_1 and V_2, we used a standard optically sensitive material with the following optical and mechanical properties in its high elasticity state:

$$T_{fr} = 130^\circ C,\quad E_{fr} = 18.3\text{ MPa},\quad \nu_{fr} = 0.5\ ,\quad \sigma_{0,fr}^{(1,0)} = 0.0337\text{ MPa}\ .$$

After "freezing" the model of the domain V_1 under a surface loading and reading the necessary experimental data, the model of domain V_2 in its natural non-deformed state was glued to the V_1-model, and the "burning" of the combined model was performed. The methods of normal translucence and the numerical integration of the equivalence equation gave the components of normal stresses at the outline and cross-sections located at the levels of foots and drainage structures of the models.

The stresses $\sigma_{ij}^{(\xi)}$ are determined through the method of "de-freezing" of forced deformations. Here a continuous field ξ obtained through the EAM (electric analogy method) method is replaced by piecewise-uniform zones ξ_i cut out from the blanks with preliminary "frozen" all-side deformations proportional to ξ_i . Further, the model of domain V_2 in its natural non-deformed state is glued to the model of domain V_1 at room temperature. The stresses in the models after the further burning correspond to the desired ones, and they are determined through the normal translucence.

According to the modeling criteria, the stress values in full-scale structures caused by volume filtration forces are determined by the formula:

$$\sigma_{fsc}^{(F)} = K_p(\bar{\sigma}_{mod}^{(p)} - \bar{\bar{\sigma}}_{mod}^{(p)}) + K_\xi(\bar{\sigma}_{mod}^{(\xi)} - \bar{\bar{\sigma}}_{mod}^{(\xi)}) + K_\sigma(\bar{\bar{\sigma}}_{mod} - \bar{\sigma}_{mod}), \tag{4}$$

where $K_p = \frac{P_{fsc}}{P_{mod}} = \frac{\gamma_в h}{P_{mod}}$; $K_\xi = \frac{E_{fsc}\xi_{fsc}}{E_{mod}\xi_{mod}} = \frac{(1-\nu_{fsc})\gamma_в h}{E_{mod}\xi_{mod}}$; $K_\sigma = \frac{\sigma_{fsk}}{\sigma_{mod}} = \frac{\gamma_в h}{\bar{\sigma}_{mod}}$;

$\bar{\sigma}_{mod}^{(p)}; \bar{\sigma}_{mod}^{(\xi)}; \bar{\sigma}_{mod}$ - the stresses in the models of domain V_1;

$\bar{\bar{\sigma}}_{mod}^{(p)}; \bar{\bar{\sigma}}_{mod}^{(\xi)}; \bar{\bar{\sigma}}_{mod}$ - the stresses in the model, combined of both models after burning.

In this problem, the stresses $\bar{\sigma}_{ij}^{(\xi)}$ are equal to zero due to the harmonicity of distribution of the piezometric head *h*.

According to the elasticity theory, the stresses with consideration of the creep coincide with the elastic stresses $\sigma_{fsc}^{(F)}$. Fig. 3 represents a dam with the head value of H=110 м and shows the loading diagrams for normal stresses caused by filtration forces.

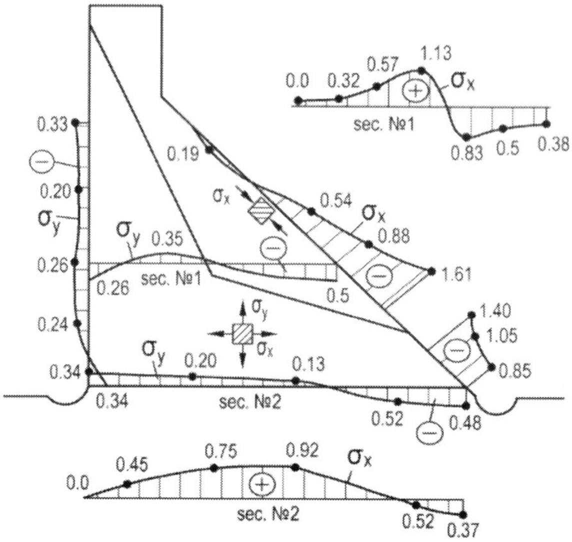

Fig. 3. The loading diagrams for normal stresses σ_x and σ_y at the cross-sections and at the outline of the dam caused by the action of filtration forces (MPa).

3 Analysis of results

For the determination of the total effect on the gravity dam, we should add the stresses caused by both the hydrostatic pressure and the own mass of the concrete to the results obtained, These results are given in [5]. Fig. 4. represents the loading diagrams for normal stresses at the cross-section at the dam foundation caused by the aforesaid loading factors as well as the loading diagrams for total stresses.

It should be noted that the action of weighing and filtration forces considerably depends on the configuration of the domain occupied by the filtration flow. For the considered configuration, the studies showed the following:

- the weighing forces cause tensile stresses at the upstream face of the dam and the compressive stresses at the downstream face;
- the action of filtration forces at the dam outline (with exception of the dam foot at the upstream face and of some part of the downstream face) causes the compressive stresses;
- the tensile stress values σ_y at the foot at the upstream dam face with the action of weighing and filtration forces approach the values of the order of 1 and 0.4 MPa, respectively, which totally equals 40 % of the value of compressive stress caused by the own mass which, in turn, equals 3.5 MPa at the foot of the upstream dam face;
- the stresses σ_y caused by the action of filtration forces become compressive ones (0.3 MPa) at the cross-section at the level of drainage structure at the upstream dam face; the stresses σ_y caused by weighing forces also have a positive value equal to 0.5 MPa; it totals

10 % of the value of compressive stresses caused by the own mass and equal to 2 MPa at the considered point.

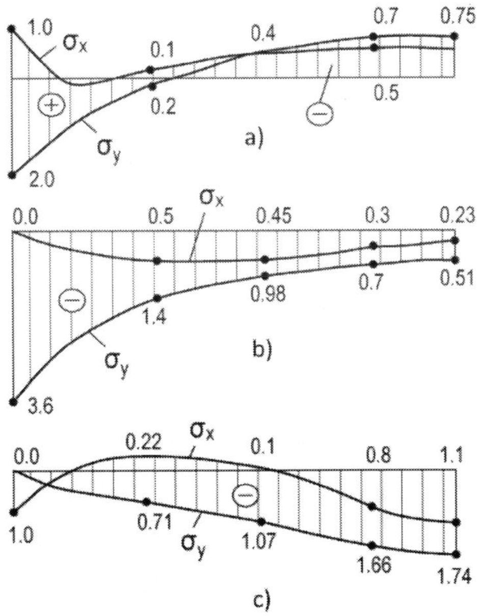

Fig. 4. The loading diagrams for normal stresses σ_x and σ_y at the cross-sections at the dam foundation (MPa): a) the effect of hydrostatic pressure along the head edge; b) the effect of the own mass of the concrete; c) the total effect of hydrostatic pressure, own concrete mass, weighing and filtration forces.

4 Conclusions

The effect of the filtration flow at the level of drainage structure is considerably weaker than that at the level of the dam foot. The action of both the weighing forces and the filtration ones with the considered domain configuration essentially reduces the compressive stresses at the foot at the upstream dam face caused by the influence of the own mass which makes worse its static work. Therefore, the action of weighing and filtration forces in such studies should always be considered.

References

1. N.S. Rozanov, Ya.G. Skomorovsky, *Model studies of the statics of hydro-engineering structures* (Moscow, Energiya, 1975)
2. E.K. Agakhanov, *Experimental mechanics, prospects of development and applications* (Moscow, 2001)
3. A.A. Khrapkov, VNIIG Proceedings **74**, 133-141 (1964)
4. G. L. Khessin, *Photoelasticity method* (Stroyizdat, Moscow, 1975)
5. G.LKhessin, V.N. Savostianov, MISI Proceedings **40**, 67-79 (1962)

DOI: 10.1051/matecconf/20168601013

Experimental assessment of some factors affecting concrete strength in a single dynamic exposure

Konstantin Tsvetkov[1,*], *Alexandra* Bazhenova[1]

[1]Moscow State University of Civil Engineering, Department of Strength of Materials, Yaroslavskoye sh. 26, 129337 Moscow, Russia

Abstract. The paper presents experiment results of testing the concrete under dynamic exposure aimed at evaluating the influence of static preloading on the dynamic strengthening coefficient.

1 Introduction

Concrete strength dependence on construction type, environment conditions, construction features and loading duration is considered in present Russian building codes [1] as coefficient of working conditions γ_b. Although [1] does not define the value of γ_b that takes into consideration the influence of dynamic exposure on concrete, it states that the coefficient has to be specified in special technical conditions. The development of special technical specifications is specified in [2] and is time-consuming and liable process. While deciding developers rely on the best practices in domestic and foreign science as well as experimental confirmation of suggested solutions. The authors suggest that the data presented in the paper might be used during the development of special technical conditions to determine the coefficient of working conditions for concrete under dynamic exposure.

It is known that strength of concrete under dynamic exposure is more than prism strength. This strength change is generally characterized by dynamic hardening coefficient [3, 4]. Influence of exposure speed (time) on the dynamic hardening coefficient is thoroughly investigated in [5, 6, 7], and these references contain dependencies mentioned above.

Although the techniques reported in the majority of the studies on definition of the dynamic hardening coefficient considered the dynamic exposure of unloaded samples. However, in real operational conditions of construction the dynamic exposure is generally an emergency load (e.g. seismic) and is applied in the addition to the operational load.

The authors attempt to specify the influence of preload on the dynamic hardening coefficient. The research was taken in two stages.

* Corresponding author: redfoxic@inbox.ru

2 Assessment of the impact of long-term static preloading degree on the dynamic hardening coefficient under axial compression

For the experiment the specimens of 1 cement : 2.8 aggregate : 4.44 sand ratio by mass on Portland cement M450 in the amount of 270 kg per cm3 with water-cement ratio of 0.67. The crushed granite fractions of 5–10 and 10–20 mm and silica sand were used as an aggregate. The hardening of the specimens had taken place in wet sawdust for 28 days, and they were stored in normal conditions for two years afterwards until the experiment was held.

The samples were divided by four groups depending on assumed level of preliminary static loading: without static loading ($\sigma_{preload} = 0$, reference samples), with static loading of $\sigma_{preload} = 0.3R_b$, $\sigma_{preload} = 0.7R_b$ and $\sigma_{preload} = 0.8R_b$.

Preloading and loading of reference samples tested for prism strength determination was held incrementally to secure the growth of stress on the value close to $0.1R_b$. The specimens were endured under design load for 360 days. A constant load was maintained and controlled, the load change did not exceed 3%.

The second stage of the experiment followed without unloading. Some of the samples from each group were brought to the failure with static load, and others – by dynamic load. Static additional load was applied incrementally with five minute interval until destruction.

Mobile dynamic module compatible with common press machines was built to perform additional loading of all the specimens at once. The testing of the specimens could be performed promptly by connecting the dynamic module and applying dynamic load to every specimen alternately, while the exposure time is almost identical. Portable dynamic module, designed and built by Igor Bezgodov, is presented on Figure 1.

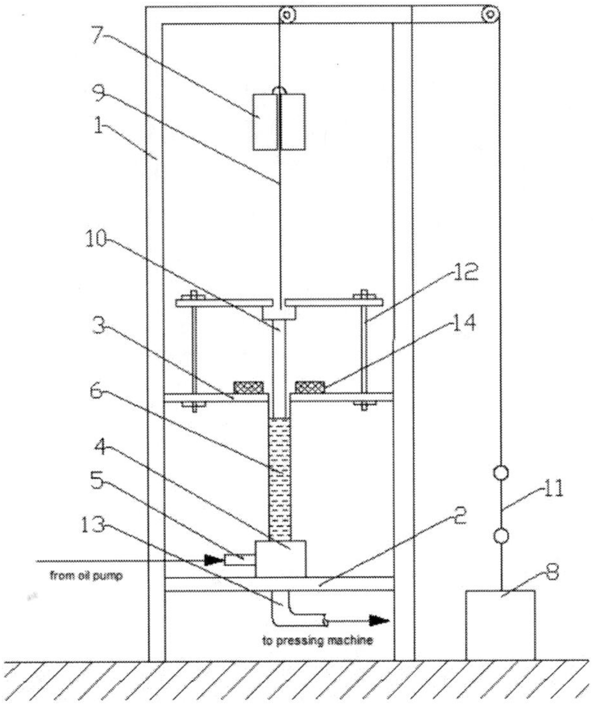

Fig. 1. Dynamic module for concrete samples testing: 1 – welded wire frame; 2, 3 – supporting table; 4 – distributor; 5 – choke; 6 – small cylinder; 7 – falling weight; 8 – counterweight; 9 – guide pin; 10 – small piston; 11 – wire; 12 – stop; 13 – power hose.

The dynamic module is a welded frame (1) that has posts with attached supporting tables (2, 3) inside. Oil is pumped into small cylinder (6) through the nozzle (5) located in the dispenser (4), which is installed on the lower table (2). Supporting table (3) limits the motion of drop weight (7), which is lifted up by counterweight (8). Drop weight (7) is directed by the guide rod (9) which is connected with the small piston (10).

Dynamic loading of specimens was carried out according by following technique. Weight (7) is lifted to certain level with counterweight (8) and is fixed by a thin wire (11). Oil is pumped into small cylinder (6) by pumping station. The piston of the small cylinder (10) comes to a specific level and is fixed by the limiter (12). The oil is supplied into small cylinder (6) and power hose (13) that is connected with press machine flat hydraulic lift jack (not shown on the figure). After turning on the recording devices, the wire (11) is cut and the falling weight (7) creates the load which is enough to destroy the specimen, and forces and deformations are measured. Porous rubber (14) laid on a supporting table (3) dampens impact load from a drop weight (7). The dynamic module could be used in facilities for axial compression testing.

The method of using a dynamic module for testing of samples that had been preloaded in conventional pressing machines with flat hydraulic jack lift has some features. Prior to dynamic loading, the power hose (13) has to be connected to the flat jack lift press. Pressure in the power hose (13) matches the pressure in the small cylinder (6) thanks to pumping station. After pressure levels equalize the flat jack lift incoming tap is opened. Shut-off valve connecting the dispenser with the pumping station is closed. Afterwards the weight (7) is discharged, and pressure enough to destroy the specimen is reached in the piston and lift jack.

Force sensor was installed between lift jack and the specimen to apply dynamic additional loading. Force sensors of all press machines were calibrated after test. Samples were tested consecutively by the dynamic module by the load of various intensity.

PC, multi-channel amplifier, analog-to-digital converter (ADC) and "ADCLab" software developed by CJSC “Rudnev-Shilyaev” (Moscow, Russia) was used to record the stress. Thanks to the use of modern amplifier, ADC and software, stresses were registered with an accuracy of 0.1 MPa. Test results are presented on Table1 and Figure 2.

Table 1.

Sample #	**Preload stress ratio $\sigma_{preload}/R_b$**	**Strength R or R_d, MPa**	**Strength R or R_d average, MPa**	**Coefficient of variation V, %**
Static tests				
15	0	34.00	34.67	3.3
18		36.00		
10		34.00		
21	0.3	35.00	35.00	0.0
3		35.00		
8		35.00		
7	0.7	35.00	35.00	0.0
13		35.00		
24		35.00		
15	0.8	36.75	35.58	2.8
19		35.00		
23		35.00		

Dynamic tests ($\dot{\sigma} \approx 500$ MPa/s, $\tau \approx 0.04$s)				
3	0	43.43	45.04	4.6
19		47.35		
23		44.33		
4	0.3	46.20	45.56	1.3
11		45.47		
22		45.00		
9	0.7	46.78	45.20	4.4
13		42.94		
16		45.88		
20	0.8	45.39	45.82	0.8
17		46.04		
5		46.04		

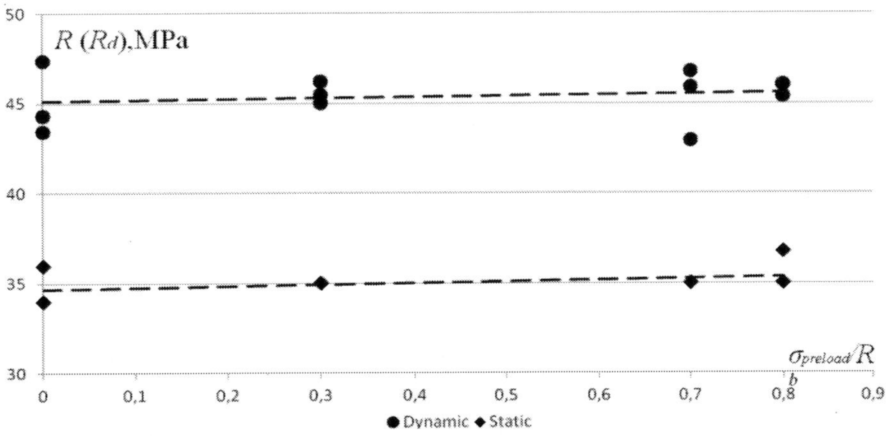

Fig. 2. Influence of long-term static preloading and its value on the dynamic strengthening factor under axial compression.

The impact of preliminary static load and its value on the dynamic strengthening factor is presented on Table 2. The dynamic strengthening factor is the ratio of dynamic strength for a particular value of static preloading to the corresponding strength under additional static loading.

Table 2.

Preload value $\sigma_{preload}/R_b$	Average static strength, R (MPa)	Average dynamic strength, R_d (MPa)	Dynamic strengthening factor
0	34.67	45.04	1.30
0.3	35.00	45.56	1.30
0.7	35.00	45.20	1.29
0.8	35.58	45.82	1.29

According to the tables and graphs the dynamic strengthening factor does not depend on static preload. Thus, it could be neglected when assigning concrete coefficient of the working conditions, and dynamic strengthening factor should be determined by axial compression tests of unloaded samples. The influence of long-term loading on concrete strength need to be considered with regard to the [1].

3 Evaluation of the static preloading value impact on the dynamic strengthening factor under biaxial compression

Concrete samples of 7x7x28 cm size, prism strength R_b = 20.41MPa and longitudinal deformations coefficient E = 28900 MPa were tested.

The equipment used allows a conventional dynamic loading with stress growth rate $\dot{\sigma}_1 = 470 - const$ MPa/s and destruction time: $\tau = 0.08$ s.

Dynamic loading was carried out smoothly from zero up to destructive values. Preliminary static load in σ_2 direction was created by the of a membrane type devices, which provide constant stress σ_2 in the subsequent dynamic loading in the direction of σ_1. Biaxial compression testing technique was preloading the sample in the direction of σ_2 to a certain prism strength ($\sigma_2 = 0.2R_b$, $0.4R_b$ or $0.6R_b$), followed by destruction of the sample by static or dynamic load in the σ_1 direction. Static and dynamic tests under uniaxial compression were also conducted. Test results for different values of stress σ_2 under static and dynamic load are presented in table 3.

Table 3.

Specimen #	Stress σ_2 (σ_2/R_b)	Strength $R_{ib}^{\sigma_2}$ ($R_{ib,d}^{\sigma_2}$), MPa	Strength average R^{σ_2} ($R_{b,d}^{\sigma_2}$), MPa	Coefficient of variation V, %
Static tests				
2	0	20.24	20.41	1.8
4		20.06		
8		20.93		
9	0.2	26.98	26.57	1.2
19		26.24		
20		26.50		
12	0.4	26.50	27.34	3.1
29		28.17		
1	0.6	29.39	28.98	1.4
32		28.57		
Dynamic tests ($\dot{\sigma} \approx 470$ MPa /s. $\tau \approx 0.08$s)				
15	0	22.80	23.97	3.4
26		25.10		
33		24.00		
5	0.2	32.02	30.89	2.6
10		30.23		
28		30.42		
11	0.4	35.30	33.97	3.0
27		32.80		
30		33.80		
14	0.6	34.68	35.04	1.0
31		35,40		

Value of preload in σ_2 direction influence on the dynamic strengthening factor was evaluated, the results are in Table 4.

Table 4. The influence of stress state and stress value σ_2 on concrete strength under static and dynamic loading.

Preload $\sigma_{2,preload}/R_b$	Average static strength, R, MPa	Average dynamic strength, R_d, MPa	Dynamic strengthening factor
0	20.41	23.97	1.17
0.2	26.57	30.89	1.16
0.4	27.34	33.97	1.24
0.6	28.98	35.04	1.21

Test results suggest the following. Stress state type and value of the lateral compression affects the dynamic strengthening factor, but to a much lesser extent than the loading rate. Dynamic strengthening factor growth on biaxial stress compared to uniaxial stress was observed at lateral compression of $0.4R_b$ and $0.6R_b$.

However, the impact of stress σ_2 on the dynamic strengthening factor change is not significant, which that can be traced from the Figure 3. The graph depicts relation between σ_2 on static exposure and loading with rate of 470 MPa/s (curves in the top) in coordinates $\frac{\sigma_2}{R_b} - \frac{R_b^{\sigma_2}}{R_b}$ for static case and $\frac{\sigma_2}{R_b} - \frac{R_{b,d}^{\sigma_2}}{R_{b,d}}$ for dynamic. Static and dynamic curves almost match.

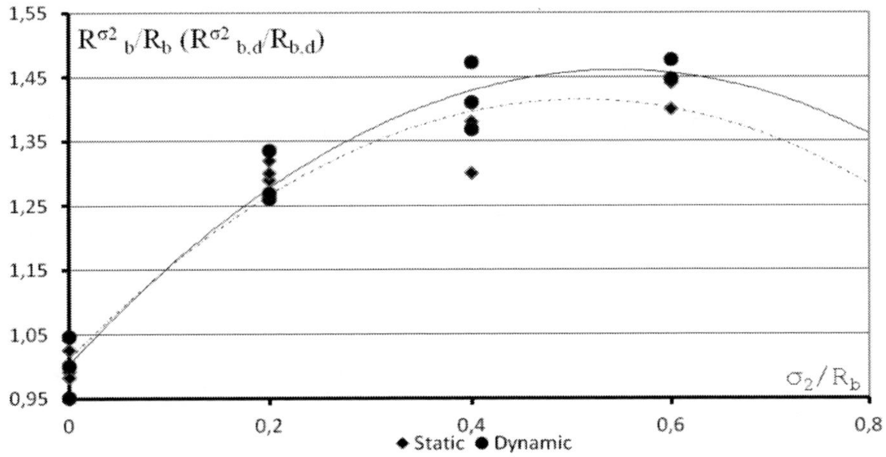

Fig. 3. Influence of static preloading on dynamic strengthening factor in biaxial compression.

At the same time it should be noted that under both static and dynamic loading, the type of stress state significantly affects concrete strength. The most significant increase in strength occurs upon the transition from uniaxial to biaxial compression, even with a small value of the second principal stress σ_2. At higher levels of lateral compression from $0.2R_b$ to $0.6R_b$ the trend of strength increase continues. Thus, when $\sigma_2 = 0.6R_b$ strength increase in both static and dynamic case increased by 40% compared to the strength un uniaxial testing with corresponding loading rate. Overall the static test results correspond to the data in [8, 9].

Thus, the evaluation of coefficient of working conditions of concrete under dynamic loading in biaxial compression can be based on data available regarding the concrete strength increase depending on the σ_2 value, by multiplying the short-term strength by dynamic strengthening factor obtained by testing the reference samples on axial compression.

4 Conclusion

Taking into account the results obtained in both directions of the study, it might be suggested that the coefficient of operating conditions of the concrete under dynamic exposure should be assumed equal to dynamic strengthening factor, obtained from the tests results on uniaxial compression, regardless of the preload duration, and stress level σ_2 under biaxial compression.

The reported study was funded by RFBR according to the research project No. 16-31-00214\16 mol_a

References

1. *Design requirements 63.13330.2012 Concrete and won concrete construction* (2012)
2. *Order of the Ministry of regional development of the Russian Federation from April 1, 2008 № 36 "On the procedure of development and approval of special technical conditions for the development of design documentation for capital construction object"* (2008)
3. Ju.M. Bazhenov, *Concrete under dynamic loading* (Moscow, Stroyizdat, 1970)
4. A.P. Kirillov, Concrete and reinforced concrete **2**, 38-39 (1987)
5. V.V.Brodskij, *Extended abstract of PhD dissertation* (Rostov-on-Don, 2001)
6. I.E. Prokopovich, V.M. Kobrinec, V.I. Polovec, I.A. Concrete and reinforced concrete **6**, 6–8 (1991)
7. A.V. Khokhlov, Problems of strength and plasticity **77(1)**, 60 (2015)
8. Ju.N. Malashkin, I.M. Bezgodov, *Evaluation of limit states of concrete under biaxial compression: the Proceedings of conferences and meetings on hydraulic engineering: Limit states of concrete and reinforced concrete structures in power plant construction* (Ehnergoizdat, 1982)
9. G.Eh. Shablinskij, O.V. Korobceva, I.M. Bezgodov, Earthquake-resistant construction. Overview **3**, 2-5 (1991)

Concrete strength evaluation using fracture mechanics technology

Valery Popov[1], *Dmitry* Popov[1], and *Anna* Davidenko[1,*]

[1]Samara State Technical University, Institute of Architecture and Civil Engineering, Molodogvardeyskaya str. 194, 443001 Samara, Russia

Abstract. This paper presents the physical phenomenon of concrete fracture process by static uniaxial compression. The analysis of strength characteristics of concrete is carried out using fracture mechanics technology, and what is more, the actual operating conditions of structures are evaluated. The expediency of using energy approach and kinetic theory is given to consider material fracture processes with a large number of defects.

1 Introduction

Nowadays, concrete and reinforced concrete are widely spread among building materials. What is more, their application is increasing. This conclusion is based on the analysis of spatial arrangement of buildings and contemporary construction. It is reflected in a considerable increase in the number of storeys, construction sites and the use of frame - monolithic structure of buildings that occur mainly in residential construction.

Moreover, the analysis of concrete properties reduces to determining the strength characteristics affecting its durability. The main concrete property is durability, that is, the capacity of concrete to resist to different external effects. But not only strength affects durability, it is also necessary to consider such characteristics as porosity, crack strength that characterizes the properties of the material itself as well as the intensity of the external load.

When we consider concrete fracture process the most effective is the application of the concept and theory of fracture mechanics. According to the proposed method the destruction of any kind of material is the result of development of the already existing defects in the body material what is suitable for concrete and takes into account the heterogeneity of its structure. After all, the concrete already has a significant number of micro- and macroscopic defects the formation of which is in accordance with the technological features of production. The control over the process of structure formation is possible only after the receipt of finished products.

[*] Corresponding author: aezg@mail.ru

2 Main Part

Thus, existing defects when subjected to external forces increase in size, develop into macro defects and reach critical disclosure in the form of main cracks. That is, they divide the material into parts destroying it. Stress redistribution, when disclosing defects, leads to the beginning of a chain reaction, and is brittle fracture of concrete.

The fracture process itself can be divided into three stages, discussed in the works [1,2], where the energy concept is applied using kinetic theory of fracture mechanics.

The energy values of crack formation, allocated in the destruction, define three stages of the considered process:

1. characterizing the accumulation of micro damages;
2. the period of preparation due to the formation of numerous microcracks;
3. the period of rapid material destruction.

At the first stage the energy amount of the material deformation is negligible and practically has no effect on the structure and characteristics of the material.

At the second stage there is more energy amount of destruction than at the first one. It results in emergence of variously oriented microcracks affecting the physical and mechanical properties of concrete. At this stage there is a destruction of the weakest links of concrete structures which are located in the places of low-strength inclusions or with a maximum stress concentration.

At the third stage, with increase in load the amount of the released destruction energy far exceeds the capacity of the first two stages. The release of such amount of energy can be justified by the process of further stress redistribution and its increase to such a critical value where the increase in length and width of cracks leads to their coalescence. Further formation and increase processes of large, main cracks absorb all the energy supplied to material and result in loss of stability and its division into parts, that is, to the destruction of material.

It should be noted that the preparation of fracture even under a static compression is extended in time and can be an indicator of durability while the process of "destroying" occurs like an avalanche. The most important influence on the duration of the preparatory destruction process has the rate of load application.

The use of energy concept makes it possible to trace the connection of the characteristics, crack strength with the proposed parameters of its evaluation.

Based on the Griffith's theory of equilibrium cracks, we can assume the development of volume damage as flat crack separation, that is, independent growth of its length and width is allowed.

The energy balance characteristic describing cracks development can be presented as:

$$\sum_{i=1}^{i=k} \frac{\pi d_i \sigma_i^2 \Delta l_i \Delta \delta_i}{2E}\left(1-\mu^2\right) = 4v\sum_{i=1}^{i=k} \Delta l_i \Delta \delta_i , \tag{1}$$

where $\Sigma\ \Delta l_i \quad \Delta\delta_i$ - the overall length and width of gaps formed in the concrete sample, m;

v – concrete surface energy, J / m^2;

E – concrete modulus of elasticity, MPa;

σ_i – working stress in i-th crack, Pa.

We simplify the formula (1) by introducing the averaged length and width of the crack, so the action of stress σ will cause the increase of the averaged crack at the specified length and width:

$$\frac{\pi\sigma^2 L\Delta L\Delta\delta}{2E}\left(1-\mu^2\right)=4v\Delta L\Delta\delta \quad (2)$$

We replace the product $\Delta L\Delta\delta = S$. The second member of equation (2) expresses the fracture or crack formation energy:

$$A=\frac{\pi\sigma^2 LS}{8E}\left(1-\mu^2\right) \quad (3)$$

Due to this, it is necessary to investigate the behavior of the formed crack depending on the stress increase in the concrete element. Wherefore we use Zhurkov's transformed equation of kinetic theory, presented by K.I. Kuznetsova [6]:

$$\frac{dl}{dt}=a_0 e^{\alpha\frac{\sigma}{R}} , \quad (4)$$

where a_0 – characteristic reflecting the tendency of the material to crack formation under certain conditions of external actions, m / s; α – characteristic reflecting the intensity of crack formation processes.

Based on the load standard conditions with linear variation of load application rate, we present the following equation:

$$\frac{d\sigma}{dt}=K \quad (5)$$

Substituting this expression into (5) we obtain:

$$\frac{dl}{d\sigma}=\frac{dl}{dt}\cdot\frac{dt}{d\sigma}=\frac{a_p}{K}exp\left(\alpha\frac{\sigma}{R_p}\right), \quad (6)$$

We replace $\beta = a_p/K$, then the expression (6) takes the form

$$\frac{dl}{d\sigma}=\beta exp\left(\alpha\frac{\sigma}{R_p}\right) \quad (7)$$

Integrating the dependence (7) we obtain:

$$l=\int_0^{\sigma}\beta exp\left(\alpha\frac{\sigma}{R_p}\right)d\sigma=\frac{\beta R_p}{\alpha}exp\left(\alpha\frac{\sigma}{R_p}\right)\Bigg|_0^{\sigma}=\frac{\beta R_p}{\alpha}\left[exp\left(\alpha\frac{\sigma}{R_p}\right)-1\right] \quad (8)$$

The equation (8) shows the dependence of the crack formation length from the level of the applied stress. Taking into account the kinetic characteristics of α and β considering various temperature – humidity conditions of concrete destruction we substitute the dependence into (3) and obtain the expression:

$$A=\frac{\pi\sigma^2 S\beta R_p\left[exp\left(\alpha\frac{\sigma}{R_p}\right)-1\right]}{8E\alpha} \quad (9)$$

Taking into account the relative level of stresses, the dependence takes the form:

$$A = \frac{\left(\frac{\sigma}{R_p}\right)^2 \pi R_p^2 S \beta R_p \left[exp\left(\alpha \frac{\sigma}{R_p} \right) - 1 \right]}{8E\alpha} \quad (10)$$

As discussed above, at the moment of destruction the release of the micro cracks formation energy exceeds to the maximum, which means that $\sigma = R_p$, and in consideration of (2) the dependence (10) takes the form:

$$\nu S = \frac{\pi S R_p^3 \beta [exp(\alpha - 1)]}{8E\alpha} \quad (11)$$

However, to simplify the descriptive part, we use the research results of V.A. Kuzmenko [5] showing that in the case of uniaxial compression the maximum tensile stresses are calculated as "reduced stresses" and are equal to $\mu\,\sigma_{com}$, that is,

$$\sigma_{tens}^{red} = -\mu \sigma_{com} \quad (12)$$

Taking into account the equivalence of "reduced stresses", we substitute the dependence (11) and derive the critical stress, which led to the destruction of the material

$$R_{com} = \sqrt[3]{\frac{8E\nu\alpha}{\pi \mu^3 \beta [exp(\alpha - 1)](1 - \mu^2)}} \quad (13)$$

The energy concept and the kinetic theory in fracture mechanics are often applied for merit rating of materials that have a significant number of micro- and macroscopic defects in the structure.

3 Conclusion

Current trends of the proposed method are the simplicity and accuracy of calculations.

The fracture mechanics methods as science based regulatory framework for evaluation the properties of concrete and reinforced concrete gained widespread with the introduction of GOST 29167-91in the 1990s [3,9]. That is the result of large-scale experiments completion conducted by RILEM (International Union of Testing and Research Laboratories for Materials and Structures) and the research results of many Russian scientists [4,7,8].

References

1. V. P. Popov, *Concrete destruction working under the cyclic freezing, Dissertation candidate of technical sciences* (MADI, Moscow, 1986)
2. P.G. Komokhov, V.P. Popov, *Energy and kinetic aspects of concrete fracture mechanics* (Publ. RIA, Samara, 1999)
3. *GOST 12730.4 – 78 Concrete. Porosity parameters determination methods* (Publishing House of Standards, Moscow, 1986)

4. V.P. Popov, A. Yu. Davidenko, Vestnik of the Russian Academy in Construction, 113-114 (2005)
5. V.A. Kuzmenko, *New deformation schemes of solid bodies* (Naukova dumka, Kiev, 1973)
6. K.I. Kuznetsova, *The laws of viscous elastic bodies destruction and some of their application features to seismology* (Nauka, Moscow, 1969)
7. V.P. Popov, A.Yu. Davidenko, Proceedings of the X scientific - technical conference "The reliability of construction objects", 34-37 (2007)
8. V.P. Popov, A.Yu. Davidenko, Vestnik of the Russian Academy in Construction, 61-62 (2009)
9. GOST 29167–91 Concrete. *Defining the characteristics of crack strength (toughness) under static loading* (Publishing House of Standards, Moscow, 1992)

DOI: 10.1051/matecconf/20168601015

Viscoelastic dynamic models of resilient elements used in railway tracks

Artur Zbiciak [1,*], *Cezary* Kraśkiewicz[1], *Wojciech* Oleksiewicz[1], and *Cezary* Lipko[2]

[1]Warsaw University of Technology, Faculty of Civil Engineering, Institute of Roads and Bridges, 16 Armii Ludowej Ave., 00-637 Warsaw, Poland
[2]TINES Capital Group SA, TINES Railway, 23 Ciepłownicza St., 31-574 Cracow, Poland

Abstract. The paper presents selected theoretical aspects concerning viscoelastic dynamic modelling of resilient elements used in railway tracks. In order to characterize the research methodology for resilient mats in railway tracks, German Standards [1-4] are used herein. The main goal of the paper is to demonstrate the procedure of insertion loss calculation for a single degree of freedom truck system containing under-ballast mats. Selected results of certain dynamic characteristics of resilient truck systems (transmissibility, Bode and Nyquist plots etc.) are also discussed. The results of calculations visualized in graphs, were obtained by using own applications written in programming language MATLAB.

1 Introduction

Resilient mats are used for vibration isolation of railway track by reducing dynamic effects of rail traffic (vertical and transverse material vibration and structure-borne noise). Under-ballast mats are used in the track superstructure of bridges, track troughs and tunnels. They increase resilience between the ballast and the track, and reduce vibrations emitted into the environment.

In ballast-less track system resilient mats have similar purpose as under ballast mats. Nonetheless, a ballast-less track system is always an individual engineering solution for a particular application. In particular, in a slab track system resilient mats are applied horizontally and vertically directly below and from the sides of the concrete track base plate. The resulting floating slab track system is effective in mitigating vibrations, especially in low frequencies. It also can achieve a significant reduction in vibrations and structure-borne noise emissions at excitation frequencies above $\sqrt{2}$ times the tuning frequency [13-15].

Detailed description of various categories of resilient mats is given in our previous paper [12]. As it was emphasized in [12], there are no European standards, which would define the requirements, methodology of research and calculation for resilient mats in railway tracks. Thus, it is substantiated to use German Standards: DIN 45673-1 [1], DIN 45673-4 [2], DIN 45673-5 [3] and DIN 45673-7 [4]. Standard [4] covers parameters related

* Corresponding author: a.zbiciak@il.pw.edu.pl

to the effectiveness of a track structure in mitigating vibrations and parameters needed for the static analysis and for the verification of track safety in floating slab track system. However, it is not possible to specify generally applicable load ranges for the resilient elements in a floating slab track as in case of under-ballast mats [3]. This is due to the fact that a floating slab track system is always an individual engineering solution for a particular application. Therefore, said standard sets out procedures for testing fitness for purpose, but does not contain requirements pertaining to the properties of resilient elements. All the substantial characteristics of resilient mats and procedures for their research are presented in studies [10-12].

In this paper we use mainly Standard [2] in order to demonstrate the procedure of insertion loss calculation for a single degree of freedom truck system containing under-ballast mats. Moreover, the results of certain dynamic characteristics of resilient truck systems (transmissibility, Bode and Nyquist plots etc.) are also discussed. Appropriate calculations were carried out by using own programs written in MATLAB.

2 Mechanical model of resilient truck elements

Rheological and kinetic properties of resilient truck elements can be characterized by linear viscoelastic dynamic models as it is proposed in Standards [1, 2]. The objective of this point is to provide a description of linear properties of an oscillatory system obeying the principle of superposition and presented in [1]. We will explain the terminology used in [1] pointing out some theoretical aspects being important for understanding the laboratory test procedures for resilient elements used in railway tracks [5-7, 9]. Moreover, this introductory part prepares basic nomenclature used in next chapter where detailed analytical calculations are carried out based on Standard [2].

Our analysis is focused on a viscoelastic single degree of freedom system containing one elastic element characterized by a spring coefficient k [N/m] and one viscous element with damping coefficient d [N s/m], linked in parallel. Moreover, effective mass of resilient elements is denoted by m_{eff} [kg]. The system is excited by a force F [N]. The response of the system is specified by displacement x [m]. The equation of motion is as follows

$$m_{eff}\,\ddot{x}(t)+d\,\dot{x}(t)+k\,x(t)=F(t) \tag{1}$$

where a superimposed dot denotes differentiation with respect to time.

Harmonic excitation of the linear system and its steady-state response can be expressed via following equations

$$F(t)=F_0\,\sin\omega t \tag{2a}$$

$$x(t;\omega)=x_0(\omega)\,\sin(\omega t-\varsigma(\omega)) \tag{2b}$$

where ω[rad/s] denotes angular frequency of excitation, F_0 and x_0 denote amplitudes of force and displacement respectively and ς is phase angle (loss angle). Angular frequency is related to ordinary frequency f [Hz] by the relation $\omega=2\pi f$.

The dynamic stiffness modulus of entire system can be defined as a ratio between amplitudes of excitation and response

$$k_{dyn}(\omega) := \frac{F_o}{x_o(\omega)} \tag{3}$$

Analysing linear viscoelastic systems excited by harmonic functions allows application of complex variables. Let us assume the following transformation of variables

$$F(t) = \operatorname{Im} F^*(t) \tag{4a}$$

$$x(t;\omega) = \operatorname{Im} x^*(t;\omega) \tag{4b}$$

The complex variables defined by Eqs. (4) are given as follows

$$F^*(t) = F_o e^{i\omega t} \tag{5a}$$

$$x^*(t;\omega) = x_o(\omega)\, e^{i\omega t - i\varsigma(\omega)} \tag{5b}$$

where i denotes imaginary unit.

Thus, it is possible to define the complex stiffness modulus containing both dynamic stiffness modulus as well as phase angle

$$k^*(i\omega) := \frac{F^*(t)}{x^*(t;\omega)} = k_{dyn}(\omega)\, e^{i\varsigma(\omega)} \tag{6a}$$

where dynamic stiffness modulus constitutes an absolute value of the complex modulus

$$k_{dyn}(\omega) = \left| k^*(i\omega) \right| \tag{6b}$$

Moreover, the complex stiffness modulus can be decomposed as follows

$$k^*(i\omega) = k_{dyn}(\omega)\cos\varsigma(\omega) + i\, k_{dyn}(\omega)\sin\varsigma(\omega) = k'(\omega) + i\, k''(\omega) \tag{7}$$

where k' denotes the storage stiffness modulus representing elastic and kinetic properties of resilient track elements while k'' is the loss modulus used to represent viscous (damping) properties.

In case of linear viscoelastic systems, it is possible to define the loss factor being equal to the tangent of the loss angle. Thus, we obtain

$$\eta(\omega) = \tan\varsigma(\omega) = \frac{k''}{k'} \tag{8}$$

Eventually, we can express the steady-state harmonic response of the system exited by sinusoidal force as follows

$$x(t;\omega) = F_o \operatorname{Im}\left\{ \frac{\exp(i\omega t)}{k^*(i\omega)} \right\} = F_o U(\omega) \sin\left(\omega t - \varsigma(\omega)\right) \tag{9}$$

where $U(\omega)$ denotes dynamic compliance defined as an inverse of dynamic stiffness (see [1]).

Consequently, $U(\omega)$ can also be interpreted as a magnitude of the complex compliance $U^*(i\omega)$:

$$U(\omega) = \left|U^*(i\omega)\right| = k_{dyn}^{-1}(\omega) \tag{10}$$

Moving back to the analysis of viscoelastic dynamic system defined via Eq. (1), the complex stiffness modulus may be expressed by the following equation

$$k^*(i\omega) = k - m_{eff}\,\omega^2 + i\omega d \tag{11}$$

while the dynamic stiffness modulus and the loss factor are as follows

$$k_{dyn}(\omega) = \sqrt{\left(k - m_{eff}\,\omega^2\right)^2 + (\omega d)^2} \tag{12a}$$

$$\eta(\omega) = \frac{\omega d}{k - m_{eff}\,\omega^2} \tag{12b}$$

The above Eq. (12) can be substituted into Eq. (10) and Eq. (9) in order to obtain the solution of equation of motion (1), under sinusoidal force excitation (steady-state response) taking additionally $\varsigma(\omega) = \operatorname{atan} \eta(\omega)$.

As it was emphasized in [12] the Standard [1] introduces more precise definitions of the spring coefficient k and dynamic stiffness modulus k_{dyn}. In case of the stiffness modulus, the following definitions are used in [1]: k_{stat} – static stiffness under a quasi-static force, $k_{stat\,0}$ – at-rest static stiffness, k_{kin} – kinetic or lower-frequency stiffness under a harmonically varying force with frequencies up to 1 Hz. Moreover, in case of the dynamic stiffness modulus, applied for the frequency range above 1 Hz, the Standard [1] distinguishes between k_{dyn1} – lower-frequency dynamic stiffness and k_{dyn2} – higher-frequency dynamic stiffness.

3 Insertion Loss and Transmissibility

In case of resilient mats hysteretic damping model is used instead of viscous damping as it was presented in Eq. (1). The equation of motion in case of single degree of freedom system is as follows

$$m_{eff}\,\ddot{x}(t) + k(1 + i\eta)x(t) = F(t) \tag{13}$$

while the complex stiffness modulus is given by

$$k^*(i\omega) = k(1 + i\eta) - m_{eff}\,\omega^2 \tag{14}$$

In case of such formulation, the loss factor is a constant value not depending of frequency of excitation. For steady-state response it is possible to establish the following relationship between the loss factor η and damping coefficient d:

$$d = \frac{\eta k}{\omega} \tag{15}$$

Let us analyse two mechanical models shown in Fig. 1 representing rail track systems: reference (Fig. 1 – left picture) and vibro-isolated (Fig. 1 – right picture). Such a model is presented in Standard [2] for calculation of insertion loss.

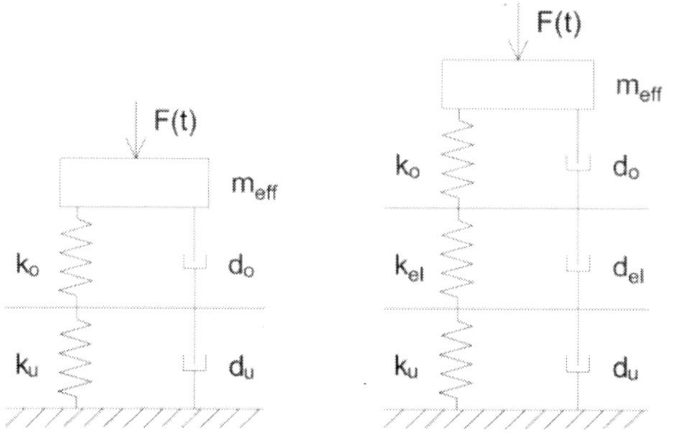

Fig. 1. Reference system (left picture) and system with Under-Ballast Mat (UBM) (right picture).

The reference system is composed of two layers. The parameters $k_o = 220$ MN/m and $\eta_o = 0.35$ represent stiffness and damping of upper layer of the system (rail, fastening, sleeper and ballast) while $k_u = 1500$ MN/m and $d_u = 1.2$ MNs/m define lower layer (track-bed). In case of isolated system, an additional layer is considered representing under-ballast mat (UBM) and characterized by parameters $k_{el} = 140$ MN/m and $\eta_{el} = 0.05$. Loss factors η_o and η_{el} can be replaced by equivalent damping coefficients d_o and d_{el} by applying Eq. (15).

Let us define the complex stiffness of entire reference system (lower and upper layer without mass)

$$k^*_{ref}(i\omega) = \left[\frac{1}{k_o + i\omega d_o} + \frac{1}{k_u + i\omega d_u}\right]^{-1} \tag{16}$$

In case of isolated system the following relation holds

$$k^*_{vib}(i\omega) = \left[\frac{1}{k^*_{ref}(i\omega)} + \frac{1}{k_{el} + i\omega d_{el}}\right]^{-1} \tag{17}$$

Thus, it is possible to define complex transfer functions for both reference system $H^*_{ref}(i\omega)$ as well as isolated system $H^*_{vib}(i\omega)$:

$$H^*_{ref}(i\omega) = \frac{k^*_{ref}(i\omega)}{k^*_{ref}(i\omega) - m_{eff}\,\omega^2} \tag{18a}$$

$$H^*_{vib}(i\omega) = \frac{k^*_{vib}(i\omega)}{k^*_{vib}(i\omega) - m_{eff}\,\omega^2} \tag{18b}$$

Absolute value of the complex transfer function can be interpreted as transmissibility of harmonically excited system.

The above definitions can be used to establish so called insertion loss parameter

$$IL(\omega) = 20 \cdot \log_{10} \left| \frac{H^*_{ref}(i\omega)}{H^*_{vib}(i\omega)} \right| \text{ dB} \tag{19}$$

The results of calculation for both reference system and vibro-isolated system with UBM are visualized in Fig. 2 and 3. Horizontal axis used in Fig. 2 represents centre frequencies of 1/3 octave bands. The plots depicted in Fig. 2 indicate a region of vibration effectiveness for approx. $f > 33$ [Hz] where $IL > 0$ dB.

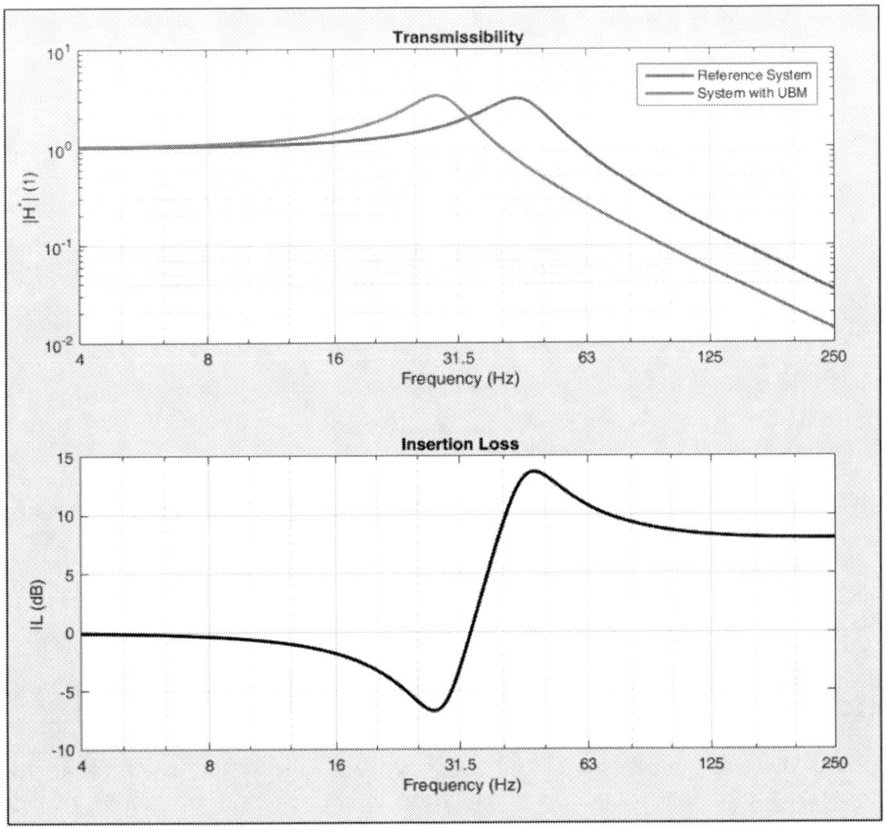

Fig. 2. Transmissibility and Insertion Loss – reference system and system with Under-Ballast Mat (UBM).

Moreover, one can define the complex compliance moduli od entire systems (reference and vibro-isolated) by the following equations

$$U^*_{ref}(i\omega) = \left[k^*_{ref}(i\omega) - m_{eff}\,\omega^2 \right]^{-1} \tag{20a}$$

$$U^*_{vib}(i\omega) = \left[k^*_{vib}(i\omega) - m_{eff}\,\omega^2 \right]^{-1} \tag{20b}$$

Figure 3 visualizes dynamic characteristics of analysed systems based on complex compliance formulations. Bode diagrams represent magnitude and phase of the complex compliance as a function of frequency. Another plots to be shown in Fig. 3 are Nyquist graph ($\operatorname{Im} U^*$ vs. $\operatorname{Re} U^*$) and Nichols graph (absolute value of U^* as a function the phase angle). Analysing Nyquist graph leads to the conclusion that even for static excitation, when $\omega \to 0$, results in energy dissipation because the imaginary part of complex modulus does not equal to zero. The reason of such an effect is the application of hysteretic damping model of UBM.

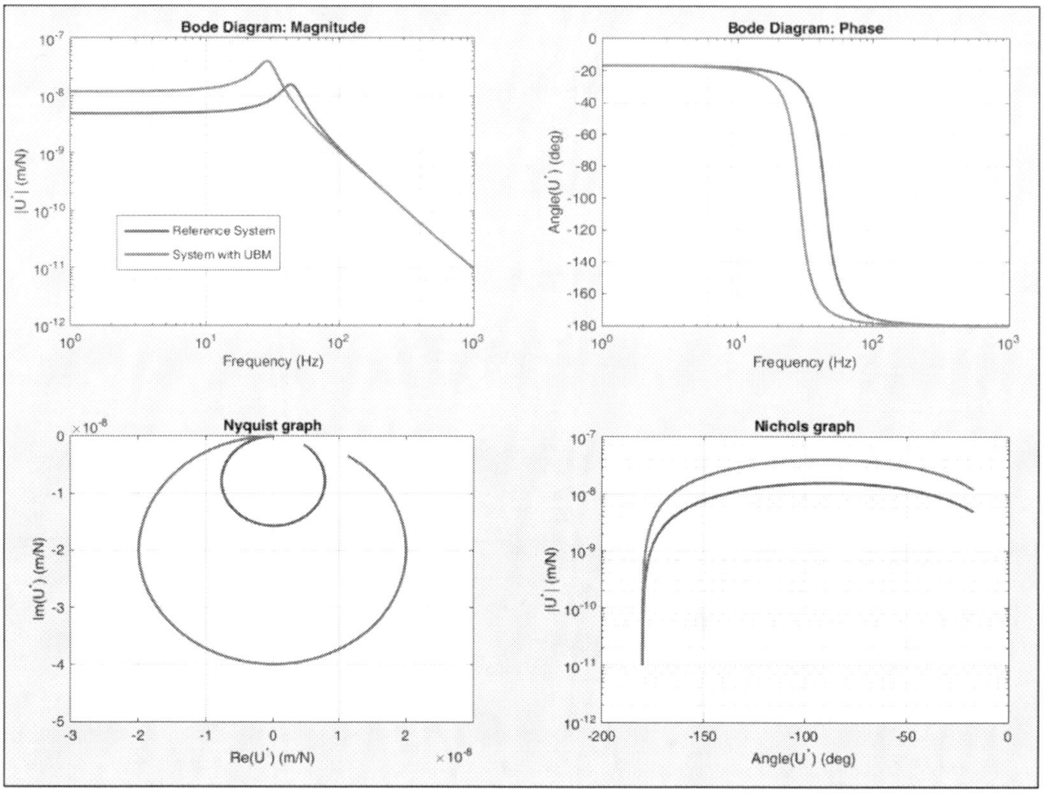

Fig. 3. Dynamic characteristics: Complex Compliance Modulus – reference rystem and rystem with Under-Ballast Mat (UBM)

4 Conclusions

Appropriate dynamic models of rail track system allow for calculation of various characteristics useful for evaluation of isolation effectiveness. The results of calculations presented in this paper were obtained by applying simple one degree of freedom system. It would be possible to use the same methodology in case of multi degree of freedom linear viscoelastic models representing rail track system with resilient elements. More general linear model should also contain non-classical viscoelastic elements described by fractional-order derivatives [8]. Such an approach would make possible to fit better experimental results.

Further explorations of problems related to the issue of modelling of resilient elements used in railway tracks by applying fractional rheological elements will be published in a subsequent paper.

References

1. *DIN 45673-1:2010-08 Mechanical vibration. Resilient elements used in railway tracks. Part 1: Terms and definitions, classification, test procedures*
2. *DIN 45673-4:2008-07 Mechanical vibration. Resilient elements used in railway tracks. Part 4: Analytical evaluation of insertion loss of mounted track systems*

3. *DIN 45673-5:2010-08 Mechanical vibration. Resilient elements used in railway tracks. Part 5: Laboratory test procedures for under-ballast mats*
4. *DIN 45673-7:2010-08 Mechanical vibration - Resilient elements used in railway tracks - Part 7: Laboratory test procedures for resilient elements of floating slab track systems*
5. C. A. Brebbia et al., *Lecture Notes in Engineering* (Springer-Verlag, Berlin,1985)
6. R. W. Clough, J. Penzien, *Dynamics of Structures* (McGraw-Hill, 1975)
7. W. Grzesikiewicz, J. Osiecki, J. Piotrowski, *Fundamentals of rail vehicle dynamics* (Publishing House of the Warsaw University of Technology, Warsaw, 1977)
8. W. Grzesikiewicz, A. Wakulicz, A. Zbiciak, International Journal of Mechanical Sciences **70**, 90–98 (2013)
9. K. G. McConnell, P. S. Varoto, *Vibration Testing: Theory and Practice, second ed.* (John Wiley and Son, 2008)
10. C. Kraśkiewicz, C. Lipko, W. Oleksiewicz, A. Zbiciak, Materiały Konferencyjne, Stowarzyszenie Inżynierów i Techników Komunikacji Rzeczpospolitej Polskiej. Oddział w Krakowie **2(106)**, 89–113 (2015)
11. C. Kraśkiewicz, C. Lipko, W. Oleksiewicz, A. Zbiciak, Przegląd Komunikacyjny **9**, 76–82 (2015)
12. C. Kraśkiewicz, C. Lipko, M. Płudowska, W. Oleksiewicz, A. Zbiciak, Procedia Engineering **153**, 317–324 (2016)
13. R. G. Wettschureck, U. J. Kurze, ACUSTICA **58** pp. 177–182 (1985)
14. R. G. Wettschureck, U. J. Kurze, ACUSTICA **58** pp. 177–182 (1985)
15. R. G. Wettschureck, M. Heim, S. Mühlbachler, Proceedings Inter-noise **97**, 577–580 (1997)

MATEC Web of Conferences **86**, 01016 (2016) DOI: 10.1051/matecconf/20168601016
IPICSE-2016

Plastic deformations in thin rotational shells

Avgustina Astakhova[1,*]

[1] Moscow State University of Civil Engineering, Yaroslavskoye sh. 26, 129337 Moscow, Russia

Abstract. The paper focuses on the method of calculation of evolution shells beyond the elastic limit. The conclusion of the basic system of differential equations is based on the linear shell theory with regard to the Hirchhoff-Lave hypothesis and on the physical equations for small elastic-plastic deformation theory using the method of elastic decisions. The boundary conditions are formulated for Cauchy problem: rigid attachment, hinged support, and free margin. The spherical shell boundary conditions in the pole are obtained from the conditions of symmetry and antisymmetry functions. The convergence of the elastic method and the method of the occurrence of superficial plastic deformations are studied. Also the stress-strain state in the spherical shell is determined and the convergence of the obtained solutions was studied. The results are presented on the symmetric load ring applied to the middle of the Meridian and on the load that can be considered as a concentrated force. The sufficient quantity of iterations is established to achieve the accuracy of 0.1%. The graphs are presented for radial displacement and for meridional bending moment as the functions that converge more rapidly and more slowly respectively.

1 State of the problem

The work is dedicated to the calculation of the isotopic shells of revolution of constant thickness under the action of symmetrical and assimetrical loads beyond the elastic limit. Long and short, closed and ring shells are considered with different hypothesis about contour intersection. The conclusion of the basic system of differential equations is based on the linear shell theory with regard to the Hirchhoff-Lave hypothesis and on the physical equations for small elastic-plastic deformation theory using the method of elastic decisions [1 - 3].

It is supposed that the shell mid-surface is the surface of rotation about the axis z without special points escluding points where the speciality is prevented. Shell generator is a quite smooth curve that curvature is too small comparing to $1/h$.

Figure 1 indicates:

θ – azimuthal angle in the horizontal plane,

S – meridian arc length,

ς – normal to the mid-surface.

---

* Corresponding author: ast956@yandex.ru

On the basis of the linear theory and Kirchhoff-Lave hypothesis the deformations are presented as two first power series expentions at ς, i.e. excluding values above the second order of smallness.

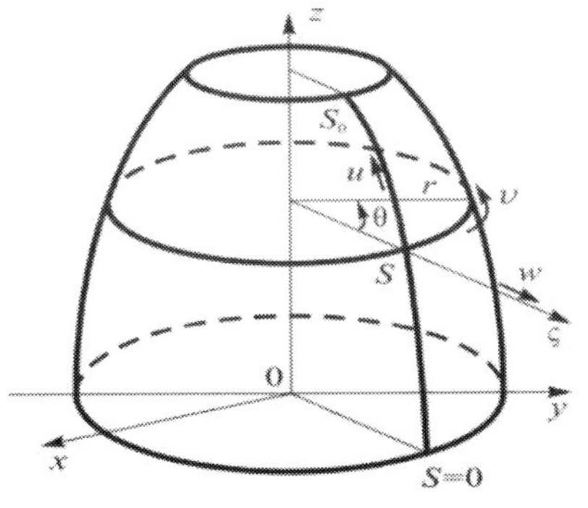

Fig. 1. Shell mid-surface.

2 The method of solution

The stress-strain state of thin shells is determined with the use of the elastic solutions method based on small elastic-plastic deformations. I. S. Tsurkov developed elastic solutions method in relation to the analysis of shells that was proposed by A. A. Ilyushin [4]. On the basis of the theory, all the generalized efforts are presented in the form of the difference of efforts that occurs in the linearly deformed structure $\mathbf{N}^0$ and incremental efforts $\mathbf{\Delta}$ that takes into account the physical nonlinearity of the material:

$$\mathbf{N} = \mathbf{N}^0 - \mathbf{\Delta}\,. \tag{1}$$

The resolving system of differential equations is derived from equilibrium equations, geometric relations of linear theory, and physical equations for small elastic-plastic deformations theory:

$$\begin{aligned} &\sigma_i = E\varepsilon_i, \quad \varepsilon_i < \varepsilon_s^0, \\ &\sigma_i = E\varepsilon_i\left[1-\lambda\left(1-\frac{\varepsilon_s^0}{\varepsilon_i}\right)\right], \quad \varepsilon_i > \varepsilon_s^0, \quad \lambda = \frac{E\text{-}E_1}{E}. \end{aligned} \tag{2}$$

All the components of the stress-strain state and displacements decompose to Fourier series. Eight resolution functions are introduced to the system. The original system of partial differential equations is transformed to the system of eight ordinary differential equations for the first order with respect to the Fourier series coefficients of resolution functions [2]:

$$\mathbf{N}^0 = \mathbf{A}(s)\times\mathbf{N} + \mathbf{f}(s) + \mathbf{B}(s)\times\mathbf{\Delta} \tag{3}$$

Here: $\mathbf{A}(s) = a_{ij}$, $\mathbf{B}(s) = b_{ik}$ – constant coefficients ($i, j = 1,2,\ldots,8$; $l, k = 1,2,\ldots,6$), $\mathbf{N} = (N_r, N_z, \hat{S}, M_s, u_r, u_z, v, \zeta_s)$, $\mathbf{f}(s) = (f_1, f_2, f_3, f_4, 0, 0, 0, 0)$, $f_1 = -q_r$, $f_2 = -q_z$, $f_3 = -q_\theta$, $f_4 = -m_s$, $\mathbf{\Delta} = (\Delta N_s, \Delta N_\theta, \Delta M_s, \Delta M_\theta, \Delta S, \Delta H)$.

Resolution functions are: N_r – radial force, N_z – force along the axis z (axial force), $\hat{S}$ – generalized tangential force, M_s – meridional bending moment, u_r – radial displacement, u_z – axial displacement, υ – tangential displacement, ζ_s – angular deflection. Furthermore, shell deformation is defined by the formulas:

$$\mathbf{E} = \mathbf{C}(s) \cdot \mathbf{N} + \mathbf{D}(s) \tag{4}$$

Here: $\mathbf{C}(s) = c_{ij}$, $\mathbf{D}(s) = d_{lk}$ – constant coefficients, $\mathbf{E} = (\varepsilon_s, \varepsilon_\theta, \varepsilon_{s\theta}, \kappa^0{}_s, \kappa^0{}_\theta, \kappa^0{}_{s\theta})$.

Elastic solutions method is based on the successive approximations method. In the first approximation incremental efforts $\mathbf{\Delta}$, that take into account physical nonlinearity of the material, are considered to be equal to zero, i.e. the elastic problem is resolved. The system of ordinary differential equations (3) should be integrated by the orthogonal sweep method to obtain n Fourier series expansions. Thus, eight resolution functions are determined by the coefficients, and the functions are used for determining six deformation components (1.4): relative stretch deformations (ε_s, ε_θ), deformations of shift ($\varepsilon_{s\theta}$), flexural deformations ($\kappa^0{}_s$, $\kappa^0{}_\theta$) and torsional deformations ($\kappa^0{}_{s\theta}$).

Four boundary conditions at each end of the generating line should be specified for the solution of Cauchy problem:

1) rigid attachment: $u_r = u_z = \upsilon = 0$, $\zeta_s = 0$;
2) hinged support: $u_r = u_z = \upsilon = 0$, $M_s = 0$;
3) free margin: $N_r = N_z = \hat{S} = M_s = 0$.

In the pole of the spherical shell boundary conditions can be obtained from the conditions of symmetry and antisymmetry functions in a neighborhood of the point $r = 0$ and from the equations that can be obtained with using passage to the limit $r \to 0$ in a geometric ratio and equilibrium equations [5]:

$u_r = \upsilon = 0$, $N_z = 0$, $\zeta_s = 0$ for $n = 0$,
$u_r + \upsilon = 0$, $u_z = 0$, $M_s = 0$ for $n = 1$,
$u_r = u_z = \upsilon = 0$, $\zeta_s = 0$ at the even n,
$M_s = 0$ at the odd n for $n > 2$.

Dimensionless loads and displacements are used in all the equations:

$(P, N_r, N_z, \hat{S}) / Eh$, $(M_s, M_\theta) / Eh^2$, $(u_r, u_z, \upsilon) / h$.

3 Results

Moreover, the intensities of the deformations on the external ε_{i2} and on the internal ε_{i1} surfaces of the shell are related to the ε_s^0 relative deformation that corresponds to the yield strength σ_T: $(\varepsilon_{i1}, \varepsilon_{i2}) / \varepsilon_s^0$.

In the result of the regional and local effects located in the thin shells of revolution, the desired efforts have large gradients. That's why to solve the problem for numerical implementation the discrete orthogonalization method was used. Direct integration was conducted by Runge-Kutta method with a stable step of integration along the shell generatrix. Load corrections Δ were only at the nodes of orthogonalization. The calculation results show that the difference in the function values is 0.5% at 50 orthogonalization points and 10 integration points at each orthogonalization step. The results coinside with an accuracy of 0.1% at 50 and 25 points of orthogonalization and at 250 integration points.

In the present work the stress-strain state in the spherical shell was determined and the convergence of the obtained solutions was studied. Also the majority of the tasks were handled with 250 integration points and with 50 orthogonalization points. As a criterion for convergence of the solution, we used the criterion where every tenth approximation is compared for all resolution functions at all N-points of orthogonalization. In the result a positive convergence of the method was discovered. It should be noted that the convergence of the solutions is too slow for the load range. The more extended the area of plastic deformations on the shell surface, the more approximations are required to achieve the same accuracy of solution.

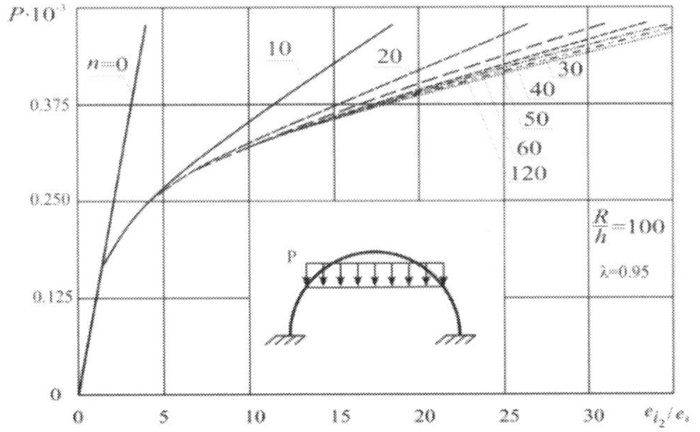

Fig. 2. Rotational shell under the action of the rotation load.

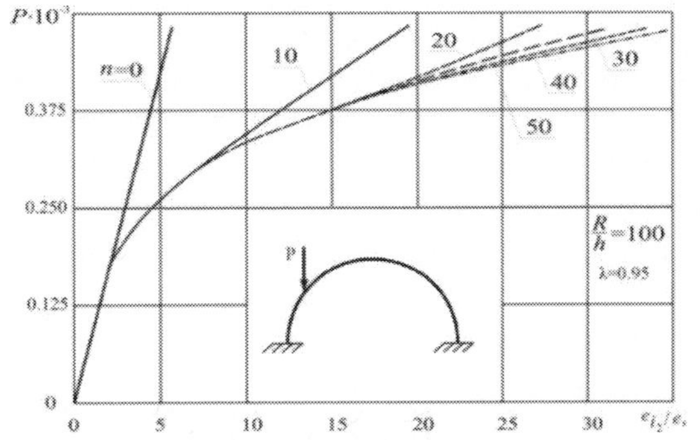

Fig. 3. Rotational shell under the action of the concentrated force.

The results presented in Fig. 2 and 3 refer to the spherical shell with the semi opening angle $\alpha = 90^{o}$ and radius ratio to the thickness $R/h = 100$. The steel chart tensile 3 approximated by the broken line with the hardening module $\lambda = 0.95$ is presented. The intensity of deformations corresponding to the beginning of yielding is equal to $\varepsilon_s^{\,0}$=0.001. Fig. 2 shows the dependence $\varepsilon_{i2} / \varepsilon_s - P$ on the symmetric load ring applied to the middle of the Meridian. Thus, the elastic solution is suitable only if the load does not exceed P/Eh= 0.15 10^{-3}. To achieve the precision of 0.1% 10 approximations with $P/Eh< 0.25\cdot10^{-3}$ and 100 approximations with $P/Eh> 0.35\cdot10^{-3}$ are required.

Fig.3 shows the dependence $\varepsilon_{i2} / \varepsilon_s - P$ on the load ring operating along the parallel arc, the corresponding angle $\theta = 0.04$ rad. The angle is small, so the load can be considered as a concentrated force. To achieve the accuracy of 0.1% at $P/Eh< 0.3\cdot10^{-3}$ it is enough to establish 10 approximations, and at $P/Eh> 0.45\cdot10^{-3}$ it is enough to establish 50 approximations.

Fig. 5. Changes in the functions of the radial displacements and the meridional bending moments.

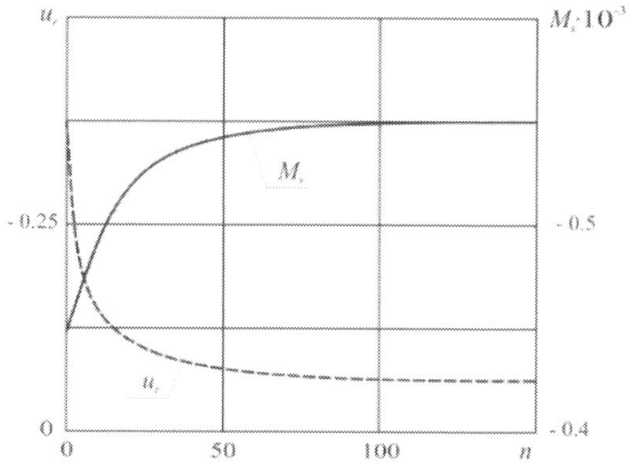

Fig. 4. Change of radial displacements functions and meridional bending moment.

Fig.4 shows the change of functions of the radial displacements u_r/h and meridional bending moment M_s / Eh^2 at the point of application of load depending on the number of approximation n under the action of load $P/Eh = 0.2862 \cdot 10^{-3}$. The solid line is the change in the meridional bending moment, and the dashed line represents the change in the radial displacement. Based on these studies with the load $P/Eh = 0.2862 \cdot 10^{-3}$ it is possible to see two results. First, the radial displacement u_r is the function that converges more rapidly where the accuracy 0.1% is achieved by the approximations with n = 10-20 iterations in the solution. Second, the meridional bending moment M_s is the function that converges slowly, that's why to achieve the accuracy 0.1% n = 80-90 iterations are required.

The studies show that the convergence of the elastic solutions method slows with the increase of the linear hardening coefficient λ, i.e. with the approaching to the Prandtl diagram. 40 iterations at λ = 0.95, 100 iterations at λ = 0.992 and 150 iterations at λ = 1 are required considering the example of the changes in the functions of the radial displacements u_r/ h and of the meridional bending moments M_s / Eh^2 at the point-of-load application (Fig.5) for $P/Eh = 0.2862 \cdot 10^{-3}$.

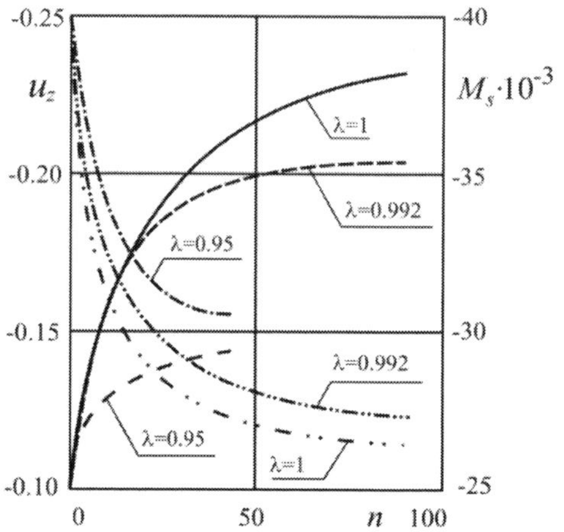

Fig. 5. Changes in the functions of the radial displacements and the meridional bending moments.

Figure 6 shows the areas of plastic deformation in the membranes of different thickness with radius to thickness ratio R/h= 200, 100, 50, 25 and with the ring load P/Eh = $0.2862 \cdot 10^{-3}$. The areas of plastic deformations are located near the point of force application.

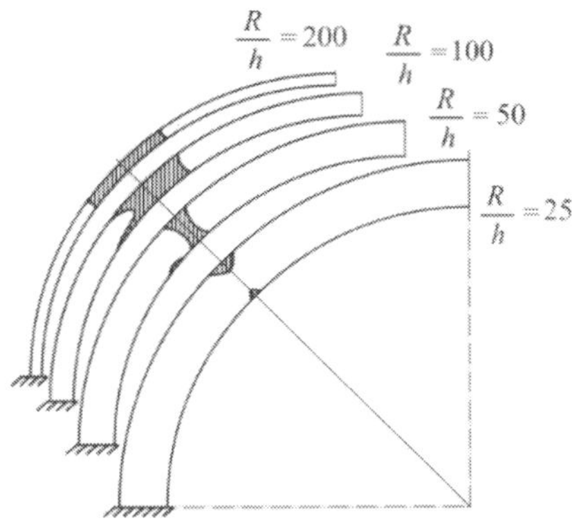

Fig. 6. Areas of plastic deformations in the membranes of different thickness.

Based on the calculation results, the dependences (Fig. 7) of the axial displacement u_z were constructed by the load for the shells with radius to thickness ratio *R/h*= 200, 100, 50, 25. Here the elastic solution is direct, the other lines are calculated considering the plastic work of the material. The graphs allow to determine the load to the quantity of which the elastic solution is true.

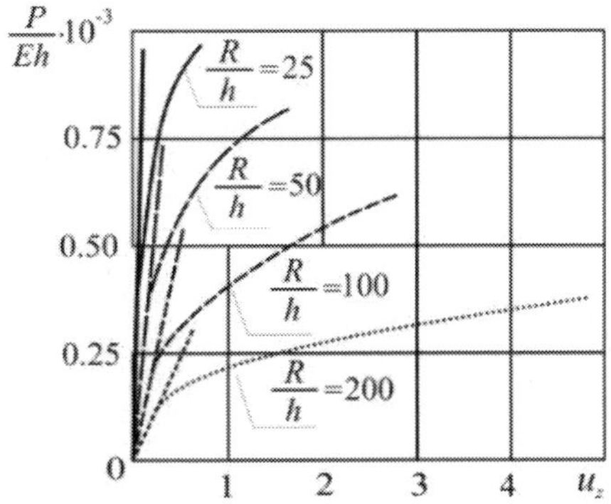

Fig. 7. Dependences of axial displacement.

4 Conclusions

The results are received on the base of the elastic solutions method. The method of the occurrence of superficial plastic deformations is studied. The results of the studying allow to define the load range at which elastic-plastic deformations appear. The areas of elastic-plastic deformations extension for different thickness ratio *R/h*= 200, 100, 50, 25 on the load ring.

References

1. V.Z. Vlasov, *Shell theory and its application to the technics* (Gostekhizdat, Moscow, 1949)
2. Y.M. Grigorenko, F.T. Vasilenko, Applied mechanics **3**, (1968)
3. A.A. Gorshkov, A.Y. Astakhova, N.Yu. Tsybin, *Fundamentals of elastic thin shells*(NIU MSUCE, Moscow, 2016)
4. I.S. Tsurkov, Collected papers of MISI named after V.V.Kuibyshev **47**, (1965)

DOI: 10.1051/matecconf/20168601016

5. A.Y. Astakhova, *Calculation of elastoplastic shells under fixed load* (MTT, Moscow, (1985)

Fundamental principles of earthquake resistance calculation to be reflected in the next generation regulations

Oleg Mkrtychev[1,*], *Guram* Dzhinchvelashvili[1]

[1]Moscow State University of Civil Engineering, Research laboratory "Safety and Seismic Resistance of Structures", 129337 Moscow, Russia

Abstract. The article scrutinizes the pressing issues of regulation in the domain of seismic construction. The existing code of rules SNIP II-7-81* "Construction in seismic areas" provides that earthquake resistance calculation be performed on two levels of impact: basic safety earthquake (BSE) and maximum considered earthquake (MCE). However, the very nature of such calculation cannot be deemed well-founded and contradicts the fundamental standards of foreign countries. The authors of the article have identified the main problems of the conceptual foundation underlying the current regulation. The first and foremost step intended to overcome the discrepancy in question is renunciation of the K_1 damage tolerance factor when calculating the BSE. The second measure to be taken is implementing the response spectrum method of calculation, but the β spectral curve of the dynamic response factor must be replaced by a spectrum of worst-case accelerograms for this particular structure or a spectrum of simulated accelerograms obtained for the specific construction site. Application of the response spectrum method when calculating the MCE impact level makes it possible to proceed into the frequency domain and to eventually obtain spectra of the accelerograms. As a result we get to know the response of the building to some extent, i.e. forces, the required reinforcement, and it can be checked whether the conditions of the ultimate limit state apply. Then, the elements under the most intense load are excluded from the design model the way it is done in case of progressive collapse calculations, because the assumption is that these elements are destroyed locally by seismic load. This procedure is based on the already existing design practices of progressive collapse calculation.

1 Introduction

About a quarter of Russian territory is located in seismic areas. Considerable funds are allocated annually for seismic safety measures. The extent of anti-seismic reinforcement is determined based on calculations of seismic resistance under the spectrum theory mandated by regulations. However, the buildings designed under the regulations in effect do not

* Corresponding author: mkrtychev@yandex.ru

ensure the level of seismic capacity required for intense earthquakes. This is corroborated by the implications of the Spitak (1988) and Neftegorsk (1995) earthquakes where the buildings were almost completely destroyed.

The recently developed design regulations allow behavior of material beyond elasticity. Non-linear behavior of structures is expected, even an assumption is made about the possible collapse of individual structural components which must be relatively accessible for investigation and repair and whose destruction does not damage the whole structure severely.

Structural designers thus have to engineer structures so that intense earthquakes inflict damage below a certain threshold, i.e. so that damage could be manageable and allowable.

Development of next generation regulations [1], which could be applied in practice by large communities of engineers, is the task of prime importance for anti-seismic engineering.

A broad implementation of such a standard presumes a high skill level of designers who master both non-linear static and dynamic methods of calculation. Besides, a certain scope of theoretical and experimental studies must be carried out in such domains as designing non-linear models for materials, non-linear computational dynamic models of reinforced concrete and steel structures while taking care of interaction with the foundation soil. Also, models of seismic impact must be elaborated for an earthquake as a nonstationary random process.

Over the past few decades, publications, reports and forum speeches produced by various scientists have stated the idea of replacing one-stage calculation with the two-stage method. This idea is based on acknowledgement of the probabilistic nature of earthquakes, whose intensity depends on their frequency (recurrence) for physical reasons. In other words, relatively weak earthquakes which occur more frequently do not dictate assumptions of considerable destruction, and the other way around, in case of intense but rare earthquakes it makes sense to plan local damage and non-elastic deformations and to assume such kind of damage albeit with no major life- and health-threatening collapse.

2 Two-level design calculation for seismic resistant structures

The engineering analysis of exposure to strong earthquakes that occurred during the validity period of SNIP P-7-81* is indicative of the fact that the regulations in force do not always ensure seismic stability of buildings and structures built in full compliance with the initial design. A range of cases demonstrate that robustness of some buildings was not ensured to the full extent [2] when the intensity of seismic impact was close to forecast or higher.

This is caused by the following circumstances.

The regulations at hand take account of non-linearity only by introducing a single integrated reduction factor †into the response spectrum method.

In fact, the calculation is done in a deterministic set-up, however, seismic impact is a highly random process and any deterministic approach will fail to determine seismic resilience of the designed buildings and structures.

Assurance of the required seismic resistance level is only possible by applying methods of the probability theory, the theory of random processes and the structural reliability theory.

† SNIP II-7-81* makes use of a magnitude which is the reverse of the reduction factor, that is the K_1 damage tolerance factor

At the moment a new conceptual framework is overdue in appearing which would make it possible to design buildings and structures protected by a certain level of seismic resistance.

Average recurrence of earthquakes of a given intensity for a particular area can be obtained from the maps of general seismic zoning [3] (Table 1).

Table 1. Average recurrence of earthquakes according to the maps of general seismic zoning (General Seismic Zoning 2012).

Map of General Seismic Zoning	Recurrence period, years
General Seismic Zoning -2012 A	100
General Seismic Zoning -2012 B	500
General Seismic Zoning -2012 C	1000
General Seismic Zoning -2012 D	2500
General Seismic Zoning -2012 E	5000
General Seismic Zoning -2012 F	10000

In Europe [4, 5] and many other countries buildings are engineered involving the maximum considered earthquake (MCE) which may occur with 0,1 probability once in 50 years. In other words, most EU countries[‡], USA, Canada, Japan etc. assume the recurrence period for MCE to be $T_{SSE} = 475$ years.

At the same time the structure of a building or facility must be engineered so that it could withstand a seismic impact which has a higher probability of occurrance than MCE, i.e. the so-called basic safety earthquake (BSE), which may occur once in 50 years with a probability of 0,5. The recurrence period of BSE is $T_{OBE} = 95$ years[§].

Below we present the approach proposed by us, which appears to be the most well-founded in our view.

2.1 Basic safety earthquake

A frequent earthquake must be expected to happen with a probability close to 1 on the specific construction site over the building's lifetime (50 - 100 years for most industrial and civil buildings and facilities).

Therefore, the seismic load of a given intensity with the recurrence period of once in a hundred years ranks similarly with, for instance, peak wind and snow loads whose recurrence periods are much shorter. Hence, the principles of defining earthquake loads for buildings and structures must be the same as for wind and snow loads.

The proposal is to make calculations for a specific intensity of seismic stress based on the ultimate limit state with no account taken of the reduction factor (by actually assuming that $K_1 = 1$), the way it is done for operational[**] loads.

Design calculations for seismic loads fitting into the serviceability limit state combined with the requirement to preserve operability of the building appear impractical. Any earthquake in any case dictates that the state of load-bearing structures of the building be examined.

[‡] The British value is $T_{SSE} = 2500$ years.

[§] Under the Japanese regulations, $T_{OBE} = 30-50$ years [6].

[**] Wind, snow etc.

Since the ultimate limit state suggests mostly elastic behavior, the response spectrum method can be considered applicable.

2.2 Maximum considered earthquake

Over the lifetime of the building on a specific construction site a 'rare' earthquake must be expected with a certain probability and a meaningfully defined recurrence period. Buildings and structures must endure earthquakes of a specified intensity without local or overall failure of load-bearing elements.

The recurrence period for the maximum considered earthquake can be determined as follows.

Reliability theory includes definition of an unconditional probability that an adverse event might happen to a person in the course of a year (severe health damage, fatal outcome). This probability accounts for about $3000 \cdot 10^{-7}$ persons-year/ (table 2) [7].

Table 2. Fatal outcome probability for a person throughout a year

Conditions and activities	10^{-7} persons/year
Car accidents	2700
Fires and explosions	400
Water bodies	280
Handling of machines	100
Air traffic	75
Electricity	51
Lightning	5.5
Public transportation	0.45
Radioactivity	0.05

Let us assume this unconditional probability of fatal outcome to be a standard maximum allowable value, i.e. $[P_{\text{fatal}}] = 3 \cdot 10^{-4}$ persons/year .

Than the design calculations must be performed based on the condition that

$$P_1 \leq [P_{\text{fatal}}], \tag{1}$$

where P_1 is the probability of a fatal outcome for a person per year for a maximum considered earthquake in the specific area of construction and the specific type of structure.

Now the P_1 probability must be determined, with the mean reference recurrence period of a maximum considered earthquake being $T_{\text{SSE}} = 500$ years.

The probability of an MCE occurring at the specific construction site throughout one year is given by

$$.P_{year} = \frac{1}{500} = 0.002 \tag{2}$$

According to a comprehensive study of implications of severe earthquakes [8] which was performed based on the lessons learned of the Russian EMERCOM, the reasonable probability of a structure's collapse P_{collapse} due to a disastrous collapse (of MCE scale) may be assumed equal to 0,1:

$$P_{collapse} = 0.1 \tag{3}$$

The probability of a person being inside the building is tentatively $P_{building} = 0.6$, while the probability of a person being hurt by the building's failure is $P_{injury} = 0.8$.

Therefore, the probability of a fatal outcome for a person per year due to MCE should be

$$P_1 = P_{year} \cdot P_{collapse} \cdot P_{building} \cdot P_{injury} = 0.002 \cdot 0.1 \cdot 0.6 \cdot 0.8 = 96 \cdot 10^{-6} \approx 1 \cdot 10^{-4} \text{persons/yea} \quad (4)$$

Hence,

$$P_1 = 1 \cdot 10^{-4} \text{ persons/year} < [P_{\text{fatal}}] = 3 \cdot 10^{-4} \text{ persons/year}. \quad (5)$$

In other words, if we assume the mean recurrence period to be $T_{\text{SSE}} = 500$ years and the probability of the building's collapse due to the maximum considered earthquake to be 0.1, than the probability of fatality for a person per year will be close to the unconditional probability and is not going to exceed it.

3 MCE impact calculation

MCE calculations must involve computation of ad-hoc limit states. Since this research is dedicated to structural states which are close to limit states and are subject to substantial non-elastic plastic strain, the non-linear nature of the structural behavior must be explicitly taken into account. The latter is only possible by using non-linear static and dynamic methods.

At the moment these methods can be utilized in design calculations made for unique buildings and facilities, for which special research and development support is provided by organizations possessing the relevant qualifications.

Design contractors who carry out design effort for standardized construction could employ the following approach for MCE calculations.

The suggestion would be to introduce the response spectrum method into the calculation of MCE impact level, as well as to involve actual worst-case accelerograms for the specific structure, while switching in such a way to the frequency domain and obtaining a spectrum of accelerograms. As a result, we get to know the response of the facility to some extent: forces, the required reinforcement, whether conditions of the ultimate limit state are fulfilled. However, the seismic impact will always deviate from the findings and differ from the impact used for the calculations. This is why it has to be checked whether another earthquake could cause the building to collapse in an avalanche-like way. The results obtained from MCE-scale earthquake calculation make it possible to identify the most intensely loaded elements which are than excluded from the computational model the way it is done for calculations of resistance to progressive collapse. Most structural designers have mastered this method successfully.

It should be noted that MCE calculation is in this case the same as progressive collapse calculation frameworked by the input seismic loads and the relevant scenarios of local failures. If the structure is not considered subject to progressive collapse, then the non-collapse criterion should be viewed as completed.

This approach is equipped to provide for the required level of seismic stability and safety of buildings and structures located in seismic areas.

4 Conclusions

This article has presented a philosophy of two-level structural design calculation for buildings exposed to earthquakes. The approach elaborated by the authors offers the possibility of designing buildings and structures with a predetermined seismic resistance and a required level of safety for human life and health.

References

1. *Proprietary standard of Moscow State University of Civil Engineering (National Research University 2015. Seismic resistance of buildings and structures. Design parameters* (Moscow State University of Civil Engineering, Moscow, 2015)
2. O. V. Mkrtychev, G. A. Dzhinchvelashvili Accounting for non-linearity in the seismic resistance theory (hypotheses and fallacies) (Moscow State University of Civil Engineering, Moscow, 2012)
3. V. I. Ulomov, L. S. Shumilina *Challenges of seismic zoning in Russia* (All-Russian Research Institute for Technical Progress and IT in Construction of the Federal agency of construction, housing and housing services of the Russian Federation, Moscow, 1999)
4. *Eurocode 8: Design of Structures for Earthquake Resistance – Part 1: General Rules, Seismic actions and Rules for Buildings* (European Committee for Standardization, Brussels, 2003)
5. M. Fardis, K. Gulvanesian *Eurocode 8: a reference for designers. Design of seismic resistant structures under EN 1998-1 and EN 1998-5. Eurocode 8: General provisions for design of seismic resistant structures, seismic impact, rules for designing buildings and retaining structures* (Moscow State University of Civil Engineering, Moscow, 2015)
6. Edoardo M. Marino, Masayoshi Nakashima, Khalid M. Mosalam Engineering Structures **27** 827–840 pp. (2005)
7. G. Augusti, A. Baratta, F. Casciati *Probabilistic methods in structural engineering* (Stroyizdat, Moscow, 1988)
8. M. A. Shakhramanyan *Seismic risk assessment and forecasting earthquake effects for rescue operations* (All-Russian Research Institute for Civil Defense and Emergencies, Moscow, 2000)

Normative approaches to structural design calculations in a non-linear framework

Oleg Mkrtychev[1], *Guram* Dzhinchvelashvili[1], and *Marina* Busalova[1,*]

[1]Moscow State University of Civil Engineering, Research laboratory "Safety and Seismic Resistance of Structures", 129337 Moscow, Russia

Abstract. This article considers the issue of taking non-linearity into account when performing design calculations for buildings and structures exposed to intense seismic impact. The existing rule codex 14.13330.2014 «SNIP II-7-81* "Construction in seismic areas" regulates the possible behavior beyond the elastic range with by introducing a damage tolerance factor. However, extensive research conducted by the authors of this article and other experts in the seismic resistance theory reveal that the values of this factor for buildings of various structural layouts must be corrected. The paper investigates the behavior of multi-element systems exposed to seismic impact, the structural layouts of them being as follows: wall-supported slab systems, reinforced concrete rigid frame buildings and buildings with a metal space frame. Calculations were made which describe the behavior of the buildings exposed to an intense earthquake modeled in the form of accelerograms standardized as 8 points on the MSK 64 scale. The calculations were performed using the LS-DYNA software package which makes use of the direct dynamic method and provides means for making all the calculations in the time domain and for taking the physical, geometric and structural non-linearity into account. The results obtained indicate that wall-supported slab buildings and rigid frame RC buildings collapse due to emerging and developing plastic deformations, while metal frame buildings lose their stability before their bearing capability is exhausted. The research carried out by the authors enabled them to pinpoint the reason for the discrepancy between the obtained values of the damage tolerance factor and the values stipulated by the regulation, as well as to draw a conclusion that these values should be refined.

1 Introduction

Seismic-resistant buildings must ensure safety of human life and preservation of tangible assets and cultural artifacts even when exposed to extremely intense earthquakes. In many cases this can be achieved without requiring that buildings or structures remain completely intact, as cracks and local damage may arise. But if human lives and the most valuable equipment are retained, such buildings and structures can be considered earthquake-resistant [1,2].

* Corresponding author: marina8busalova@gmail.com

When structures should be capable of resisting seismic impact beyond elasticity limits, it is usually acceptable that their resistance to seismic stress can be designed to be less than the stress corresponding to the linear elastic response.

Analysis of non-elastic behavior of structures should be left out from the design effort. For this, the current design regulation takes account of the capability of a structure to dissipate energy, which is driven by mainly elastoplastic behavior of its elements and other phenomena. Specifically, the regulation provides for an elasticity analysis based on the design response spectrum and introduces a reduction factor for the elastic response spectrum.

The effort of numerous researchers resulted in methods of using the above mentioned framework for conducting performance calculations of non-linear oscillations of buildings exposed to seismic impact. To determine the seismic design loads, most countries' regulations make use of reduction methodology when considering nonlinear effects. This approach is generally accepted, however, the latest research in seismic resistant construction has discovered a possibly deficient seismic capacity of the buildings and structures designed under this methodology.

2 The rationale behind the damage tolerance factor

The generally accepted design calculation of seismic resistance is based on matching the forces or strains driven by external load with maximum permissible values of internal forces (or strains) which can manifest themselves in the relevant sections of the structure. The values of maximum allowable stress in inside dimensions of a structure vary depending on whether we consider solely elastic behaviour of the material or also take account of its plastic deformations.

In fact, it is impossible to say anything certain about seismic resilience of a designed structural system. Hence, we obtain a structure with an unknown degree of seismic capacity.

Structural design which accounts for plastic and non-linear behavior or even for destruction of individual structural elements requires usage of more complex mechanical and mathematical models and theories, i.e. assumptions of controlled damage. Also, regard must be had to the redistribution of forces within the structure, the cause of which are the emerging zones of plastic deformations. But the principle of matching external and internal forces remains a criterion for the assessment of a structure's bearing capacity in all cases.

The regulations [3] allow for development of non-elastic (plastic) deformations for most buildings and structures. Non-linear behavior is taken care of by multiplying the seismic forces by the K_1 factor which absorbs the allowable damage. For instance, K_1=0.22 is used for monolithic wall-supported slab buildings whose structures can tolerate residual deformations and damage detrimental to normal operation but still provide safety for human life and equipment. Numerous experts in seismic resistant construction deem the values of K_1 presented in Table 3 in [3] ill-founded.

The necessity to reconsider the K_1 factor manifests itself, firstly, when data are analyzed for calculating the seismic capacity of a building using the regulations currently in effect. Research reveals that misuse of the damage tolerance factor may result in reduced seismic resilience of structures and their possible overall collapse [4]. Secondly, the authors have done research of buildings with different types of structures in a non-linear dynamic framework and arrived at the conclusion that the K_1 factor must assume values other than those stipulated in the regulation [3].

The findings of this research are presented below.

2.1 Multi-element wall-supported slab system

Let us consider a 27-storey monolithic tower block H=81.6 m high with an elevator shaft at the center (Figure 1). The building was designed to be erected in a seismic area. The calculation under the response spectrum theory [3] indicates that the seismic capacity of the building is 9 points on the MSK-64 scale.

Fig. 1. Design diagram of the multistorey monolithic building.

Now the building is exposed to an external impact along the horizontal axis X which is set by the accelerogram presented in Figure 2.

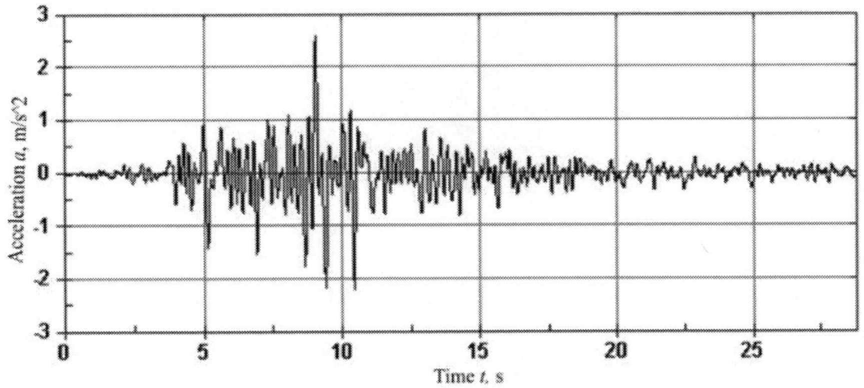

Fig. 2. Accelerogram of the original impact.

Figure 3 displays the motions of the point w at the level of the upper floor slab for a problem in a linear (curve 1) and nonlinear (curves 2 and 3) setups. The non-linear setup presumes that the material deforms under the Prandtl diagram with the initial modulus of elasticity being $E=3\cdot10^4$ MPa. Further, options with incremental reduction of the yield point of the material σ_T, are considered, and this is the way how increased intensity of non-linear effects is modeled. The yield point σ_T = 20 MPa is reflected by curve 2, while σ_T=13 MPa is depicted by curve 3.

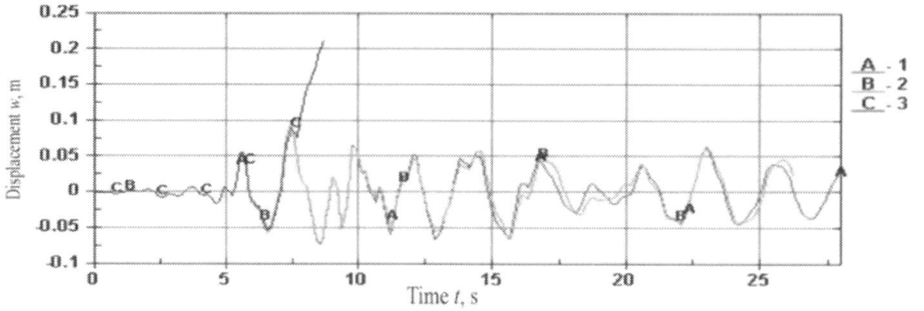

Fig. 3. Motion of the *w* point at the level of the upper floor slab.

σ_T=13 MPa at the time point t=7.6 s triggers progressive collapse of the building (see curve 3).

Figure 3 displays that design calculations made for multistorey monolithic buildings in a non-linear setup do not demonstrate any reduction in the displacement amplitude of the structure's characteristic points. The reason for this is that plastic deformations can be spotted in elements insignificant in number compared to the total number of elements (Figure 4). On the whole, the stiffness of the structure barely changes till collapse.

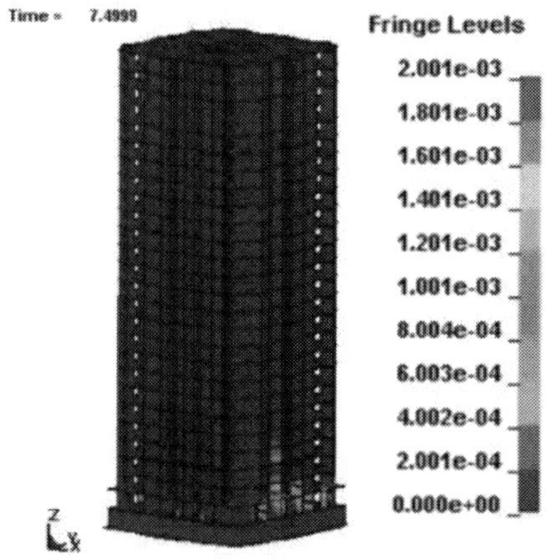

Fig. 4. Isofields of plastic deformations prior to collapse (t=7.6 s).

Seismic loads assigned to the type of buildings under consideration must not be scaled down, and the K_1 factor must not be applied.

If damage tolerance is taken into account in a linear calculation, seismic impact may be reduced only in case a surplus of bearing capacity can be proven and stress can be redistributed between the elements. E.g., design regulations of foreign countries provide for load reduction (to a much lower extent) only for buildings and structures of specific structural layouts after an in-depth analysis of structural behavior.

Ultimately, the buildings of specific structural layouts designed in accordance with the rule codex 14.13330.2014 SNIP II-7-81* "Construction in seismic areas" may exhibit deficiency of seismic capacity up to 2 points. For instance, Russian 9-point zones are home to buildings able to tolerate 7 points only.

2.2 Reinforced concrete rigid frame building

Let is consider a 5-storey reinforced concrete monolithic building, the structural layout of which is a complete space frame (exterior walls non-bearing) with column spacing being 6 meters in both directions (Figure 5). Column section is 400×400 mm, beam section is 500×300 mm.

Fig. 5. Design model of the rigid frame building.

The building is exposed to an external impact set by the accelerogram standardized as 8 poins on the MSK-64 scale (see Figure 2).

Figure 6 displays the motions of the *w* point on the roofing level in a linear and non-linear setup. The non-linear setup presumes that the material deforms under the Prandtl diagram with the initial modulus of elasticity being E=3·10^4 MPa. The maximum allowed value of plastic deformations in introduced into the calculation [5,6,7]. Further, options with incremental reduction of the yield point of the material σ_T are considered, and this is the way how the increasing intensity of non-linear effects is modelled. Curve 1 denotes elastic behavior, curve 2 – σ_T=40 MPa yield point; curve 3 – σ_T=30 MPa yield point; curve 4 – σ_T=29 MPa yield point.

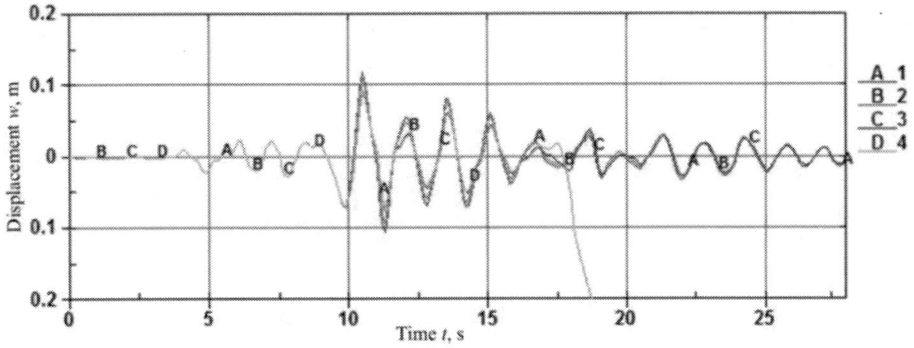

Fig. 6. Motions of the *w*–point on the roofing level.

σ_T=29 MPa at the time point of 17 seconds triggers progressive collapse of the building (see curve 4). Figure 7 displays the building at the time point t=18 s along with the isofields of maximum stress intensity σ_i in the elements.

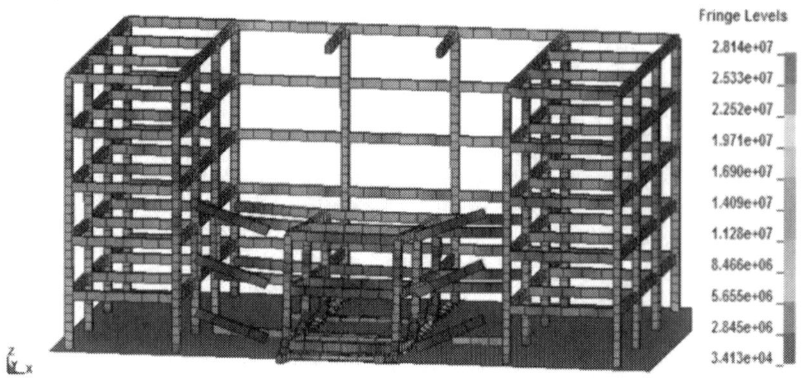

Fig. 7. Collapse of the building.

The research undertaken enables us to draw a conclusion that this type of reinforced concrete rigid frame buildings dictates that the reduction factor to take account of non-linear behavior of structures be considered equal to 1.3. The factor which absorbs the allowable damage to buildings and structures (damage tolerance factor) must be set to 0.77.

2.3 A building with a space metal frame

Now we are going to consider the same seismic impact and the response of a 5-storey building to it. The building's structural layout is a complete rigid space metal frame (exterior walls nonbearing) with the identical 6 m spacing between the columns in both directions (see Figure 5). The columns and beams are I-shaped.

Figure 8 displays the motions of the point *w* at the roofing level for a problem in a linear and non-linear setups. The non-linear setup presumes that the material deforms under the Prandtl diagram with the initial modulus of elasticity being E=2.1 MPa. The limit value of intensity of plastic deformations is assumed to be ε_i= 0.08. Options with incremental reduction of the yield point of the material σ_T are considered. Curve 1 denotes elastic behavior; curve 2 – yield point σ_T=300 MPa; curve 3 – yield point σ_T=200 MPa; curve 4 – yield point σ_T=150 MPa.

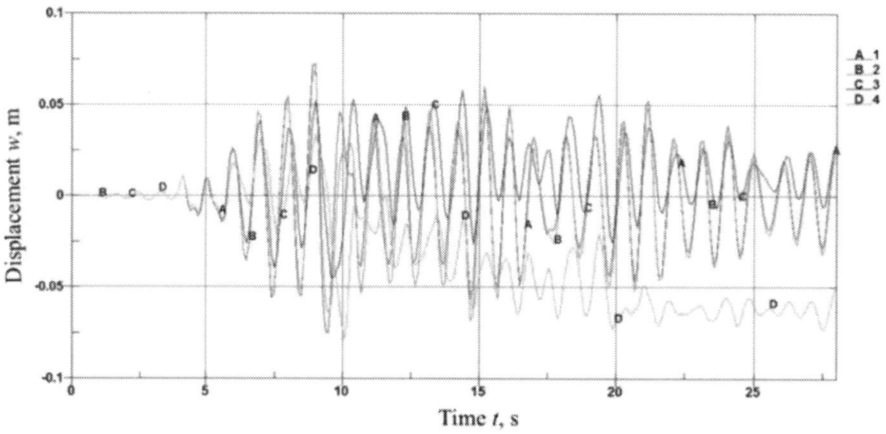

Fig. 8. Motions of the *w* point on the roofing level.

It should be noted that the load-bearing metal frame of the building exposed to an earthquake may lose stability before the bearing capacity of the system is exhausted as plastic deformations reach their limtis. Thus, if σ_T=130 MPa the structure will lose stability altogether at the time point of 13 seconds. Figure 9 depicts the buckling and the isofields of motions along the *Y* axis.

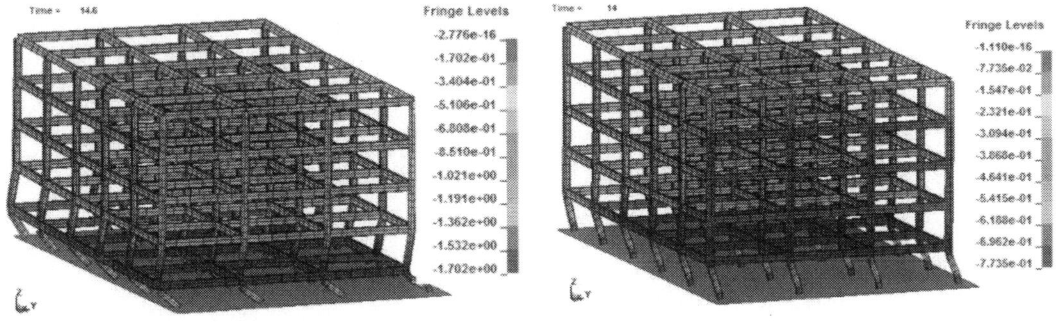

Fig. 9. Overall failure of the building: a) t=14 s; b) t=14.6 s.

The output evaluation goes to show that the building in question with a complete steel frame (exterior walls nonbearing) dictates that the reduction factor to take account of non-linear behavior of structures should be set to 5. The coefficient which absorbs the allowable damage to buildings and structures (damage tolerance factor) is equal to 0,2.

3 Conclusions

The authors used the output of calculations made for the buildings of different structural layouts to conduct additional research, which enabled them to substantiate why the values of the K1 factor are different from the regulatory values (Table 1) [3,8].

Table 1. K_1 comparative table.

	Structural layout	**K_1 factor (SNIP II-7-81*)**	**Values obtained from a direct nonlinear analysis**	**Overestimation of bearing capacity, times**
1	Rigid steel framework buildings	0.25	0.2	0.8
2	Buildigs with a monolithic RC rigid framework	0.35	0.77	2.2
3	Monolithic wall-supported slab buildings	0.22	1	4.5

Consequently, the reduction factor is determined unreasonably in most cases. More research in this domain along with the use of non-linear static and dynamic methods is necessary. The findings of such research should form the basis for the next generation regulations.

References

1. A. N. Akhvelediani, *Theorems on logical collapse in terms of the theory of limit equilibrium of plastic systems* (Russika encyclopaedic foundation, Saint Petersburg, 2011)
2. O. V. Mkrtychev, G. A. Dzhinchvelashvili, Science and transportation, E **2**, 8 (2002)
3. *SP 14.13330.2014 SNIP II-7-81. Construction in Seismic Areas* (Analitik, Moscow, 2014)

4. O. V. Mkrtychev, V. B. Dorozhinskiy, D. S. Sidorov, Proceedings of Moscow State University of Civil Engineering **12**, 10 (2015)
5. O. V. Mkrtychev *Safety of buildings and facilities exposed to seismic and emergency impact* (Moscow State University of Civil Engineering, Moscow, 2010)
6. V. A. Rzhevsky *A research on non-stationary elastoplastic systems exposed to complex seismic impact* (Stroyizdat, Moscow, 1988)
7. O. V. Mkrtychev, G. A. Dzhinchvelashvili, M. S. Busalova, Proceedings of Moscow State University of Civil Engineering **12**, 7 (2013)
8. O. V. Mkrtychev, G. A. Dzhinchvelashvili *Accounting for non-linearity in the seismic resistance theory (hypotheses and fallacies)* (Moscow State University of Civil Engineering, Moscow, 2012)

Mathematical simulation of two-phase gas-liquid flow for jet aeration structures

Vadim Akhmetov[1,*], *Valentina* Akhmetova[1]

[1]Moscow State University of Civil Engineering, 26, Yaroslavskoe Shosse, Moscow, 129337, Russia

Abstract. The problem of the motion of an aerated submerged jet into stationary fluid mass is investigated. The equations of balance of mass, momentum and forces of interface interaction are solved numerically by finite-difference method. Jet propagation path and maximum depth of submergence of the air bubbles are calculated for various Froude numbers. The results are compared with available experimental data.

1 Introduction

The polluted state of natural and artificial water bodies, settling ponds, and sludge ponds in the biological and chemical industries represents an increasing danger for the environment. In connection with this the acute need arises to construct highly productive treatment and aeration plants. The artificial aeration method used at them by means of an aerated jet of water is rather simple to realize and is quite effective for enriching fluid masses with dissolved oxygen. By virtue of this the problem of the propagation of an aerated submerged jet is extremely important and urgent.

2 Formulation of the problem and numerical procedure

Let us examine on the basis of the method of integral relations the general statement of the problem of propagation of a submerged jet containing uniformly distributed air bubbles and issuing with velocity U_0 from a round pipe with diameter D_0 into a fluid mass at angle θ_0 ($0 \le \theta_0 \le 90^0$) to its surface (Fig. 1). Here we will assume that the air bubbles in the investigated jet have the same size and a regular spherical shape, the radius of curvature for the path of the central line of the jet is much greater than the diameter of the bubbles, and phenomena of disintegration, merger, and formation of new bubbles and phase transitions are absent. We introduce curvilinear coordinates s, n, θ, where s is the coordinate along the central line of the jet; n is the normal to it; θ is the angle between the surface of the water body and the central line of the jet. We will consider that the axial component of the velocity in each cross section is a constant, i.e., $U = U(s)$. Then the equations of balance of mass and momentum for a control volume of the aerated jet in accordance with the method of integral relations can be written as follows:

* Corresponding author: vadim.akhmetov@gmail.com

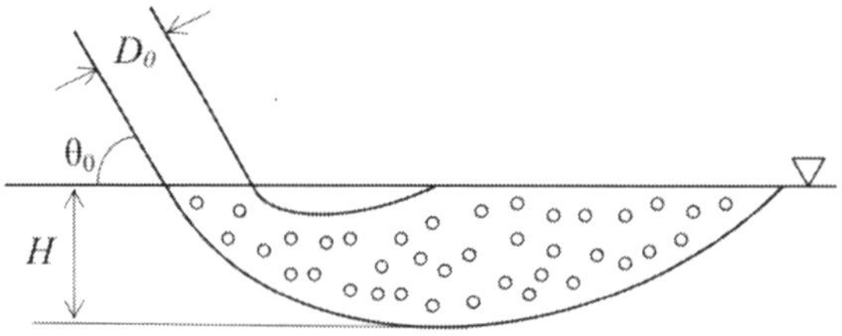

Fig. 1. Schematic of the problem.

$$\frac{d}{ds}[\rho_L U_L(1-\alpha)A] = E \quad (1)$$

$$\frac{d}{ds}\left[\rho_L U_L^2(1-\alpha)A\right]+\frac{d}{ds}\left[\rho_B \alpha U_B^2 A\right]=(\rho_L-\rho_B)\alpha Ag\sin\theta \quad (2)$$

$$\frac{d}{ds}\left[\rho_B U_B \alpha A\right]+\rho_B V_B \alpha D = 0 \quad (3)$$

$$A\left[\rho_L(1-\alpha)U_L^2+\rho_B\alpha U_B^2\right]\frac{d\theta}{ds}=\alpha(\rho_L-\rho_B)Ag\cos\theta+\rho_B\alpha DV_B^2 \quad (4)$$

where V is the transverse velocity component; D is the jet diameter; p is the density; A is the cross-sectional area of the jet; α is a parameter characterizing the ratio of the volume of air to a unit of the mixture; g is the acceleration of gravity; the indices L and B pertain respectively to the liquid and bubble phases. The quantity E characterizes the ejection properties of the jet during its propagation and is proportional to the perimeter of the local cross section of the jet, the velocity, and the density. The equations of motion of the bubbles can be obtained from an examination of the balance of forces of interface interaction, including friction (Stokes force), buoyant, and gravitational forces and the force related to interaction of the apparent additional masses. Writing these relations in projections onto the axial and transverse directions, we will have

$$(\rho_B+k\rho_L)U_B\frac{dU_B}{ds}-k\rho_L U_B\frac{dU_L}{ds}=\frac{3}{4}c_{Ds}\rho_L(U_L-U_B)\left|U_L-U_B\right|\frac{1}{d_B}+(\rho_L-\rho_B)g\sin\theta \quad (5)$$

$$\rho_B U_B\frac{dV_B}{ds}=-\frac{3}{4}c_{Dn}V_B\left|V_B\right|\frac{1}{d_B}+(\rho_L-\rho_B)g\cos\theta-k\rho_L U_B\frac{dV_B}{ds} \quad (6)$$

Here c_{Ds}, c_{Dn} are the drag coefficients of the bubbles in the axial and transverse directions; d_B is the diameter of the bubbles; k is the coefficient of the apparent additional mass. Supplementing these equations with relations for determining the Cartesian coordinates x_c, y_c of the central line of the jet

$$\frac{dx_c}{ds}=\cos\theta, \qquad \frac{dy_c}{ds}=\sin\theta \quad (7)$$

we obtain a closed system of equations (1)-(7) for the unknowns U_B, V_B, θ, α, D, x_c, y_c.

Changing to dimensionless variables, having determined them by means of the initial values D_0, α_0, U_{L0} and V_{B0}

$$\xi = \frac{s}{D_0}, \quad \widetilde{D} = \frac{D}{D_0}, \quad \widetilde{A} = \frac{A}{A_0}, \quad \widetilde{U}_L = \frac{U_L}{U_{L0}}, \quad \widetilde{U}_B = \frac{U_B}{U_{B0}}$$

$$\widetilde{E} = \frac{E}{U_{L0}D_0}, \quad \widetilde{V}_B = \frac{V_B}{V_{B0}}, \quad \widetilde{\alpha} = \frac{\alpha}{\alpha_0}, \quad \widetilde{x}_c = \frac{x_c}{D_0}, \quad \widetilde{y}_c = \frac{y_c}{D_0}.$$

we can write system (1)-(7) in the form of eight ordinary differential equations (omitting the tilde sign):

$$\frac{1}{A}\frac{dA}{d\xi} - \frac{\alpha_0}{1-\alpha\alpha_0}\frac{d\alpha}{d\xi} = \frac{4}{\pi}\frac{E}{U_L A(1-\alpha\alpha_0)} - \frac{1}{U_L}\frac{dU_L}{d\xi} \tag{8}$$

$$U_B \alpha \rho_R U_R^2 \alpha_0 \frac{dU_B}{d\xi} + U_L(1-\alpha\alpha_0)\frac{dU_L}{d\xi} =$$

$$= -\frac{4}{\pi}\frac{U_L}{A}E + \frac{4}{\pi}\frac{U_B}{D}\rho_R U_R^2 \alpha_0 V_B \alpha + (1-\rho_R)\alpha\alpha_0 \frac{1}{\mathrm{Fr}}\sin\theta \tag{9}$$

$$\left[(1-\alpha\alpha_0)U_L^2 + \rho_R \alpha\alpha_0 U_B^2 U_R^2\right]\frac{d\theta}{d\xi} = (1-\rho_R)\alpha\alpha_0 \frac{1}{\mathrm{Fr}}\cos\theta + \frac{4}{\pi}\frac{\rho_R \alpha\alpha_0 V_B^2 U_R^2}{\sqrt{A}} \tag{10}$$

$$\frac{1}{A}\frac{dA}{d\xi} + \frac{1}{\alpha}\frac{d\alpha}{d\xi} = -\frac{4}{\pi}\frac{V_B}{U_B D} - \frac{1}{U_B}\frac{dU_B}{d\xi} \tag{11}$$

$$(k+\rho_R)U_B U_R^2 \frac{dU_B}{d\xi} - kU_B U_R \frac{dU_L}{d\xi} =$$

$$= \frac{3}{4}c_{Ds}(U_L - U_B U_R)\left|U_L - U_B U_R\right|\frac{1}{d_R} + (1-\rho_R)\frac{1}{\mathrm{Fr}}\sin\theta \tag{12}$$

$$U_B \frac{dV_B}{d\xi} = \frac{1}{k+\rho_R}\left[-\frac{3}{4}c_{Dn} V_B |V_B| \frac{1}{d_R} + (1-\rho_R)\frac{1}{\mathrm{Fr}U_R^2}\cos\theta\right] \tag{13}$$

$$\frac{dx_c}{ds} = \cos\theta, \qquad \frac{dy_c}{ds} = \sin\theta \tag{14}$$

where $Fr = U_{L0}^2 / gD_0$, $U_R = U_{B0}/U_{L0}$, $d_R = d_B / D_0$, $\rho_R = \rho_B / \rho_L$.

The Cauchy problem is examined for system (8)-(14). We will seek the solution of the initial problem for various values of the Froude numbers Fr, initial concentration α_0 and slope angle of the θ_0 ,considering $\rho_R = 0.00129$ and $d_R = 0.01$ to be constant. The system of equations (8)-(14) was integrated numerically by the Runge-Kutta method. The integration step was 0.01-0.05.

3 Numerical results

Let us examine the main properties of the solutions obtained. One of the most important characteristics is the depth H of development in the water body, as which the maximum depth of submergence of the air bubbles is taken here. Figure 2,a shows the dependences of H referred to the initial jet diameter D_0 on the Fr number for slope angles of the jet to the surface of the water body $\theta_0 = 90^0$ and fixed values $\alpha_0 = 0.1 - 0.6$. For a small value $\alpha_0 = 0.1$ the depth H rapidly increases with the Froude number and for large Fr numbers approaches the limiting value $H \approx 60$. This value of H corresponds to Abramovich's analytical calculation [1] for an unaerated jet, and at this depth the axial velocity of the jet is about 10% of the initial value. However, with such a small initial value α_0 the air concentration at large values of H is quite small. With an increase of α_0 the depth H increases noticeably more slowly, but the character of the dependence is preserved qualitatively.

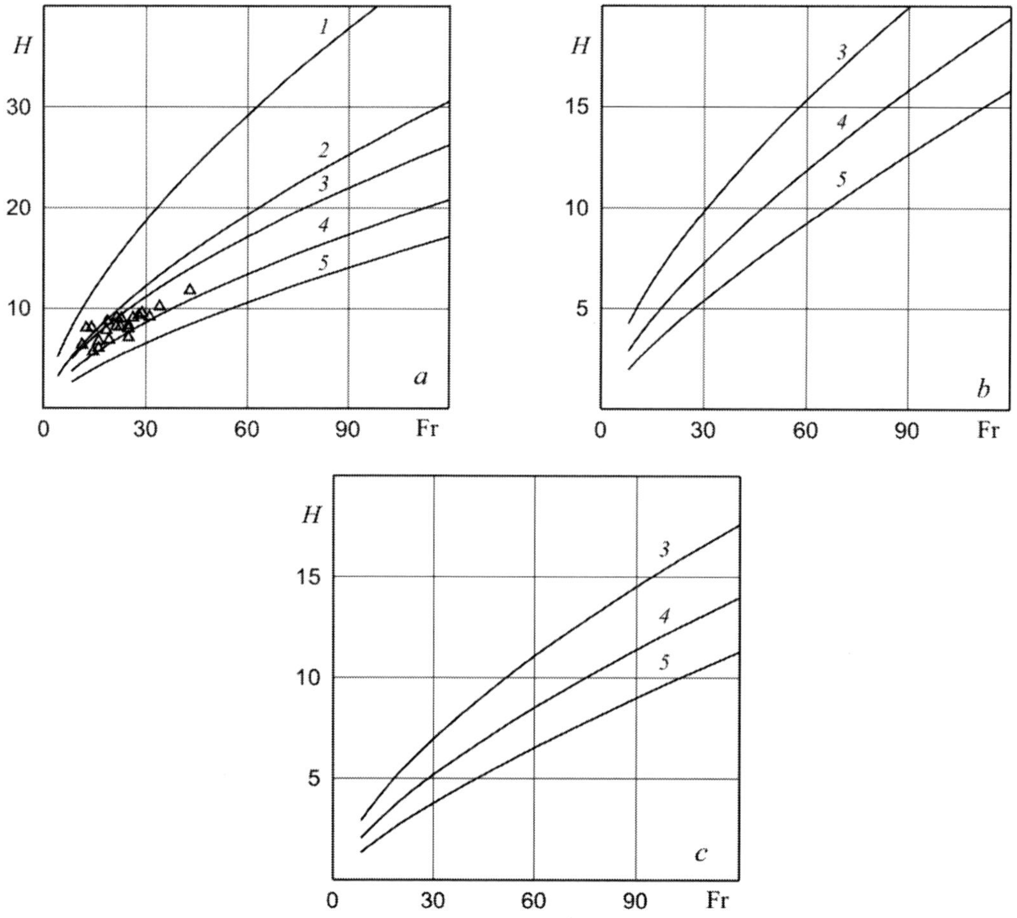

Fig. 2. Dependence of the depth H on the Froude number Fr for slope angles of the jet $\theta_0 = 90^0, 60^0, 45^0$ (a-c) for $\alpha_0 = 0.1, 0.3, 0.4, 0.5, 0.6$ (curves 1-6).

Let us compare the obtained functional dependences $H = H(Fr)$ for $\theta_0 = 90^0$ with the experimental data [2-4]. Hydraulic investigations of the propagation of an aerated jet were conducted on a laboratory device in Moscow State University of Civil Engineering. The working process in it is organized in the following way. Two coaxially swirled flows rotating in opposite directions are formed by means of spiral swirlers. A vacuum occurs on the axis of the flow, promoting sucking of air from the atmosphere into the flow. As a result of interaction of the swirled flows, the air is broken into small bubbles and the intensity of swirling of the interacting flows rapidly decreases. Thus, at the exit of the aerator nozzle there is an axial flow containing air bubbles rather uniformly distributed over the cross section of the jet. The main parameters of the experimental device have the values: $D_0 = 0.126, 0.168$ and 0.168 m; water discharge up to 0.1 m^3/sec; $\alpha_0 = 0.43 - 0.58$.

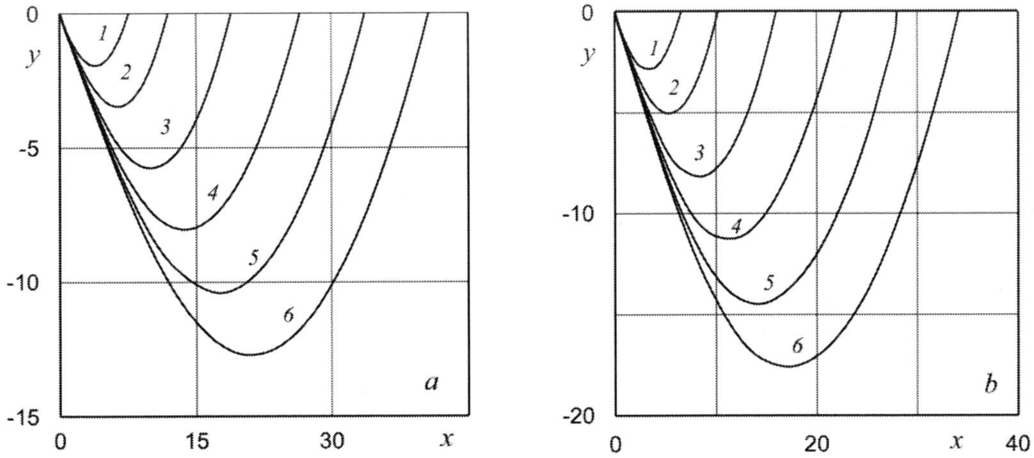

Fig. 3. Paths of the axial line of the propagation of a jet with slope angle $\theta_0 = 45^0$ (a) and $\theta_0 = 60^0$ (b) for Fr = 8.16, 16.33, 32.65, 54.4, 85.03, 122.45 (curves 1-6).

The results of these investigations characterizing the maximum depth of submergence H of the air bubbles for $\theta_0 = 90^0$ are plotted as asterisks in Fig. 2,a. It is seen that the experimental points in this case agree rather well with the theoretical curves and that the mathematical model reflects the essence of the phenomenon.

Figure 2,b,c shows the analogous dependences of H on the Fr number for slope angles of the jet $\theta_0 = 60^0$ and $\theta_0 = 45^0$ respectively. The character of these curves remains qualitatively the same as in the preceding case. Here the value of H for the same values of α_0 and Fr and a smaller value of θ_0 becomes somewhat smaller. However, the slope of the jet makes it possible to increase the surface area of aeration.

Figure 3,a,b show the calculated paths of the axial line of the jet during its propagation with slope angles to the surface of the water body $\theta_0 = 45^0$ and $\theta_0 = 60^0$ respectively, for the value $\alpha_0 = 0.5$ and various Froude numbers. The Cartesian coordinates x, y are read from the point of the surface of the water body to which the aerated jet is delivered and are referred to the outlet diameter D_0. The maximum depth to which the air bubbles are transported is clearly seen in Fig. 3,a,b. On the other hand, by means of these dependences we can estimate the distance from the site of installing the aerator at which effective mixing of air bubbles with the aerated mass occurs. Comparing the dependences in Fig. 3,a and 3,b we see that for the same values of the Froude number and a large slope angle ($\theta_0 = 60^0$)

the depth of the bubble zone is greater but the surface area is smaller. To all appearances, the optimal variant here is the installation of the aerator with an angle of delivery of the jet

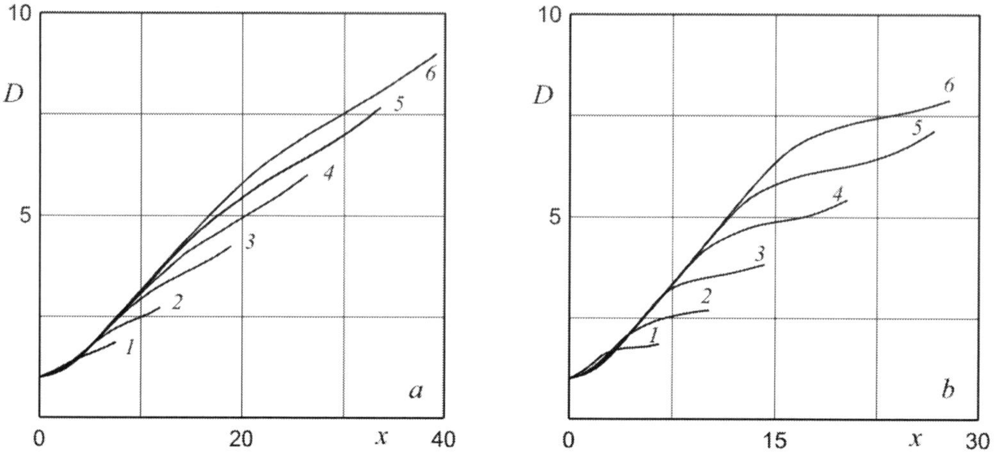

$\theta_0 = 45^0$ to the surface of the water mass.

Fig. 4. Change in the diameter of the jet during its propagation with slope angle $\theta_0 = 45^0$ (a) and $\theta_0 = 60^0$ (b) for $Fr = 8.16, 16.33, 32.65, 54.4, 85.03, 122.45$ (curves 1-6).

Figure 4,a,b shows the calculated change in the relative diameter of the jet during its propagation for angles $\theta_0 = 45^0$ and $\theta_0 = 60^0$ for the same values of the Froude numbers as in Fig. 3. It is seen that with the exception of the initial and end segments the change in the diameter is a linear function of x, which agrees with Abramovich's theory.

Conclusions

Thus the investigations permit calculating the main parameters of the process of propagation of the aerated jet, and the results obtained can be used when designing various types of jet aeration systems.

References

1. G. N. Abramovich, *Theory of Turbulent Jets* (Nauka, Moscow, 1984)
2. V.K. Akhmetov , V.Ya. Shkadov, *Numerical Simulation of Viscous Vortex Flows for Applications to Technics* (ASV Publishing House, Moscow, 2009)
3. V.K. Akhmetov , V.V. Volshanik, A.L. Zuikov, G.V. Orekhov, *Modelling and calculation of counter vortex flows* (ASV Publishing House, Moscow, 2012)
4. V.K. Akhmetov, *Structure and hydrodynamic stability of swirling flows with recirculation zones* (MGSU, Moscow, 2009)

Experimental and theoretical studies of strength and stress-strain properties of the concrete under conditions of complex stress-strain states with presence of tensile stresses from the effect of a single-shot dynamic load of the general form

Konstantin Tsvetkov [1,*]

[1]Moscow State University of Civil Engineering, Department of Strength of Materials, Moscow, Russia

Abstract. The publication examines the generalization of results of experimental and theoretical studies of concrete under static and single-shot dynamic loading under conditions of single-axial, two-axis, three-axis stress states with the presence of tensile stresses. Dependences for determining concrete strength, concrete stress-strain modulus, concrete ultimate strains. Research results extend the idea of concrete behaviour at various types of force action and can be used in the calculation of structures for special combination of loads, taking into account the actual concrete properties.

1 Formulation of the problem

In order to improve the calculations of building structures for the special combination of loads, as well as when carrying out check calculations to assess the consequence of the emergency dynamic impacts, it is necessary to have the reliable information on strength and stress-strain characteristics of structural materials, taking into account the type of stress-strain state (SSS) and loading rate.

Emergency action of the dynamic load can lead to appearance of various SSS in the construction, including different to accepted in calculations for the basic combination of loads, at that the most dangerous in terms of the concrete behaviour will be zones with the presence of tensile stresses.

A limited number of studies is devoted to the assessment of the joint influence of the stress state type and increased loading rate on properties of the concrete. Carried out studies concern mainly issues of strength under biaxial and triaxial compression, at that the data on concrete behaviour under dynamic loading under conditions of complex stress states with the presence of tensile stresses are practically absent. Experimental and theoretical studies

* Corresponding author: sopromat@mgsu.ru

of the concrete under conditions of complex SSS with the presence of tensile stresses from the action of the single-shot dynamic load of the general type expand notions about the behaviour of concrete at various power impacts. Obtained results can be used when developing the theory of concrete dynamic strength, as well as after carrying out additional studies on a larger number of samples, when varying the loading rate can be used in calculation of structures for the special combination of loads, taking into account SSS factor and the loading rate.

With account of the above it was important to carry out experimental studies of strength and stress-strain characteristics of concrete under single-axis tension and "compression-tension" stress state at static and dynamic loading.

2 Methodology of experimental research

Concrete prisms of 10x10x40 mm were tested. 2 series of samples were tested. Samples of the first series (P-1) were tested at "compression-tension" stress state. Samples of P-2 series were tested at "two compressions - tension" stress state. Besides, samples of both series were tested for the single-axis tension and compression.

Strength of P-1 series concrete under single-axis compression was R_b = 31.5MPa, under single-axis tension R_{bt} = 2.95 MPa. P-2 series concrete had the strength for compression R_b = 16 MPa, for tension R_{bt} = 1.45 MPa.

When carrying out tests the set-up was used similar to that described in [1]. The difference was in the application of the reversing principle - the conversion of compressive forces into tensile, that was achieved through the use of the special device.

Fixing of samples in the set-up was carried out with the use of studs, which were screwed into heads of samples and fixed in hinged system of set-ups. Use of the hinged system allowed centring samples along the physical axis with the trial load.

The set-up allowed realization of the dynamic loading at a constant rate of stress increase. In our experiments $\dot{\sigma}_1 \approx$ 70MPa/s, the destruction time was $\tau \approx$ 0.05s. Static tests were carried out with the same set-up.

Creation of compressive stresses σ_2 and $\sigma_2 = \sigma_3$ was fulfilled with the use of membrane-type set-ups, which provide the hydrostatic method of load transmission.

Experimental studies at the stress states of "compression - tension" and "two compressions - tension" were carried out according to the scheme of complex loading. At the first stage, the static compressive load was applied to the sample to a certain level from the prism strength (in our experiments: $\sigma_2 = 0.15R_b$, $\sigma_2 = 0.3R_b$, $\sigma_2 = 0.45R_b$ or $\sigma_2 = 0.6R_b$ – at "compression-tension" stress state and $\sigma_2 = \sigma_3 = 0.3R_b$ – at «two compressions - tension») stress state, which value was maintained constant at the second stage of loading. At the second stage, loading of the sample was performed with static or dynamic tensile load, acting in the direction of σ_1 and increasing from zero to the destructive values.

At all stages of loading, forces and longitudinal strains were synchronously recorded using strain gauges, modern electronic equipment and special software.

For each level of stress σ_2 was tested from two to five samples at static and dynamic loading. At that, for experiments carried out under the same conditions, close results were obtained (variation coefficient did not exceed 10%).

3 The main results of experiments

Main results were obtained for samples of P-1 series.
Strength experimental results of P-1 series samples are given in Fig. 1.

Experimental results showed that the destruction at "compression-tension 'stress state takes place as a result of tensile stress achievement of a certain limit value, at that the presence of the compressive stress σ_2 influences the strength value in dependence on σ_2/R_b.

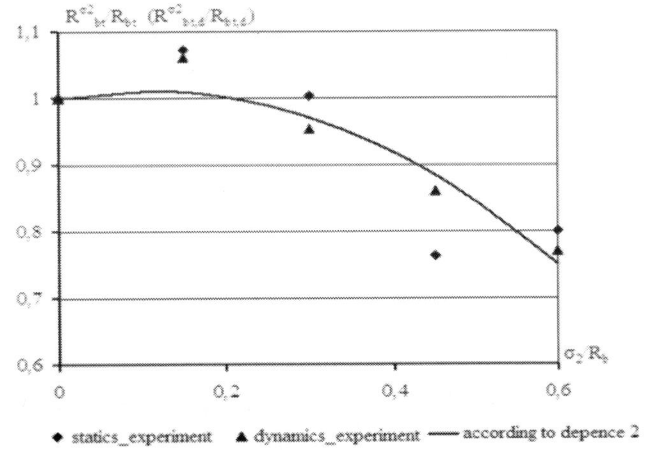

Fig. 1. The influence of the stress state type and the stress σ_2 at static and dynamic loading under conditions of single-axis tension and "compression – tension" stress state.

Both at statics and dynamics the following influence of σ_2/R_b level on the concrete strength is observed. When applying compressive stresses $\sigma_2 = 0.15R_b$, a certain small increase of the concrete strength (up to 7%) takes place. At $\sigma_2 = 0.3R_b$ the strength value becomes again close to the strength at the single-axis tension. Further increase of the stress level σ_2 keeps the tendency of the strength decrease. Obtained results as applied to the static loading coincide with the data [2].

Type of the stressed state and the level of the edging draft do not significantly influence the dynamic hardening factor, in other words, the change in the concrete strength, depending on the growth rate of stresses when "compression-tension" stress state with different levels of σ_2 takes place proportionally to the dynamic hardening factor corresponding to the single-axis tension.

The value of the dynamic hardening factor for the rate accepted in our studies $k_{bt,v}$ was 1.32–1.39.

Based on the experimental studies, the attempt was made to write the concrete strength criteria under single-axis tension and "compression-tension" stress state, taking into account the type of stress state, stress level σ_2 and stress growth rate (1).

$$\begin{cases} \sigma_2 \le R^{\sigma_2}_{bt,v} = k_{bt,v} F^{\sigma_2}_{bt} R_{bt} \\ k_{bt,v} = 1.08(\dot{\sigma}_1)^{0.05} \\ F^{\sigma_2}_{b} = 1 + 0.22\left(\frac{\sigma_2}{R_b}\right) - \left(\frac{\sigma_2}{R_b}\right)^2 \\ 0 \le \frac{\sigma_2}{R_b} \le 0.8 \\ 10^0 \le \dot{\sigma}_2 \le 10^2,\ \text{MPA/s} \quad \left(10^{-4} \le \dot{\varepsilon}_2 \le 10^{-2}\ \text{url/s}\right) \\ 10^{-1} \le \dot{\sigma}_1 \le 10^4,\ \text{MPA/s} \quad \left(1\cdot 10^{-1} \le \dot{\varepsilon}_1 \le 10^{-4}\ \text{url/s}\right) \\ \dot{\sigma}_2 = const \end{cases} \quad (1)$$

In expression (1):

$R^{\sigma_2}_{bt,v}$- concrete strength with account of the type of the stress state and loading rate;

$k_{bt,v}$ - dynamic hardening factor of the concrete at single-axis tension ($\dot{\sigma}$- stress growth rate, MPa/s. Expression for $k_{bt,v}$ is written with the use of the data [3];

$F^{\sigma_2}_{bt}$- function characterizing the influence of the stress level σ_2 on the change in the concrete strength under static loading. It is recorded according to the results of these studies. Factors are selected using MS Excel means;

R_{bt}- concrete strength at the single-axis tension under static loading;

σ_1 – tensile stress;

σ_2 – compressive stress;

Limits of applicability (1) are given in accordance with the classification of loading modes, proposed in [4].

Based on measurement results of longitudinal strain values, the evaluation of the influence of stress state type, stress level σ_2 and the rate of loading on the concrete strain modulus and value of ultimate strains, determined with account of the first stage of loading $\varepsilon_{1,u}$ and without account of the first stage (maximum strains from action of tensile stresses) $\varepsilon_{1,u,1}$.

At static loading, the type of stress state and level σ_2 do not have a tangible influence on the value of concrete strain initial modulus, calculated with account of the strain curve. At the same time, with increase of the loading rate the value of the dynamic modulus decreases with the growth of σ_2/R_b.

For dynamic loading with 70 MPa/s rate there were recorded the following dependences:

for initial modulus:

$$E^{\sigma 2}_{o,d} = E_{o,d}\, e^{-0.4\sigma_2/R_b} \quad (2)$$

for modulus calculated with account of the entire strain curve:

$$E^{\sigma 2}_{o,d} = 0.88\, E_{o,d}\, e^{-0.4\sigma_2/R_b} \quad (3)$$

Factors to the power *e* are adjusted depending on the loading rate.

In our experiments for the single-axis tension it was obtained $E_{0,d}/E_0 = 1.21$. In order to determine the dynamic initial modulus at rates which are significantly different from realized in these studies, carrying out of additional experimental studies is required.

Data on strains $\varepsilon_{1,u,t}$ and $\varepsilon_{1,u}$ are shown in Table 1.

Table 1. Ultimate strains and maximum strains from the action of the tensile load at single-axis compression and "compression-tension" stress state.

Type of the strain	σ_2/R_b				
	0	0.15	0.3	0.45	0.6
	Statics				
$\varepsilon_{1,u}$, url × 10^5	9.3	12.5	14.0	14.3	16.8
$\varepsilon_{1,u,t}$, url × 10^5	9.3	10.2	9.6	7.5	8.3
Dynamics ($\dot{\sigma}$=70 MPa/s)					
$\varepsilon_{1,u}$, url ×10^5	10.3	14.0	14.6	16.6	18.1
$\varepsilon_{1,u,t}$, url ×10^5	10.3	11.7	10.2	9.8	9.4

Strains $\varepsilon_{1,u,t}$ little depend on the stress state type and level σ_2/R_b. So, at σ_2/R_b=0.15 a slight increase in the value of strain $\varepsilon_{1,u,t}$ is observed (by 10–13% compared to the single-axis tension), at further increase of σ_2 value the tendency to a slight decrease in strains $\varepsilon_{1,u,t}$ is observed (at σ_2/R_b=0,6 $\varepsilon_{1,u,t}$ less the similar strains calculated at single-axis tension by 10%).

Such state is true for both static and dynamic loading.

Full ultimate strains $\varepsilon_{1,u}$ increase with the growth of σ_2/R_b. At that, increase of $\varepsilon_{1,u}$ takes place, mainly, due to strains from the compressive load action at the first stage of loading, share of which in the full ultimate strain, while growth of σ_2/R_b, increases and at the level of $\sigma_2/R_b = 0.6$ even exceeds the value of strains $\varepsilon_{1,u,t}$.

When analysing the loading rate effect on ultimate strains, it is noted that strains $\varepsilon_{1,u,t}$ and $\varepsilon_{1,u}$ at dynamics turn out to be higher by 10–15% ,than similar strains at statics, both at single-axis tension and at "compression-tension" stress state with different level σ_2/R_b.

When analysing the results of P-2 series, first of all it is necessary to point out the following methodical difficulties that have appeared at the implementation of researches at "two compression-tension" stress state. As a result of the direct action on membrane device surface sensors their pressing in the concrete was observed, because of that it was impossible to record strains at the first stage of loading. In tension, it was necessary to record strains directly after the load increase at the stage. Further changes in the equipment readings, recording the growth of strains at the stage was connected with resistive strain gauges pressing in the concrete surface. These readings were excluded from the total deformation.

Besides, it is necessary to pointed out the fracture pattern of samples under dynamic loading. Fracture surface was formed at the top part of the sample in close proximity to the head. This is the zone of the single-axis tension, since due to design considerations the membrane unit is a few centimetres shorter than the sample. Thus, recorded values recorded of the strength loss can not be taken for the strength values at modelled SSS and hence to draw conclusions about the impact of the stress state type on the strength at the dynamic loading.

Indicated effects did not appear at the static loading. From what it is possible to conclude that the increase in the loading rate increases the friction along the sample surface. Hereinafter it is needed to pay the special attention to procedures to reduce the friction.

Strength results are specified in Table 2.

Table 2. Results of strength determination at static and dynamic loading under conditions of single-axis tension and "two compressions - tension" stress state (P-2 series).

Sample No.	**Stress level** σ_2 (σ_2/R_b)	**Strength** $R^{\sigma 2}_{ibt}$ $(R^{\sigma 2}_{ibt,d})$, **MPa**	**Strength average value** $R^{\sigma 2}_{bt}(R^{\sigma 2}_{bt,d})$,**MPa**
1	2	3	4
Static tests			
O4	0	1.4	1.45
O5		1.5	
O9	0.3	1.9	1.78
O12		1.8	
O13		1.65	
Dynamic tests ($\dot{\sigma}$ =70 MPa/s, τ=0.05 s)			
O6	0	2.3	2.23
O7		2.2	
O8		2.2	
O10	0.3	2.2	2.2

Double-sided drafting at σ_2=σ_3=0.3R_b increases the strength at the static loading by about 20%. At that the dynamic strength is recorded at the strength level at the single-axis tension, however, taking into account described above fracture pattern, one should not be guided by these values as the final ones.

Diagrams of series P-2 concrete strain under static and dynamic loading at single-base tension and "two compression-tension" stress state are shown in Fig. 2.

Taking into account the limitations of obtained strain data, it is impossible to give a numerical estimate of longitudinal strain modulus values. Analysis of the available data does not allow speaking about an increase in the initial strain modulus with increasing loading rate, as it was in the case with P-1 series samples.

At the dame time, it is possible to speak confidently about the reducing tension strain modulus with the growth of the stresses $\sigma_2 = \sigma_3$ at both static and dynamic loading. Such modulus decrease is about 25%.

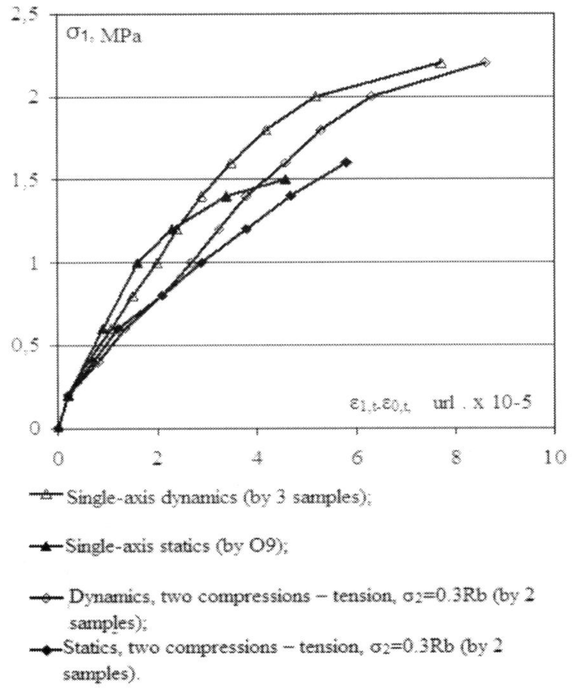

Fig. 2. Series P-2 concrete strain under static and dynamic loading at single-base tension and "two compression-tension" stress state.

Tendency of increasing ultimate longitudinal strains with the growth of the loading rate, which was observed in a testing of P-1 series samples, remained for P-2 series samples both at single-axis tension and at 'two compression-tension" SSS. When comparing longitudinal strain values at 0.9R level, it is possible to note that strains at dynamic impact exceed strains at statics by 35% at single-axis tension and by 10% at "two compression - tension" stress state at з $\sigma_2 = \sigma_3 = 0.3R_b$.

Conclusion

1. Strength and strain characteristics (strain modulus and ultimate strains) of concrete at static and dynamic loading under conditions of single-axis tension and "tension-compression"stress state were determined experimentally;

2. Regularities of the influence of stress state, stress level σ_2 and loading rate on the concrete strength were revealed. Criterion of the concrete strength at single-axis tension and "compression-tension" stress state was recorded, which takes into account these factors influence on the strength;

3. Initial modulus of strains and modulus, calculated taking into account the entire strain curve increases with increase of the loading rate (according to results of P-1 series tests);

4. Ultimate longitudinal strains increase with the growth of stress σ_2 for the value of strains from compressive load action at the first stage of loading. Maximum longitudinal strains from only tensile load action little depend on σ_2;

5. Ultimate longitudinal strains and maximum longitudinal strains from only tensile load action at dynamics by 10-15% exceed the similar strains at statics.

6. In the course of testing at "compression-tension" 'stress state, methodical difficulties on modelling the specified SSS, connected with a significant increase of surface friction forces at the dynamic loading. In the future, it is necessary to solve the issues of improving the techniques of carrying out dynamic tests under conditions "two compression - tension" stress state.

The work is carried out pursuant to the grant of the Russian Academy of Architecture and Construction Science for young scientists

References

1. Y.N. Malashkin, I.M. Bezgodov, K.A. Tsvetkov, Natural and Technical Sciences.-, **1**, 182-190 (2007).
2. N.F. Eremin, I.S. Tsurkov, Y.N. Malashkin, B.V. Tyablikov, I.M. Bezgodov, *Study of concrete mechanical characteristics with account of its actual operation in hydraulic works*, 62 (1978)
3. N.N. Popov, B.S. Rastorguev, A.V. Zabegaev, *Calculation of structures for dynamic special loads*, 319 (1992)
4. V.A. Kopnov, A.N. Voronin, Building mechanics of engineering structures and constructions. – **6**, 7-11 (1996)

Buckling resistance assessment of thin-walled open section element under pure compression

Andrzej Piotrowski[1], *Łukasz* Kowalewski[1*], *Radosław* Szczerba[1], *Marcin* Gajewski[1] and *Stanisław* Jemioło[1]

[1]WUT, Faculty of Civil Engineering, Armii Ludowej 16, 00-637 Warsaw, Poland

Abstract. The paper deals with the problem of buckling resistance determination in case of perforated thin-walled elements. Depending on element slenderness the limit load in compression mode may be determined taking into account elastic (LBA analysis), elasto-plastic and plastic material properties (GMNA analysis). The limit load value is determined on the basis of thin-walled open section beam theory in case of elastic buckling and using Johnson-Ostenfeld approximation for elasto-plastic buckling. Obtained results are compared with finite element solutions derived using shell modelling and large deformation theory with Huber-Mises elasto-plasticity constitutive model without any geometrical imperfections. Some results of the carried out experiments are also shown.

1 Introduction

In the paper the buckling resistance of the perforated thin-walled bar (without any geometrical imperfections) is under investigation. Analysed elements with the sigmoidal cross-section are used as a vertical support element for storage systems. Perforations are used as a supports for shelfs mounting elements and also as a mounting places for structure stiffening metal tie cables.

Depending on the distance between shelfs the slenderness for the analysed element is changing, leading to the elastic buckling or elasto-plastic load capacity problems. The limit load analysis was conducted with the use of thin-walled beam theory and compared with the results from the FEM analysis (shell models). In the case of analytical modelling with thin-walled beam theory the Johnson-Ostenfeld approximation has been used for the elastic-plastic range. In large deformation finite element shell models the material description is done in the frame of elasto-plasticity with flow rule associated with Huber-Mises yield condition and isotropic hardening. The main purpose of the paper is to compare the analytical solution for the perforated bar, for which perforations are taken into account through the concept of thinening of the bar wall (assuming equivalence of the material volume), with numerical solutions. Both analysis are done for perfect members, and finaly presented against some experimental tests done on imperfect samples with limited length (limitations of the testing systems). In the case of tested samples the geometrical and boundary type imperfections were identified.

* Corresponding author: l.kowalewski@il.pw.edu.pl

2 Material and geometry

The storage system elements are made from steel, which is treated as homogenous and isotropic material. The material parameters were determined based on uniaxial tensile tests carried out on three specimens cut out of the analysed structure in accordance with the standard [1]. The samples of 12mm width and 2mm thickness were tested using the Instron material testing system. The parameters determined on the basis of these tests are shown in Table 1. Such parameters are needed for analytical modelling (Johnson-Ostenfeld approximation). In case of parameters for Huber-Mises elasto plasticity additionaly the Poisson coefficient is assumed as equal to 0.3 and uniaxial tension test results are reproduced using elasto-plasticity model with isotropic isotropic strain hardening in wide range of strains. The dimensions of analysed element and its cross-section are shown in Figure 1.

Table 1. Material parameters.

Young's modulus E [GPa]	**Proportionality limit R_H [MPa]**	**Plasticity limit R_e [MPa]**
230.0	382.2	453.5

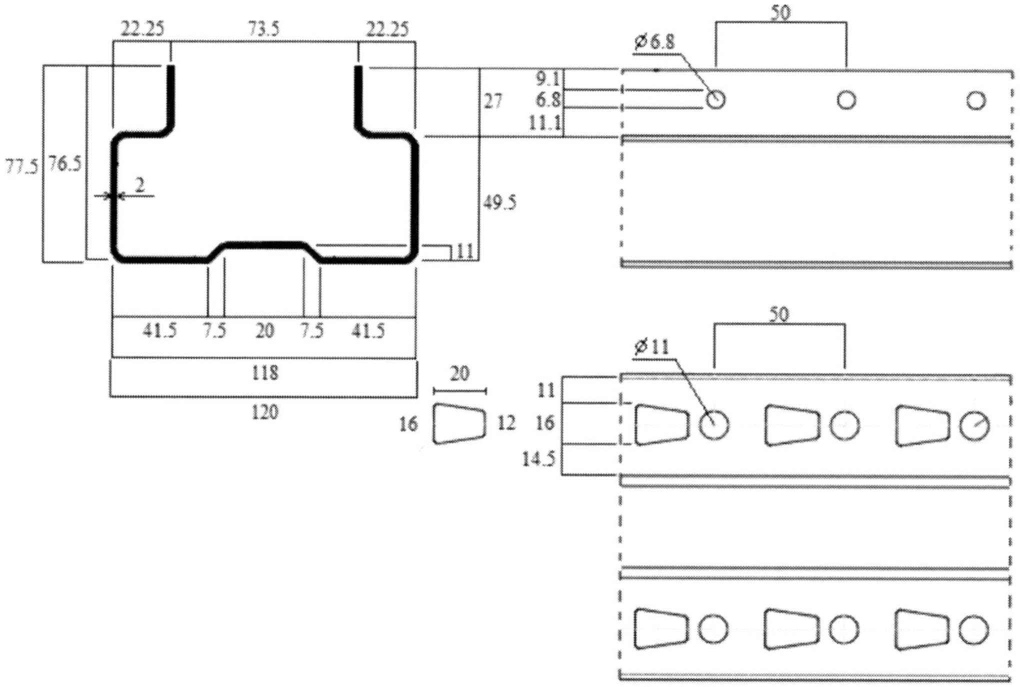

Fig. 1. Dimensions of analysed perforated element.

3 Determination of the compressive load capacity by the thin-walled beam theory

The perforated element shown in Figure 1 was replaced – according to [2] (chapter 8.3) – with the bar with equivalent cross-section, where the walls have been locally thinned in a way so as to keep the material volume (Figure 2).

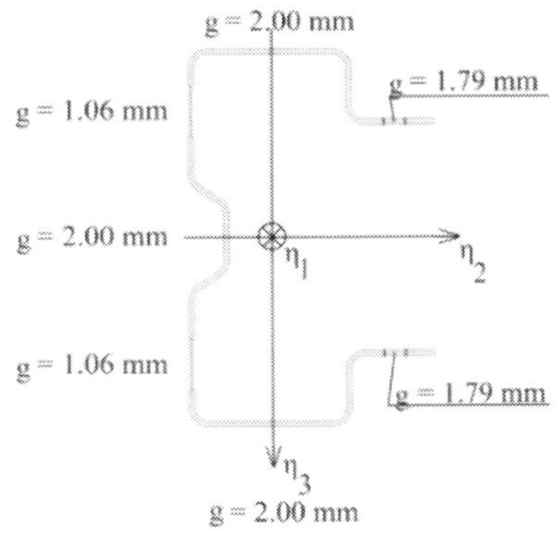

Fig. 2. Cross-section with thinned walls and rounded corners (g stands for wall thickness).

The computations were carried out for four types of element's cross-section: full cross-section (with filled perforations) and either rounded corners or simplified not-rounded corners (the latter one refered to as 'max' later on) or thinned cross-section (with perforations), again with either rounded ('min' later on) or not-rounded corners. In case of rounded corners – according to [3] – the geometrical cross-section characteristics are derived as:

$$A \approx A_k(1-\delta), \quad I \approx I_k(1-2\delta), \quad I_\omega \approx I_{\omega;k}(1-4\delta), \tag{1}$$

Where „k" index stands for characteristics of a cross-section with not-rounded corners. The δ parameter can be determined from:

$$\delta = 0.43\left(\sum_j R_j \tfrac{\varphi_j}{90^\circ}\right) / \left(\sum_i b_i\right), \tag{2}$$

Where: R – internal radius of rounding, φ – angle between the coincident walls (in degrees), b – the distance between the roundings' middle points.

The calculated geometrical characteristics are presented in Table 2, where No.1 stands for full and not-rounded, No.2 for full and rounded, No.3 for thinned and not-rounded and No.4 for thinned and rounded cross-section.

Table 2. Geometrical characteristics.

No.	η_{2c} [mm]	η_{2A} [mm]	A [mm^2]	I_2 [mm^4]	I_3 [mm^4]	I_ω [mm^6]	I_s [mm^4]
1	25.7	-57.1	654	1 321 104	388 392	1 952 013 679	872
2	25.2	-55.9	625	1 203 350	353 773	1 604 034 794	795
3	26.8	-59.4	621	1 268 536	363 541	1 848 685 312	789
4	26.2	-58.1	594	1 155 467	331 137	1 519 126 426	719

To determine the Kirchhoff modulus G = 88.46 MPa, the torsional stiffness $K_s = GI_s$ and the modified Young modulus E_1 = 252.7 GPa (which is used in the thin-walled beam theory because of the assumption of the beam wall contour constant length) the results from Table 1 were combined with the Poisson ratio assumed based on the [4] as 0.3.

The critical load was computed as the smallest root of the polynomial shown in equation (3), see also [2, 5]. The formula on the left hand side of the equation is the determinant of the stiffness matrix created by the 2nd order theory with the assumption of axial compression of the bar and the symetry of the cross-section:

$$P^3\left(r^2-\eta_{2A}{}^2\right)-P^2\left(\left(P_\omega+P_2+P_3\right)r^2-P_3\eta_{2A}{}^2\right)+ \\ +Pr^2\left(P_2P_3+P_2P_\omega+P_3P_\omega\right)-P_2P_3P_\omega r^2=0 \quad (3)$$

Where the symbols were adopted as in [2].The following types of boundary conditions were analysed (Table 3):

1. bar fixed at each end ($\mu_1 = \mu_2 = \mu_3 = 0.5$),
2. elastic boundary conditions at each end according to [2, 6] ($\mu_1 = \mu_2 = \mu_3 = 0.65$),
3. mixed boundary conditions: fixed rotation about the axis η_1 (longitudinal), allowed rotation about the η_3 axis, mixed boundary conditions for the η_2 axis: $\mu_1 = 0.5$, $\mu_2 = 1.0$, $\mu_3 \in [0.5,1.0]$,
4. hinged boundary condition at each end ($\mu_1 = \mu_2 = \mu_3 = 1$).

Table 3. Boundary conditions.

Case 1. fixed			**Case 2. elastic**			**Case 3. mixed**			**Case 4. hinged**		
μ_2	μ_3	μ_1	μ_2	μ_3	μ_1	μ_2	μ_3	μ_1	μ_2	μ_3	μ_1
0.5	0.5	0.5	0.65	0.65	0.65	1.0	0.7	0.5	1.0	1.0	1.0

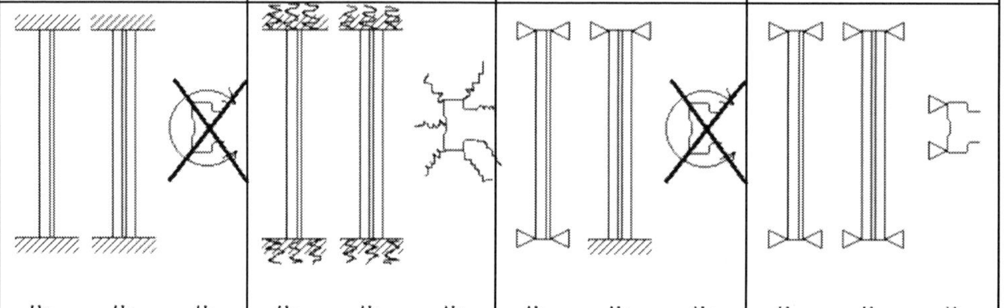

It was noted that with mixed boundary conditions the value of μ_3 when taken from interval [0.5, 1] has no influence on the results. In the elastic-plastic buckling range the load capacity of analyzed bar was approximated with the Johnson-Ostenfeld parabola:

$$P_{kr} = A\left(R_e - \frac{R_e - R_H}{l_{gr}^{\ 2}} l^2\right), \quad (4)$$

Where the limit length (l_{gr}) value can be found by intersecting the linear buckling curve with the value of $P_H = R_H A$. The obtained limit lengths are shown in Table 4, the values of critical load with respect to bar's length are shown in Figure 3.

Table 4. Limit lengths [cm].

Thinned, not-rounded corners				**Thinned, rounded corners**			
fixed	hinged	mixed	elastic	fixed	hinged	mixed	elastic
355.1	177.5	205.3	273.1	338.3	169.1	200.4	260.2
Full, not-rounded corners				Full, rounded corners			
fixed	hinged	mixed	elastic	fixed	hinged	mixed	elastic
345.5	172.8	196.9	265.8	329.3	164.6	192.2	253.3

By comparing the minimal results (obtained for the bar with thinned cross-section and rounded corners) with the maximal results (obtained for the filled bar with not-rounded corners) it can be observed that the difference between both results reach up to 15% of maximal value. Analysing the results with respect to boundary conditions shows that the difference between mixed and hinged boundary conditions disappears with increasing length of the bars – which can be justified by reducing the influence of torsion in the cases of flexural-torsional buckling (the buckling mode of the bar with mixed boundary conditions is determined as flexural, while for the hinged support as flexural-torsional).

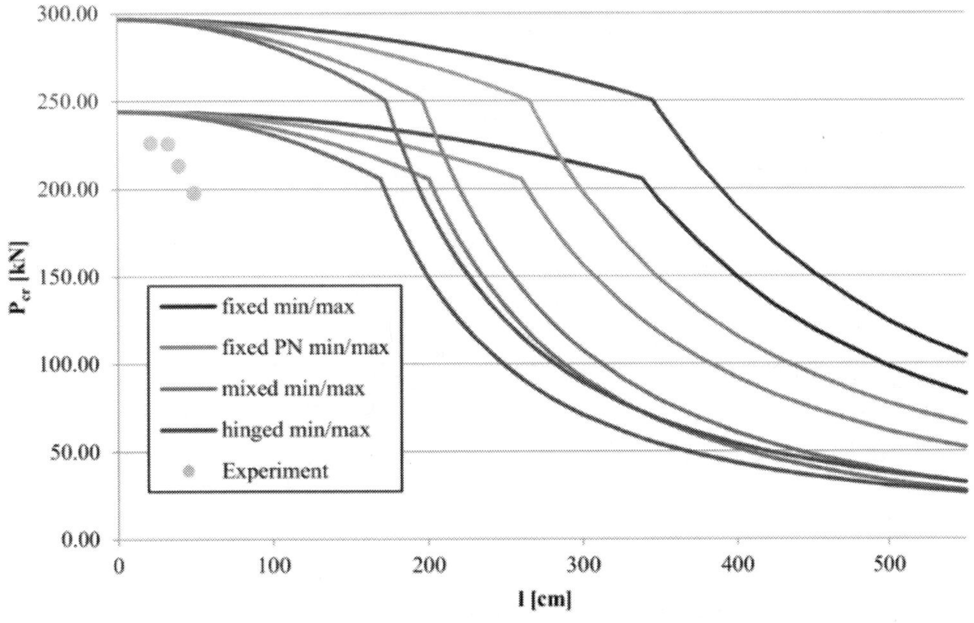

Fig. 3. Limit load for elastic-plastic buckling with respect to element's length.

4 FEM analysis description

As a verification of theoretical approach multiple FEM models, representing bars of multiple lengths, were created, and a parametric study was carried out in order to create the function of compressive bearing capacity of a bar with respect to its length. Only the hinged boundary conditions (compare Table 3 – Case 4) were taken into account.

4.1 FEM model

Each FEM model consisted of multiple repeated modules of 5 cm length joined together by merging the coincident nodes. Total length of the bars used for parametric study varied from 5 to 500 cm - 32 models were analysed.

Each module consisted of 5080 linear shell elements S4R and 82 linear shell elements S3 [7, 8]. The number of variables in the models (DOFs) varied from 32k to 3160k. The material used for constitutive relationship was a linear-elastic material with Huber-Mises plasticity combined with isotropic hardening [9], the parameters of which were adjusted based on the laboratory uniaxial tensile test.

At each end of the bar the boundary conditions were prescribed by applying the displacement load to the point obtained by projecting the bar's center of mass onto the extreme cross-section (see points C-1 and C-2 in Figure 4). All of the nodes belonging to the extreme cross-section were then connected with corresponding point via MPC beam constraint, thus creating a rigid body. For the FEM analysis only the hinged boudary conditions where applied – each end of the bar had the translations in x, y and z direction, as well as the rotation alongside z axis, restricted (with the displacement load in the z direction). The FE mesh of a module with the location of aforementioned points is shown in Figure 4.

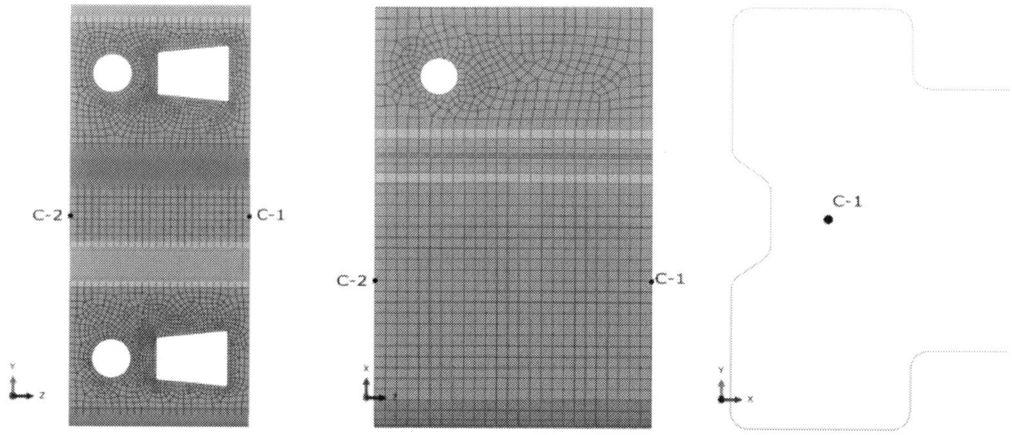

Fig. 4. FEM mesh of the single module with the position of center of mass.

4.2 Analysis

Two sets of analysis were performed: a set of eigenvalue buckling analysis (LBA), where plasticity was not taken into account and only the first critical loads were measured; and a set of quasi-static analysis with elasto-plastic constitutive relationships (GMNA) and the displacement load described in section 4.1. In the latter analysis the maximum force carried by the bar was calculated as the compressive bearing capacity. In each analysis the large displacement theory was used (NLGEOM option, see [8, 9]).

5 Comparison of numerical and analytical results

All the results presented in this section are for the bars with boundary conditions described as Case 4. Graph in Figure 5 shows the comparison of the buckling curves (limit load with respect to the length of the bar) obtained from the eigenvalue buckling FEM analysis with the one calculated analytically as shown in section 3. Bigger diversity can be seen for bars of smaller slenderness – as the use of shell elements in FEM analysis allows to consider local buckling in the walls of the structure.

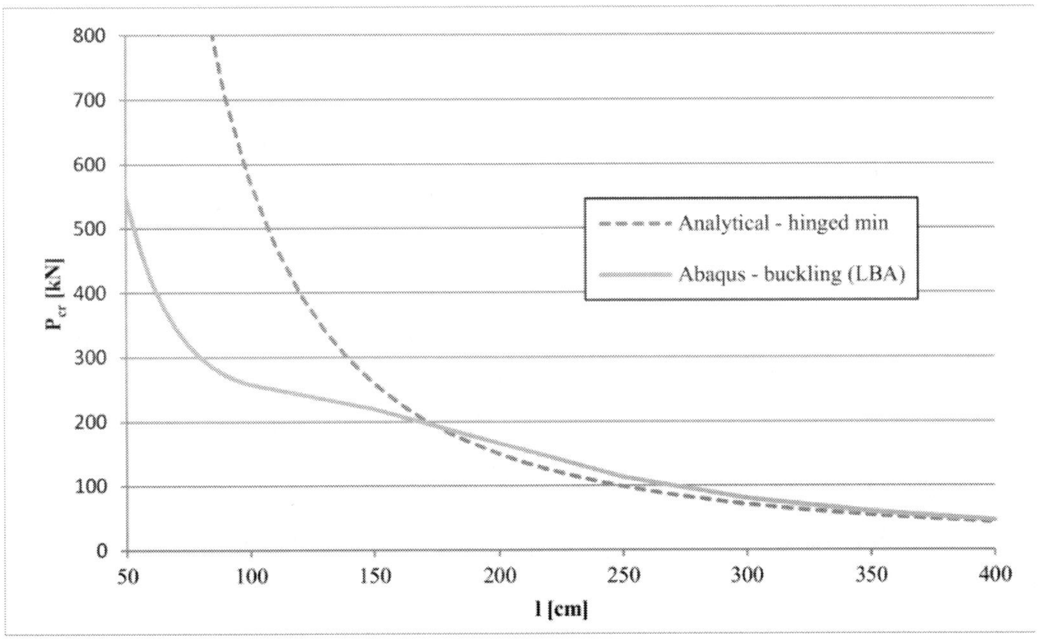

Fig. 5. Results from FEM (LBA) analysis compared to the theoretical results.

Figure 6 shows the load capacity of the bars with respect to their lengths based on the results of quasi-static FEM analysis and compares it to the analytical and experimental results. Here both the elastic and plastic material properties were taken into account. The analytical and numerical results show fine convergence, especially for bars of larger slenderness. Comparison with the results of eigenvalue buckling analysis shows the length for which the load capacity can be determined by linear buckling – here around 3.5m.

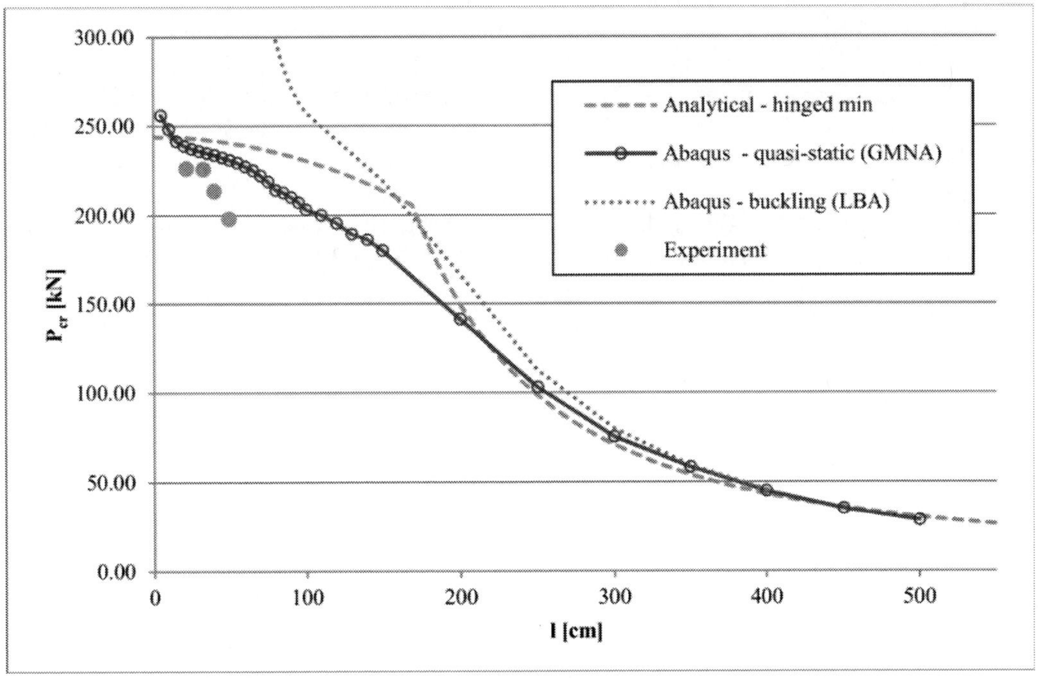

Fig. 6. Results from FEM analysis compared to the theoretical results in case of elastic – plastic buckling.

6 Final remarks and conclusion

The results obtained from FEM solutions showed good convergence with the analytical results by thin-walled beam theory proving that the applied concept of taking into account perforations through thinning of the element wall is rational and leads to proper results. The main divergence between both results appeared for the bars of smaller lengths, as for such lengths the local buckling of the structure walls was more important than global one, and the shell theory would be more appropriate for analytical approach – this conclusion can be taken from both elastic and elastic-plastic analysis. All analysis were made for elements without any imperfections (geometrical shape or boundary conditions imperfections) so the experimental results obtained for tested imperfect models for quite low slenderness ratios should be located below the obtained analytical and numerical results what is visible in graph presented in Figure 6, cf. also [10, 11]. By comparing the buckling load (in the range where the global buckling mode is dominant) obtained from the thin-walled beam theory with the buckling load obtained for shell elements in FEM analysis it is clear that thin-walled beam theory gives smaller values of the critical load. What's more these results are closer to the results obtained from quasi-static FEM analysis done for elasto-plastic material.

References

1. PN-EN ISO 6892-1,Polish Standarization Committee, (2010)
2. J.B. Obrębski, Warsaw University of Technology Publishing House, (1991)
3. PN-EN 1993-1-3: 2008 Eurocode 3, (2008)
4. PN-90/B-03200, Polish Committee for Standarization Measures and Quality, (2008)
5. J. Mutermilch, A. Kociołek, Warsaw University of Technology Publishing House, Warsaw, (1964)
6. PN-80/B-03200, Polish Committee for Standarization Measures and Quality, (2010)
7. ABAQUS/Standard User's manual, Hibbitt, Karlsson & Sorensen, Inc., Pawtucket, (2000)
8. ABAQUS Theory manual, Hibbitt, Karlsson & Sorensen, Inc., Pawtucket, (2000)
9. S. Jemioło, M. Gajewski, Warsaw University of Technology Publishing House, (2014)
10. M.A Giżejowski, R. Szczerba, M. Gajewski, Z. Stachura, Open Access Journal Elsevier **111**, 254-261 (2015)
11. R. Szczerba, M. Gajewski, M.A. Giżejowski, Journal of Civil Engineering, Environment and Architecture **JCEEA-62-03-t2**, 425-438 (2015)

Mathematical model of innovative sustainability "green" construction object

Michail Slesarev[1*], *Evgeniy* Pankratov[2] and *Viktor* Fedorov[1]

[1]Moscow State University of Civil Engineering, 129337, Yaroslavskoye Shosse 26, Moscow, Russia
[2]Plekhanov Russian University of Economics, 117997, Stremyanny lane, 36, Moscow, Russia

Abstract. The paper addresses the issue of finding sustainability of "green" innovative processes in interaction between construction activities and the environment. The problem of today's construction science is stated as comprehensive integration and automation of natural and artificial intellects within systems that ensure environmental safety of construction based on innovative sustainability of "green" technologies in the life environment, and "green" innovative products. The suggested solution to the problem should formalize sustainability models and methods for interpretation of optimization mathematical modeling problems respective to problems of environmental-based innovative process management, adapted to construction of "green" objects, "green" construction technologies, "green" innovative materials and structures.

1 Introduction

For years now, the concept of "smart management" has been widely used in the systems theory; the concept shifts the focus on heuristic (expert-based, adaptive, fuzzy, etc.) correction of formal mathematical description of a control object in order to simulate the life environment's dynamics. The known gamut of science and engineering instruments along with effective regulation on environmental research management fail to sufficiently validate the grounds underlying managerial decisions in environmental safety. Numerous issues of the taxonomy of concepts, methods and means related to environmental safety management remain without attention; this includes issues of rational selection of a scientific approach used to describe the processes of environmental safety management in construction, rational choice of modeling and management approach, issues of concurrent use and integration of the results of environmental monitoring along various vectors.

Priority analysis of the growing construction sector of economy seen through the paradigm of construction organization and innovative sustainability based on the principles of innovative governance helped to work out certain approaches to model building for control systems of construction and environmental safety of the living environment. These include: approaches to mathematical interpretation of optimization models for construction and environmental safety management of objects, and models of their innovative sustainability.

Because methods of calculating optimal programs (or trajectories) are well designed by

* Corresponding author: slesarev@mgsu.ru

now, to ensure optimal allocation of resources, it is sufficient to have a model of the innovation process – its description in the mathematical language [1, 2]. To compensate for possible interferences and departures from the script, there must be a feedback-generating mechanism of adjusting innovative actions that keeps the innovation process in its preset path [5].

Green technology is comprehensive engineering knowledge, structured with standards of design, manufacturing and construction. The level of its development depends directly on advances of science and technology, activity of industrial engineers, and public awareness of the environmental principles.

In the construction sector, the purpose of the *green standards* is to speed up the transition from conventional design and construction of buildings and structures towards sustainable development that adheres to the green principles:

- safety and wholesome favorable conditions of human life;
- curbing negative impacts on the environment;
- taking the interests of future generations into account.

Green standards are designed to regulate a robust approach to building and to assess compliance of life environment technologies with the green principles [4].

Development and enforcement of *green construction* standards creates incentives for business growth on the basis of technical regulation on the market of innovation technologies and products, it also raises the quality of public life and condition of the environment [3].

The **hypothesis** of innovation-supported environmentally safe development of an object, for example, a thermal or nuclear power plant, understands interpretation of environment parameters under the influence of the object in the form of homogeneous differential equation describing the trajectory of the system "object - environment" traveling through a multiple-parameter information space in the life environment. The required accuracy of research is ensured by known instruments: mathematical optimization models and mathematical tools of motion stability according to the classical automatic control theory.

A fragment of the information space in arbitrary coordinates (*materials, components, energy*) needed to build an innovative development path for a thermal and nuclear power plant can be illustrated graphically as represented in Fig. 1.

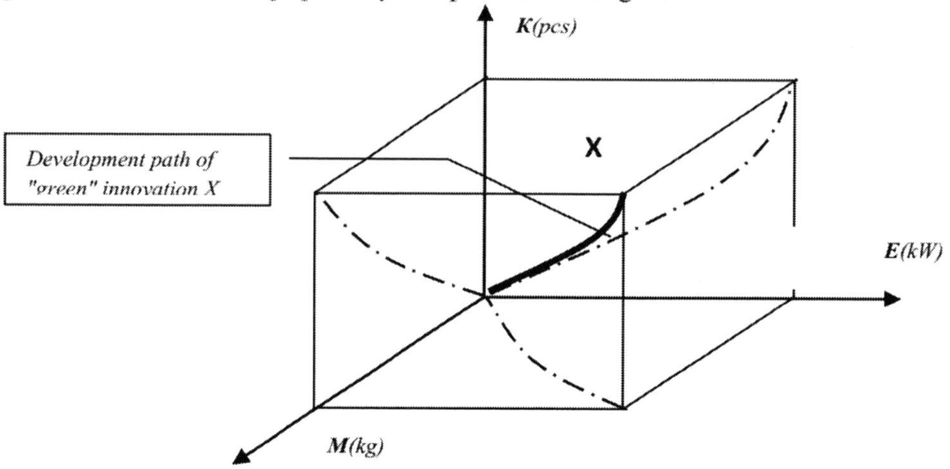

Fig. 1. Graphical interpretation of the development path of an innovative technical object (dynamics of point X) in the information coordinates is seen as: K (number of components in the object); E (energy spent during the object's life cycle); M (material mass consumed during the object's life cycle).

To build the development path for a local object of engineering, one has to use data from environmental monitoring parameters of the life environment in the object's particular location.

In general, the equation of the uninhibited development path of a "green" construction within the space of environmental parameters can be represented as:

$$a_0 \frac{d^n X}{dt^n} + a_1 \frac{d^{n-1} X}{dt^{n-1}} + \cdots + a_{n-1} \frac{dX}{dt} + a_n X = 0 \quad (1)$$

or symbolically as

where $a_0, a_1, \ldots a_n$ — coefficients of development path, factored by the life environment parameters, and X — relative departure of the control variable of the environment parameter.

As analysis of a system's environmental stability examines unimpeded development of the life environment, the starting conditions of the control system's travel along the development path in the most general case can be represented as follows:

In this case, solution to the homogeneous equation

$$a_0 \frac{d^n X}{dt^n} + a_1 \frac{d^{n-1} X}{dt^{n-1}} + \cdots a_n X = 0, \quad (2)$$

that describes the object's motion through the information space of the life environment parameters, is expressed as:

$$X(t) = C_1 e^{p_1 t} + C_2 e^{p_2 t} + \cdots + C_n e^{p_n t} = \sum_{k=1}^{n} C_k e^{p_k t}, \quad (3)$$

where $p_1, p_2, \ldots p_n$ — are roots of the characteristic equation

and $C_1, C_2, \ldots C_n$— are constants determined by the initial conditions.

To ensure that the object in the information space of life environment parameters remains sustainable on terms of innovation, the solution must meet the requirement:

$$\lim_{t \to \infty} X(t) = 0. \quad (4)$$

This is obviously possible when

$$\sum_{k=1}^{n} C_k e^{p_k t} \to 0. \quad (5)$$

Let us know the restrictions imposed on roots $p_1, p_2, \ldots p_n$, under which the life environment remains ecologically stable.

From among the roots p, let us separate those that are real α_0 from those that are complex $\alpha_k \pm j\omega k$. Here, indexes n and k can be different integers.

Now the solution $X(t)$ can be written out as:

$$X(t) = \sum_{v=1}^{S} C_v e^{\alpha_v t} + \sum_{k=1}^{n-S} C_k e^{\alpha_v t} sin(\omega_k t + \Psi_k) \quad (6)$$

(S — number of real roots).

First we assume that all roots p are simple. Apparently,

$$\lim_{t \to \infty} X(t) = 0, \tag{7}$$

if each component in the right-hand part tends towards zero. The latter occurs where $\alpha < 0$, i.e. the real parts of all roots in the characteristic equation are negative values.

If the real or complex root has the multiplicity of r, then the components that form the solution $X(t)$ will have members of the type:

$$C_1' + C_2't + \cdots + C_r't^{r-1})e^{\alpha_v t} \tag{8}$$

and

$$\left(C_1'' + C_2''t + \cdots C_r''t^{r-1}\right) \cdot e^{\alpha_k t} sin(\omega_k t + \Psi_k). \tag{9}$$

For negative real parts of the roots in a characteristic equation ($\alpha < 0$) the presence of polynomial factors of t will bear no influence on the final result, because $e^{\alpha t}$ at $t \to \infty$ decreases faster than the polynomial of t increases:

$$\lim_{t \to \infty} = \left(C_1' + C_2't + \cdots + C_r't^{r-1}\right)e^{\alpha_v t} = 0 \tag{10}$$

Next we consider the case of zero roots and purely imaginary roots when the array of α contains.

Here, there are two ways to address the issue of environmental sustainability. For a simple zero root, the general solution will have a constant component that determines the stationary error.

Conclusion

If there is at least one root with positive real part among all roots, the "green" technology control system in the life environment and/or "green" innovative products will be unstable.

The prerequisite and sufficient condition for the stability of the control system for innovative technologies and innovative products is that the real parts of roots in the characteristic equation should be negative .

References

1. V.I. Telichenko, M.Yu. Slesarev, *Construction Materials, Equipment and Technologies of XXI Century*, **2**. 6-7 (1999)
2. M.Yu. Slesarev, *Abstract of dissertation for the degree of Doctor of Technical Sciences,* 43 (Moscow, FGBOU VO NIU MGSU, 2007)
3. K.Yu. Korolevskiy, M.Yu. Slesarev, Ind. and Civil Constr., **4**, 55-57 (2008)
4. A. Zhukov, Y. Bobrova, D. Zelenshchikov, R. Mustafaev, A. Khimich, Adv, Mat, Res., **1025-1026**, 1031-1034 (2014)
5. A. Volkov, A. Sedov, P. Chelyshkov, A. Pavlov, L. Kievskiy, Int. Jour. of App. Eng. Res., **11**, 1645-1648 (2016)
6. V.G. Borkovskaya, App. Mech. and Mat., **467**, 287-290 (2014)

Assessing destructive capacity of impulse loads

Alexander Komarov[1*], *Nikolay* Gromov[1] and *Yulia* Sokolova[2]

[1]Moscow State University of Civil Engineering, 129337, Yaroslavskoye Shosse 26, Moscow, Russia
[2] NOU INEP, 115477, Bekhtereva ul, 13, korp 3, Moscow, Russia

Abstract. The paper gives an overview of the methodology to find dynamic coefficients of impulse loads typical of various emergencies (explosive blast, massive body falling, hydraulic shock, etc.). Criteria are described under which dynamic coefficients are determined by impulse impact alone irrelevant of its form. Correlations are quoted to find dynamic coefficients for impact loads that accompany such emergencies.

1 Introduction

It is the accepted practice of construction design that non-static loads are reduced to equivalent static loads. Calculations of dynamic impact on building structures use the following methodology. Dynamic load is substituted with its equivalent static load found by this formula [1-3]:

where $P\ max_c$ – maximum dynamic load value; K_d- dynamic coefficient equal to maximum value of dynamic function *T(t)* that describes motion of the structure (its sections shifting) in time.

As is apparent from (2.2.1), to substitute dynamic load with its equivalent load we need to know dynamic factor K_d.

For example, dynamic coefficient *Kd* for a triangular-shaped load is frequently known from diagrams (Fig. 1) available in many reference manuals on strength calculation of structures, where θ and θ_l – duration of dynamic load and the time it takes to peak; ω_K - frequency of main oscillation pitch in the structure, found using the methods of construction mechanics.

The diagram means that if θ_l tends to zero (saw-tooth load form), then the dynamic coefficient's numeric value approaches 2. Considering this, designers double any sawtooth load when they do static calculations. However, this approach is wrong for short-term (impulse) loads. This is the issue discussed in this paper.

* Corresponding author: KomarovAA @mgsu.ru

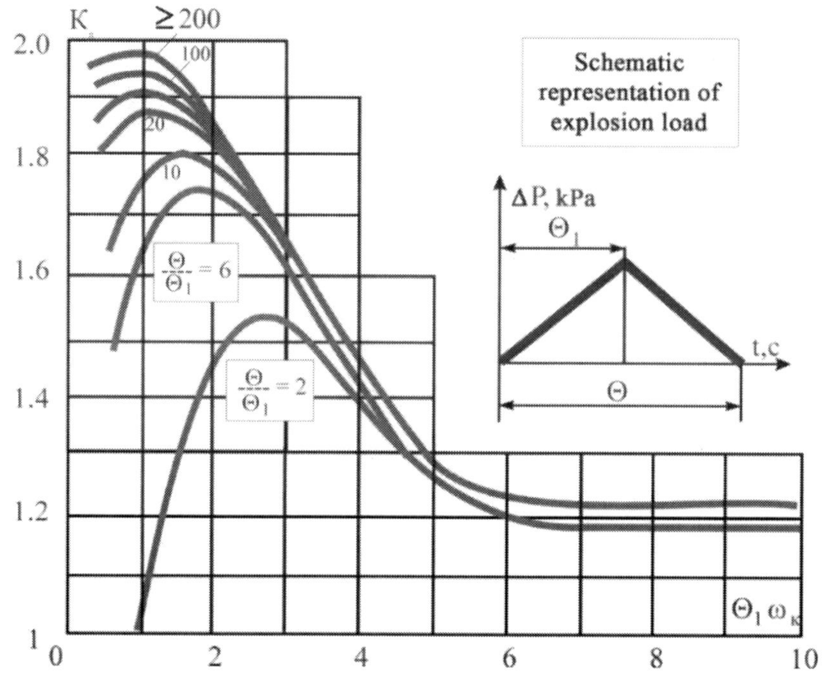

Fig. 1. Diagrams of dynamic coefficient Kd

Emergency situations create momentary loads rather often. Discharge of explosive material or impact of a massive body falling (e.g. transport containers falling in a cooling pool of a nuclear power plant) or hydraulic shock etc. creates loads that last for milliseconds at very high pressures.

Let us examine the dynamic factor methodology for such loads. This discussion will adhere to the approaches accepted in structural dynamics.

We know that the structure's movement at the elastic deformation stage is described by the equation:

$$\frac{d^2T}{dt^2} + \omega^2 \cdot T = \omega^2 \cdot f(t), \tag{1}$$

where ω - circular frequency of the structure's own vibrations круговая частота; *f(t)* – a function to describe changes in time of dimensionless $f(t) = \dfrac{P(t)}{P_{\max}}$ dynamic load; $T(t)=S(t)/S_{MAKC}$ – relative movement of the structure, or dynamic function; *Smax* – offset of the structure responding to static load – *Pmax*.

The solution to equation (1) with zero initial conditions appears as:

$$T(t) = \omega \cdot \int_0^t f(u) \cdot \sin\big(\omega \cdot (t-u)\big) \cdot du = \omega \cdot (\sin \omega \cdot t \cdot \int_0^t f(u) \cdot \cos(\omega \cdot u) \cdot du -$$

$$- \cos \omega \cdot t \cdot \int_0^t f(u) \cdot \sin(\omega \cdot u) \cdot du). \tag{2}$$

Numerical integration of expression (3) produces a function of the structure's movement in time responding to force *f(t)*. The maximum value of displacement function *T(t)* is dynamic coefficient K_d.

To find the dynamic coefficient we need to know the frequency of the structure's own oscillation pitch - ω, expressed with formula [4]:

$$\omega = \frac{\alpha}{L^2} \cdot \sqrt{\frac{B}{M_k}}, \tag{3}$$

where L – structure's span; B – bend rigidity: $B = E \cdot I$, where E - dynamic modulus of elasticity, I - axial moment of inertia, M_K – linear mass of structure: $M_K = \frac{q}{g}$, where q – linear load from explosion, plus load of the structure's own weight, α- factor quoted from reference literature (e.g., [4]), g – free fall acceleration.

Equation (1) can be solved by solving numerically this system of differential equations of the first order:

$$\frac{dU}{dt} = -\omega^2 \cdot T + \omega^2 \cdot f(t)$$
$$\frac{dT}{dt} = U \tag{4}$$

This method has the following disadvantage. Calculations assume that the process of each element's oscillation uses a certain frequency (typically that of the pitch), although in reality oscillations occur across the entire spectrum of natural frequencies of the elastic element. Moreover, arbitrary limiting conditions can be set with approximation in this method, but to calculate strength of building structures in emergency, when maximum possible loads are assumed as inputs, even though they are unlikely, this approach to calculate K_d is justified to an extent.

Equation (1) was integrated for several load types and for different ratios of load duration (T_{load}) to the structure's own oscillation period (T_0).

Fig. 2 gives values of dynamic coefficients for different ratios of load duration (T_{load}) to the structure's own oscillation period (T_0), and for five load types. Evidently, the temporal nature of the load is an influence on the dynamic factor only for load duration to oscillation period ratios greater than $0.5*T_0$. This fact is actually illustrated in Fig. 1, plotted based on rather lengthy loads (relative to the structure's own oscillation period), but not on impulse loads.

Fig. 3 gives values of dynamic coefficients for different load duration to oscillation period ratios, K=0 for shock load, and K=1 for a load type that only has a buildup phase. Conditionally these can be regarded as *extreme* forms: shock load and smooth load.

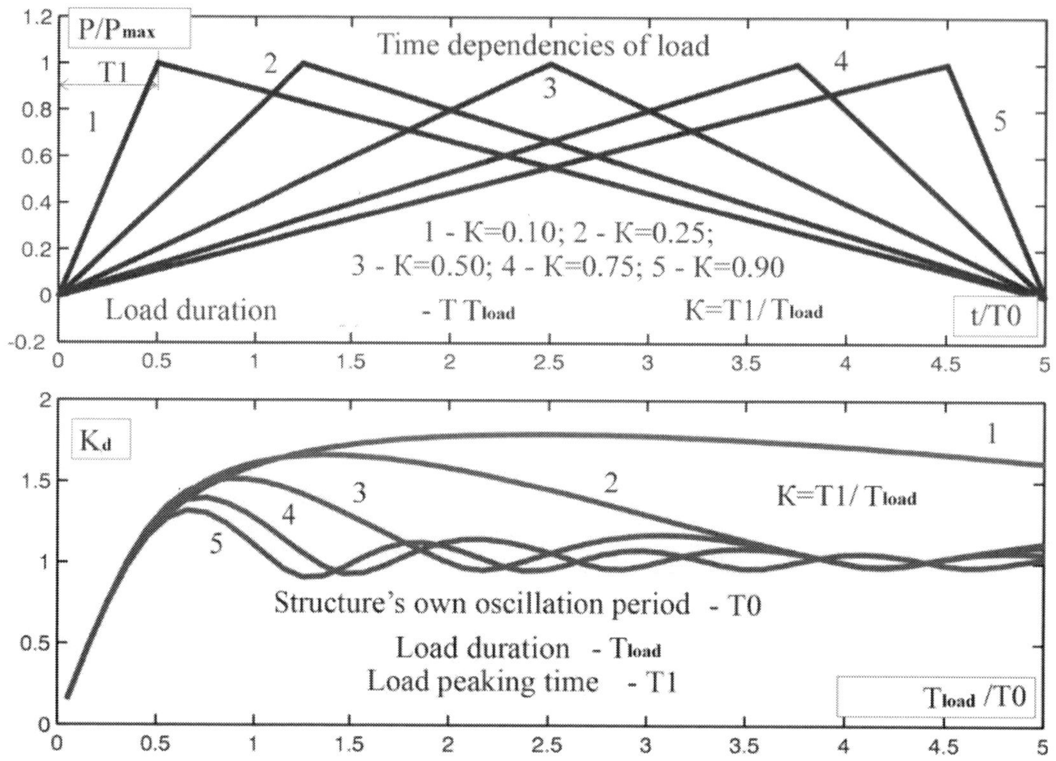

Fig. 2. Dynamic coefficients for five load types

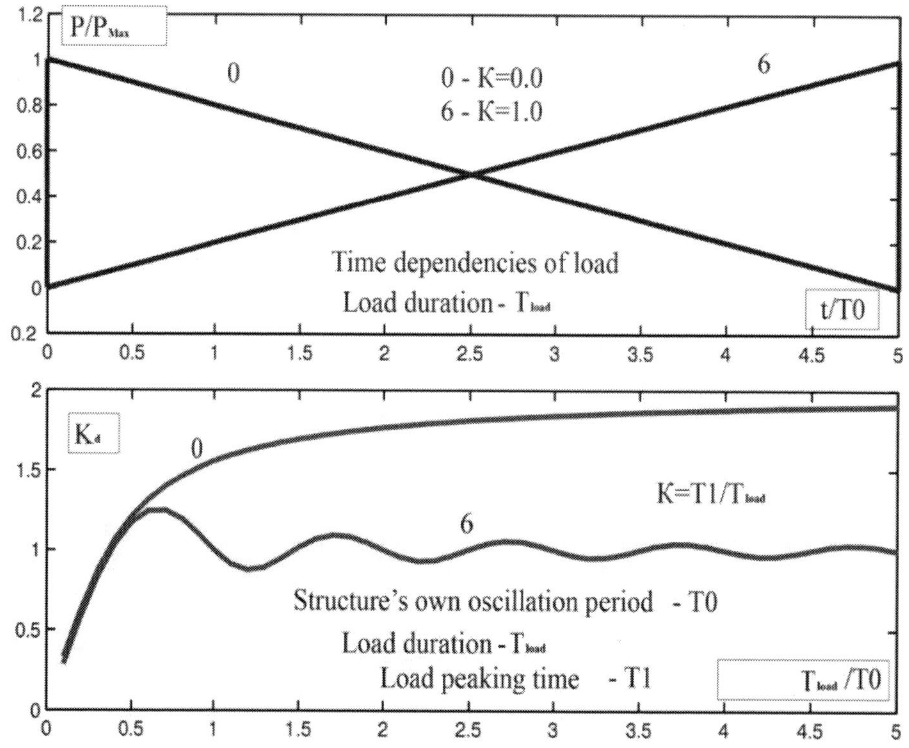

Fig. 3. Dynamic coefficients for two extreme load types

As stated above (see Fig. 2), for ratios $T_{НАГРУЗКИ}/T_0<0.5$, the form of load does not affect the numeric values of dynamic coefficients. They are mainly influenced by pressure impulse: $I = \int P(t)dt$. This is seen from the results in Fig. 4 that quotes dynamic coefficients for triangle-shaped loads (*I=0.5*Pmax*Tlaod*) and for rectangle-shaped loads (*I=Pmax*Tload*).

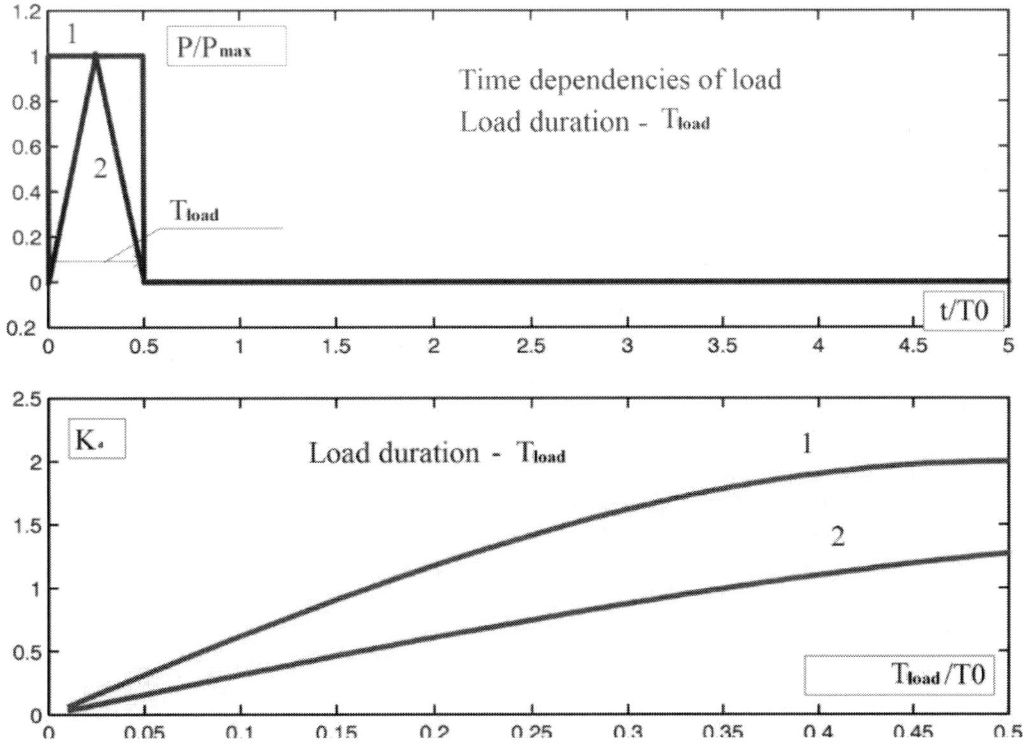

Fig. 4. Dynamic coefficients for ratios $T_{НАГРУЗКИ}/T_0<0.5$ and for triangle- and rectangle-shaped loads

Calculation demonstrated that dynamic coefficient for brief loads ($T_{load}/T_0<0.5$) can be approximately found using this formula:

$$Kd = \frac{2 \cdot T_{load}}{T_0} \cdot \left(2 - 1.5 \cdot \left(1 - \frac{I}{P_{\max} \cdot T_{load}}\right)\right), \tag{5}$$

where I – pressure impulse - $I = \int P(t)dt$.

For example, in case of a triangle-shaped load (*I=0.5*Pmax*Tload*), the dynamic coefficient will approximately equal:

$$Kd = 1.25 \cdot \frac{2 \cdot T_{load}}{T_0} \tag{6}$$

and for a rectangle-shaped load (*I=Pmax*Tload*) the dynamic coefficient is:

$$Kd = 2.0 \cdot \frac{2 \cdot T_{load}}{T_0} \tag{7}$$

Conclusions

The paper discusses specifics of calculating dynamic coefficients for impulse loads typical of many emergency situations.

Criteria are described under which dynamic coefficients depend only by shock load impulse but not by its shape form.

The paper gives ratios needed to calculate dynamic coefficients for shock loads that accompany emergency situations (discharge of explosive materials, massive body falling, hydraulic shock etc.)

References

1. N.N. Popov, B.S. Rastorguyev, *Dynamic Calculation for Reinforced Concrete Structures* (Moscow, Stroyizdat, 1974)
2. B.S. Rastorguyev, *Methodological Guidance on Design and Examination of New and Existing Structures in Explosive Production Facilities (Version 1)* (Moscow, 1996)
3. SNIP 2.01.07-85. Loads and Impacts. (Supplements. Section 10. Deflection and Displacement).
4. *Project Designer's Reference Book. Dynamic Calculation of Special Factors in Construction* (Moscow, Stroyizdat, 1981)
5. A.A. Komarov, *Forecasting Loads and Estimating Consequences of Their Impact on Buildings and Structure.* (Moscow State Construction University (National Research University), Moscow, 2001)
6. A.A. Abrosimov, A.A. Komarov, Oil, Gas and Bus., **6**, 58-61 (2002)

Method of Calculation of Stresses in the Layered Elastic-Creeping Arrays

Tatiana N. Bobyleva[1,*]

[1]Moscow State University of Civil Engineering, 129337, Yaroslavskoye sh. Moscow, Russia

Abstract. Stress calculation problem in an elastically creeping layered array is solved by the method of homogenization. Exponential function of time is taken as the kernel of creep of each layer. Experiments prove that such functions are the best way to describe the creep of mountain ranges over long periods of time. Integrals of convolution type with such exponential kernels are used to describe the creep process. The article contains formulas for calculating effective elastic-creep coefficients. These formulas contain exponential functions with both initial parameters and with new ones, which are also negative real numbers. As a computational example, kernel creep graph is shown for one of the average environment's effective modulus. Obtained solution can be used in underground construction.

1 Introduction

Most of the existing in nature and man-made materials are characterized by a heterogeneous composition. Numerous experimental studies show that the properties of structurally inhomogeneous materials (e.g., rocks, composite materials) can considerably differ from the properties of the individual components included in their structure. These issues are important in solving practical problems that arise during the construction and exploitation of buildings, the operation of composite materials. The construction of structures on rock foundations and inside rocks is an actual problem, which requires knowledge of the properties of rocks and their reaction to those or other impacts. Partial differential equations that describe the behavior of such heterogeneous materials contain rapidly changing coefficients. The method of asymptotic averaging can be used to solve such problems. This method allows to find the effective mechanical properties of heterogeneous material using the values of the mechanical characteristics of its components and their geometry. Some model structure (the periodic cell) is allocated in the array. As a result, the investigated area is considered as a periodic system with known characteristics of the individual components. To solve this problem, we can consider the environment as a macro-homogeneous, which obeys the averaged equations with constant coefficients. These coefficients are called effective. In general, the obtained homogeneous medium is anisotropic, even if the original layers were isotropic. This method of averaging is developed in the works [1-9]. In this article the characteristics of the original array (elastic

* Corresponding author: tatyana2211@outlook.com

modulus, coefficients of creep kernels, specific weight) are the step functions, abruptly changing from layer to layer. Problems for layered materials with continuously changing characteristics have been solved in [10 -12].

2 Problem specification and decision

We use the method of homogenization to calculate the coefficients in the constitutive stress-strain relations of elastic-creeping array. The array has a plane parallel pairs of alternating layers (Fig.1).

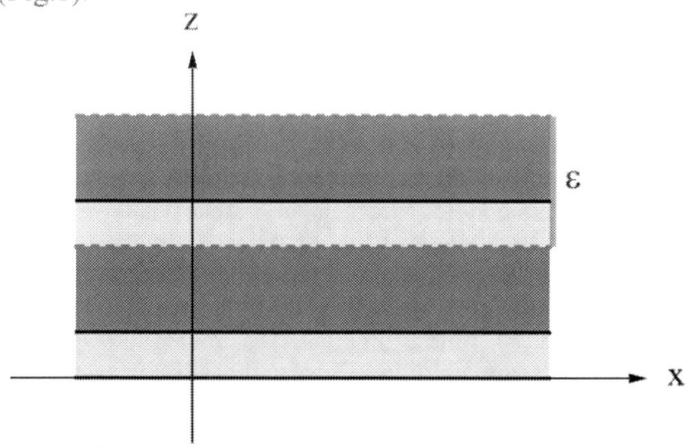

Fig. 1. Layered elastic-creeping array.

Each layer consists of isotropic elastic-creeping material.

The thickness ratio of one layer to the cell period ε is denoted by h. Consequently, the ratio of the thickness of the other layer to the cell period will be $1-h$. $(0<h<1)$.

The original constitutive relations are [13]:

$$-\frac{\partial}{\partial x_k}\left(a_{ijkh}\delta(t)+d_{ijkh}\right)*\frac{\partial u_j}{\partial x_h}=f_i(x,t),\ (i,j,k,h=1\div 3). \quad (1)$$

In (1) the symbol * denotes the convolution operator:

$$\left(d_{ijkh}(x,t)*\frac{\partial u_j}{\partial x_k}\right)=\int_0^t d_{ijkh}(x,t-\tau)\frac{\partial u_j(x,\tau)}{\partial x_k}d\tau \quad (2)$$

$u_j\ (j=1,2,3)$ are components of the displacement, $\delta(t)$ is Dirac-delta function, operation $(a_{ijkh}\delta(t))*$ is equivalent to multiplication by a_{ijkh}. (Einstein convention for repeated indices is used.) We are considering isotropic materials therefore the tensors in (1) can be expressed in the form [1 - 3, 13, 14]:

$$a_{ijkh}=\lambda\delta_{ij}\delta_{kh}+\mu(\delta_{ik}\delta_{jh}+\delta_{ih}\delta_{jk}), \quad (3)$$

$$d_{ijkh}=-(G_v(t)-\frac{1}{3}G_s(t))\delta_{ij}\delta_{kh}-\frac{1}{2}G_s(t)(\delta_{ik}\delta_{jh}+\delta_{ih}\delta_{jk}). \quad (4)$$

We denote here by G_s and G_v the regular part of the shear and the bulk relaxation respectively, by δ_{ij} Kronecker symbol. Let us assume that the amplitude of the bulk creeping kernel is proportional to the shear amplitude with a coefficient of proportionality k_i for each layer that is: $(G_v)_i = k_i(G_s)_i$, $k_i = \text{const}$, $k > 0$, $(i = 1,2)$. Further, G_s is denoted by G.

As usual we denote by λ_i, μ_i Lame parameters, by γ_i specific weight, by G_i creeping kernel for each layer $(i = 1,2)$.

In this problem, all elastic modulus, specific weight and creeping kernels are periodic functions of the coordinate $\xi = \frac{z}{\varepsilon}$ (ε is the cell period) and are piecewise constant functions of this variable, i.e., elastic modulus, specific weight and creep kernels have the form:

$$\lambda_i = \begin{cases} \lambda_1, \xi \in [0;h], \\ \lambda_2, \xi \in [1-h;1], \end{cases} \quad \mu_i = \begin{cases} \mu_1, \xi \in [0;h], \\ \mu_2, \xi \in [1-h;1], \end{cases} \quad \gamma_i = \begin{cases} \gamma_1, \xi \in [0;h], \\ \gamma_2, \xi \in [1-h;1], \end{cases}$$

$$G_i = \begin{cases} G_1, \xi \in [0;h], \\ G_2, \xi \in [1-h;1], \end{cases} \quad 0 < h < 1,\ i = 1,2. \tag{5}$$

Let us choose creep kernel of exponential type for each layer: $G_i = g_i e^{-\alpha_i t}$, where α_i, g_i are the constants, $\alpha_i > 0$, $g_i > 0$, $(i = 1,2)$, t is the variable that specifies the time.

We apply the Laplace transform in the time domain to the equations (1)

$$\tilde{f}(p) = \int_0^\infty f(t) e^{-pt} dt \tag{6}$$

The result is the system of elasticity theory with a complex parameter p. We apply to this system the homogenization method described in [1, 2]. After that we use the inverse Laplace transform and obtain the following symmetric matrix of effective coefficients for the elastic creeping array:

$$\begin{pmatrix} <c_{11}> & <c_{12}> & <c_{13}> & 0 & 0 & 0 \\ & <c_{22}> & <c_{23}> & 0 & 0 & 0 \\ & & <c_{33}> & 0 & 0 & 0 \\ & & & <c_{44}> & 0 & 0 \\ & & & & <c_{55}> & 0 \\ & & & & & <c_{66}> \end{pmatrix} \tag{7}$$

In the next formulas we have denoted for brevity:

$$L_i = \lambda_i + 2\mu_i,\ G_i = g_i\left(k_i + \frac{2}{3}\right),\ j_i = g_i\left(k_i - \frac{1}{3}\right), i = 1,2.$$

We have got a transversely isotropic homogeneous medium with the next effective modulus:

$$<c_{11}>=<c_{22}>=-\frac{(1-h)h}{L}\left\{\frac{e^{p_1t}[(\alpha_1+p_1)(j_2-\lambda_2(\alpha_2+p_1))-(\alpha_2+p_1)(j_1-\lambda_1(\alpha_1+p_1))]^2}{(\alpha_1+p_1)(\alpha_2+p_1)(p_1-p_2)}+\right.$$

$$+\frac{e^{p_2t}[(\alpha_1+p_2)(j_2-\lambda_2(\alpha_2+p_2))-(\alpha_2+p_2)(j_1-\lambda_1(\alpha_1+p_2))]^2}{(\alpha_1+p_2)(\alpha_2+p_2)(p_2-p_1)}+\frac{e^{-\alpha_1t}j_1^2(\alpha_2-\alpha_1)}{(\alpha_1+p_1)(\alpha_1+p_2)}+$$

$$\left.+\frac{e^{-\alpha_2t}j_2^2(\alpha_1-\alpha_2)}{(\alpha_2+p_1)(\alpha_2+p_2)}+(\lambda_1-\lambda_2)^2\delta(t)\right\}+(L_1+L_2)\delta(t)-e^{-\alpha_1t}G_1h-e^{-\alpha_2t}G_2(1-h). \quad (8)$$

Here p_1, p_2 are the roots of the next quadratic equation:

$$Lp^2+[L(\alpha_1+\alpha_2)-((1-h)G_1+hG_2)]p+[L\alpha_1\alpha_2-G_2h\alpha_1-G_1(1-h)\alpha_2]=0. \quad (9)$$

This equation has a positive discriminant:

$$D=[G_2h-G_1(1-h)+L(\alpha_1-\alpha_2)]^2+4G_1G_2h(1-h), \quad (10)$$

consequently the roots p_1, p_2 will always be real. (We denote in (10): $L=(1-h)L_1+hL_2$).

$$<c_{12}>=-\frac{(1-h)h}{(1-h)L_1+hL_2}\left\{\frac{e^{p_1t}[(\alpha_1+p_1)(j_2-\lambda_2(\alpha_2+p_1))-(\alpha_2+p_1)(j_1-\lambda_1(\alpha_1+p_1))]^2}{(\alpha_1+p_1)(\alpha_2+p_1)(p_1-p_2)}+\right.$$

$$+\frac{e^{p_2t}[(\alpha_1+p_2)(j_2-\lambda_2(\alpha_2+p_2))-(\alpha_2+p_2)(j_1-\lambda_1(\alpha_1+p_2))]^2}{(\alpha_1+p_2)(\alpha_2+p_2)(p_1-p_2)}+\frac{e^{-\alpha_1t}j_2^2(\alpha_2-\alpha_1)}{(\alpha_1+p_1)(\alpha_1+p_2)}+$$

$$\left.+\frac{e^{-\alpha_2t}j_2^2(\alpha_1-\alpha_2)}{(\alpha_2+p_1)(\alpha_2+p_2)}+(\lambda_1-\lambda_2)^2\delta(t)\right\}+(\lambda_1+\lambda_2)\delta(t)-e^{-\alpha_1t}j_1h-e^{-\alpha_2t}j_2(1-h). \quad (11)$$

$$<c_{33}>=\frac{1}{L}\left\{\frac{e^{p_1t}[G_1-L_1(\alpha_1+p_1)][G_2-L_2(\alpha_2+p_1)]}{p_1-p_2}-\right.$$

$$\left.-\frac{e^{p_2t}[G_1-L_1(\alpha_1+p_2)][G_2-L_2(\alpha_2+p_2)]}{p_1-p_2}+L_1L_2\delta(t)\right\}. \quad (12)$$

$$<c_{13}>=<c_{23}>=\frac{1}{L^2(p_1-p_2)}\left\{L[(1-h)j_2L_1+hL_2(j_1-\lambda_1(\alpha_1-\alpha_2))][e^{p_1t}(\alpha_1+p_1)-e^{p_2t}(\alpha_1+p_2)]-\right.$$

$$\left.-L[G_2hj_1+G_1(1-h)j_2]+h(1-h)(\lambda_1-\lambda_2)[G_2L_1(\alpha_1+p_1)-G_1L_2(\alpha_2+p_1)](e^{p_1t}-e^{p_2t})\right\}+$$

$$+\frac{L_1\lambda_2(1-h)+h\lambda_1 L_2}{L}\delta(t). \qquad (13)$$

In (11-13) p_1, p_2 are the roots of the equation (9).

$$<c_{44}>=<c_{55}>=\frac{1}{[(1-h)\mu_1+h\mu_2](p_1-p_2)}\Big\{e^{p_1 t}[g_1-\mu_1(\alpha_1+p_1)][g_2-\mu_2(\alpha_2+p_1)-$$

$$-e^{p_2 t}[g_1-\mu_1(\alpha_1+p_2)][g_2-\mu_2(\alpha_2+p_2)]\Big\}+\frac{\mu_1\mu_2\delta(t)}{(1-h)\mu_1+h\mu_2}. \qquad (14)$$

Here p_1, p_2 are the roots of the next quadratic equation:

$$Mp^2+[M(\alpha_1+\alpha_2)-((1-h)g_1+hg_2)]p+[M\alpha_1\alpha_2-g_2h\alpha_1-g_1(1-h)\alpha_2]=0, \qquad (15)$$

which also has a positive discriminant. (We denote in (14): $M=2[(1-h)\mu_1+h\mu_2]$).

$$<c_{66}>=\frac{1}{2}(c_{11}-c_{12})=-e^{-\alpha_1 t}g_1h-e^{-\alpha_2 t}g_2(1-h)+[h\mu_1+(1-h)\mu_2]\delta(t). \qquad (16)$$

The calculation takes the following values of Lamé constants and the parameters of the creep kernels for each layer [15, 16]:
$\lambda_1=288{,}46$ MPa, $\lambda_2=2019$ MPa, $\mu_1=192{,}3$ MPa, $\mu_2=1346$ MPa, $k_1=k_2=0{,}7$, $g_1=17{,}3\times10^{-4}$ (hour)$^{-1}$, $g_2=3{,}45\times10^{-4}$ (hour)$^{-1}$, $\alpha_1=0{,}003$, $\alpha_2=0{,}0015$.

As a result, for example, we plotted creep kernel g_{11} containing exponential functions depending on time t (hours) for $t\in[0;5000]$ (Fig. 2). This creep kernel is included in the average modulus $<c_{11}>$.

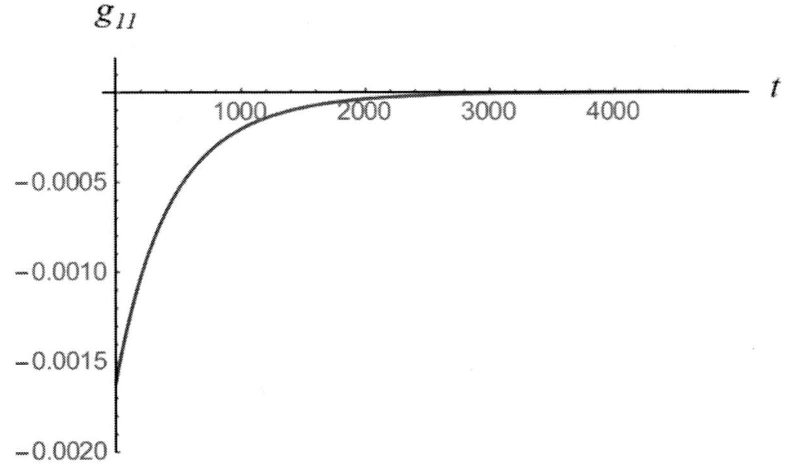

Fig. 2. Creep kernel g_{11} after homogenization.

3 Conclusion

Received formulas give the explicit analytical solution of the problem of calculating displacements in elastic creeping layered array. These explicit expressions can be useful in structural mechanics for example in underground construction and in problems of the science of materials, in particular, for the deformation analysis of layered polymers over large intervals of time.

Considered convolution kernels are sums of exponential functions. Such kernels are sometimes called spectral. In this paper the spectral kernels containing only one's own exponential function describe each of the two original materials. By the methods of this work it is easy to prove that if the kernel of creep of each material was set with sums of a finite number of exponential functions, then effective kernels of creep for the averaged material were also represented by the sums of a finite number of exponential functions. In this case indicators of exponential functions are roots of some polynomial, and the coefficients of this polynomial can be explicitly calculated by using the characteristics of the original two materials and the geometrical parameters of the layers.

References

1. O.A. Oleynik, A.S. Shamaev, G.A. Yosifian, *Mathematical Problems in Elasticity and Homogenization* (1992)
2. D.I. Bardzokas, A.I. Zobnin, *Mathematical Modelling of Physical Processes in Composite Materials of Periodical Structures* (2003)
3. B.E. Pobedrya, *Mechanics of Composite Materials* (1984)
4. A.S. Shamaev, V.V. Shumilova, *Doklady Physics* **56** (1) 43 (2011) doi: 10.1134/S1028335811010083
5. A.S. Shamaev, V.V. Shumilova, Fluid Dynamics **46** (2), 250 (2011) doi: 10.1134/S0015462811020074
6. E. Sanchez-Palencia, *Non-Homogeneous Media and Vibration Theory* (1980)
7. A.S. Shamaev, V.V. Shumilova, Proceedings of the Steklov Institute of Mathematics **295**, 23 (2016)
8. A.S. Shamaev, V.V. Shumilova, Proceedings of the Steklov Institute of Mathematics **295**, 35 (2016)
9. T. Bobyleva, Procedia Engineering **153**, 103 (2016) doi: 10.1016/j.proeng.2016.08.087
10. V.I. Andreev, D.A. Kapliy, Procedia Engineering **111**, 36 (2015) doi: 10.1016/j.proeng.2015.07.032
11. V.I. Andreev, *Some problems and methods of mechanics inhomogeneous bodies* (2002)
12. N.Yu.Tsybina, R.A.Turusova, V.I. Andreev, Procedia Engineering **153**, 51 (2016) doi: 10.1016/j.proeng.2016.08.079
13. A.A. Ilyushin, B.E. Pobedrya, *Foundations of the mathematical theory of thermovisco-elasticity* (1970)
14. Yu.N. Rabotnov, *Creep of Structural Elements* (1966)
15. S.A. Konstantinova. V.M. Pestrenin. I.V. Pestrenina, *Proceedings of higher educational institution,* Mining journal **4,** 92 (2007)
16. M. H. Gabdrahimov, M.N. Dedyukin, *Research on mining.* The proceedings of the conference on the results of scientific research (1968)

 DOI: 10.1051/matecconf/20168601025

The research of vibrations of the beam with the attached system of mass-spring-damper using functions Tymoshenko

Ivanov Pavel[1,*]

[1]Moscow State University of Civil Engineering, Yaroslavskoye Shosse, 26, Moscow, 129337, Russian Federation

Abstract. In the article [1], is considered a classic method for analyze the dynamics of the loaded beam, as a result leading to the solution of a system of transcendental equations for the eigenvalues. In this paper, shall show that perhaps the solution using functions Tymoshenko arising in the analysis of the free beam vibrations. As a result, the problem will be reduced to infinite dimensional system of differential equations satisfying the conditions for the approximate solution.

1 Introduction

The study of the beam vibrations continues to be actual over the years and finds wide application in engineering practice. The authors [4-6] studied this problem is performed analysis to find eigenfrequencies and eigenmodes. This approach is useful when considering the free vibrations of the beam. If mass or spring-mass attach to a beam, having difficulty with the calculation of eigenfrequencies. In this paper, we will show how, without changing the system of eigenfunctions, explore the fluctuations in the beam with a system of mass-spring-damper.

2 Equations of motion

Consider the free vibrations of the simply-supported uniform beam (length l, linear density ρ, stiffness EI and area of cross section F) with system of mass-spring-damper on на interior point x_0 (Fig. 1.). The stiffness of the spring g, a viscosity damper c and mass weight M. For this system is well-known equation of motion [2]:

$$EI\frac{\partial^4 U}{\partial x^4} = -\rho F\frac{\partial^2 U}{\partial t^2} \tag{1}$$

where $U(x,t)$ - function of a displacement perpendicular to the axis of the beam.

* Corresponding author: ar1on@mail.ru

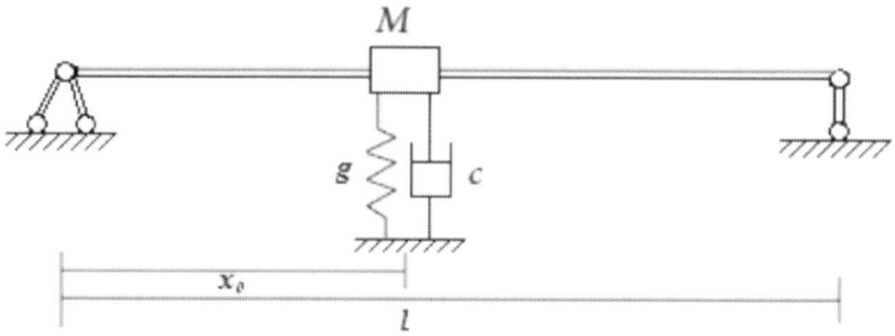

Fig. 1. Beam with system of mass-spring-damper.

The boundary conditions are of the form

$$\begin{cases} U\big|_{x=0}=0 \\ \left.\dfrac{\partial^2 U}{\partial x^2}\right|_{x=0}=0 \end{cases} \qquad \begin{cases} U\big|_{x=l}=0 \\ \left.\dfrac{\partial^2 U}{\partial x^2}\right|_{x=l}=0 \end{cases} \tag{2}$$

Influence of a system of mass-spring-damper is determined by the coupling conditions at the point x_0 :

1) Continuity condition: $U\big|_{x=x_0-0}=U\big|_{x=x_0+0}$;

2) Conditions smoothness: $\left.\dfrac{\partial U}{\partial x}\right|_{x=x_0-0}=\left.\dfrac{\partial U}{\partial x}\right|_{x=x_0+0}$;

3) Equilibrium of moments: $EI\left.\dfrac{\partial^2 U}{\partial x^2}\right|_{x=x_0-0}=EI\left.\dfrac{\partial^2 U}{\partial x^2}\right|_{x=x_0+0}$; (3)

4) Equilibrium of forces:

$$EI\left(\left.\frac{\partial^3 U}{\partial x^3}\right|_{x=x_0-0}-\left.\frac{\partial^3 U}{\partial x^3}\right|_{x=x_0+0}\right)=M\left.\frac{\partial^2 U}{\partial t^2}\right|_{x=x_0}+c\left.\frac{\partial U}{\partial t}\right|_{x=x_0}+gU\big|_{x=x_0}.$$

3 The solution using functions Tymoshenko

Apply the method of separating variables, i.e. $U(x,t)=X(x)\cdot T(t)$.

If consider only (1) and (2), the eigenfunctions would constitute functions Tymoshenko:

$$X_n(x)=\cosh k_n x-\cos k_n x-\frac{\cosh k_n l-\cos k_n l}{\sinh k_n l-\sin k_n l}(\sinh k_n x-\sin k_n x),\ \text{where}\ k_n=\frac{\pi n}{l}.$$

The orthogonality condition

$$\int_0^l X_n(x)X_n(x)dx=\delta_{nm},$$

where δ_{nm} - Kronecker delta function.

Let us assume that the solution of the current problem can be represented in the form:

$$U(x,t)=\sum_{n=1}^{\infty}a_n(t)X_n(x) \tag{4}$$

And also rightly:

$$\frac{\partial^4 U}{\partial x^4}=\sum_{n=1}^{\infty}b_n(t)X_n(x) \tag{5}$$

To reduce the size of the calculations, we agree that strokes or numbers in brackets at the upper right index indicates the order derivatives of x, and the points on functions derivative of t.

Establish the relation between $\{a_n(t)\}_{n=1}^{\infty}$ и $\{b_n(t)\}_{n=1}^{\infty}$. For this use the representation $a_n(t)$, as the Fourier coefficients and 4 times apply integration by parts and taking into account $X_n^{(IV)}(x)-k_n^4\cdot X(x)=0$:

$$a_n(t)=\int_0^l U(x,t)X_n(x)dx=\frac{1}{k_n^4}\int_0^l U(x,t)X_n^{(IV)}(x)dx=\frac{1}{k_n^4}\Big(U(x,t)X_n'''(x)\Big|_0^l-$$

$$-\int_0^l U'(x,t)X_n'''(x)dx\Big)=\frac{1}{k_n^4}\Big(U(x,t)X_n'''(x)\Big|_0^l-U'(x,t)X_n''(x)\Big|_0^l+U''(x,t)X_n'(x)\Big|_0^l-$$

$$-\int_0^l U'''(x,t)X_n'(x)dx\Big)=\frac{1}{k_n^4}\Big(U(x,t)X_n'''(x)\Big|_0^l-U'(x,t)X_n''(x)\Big|_0^l+U''(x,t)X_n'(x)\Big|_0^l-$$

$$-\left(\int_0^c U'''(x,t)X_n'(x)dx+\int_c^l U'''(x,t)X_n'(x)dx\right)\Bigg)=\frac{1}{k_n^4}\Big(U(x,t)X_n'''(x)\Big|_0^l-U'(x,t)X_n''(x)\Big|_0^l+$$

$$+U''(x,t)X_n'(x)\Big|_0^l-U'''(x,t)X_n(x)\Big|_0^l+U'''(x,t)X_n(x)\Big|_{x_0-0}^{x_0+0}+\int_0^l U^{(IV)}(x,t)X_n(x)dx\Big).$$

In accordance with the boundary conditions, the first four expressions in brackets are zero. We obtain

$$a_n(t)=\frac{1}{k_n^4}\left(\big(U'''(x_0+0,t)-U'''(x_0-0,t)\big)X_n(x_0)+\int_0^l U^{(IV)}(x,t)X_n(x)dx\right).$$

As can be seen, the integral in the brackets is the coefficient $b_n(t)$.

From (4) follows that

$$\dot U(x,t)=\sum_{n=1}^{\infty}\dot a_n(t)X_n(x) \qquad \ddot U(x,t)=\sum_{n=1}^{\infty}\ddot a_n(t)X_n(x)$$

Substitute the expression for $a_n(t)$, have

$$a_n(t)=\frac{1}{k_n^4}\left(-\frac{1}{EI}\left(M\frac{\partial^2 U}{\partial t^2}\bigg|_{x=x_0}+c\frac{\partial U}{\partial t}\bigg|_{x=x_0}+gU|_{x=x_0}\right)X_n(x_0)+b_n(t)\right).$$

Apply (4) and obtain the final formula

$$a_n(t)=\frac{1}{k_n^4}\left(-\frac{X_n(x_0)}{EI}\sum_{s=1}^{\infty}\left(M\ddot{a}_s(t)+c\dot{a}_s(t)+ga_s(t)\right)X_s(x_0)+b_n(t)\right)$$

or

$$b_n(t)=k_n^4 a_n(t)+\frac{X_n(x_0)}{EI}\sum_{s=1}^{\infty}\left(M\ddot{a}_s(t)+c\dot{a}_s(t)+ga_s(t)\right)X_s(x_0). \qquad (6)$$

The formula is clearly seen the main difference of this method from the classic separation of variables - additional load is taken into account in the Fourier coefficients, rather than in orthogonal conditions eigenfunctions. Also note that when $M=k=c=0$, obtain $b_n(t)=k_n^4 a_n(t)$, which is consistent with the results for the free vibration.

Write the equation (1) substituting (4) and (5)

$$EI\sum_{n=1}^{\infty}b_n(t)X_n(x)=-\rho F\sum_{n=1}^{\infty}\ddot{a}(t)X_n(x)$$

$$\sum_{n=1}^{\infty}\left(EI\cdot b_n(t)+\rho F\cdot\ddot{a}_n(t)\right)X_n(x)=0.$$

$$EI\cdot b_n(t)+\rho F\cdot\ddot{a}_n(t)=0,\ \text{где}\ n=\overline{1,\infty}.$$

Using resulting formula (6) in the last equation

$$k_n^4 EIa_n(t)+X_n(x_0)\sum_{s=1}^{\infty}\left(M\ddot{a}_s(t)+c\dot{a}_s(t)+ga_s(t)\right)X_s(x_0)+\rho F\cdot\ddot{a}_n(t)=0,\ \text{где}\ n=\overline{1,\infty}.$$

or

$$\left(MX_n^2(x_0)+\rho F\right)\ddot{a}_n(t)+cX_n^2(x_0)\dot{a}_n(t)+\left(k_n^4 EI+gX_n^2(x_0)\right)a_n(t)=$$

$$=-X_n(x_0)\sum_{\substack{s=1\\ s\neq n}}^{\infty}\left(M\ddot{a}_s(t)+k\dot{a}_s(t)+ca_s(t)\right)X_s(x_0),\ \text{где}\ n=\overline{1,\infty}.$$

For practical application, the system can be approximately solved using numerical methods and reduction method [3], because $\left\{X_n^2(x_0)\right\}_{n=1}^{\infty}$ limited and $\lim_{n\to\infty}a_n(t)=0$.

4 Conclusion

In this article, we have been shown a method which does not require calculation of eigenfunctions and eigenfrequencies, and reduces the problem to the infinite dimensional system of differential equations. This system can easily be solved using numerical methods and method of reduction, since It satisfies the necessary conditions.

Since the solution uses functions Tymoshenko system it makes it easier the problem of convergence of Fourier series, and also allows a comparative analysis with non-loaded beam.

It is worth noting that all of the solution was performed for free-supported beam, but it is similar to any combination of the classic fixings. Enough to take corresponding functions Tymoshenko. This solution can be generalized to multiple systems of mass-spring-damper.

References

1. O. Barry, D.C.D. Oguamanam, J.W. Zu, Shock and Vibration, **2014**, 9 (2014)
2. S.P. Tymoshenko, *Vibration problems in engineering*, Mashinostroenie (1985)
3. L.V. Kantorovich, G.P. Akilov, *Functional Analysis*, Phys.-Mat. Lit. (1962)
4. G. Chen, L. Qian, Q. Yin, Shock and Vibration, **2014**, 12 (2014)
5. H.-Y. Lin, Y.-C. Tsai, Journal of Sound and Vibration, **302**, 442-456 (2007)
6. M.H. Ghayesh, F. Alijani, M.A. Darabi, Journal of mechanical Science and Technology, **25**, 1915-1923 (2011)

The rod as super-element of the subsystem

Oleg Kovalchuk [1,*]

[1]Moscow State University of Civil Engineering, 129337 Yaroslavskoe Sh.26, Moscow, Russia

Abstract. The article reviews some aspects of simulation of static and dynamic conditions of building bearing structures due to its necessity for durability estimation. The main method of this simulation is Finite Super-Element Method (FSEM), which is characterized by its main principle «One rod - one super-element». This element, called a super-element, may be a part of a larger multi-body system or it may be an entire mechanical system. The main algorithm of FSEM modelling focused on possible variation of bearing structure is given.

1 Introduction

For the simulation core supporting structure the most suitable method of super-elements (FEM) in the variant of the method of the displacements. [1, 2] In this case, due to the use of the Lagrange&D'Alembert variational principle and arbitrary functions of shape the resolving set of equations has the meaning of static or dynamic equations of equilibrium with symmetrical stiffness and mass matrices. Every finite element is described with its own stiffness matrix, load vector and mass matrix (for dynamic analysis only). The structure model is a result of the summation of lines of all elements stiffness matrices, representing internal forces projections and the summation of vector load components, representing external forces projections on global coordinates system, acting on the nodes of one element.

The feature of the integration process is that all external and internal forces, acting from the all the finite elements, containing one common node, are summed up for this node.

An integration of all nodal unknown variable (node displacements) for each node is the most rational. For rod systems it is attributed six degrees of freedom to each node, three of them are linear displacements and three other are angle of rotation relative to the global coordinate axes.

2 The algorithm of condition estimation of rod bearing structure

If the number of the node is known, its degree of freedom position of node displacement vector is estimated as in the following equation:

$$I_{cc} = 6(Nnode - 1) + i_{cc} \tag{1}$$

where I_{cc} – the degree of freedom index of node displacement vector, icc – the degree of

[*] Corresponding author: Kovalchuk@mgsu.ru

freedom index of node displacement vector of the single node.

The algorithm submitted is effective in the simulation of two- and three-dimensional bodies. For example, it is implemented, for example, in ANSYS.

At the same time, the use of this algorithm is complicated for modelling rod systems, where you can change the structure: extra degree of freedom appear and rearrangement of the matrix stiffness of an ensemble is required; moreover, it is required when a defect occurs in one or more rods. [4, 5] The paper [6] proposed one of the methods of formulation ratios for FE using analytical approaches.

A version of FSEM, suitable for durability of bearing system analysis problems with features given above, is shown below. The meaning of this version is equivalent to the method of joints, for example, in [7, 8]. The meaning of its resolving set of equations is the same as for classic FEM or FCEM. The difference of the proposed method from the classic versions of FEM and FCEM is formulated.

Assume that Stiffness Matrix of all the FE (or SE) is known, for example, in the form of algorithms. Imagine the vector of the nodal displacements arranged in the rods:

$$\begin{aligned} &\mathbf{U}=\{\mathbf{U}_1 \quad \mathbf{U}_2 \quad \dots \quad \mathbf{U}_N\} \\ &\mathbf{U}_k=\{\mathbf{U}_{1k} \quad \mathbf{U}_{2k}\} \; k=1,2,3..N_{Rods} \\ &\mathbf{U}_{ik}=\{u_{xik} \quad u_{yik} \quad u_{zik} \quad \theta_{xik} \quad \theta_{yik} \quad \theta_{zik}\}, \; i=1,2 \end{aligned} \tag{2}$$

Here k is the number of terminal, i=1 – the beginning of the k-th shaft, i=2 – the end of the k-th terminal, N_{Rods} – the number of rods in the system. The same structure, suppose for the vector of the right side. If members are Autonomous, not linked to each other, then the resolving system of equations has a block-diagonal matrix composed of stiffness matrix finite elements (SMFE) in the order of their numbering:

$$\begin{aligned} &\mathrm{K}\mathbf{U}=\mathbf{R} \\ &\mathrm{K}=\begin{bmatrix} \mathrm{K}_1 & 0 & \dots & 0 \\ 0 & \mathrm{K}_2 & \dots & 0 \\ \dots & \dots & \dots & \dots \\ 0 & 0 & \dots & \mathrm{K}_{N_{Rods}} \end{bmatrix} \quad \mathbf{R}=\begin{Bmatrix} \mathbf{R}_1 \\ \mathbf{R}_2 \\ \dots \\ \mathbf{R}_N \end{Bmatrix} \end{aligned} \tag{3}$$

the dimensions of the matrix K – $12\cdot N_{Rods}\times 12\cdot N_{Rods}$.

The assume that M_1-number of constraints of type

$$u_{xik}-u_{xjn}=0, \;\; k,n\in[1,2..N], \;\; i,j\in[1,2] \tag{4}$$

are set on the system, which Express the conditions of equal displacement u_{xik} and u_{xjn}. Let us add M_2-number of constraints to primary set to represent displacement boundary conditions

$$u_{\alpha ik}=\overset{*}{u}_{\alpha ik}(t), \;\; \theta_{\alpha ik}=\overset{*}{\theta}_{\alpha ik}(t), \;\; k\in[1,2..N], \;\; i\in[1,2], \;\; \alpha=x,y,z \tag{5}$$

here $\overset{*}{u}_{\alpha ik}(t)$, $\overset{*}{\theta}_{\alpha ik}(t)$ – prescribed displacements and rotation angles of the sections.

Due to linearity of relations (4), (5), they can be represented as the matrix type:

$$\mathrm{C}\mathbf{U}=\mathbf{U}^*(t) \tag{6}$$

here matrix relating nodal dof to structure dof (matrix links) C has a size of (M_2+

M_2)×(12·N_{Rods}).This matrix is in General rectangular and rarefied: it has two element in the line (1 и -1) – connection type (5) and one element in the string (1) – connection type (6).

We introduce the functional of the fine:

$$S = \boldsymbol{\lambda}^T\left(\mathbf{CU} - \mathbf{U}^*\right) \tag{7}$$

where $\boldsymbol{\lambda}$ – Lagrange multipliers, all – greater than zero.

We introduce the functional F for which conditions of minimum are the equations (3):

$$F = \mathbf{U}^T(\mathbf{KU} - \mathbf{R}) \tag{8}$$

The functional F is the sum of works of external and internal forces to nodal displacements; the functional S a can be interpreted as the finefor the violation of boundary conditions and conditions of jointness of cores in the design.

Here require the minimum sum $F + S$:

$$\begin{aligned} \frac{\partial(F+S)}{\partial \mathbf{U}} &= \mathrm{K}\mathbf{U} - \mathbf{R} + \mathrm{C}^T\boldsymbol{\lambda} = 0 \\ \frac{\partial(F+S)}{\partial \boldsymbol{\lambda}} &= \mathrm{C}\mathbf{U} - \mathbf{U}^* = 0 \end{aligned} \tag{9}$$

In order to define node displacements a system of linear equations with square size matrix and inhomogeneous right side is obtained:

$$\begin{aligned} &\mathrm{K}_\lambda \mathbf{U}_\lambda = \mathbf{R}_\lambda, \\ &\mathrm{K}_\lambda = \begin{bmatrix} \mathrm{K}_{cb} & \mathrm{C}^T \\ \mathrm{C} & 0 \end{bmatrix};\ \mathbf{U}_\lambda = \begin{Bmatrix} \mathbf{U} \\ \boldsymbol{\lambda} \end{Bmatrix};\ \mathbf{R}_\lambda = \begin{Bmatrix} \mathbf{R} \\ \mathbf{U}^* \end{Bmatrix} \end{aligned} \tag{10}$$

This algorithm makes it easy to model systems under variable structure with beyond design impacts such as changes in the structure: the destruction of individual links. [9, 10] For this you need to change the structure of the matrix links C, excluding the destroyed connection. When are appearance of defects (cracks) in a separate rod, it is sufficient to change one stiffness matrix in quasidiagonality the matrix K corresponding to this rod. Of course, to obtain the decision values (nodal displacements) is required after each change of the structure of the rod system.

Thus, to solve one problem (for a given structure of the system of rods), you need to build expressions for the nodal forces and the nodal displacements and set the matrix C that defines the system structure and kinematic boundary conditions

3 The principle «One rod - one super element»

As shown in Ch.**1** the required condition of making FEM model of the rod system is an availability of expression allowing the estimation of node forces through node displacements. Vector of initial conditions consist of kinematic and force part:

$$\begin{aligned} \mathbf{y}_0 &= \{\mathbf{y}_{0C} \quad \mathbf{y}_{0F}\} \\ \mathbf{y}_{0C} &= \{u_0 \quad v_0 \quad w_0 \quad \theta_{x0} \quad \theta_{y0} \quad \theta_{z0}\} \\ \mathbf{y}_{0F} &= \{M_{x0} \quad M_{y0} \quad M_{z0} \quad N_0 \quad Q_{y0} \quad Q_{z0}\} \end{aligned} \tag{11}$$

Then the expression for the static condition can also be represented in the form of a parts:

$$\begin{aligned}
&\mathbf{y}_C(x) = \mathrm{V}_{CC}^{st}(x)\mathbf{y}_{0C} + \mathrm{V}_{CF}^{st}(x)\mathbf{y}_{0F} - \mathbf{F}_{qC}(x)\\
&\mathbf{y}_F(x) = \mathrm{V}_{FC}^{st}(x)\mathbf{y}_{0C} + \mathrm{V}_{FF}^{st}(x)\mathbf{y}_{0F} - \mathbf{F}_{qF}(x)\\
&\mathrm{V}_{CC}^{st}(x) = \left[\mathrm{V}^{st}\right]_{1..6,1..6};\ \mathrm{V}_{CF}^{st}(x) = \left[\mathrm{V}^{st}\right]_{1..6,7..12}\\
&\mathrm{V}_{FC}^{st}(x) = \left[\mathrm{V}^{st}\right]_{7..12,1..6};\ \mathrm{V}_{FF}^{st}(x) = \left[\mathrm{V}^{st}\right]_{7..12,7..12}
\end{aligned} \tag{12}$$

Eliminate the force factors at the beginning of the rod through the displacement of its end, recording the first equality of (12) to the end of the rod in case $x=L$:

$$\begin{aligned}
&\mathbf{y}_C(L) = \mathbf{y}_{kC} = \mathrm{V}_{CC}^{st}(L)\mathbf{y}_{0C} + \mathrm{V}_{CF}^{st}(L)\mathbf{y}_{0F} - \mathbf{F}_{qC}(L) \Rightarrow\\
&\Rightarrow \mathbf{y}_{0F} = \mathrm{V}_{CF}^{st}(L)^{-1}\left\{\mathbf{y}_{kC} - \mathrm{V}_{CC}^{st}(L)\mathbf{y}_{0C} + \mathbf{F}_{qC}(L)\right\}
\end{aligned} \tag{13}$$

Now substitute the result in (12) and obtain the expression state at any point of the rod using the nodal displacements:

$$\begin{aligned}
&\mathbf{y}_C(x) = \mathrm{V}_{CC}^{st}(x)\mathbf{y}_{0C} +\\
&+ \mathrm{V}_{CF}^{st}(x)\mathrm{V}_{CF}^{st}(L)^{-1}\left\{\mathbf{y}_{kC} - \mathrm{V}_{CC}^{st}(L)\mathbf{y}_{0C} + \mathbf{F}_{qC}(L)\right\} - \mathbf{F}_{qC}(x) =\\
&= \left\{\mathrm{V}_{CC}^{st}(x) - \mathrm{V}_{CF}^{st}(x)\mathrm{V}_{CF}^{st}(L)^{-1}\mathrm{V}_{CC}^{st}(L)\right\}\mathbf{y}_{0C} + \mathrm{V}_{CF}^{st}(x)\mathrm{V}_{CF}^{st}(L)^{-1}\mathbf{y}_{kC} -\\
&- \mathbf{F}_{qC}(x) + \mathrm{V}_{CF}^{st}(x)\mathrm{V}_{CF}^{st}(L)^{-1}\mathbf{F}_{qC}(L)
\end{aligned} \tag{14}$$

$$\begin{aligned}
&\mathbf{y}_F(x) = \mathrm{V}_{FC}^{st}(x)\mathbf{y}_{0C} +\\
&+ \mathrm{V}_{FF}^{st}(x)\mathrm{V}_{CF}^{st}(L)^{-1}\left\{\mathbf{y}_{kC} - \mathrm{V}_{CC}^{st}(L)\mathbf{y}_{0C} + \mathbf{F}_{qC}(L)\right\} - \mathbf{F}_{qF}(x) =\\
&= \left\{\mathrm{V}_{FC}^{st}(x) - \mathrm{V}_{FF}^{st}(x)\mathrm{V}_{CF}^{st}(L)^{-1}\mathrm{V}_{CC}^{st}(L)\right\}\mathbf{y}_{0C} + \mathrm{V}_{FF}^{st}(x)\mathrm{V}_{CF}^{st}(L)^{-1}\mathbf{y}_{kC} -\\
&- \mathbf{F}_{qF}(x) + \mathrm{V}_{FF}^{st}(x)\mathrm{V}_{CF}^{st}(L)^{-1}\mathbf{F}_{qC}(L)
\end{aligned} \tag{15}$$

Let us write the expression (15) for the start ($x=0$) and the end ($x=L$) rod:

$$\begin{aligned}
&\mathbf{y}_F(0) = \left\{\mathrm{V}_{FC}^{st}(0) - \mathrm{V}_{FF}^{st}(0)\mathrm{V}_{CF}^{st}(L)^{-1}\mathrm{V}_{CC}^{st}(L)\right\}\mathbf{y}_{0C} + \mathrm{V}_{FF}^{st}(0)\mathrm{V}_{CF}^{st}(L)^{-1}\mathbf{y}_{kC} -\\
&- \mathbf{F}_{qF}(0) + \mathrm{V}_{FF}^{st}(0)\mathrm{V}_{CF}^{st}(L)^{-1}\mathbf{F}_{qC}(L);\\
&\mathbf{y}_F(L) = \left\{\mathrm{V}_{FC}^{st}(L) - \mathrm{V}_{FF}^{st}(L)\mathrm{V}_{CF}^{st}(L)^{-1}\mathrm{V}_{CC}^{st}(L)\right\}\mathbf{y}_{0C} + \mathrm{V}_{FF}^{st}(L)\mathrm{V}_{CF}^{st}(L)^{-1}\mathbf{y}_{kC} -\\
&- \mathbf{F}_{qF}(L) + \mathrm{V}_{FF}^{st}(L)\mathrm{V}_{CF}^{st}(L)^{-1}\mathbf{F}_{qC}(L)
\end{aligned} \tag{16}$$

In the left part (16) are the nodal moments and forces. Combine them into the vector of nodal power factors:

$$\mathbf{F}_{node} = \left\{\mathbf{y}_F(0)\quad \mathbf{y}_F(L)\right\} \tag{17}$$

Similarly dealing with the movements of the ends of the rod, combining them into a vector of nodal displacements:

$$\mathbf{U}_{node} = \left\{\mathbf{y}_{0C}(0)\quad \mathbf{y}_{kC}\right\} \tag{18}$$

Then (16) can be written compactly:

$$\mathbf{F}_{node} = \mathbf{K}\mathbf{U}_{node} - \mathbf{F}_q^{node} \tag{19}$$

Here are introduced the stiffness matrix of the rod:

$$\mathrm{K}=\begin{bmatrix} \mathrm{V}_{FC}^{st}(0)-\mathrm{V}_{FF}^{st}(0)\mathrm{V}_{CF}^{st}(L)^{-1}\mathrm{V}_{CC}^{st}(L) & \mathrm{V}_{FF}^{st}(0)\mathrm{V}_{CF}^{st}(L)^{-1} \\ \mathrm{V}_{FC}^{st}(L)-\mathrm{V}_{FF}^{st}(L)\mathrm{V}_{CF}^{st}(L)^{-1}\mathrm{V}_{CC}^{st}(L) & \mathrm{V}_{FF}^{st}(L)\mathrm{V}_{CF}^{st}(L)^{-1} \end{bmatrix} \tag{20}$$

and the vector of nodal loads from the external loads on the rod:

$$\mathbf{F}_q^{node}=\left\{\mathbf{F}_{qF}(0)+\mathrm{V}_{FF}^{st}(0)\mathrm{V}_{CF}^{st}(L)^{-1}\mathbf{F}_{qC}(L) \quad \mathbf{F}_{qF}(L)+\mathrm{V}_{FF}^{st}(L)\mathrm{V}_{CF}^{st}(L)^{-1}\mathbf{F}_{qC}(L)\right\} \tag{21}$$

The expression (19) the form is the same as analogous equation of the FEM and, therefore, it is possible to apply a known algorithm of the FEM in form the displacement method.

Note that for the spectral problem you can do the same calculations as above and get the result the same as in (19), with the exception of equality to zero nodal forces $\mathbf{F}_q^{node}$:

$$\mathbf{F}_{node}(\omega_n)=\mathrm{K}(\omega_n)\mathbf{Y}_{node}(\omega_n) \tag{22}$$

here $\mathbf{Y}_{node}$ – vector of nodal displacements corresponding to the natural frequency ωn.

The result obtained is FE or SE? [11] Apparently, it is the super-element of a straight rod, as its original ratio do not contain internal degrees of freedom. He models an element of a subsystem – one rod with its structural features. To interpret the results compute nodal displacements should be used (14), (15), rewriting them in matrix form:

$$\mathbf{y}(x)=\mathrm{D}(x)\mathbf{U}^{node}-\mathbf{F}(x) \tag{23}$$

where

$$\begin{gathered} \mathrm{D}(x)=\begin{bmatrix} \mathrm{V}_{CC}^{st}(x)-\mathrm{V}_{CF}^{st}(x)\mathrm{V}_{CF}^{st}(L)^{-1}\mathrm{V}_{CC}^{st}(L) & \mathrm{V}_{CF}^{st}(x)\mathrm{V}_{CF}^{st}(L)^{-1} \\ \mathrm{V}_{FC}^{st}(x)-\mathrm{V}_{FF}^{st}(x)\mathrm{V}_{CF}^{st}(L)^{-1}\mathrm{V}_{CC}^{st}(L) & \mathrm{V}_{FF}^{st}(x)\mathrm{V}_{CF}^{st}(L)^{-1} \end{bmatrix} \\ \mathbf{F}(x)=\begin{Bmatrix} \mathbf{F}_{qC}(x)-\mathrm{V}_{CF}^{st}(x)\mathrm{V}_{CF}^{st}(L)^{-1}\mathbf{F}_{qC}(L) \\ \mathbf{F}_{qF}(x)-\mathrm{V}_{FF}^{st}(x)\mathrm{V}_{CF}^{st}(L)^{-1}\mathbf{F}_{qC}(L) \end{Bmatrix} \end{gathered} \tag{24}$$

4 Conclusion

Similarly, should seek own of the condition.

The interpretation of the results is the most time-consuming part of the computational work due to the necessity of calculating the influence matrices to determine the weak sections of the rod. At the same time, for typical structural elements possible to perform cataloging stiffness matrix K and interpretation matrix D for use in calculating of other structures, possessing the same elements. [12]

Solution of inhomogeneous dynamic problems for the entire load-bearing structure should be carried out after determining its spectrum.

References

1. O.A. Kovalchuk, Natural and technical Sciences, **11-12**, 344-348 (2014)
2. R. W. Clough, Computers and Structures, **12**, 361-370 (1980)
3. E.I. Sorokina, K.M. Melikhov, L.N. Makovkina, Technical science in Russia and abroad, **4**, 134-137 (2015)
4. E.E. Krasnovsky, A.S. Shadskiy, Bulletin of Moscow state technical University N. Uh. Bauman, **S**, 169-176 (2011)

5. Kai-Liang Lu, Yuan Liu, Wei-guo Zhang, Hui-qing Qiu, Wei-jian Mi, Procedia Engineering, **16**, 685-694 (2011)
6. O.A. Kovalchuk, Building: Science and Education, **1**, 1-6 (2012)
7. S.S. Law, T.H.T. Chan, D. Wu, Journal of Sound and Vibration, **239-1**, 19-39 (2001)
8. A.G. Tamrazyan, O.A. Kovalchuk, National Research Center Bulletin, **3-4**, 120-130 (2011)
9. V.M. Bondarenko, V.I. Kolchunov, E.D. Vorobyov, The Bulletin of the construction technology, **1**, 8-11, (2004)
10. V.D. Rayzer, Construction mechanics and calculation of structures, **5**, 77-78, (2012)
11. O.A. Kovalchuk, Industrial and Civil Engineering, **8**, 70-73 (2016)
12. I. Cismasiu, J.P. Moitinho de Almeida, Advances in Engineering Software, **38**, 11-12, 750-762 (2007)

Blast load estimation using Finite Volume Method and linear heat transfer

Michał Lidner[1,*] and *Zbigniew* Szcześniak[1]

[1]Military University of Technology, Faculty of Civil Engineering and Geodesy, ul. Gen. S. Kaliskiego 2, 00-908 Warsaw, Poland

Abstract. From the point of view of people and building security one of the main destroying factor is the blast load. Rational estimating of its results should be preceded with knowledge of complex wave field distribution in time and space. As a result one can estimate the blast load distribution in time. In considered conditions, the values of blast load are estimating using the empirical functions of overpressure distribution in time ($\Delta p(t)$). The $\Delta p(t)$ functions are monotonic and are the approximation of reality. The distributions of these functions are often linearized due to simplifying of estimating the blast reaction of elements. The article presents a method of numerical analysis of the phenomenon of the air shock wave propagation. The main scope of this paper is getting the ability to make more realistic the $\Delta p(t)$ functions. An explicit own solution using Finite Volume Method was used. This method considers changes in energy due to heat transfer with conservation of linear heat transfer. For validation, the results of numerical analysis were compared with the literature reports. Values of impulse, pressure, and its duration were studied.

1 Introduction

The need to accurately quantify the blast overpressure loadings is important because detonations represent a common threat for the security design of building (i.eg. progressive collapse [1]) as a result of big pressures and also fire [2].

For several years great effort has been devoted to the study of blast response of structural elements such as columns [3], slabs [4], RC walls [5], brick walls [6] or prestressed elements [7]. In those papers and in related references, however, the blast load overpressure is estimated with simplified triangular distribution in time. It is well-known that exponential function approximates the values of the overpressure distribution in time more closely [8,9]. As mentioned in [10], this indicates changes in the pressure-impulse diagrams ($\Delta p(I)$) proposed in [3,4].

The paper presents a method of numerical analysis of the phenomenon of the air shock wave propagation that can be used to make the $\Delta p(t)$ functions more realistic. This method describes an explicit own solution using Finite Volume Method (FVM) and considers changes in energy due to linear heat transfer. For validation, the results of numerical

* Corresponding author: michal.lidner@wat.edu.pl

analysis were compared with the literature reports. The free field explosion and also one-dimensional flow (an explosion in the pipe) and three-dimensional flow (explosion within the compartment) of the shock wave were analyzed. Values of impulse, pressure, and its duration were studied. Finally, an overall good convergence of numerical results with experiments was achieved.

2 Explicit solution using Finite Volume Method

The unknown *Δp(t)* function reflects the thermodynamic state in each point of disturbed gaseous medium, and also reflects the influence of the boundary conditions. The most advanced models available nowadays for Computational Fluid Dynamics are based on Finite Volume formulations, in which the governing equations for the fluid domain (equations for a compressible inviscid fluid, expressing the conservation of mass, momentum and energy) are formulated and solved in conservative form. The Finite Volume Method (FVM) is one of the best methods of solving issues related to gas flow [11]. Thus conservation of mass, momentum and energy can be written as follows [12]:

$$\frac{\partial}{\partial t}\iiint_V \rho dV + \iint_S \rho v_n dS = 0 \tag{1}$$

$$\frac{d}{dt}\left(\iiint_V \rho \vec{v}\, dV\right) = \iint_S p_n dS + \iiint_V \rho F_m dV \tag{2}$$

$$\frac{d}{dt}\left[\iiint_V \rho\left(c_v T + \frac{v^2}{2}\right) dV\right] = \iint_S p_n v_n dS + \iiint_V \rho \vec{F}_m \vec{v}\, dV + \iint_S \dot{q}_n\, dS + \iiint_V \dot{q}_m\, \rho dV \tag{3}$$

where : t - time, ρ - gas density, V - finite volume, S - surface area, v_n - flow rate of the gaseous medium by the surface S, $\vec{v}$ - velocity vector consisting of components $[u,v,w]$ in each of the three orthogonal directions, p_n - pressure by the surface S, $\vec{F}_m$ - unit vector of internal forces, c_v - specific heat of the gaseous medium at constant volume, T - temperature, $\dot{q}_m$ - heat flux density related to the unit mass of the gas, $\dot{q}_n$ - surface density of the heat flux.

When analyzing the state of energy, the change of kinetic energy in heat energy was considered, taking into account the linear heat transfer [13]. It was assumed that the thickness of the shock wave front is equal to three cell sizes. If the particle velocity in the first cell is equal u, then the particle velocities in second and third are equal to $0.67u$ and $0.33u$ respectively. The temperature decrease was assumed in the same way. Before arrival of the air shock wave the particle velocity is equal to 0 and the temperature is equal to the ambient gas temperature [14].

For further analysis finite difference scheme was applied. In the fluid domain a node-centered Finite Volume scheme is adopted and uses the classical explicit Finite Difference time integration scheme. Finite Volume Method on a full unstructured grid is introduced to treat the fluid domain. Equations (1) to (3) have the following difference solution in each of the three orthogonal directions (to compare the 1D case see [11]):

$$\begin{aligned} M_l^{n+1} = M_l^n &- \rho^n \Delta t \Delta z \Delta y\left(U^n_{x,l-1/2} - U^{n+1}_{x,l+1/2}\right) - \rho^n \Delta t \Delta z \Delta x\left(U^n_{y,k-1/2} - U^{n+1}_{y,k+1/2}\right) - \\ &- \rho^n \Delta t \Delta x \Delta y\left(U^n_{z,m-1/2} - U^{n+1}_{z,m+1/2}\right) \end{aligned} \tag{4}$$

$$F_{x,l+1/2}^{n+1} = F_{x,l-1/2}^{n} + \frac{\left(M_l^{n+1} - M_l^{n}\right)\left(U_{x,l-1/2}^{n} - U_{x,l+1/2}^{n+1}\right)}{\Delta t} \tag{5}$$

$$Fy_{k+1/2}^{n+1} = Fy_{k-1/2}^{n} + \frac{\left(M_l^{n+1} - M_l^{n}\right)\left(Uy_{k-1/2}^{n} - Uy_{k+1/2}^{n+1}\right)}{\Delta t} \tag{6}$$

$$F_{z,m+1/2}^{n+1} = F_{z,m-1/2}^{n} + \frac{\left(M_l^{n+1} - M_l^{n}\right)\left(U_{z,m-1/2}^{n} - U_{z,m+1/2}^{n+1}\right)}{\Delta t} \tag{7}$$

$$\begin{aligned}
&\frac{\left(M_l^{n+1} - M_l^{n}\right)}{\Delta t}\left[c_v \Delta T^n + 0.5\left(U_{x,l+1/2}^{n+1} - U_{x,l-1/2}^{n}\right)^2 + 0.5\left(U_{y,k+1/2}^{n+1} - U_{y,k-1/2}^{n}\right)^2 + 0.5\left(U_{z,m+1/2}^{n+1} - U_{z,m-1/2}^{n}\right)^2\right] = \\
&= \left(F_{x,l+1/2}^{n+1} - F_{x,l-1/2}^{n}\right)\left(U_{x,l+1/2}^{n+1} - U_{x,l-1/2}^{n}\right) + \left(F_{y,k+1/2}^{n+1} - F_{y,k-1/2}^{n}\right)\left(U_{y,k+1/2}^{n+1} - U_{y,k-1/2}^{n}\right) + \left(F_{z,m+1/2}^{n+1} - F_{z,m-1/2}^{n}\right)\cdot \\
&\cdot\left(U_{z,m+1/2}^{n+1} - U_{z,m-1/2}^{n}\right) + \frac{\left(M_l^{n+1} - M_l^{n}\right)}{\Delta t}\left(U_{x,l-1/2}^{n}U_{x,l+1/2}^{n+1} + U_{y,k-1/2}^{n}U_{y,k+1/2}^{n+1} + U_{z,m-1/2}^{n}U_{z,m+1/2}^{n+1}\right) + \Delta q S
\end{aligned} \tag{8}$$

where: M_l^n - finite volume mass in cell l and in time step n, Δt - time step, $\Delta x, \Delta y, \Delta z$ - cell size in each of the three orthogonal directions, $U_{x,l}^n, U_{y,k}^n, U_{z,m}^n$ - particle velocities in each of the three orthogonal directions in cell $l(k;\ m)$ and in time step n, $F_{x,l}^n, F_{y,k}^n, F_{z,m}^n$ - external forces, acting on the surface of the finite volume, in each of the three orthogonal directions Δq - heat flux.

The energy of the system is conditioned by the state of the kinetic energy (velocity dependent) and heat. For the purposes of this paper the simplest model of heat transfer was applied (namely the linear temperature decrease) which is expressed as follows [13]:

$$\Delta T^n = \frac{\Delta q \Delta t}{\Delta x \Delta y \Delta z \rho^n c_p} \tag{9}$$

where: c_p – thermal conductivity coefficient of gaseous medium.

This strategy requires careful consideration and proper synchronization of the time integration schemes used in the two subdomains. The error introduced should be negligible given the shortness of time increments in explicit time stepping. This paper presents also an extended version of equation of the maximum time step in the 3D as given:

$$\Delta t \le \frac{C_{FL} \min\{\Delta x, \Delta y, \Delta z\}}{a_1} \tag{10}$$

where: C_{FL} – non dimensional parameter at values from 0 to 1, a_l – sound propagation velocity in gaseous medium.

The system of equations (4) to (9) expresses the gaseous medium flow in the free field region. Graphical interpretation is presented in Fig. 1. The considered volume of gaseous medium is outlined by thick lines and the adjacent volumes by dotted lines. It is assumed that the two parallel sides are being displaced with velocities u^n and u^{n+1} due to changes in energy. This results in displacement of the mass (M) of gaseous medium to a finite volume, which is highlighted by thick line, and hence the change of density and mass of this volume. Weight increase is associated with a pressure change (p). Consequently, there is also a change in energy. Next, a loop is performed over all finite volumes for the following time steps in order to compute the internal forces. Assume that a complete solution, i.eg. all discrete quantities related to the gas state, are known at time t_n and one wants to find the solution at t_{n+1}. First, the velocity from equation (9) is computed. Therefore, the new mass is evaluated via equation (4). Then, the internal forces are computed via equations (5) to

DOI: 10.1051/matecconf/20168601027

(7). Divide values of computed external forces by surface areas to give the pressure values, then subtract the value of ambient air pressure from the external forces to compute the overpressure values ($\Delta p(t)$).

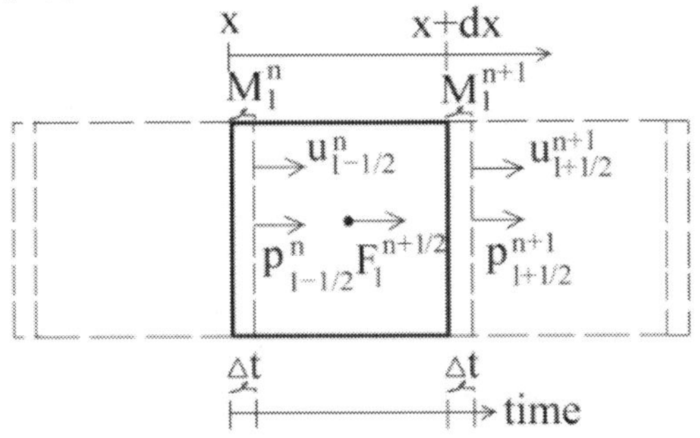

Fig. 1. Diagram of gas flow in Finite Volume Method.

The system of equations must be supplemented by the boundary conditions at the interface of building compartments with adjacent finite volumes and at the point of detonation of condensed explosive. When assuming boundary conditions one can simulate the inhibition of gaseous flux by the building compartments. The assumption of the compartment velocity equal to 0 (boundary condition) in case of high mass of compartments (concrete, RC) is reasonable, because the blast loading is completed before the compartment deformation started [15]. When considering light compartments (made of steel or glass) a suitable coupling strategy must be chosen. One of the best is the strategy based upon suitable kinematic constraints on the velocities of the fluid and of the structure along the fluid-structure interface [16]. In the point, when the shock wave region starts, some boundary condition should be known. These are the particle velocity $u_{1/2}^{1}$, the shock wave pressure $p_{1/2}^{1}$, the density of post-explosion gases ρ_1^{1} and the temperature of post-explosion gases. They can be computed based on the rules presented in [17] and as follows below.

This study examines also numerical modeling of detonations of spherical and cubical high explosives and characterizes their effects in the near field, called the region of post-explosion gases. This is the region defined by a distance equal to approximately from 10 to 15 times the charge radius, within which the shock wave is affected by local phenomena, including expansion of the detonation products. In case of spherical charge gasses has the shape of a sphere, and in case of cubical charge the shape of an octahedron [18].

Then the procedure based on [17] is applied. In the first step the volume of the post-explosion gases should be known. When considering the discrete space the quantity of cells is the main factor (s). In case of a sphere, the finite volume at coordinates ($l; k; m$) is inside the region of post-explosion gases in n^{th} time step if the following inequality is satisfied:

$$\sqrt{l^2 + k^2 + m^2} \le n \tag{11}$$

and in case of an octahedron (the plane equation):

$$l + k + m \le n. \tag{12}$$

Note that in the time-stepping procedure the gas density is obtained first. Then, the particle velocity is solved on the current configuration. Finally, the gas pressure is obtained as the last result as follows:

$$\rho_f = \frac{m}{\Delta x \Delta y \Delta z s} \tag{13}$$

$$v_f = \left(1 - \frac{\rho_1}{\rho_f}\right) D \tag{14}$$

$$p_f = \rho_1 v_f D + p_1 \tag{15}$$

where: ρ_f - density of post-explosion gases, m - mass of explosive charge, s -quantity of cells, v_f - particle velocity of post-explosion gases, ρ_1 - ambient air density, ρ_f - ambient air density, D - detonation velocity of considered explosive material, p_f - pressure of post-explosion gases, p_1 - ambient air pressure.

Then the density of post-explosion gases, computed from equation (13) (assume that the cell size is equal to the charge dimension) is equal to the air density in 7th time step in case of spherical charge (2 x 6,5 = 13 – the medium between 10 and 15) and till this point the shock wave begins to propagate. In case of a cubical charge the region of post-explosion gases finishes in 9th time step. This difference is due to bigger volume of a sphere than an octahedron entered into this sphere. Discrete regions of post-explosion gases in case of spherical charge (7th time step, squares) and cubical charge (9th time step, dots) are presented in Fig. 2.

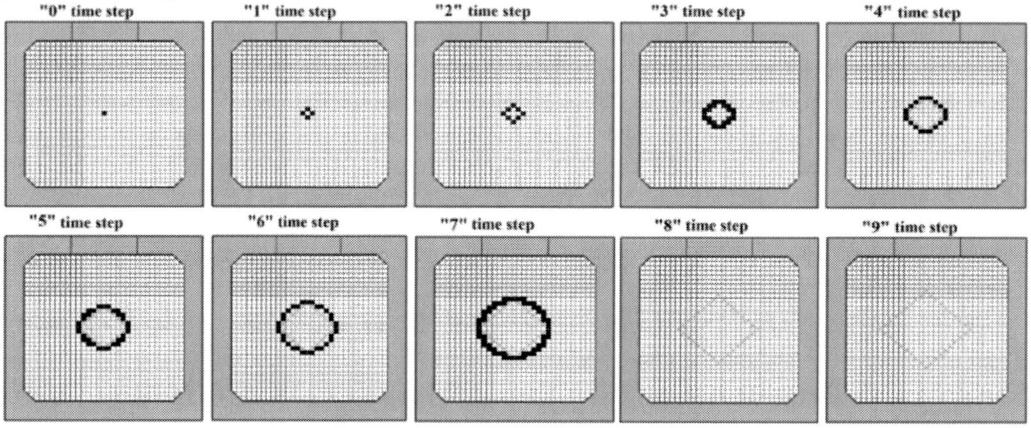

Fig. 2. Discrete spheres (drawn using squares) and octahedrons (drawn using dots) in the following time steps

When calculating the overpressure values in the shock wave region, assume that some essential boundary conditions are imposed, which can be expressed by the initial velocity ($u_{1/2}^{1}$), the shock wave pressure ($p_{1/2}^{1}$) and the density (ρ_1^{1}) equal to the particle velocity (v_f), the pressure of post-explosion gases (p_f) and the gas density (ρ_f) respectively.

3 Comparisons with the literature reports

The first test problem is the classical free field detonation of 1 kg TNT at a standoff distance of 5 m, for which an empirical solution is available. Nowadays scientists do not carry on research in such a simple configuration (without compartments reflecting the shock wave), so the empirical equations based on [17] were applied. The relative overpressure proposed by Sadowski ($\Delta p_{f1}/p_1$) or the relative overpressure proposed by Brode ($\Delta p_{f2}/p_1$), the overpressure duration (τ in *ms*), the exponential parameter (n) and the overpressure distribution in time (Δp) are given by:

$$\frac{\Delta p_f}{p_1} = \begin{cases} \dfrac{\Delta p_{f1}}{p_1} = 0.806\dfrac{\sqrt[3]{m}}{r} + 2.56\dfrac{\left(\sqrt[3]{m}\right)^2}{r^2} + 6.64\dfrac{\left(\sqrt[3]{m}\right)^3}{r^3} \\ \dfrac{\Delta p_{f2}}{p_1} = -0.02 + 0.78\dfrac{\sqrt[3]{m}}{r} + 1.40\dfrac{\left(\sqrt[3]{m}\right)^2}{r^2} + 5.52\dfrac{\left(\sqrt[3]{m}\right)^3}{r^3} \end{cases} \tag{16}$$

$$\tau = 1.219\sqrt{\sqrt[3]{m}r} \tag{17}$$

$$n = 1.9\sqrt{\Delta p_f / p_1} \tag{18}$$

$$\Delta p = \Delta p_f \left(1 - \frac{t}{\tau}\right)^n \tag{19}$$

where: r - standoff distance between the considered point and the detonation point. The overpressure diagram is presented in Fig. 3a.

The second case considers a steel cube with insusceptible walls. One cube wall was joined to a tube of 16.8 cm in diameter and 128 cm in length, which was in turn connected to another tube at some distance (see Fig. 3b) [19]. A TNT explosive charge with mass of 18.5 g was detonated at the joint of the cube with the tube. Graphs of the overpressure recorded by sensors located in points 1 and 2 provide the best illustration of the assumptions for the 1D model. The overpressure diagram is presented in Fig. 3c.

To validate correctness of three-dimensional solution, blast pressure distribution from a detonation of 1 kg TNT charge in the center point of a vented room was examined [20]. The room was a composite steel and concrete structure of a horizontal square projection with a side of 2.9 m and height 2.7 m (internal dimensions) with a hole in the roof of 1.20 m in diameter (see Fig. 3d). Nine pressure gauges were installed on one of the walls (G1 to G9). The overpressure diagram is presented in Fig. 3e.

Table 1 presents initial and boundary conditions. The cell size should not be bigger than the charge dimension. The compartment velocity is equal to 0. The time step value (Δt) is equal to $\Delta x/7000$. This is the time to reach the distance of first cell by the shock wave (the TNT detonation velocity – 7000 m/s [21]).

Table 1. Initial and boundary conditions.

	ρ_1^1 (*kg/m*^{3})	Δx (*m*)	Δt (*ms*)	C_{FL} (-)	$u_{1/2}^1$ (*m/s*)	$p_{1/2}^1$ (*MPa*)	c_p (J/(kg · °C))	c_p (J/(kg · °C))	T_1^1 (°C)
1st case	1560 [21]	0.20	0.029	0.05	690	6.7	718	1005	2800 [21]
2nd case	1560 [21]	0.022	0.003	0.05	690	6.7	718	1005	2800 [21]
3rd case	1560 [21]	0.085	0.012	0.05	690	6.7	718	1005	2800 [21]

4 Results and discussion

Fig. 3a, 3c and 3e present graphs with an overpressure versus time in 1st case, 2nd case and 3rd case respectively. As can be seen, numerical results reflect well the time to reach the shock wave and the duration of the shock wave. It can also be observed that when the overpressure obtained in the tests increases, the overpressure obtained numerically also increases. The same applies to the overpressure decrease. The values of maximum overpressures and impulses (area under the overpressure graph) obtained numerically and those from literature reports are similar.

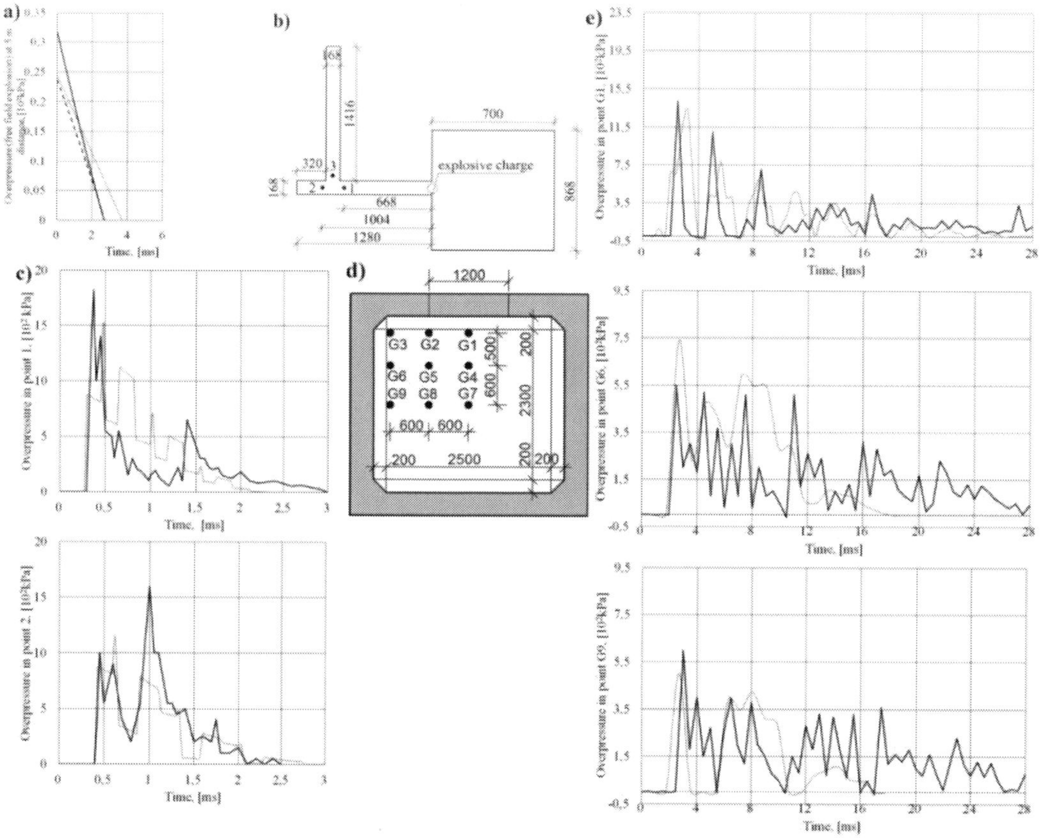

Fig. 3. a) overpressure distribution in time (1st case), thick line-Sadowski, dotted line-Brode, thin line-numerical analysis; b) test stand in 2nd case (mm); c) overpressure distribution in time (2nd case); thick line-research, thin line-numerical analysis; d) test stand in 3rd case (mm); e) overpressure distribution in time (3rd case); thick line-research, thin line-numerical analysis.

As mentioned in [10] this results in the change of pressure-impulse diagram location. In consequence, the point at coordinates (I,p_{max}), which expresses the detonation parameters, could be beneath $\Delta p(I)$ diagram location. This means the possibility of surviving the element during the detonation, which normally should be destroyed. (see Fig. 4).

The method can also be used for analyzing internal explosion, which tends to be difficult to determine in the case of using empirical equations [9].

5 Conclusion

DOI: 10.1051/matecconf/20168601027

The paper presents the solution that uses Finite Volume Method to solve the governing conservation equations at the base of the transient explicit formulation for the fluid domain. A verified and validated computational fluid dynamic code could be used to predict the overpressure histories. Calculations were performed to verify the code in the 1D and 3D,

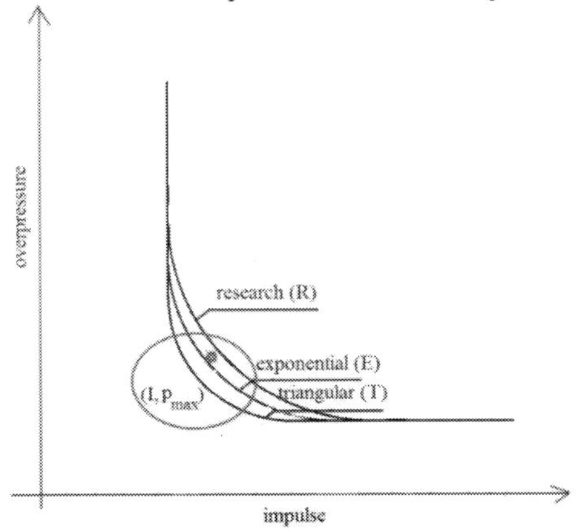

Fig. 4. Influence of change of Δp(I) diagram location on assessment of bearing capacity of element.

predict incident overpressures and impulses and provide guidance on the use of reflecting boundaries. The algorithm can be used to analyze the internal explosions, where the wave field is more complex.

The paper is the result of research tasks carried out under the research RMN No. 800/2016, implemented in the Faculty of Civil Engineering and Geodesy in Jaroslaw Dabrowski Military University of Technology.

References

1. M. Ghahremannejad, Y. Park, Eng. Struct. **111**, 11 (2016)
2. Z. Ruan, L. Chen, Q. Fang, J. Loss Prev. Process Ind. **34**, 10 (2015)
3. Y. Shi, H. Hao, Z.X. Li, Int. J. Impact Eng. **35**, 1213 (2008)
4. J. Xu, C. Wu, Z.X. Li, Int. J. Impact Eng. **69**, 136 (2014)
5. X.Q. Zhou, H. Hao, Adv. Struct. Eng. **11**, 355 (2008)
6. F. Parisi, C. Balestrieri, D. Asprone, Eng. Struct. **113**, 233 (2016).
7. N. Cramsey, C. Naito, PCI J. **11**, 67 (2007)
8. J. Siwiński, A. Stolarski, Bulletin MUT. **64**, 173 (2015)
9. J. Siwiński, A. Stolarski, Bulletin MUT. **64**, 197 (2015)
10. T. Krauthammer, S. Astarlioglu, J. Blasko, T.B. Soh, P.H. Ng, Int. J. Impact Eng. **35**, 771 (2008)
11. F. Casadei, N. Leconte, Int. J. Numer. Methods Eng. **86**, 1 (2011)
12. M. Frant, *Numerical analysis of aerodynamics of complex objects using Finite Volume Method, Ph. D. Thesis* (WAT, Warsaw, 2009)
13. M. Owczarek, W. Radzikowska-Juś, Bulletin MUT. **64**, 149 (2015)
14. A.I. Shestakov, Phys. Fluids **11**, 1091 (1999)
15. J.T. Baylot, T.L. Bevins, Comput. Struct. **85**, 891 (2007)
16. T. Ngo, D. Mohotti, A. Remennikov, B. Uy, J. Constr. Steel Res. **105**, 151 (2015)
17. E. Włodarczyk, Bulletin MUT. **39**, 131 (1980)
18. L. Mazurkiewicz, J. Malachowski, P. Baranowski, Eng. Struct. **104**, 107 (2015)
19. A.M. Benselama, M.J.P. William-Louis, F. Monnoyer, J. Comput. Phys. **228**, 6796 (2009)
20. V.R. Feldgun, Y.S. Karinski, D.Z. Yankelevsky, Int. J. Imp. Eng. **38**, 964 (2011)

21. R. Krzewiński, R. Rekucki, *Building work using explosive materials* (Polcen, Warsaw, 2005)

Numerical simulation of fluid-structure interaction between elastic thin-walled structure and transient fluid flow

Evgeny I. Boznyakov [1,3*], *Irina N.* Afanasyeva[1,2], *and Alexander M.* Belostotsky [1,2]

[1]Moscow State University of Civil Engineering, Research&Educational Center of Computational Simulation, 26 Yaroslavskoye Shosse, Moscow, 129337, Russia
[2]Research & Development Centre StaDyO Ltd, Office 810, 18, 3-ya Ulitsa Yamskogo Polya, Moscow, 125040, Russia
[3]«PIK-Proekt» Ltd., Moscow, Russia

Abstract. The present paper deals with the verification of a numerical simulation technique of fluid-structure interaction (FSI) problems. The configuration of the task consists of viscous incompressible fluid around an elastic cylindrical thin-walled structure where the external flow is fully turbulent (Reynolds numbers $Re=1.2 \cdot 10^7$). The verification is divided into three stages: 1) verification of the mechanical finite element model, 2) verification of the fluid finite volume model and 3) verification of the Fluid-Structure interaction model. Within the stage 1 the comparison of the obtained results with the full-scale measurements and the results of the alternative numerical simulations are presented below.

1 Introduction

Structures of cylindrical thin-walled shells with a certain combination of a specific geometrical parameters, physical, mechanical and dynamic properties of the system and the environment are sensitive to wind effects. Because of the interaction between consider structures and wind flow certain aeroelastic phenomena appeared due to the vortex shedding can be observed (oscillations of a thin-walled cylindrical shell under the flow excitation).

Considering dissipative dynamic system "flow–structure", it should pay attention to the range of natural frequencies and modes of structure and the main frequency of the vortex shedding. In the case of their proximity it leads to significant displacement with increasing amplitude of thin-walled shell oscillations which leads to a change of boundary conditions.

An assessment of the mutual influence of wind gusts on the structure and the deformed structure on the spatial and temporal characteristics of the flow is possible only using a modern numerical models and methods that lets to solve multidisciplinary coupled problems.

* Corresponding author: evgenyboznyakov@gmail.com

2 Problem statement

In the development of the numerical simulation technique for solving three-dimensional aerohydroelasticity problems in civil engineering [4] as well as to examine in detail the nature of aeroelastic vibrations of thin-walled cylindrical structures the problem of numerical simulation of the empty silo oscillations induced by external flow was selected. This verification example is based on the fact that in October 2002 at the port of Antwerp (Belgium) during the storm the oscillations of several empty silos from a group of forty ones was observed [1, 2].

The verification task is divided into three stages. The first stage is the verification of finite element models of a thin-walled cylindrical tank based on a modal analysis and comparison of the obtained results with full-scale measurements and the results of alternative numerical simulations. The second one is the verification of the fluid finite volume model based on the numerical solution of the equations of the Navier-Stokes equations for incompressible Newtonian fluid using a turbulence model URANS SST. Comparison of the obtained results is occurred with data from alternative numerical simulations [1, 2]. The third stage – the verification of the aeroelastic model, involves the determination of the spatial and temporal characteristics of the fluid flow and the stress-strain state of structures in time using FSI technology (Fluid-Structure Interaction). Within this stage the obtained results compared with data from [1, 2].

As a primary tool for performing a numerical simulation of the fluid-structure interaction problems the specialized multidisciplinary software ANSYS (Mechanical and CFX) was chosen.

3 Stage 1: Verification of the mechanical finite element model

3.1 Description of the computational models

The object of study is one of the forty silos located in 5 rows of 8 each. The distance between them is 30 cm (Fig. 1). Cylindrical shell with a diameter of 5.5 m and a height of 25 m consists of 10 aluminum plates with a height of 2.5 m and a thickness decreasing with a height (Fig. 1, Table 1).

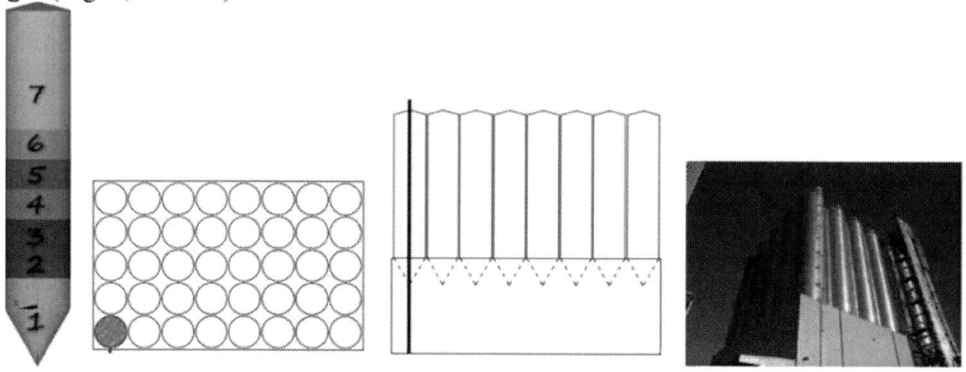

Fig. 1. Geometrical model of the silo (ANSYS Mechanical, different shell thickness marked with color and number), plan, lateral view and picture of the silo group in the port of Antwerp (Belgium).

Table 1. Variation of the shell thickness of the cylindrical silo along the height

Height, [m]	Thickness, [mm]	Numbers matching (Fig.1)
0.0-2.5	10.5	(1)
2.5-5.0	9.0	(2)
5.0-7.5	8.5	(3)
7.5-10.0	7.5	(4)
10.0-12.5	7.0	(5)
12.5-15.0	6.5	(6)
15.0-25.0	6.0	(7)

The parametric three-dimensional shell finite-element (FE) model of the silo was created in the ANSYS Mechanical preprocessor. Type element SHELL181 was used that is dictated by the structure geometry. It is four-node element with six degrees of freedom at each node that implements Mindlin-Reissner shell theory.

The cones at an angle of 15° and 60° are welded to the top and bottom of the cylinder respectively. Perpendicular beam-pillars support the lower part of the cylinder and fastened at four points on the circle. For which of them the appropriate boundary conditions in the FE model are set: all linear (UX=UY=UZ=0) and angular (ROTX=ROTX=ROTZ=0) displacements are limited at the four nodes where the silo connects with support beams.

Table 2. FE computational grids characteristics

Model	Elements number	Nodes number	Elements number along the ccircumference	Vertical elements number	Ratio of the recommended h_e to factual one
D1	1686	1680	60	20	48.64
M1	4142	4142	60	52	11.15
M2	15000	15002	120	80	2.23
M3	60000	60002	240	160	1.00

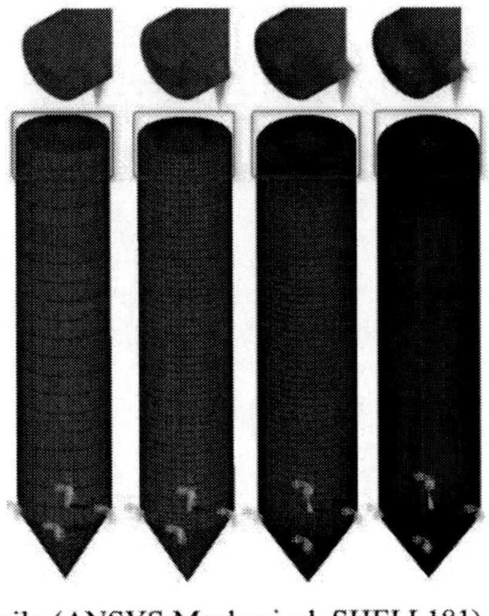

Fig.2. FE models of the single silo (ANSYS Mechanical, SHELL181) with an indication of the boundary conditions (Models D1, M1, M2, M3).

To study the grid convergence for four types of the computational grid (Fig. 2, tab. 2) with a successive increasing of an elements number were tested. In the areas of the shell edge effect (the joints "floor-wall", "wall-cover") FE vertical size h_e was chosen according the condition of an adequate definition of the torque components:

$$h_e \leq 0{,}2\sqrt{R \cdot th} \tag{1}$$

where R=2.75 [m] – the radius of the silo; th=0.006 [m] – wall thickness of the silo.

3.2 Simulation parameters

The following physical and mechanical properties of FE models were set: Young's modulus E = 6,76 10^4 MPa, density ρ=2700 kg/m^3 and Poisson's ratio ν=0,35 corresponding to the material (aluminum) of the structural elements of the silo.

Modal analysis was conducted in ANSYS Mechanical using Block Lanczos method [3]. Sparse direct solver is used for solving the linear algebraic equations.

Mode shapes are characterized by a pair of parameters (m, n), where m is a number of half waves in a longitudinal (along a height) direction of the shell, n is a number of circumferential waves of the shell. Data of the field tests are taken from the electronic source [1]. Vibration measurements were conducted on the corner silo, where the maximum amplitude of vibration under the strong wind excitation was detected. The sensors were located at various points along the height. The radial acceleration was measured at 10 points.

3.3 Results and conclusions for the stage 1

The Figure 3 illustrates the numerical results of the stage 1: natural frequencies and mode shapes of a single silo in comparison with numerical and experimental reference data [1, 2] and the relative errors calculated according to the formula (2):

$$\varepsilon = \frac{\left|x_{ref} - x_{calc}\right|}{x_{ref}} \cdot 100\%, \tag{2}$$

where x_{ref} – reference data [1, 2] and x_{calc} – the results of numerical calculations performed by the authors of the present article.

The grid convergence for four types of the computational grid is demonstrated.

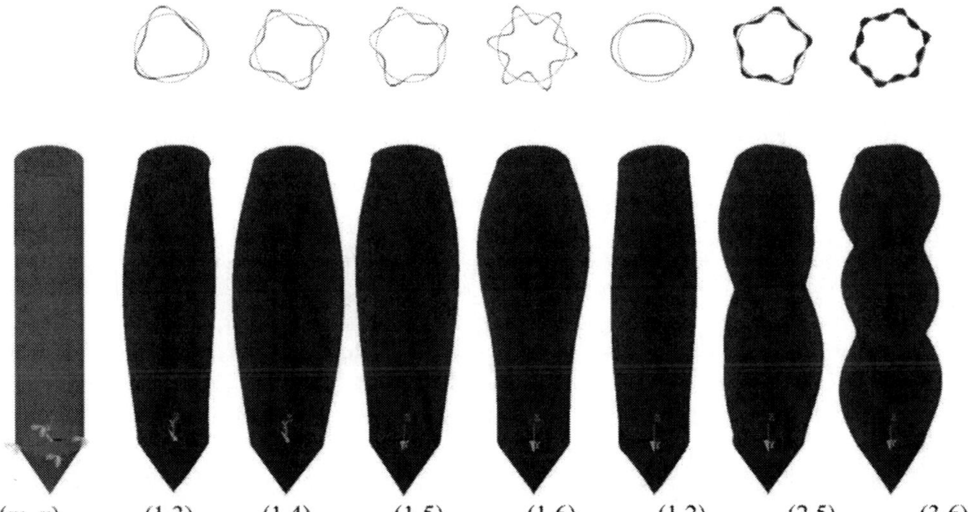

(m, n)	(1,3)	(1,4)	(1,5)	(1,6)	(1,2)	(2,5)	(3,6)
$f_{calc\ M3}$, Hz	3.99	3.93	5.31	7.40	7.81	8.03	12.04
$f_{calc\ [1,2]}$, Hz	3.93	3.93	5.25	7.37	7.75	8.01	12.00
$f_{exp\ [1,2]}$, Hz	3.91	3.91	5.31	7.49	7.87	-	12.00
ε_{calc}, %	1.5	0.0	1.1	0.4	0.8	0.2	0.3
ε_{exp}, %	2.0	0.5	0.0	1.2	0.8	-	0.3

Fig.3. The mode shapes and corresponding natural frequencies of the single silo with an indication of the relative error ε. First row – reference numerical results [1, 2], second row – obtained numerical results (Model M3).

The comparison with experiment shows the expected smallness of error for the best series of experiments. The discrepancy is convincingly explained by errors of the experiment (first of all by the imperfection of the cylindrical shape). At the same time the high proximity of the obtained numerical results to the numerical results presented in reference [1, 2] is observed.

Taking into account the numerical results of aerodynamic studies from [1, 2] and data from domestic codes [5], we can estimate the main vortex shedding frequency (f_{vs}) for cylindrical structures is in the range from 1.56 Hz to 2.13 Hz. As shown by numerical simulations and experiments the closest natural frequencies of considering cylindrical structure to these range of frequencies are those that correspond to m=1, $n=3$ and $n=4$ ($f(1,3)$=4.06 Hz, $f(1,4)$=4.00 Hz). The effect of other modes (the nearest one $f(1,5)$ = 5.50 Hz) can be neglected.

For further FSI investigations the "coarse" FE model *D1* is chosen because it provides an acceptable accuracy and computational efficiency so necessary for solving nonlinear dynamic coupled problems.

3 Summary

Within the stage 1 the comparison of the obtained results with the full-scale measurements and the results of the alternative numerical simulations [1, 2] were carried out. For further FSI investigations the main FE-model are chosen.

As a next step of study it is planning to perform the verification of the fluid finite volume model (stage 2). Outcomes of 2D- and 3D-simulations will present. The grid convergence for several types of the computational grid will demonstrate. The comparison with the alternative numerical simulations [1, 2] will conduct. For further FSI investigations the main FV-model will chosen.

The third stage will be devoted to the numerical simulation of the thin-walled silo behavior under transient wind acting. The values of radial displacements at selected points located halfway up the height of the cylindrical part of the silo will compare with the alternative numerical simulations [1, 2]. The assessment of the mutual influence of wind gusts on the structure and the deformed structure on the spatial and temporal characteristics of the fluid flow will study.

The Reported study was funded by Ministry of Education and Science of the Russian Federation (RF President Grant, agreement №14.Z56.16.8493-MK).

References

1. D. Dooms, G. Degrande, De G. Roeck and E. Reynders, *Wind induced vibration of thin-walled cylindrical structures*, Proceedings for ISMA2004, 781-796 (2004).
2. D. Dooms, *Fluid structure interaction applied to flexible silo constructions*, PhD thesis (2009).
3. Mark T. Jones, Merrell L. Patrick, *The use of Lanczos's method to solve the large generalized symmetric definite eigenvalue problem*, NASA Contractor report 181914 ICASE Report **89-69** (1989).
4. I.N. Afanasyeva, *Adaptivnaya metodika chislennogo modelirovaniya trekhmernikh dinamicheskikh zadach stroitelnoy aerogidrouprugosti,*Dissertaciya na soiskanie uchenoy stepeni kandidata tekhnicheskikh nauk (2015).
5. Eurocode 1: Basis design and action on structures. Part 2–4: Wind action. ENV 1991-2-4, CEN (1994).
6. ANSYS 15.0 Help.

Comparative analysis of analytical and experimental results of the strength of compressed reinforced concrete columns under special combinations of loads

Ashot Tamrazyan[1], *and Levon* Avetisyan[1,*]
[1]Moscow State University of Civil Engineering, Department of Reinforced Concrete and Masonry Structures, 26 Yaroslavskoye Shosse, Moscow, 129337, Russia

Abstract. The article presents a comparison of results of experimental and analytical calculations of the strength of the central and eccentrically compressed concrete elements working in the conditions of dynamic loads under fire exposure. The diagram shows a coefficient of dynamic strength depending on the dynamic loading under high temperature obtained by experimentation way. It is shown, that accounting the dynamic effects in fire condition reduces the load carrying capacity of columns by 40%. Therefore, it is recommended to check the possibility of progressive collapse of buildings and, results of which, appearing the character of dynamic loads on the structure, in calculating load carrying capacity of the structure in fire testing.

Introduction

Using the concept of a dynamic analysis of structures for fire resistance makes it possible to precise the valuation of resistance of buildings to the progressive collapse, survivability.

In [1], it is presented a calculation of the fire resistance of the pylon of the 1st floor of 59-storey residential buildings in standard fire condition. Loss of the load carrying capacity of the carrier element in fire impacts is determined by the formula.

$$N_2 \leq N_1 \text{ (1)} \tag{1}$$

$$N_1 = \varphi(\gamma_{bt} R_{bn} A_b + \gamma_{st} R_{sn} A_s), \tag{2}$$

N_1-Load carrying capacity of pilon, N_2- acting load on the pylon.

As experimental studies devoted to the dynamic work of reinforced concrete elements operating in different temperature conditions were absent [2,3], the following experimental studies were performed by the authors.

Reinforced concrete specimens were prepared and tested both under central compression and at different eccentricities: 3 cm, 3.5 cm, 4.0 cm. Experimentally the values

* Corresponding author: avetisyanlevon@inbox.ru

of the strength of the elements came after the static and dynamic tests at different eccentricities, taking into account the change of the loading speeds.

For obtaining the values of the dynamic strength of compressed reinforced concrete elements in different temperature conditions at different speeds of dynamic impact, the experimental researches were carried out by sample loading rate - 1sec and 0.4 seconds. In normal conditions all previously known results were confirmed. After receiving all the results in normal conditions the samples were divided by series and heated to T = 300°C, 500°C, 700°C, 900°C in the test center "TPB TEST" with the plant (Oven Inv. No. 002) for fire tests of small samples of beam structures, according to the standard temperature regime ISO 834.

In order to identify the impact of high temperatures on the properties of concrete under static and dynamic loadings concrete cubes and prisms were tested in different time of load at different temperatures of fire.

Coefficient of dynamic strength of concrete is defined as the ratio of the dynamic to the static strength of concrete $K_{b,d} = R_d / R_{st}$. Coefficient of dynamic strength is a dimensionless relative value and allows the best way to characterize the comparative influence of various factors on the dynamic strength of concrete.

It was revealed that the dynamic hardening coefficient was reduced to $K_{b,d} = 0.441$.

Coefficient of dynamic strength is less than unity, because the thermal effects on concrete and reinforced concrete structures are considerably stronger, than the change of speed of application load. The graph of experimental curves (Figure 1) $K_{b,d}$ depending on the heating temperature is found.

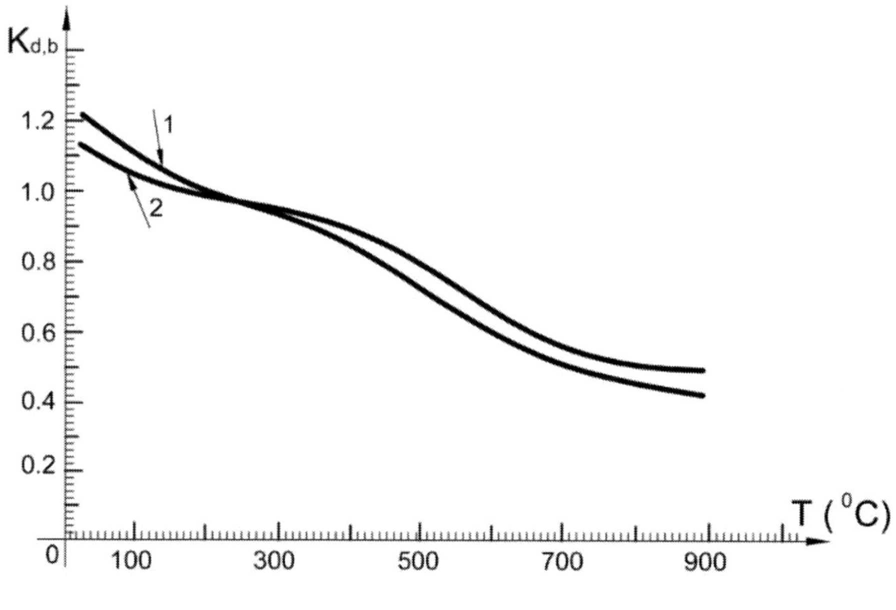

Fig. 1. Dependence $K_{b,d}$ on temperature 1) at the loading time t = 0.4 sec and 2) at the loading time t = 1.0 sec

The following experiments were carried out to study the changes of physical and mechanical properties of reinforced concrete columns after the temperature exposure [4] and to identify the value of the coefficient of dynamic strength after fire exposure in a single dynamic loading with different eccentricities at different speeds (Figure 2,3).

Fig. 2. Central compression (after heating up $900^{\circ}C$)

Fig. 3. Eccentric compression, eccentricity $e_0 = 3.5cm$

As a result of experimental research the following results of strength of reinforced concrete columns (Table 1;2) were found.

Table 1. Strength of reinforced concrete columns under static loading, with different eccentricities after different temperature influences

$T(^{\circ}C)$	Static central loading (kN), $e = 0.00$	Static eccentrically loading (kN), $e = 3.00cm$	Static eccentrically loading (kN), $e = 3.5cm$	Static eccentrically loading (kN), $e_0 = 4.0cm$
$20^{\circ}C$	396.60	278.30	205.40	129.60
$300^{\circ}C$	356.44	244.90	182.50	116.80
$500^{\circ}C$	317.00	221.50	163.60	103.00
$700^{\circ}C$	249.80	183.80	131.70	81.10
$900^{\circ}C$	183.00	128.25	96.80	61.20

Table 2. Results of the strength of reinforced concrete columns at different speeds of dynamic loading with different eccentricities after different temperature influences

$T(^{\circ}C)$	Dynamic loading central loading (kN), t=1.0sec	Dynamic loading central loading (kN), t=0.4sec	Dynamic ecc. loading (kN), e=3.0cm, t=1.0sec	Dynamic ecc. loading (kN), e=3.0cm, t=0.4 sec	Dynamic ecc. loading (kN), e=3.5cm, t=1.0sec	Dynamic ecc. loading (kN), e=3.5cm, t=0.4sec	Dynamic ecc. loading (kN), e=4.0cm, t=1.0sec.	Dynamic ecc. loading (kN), e=4.0cm, t=0.4 sec.
$20^{\circ}C$	441.68	452.40	315.60	322.40	237.50	238.15	150.2	152.66

$300^{\circ}C$	404.74	405.10	274.50	260.80	212.66	202.33	133.0	126.30
$500^{\circ}C$	367.94	370.50	237.90	214.10	178.00	161.20	103.9	88.74
$700^{\circ}C$	287.88	285.70	198.70	169.55	142.00	122.90	82.50	67.95
$900^{\circ}C$	216.00	210.72	132.60	113.47	106.87	84.89	60.48	47.28
	Average deviation 2.45%		Average deviation 14.45%		Average deviation 20.5%		Average deviation 21.8%	

Experimentally, the values of the coefficient of dynamic strength at different eccentricities and at different speeds of dynamic loading came after fire exposure (Table 3).

Table 3. The values of coefficient of dynamic strength $K_{d,T}$ at different speeds of dynamic loading with different eccentricities

T(°C)	e=0.0cm, t=1.0sec	e=0.0cm, t=0.4sec	e=3.0cm, t=1.0sec	e=3.0cm, t=0.4 sec	e=3.5cm, t=1.0sec	e=3.5cm, t=0.4sec	e=4.0cm, t=1.0sec.	e=4.0cm, t=0.4 sec.
20	1.130	1.150	1.134	1.150	1.156	1.159	1.158	1.177
300	1.135	1.136	1.120	1.060	1.165	1.108	1.130	1.081
500	1.160	1.168	1.074	0.960	1.088	0.985	1.008	0.861
700	1.152	1.143	1.081	0.920	1.078	0.933	1.017	0.838
900	1.18	1.150	1.033	0.884	1.104	0.876	0.987	0.789

According to the results, shown in Table 1.2, it becomes clear that under static loading reduction the strength is up to 54% (at temperatures T=900°C). Almost the same regularity is kept during the experiments at different eccentricities and temperatures. For a single dynamic impact, with different speeds, the situation obviously changes. Coefficient of dynamic strength, when the experiments is performed in normal conditions varies from 1.1 to 1.2, and the greater the impact speed, the closer K_d to 1.2. After fire explosions the coefficient of dynamic strength at the central dynamic loadings over time 1.0 sec shows values closer to unit, and at the loading time 0.4 sec it decreases to 0.95. With the increase of the eccentricity the coefficient of dynamic strength is reduced to 0.8. Research of influence of the loading speed after fire explosions at different eccentricities shows that after heating over 500 °C with increasing of eccentricity the increase of speed of dynamic loading can lead to a reduction of $K_{d,T}$ to 21.8 % (Table 2). This is explained by brittle fracture of reinforcing steel, because after fire influences the load carrying capacity of reinforcing steel is reduced to 80%.

This is due to the increased share of participation of work reinforcement steel in a dynamic deformation of eccentrically compressed elements [5,6,7].

In the calculations of eccentrically compression elements of construction under dynamic loading, the calculations must be carried out not only in the elastic stage, but also in the plastic one. In the calculation of particular elements, it needs to get solution of the problem based on the deformed schemes of structure. This circumstance, together with the accounting of the plastic work of elements, allows to reveal the real work of construction and thereby to reveal the significant reserve of their dynamic deformation, since a large proportion of input external energy is spent on the work of bending element and plastic deformations before construction collapses.

For comparison, we define the load carrying capacity of reinforced concrete specimens (columns) under static and dynamic loading in standard fire condition.

Initial data: sizes $600 \times 100 \times 100\,mm$; concrete class B30; the average density in a dry condition (granite crushed stone) $\rho_{oc} = 2485 kg/m^3$; gravimetric humidity $w = 4.6\%$; reinforcing steel 4Ø8 A400; thickness of the protective layer $a_0 = 12mm$.

Here, the destruction begins from achievement of the yield point of reinforcement steel in a stretched rebar, and ends by crushing the concrete of compression zone during the development of plastic deformations in the stretched reinforcement rebar.

One of the most important tasks of calculation of fire resistance of building structures is determination of the temperature fields for rectangular columns at the quadripartite fire exposure according to fire standard ISO 834

Calculation of temperature fields of reinforced concrete structures for fire resistance is based on the solution of the boundary value problems of non-stationary thermal conductivity of capillary-porous bodies in standard temperature conditions.

As a result of the analytical solution of heat engineering tasks the temperature of reinforcement rebar for specific intervals time of fire has come, and the cross section area of the core bounded computational(critical) temperature (Table 4) has been determined.

Table 4. The results of thermo-technical calculation

	15min	30min	40min	50min
Reinforcement steel (°C)	342.18	736.92	790.36	825.9
The core area (mm)	98.2	96.0	88.1	79.0

The graph is found on the basis of the data of analytical calculations and experimental research for the load carrying capacity of reinforced concrete columns with central and eccentric compression, at the different thermo-power loadings (Figure 4 a, b).

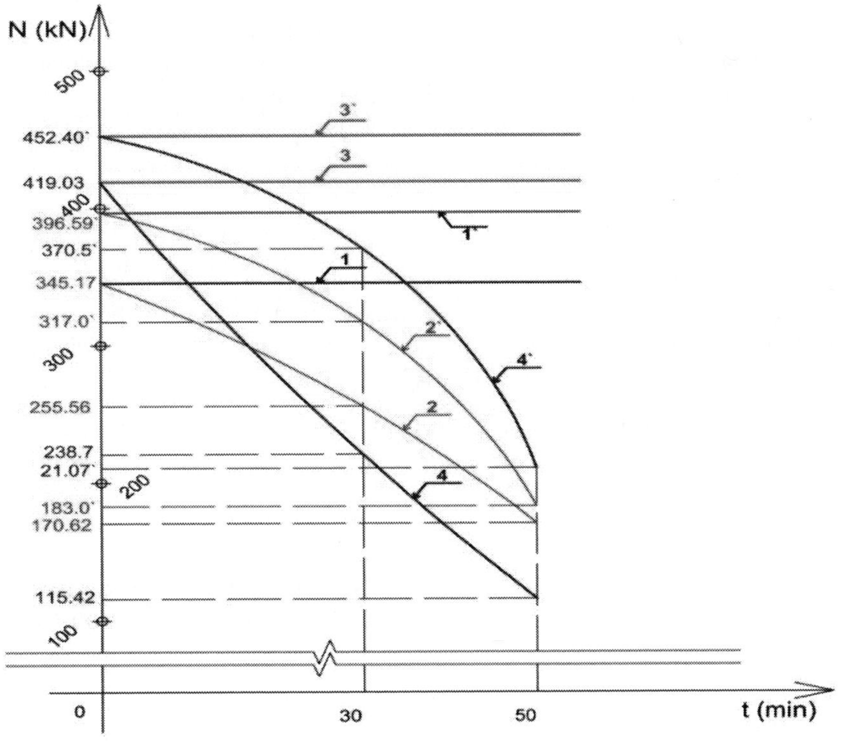

a) Central compression

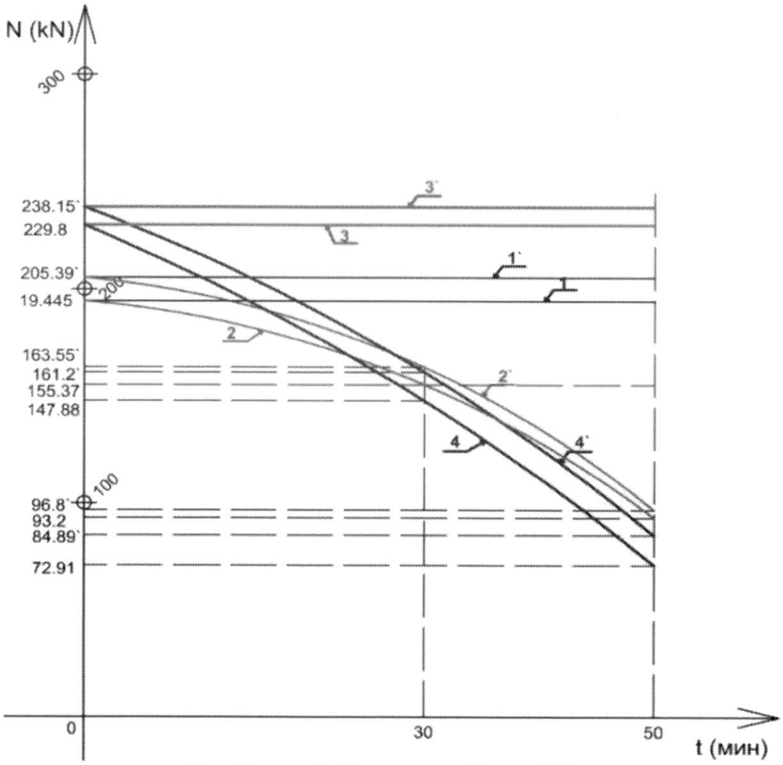

b) Eccentrically compression e=3.5cm.

Fig. 4. Comparison of the analytical and experimental results of the change of the load carrying capacity of reinforced concrete columns under conditions of thermo-dynamic loadings

Calculation results	Experimental results
1. Static loading	1`. Static loading
2. Static loading in fire condition	2`. Static loading in fire condition
3. Dynamic loading	3`. Dynamic loading
4. Dynamic loading in fire condition	4`. Dynamic loading in fire condition

Conclusions

1. Static strength of reinforced concrete columns after fire exposure is reduced to 55%, coefficient of dynamic strength reaches 1.2 depending on the speed of dynamic loading in normal condition. After heating above 500 ° C, with increasing of eccentricity the increase of the speed of dynamic loading leads to the reduction of $K_{d,T}$ to 21.8%.

2. The load carrying capacity of eccentrically compressed reinforced concrete columns (e=3.5cm) under static loading in fire condition (900 ° C) is reduced by 48%.

3. The load carrying capacity of eccentrically compressed reinforced concrete column (e=3.5cm) under dynamic loading (t = 0.4s) in fire impact condition (900 ° C) is reduced by 70%.

4. Coefficient of dynamic strength depending on the temperature decreases by linear dependence at: . T=300°C- $K_{b,d,T} = 0.96$, T=500°C- $K_{b,d,T} = 0.844$, T=700°C- $K_{b,d,T} = 0.59$, T=900°C- $K_{b,d,T} = 0.441$.

References

1. A. G. Tamrazyan, *ICSMIM 2013, 2nd International conference on sensors, measurement and intelligent materials,* 1563-1566 (2013).
2. T.T. Lie, R.J. Irwin, ACI Structural Journal **90 (1)**, 52-60 (1993).
3. J. Xie, A.E. Elwi, J.G. MacGregor, Structural Journal **93 (4)**, 449-461 (1996).
4. A.G.Tamrazyan, L.A. Avetisyan, Applied Mechanics and Materials **638-640,** 62-65 (2014).
5. T.T. Lie, T.D. Lin, D. E. Allen, M.S. Abrams, National research Council Canada Division of Building research (1984).
6. T. Majevski, J. Bobinski, Engineering Structures **30(2),** 300-317 (2008).
7. M. Mohamed Bikhiet, F. El-Shafey Nasser , M. El-Hashimy Hany, Alexandria Engineering Journal **53**, 643–653 (2014).

Influence of Strip Bite Time in Work Rolls Gap on Dynamic Loads in Strip Rolling Stands

Viktor Artiukh[1,*], *Vladlen* Mazur[2] and Elena Pokrovskaya[3]

[1]Peter the Great St. Petersburg Polytechnic University, Polytechnicheskaya, 29, Saint-Petersburg, 195251, Russia
[2]LLC "Saint-Petersburg Electrotechnical Company", Pushkin, Parkovaya, 56, Saint-Petersburg, 196603, Russia
[3]Moscow State University of Civil Engineering, 129337, 26, Yaroslavskoye Shosse, Moscow, Russia

Abstract. Reasons of increase of dynamic coefficients of main drive lines are discussed. Methods of their calculation and technical solutions for their reduction are offered by means of increase of elastic torsion angle of main drive lines during unsteady rolling when metal bites happen. According to results of calculations of elastic shaft-energy accumulator it was recommended to install it in motor section of #4A rolling stand of hot rolling mill 1700 of PJSC 'ILYICH iron and steel works' (Mariupol city, Ukraine).

Introduction

Productivity of rolling stands depend on speeds of their acting parts and minimal technological time. Due to it there is a 'speed problem' in rolling stands equipment design which is in constant solve with durability and stability of equipment [1]. To the point, reduction of dynamics in rolling stands is important for production of new flat products.

For characteristics of dynamic loads in rolling stands main drive lines dynamic coefficient is used [2, 3]:

$$k_d = \frac{T_d}{T_{st}}, \tag{1}$$

where T_d is rolling torque during unsteady rolling, N·m;
T_{st} is rolling torque during steady rolling, N·m.
Value of k_d is not constant for single main drive line. It is proved that at bigger reduction of metal values of k_d and T_d are less compared to ones at smaller reductions. In addition, when width of strip is increased by 1.5 times k_d increases by 10...20% . Also, it is important to consider:

- difference between speed of metal strip entering work rolls (WRs) gap and circumferential speed of WRs;
- values of radial gaps in elements of main drive lines;

[1] Corresponding author: artiukh@mail.ru

- forms of front parts of metal sheet;
- time of metal strip bite t_b.

This paper is devoted to consideration of t_b influence on increase of T_d.

For characteristics of work of metallurgical and rolling equipment a term 'Quality' Q is equal to technological load divided by maximum load. Q = 0.50…0.55 during metal strip bite. It means that k_d = const can be used for preliminary evaluation of dynamic loads acting in main drive lines during particular periods of time.

For calculation of maximum values of dynamic coefficients $k_d^{'}$ at interval $0.01\text{sec} \leq t_b \leq 0.03\text{sec}$ when radial gaps are closed it is reasonable to use method which was modified by authors of this paper by means of modern mathematical software and data given in particular scientific work reports :

$$k_d^{'} = \lambda + \frac{A}{t_b}, \tag{2}$$

where λ is a coefficient that takes into account dynamics due to t_b during hot strip rolling and defined from graph on Fig. 1;

A is a coefficient that takes into account parametric characteristics of main drive line and its flexibility , sec.

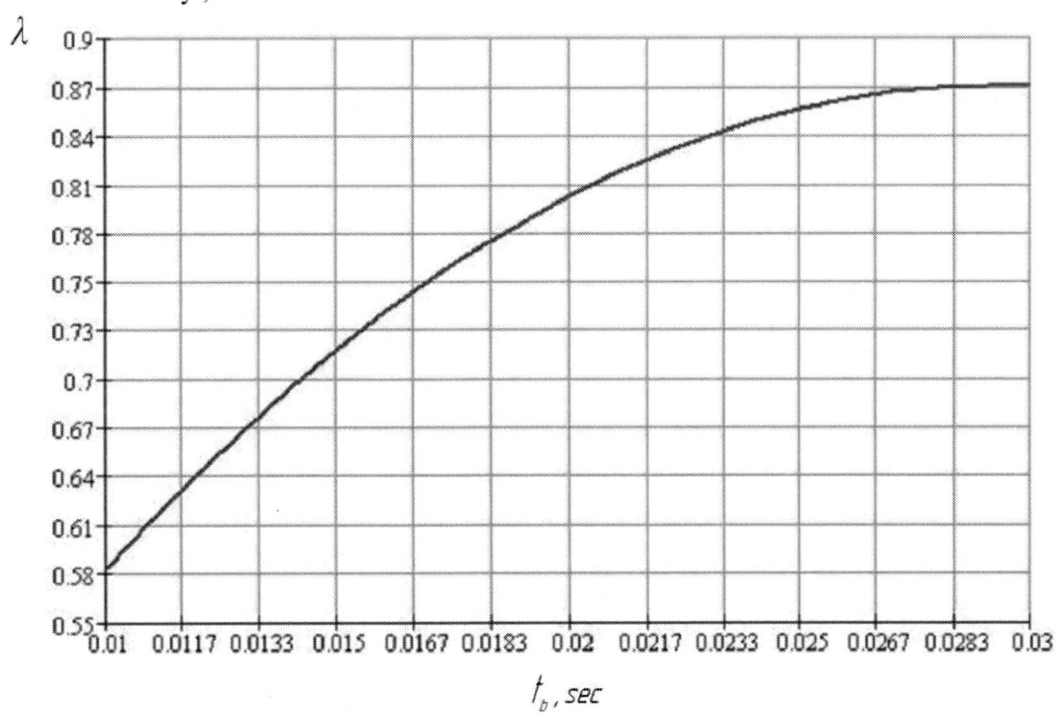

Fig. 1. Graph of coefficient definition which takes into account dynamics due to t_b

For approximate calculation of maximum value of dynamics coefficient $k_d^{''}$ during thick sheet rolling at intervals $0.006\text{sec} \leq t_b \leq 0.5\text{sec}$ and $1.0 \leq k_d \leq 3.0$ it is reasonable to use method described as linear function, modified by authors of this paper:

$$k_d^{''} = 1{,}8 + m_1 \cdot e - m_2 \cdot t_b, \tag{3}$$

where $m_1 = 0.3...0.35mm^{-1}$ is a coefficient that takes into account technological features of use of thick sheet rolling stands, $m_1 = 0.3$ is for roughing rolling stands, $m_1 = 0.35$ is for finishing rolling stands;

e is gap in main drive line within interval $0.0mm \le e \le 6.0mm$ which opens and closes during metal strip bite by WRs, mm;

$m_2 = 10...20sec^{-1}$ is a coefficient that takes into account speed of metal strip bite by WRs; $m_2 = 10$ is for high-speed main drive lines with speeds of metal strip bite equal to 2m/s and more, $m_2 = 20$ is for low-speeds main drive lines (low speeds of metal strip bite by WRs).

Based on proposed methods of dynamics coefficients calculations at hot sheet/strip rolling it is possible to theoretically define dynamics coefficients relations to times of bites by WRs for different types of rolling stands. Initial data for calculation and its results are given in Tab. 1, where $k_{d\ theor.}$ is theoretically calculated dynamics coefficient at particular time of bite as per methods (2) and (3), $k_{d\ exper.}$ is experimentally calculated dynamics coefficient at particular time of bite, Δ is calculation error of $k_{d\ theor.}$.

Analysis of graphs of theoretical relation of dynamics coefficient and time of strip metal bite for main drive lines of different sheet rolling stands (refer to Fig. 2 – Fig. 6) and results of dynamics coefficients calculations bring us to conclusions:

- on practice, for preliminary calculation of elements and details of hot strip stands it is possible to use methods (2) and (3) because calculation error of $k_{d\ theor.}$ is not more than 5%;
- according to theoretical calculation results, depending on type of rolling stand increase of time of metal bite by 30% can reduce dynamics coefficient by 5...20%; moreover, reduction of dynamics coefficient results in reduction of amplitude of T_d oscillations at unsteady rolling that decreases thickness deviation of finish strip

Table 1. Initial data and results of theoretical calculations of relation between dynamics coefficient and time of metal strip bite

Parameter / Rolling stand	t_b, sec	$k_{d\ theor.}$	A, sec	$k_{d\ exper}$	m_1, mm^{-1}	e, mm	m_2, sec^{-1}	Δ, %	$k_d(t_b)$
750/350/750–800 3-Hi rolling stand of JSC 'Metallurgical Plant 'Electrostal' (Electrostal city, Russia)	0.0162	1.36	0.01	1.40				2.85	Refer to Fig. 2
	0.023	1.28	0.01	1.26				1.56	
#4A rolling stand of rolling mill 1700 of PJSC 'ILYICH iron and steel works' (Mariupol city, Ukraine)	0.02		0.078	4.71					Refer to Fig. 3
#1 rolling stand of rolling mill 1700 of PJSC 'ILYICH iron and steel works'	0.128	2.26		2.36	0.3	6.0	10.5	4.24	Refer to Fig. 4
Roughing stand of thick sheet rolling mill 3000 of PJSC 'ILYICH iron and steel works'	0.058	2.56			0.3	4.5	10.2		Refer to Fig. 5
Finishing stand of thick sheet rolling mill 3000 of PJSC 'ILYICH iron and steel works'	0.04	2.55			0.35	3.3	10.0		Refer to Fig. 6

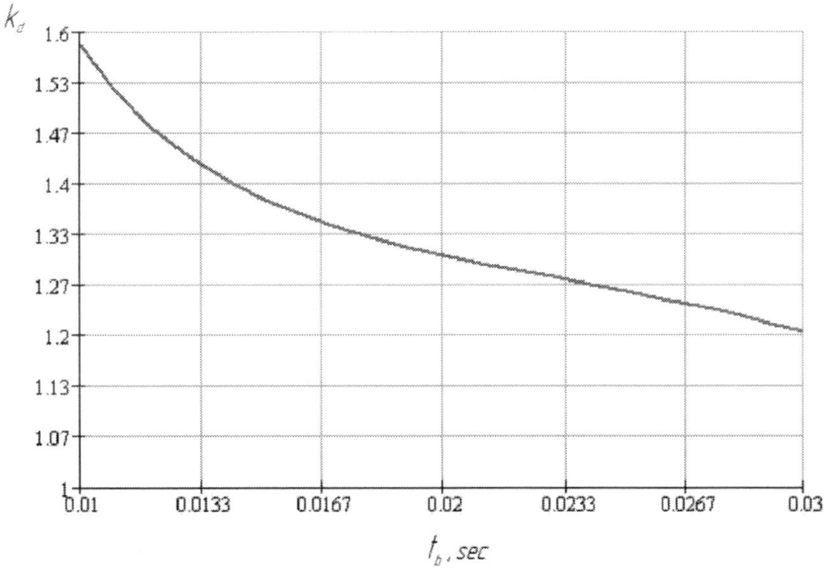

Fig. 2. Graph of relation between dynamics coefficient and time of metal strip bite for main drive line of 3-Hi rolling stand 750/350/750-800

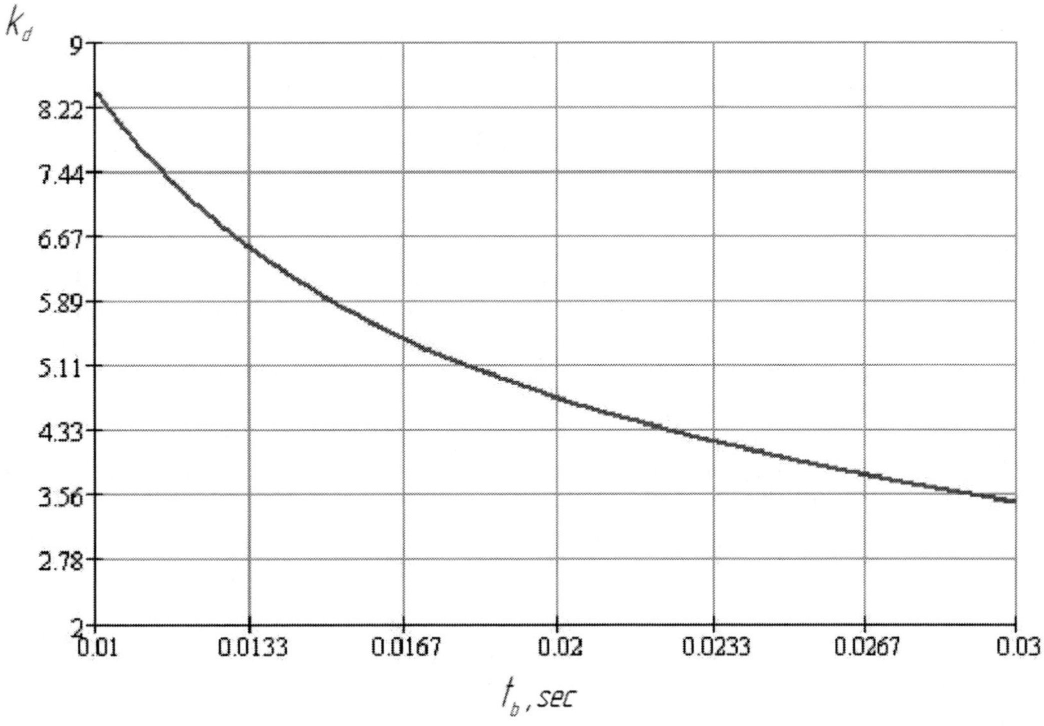

Fig. 3. Graph of relation between dynamics coefficient and time of metal strip bite for main drive line of rolling stand #4A of rolling mill 1700

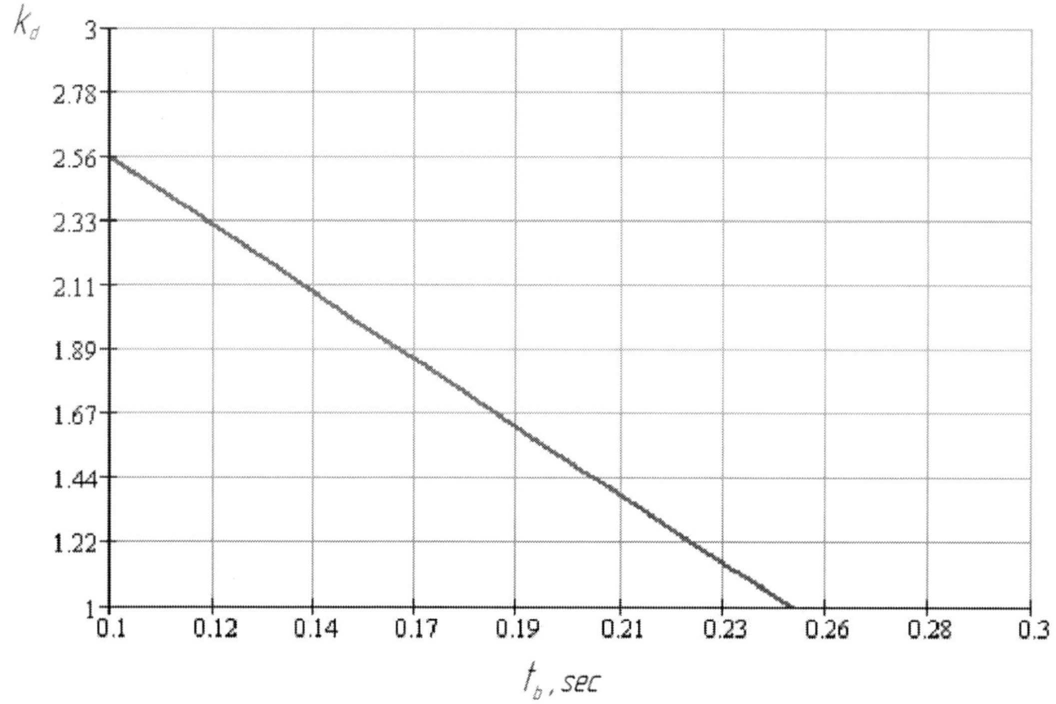

Fig. 4. Graph of relation between dynamics coefficient and time of metal strip bite for main drive line of rolling stand #1 of rolling mill 1700

For reversing rolling stands where WRs are driven by DC motors there is a possibility to regulate speed of metal bite. For rolling stands where WRs are driven by synchronous AC motors there is possibility to regulate speed of metal bite by change of front form of metal sheet/strip. There are numerous proposals to make cuts of front form of metal sheets/strips by angular, chevron and trapezoidal contours. Experimental results of length evaluation of such contours within one rolling campaign on hot rolling mill 1680 of PJSC 'Zaporizhstal' (Zaporizhya city, Ukraine) after roughing group were 0.05…0.2m and after finishing group were 0.6…1.5m. Similar experiments were done on roughing group of hot rolling mill 2000 of PJSC 'NLMK' (Lipetsk city, Russia) where dynamics coefficients were reduced by 8...13% by usage of contours length equal to 0.22m. On practice method of use of angular, chevron and trapezoidal contours is not popular because of uniform further deformation (during reduction) of sheets/strips front parts.

Based on information given above authors of this paper propose new approach to increase time of metal bite by WRs and to reduce oscillations of T_d by means of usage of flexibility of main drive line and increase of its energy capacity due to use of high-torque elastic shaft in main drive line and elastic shaft-energy accumulator which has elastic torsion angle φ_{oi} (given to WRs) comparable to bite angle α_o. WRs during metal bite will be rotating by angle (refer to Fig. 7):

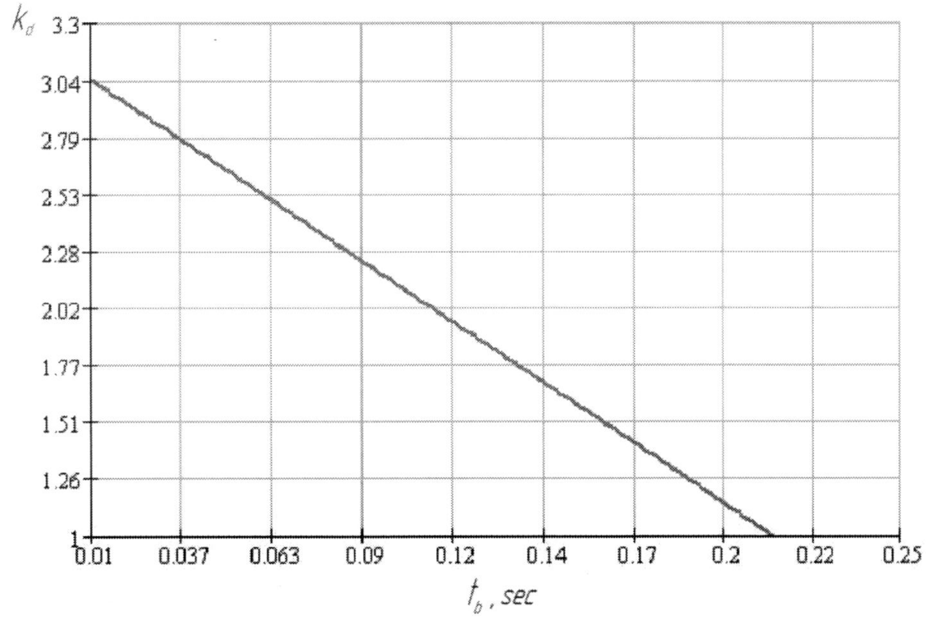

Fig. 5. Graph of relation between dynamics coefficient and time of metal bite for main drive line of roughing rolling stand of thick sheet rolling mill 3000

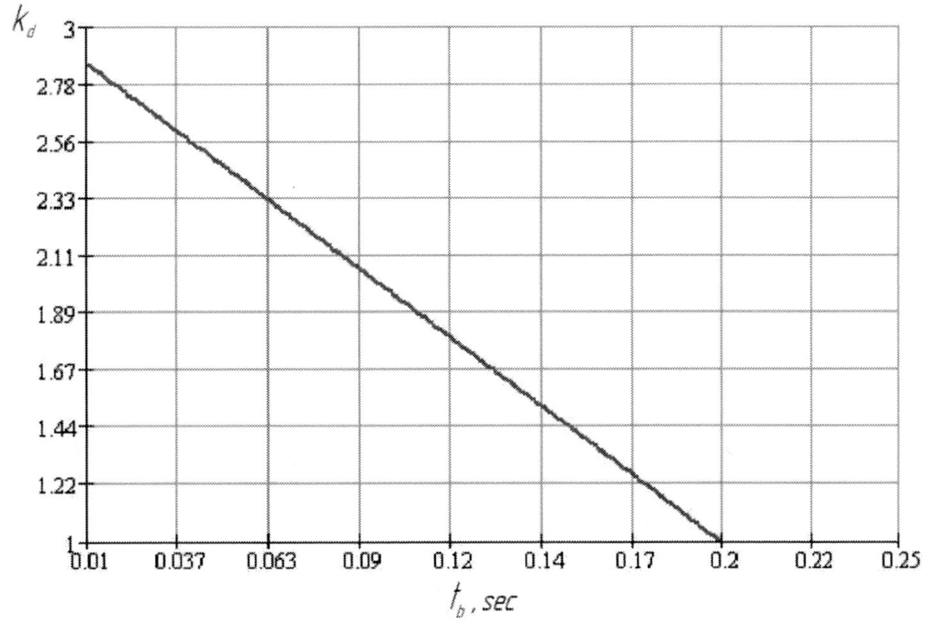

Fig. 6. Graph of relation between dynamics coefficient and time of metal bite for main drive line of finishing rolling stand of thick sheet rolling mill 3000

$$\alpha = \alpha_o + \varphi_{elast}, \text{deg} \quad (4)$$

where φ_{elast} is angle of elastic torsion of main drive line, deg.

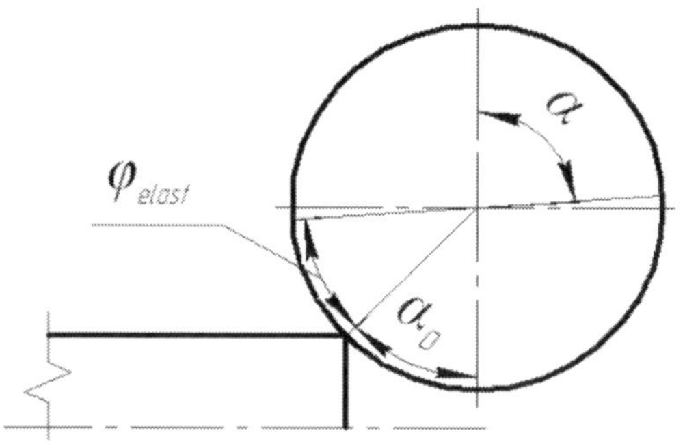

Fig. 7. Scheme of WR rotation angle calculation

Experimental measurements results of T_d and T_{st} show that dynamics coefficients of motor sections mostly are higher than dynamics coefficients of spindle sections of main drive lines with reducer. It means that it is reasonable to install elastic energy-accumulator shaft in motor section.

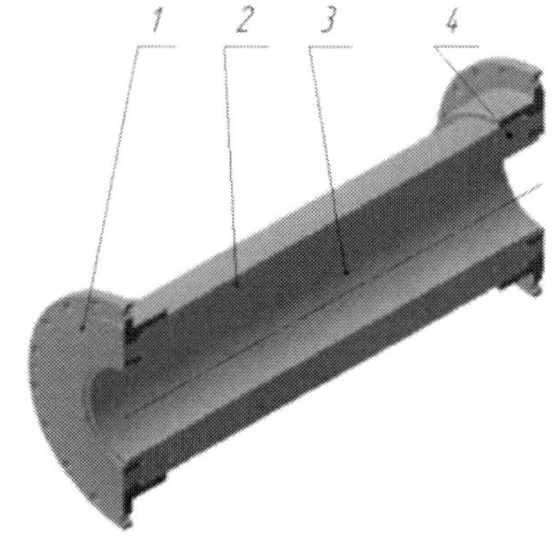

Fig. 8. General view of elastic shaft-energy accumulator: 1 – half-clutch; 2 – polyurethane elastic shaft element; 3 – cylindrical tube; 4 – pin.

Theoretical calculation shows that during steady rolling elastic shaft-energy accumulator (installed instead of initial shaft in motor section) has elastic torsion angle which does not significantly affect productivity of rolling stand.

Experimental shaft, which can provide dynamic rolling torque up to 250 kN·m and torsion angle up to 180deg will be installed soon in the main drive line of hot strip rolling stand #4A of rolling mill 1700 of PJSC 'ILYICH iron and steel works'. It is planned to increase metal bite time by 2...2.5 times due to usage of polyurethane elastic shaft element pos. 2 (refer to fig. 8) with damping coefficient equal to about 0.3 (was established experimentally). Moreover, damping can be significantly increased by usage of additional friction between polyurethane elastic shaft element pos. 2 and cylindrical tube pos. 3.

FEM calculation results of elastic torsion angle of elastic shaft-energy accumulator during its loading by torque equal to 1N·m are shown on Fig. 9, where UY states for movements (in meters) of particular points relative to axis Y. It means that when it comes to drive with T_d = 217 kN·m [20] elastic torsion angle of elastic shaft-energy accumulator will be around 25.35deg and its value can be changed by changes in design of elastic shaft-energy accumulator and/or by usage of elastic shaft element made from another material.

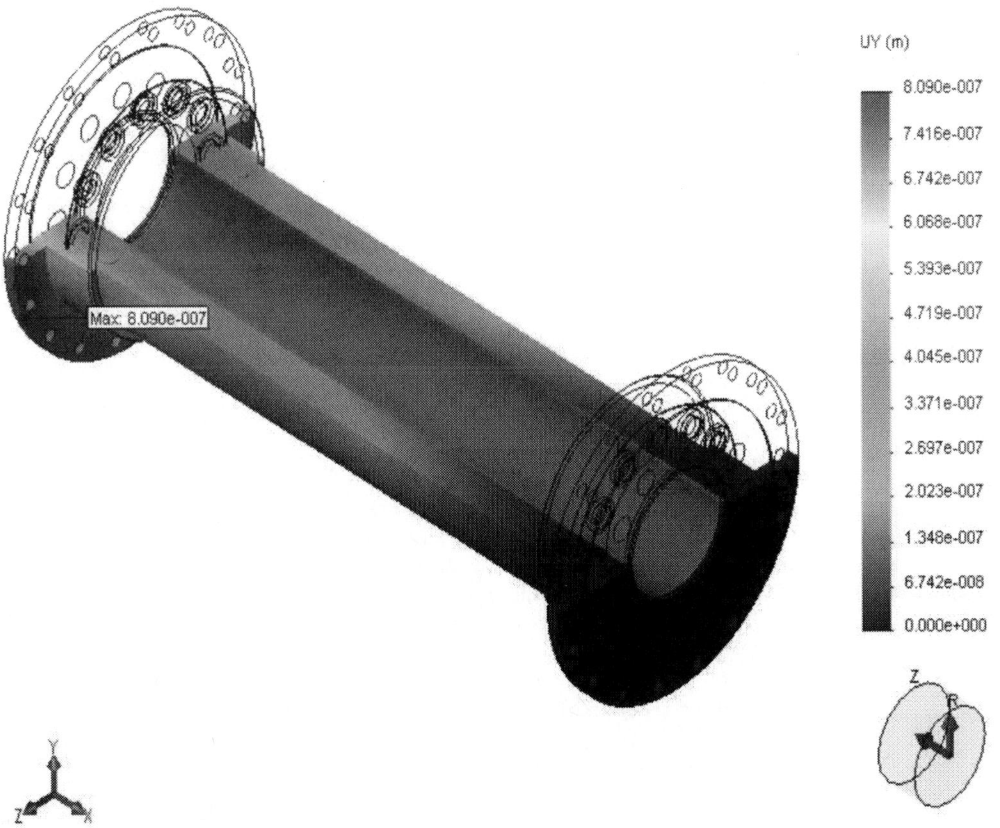

Fig. 9. FEM calculation results of elastic shaft-energy accumulator

The reported study was funded by RFBR according to the research project №16-08-00845a «Verification and development of models of inelastic deformation at the passive loading». The authors declare that there is no conflict of interest regarding the publication of this paper.

Conclusions

1. Almost all hot sheet/strip rolling mills have low 'Quality' of their main drive lines that results in low durability and stability of their equipment.
2. One of main parameters which provokes inertia forces to arise during metal bite by WRs is time of metal bite which is reasonable to increase.
3. Theoretical methods of relation between dynamics coefficient and time of metal bite for main drive line are given.
4. Innovative approach to increase time of metal bite for main drive lines with synchronous AC motors is proposed.

References

1. E. Sorochan, V. Artiukh, B. Melnikov, T. Raimberdiyev. MATEC Web of Conferences, **73**, 04009 (2016) .
2. V. Artiukh, V. Mazur, R. Prakash. Solid State Phenomena, **871**, pp. 3-8, (2016).
3. V. Mazur, V. Artyukh, G. Artyukh, M. Takadzhi. Eng. Designer, **37(1)**, 26–29 (2012)

The problem 4 of placement triangular geometric line field

Vladimir Iljich Travush[1] , *Vasilij Dmitrievich* Antoshkin[2,*], *Vladimir Trofimovich* Erofeev[2]

[1] *RAASN, Bolshaya Dmitrovka 24-1,Moscow,107031, Russia*
[2] *Mordovian State University, Bolshevistskaya 68, Saransk,430005, Russia*

Abstract: One of the a method of formation of triangular networks in the field is investigated. Conditions the problem of locating a triangular network in the area are delivered. The criterion for assessing the effectiveness of the solution of the problem is the minimum number of sizes of the dome elements, the possibility of pre-assembly and pre-stressing. The solution of the problem of one embodiment of a triangular network of accommodation in a compatible spherical triangle and, accordingly, on the sphere. Optimization of triangular geometric network on a sphere on the criterion of minimum sizes of elements can be solved by placing the system in an irregular hexagon inscribed in a circle of minimal size, maximum regular hexagons.

1 Introduction

Placing on the area of regular and irregular hexagons, inscribed in a circle, ie, flat figures or composed in turn of spherical triangles (see Fig. 1) with minimum dimensions of the ribs, is an effective solution in the form of a network formed by circles of minimum radii, ie, circles on a sphere obtained at the touch of three adjacent circles whose centers are at the shortest distance from each other [1-3. 5-8]. Optimization of triangular geometric network on a sphere by the criterion of minimum sizes of elements may be presented and solved by placing the system of irregular hexagon, inscribed in the minimum size of the circle, the maximum regular hexagons, for example, compatible spherical triangles (segments) icosahedron with the scheme of cutting, as shown in Figures 1 and 2. In the application options of the symmetry properties of the main and parallel lines of circles spheres, in such cutting must be realized the possibility of central symmetry of circles [1-4].

The problem 4 of placement triangular geometric line field.

The diagrams Figures 1 and 2 shows the placement of the circumscribed circle of the hexagons in a spherical triangle (compatible spherical segment of the icosahedron) with

* Corresponding author: antovd@mail.ru

internal angles of 36, 90 and 60 °. When optimizing the network using a triangular symmetry of circles and spheres of major lines can be identified as one of the intermediate objectives - determining the position of the first rows of hexagons centers O_1, O_2, O_3, O_4 (Fig. 2). Accommodation centers of circles describing irregular and regular hexagons, are conducted in the cutting of a 3920-polyhedron (Fig. 1). So, at the first stage we determine the position O_1 centers, O_2, O_3, O_4hexagons equal to the radius in the spherical triangle (Fig. 2 shows part of the triangle-segment have interior angles of 60 and 90 degrees). There is a rectangular spherical triangle with internal angles of 90 and 60 degrees and the leg equal to a. Required to determine the values of the radii between the centers O_1 and O_2, O_3 and O_1, O_1 and O_4.

To solve this problem, we consider a half regular hexagon $O_1, O_2, O_3, O_4$and the right-angled triangle $O_2 O_1 M$ with the leg r_0 and the hypotenuse is equal to the total radius $r_1 + r_2$ and right triangle $O_2 O_0 M$, legs r_0 and x, and

$$x = a - 2r_0.$$

Based on the fundamental formulas of spherical trigonometry, we get the system of equations (1 and 2).

Applying the theorem of cosines and sines to the triangle $O_2 O_0$ M we get

$$\operatorname{tg} y = \frac{\operatorname{tg}(a - 2r_0)}{\cos 60°},$$
$$\cos y = \cos(a - 2r_0)\cos r_2,$$
$$\sin y = \frac{\sin r_2}{\sin 60°}. \quad (1)$$

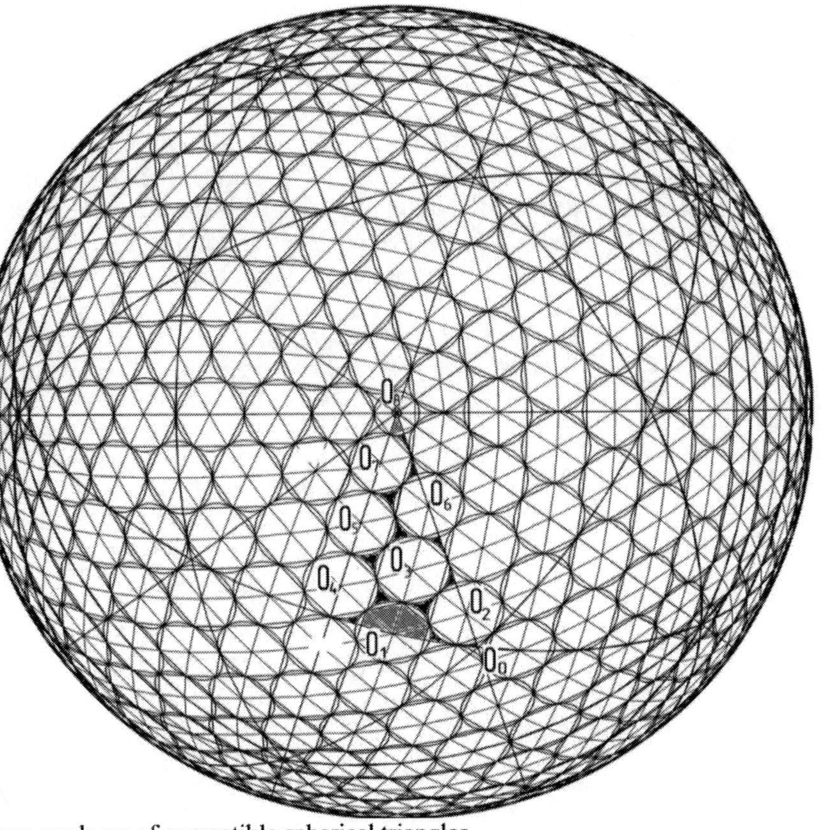

Fig. 1. A sphere made up of compatible spherical triangles

A sphere made up of compatible spherical triangles (segments, representing 1/6 of the brink of the icosahedron), by cutting a triangular network of 3920-polyhedron having sizes of 20 elements or mounting elements 7 sizes (all sizes -15); 0_0 - center of the face of the icosahedron and the panel in the form of an equilateral triangle on the sphere; $0_1, 0_2, 0_3, 0_4$ be the centers of the mounting panels in the form of flat hexagons and pentagons in a segment of the internal angles of 36, 90, 60 ° for solving the optimization problem of the triangular network
Whence

$$\frac{\operatorname{tg}(a-2r_0)}{\cos 60^\circ}=\frac{\sin r_2}{\sin 60^\circ \cos(a-2r_0)\cos r_2}.$$

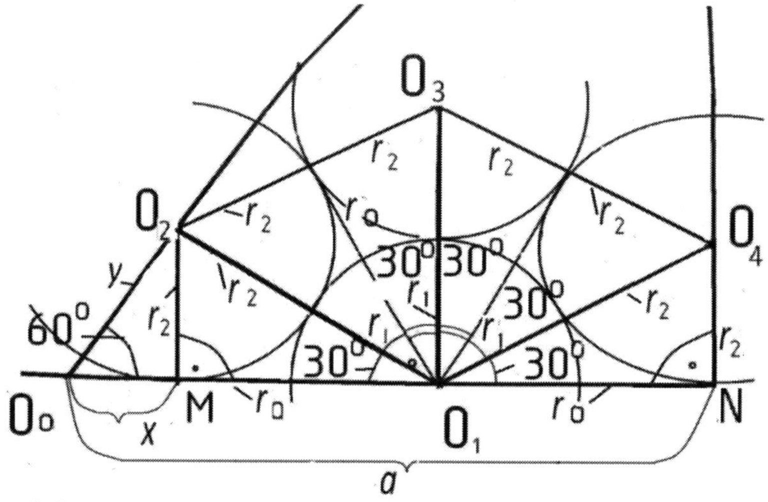

Fig. 2. Scheme of placement

Scheme of placement of the first four flat hexagonal in segment with the inner corners 36, 90, 60 ° for solving the optimization problem of the triangular network using the symmetry properties of circles and spheres of major lines (main lines are axes within a spherical segment): $0_1, 0_2, 0_3, 0_4$ - the centers of the panels in the form of flat hexagons

Applying the theorem of cosines and sines to the triangle $0_2 0_1$ M we get

$$\operatorname{tg}(r_1+r_2)=\frac{\operatorname{tg} r_0}{\cos 30^\circ},$$
$$\cos(r_1+r_2)=\cos r_0 \cos r_2,$$
$$\sin(r_1+r_2)=\frac{\sin r_2}{\sin 30^\circ}; \qquad (2)$$

whence

$$\frac{\operatorname{tg} r_0}{\cos 30^\circ}=\frac{\sin r_2}{\sin 30^\circ \cos r_0 \cos r_2}.$$

Through transformation of the equations (1, 2), we obtain the equation

$$\operatorname{tg} r_2=\cos 30^\circ \sin 60^\circ \sin(a-2r_0)=\cos 60^\circ \sin 30^\circ \sin r_0, \qquad (3)$$

From equations (3) we obtain

$$3\sin(a-2r_0)=\sin r_0, \qquad (4)$$

through transformation of the equation (4), we obtain the equation

$$3\ (\sin a \cos 2r_0\ - \cos a \sin 2r_0\ ) = \sin\ r_0,$$
$$(3 \sin a - \sin r_0) - 6 \sin a \sin^2 r_0)^2 = 36\cos^2 a(1 - \sin^2 r_0)\ \sin^2 r_0\ ,$$
$$9\sin^2 a - 6 \sin a \sin r_0 + \sin^2 r_0 - 12 \sin a \sin^2\ r_0\ (3\sin a - \sin r_0) + 36 \sin^2 a \sin^4 r_0 =$$
$$= 36\cos^2 a \sin^2 r_0 - 3\ 6\cos^2 a \sin^4\ r_0$$

So we get the equation

$$9\sin^2 a - 6 \sin a \sin r_0 +\ \sin^2 r_0 - 36 \sin^2 a \sin^2\ r_0 + 12\sin a \sin^3 r_0 + + 36\sin^2 a \sin^4 r_0$$
$$= 36\cos^2 a \sin^2\ r_0 - 36\cos^2 a \sin^4 r_0,$$
$$9\sin^2 a - 6 \sin a \sin r_0 +\ (1 - 36 \sin^2 a - 3\ 6 \cos^2 a) \sin^2\ r_0 + 12\sin a \sin^3 r_0$$
$$+ 36\sin^2 a \sin^4 r_0 + 3\ 6\cos^2 a \sin^4 r_0 = 0.$$
$$9\sin^2 a - 6 \sin a \sin r_0\ \ - 35\ \sin {r_0}^2 + 12\sin a \sin^3 r_0 + 36 \sin^4\ r_0 = 0,$$
$$36 \sin^4 r_0 + 12\sin a \sin {r_0}^3 - 35\ \sin^2 r_0 - 6 \sin a \sin r_0\ + 9\sin^2 a = 0. \qquad (5)$$

Introducing auxiliary notation

$$k = \sin\ a, \qquad t\ = \sin r_0,$$

and getting rid of irrationality, we turn up to the equation

$$36t^4 + 12kt^3 - 35t^2 - 6kt + 9k^2 = 0\,. \qquad (6)$$

Spending estimates for the value of leg $a\ = 20.9051628°$, we get

$$36t^4 + 4{,}281866124t^3 - 35t^2 - 2{,}14093306t + 1{,}145898594 = 0\,.$$
$$\sin r_0 =\ 0{,}15581464.$$
$$r_0 = 8{,}9640458.$$

Then the relations connecting radii r_1 and r_2 with parameter r_0:

$$\mathrm{tg}\, r_2 = \cos 60 \sin°30° \sin\ r_0,$$
$$\sin(\ r_1 + r_2) = \frac{\sin r_2}{\sin\ 30°};$$

We obtain the required values $r_1\ = 5.1824133°$ and $r_2 = 5.1404432$.
To solve the equation (6) we ve used free software Sclab 5.4.1 - The free platform for Numerical Computation (analogue Matlab) [9].

For the got position of centers of the first rows of the hexagons entered in circumferences, for them corners, sizes of parties of hexagons, are already certain. Accepting the location of cathetus with the size of arc, equal a, on the equator of sphere, we use for the construction of circumferences for other hexagons symmetry of main circumferences (lines) spheres that with a step, equal r 0, we range athwart to this equator [10-12]. When approaching the Pentagon internal symmetry is broken and the circumference is slightly shifted relative to these axes. Calculating the position of the centres of circles in these cases, produced by the method of triangulation. The changes of corners of internal corners of ribs-parties, and and optimization, are possible here for hexagons with centers $O_5 - O_7$ [10-12].

Conclusions

These solutions allow to implement the scope of approximation algorithms triangular geometric network with the maximum number of regular hexagons and prepare options for optimization of cutting sphere.

References

1. V.D. Antoshkin, V.T. Erofeev, V.I. Travush , V.I. Rimshin, V.L. Kurbatov , Modern Applied Science. 9(3), 46-50 (2015)
2. V.D. Antoshkin, Regional architecture and engineering. 3 (24), 112-121 (2015)
3. C.A. Bretschneider. Arch. Math. 2. 225-261 (1842)
4. G. A. Korn *Mathematical handbook: For scientists and engineers*, (1968)
5. W. Lienhard, Elem. math. 66(2) 74-82 (2011)
6. W. J. M'Clelland, T. Preston. *A treatise on spherical trigonometry with application to spherical geometry and numerous examples.* P. II. (London: Macmillian and Co. 1886)
7. B.V. Miryaev B.V. Regional architecture and engineering, 3, 122—125 (2012)
8. V. F. Manukhov, A. S. Turchin , *Glossary of geodetic terms: proc. manual* (Publishing Mord. Univ. Saransk, 2005)
9. Software Scilab 5.4.1 - The free platform for Numerical Computation/ Date Views 06.17.20
10. V. I. Travush, V.D. Antoshkin, I.V. Yerofeyeva, D.V. Antoshkin. TEAM spherical shell Patent RU № 2564545 07/28/2014 year
11. V. I. Travush, V.D. Antoshkin, V.T. Erofeev, S.S. Gudozhnikov. Construction and reconstruction. 2012. № 6 (44). pp. 45-55.

The problem 7 forming triangular geometric line field

Vladimir Iljich Travush[1] , *Vasilij Dmitrievich* Antoshkin[2,*], *Vladimir Trofimovich* Erofeev[2]

[1]RAASN, Bolshaya Dmitrovka 24-1,Moscow,107031, Russia
[2]Mordovian State University, Bolshevistskaya 68, Saransk,430005, Russia

Abstract. Investigated a method of formation of triangular networks in the field. Delivered conditions the problem of locating a triangular network in the area. The criterion for assessing the effectiveness of the solution of the problem is the minimum number of sizes of the dome elements, the possibility of pre-assembly and pre-stressing. The solution of the problem of one embodiment of a triangular network of accommodation in a compatible spherical triangle and, accordingly, on the sphere. Optimization of triangular geometric network on a sphere on the criterion of minimum sizes of elements can be solved by placing the system in an irregular hexagon inscribed in a circle of minimal size, maximum regular hexagons.

1 Introduction

Placing on the area of regular and irregular hexagon inscribed in a circle, ie, flat figures or composed in turn of spherical triangles (see Fig. 1) with minimum dimensions of the ribs, is an effective solution in the form of a network formed by circles of minimum radii, ie, circles on a sphere obtained at the touch of three adjacent circles whose centers are at the shortest distance from each other [1-3. 5-8].

Optimization of triangular geometric network on a sphere on the criterion of minimum sizes of elements may be presented and solved by placing the system of irregular hexagon, inscribed in the minimum size of the circle, the maximum regular hexagons, for example, compatible spherical triangles (segments) icosahedron with the scheme of cutting, as shown in Figures 1 and 2. in the application options of the symmetry properties of the main and parallel lines of circles spheres, in such cutting must be realized the possibility of central symmetry of circles [9, 10].

2 The Main Part

The diagrams Figures 1 and 2 shows the placement of the circumscribed circle of the hexagons in a spherical triangle (compatible spherical segment of the icosahedron) with

* Corresponding author: antovd@mail.ru

internal angles of 36, 90 and 60°. When optimizing the network using a triangular symmetry of circles and spheres of major lines can be identified as one of the intermediate objectives - determining the position of the first rows of hexagons centers O_1, O_2, O_3, O_4 (Figure 2). Accommodation centers of circles describing irregular and regular hexagons, are conducted in the cutting of a 3920-polyhedron (Figure 1).

So, at the first stage determines the position O_1 centers , O_2, O_3, O_4 hexagons equal to the radius in the spherical triangle (Fig. 2 shows part of the triangle-segment have interior angles of 60 and 90 degrees). There is a rectangular spherical triangle with internal angles of 90 and 60 degrees and the leg equal to a. Required to determine the values of the radii between the centers O_1 and O_2, O_1 and O_3, O_1 and O_4.

For a decision the set problem we will consider the fourth of correct hexagon, having centers O_1, O_4, O_6 and half of correct hexagon, having centers O_2, O_3, O_5, O_4, and also rectangular triangle O_2O_3E, with the cathetuses of r_2, r_0 and hypotenuse equal to the total radius of $r_1 + r_2$, and rectangular triangle O_3BE, also with the cathetuses r_2 and x, thus

$$x = a - 3r_0$$

Based on the fundamental formulas of spherical trigonometry, we get the system of equations (1 and 2). Applying the theorem of cosines and sines to the triangle O_3BE, we get

$$\operatorname{tg} y = \frac{\operatorname{tg}(a - 3r_0)}{\cos 60°},$$

$$\cos y = \cos(a - 3r_0) \cos r_2,$$

$$\sin y = \frac{\sin r_2}{\sin 60°}; \qquad (1)$$

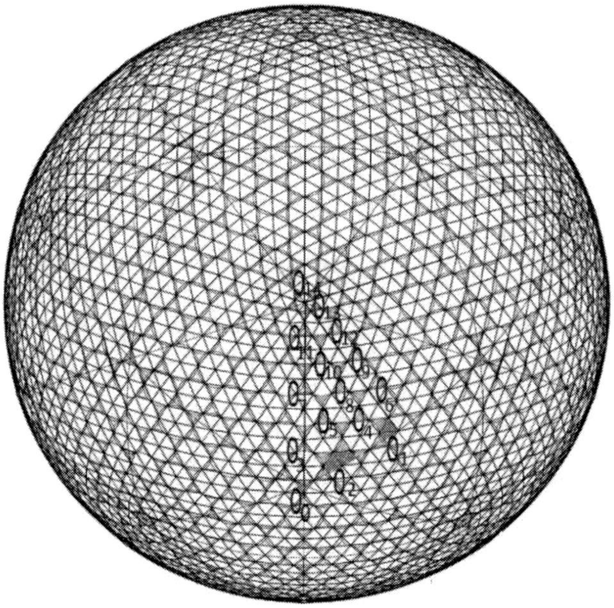

Fig. 1. Is a sphere made up of compatible spherical triangles (segments, representing 1/6 of the brink of the icosahedron), by cutting a triangular network of 7940-polyhedron having sizes of 41 elements or mounting elements 11 sizes (all sizes -27); Oo - center of the face of the icosahedron and the panel in the form of an equilateral triangle on the sphere; $O_1, O_2, O_3, O_4, \ldots$ – be the centers of the

mounting panels in the form of flat hexagons and pentagons in a segment of the internal angles of 36, 90, 60 ° for solving the optimization problem of the triangular network.

Whence

$$\frac{\mathrm{tg}(a-3r_0)}{\cos 60°}=\frac{\sin r_2}{\sin 60° \cos(a-3r_0)\cos r_2}.$$

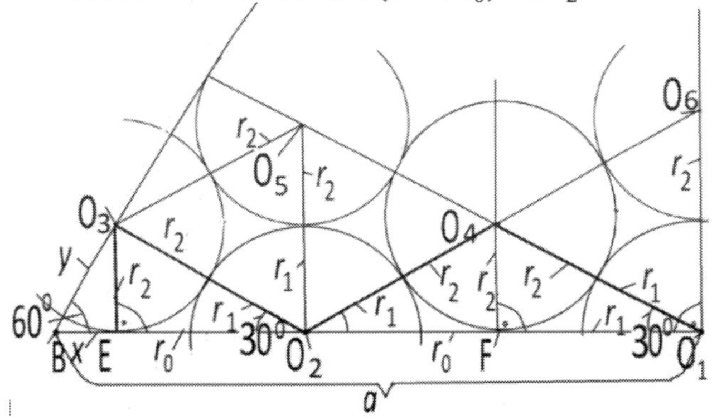

Fig. 2. Scheme of placement of the first four flat hexagonal segment with the inner corners 36, 90, 60 ° for solving the optimization problem of the triangular network using the symmetry properties of circles and spheres of major lines (main lines are axes within a spherical segment): O_1, O_2, O_3, O_4 - be the centers of the panels in the form of flat hexagons

Applying the theorem of cosines and sines to the triangle O_2O_3E, we get

$$\mathrm{tg}(r_1+r_2)=\frac{\mathrm{tg} r_0}{\cos 30°},$$

$$\cos(r_1+r_2)=\cos r_0 \cos r_2,$$

$$\sin(r_1+r_2)=\frac{\sin r_2}{\sin 30°}; \qquad (2)$$

Whence

$$\frac{\mathrm{tg}\, r_0}{\cos 30°}=\frac{\sin r_2}{\sin 30° \cos r_0 \cos r_2}$$

Through transformation of the equations (1, 2), we obtain the equation:

$$\mathrm{tg}\, r_2=\cos 30° \sin 60° \sin(a-3r_0)=\cos 60 \sin° 30° \sin r_0, \qquad (3)$$

From equations (3) we obtain

$$3\sin(a-3r_0)=\sin r_0, \qquad (4)$$

Through transformation of the equation (4), we obtain the equation

$$3(\sin a \cos 3r_0-\cos a \sin 3r_0)=\sin r_0,$$

$$3\sin a \cos^3 r_0-9\sin a \cos r_0 \sin^2 r_0-9\cos a \cos^2 r_0 \sin r_0 + 3\cos a \sin^3 r_0)=\sin r_0,$$

$$3\sin a(\cos^2 r_0-3\sin^2 r_0)\cos r_0-3\cos a(3\cos^2 r_0-\sin^2 r_0)\sin r_0=\sin r_0,$$

$$3 \sin a\,(1-4\ \sin^2 r_0)\cos r_0 = 3 \cos a\,(3-4 \sin^2 r_0)\sin r_0 + \sin r_0,$$

$$3 \sin a\,(1-4\ \sin^2 r_0)\cos r_0 = (3 \cos a\,(3-4 \sin^2 r_0) + 1) \sin r_0.$$

Substitute $1+\mathrm{tg}^2 x = 1/\cos^2 x\,; 1+\mathrm{ctg}^2 x = 1/\sin^2 x$:

$$9 \sin^2 a\,(1-\frac{4}{1+\mathrm{ctg}^2 r_0})^2 \mathrm{ctg}^2 r_0 = (3\cos a\,(3-\frac{4}{1+\mathrm{ctg}^2 r_0}) + 1)^2,$$

$$9 \sin^2 a\,(\mathrm{ctg}^2 r_0 - 3)^2 \mathrm{ctg}^2 r_0 = (3 \cos a\,(3 + 3\ ctg^2 r_0 - 4) + 1 + \mathrm{ctg}^2 r_0)^2,$$

$$9 \sin^2 a\,(ctg^2 r_0 - 3)^2 \mathrm{ctg}^2 r_0 = ((-3\ \cos a\ + 9\ cosa\ \mathrm{ctg}^2 r_0 + 1 + \mathrm{ctg}^2 r_0)^2.$$

How do we arrive at the equation

$$9 \sin^2 a\,(\mathrm{ctg}^4 r_0 - 6\mathrm{ctg}^2 r_0 + 9)\mathrm{ctg}^2 r_0 = ((1-3\cos a) + (9\ \mathrm{cos}a + 1)\ \mathrm{ctg}^2 r_0)^2,$$

$$9 \sin^2 a\,(\mathrm{ctg}^6 r_0 - 6\mathrm{ctg}^4 r_0 + 9\mathrm{ctg}^2 r_0) =$$

$$= (1-3\cos a)^2 + 2(1+9\ \cos a)(1-3\cos a)\mathrm{ctg}^2 r_0 +$$

$$+(1+9 cos\ a)^2\ \mathrm{ctg}^4 r_0.$$

$$9 \sin^2 a\, \mathrm{ctg}^6 r_0 - 54\ \sin^2 a\, \mathrm{ctg}^4 r_0 + 81\ \sin^2 a\, \mathrm{ctg}^2 r_0) =$$

$$= (1-3\cos a)^2 + 2(1+9\ \cos a)(1-3\cos a)\mathrm{ctg}^2 r_0 +$$

$$+(1+9 cos\ a)^2\ \mathrm{ctg}^4 r_0.$$

$$9 \sin^2 a\, \mathrm{ctg}^6 r_0 - (54\, sin^2\, a + (1+9\cos a)^2)\ \mathrm{ctg}^4 r_0 +$$

$$+(81 \sin^2 a - 2(1+9\ cos\ a)(1-3\cos a\,))\mathrm{ctg}^2 r_0 -$$

$$-(1-3\ \cos a)^2 = 0\,.$$

Substitute $k = \mathrm{ctg}^2 r_0$.

$$9 \sin^2 a\, k^3 - (54 \sin^2 a + (1+9\cos a)^2)\ k^2 +$$

$$+(81 \sin^2 a - 2(1+9\ \cos a)(1-3\cos a))k - (1-3\ \cos a)^2 = 0. \qquad (6)$$

Spending estimates for the value of leg $a = 20.9051628°$, we get

$$1.14589859422x^3 - 95.3774060787x^2 + 44.22762787509x - 3.24906745199 = 0;$$

$$k = \mathrm{ctg}^2 r_0 = 82{,}76801,$$

$$r_0 = 6{,}27265567.$$

Then from the relations for the radii r_1 and r_2 parameter r_0:

$$\mathrm{tg}\,(r_1 + r_2) = \frac{\mathrm{tg}\, r_0}{cos\ 30°} = \frac{\sin r_2}{\sin 30° \cos r_0 \cos r_2}.$$

We obtain the required values $r_1 = 3{,}62393°$ and $r_2 = 3{,}60950°$. To solve the equation (6) to use free software Sclab 5.4.1 - The free platform for Numerical Computation (analogue Matlab) [9].

For the got position of centers of the first rows of the hexagons entered in circumferences, for them corners, sizes of parties of hexagons, are already certain. Accepting the location of cathetus with the size of arc, equal a, on the equator of sphere, we use for the construction of circumferences for other hexagons symmetry of main circumferences (lines) spheres that with a step, equal r 0, we range athwart to this equator [10-12]. When approaching the Pentagon internal symmetry is broken and the circumference is slightly shifted relative to these axes. Calculating the position of the centres of circles in these cases, produced by the method of triangulation. The changes of corners of internal corners of ribs-parties, and optimization, are possible here for hexagons with centers $O_5 - O_{13}$ [10-12].

Conclusions

These solutions allow to implement the scope of approximation algorithms triangular geometric network with the maximum number of regular hexagons and prepare options for optimization of cutting sphere.

References

1. V.D. Antoshkin, V.T. Erofeev, V.I. Travush, V.I. Rimshin, V.L. Kurbatov, Modern Applied Science **9 (3)**, 46-50 (2015)
2. V.D. Antoshkin, Regional architecture and engineering **3 (24)**, 112-121 (2015)
3. D.V. Antoshkin, S.S. Gudozhnikov, O.I. Perfilieva, I.V. Erofeeva, Proceedings of the Thirteenth International scientific-technical conference **2**, 4-15 (2014)
4. V.D. Antoshkin, V.I. Nikonov, Fundamental research **11-8**, 1669-1673 (2014)
5. E.F. Ezhov, Y.V. Yurkin, D.V. Antoshkin., E.E. Yezhov, Materials of all-Russian scientific and technical conference **23**, 160-164 (2003)
6. V.F. Manukhov, A.S. Turchin, Publishing Mord. Univ., 44 (2005)
7. B.V. Miryaev, Regional architecture and engineering **3**, 122—125 (2012)
8. Software Scilab 5.4.1, The free platform for Numerical Computation (2008)
9. V.I. Travush, V.D. Antoshkin, I.V. Yerofeyeva, D.V. Antoshkin, patent RU № 129534 (2013)
10. V.I. Travush, V.D. Antoshkin, I.V. Yerofeyeva, D.V. Antoshkin, patent RU № 2564545 (2013)
11. V.I. Travush, V.D. Antoshkin, V.T. Erofeev, S.S. Gudozhnikov, Construction and reconstruction **6 (44)**, 45-55 (2012)
12. V.I. Travush, V.D. Antoshkin, V.T. Erofeev, Construction and reconstruction **6 (50)**, 36-48 (2013)

MATEC Web of Conferences **86**, 01033 (2016) DOI: 10.1051/matecconf/20168601033
IPICSE-2016

Optimization of the thin-walled rod with an open profile

Vladimir Andreev[1,*] *Elena* Barmenkova[1,2]

[1]Moccow State University of Civil Engineering, Yaroslavskoye sh., Moscow 129110, Russia
[2] Peoples' Friendship University of Russia, Miklukho-Maklaya str., Moscow 117198, Russia

Abstract. The purpose of this study is to develop optimization of the cross-section of the inhomogeneous thin-walled rod with an open profile compared to the traditionally used homogeneous. The idea of the developed method of optimization is to determine the law of variation of the elasticity modulus, at which the stress across the cross-section "aligned" and approaching to a constant value. Thus, it is possible to use all material of rod, which leads to optimization of its structure, i.e. an increase efficiency of rod.

1 Introduction

Due to the widespread of thin-walled structures in the construction practice, which characterized by reduced material consumption and weight, we substantiate the optimization method of such structures. A number of works on optimization of thin-walled open profile rod provides the variation of different parameters subject to the conditions of the problem of optimal design (for example, conditions of strength) [1-5]. In this paper we consider optimization method, which based on solving the inverse problem of the theory of elasticity of inhomogeneous bodies, the essence of which is to determine the law of changing the modulus of elasticity on the rod's height for which stress state will be given.

2 Strength analyses

Below are the results of the strength calculations of the roof run. The design scheme is presented in the form of simply-supported at its ends thin-walled rod (Fig. 1). As the profile of the rod is considered the steel channel, because it is the most profitable for the roof runs, the most desirable structurally and in terms of the convenience of attaching it to a belt truss (Fig. 2).

The initial data:

Length of the roof run: l = 600 cm. Roof pitch: $\alpha = 5°$.

Design load on the run: q = 300 kg/m.

Geometric characteristics of the profile: №16P by GOST 8240-89:

* Corresponding author: asv@mgsu.ru

h = 16 cm, b = 6.4cm, d = 0.5 cm, t = 0.84cm, x_0 = 1.97 cm, Wx = 93.8 cm^3, $Wy_{\min}$= 16.4 cm^3, I_x = 750 cm^4, I_y = 72.8 cm^4.

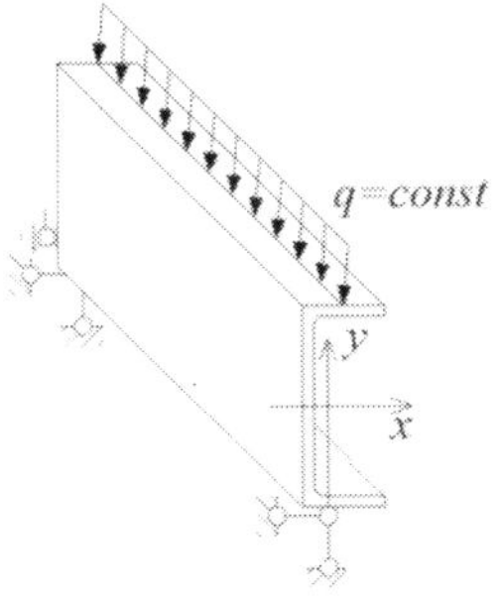

Fig.1. Design scheme.

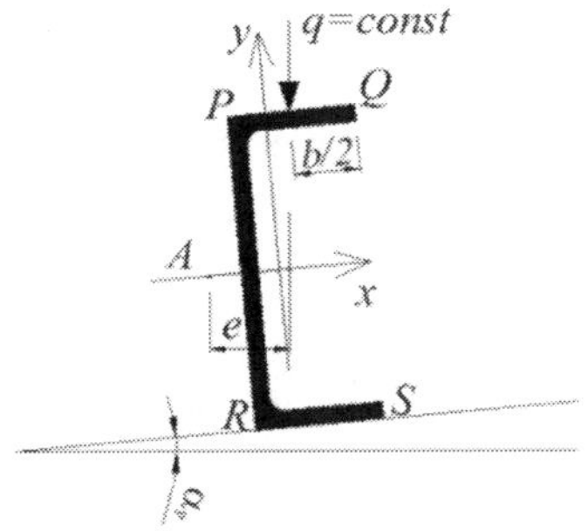

Fig.2. Channel cross section.

Characteristics of the material steel: R_y = 240MPa, E = 2.1·10^{5}MPa, G = 0.8·10^{5}MPa.

The roof run, resting on an inclined plane and being under the influence of vertical loads, should be calculated on the combined effect of bending with torsion. So it is necessary to carry out the calculation of the strength of elements under the action of moments (M_x, M_y) in the two principal planes and the presence of bimoment (B_ω) (paragraph 8.2.1 SP 16.13330.2011).

Bending moments in the two principal planes:

$$M_x = \frac{q_p \cdot l^2}{8} \cdot \sin\alpha_p = 117.66 \text{ kHcm}\,; \quad M_y = \frac{q_p \cdot l^2}{8} \cdot \cos\alpha_p = 1345 \text{ kHcm}\,.$$

To determine the bimoment calculate the sectorial geometric section properties.

Sectorial coordinate:

$$\omega_0 = 2 \cdot \frac{1}{2} \cdot \left(\frac{h}{2} - \frac{t}{2}\right) \cdot \left(b - \frac{d}{2}\right) = 2 \cdot \frac{1}{2} \cdot 7.58 \cdot 6.15 = 46.62 \text{ cm}^2\,.$$

Coordinate of the torsion center:

$$\alpha_x = -\frac{I_{\omega y}}{I_x} = -2.43 \text{cm}\,,$$

Here $I_{\omega y} = 2 \cdot \left(\frac{1}{2} \cdot \omega_0 \cdot \left(\frac{h}{2} - \frac{t}{2}\right) \cdot \left(b - \frac{d}{2}\right)\right) \cdot t = 1825 \text{ cm}^5$ – sector-linear moment of inertia (static moment).

The values of the main sectorial coordinate in the characteristic points of the cross section (Fig. 2):

lower the leftmost point of the profile

$$\omega_R = -2 \cdot \frac{1}{2} \cdot \left(\frac{h}{2} - \frac{t}{2}\right) \cdot |\alpha_x| = -18.45 \text{ cm}^2\,.$$

lower rightmost point of the profile

$$\omega_S = -2 \cdot \frac{1}{2} \cdot \left(\frac{h}{2} - \frac{t}{2}\right) \cdot |\alpha_x| + \left(\frac{h}{2} - \frac{t}{2}\right) \cdot \left(b - \frac{d}{2}\right) = 28.17 \text{ cm}^2\,.$$

Sectorial moment of inertia:

$$I_\omega = 2 \cdot \left[t \cdot \frac{\left(b - \frac{d}{2}\right)}{6} \cdot \left(2 \cdot \omega_S^2 + 2 \cdot \omega_R^2 - 2 \cdot \omega_R \cdot \omega_S\right) + d \cdot \frac{\left(\frac{h}{2} - \frac{t}{2}\right) \cdot \omega_R}{2} \cdot \frac{2}{3} \omega_R \right] = 2975 \text{ cm}^6 .$$

Moment of inertia at free torsion considering factor for channel $\beta = 1.12$:

$$I_d = \beta \frac{1}{3}\left(2 \cdot b \cdot t^3 + h \cdot d^3\right) = 3.58 \text{ cm}^4 ;$$

Eccentricity (Fig. 2):

$$e = \left(|\alpha_x| + \frac{b}{2} - \frac{h}{2} \cdot tg\alpha_p \right) \cdot \cos\alpha_p = 4.92 \text{ cm} .$$

Bimoment taking into account the design scheme:

$$B_\omega = \frac{q_p \cdot e \cdot \left(\text{ch}\left(\frac{kl}{2}\right) - 1 \right)}{k^2 \cdot \text{ch}\left(\frac{kl}{2}\right)} = 320.71 \text{ kHcm} ,$$

where $k = \sqrt{\frac{G \cdot I_d}{E \cdot I_\omega}} = 0,021$ – flexural-torsional characteristics.

Figure 3 shows the following diagrams of normal stresses: $\sigma(M_x)$, $\sigma(M_y)$– diagrams of bending in the two principal planes, σ_ω - sectorial stresses diagram, σ - diagram of total stresses.

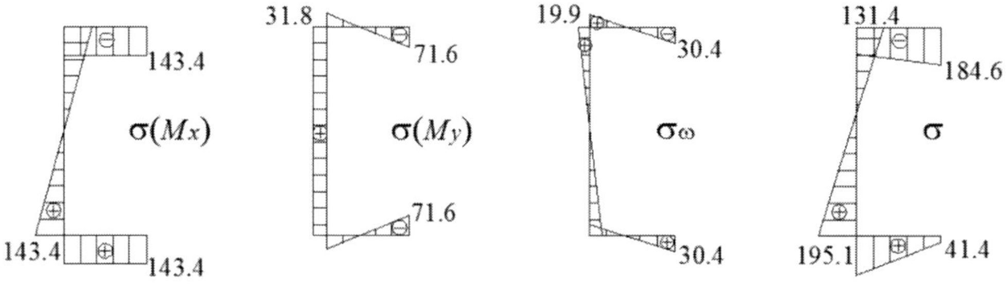

Fig. 3. The diagrams of normal stresses $\sigma(M_x)$, $\sigma(M_x)$, σ_ω, σ, [MPa].

Thus, for the most strenuous point the value of total stress is: σ_{max}=195.1 MPa

The analysis of strength:

$\sigma_{max} = 195.1 \text{ MPa} < R_y \cdot \gamma_c = 240 \cdot 0.9 = 216 \text{ MPa}$.

Strength is provided.

As can be seen from the results, for run channel profile account torsion gives reduction of normal stresses that is actually bimoment downloads profile, and it will be a reserve of strength. Thus, *the strength analysis* showed the strength *of the rod is provided taking into account the unloading of profile of due to sectorial stress.*

3 Stress state optimization method

To optimize thin-walled open profile rods on torsion with bending will use the iterative method. As the initial solution, we take the solution for a homogeneous material, represented in the above paragraph 2. According to the presented stress diagram (Fig. 2) for a homogeneous rod, obviously, for equal-stress profile it is necessary to reduce the modulus of elasticity should be reduce as compared with a homogeneous material on the lower and upper fibers and increase at the neutral layer.

As a first approximation we consider the stress state of the rod, when the modulus of elasticity varies linearly:

$$E^{(1)}(y) = E_1 + \frac{2(E_1 - E_2)}{h} y,$$

where $E_1 = 1.47 \cdot 10^6$MPa (the elastic modulus at $y = 0$), $E_2 = 2.1 \cdot 10^5$MPa (the elastic modulus at $y = -h/2$). Dependence of Young's modulus of the coordinate y is searched in the range $(-h/2, 0)$, this is caused by the symmetry of the cross section and presented in point 2 *strength analysis*, according to which the lower-left point of the profile is the most strenuous.

Normal stresses $\sigma^{(1)}$ can be calculated by the formula:

$$\sigma^{(1)}(y) = E^{(1)}(y)\left(\frac{(-M_x)}{[EI_x]_1} y + \frac{M_y}{\left[EI_y\right]_1} x - \frac{(-B_\omega)}{[EI_\omega]_1} \omega(y)\right).$$

Here

$$[EI_x]_1 = 2\left(\int_{-\left(\frac{h}{2}-t\right)}^{0} d \cdot E^{(1)}(y) \cdot y^2 dy + \int_{-\frac{h}{2}}^{-\left(\frac{h}{2}-t\right)} b \cdot E^{(1)}(y) \cdot y^2 dy\right) = 2.5 \cdot 10^7 \text{ kN} \cdot \text{cm}^2,$$

$$\left[EI_y\right]_1 = 2\left(\int_{-\left(\frac{h}{2}-t\right)}^{0} E^{(1)}(y) \cdot \left[\int_{-x_0}^{-(x_0-d)} x^2 dx\right] dy + \int_{-\frac{h}{2}}^{-\left(\frac{h}{2}-t\right)} E^{(1)}(y) \cdot \left[\int_{-x_0}^{b-x_0} x^2 dx\right] dy\right) = 3.4 \cdot 10^6 \text{ kN} \cdot \text{cm}^2 -$$

the reduced bending stiffness,

$$[EI_\omega]_1 = 2\left[\int_{-\left(\frac{h}{2}-\frac{t}{2}\right)}^{0} d \cdot E^{(1)}(y) \cdot \left(\frac{\omega_P}{\frac{h}{2}-\frac{t}{2}} y\right)^2 dy + \cdot \frac{\left(\mathrm{b}-\frac{\mathrm{d}}{2}\right)}{6} \cdot \left(2 \cdot \omega_S{}^2 + 2 \cdot \omega_R{}^2 - 2 \cdot \omega_R \cdot \omega_S\right) \int_{-\frac{h}{2}}^{-\left(\frac{h}{2}-\frac{t}{2}\right)} E^{(1)}(y) dy\right] =$$

$= 1.1 \cdot 10^8 \text{ kN} \cdot \text{cm}^4 -$ the sectorial stiffness,

On the next step of iteration process using condition $\sigma = const$ we define the function $E^{(2)}(y)$:

$$E^{(2)}(y) = \frac{\sigma}{\frac{(-M_x)}{[EI_x]_1} y + \frac{M_y}{\left[EI_y\right]_1} x - \frac{(-B_\omega)}{[EI_\omega]_1} \omega(y)}.$$

Thus, we have obtained the second distribution of the elastic modulus on height of cross-section. Distribution $E^{(2)}(y)$ represented in the Table 1.

Now we can define the stresses $\sigma = \sigma^{(2)}(y)$ by using formula for $E(y) = E^{(2)}(y)$

$$\sigma^{(2)}(y) = E^{(2)}(y)\left(\frac{(-M_x)}{[EI_x]_2}y + \frac{M_y}{[EI_y]_2}x - \frac{(-B_\omega)}{[EI_\omega]_2}\omega(y)\right)$$

Stiffness are increased in comparison with the values on the first step of iteration process:

$$[EI_x]_2 = 2\left(\int_{-\left(\frac{h}{2}-t\right)}^{0} d\cdot E^{(2)}(y)\cdot y^2 dy + \int_{-\frac{h}{2}}^{-\left(\frac{h}{2}-t\right)} b\cdot E^{(2)}(y)\cdot y^2 dy\right) = 3.2\cdot 10^7 \text{ kN}\cdot\text{cm}^2,$$

$$[EI_y]_2 = 2\left(\int_{-\left(\frac{h}{2}-t\right)}^{0} E^{(2)}(y)\cdot\left[\int_{-x_0}^{-(x_0-d)} x^2 dx\right]dy + \int_{-\frac{h}{2}}^{-\left(\frac{h}{2}-t\right)} E^{(2)}(y)\cdot\left[\int_{-x_0}^{b-x_0} x^2 dx\right]dy\right) = 4.22\cdot 10^6 \text{ kN}\cdot\text{cm}^2$$

$$[EI_\omega]_2 = 2\left(\int_{-\left(\frac{h}{2}-\frac{t}{2}\right)}^{0} dE^{(1)}(y)\cdot\left(\frac{\omega_P}{\frac{h}{2}-\frac{t}{2}}y\right)^2 dy + \frac{\left(b-\frac{d}{2}\right)}{6}\left(2\omega_S^2 + 2\omega_R^2 - 2\omega_R\omega_S\right)\int_{-\frac{h}{2}}^{-\left(\frac{h}{2}-\frac{t}{2}\right)} E^{(1)}(y)dy\right) =$$

$$= 1.3\cdot 10^8 \text{ kN}\cdot\text{cm}^4.$$

The function $E^{(3)}(y)$ is given by:

$$E^{(3)}(y) = \frac{\sigma}{\frac{(-M_x)}{[EI_x]_2}y + \frac{M_y}{[EI_y]_2}x - \frac{(-B_\omega)}{[EI_\omega]_2}\omega(y)}.$$

The expression for $\sigma^{(3)}(y)$ takes the following form:

$$\sigma^{(3)}(y) = E^{(3)}(y)\left(\frac{(-M_x)}{[EI_x]_3}y + \frac{M_y}{[EI_y]_3}x - \frac{(-B_\omega)}{[EI_\omega]_3}\omega(y)\right).$$

The calculation results are shown in Table 1. According results, it can be argued that in the considered problem will be enough execute three approximations.

Table 1. The values of the modulus of elasticity $E^{(i)}$ and the total stresses $\sigma^{(i)}$ for the stages of the iterative process.

***y*, m**	$E^{(0)}$	$\sigma^{(0)}$	$E^{(1)}(y)$	$\sigma^{(1)}$	$E^{(2)}(y)$	$\sigma^{(2)}$	$E^{(3)}(y)$	$\sigma^{(3)}$
	MPa							
0	$2.4\cdot10^5$	31.2	$1.47\cdot10^6$	100.3	$3.16\cdot10^6$	173.8	$3.93\cdot10^6$	169.5
-2		72.8	$1.16\cdot10^6$	222.0	$1.12\cdot10^6$	169.9	$1.43\cdot10^6$	168.9
-4		114.3	$8.4\cdot10^5$	265.5	$6.83\cdot10^5$	169.1	$8.73\cdot10^5$	168.8
-6		155.9	$1.16\cdot10^6$	231.1	$4.91\cdot10^5$	168.7	$6.28\cdot10^5$	168.9
-8		195.1	$2.1\cdot10^5$	118.5	$3.83\cdot10^5$	168.5	$4.91\cdot10^5$	168.8

As seen from the results, taking into account inhomogeneity of the material leads to a substantial redistribution of the stresses in the structure. This redistribution has quantitative and qualitative character.

In the case of a homogeneous material stresses in the extreme fibers differ more than by 6.0 times, whence it follows that the profile is generally underloaded. In the inhomogeneous rod for the case $E(y) = E^{(3)}(y)$ stresses are constant over the cross section, that allows using all material of rod, which generally increases its efficiency.

To determine the effect of an optimized model of an inhomogeneous structure, we introduce the efficiency ratio of work of inhomogeneous beam: $\beta = \sigma_{max}^{(hom)} / \sigma_{max}^{(inh)}$ [7-11]. In formula $\sigma_{max}^{(hom)}$ и $\sigma_{max}^{(inh)} = const$ are respectively the maximum stresses in the homogeneous and inhomogeneous rods. In the above case, this factor is: $\beta = \sigma_{max}^{(0)} / \sigma_{max}^{(3)} = 1.2$ (see Table 1).

The question naturally arises how to create a beam with a continuous change of its stiffness on height cross-section. We can offer a way to create a piecewise homogeneous structure, as was done in [7-11]. Thus, we come to the problem of multilayered structures. Mechanical characteristics of each layer of the multilayer structure are assigned from the solution of the inverse problem of determining the dependence of *E(y)*.

In this case, naturally, the coefficient β for multilayer-rod will be smaller than for the beam with continuous inhomogeneity, but with an increasing number of layers it will be closer to the above value, which will give economic effect.

4 Conclusions

The obtained solution of problem for the *channel* profile at combined effect of *bending* with *torsion* gives justification to use the developed optimization method for thin-walled rod of various profiles.

The used method allows to improve the structure of thin-walled rod with the open profile under *bending* with *torsion*. The developed method of optimization leads to a significant reduction of stresses in inhomogeneous rod compared to the traditionally used homogeneous and the stresses over the cross section are approaching to a constant value. Thus, it is possible to use all material of rod, which leads to optimization of its structure, i.e. an increase efficiency of rod. The reduction in stresses in comparison with the homogeneous rod can increase the loads, reduce the size of the rod, so it will allow to receive some economic effect.

Acknowledgment

This work was financially supported by the Ministry of Russian Education (state task #2014/14).

References

1. *D.V. Bychkov, Structural Mechanics of thin-walled structures,* (1962)
2. G.I. Grebenyuk, A.A. Gavrilov, E.V. Yankov, Proceedings of the universities. Building **7**, 3-11 (2013)
3. T.L. Dmitrieva, Vestnik of Irkutsk State Technical University **5**, 88-94 (2010)

4. A.V. Izhendeev, International Research Journal **4(46)**, (2016)
5. Z. Freidman, M.D. Fuchs, Comput. and Structures **3(25)**, 405-414 (1987)
6. *V.I. Andreev, Some problems and methods of mechanics of inhomogeneous bodies,* (2002)
7. V.I. Andreev, I.A. Potekhin, Bulletin of the Department of Civil Engineering of the Russian Academy of Architecture and Building Sciences **11**, 48-52 (2007)
8. V.I. Andreev, E.V. Barmenkova, A.V. Matveeva, Vestnik of Moscow State University of Civil Engineering **1**, 25-32 (2014)
9. V.I. Andreev, E.V. Barmenkova, A.V. Matveeva, Applied Mechanics and Materials **501-504**, 645-648 (2014)
10. V.I. Andreev, E.V. Barmenkova, Procedia Engineering **91**, 20-25 (2014)
11. V.I. Andreev, E.V. Barmenkova, I. A. Potekhin, Procedia engineering **153**, 37-44 (2016)

International style interior in the historical steam power plant in Žilina

Zuzana Grúňová[1,*], and *Miloslava* Borošová Michalcová[2]

[1]Faculty of Civil Engineering, University of Žilina, Univerzitná 8215/1, 010 26 Žilina, Slovak Republic
[2]Faculty of Education, Matej Bel University in Banská Bystrica, Ružová 13, 974 11 Banská Bystrica, Slovak Republic

Abstract. The article deals with one of the examples of good interior design in Slovak functionalist architecture uncovered during recent restorational research (2016). It points out that even provincial, small but growing town, which Žilina in the northern Slovakia region was at that time, can be a place of high-quality implementation of aesthetically valuable and structurally developed architectural design. The high level of composition, choice of color scheme, quality of materials and attention to detail prove the fact, that the architect František Bednárik mastered to apply international style ideas also to a small extent reconstruction of the original steam power plant in Žilina.

1 International style in the world of interior design

Classically cited feature of international style buildings is primarily their declared absence of ornament. The ornament was declared a crime by Loos - it causes decorated work to became obsolete mentally, outdated and therefore rejected, buildings rebuilt a far before they reach the limit of their physical life. This memento is valid especially the ecological nowadays, when the emphasis is on quality product with the longest possible both physical and mental longevity.

All theoretic ideas find their reflection in a precise logic, composition, quality of the materials and above all high-quality, aesthetics and functionality of each seemingly insignificant detail. Composition and materials (color, gloss, texture, quality) all that bears a strong second plan of the purely aesthetic forms. The concept of the built-in furniture into the volume of interior partitions, walls, bulkheads as compositional elements as well as functional elements, necessary equipment is the main characteristic feature in the large repertoire of the international style's interior design.

2 Purism and functionalism in Žilina

The last years of the 19th century are closely connected with a search for new forms also in the architecture of Slovakia. The architects of the new century were often the students of

[*] Corresponding author: grunova@fstav.uniza.sk

Budapest or Vienna technical and art schools, where eclecticism or historicism still persisted. Žilina, a small town in the northern Slovakia, has a rich collection of these styles buildings and considerable amount of the ornamental Art Nouveau buildings. War, the collapse of the monarchy and the establishment of the new Czechoslovak Republic of changed the social structure. A large part of the German and Hungarian bourgeoisie left the new republic and the new Slovak middle class is susceptible to change. This is also reflected in the architecture - a new one, purist based. Architects and investors of Jewish origin were very close to its economic efficiency and pragmatism [4]. With the establishment of an independent state comes the logical refusal of the eclectic forms, particularly of the Hungarian provenance. Architects are more oriented towards the German environment ideas or the Czech schools. The first functionalist ideas came at that time from Vienna. Adolf Loos and others were the inspiration.

Žilina was in this period an important place – even a capitol of the new republic for the first two months in December 1918 and January 1919. The first major international architectural competition is held for building of the Neologic synagogue. Among the participants the famous names could be found (Hoffmann, Scheer).

Fig. 1. Peter Behrens, Neologic synagogue in Žilina, built 1928-1931.

Proposal of the Berlin architect Peter Behrens won and synagogue was built shortly afterwards in 1928-31. (fig.1.) At the end of the 20´s the scene of the Žilina´s progressive architecture was rich with a large number of authors, like a celebrity architect Fridrich Weinwurm, Ignác Vécsei, Julius Stein and famous architect Michal Maximilián Scheer. [3]

2.1 Architect František Bednárik and his works in Žilina

František Bednárik comes from a family of Moravian builders. After studies he returns to work in his father's company and to work on the design of the residential cooperative houses of the Czechoslovak Legionaries in Žilina. A row of the houses (then on the outskirts of town) bears traces so called rondo-cubism with a powerful tectonic facades. Bednárik was settled in Žilina afterwards. In 1925 he founded his own construction company [4]. Bednárik´s wife sister was a wife of his colleague a young architect Ferdinand Čapka, thus a long-term cooperation was born. Academic architect Ferdinand Čapka was a native of Vienna, Scheer got him to work in Žilina - in 1933-1935 Bednárik´s company, then he became independent. Bednárik-Čapka joint work include Žilina's Railway Station (1939), Representative House - now the Municipal Theatre (Reichner, Čapka, Bednárik, 1940). Bednárik´s sister was the wife of the power plants company director Ing. František Reich. Family ties logically led Bednárik to work on several contracts for the power plants company. In 1941, Bednárik and Čapka in collaboration with architect Paulik designed dynamically expressive functionalist conception for the new administration building of the power plants company, with rounded curves of the facade and entrance accented by distinctive awning [2].

3 Žilina in early 20th century and the industrial boom

Žilina experienced a period of rapid expansion in the late 19th century. From the provincial town it became the new, agile center of the region. The after-fire recovery meant extensive building activities in the city, along with demolition of fortification walls. Further, more important impetus was the construction of the Košice-Bohumín railway from 1870 to 1872. Population was growing as well as Žilina has devolved into a transport hub and commercial center to more than 5000 inhabitants. The late 19th century brought the construction of new industrial plants, using natural resources and water. Cellulose, tar, matches, cloth and chemicals were the main production.

Growing industry needed more and more abundant energy sources. The first choice logical was hydropower. In 1893 was built a small hydroelectric power plant Helios at Rajčianka stream. But it soon ceased to be enough to cover growing demands. In Žilina there was already at that time progressive gas lighting. The face of the town was transformed into a modern urban structure, for which was intended to build an electric lighting. Already in 1903 the city began negotiations with the company Ganz és Társa on building a new power plant. The newly founded Žilina electric plant company (Zsolnai Villamos Muvek részvénytársaság) concluded in 1906 an agreement on free use of public spaces for the distribution of electricity mainly for lighting and industrial purposes. In 1908 the company builds steam power plant, Ganz owned and operated technology up to 1923. [1]

Steam power plant was built at the outskirts of the town, near the slaughterhouse. Historicist industrial architecture, arch lined frieze and tall windows with rounded wedges was applied to modern two-hall plant with the center saddled skylights. Chimney was detached and a small, single-storey residential building was built in the street line. In the halls of power plant was the boiler room with steam boiler, coal shed and room for pullets. It was located next to the engine room with two steam engines, DC dynamos and reserve for another machine and dynamo. [1]

Fig. 2. Steam Power Plant area, Žilina, around 1910.

The Northwest Slovak Joint Power Plant Company (SESzS) was established in 1922 in Žilina. Company rebuilt hydroelectric project on Rajčianka stream and delivered energy to town and surrounding villages (Bánová, Gbeľany, Nededza ...) and growing businesses in and around the town. Probably in January 1926 SESzS bought also the steam power plant - a supply of the direct current. In 20´s the network was gradually reconstructed for alternating current, older solution had less and less coverage. Conversion of the technology and a steam power to a new trend would not be cost-effective, electricity production ceased around 1928. [1] The chimney was demolished in 1930 and the courtyard building of the former steam plant was converted to administrative duties and technical premises.

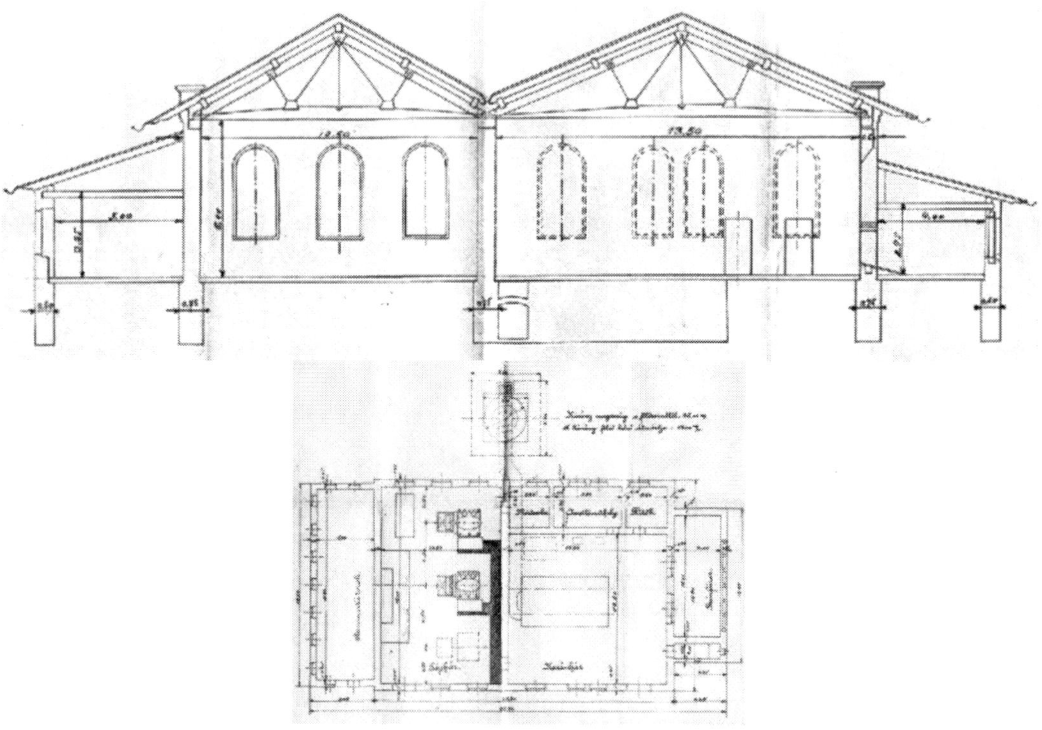

Fig. 3. Steam Power Plant drawing, section and a ground plan. Source: State Archives Žilina.

The first Bednárik´s project from 1930 documents the rebuilding of the left, northwest side of the building. The large space of the engine hall was divided by three floors with a new entrance, corridors and staircase built and given a new administrative function.

In 1935 Bednárik created another project for the renovation of the former steam plant, this time more complex. Again was generous indoor space divided to three floors with the preservation of industrial truss roof structure. The space for installers and warehouse, the technical areas for machine shop and forge was proposed in the other steam plant hall. The second floor is accessible from the staircase already implemented the first reconstruction of 1930. The interior is a coherent compositional and artistic solution in the spirit of functionalist architecture - an elegant office space with entrance hall, offices, divided glazed partitions and large openings with wooden (probably glazed) doors. There is a generous meeting room, followed by less luxuriously designed archives, store and sanitation facilities shared by the whole floor. [1]

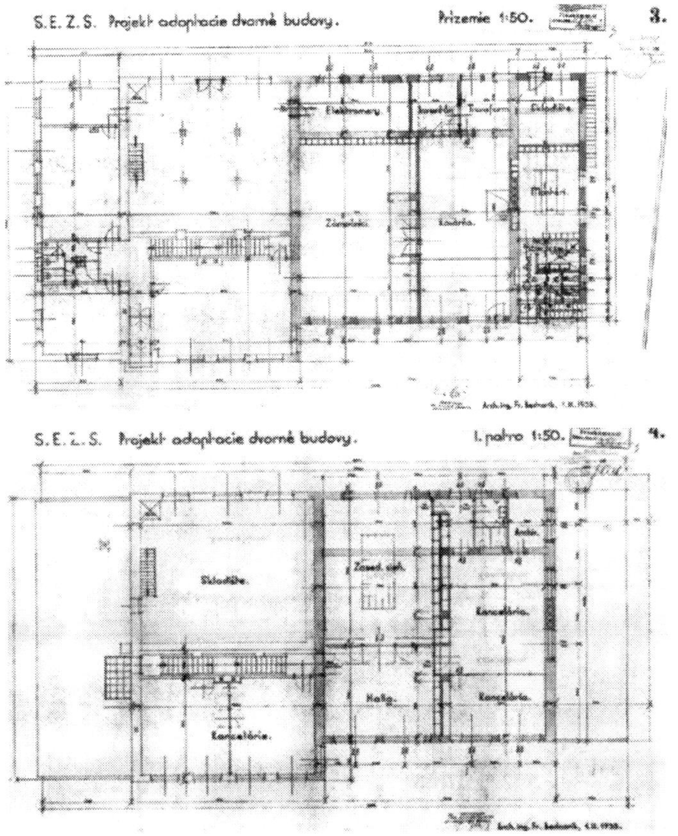

Fig. 4. Adaptation plans. Left ground floor, right newly proposed 1st floor plan. Bednárik, 1935. Source: State Archives Žilina.

The walls were covered with quality terracotta colored wallpaper (lately painted in the same color tone) that was glued also to parts of the ceiling. Area has the very carefully handled seams of the each wallpaper strip. From the each corner of the office seam goes diagonally at precise 45 degree angle. The gap was almost invisible - only a soft texture and color variation of the wallpaper shows direction of gluing.

Fig. 5. Restorational research. Left: soundings – veneer applied to window frames. Right: Entrance and stairs with skylight. Photo: Grúňová, 2016.

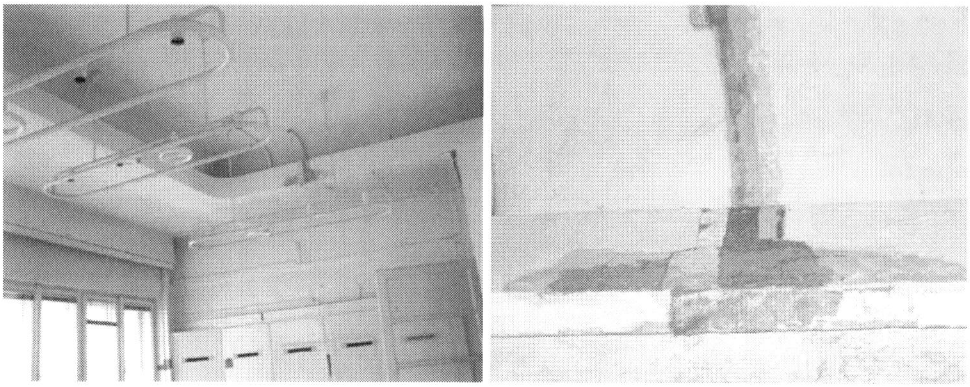

Fig. 6. Restorational research. Office with built-in storage and recessed ceiling. Soundings with wallpaper fragments. Photo: Grúňová, 2016.

Lighting was designed in recessed ceiling as indirect illumination area. The niche for a tubular bulbs was decorated with light olive wallpaper that also as the narrow strip lined niche around. Lightning niche has gently rounded, soft molded edge. Walls between office spaces absorbed load-bearing concrete columns, the volume is used to create a storage space. Veneered doors of the cabinets are dark brown, complemented by fine acting technical detail - thin chrome strips. The floor is wooden parquet.

Fig. 7. Visuals of the meeting room based upon restorational research findings. Grúňová, 2016.

High quality of the details is shown also in the construction lining offices windows. The design is the embedded system with wooden blinds, covered by dark brown veneer in combination with striped exotic wood veneer, the same as in the case of the doors. One of the original glass partitions is now lined on both sides, only fragments of fixing profiles are preserved.

Fig. 8. Visuals of the office based upon restorational research findings and ground plan, partly analogical reconstruction. Grúňová, 2016.

Fig. 9. Visuals of the lobby hall, based upon restorational research findings. Grúňová, 2016.

CaD visuals of the spaces (preserved and partly based on the Bednárik´s ground plan) show a real elegance, clear, pure beauty of high quality design. Bednárik created interior space, which is fully comparable to the level of the mainstream international style architecture in Europe.

References

1. P. Groma, M. Novák, M. Pfliegel, P. Štanský, *Žilina in the lamp light – (not)forgotten history of energetics* (GEORG, Žilina, 2015)
2. *Elektrárne. Power plant company* (Istrofinal KN, s.r.o., 2015)

3. H. Moravčíková, M. Dulla, J. Pohaničová, D. Šoltésová, M. Kvasnicová, et all., *Architektúra na Slovensku - Stručné dejiny, Architecture of Slovakia – short history* (Slovart, Bratislava, 2005)
4. M. Dulla, H. Moravčíková, *Architektúra Slovenska v 20. Storočí. Architecture of Slovakia in the 20th cent.* (Slovart, Bratislava, 2002)

The real static behaviour of historic bridge in town Krásno nad Kysucou in Slovakia

Miroslav Strieška [1,*], *Peter* Koteš [1] , and *Petra* Bujňáková [1]

[1]University of Žilina, Faculty of civil engineering, Univerzitná 8215/1, 010 26 Žilina, Slovakia

Abstract. The paper is focused on real static behaviour of the oldest reinforced concrete arch bridge in middle Europe, which is located in the town Krásno nad Kysucou over the river Bystrica. This bridge was diagnosed and the task was to determine the load-carrying capacity of the bridge. Documentation of the object has not been available, so, it was not clear what the static schema of the bridge is – whether it is fixed arch or two-hinged arch and whether the spandrel walls located on the both sides in the longitudinal direction are cooperating with the arch. The paper is focused on the static calculation and the load-carrying capacity calculation in order to compare various static schemas.

1 Introduction

There are situated many old bridges in Slovakia, some of them have historical significance, but one of the most important bridges in middle Europe is the oldest reinforced concrete arch bridge, which is located in the town Krásno nad Kysucou [1]. It is a two span Monier type bridge structure, which has operated since 1892. In the frame of research activities of the Department of structures and bridges, University of Zilina, the bridge was diagnosed and, consequently, it was needed to verify its real static behaviour and to determine load-carrying capacity [2,3]. It is not known the exact (real) static schema of the bridge and also it is not known how the spandrel walls are cooperating with the arch concerning the very old bridge structure (more than 124 years). Because the static schema of the bridge and the spandrel walls cooperating with the arch have considerable impact on the load-carrying capacity, that paper is focused on the most precise calculation in order to determine the real behaviour of bridge structure.

2 Description of the real state of the construction

The bridge consists of two reinforced concrete arches, which are built on two stone abutments and one stone pier. Its two spans have clear span as much as 16.8 m [4,5]

The thickness of the arch changes in the transverse direction as well as the longitudinal direction. The thickness in the longitudinal direction is changed from 400mm in the arch spring to 200 mm in the middle of the first arch (arch top) and 190 mm in the arch top of the second arch. The thickness of 400mm is the same in the springs of the arches, but the

* Corresponding author: miroslav.strieska@fstav.uniza.sk

thickness in the top of the arches (middle span) in the transverse direction is variable from 250 mm to 150 mm in the middle of first arch and from 250 mm to 130 mm in the middle of second arch. This thickness variability in the transverse direction is gradually disappeared from the middle to the quarter of the arch. There is concrete overlay with plain concrete above the base arch, which has reinforcement in both surfaces. The overlay reaches a thickness up to 600mm in the arch springs and decreases to the 1/3 of span about [4,5].

During the diagnostic of the bridge [6,7], crack was found in the middle of the second arch through its height and width of cross-section, the cracks were also located between the arch and the spandrel walls through the whole longitudinal direction.

3 Static calculation – load-carrying capacity

The paper is focused on the load-carrying capacity calculation of the bridge structure [8] according to the provision USM 01/2012 [9], which was valid at the time of diagnostic and calculation. Subsequently, percentage of load capacity was recalculated according to current code STN EN 1991-2 [10].

The arches were simulated in Scia Engineer software as 3D slab-wall model. Measured geometric parameters of arches and spandrel walls were applied in the model simulation.

In order to analyse the real static behaviour of the bridge and be able to prove the crack formation and development in the middle of the second arch (formation of plastic hinge), it was necessary to create several models to see, which one describes the best the real bridge structure [11].

The first simulations considered fixed arch. To compare the cooperation of spandrel walls with the arch, there were created three models of fixed arch (Fig. 1), where the spandrel walls were connected to the arch as:

- fixed spandrel walls to the arch,
- hinge connection to the arch,
- model without spandrel walls (the spandrel walls were simulated as a load).

In other models, there was considered the hypothesis of two hinged arch with hinge connection of spandrel walls to the arch or model of two hinged arch without spandrel walls, and it was considered in both cases:

- without concrete overlay,
- with concrete overlay.

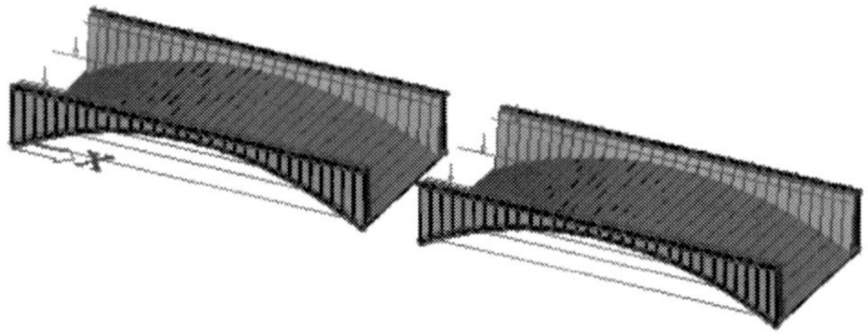

a) The model - spandrel walls cooperate with the arch

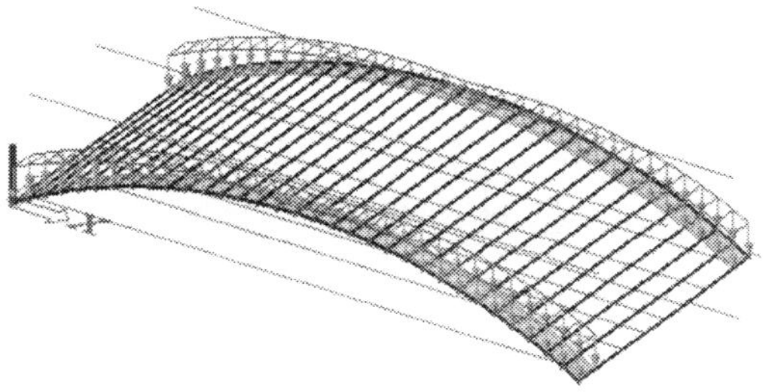

a) The model without spandrel walls (the spandrel walls were modelled as a load)

Fig. 1. The models of arches.

Altogether, it was seven basic models (Fig. 1). In first three models (fixed arch), the most critical cross section were in the arch springs. These models could not show formation of crack in the middle of the arch but it proved that the arch is two hinged arch. Model without spandrel walls, where spandrel walls were simulated as load (fixed arch model), showed that cross section in the spring arch cannot even carry dead weight as well as two hinged arch model without the spandrel walls cannot even carry dead weight in the middle of the arch. In means that spandrel walls have to cooperate with the arches (Tables 1 and 2). As there was found crack between the arch and spandrel walls during the diagnostic of the bridge, it can be assumed that the spandrel walls are hinge connected to the arches.

Considering that it is variable thickness of the arch and also that the spandrel walls cooperate with the arch, it was appropriate to calculate buckling length by the stability calculation in Scia Engineer software. It means that it is a stability problem [12,13,14].

Because it is combination of bending moment and normal (compressive) force, it was needed to make analysis of cross-section by iteration method (step by step procedure) in such a way that the load was simulated in successive steps by 10% of variable load according to USM 01/2012 [9] as well as STN EN 1991-2 [10]. Subsequently, the cross-section in the middle, in the quarter and in the arch spring were verified by iterative calculation by changing the percentage of variable load for each cross section using interactive diagram.

3.1 Variable load according to provision USM 01/2012

According to the provision USM 01/2012 [9], it was necessary to consider the old load models in accordance with the expired Slovak standard – ZZI, ZZII and four axle vehicle. The examples of the critical load ZZI location and four axle vehicle location for normal load-carrying capacity and exclusive load-carrying capacity are show in Fig. 2. The final load-carrying capacity of individual models are shown in Tables 1 and 2 (normal Z_N and exclusive Z_E load-carrying capacity).

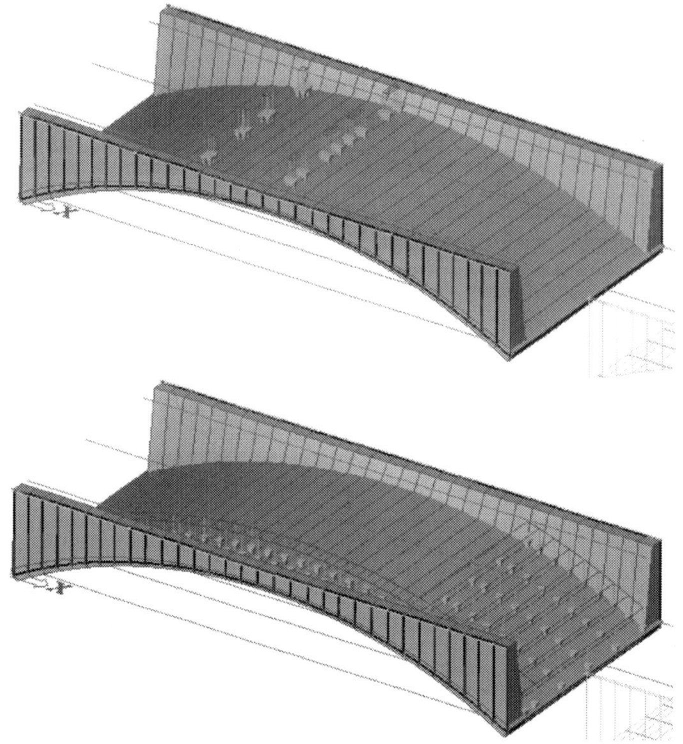

a) Load: ZZI in the middle of the arch

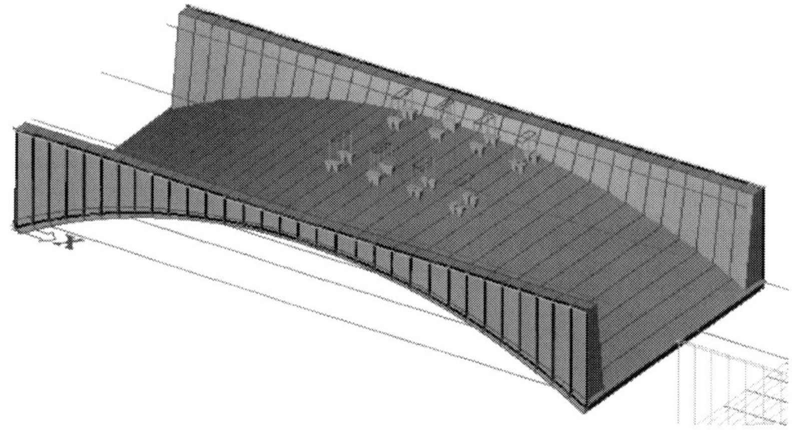

b) Load: four axle vehicle in the middle of the arch

Fig. 2. Examples of load location according to the guideline USM 01/2012 [9].

The fixed arch model with hinge connection of spandrel walls to the arch has normal load-carrying capacity of $Z_N = 19.2$ tons and exclusive load-carrying capacity of $Z_E = 24.0$ tons, which are relatively small values. This load-carrying capacity Z_N would be exceeded during the service life. This means that the crack would be created in the arch spring (formation of plastic hinge – model of two hinged arch). The model of two hinged arch with hinge connection of spandrel walls to the arch without overlay has the most critical cross-section in the quarter of the span, but not in the middle of the arch (in the middle of the span). Concerning that the crack was found in the middle of the second arch and not in the quarter of the arch, it can be concluded that the concrete overlay with plain concrete above the base arch is cooperating with the arch. It indicates that the model, which describes the best the actual construction is the model of two hinged arch with hinge connection of spandrel walls to the arch with concrete overlay, because, in this case, critical cross section is in the middle of the arch, which can leads to the crack (formation of plastic hinge). The normal load-carrying capacity of the first arch (without the crack in the middle of the arch) is Z_N = 22.4 tons and exclusive load-carrying capacity is Z_E = 48.0 tons.

Table 1. Normal load-carrying capacity ZN in tones [t].

Static schema	arch spring	quarter of the arch	middle of the arch
fixed arch model, fixed spandrel walls to the arch	32.0t	32.0t	25.6t
fixed arch model, hinge connection of spandrel walls to the arch	19.2t	32.0t	22.4t
fixed arch model, without spandrel walls	0.0t	19.2t	19.2t
two hinged arch model, without spandrel walls	32.0t	0.0t	0.0t
two hinged arch model, without spandrel walls with concrete overlay	32.0t	9.6t	0.0t
two hinged arch model, hinge connection of spandrel walls to the arch	32.0t	12.8t	22.4t
two hinged arch model, hinge connection of spandrel walls to the arch with concrete overlay	32.0t	28.8t	22.4t

Table 2. Exclusive load-carrying capacity ZE in tones [t].

Static schema	arch spring	quarter of the arch	middle of the arch
fixed arch model, fixed spandrel walls to the arch	48.0t	80.0t	64.0t
fixed arch model, hinge connection of spandrel walls to the arch	24.0t	80.0t	56.0t
fixed arch model, without spandrel walls	0.0t	48.0t	48.0t
two hinged arch model, without spandrel walls	48.0t	0.0t	0.0t
two hinged arch model, without spandrel walls with concrete overlay	48.0t	16.0t	0.0t
two hinged arch model, hinge connection of spandrel walls to the arch	80.0t	24.0t	48.0t
two hinged arch model, hinge connection of spandrel walls to the arch with concrete overlay	80.0t	64.0t	48.0t

Subsequently, two other models of second arch were created in Scia Engineering software, which represented the arch with crack (in the middle of the arch). The crack was simulated like hinge. This means that static schemas were three-hinged arch with hinge connection of spandrel walls to the arch (one model with concrete overlay and another one without concrete overlay). Moreover, the load-carrying capacity of the second arch was calculated. The results of load-carrying capacity of these models are shown in Table 3.

Table 3. Load-carrying capacity.

Static schema	arch spring	quarter of the arch	middle of the arch
Normal load-carrying capacity Z_N			
three hinged arch model, hinge connection of spandrel walls to the arch	28.8t	6.4t	32.0t
three hinged arch model, hinge connection of spandrel walls to the arch with concrete overlay	28.8t	22.4t	32.0t
Exclusive load-carrying capacity Z_E			
three hinged arch model, hinge connection of spandrel	64.0t	8.0t	56.0t

walls to the arch			
three hinged arch model, hinge connection of spandrel walls to the arch with concrete overlay	64.0t	48.0t	56.0t

3.2 Variable load according to the code STN EN 1991-2

In this case, the bridge was loaded according to current Eurocode STN EN 1991-2 [10], specifically load models: LM1, LM2 and LM4. The load-carrying capacity of the cross section was recalculated on the most efficient loads locations. This Eurocode was not intended for calculation of the load-carrying capacity of the bridge in time of calculation. This is reason why the percentage of variable load of each cross section was used and needed to compare (the cross section in the middle of the arch, in the quarter of the arch and in the arch spring). The results of percentage load-carrying capacity of these models are given in Table 4. For example, the most efficient location load (load model LM1) is shown in the Fig. 2.

Table 4. Load-carrying capacity of the arch recalculated according to STN EN 1991-2 [6].

Static schema	arch spring	quarter of the arch	middle of the arch
fixed arch model, fixed spandrel walls to the arch	70%	100%	80%
fixed arch model, hinge connection of spandrel walls to the arch	20%	80%	50%
fixed arch model, without spandrel walls	0%	50%	40%
two hinged arch model, without spandrel walls	60%	0%	0%
two hinged arch model, without spandrel walls with concrete overlay	60%	20%	0%
two hinged arch model, hinge connection of spandrel walls to the arch	100%	30%	50%
two hinged arch model, hinge connection of spandrel walls to the arch with concrete overlay	100%	60%	50%

Also, in this case (according to STN EN 1991-2 [10]), the static behaviour of the bridge was similar to the previous case (USM 01/2012 [9]). The results of simulated models show that model of two hinged arch with hinge connection of spandrel walls to the arch with concrete overlay describes the best the real static behaviour of the first arch. And the model of three hinged arch with hinge connection of spandrel walls to the arch with concrete overlay describes the best the real static behaviour of the second arch (where crack was formed in the middle of the arch).

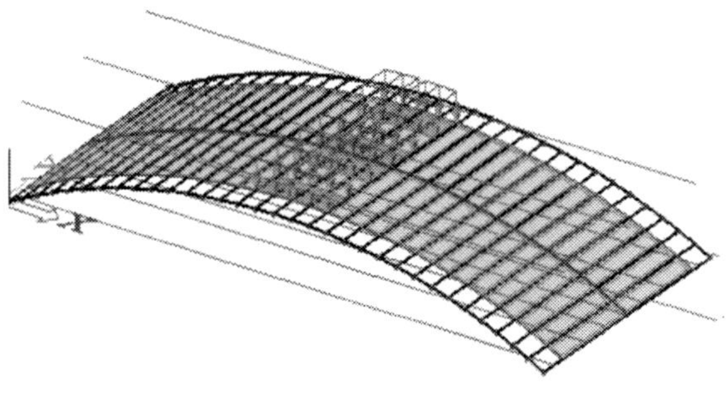

Fig. 3. Example of location of load model LM1 in the middle of the arch according to the code STN EN 1991-2 [10]

4 Conclusion

The simulated models showed the real static behaviour of the oldest reinforced concrete arch bridge in the middle of Europe. The paper presents the results of numerical models.

This paper demonstrated that the real static behaviour of the bridge is similar in both calculation, according to the provision USM 01/2012 as well as according to the current code STN EN 1991-2. Moreover, it was also shown that the model of two hinged arch with hinge connection of spandrel walls to the arch with concrete overlay proves the crack formation and development in the middle of the second arch (formation of plastic hinge).

Recalculation of load-carrying capacity was based on these static schemas and the results are as follows:

- the first arch: the static schema is two hinged arch with hinge connection of spandrel walls to the arch with concrete overlay (normal load-carrying capacity is 22.4 ton and exclusive load-carrying capacity is 48.0ton),
- the second arch: the static schema is three hinged arch with hinge connection of spandrel walls to the arch with concrete overlay (normal load-carrying capacity is 22.4 ton and exclusive load-carrying capacity is 48.0ton).

The research is supported by the Slovak Research and Development Agency under contract No. APVV-14-0772, and by Research Project No. 1/0566/15 of Slovak Grant Agency.

References

1. P. Paulík, *Bridges in Slovakia*, JAGA, Bratislava, **258** (2012)
2. J. Antucheviciene, Z. Kala, M. Marzouk, E.R. Vaidogas, Math. Prob. in Eng. **16(2015)**, 362579 (2015)
3. J. Antucheviciene, Z. Kala, M. Marzouk, E.R. Vaidogas, Math. Prob. in Eng. **3(2015)**, 160569 (2015)
4. P. Paulík, M. Bačuvčík, M. Brodňan, P. Koteš, Beton TKS **5**, 30-34 (2015)
5. P. Paulík, P. Koteš, J. Vičan, M. Bačuvčík, 9th sem. „Sanation of Concrete Structures“, 73-78 (2015)
6. I. Holly, J. Bilcik, In App. Mech. and Mat., **206-209** Svratka, ČR (2015)
7. J. Bilcik, I. Holly, Proc. Eng. **65**, 248-253 (2013)
8. M. Strieška, *Recalculation and reconstruction of reinforced arch bridge in Krásno nad Kysucou*, diploma thesis, Žilina, **268**, (2016)
9. *USM 01/2012, Load-carrying capacity of bridges* (2012)
10. *STN EN 1991-2, Eurocode 2: Design of concrete structures - Part 1-1: General rules and rules for buildings, National Annex* (SÚTN, Bratislava, 2006)
11. V. Benko, Ľ. Fillo, P. Kendický, V. Knapocová, In App. Mech. and Mat., **747-752**, Svratka, ČR (2015)
12. J. Vican, J. Gocal, J. Odrobinak, R. Hlinka, Communications **15/4**, 27-31 (2013)
13. J. Vican, J. Odrobinak, J. Gocal, R. Hlinka, *Bridges in Danube basin Bridges*, **267-278** (Springer Fachmedien, Wiesbaden, 2013)
14. J. Vican, J. Odrobinak, J. Gocal, Communications **18/4**, 3-9 (2016)

Existing bridge evaluation using deficiency point method

Josef Vičan[1,*], *Jaroslav* Repa[1] , and *Peter* Koteš[1]

[1]University of Žilina, Faculty of Civil Engineering, Department of structures and bridges, Univerzitná 8215/1 Žilina, Slovakia

Abstract. In the transforming EU countries, transportation infrastructure has a prominent position in advancing industry and society. Recent developments show, that attention should be moved from the design of new structures towards the repair and reconstruction of existing ones to ensure and increase their satisfactory structural reliability and durability. The problem is very urgent because many construction projects, especially transport infrastructure, in most European countries are more than 50-60 years old and require rehabilitations based on objective evaluations. Therefore, the paper presents methodology of existing bridge evaluation based on reliability concept using Deficiency Point Method. The methodology was prepared from the viewpoint to determine the priority order for existing bridge rehabilitation.

1 Introduction

Bridges are considered an inseparable and strategic part of transport infrastructure, and they should not become the limiting components of communication capacity and traffic reliability. In the past, bridge maintenance, repair and rehabilitation activities were performed on an "as-needed" basis. This changed in the late 1960s, when a series of bridge failures focused public attention on the deterioration of existing bridges and motivated governments to initiate standardized bridge inspection and evaluation procedures. Data collected through these inspection activities formed the basis for computer-based bridge management systems (BMS), [1-3].

Evaluation of existing bridge structures is the most important process in the global BMS and provides basic information about existing bridges for decision-making processes related to the optimal bridge maintenance and rehabilitation strategy. The Department of Structures and Bridges prepared the new methodology based on the concept of reliability evaluation respecting the actual bridge load-carrying capacity and its real condition. The complex methodology of existing highway bridge evaluation was based on the Deficiency Point Method. Using the developed methodology, the priority order for existing bridge rehabilitation could be determined.

[*] Corresponding author: vican@fstav.uniza.sk

2 Methodology of existing bridge evaluation

To determine the priority order for existing bridge rehabilitation, the development of the methodology for bridge complex evaluation shall be assumed. It was a long-term problem of the research activities of the Department of Structures and Bridges in frame of research supported by Slovak Research and Development Agency [4] based on the previous activities in frame of the applied research for Slovak Road Administration (SRA) [5] or Slovak Railways (SR) [6]. The idea of the bridge reliability evaluation was presented in this research that prefers the existing bridge reliability as the main parameter of the bridge evaluation. The safety is the basic and the most significant attribute of the structural reliability. In the case of existing bridge structure, the reliability is quantitative expressed by its load-carrying capacity taking into account the bridge actual condition. Therefore, the unified methodology of the bridge load-carrying capacity calculations for existing highway bridges [5] and railway bridges [6] based on the partial safety factor method were developed in frame of applied research activities for SRA and SR. Nevertheless, the bridge evaluation using its load-carrying capacity only does not enable allowing for all existing bridge deficiencies which could influence another reliability attributes, especially parameters related to the bridge serviceability. Among those deficiencies, e.g. the insufficient spatial arrangement on or under the bridge should be included. The influence of those parameters shall be solved separately, so that the evaluation objectivity is reduced due to the need to determine weighting factors for both or more evaluating parameters affecting existing bridge evaluation. From this viewpoint, we have tried to develop the most complex bridge evaluation in frame of above mentioned research project [4] based on the concept introduced in previous research activities [5, 6].

The methodology proposal was based on Deficiency Point Method taking into account four basic evaluation parameters - actual bridge load-carrying capacity, actual bridge deck width, actual bridge vertical clearance and actual bridge age. Two secondary parameters – road traffic intensity and detour length due to insufficient load-carrying capacity or inconvenient bridge arrangement were also considered in the first methodology proposal. Respecting results of the project entrance disputation and actual approaches to the determination of highway bridge load-carrying capacity, where simplified methods and estimations are preferred, the number of evaluating parameters had to be extended taking into account the effect of actual bridge condition determined by regular bridge inspections. At present, this parameter is considered as the main evaluating parameter in the case of highway bridges and decision-making processes related to maintenance strategy and bridge rehabilitation are mostly based on this parameter.

Due to questionable determination of the detour length in the case of insufficient bridge load-carrying capacity or bridge arrangement, this secondary parameter was replaced by other parameter allowing for the actual bridge length. Thus, the resulting number of the main evaluating parameters was increased to five, as follow:

- Actual bridge load-carrying capacity (B_{LC}).
- Actual bridge technical condition evaluated according to [7] (B_{TC}).
- Actual bridge deck width (B_{DW}).
- Actual vertical bridge clearance over and under the observed bridge (B_{VC}).
- Actual bridge age and its planned remaining lifetime (B_{RL}).

Two secondary parameters complete the bridge significance from the transportation and building viewpoint:

- Parameter of road traffic intensity replacing bridge categorization according to road classification.
- Parameter of the bridge length underlining bridge significance from the building and economical viewpoint.

3 Deficiency point method

General mathematic formulation of the deficiency points calculation of the bridge evaluated is given by the following equation

$$DP = \Sigma\left[W_i \cdot f_i(a,b,c,d,e) \cdot W_j f_j(\mathrm{f},\mathrm{g})\right] \quad (1)$$

W_i are the weighting factors defined for main individual evaluating parameters according to its significance.
$f_i\,(a,b,c,d,e)$ expresses mathematic interpretation of the individual main parameter influence.
a, b, c, d, e are the individual main evaluating parameters.
W_j are the weighting factors defined for secondary individual evaluating parameters.
$f_i\,(g,f)$ represents the mathematic formulation for influence of the individual secondary evaluating parameters.
g,f are the secondary evaluating parameters.

In the evaluation model presented above, the general equation (1) can be written in following form

$$DP = W_{LC}B_{LC} + W_{TC}B_{TC} + W_{DW}B_{DW} + W_{VC}B_{VC} + W_{RL}B_{RL} \quad (2)$$

DP is the resulting number of deficiency points of the evaluated bridge.
B_{LC} is the number of deficiency points due to actual bridge load-carrying capacity.
B_{TC} is the number of deficiency points taking into account the bridge actual technical condition.
B_{DW} is the number of deficiency points allowing for actual bridge deck width.
B_{VC} is the number of deficiency points allowing for actual bridge vertical clearance over and under the bridge observed.
B_{RL} is the number of deficiency points from the bridge age viewpoint.
W_j are the weighting factors of individual main evaluating parameters from the viewpoint of their influence on the complex bridge evaluation.

Mathematic interpretations of the individual evaluating parameters are introduced in following chapters.

3.1 Effect of the actual bridge load-carrying capacity

Effect of the actual bridge load-carrying capacity should be expressed by following formula

$$B_{LC} = \left(\frac{V_{n,\max} - V_n}{V_{n,\max}}\right)\left(W_T K_T + W_L K_L\right)^{n_1} \quad (3)$$

but $B_{LC} \geq 0$ and
$V_{n,max}$ is the maximum characteristic value of the bridge normal load-carrying capacity expressed in tons.
V_n is the bridge actual normal loading capacity in tons.
n_1 is the exponent allowing for classification of the highways.
W_T is the weighting factor of the bridge road traffic intensity effect.
W_L is the weighting factor of the bridge length effect.

$$W_T + W_L = 1.0 \quad (4)$$

K_T is the factor taking into account the effect of the bridge traffic intensity.

$$K_T = n_2\left(\frac{ADI - ADTI}{\max DI}\right) + n_3\left(\frac{ADTI}{\max DI}\right) \tag{5}$$

ADI is the average daily highway traffic intensity on the observed bridge (vehicles per day).
ADTI is the average daily truck traffic intensity on the observed bridge.
max DI is the maximum daily highway traffic intensity in the evaluated region.
n_2, n_3 are factors allowing for effects of the automobile or truck traffic, $0 \le n_2 \le 1.0$ and $0 < n_3 \le 1.0$.
K_L is the factor taking into account the effect of the bridge length and should be determined according to formula

$$K_L = \frac{BL}{\max BL} \tag{6}$$

BL is the actual bridge length.
max BL is the maximum bridge length in the evaluated part of highway, region, etc.

3.2 Effect of the bridge technical condition

Effect of the bridge actual technical condition should be expressed by following formula respecting the system of evaluating bridge technical condition in accordance with standard [7]

$$B_{TC} = \left(1 - \frac{7 - TC}{4}\right)\left(W_T K_T + W_L K_L\right)^{n_1} \tag{7}$$

but $B_{TC} \ge 0$ and $TC > 3$,
TC is the evaluation of the bridge actual technical condition according to [7] defined by numbers from 1 till 7.

In compliance with [7], the parameter B_{TC} respects only bridges classified by $TC > 3$, whose deficiencies could affect bridge load-carrying capacity.

3.3 Effect of the bridge arrangement

In this case, the effect of the bridge deck width and bridge vertical clearance should be distinguished. The number of the deficiency points related to the bridge deck width can be calculated according to formula

$$B_{DW} = \left(\frac{SDW - DW}{SDW}\right)\left(W_T K_T + W_L K_L\right)^{n_1} \tag{8}$$

but $B_{DW} \ge 0$ and
SDW is the standard bridge deck width in accordance with standards [8, 9].
DW is the actual bridge deck width.

The ranking formula for bridge vertical clearance is in two components to account for traffic over and under the bridge and can be written in following form

$$B_{VC} = 0.5(VC_0 + VC_u)(W_T K_T + W_L K_L)^{n_1} \tag{9}$$

but $B_{VC} \geq 0$ and
VC_o expresses effect of the bridge vertical overclearence.
VC_u expresses effect of the bridge vertical underclearence.
Effect of the bridge vertical overclearence could be determined in accordance with formula

$$VC_o = \frac{SVOC - VOC}{SVOC} \tag{10}$$

but $VC_o \geq 0$ and
$SVOC$ is the standard bridge vertical overclearence according to [8].
VOC is the actual bridge vertical overclearence.
Effect of the bridge vertical underclearence could be determined in accordance with the following equation

$$VC_u = \frac{SVUC - VUC}{SVUC} \tag{11}$$

but $VC_u \geq 0$ and
$SVUC$ is the standard bridge vertical underclearence according to [8].
VUC is the actual bridge vertical underclearence.

3.4 Effect of the bridge age

From the viewpoint of this evaluation parameter, the aim was to allow for bridge remaining lifetime. Bridge remaining lifetime is the very significant parameter from the viewpoint of the existing bridge rehabilitation. Because of complicated calculation of the bridge remaining lifetime due to various type of material degradation, currently not knowing, the actual bridge age was finally considered for this parameter only. Then, the effect of the bridge age should be expressed by formula

$$B_{RL} = \left[1 - \left(\frac{BC + T_D - BE}{T_D}\right)\right](W_T K_T + W_L K_L)^{n_1} \tag{12}$$

but $B_{RL} \geq 0$ and
T_D is the design bridge lifetime considered as *100* years.
BC is the year of the bridge construction or reconstruction.
BE is the year of the bridge evaluation.

3.5 Calibration of the parameter weights

The methodology presented above was processed for computational utilization. The first software version was prepared using table processor Excel, in which the methodology was tested and the optimal choice of the weighting main and secondary factors was found by means of the parametric studies. In frame of the parametric studies, the data files of the 96 bridges from Žilina region, 72 bridges from the region of Banská Bystrica and 16 bridges from Košice region were tested. The following value of the weighting factors were recommended for practical use in frame of Slovak Road Administration as the outputs of the performed parametric studies

- Effect of the bridge loading capacity $W_{LC} = 45$
- Effect of the bridge actual technical condition $W_{TC} = 30$
- Effect of the bridge deck width $W_{DW} = 10$
- Effect of the bridge vertical clearance $W_{VC} = 5$
- Effect of the bridge age $W_{RL} = 10$
- Effect of the highway traffic intensity $W_T = 0.8$
- Effect of the bridge length $W_L = 0.2$
- Effect of the highway classification $n_1 = 0.5$
- Effect of the highway truck traffic intensity $n_2 = 1.0$
- Effect of the highway automobile traffic intensity $n_3 = 1.0$

4 Conclusions

The contribution presents results of the research project elaborated in frame of applied research related to existing bride evaluation. Methodology of the complex existing bridge evaluation was developed based on the Deficiency Points Method. Using the methodology, the priority order for existing bridge rehabilitation could be determined.

The paper presents results of the research project APVV-14-0772 supported by the Slovak Research and Development Agency and the project 012ŽU-4/2016 supported by the Cultural and Education Agency of Slovak Republic.

References

1. J. Lauridsen, J. Bjerrum, N.H. Andersen, B. Lassen, Struc. Eng. Inter. **3**, 216-220 (1998)
2. P.D. Thompson, E.P Small, M. Johnson, A.D. Marshall, Struc. Eng. Inter. **4**, 303-308 (1998)
3. J. Vičan, *Bridge Management System – Methodology of the Existing Road Bridge Evaluation, Final Report for Slovak Road Administration* (University of Žilina, Žilina, 1998)
4. J. Vičan at all, *Evaluation methodology of existing bridges, Project of the SRDA APVV-0106-11* (University of Žilina, Žilina 2015)
5. J. Vičan at all, *Bridge Management System* (Ministry of Transportation of Slovak Republic, Bratislava, 2003)
6. J. Vičan at all, *Program of planning maintenance and repair of infrastructure for project of the transformation and restructuralization of Slovak Railways* (University of Žilina, Žilina, 2001)
7. OTN 73 6220, *Bridge evidence on the motorways, roads and local communications* (SOSMT, Bratislava, 1993)
8. STN 73 6201, *Standard specifications for bridges* (SOSMT, Bratislava, 1999)

Preconditions of emergence of large-span buildings in the world

Elena Sysoeva[1,*]

[1]Moscow State University of Civil Engineering, Department of Design of Buildings and Structures, Yaroslavskoe sh. 26, 129337 Moscow, Russia

Abstract. The article is devoted to the design and construction of Olympic facilities in the world, preconditions of the emergence of large-span buildings and structures. Every example includes information about the building structural systems, names of architects and engineers, some interesting facts about the construction technology. There is a review about several Olympic stadiums in the world, including the stadium "Friendship", Small Olympic Palace in Rome, indoor ice arena Makomanai in Sapporo (Japan), the Olympic stadium in Montreal, Kanteerava Indoor Stadium (India) etc. The article includes opinions of scientists on the development of long-span structures in the coming years.

1 Introduction

Public buildings are often architectural dominants of major cities. Bright expression and the whole image of residential areas attract tourists from all over the world. The design and architecture of public buildings will be given significant attention. Building material and the coating form of such buildings can play an important visual role in the perception of the building by any person. Before the end of the eighteenth century the main construction kinds of material were stone, wood and ancient concrete but from the end of the twentieth century besides reinforced concrete and glued wood synthetic material have been often used, such as ETFE, for example.

In ancient times there was a need in the construction of arenas for performances of gladiators (for example, The Coliseum) and a sport competition (the first athletic building was built by architect Leonid in the fourth century BC "Leonidaion"). But due to the prohibition of gladiatorial combats and the cessation of the sports of the Olympics basically religious buildings had been built until 1896 (churches, temples, cathedrals, etc.). And only at the end of the nineteenth century a new surge of the Olympic movement began. It led to the need of the construction of buildings with long spans. To this moment, in addition to the need for such buildings, there are new great opportunities – new building material – reinforced concrete (the invention of Monier in 1867).

An opening of a new building material (reinforced concrete), development of the theory of reinforced concrete as a science, the engineering school in the world which led to the

* Corresponding author: sev4279@yandex.ru

emergence of a galaxy of scientists all around the world, became the prerequisites for the formation of thin-walled reinforced concrete long-span buildings.

2 The modern Olympics

The revival of modern Olympic summer games at the end of the nineteenth century and in 28 years the winter Olympic Games has led to the need of construction of buildings with large unsupported spaces for stadiums, swimming pools, race tracks, etc. Scientists state that since 1960s Olympic Games are reviving the continued international interest in the creation of large span buildings [1].

The tradition of holding of the Olympic Games leads to the necessity of designing and building of stadiums, concert halls and other large-span complexes in the cities to host the Olympics. From 1955 to 1956 it was the designing and construction of "Central stadium named after V. I. Lenin" (now a sport hall "Druzhba") with dimensions in plan 85x85 m, height 18 m and a capacity of 4,000 seats. The central part of the roof is ribbed thin concrete coating supported by 28 reinforced concrete panels with complex geometric shapes (Fig. 1) [2].

The authors of the project were architects A. V. Vlasov, I. E. Rozhin, and A. F. Boars, N. N. Ullas, and engineers V. N. Nasonov, N. M. Reznikov, V. P. Polikarpov [4]. The sports complex was built in record time of 450 days. By 1980, the building had undergone renovations to prepare for the Moscow Olympics.

Fig. 1. The «Druzhba» Multipurpose Arena (photo credit: http://www.vashdosug.ru/msk/concert/place/7664/).

But if the first Games was held at the stadium Panathinaikos restored in the mid-nineteenth century and for the next fourteen games only opened stadiums were built, for the Olympic games in Rome in 1960 an Italian architect and engineer Luigi Nervi [3] and his friend and architect Annibale Vitellozzi Small and Large Olympic stadiums in Rome with coating of ribbed concrete domes were designed. During the construction of Small Olympic Hall (Fig. 2) Nervi used a caisson cover, reinforced with thin steel mesh. The diameter of the dome is 61 m.

Fig. 2. Small Olympic Hall, Rome, Italy (photo credit://www.nashi-progulki.ru).

In 1972 indoor ice arena Makomanai located in Sapporo (Japan) was built. Its capacity is 6020 seats (Fig. 3, 4). The diameter of the building is 109 meters. The building structural system consists of 48 concrete columns supporting a steel frame, which is a support for 24 metal trusses. «The rink was surrounded plastic fence to give the spectators an unobstructed view of the action at all times» [4].

Fig. 3. Makomanai Sekisui Heim Ice Arena, Sapporo, Japan.

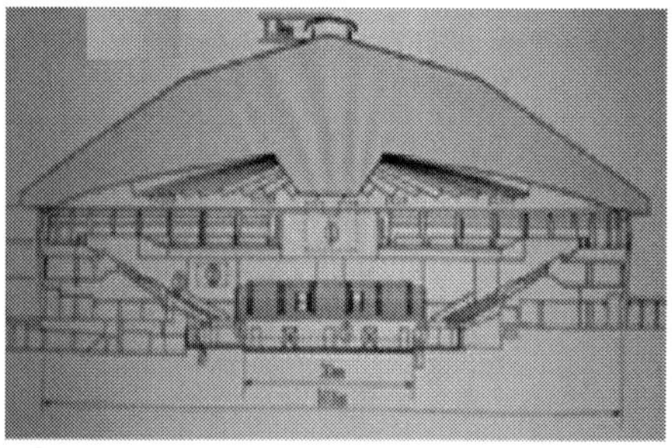

Fig. 4. Cross sectional view.

In 1976, architect Roger Taillibert and his team the Olympic stadium in Montreal was designed. The stadium consisted of a slanted tower (at that time the highest in the world) with a retractable membrane roof (Fig. 5).

Fig. 5. Concrete frame of the Olympic Stadium in Montreal.

It was the stadium which was designed and made of only concrete and consisted of 12000 prefabricated elements, «adjusted and assembled to the others with the utmost precision »[4].

Building structural system was created of 34 reinforced concrete curved consoles connected with a central concrete ring creating a single monolithic system. Technical equipment was installed on two floors. A large part of the lighting equipment was installed on the ground floor. The second floor is occupied by technical ventilation systems. More than 10,000 workers participated in the construction of this unique building, constantly working from August design is the Stadium's Tower. It is 165 meters high and has a 45 degree incline.

For the Olympic Games in 1980 one of the largest stadiums in the world "Olympic" at Prospect Mira in Moscow (Russia) was built (Fig. 6).

Fig. 6. Olympic sport Stadium "Olympic" (photo credit: https://ru.wikipedia.org/wiki).

Architects of the project were M. V. Posokhin, B. I. Thor, HP, Aranauskas and engineers Lvov Y. P., Chertkov L. I., Eremeev P. G., Trofimov V. I. The design of the coating is a metal hanging membrane with positive curvature Gassoway with the sag of 12.5 m and 5 mm thick, dimensions in plan are 224x183м. Membrane design has been supported by 64 steel wire ropes. The hall's capacity is 45 000 [1].

Kanteerava Indoor Stadium (Fig. 7) is an indoor sporting arena in Bangalore, India (1998). The capacity of the arena is 4,000 people. Elliptical structural system consists of 120 precast folded plates spanning between a tension ring at its base and a compression ring at its crown [1].

Fig. 7. Kanteerava Indoor Stadium, Bangalore, India.

One of the most interesting works of an engineer Christiansen was the Seattle Kingdome (completed in 1976 and demolished in 2000, Fig. 8). The building construction consists of 40 triangle hyperboloid paraboloid segments, stiffened by ribs. Its maximum span is 201,5 m.

Fig. 8. Kingdome, Seattle, 1976.

Indoor sports arena in Raleigh (North Carolina, USA) has a capacity of 5500 seats (Fig. 9). The building is covered hanging saddle-like shape. The basis for the structural system of the arena in Raleigh is the two concrete arch cross sections which are inclined to the horizon at an angle 22^0. Arches attached to the cables with a pitch of 1.8 m for the formation of saddle-shaped rigid surface.

Fig. 9. Sports arena in Raleigh, North Carolina, USA.

Rungnado 1st of May Stadium is located on a sandy island Ninnada in Pyongyang (North Korea). The stadium capacity is 150,000 spectators. The construction finished in 1989 (Fig. 10).

Fig. 10. Rungnado 1st of May Stadium (North Korea).

Concrete covering of the stadium consists of 16 shell segments. Building has a height of 60 m. The construction lasted about 2,5 years.

3 The prospects for further development

For many decades, designers and architects have been looking for the perfect geometry, shape and static conditions, which correspond to the most profitable use of the material and the smallest expenditure of labor. Calculator engineers are engaging of the development of numerical methods for detailed qualitative and numerical analysis of the stress-strain state of an arbitrary shell. All the efforts of scientists and professionals of all areas in the structural design of long-span public buildings aimed at finding new opportunities to cover the unsupported spaces [5, 6]. But if in the previous century the construction of buildings of this type was related to preparation for the Olympics, the emergence of theatres, circuses, temples, nowadays large spans are needed to create a comfortable stay in undeveloped areas with an unfavourable climate, comfortable work of builders in areas with very low or very high temperatures, for the frame tent and pneumatic building used for sports, shopping facilities, etc. Therefore, the attention of designers is drawn to new structural types of long-span buildings and new numerical methods for their calculation.

References

1. J.F. Abel, Fifty Years of Progress for Shell and Spatial Structures, 485 (2011)
2. P.G. Eremeev, *The modern design of coverings over the stands of stadiums* (Publishing house ASV, Moscow, 2015).
3. L. Nervi, Proceedings of IASS **2**, 19 (1969)
4. Official Report of the XIth Olympic Winter Games, Sapporo (1972)
5. M.S. Tupolev, Proceedings of IASS **1**, 747 (1969)
6. G.A. Karavaev, Proceedings of IASS **2**, 403 (1969)

"Markhi" spatial design structure: numerical study of its work under static load

Vadim Alpatov[1,*]

[1]Samara State Technical University, Institute of Architecture and Civil Engineering, Molodogvardeyskaya str. 194, 443001 Samara, Russia

Abstract. There is a problem of internal stress volume existing for some types of spatial structures and their joint connections. The problem occurs when a massive body is used as a joint connector. It is quite simple to determine tension on this joint connector surface using electric resistive tensometry method. It is not simple though to empirically determine internal tension in the massive body of the connector. To determine internal tension we can use modern calculation systems, such as Ansys, Abaqus, CosmosWorks, Nastran, Autodesk Inventor, Robot Structural Analysis, Bentley STAAD, CSI SAP2000; etc: Internal tension analysis in a massive joint connector makes possible to select both surplus stock parts and shortage stock parts. In this paper the authors base their analysis on both surface and internal tension of MARKHI connector and come up with solutions for its improvement.

1 Introduction

Many elements converge in joints of spatial lattice design structures (SLS). The peculiarity of the SLS is that each element here is involved in receiving of external load and transferring this load to the foundations [1-5, 20, 21]. Each item of the SLS also possesses some internal force which it transfers to the joint connection. Joint connection in this case accepts many differently directed loads coming from adjoining elements. Thus, the joint work here can be described as combined stress state [2, 4-11].

2 Main Part

Traditionally, experimental methods are used to study stress-strain behaviour (SSB) of joint connections. The most popular method here is the electric resistive tensometry method. Experimental methods make it possible to determine the level of tension only on the surface of the item tested. It is not possible to determine internal stress by experimental methods. In practice the internal stresses in the tested item are determined when judged by stresses on its surface. This method proved valid for thin-walled elements. Internal stress for massive bodies can be calculated that easy. As for massive bodies with complex internal geometry,

* Corresponding author: avu75@mail.ru

cavities and holes, it is almost impossible to assess their internal stress judging by their surface stresses.

Modern calculation software systems allow to calculate internal stress in almost all the models. While analyzing SSB of commuter models, modern calculation systems make it possible to peep inside these models and to build any cross-section with a distribution level of internal stresses [6, 8, 9].

When SLS is designed, the method of material concentration in places with high levels of stress and in places requiring increased reliability is widely used. SLS joins are often quite massive as they are made of steel with high or increased strength So, this is the way how the reliability of responsible joints is increased in practical conditions. It also helps to prevent "uncertainty" of internal stresses.

The most well-known joints if spatial structures (Mepo, Orona, Cubotto, Veymar, Markhi, Kislovodsk) have either a solid ball or a holed polygon as their basis. Up to 16 cored can be jointed together in such systems. R.I. Khisaimov showed in his research that joints weight ratio can reach up to 40% of the whole SLS weight [11]. In terms of the theory of structures optimal design, joint elements are referred to auxiliary elements, unlike rods which are referred to the essential elements [12, 13]. One of the commonly accepted ways to optimize designs is to reduce the size and number of support elements. Chamferless trusses made of bend-welded profiles can serve as an example of successful implementation of this principle. In these trusses the principle of refraining from joint connectors (supporting elements) is implemented to the fullest extent. As a result, the weight of such trusses is considerably smaller than the weight of gusset plates. Thus, reducing the weight of joint connectors can help reduce steel spread up to 40%.

The authors are sure that it is possible to reduce materials consumption of SLS while studying the reserves of joint connectors load bearing capacity by computer modeling. SLS with joints made on the basis of filled-in massive details of MARKHI type. The aim of this research is to study the distribution of internal stress in MARCHI joint connection and to find ways to reduce its materials consumption. To achieve this goal the authors used a heel joint of MARKHI spatial structure as the object of study.

The authors took all necessary input data for their study from a famous project. A heel joint was selected for analysis because of the following reasons. Firstly, this joint is the most loaded in the structure, secondly, the way this joint is fixed to the column is the most typical for describing clearly the boundary conditions of fixing, thirdly, demands of maximum reliability are usually applied to support members due to the possible maximum damage in the development of progressive destruction [14, 15].

Modeling of the joint geometry was done in SolidWorks software system. While creating the model of this joint connector, the authors did their best to take into account the most essential features of its design, such as: internal and external geometry of the joint connector, adjacent coupling (nuts) and bolts, the difference of physical and mechanical characteristics of the materials (due to the use of different steels). Some of the features of the joint connections had to be disregarded because of the complexity of the model. Their influence on the final result was considered as non-essential. Thus, the research simplified: screwed connection as well as conditions under which the connector bears against the column. Screwed connection was substituted with a solid contact on surfaces of the bolt and the connector.

The way the connector is supported is in practice a relatively "free" notion, implemented through the centralizer and safety angle stops installation. Vertical forces from the connector to the column are transmitted along the lower side of the connector. Horizontal loads are perceived by the friction forces, and for an emergency case safety angle stops are applied. The model under analysis excludes the column and its structural elements (that is centralizer, angle stops, etc). The way the joint is supported was modelled by means of rigid fixing of the lower side of the connector. It corresponds to a situation in which the force of friction on the surface of the lower side of the connector holds all possible horizontal efforts, i.e. the situation of regular operating mode.

As a result of calculation of the joint connection model the authors determined tension both on the surface of the connector, and inside it. This computational model makes it possible to display stress fields of cross section. Figure 1 shows internal tension in a joint connector.

As expected, tension at the connector surface does not exceed proof stress for the material. It means that externally connector is in the elastic stage. Exceptions here are the areas under adjoining nuts. In an assembled joint these areas are hidden and they have to therefore be attributed to inside areas of the assembled unit.

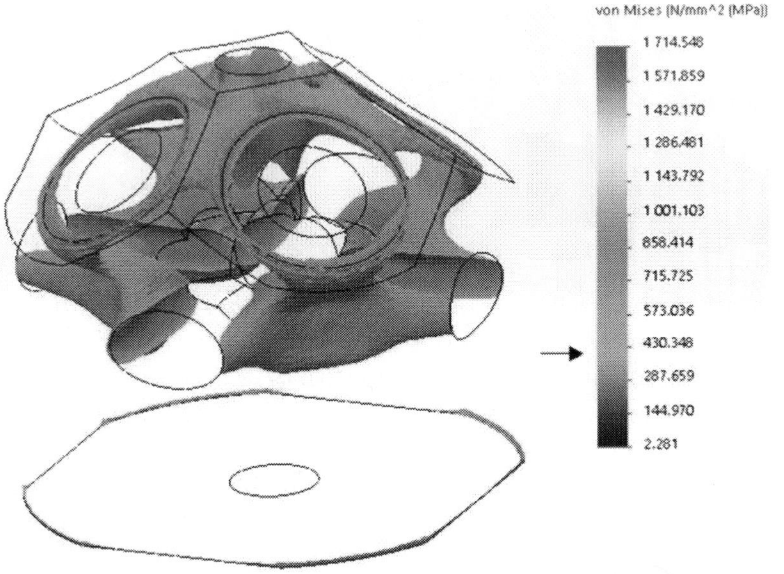

Fig. 1. Internal tension in a joint connector.

Inside the connector the researchers found come small areas with tension exceeding that of nominal yield strength steel as well as smaller areas with tension exceeding greater than the tensile strength of steel (fig. 1). The latter should be referred to the so-called "emissions", characteristic for the calculation models done by software systems and working on the basis of the finite element method. These areas are affected by a well-known fact stating that there are usually sections with peak values of calculation results in the areas of joint (dot) load application and in the areas of joints fixing (boundary conditions). Peak values should be neglected when determining the scope of reliable solutions empirically. Areas with peak tension values occur along the edge of the supporting surface of the connector, on the edge of the nut and on the bolt and connector contact surfaces. If we exclude these areas form the analysis, we can conclude that the whole volume of the connector massive body is within the scope of tension which is within the strength limit. Inside the volume there are small areas with the scope of tension greater than nominal yield stress of steel (that is in case the worst possible combination of loads is taken into account).

For the connector geometry under analysis the distribution of the material by its volume turned out to be rational. It proved impossible to reduce its mass by simply reducing of its size.

Internal stress distribution in a massive body with inside holes (made as cross-holes for bolts) is characteristic for crystallographic structures. Tensions in such a massive body are concentrated in accordance with stress application and are distributed onto the supporting structure through shortcuts. Porosity in the upper part of the massive body (bolt installation place) causes concentration of tension in "narrow" parts. The level of tension declines in the lower side of the connector massive body (closer to the supporting surface). There's more material and less porosity in this part of the body. Surface areas of the connector (which are comparatively far from the borders of contacts connector-nut and bolt connector) are involved in the work of the joint comparatively seldom. To these areas we refer outer edges of the faces of the polyhedron connector. In these areas the material can be removed completely , with the result that the from of the connector will become a ball characteristic for a Mero joint (out of a polyhedron) (see Figure 1).

It is also possible to reduce the consumption of steel used for a massive joint connector by changing its design and form. For example, the joint can be formed as a hollow spherical shell of high-strength steels. In this case, there will be no internal stresses typical for massive connectors [16-18]. Power lines here will be distributed along the surface of the shell causing surface tension only. They can be easily verified by using electric resistive tensometry method. SDC, Nodus, NS, Oktaplatta, Vestrut and other systems are examples of such joints.

3 Conclusion

The research yielded the following conclusions:

1. Tensions at the connector surface coming from calculated loads do not exceed that of nominal yield strength steel.
2. The maximum tension at the connector surface occurs on the supporting surface and in the parts where nuts and connector contact, i.e. under the nuts.
3. There are some small areas with tension exceeding that of nominal yield strength steel.
4. Areas with high internal stresses are located in the bolt and connector contact surfaces.(screw connection area) and inside of the connector in the areas serving as cross-holes for bolts.
5. For the connector geometry under analysis the distribution of the material by its volume turned out to be rational. It proved impossible to reduce its mass by simply reducing of its size.
6. It is possible to reduce the consumption of steel used for a massive joint connector by changing its design and form. For example, the joint can be formed as a hollow shell.

References

1. I.S. Kholopov, M.I. Balzannikov, V.Yu. Alpatov, Vestnik VolgGASU **28,** 225-232 (2012)
2. V.N. Trofimov, G.B. Begun, *Structural Constructions* (Stroiizdat, Moscow, 1972)
3. I.S. Kholopov, V.Yu. Alpatov, V.N. Mochalnikov, N.N. Moiseev, V.Yu. Vetschin, Modernization and development of metal, timber and plastic constructions in civil engineering and on transport actual problems: International scientific and technical conference proceedings, 199-206 (2002)
4. V.K. Faibishenko, *Metal cross-bar spatial structures of roofs* (VNIINTPI. Moscow, 1990)
5. A.V. Perelmuter, V.V. Yurchenko, Structural Mechanics and constructions calculations **6**,18-25 (2012)

6. V.Yu. Alpatov, I.S. Kholopov, A.V. Solovjev, Effective constructions, materials and technologies in architecture and civil engineering: International conference proceedings, 122-127 (2009)
7. E.V. Gorokhov, V.F. Muschanov, I.V. Romenskiy, A.V. Muschanov, Metal Structures,**4,** 191-206 (2015)
8. N.N. Halobyta, Herald of Brest State Technical University, 1, 94-102 (2008)
9. N.A. Buzalo, S.A. Alekseev, N.G. Tzaritova, Civil Engineering-2014: Actual problems of industrial and civil engineering: International scientific and pratical conference proceedings, 215-216 (2014)
10. A.V. Muschanov, V.F. Muschanov, I.V. Romenskiy, Unique buildings and constructions, **2,** 18-29 (2016)
11. R.I. Khisaimov, L.A. Isaeva, *Calculating technical-and-economic index of space frames* (Kazan institution of civil engineering, Kazan, 1979)
12. K.I. Mazhid, *Constructions optimal design* (High school, Moscow, 1979)
13. V.Yu. Alpatov, *Metal constructions optimal design* (Candidate of Engineering Sciences dissertation, Samara, 2002)
14. V.Yu. Alpatov, A.O. Lukin, S.M. Petrov, Industrial and civil engineering **3**, 47-51 (2014)
15. T.Ye. Gordeeva, N.S. Belomytzeva, Architecture and civil engineering: traditions, and innovations, Natural sciences and technospheric safety. Collected works of 72th all-Russia scientific technical conference, 406-410 (2015)
16. A.V. Tur, V.I. Tur, I.S. Kholopov, *Joint connections of spatial structures thin-walled bars* (Patent 2467133 RF,2011)
17. S.A. Selin, A.V. Tur, V.I. Tur, *Joint connections of spatial structures thin-walled bars* (Patent 2468157 RF,2012)
18. I.S. Kholopov, A.V. Tur, V.I. Tur, *Higher Education Institutions Proceedings. Civil Engineering* **4**, 104-111 (2012)
19. M.I. Balzannikov, V.Yu. Alpatov, I.S. Kholopov, A.A. Saharov, A.O. Lukin, MATEC Web Conf. **73**, 1-6 (2016)
20. I.S. Kholopov, M.I. Balzannikov, V.Yu. Alpatov, A.V. Soloviev, Procedia Engineering **153**, 277-282 (2016)

Research of flow parameters in a blow off of a hydraulic installation

Mikhail Balzannikov[1,*], *Semyon* Piyavsky[1], and *Andrey* Kurmanayev[1]

[1] Samara State Technical University, Institute of Architecture and Civil Engineering, Molodogvardeyskaya str. 194, 443001 Samara, Russia

Abstract. The paper deals with conditions and parameters of water flow in a blowoff of a hydraulic installation and adjacent area of a discharge channel. The authors point out the significant impact of forms and geometric sizes of outlet structures, as well as stilling structures on the linkage parameters of outlet water flow. The article presents results of research of the impact of a baffle on the water flow in discharge outlet and outlet canal and shows a possibility of hydraulic conditions improvement in discharge canal due to water-diverting wall construction. The authors conclude that deletion of a water-diverting wall in an outlet section of discharge outlet should be proved by technical and economic estimation.

1 Introduction

Waste water facilities are always considered as a part of hydraulic engineering units, constructed at natural water flows. These constructions are used for water discharge into the tail bay in the high water period, and design water flow in low-flow period [1-4]. If hydraulic water system has energy application, then part of water consumption is transmitted into the tail bay and also through the water conducting conduit of hydraulic units [5-8]. Water sluice constructions should meet high-level requirements, concerning their work safety and specific water flow design. Besides, water flow facilities should provide fast termination of water pass in case of emergency situations or their threat, and provide favorable hydraulic conditions of water flow, including output area into discharge channel [9-13].
Analysis of the results of studies of hydraulic structures and their water flow facilities allows to conclude that hydraulic conditions of water flow in the outlet channel are mostly influenced by the kinematic characteristics of the water flow and the nature of the velocity distribution throughout the height of the outlet section of a water flow facility, which are in turn determined by the shape and geometric parameters of the outlet area of the conduit [14-18]. It should be noted that the values of bottom velocity flow and, consequently, the parameters (firstly, thickness) of anchoring the bottom of outlet channel and its construction expenses are dependent on the matching conditions of output flow and tail bay. Moreover, the flow matching conditions and the characteristics of the flow stream is greatly influenced by the presence and form of stilling structures that provide flow deviation [19]. All

[*] Corresponding author: sgasu@samgasu.ru

mentioned conditions and parameters largely determine the design solutions in the justification of the constructions of these elements in hydropower facilities [20-22].

Thus, research of hydraulic conditions of output flow in blow off, that helped to work out the recommendations on selection of rational shapes and geometric dimensions of water outlet, is very important, relevant and has a significant economic application.

The research objective – is research of hydraulic conditions of flow in blow off of hydraulic structures and outlet channel.

The main tasks that are necessary to achieve are the following:

- select typical constructive solutions of a water-discharge facility for the research sake;
- development of a mathematical model of outlet facility and outlet channel;
- implementation of the mathematical model description of outlet facility;
- application of the methodology of conducting model research and the rationale for the experiment boundary conditions related to the developed model;
- conducting of design-basis experiment and obtaining kinematic parameters of the flow in the water outlet facility and the outlet channel;
- analysis of the obtained results;
- formulation of conclusions and elaboration of recommendations to improve the hydraulic conditions of flow in the outlet facility of hydraulic structures and outlet channel.

Research of flow parameters in a blow off of a hydraulic installation

Upstream blow offs, in which water from the reservoir flows via the penstock and further goes under the level of a tail bay, are mostly widespread in low- and medium-head water power development. Structurally, the discharge outlet should include a flared part of rectangular cross section. The flared shape is necessary for reduction of water flow speed at the outlet of the discharge channel, and the rectangular cross-sectional shape is most efficient for its location in the output zone of the plain gate of rectangular shape.

Baffles of various types and constructions are used in order to reduce the section length of outlet canal that undergoes significant water flow bottom velocity. Stilling baffles and wells are often used as baffling facilities. These elements are designed to reduce flow velocities in the bottom section of the outlet channel and its scoring action due to deflection of water jets of water flow to the surface in the outlet channel in a tail bay.

However, if the water conduit is used to divert water from hydraulic units, providing work not only in the turbine, but in the pump modes, such stilling facilities may significantly and negatively affect the hydraulic conditions of reverse flow and, consequently, reduce the energy output of the power unit. The authors suggested the outlet structure with a movable water-diverting wall to ensure good working conditions of the outlet facility in both modes of possible operation of the power plant [23]. This structure provides protection of bottom anchoring of outlet channel from destructing effect of the flow, discharged from the water conduit. At the same time, the water-diverting wall is able to automatically clean the bottom of the pit and thereby eliminate the obstacle for the incoming flow, working in the water pump mode. This design was applied in the main research, based on the mathematical model.

A mathematical model of the outlet facility is designed in the software "ANSYS". The walls of the water conduit as well as the water level in tail bay are modeled as high-velocity contrasts.

However, all the main geometric parameters of the model of the outlet facility and the depth of water in tail bay in the model varied significantly during the experiment. The length of the flaring section changed particularly from 1H to 7H, where H – is the height of the pressure conduit in front of the output section. The incidence angle of the top in the flaring section site varied from 0^0 to 40^0. Removal of water-diverting wall from outlet section of the conduit ranged from 1H to 9H. The water depth in the outlet channel was calculated in the range from 3H to 7H. The incidence angle of water-diverting wall was from 0^0 to 45^0 in vertical axis. The flow velocity in a characteristic cross-section in the discharge outlet was taken to be equal to 4 m/s for this experiment.

The studies of the mathematical model resulted in flow characteristics in the operational part of the outlet facility and the outlet channel of a hydraulic structure.

The following figures show the results of calculations of flow rates for the water outlet facilities in hydraulic structures at varying distances of water-diverting wall from outlet section exit of a water outlet. The deviation angle of the wall from a vertical plane remained constant - 15^0.

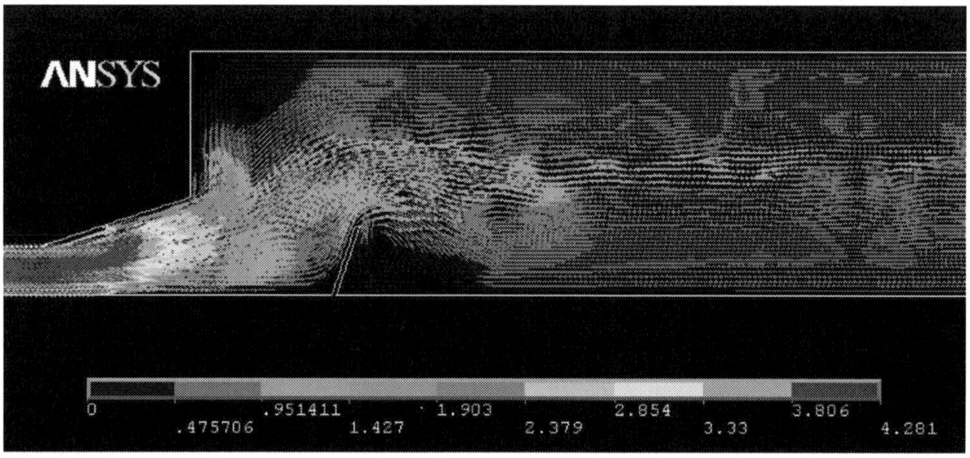

Fig. 1. The result of flow rates calculation in the process of removing water-diverting wall from outlet section equal to 3H.

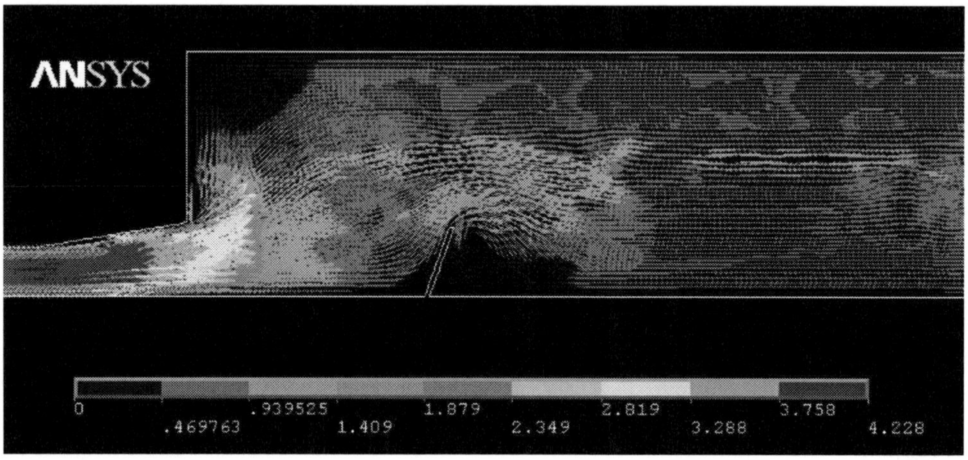

Fig. 2. The result of flow rates calculation in the process of removing water-diverting wall from outlet section equal to 5H.

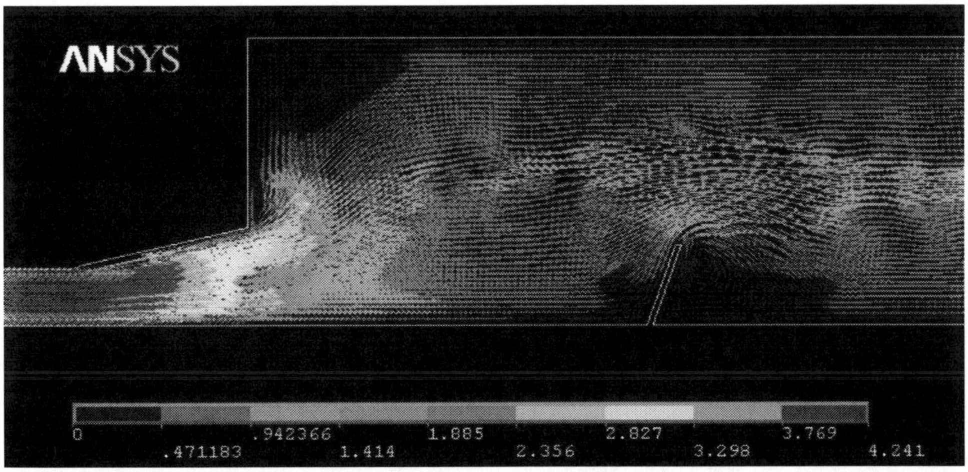

Fig. 3. The result of flow rates calculation in the process of removing water-diverting wall from outlet section equal to 7H.

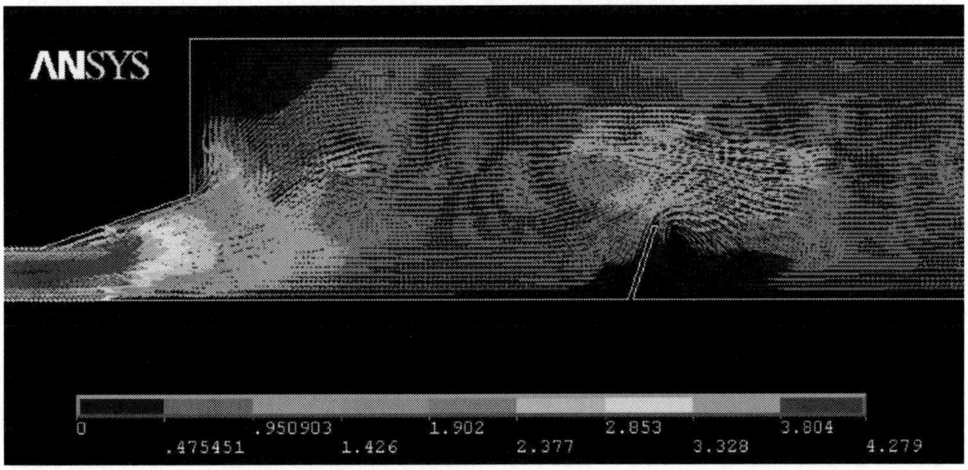

Fig. 4. The result of flow rates calculation in the process of removing water-diverting wall from outlet section equal to 9H.

The results of the study show that the flaring section of an outlet facility in all considered variants plays a significant role in ensuring the reduction of velocity of water flow in the outlet section. The average speed in this area decreases from 4.0 m/s to 2.0-3.0 m/s depending on the length of the section and the angle of incidence of the ceiling.

Then, after the flaring section, the flow goes to the outlet channel in the form of volatile widening transitional jet. Stagnant and whirling zones are clearly observed throughout the height of the discharge channel: at the top near the outlet section and in the bottom areas in front of a water-diverting wall and behind it.

The parameters (location and size) of the first stagnation zone for the considered variants do not differ from each other. The parameters of the stagnant zone in front of water-diverting wall change. In particular, the closer the wall is to the output section, the smaller size is formed in front of its area with small flow rates. However, the water flow is more actively dropped out in the upper layers, i.e. the locus of the streamlines becomes highly curved. It is important to consider that a significant change of the velocity vector of the transit flow will lead to an increase of pressure loss, which is undesirable for power water conducting system. In all cases, water-diverting wall of the bottom flow rate is much less than in front of the wall. Their values vary from zero value (at the distance of 2.0-2.5 of parameter H), up to 0.3-0.4 m/s – at a greater distance.

Conclusion

Thus, the analysis of the results showed that the presence of water-diverting wall in the outlet channel of the water conduit has a significant impact on the velocity distribution coming out of the water flow on the depth. In particular, a zone with very low velocities of water flow emerge almost immediately behind such wall, which, in turn, will allow to avoid the facility of considerable thickness of the bottom (usually concrete) stilling and reduce the amount of expensive material which will be beneficial from economic point of view.

The following two opposing trends should be considered in the process of designing location of a water-diverting wall.

On the one hand, fitting water-diverting wall to the outlet section of the conduit will reduce the length of the required solid concrete for mounting the bottom channel and bring down lump-sum costs (this is a positive effect). On the other hand, this solution will require strengthening of bearing elements of the used wall for increasing the pressure of the water flow and will increase the head losses due to the worsening of hydraulic conditions of the output flow (kinematics). The second factor will cause the need to increase non-recurring costs (cost of the wall) and other operational losses. Finally the location of water-diverting wall should be designed after conducting economic and technical research of possible options and identify the best economic solution.

References

1. V.V. Elistratov, V.I. Maslikov, G.I. Sidorenko, Water engineering **11**, 11-15 (2014)
2. M.I. Balzannikov, V.M. Yurov, A.A. Mikhasek, Scientific survey **3**, 101-108 (2015)
3. F. Switala, S.V. Evdokimov, Yu.M. Galitskova, Industrial and civil engineering **12**, 87-90 (2014)
4. M.I. Balzannikov, M.V. Rodionov, V.A. Seliverstov, Vestnik SGASU. Town planning and architecture **1**, 100-105 (2011)
5. M. I. Balzannikov, Vestnik SGASU. Town planning and architecture **4**, 86-92 (2015)
6. Y.S. Vasiliyev, V.V. Elistratov, *Hydropower installation* (Publishing house of St. Petersburg state polytechnical university, Saint Petersburg, 2011)
7. M.I. Balzannikov, V.M.Yurov, A.A. Mikhasek, Procedia Engineering **111**, 82-88 (2015)
8. Y. S.Vasiliyev, G. I.Sidorenko, V. V. Frolov, Scientific-technical Bulletin of Saint-Petersburg state polytechnic university **147-1**, 76-84 (2012)
9. S.V. Evdokimov, Vestnik SGASU. Town planning and architecture **2(6)**, 68-72 (2012)
10. S.A. Piyavskiy, S.V. Evdokimov, Proceedings of higher educational institutions. Construction **6 (643)**, 36-42 (2012)
11. M.I. Bal'zannikov, V.A. Zubkov, N.V. Kondrat'eva, V.A. Khurtin, Power Technology and Engineering **4(47)**, 267-272 (2013)
12. M. I. Balzannikov, B.G. Ivanov, A.A. Mikhasek, MGSU Bulletin **7**, 119-124 (2012)
13. S.V. Evdokimov, T.V. Dormidontova, Vestnik SGASU. Town planning and architecture **1(5)**, 64-68 (2012)
14. M.I. Bal'zannikov, V.V. Elistratov, Power Technology and Engineering **12(28)**, 720-725 (1994)
15. V.V. Elistratov, I.G. Kudryasheva, Yu.A. Miroshnikov, Hydraulic engineering **11**, 22-27 (2014)
16. M.I. Balzannikov, V.A. Seliverstov, Industrial and civil engineering **8**, 17-19 (2010)

17. M.I. Bal'zannikov, V.A. Seliverstov, Power Technology and Engineering (Springer New York Consultants Bureau) **3(46)**, 210-214 (2012)
18. G. S. Konigh, N.V. Kasatkin, A.M. Weinstein, A.G. Shkuropat, Proceedings of all-Russian research Institute of hydraulic engineering in honour of B. E. Vedeneeva **259**, 3-10 (2010)
19. G. K. Deryugin, Proceedings of all-Russian research Institute of hydraulic engineering in honour of B. E. Vedeneeva **275**, 10-23 (2015)
20. Y.S. Vasiliyev, L.I. Kubyshkin, Hydraulic engineering **7**, 2-8 (2014)
21. M.I. Bal'zannikov, V.A. Seliverstov, Power Technology and Engineering **1(49)**, 22-26 (2015)
22. S.V. Evdokimov, Industrial engineering **8**, 35-38 (2010)
23. M.I. Balzannikov, A.L. Kurmanaev, Eastern European Scientific Journal **2**, 282-286 (2014)

The reliability control of the existing buildings

Alexander Frolov[1], *Irina* Frolova[1], and *Olga* Kotkova[1,*]

[1]Samara State Technical University, Institute of Architecture and Civil Engineering, Molodogvardeyskaya str. 194, 443001 Samara, Russia

Abstract. Change control of enclosure structures state includes several tasks depending on a structural concept. To test the performance of thermal protective functions the distribution uniformity of the thermal field, possible losses or protection quality are controlled. Defects can emerge in bearing wall constructions, such as the material structure violation and it reduces the walls safety. Cracks, loose layer, loss of individual elements, the disclosure of stitches and so forth also lead to the change of the thermal field. Thermovision research based on the developed methodology helps to monitor emerging defects in the materials of buildings wall structure. Most of the detected thermal protection defects are not critical for the bearing capacity of building structures in general, but they are important for the microclimate of separate rooms.

1 Introduction

Vertical walls of the building can be load bearing and non-load bearing. During the operation they obtain various defects affecting their safety or just reducing operational characteristics. Defects can be visible and hidden.

Heat monitoring method allows us to control the thermal protection quality of single-layer and multi-layer structures and their connections. It also helps to detect under natural conditions sections of walls slated for repair and with hidden defects. There are two ways of applying this method: assessment of thermal insulation characteristics of the enclosing structures and defective zones establishment of wall constructions [1].

2 Heat Monitoring Implementation

Thermal insulation control method is performed by using thermographic camera. It is based on remote measurement of the surface enclosing structure temperature field. Temperature anomalies are visualized on the defective areas as areas of high temperatures.

Temperature fields of enclosing structures are obtained in the form of false color image accompanied by the appropriate scale. The color boundaries correspond to different temperatures. When there is a monochrome image the temperature gradations are determined by the image brightness that is set by the temperature scale that accompanies the image. It is recommended to carry out heat monitoring in spring and autumn when the

[*] Corresponding author: kotkovaon@yandex.ru

temperature difference between the outside and indoor air is equal to 10-15 ° C. The higher the differential temperature, the more accurate analyses.

It is acceptable to carry out thermovision measurements when the differential temperature between the outdoor and indoor air exceeds Δt_{min}°C:

$$\Delta t_{min} = 2\Delta\theta R_0^n \cdot(\alpha\cdot\tau)/(1-\tau) \qquad (1)$$

$\Delta\theta$ – temperature sensitivity limit, °C (in this case $\Delta\theta = 2$°C);
α – heat-transfer coefficient accepted for face of walls depending on wind speed:
1; 3; 6 m/s – 11; 20; 30 W/°C·m^2, correspondingly;
R_0^n – design value of heat transmission resistance, °C;
τ – relative resistance to heat transfer of the investigated fence zone.

In practice the deferential of indoor and outdoor air not less than 10 -15 °C is sufficient.

The monitoring, according to [2], is divided into three types. The first one provides the examination of the object to form the overall characteristics and to identify areas to be further thermographed. The second one represents an overview thermography of wall surfaces including large-scale overview thermography. Based on these two types of inspection fiducial or base areas and shooting points for the detailed thermography are selected. The third type is the detailed thermography (if necessary) - depending on the task.

The results of the review and detailed thermography are divided into qualitative and quantitative. Qualitative are limited to information contained in thermograms and used for surveys aimed at identifying defects without quantitative calculation of local heat transfer resistance. Quantitative results of thermography can be used in further calculations for the recovery of enclosing structures heat-shielding properties.

In order to obtain qualitative results one or two fiducial (base) areas are chosen in the thermogram: temperature difference there exceeded several times the accuracy of the temperature measurement by non-contact method and the thermographic camera sensitivity.

When performing the first stage of the object inspection with a thermographic camera the fiducial point is not required. The inspection is carried out to create a common quality characteristic and the identification of heat loss defects.

During the inspection the base area is determined (marked on the design by any means); its linear dimensions should exceed two thicknesses. The area is selected in the temperature zone close to the minimum at the surface.

Before and after thermography humidity and temperature of outdoor and indoor air are measured. During the survey the parameter checkout is carried out in 15-30 minutes. Additionally, the wind speed is measured at the external survey. If the thermogram is created from individual sections then in order to obtain a single image, thermography is produced with an overlap of each adjacent section by 10%.

It is recommended to expose corner sections and joints to additional thermography. The surface survey should not be in the irradiation zone within 12 hours prior to measurement. The survey points are selected so that the surface of the object measurement is in direct line of sight at an observation angle of not more than 60 °C.

3 The Results Processing of the Survey

The aim of the thermogram processing is the transition from radiation temperatures, recorded by the thermographic camera, to actual temperatures of the surface. Processing is carried out by software of the used device. It is not required to conduct any additional research using a computer if the obtained temperature values at the base areas coincide with the temperatures measured by the contact method.

 DOI: 10.1051/matecconf/20168602007

The thermogram processing is not carried out if the surveys are conducted in a simplified form at the qualitative level, for example, to monitor the condition of enclosure structures. In this case the defects with different temperatures are defined in order to make a decision on the enclosure defectiveness in terms of thermal protection. In this way, the internal structures defects are detected.

At the qualitative level the thermography is carried out under the quality monitoring of thermal insulation implementation. In this case, the heat loss is calculated. Based on the data that characterize the thermal protection quality of the enclosure structures, we can decide on carrying out additional works to reduce the heat loss of the building [3].

When interpreting the thermograms we should pay attention to the following aspects:

1. On the thermogram taken indoors, the interest area is with a lower temperature. At the color palette these are blue, purple and black, accordingly. As the thermography is a qualitative evaluation method then it is necessary to pay attention to the irregularity of temperature distribution at the areas where, according to the technical design, there are no reasons for the anomaly.
2. On the thermogram taken outdoors, the situation is reversed and defects are considered to be the areas with high temperatures, usually white, yellow and red colors.
3. The temperature value plays a role in assessing the risk of condensing - dew point, which is calculated from the temperature and relative humidity. Also for the condensate formation on the surface, the lack of air exchange is required. It should be taken into consideration that the error of the thermographic camera is +/- 2°C.

Figure 1 shows the thermogram of enclosure structure areas with defects.

a)

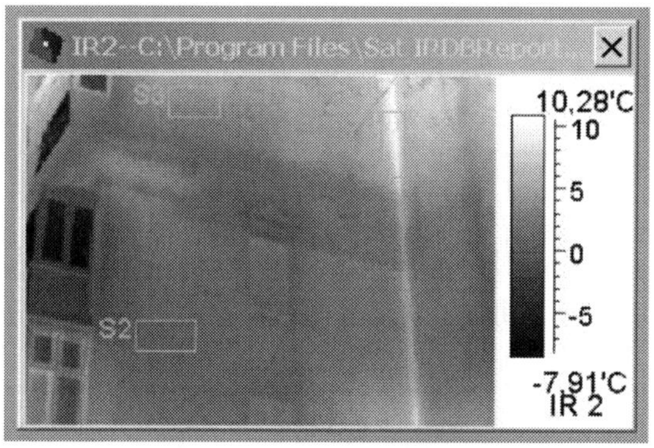

b)

Fig. 1. The thermogram of enclosure structure areas with defects: a – through crack in the wall; b – steeping the wall basement with surface deterioration.

4 Conclusion

Most of the detected thermal protection defects are not critical for the bearing capacity of building structures in general but they are important for the microclimate of separate rooms [4-7]. The low level of enclosure structures thermal protection and local thermal protection defects are the reason for comfort criteria violations [8-10]. As shown above, heat monitoring helps to detect and remove defects of buildings thermal protection [11].

References

1. V.D. Tulukanov, News Bulletin "Energosvet". 3rd (3) Edition, 14-17 (2009)
2. *GOST P 54852-2011. Thermal insulation quality control method of walling. M.: Russian Ministry of Regional Development* (2011)
3. *Design and Construction Specifications 50.13330.2012. Buildings Heat Insulation. SNiP revised edition 23-02-2003, M.: Russian Ministry of Regional Development* (2013)
4. *GOST 31937-2011. Buildings and structures. Examination and monitoring rules of technical condition, M.: Russian Ministry of Regional Development* (2011)
5. M.V. Yakovleva, E.A. Frolov, A.E. Frolov, Science and Technosphere safety: collection of articles, 394-398 (2015)
6. M.V. Yakovleva, E.A. Frolov, Traditions and innovations in construction and architecture: proceedings of the 66th All-Russian science technical conference according to the R&D, (2009)
7. M.V. Yakovleva, E.A. Frolov, Traditions and innovations in construction and architecture: Proceedings of the 68th All-Russian science technical conference according to the R&D,m935-936 (2011)
8. T.E. Gordeeva, Science Vector of Togliatti State University **1(23)**, 104-107 (2013)
9. E.A. Frolov, A.E. Frolov, Traditions and innovations in construction and architecture: Proceedings of the 71st All-Russian science technical conference according to the R&D, 913-914 (2014)
10. V.M. Balkin, Traditions and innovations in construction and architecture: Proceedings of the 71st All-Russian science technical conference according to the R&D, 917-919 (2014)
11. I.G. Frolova, Traditions and innovations in construction and: Proceedings of the 71st All-Russian science technical conference according to the R&D, 926-927 (2014)

Corrosive wear forecasting of steel elements on the basis of mathematical modeling methods

Alexey Soloviev[1,*], *Larisa* Rybakova[1], and *Olga* Kostakova[1]

[1]Samara State Technical University, Institute of Architecture and Civil Engineering, Molodogvardeyskaya str. 194, 443001 Samara, Russia

Abstract. Life Extension and resistance increase of metal materials and constructions to the corrosion destruction processes is the most important scientific and technical problem. To solve this problem it is necessary to develop complex scientific research to study the corrosion phenomena, along with practical actions against corrosion directed to selecting new corrosion resistant metal materials and methods of their protection. This research is carried out for searching mathematical model which could predict corrosive wear in metal constructions with a certain accuracy taking into account design and the type of corrosion process.

1 Introduction

Damage assessment of bearing constructions and determination of their residual resource is an important task [1,2]. Corrosion destruction is one of basic reasons of steel structures durability decrease. Researches in the field of durability increase and improvement of anticorrosive protection of building constructions are conducted in complex and include: on-site investigations, experimental and theoretical developments.

When studying development results of corrosion processes it is necessary to analyse a large number of factors. Handling of considerable amount of information and determination of common factors of damage formation due to corrosion require the use of modern computing devices and corresponding mathematical models [3]. Mathematical models development is impossible without clear idea of process mechanism, experimental data characterizing the influence of different factors on process kinetics and reliability control of the forecast methodology in natural conditions. It particularly concerns corrosion processes which are multistage. Stages can proceed in consecutive, as well as in parallel in case of various combinations of internal and external factors. Chemical composition of metal and alloy, metal surface condition, protective coatings type and condition, operation mode, hostile environment characteristics belong to such factors [4, 5-8].

2 Main Part

In the pilot study process of corrosive wear of different metalwork it is necessary to carry out the analysis of corrosion damage nature, as well as quantitative indices of wear in long

* Corresponding author: savsmr@rambler.ru

operational conditions. In this paper the installation for hydrochloric acid acceptance and pouring of the galvanic shop was chosen as a research object. The installation represents three level metal framework consisting of 8 racks (a square pipe 200x7) connected by platform beams (channel No. 14) at which boards from corrugated steel are put. On the upper platform (mark. +11.000) there were tanks with hydrochloric acid. All the constructions are made of S 235 steel (Russian standard - VSt3kp2 GOST 380-71*). Installation service life is 42 years in conditions of moderately aggressive environment [5]

During on-site investigation residual elements thickness was measured: i.e. stay braces (4 pieces) and beams (4 pieces). Forty measurements in various section points were performed on each element. The total number of measurements was 320. Survey allowed to determine the type and nature of corrosion damages. Common uneven corrosion was inherent for all the elements. In particular, pitting and localized were found on certain sites. Hence, it is possible to state that corrosion process caused the damages formation of chaotic character on the surface of studied elements. And they can be described statistically. When analyzing statistical data, the measured values are most likely, to submit to lognormal distribution law as corrosion destructions happen owing to a large number of arbitrary factors. It means that accidental process is stationary, normal and can be described by means of the least significant moments of probabilities distribution. Mean square deviation value and points deviation dispersion from the average line belong to these very probabilities. These two values describe distribution function of geometrical sizes of construction elements damaged by corrosion (thickness). Long-term practice of scientific research conducted in the field of studying of corrosion wear influence on construction showed that a true mathematical wear size distribution model of steel elements of a framework is lognormal distribution [7].

To detect distribution nature of spreading depth on the elements surface and to determine construction wear value let us carry out the statistical analysis to draw distribution curves of wear values. (Fig. 1)

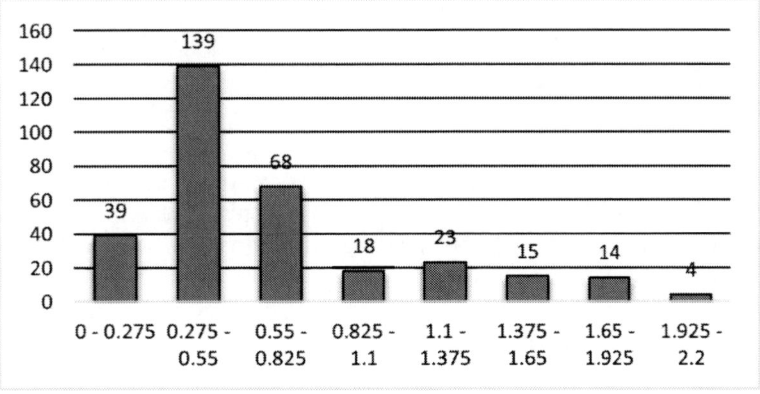

Fig. 1. Histogram of results distribution of corrosion wear measurement of elements thickness -Δt, mm. Δt=t0-t, where t0 – the initial thickness of elements, mm; t – the measured thickness of elements, mm.

Corrosion wear was determined as difference between the initial thickness of an element and residual one according to measurements. After visual assessment of this distribution it is possible to make the assumption of lognormal distribution since form and outline quite correspond to values distribution according to Gauss's law. There are several methods of justifying assumptions about distribution law of random variable. Pearson consent criteria used in case of large amount of test samples is the most widespread and it was considered in this paper. Non-negative random variable is known to be distributed logarithmically normal if the decimal logarithm of this value is distributed according to the normal law. Values of logarithmic and normal random variables are formed as "accidental misstatements" of some "true value". The last, eventually, acts not as an average value, but as a median. Let us take the logarithms of residual thickness and draw the histogram for the given values (Fig. 2,3).

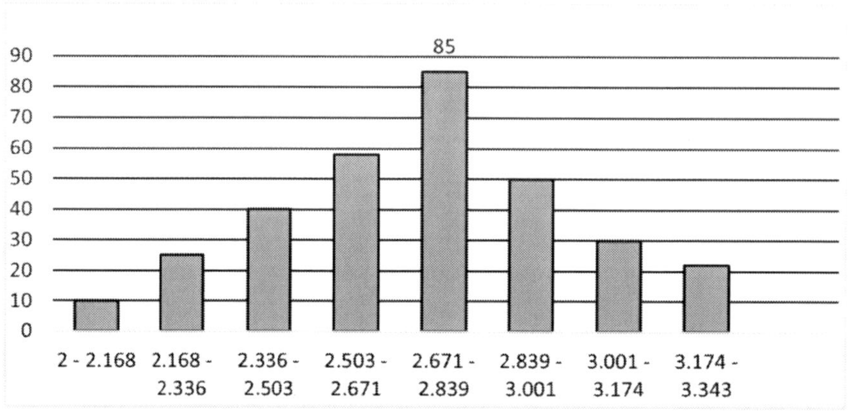

Fig. 2. Distribution type of corrosion wear values for stay braces.

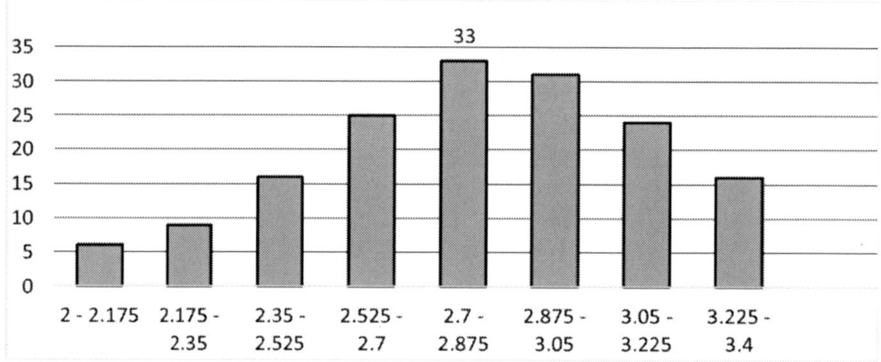

Fig. 3. Distribution type of corrosion wear values for beams.

The results received showed that values of corrosion losses of steel elements of this installation (a beam, a stay brace) are of lognormal distribution [3].

Let's figure mathematical distribution models of corrosion wear values of steel elements on the example of steel stay braces (Fig. 4).

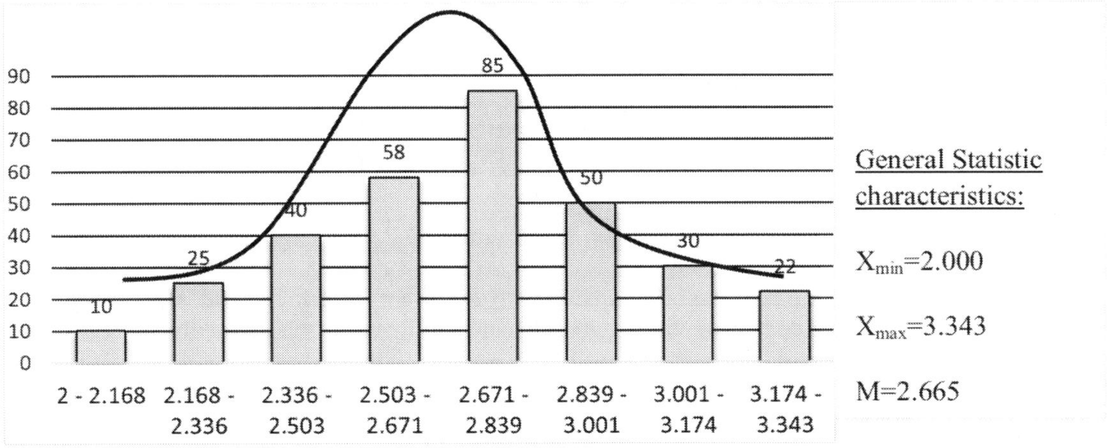

Fig. 4. A mathematical distribution model of corrosion wear of steel elements – lognormal distribution.

Lognormal distribution has its own regularities with the use of which, it is possible to define elements average wear, and also the greatest and smallest probable wear values of constructions in the given operating conditions, with a certain accuracy. Let's consider the maximum corrosion damage depth with the probability of 0,95 which is quite accepted for technical calculations. In case of lognormal distribution it will be:

$$t_{max} = 10^{m+1.65\sigma} \tag{1}$$

where m – logarithms mean value of corrosion damage depth of test samples.

For the measured stay braces:

$$m = \frac{\Sigma x_i}{N} = \frac{852.809}{320} = 2.665 \tag{2}$$

For the measured beams:

$$m = \frac{\Sigma x_i}{N} = \frac{443.81}{160} = 2.774 \tag{3}$$

where x_i – the logarithms values amount of corrosion damage depth of constructions; N – a number of measurements (N=320 – for stay braces, N=160 – for beams).

σ – root mean squared deviation of test samples logarithms is defined:

- For the measured stay braces:

$$\sigma = \sqrt{\frac{\Sigma(m-x_i)^2}{N-1}} = 0.462 \tag{4}$$

- For the measured beams:

$$\sigma = \sqrt{\frac{\Sigma(m-x_i)^2}{N-1}} = 0.409 \tag{5}$$

Thus, the maximum depth of corrosion damages, according to the statistic analysis of lognormal distribution, is equal:

For the measured stay braces:

$$t_{max} = 10^{m+1.65\sigma} = 10^{2.665+1.65\cdot 0.462} = 2680\ mkm = 2.68\ mm \quad (6)$$

For the measured beams:

$$t_{max} = 10^{m+1.65\sigma} = 10^{2.774+1.65\cdot 0.409} = 2818\ мкм = 2.82\ mm \quad (7)$$

Comparing these data we see that:

$$t^{st}{}_{max} = 2.68\ mm < t^{b}{}_{max} = 2.82\ (mm) \quad (8)$$

Difference in thickness value of corrosive wear of beams and stay braces shows a significant influence role of a geometrical arrangement of elements in space. Smaller corrosion losses in stay braces can also be explained by successfully chosen section form – the closed square profile since corrosion process takes place on open profiles more intensively (corners, channels, I-beams, etc.).

As a result of the conducted research it is stated that corrosion wear process of various elements is described with the help of exponential function quite accurately by means of which it is possible to determine wear value of metal elements from various constructive forms at the definite operation moment in strong and the moderately aggressive medium, thus, predicting their durability, appointing the optimum inter-repair periods and different anticorrosive actions. The law of lognormal distribution allows to view processes, to predict them with high precision and that can be used in case of real inspection of the constructions subjected to corrosion. The application possibility of this mathematical model for the long-term forecast of corrosion process can be performed by indirect verification, i.e. comparison of model and the research results of similar processes reviewed in literature.

References

1. N.P. Shalamov, *Inspection results analysis of the plant buildings operated in hostile environment, Works concrete and ferroconcrete institute, Release 2* (Gosstoyizdat, 1958)
2. L.Y. Rybakova, O.M. Kostakova, Section form influence on the speed of corrosion process, Student's science, Research in the field of architecture, constructions and environmental protections: theses of reports of the 33rd interuniversity student's scientific and technical conference in 2013 (2014)
3. G. Kramer, *Mathematical methods of statistics* (Moscow, 1948)
4. J. Aitchison, J. A. C. Brown, *The lognormal distribution* (Camb., 1957)
5. A.I. Kikin, etc. *Increase in durability of steel structures of plant buildings* (Stroyizdat, 1969)
6. L.Yu. Rybakova, O.M. Kostakova, Materials XVI of the International scientific and practical conference, 156-162 (2015)
7. L.Yu. Rybakova, O.M. Kostakov, Theoretical and practical problems of modern science development: collection of materials VIII of the International scientific and practical conference, 44-47 (2015)
8. N.S. Piskunov, *Differential and integrated calculations for technical colleges, v. 2: The education guidance for technical colleges* (Science, 1985)
9. V.V. Shabanin. *Features of metal constructions designing and operation in the hostile environment. Education guidance* (Kuybyshev State University, Kuibyshev, 1985)
10. N.A. Azarenkov, S.V. Litovchenko, I.M. Neklyudov, P.I. Stoyev, *Corrosion and protection of metals, Part 1, Chemical corrosion of metals. Education guidance* (HENNA, Kharkov, 2007)

11. *Set of Rules 28.13330.2012, Protection of building constructions against corrosion, Norms and Regulations II-28-73*, Ministry of Regional Development of the Russian Federation* (2012)
12. M.D. Mosesov, A.V. Solovyov, I.S. Holopov, Traditions and innovations in construction and architecture. the Electronic resource: materials of the 71st All-Russian scientific and technical conference following the results of NIR of 2013, 840-841 (2014)
13. V.Yu. Alpatov, A.V. Solovyov, Metal constructions **1(15)**, 39-46 (2009)

Reliability theory application for building structures reliability determination

Nadezhda Chiganova[1,*]

[1]National research Moscow state university of civil engineering, Yaroslavskoye sh. 26, 129337 Moscow, Russia

Abstract. Nowadays the determination of building structures and their components reliability in the variable mode is an important task. These tests are as follows: the batch of products intended for test are first tested in the first mode, and then in the second mode. A certain moment of time for mode switching is selected randomly by a specified law. The product is tested in the variable two-stage mode. During testing, moments of time of product operation in the first and second modes are registered. Two experimentation programs that differ in the way of mode switching are proposed: program tests and dynamic tests. Examples of the methodology application to the determination of the product failure rate were discussed. Besides, the approximate calculation method of the reliability characteristics in the second mode was discussed.

1 Introduction

The solution to many of the reliability theory problems [8] requires building structures and their components (hereinafter – “products”) testing in the variable mode. Such problems include, for example, stress test problems, reliability calculation in the variable mode [1-4], life utilization principles validation of the products [5], replacement of the operation modes with laboratory ones [6].

Tests in variable regime are as follows. The batch of products intended for test are first tested in the mode ε_1, and then in the mode ε_2. Moment τ of time for mode switching is selected randomly by a specified law $H(t)=P(t<\tau)$. The product is tested in the variable two-stage mode:

$\varepsilon = \begin{cases} \varepsilon_1, 0 \le t < \tau \\ \varepsilon_{2,}\tau \le t < \infty \end{cases}$, where τ – random variable distributed according to the law $H(t)$.

The article studies the problem of determining the building structures and their components reliability performance in mode ε_1 and ε_2 based on the results of the tests described. Let us set ξ_i and $\tilde{\xi}$ product failure time in the modes ε_i and $\tilde{\varepsilon}$, respectively and through F_i and $\tilde{F}$ their distribution function, i.e. $F_i(t) = P(\xi_i < t)$, i=1.2, $\tilde{F}(t) = P(\tilde{\xi} < t)$.

* Corresponding author: chiganovanm.mgsu@gmail.com

2 Determination of the building structures and their components reliability performance

Note that there is the following relationship between random variables θ_1, θ_2 and ξ_i, $\tilde{\xi}$: $\theta_1 = min\{\xi_i, \tau\}$, $\theta_2 = max\{0, \tilde{\xi} - \tau\}$.

As the characteristics θ_1, θ_2 and $\tilde{\xi}$ are observed during the test, after the experimental test distribution functions $\tilde{F}(t), \theta_i(t) = P(\theta_i < t), i = 1.2$ and $S(t) = P(\theta_2 > t/\xi_i > \tau)$ can be restored. S(t) function is the conditional probability that the product not failed in the first mode during its testing on the program $\tilde{\varepsilon}$, will operate in the second mode for no longer than t. Subsequently we distinguish between two experimentation programs that differ in the way of mode switching.

A) Program tests. They consist in the fact that before the experiments start they are specified by the distribution function H(t). Further, n random values $\mu_1, \mu_2, \ldots, \mu_n$, evenly distributed by the law are determined by the random number tables corresponding to the number n of test products. Values $\tau_i = H^{-1}(\mu_i), i = 1,2, \ldots, n$ are calculated, where H^{-1} – inverse H function. Product number i is tested first in the mode ε_1 for a time τ, then, if it does not fail – in the mode ε_2. Any function H that has the probability density h(t)>0 at all t> 0 can be chosen.

B) Dynamic tests. In these tests, the mode switching is carried out only in the product failure times. Briefly the test sequence will be as follows. At first, all products are tested in the mode ε_1. Let's assume that the first failure occurred in the time t. Then randomly selected not failed product is switched to the second mode, and the remaining n-2 continue to be tested in the mode ε_1 as long as one of them fails. Assume that it happens in the moment of time τ_2. Again, randomly selected not failed product is switched to the second mode, and the remaining n-4 continue to be tested in the first mode, etc. It is evident that in this way each product will be tested in the variable mode of (I) type.

So, we will proceed from the fact that similar tests were carried out and, based on their results, distributions Q_i, S were obtained. First let us set a goal to find the distribution of products failure-free operation in the first mode $\bar{F}_1(t) = 1 - F_1(t)$ by received Q_i, S .

1. Let the experiment be carried out under program A) and based on its results function H, Q_i, S are obtained. It follows from the definition of random variable θ_1 that

$$\begin{aligned}\theta_1(\mathrm{e}) &= P(\theta_1 < t) = P(min\{\xi_1, \tau\} < t) = 1 - P(\xi_1 > t, \tau < t) \\ &= 1 - [1 - F_1(t)][1 - H(t)] = 1 - \bar{F}_1(t)\bar{H}(t), \bar{Q}_1(t) \\ &= \bar{F}_1(t)\bar{H}(t)\end{aligned} \quad (1)$$

From formula (2) we obtain the reliability function in the first mode

$$\bar{F}_1(t) = \frac{\bar{Q}_1(t)}{\bar{H}(t)}, \bar{F}_1(t) = 1 - F_1(t) = P(\xi_1 > t) \quad (2)$$

We shall write the numerical characteristics of this distribution: mathematical expectation and variance of random variable:

$$M(\xi) = \int_0^\infty \bar{F}_1(t)dt = \int_0^\infty \frac{\bar{Q}_1(t)}{\bar{H}(t)}dt, D(\xi_1) = M(\xi_1^2) - M^2(\xi)$$

It should be noted that

$$M(\xi_1^2) = 2\int_0^\infty t\,\bar{F}_1(t)dt = 2\int_0^\infty \frac{\bar{Q}_1(t)}{\bar{H}(t)}dt$$

2. Let the tests be carried out under the programs B). In this case the switching distribution H (t) will not be known because it coincides with $F_1(t)$. Applying $H=F_1$ in (2), we obtain

$$F_1(t) = \sqrt{\bar{Q}(t)} \tag{3}$$

Formulas (2) and (3) show that in order to determine the product reliability characteristics in the first mode it is sufficient to find the lifetime distribution Q_1 based on the test conducted results θ_1.

Although formulas (2) and (3) are simple, their use in practice may give rise to some difficulties. In fact, after the experimental study failure time will be known, and it is difficult to determine the distribution type by this data.

Therefore, Q_1 functions for a given failures distribution of F_1 and H are a matter of interest.

3 Examples

1. Let the products fail in the first mode exponentially $F_1(t) = 1 - e^{-\lambda_1 t}$ and switching points are also set exponentially $H(t) = 1 - \exp(-\lambda_2 t)$. Then $\bar{Q}_1(t) = \bar{F}_1(t)H(t) = exp[-(\lambda_1 + \lambda_2)t]$.

$$M(\theta_1) = \int_0^\infty \bar{Q}_1(t)dt = \int_0^\infty exp[-(\lambda_1 + \lambda_2)t]dt = \frac{1}{\lambda_1 + \lambda_2}$$

We deduce from the last equation the failure rate in the first mode:

$$\lambda_1 = \frac{1 - \lambda_2 M(\theta_1)}{M(\theta_1)}$$

2. Failures time is distributed by Poisson law: $F_1(t) = t^n e^{-t}/n!$ and switching times are chosen exponentially: $H(t) = 1 - \exp(-\lambda t)$. In this case:

$$\bar{Q}_1(t) = e^{-\lambda t}\left(1 - \frac{t^n e^{-t}}{n!}\right), M(\theta_1) = \int_0^\infty \left[e^{-\lambda t} - \frac{t^n e^{-t(1+\lambda)}}{n!}\right] dt = \frac{1}{\lambda} - \frac{1}{(1+\lambda)^{n+1}},$$ where

$$(1 + \lambda)^{n+1} = \frac{\lambda}{1-\lambda M(\theta_1)},\ n = log_{(\lambda+1)}\left[\frac{\lambda}{1-\lambda M(\theta_1)}\right] - 1.$$

4 Determination of the failures distribution in dynamic regime

Let us now determine the failures distribution F_2 in the second mode. Note that the F_2 function can not be clearly determined according to Q_i and H data obtained after the tests. It is necessary to have additional information on the products life utilization in the variable mode.

Currently, many product functional loss models were proposed [1-5]. We select the most widespread model proposed by N.M. Sedyakin [6]. According to this model, the products reliability depends on its life span value in the past and does not depend on the manner (mode, time) of the life span end. N.M. Sedyakin proposes to select the cumulative failure rate as a life span measure. Sedyakin principle mathematically may be written as follows.

If $P(t_1, \varepsilon_1) = P(t_2, \varepsilon_2)$, then the equality $P(t, \varepsilon_2/t_1, \varepsilon_1) = P(t, \varepsilon_2/t_2, \varepsilon_2)$ is valid. Here $P(t, \varepsilon_2/t_i, \varepsilon_i), i = 1, 2$, it means the probability of failure-free operation for a time t in the mode ε_2 for such products which have been already operated in the mode ε_i without a failure for a time t_i.

Based on Sedyakin principle it is easy to solve the inverse problem of determining distribution functions F_1, F_2 and H $S(t) = P(\theta_2 > t/\xi_1 > \tau)$. In fact, $S(t/x) = P(\theta_2 > t/\xi_1 > \tau, \tau = x)$.

By the total probability formula

$$S(t) = P(\theta_2 > t/\xi_1 > \tau) = \int_0^\infty P(\theta_2 > t/\xi_1 > \tau, \tau = x)\, dH(x) = \int_0^\infty S(t/x) dH(x) \quad (4)$$

The conditional distribution $S(t/x)$ is uniquely adjusted by F_1 and F_2 functions based on Sedyakin principle. Let us denote $\hat{x}$ as the solution of equation

$$F_1(x) = F_2(\hat{x}) \quad (5)$$

According to Sedyakin principle, product testing in the mode ε_1 for a time x equivalent to test in the mode ε_2 and duration $\hat{x}$. Therefore,

$$S(t/x) = P(\xi(\bar{\varepsilon}) > t + x/\xi_1 > x) = P(\xi_2 > t + \hat{x}/\xi_2 > x) = \frac{\bar{F}_2(t+\hat{x})}{\bar{F}_2(\hat{x})} \quad (6)$$

After substituting (5) and (6) in (4) we obtain

$$S(t) = \int_0^\infty \frac{\bar{F}_2(t+\hat{x})}{\bar{F}_2(\hat{x})} dH(x)$$

Or by replacing $\hat{x} = \bar{F}_2^{-1}(\bar{F}_1(x))$ in accordance with (6)

$$S(t) = \int_0^\infty \{\bar{F}_2[t + \bar{F}_2^{-1}(\bar{F}_1(x))]/\bar{F}_1(x)\} dH(x) \quad (7)$$

Thus, the distribution $\bar{F}_2$ is defined as the solution of the integral equation (7).

In case of dynamic tests $H = F_1$ formula (7) can be simplified by introducing a new variable of integration $\xi = \bar{F}_1(x)$, that enables to reduce (7) to the form

$$S(t) = \int_0^1 \{\bar{F}_2(t + \bar{F}_2^{-1}(\xi)/\xi\} d\xi$$

The solution of this integral equation poses a challenge, because the element of integration includes the inverse function $\bar{F}_2^{-1}$. This disadvantage can be eliminated if instead of distribution F_2 the failure rate $\lambda_2(x) = -\frac{d ln F_2}{dx}$ will be found. It is easy to verify, if we pass to a new variable of integration $\eta = \bar{F}_2^{-1}(\xi)$.

Then $\xi = \bar{F}_2(\eta)$, $d\xi = -\lambda_2(\eta)\bar{F}_2(\eta)d\eta$ and, hence,

$$S(t) = \int_0^\infty exp\left[-\int_0^{t+\eta} \lambda_2(x)dx\right] \lambda_2(\eta)d\eta$$

In general, the solution of equation (7) may be carried out by a computer.

Let us consider the approximate calculation method of the reliability characteristics in the second mode by the known functions S, Q_i and H. Let $Q(t/x) = P(\xi_2 < t/\xi_1 = x)$ – conditional distribution of failures in the mode ε_2 of product, which in the first mode would have operated exactly the time x. By setting the function Q distributions of F_2 and S are clearly determined. Indeed,

$$F_2(t) = \int_{-\infty}^{\infty} Q(t/x)dF_1(x) \quad (8)$$

Assuming Q as a constant during the products tests in the mode ε_1, we can write

$$S(t/x) = P(\xi_2 > t/\xi_1 = x) = \int_0^{\infty} [1 - Q(t/y)]\frac{dF_1(y)}{F_1(x)} = \frac{1}{F_1(x)}\int_0^{\infty} dF_1(y) \int_0^{\infty} \psi_t(z)dQ(z/x)$$

$$\text{where } \psi_t(z) = \begin{cases} 0, z < t, \\ 1, z \geq t. \end{cases}$$

By inserting $S(t/x)$ into (4) we obtain

$$S(t/x) = \int_0^{\infty} \frac{dH(x)}{F_1(x)} \int_x^{\infty} dF_1(y) \int_0^{\infty} \psi_t(z)dQ(z/x)$$

Or, changing the order of integration

$$S(t/x) = \int_0^{\infty} dF_1(y) \int_0^{y} \frac{dH(x)}{F_1(x)} \int_0^{\infty} \psi_t(z)dQ(z/x) \quad (9)$$

Let us set $\Pi(x) = \int_0^{\infty} dF_1(y) \int_0^{y} \frac{dH(\xi)}{F_1(\xi)}$, $I_Q^t(x) = \int_0^{\infty} \psi_t(z)dQ(z/x)$.

Then (9) can be presented as

$$S(t) = \int_0^{\infty} I_Q^t(x)\, d\Pi(x) \quad (10)$$

Since the bivariate function Q is not known and can not be determined in principle based on the conducted tests results (because the simultaneous failure time measuring of the same product in different modes is not possible), then F_2 distribution is ambiguously determined by setting the function S. In this regard, let us find the lower F_{2*} and upper F_2^* boundaries for F_2 under condition (10), considering F_2 as functional of Q. Let $M(F_1, F_2)$ and $R(S, \Pi)$ are set of all distributions Q that satisfy equations (8) and (10), respectively. In addition, let us introduce the set L_t of all conditional distributions for which $I_Q^t(x)$ is a non-decreasing function x at any fixed t.

Let's assume that $Q \cap_{t \in x} L_t \equiv L$. Physically, it means that the longer the product operates in the first mode, the more reliable in the mode ε_2 it is.

From the mathematical point of view, the evaluation of F_2 distribution is reduced to the variational problem associated with the determination of absolute minimum $F_{2*}(t) = \inf_{Q \in X} F_2(t)$ and absolute maximum $F_2^*(t) = sup_{Q \in X} F_2(t)$ of F_2 functional on the set $X = L \cap M$. Because functionals F_2, I_Q^t, $S(t)$ are linear in Q, the duality principle is valid. According to this principle

$$\left[\int Q_* dF_1 = \inf_{Q \in X} \int Q dF_1\right] <=> \left[\int I_Q d\Pi = \inf_{Q \in X} \int I_Q d\Pi\right],$$
$$\left[\int Q^* dF_1 = sup_{Q \in X} \int Q dF_1\right] <=> \left[\int I_Q d\Pi = sup_{Q \in X} \int I_Q d\Pi\right],$$
$$Q_*, Q^* \in X \cap S.$$

This principle allows the functional extremums $S(t)$ to be found instead of the variational problem on the set $M \cap L$. Therefore, let us find $S_*(t)$:

$$S_*(t) = \inf_{Q \in X} \int_0^{\infty} I_Q^t(x)\, d\Pi(x),\ S^*(t) = \sup_{Q \in X} \int_0^{\infty} I_Q^t(x)\, d\Pi(x).$$

Decision algorithm for this variational problem is given in [9]. According to this algorithm, let us form the probabilities ratio

$$\gamma(x) = d\Pi\,(x)/dF_1(x) = \int_0^x (1/\bar{F}_1(\xi))dH(\xi)$$

Because $\gamma(x)$ is a non-decreasing function, then in virtue of Theorem (3) [9]

$$Q_*(t/x)\begin{cases}0, t < t_*(x)\\ 1, t \geq t_*(x)\end{cases}, Q^*(t/x) = F_2(t) \tag{11}$$

Where $t_*(x)$ – solution of equation $F_2(t_*) = F_1(t)$.

By inserting (11) into (10) we obtain:

$$S(t) = \int_0^\infty dF_1(y)\int_0^y \frac{dH(x)}{\bar{F}_1(x)} x[t_*(x) - t], x(t) = \begin{cases}0, t \leq 0\\ 1, t > 0\end{cases} \tag{12}$$

$$S(t) = \int_0^\infty dF_1(y)\int_0^y \frac{dH(x)}{\bar{F}_1(x)} \bar{F}_2(t) \tag{13}$$

Because $\int_0^\infty dF_1(y)\int_0^y \frac{dH(x)}{\bar{F}_1(x)} = 1$, from (13) it follows that $\bar{F}_2^*(t) = S(t)$.

By inserting $t_*(x) = F_2^{-1}[F_1(x)]$ into (12) we obtain

$$S(t) = \int_0^\infty dF_1(y)\int_0^y \frac{dH(x)}{\bar{F}_1(x)} xF_2^{-1}[F_1(x) - t]$$

It is evident that $F_2^{-1}[F_1(x)] \geq t$.

Then $x \geq F_2^{-1}[F_1(x)]$, $S(t) = \int_{F_1^{-1}[F_2(t)]}^\infty dF_1(y)\int_{F_1^{-1}[F_2(t)]}^\infty (1/F_1(x))dH(x)$.

Let us change the order of integration.

$$S(t) = \int_{F_1^{-1}[F_2(t)]}^\infty (1/F_1(x))dH(x)\int_x^\infty dF_1(y) = \int_{F_1^{-1}[F_2(t)]}^\infty dH(x)$$

It follows that $S(t) = \bar{H}[F_1^{-1}(F_2(t))]$ and $F_{2*}(t) = F_1[H^{-1}(S(t))]$. In case of dynamic tests $H = F_1$ the formula will be $F_{2*}(t) = S(t)$.

Thus, the distribution of failure-free products operation in the second mode is within the boundaries

$$\bar{F}_1[H^{-1}(\bar{S}(t))] \leq \bar{F}_2(t) \leq S(t) \tag{14}$$

Function $\bar{F}_1$ included in (14) is determined by formulas (2) and (3).

5 Conclusions

In the article of I.A. Ushakov "Is reliability theory still alive?" [8] the definite answer is given – it is alive. Today the reliability theory becomes increasingly important because the quality and reliability determination in the variable mode for building structures and their components is frequently required.

The results obtained in this article can simplify the determination of the quality and reliability of building structures and their components during testing modes switching.

References

1. B.V. Gnedenko, J.K. Beljaev, A.D. Solov'ev, *Mathematical methods in reliability theory* (Nauka, 2010)

2. N.M. Chiganova, Natural and Technical Sciences **6**, 14–17 (2015)
3. V. Medvedev, A. Pustovgar, Applied Mechanics and Materials **725-726**, 337–382 (2015)
4. G.D. Kartashov, N.M. Chiganova, Journal of Mathematical Sciences **39**, 2578–2588 (1987)
5. G.D. Kartashov, *Elements and schemes accelerated testing (monograph)* (Knowledge, 1990)
6. N.M. Sedjakin, Technical Cybernetics - USSR Academy of sciences proceedings **3**, 80-87 (1968)
7. A.I. Kubarev, I.Z. Aronov, O.I. Teskin et al., *Methodical guidelines. The reliability of the technique. Methods for assessing the reliability of indicators based on experimental data* (Standards Publishing House - USSR Gosstandart, 1991)
8. I.A. Ushakov, Reliability: Theory & Applications **2**, 6–19 (2007)
9. G.D. Kartashov, Theory of Probability and its Applications **14(4)**, 595–611 (1969)

Modal properties of suspension electricity pylons

Zofia Kozyra[1] , *Sławomir* Dudziak[1,*]

[1] Warsaw University of Technology, Faculty of Civil Engineering, al. Armii Ludowej 16, 00-637, Warsaw, Poland

Abstract. The paper presents the results of the modal analysis of suspension electricity pylons (S12P, S24P, Y52P). The results were obtained from FEM models built in the Abaqus software. On this basis, there were proposed formulas to predict the first two natural frequencies for pylons S12P and S24P with good accuracy.

1 Introduction

Power lines are essential for the energetic security of each country. To increase the security of transmission a recorder has been constructed as a result (as an outcome) of the SDZP project [The System of Dynamic Management of Transmission]. The recorder monitors the deformation of the lines in real-time taking into account the electricity pylons – conductors interaction.

To determine the influence of this interaction 17 electricity pylon models were prepared by the team from the Warsaw University of Technology - the Faculty of Civil Engineering. Static and modal parameters were calculated for each model. The functions from which the first two natural frequencies can be obtained approximately for suspension pylons were proposed on the basis of the data gathered in the project.

2 The aim of natural frequencies prediction

In linear dynamics the dynamic response of a structure is a linear combination of eigenvectors with weights, which depend on the corresponding eigenvalues. The higher the number of the eigenvalue, the smaller the weight is in most cases (when the resonance or beat phenomenon does not occur). Therefore, the first few eigenmodes are sufficient to describe vibrations of a structure with the required accuracy [1]. Many methods of solving the linear dynamics problem use this fact because of their low numerical cost compared with the direct integration of motion equations methods [1 - 4].

For engineering applications the first natural frequency is often the most important, so many standards contain formulas to predict it. The first example is standard [5], in which the building's natural period of vibrations is associated with its height and the coefficient taking into account the type of construction. The second example is standard [6], in which

* Corresponding author: s.dudziak@il.pw.edu.pl

the fundamental frequencies of different structures are related to their parameters such as height, mass, dimensions of their cross-sections.

The quoted examples show that engineers aim for formulas in which the prediction of the first eigenvalue is associated with the basic geometric parameters of the structure. In this paper, the formulas linking the first two natural frequencies with height for suspension towers are proposed. In paper [7] it is proved that dynamic effects cannot be omitted if a precision device is planned to be mounted on the cross arm of a pylon.

3 Electricity pylons and their numerical models

The paper concerns only suspension pylons, since the formulas to predict the fundamental frequency for tension towers cannot be found easily due to the variety of their design solutions, e.g. dimensions of the column projection, number of transverse ties.

Three suspension pylons of a single power line in different height versions are analysed in this paper: S12P, S24P, Y52P. All of the pylons were designed as 3D lattice frames made of steel angles with equal and non-equal sides. They consist of: columns, cross arms and arms supporting ground wires. The basic information about each pylon is summarised in Table 1.

Table 1. Characteristics of poles.

Pylon	Height version	Total height [m]	Arrangement of conductors	Shape
S12P	Basic	22.8	Triangular	T-shaped
	+2.5	25.3		
S24P	-2	20.1	Triangular	T-shaped
	Basic	22.1		
	+5	27.1		
	+10	32.1		
Y52P	-4	27.3	Flat	Y-shaped
	Basic	31.3		

Pylon models were built in the Abaqus FEM software using b33 elements (Euler-Bernoulli beam [2-3]) as 3D frames with rigidly connected bars. The linear elastic material model of steel was assumed in the analysis. The material parameters: Young modulus E = 200 GPa, Poisson's ratio μ = 0.3. In paper [8] the authors proved that the introduction of pinned joints in the pylon models changes the first two natural modes insignificantly. The tie rods were modelled as truss elements (symbol t3d2) made of a non-linear material which has no compressive stiffness. All the structure models have pinned supports. The models in the basic height version of each analysed pylon are shown in Figure 1.

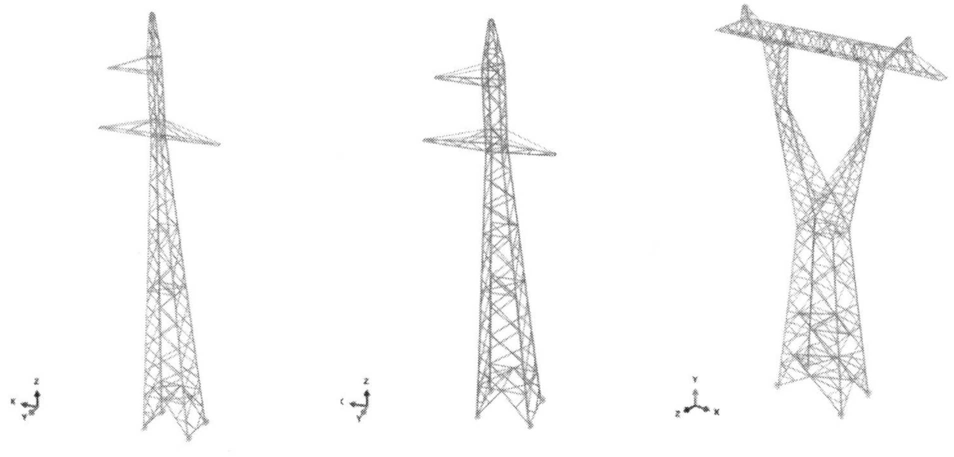

Fig. 1. The numerical models of the pylons: S12P, S24P, Y52P (left to right).

4 Modal analysis

Every linear FEM model used in dynamic analysis is described by two symmetric and positive definite matrices: the stiffness matrix - **K** and the mass matrix – **M** [1]. The structural eigenvalue problem consists in finding the pairs of eigenvalues - ω and eigenvectors **a**. It is described by the following formula:

$$(\mathbf{K} - \omega^2 \mathbf{M})\,\mathbf{a} = \mathbf{0}. \tag{1}$$

The relation between the angular frequency ω and the normal frequency f is as follows:

$$f = \frac{\omega}{2\pi}. \tag{2}$$

Some lowest eigenvalues and corresponding eigenvectors were extracted for the analyzed pylon models using the subspace iteration method implemented in the Abaqus software [2-3].

The plots which show the relation between the first three (the lowest) natural frequencies and the height of the pylons are presented in Figures 2 - 4.

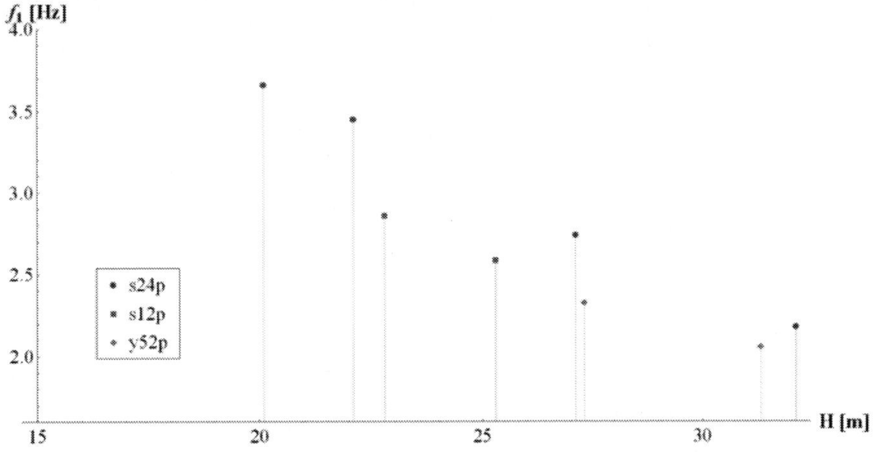

Fig. 2. The plot of relation between the first natural frequency and the height of the pylon.

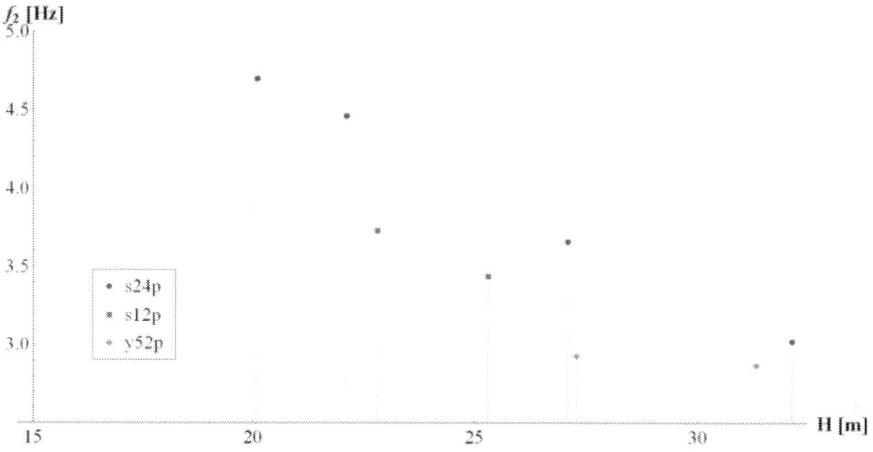

Fig. 3. The plot of relation between the second natural frequency and the height of the pylon.

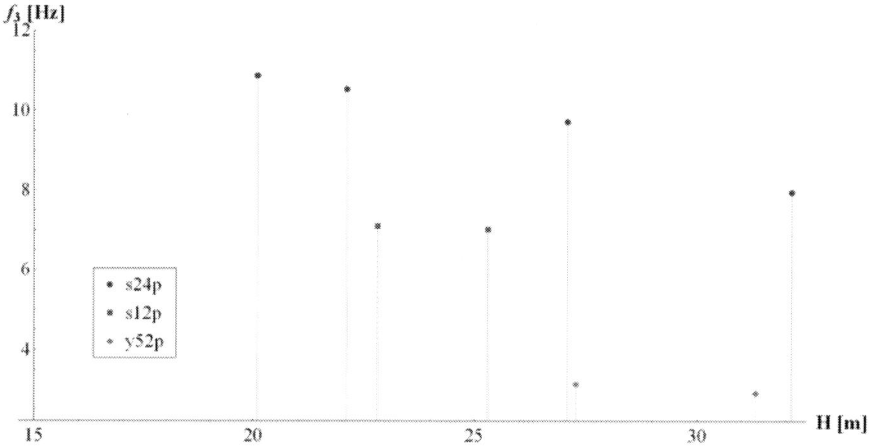

Fig. 4. The plot of relation between the second natural frequency and the height of the pylon.

Figures 5-7 present the first three eigenvectors for the basic height versions of the analysed pylons.

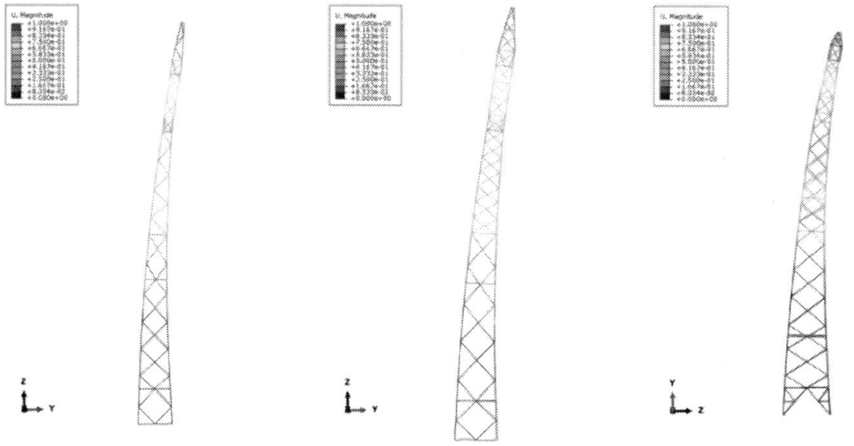

Fig. 5. The first eigenvector for the pylons: S12P, S24P, Y52P (left to right).

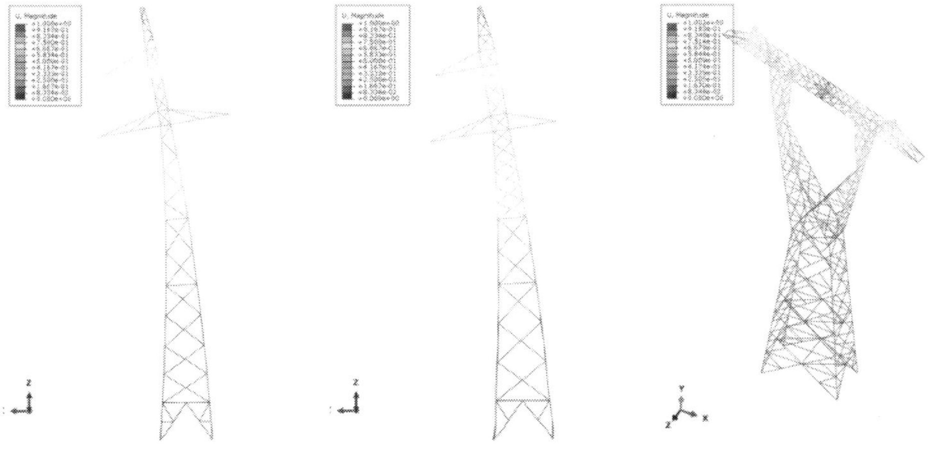

Fig. 6. The second eigenvector for the pylons: S12P, S24P, Y52P (left to right).

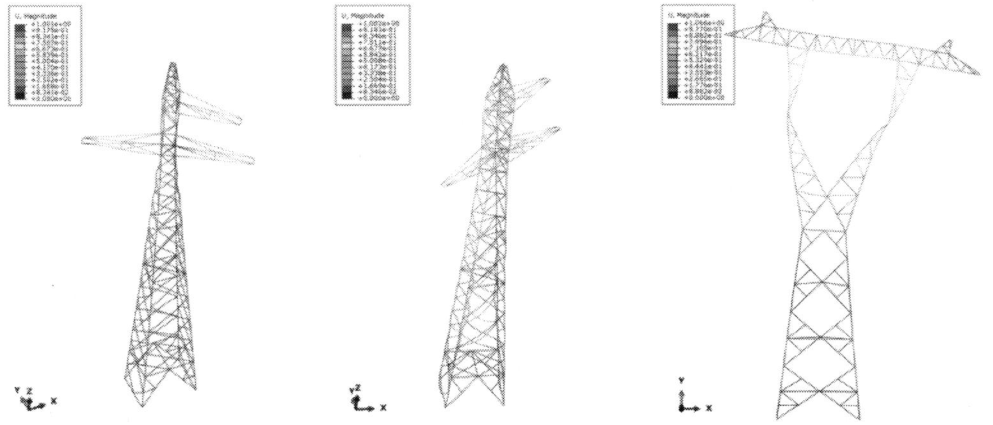

Fig. 7. The third eigenvector for the pylons: S12P, S24P, Y52P (left to right).

5 Formulas for the approximate natural frequencies

On the basis of the analysis of Figures 2 – 4 the following conclusions can be drawn. The relation between the first two natural frequencies (numbers $i = 1,2$) and the height (H) for the S24P pylon is linear so it can be written as:

$$f_i(H) = A \cdot H + B. \quad (3)$$

Coefficients A and B were found using the method of least squares. After substituting them in the formula (3) the first two natural frequencies can be approximated from the formulas:

$$f_1(H) = -\,0.126 \cdot H + 6.198 \quad (4)$$

$$f_2(H) = -\,0.143 \cdot H + 7.578. \quad (5)$$

The height (H) must be substituted in metres [m] and greater than 20 m in the formulas above. Their results have unit [Hz].

The tangent of the angle of the line drawn through the points for the S12P pylon is similar to coefficient A found for the S24P pylon. Hence, it is proposed to conduct the following algorithm for the S12P pylon:

- to substitute coefficient A calculated for the S24P pylon,
- to find coefficient B from the linear equation formed by assuming the height and the natural frequency for one height version,
- to compare the values for the other height version obtained from the FEM model and the proposed formula.

This approach produced the following result for the S12P pylon:

$$f_1(H) = -0.126 \cdot H + 5.732 \tag{6}$$

$$f_2(H) = -0.143 \cdot H + 6.990. \tag{7}$$

The results of the considerations above are shown in Figures 8 – 9. The error of approximated formulas for the S12P pylon is 0.05 Hz for the fundamental frequency and 0.03 Hz for the second natural frequency.

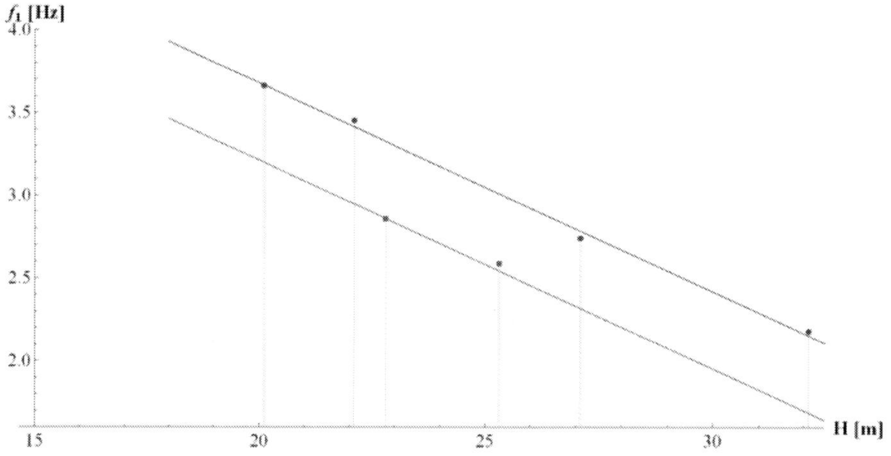

Fig. 8. The fundamental frequency for the S24P and S12P pylons obtained from FEM models and the proposed formulas.

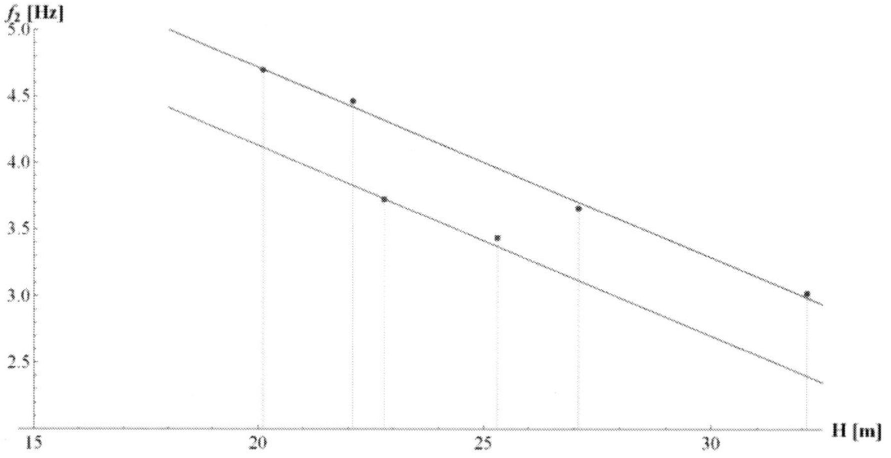

Fig. 9. The second natural frequency for the S24P and S12P pylons obtained from FEM models and the proposed formulas.

The proposed formulas provide good estimation accuracy of the first two eigenvalues. The first two modes have bending character. On the basis of the Figure 4 analysis it can be concluded that the third eigenvalue (for the torsion mode) is not associated with the height of the pylon, but probably with the geometric properties of the cross arms.

6 Conclusions

On the basis of the results for two height versions of a suspension pylon, there can be found the formula to estimate the first two natural frequencies of the other versions. Good accuracy was obtained for the exemplary two towers – S12P, S24P by using the linear functions (4) - (6), which have the same directional factor.

The functions for tension towers would be more complex and have more variables. The height cannot be the only factor taken into account. The influence of properties like the cross arm span or proportions of the tower should be analysed as well.

The paper is a result of the SDZP project [System of Dynamic Management of Transmission], which was financially supported by the National Centre for Research and Development and the National Fund for Environmental Protection and Water Management in the GEKON contest (NCBR/214108).

References

1. R. Lewandowski, *Dynamic of building structures* (Poznań, 2006, in Polish)
2. *Abaqus Analysis User's Manual* (Dassault Systèmes, 2012)
3. *Abaqus Theory Manual* (Dassault Systèmes, 2012)
4. K.J. Bathe, *Finite element procedures* (Prentice Hall, 1996)
5. *EN 1998-1 Eurocode 8: Design of structures for earthquake resistance - Part 1 : General rules, seismic actions and rules for buildings*
6. *EN 1991-1-4 Eurocode 1: Actions on structures - Part 1-4: General actions - Wind actions*
7. S. Dudziak, Z. Kozyra, TTS **12**, 2463-2467 (2015)
8. S. Dudziak, Z. Kozyra, *Theoretical Foundations of Civil Engineering*, vol. **23**, W. Szcześniak, ed., 79-88 (2015)

Numerical estimation of concrete beams reinforced with FRP bars

Kostiantyn Protchenko[1,*], *Krystian* Młodzik[2], *Marek* Urbański[1], *Elżbieta* Szmigiera[1] and *Andrzej* Garbacz[1]

[1]Warsaw University of Technology, Department of Civil Engineering, Al. Armii Ludowej 16, 00-637 Warsaw, Poland
[2]MK Projekt – Construction and Design Office, Plac Wolności 12, 46-100 Namysłów, Poland; KGE LLC, ul. Krokwi 32/25, 03-114 Warsaw, Poland

Abstract. This paper introduces numerical investigation on mechanical performance of a concrete beam reinforced with Fibre Reinforced Polymer (FRP) bars, which can be competitive alternative to steel bars for enhancing concrete structures. The objective of this work is being identified as elaborating of reliable numerical model for predicting strength capacity of structural elements with implementation of Finite Element Analysis (FEA). The numerical model is based on experimental study prepared for the beams, which were reinforced with Basalt FRP (BFRP) bars and steel bars (for comparison). The results obtained for the beams reinforced with steel bars are found to be in close agreement with the experimental results. However, the beams reinforced with BFRP bars in experimental programme demonstrated higher bearing capacity than those reinforced with steel bars, which is not in a good convergence with numerical results. Authors did attempt to describe the reasons on achieving experimentally higher bearing capacity of beams reinforced with BFRP bars.

1 Introduction

In occasion, design service life of structures is drastically affected by highly corrosive environment. The corrosion of steel bars is a material problem rather than structural one. Instead of using common practices to eliminate corrosion, more durable materials can be used.

The premise of fibre reinforced polymer FRP composites lies in their high-strength, lightweight, noncorrosive, nonconducting, and nonmagnetic properties. In addition, FRP manufacturing, using various cross-sectional shapes and material combination, offers unique opportunities for the development of shapes and forms that would be difficult or impossible with conventional steel materials [1]. The widespread implementation of FRP as a reinforcement for reinforced concrete elements requires: a comprehensive understanding of how each of these materials behaves alone as well as the behavior of the complete structural system [2].

* Corresponding author: k.protchenko@il.pw.edu.pl

In a certain circumstances, using of FRP bars can be considered as the substitute material to conventional steel in concrete detailing. If FRP is utilized as the internal reinforcement for concrete structures, it can provide high specific strength compared to the steel reinforcement [3, 4].

Current situation on the market shows that using of FRP will be confined to applications where their unique characteristics will be the most appropriate. Exist common practice to use FRP bars in a range of structural applications, such as: prestressed concrete structures, foundations, road surfaces, parking, and bridges.

The selection of the fibres is primarily based on consideration of characteristics on strength, stiffness, durability, and cost. Nevertheless, BFRP bars characterised as a new variety of material which mechanical properties are not completely investigated.

2 Mechanical Behaviour of the FRP Bar

Mechanical properties of the bar are mostly characterised by properties of their constituents i.e. fibres and matrices (described in the Figure 1). Fibres can be characterized as continuous, non-intersecting, directionalised, non-metallic and fire resistant. Basalt fibres have a longitudinal/transverse modulus ratio close to 1. Epoxy resin has viscoelastic properties and it is assumed to use it as a polymer matrix. According to test results in [5] FRP bars with epoxy resin demonstrated better physical and mechanical characteristics and lowest degradation rate after conditioning in alkaline solution than bars with other tested types of matrices.

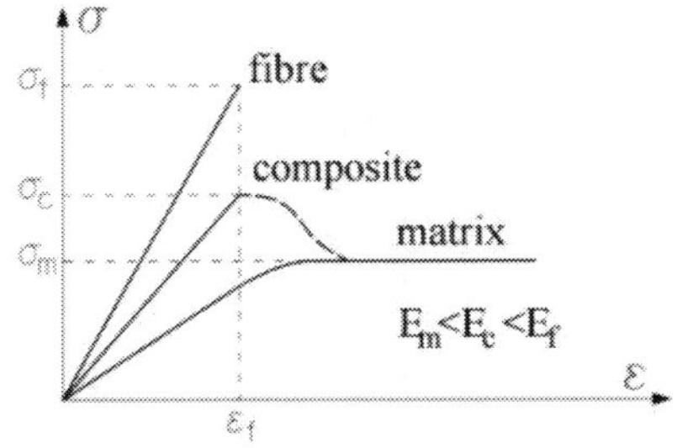

Fig. 1. Stress-strain behaviour of constituents and composite materials.

The BFRP bar can be produced in different arrangements of fibres, which then are bounded intimately in polymer matrix. Configuration of fibres defines the number of material` constants, and for the BFRP bars is assumed as hexagonal one, which can be related to the case of transverse isotropy. Detailed mechanical parameters of different bar constituents are described in [6, 7, 8, 9].

It is verified from the experimental data that for the analysis of FRP properties in the longitudinal direction (direction of fibres) the Voigt Model can be used [10, 11, 12] and for the transverse direction (perpendicular direction to direction of fibres) semi-empirical models, such as Halpin-Tsai model, which takes into account experimental data regarding bond behaviour between fibres and matrices, can be used [12].

3 The Experimental Programme

The experimental tests were performed in the laboratory of Warsaw University of Technology and described in detail in [13]. An experimental programme consisted of six beams made of concrete class *C30/37* with the dimension of $b \times h \times L = 80 \times 140 \times 1200$ *mm*, which were subjected to flexural tests. Figure 2 shows the beam structure. The actual average strength of concrete was tested on cubic samples - $f_{ck,cube} = 41.02$ *MPa*. Three of them were reinforced at the bottom with BFRP bars of *8 mm* diameter. For comparison, three reference RC beams with bottom steel reinforcement of the same diameter were prepared. Modulus of elasticity of basalt bars was determined based on experimental data $E_f=39,05GPa$ [5] and Modulus of elasticity for steel bars was assumed as *200 GPa*.

The support regions of the beams were reinforced at top by two steel bars of *8 mm* diameter and steel stirrups, however the middle part of beams were without stirrups and top reinforcement. In the beams, which were reinforced with BFRP bars, the central bottom basalt bar was made to protrude on both ends to enable the measurement of the slip in the process of loading. Clear cover were assumed as *20 mm*. The 7 pairs of bench-marks, with spacing *20 mm*, were arranged on the side surface of the beam. Registration of concrete strains was made with a mechanical extensometer with a measuring length of *100 mm*.

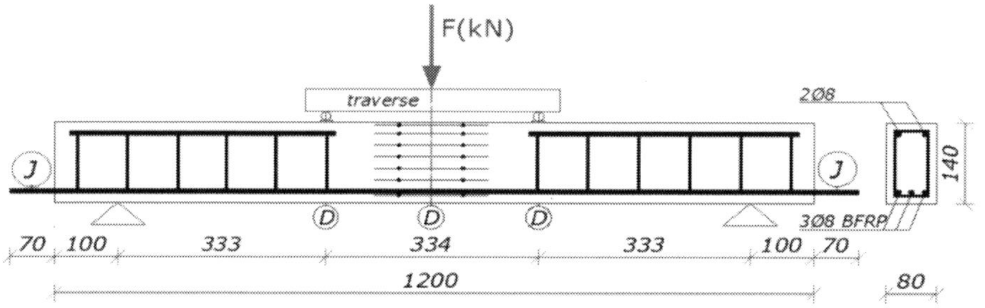

Fig. 2. The structure of tested beams with BFRP reinforcement at the bottom, J - slip measurement sensor, D - deflection gauge, dimensions in mm [13].

Loading was placed in the four point system made of steel traverse, respectively in the third and two thirds on the beam span. Loading was performed in several cycles. During the first cycle of loading, the beams were subjected to load equal to *10 kN* and then the load was reduced to *5 kN*. In each following cycle loading was increased by *10 kN*, and then reduced again to *5 kN*, till the failure of the structure.

Table 1 describes the deflections in the mid-span depending on the level of loading.

Table 1. Deformational characteristics of tested beams obtained during the experiments [13].

Applied force, kN	**Deflections, mm**					
	SRC1	SRC2	SRC3	BFRP1	BFRP2	BFRP3
5	0.97	1.81	1.61	3.21	3.04	3.10
10	1.71	2.23	2.06	4.35	4.11	4.23
20	2.45	3.30	3.14	7.33	6.99	7.21
30	3.48	4.43	4.24	12.63	11.58	12.01
40	-	-	-	-*	19.54	19.03

*The failure of the beam due to shear

Experimental results show that beams reinforced with BFRP bars have greater deflections than beams, which are reinforced with steel bars. That is mainly because the BFRP bars have significantly lower modulus of elasticity.

The Figure 3 shows the final phase of the beam "BFRP3" with the load $F_U=45$ *kN*. It is noteworthy that there was no rupture of the flexural basalt bars, and therefore they have not reached tensile strength. The destruction of beam took place only by shear in support zone and it had brittle nature.

Fig. 3. The destruction phase of the beam reinforced with BFRP (BFRP3) [13].

Table 2 describes the flexural capacity for beams with different reinforcement, namely SRC for the beams with steel reinforcement at the bottom, BFRP for the beams reinforced with BFRP bars at the bottom.

Table 2. Beam flexural capacity achieved with BFRP bars (BFRP) and steel bars (SRC) [13].

Parameters	**Steel reinforced beams**			**BFRP reinforced beam**		
	SRC1	SRC2	SRC3	BFRP1	BFRP2	BFRP3
F_u, kN	37.5	35.0	40.5	47.5	47.5	45.0
$F_{u,ave}$, kN	37.6			46,7		
ε_1 , ‰	-1.58	-2.17	-2.02	-1.78	-2.60	-3.25
ε_1 , ‰	4.18	5.69	6.52	9.43	13.60	7.76
$M_{R,fl}$, kNm	6.3	5.8	6.8	7.9	7,9	7.5
$M_{R,fl,ave}$, kNm	6.3			7.8		

4 The Numerical Model

Figure 4 represents the FEA model, which was performed in the Abaqus/CAE Software. The materials, which were used for the model are the same as in the experimental programme. FEA model for the concrete beam was designed by the use of grid, which consist of *660 elements* of *C3D8R* type. For the modelling of linear objects, such as: BFRP bars, steel bars and stirrups *300 elements* were used of *T3D2 type*.

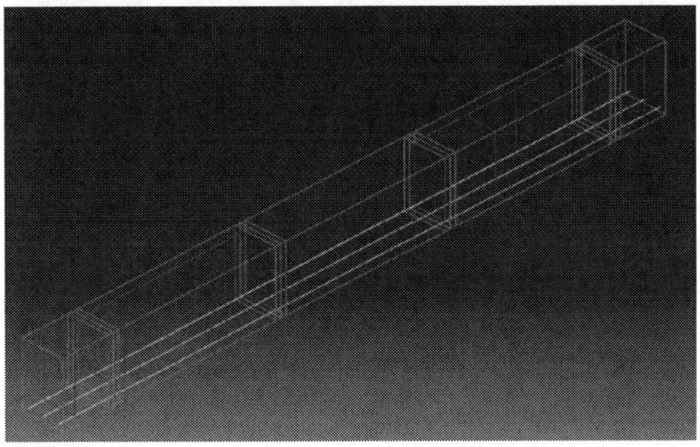

Fig. 4. The FEA model of tested beam.

Figure 5 shows the placing of the loading, which was applied in the same manner as in the experimental tests, in the form of strips with the sizes *2.0 cm* and *width of the beam*.

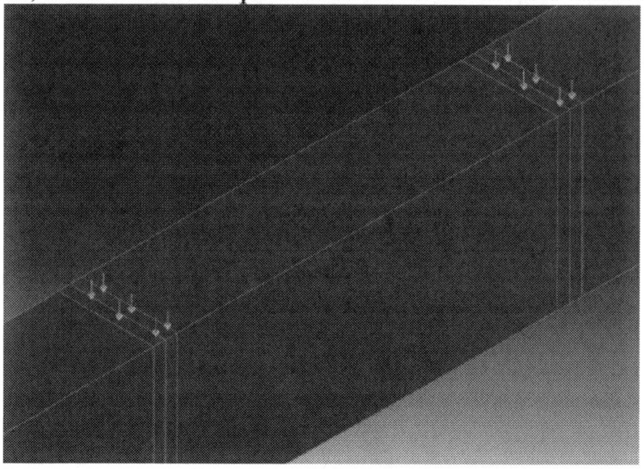

Fig. 5. The placing of loading.

The non-linear model was used with the gradual accumulation, $S=0.002$, and the beam was planned to load force, $F_{u,pl} = 20\ MN/m2$. Assuming the load transfer by the strip with dimensions, $b \times h=0.02 \times 0.08\ m$, the final applied loading for the point force was defined according to expression 1:

$$F= F_{u,pl} \times b \times h = 20.00 \times 0.02 \times 0.08 =32.00\ kN \tag{1}$$

The total force which was planned to apply to the beam is equal to twice the point force, which corresponds to the value of *64.00 kN*. This value exceeds the average strength obtained from experimental tests by about *37%*.

The Figure 6 shows that the program stops computation, when the element is subjected to *0.336* of the planned force. At this point, the element is destroyed.

Step	Increment	Att	Severe Discon Iter	Equil Iter	Total Iter	Total Time/Freq	Step Time/LPF	Time/LPF Inc
1	165	1	0	2	2	0.33	0.33	0.002
1	166	1	0	1	1	0.332	0.332	0.002
1	167	1	0	1	1	0.334	0.334	0.002
1	168	1	0	1	1	0.336	0.336	0.002
1	169	1	0	6	6	0.336	0.336	0.002

Fig. 6. The applying of loading at the moment of element failure.

Obtained strength, $F_{u,FEA}$, of the element is defined according to expression 2:

$$F_{u,\ FEA}= S \times F = 0.336 \times 64 =21.50\ kN \tag{2}$$

Figure 7 and Figure 8 show distribution of stresses in the concrete and in BFRP bars, respectively, just before the destruction of the element.

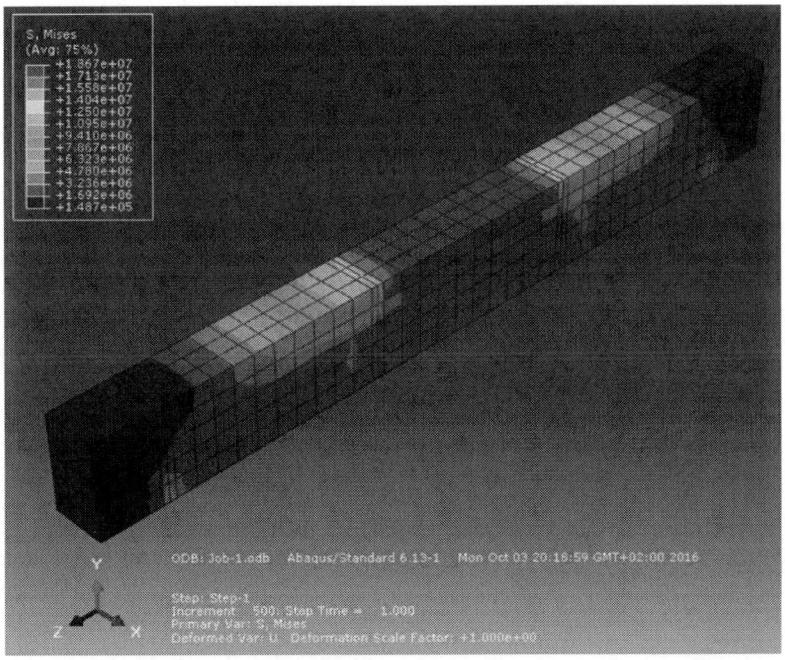

Fig. 7. Distribution of stresses in concrete element, units in Pa.

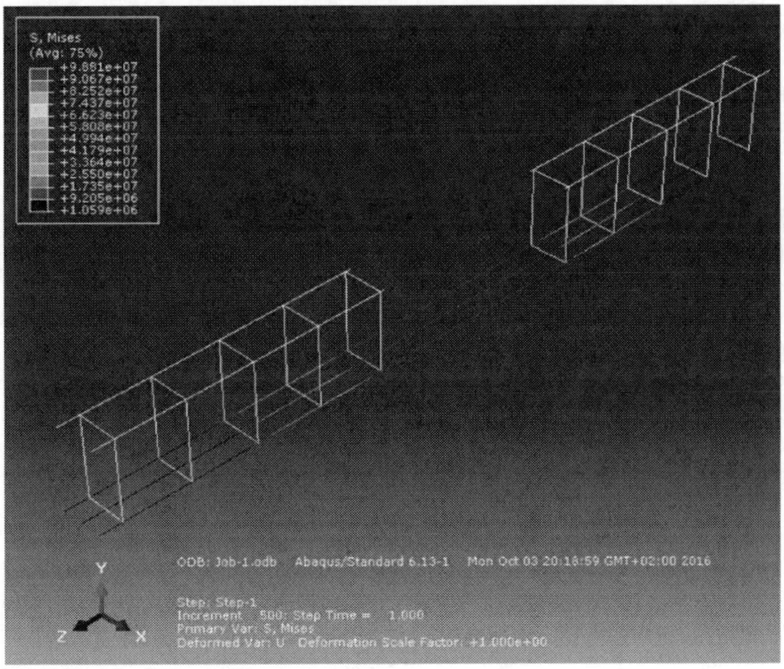

Fig. 8. Distribution of stresses in FRP reinforcement, units in Pa.

Figure 9 describes the distribution of strains for the whole element.

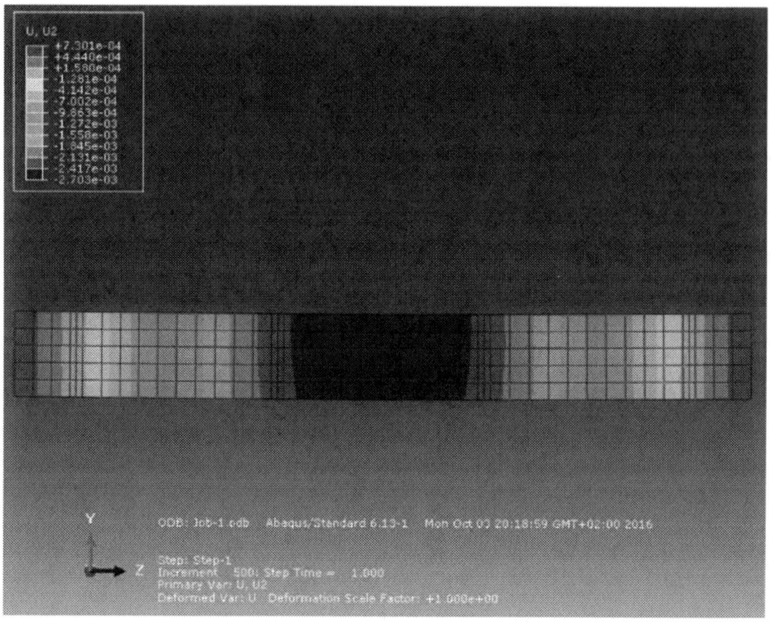

Fig. 9. Distribution of strains in the element, units in Pa.

The bearing capacity of the beams reinforced with steel bars was calculated in similar manner and found to be *44.15 kN*, which is *17%* higher than experimental results.

5 Results and Discussion

Experimental as well as numerical deflections of the beams reinforced with BFRP bars were significantly greater than the deflections for beams reinforced with steel bars, that can be caused due to the significantly lower elasticity modulus of BFRP bars than that one for steel bars.

Additional aspect that demonstrate the convergence between experimental and numerical programmes was that in both studies the concrete regions were destroyed, rather than BFRP or steel bars. It is also observed that with an increase in tensile strength of concrete, the carrying capacity of the beam increases. That can be considered as one of the major aspects for estimation of the strength capacity of elements.

The difference between experimental and numerical outcomes for bearing capacity can be described by the presence of one or several factors. Initially, it can be explained by different bond characteristics between materials. Some scientific papers confirmed that FRP bar-to-concrete bond can be considered as competitive [14]. The test results obtained in [15] for FRP bars support that bond strength is proportional to the square root of the development length. The mechanism failure and high resistance to cyclic loadings can be interpret as important factor comparing with working of steel bars in this case. These issues indicates on the possibility to use BFRP bars in order to achieve balanced ratio, which is defined as the reinforcement ratio that simultaneously results in the rupture of the bars and crushing of concrete. Balanced ratio can be the reason of higher bearing capacity of the element.

Authors want to conclude, that for the investigation of a new materials, numerical modelling is insufficient and experimental work is a vital issue.

The authors wish to acknowledge the financial support of the project – "Innovative Hybrid - FRP composites for infrastructure design with high durability" NCBR: PBS3/A2/20/2015.

References

1. *ACI 440.4R-04: Prestressing Concrete Structures with FRP Tendons (Reapproved 2011), American Concrete Institute, the USA* (2011)
2. K. Protchenko, M. Wlodarchyk, E. Szmigiera, Proc. Eng. **111C**, 679-686 (2015)
3. A. Lapko, M. Urbanski, Materiały Budowlane **2013/3**, (2013)
4. M. Urbański, A. Łapko, Budownictwo i Architektura **13(3)**, 201-208 (2014)
5. B. Benmokrane, A.H. Ali, H.M. Mohamed, M. Robert, conference paper: 7th Intern. Conference on Advanced Comp. Materials in Bridges and Structures (2016)
6. K. Protchenko, J. Dobosz, M. Urbański, A. Garbacz, Journal of Civil Engineering, Environment and Architecture **XXXIII - 63**, 149-156 (2016)
7. N.M. Chikhradze, L.A. Japaridze, G.S. Abashidze, Composites and Their Applications, 221-246 (2012)
8. A. Dorigato, A. Pegoretti, Journal of Composite Materials **48**, 1121-1130 (2014)
9. R. Parnas, M. Shaw, Q. Liu, *Basalt Fiber Reinforced Polymer Composites* (Technical Report, Storrs, 2007)
10. R. Piekarski, Wydawnictwo Politechniki Częstochowskiej, Kompozyty **1(6)**, 26-31
11. E. Kormanikova, K. Kotrasova, Chem. Listy **105**, 758-762 (2011).
12. M. Abdel Gfahaar, A.A. Mazen, N.A. El-Mahallawy, Journal of Engineering Sciences, **34**, 227-236 (2006).
13. M. Urbanski, A. Lapko, A. Garbacz, Procedia Engineering **57**, 1183–1191, (2013)
14. A. Garbacz, M. Urbański, A. Łapko, Advanced Materials Research **1129**, 233-241, (2015)
15. C. P. Mosley, A. Koray Tureyen, R J. Frosch, ACI Structural Journal **105-S60**, 634-642 (2008).

Experimental study of wind loads on unique buildings and structures in Russia

Olga Poddaeva[1], *Pavel* Churin[1,*] and *Ilya* Dunichkin[1]
[1]Moscow State University of Civil Engineering, Yaroslavskoe sh. 26, 129337 Moscow, Russia

Abstract. Design and construction of unique buildings and structures (sports arenas, airport complexes, business centres, etc.) from an engineering point of view is a very difficult task as in most cases these facilities have an original architectural form. Therefore, consideration of wind loads is an important part of the design. The paper presents the definition of wind load for two complex of airport. Researches was applied the combined calculation an experimental method. During the experimental study a wind tunnel architectural and construction type NRU MSUCE was used. Numerical simulations were performed using the software package ANSYS. The result of research on each object are integral aerodynamic loads on the object (coefficients Cx, Cy, Cmz) and picture of the distribution of aerodynamic pressure coefficient Cp obtained in the numerical simulation. In conclusion, we discuss the possible formation of deposits of snow and recommendations to eliminate them from the roof of researched objects.

1 Introduction

Modern Russian policy is aimed at promoting the different types of sports. Since 2014 our country carried out many international sporting events, including the Olympic Games in Sochi. At present, we are actively preparing for the FIFA World Cup 2018. In this regard is actively built a unique building objects - sports arenas, football stadiums, airport complexes, and others.

The Russian and foreign documents, regulating the accounting of various loadings and impacts on constructions, for unique buildings and constructions recommend carrying out researches of wind loading in specialized wind tunnels [1,2,3]. In 2013, on the basis of the Moscow State University of Civil Engineering was put into operation the first Russian wind tunnel for buildings and constructions corresponding to modern requirements.

2 Research method

For the purpose of minimization of labor costs and time of carrying out researches was applied the combined calculation an experimental method. Calculation and experimental approach combines the advantages of physical and numerical modeling, and is the most optimal and efficient in solving problems of building aerodynamics.

* Corresponding author: pashok_@inbox.ru

DOI: 10.1051/matecconf/20168602012

This method involves experimental simulation in the wind tunnel and numerical modeling in a specialized software package (in this case, was used ANSYS) [5,8]. In the process of experimental simulation are defined integral aerodynamic forces, moments, and aerodynamic pressure coefficients at the control points on the surface of the models. In the process of numerical simulation was defined the complete picture of the distribution of the aerodynamic pressure coefficients on the surfaces of the models. This approach allows for the verification of the numerical model with the results of more reliable experimental studies, and at the same time to obtain the most complete information about the wind loads.

3 Objects of study

As objects of research are chosen the projected airport complexes in the large regional centers - Rostov-on-Don and Perm.

Fig. 1. Air terminal complex in Perm and in Rostov-on-Don.

4 Prototyping

One of the most difficult stages of carrying out such researches is prototyping. This is related to the necessity of accurately reproduce complex spatial geometric shape of the roof of the investigated object. Currently, this problem is solved by creating a 3D model of the object under study and subsequent printing on a 3D printer.

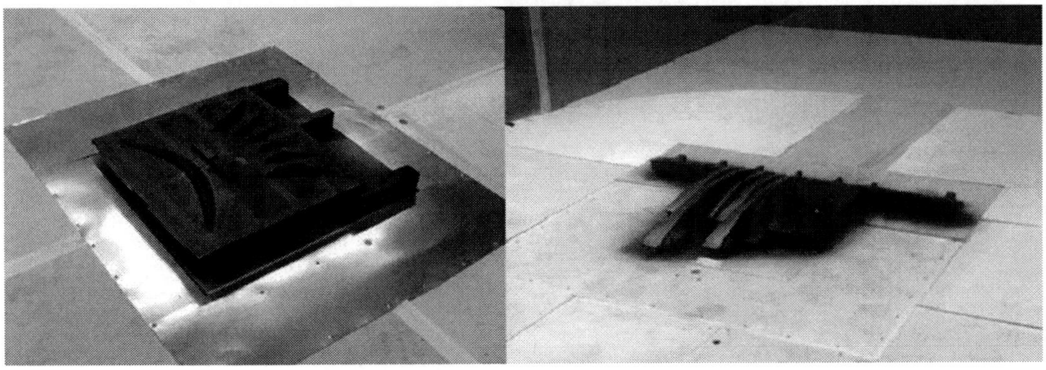

Fig. 2. Models of air terminal complex The process of research.

5 The process and results of research

In the process of research, a scale model airport in the wind tunnel to match full-scale Reynolds number ($Re_{nat} \approx 2*10^7$) is unattainable. However, for this class of bodies (bluff) is characterized by the existence of "self-similarity on Reynolds's number" range when with increasing Re dimensionless characteristics of a flow around a body practically cease to depend on this parameter [4,6].

To determine the self-similarity range for the investigated object model were performed auxiliary weight test by gradually increasing the flow velocity in the tunnel from $V\infty = 5$ to 30 m/s. The dependences obtained allows us to conclude that for such models at $V\infty \geq 20$ m/s ($Re \geq 0.29*106$) we have the self-similarity of the Reynolds number.

Measured using tensor-pressure sensors on the surface of airports models in drainage points gives an indication of the distribution of the facades and roof of aerodynamic pressure of the object at an air flow attack angles from 0 to 360 degrees with a pitch of 15 degrees.

Spectral analysis of pressure fluctuations in the drainage points allowed verifying the numerical simulation. These data are reusable. In addition to the definition of pressure on the building envelope airports is possible to perform an assessment of bioclimatic comfort of the adjacent space, which is important for the comfort and safety of the occupants of the buildings near airports [6, 8].

Using six-component force-torque sensors and ATI DAQ F/T software defined Fx, Fy, Fz, Mx, My and Mz - the total force and torques along the axes X, Y and the Z, respectively, and calculated the Cx, Cy, Cz, CMx, CMy and CMz- total aerodynamic coefficients [9].

The result of research on each object are integral aerodynamic loads on the object (coefficients Cx, Cy, Cmz), and picture of the distribution of aerodynamic pressure coefficient Cp obtained in the numerical simulation.

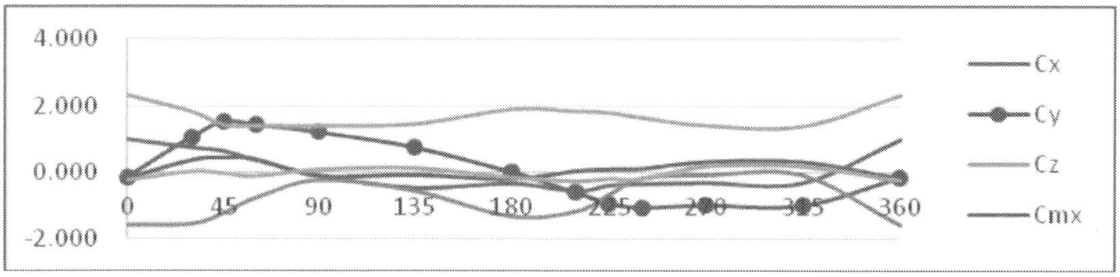

Fig. 3. The schedule of aerodynamic coefficients dependence Cx, Cy, Cz, Cmx, Cmy, Cmz at change of attack angle from 0° to 360° (Rostov-on-Don).

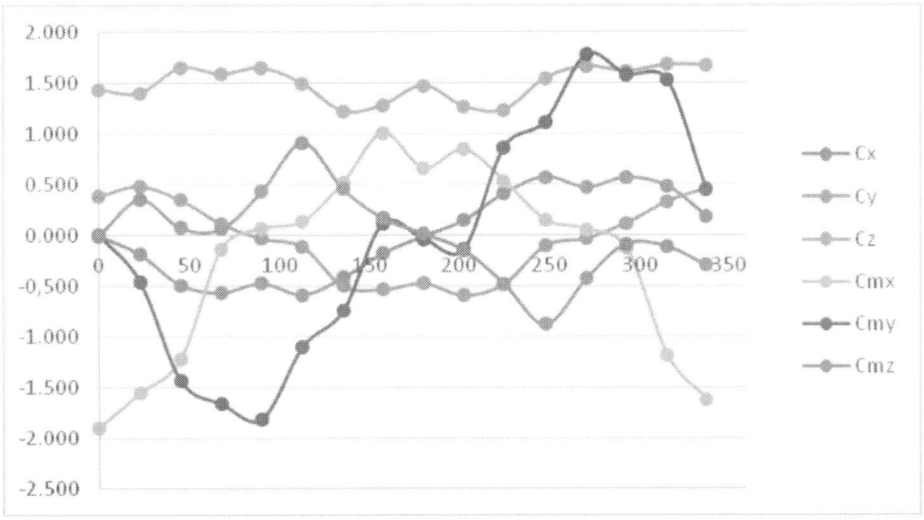

Fig. 4. The schedule of aerodynamic coefficients dependence Cx, Cy, Cz, Cmx, Cmy, Cmz at change of attack angle from 0° to 360° (Perm).

For the airport of Rostov-on-Don max load occur at angles blowing 0°, 30° and 210° (Cx = 0.968,Cx = 0.726 and Cx =-0.569, respectively) and at angles of airflow in 45° and 240° (Cy = 1.535 and Cy = -1.04 respectively).

For airport of Perm maximum loads occur when blowing corners 90°, and 270° (Cx = -1.089 and Cx = 1.781, respectively).

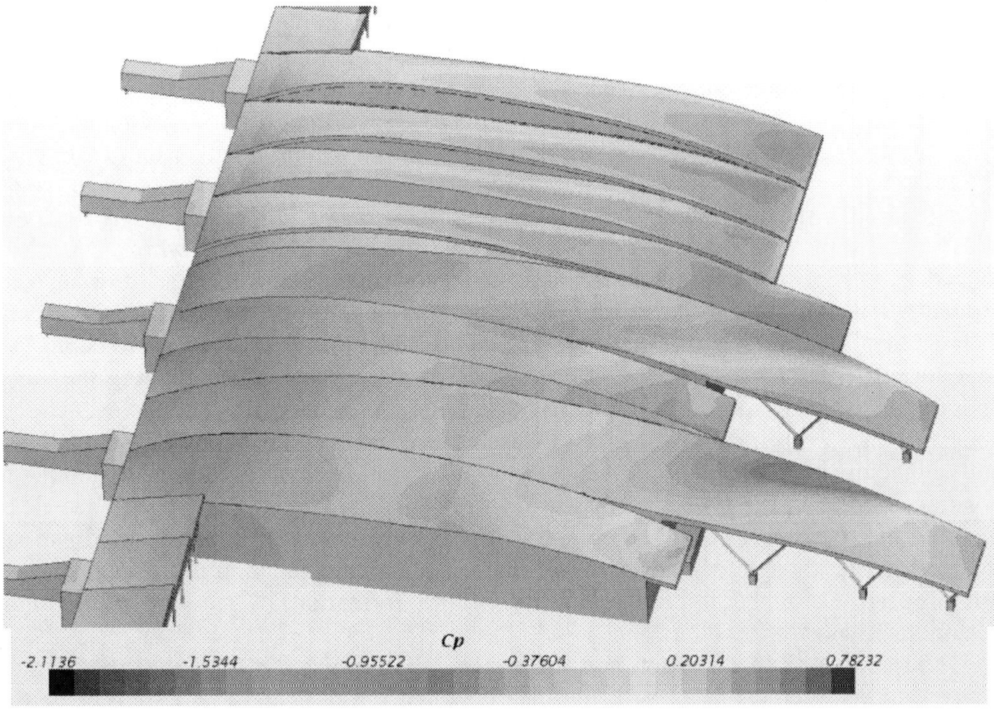

Fig. 5. Distribution of aerodynamic coefficient of pressure values on the surface (Rostov-on-Don).

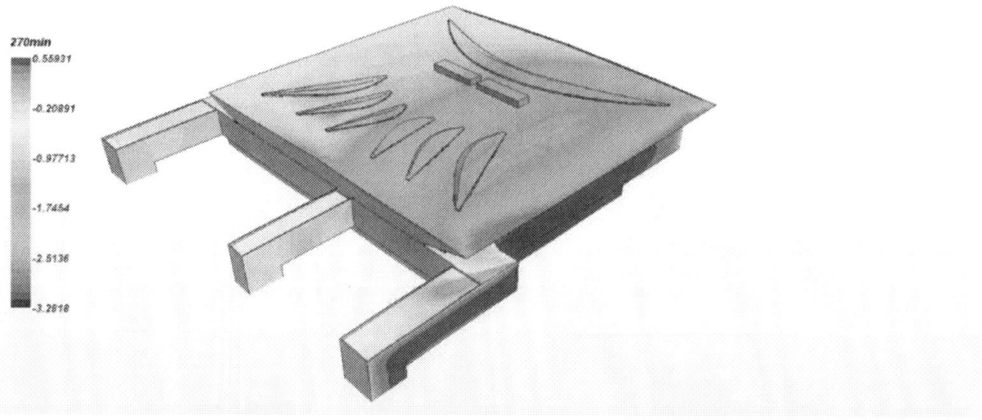

Fig. 6. Distribution of aerodynamic coefficient of pressure values on the surface (Perm).

Bioclimatic comfort space near the airport is connected not only with the speed of the wind, but also with deposition of snow and necessary maintenance works on clearing of snow and ice. Wind effects and snow are subjected to the passengers during the boarding and disembarking from the domestic airport transportation, as well as airport staff, who spends a long time in the open space. In this connection, obtained by experiment results help to identify problem areas on bioclimatic discomfort and avoid their placement embarkation and disembarkation of passengers in domestic vehicles airport, as well as to optimize the allocation of posts and technological platforms on which it is supposed finding of the airport staff.

Criterion of wind speeds comfort for the evaluation human presence close to the airport was used for the winter, and is in the range from calm to 3m /s. For evaluation the short stay on street (10 minutes) in relative discomfort it was assumed the value of wind speed is 7 m/s, based on the recommendations given P.P. Kovalenko and L.N. Orlova in the chart the effects of wind and temperature in the environment [10].

It should also be noted, that the most vulnerable structures of the building in this case are telescopic gangways for boarding and roofs, or rather the airports terminals coating systems. The telescopic ladder is the mobile part of the building for the passage from the aircraft in the airport building, without an intermediate exit to the street. His device is a telescopically moved apart corridor consisting of several modules, which can also be moved relative to the aircraft: horizontally on the ground using a wheel and suspended vertically with electro- or hydro-lifts as well as the position of the entrance doors at the different types of aircraft different. During using the parking lot with a telescopic ladder first aircraft usually rides to park yourself under own engines. After stopping the aircraft, operator telescopic ladder set it to the front door of the aircraft. Thereafter, telescopic gangway must keep track of the aircraft fuselage vertical displacement (due to its loading and unloading) himself and move vertically to the extent necessary, thereby maintaining its position relative to the door. In connection with this, used structure should have flexible metallic insert parts of variable geometry, which is very sensitive to the wind and snow loads. In winter the telescopic ladder can freeze and damage the aircraft, if dock space was not properly examined to the possibility of the formation of ice due to melting of accumulated snow.

The question of specification and snow cleaning optimization of airport roof is also in many respects interconnected with the data obtained during experiment study. Due to the winter temperature drops at the airport roofs and canopies a frost is likely will be formed. Along with snow, hanging from the roof and icicles edges, it can damage the roof and gutter system. It is not only the risk of damage to the airfield equipment, but also a real threat to the health and lives of people passing by. To reduce risks and protect the surrounding area, the results of the experiment must be considered in the development of regulation works on the snow and ice roof clean, including the cleaning of the pitched roof areas of snow on the width of 1.5 m. Such work of the roof cleaning is recommended to avoid the risk of falling snow masses from the roof and it is used under extreme weather conditions. Cleaning roofs from snow and ice belongs to the category of particularly dangerous works carried out at height, require compliance with safety rules and taking into account the weather conditions. At differences of external air temperatures with transition on positive on metal cattle sites of roofs (except for a roofing covering) the roofing covering, trenches and covers trays established on water reception funnels have to be completely cleared. This is necessary to avoid the formation of ice dams.

It is recommended to clean the roof of the airport in Perm in extreme weather conditions, partly for supporting the work of drainage funnels and testing the thickness of the snow cover at the light and aeration lanterns. The airport building in Rostov-on-Don has largely pitched roof, which requires complete removal of snow around the edges in case of extreme weather conditions due to the danger of icing. In both buildings, telescopic ladders receive considerable wind load without serious snow deposits. Thus, the obtained results allow us to estimate not only the necessary requirements for snow removal of the roof and the load on the roof structure, but also to predict the snow load on the cover and flexible insert of telescopic ladders.

This work was financially supported by the Ministry of Education and Science of the Russian Federation within the framework of the state order to MGSU №2014/107 project "Fundamental studies of wind effects (including extreme) to the unique building and bridge structures".

References

1. *SP 20.13330.2011, The updated edition of SNiP 2.01.07-85 * "Loads and effects.* (2011)
2. *American Society of Civil Engineers. Minimum design loads for buildings and other structures. ANSI/ASCE* (1998)
3. *MDS 20-1.2006. Temporary regulations for loads and impacts of destination, acting on a multifunctional high-rise buildings and complexes in Moscow* (2006)
4. E. Simiu, R. Skanlan, *The impact of wind on buildings and structures* (1984)
5. V.G. Gagarin, S.V. Guvernyuk, A.S. Kubenin, A.A. Sinyavin, Bulletin of the Department of Civil Engineering of the Russian Academy of Architecture and Building Sciences **18**, 151 (2014)
6. O.I. Poddaeva, G.V. Orekhov, I.V. Dunichkin, O.A. Kochanov, Proc. Integration, Partnership and Innovation in Construction Science and Education Scientific publication, 133-138 (2012)
7. O. Flamand, Czasopismo Techniczne **12**, 51-61 (2015)
8. A. Zasso, D. Rocchi, P. Schito, EACWE **5** (2009)
9. P.S. Churin, J.S. Buslaeva, International Journal of Applied Engineering Research **10**, №20, 40841-40843 (2015)
10. P.P. Kovalenko, L.N. Orlova, Urban climatology, 142 (1993)

Determining safety criteria for reinforced concrete structures of power plants taking the example of a nuclear power plant with RBMK

Valery Nikolayev[1], *Dmitriy* Nikolayev[1] and *Anton* Yakushev[1*]

[1]AO Atomenergoproyekt, 107996, ul. Bakuninskaya, 7 p. 1, Moscow. Russia

Abstract. The paper shows how safety criteria of nuclear power plants with reactor RBMK can be defined based on analytical, numerical and mixed calculation methods using data about strength characteristics of materials with the course of time.

1 Introduction

Buildings and facilities for use of nuclear power are qualified as super-critical structures according to [1]; this means that they call for extreme safety requirements. Monitoring of the condition of building structures is highly important for assurance of safety. Monitoring includes not merely registration but also analysis of measured physical values. Such analysis compares recorded values with respective permissible high or criterion. Criteria values are thresholds that trigger the near-emergency or emergency status according to [2-4] and [5]. At this time, two levels of criteria values are assigned: alert and alarm. The alerting level of safety criteria leaves room for the control parameter being incremented by a value that corresponds to a particular load; whereas the alarm level switches the structure to the emergency (near-emergency) state. In other words, the alerting criterion is the emergency criterion diminished by a value expected from specific impact (seismic shock, blast wave, falling aircraft, etc.) [6-8]

For concrete structures, monitoring is carried out to control opening of cracks, rebar stress and, if need be, also structural material temperature and compressive loads inside concrete, focusing on stress in compressed rebar assuming they undergo deformation together [9]. Stress of rebar material and opening fissures are typically measured by means of string devices. Each specific sensor is assigned its own specific safety criterion to take into account the stress of the structure around the sensor, plus the geometry and dimensions of the cross-section and characteristics of concrete and rebar material. If the main crack opens in zones not covered by a sensor, stress can be measured by the extent of the opening crack [2-4]; also, the unloading method can be used for greater accuracy when stress level boarders on the criterial values [10].

* Corresponding author: Yakushev_AA@aep.ru

DOI: 10.1051/matecconf/20168602013

Building structures of power plants and facilities frequently have areas where access to the most stressed aspects of the structure is either limited or impossible. For example, consider rooms with high level of ionizing radiation or a structure with one side exposed to water or soil. In such cases, stress level can be measured indirectly by rebar force gage installed wherever accessible on the rear side.

The resulting monitoring criteria (comparing major factors of technical condition with criterial values) [10] were applied to the "hot" boxes and rooms in a nuclear power plant (NPP) with a RBMK (Power Units One and Two of Smolensk NPP, and Power Units Three and Four of Kursk NPP), where cooling system for concrete became unusable due to lost tightness in water screen pipes.

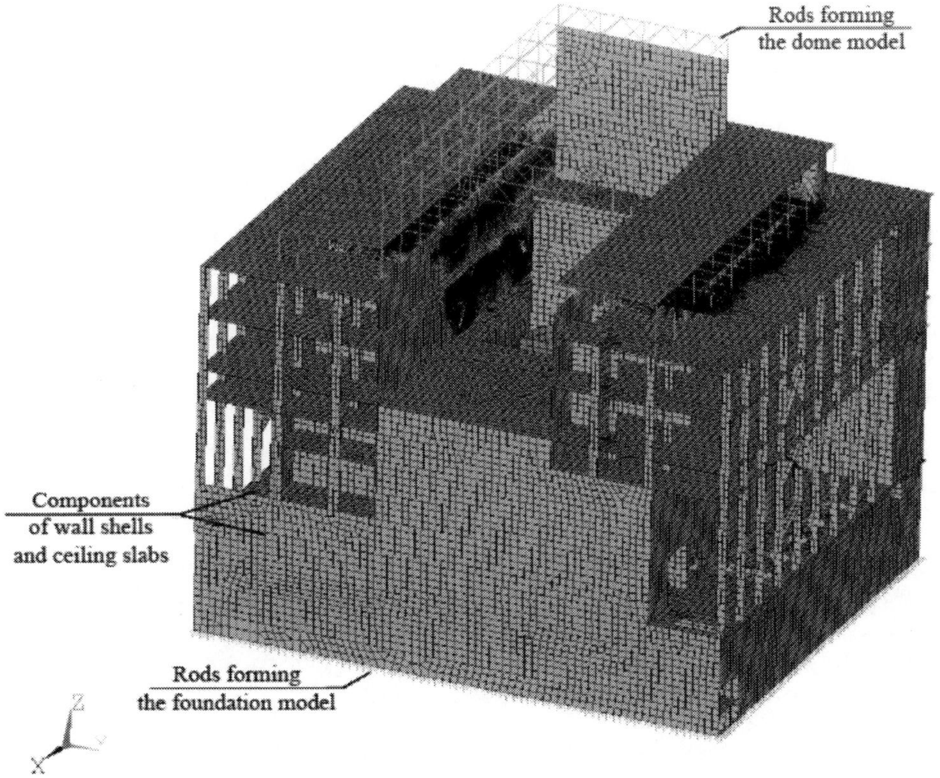

Fig. 1. Standard calculation model of a RBMK unit

Fig. 2. Rebar panels layout model

Power unit structures undergo considerable stress caused by thermal impacts beyond their designed level. Stress inside the walls and ceilings of the power units is monitored scores of rebar material dynamometers, each installed on the outer side of hot rooms (see Fig. 1). To measure stressed conditions in a power unit, an associated finite element model was developed (see Fig. 2). The power unit's walls were designed to be built as precast-monolithic. As prefabricated components, 3x6 m (two-directional) ribbed reinforcement panels are used, with maximum rib height of 200 mm (see Fig. 2). Longitudinal working rebar is placed inside the ribs.

$$\sigma_s = E_s \cdot \frac{R_{b,m}}{E_{b,\tau,m} \cdot k_n}, \tag{1}$$

where σ_s – stress in reinforcement steel;

E_s – modulus of elasticity of reinforcement steel;

$R_{b,m}$ – calculated resistance of filling concrete to axial compression;

$E_{b,\tau,m}$ – continuous deformation module of filling concrete;

k_n – irregularity factor for compressive stresses on the structure's faces, caused by one-sided heating of surfaces and emergency pressure (2):

$$k_n = \sigma_{t,hot.} + \sigma_{press}\sigma_{t,armdyn} - \sigma_{press}, \tag{2}$$

where $\sigma_{t,\text{armdyn}}$, $\sigma_{t,\text{hot.}}$ – normal stress in reinforcement steel from temperature on the side of the rebar dynamometer and on the side of the hot room, respectively;

σ_{seism} – stress in reinforcement steel caused by seismic factor;

σ_{press} - stress in reinforcement steel caused by emergency pressure;

As monolithic layer of concrete shrinks it gets stressed due to hanging on rebar panels (Fig. 3). Tensile stresses are described with this equation (3):

$$\sigma_m = \frac{0.5\alpha_y}{h_m \frac{1}{2h_p E_{b,p}} + \frac{1}{E_{b,m}}\left(1 + \frac{\varphi_{b,cr,m}}{4}\right)}, \tag{3}$$

where σ_m – shrinkage stress in the filling concrete;

α_y – creep of concrete;

h_m –filling concrete section height;

h_p – reinforcement panel section height;
$E_{b,p}$ – initial modulus of elasticity of the filling concrete on compression;
$E_{b,m}$ – initial modulus of elasticity of the filling concrete on compression;
$\varphi_{b,cr,m}$ – filling concrete creep ratio

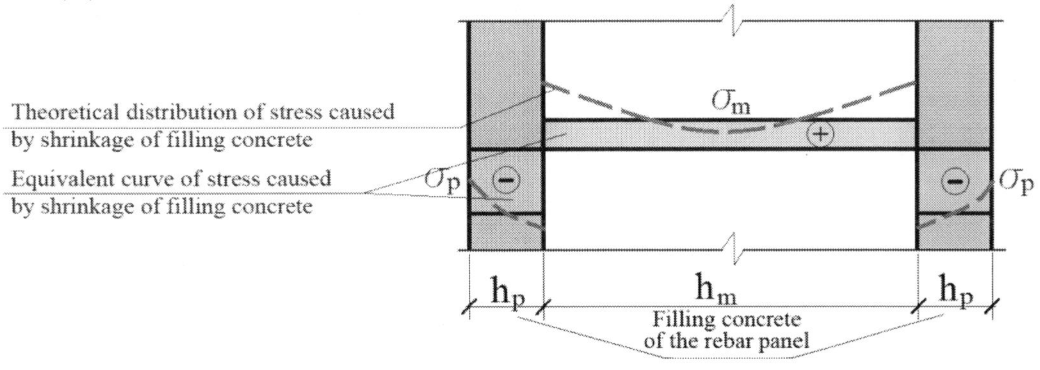

Fig. 3. Illustration of measured stress caused by shrinkage of the filling concrete

Stress in precast concrete caused by shrinkage taking into account the creep of in-situ concrete is described with equation (4):

$$\sigma_p = \sigma_m \cdot \frac{h_m}{2 \cdot h_p} \tag{4}$$

Stress in reinforcement steel caused by shrinkage taking into account the creep is described with equation (5):

$$\sigma_{sp} = \sigma_p \cdot \frac{E_s}{E_{b,p}} \tag{5}$$

Sum total σ_s of stress in reinforcement steel and total shrinkage stress taking into account creep σ_{sp} less stress caused by specific loads σ_{seism} and σ_{press} qualify as the alerting safety criterion (6):

$$\sigma_{ss} = \sigma_s + \sigma_{sp} - \sigma_{seism} - \sigma_{press} \tag{6}$$

The monolithic structures, the approach to the definition of the criteria does not change fundamentally: during shrinkage, reinforcement steel rather than precast structure acts as the decelerating element.

If the section has a compressed and extended reinforcement steel, the height of the compressed zone of concrete (x) and stress in the extended reinforcement steel (σ_s) should be found from the equilibrium equations (7) and (7a). Torque (M) and longitudinal force (N) should be found in advance through calculation based on the associated model. The strength of concrete R_b is assumed based on varying strength of precast and in-situ layers of concrete, in proportion to the area of the compressed zone curve diagram (Fig. 4). The solution is iterative, and we check the height of the compressed zone and respective concrete strength after each iteration step. The resulting stresses σ_s less seismic load and emergency pressure load will be in fact the alerting criterion.

$$\begin{cases} \mathrm{N} = \mathrm{N_b} + \mathrm{N'_s} - \mathrm{N_s} \\ \mathrm{M} = \mathrm{M_b} + \mathrm{M'_s} + \mathrm{M_s} \end{cases} \tag{7}$$

$$\begin{cases} N = \frac{R_b \cdot b \cdot x}{2} + R_b \cdot \alpha_s \cdot A'_s - \sigma_s \cdot A_s \\ M = \frac{R_b \cdot b \cdot x}{2} \cdot \frac{2 \cdot x}{3} + R_b \cdot \alpha_s \cdot A'_s \cdot (x - a') + \sigma_s \cdot A_s \cdot (h_0 - x) \end{cases} \quad (8)$$

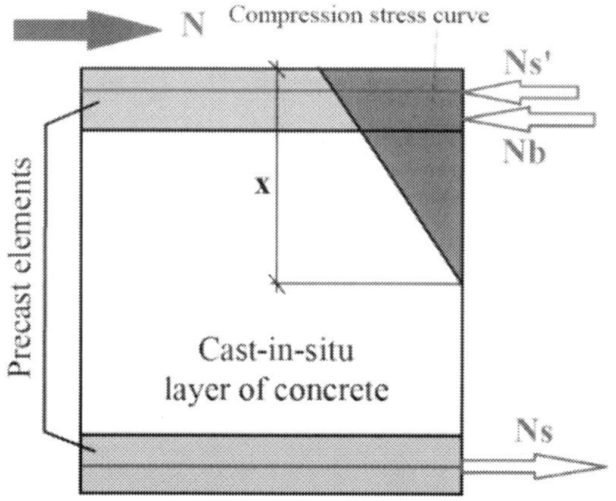

Fig. 4. Illustration of criterion calculation to control expansion stress

If the structure has any cracks or other discontinuities commensurable with the resolution threshold of the numerical method adopted, such flaws should be incorporated in the calculated model [2].

As inputs used to determine the safety criteria, we can use not only the forces existing inside the structures, but also directly stress in core elements that simulate reinforcement steel.

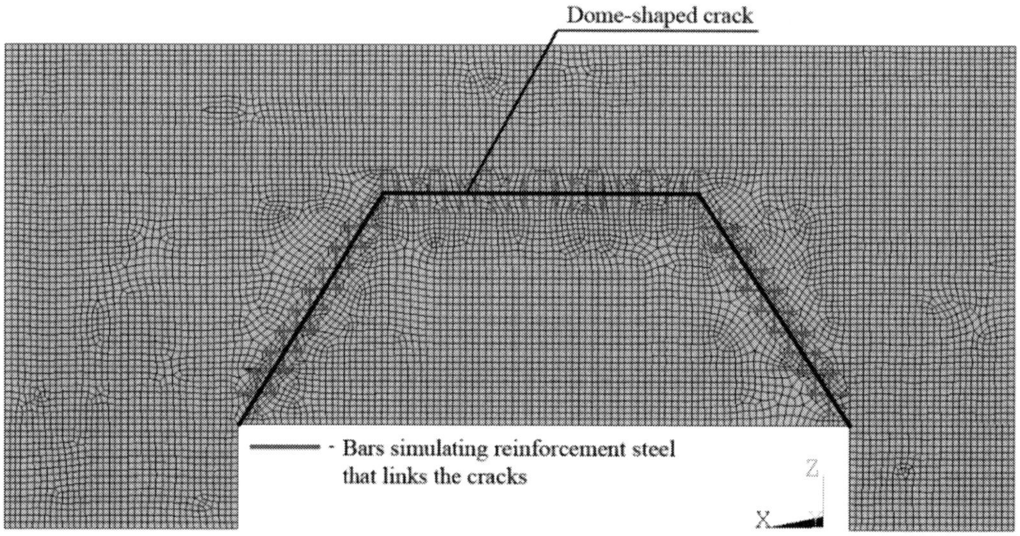

Fig.5. Simulation of cracks in the Central Hall

For example, in the course of operation of the aforesaid NPP units, the walls of the Central Room developed cracks opening in excess of the regulated limits. Existence of a special 24-m long service embrasure predetermined the character of cracks emerging in the walls, as in a wall beam. The areas around supports register oblique cracks created by impacting vertical loads that tend to meet under the LUM console, which results in the formation of the so-called dome-shaped crack, setting out a wall segment connected to the rest of the wall virtually by reinforcement steel alone. In the mathematical model of the structure, shell elements have gaps in the cracked areas and are connected with reinforcement bars that simulate reinforcement steel (see Fig. 5).

Conclusion

We have developed a methodology to address the key safety criterion: stressed condition of reinforcement steel and concrete in a reinforced-concrete structure. Safety criteria for power plants can be defined based on analytical, numerical and mixed calculation methods; they also rely on field data and take into account compromised strength characteristics of materials with the course of time.

References

1. Town Plan. Code of the Russ. Fed. (2016)
2. GOST 31937-2011 *Buildings and Structures. Rules for Examination and Monitoring of Technical Condition. Minregion of Russia* (Moscow, 2011)
3. A.V. Korgin, Y.I. Kudishin, V.A. Ermakov, M.V. Emelianov, L.Z. Zeid Kilani, Int. Jour. of App. Eng. Res., **11**, 1680-1686 (2016)
4. Valintinovich, K.A., Uriy Ivanovich, K., Leys Zeidovich, Z. Int. Jour. of App. Eng. Res., **10**, 43139-43142 (2015)
5. RD 153-34.2-21.342-00 *Methods to Determine Safety Criteria of Hydraulic Facilities* (Moscow, RAO UES of Russia, 2001)
6. F.F. Bryukhan', A.Y. Vinogradov, A.A. Lavrusevich, At. En., **118**, 365-368 (2015)
7. F.F. Bryukhan', A.D. Potapov, At. En., **115**, 346-350 (2014)
8. V.V. Potapov, A.M. Loginov, V.A. Il'in, V.G. Kolomiets, A.P. Pustovgar, At. En., **119**, 380-383 (2016)
9. B.K. Bylkin, A.N. Kozhevnikov, I.A. Engovatov, At. En., **118**, 436-441 (2015)
10. RD EO 1.1.2.99.0007-2011 *Standard Manual on Operation of Production Buildings and Facilities for Nuclear Power Plants*.
11. RD EO 0462-03 *Methodology for Validation of Service Life of Structures in Buildings and Facilities for Nuclear Power Plants*.

Aspects of designing construction of balcony parapets with photoelectric cells

Ilya Dunichkin[1,*]
[1] Moscow State University of Civil Engineering, 129337 Yaroslavskoye sh. 26, Moscow, Russia

Abstract. The aim of the work is to develop a modular system design of the balcony parapets of photoelectric panels "photoelectric barrier" (the "PB"). As a result of research conducted it has been identified the dimensions of the module "photoelectric barrier" ("PB"), height 1200 mm and width 1000 mm. The Parapet design parameters have been determined and technological problems have been posed. The design of the module "PB" has been explored, elements that make up the module design have been developed. Subject to the requirements, objectives and constraints there were designed sketches and design drawings that were evaluated by physical influences. While determining the design parameters it was used in the calculation of the energy potential and the designation of the optimal size of the modules "photoelectric barrier". It is possible to identify the optimum size of the module design and the energy potential of the project. The article gives full view into the scientific work of its prospects and applications of developing design "PB" module, as well as forms the foundation for the development and research of the issues of management and organization in the implementation of the design in the real construction

1 Introduction

The urgency of this development and its advantages include the innovative use of renewable energy technologies in the balcony railings, renewable energy technology deployment opportunities in any object of the construction industry in new construction and renovation, the formation of a new quality of exterior and energy supply of buildings and structures, ease of implementation in industrial and construction production run spread with large-scale production through its retail sales network and use as a household appliance.

2 The calculation of the energy potential and defining the optimal size of the "photoelectric barrier" modules

To determine the parameters of the module "PB" is necessary to submit the basic model. Photoelectric panel module "PB" is made up of solar cells, the type of which will influence on the energy potential of the module. Consequently, the parameters of the module size and its energy potential are interdependent.

* Corresponding author: ecse@bk.ru

According to the materials presented in the research works of D.S. Strebkov and his team to efficiently convert solar energy into electricity photo converter plane must be oriented perpendicular to the direction of radiation. In stationary mode photo converter must be oriented it at an angle equal to the latitude of the area, taking into account seasonal correction for tilt of Earth's axis. [1]

The scope of application of the balcony fences "PB" are balconies of residential and public buildings. The development can be applied both in new construction, the mass and the individual, and in the reconstruction. To use the "PB" on the territory of the existing centralized power network its required an economic analysis, which will be executed in subsequent phases of research. Therefore, there will be inspected the territory, where the application module "PB" will be competitive, [2]. Large areas with favorable conditions take place in areas with low-density development. In view of this it is advisable to orient the use of the module on new construction.

The question of application of photoelectric panels in the new energy efficient construction was considered by the author in the pilot project "eco-settlement" GENOME "and presented on the World Renewable Energy Forum (WREF 2012) and Including World Renewable Energy Congress XII and Colorado Renewable Energy Society (CRES), [3]. The result of the study is the conclusion that the most effective use of the modules "PB" in the new building must be designed with an extended front of the building or special plastic south facade.

Studying urban development for the application module "PB", exactly, the orientation of the buildings on the sides of light, it can be concluded that a significant part of the buildings in Russia are located without orientation relative to the cardinal, considering many other factors shaping development. The norms in residential buildings is permitted according on the number of rooms, insolation values equal to one and a half to two hours [4]. Therefore, relying on sun exposure balcony railings throughout the day in dense urban areas is not advisable.

Based on this materials, you can define the measure for the development of the module above "PB". Photoelectric panels in the module "PB" should be able to change its position according to:

1. The horizontal axis of rotation, as depending on the geographical latitude of its installation, the panel must be adapted to the angle to the horizontal.
2. The vertical axis of rotation, for use in buildings that are not oriented in a southerly direction.

To determine the optimal size of the module, and as a consequence its efficiency it is necessary to develop a conceptual model of the fence construction.

GOST 25772-83 "Fences of stairways, balconies and roofs of steel" sets the normative parameters of protections. Fences must withstand the loads specified in SNiP 2.01.07-85 * "Loads and effects", including wind loads.

In addition, it should be noted that the application of these modules on buildings and structures of high level of responsibility, must take into account their impact on the distribution of the average and peak values of wind loads on the facade. Current regulations for these building structures [5], regulate carrying out of experimental studies in specialized wind tunnels. The methodology of these tests is described in detail in [6, 7]. Determining the above criteria for the development of the module "PB", you can make of the fence design concept model. Since solar photoelectric should be located at a certain angle to the supporting framework fence, depending on the geographical location of the place of its application, the panels of solar cells should be articulated at the upper point of attachment "A" and the value of "b" should be changed to adjust the tilt angle panel «α». To orient the panel in a southerly direction it is necessary to provide rotation about point "A", Fig.1.

While developing the module "PB" solved the problem of placing the maximum number of solar cells in the photoelectric panels for the most efficient work. The width of photoelectric panels will be equal to the module width, 1000 mm, and the panel height will be equal to 1090 mm. Due to the placement of the module "PB" on the balcony of the building it must take into account the ratio of emerging wind load and mechanical strength of the panel and enclosure design. Therefore, taking into account the maximum size of the photoelectric panels, to reduce the impact of dynamic loads of the environment it is proposed to reduce the linear dimensions of the panels photoelectric module without reducing the total area of "PB". In this regard, in the module "PB" must be reduced two times a linear panel dimensions. On the basis of the above-defined parameters it will created a sketch of the module base model "PB", Fig. 2. Fig. 3 shows the geometry of the light beam falling on the photoelectric panel module "PB".

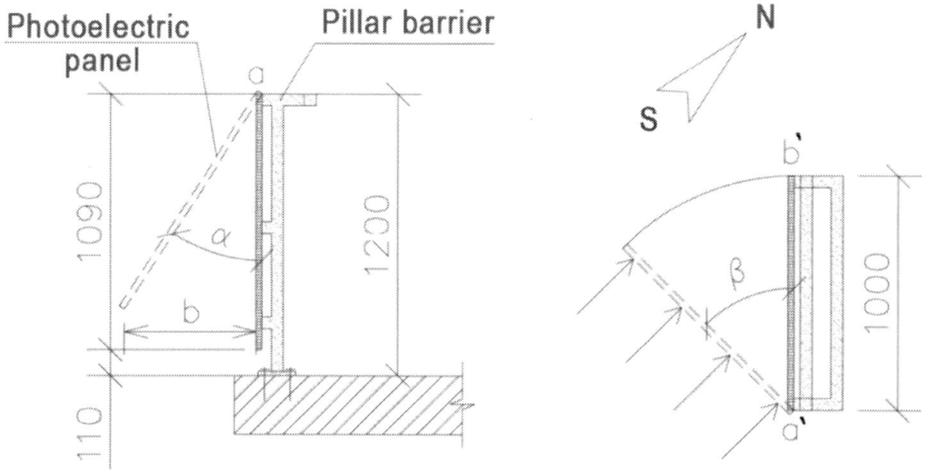

Fig. 1. Driving orientation of photoelectric solar panels

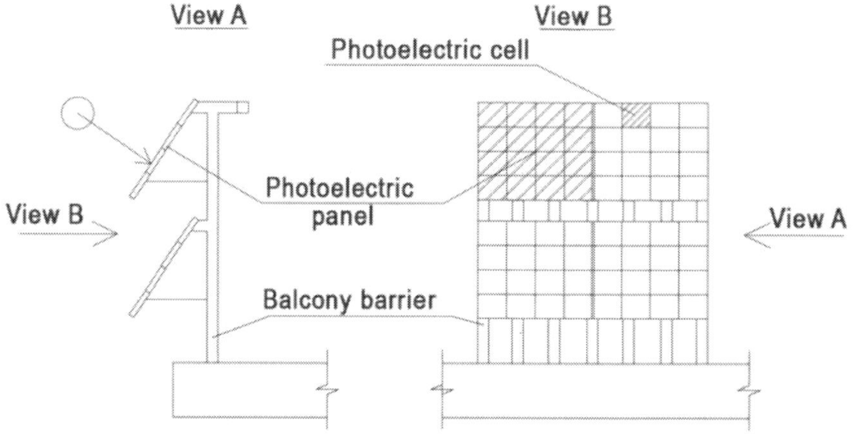

Fig. 2. Basic module model "PB"

We define the effective area of a panel of solar cells:

$$S2= S - S1=h * (L - L1) \quad (1)$$

where: S-area panels, S1-shaded area, S2- effective area, h-panel width.

Substituting (1) into (2) we get:

$$S2=h * (L - (L * \sin \alpha * \tan \beta - b)/(\cos \alpha + \sin \alpha * \sin \beta)), \quad (2)$$

With increasing angle shading area increases, which will affect the efficiency of the module "PB", and will reduce the amount of generated electric power module. Because of the problem, one of the questions that should be solved in the development of design drawings - is to minimize the area of the shading of photoelectric panels. Therefore, in determining the capacity of the power module "PB" is should be neglected area shading, considering that the entire area of the photoelectric panels module "PB" insolates throughout the daylight hours.

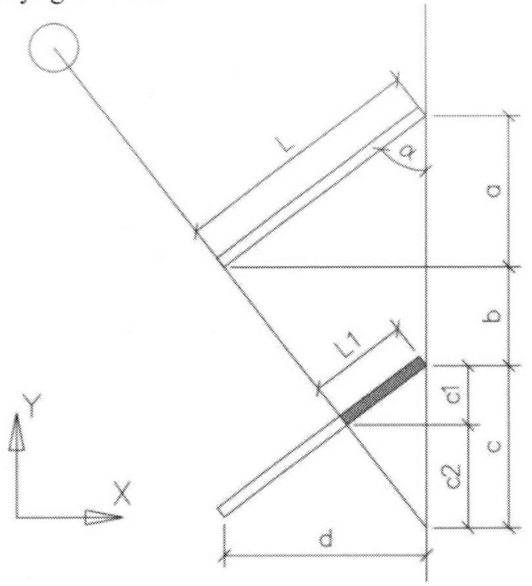

Fig. 3. Scheme of photoelectric panels in the module "PB" in the horizontal plane.

With increasing angle shading area increases, which will affect the efficiency of the module "PB", and will reduce the amount of generated electric power module. Because of the problem, one of the questions that should be solved in the development of design drawings - is to minimize the area of the shading of photoelectric panels. Therefore, in determining the capacity of the power module "PB" is should be neglected area shading, considering that the entire area of the photoelectric panels module "PB" insolates throughout the daylight hours.
Exploring the energy potential of the module "PB" on the example of the city of Krasnodar.

The annual amount of total solar radiation on a stationary platform oriented to the sun at an angle of 35 is 1433 kWh / m2. [2] Consequently, the energy potential of the "PB" module will be equal to:

$$P= (EFFICIENCY) \times R \times S, \quad (3)$$

Where P-Power, R-total solar radiation on 1 m^2, S-area module panel [8] Substituting the data we obtain: P = 286,6 kW per year Check the results. We choose the basic element for the solar panel with the approximate performance efficiencies and consider the duration of sunshine in the territory of Krasnodar. Knowing these parameters, we can calculate the energy potential of the module "PB". The energy potential of "PB" module will be equal to:

$$P= Pi * n * t * k, \quad (4)$$

Where p - power, n- number of elements in the module, t-sunshine duration, k-correction factor for the visible spectral range, perceived photodetector equal to ~ 0.7. [9] Substituting the data we obtain: P = 274,12 kWh per year. The difference in results is 4%, which is slightly due to the averaging of original data. One module "PB" can produce about 280 kWh per year if it is used in Krasnodar. Therefore, depending on where the application module in the territory of the Russian Federation of its energy potential will be taking into account the constraints of territories, suitable for its application, the duration of sunshine from 200 kWh per year to 290 kWh per year.

3 Development of design and construction drawings modules "photoelectric barrier"

As a result it was identified the criteria that influence the size of the module parameters and its energy potential, as well there were determined the size of the module "photoelectric barrier", height 1200 mm and width 1000 mm, and the parameters of its parts. The module consists of a frame fence and 4 attached on it photoelectric panels 2 having an axis of rotation (2 degrees of freedom). As a basic element for the construction of photoelectric panel sized 500h545mm took 125h125mm photoelectric cells [10]. These options take the starting point for the development of the fence construction. Due to the problem of shading of adjacent photoelectric panels, it is necessary to develop a design that can minimize area of the shading of photoelectric panels

In this case, the two vectors of fences constructive scheme can be considered:

1. Monolithic module design, without fractional structure that leads to its rise in price. In this case, it needs economic analysis of the appropriateness of the design type with the possible reduction of the energy efficiency of the module "PB"

2. Developing fractional structure design, with technological improvements module design "PB" to solve the problem of shading of solar panels, Fig. 2.

To develop the most efficient design in terms of construction work and design aspects of economic criteria should be introduced, which will adhere to the further study, and that will build in the development of the module design "PB":

1. Using of rolled profiles and cost-effective materials in the development of elements of "PB";

2. Using a unified standard construction elements;

3. Developing the module construction "PB" and its elements based on existing regulations;

4. Considering the operating conditions of the design performance of the materials in the design and calculation;

5. Design a structure with minimizing labor costs in its manufacture;

6. Select the optimum technical and economical production of circuit elements

It is necessary to consider the requirements of regulatory documents. Fencing design should withstand the regulatory burden [5].

The characteristic values of horizontal loads on the rails of railing balconies and loggias should be equal to:

a) for residential buildings, kindergartens, rest homes, health centers, hospitals and other medical institutions - 0.3 kN / m (30 kg / m);

b) for the stands and gyms - 1.5 kN / m (150 kgf / m);

c) for other buildings and premises in the absence of specific requirements - 0.8 kN / m (80 kg / m).

According to SNiP 2.08.01-89 * fencing loggias and balconies in buildings with three floors and more should be made of non-combustible materials.

Based on the regulations and the economic aspect, it should be used steel profiles for fence structure.

SNiP II-23-81 * provides guidance for the design of metal constructions, which coincides largely with the proposed criteria for the development of the above structures.

To develop design drawings it should be considered components of the module "PB". The construction of the fence can be divided into three main parts:

1. Ball fence frame (rack and rail)
2. Filling the frame fence
3. Photoelectric panels with fastening

The front is the main carrier of the module "PB", which are attached to the brackets with photoelectric panels, mounted handrail and fence filling. Racks must ensure the safe operation of the module and accept all major loads applied to the module. Racks are mounted vertically and are fastened to the base of the balcony, set the parameters of the height of the fence, which normalized.

Handrail parapet is the most important element to ensure human security. The handrail is attached horizontally to the fence uprights with handrail bracket. Brackets can be in one piece and hinged. Articulated provide the ability to install the rail fence at an angle. By the racks and the rail bracket can be attached by welding or by using prefabricated compound.

Filling the frame fence has a safety feature, as well as decorative. To ensure the reliability and solidity of construction it is proposed to use the module "PB" through filling [11]. This frees up rack enclosure for mounting on the plane her arms design with photoelectric panels.

Photoelectric panel mounting bracket has carrier function, it is fixed to the solar panel, as well it has adaptive function, that gives opportunity to change the angle of the solar panel, depending on the operating conditions. Solar panel mounting brackets can be single- and dual-axis. Scheme of the device one axle brackets are more economical and is showed on Figure 4.

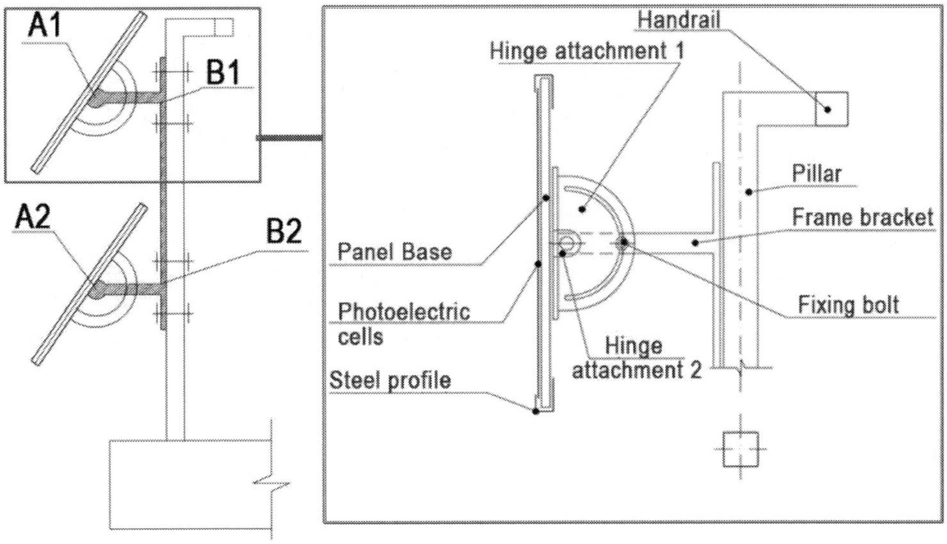

Fig. 4. Mounting photoelectric panels design module "PB".

Single-axis bracket module has a total base of the upper and lower mounting solar panels to provide rigidity. Horizontal guide statically mounted on a common base at the points B1 and B2. At the points A1 and A2 solar panel pivotally mounted on a horizontal rail. To solve the problem of shading upper bottom panel A2-B2 bracket should have a telescopic design or a predetermined extended length.

To solve the problem of the module for orientation of photoelectric panels further rotation associated with not only their axis on a horizontal axis but also about a vertical axis for orientation module panels towards the sun for more efficiency, this design can be implemented by uniaxial bracket. During detailed design, dimensions can be changed.

4 "Photoelectric barrier" in the scheme of power supply

For the energy conversion process module into electrical light, which can be used for supply of domestic users, its needed power supply system, comprising a module "PB".

Autonomous power supply system with "PB" module consists of the following components:

1. Module "PB" with necessary power;
2. The battery charge controller, which prevents harmful to the battery deep discharge and overcharging;
3. Accumulator battery (than AB);
4. The inverter, which converts direct current into alternating (for 220 V);
5. Energy efficient continuous load and / or alternating current

Power supply system using the modules "PB" and the use of the electricity grid should be supplemented by automatic switching unit to the electricity grid.

Batteries determine the reliability and efficiency of the electrical system. You must use the rechargeable battery, which can be a long time to work in severe cyclic conditions. Charging AB is small currents. Tanks battery at an autonomous power supply system should be enough to ensure the load for a few days, so you need to choose the type and capacity of the battery corresponding to the task. These measures correspond to lead-acid batteries. Another kind of used batteries - lithium-ion batteries, but due to uneconomical almost do not apply to stand-alone power supply system's [12].

An important element of the solar system is a charge controller / battery discharge. The correct choice of controller can affect system performance and longevity of the AB [12]. There are different types of controllers - from the simplest, in which when a certain threshold voltage is shut down solar modules or load to the microprocessor, which made tracking the maximum power point of the solar modules and pulse width modulation (hereinafter PWM) charging current. PWM AB allow you to charge up to 100% (the simplest controllers typically charge 60-70% AB; systematic undercharging battery leads to rapid failure of its failure).

To ensure reliable power supply is needed the backup power supply. When independent power source such as a wind turbine or can be a small petrol or diesel electric generator [13, 14]. When centralized power supply module with "PB" to reduce the costs of energy supply as a backup source can be a centralized power supply network. When battery power battery controller using automation unit can either transfer the load on centralized power network, or turn on the charger for a quick battery charge.

5 Conclusion

The reliability of the module structure "Photoelectric barrier" depends on its structural rigidity and mounting, quality, reliability and number of elements used. Prefabricated structural components type compound can occupy a small volume for storage and transportation of the fence, which reflects its universality. Welded components are large compared to the size of the teams of the elements due to the impossibility to ensure the welding process at the site of application design and provide the necessary strength properties of the compounds, and aesthetic properties [14].

Therefore the optimal design scheme should include the largest possible number of elements manufactured at the plant are connected by welds, and to ensure the mobility and versatility of the design elements have mounted directly onto the application site prefabricated construction method.

Application design modules "PB" is acceptable for both stand-alone systems, backup and auxiliary power supply to consumers and the design parameters obtained allow to install the module on the most southern of balconies and loggias, south-west, south-east orientation.

This work was financially supported by the Ministry of Education and Science of the Russian Federation within the framework of the state order to MGSU №2014/107 project "Fundamental studies of wind effects (including extreme) to the unique building and bridge structures".

References

1. D.S. Strebkov, Y.D. Arbuzov, V.M. Evdokimov, V.A. Mayrov, V.I. Polyakov, *Photovoltaic modules concentrate the radiation on the basis of the matrix of silicon solar cells Low energy*, Recurrent scientific and technical journal **1-2** (2011)
2. Y.Y. Rafikova, S.V. Kiseleva, L.V. Nefedova, *The use of renewable energy GIS technologies: foreign and domestic experience,* International Journal of Alternative Energy and Ecology **12 (152)**, 96-106 (2014)
3. O. Shepovalova, D. Strebkov, I. Dunichkin, *Energetically Independent Buildings of the Resort-improving and Educational-recreational Complex in Ecological Settlement GENOM,* Colorado Renewable Energy Society (CRES) Annual Conference, 3767-3772 (2012)
4. *The Sanitary Regulations and Norms (SanR and N) 2.2.1/2.1.1.1076-01 Hygienic requirements for sun exposure and sun protection facilities of residential and public buildings and areas* (2002)
5. *SP 20.13330.2011 The updated edition of SNiP 2.01.07-85 * Loads and effects*, JSC "LAC" (2011)
6. O. Poddaeva, S. Trushin, P. Churin, A. Zubkov, *Experimental study of extreme wind loads on tall buildings and facilities*, International Journal of Applied Engineering Research **21**, 42797-42799 (2015).
7. P.S. Churin, J.S. Buslaeva, *Experimental Study Of Wind And Snow Influence On The Unique Buildings And Facilities*, International Journal of Applied Engineering Research **20**, 40841-40843 (2015)
8. A.A. Soloviev, S.D. Varfolomeyev, P.P. Bezrukikh, O.S. Popel, A.B. Tarasenko, E.V. Antipov, G.V. Tomarov, M.V. Slipenchuk, E.I. Golubeva, S.V. Kiseleva, V.G. Nikolaev, Y.I. Blyashko, G.V. Ermolenko, I.Y. Yegorov, N.I. Chernova, *Renewable energy: Lectures. Release 8* (2015)
9. D.S. Strebkov., A.E. Irodionov, *Global solar power system,* Eurosun – 2004 **2,** 336–343 (2004)
10. *Renewable energy technologies and energy-saving equipment. Catalog of technologies and products, developed within GNU VIESH,* GNU VIES (2009)
11. *GOST 25772-83 Fences of stairways, balconies and roofs of steel* (1983)
12. A.P. Kashkarov, *Wind turbine, solar panels and other useful design* (2011)
13. E. Duraeva, *Renewable energy in Russia from possibility to reality,* OECD / IEA (2004)
14. I.A. Polyakov, S.V. Ilvitskaya, *Thesaurus architectural mentality of the XXI century*, Russian Architecture and Building **1-2,** 166-167 (2016)

Trial design of a dome roof for an church

Vadim Yu. Alpatov[1,*], *Oleg Yu.* Veremeenko[1], *Andrei A.* Sakharov[1] and *Vyacheslav S.* Shirokov[1]

[1]Samara State Technical University Institute of Architecture and Civil Engineering, Molodogvardeyskaya st. 194, 443001 Samara, Russia

Abstract. The paper describes some construction solutions for a metalic dome roof of an orthodox church. The authors describe three alternative variants (with different structural diagrams) of designing its dome framework. The paper introduces ribbed ring-shaped, grid and webbed geodesic dome constructions which are aimed to be used in the same operation conditions. The researchers calculated structural design, strength, proportioning of members and weight index as well as geometrically arranged elements of each dome design structure. Then their general total economic parameters were compared.

1 Introduction

Dome structures are often used successfully in round buildings [1-6, 11]. Dome roofs are usually designed for buildings of social entertainment (such as circuses and stadiums) and for religious buildings (such as orthodox churches). Dome roofs can be classified according to different characteristics, such as their purpose, design shape, size, rising height, materials used, etc. According to the dome construction, they are divided into ribbed, ribbed ring-shaped, webbed, radial beam and grid domes [3, 7].

Dome design mostly depends on technical-and-economic properties of its coating [7-9]. Complexity / simplicity of a dome calculations, manufacturing, final fitting-up and so on depend, in their turn, on this dome design. Above all, weight and price of a dome depend on this dome design. We find it possible to estimate efficiency of different dome constructions (being used in certain operation conditions) while applying optimal design theory. For that, we have to set out to calculate the minimum weight of the construction as a parameter optimization problem and to find a solution to the problem. Trial design is most frequently used to point out the most ergonomic dome structure in real life. To do that we need to work out several alternative variants of the construction in question and then to choose the best one taking into account a definite criterion [5-8, 10].

[*] Corresponding author: avu75@mail.ru

DOI: 10.1051/matecconf/20168602015

2 Calculate economic efficiency of the designed dome structures

The authors applied the trial design method when constructing a dome roof of the Orthodox Church. While designing this roof the authors chose ribbed ring-shaped, grid and webbed geodesic domes for comparative analysis. The researchers calculated structural design, strength, proportioning of members and weight index as well as geometrically arranged elements of each dome design structure. Geometrical arrangement of the dome different design structures was made while using AutoCAD and SolidWorks software systems. Static analysis of the dome different design structures was made while using LIRA software system. Joint connection deign was made by SolidWorks software system. Dome elements weight and its whole weight was calculated in Excel application. Dome different design structures made by SolidWorks software system can be seen in Figures 1-3.

Fig. 1. Ribbed ring-shaped dome.

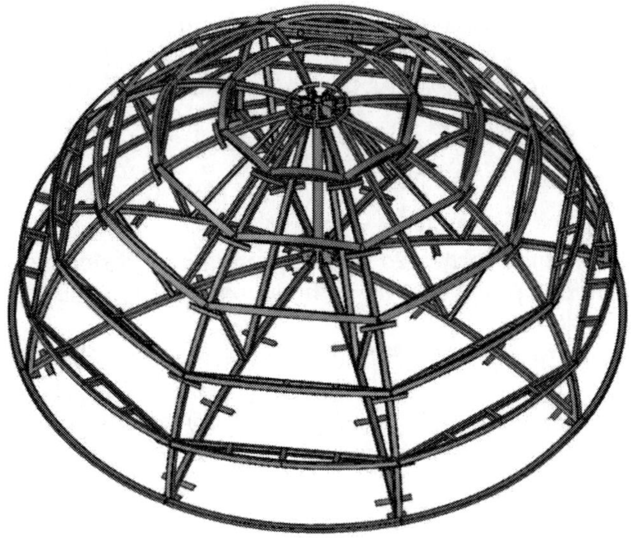

Fig. 2. Grid dome.

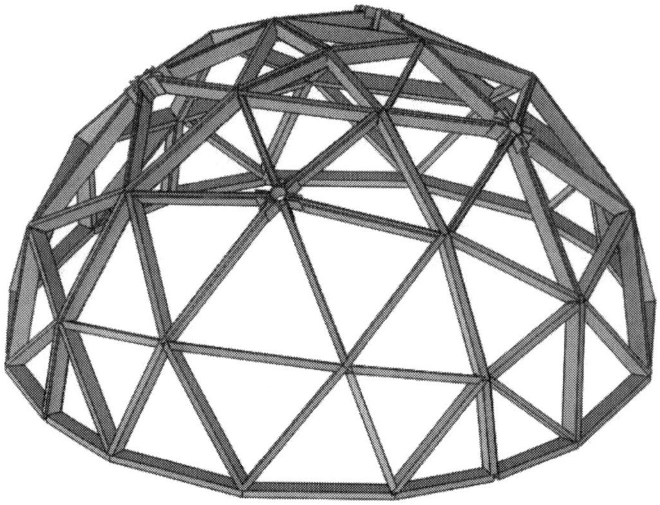

Fig. 3. Webbed geodesic dome.

To calculate economic efficiency of the designed dome structures the authors used Ya.M. Likhtarnikov's methodology [10]. According to this methodology to determine the cost of the construction one has to calculate the construction weight and its elements and units weight , labour hours, transport operation and construction assembling:

$$T = T^{prdct} + T^{trnsp} + T^{inst} \quad (1)$$

To calculate labour hours we need to define the necessary amount of carving, bending, drilling, welding, etc. All these calculations depend on the roof elements type (if its rolled stock or sheet), weight and amount. That is why three dome different design structures were calculated and their technical-and-economic properties were compared.

To calculate transport operations it was necessary to decide on the type of assembling, the type of deliverable assemblies and their quantity, to heft assemblies and choose the mode of transportation. To calculate assembling costs it was necessary to decide on the erection method and assembling vehicles and machinery.

While designing these three dome structures the researches did not take into account transport operation and assembling labour input. They found it possible to rely on the cost of materials required for the construction and the cost of its manufacturing (which was based on the required labour hours).

3 Conclusion

1. Taking into account the calculated dome different design structures and comparing their technical-and-economic properties the authors come to the conclusion that a ribbed ring-shaped dome is the best option.
2. The total cost of assembling a grid dome is 7% more than that of a ribbed ring-shaped dome.
3. The total cost of assembling a webbed geodesic dome is twice as much compared to that of a ribbed ring-shaped dome.

References

1. I.S. Kholopov, V.Yu. Alpatov, V.N. Mochalnikov, N.N. Moiseev, V.Yu. Veschin, *Actual problems of metal, timber and plastic constructions developing and modernizing (Civil Engineering and Transport): International Scientific-Technological Conference Materials* (SamGASA, Samara, 2002)
2. V.Yu. Alpatov, I.S. Kholopov, A.V. Solovjev, *Stroy-info: news bulletin* **22(214)**, 7–10 (2003)
3. A.V. Tur, I.S. Kholopov, M.D. Mosesov, *Traditions and innovations in architecture and civil engineering, 70th anniversary scientific and technical conference proceedings* (SGASU, Samara, 2013)
4. V.Yu. Alpatov, N.R. Ratzev, A.R. Ratzev, *Materialy IX Miedzynarodowej naukowi-praktycznej konferencji «Wschodnie partnerstwo – 2013»*, Budownictwo i architektura: Przemysl. Nauka i studia **32**, 40-43 (2013)
5. V.S. Shirokov, V.Yu. Alpatov, *Scientific electronic archival depository. URL: http://econf.rae.ru/article/7097 (Reference date: 19.11.2012)* (2012)
6. *XXI century: building materials, equipment, technologies*
7. V.Yu. Alpatov, O.V. Alpatova, A.V. Alpatov, A.A. Belyakova, Scientific Survey **24**, 51-56 (2015)
8. V.Yu. Alpatov, S.S. Igoshin, *Traditions and innovations in architecture and civil engineering, 67th scientific and technical conference proceedings* (SGASU, Samara, 2010)
9. V.Yu. Alpatov, M.I. Balzannikov, I.S. Kholopov, A.A. Lukin, Procedia Engineering **111**, 72-81 (2015)
10. Ya.M. Likhtarnikov, *Metal structures. Methods of technical economical analisys in design* (Stroyizdat, Moscow, 1968)
11. I.S. Kholopov, M.I. Balzannikov, V.Yu. Alpatov, Vestnik SGASU, Town Planning and Architecture **28(47)**, 225-232 (2012)

Coffered slabs as a perspective type of the reinforced concrete structures

Anton Kibkalo[1], *Mikhail* Volkov[1,*], *Anna* Vodolagina[1], *Vera* Murgul[1]

[1] Peter the Great Saint-Petersburg Polytechnic University, 195251 St. Petersburg, Russia

Abstract. The article discusses coffered slabs. In this paper considered the technology of arrangement of this slabs. Cast-in-place and precast ways of construction of coffered slab are reviewed. Cast-in-place and precast coffered slabs has been analysed in this article. Among other things construction of coffered slabs has an economical and technical advantages.

1 Introduction

Search of optimal types of structure is one of the motivations of science development in constructing which could save time and money and also don't give strength characteristics to its analogues.

One of the most important problems of constructing branch is reduce of building materials and weight of erected parts of building.

It is possible due to:

- applying new effective structures;
- use of prestressed elements of building;
- intensive introducing light weight concretes [1].

The introducing of effective cast-in-place structures only starts now in civil and industrial building in our country. Their introduction with no doubt will reduce cost price of erection and reconstruction of buildings. It depends of methods of design of such structures in domestic building branch that are bad developed.

2 The effective forms of coffered slabs

The effective forms of slabs are applied in many European countries in the construction industry, to reduce the weight of slabs constructed of heavy cast-in-place concrete slab.

Single-span reinforced concrete slabs with a solid cross section is the most economical to the length of span to 4,5 m. The span can be increased due to supporting the plate on the contour. In this case, it is possible to overlap a large area. In order to reduce consumption of concrete in the construction of floors, and consequently reducing loads on the supporting structures and foundations, it is advisable to use a cast-in-place ribbed ceiling [2].

Coffered slab (Fig. 1) is a kind of ribbed slab, it is also called waffle reinforced concrete floor.

* Corresponding author: w2olf@mail.ru

Fig. 1. Coffered slab.[3]

Coffered slab is a ribbed structure with reinforced ribs in the lower zone which have mutually perpendicular arrangement. In comparing with plate slab or cast-in-place slab with profiled sheet formwork, application of cast-in-place ribbed slab reduces the consumption of concrete for the construction of floors and vertical load bearing walls, foundations of buildings. Hence there is a reduction of costs for its construction. There is a reduction of its own weight of overlap - from 40% to 60%, while rigidity of the slab is increased, which allows to arrange large spans without intermediate supports devices. The resulting thickness of the slab is much less flat.

Fig. 2. Coffered slab in parking construction.[4]

Coffered slab consist of slabs supported on a contour of a system mutually perpendicular beams. Slabs have a rigid conjugation with beams and joint bearing on walls.

The optimal side ratio for rooms with coffered ceilings in the range 1 - 1.5. Beams can be placed either perpendicular or at an angle of 45 ° (a rare variant) to the sides of the overlapped areas. The normalized height of the beams in both directions should be the same and not less than 1/20 of the span. The thickness of the plates overlap coffered is 6 - 7 cm, the thickness of the ribs varies from 10 to 20 cm. [5, 6]

For devices coffered cast-in-place slab are used formwork modular system consisting of telescopic racks, lathing which is located with the size of formwork. The most common type of coffered self-supporting formwork system type «Skydome» (Fig. 3). All elements of the formwork is prefabricated; system does not require reinforcement or special protective devices. Boxes, which form voids between slabs, laid on the lathing. They have not a high adhesion to concrete and can be easily removed after a set of concrete required strength. [6]

Fig. 3. Coffered formwork system type «Skydome».[7]

Comparison of the technical and economic indicators of a cast-in-place coffered floor and solid cast-in-place slab under the same loads and boundary conditions shows that by reducing the weight of the first possible to increase the thickness of the floor up to 40% while saving 15% of the concrete by removing it from the stretched bottom zone. [8]

Table 1. Technical and economic indicators overlap with the device span 12 m, and short-term regulatory load 6 kN/m^2.

Characteristic	Thickness, cm	Concrete consumption, m^2	Consumption of the working reinforcement, kg / sq.m.
Solid cast-in-place	25.0	0.25	63.84
Coffered cast-in-place	46.0	0.19	24.2
Cost of materials, %	-	24	62.1

3 Types of manufacturing coffered

By way of manufacturing coffered floors are divided into cast-in-place and precast-monolithic. Precast-monolithic coffered floor includes prefabricated hollow concrete blocks the size from 0.2x0.2x0.6 to 0.3x0.3x0.8 m to m and in situ concrete. The blocks form geometric shape are closed on all sides. The layer thickness of the monolithic concrete under blocks - 5 - 6 cm. [9] Concrete blocks act as permanent formwork, they are left in the ceiling. Thus cast-in-place concrete forms vertical ribs in the tension zone and a solid floor in the compressed zone of overlap. In the area under supports (the intersection with columns) structure is arranged as a continuous cast-in-place slab. Reinforcing mesh is placed on top of blocks and the ends of reinforcing bars are set on the blocks in the area of solid cast-in-place plate. (Figure 4).

Fig. 4. Precast-monolithic coffered floor.[10]

In the manufacture of concrete blocks with a wall thickness of 2 cm is required at least 62-65% of the concrete mix than solid block similar size. Frame buildings with cast-in-place columns and precast-monolithic coffered floors much lighter, compared to the like of cast-in-place columns and solid monolithic slabs.

In the calculation of technical and economic parameters that characterize the consumption of the materials to device solid cast-in-place and precast-monolithic coffered ceilings it concludes that for the same consumption of concrete on the slab length of 6 m, and the same regulatory burden savings of the working reinforcement is about 48% (Table. 2). [11-16]

Table 2. Technical and economic indicators overlap with the device span 6 m, width 1.2 m, and short-term regulatory load 6 kN/m^2.

Characteristic	Thicknes, cm	Concrete consumption, m^2	Consumption of the working reinforcement, kg.
Solid cast-in-place	16.0	1.15	45.15
Coffered precast-monolithic with concrete blocks	25.0	1.15	28.8
Cost of materials, %	-	0.0	48.7

4 Conclusion

Therefore, we can conclude that the use of cast-in-place coffered slabs and precast-monolithic slabs allows to:

- to reduce the construction time by eliminating the need for installation of additional columns and foundations;
- reduce the total thickness of the floor compared to the beam structures;
- significantly reduce the consumption of material compared with a solid

monolithic slabs;
- use open space for free layout.

References

1. I. S. Loskutov, D. A. Glotov, (2015) Information on: http:///gbk.ru/concrete/ simulation /ing_fea/kessons_sim/kessons_sim.php
2. A.N. Malakova,Vestnik MGSU **1**, 79-86 (2013)
3. Information on http://www.rdconsultantsmw.com/?p=93
4. Information on https:///www.flickr.com/photos86792135/@N0410784419463/ inphotostream
5. P.F. Vakhnenko, *Calculation and design of parts of residential and public buildings* (Designer directory, Kiev, 1987)
6. *Manual for the design of concrete and reinforced concrete structures of heavy concrete without prestressing reinforcement (to set of rules 52-101—2003)* (CNIIPromzdanii, Moscow, 2005)
7. Information on http://www.calizausa.com/en/products/slabs/skydome
8. Information on http://www.progressbuild.ru/?page_id=159
9. *SP 52-101—2003. Concrete and reinforced concrete constructions without prestressing reinforcement* (Moscow, 2005)
10. Information on http://forums.justcommodores.com.au/general/248098-waffle-slab.html
11. Information on http:///gbk.ru/concrete/simulation/ing_fea/kessons_sim/ kessons_sim.php
12. A. Kibkalo, M. Lebedeva, M. Volkov, MATEC Web of Conferences **53**, 01051 (2016)
13. S. Ognjenovic, T. Sazonova, R. Ristov, I. Nedevska, A. Kibkalo, MATEC Web of Conferences **73**, 07002 (2016)
14. V. Murgul, Procedia Engineering **117**,813-823 (2015)
15. V. Korsun, N. Vatin, A. Franchi, A. Korsun, P. Crespi, Mashtaler S. Procedia Engineering **117**, 975-984 (2015).
16. V. Pukhkal, M. Tanić, N. Vatin, V. Murgul, Procedia Engineering **117**, 864-869 (2015)

The problems of calculating the load-bearing structures made of light steel thin-walled profiles

Vera Roy[1,*], *Oxana* Izvekova[1], and *Issa Togo*[1]

[1]Peter the Great St. Petersburg Polytechnic University, 29 Politechnicheskaya St., St. Petersburg, 195251, Russia

Abstract. The article presents the results of a study of bearing capacity of thin-walled cold-formed steel beam of the guide profile. Such profiles have a small thickness and complex cross-sectional shape. Bending deformation develops in the cross-sectional plane under the influence of loads in beam. In addition, deformation of constrained torsion and warping arise. These deformations influence the stress distribution at the points of the cross-section of the beam and thereby determine its load-bearing capacity.

1 Introduction

Today in Russia buildings with load-bearing frame of thin-walled cold-formed sections are widely used [1]. They have a number of technological and operational advantages (lightness, high speed of erection, etc.) [2].

As practice shows, often due to errors in the design of buildings of light steel thin-walled structures (LSTS), frequent cases of emergency [3]. Sustaining the work of such structures is a major problem to the widespread introduction of LSTS in massive construction. Obviously, the thin-walled structures at the design stage requires a thorough analysis of their bearing capacity. Not paying attention to the growing popularity of such structures and a large amount of research on all LSTS stage of development, today in Russia there are significant deficiencies in the prescriptive, methodological and computational databases of design of LSTS which does not make it easy and clear to make their calculation.

Theory of calculation based on the hypothesis of plane sections, are not applicable to the thin-walled bars, due to their small thickness and mismatch the centers of gravity and bending. To solve engineering problems of calculation of elements of thin-walled structures can distinguish two groups of ways to calculate: based on the modeling of shells and rods. The first group of methods is associated with the representation of a thin-walled rod in the form of a shell and further numerical calculation, as a rule, using the finite element method, in the program complexes SCAD, Lira, SOFiSTiK etc. Such methods of calculation are accurate enough, but very time-consuming in the engineering design activities especially in terms of the complex structure analysis. In the second group of methods can be

* Corresponding author: veraroya@gmail.com

distinguished analytical and numerical methods for calculation of thin rods associated with the introduction of additional "seventh" degree of freedom and the corresponding power factor – bimoment [2, 4, 5].

Study of the bearing capacity, deformability and manufacturing technology of steel structures of cold-formed sections are devoted to [6-21]. In [6, 11, 12] are considered the possibility of accounting the seventh degree of freedom in the creation of a thin-walled finite element.

2 Statement of the research problem

Let us analyze the degree of influence of component of warping on the stress-strain state of the thin-walled bar.

We will lead the calculation for cantilever balcony beam with cross section GP 254-1.5 [22] like for beam having strain of warping.

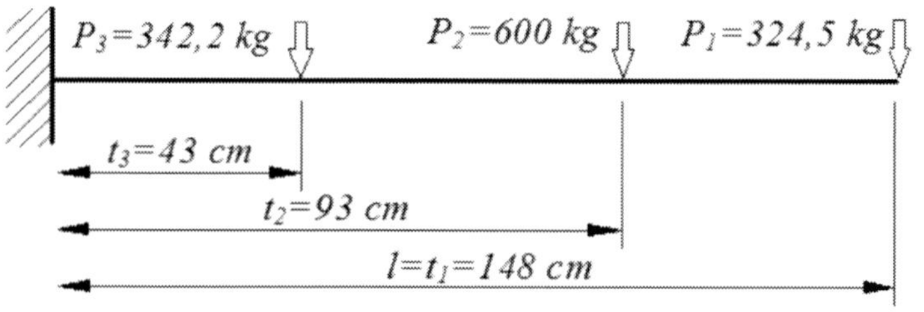

Fig. 1. Design scheme.

2.1 Calculation of the strength of solid section bent elements

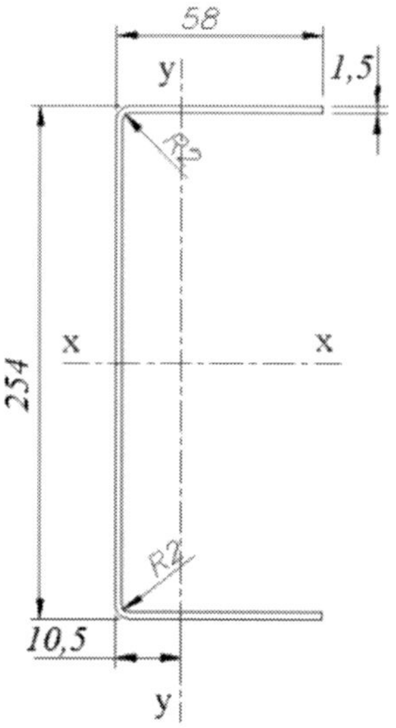

Fig. 2. Cross section [22].

Geometric characteristics in calculation are taken the lower of two: by assortment [22] and by calculation in the program TONUS (PC SCAD Office) – table 1.

Table 1. Geometric characteristics of cross section GP 254-1.5.

	Parameter	**Value according to STO[22]**	**Value according to TONUS**	**Units**
A	Cross sectional area	5.46	5.458	cm^2
I_t	Moment of inertia at free torsion	-	0.041	cm^4
I_w	Sectorial moment of inertia	-	1669.71	cm^6
W_{u+}	The maximum moment of resistance about an axis U	37.99	36.8	cm^3
W_{v+}	The maximum moment of resistance about an axis V	3.28	2.973	cm^3
I_u	The maximum moment of inertia	484.35	466.44	cm^4
I_v	The mininmum moment of inertia	16.53	14.34	cm^4
Y_b	Coordinate of the center of the bend on the x axis	-	-1.672	cm
Z_b	Coordinate of the center of gravity in the x direction	1.05	0.906	cm

Flexural-torsional characteristics (it is a physical constant of cross section and mathematically representing a characteristic number of differential equation of equilibrium 4th order) [23] amount to:

$$k = \sqrt{\frac{GJ_t}{EJ_\omega}} = 0.0030775cm^{-1} \tag{1}$$

Upon application of a uniformly distributed load and bringing it up to the maximum (destructive) value due to a sufficiently high ductility of the top shelf, which will be significantly deformed, the profile loses its original geometric shape. Load, in most cases being a "stamping" that is more rigid than the section by its nature, will not have tightly fit to the shelf. To indirect accounting for this geometric nonlinearity we will apply the load is not uniform along the shelf, but on the law of the triangle, which is a fairly common method. Then, the resultant load vector will pass through the center of gravity of the load diagram, lying in the intersection of the medians of the triangle, that is, at a distance (b/3) from the wall edge [4].

As can be seen from Fig. 3 full eccentricity of load application will consist of the eccentricity, due to the mismatch between the center of gravity and the center of the bending cross section of α_x, (equal Y_b- coordinate of shear center y-axis) and the direct eccentricity of application of the load equal:

$$e = \alpha_x + \frac{b}{3} = 1.672 + \frac{5.8}{3} = 3.605\,cm \tag{2}$$

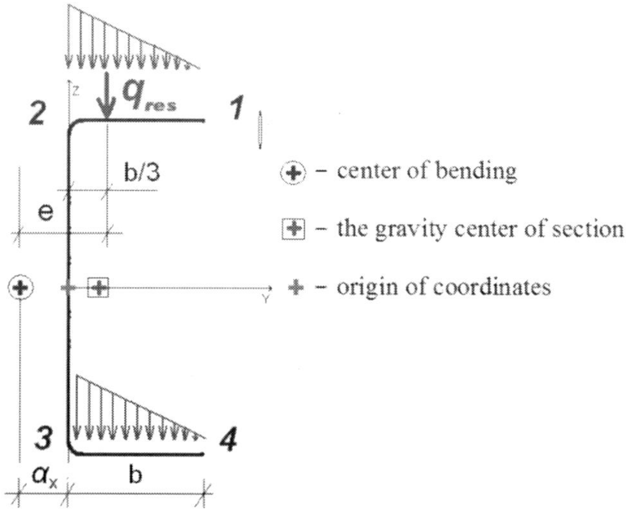

Fig. 3. Scheme of axes and load application, points numbering of σ_y.

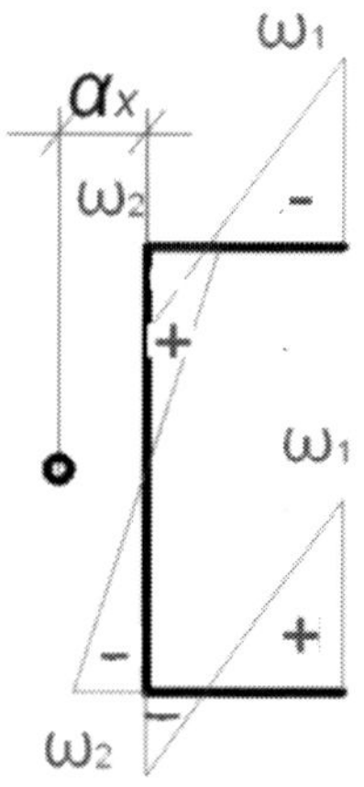

Fig. 4. Diagram of sectoral coordinates [2].

Diagram of sectoral coordinates of section is shown in Fig. 4. The maximum sectoral coordinates are necessary for detecting the stress from bimoment; they comprise:

$$\omega_1 = (\alpha_x - b)\frac{h}{2} = (1.672 - 5.8)\frac{25.34}{2} = -52.30\,cm^2 \tag{3}$$

$$\omega_2 = \alpha_x \frac{h}{2} = 21.19\,cm^2 \tag{4}$$

According to p. 8.2.1. SR 16.13330-2011[24], calculation of the strength of the beams should be carried out under the action of moments in the two principal planes (and bimoment available) according to the formula

$$\frac{M_x}{I_{xn}R_y\gamma_c}y \pm \frac{M_y}{I_{xn}R_y\gamma_c}x \pm \frac{B_\omega}{I_\omega R_y\gamma_c} \le 1 \qquad (5)$$

where x and y – the distance from the main axis to the point of the section under consideration; while the action in the wall of the beam moment and shear forces.

The bending moment in the reference section:

$$M = -P_1 t_1 - P_2 t_2 - P_3 t_3 \qquad (6)$$

$$M = -(324.5 \cdot 1.48 + 600 \cdot 0.93 + 342.2 \cdot 0.43) =$$
$$= 480.3 + 558.0 + 147.4 = 118540\ kg \cdot cm^2 \qquad (7)$$

where t – the distance from the end of beam to the point of load application,
P – applied load

Assuming that the bending occurs only in one plane, taking $M_y = 0$

The formula for the bimoment in the case of a cantilever beam, according to the technical literature [2]:

$$B = -\frac{1}{k}\frac{Pe}{ch(kl)}\left[sh(k(l-t)) - sh(kl)\right] \qquad (8)$$

$$B_1 = 162077.9\ kg \cdot cm^2 \qquad (9)$$

$$B_2 = 191558.4\ kg \cdot cm^2 \qquad (10)$$

$$B_3 = 51702.35\ kg \cdot cm^2 \qquad (11)$$

Bimoment in the reference section obtain as a superposition of moments of individual forces:

$$B = B_1 + B_2 + B_3 \qquad (12)$$

$$B = 162077.9 + 191558.4 + 51702.35 = 405339\ kg \cdot cm^2 \qquad (13)$$

The maximum normal stress σ_1, σ_2, σ_3, σ_4 occurring in the most tense cross-section, located in the middle of the span, will be different in the four characteristic points (1, 2, 3 and 4, respectively) of the cross-section will be:

$$\sigma_1 = \sigma_{x1} + \sigma_{\omega 1} = -\frac{M_x}{W_x} + \frac{B_\omega}{I_\omega}\omega_1 \qquad (14)$$

$$\sigma_1 = \frac{118540}{36.8} - \frac{405339}{1669.71} \cdot 52.30 = +3221.2 - 12696.4 = -9475.2\ kg/cm^2 \qquad (15)$$

Similarly, we calculate the value of the normal stress σ_2, σ_3, σ_4:

$$\sigma_2 = -8365.3\ kg/cm^2 \qquad (16)$$

$$\sigma_3 = -8365.3\ kg/cm^2 \qquad (17)$$

$$\sigma_4 = -9472.2\ kg/cm^2 \qquad (18)$$

Allowable stress:

$$\sigma_{per} = R_y \cdot \gamma_c = 3200 \cdot 0.9 = 2880\ kg/cm^2 \qquad (19)$$

2.2 The calculation using the finite element method (analysis of stress-strain state)

According to p. 8.2.1. SR 16.13330-2011 [24], values of stress in the wall of the beam must be checked according to the following formula in the two principal planes of bending:

$$\frac{0.87}{R_y\gamma_c}\sqrt{\sigma_x^2 + \sigma_y^2 - \sigma_x\sigma_y + 3\tau_{xy}^2} \le 1 \qquad (20)$$

$$\frac{\tau_{xy}}{R_y\gamma_c} \leq 1 \qquad (21)$$

where σ_x – normal stress in the median plane of the wall, parallel to the longitudinal axis of the beam;

σ_y – same, perpendicular to the longitudinal axis of the beam;
τ_{xy} – shear stress in the wall.

$$\tau_{xy} = \frac{Q \cdot S}{I \cdot t_w} \qquad (22)$$

Stress σ_x and σ_y, taken in formula (20) with their signs and τ_{xy}, should be determined in the same point of the beam wall.

For this, we use the finite element method in a static statement and solve the problem in the software package SCAD Office.

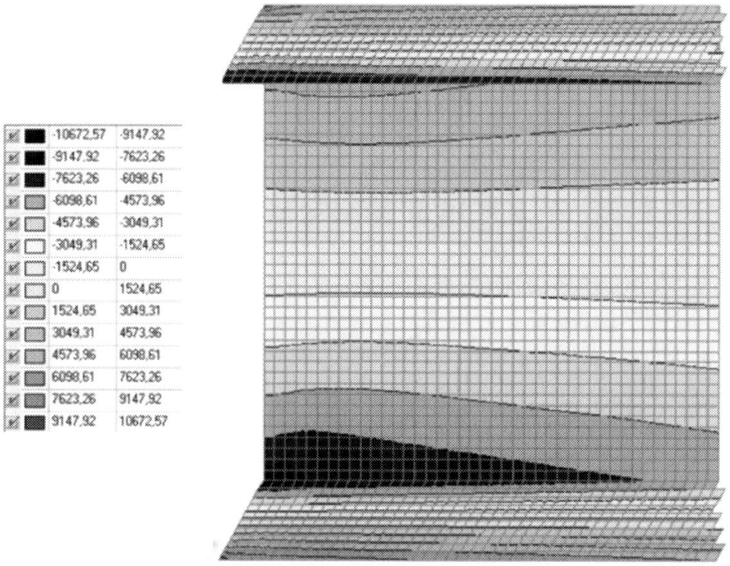

Fig. 5. Normal stress σ_y. Minimum value *–10672.57 kg/cm²*, Maximum value *10672.57 kg/cm²*

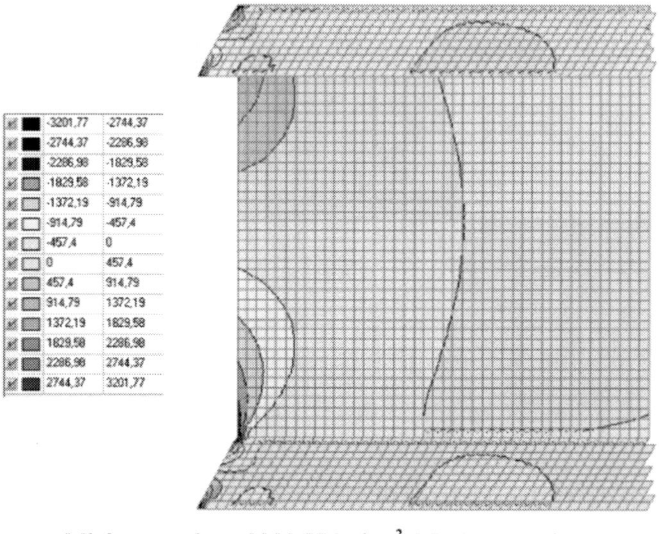

Fig. 6. Normal stress σ_x. Minimum value *–3201.77 kg/cm²*, Maximum value *3201.77 kg/cm²*

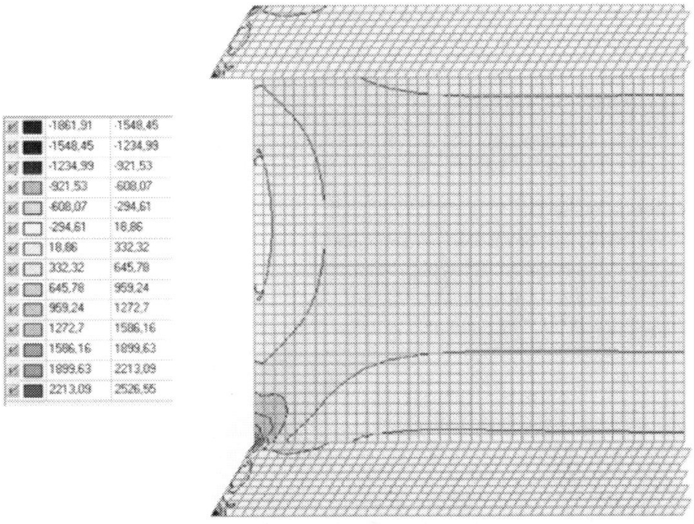

Fig. 7. Shear stress τ_{xy}. Minimum value *–1861.91 kg/cm²*, Maximum value *2526.55 kg/cm²*

For this, we use the finite element method in a static statement and solve the problem in the software package SCAD Office.

$$\frac{0.87}{3200\cdot 0.9}\sqrt{10672^2+3202^2-10672\cdot 3202+3\cdot 1862^2}= \\ =\frac{0.87}{3200\cdot 0.9}\cdot 10010=3.02\,kg/cm^2>1 \tag{23}$$

3 Conclusions

According to the calculation results, we can conclude that the curvature of the cross section of the thin-walled cold-formed beams significantly affects their stress-strain state. Contribution of bimoment's component is an essential.

In the analysis of the stress-strain state revealed that the strength condition is not met, under a given load the beam of will lose the bearing capacity.

References

1. N.I. Vatin, A.S. Sinel'nikov, Construction of Unique Buildings and Structures **1**, 47-53 (2012)
2. V.A. Rybakov, *Osnovy stroitel'noj mehaniki legkih stal'nyh tonkostennyh konstrukcij: ucheb. Posobie* (SPbGPU, SPb, 2011)
3. M.R. Garifullin, N.I. Vatin, Construction of Unique Buildings and Structures **6 (21)**, 32-57 (2014)
4. V.A. Rybakov, O.S Gamayunova, Construction of Unique Buildings and Structures **17**, 79-123 (2013)
5. V.A Rybakov, *Application of half-moving theory VI Slivkera for analysis of stress-strain state of thin-walled systems* (2012)
6. E.L. Ayrumyan, Erecting and Special Works in Construction **3**, 2-7 (2008)
7. V.V. Lalin, V.A Rybakov, A. Sergey, Applied Mechanics and Materials **578-579**, 858-863 (2014)
8. A.I. Alkhimenko, N.I. Vatin, V.A. Rybakov, *The light steel constructions techology* (SPbGPU, SPb, 2008)

9. V.A Rybakov, O.S Gamayunova, Stroyyprofil **1**, 128-130 (2008)
10. D.A. Trubina, D. Abdulaev, E. Pichugin, V.A. Rybakov, Applied Mechanics and Materials **633-634**, 982-990 (2014)
11. D.A. Trubina, D. Abdulaev, E. Pichugin, V.A. Rybakov, Applied Mechanics and Materials **633-634**, 1133-1139 (2014)
12. D.A. Trubina, E. Pichugin, V.A. Rybakov, M.R. Garifullin, O. Sokolova, Applied Mechanics and Materials **725-726**, 752-757 (2015)
13. V.A. Rybakov, A. Panteleev, G. Sharbabchev, E. Epshtein, Applied Mechanics and Materials 584-586, 1876-1880 (2014)
14. V.A. Rybakov, A. Sergey, Applied Mechanics and Materials 725-726, 746-751 (2015)
15. V.A. Rybakov, N. Molchanova, V. Laptev, A. Suslova, A. Sivokhin, MATEC Web of Conferences **53**, 01017 (2016)
16. N.I. Vatin, V.A. Rybakov, Stroyyprofil **2**, 60-64 (2007)
17. A.R. Tusnin, O.A. Tusnina, Industrial and Civil Engineering **8**, 62-65 (2012)
18. A.O. Gordeeva, N.I Vatin, Engineering and Construction Journal **3 (21)**, 36-46 (2011)
19. V.V. Lalin, V.A. Rybakov, S.A Morozov, Engineering and Construction Journal **1(27)**, 53-73 (2012)
20. V.V. Lalin, V.A. Rybakov, Engineering and Construction Journal **8 (26)**, 69-80 (2011)
21. A.R.Tusnin, O.A. Tusnina, Industrial and Civil Engineering **8**, 62-65 (2012)
22. *STO 96199376-002-2008 Standard Organization. Steel bent profile, for production of buildings and structures on the LSTC technology* (2008)
23. V.Z. Vlasov, *Tonkostennye uprugie sterzhni* (Strojizdat, Moscow, 1940)
24. *SP 16.13330.2011 Steel structures. Updated edition SNIP II-23-81*(2011)

Features technologies calculation of constructions with concrete-filled steel tubes

Ksenia Bolotova[1], *Sergey* Lukichev[1,*], *Vera* Murgul[1]

[1] Peter the Great Saint-Petersburg Polytechnic University, 195251 St. Petersburg, Russia

Abstract. In modern construction industry around the world constructions with concrete-filled steel elements are used more and more frequently. Usage of such structures allows substantially reduce the weight and cost of construction works (about 3 times) comparing with similar structures made of reinforced concrete. Labor costs are reduced in 5 times. Comparing with metal structures cost reduces significantly (up to 40%). Also the consumption of steel with a slight weight gain reduces in 3 times. In our country currently usage of concrete-filled steel structures is not widely used due to the lack of studies of such structures under different influences. Now there is no the normative base offering a calculation methodology of concrete-filled steel tubes, taking into account the modulus of elasticity in bending metal and concrete because at some point collaboration of materials is lost. This article presents the main trends in the usage of concrete-filled steel tubes in the construction. The advantages and disadvantages of such constructions are given, main constructive solutions are shown. The basic technologies of calculation of buildings with concrete-filled steel tubes are reviewed and analyzed.

1 Introduction

A good example of when concrete and steel are mutually help resist the force effect, which leads to an increase in the bearing capacity of the element, are the pillars of steel tubes filled with concrete. Working pipe-concrete rod elements has a number of significant features, compared with conventional reinforced concrete elements, reinforced rod fitting and also compared to the rigid concrete elements with an open structure reinforcement arranged inside the concrete section [1].

Features the work of compressed and compressed-bent concrete-filled steel elements are described in a number of works. In many ways, these descriptions are the same.

The main features of the concrete-filled steel tubes:

- High load-bearing capacity;
- Steel ferrule performs the role of longitudinal and transverse reinforcement;
- High reliability while using high-strength concrete kernel;
- High resistance of a building to seismic influences;
- Increased fire resistance of bearing elements comparing to steel frame structures;

* Corresponding author: serj_lu@mail.ru

- High durability and weather resistance of concrete;
- Lower frame weight [2]

Ultimately, all of this leads to a reduction in the consumption of material and money resources to manufacture concrete-filled steel elements. Concrete-filled steel are using effectively as highly stressed structural members and provide high security of buildings. Widespread using of concrete-filled steel tubes in our country is constrained by lack of domestic rules on their design and calculation. There is still no reliable calculation model for practical use of concrete-filled steel section in limit state, reflecting its specific characteristics. In view of this, it may be assumed that further researches in this area are promising. The idea of using metal pipes filled with concrete for manufacture of load-bearing columns of buildings was proposed by engineer John Lally. At present, concrete-filled metal pipes used as columns of the high-rise buildings or fencings of pits. It may appear that the concrete-filled steel constructions have no weaknesses. However, it's falce. For example, these constructions are very difficult to work together the metallic tubular casing and beton core at all stages of the operation. Tube starts working as ferrule with loads under which starts cracking [3-5].

2 Material and methods

Metal pipe in concrete-filled steel structures simultaneously performs the role of shell and the longitudinal and transverse reinforcement. It is able to perceive emerging efforts from different angles and in all directions. Concrete in tube-shell, being under the Triaxial compression withstands stress well above cube-strength and prism strength. Pipe prevents the formation of microcracks in concrete core, due to fullness of concrete probability of local buckling of pipes' walls and a total loss of stability in general is significantly reduced[6-7].

The existing standards and researches offer different methods for calculating steel shells, filled with concrete, off-center and axial compression. Widely known dependence for evaluation of bearing capacity of concrete-filled steel central compressed columns of circular cross-section:

$$N = cR_bA_b + \alpha R_cA_c \tag{1}$$

where c - the coefficient of the joint work of concrete and steel; α - the coefficient of hardening; R_b, A_b - tensile strength and cross-sectional area of concrete, respectively; R_c, A_c -yield point steel and cross-sectional area of the pipe, respectively.

Formula (1) does not fully reflect the physical essence of the bearing capacity of the concrete-filled steel tube, because the concrete compressed shell core is hardened, not the steel shell. Compression level depends on the ratio of the shell wall thickness and diameter of the concrete core. Most appeared later dependencies can be presented in the summarized form:

$$N = (cR_b + d)A_b + \alpha R_cA_c, \tag{2}$$

where c, d, α- coefficients.

Conditions for limit state:

- Achieving normal strains of the axial direction in the concrete core values of concrete strength under tri-axial compression $\sigma_{bz} = R_{b3}$;
- Achievement of the stress intensity in the most compressed fiber steel shell physical or adverse yield strength $\sigma_{pi} = \sigma_{p,y}$;
- Achieving the axial direction by normal strains in most stretched fiber steel shell yield $\sigma_{pz} = \sigma_{p,y}$.

Considering the internal static undetectable concrete-filled steel construction the first condition of occurrence of limit State should be performed in conjunction with the second or third (fig.1)

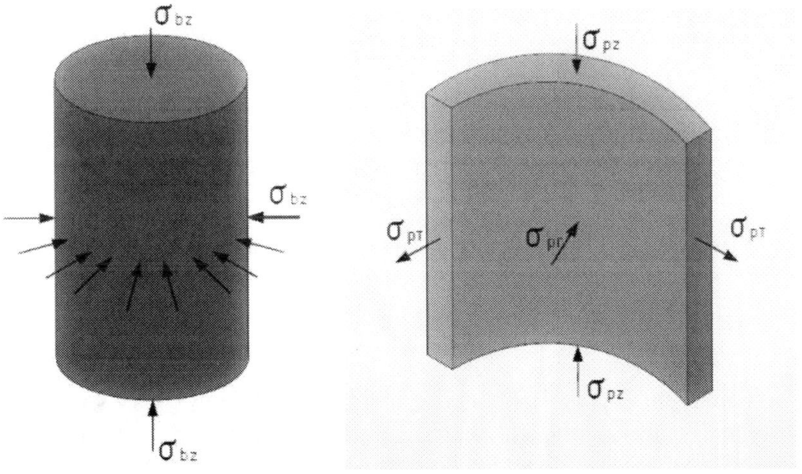

Figure. 1. Stress state of the concrete core and a steel shell in axial compression of concrete-filled steel tubes.

To ensure the serviceability of the columns under the action at design loads of magnitude of steel shell deformations should be limited to appropriate values [8-12].

In this regard, the strength calculation of normal sectional concrete-filled steel tubes in the general case of eccentrical compression should be made on the basis of nonlinear deformation of reinforced concrete model taking into account the peculiarities of deformation of the concrete core and a steel shell in terms of volumetric stress state. Deformation ability of materials and changing the coefficients of shear deformations in the concrete core and a steel shell stress level increases should also be considered for a more accurate calculation [13-15].

A common system for calculating the ratio of normal physical sections of concrete-filled steel tubes on strength, as in conventional deformation models take from the joint review:

- equations of equilibrium of external forces and inner forces in the normal section of element;
- equations, establishing the distribution of axial deformation of concrete and reinforcement for normal cross section, on the basis of a flat turn and flat bias section;
- dependencies linking stress and relative longitudinal deformation of concrete, steel cages and longitudinal reinforcement.

3 Results and Discussion

Consider the existing method for calculating pipe-concrete structures with metal shell. Consider the example of column height has a constant cross section and a steel tube of C235 steel with an outer diameter of 490 mm and a wall thickness of 8 mm, class B25 filled with concrete. The estimated height of the column is 3 m.

Column has the following loads:

- the longitudinal compressive force N = 4168 kN;
- bending moments M_y = 16.6 kNm, M_z = 51 kNm;
- shear force Q_y = 24 kN.

Let's calculate concrete-filled steel constructions using different approaches:

1) Reduction of concrete filled tube section to the steel

Under this approach, the composite cross section of pipe-concrete column is one material - steel, taking into consideration that in addition to the longitudinal force acting on the column and bending moments, so the column is calculated as the eccentric compression. And the calculation is carried out in two variants - considering only the elastic properties and taking into account the plastic deformation. And the calculation is performed in two version: with only elastic properties and considering plastic deformation.

a) While calculating the elastic stage area steel square is accepted A_s =117.8 cm^2, square concrete A_b = 1865 cm^2. The calculated resistance steel R_s = 235 MPa concrete R_b = 14.8 MPa. Reduction coefficient n = Eb/Es = 0.13.

The calculation is made according to the formula

$$\frac{N}{A_n} \pm \frac{M_y}{I_n} \leq R_n \gamma \quad (3)$$

With that in the margin of safety was thought that during the eccentrical compression the bending moment is perceived only a steel pipe. Working stresses were equal 142,5 MPa that substantially less of the calculated resistance R_s = 235 MPa. Factor of safety k = 235/142.5 = 1.65.

b) In the calculation taking into account the plastic deformation of concrete was taken into account the concrete elastic modulus decrease administration coefficients β = 0.45 and consequently reduction coefficient n = βE_b / E_s = 0,06. Operating voltage were equal to 201.9 MPa, which is also less than the calculated resistance of R_s = 235 MPa. Factor of strength with respect to k = 235 / 201.9 = 1.16.

2) The calculation of a concrete structure with a rigid reinforcement

In this case, the marginal bearing capacity of pipe-concrete column section, which was found to be N_{pr} = 5877 kN, which is higher than the force of N = 4168 kN, the margin of safety is k = 5877/4168 = 1.41.

3) The project jodnogo normative document, load-bearing capacity of the compressed pipe-concrete column is given by:

$$N \leq N_{per} = \varphi_{\gamma bs} (R_b A_b + \gamma_{s2} R_s A_s) \quad (4)$$

Here φ – buckling coefficient, R_b-concrete design resistance in the pipe-concrete:

$R_b = 0{,}65B(1 + \mu_{pb}\alpha)$

where α = 0,35; The class of concrete strength, μpb- reinforcement ratio, and μ_{pb} = 117.8 / 1865 = 0.063. In this case, Rb = 19.9 MPa.

The result is a N_{per} = 5543.4 kN, ie the margin of safety is k = 5543.4 / 4168 = 1.33.

4) Verification methodology using Eurocode 4

a) Calculation of pipe-concrete column axial compression according to international regulations Eurocode 4 gives the following meaning bearing capacity column N_{osszh} = 5476.72kN, ie margin of k = 5476.72 / 4168 = 1.29.

b) Calculation of pipe-concrete columns on the eccentric contraction accordance with international regulations Eurocode 4 gives the following meaning bearing pillar ability Nvnetsszh = 5335 kN, ie the margin of safety is k = 5335/4168 = 1.28.

We reduce the results in Table 1.

Table 1. Comparison of methodologies.

Methods	1a	1б	2	3	4a	4б
The bearing capacity of pipe-concrete columns, kN	-	-	5877	5543.4	5476.72	5335
Factor of safety k	1.65	1.16	1.41	1.33	1.29	1.28

4 Conclusion

Concrete-filled steel is a concrete enclosed in a metal pipe of round or a more complex cross-section. Concrete in such column is under triaxial compression and because the bearing capacity and deformability of concrete-filled steel in compression is highly increasing. Comparing the concrete-filled steel tubes with metal column with the same bearing capacity and weight reduction of metal consumption in case of concrete-filled steel can achieve the 50%. If reinforced concrete columns with the same amount of metal as a tubular sheath, it appears that the pipe-concrete section of the column at the same load bearing capacity can also be reduced by up to 50%.

Now filled with concrete constructions can be made not only on the basis of circular tubular membranes. It is possible to use polygonal tubular membranes of cold forming with different numbers of faces that are already used in some constructions. This promising modification of concrete-filled steel constructions allows to expand range of design solutions of steel-concrete composite construction that enhances their aesthetic qualities.

Comparative analysis of 4 calculation methods of strength shows that the calculation of concrete-filled steel tubes by using the second, third and fourth methodologies give similar results, while the calculation using the transformed section gives a significantly different results, moreover record of plastic deformations leads not to increasing but to reduction of safety factor in strength.

References

1. A. L. Krishan, A. I. Sagadatov, I. V. Atkishkin, UGTU-UPI, 108-110 (2004)
2. A. L. Krishan, AOPICUD, 725–733 (2005)
3. G. V. Nesvetaev, I. V. Rezvan, FR **12**, 580-583 (2011)
4. V. P. Mitrofanov, N. Dergam, PoltNTU, 91 (2008)
5. A. L. Krishan, V. V. Remnev, ICE **10**, 22–24 (2009)
6. I. M. Garanzha, MC **20**, 45-53 (2014)
7. I. A. Duvanova, I. D. Sal'manov, CUBS **21**, 89-103 (2014)
8. L. I. Storozhenko, A. V. Semko, UC **63**, 59-67 (2005)
9. I.I. Ovchinnikov, I.G. Ovchinnikov, G.V. Chesnokov, E.S. Mikhaldykin, IJN, 1-20 (2015)
10. J. Seo, H. Park, ASCE, 716-726 (2015)
11. U.K. Zolkafli, Z. Yahya, N. Zakaria, F.W. Akashah, A.S. Ali, SS **33**, 309-321 (2015)
12. C.R. Liñán, M.J. Morales Conde, P.R. De Hita, F.P. Gálvez, IJAH **9**, 324-340 (2015)
13. B. Carlo, C. Pietro, D. Vincenzo, F. Nora, IAARC, 1-8 (2015)
14. I. Stefanou, M. Fragiadakis, I.N. Psycharis, CMAS **37**, 61-82 (2015)
15. *SP 63.13330.2012, Concrete and won concrete construction. Design requirements,* (2013)

Determination of stress-strain state of the wooden church log walls with software package

Anastasia Chulkova[1], *Sergey* Lukichev[1,*] , *Marina* Romanovich[1]

[1] Peter the Great Saint-Petersburg Polytechnic University, 195251 St. Petersburg, Russia

Abstract. The restoration of architectural monuments is going on all over the world today. The main aim of restoration is the renewal of stable functioning of building constructions in normal state. In this article, we have tried to figure out with special software the bearing capacity of log cabins of the Church of Transfiguration on Kizhi island. As shown in research results, determination of stress-strain stage with software package is necessary for the bearing capacity computation as well as field tests.

1 Introduction

Kizhi is an island nearby the lake Onega. It is one of the greatest open-air museums in Russian Federation since 1960. Kizhi is a unique historical and cultural complex, which is considered to be a particularly valuable object of Russian cultural heritage. The architectural ensemble "Kizhi Pogost" belongs to UNESCO World Heritage site. The central object of the museum is the Church of Transfiguration, which was built in 1714. This building is unique because there are no nails in the whole construction [1].

The purpose of the study is research of the Church of Transfiguration wooden constructions, which are over the age of 300 years. The study was also expected to reason a need of construction reinforcement and to give some information for working up of the optimal reinforcement project. It is essential to build up a design model, which would work excellently, for using it while designing the reinforcement project of restoration.

Following theoretical and design problems are solving in the process:

- studying of the log cabins imperfections influence on a bearing capacity;
- studying of bearing zones and contact zones of logs and its intermediate elements;
- optimisation of diametres and spacing of the intermediate elements.

2 Material and methods

The constructive basis of the church is a pyramid. The basis of constructive volume of the church - a pyramidal shape. There are several design types in the building that are connected to each other shaping a unified form. A basement of the church is made of separate boulders. The bearing constructions above the basement are walls made of

* Corresponding author: serj_lu@mail.ru

horizontal logs – log cabins. These are vertical panels, which thickness is determined by diameters of the logs (20-30 cm). The panels are arranged to a plan, that consists of rectangles and octagons. The main octagon is connected to the basement by four walls. This octagon supports upper constructions, such as smaller octagons and rectangles and 22 domes. There are rectangles of log panels in each octagon for bearing upper smaller octagon [2-3].

Quadrangle are square in terms of intersecting walls. Each wall is made of several beams arranged in the same vertical plane one above the other with a gap, which is determined by the diameter of the log of the orthogonal faces. Logs quadrangle walls in areas of mutual intersection has no cuttings and freely reach the poles (diagonal walls), creating girder design. Collaboration logs in the quadrangle provides pads in the gaps. The building is middle and upper quadrangular create a support for the middle and upper octagons. This is the most crucial elements of the building, since they represent a large span structure made of conventional logs. This transferred the load of its own weight upstream structures are also snow and wind loads, the intensity of which increases with height in the middle and the upper quadrangle.

All of the above various types of design are combined in a single volume, the calculated scheme is very difficult to interpret. These circumstances, together with the natural imperfections of the wood, have a negative impact on the reliability of settlement schemes and require a cautious approach to the results of calculations [4-7].

The most problematic part of the church is the average bushel with the side faces approximately 9.6 m due to its design features and a large number of perceived stress. Deflections of beams medium quadrangular reach 14 cm (1/17 of flight), which exceeds the allowable.

The basic design scheme chosen the northern wall of the middle of the quadrangle. The following conditional premises were adopted:

Panel 5 consists of a timber with a length of 9.6 m, diameter 30 cm originally adopted;

load acting only in the vertical direction;

in spans load beams must pass each other through the spacers, which is now lost. The gaskets are included in the calculation scheme, adopted by their step 2 m from the mean;

laying modeled vertical links with stiffness corresponding to the logs;

The natural curvature of the timber is taken into account the position deviation of the average units of the vertical plane.

Loads timbered panel have been adopted according to LLC "Stroyrekonstruktsiya". It is uniformly distributed load from the overlying structures q = 5.3 T / m, attributable to the middle part of the top of the log. Also according to LLC "Stroyrekonstruktsiya" accepted deviations from the middle of the logs to the longitudinal axis (natural bending); to the north wall is:

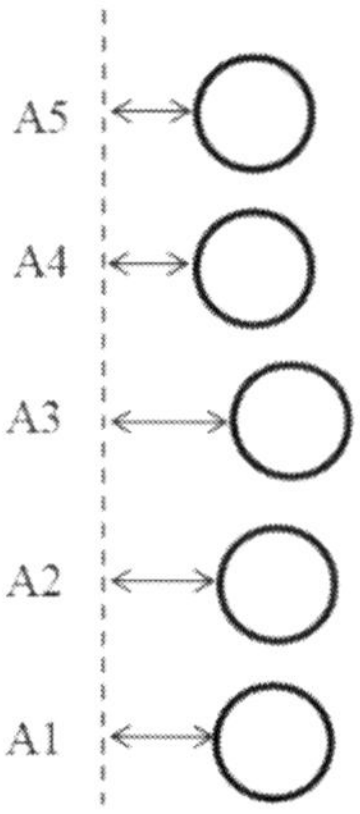

Fig. 1. Deviations logs.

Table 1. Setting Word's margins.

Marking	Value
A5	-
A4	0
A3	-15.5
A2	-9
A1	-8

For modeling of design scheme FEM tools were used. Design scheme was simulated as a truss model in the Lira SAPR software. Stiffness for truss elements were modelled by means of numerical descriptions: EF is a longitudinal stiffness, where $F = \pi D^2/4$ is the area of element; $EI_z = EI_y$ is a bending stiffness, where $I_z = I_y = \pi D^4/64$ are the moments of inertia. All these characteristics for logs and intermediate elements were calculated and provided in Table 1. Stiffness pads is taken to be the stiffness of logs, because they fill the distance between the logs equal to their diameter [8-11].

Table 2. Setting Word's margins.

Argument	Value
$EF,\ [T]$	70650
$EI_z = EI_y,\ [T \cdot m^2]$	397.4

3 Results and Discussion

The log ending junctions were taken into consideration by putting braces in ending joints of truss elements. There are prohibitions of straight displacements by the main axes (X, Y, Z) and rotation displacements by the X (longitudinal) axis on the one ending, and prohibition of straight displacement by Y and Z (vertical displacement) axis on the other ending. It was made for correct calculation and output data. The other joints are not fixed in space, but are stiff and provide rigid connections between longitude truss elements.

In contact zones of logs and its intermediate elements there are hinged ends. A hinge means the removal or restriction of bend stiffness of a truss bar ending and a joint of elements. A hinge may be set on the one ending of an element or on both by some degree of freedom. It is allowed to set angled (around the main axes) or linear hinges (along the axes). Appropriate stiffness should be set to hinges. If stiffness of a hinge is zero, it is called the straight hinge. If stiffness is set by a huge number, than a rigid bending between an element and a scheme joint. By varying these parameters correct design model working is attained. It should match the real construction behavior [12-15].

The most realistic functioning of the design scheme was achieved with following parameters of hinges: X1, Y1, Z1 = 0 (for displacements along the axes), Ux, Uy, Uz =1000 ton·m (for displacements around the axes) on both endings of an intermediate element.

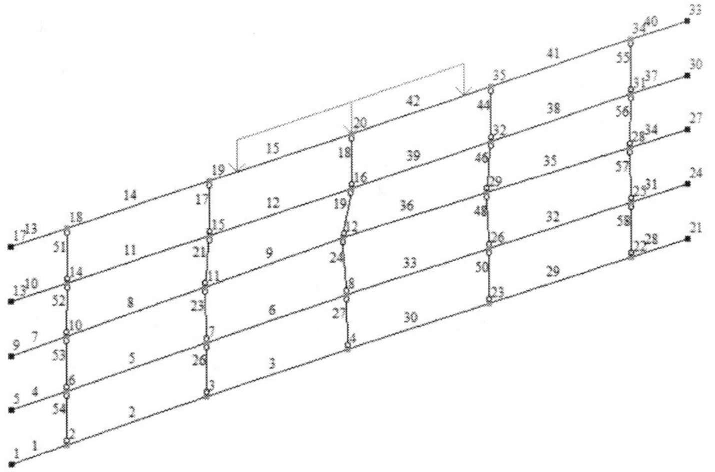

Fig. 2. The design scheme of a log wall.

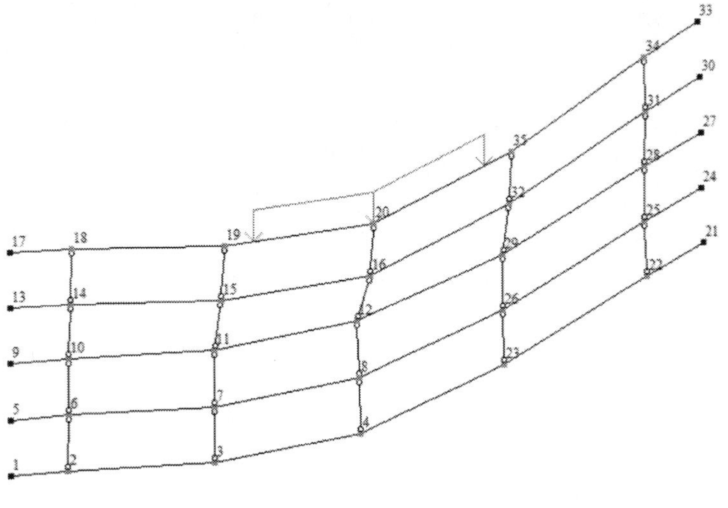

Fig. 3. The strain state.

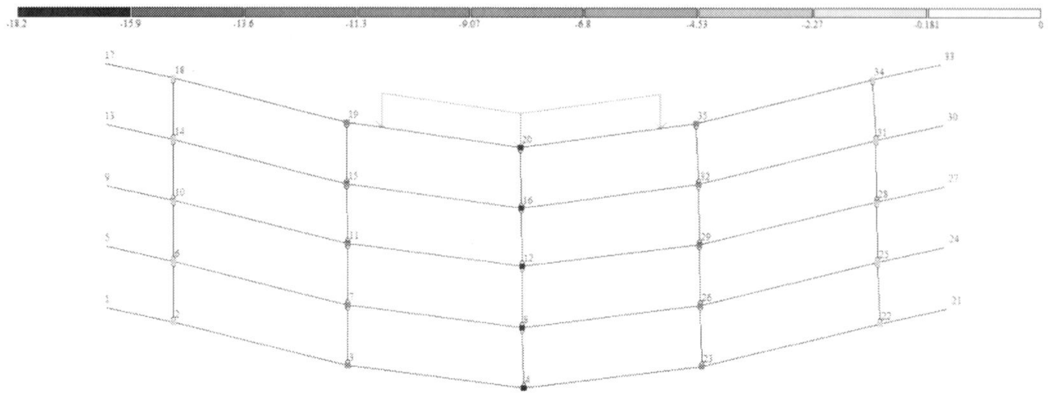

Fig. 4. Vertical displacements.

Due to considering of even spacing of the intermediate elements in the design model, the deflection of logs have decreased (1.82 cm instead of 14 cm according to fig. 4). In the central part of a log wall considerable internal forces emerge. By setting the intermediate elements it became possible to reallocate loads to another elements, and the forces decrease slightly (from -9.27 tons in the 18th element to -1.68 tons in the 27th one).

Table 3. The internal forces.

Element №	N (tons)
27	-1.68
24	-4.08
19	-6.52
18	-9.27

4 Conclusion

During the study process, the following conclusions were made:

- The construction of the log wall was analyzed accurately. The received data contains information about wall geometrical proportions, such as length, diameters, the logs ideal position deviation (original curvature). The log ending junctions and contact zones of logs and its intermediate elements were also taken in consideration;
- Each log in a wooden wall was modelled by using FEM tools, optimal fixing conditions and bonds of elements were picked over and appropriate loadings was set on elements of a design scheme;
- Special methods of accounting the intermediate elements between logs were worked out in design scheme;
- As a calculation result, the information about stress-strain state were recieved

The study shows that it is necessary to compare the received data with results of field tests and standart values. The conclusions of this comparison will be used for proving (or rejecting) a need of construction reinforcement in the restoration of the Church of the Transfiguration process. Furthermore, it can be used for restoration of other similar wooden buildings.

References

1. A. Onen, J. Jung, M. Dilek, D. Cheng, R. Broadwater, C. Scirbona, G. Cocks , S. Hamilton , X. Wang, EPCS **44**, 495-505 (2016)
2. G. Qiao, T. Li, Y. Frank Chen, CBM **105**, 435-447 (2016)
3. D. De León, C.A. González-Pérez, S. Díaz, D. Delgado, J.C. Arteaga, CRC, 2225-2229 (2012)
4. J. Seo, H. Park, ASCE, 716-726 (2015)
5. B. Serrano-Lanzarote, E. Fenollosa Forner, F. Arnau Paltor, IC **541**, (2016)
6. U.K. Zolkafli, Z. Yahya, N. Zakaria, F.W. Akashah, A.S. Ali, SS **33**, 309-321 (2015)
7. G.R. Petrozzi, F. Carbajal, C.J. Schexnayder, PPSDC **20**, (2015)
8. J.L. Yuan, Y.D. Yan, D. Huang, C. Du, AMM **694**, 95-101 (2014)
9. T. Tribulová, P. Kotlík, IJAH **9**, 352-366 (2015)
10. C.R. Liñán, M.J. Morales Conde, P.R. De Hita, F.P. Gálvez, IJAH **9**, 324-340 (2015)
11. E. Rodríguez-Mayorga, E. Yanes-Bustamante, A. Sáez-Pérez, IC **67**, (2015)
12. B. Carlo, C. Pietro, D. Vincenzo, F. Nora, IAARC, 1-8 (2015)
13. I. Stefanou, M. Fragiadakis, I.N. Psycharis, CMAS **37**, 61-82 (2015)
14. T. Aven, L.M. Pedersen, RESS **124**, 165-170 (2014)
15. *SP 64.13330.2011, Timber structures,* (2011)

Metal structural roofing with a flat truss with continuous chords as an enlarged assembly unit

Vadim Alpatov[1,*]

[1]Samara State Technical University Institute of Architecture and Civil Engineering, Molodogvardeyskaya st. 194, 443001 Samara, Russia

Abstract. When building a new block of a health and recreation center, designer metal structures were used in the roofing coverage. The over covered three storey building consists of two blocks. It has a variable height and a mixed framework. Both building blocks were over covered with structural lattice roofing of similar design. When designing spatial structural roofing, the authors proposed a new design concept. The novelty of the design lies in the fact that a modified flat truss is used as the basic assembly unit of a spatial structure. To develop a spatial structure design the roof-girder trusses with continuous chords from rolled I-beams and a lattice from roll-welded sections with non gusset joists were used. They are well-known and work well in practice. When designing structural roofing, previously known trusses of the mentioned design were slightly modified. In particular, trusses with parallel chords were given greater camber (2.5% instead of 1.5%), the design of a supporting truss joist was changed and fixing elements for chord and lattice members joining trusses into a spatial structure were added to truss joists.

1 Introduction

Metal structures are well-known for their architectural expressiveness. This advantage determines in many respects the popularity of their application as roof-covering structures of public entertainment objects which are visible. Sports facilities, fitness centers, shopping centers and other buildings are in the sphere of application of structural roofing in civil engineering.

2 Main Part

When constructing the new building of one of the city's health and recreation centers in Samara, designer metal structures were used in the roofing coverage. The over covered three storey building consists of two blocks (24x30 m and 36x18m). It has a variable height and a mixed framework. Both building blocks were over covered with structural lattice roofing of similar design.

[*] Corresponding author: avu75@mail.ru

When designing spatial structural roofing, the authors proposed a new design concept. The novelty of the design lies in the fact that a modified flat truss is used as the basic assembly unit of a spatial structure. To develop a spatial structure design the roof-girder trusses [1] with continuous chords from rolled I-beams and a lattice from roll-welded sections with non gusset joists were used. They are well-known and work well in practice.

The trusses of mentioned design have a number of advantages:

1) the usage of I-beam cross-section as a truss chord allows the application of such structures in roofing without roof-bearers;

2) continuous chords have smaller effective lengths;

3) the usage of effective tubes as a truss lattice allows to achieve saving metal;

4) the joist connections without sheet gussets for combining a lattice and chords saves metal, which according to various sources can be from 10 to 20% of the whole truss weight for roof girder trusses. [2]

Continuous truss chords used as an assembly unit of spatial structure allows to use the advantages of factory precast structures in full. In particular these are the well-known advantages of reducing the labor intensivity when manufacturing structures at the factory before assembling them at the installation site; the use of the advantages of mass and in-line manufacturing, and reduction of the labor intensivity of pre-assembly works, etc. Economic and technical advantages of assembling structures from large size assembly units were mentioned in lots of publications, including publications by the authors [3] - [7].

When designing structural roofing, previously known trusses of the mentioned design were slightly modified. In particular, trusses with parallel chords were given greater camber than in the prototype (2.5% instead of 1.5%), the design of a supporting truss joist was changed and fixing elements for chord and lattice members joining trusses into a spatial structure were added to truss joists.

The covering without roof bearers as an assembly unit is among the advantages of the designed structure solutions. To reduce the weight, the chord lattices of the structure are made thinned. The chord lattices contain only the minimum number of elements that join together flat trusses. The required quantity of joining chord elements is defined by the requirement to ensure geometric invariability of the structure in general.

The structure is intended for usage in a building with variable height, wherefore the possibility of accumulating a snow bag on the roofing was taken into account in the design. Anti-symmetrical load of considerable intensity due to the formation of a snow bag resulted in refusal to use the complete unification of assembly units. It was decided to design two types of flat trusses, which differ from each other in effective cross-sections of elements - trusses that happen to be in the area of a snow bag and trusses that do not happen to be in the area of a snow bag. Taking into account significant snow load with a higher value of the snow accumulation coefficient in the snow bag area caused considerable increase in material consumption. Despite this, the previously mentioned advantages of the proposed design solution helped to retain acceptable technical and economic parameters of the structure. Weight parameters of proposed structure can be comparable to parameters of well-known standard structural designs.

3 Conclusion

The building with the structure proposed by authors consists of two thermal blocks of various height. Both blocks are over covered by the proposed structures. General view of the building with roofing in the form of spatial structures is presented in Figures 1 and 2. Projection views of structural design are presented in Figure 3. Average consumption of the metal for manufacturing metal roofing was 25 kg/m^2.

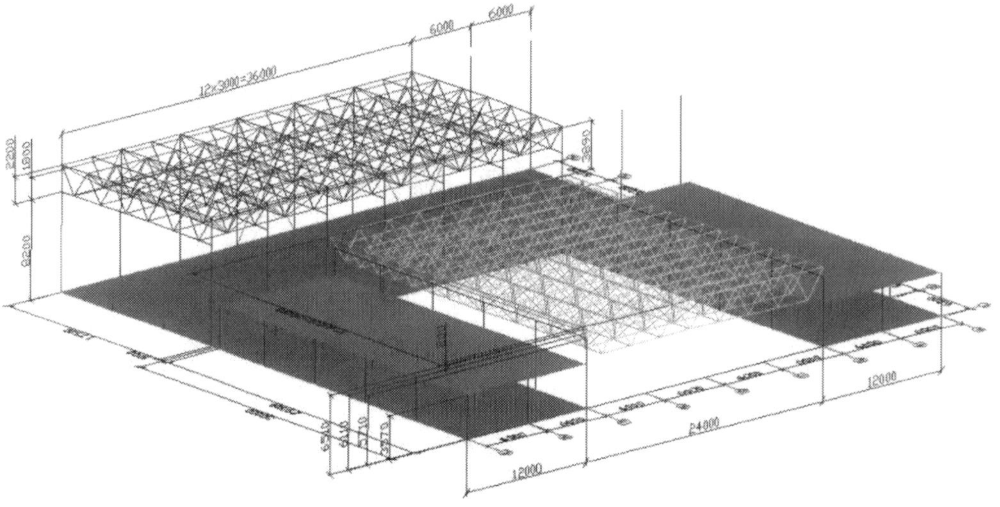

Fig. 1. General view of the building.

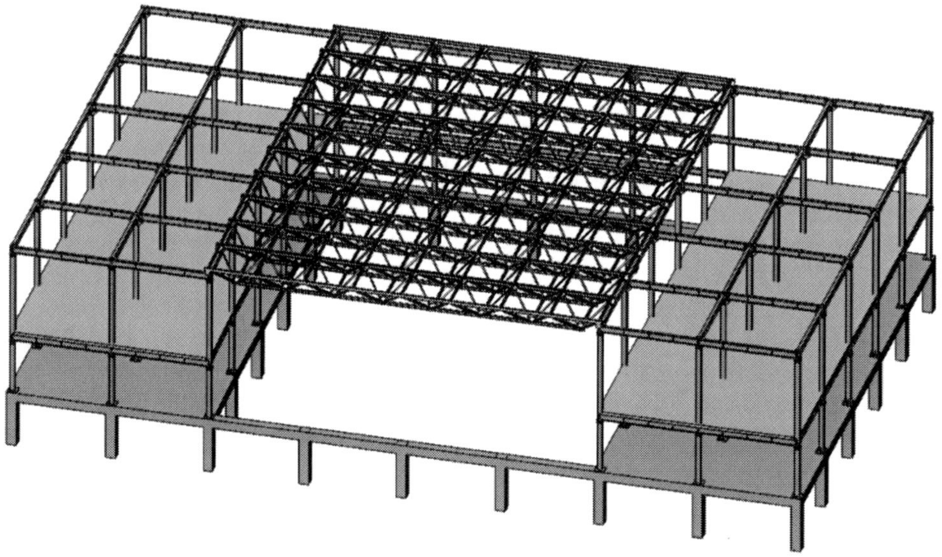

Fig. 2. Frame of the building block with structural roofing sized 24x30 m.

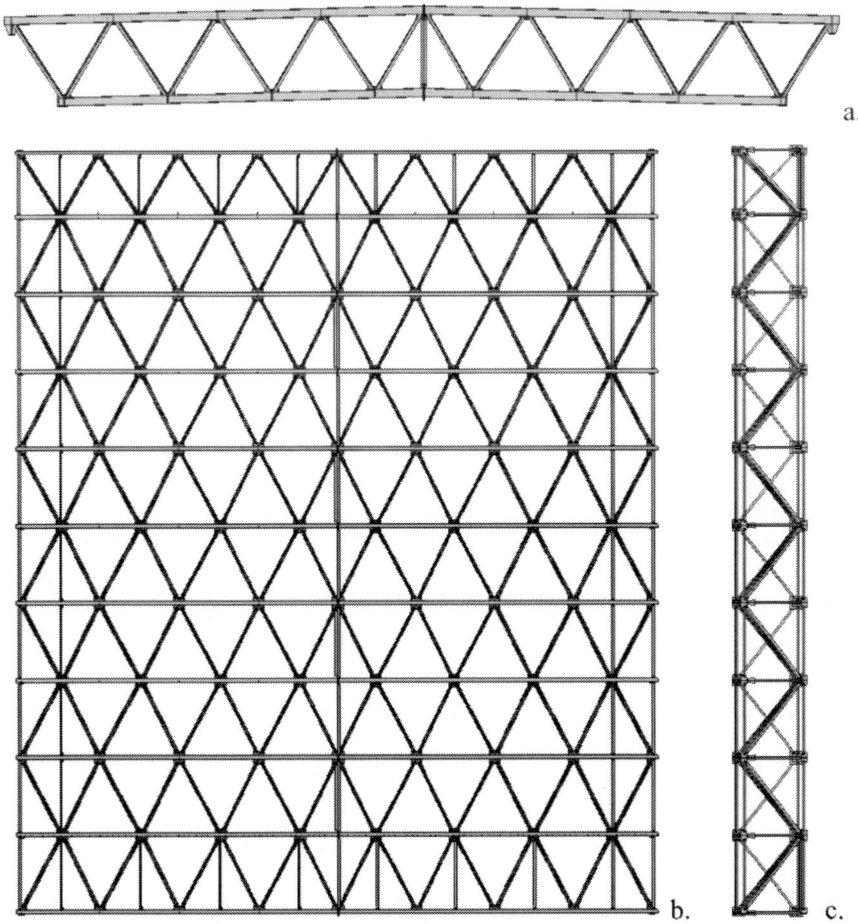

Fig. 3. Projection views of structural design.
a. - General view;
b. - Top view;
c. - Side view.

References

1. I. V. Levitanskiy, F.F. Kuklin, Sbornik nauchnyih trudov TsNIIproektstalkonstruktsii im. Melnikova, 52–61 (1984)
2. I.I. Ischenko, E.G. Kutuhtin, V.M. Spiridonov, Yu.N. Hromets, Light metal work of single-storey industrial buildings. Designer's Reference Book (Stroyizdat, Moscow, 1979)
3. V.Yu. Alpatov, I.S. Holopov, *Framed structure from long-sized elements [Patent]: 33381: A utility model patent. - Russian Federation* (2003)
4. V.Yu. Alpatov, *Spatial frame structure [Patent]: 2213185: patent for the invention - Russian Federation* (2007)
5. V.Yu. Alpatov, I.S. Holopov, A.V. Solovev, Stroy-info. Informatsionniy byulleten **22(214)**, 7–10 (2003)
6. M.I. Balzannikov, V.Yu. Alpatov, I.S. Kholopov, A.A. Saharov, A.O. Lukin, MATEC Web Conf. **73**, 1-6 (2016)
7. I.S. Kholopov, M.I. Balzannikov, V.Yu. Alpatov, A.V. Soloviev, Procedia Engineering **153**, 277-282 (2016)

MATEC Web of Conferences **86**, 02021 (2016) DOI: 10.1051/matecconf/20168602021
IPICSE-2016

Bar optimal design with the use of out-form

Vadim Yu. Alpatov[1,*], *Andrei A.* Sakharov[1], *Oleg Yu.* Veremeenko[1] and *Alexey A.* Lukin[1]

[1]Samara State Technical University Institute of Architecture and Civil Engineering, Molodogvardeyskaya st. 194, 443001 Samara, Russia

Abstract. The paper describes steel girders optimal design. Steel girders optimal design is based on the strength theory introduced by M. Huber, R. Mises, H. Hencky and on the theory of mirror functions. With the help of out-form volume we can either calculate the optimal form for a girder with variable cross-section or the optimal height for a girder of uniform cross-section. The paper introduces girders calculations according to the authors' methodology. For use this methodology effect of steel saving is about 10-20% compared with a girder or a beam with uniform cross-section. An example of determining optimal shape of a compressed-bend girder the paper describes.

1 Introduction

There are several well-known methods of building structures optimal design [1, 2, 5, 6, 7, 8] that can be applied while designing building structures cross-beams or girders. In this paper, the authors present an example of determining optimal shape of a compressed-bend. The paper introduces girders calculations according to the authors' methodology.

2 Bar optimal design with the use of out-form

According to theory [3], strength in this case is provided on condition that:

$$(1+\mu)\cdot\frac{R_y^2}{3}\cdot E \qquad (1)$$

For a plane stress problem, direct stress in i point of the bar can be determined from the formula:

$$\sigma_{ic}=\frac{N_{ic}}{F_{ic}}+\frac{M_{ic}}{W_{ic}} \qquad (2)$$

Specific potential energy of the shape change in i point of the bar is

$$\frac{1+\mu}{3\cdot E}\cdot\left(\frac{N_{ic}}{F_{ic}}+\frac{M_{ic}}{W_{ic}}\right)^2+\frac{1}{2\cdot G}\cdot\frac{Q_{ic}^2\cdot S_{ic}^2}{I_{ic}^2\cdot\sigma_{ic}^2}\leq[u]_{\Phi} \qquad (3)$$

---

[*] Corresponding author: avu75@mail.ru

As a result of transformations we can introduce a new equation from the strength condition. This equation is mathematically determined as non-linear modular form i bar capacity [4-6]:

$$V = \frac{l_i}{R_i \cdot \sqrt{[N_{ic}^2 + 2 \cdot N_{ic} \cdot M_{ic} \cdot \alpha_{ic} + M_{ic}^2 \cdot \alpha_{ic}^2 + 3 \cdot Q_{ic}^2 \cdot \beta_{ic}^2]}} \tag{4}$$

where $\alpha_{ic} = a_c / r_c^2$ and $\beta_{ic} = S_{ic}/(r_c^2 \delta_c)$.

Shape coefficients αi и βi are cross sections accessible characteristics and can be easily derived from gauge material or calculated. With given cross-sectional height, cross-section form coefficients α and β are taken as constant.

For a statically indeterminate bar with uniform cross-section the volume function can be presented as it follows:

$$V_i = l_i \sqrt{\frac{0.433}{v_i [U]_\phi E} \begin{bmatrix} (N_{ip} + \sum N_{ij} x_j)^2 + 2 \cdot (N_{ip} + \sum N_{ij} x_j)(M_{ip} + \sum M_{ij} x_j)\alpha_i + \\ + (M_{ip} + \sum M_{ij} x_j)^2 \alpha_i^2 + 3 \cdot (Q_{ip} + \sum Q_{ij} x_j)^2 \beta_i^2 \end{bmatrix}} \tag{5}$$

In general terms, Formula 5 can be given as

$$V_i = F_i l_i = \sqrt{\sum_{j=1}^{n} \sum_{k=1}^{n} A_{ijk} x_j x_k + 2 \sum_{j=1}^{n} B_{ij} x_j + C_i} \tag{6}$$

where

$$A_{ikj} = \frac{l^2 \cdot 0.433}{v_i [U]_\phi E} \left[(N_{ik} N_{ij} + N_{ik} M_{ij} \alpha_i + M_{ik} N_{ij} \alpha_i + M_{ik} M_{ij} \alpha_i^2) + 3 Q_{ik} Q_{ij} \beta_i^2 \right] \tag{7}$$

$$B_{ij} = \frac{l^2 \cdot 0.433}{v_i [U]_\phi E} \left[(N_{ip} N_{ij} + N_{ip} M_{ij} \alpha_i + M_{ip} N_{ij} \alpha_i + M_{ip} M_{ij} \alpha_i^2) + 3 Q_{ip} Q_{ij} \beta_i^2 \right] \tag{8}$$

$$C_i = \frac{l^2 \cdot 0.433}{v_i [U]_\phi E} \left[(N_{ip}^2 + 2 N_{ip} M_{ip} \alpha_i + M_{ip}^2 \alpha_i^2) + 3 Q_{ip}^2 \beta_i^2 \right] \tag{9}$$

If this statically determinate bar is variable in its cross-section along its length, we can find the optimal shape or height for a uniform cross-section bar while using the form coefficient α and β. For that we need to find the dependence α and β from the cross-sectional height h [9]. If we take I-beams (according to All Union State Standard - GOST - 8239-89) and compare all the coefficients α and β for outer fibers, fibers on flange and wall junction and neutral fibers we'll see that there is an approximate hyperbolic dependence between 3 and α and linear dependence between h and β (as Fig. 1 shows).

Paper [7] introduces statistical research proving that the radius of cross-section gyration of a defined height h-const in flexural and compressed flexural elements does not depend on their other characteristics if their construction and calculation constraints are taken into account. The radius of gyration changes only if cross-sectional height is changed. We can see almost the same dependence in respect of the width of (flange) section b.

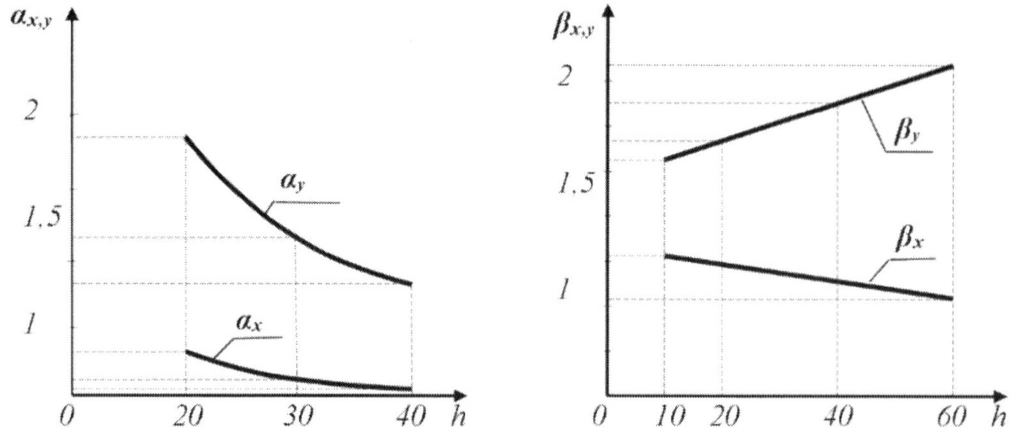

Fig. 1. Dependency diagrams of cross-section form coefficients for I-beams (according to All Union State Standard (GOST 8239-89) α_x, α_y, β_x, β_y from cross-sectional height h.

Tables 1 and 2 show α, β coefficients for rolled sections of different types.

Table 1. Cross-section form coefficients α_x, α_y.

Cross-sectional height *h* (cm)	**All Union State Standard (GOST) 26020-83**	α_x **(1/cm)**	α_y **(1/cm)**	**All Union State Standard (GOST) 26020-83**	α_x **(1/cm)**	α_y **(1/cm)**	**All Union State Standard (GOST) 8239-89**	α_x **(1/cm)**	α_y **(1/cm)**
10	–	–	–	–	–	–	I 10	0.303	1.848
20	–	–	–	20SH1	0.146	0.576	I 20	0.146	1.167
30	30B1	0.098	0.748	30SH2	0.093	0.435	I 30	0.100	0.939
	30B3	0.095	0.692	30SH4	0.092	0.426	–	–	–
40	40B1	0.076	0.674	40SH2	0.068	0.281	I 40	0.077	0.844
	40B3	0.072	0.611	40SH4	0.068	0.279	–	–	–
50	50B1	0.061	0.562	50SH2	0.056	0.299	I 50	0.063	0.815
	50B3	0.058	0.514	50SH4	0.057	0.306	–	–	–
60	60B2	0.049	0.447	60SH2	0.048	0.291	I 60	0.054	0.758
	–	–	–	60SH4	0.051	0.317	–	–	–
70	70B2	0.043	0.461	70SH2	0.043	0.307	–	–	–
	70B4	0.041	0.400	70SH4	0.044	0.314	–	–	–
80	80B4	0.038	0.408	–	–	–	–	–	–
100	100B2	0.031	0.361	–	–	–	–	–	–
	100B4	0.031	0.344	–	–	–	–	–	–

Cross-section form coefficients for rolled sections, fabricated sections of simple and complex types, U-sections, etc can be approximate by functions like the following:

$$\alpha = a_1 + a_2 / h \qquad \beta = c_1 + c_2 \cdot h \tag{10}$$

For example, the following approximations correspond to wide-flange I-beam gauge:

$$\begin{aligned} \alpha_x &= 0.00318213 + 2.6285/h \\ \beta_x &= 4.302782 - 0.019985 \cdot h \\ \alpha_y &= 0.194298 + 16.4660/b \\ \beta_y &= 3.9211329 - 0.44453 \cdot b \end{aligned} \tag{11}$$

Let us put functions (10) into the strength condition (3). Here we get a modular form of the volume as regard to h in k point of the bar.

$$V_{ic}(h)=F_i l_i = l_i / R\sqrt{[N_{ic}^2+2\cdot N_{ic}M_{ic}(a_1+a_2/h)_{ic}+M_{ic}^2(a_1+a_2/h)_k^2+3\cdot Q_{ic}^2(c_1+c_2h)_{ic}^2]} \quad (12)$$

For a part of the bar with d_z length the elementary capacity can be presented as an out-form for n fiber points:

$$dV = \| V_1(h) \| V_2(h) \| V_3(h) \| .....V_n(h) \| \, dz \quad (13)$$

Table 2. Cross-section form coefficients β_x, β_y.

Cross-sectional height h (cm)	All Union State Standard (GOST) 26020-83	β_x (b/r)	β_y (b/r)	All Union State Standard (GOST) 26020-83	β_x (b/r)	β_y (b/r)	All Union State Standard (GOST) 8239-89	β_x (b/r)	β_y (b/r)
10	–	–	–	–	–	–	I 10	3.101	5.080
20	–	–	–	20SH1	3.45	2.15	I 20	2.917	5.834
30	30B1	2.724	5.233	30SH2	3.714	4.349	I 30	2.725	6.296
	30B3	2.929	4.880	30SH4	3.656	4.297	–	–	–
40	40B1	2.538	5.576	40SH2	3.934	4.211	I 40	2.503	6.542
	40B3	2.776	5.063	40SH4	3.836	4.209	–	–	–
50	50B1	2.493	5.615	50SH2	3.467	4.489	I 50	2.321	6.925
	50B3	2.748	5.126	50SH4	3.174	4.642	–	–	–
60	60B2	2.732	5.145	60SH2	3.209	4.662	I 60	2.321	7.202
	–	–	–	60SH4	2.686	5.165	–	–	–
70	70B2	2.519	5.486	70SH2	2.925	4.920	–	–	–
	70B4	2.603	5.233	70SH4	2.647	5.136	–	–	–
80	80B4	2.434	5.548	–	–	–	–	–	–
100	100B2	2.372	5.772	–	–	–	–	–	–
	100B4	2.424	5.562	–	–	–	–	–	–

3 An example of determining optimal shape of a compressed-bend girder

Let us analyze Figure 2, which shows the process of determining optimal shape of a hinge-supported compressed-bend girder of a hydraulic gate [11, 12, 13] with 12 m span. Steel design resistance is R=240 MPa = 24 kN/cm^2.

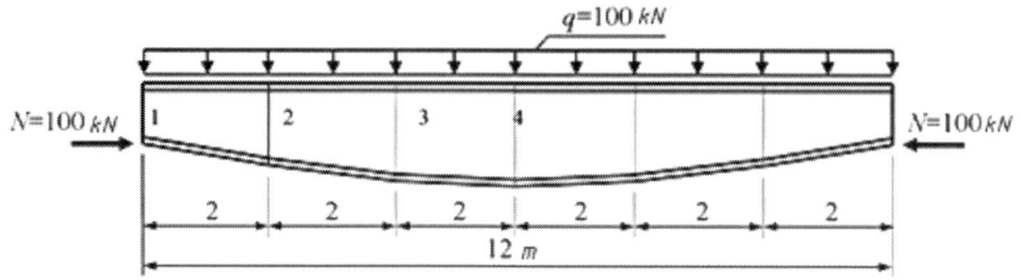

Fig. 2. Compressed-bend girder.

Axial force in a hinge-supported girder is usually applied on the pier level. As Force N for a beam with uniform cross-section changes the bending moment in the middle of each cross-section, let us determine cross-sectional height in the middle of the span according to strength condition:

$$W_x = M_{\max} / R = (ql^2/8)/R = (50 \cdot 12^2/8)/24 = 900/4 = 3750\ cm^3 \tag{14}$$

Gauge material for such a moment corresponds to cross-section 60SH2, with W_x=4490cm^3.

From the condition of deflection limitation in the middle equal to 1/400L=3 cm we have:

$$J_x = (5/48) M''_{max} l^2 / 3E = (5/48)(50 \cdot 12^2/8)/(1.2 \cdot 2.05 \cdot 10^4) = 175609\ cm^4 \tag{15}$$

It corresponds to I-beam 70 SH2.

So, from now on we are going to make use of wide-flange I-beams and use the following heights: H=(70, 60, 50, 40, 30, 20). According to H we'll calculate cross-section form coefficients α_{x1}, α_{x2}, α_{x3}, β_{x1}, β_{x2},, β_{x3}, and then put them together in Table 3. Here we use gauge material according to GOST 26020-83. Points 1, 2, 3 consequently correspond to outer fibers, fibers on flange and wall junction and neutral fibers.

Table 3. Cross-section form coefficients for the line *h*.

Height *h* **Coef.**	**70**	**60**	**50**	**40**	**30**	**20**
$\alpha_{x,1}$	0.043	0.042	0.058	0.069	0.093	0.146
$\beta_{x,1}$	0	0	0	0	0	0
$\alpha_{x,2}$	0.0395	0.038	0.054	0.065	0.096	0.1414
$\beta_{x,2}$	1.976	1.986	2.019	2.796	2.747	3.15
$\alpha_{x,3}$	0	0	0	0	0	0
$\beta_{x,3}$	2.742	2.71	2.816	3.49	3.68	3.45

We take into account the symmetry of the girder and analyze cross-sections spaced at intervals of 2 m.

1) Optimal design of a hinged beam with variable cross-section starts with determining its bearing cross-section. In bearing cross-section 1 we have $N_x=100$ *kN*; $M_x=N_x \cdot h/2=100 \cdot h/2$; $Q_y=300$ *kN*. Let us take $h=30cm$ cross-sectional height. Then $M_x=100 \cdot 15=1500$ *kN·cm*. The area we need is calculated according to equations (4).

$$F_{11} = 1/24\sqrt{\left[100^2 + 2 \cdot 100 \cdot 1500 \cdot 0.093 + 1500^2 \cdot 0.093^2\right]} = 9.98cm^2$$

$$F_{12} = 1/24\sqrt{\left[100^2 + 2\cdot 100\cdot 1500\cdot 0.096 + 1500^2\cdot 0.096^2 + 3\cdot 300^2\cdot 2.747^2\right]} = 65.93\,cm^2$$

$$F_{13} = 1/24\sqrt{\left[100^2 + 2\cdot 100\cdot 1500\cdot 0 + 1500^2\cdot 0 + 3\cdot 300^2\cdot 3.68^2\right]} = 79.78\,cm^2$$

We take $h=30cm$ and area $F_1=79.78cm^2$ along neutral fiber where both shearing force and axial force are quite strong.

2) Let us analyze the cross-section at the distance of *2 m* from its supporting structure. Let us take $h=40cm$ cross-sectional height. There are forces $N_x=100\ kN$; $M_x=(300\cdot 2-50\cdot 2\cdot 1)-100\cdot 0.1=400\ kN\cdot m=40000\ kN\cdot cm$; $Q_y=300-100=200\ kN$ in the cross-section. The area we need is

$$F_{21} = 1/24\sqrt{\left[100^2 + 2\cdot 100\cdot 40000\cdot 0.069 + 40000^2\cdot 0.069^2\right]} = 117.8\,cm^2$$

$$F_{22} = 1/24\sqrt{\left[100^2 + 2\cdot 100\cdot 40000\cdot 0.065 + 40000^2\cdot 0.065^2 + 3\cdot 200^2\cdot 2.796^2\right]} = 119.52\ cm^2$$

$$F_{23} = 1/24\sqrt{\left[100^2 + 2\cdot 100\cdot 40000\cdot 0 + 40000^2\cdot 0 + 3\cdot 200^2\cdot 3.68^2\right]} = 53.3\,cm^2$$

It is clear that in cross-section 2 its area depends on the bending moment and on the tension at flange and wall junction.

Let us take $h=50cm$ cross-sectional height. Let us determine the forces by eccentricity. Here we have $N_x=100\ kN$; $M_x=(300\cdot 2-50\cdot 2\cdot 1)\ -\ 100\cdot 0.05=\ 450\ kN\cdot m=\ 45000\ kN\cdot cm$; $Q_y=300-100=200\ kN$.

$$F_{21} = 1/24\sqrt{\left[100^2 + 2\cdot 100\cdot 45000\cdot 0.058 + 45000^2\cdot 0.058^2\right]} = 112.9\,cm^2$$

$$F_{22} = 1/24\sqrt{\left[100^2 + 2\cdot 100\cdot 45000\cdot 0.054 + 45000^2\cdot 0.054^2 + 3\cdot 200^2\cdot 2.019^2\right]} = 109.4\ cm^2$$

$$F_{23} = 1/24\sqrt{\left[100^2 + 2\cdot 100\cdot 45000\cdot 0 + 45000^2\cdot 0 + 3\cdot 200^2\cdot 2.816^2\right]} = 40.85\,cm^2$$

We tale $h=50cm$ and the area which is $F_2=minmax=112.9\ cm^2$.

3) Let us analyze the cross-section at the distance of *4 m* from its supporting structure. Let us take $h=60cm$ cross-sectional height. There are forces $N_x=100\ kN$; $M_x=(300\cdot 4-50\cdot 4\cdot 2)=800\ kN\cdot m=\ 80000\ kN\cdot cm$; $Q_y=300-200=100\ kN$ in the cross-section. The area we need is

$$F_{31} = 1/24\sqrt{\left[100^2 + 2\cdot 100\cdot 80000\cdot 0.042 + 80000^2\cdot 0.042^2\right]} = 144.2\,cm^2$$

$$F_{32} = 1/24\sqrt{\left[100^2 + 2\cdot 100\cdot 80000\cdot 0.038 + 80000^2\cdot 0.038^2 + 3\cdot 100^2\cdot 1.986^2\right]} = 131.6\ cm^2$$

$$F_{33} = 1/24\sqrt{\left[100^2 + 2\cdot 100\cdot 80000\cdot 0 + 80000^2\cdot 0 + 3\cdot 100^2\cdot 2.71^2\right]} = 20\,cm^2$$

Let us take $h=70cm$ cross-sectional height. Let us determine the forces by eccentricity. Here we have $N_x=100\ kN$; $M_x=(300\cdot 4-50\cdot 4\cdot 2)\ +\ 100\cdot 0.05=\ 805\ kN\cdot m=\ 80500\ kN\cdot cm$; $Q_y=300-200=100\ kN$.

$$F_{31} = 1/24\sqrt{\left[100^2 + 2\cdot 100\cdot 80500\cdot 0.043 + 80500^2\cdot 0.043^2\right]} = 148.4\,cm^2$$

$$F_{32} = 1/24\sqrt{\left[100^2 + 2\cdot 100\cdot 80500\cdot 0.0395 + 80500^2\cdot 0.0395^2 + 3\cdot 100^2\cdot 1.976^2\right]} = 137.4\ cm^2$$

$$F_{33} = 1/24\sqrt{\left[100^2 + 2\cdot 100\cdot 80500\cdot 0 + 80500^2\cdot 0 + 3\cdot 100^2\cdot 2.742^2\right]} = 20.2\,cm^2$$

We take $h=60\ cm$ and the area $F_3=144.2\ cm^2$.

4) Let us analyze the cross-section at the distance of *6 m* from its supporting structure (the middle of the span). Let us take $h=70cm$ cross-sectional height. Let us take $h=70cm$ height. Let us determine the forces by eccentricity. Here we have $N_x=100\ kN$; $M_x=(300\cdot 6-50\cdot 6\cdot 3)\ +\ 100\cdot 0.05=\ 905\ kN\cdot m=\ 90500\ kN\cdot cm$; $Q_y=300-300=0\ kN$.

$$F_{41} = 1/24\sqrt{\left[100^2 + 2\cdot100\cdot90500\cdot0.043 + 90500^2\cdot0.043^2\right]} = 162.62\,cm^2$$

$$F_{42} = 1/24\sqrt{\left[100^2 + 2\cdot100\cdot90500\cdot0.0395 + 90500^2\cdot0.0395^2 + 0\right]} = 153.1\,cm^2$$

$$F_{43} = 1/24\sqrt{\left[100^2 + 2\cdot100\cdot80500\cdot0 + 80500^2\cdot0 + 0\right]} = 4.2\,cm^2$$

Let us take $h=80cm$ cross-sectional height. Let us determine the forces by eccentricity. $N_x=100\ kN$; $M_x=(300\cdot4-50\cdot4\cdot2) + 100\cdot0.1=910\ kN\cdot m=91000\ kN\cdot cm$; $Q_y=0\ kN$.

$$F_4 = 1/24\sqrt{\left[100^2 + 2\cdot100\cdot91000\cdot0.041 + 91000^2\cdot0.041^2 + 0\right]} = 159.62\,cm^2$$

We take $h=80\ cm$ and the area $F_4=59.62\ cm^2$.

5) The girder integral capacity with account of the symmetry is:

$$V=[(79.78+112.9)/2\cdot200+(112.9+144.2)/2\cdot200+(144.2+159.62)/2\cdot200]=150720\ cm^3$$

The girder with uniform cross-section integral capacity is
$V=(159.62\cdot600)\cdot 2=191544\ cm^3$.
Theoretically we save $|191544-150720|/191544\cdot100\% = 21.3\%$ of steel.

4 Conclusion

In real life, the construction also requires the account of the cross-section flanges and walls local buckling resistance, bending limitations, limitations connected with width, depth and size unification of flanges and walls. Still if we take all these limitations into consideration, the whole effect of steel saving is about 10-20% compared with a girder or a beam with uniform cross-section.

References

1. Chang-Yu Cui, Bao-Shi Jiang, Engineering Structures **77**, 109-118 (2014)
2. L.E. Mavromatidis, Energy and Buildings **106**, 192-202 (2015)
3. I.I. Goldenblat, V.A. Kopnov, *Strength and plasticity criteria of construction materials* (Mashinostrojenie, Moscow, 1968)
4. Ya.I. Olkov, I.S. Kholopov, *Steel pre-stressed trusses optimal design* (Stroyizdat, Moscow, 1985)
5. I.S. Kholopov, *Bar systems optimization as applied to SAE (computer-aided engineering systems), Doctoral dissertation* (Moscow, 1972)
6. I.S. Kholopov, *Mathematical modeling of bar systems volume by use of modular forms (textbook)* (Kuibyshev State University ed., 1982)
7. Yu.A. Radtsig, *Statically indeterminate trusses of minimum weight* (Kazan University ed., 1969)
8. V.Yu. Alpatov, A.A. Lukin, Procedia Engineering **111**, 20-29 (2015)
9. I.S. Kholopov, Building Mechanics and Structures Calculation, 14-17 (1981)
10. I.S. Kholopov, I.V. Loseva, Architecture and Civil engineering **5**, (1987)
11. I.S. Kholopov, M.I. Balzannikov, A.O. Lukin, A.V. Solovjev, Vestnik of MGSU **11**, 34-41 (2013)
12. I.S. Kholopov, I.V. Loseva, Higher education establishments news: Architecture and civil engineering **1,** 54-58 (1980)
13. M.I. Balzannikov, B.G. Ivanov, A.A. Mikhasek, Vestnik MSSU, Scientific-Technological Journal **7**, 119-124 (2012)

DOI: 10.1051/matecconf/20168602021

14. N.V. Efimov, E.R. Rozendorn, *Linear algebra and multidimensional geometry* (Moscow, 1970)

MATEC Web of Conferences **86**, 03001 (2016) DOI: 10.1051/matecconf/20168603001
IPICSE-2016

Theoretical basics of applying the one dimensional problem of soils compression seal theory to large foundation plates calculation

Veniamin Isaev[1], *Andrei* Maltsev[1], and *Andrei* Karpov[1,*]

[1]Samara State Technical University, Institute of Architecture and Civil Engineering, Molodogvardeyskaya str. 194, 443001 Samara, Russia

Abstract. Calculation of compressed footings settlement is one of the most vital tasks of soil mechanics. The calculation method of layer-by-layer addition, which recommended by current regulations and used in structural engineering practice, is most suitable for determining the settlement of foundations with an area of less than 50 m^2. The authors prove that it's possible to apply the one-dimensional problem of soils compression seal theory to the calculation of the settlement of large foundation plates. The proposed method of determining the settlement of large foundation plates makes the calculations simpler. There are examples of comparable calculations using the existing and proposed methods.

1 Introduction

Currently in design practice, to calculate the deformations of soil footing, in most cases the theory of linearly deformed half-space is used. It is based on the assumption that the stress propagation zone deep into the soil body from the load applied to the surface is not limited and goes to infinity. However, at great depths the stress of the overlying soil layers' own weight significantly exceeds the additional stress of external load. Therefore, we can assume the following: at a certain depth the soil is already so much compressed by the load of its own weight, that a slight increase in total stress due to the load of a building practically does not cause any additional soil compression. For this reason, in the majority of methods for calculating settlement the thickness of the compressed soil body under the foundation is limited by the layer, under which soil deformations are considered to be minor, and therefore they may not be taken into account. In such a case the compressed width is regarded as the active zone of the soil body under the foundation base, in which the most significant share of the settlement deformation takes place.

* Corresponding author: andrkarpow@yandex.ru

2 Analysis of existing methods of determining compressed width's depth

Notation system:

P_0 compacting pressure, kPa
P pressure of the building, kPa
γ soil's specific gravity, кH/m^3
d foundation depth, m
z soil body thickness, m
σ_{zpi} stress in the soil body caused by the pressure of the building, кPa
σ_{zgi} stress caused by the soil's own weight, kPa
R width/length of the rectangular foundation, m
r «floating» coordinate, m
H_c compressed width's depth, m
β non-dimensional coefficient, equal to 0.8
E soil deformation modulus, MPa
S foundation settlement, m

An analysis of different methods for calculating the compressed width's depth, based on the theory under consideration, was made by Professor P.A. Konovalov [1]. The resulting dependences of the changes in the depth of the active zone's bottom boundary on the foundation's width, obtained using several different calculation methods, are shown on

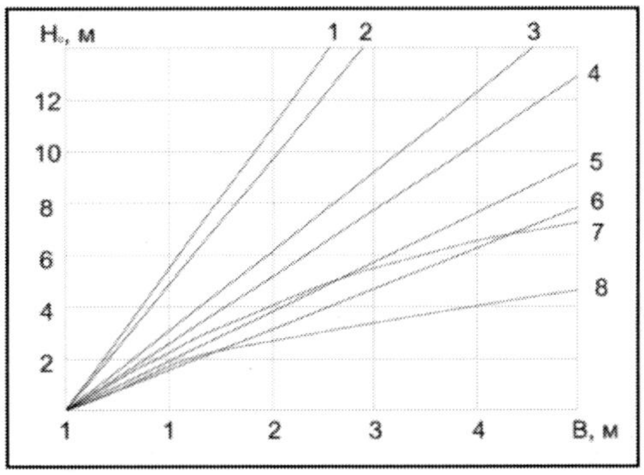

Fig 1. Dependence of the compressed soil width's depth on the size of the square foundation, determined according to different methods: 1- by V.A. Florin; 2- by PlagemanLanger; 3-by N.A. Tsytovich;4- by I.A. Rosenfeld; 5-American method; 6-by E.F. Vinokurov; 7-SNIP 2.02.01-83*;8-SNIP II-16-76.

As seen from the graphs presented in Figure 1, the results of the calculation of the compressed width's depth very much, and it's impossible to choose one particular method. Therefore, let us consider in more detail the methods described in SP 22.13330.2011 [2], which is the current normative document.

According to the current normative document [2], it is recommended to calculate foundation settlement, considering the footing as a linearly deformed half-space, here it is meant that a certain layer is compressed (referred to below as the method by K.E. Egorov [3]).

Also settlement calculation is done using the layer-by-layer addition method (referred to below as the method by N.A. Tsytovich [4]), which has got the following assumptions:

• the calculation of settlement is made for the central point of the foundation, taking into account axial compression stresses;

• the compressed layer is defined as thė distance from the foundation base to the point in the soil body, at which the certain defined ratio of compacting and overburden pressure values is met;

• the compressed layer is divided into elementary layers, and the compression seal principle is applied to each of these layers;

• the compression of the elementary layer of soil takes place under the pressure of the building, net of the existing overburden pressure.

The term "compacting pressure", or "overpressure", applies in case the foundation has got some laying depth. When this occurs the value of the pressure of own weight of the soil excavated to the foundation base is deducted from the actual pressure of the building on the foundation base. This is expressed by the formula (1)

$$P_0 = P - \gamma \cdot d \tag{1}$$

3 Justification of the possibility to apply the one dimensional problem of soils compression seal theory to large foundation plates calculation

From the works [3-10] it's known that the layer-by-layer addition method is most suitable for determining the settlement of foundations with an area of less than 50 m^2.

To apply the principle of one-dimensional soils compression seal to calculate the settlement of large foundation plates it's necessary to prove the following:

• under the plates there's compressed width of insignificant depth;

• compacting pressure is uniformly distributed along the layer height over the entire area;

• uniformly distributed load under the plate extends sideways to infinity.

The size of the compressed layer, according to N.A. Tsytovich [4], is defined as the distance from the surface of load application to the point (horizon), where such load is equal to zero.

According to the theory of linearly deformed half-space, the stresses in the half-space attenuate at infinity.

However, in reality, the impact of the load of the building is over at the depth (on the horizontal plane), where such load is balanced by the weight of excavated soil, which is expressed by the formula:

$$P_0 = \gamma \cdot z \tag{2}$$

If the foundation has got some definite laying depth, the formula changes to:

$$P_0 = \gamma \cdot (z - d) \tag{3}$$

Thus, the soil body the width of which is z will experience the compression caused by the actual load P applied to the surface, and in this case the width z can be called the compressed width.

Since all foundations are designed with the laying depth d given in advance, the value of the compressed width is found from the expression:

$$H_c = z - d = \frac{P}{\gamma} - d \tag{4}$$

To prove the uniform distribution of stresses along the height of the linearly deformed layer, let's consider the well-known solution of the theory of elasticity for stresses under round rigid stamp [4]:

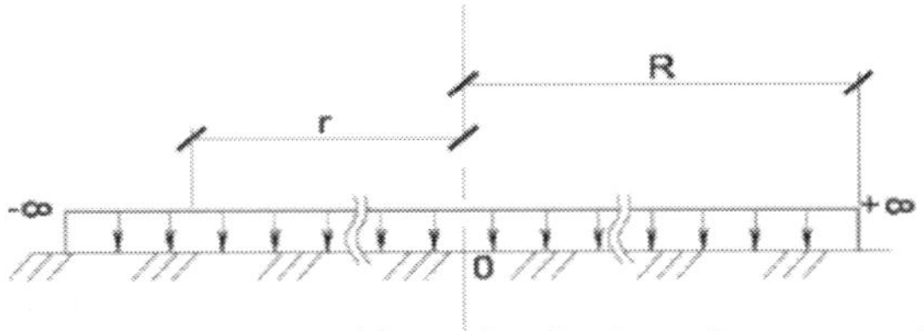

Fig 2. Design diagram for determining the uniform distribution of stresses under very big foundations.

$$\sigma_{zpi} = \frac{P}{2\cdot\sqrt{1-(\frac{r}{R})^2}} \tag{5}$$

Where at r=0 (under the foundation's centre) the formula changes to:

$$\sigma_{zpi} = \frac{P}{2\cdot\sqrt{1-(\frac{0}{R})^2}} = \frac{P}{2} \tag{6}$$

For distributed load, which extends sideways to infinity and passes through a large foundation plate, the latter's width in the calculation is taken as an infinite value. After substituting R=∞в in formula (5) we get the stress values, similar to formula (6):

$$\sigma_{zpi} = \frac{P}{2\cdot\sqrt{1-(\frac{0}{R})^2}} = \frac{P}{2} \tag{7}$$

The comparison of formula (6) and formula (7) demonstrates the uniform distribution of compacting pressures along the layer height under the entire area of the plate, which proves one-dimensional soils compression seal of the compression width.

The above-given proof allows using the formula of the one-dimensional problem for large plates [4]:

$$S = H_e \cdot P \cdot \frac{\beta}{E} \tag{8}$$

4Examples of calculations

As an example of using the one-dimensional problem of soils compression seal theory for large foundation plates calculation, below we give the calculations of settlement of two foundation plates under a 10-storey apartment building in the city of Samara.

Example №1. The foundation is a plate of square shape, side length = 18.0 m, actual soil bearing pressure P = 176 kPa, loam, width = 12.5 m, specific gravity γ = 19.8 kH / m^3, deformation modulus E = 10 MPa. Laying depth of plate d = 4.0 m

The calculation results according to the proposed method:
-compressed width from formula (4)

$$H_c = \frac{176}{19.8} - 4.0 = 4.9m \tag{9}$$

- settlement according to formula (8):

$$S = 176 \cdot 4.9 \cdot \frac{0.8}{10000} = 0.069m \tag{10}$$

For comparison:

Calculation results according to the method by K.E. Egorov: H_c=10.42 м, S=0.069 m.

Calculation results according to the layer-by-layer addition method: compressed width's depth at σ_{zpi}=85 kPa, σ_{zgi}=170 kPa - H_c=4.3 m; settlement at P_0=96.8 kPa -S=0.03 m.

Example №2. The foundation is a plate with the side length of 19.2 m, the actual pressure under the foundation base is P = 156 kPa. Soil conditions are similar to example №1.

The calculation results according to the proposed method:

-compressed width from formula (4)

$$H_c = \frac{156}{19.8} - 4.0 = 3.9\text{м} \tag{11}$$

- settlement according to formula (8):

$$S = 156 \cdot 3.9 \cdot \frac{0.8}{10000} = 0.05\text{м} \tag{12}$$

For comparison:

Calculation results according to the method by K.E. Egorov: H_c=10.21 м, S=6.49cm.

Calculation results according to the layer-by-layer addition method: compressed width's depth at σ_{zpi}=70 kPa, σ_{zgi}=140 kPa - H_c=3.2 m; settlement at P_0=76.8 kPa -S=0.02 m.

5 Conclusion

Finally, it can be concluded that the proposed method of determining the settlement of large foundation plates makes the calculations simpler. In the future, the authors intend to specify the field of application of the method (plate size) - of the one-dimensional problem of soils compression seal, as well as to confirm the proposed theory by experiment.

References

1. P.A. Konovalov, *Footings and foundations of renovated buildings, 4th edition* (VNIINTPI, Bumazhnaya Galereia Publisher, Moscow, 2000)
2. SP 22.13330.2011 *Footings of buildings and structures* (Ministry of Regional Development of the Russian Federation, Moscow, 2011)
3. K.E. Egorov, *On deformation of finite footings: footings, foundations and soil mechanics* (Moscow, 1961)
4. N.A. Tsytovich, *Soil mechanics, comprehensive study, 5th edition* (Lenand Publisher, Moscow, 2014)
5. V.I. Isaev, O.M. Kakutina, A.V. Maltsev, *On sealing of water-saturated clay soils, Vital problems in civil engineering and architecture, Education, science, practice:*

Proceedings of the 66th all-Russian scientific and technical conference on R&D results in 2008, part II (SGASU, Samara, 2009)
6. A.V. Maltsev, T.O. Klokova, *Analysis of areas for efficient application of methods to calculate linearly deformed footings settlement, Vital problems in civil engineering and architecture, Education, science, practice: Proceedings of the 66th all-Russian scientific and technical conference on R&D results in 2008, part II* (SGASU, Samara, 2009)
7. A.V. Maltsev, D.G. Skopintsev, Proceedings of the 59th regional scientific and technical conference in April **2**, 412-414 (2002)
8. A.V. Maltsev, D.G. Skopintsev, Vital problems in civil engineering and architecture, Education, science, practice: Proceedings of the 60th all-Russian scientific and technical conference on R&D results in 2002, part II, 230-234 (2003)
9. A.A. Averkiev, O.V. Zayats, Tradition and innovation in civil engineering and architecture: Proceedings of the 69th all-Russian scientific and technical conference on R&D results in 2011, part 2, 422-423 (2012)
10. A.V. Maltsev, N.O. Tolmachev, I.V. Tikhonov, Tradition and innovation in civil engineering and architecture: Proceedings of the 70th anniversary all-Russian scientific and technical conference on R&D results in 2012, part 2, 378-379 (2013)

Algorithm development of structural variables for slope reinforcement of hydrotechnic structures with synthetic materials

Andrey Mikhasek [1,*] *Maxim* Rodionov[1], *Vladimir* Seliverstov[1], and *Yury* Senitsky[1]

[1]Samara State Technical University, Institute of Architecture and Civil Engineering, Molodogvardeyskaya str. 194, 443001 Samara, Russia

Abstract. The construction of hydrotechnic structures out of watercourse area is usually done with grass seeding above a vegetative soil layer. For stability of slope protection, during and after construction, it is suggested that vegetated soil is reinforced with cubic geosynthetic mats. To choose a geosynthetic material the algorithm is offered, taking into account conditions that provide resistance of vegetated soil layer, and depending on forces that act upon the slope and considering changes of material properties in course of time technological peculiarities of its placement.

1 Introduction

In the process of hydrotechnic structures construction with reinforced slopes we suggest to distinguish 3 vertical elements: a lower (underwater), a middle (in the area of variable level) and an upper one. Based on structural-technological analysis of the already existed bank protections and slope reinforcements of hydrotechnic structures and solutions involving modern materials [1-6] we come to the following construction solution (Fig. 1): the lower element - в виде банкет из rock mass, where a crest and a higher point of the slope are fastened with a quarry fill which sizes and mass are determined according to calculations of resistance to ice and wave impacts. The inner slope of the bank is protected with a breakstone filter of 20-40 мм and 20 см thick, on which a geotextile layer with density 350 gr/m^2 is placed, and a foundation of the existed slope is filled with sand drainage) 0.5 m thick to prevent suffusion carry-over. The lower side of the bank is constructed with sand backing. The middle one is a slope with filling 1:2.5÷3, which is fastened to concrete mats, placed on a geotextile layer. This solution is quite technological as well as economical and time-saving.

The construction of the upper element, which we recommend to construct up to the natural boundaries of a higher slope part is grass seeding on the vegetated soil layer with reinforcement of synthetic erosion-preventive material.

There are a lot of different types of geosynthetic materials which are produced in Europe, Asia region, Russia and can be used for slope protection of the upper element.

* Corresponding author: andremixas@mail.ru

But nevertheless there is no method for structural variables, that`s why algorithm development for choosing geosynthetic material for slope reinforcement of hydrotechnic structure is of great importance.

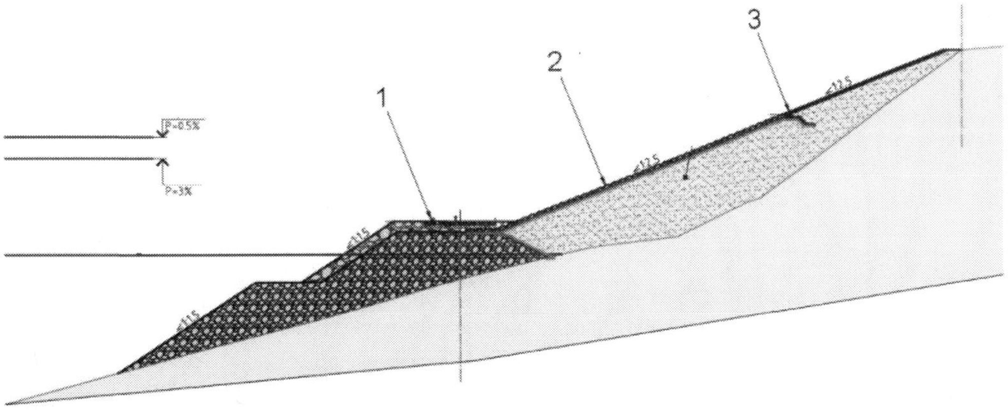

Fig. 1. Reinforcement option using composite and synthetic materials 1 – lower reinforcement element, 2 – middle, 3 – upper.

2 Research

The requirements for the material which is placed in the upper element of reinforcement are the following:

- slope protection from wind erosion;
- protection from water erosion.

These tasks are maintained by plurannual grass root system, planted along the slope. But from construction period up to strengthening of the root system additional protection is required, which can be achieved by adding a reinforcement layer made from geosynthetic materials.

It is recommended to use a cubic geosynthetic mat which allows to increase adhesion area with vegetated soil as well as surface resistance to rain and ice water. Considering the conditions of construction work and reinforcement slope stability it is recommended to place 5 cm vegetated soil layer just above and below the mat , and plurannual grass seeding - about 2 cm from the upper layer boundary.

As there is no method for choosing geosynthetic material, we offer to do it according stiffness properties. Calculated scheme for load determination is shown in (Fig. 1), calculated stiffness of surface reinforced material Fs is calculated [1]:

$$F_s = L \cdot (H \cdot \gamma_s + S_d \cdot \cos\beta) \cdot (\sin\beta - tg\delta \cdot \cos\beta) \quad (1)$$

L – actual slope lenght, m;
H – thickness of geomat and vegetated soil layer, m;
γ_s – calculated relative density of soil, кN/m^3;
S_d – calculated total load on the slope (including snow load), кN/m^2;
β – slope angle, °;
δ – friction interaction angle «geomat - soil» (depends on internal friction soil angle and a type of reinforced material), °.

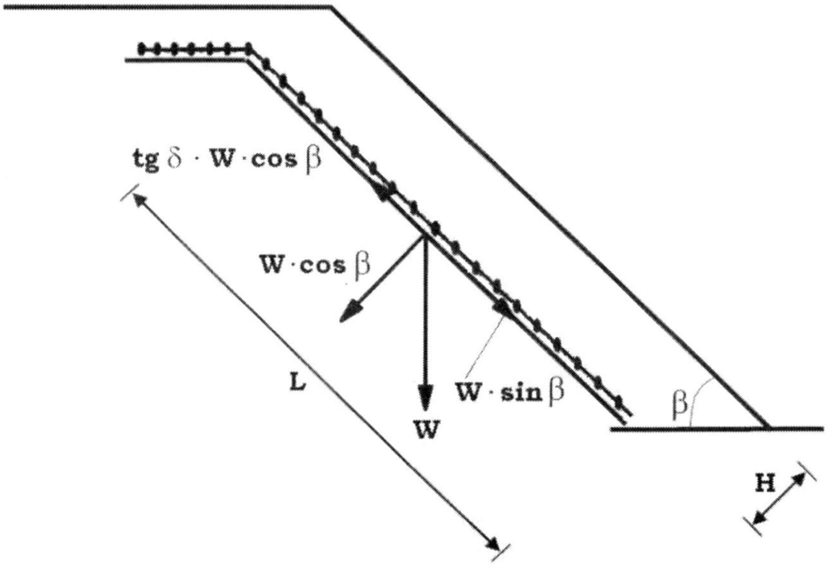

Fig. 2. Calculated scheme.

For the stability of slope surface the fulfillment of the condition is required:

$$F_d \geq F_s, \tag{2}$$

F_d – calculated (designed) hardness of reinforced material, estimated with a formula:

$$F_d = F/(A_1 \cdot A_2 \cdot A_3 \cdot A_4 \cdot \gamma_B), \tag{3}$$

F – short-term hardness of reinforced material (tensile strength);
A_1 – decreasing coefficient of long-lasting material stiffness, directly connected with material creep and structure durability [8, 9];
A_2 – increasing coefficient of material in the course of handling, placement and soil compaction, depends on the type of geosynthetic material and soil type [8, 9];
A_3 – decreasing coefficient of material stiffness if there are joints (for calculations A_3 = 1.00 with the absence of joints towards load acting) [8, 9];
A_4 – decreasing coefficient of material stiffness for outer effects resistance – resistance to microorganisms, X-rays, chemical environment, temperature change [8, 9];
γ_B – general reserve coefficient of geosynthetic material construction, is taken according to the recommendations [8].

Variables A_1, A_2, и A_4 for synthetic materials are provided by manufacturing plant and determined by the results of independent lab tests and must be certified.

Base on the formulae given above, a short-term material stiffness for surface reinforcement is estimated by the following way:

$$F \geq L_e \cdot (H \cdot \gamma_s + S_d \cdot \cos\beta) \cdot (\sin\beta - \mathrm{tg}\delta \cdot \cos\beta) \cdot A_1 \cdot A_2 \cdot A_3 \cdot A_4 \cdot \gamma_B \cdot \tag{4}$$

3 Conclusion

The research yielded the following conclusions:

- for the protection of upper part of bank stabilization as hydrotechnic structure which is not affected by watercourse , it is necessary to use flexible reinforcements like grass seeding on a vegetated soil layer including a cubic geosynthetic mat as a reinforced element;
- there is no method for choosing geosynthetic material for slope reinforcement of hydrotechnic structure, that`s why algorithm development is quite an important;
- the choice of geosynthetic material should be done taking into account the resistance of vegetated soil layer, forces that act upon the slope and changes of material properties in course of time and technological peculiarities of its placement.

References

1. A.A. Mikhasek, A.A. Smivalov, Scientific review **14**, 102-109 (2015)
2. M.I. Balzannikov, A.A. Mikhasek, M.V. Rodionov, Scientific review **12**, 471-475 (2014)
3. S.A. Pyavsky, M.V. Rodionov, I.S. Kholopov, MSAU Newsletter **6**, 54-61 (2012)
4. M.I. Balzannikov, A.A. Mikhasek. Procedia Engineering **91**, 183-187 (2014)
5. A.A. Mikhasek, M.V. Rodionov Procedia Engineering **91**, 82-88 (2015)
6. Y.A. Senitsky, A.A. Mikhasek, A.V. Kozlov, Scientific review **23**, 64-70 (2015)
7. V.A. Seliverstov, A.A. Orlova, Scientific review **3**, 109-115 (2015).
8. *RIMD Recommendations on geosynthetic material usage in road building and road reconstruction* (2010).
9. *EBGEO (Empfehlungen für den Entwurf und die Berechnung von Erdkörpern mit Bewehrungen aus Geokunststoffen)* (2010)

MATEC Web of Conferences 86, 03003 (2016) DOI: 10.1051/matecconf/20168603003
IPICSE-2016

Modeling of Fine Migration in a Porous Medium

Yuri Galaguz[1], and *Galina* Safina[1,*]

[1] Moscow State University of Civil Engineering, 129337 Moscow, Russia

Abstract. The structure of underground waters and water permeability of rocks must be taken into account when choosing the location of construction objects, the design of tunnels, hydraulic and underground structures. The model of filtration for suspension in solid porous medium during its displacement with clean water is considered. The numerical calculation of boundary of two phases is carried out and concentrations of suspended and retained particles are calculated for different values of porosity and permeability of the porous medium.

1 Introduction

The structure of underground waters and water permeability of rocks must be taken into account when choosing the location of construction objects, the design of tunnels, hydraulic and underground structures [1]. Underground waters significantly effect the construction properties of soils. Intensive filtration of water flows leads to soil loosening and reducing its strength. Solution of filtration problems allows forecasting flooding of multistory buildings and estimates the long-term sustainability of building structures.

Underground waters passing through the porous rocks and soil transport a lot of small solid particles. The fluid of particles in liquid is called the suspension. Fine migration – one-dimensional problem of particle filtration in the solid porous medium during its displacement with pure water is considered [2,3]. The pores in the filter form intersecting channels of different lengths and shapes. In this paper the mathematical model of the particles motion in the filter based on the mechanical-geometric interaction of particles with porous medium is considered. Assume that all the particles are solid balls of equal size, and the channels have circular cross-section of constant diameter throughout its length. When constructing the particle diffusion models the influence of fluid viscosity and electrical interaction of the particles with the channel walls are neglected. It is assumed that the solid particles pass freely through the large pores and get stuck in the throat of small pores which are smaller than the particle diameter [4,5]. The retained particles cannot be driven out by the fluid flow or other particles. Deep bed filtration is described by the equation of mass balance of suspended and retained particles of suspension and the kinetic equation for deposit growth [6]. During long-time filtration the number of free small pores is significantly reduced, and the porosity and permeability of the porous medium are changed. To take into account this phenomenon, in contrast to the classical equations of filtration the dependence of coefficients of the mass balance equation on deposit concentration $S(x,t)$ is

* Corresponding author: minkinag@mail.ru

introduced [7]. It is assumed that the deposit growth is proportional to the concentration $C(x,t)$ of suspended particles. The proportionality coefficient depends on the deposit concentration and is called the filtration coefficient [8].

The solution of the problem of particle diffusion in the filter describes the process of particles transport through the pores of the filter and the dynamics of precipitation. Exact and asymptotic solutions of filtration problems are found for some important special cases [6, 9-15], but in general, the problem does not have an analytical solution.

In problem considered at the initial moment the porous medium is filled with suspension with given concentrations of suspended and retained particles. At the filter inlet the clean water begins to flow. It is gradually displacing the suspension and filling the porous medium. In a part of the porous medium, which is after water front, filtration of the suspension is terminated, the concentration of suspended particles becomes zero, and the concentration of retained particles is constant.

2 Mathematical model

In the domain $\Omega = \{0 < x < 1, t > 0\}$ the suspended and retained particles concentrations $C(x,t)$, $S(x,t)$ satisfy the equations

$$\frac{\partial}{\partial t}(g(S)C) + \frac{\partial}{\partial x}(f(S)C) + \frac{\partial S}{\partial t} = 0; \tag{1}$$

$$\frac{\partial S}{\partial t} = \Lambda(S)C; \tag{2}$$

with border and initial conditions

$$x = 0: C = 0; \tag{3}$$

$$t = 0: C = C_0(x),\ S = S_0(x). \tag{4}$$

Here porosity $g(S)$, permeability $f(S)$, the filter coefficient $\Lambda(S)$ and initial concentrations $C_0(x)$, $S_0(x)$ are continuous positive functions.

Equations (1), (2) form a quasi-linear hyperbolic system of the first order. The characteristic curve emerging from the origin is the boundary Γ of two phases, it divides the domain Ω into two subdomains Ω_w and Ω_s containing water and suspense, respectively (see Fig. 1). In Ω_w the suspended particles concentration is zero, the retained particles concentration does not depend on time; in Ω_s the concentrations $C(x,t)$, $S(x,t)$ are variable и positive. According to the method of characteristics since the conditions (3) and (4) disagree at the origin, the solution $C(x,t)$ has a strong discontinuity on the boundary Γ. The solution $S(x,t)$ is continuous in Ω and has a weak break on the water front Γ.

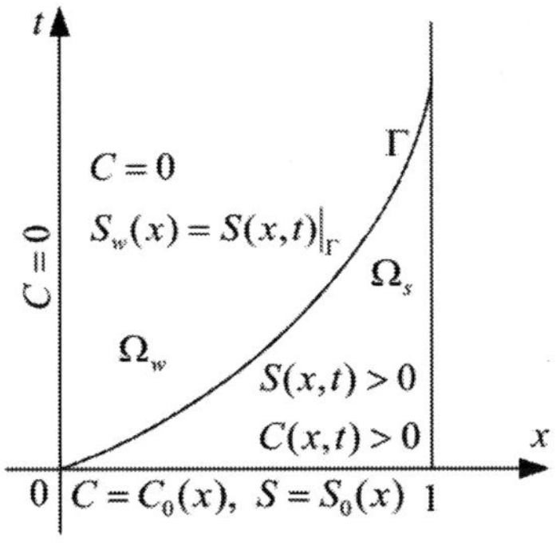

Fig. 1. A schematic solution of the problem (1)–(4).

If the functions included in the equation (1) are constant and positive

$$g(S) = g_0 > 0,\ f(S) = f_0 > 0\ , \tag{5}$$

then the boundary Γ of two phases and the characteristic of the equation (1) is a straight line

$$t = \alpha x,\ \ \alpha = g_0/f_0\ . \tag{6}$$

For non-permanent functions $g(S)$, $f(S)$ the boundary Γ is a curved line, for which there is no analytical expression.

In general, the problem (1)-(4) does not have an analytic solution. The results of numerical calculation by finite differences method as in [16] are given below.

3 Calculation of fine migration process

Calculation of boundary of two phases is performed for the filter coefficient $\Lambda(S) = 2 - S$, porosity $g(S) = 1 + 3S$ and constant initial conditions $t = 0: C = 1,\ S = 0.5$ at different values of permeability: $f(S) = 1 + 2S$ (Fig. 2, a) and $f(S) = 1 + 0.1S$ (Fig. 2, b).

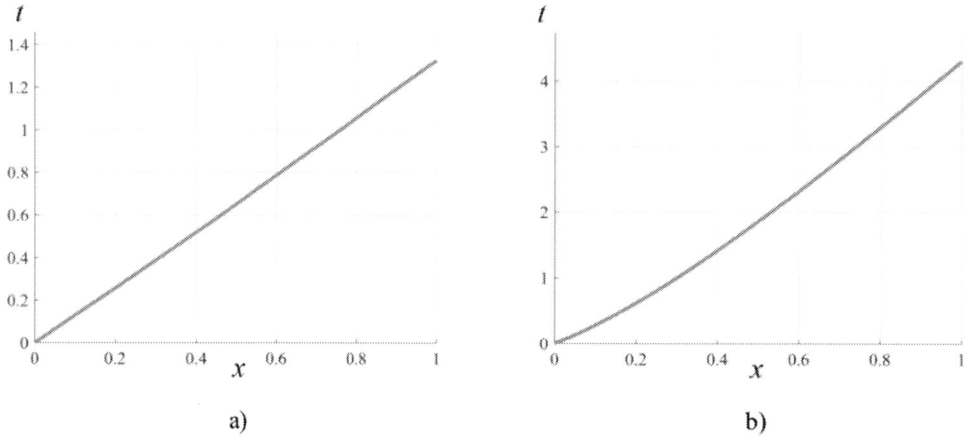

Fig. 2. The boundary between water and suspence.

Calculation of filtration problem is performed by finite difference method for the explicit scheme. The relationship between step τ in time and step h in coordinate x is selected from the Courant condition $\tau \le \min\limits_{x,t}(g(S)/f(S)) \cdot h$. However, the best option is reached at $\dfrac{\tau}{h} = 4$, when the numerical solution reflects well the behavior of solutions in discontinuity points. Numerical solution of the problem (1)-(4) is obtained for the coefficients of the equations and constant initial conditions mentioned above for Fig. 2, b) (blue solid line is in Fig. 3-6). Red dotted line represents the solution of the problem for the permanent porosity and permeability $g(S) = f(S) = 1$.

DOI: 10.1051/matecconf/20168603003

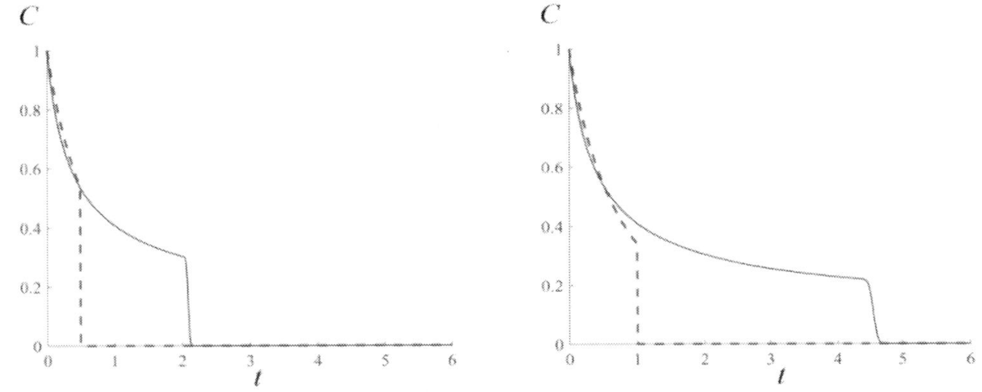

a) b)

Fig. 3. Suspended particles concentrations for a) $x = 0.5$; b) $x = 1$.

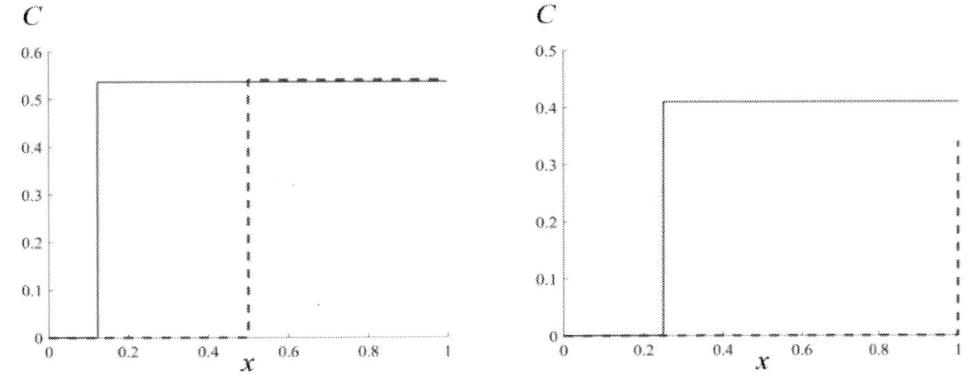

a) b)

Fig. 4. Suspended particles concentrations for a) $t = 0.5$; b) $t = 1$.

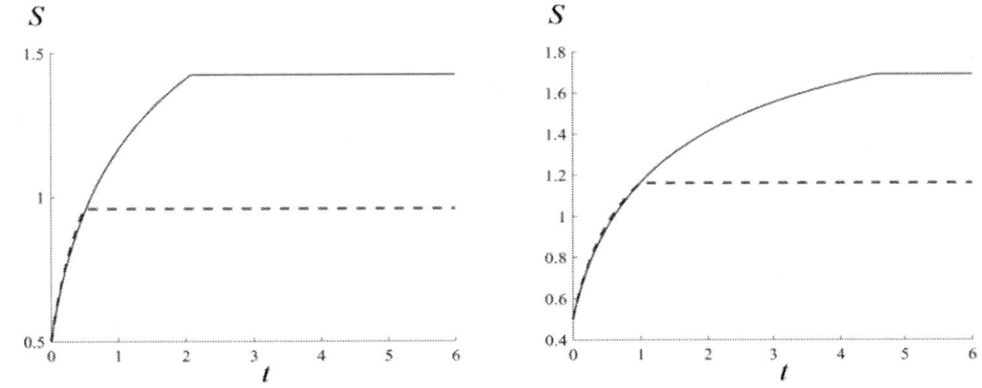

a) b)

Fig. 5. Retained particles concentrations for a) $x = 0.5$; b) $x = 1$.

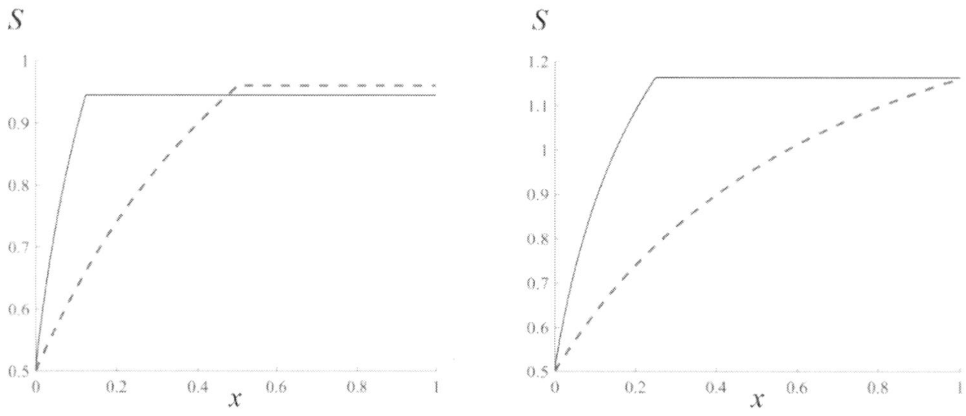

a) b)

Fig. 6. Retained particles concentrations for a) $t=0.5$; b) $t=1$.

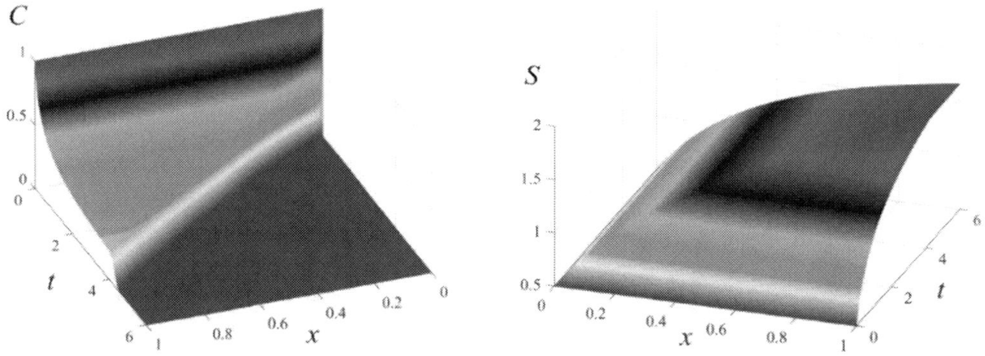

a) b)

Fig. 7. 3D-graphs a) Suspended particles concentration; b) Retained particles concentration.

4 Conclusion

In this paper a numerical solution of the problem of the displacement of a suspension by clean water flow in porous medium is obtained. The boundary of two phases is found, three-dimensional graphics of the concentration of suspended and retained particles and their two-dimensional cross-sections at a fixed time and at a given distance from the filter inlet are constructed.

Calculations show that the boundary of two phases depends on the coefficients of the equations (1) and (2). If $g(S)=f(S)$ then the boundary is a straight line segment. The greater the ratio $g(S)/f(S)$ differs from the constant, the more the curve Γ deviates from a straight line.

The introduction of variable porosity and permeability significantly changes the relationship of suspended and retained particles concentrations depending on the time and location. It follows from Fig. 3, 4, 6 that the filtering model with constant functions of porosity and permeability can not serve as a linear approximation of the general non-linear model.

For the numerical calculation the blocking filtration coefficient $\Lambda(S)=2-S$ is selected. In this case for an unlimited time of filtration the retained particles concentration

$S(x,t) \to 2$ and the suspended particles concentration $C(x,t) \to 1$. However water forces out the suspension from a porous medium and prevents particles concentrations to achieve limit values (Fig. 3, 5).

References

1. K.R.Arora, *Soil mechanics and foundation engineering* (Delhi, 2004)
2. M.M. Sharma, Y.C. Yortsos, AIChE Journal **33**, 10 (1987)
3. K.C. Khilar, H.S. Fogler, *Migrations of Fines in Porous Media* (Dordrecht, 1998)
4. A. Badalyan, Z. You, K. Aji, P. Bedrikovetsky, T. Carageorgos, A. Rev. Sci. Instrum. **85**, ID 015111 (2014)
5. Z. You, A. Badalyan, P. Bedrikovetsky, SPE Journal. **18**, 4 (2013)
6. J. P. Herzig, D. M. Leclerc, and P. Legoff, Ind. Eng. Chem. **62**, (1970)
7. P. Bedrikovetsky, Transp. in Por. Med. **75**, 3 (2008)
8. C. Tien, B.V. Ramarao, *Granular filtration of aerosols and hydrosols* (Elsevier, Amsterdam, 2007)
9. E.A. Vyazmina, P.G. Bedrikovetskii, A.D. Polyanin, Theor. Found. of Chem. Eng. **41**, 5 (2007)
10. Z. You, P. Bedrikovetsky, L. Kuzmina, Abs. and App. Anal., 680693 (2013)
11. Z. You, Y. Osipov, P. Bedrikovetsky, L. Kuzmina, Chem. Eng. J., 258 (2014)
12. L.I. Kuzmina, Yu.V. Osipov, Proc. Eng. **111**, (2015)
13. L.I. Kuzmina, Yu.V. Osipov, Proc. Eng. **153**, (2016)
14. L.I. Kuzmina, Yu.V. Osipov, Vestnik MGSU. **1**, (2015)
15. L.I. Kuzmina, Yu.V. Osipov, Int. J. for Comp. Civ. and Str. Eng. **11**, 1 (2015)
16. Yu.P. Galaguz , G.L. Safina, Proc. Eng. **153**, (2016)

Subgrade strengthening of roads and railroads as planar geotextile reinforcement

Krzysztof Gradkowski[1,*]

[1]Warsaw University of Technology, Civil Engineering Department, Al. Armii Ludowej 16, Poland

Abstract. The proper quality of construction of roads and railroads can be achieved by strengthening the of the foundation bed. Measurements and tests the number and order of the layers of geotextile layer of ground indicate the effectiveness of technical solutions involving the reinforcement laid horizontally. The offer technical use of the geotextile reinforcement in the ground layer substrates of roads indicates the possible effectiveness of the strengthening of the subsoil of roads and is also included in this article.

1 Introduction

In practice, the implementation of the sections of the transport infrastructure is the need to improve and strengthen the ground layer of pavement surfaces of roads and railroads. In the case of semi-rigid surface highways are layers adjacent to the substructure. At the European railway routes it is envisaged building the so-called. ground the protective layer. In both cases, the use of geotextile layers into ground layers is an alternative to the widespread use of hydraulic binders (cement, lime). By installing geotextile layers in the subsoil of roads and railroads also we obtain as well;

- proper strengthening layer substrate surface,
- favourable changes in soil moisture in the surface layers of the substrate,
- reducing the cost of construction in the area of substrates surface.

These effects are due to the reinforcing properties of geotextiles interaction with the soil substrate. The scope of strengthening the structural layer of the ground floor surface of roads depends on several factors. Some of them have been identified on the basis of laboratory testing measurements.

2 Tests measuring reinforced ground with geotextiles

Since the beginning of the twentieth century saw rapid growth applications of geosynthetic materials in infrastructure construction. The use of geosynthetic materials and geotextile in round constructions increases the stability and durability and stability, especially road infrastructure construction [1]. Different types of geotextile expose the different properties, whose effective use in the construction of roads is the subject of specific research programs

* Corresponding author: k.gradkowski@il.pw.edu.pl

[2, 3]. Commonly known and used are properties of filtration and separation geotextiles. Effective and tangible research results obtained in the measurement chamber, beyond the actual sections of roads as well. Laboratory research programs on the use of geotextiles are economical and indicate the direct technical solutions [4].

2.1 Preliminary oedometer test

As part of the preliminary measurements of signal preparation for wider research, made several signalling series, standard measurements oedometer test, according to two schemes fulfilling the oedometer barrel, fig.1. A series of measurements performed for medium granular sand (MSA) of natural humidity of about 10%. The study, performed by Polish standards PN-88 / B-04481, is used laboratory oedometer test with a range of loads $\boldsymbol{p}$ and $\boldsymbol{p_z}$ from 12.5 to 400 kPa and a sample height of 20cm.

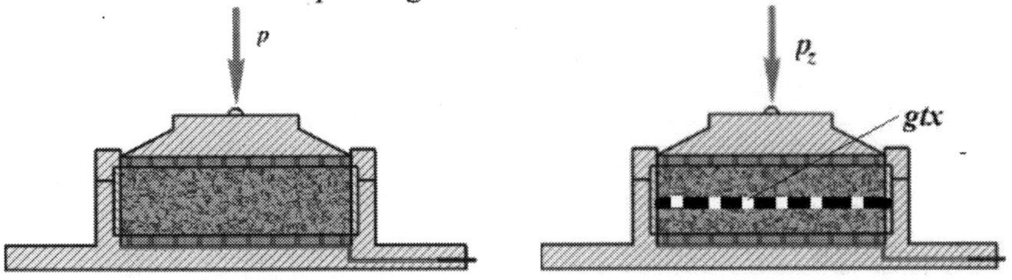

Fig.1. Schemes of fulfil and oedometer loading.

Repeated sequences of oedometer measurements and statistical results approximations, allowed us to obtain the graph of figure 2.

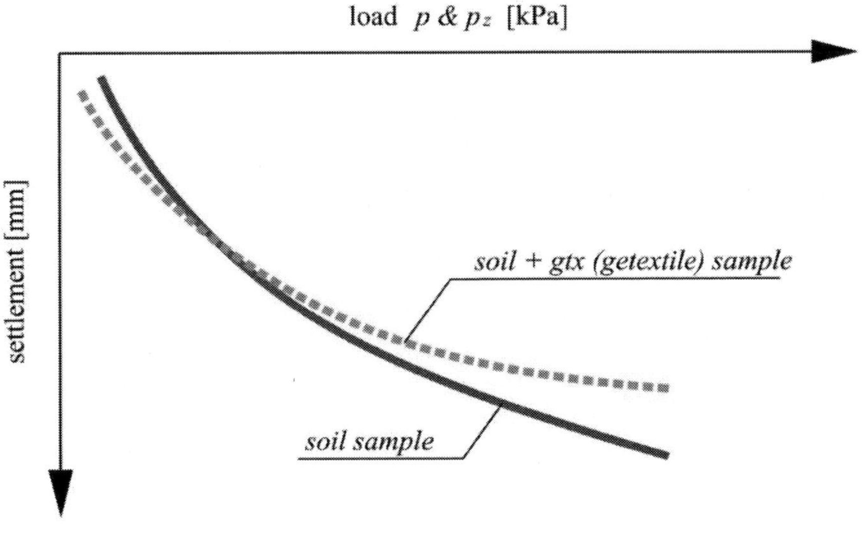

Fig.2. Statistical performance the measurements of two ways of filling oedometer barrel.

The graph in figure 2 clearly shows a tendency to strengthen the soil through the geotextiles reinforcement with one layer. The course of the curves is an approximations not less enhancement effect by the geotextile is very clear.

2.2 Oedometer test program

In order to perform direct testing laboratory prepared samples appropriate forms to fill different structures stratification geotextiles, one type of sandy soil with known physical parameters. Land filling was a medium sand, ρ = 1.78 g / cm^3, solid interior humidity = 5% and the particle size uniformity index C_U = 5. Methods roller made of non-deformable sections of the tube height H = 25 cm and an internal diameter d = 10 cm, that is, with a coefficient of proportionality, the shape of the sample of 2, 5. Used in the research commonly used in road geotextile having a basis weight of 200 g / m^2. In the six forms of the samples were performed four series fills the ground with different structures stratification geotextile in each sample, Figure 3. In the next series A, B, C and D were both variable number of layers of geotextile N, and the distance between the layers of geotextile. In each series was made a comparative sample only filled with soil. The series C and D geotextile used in a double arrangement, thereby obtaining a basis weight of 400 g / m^2. Soil-geotextile sample preparation system shown in figure 3

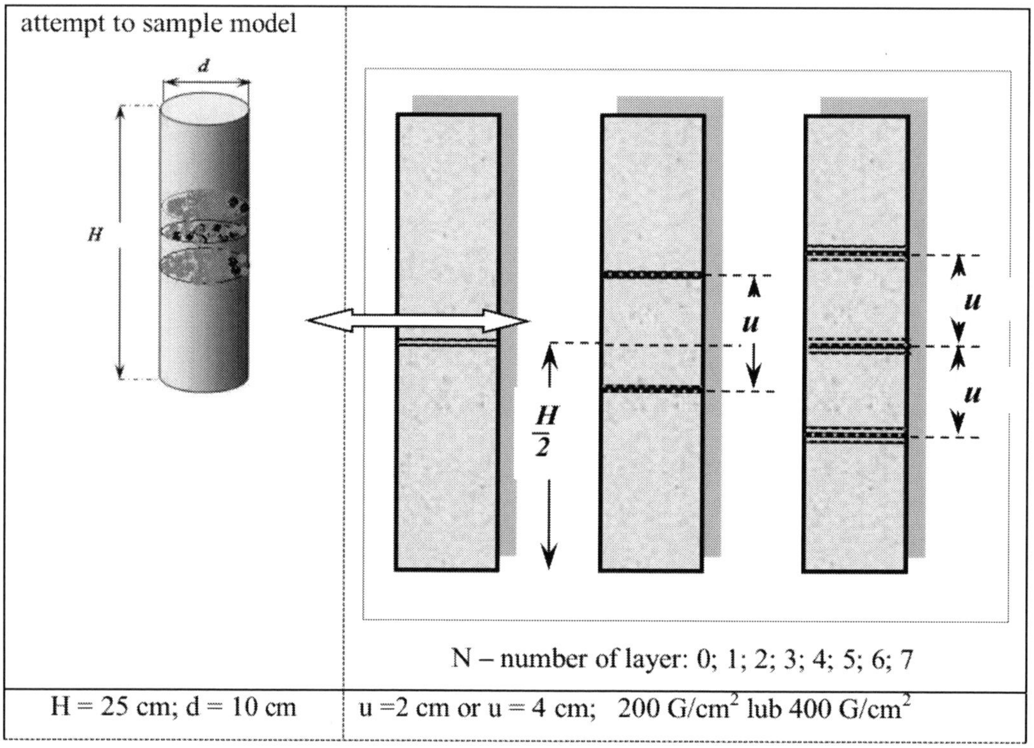

Fig. 3. The structures of preparing a series of samples having a variable number of N layers at distances of u.

Determine the extent and scale of load and displacement were made on a single sample zero, filled with the soil compacted layers in the manner adopted for all samples. Before you can record movements, in order to harmonize the conditions for direct measurement and maximum elimination of differences in the density of filling the ground, each of the samples were subjected to pre-load of 100 kPa. The load was single, and after unloading the sample was applied as the initial load. The expected layout changes susceptibility of the tested samples, depending on the increase in the number of layers of geotextiles. Susceptibility to deformation of the sample that is, changing the height of the sample is defined as the reciprocal of the spring constant of the sample and may be a measure of the compressive strength of the laminar structure of the ground geotextile - layer. Testing

DOI: 10.1051/matecconf/20168603004

machine is equipped with a digital recorder and monitor graphical image measurement of strains and loading forces fig.4, 5.

Fig. 4. Uniaxial load filled with samples.

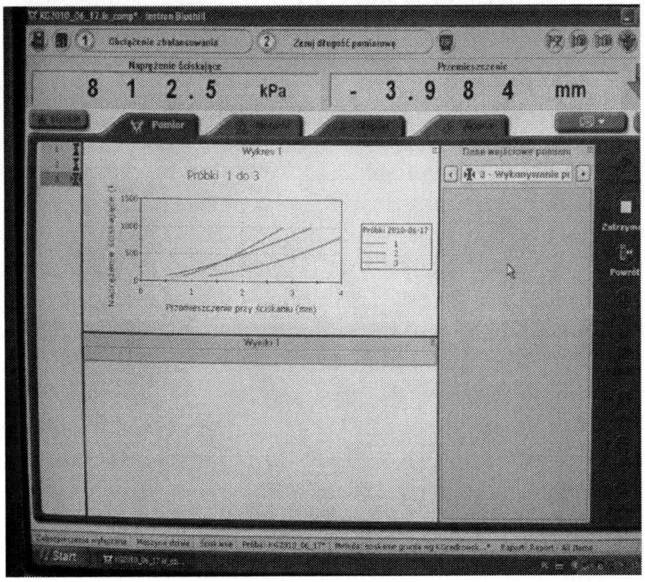

Fig. 5. Screen registrar.

Summary of results and presented in graphs the results of direct measurement of susceptibility samples of each series. And a graphical summary of these results fig. 6 is of some importance to determine the appropriate way to strengthen planar geotextile ground reinforced. The analysis results were not taken into account the extreme that could arise as a result does not maintain uniform conditions for the preparation of all samples or difficult to avoid errors in the care of the samples.

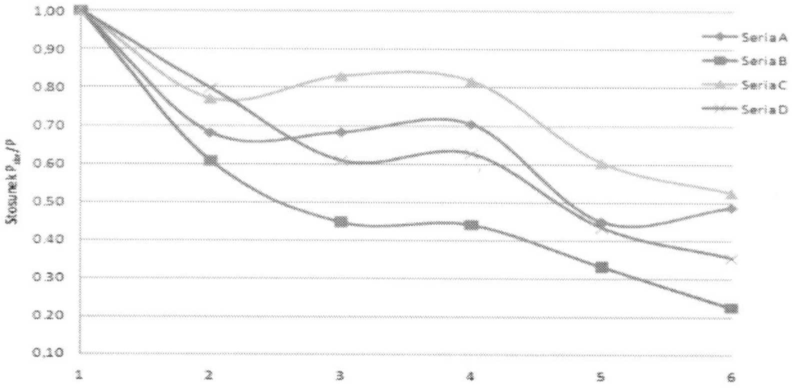

Fig. 6. Graphical comparison of the settlement susceptibility of the samples series A, B, C, D.

3 Conclusions and scope of applications

Generalizations of the specific proposals from the conducted test sequence and measurements referred to from [4], lead to the following statements:

- Subgrade and subbase ground layers can be reinforced with geosynthetics including geotextiles and geogrids. The use of these give a significant strengthening effect of the substrate road surface, a larger modulus of the multilayer substrate surface and reduces the susceptibility to permanent deformation and settlement.
- The use of several layers of geotextiles resulting in a material gain range of multilayer substrate surface highways and rail roads, and does not limit their functions separation and filtration in a multi-layer system, ground substrates foundation of roads and railroad.
- Single-layer geotextiles reinforcement is not a sufficient reinforcement of the substrate surfaces of highways, only retain their function layer separation and filtration function circuit substrate. Number and order of geotextile layers in the soil-ground mobilizes interaction and friction of the ground layer strengthening.
- An effective form of reinforcement layers of soil substrate surface is to use two or three layers of geotextiles, the total thickness of the reinforced layer of primer should not exceed 25 cm significantly greater enhancing effect layers can be obtained by replacing the geogrid of one of the layers of geotextile.
- The appropriate thickness of the soil layer between the sandy layers of geotextiles depends on the parameters of the quality of the land and should not exceed 10 cm while maintaining the required densities.
- Strengthening and reinforcement geosynthetics ground floor paving of roads is an alternative design for the reinforcement and improvement of ground surface using hydraulic binders and can be economically competitive.

References

1. J.G. Zamberg, R. Gupta, 9th International Conference on Geosynthetics, Brazil, 380-398 (2010)
2. M.S.N. Tafreshi, A.R. Dawson, Geotextiles and Geomembranes **28**, 72-84 (2010)
3. E. M Palmira, Geotextiles and Geomembranes **27**, 368-390 (2009)
4. K. Gradkowski, *Testing ground subgrade reinforced with geosynthetics*. (Publishing House WUT, Warsaw, 2013)

The modern evaluation method of fracture toughness of basaltplastic lining for engineering collectors

Olga Lyapidevskaya[1*], and *Andrey* Sherstnev[1]

[1]Moscow State University of Civil Engineering, Yaroslavskoe sh. 26, 129337 Moscow, Russia

Abstract. The estimation methods of fracture toughness of reinforced blocks are considered for construction for engineering collectors. There are presented the methodical accesses that make possible to numerical describe all studies of destruction process of polymer lining with regard to simultaneously action the whole complex of exploitation factors. It is stated the data of fracture toughness of the composites on organic and nonorganic matrix in normal condition in comparison with basaltplastic lining. It is reported the dates of safety and critical coefficients of composite strains intensity on organic and nonorganic base in normal conditions.

1 Methods of evaluation of fracture

The fracture toughness is the one of the most important characteristic of safety of polymer lining.

Processing from the fracture toughness condition under action of the exploitation load it is determined the thickness of polymer lining. It is a number of evaluation methods of estimation of the fracture toughness that permit to determine the possibility of material to be resistant to developing of stress (to failure).

This paper is presented the modern method of estimation of the fracture toughness and comparative analysis of composites on the base of organic and nonorganic matrix with basaltplastic.

The basaltplastic lining is the special multi-layered lining consisted from nano modified thermosetting resins, chopped basalt fibers, metal reinforcing elements for coupling of lining with concrete that is designed to protect the inner surfaces of concrete blocks from possible mechanical, thermal, physical and chemical damage.

For the formulation and solution of this task the plane of polymer lining was combined with the plane of a Cartesian coordinate system with respect to which it was set orientation and location ordered the cyclically symmetric system of fibers in composite (fig. 1).

* Corresponding author : olga.lyapidevskaya@inbox.ru

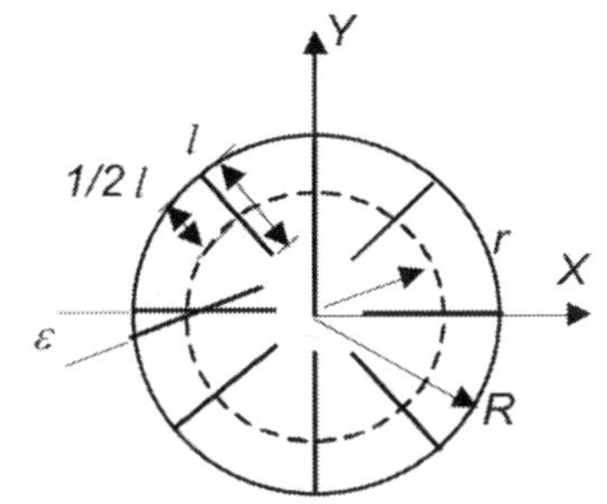

Fig. 1. The model of randomly reinforced basaltplastic by discrete fibers: l- the length of rigid inclusion, R - the radius of the composite, ε - the angle of rotation of the rigid inclusion.

On the basis of application of the method of mechanical quadrature to calculate the ratio of intensity stresses in the work [1] it was established that giving the length of the fibers L and values K_o and K_q can be found the values of σ_o and σ_k:

$$\sigma_0 = \frac{K_0}{0.0604\sqrt{l}}; \sigma_k = \frac{K_q}{0.0604\sqrt{l}} \quad (1)$$

where σ_o – safety stress, σ_k - critical stress, that are substantial data when assessing bearing capacity of the material. Therefore, the values σ_o, σ_k, K_o and K_q are connected by the next ratio:

$$\frac{\sigma_0}{\sigma_q} = \frac{K_0}{K_q} \quad (2)$$

The legitimacy of the use of methods of linear mechanics of destruction for the investigation of fracture toughness of polymer and composites on its base is proven by the results of a number of works.

The paper [2] presents the results of experimental determination fracture toughness of epoxy resin. Straightforward character typical chart "stress - offset", and also the consistency of critical factor of intensity stresses K1c evidence of the legality of relevant application of linear fracture mechanics.

In the work [3] it is analysed the possibility of using the classical methods of destruction for investigation of cracks developing processes in composite materials and showed that at the macro level to composites may be applied such concepts as the coefficient of stress intensity, the rate of release of strain energy, he critical crack opening, J – integral. To determine the strength properties of composites with cracks the authors of the works [4, 5] suggest to use well-developed methods of fracture mechanics of homogeneous anisotropic body.

However linear fracture mechanics may be used to estimate the fracture toughness only to composites having linear elastic properties until the destruction. Quite often the destruction of the material is preceded by the formation of a prefracture zone in the vicinity of crack tip (tears and elongation of individual fibers, loosening of the matrix and violation of adhesion on the interface) that starts with a load equal to 80% of the critical [6]. When this zone is small compared to the size of the defect which cause the destruction of the composite it is possible to apply methods of linear mechanics to such materials [3], and in particular, the concept of the K coefficient, introducing an amendment to Irvine [7], lengthening the initial crack at a certain amount:

$$r_y = \frac{1}{2\pi}(\frac{K_{1c}}{\sigma_{0,2}})^2 \tag{3}$$

where $\sigma_{0,2}$ - the yield strength of the material.

Effective fracture length – $L_{ef}=L_0+r_y$.

The authors of the work [8] suggested for estimation of crack resistance of randomly reinforced fiberglass on the base of epoxy and polyester resins to use the method R-curve, which is widely used to describe the behaviour of metallic materials at destruction [6]. The essence of the method of R-curve lies is to estimate the increment of crack length at the time of its unstable distribution. The meaning of this is analogous to the reception of Jones and Brown's in determining coefficient K_q corresponding to the two per cent tolerance on stable growth of crack when tested to destruction. The plate specimens with edge crack were tested. The coefficient of stress intensity was determined [9] by the formula:

$$K = \frac{Pa^{0.5}}{tH}Y \tag{4}$$

where Y- the shape factor of the sample coefficient, a – the length of crack corresponding to the load P; t and H – the thickness and the width of the test sample respectively.

The results of the work [6] showed that the method R-curve can be successfully applied to composite chaotic reinforced materials with developed prefracture zone.

2 The modern method of fracture assessment

The above experimental approaches to the evaluation of performance of composites from the position of linear fracture mechanics do not give total information about the parameters of fracture toughness of composite materials as in normal condition as in terms of the impact of operational factors (temperature , aggressive environment , load and so on), since the fracture toughness is evaluated only at the stage of propagation of the macrocrack, which is the second final part of a long process of destruction.

Thus, it is obvious that for modern composite materials it is necessary the methodological approaches to quantitatively describe all the stages of the process of destruction of the material subject to the simultaneous exposure of the whole complex of operational factors.

Thus, the main fundamental characteristic of composites is the coefficient of stress intensity. Knowing the latest it is possible to estimate the stress-strain state in the vicinity of the most dangerous areas of the composite, its load capacity (since most of the criteria for the occurrence of the threat condition are expressed through the components of the stress tensor), and to establish the effect of operational factors.

The basis for determining the critical stress intensity factor K for the case of generalized plane stress state the condition defined at the moment of breakaway of a crack-cuts were based on the methodology outlined in this paper 9.

3 The results of laboratory fracture tests of composite materials using modern method of its evaluation

The choice of this technique is predefined with the shape of test specimen that is a wedge-shaped sample of type a double console beam, allows long to exhibit it in terms simultaneous exposure to different aggressive spheres, load and temperature (fig. 2).

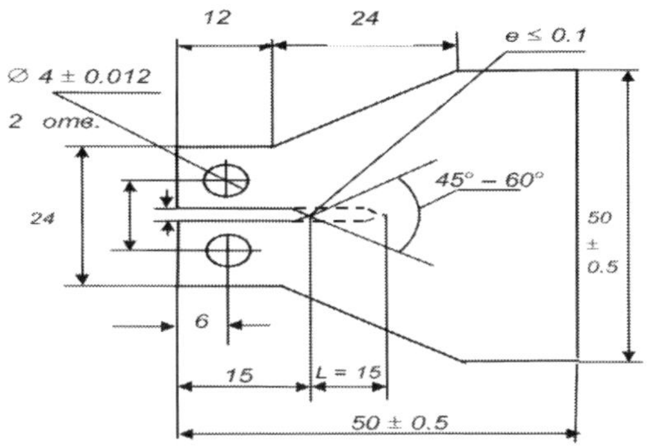

Fig. 2. A wedge-shaped sample of type a double console beam: L- the valid range of initial crack-incision n lengths, e - its width.

A wedge-shaped sample was stretched on a tensile testing machine at a speed of the active capture 1 – 0,5 mm/min. The error in determining the power does not exceed 10% of the measured value [10].

To determine the moment of breakaway of a crack-cut it was proposed simultaneously with the recording of the curve "force-displacement" record in real time the differential and integral parameters of acoustic emission (fig. 3).

Experimentally it was found that the minimum pulse amplitude acoustic emission associated with the beginning of the micro fractures of the tested components is 50 μv. Pulses with the amplitude of 10 μv relate to fluctuation noise. The acoustic emission signals with an amplitude less than 100 μv reflect the kinetics of accumulation of submicron cracks in the vicinity of crack tip of the notch and readiness to start.

For the moving crack front-cut is taken as the time of occurrence of the characteristic "jump" in the diagram "stress-displacement" X [11]. In the absence of a specified characteristic "jump" for the moving is taken as the time of appearance of acoustic emission signals exceeding 150 μv.

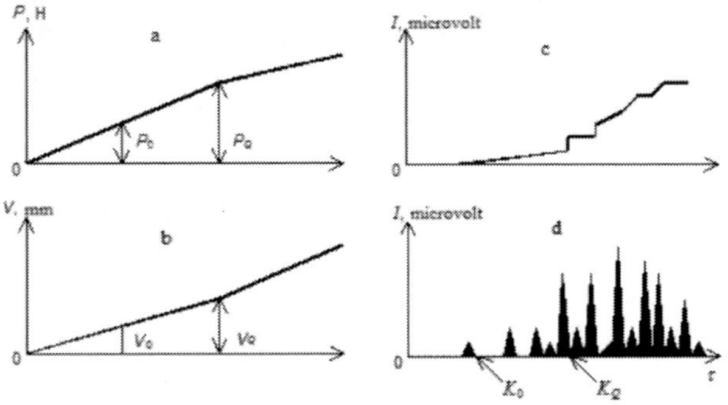

Fig. 3. Synchronized recording: a - stress P, b- displacement V, c - the amount of acoustic emission pulses, d - the acoustic emission signals.

Believing that the process of microfractures of a composite is the process of nucleation of microcracks we denote the stress intensity factor corresponding to this process, -K_o.

In establishing the diagram "force-displacement" at the outbreak of the first acoustic emission signals and the pushing-notch fracture the amount of K_o and K_g are determine by the relations:

$$K_0 = \sqrt{\frac{P_0 E V_0}{2bl}}; K_q = \sqrt{\frac{P_q E V_q}{2bl}} \quad (5)$$

where P_0 and P_q - load, respectively at the time of the first signal of acoustic emission and breakaway crack; V_0 and V_q - disclosure of the crack, respectively, at the time of the first acoustic emission signal cracks and breakaway; b - the thickness of the sample in the vicinity of the top of the notch-cracks, l- the original length of the crack notch.

Table 1. The fracture toughness data based composites of inorganic and organic matrix in normal conditions.

Resin, composite	The content of the reinforcing component, max. %	K_q, N/mm $^{3/2}$	The coefficient of variation, %
PN -15	-	19.1	9.0
PN -15+ lavsan	10	28.0	12.1
PN -15+MPS	10	30.5	11.9
PN -15+LBB-SP	10	32.3	13.6
PN -15+UTM-8	50	110.2	14.0
PSK -15-SX	50	101.0	14.8
PPM -15- SX	50	127.3	15.1
UNIPEK+Basalt	30	67.5	14.3

Table 1 shows the fracture toughness of the composites based on organic and inorganic matrices obtained by the method described above with and without the use of acoustic emission (acoustic emission characteristic of recording signals in the process of loading the samples shown in fig. 4).

Comparison of the fracture toughness of composites parameters shows that the estimate K_Q without the use of acoustic emission method materials with elasto-plastic properties which remain virtually unchanged at the start-notch fracture leads to an overestimation of the results by 6-11% compared with the data obtained using the acoustic emission method.

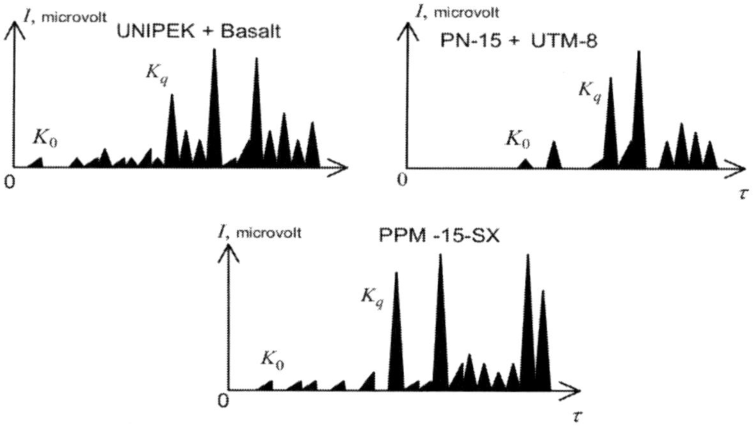

Fig. 4. The characteristic acoustic emission signals during the destruction of samples of composite materials.

The K_0 and K_q values of composite materials based on polyester resins PN-15 are presented in Table 2.

Table 2. Values K_0 and K_q of composites based on organic and inorganic matrix in normal conditions.

Resin, composite	K_0, N/mm $^{3/2}$	The coefficient of variation, %	K_q, N/mm $^{3/2}$	The coefficient of variation, %

PN-15	19.1	9.0	19.1	9.0
PN-15+lavsan	28.0	12.1	28.0	12.1
PN-15+MPS	16.0	12.3	30.5	11.9
PN-15+DBB-SP	13.3	9.2	32.3	13.6
PN-15+U TM-8	47.0	11.1	110.2	14.0
PPM-15-SX	65.0	14.4	127.3	15.1
UNIPEK+basalt	9.4	11.4	67.5	14.3

Table 3 shows the values of the parameters of fracture toughness and solidity ratio of randomly reinforced fiberglass.

Analysis of the data in Table 3 leads to the conclusion about clear symbate connection M_x, K_0, and K_q.

Table 3. Values Mx, K0, and Kq of randomly reinforced fiber glass.

Composite	**K_0 N/mm $^{3/2}$**	**K_q,N/mm $^{3/2}$**	**M_x**
PN-15+DBB-SP	13.3	32.3	0.40
PN-16+DBB-SP	9.1	23.6	0.29
PN-10+DBB-SP	5.8	15.2	0.25
EDT-10+DBB-SP	19.5	40.0	0.56
UNIPEK+basalt	14.5	33.1	0.47

4 Conclusion

Solidity ratio of randomly reinforced fiber glass is indicative of not only the strength of these composite materials and also the crack resistance. Thus, the monolithic conditions set forth above may be used to produce high strength and fracture toughness at random discrete basalt fiber-reinforced composite materials.

References

1. I. T. Berejnitsky, (Nauk. dumka, Kiev, 1983)
2. G. Korten, *Micromechanics and character of compositions breaking: Modern composite materials* (Mir, Moscow, 1973)
3. G.P. Zaitsev, Problem of strength **6**, 78-83 (1977)
4. N. Cornish, I. Swerdlow, T.A. Cruse, Rocketry and Astronautics **1**, 45-50 (1974)
5. N. Cornish, I. Swerdlow, T.A. Cruse, J. Composite Mater. **6**, 114-124 (1972)
6. C.C Poe, Eng. Fract. Mech. **2(17)**, 153-171 (1983)
7. Proc. of Seventh Ordnance Material Research Conf. Sagamore, 1961, P.63.
8. S. Gaggar, L.G. Broutman, Ibid. **9**, (1975)
9. P. Crosly, A. Ripling, Fracture Mechanics **25**, 199-221 (1981)
10. B.I. Bulgakov, T. V. Lum., Industrial and Civil Engineering. **8**, 45-48 (2016)
11. E.B. Tkach, Concrete technology **7-8**, 44-47 (2011)

The influence of hydrodynamic factors, membrane surface properties and channel geometries on membrane performance and fouling mechanisms

Alexey Pervov[1,*]

[1]Moscow State University of Civil Engineering, Department of Water Supply and Water Disposal, Yaroslavskoye sh. 26 , 129337 Moscow, Russia

Abstract. Modern theoretical understanding of colloidal and suspended matter membrane fouling mechanisms are presented and discussed. State-of-the-art simulation models of concentration polarization calculations for different channel conditions are described and influence of the fouling layers on the flux and rejection decrease are evaluated. Results of experimental investigations are presented that suggest a quantitative evaluation of fouling rates and membrane flux prognosis due to colloidal fouling with time. The influence of channel geometry on fouling is demonstrated and discussed. The main disadvantage of spiral wounded membrane modules which is attributed to the presence of a separation spacer mesh in the feed channel is discussed.

1 Introduction

The problem of fouling emerged at the beginning of industrial application of membranes to treat natural water for drinking and technical purposes. Theoretical and experimental investigations of membrane fouling processes reveal that processes of membrane transport, flow hydrodynamics, and chemical interactions that occur in membrane modules are very closely connected. Many modern research reports present results of membrane fouling investigations and develop fouling theories without accounting this relationship. Each process is very important to create a broad understanding of fouling.

Natural surface water contains suspended and colloidal matter and dissolved organics that have different molecular weight. Fouling mechanisms are different for different foulants. Suspended and colloidal matter can sediment on membrane surface, adhere to membrane, and/or coagulate. They form fouling layers that decrease membrane flux. Organic molecules adsorb on membrane surface or on foulant particles.

The first step to solving this problem is to understand that fouling and membrane performance deterioration was completely attributed to hydrodynamic factors (e.g., crossflow velocity, flux, etc.) and development of concentration polarization effect on membrane surface was due to uneven velocities distribution. Sorption and adhesive

* Corresponding author: ale-pervov@yandex.ru

processes were not diligently accounted for. The main techniques to control fouling were based on creation of suitable flow conditions in membrane channels. Meanwhile, further experience demonstrated that hydrodynamic tools to control fouling have been exhausted. Other important factors to be taken into consideration are organic adsorption and the adhesion of particles and bacteria to membrane surface. Membrane material properties (e.g., charge, hydrophility, and adsorption characteristics) are decisive for fouling intensiveness and selection of these properties becomes an essential tool to control colloidal, organic and bacterial fouling. Low fouling membranes with a modified surface are developed and introduced into practice [1]. But it is not only flow hydrodynamics and surface properties that completely solve the problem.

Membrane channel configuration also plays important role in fouling. For a long time it was considered that spiral wound construction has high fouling propensities and requires very careful pretreatment. Currently, to remove colloidal and organic matter pretreatment uses ultrafiltration membranes tailored by tubular or capillary membrane channels that are considered more reliable than spiral wound configurations as they have lower hydraulic resistance and less membrane backwashing [2].

Spiral wound configuration is considered very efficient because of its compact design and high membrane surface per unit volume of the module. The main disadvantages of spiral wound modules are attributed to the presence of separation spacer mesh in the feed channel as it traps fouling particles and increases crossflow resistance [3, 4]. Elimination of said spacer mesh from the feed channels could be a decisive step in solving the fouling problem and developing efficient surface water and wastewater treatment techniques.

2 Simulation models and analysis of colloidal fouling influence on reverse osmosis (RO) membrane mechanisms

Prediction of product flow decrease due to membrane fouling remains one of the most important problems of RO process design and operation. Such prognostic techniques are based on the understanding of fouling mechanisms by authors who developed them.

A review of existing fouling theories demonstrated different views on the process. Filtration theory is developed and discussed in many publications presented by G. Belfort [5-9]. The described mechanism follows "cake filtration" mechanism. Belfort presents a filtration theory as a main tool to describe fouling of microfiltration membranes both in a "dead end" and cross-flow modes [7] that correspond to flow conditions in RO modules.

In [8], Belfort and co-authors developed a filtration theory that accounts for trajectories of particles in the cross-flow mode depending on their mass and velocity. As it was shown in [8], filtration theory often predicts stronger product flow decrease than was observed during experiments. Therefore, conventionally used filtration mechanism was improved in respect to determination of force that influences the particles in the flow. Calculations were made for laminar flow conditions with a parabolic distribution of velocities in the channel. The particle near membrane surface is influenced by convective and back diffusion forces and also by the lifting force that is formed due to difference of various layers velocities.

Theory of stability of colloidal solutions at membrane surface, as described by Brunelle in [10] and later developed in [11], combines concentration polarization effect with coagulation of particles on membrane surface. Coagulation occurs due to the increase of salt concentration in the boundary layer and the loss of colloidal solution stability.

Gel formation theory, or gel-polarization theory [12], describes coagulation or gel formation in high molecular organic solutions at membrane surface due to the concentration polarization effect. According to this theory, gel is formed after the maximum concentration value is reached near membrane surface.

Colloidal fouling model based on adsorption mechanisms is presented by Jackson and Landolt [13] in respect to iron dioxide colloidal solutions. According to their suggested mechanism, membrane fouling occurs in two stages: nucleation and foulant growth. To describe foulant growth process, the authors use colloidal stability theory [10]. To describe and predict product flow decrease due to membrane fouling, cake filtration mechanism is widely used. The influence of accumulated fouling layers on membrane behavior is discussed below.

3 The influence of fouling layers on RO membrane performance: product flow decrease and loss of rejection

Different understandings also exist about the process of foulant influence on membrane performance. According to Kimura [14], accumulation of foulant on membrane surface leads to product flow decrease due to hydraulic resistance growth of the fouling layer. The decrease of product flow also decreases concentration polarization level, therefore membrane rejection value increases (Figure 1). This is not confirmed by the operational experience, as fouling always leads to loss of rejection.

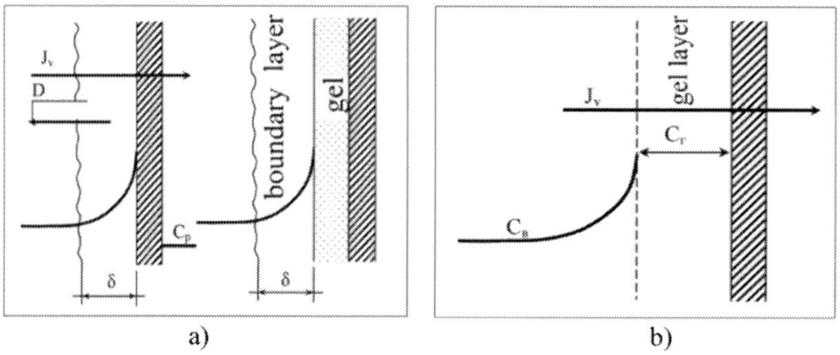

Fig. 1. Increase of salt concentration near membrane surface due to layer formation: Kimura model (a) and gel formation model (b).

In [8], Belfort claims that rejection decrease (product quality deterioration) is connected with decrease of diffusion rate from membrane surface to the bulk solution due to the presence of the porous foulant layer on membrane surface. According to Merten's model, fouling layer thickness increase membrane boundary layer (where concentration polarization occurs), as shown in Figure 2. Therefore, reduction of membrane flux can be explained not by the increase of flow resistance (transmembrane pressure), but by the increase of concentration polarization and decrease of active pressure value that is described as $(P-\pi)$, where π is osmotic pressure of solution on membrane surface.

The increase of flow resistance in membrane channels can significantly influence membrane flux as active pressure value decrease through the module length. It is well known that this resistance (called also "delta pressure") can reach high values, producing a "telescoping effect" of spiral wound membranes. Mechanism of flow resistance increase is greatly attributed to membrane channel configuration. Autopsies of fouled spiral wound membranes reveal that spacer mesh, mainly the mesh bundles are covered by foulant more than other membrane surfaces, trapping foulant particles [15]. Thus, flux reduction is attributed to the concentration polarization growth and active pressure value loss.

Hydraulic resistance of the fouling layer obviously does not influence RO membrane flux rate. Membrane autopsies show that fouling layers have very soft structure and its hydraulic resistance is negligible as compared to membrane resistance. Thus, calculations of fouling layers flow resistance do not have practical significance. In practical observation, flux decrease is always accompanied by an increase of product TDS. This drop of product flux due to the increase of foulant resistance does not explain the decrease of rejection values.

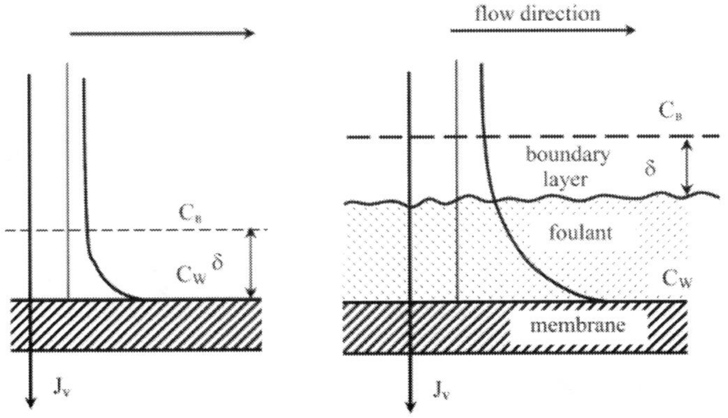

Fig. 2. Concentration profile near RO membrane surface in turbulent flow with fouling and without fouling layer influence.

Concentration polarization is thus recognized as the main reason of product flux decrease, and the thickness of the fouling layer is recognized as the main factor that affects membrane performance. When concentration polarization level is negligible (for the cases where tangential flow is high), the thickness of boundary layer is equal to the thickness of the fouling layer, and concentration polarization level in the fouled membrane channel is determined according to equation (1):

$$C_w / C_b = \frac{exp(v\delta / D)}{R + (1 - R)exp(v\delta / D)}, \tag{1}$$

where C_w, C_b – salt concentration at the membrane wall and in bulk solution, D – diffusion coefficient, R – salt rejection o the membrane ($R = (C_b - C_p)/C_b$, C_p – salt concentration in permeate), v – convective velocity, δ – thickness of boundary layer

The model that describes influence of the fouling layer on membrane rejection used the Nernst model equations. According to this model, a thin boundary layer exists at membrane surface (a stagnant film with a thickness δ). Dissolved salt transport in this case is limited by molecular mechanism – diffusion that is determined by coefficient D. When fouling layer is formed on membrane surface, the thickness of boundary layer is increased by the thickness of the fouling layer, and salt transport through the layer from membrane to the bulk flow is limited as compared to the case where is no foulant (Figure 3).

Density of layer's structure can also influence membrane performance. Loose layer structure creates a lower thickness of the boundary layer than does a firm, dense structure.

Under laminar flow conditions a special velocity value profile is formed along with concentration profile in the boundary layer [16]. The concentration values profile is formed due to convection and diffusion. Concentration profile is formed even at the channel inlet. If the fouling layer is formed on membrane surface, concentration gradient is formed in the channel above this layer. The velocities' profile does not influence salt transport within fouling layer thickness. At the end of the channel, a new concentration gradient is built-up, which depends on velocities' profile in the bulk flow and within the fouling layer thickness, as shown on Figure 3.

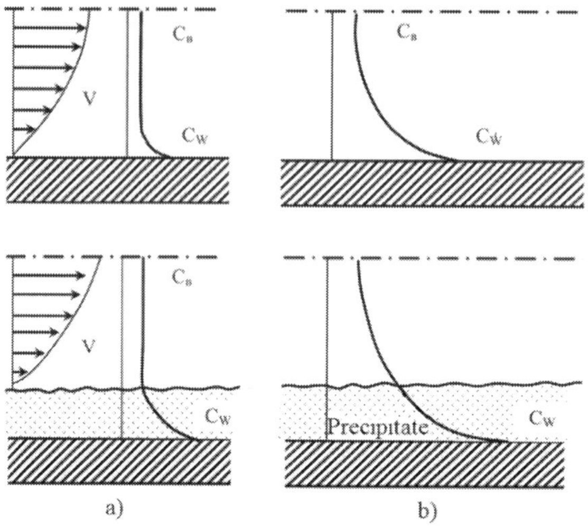

Fig. 3. Development of concentration polarization with fouling under laminar flow conditions: at the channel inlet (a) and in the end of the channel (b).

4 Colloidal and organic fouling: the influence of "shear" forces

Cake filtration model is widely used to describe membrane fouling by suspended particles when water is filtered through microfiltration or ultrafiltration membrane in "dead-end" mode. As was already mentioned in [14], this model does not efficiently describe fouling layer formation under "cross-flow" conditions. In the "cross-flow" mode after the fouling layer is formed, the subsequent fouling does not occur due to "tangential" flow shear forces.

When surface water is treated with membranes, suspended particles, colloidal particles, and large organic molecules (e.g., humic acids) foul membrane surfaces. Colloidal particles differ by size and their diffusion characteristics vary. In the beginning of the process, a fouling layer forms on membrane surface and product flux decreases. The foulant growth stops after product flux decreases and concentration polarization level is reduced. Thus, cake filtration theory also can be used to describe fouling under "cross-flow" conditions. According to "cross-flow" model, fouling process consists of two steps:

1) accumulation of the fouling layer that cannot be sheared-off;

2) shearing rate and fouling rates reach equilibrium.

Research specialists often share their experience – that this equilibrium is reached usually after membrane unit is completely fouled-up and product flux has significantly been decreased [14]. Many technological tools to control colloidal fouling are recommended based on hydrodynamic understanding of the problem: to use flocculation to increase particle size and thus decrease fouling layer resistance; to increase "cross-flow" velocities and "shear-off" effect, to apply backwashes (for UF applications) and membrane flushing.

Fouling control by cross-flow velocities requires development of the "back transport" models. There are three models that describe shearing mechanisms: Brown's diffusion model for low molecular organics, "shear" diffusion for macromolecules, and migration model for colloids. These "back transport" models provide different evaluation of fouling potential for colloidal particles and humic substances. For small particle size and for low molecular weight substance, Brown's diffusion coefficient value is higher and "shear" diffusion coefficient value is lower than for large particles and high molecular organics. When organic solution is separated, concentration of organics in the boundary layer increases. Large organic molecules with smaller diffusion abilities are more likely to form fouling layer due to low solubility and higher concentration polarization level [17].

Product flux decrease when organic solution is separated is attributed to gel formation and sorption processes that occur between organic molecules and membrane material [18] and described by equation (2):

$$J = \frac{P}{\mu(R_m + R_{ads1} + R_{ads2})}, \quad (2)$$

where: P – pressure drop through membrane; R_{ads1} – flow resistance of the gel layer plus additional resistance of adsorbed organic matter; R_{ads2} – additional resistance attributed to interactions of adsorbed organics and dissolved organics.

Formation of gel fouling layer is described by the following equation (that is modified equation 2):

$$\frac{P}{\mu J} = R_o + R_c = K_c \frac{V}{A} + \frac{P}{\mu J_o}, \quad (3)$$

where: J_0 – initial membrane flux velocity, R_0 – membrane resistance; V – permeate volume; A – membrane surface; K_c – a unit gel resistance coefficient; R_c – gel resistance.

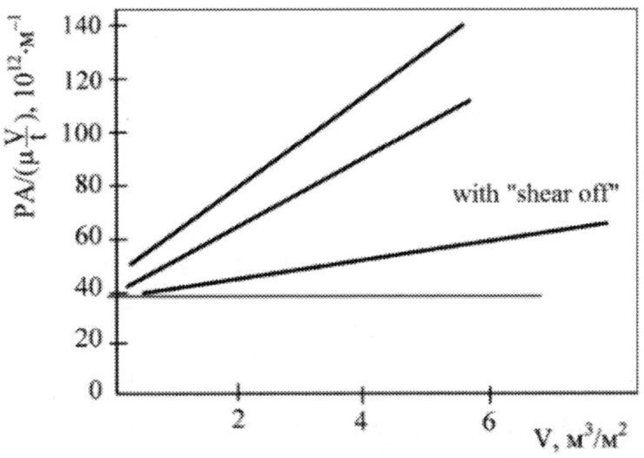

Fig. 4. Dependence of *PA/(μV/t)* value on filtered volume for various cross-flow velocities [19].

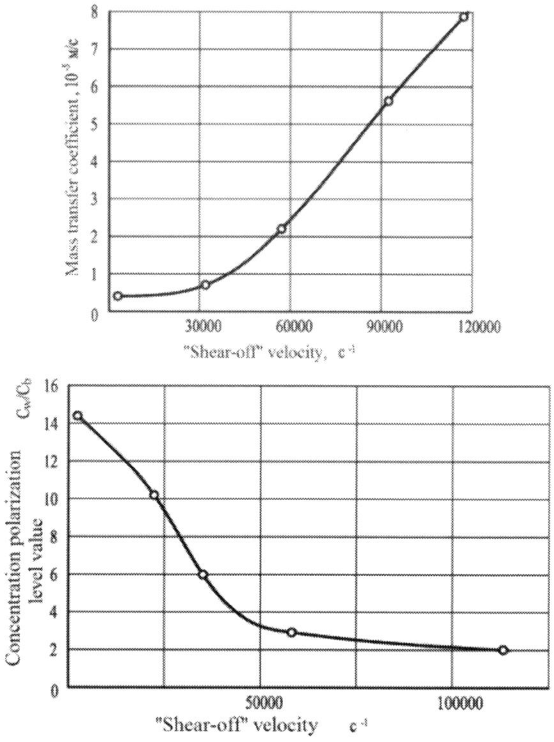

a) b)

Fig. 5. The influence of "shear-off" velocity values on mass transfer coefficient and concentration polarization effect [19].

Figure 4 shows dependence of the *PA/(μV/t)* value on the volume *V* of water passed through membrane filter ($V/t = J_t$, where J_t – membrane flux at time *t*). The change of slope can be explained by particle "shear-off" effect, decrease of their concentration, and gel layer resistance [19]. Figure 5 demonstrate the influence of cross-flow velocity values on concentration polarization level and mass transfer coefficient value.

5 Increase of hydraulic resistance in membrane channels during operation

The main disadvantages of spiral wound modules are attributed to the presence of separation spacer mesh in the feed channel as it traps fouling particles and increase a crossflow channel resistance [4, 20]. The places (spots) where mesh contacts the membrane surface provide "dead" areas without cross-flow, resulting in high concentration increase at the membrane surface that initiates formation of colloids and crystals of sparingly soluble salts.

When water is filtered through a sand bed, suspended and colloidal particles adhere to the sand particles and sediment in the pore volume forming "bridges" in pore channels that increase hydraulic resistance for the flow. Experimentally obtained relationships of hydraulic resistance versus time, flow and suspended matter concentrations demonstrate similar process for spiral wound membrane channel and sand filter bed (Figure 6).

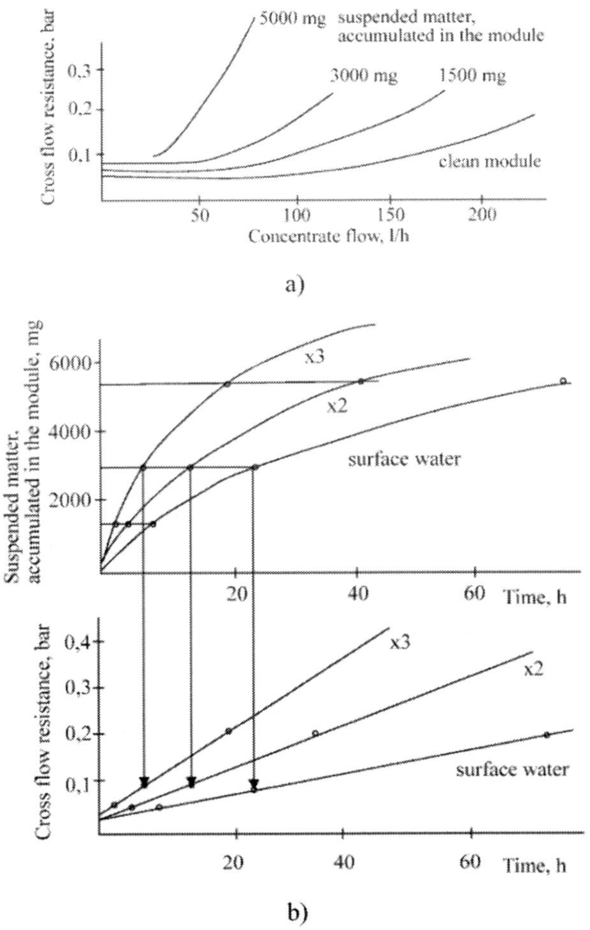

Fig. 6. Example of flow resistance increase in spiral wound membrane module.

Figure 6 (a) shows the increase of cross-flow resistance versus cross-flow value at different stages of fouling (for different amounts of foulant, accumulated in the spiral wound module 1812 standard). Determination of fouling rate values at different stages of fouling (corresponding to different amounts of accumulated foulant) enabled us to predict amount of foulant sedimented on membrane surface depending on operation time and feedwater composition. Prognosis of foulant accumulation is demonstrated in Figure 6 (b), where feed water from the river was concentrated by RO twice and by three times and these solutions were tested. Based on the data presented in Figure 6 (b), each foulant amount corresponds to the operation time and to hydraulic resistance value (Figure 6 (c)).

6. Conclusions

Colloidal fouling rate values are very dependent on the cross-flow velocities: the higher cross-flow velocity, the less is foulant accumulates on membrane surface. Mechanism of suspended and colloidal fouling is mainly attributed to particle adhesion on membrane surface due to convective flow. Colloidal fouling varies depending on membrane properties such as surface charge and hydrophility. Channel geometry has decisive influence on fouling process. During foulant accumulation in spiral wound modules, cross-flow resistance grows. The product flux drop is directly dependent on hydraulic resistance increase as active pressure above membrane surface in the channel decreases.

Future work will show that development of a new module with "open channel" with low cross-flow resistance enables us to suggest a reliable membrane water treatment technique that does not require pretreatment due to efficient channel configuration, modified membrane material and hydrodynamic mode.

References

1. J. Gilron, S. Belfer, P. Vaisanen, M. Nystrom, Desalination **140**, 167-179 (2001)
2. X. Shi, G. Tal, N.P. Hankins, V. Gitis, J. of Water Process Eng. **1**, 121-138 (2014)
3. C.P. Koutsou, A.J. Karabelas, J. of Membrane Sci. **488**, 129-142 (2015)
4. A.H. Haidari, S.G.J. Heijman, W.G.J. Meer, Water Res. **106**, 232-241 (2016)
5. G. Belfort, G. Alexandrowisz, B. Marx, Desalination **28**, 13-30 (1979)
6. G. Belfort G., Desalination **34**, 159-169 (1980)
7. R.H. Reed, G. Belfort, Water Sci. Tech. **14**, 499-522 (1982)
8. G. Belfort, R.J. Weigand, R.I. Mahar, *Reverse Osmosis and Ultrafiltration Symp. 188th Meet. Amer. Chem. Soc.* (Philadelphia, Washington, 1985).
9. G. Green, G. Belfort, Desalination **35**, 129-147 (1980)
10. M.T. Brunnele, Desalination **32**, 127-135 (1980)
11. J.W. Kaakinen, C.D. Moody, *Reverse Osmosis and Ultrafiltration Symp. 188th Meet. Amer. Chem. Soc.* (Philadelphia, Washington, 1985).
12. S.-I. Nakao, T. Nomura, S. Kimura, AICHE Journal **25**, No4, 615 (1979)
13. J.M. Jackson, D. Landolt, Desalination **12**, 361-378 (1973)
14. D.E. Potts, R. Alert, S. Wang, Desalination **36**, 235-264 (1981)
15. A.I. Shafer, A.G. Fane, T.D. Waite, Desalination **131**, 215-224 (2000)
16. J. Cho, G. Amy, Y. Yoon, J. Sohn, Desalination **142**, 245-255 (2002)
17. J.G. Grespo, K.W. Buddeken (ed.) *Membrane processes in separation and purification* (NATO ASI Series, 1994)
18. T.L. Champlin, *Proc. of the Conf. on Membranes in Drinking and Industrial Water Production* (Paris, France, 2000)
19. R. Bian, K. Yamamoto, Y. Watanabe, *Proc. of the Conf. on Membranes in Drinking and Industrial Water Production* (Paris, France, 2000)
20. A. Pervov, A. Andrianov, Desalination and Water Treatment **35**, 2-9 (2011)

Research dedusting efficiency of the inertial hub with adjustable parameters

Rashid Sharapov[1,*], *Pavel* Kapyrin[1], *Svetlana* Lozovaya[2], *Valentina* Yadykina[2], *Aleksandr* Agarkov[2]

[1]Moscow State University of Civil Engineering, Moscow, Russia
[2]Belgorod State Technological University named after V.G. Shukhov, Belgorod, Russia

Abstract. Production of construction materials every year production increases, thus increasing emissions pollutants into the atmosphere subjected to cleaning. At the enterprises of building materials production equipment becoming outdated both morally and physically exhausted their useful life. Technological design standards in the operation of purification equipment has undergone a number of changes. There are a large number of works devoted to the theory of cyclone separation. They examined calculation method and to determine the minimum size of particles is separated completely from the air stream in a cyclone. In theory, the high efficiency cyclones are relatively small separation of particles from the air stream. The experiments show that certain fractions of full capture occurs from the cyclones and out relatively large particles. Designed dust device hubs. They capture dust particles with low specific density and smaller than 5–10 microns. Rational use of hubs is achieved in the aspiration systems high performance. They are placed in front of or between the filter and the cyclone filter. Hub much reduces the amount of dust-laden air directed onto a thin cleaning. This reduces energy consumption for manufacturing as a whole, as well as, due to a possible decrease in performance fine filters, reducing their initial cost, metal consumption, maintenance costs and reduce the area occupied by them.

1 Introduction

The rapid growth of industrial production, the accumulation of huge amounts of waste, significant energy consumption in various industries contributes pollution of our planet [1-6].

The catastrophic increase in air pollution due to the sharp disparity between the development of industrial production and equipment cleaning, in particular of air dust, which has not been able to solve the problems posed to the industry. In the future, dedusting equipment has evolved significantly, and the growth of air pollution in some regions has been suspended.

* Corresponding author: ptdm_zavkaf@mail.ru

At the same time successes are local, and air pollution is increasing. This is indirectly saying and reducing life expectancy in selected industrial areas in different countries and a global average temperature increase of our planet [7].

One of the most important tasks for the prevention of the consequences of air pollution is the use of modern dust removal equipment in all process stages of technological processes related to dust emission. This equipment can be used not only in the last stage dedusting process, but also in the early stages, reducing the load on the finishing dedusting.

The aim of this work is the development of inertial dust hub, providing more efficient clearing of coarse dusty process gases above 90% while reducing its hydraulic resistance.

2 Development of the concept of inertial dust hub with adjustable parameters

A comparative analysis conducted on the basis of contemporary sources, leads to the conclusion that for effective dust cleaning is necessary to develop and use inertial dust hubs.

There is a need for further development of the scientific field and to create a universal equipment with greater efficiency, increased reliability and more features.

In order to implement our proposed destinations are encouraged to use the following technical solution, as reflected in the patent of the Russian Federation for useful model (Fig. 1).

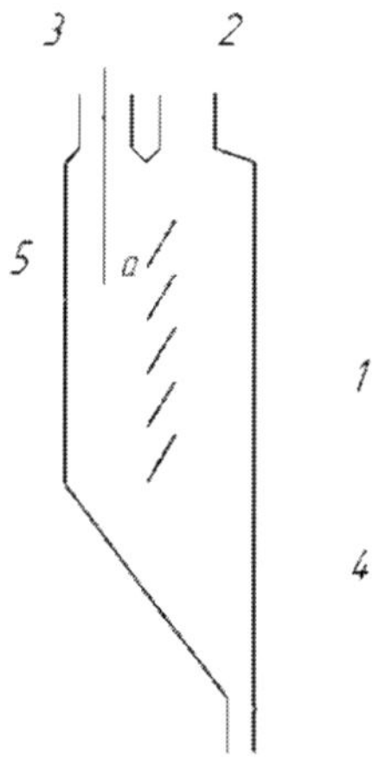

Fig. 1. Inertial dust hub: 1 – body, 2 – inlet pipe, 3 – output thin tube material, 4 – pipe the output of the coarse fraction, 5 – louver elements.

Inertial dust concentrator consists of a housing 1. The upper housing inlet pipe 2 arranged starting material and output 3 fine material outlet. On the opposite side of the housing in its lower part located output nozzle 4 of the coarse fraction. In the middle part of the housing relative to its vertical axis mounted adjustable louver system elements 5. Louvered fixed elements, such as axles, shafts, which are mounted on the side walls of the housing. Elements cinematically connected to each other, for example, traction, for simultaneous rotation about their axes and horizontal displacement. For this purpose, made in the body bore. Louvered elements are rotatable about their axes from 0° to 180° and move across the housing from 0 mm up to a.

Rotation is provided by the fact that on one of the axes, for example, the top mounted actuator or electro-mechanical or manual, and an axis between cinematically connected with the possibility of simultaneous rotation. Also, simultaneous movement of louver elements provided across the body from 0 mm to a. Varying quantities and parameters [alpha] and can widely alter the dispersibility of product exiting the dust concentrator. To control the process of separation of dispersed particles in a stream of air louver elements are rotated about their axes 0° to 180°.

Inertial dust concentrator operates under discharge generated by an external fan. Dusty air enters the housing 1 on the hub inertial dust inlet pipe 2 and then passes to zone louver installation elements 5. Due to the vacuum created by an external fan, the air flow with small turns the dust particles passing through 180° louver elements and out through output pipe 3 of a dust concentrator. Large particles in the air stream by inertial forces trying to keep the trajectory of motion in a straight line in the output drop tube 4 and are removed from the hub.

Due to the fact that the louvre elements are able to rotate around their axes and to move horizontally, there are conditions for the control process of separating the dust-laden flow required particle size. You can also select such modes of his work in which, depending on the volume of pumping dusty air and dust can reduce the flow resistance of the device to a minimum.

This dust concentrator can be used in complex systems aerodynamic dedusting separation processes or fine powders, allowing the whole or improve dust collection efficiency of the separation.

3 Strike the particles of the shoulder blade

Upon impact, the spherical particles of the shoulder blade must change the center of mass of the particle velocity. Let the mass m_1 of a particle moving at a certain speed V, hits the vane mass m_2 at an angle [gamma] to the axis x. The surface of the blade body and consider it absolutely smooth.

The projections of the center of gravity of the speed of a falling body on the axis $\vec{n}$ and $\vec{\tau}$ (Fig. 2) are of the form:

$$V_n = V\cos\delta,\ V_\tau = V\sin\delta, \tag{1}$$

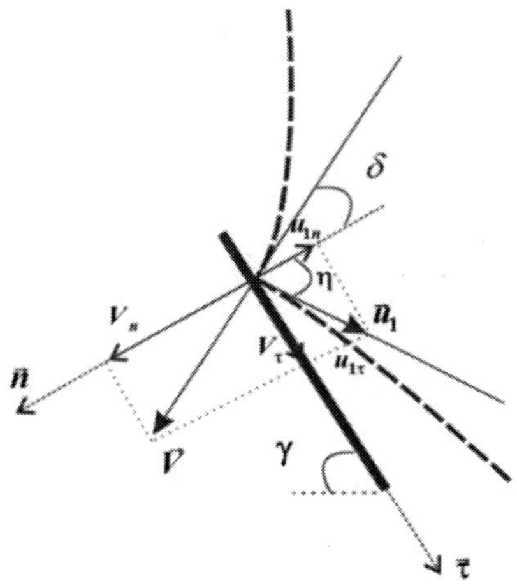

Fig. 2. Strike the particles of the shoulder blade to the hub.

We consider an oblique partially elastic collision. Consideration of the impact carried out in two stages [8]. The first phase takes place a deformation of the colliding bodies. At the end of this phase velocity of the particle's center of gravity is given by

$$u_n = \frac{m_1 V_n}{m_1 + m_2} = \frac{m_1/m_2}{m_1/m_2 + 1} V_n. \tag{2}$$

This rate is the rate of inelastic collision.

During the second stage a partial recovery of non-deformed state. At the beginning of the second phase particle has the same speed as when the first closure. Projection particle velocity of the center of gravity at the end of the second stage (after impact) equal to

$$u_{1\tau} = V_\tau = V \sin\delta, \tag{3}$$

$$u_{1n} = u_n + k(u_n - V_n) = \frac{m_1/m_2}{m_1/m_2 + 1} V\cos\delta + k\left(\frac{m_1/m_2}{m_1/m_2 + 1} V\cos\delta - V\cos\delta\right), \tag{4}$$

where – coefficient of restitution ($0 < k < 1$). After reduction of formula (4) can be written

$$u_{1n} = u_n + k(u_n - V_n) = \frac{m_1/m_2 - k}{m_1/m_2 + 1} V\cos\delta, \tag{5}$$

Module of the particle velocity in the center of gravity in view of the end of a pin (3) – (5)

$$u_1 = \sqrt{u_{1\tau}^2 + u_{1n}^2} = \sqrt{(V\sin\delta)^2 + \left(\frac{m_1/m_2 - k}{m_1/m_2 + 1} V\cos\delta\right)^2}. \tag{6}$$

Reflection of the particles will be at an angle η [eta], the tangent of which is equal to

$$tg\eta = \left|\frac{u_{1n}}{u_{1\tau}}\right| = \left|\frac{V\sin\delta}{\frac{m_1/m_2 - k}{m_1/m_2 + 1} V\cos\delta}\right| = \left|\frac{tg\delta}{\frac{m_1/m_2 - k}{m_1/m_2 + 1}}\right| = \left|tg\delta \cdot \frac{m_1/m_2 + 1}{m_1/m_2 - k}\right|. \tag{7}$$

Because the particles are considered mass not exceeding 1.3×10^{-5} kg, it $m_2 >> m_1$. Then we can assume $m_1/m_2 \approx 0$ and taking into account the conditions of the formula (5), (6) and (7) take the form

$$u_{1n} = -kV\cos\delta, \tag{8}$$

$$u_1 = \sqrt{V^2 \sin\delta^2 + k^2 V^2 \cos\delta^2}, \tag{9}$$

$$tg\eta = \left|\frac{tg\delta}{-k}\right|. \tag{10}$$

In the case of an absolutely elastic impact k = 1 and

$$u_{1n} = 2u_n - V_n = -V\cos\delta, \tag{11}$$

$$u_1 = V, \tag{12}$$

$$tg\eta = tg\delta. \tag{13}$$

These formulas make it possible to account for changes in the particle velocity vector in the reflection of the blades in the case of elastic or partially elastic impact.

4 The particle deposition efficiency

The acceleration of the vertical motion of the particles is determined by the resultant force of gravity, Archimedes and environmental resistance [9, 10]. In the case of gas Archimedes force can be neglected. Through elementary transformations [11], we obtain

$$C_X \operatorname{Re}^2 = \rho \frac{(\rho_p - \rho)d^3}{3\mu\beta \frac{S_p}{V_p}} g \tag{14}$$

How do we get the speed of the particles of hovering considering reflection particles from the blade (13)

$$v_B = \frac{(\rho_p - \rho)d^2}{18\mu\beta \frac{S_p}{V_p}\cos(\pi - 2\gamma)} g \tag{15}$$

Then calculate the cleaning parameters x1 and x2 (partial parameters of the distribution function F (x)) by the formulas [8]:

$$x_1 = \frac{H + h - L\frac{v_B}{v}}{\sqrt{2D_t - \frac{l}{v}}}, \quad x_2 = \frac{H - h - L\frac{v_B}{v}}{\sqrt{2D_t - \frac{l}{v}}}, \tag{16}$$

where h – distance from the camera of the ceiling; Dt – coefficient of turbulent particle diffusion.

Next on the values of x1 and x2, we find the probability integral Ф (x1) and Ф (x2), and calculate the ratio of the concentration of particles in the size calculated at the outlet section of the chamber to their concentration in the inlet section Ni

$$N_i = \Phi(x_1) + \Phi(x_2) - 1, \tag{17}$$

We average value of N in the cross section, calculating it as a simple average for the height of the section:

$$N_{av} = \sum_{i=1}^{k} \frac{N_i}{k}. \tag{18}$$

Next, we consider the average ratio of the partial deposition of the considered particle size:

$$\varepsilon_{n.av} = 100 - N_{av} \tag{19}$$

After this we find the full rate treatment as the sum of shares of the particles of the appropriate fractions to their fractional purification coefficients

$$\varepsilon = \sum \varepsilon_n \frac{N}{100} \Delta d. \tag{20}$$

To validate the above expressions, we have performed a calculation of the concentrator with the following parameters: the volume of air aspirated V in inertial dust hub, width of a of the working area chamber, concentration [beta] of dusty air and the angle [alpha] tilt blade inertial dust hub.

Fig. 3 shows the influence of the volume of air aspirated V dust collecting efficiency ε for different values of [beta] concentration of dust-laden air chamber working area of constant width of a = 0.36 m and a tilt angle blades [alpha] = 40 degree.

Fig. 3 shows that all the dependences are extreme character in the investigated range of varying factors.

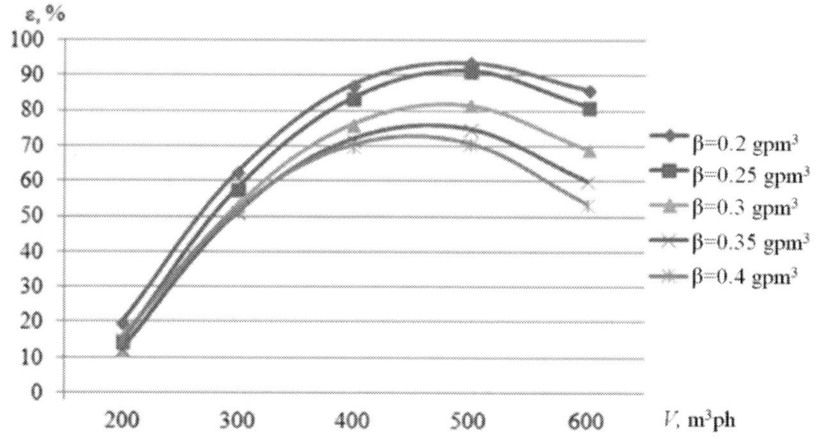

Fig. 3. Dependence of the efficiency of dust collection ε aspirated air volume V for various values of [beta] concentration of dust-laden air, a constant chamber width of the working area a = 0.36 m and a tilt angle of the blades [alpha] = 40 degree.

When analyzing the graphs shown in Fig. 3, it can be concluded that maximum dust separation efficiency is achieved when ε aspirated air volume V = 500 m^3ph, the concentration [beta] of dust-laden air [beta] = 0.2 gpm3 and is [epsilon] = 93.6 %.

5 Conclusions

The conducted analyzes allowed to determine conditions conducive to increasing the efficiency of dust collection is due to changes in the design parameters, and by changing the process parameters, and also showed the need to clarify the experimental fractional composition of the output product.

Improving the efficiency of dust collection inertial concentrator can be achieved both by varying the angle of inclination of the blades and due to increase to a certain value of air volume aspirated. Moreover, the increase in the volume aspirated air above a certain decline in the efficiency of dust collection entails.

As a result of experimental studies have provided updated according to the inertial dust concentrator, allowing to calculate the aerodynamic parameters of aspiration systems.

References

1. R.R. Sharapov, A.M Agarkov, World Applied Sciences Journal **24(10)**, 1399–1403 (2013)
2. N.S. Sevryugina, Construction and road machines **7**, 28-29 (2007)
3. A.M Agarkov, Proceedings of the International scientific and technical conference. Kazan State Architectural University. Kazan., 7–11 (2015)
4. A.M Agarkov, L.M. Isanov, Y.I. Yalamov, The era of science **4**, 89 (2015)
5. A.J. Waldberg. *Theoretical foundations of air protection from pollution by industrial aerosols* (MP "NIIOGAZ filter", Saint-Petersburg, 1993)
6. W.H. Duda. *Cement-data-book. Band 1 – Internationale Verfahrenstechniken der Zementindustrie, Wiesbaden* (Bauverlag, Berlin, 1985)
7. N.S. Sevryugina, V.M. Babin, Building and road machines **9**, 49-53 (2007)
8. A.G. Vetoshkin. *Processes and devices of dust cleaning. Tutorial* (Penz. state University Press, Penza, 2005)
9. M.I. Bach, G.Y. Dzhanelidze, A.S.Kelzon. *Theoretical Mechanics in examples and problems* (Science, Moscow, 1973)
10. E.V. Kharlamov, R.R. Sharapov, M.A. Stepanov, Mechanization of construction **6(840)**, 28-32 (2014)
11. E.V. Kharlamov, R.R. Sharapov, V.G. Shaptala V.G., V.V. Shaptala, Bulletin of BSTU named after V. G. Shukhov **4**, 91-95 (2013)
12. R.R. Sharapov, I.P. Boychuk, A.M Agarkov, V.S. Prokopenko, Bulletin of BSTU named after V. G. Shukhov **5**, 175-178 (2015)
13. R.R. Sharapov, V.S. Prokopenko, A.M Agarkov, Bulletin of BSTU named after V. G. Shukhov **2**, 84–89 (2015)
14. R. Schnatz, H.-G. Ellerbrock, S. Sprung, Zement-Kalk-Gips **48(2)**, 63-71 (1995)

Numerical simulation of the process of geothermal low-potential ground energy extraction in Perm region (Russia)

Andrey Ponomaryov[1,*] and *Aleksandr* Zakharov[1]

[1]Perm national research polytechnic university, Komsomolsky prospect 29a, 614990 Perm, Russia

Abstract. The aim of our research is to study the interaction of energy foundations with the ground mass and to develop methods for their construction on the example of the city of Perm. Field studies of ground were carried out in a specially chosen pilot site to determine temperature distribution in the ground mass, change of ground-water level and physical-mechanical and thermal-physical characteristics of the ground mass. The diagrams of depth temperature distribution in the ground and its seasonal variations were obtained on the results of monitoring, and also the average groundwater level. To carry out numerical simulation, software-complex "GeoStudio" was selected. Its basic differential equation is the fundamental heat conduction equation with an internal heat source. The purpose of the numerical simulation was quantitative evaluation of the thermal energy extracted from different energy foundations under soil conditions in the city of Perm. By results of the spent numerical experiments the equations of regress and nomographs dependences of size of received thermal energy on geometrical parameters of the projected power bases to hydro-geological and climatic conditions of the Perm region are constructed.

1 Introduction

One of the ways of increasing energy consumption efficiency when heating buildings is the use of renewable (alternative) energy sources. In developed countries ground thermal energy makes up a considerable proportion of energy used for heating.

Although studies of this problem have taken much time, technologies based on them are rather young. Nowadays, these technologies are widely used in many countries, such as Canada, Australia, the United States, and most European countries. Energy geothermal systems together with important environmental aspect have a great number of advantages [1-7]:

- They allow reduced energy consumption for heating buildings by 50-70 %.
- The use of foundations as ground heat exchangers necessary from the structural point of view becomes possible.
- They have fewer current costs in operation.

* Corresponding author: spstf@pstu.ru

Technologies that use geothermal energy have been used very rarely in Russia so far. Taking into account the advantages of the technology given and the state policy in the area of energy consumption we think that the problem of ground thermal energy investigation when laying foundations and building underground structures is pertinent.

2 Experimental investigations

The aim of our research is to study the interaction of energy foundations with the ground mass and to develop methods for their construction based on the example of the city of Perm.

Heating system using ground thermal energy consists of three main parts: the system of pipelines embedded in the ground mass or in contact with the ground (primary circuit); the system of pipelines intended for heating or conditioning (secondary circuit) and a heat pump combining these pipeline systems [5-8].

The primary circuit is used to generate ground thermal energy and is located in the body of energy foundations. Piles, foundation plates, "slurry walls", diaphragms, anchors, walls of the underground floors and other constructions being in contact with ground can be used as energy foundations.

The advantage of energy foundations is that these structures (piles, foundation plates, etc.) are required for the conditions of constructional safety (that is, to ensure bearing capacity and deformability). Correspondingly, there is no need in their additional construction. Therefore, they are double-purpose structures acting as load-bearing elements and ground heat exchangers.

The secondary circuit is a closed heating system in the walls and slabs of a building. The functions of the heat pump are to increase the temperature of the primary circuit heat carrier to the necessary one.

Ground is a multiphase system with a complex mechanism of heat transmission, which includes [1, 8, 9]: conductivity, convective transfer (convection), the processes of evaporation and condensation (latent heat transfer), heat radiation, ion exchange, freezing – thawing processes.

If the size of soil particles and pores is significantly small in relation to the total soil, the complex process of heat transfer in the ground can be reduced only to conductivity, which dominates in the case of energy foundations.

If there is an internal heat source (internal heat generation) in the concerned ground quantity, the basic equation of heat conductivity is as follows [1, 11-13]:

$$\frac{\partial t}{\partial \tau} = \alpha \nabla^2 t + \frac{q_\upsilon}{\rho c} \tag{1}$$

where, α is thermal diffusivity, $\nabla^2 t$ is Laplace operator, q_υ is power of internal heat sources, c is specific thermal capacity and ρ is density of solid medium.

Differential equations of heat conductivity show a character of the process and have many solutions. To obtain the solution of a specific task it is necessary to have initial and boundary conditions. Because of mathematical difficulties, the analytical solutions of these equations are possible only for simple cases. At present, a number of software packages that solve the problem of heat transfer in soils, including calculations of energy foundations have been developed.

Field studies of ground were carried out at a specially chosen pilot site (Site 1) to determine temperature distribution in the ground mass and change physical-mechanical and thermal-physical characteristics of the ground mass [6, 13, 14]. The monitoring system is installed at the Civil Engineering Faculty PNRPU in the Sverdlovsky district of Perm (left bank of the Perm city). Site 1 is characterized by dense urban development. Development age is more than 50 years. The distance from the observation well to the nearest building is about 3 m.

Geological structure of the site 1 was defined by the results of the research done. Geologically, the experimental platform is composed of Quaternary alluvial-dluvial clay soils, at the base with pebbles up to 60-70 % of the total thickness of 11.6m, overlapped by the thickness of filled-up ground of 6.0m thick. Bedrock is argillites, uncovered at a depth of 17.6m.

The following physical-mechanical and thermal-physical characteristics [15-18] of the experimental site ground were obtained on the results of laboratory work:

Table 1. Characteristics of the experimental site 1 ground.

№	Soil classification	Deep (m)	ρ (t/m3)	ω	e	c (kJ/kg°C)	λ (W/m°C)
1	Filled-up ground	0-6	1.91	0.25	0.73	1.27	1.33
2	Low plasticity loam	6-13	1.92	0.22	0.8	1.25	1.21
3	Gravel ground	13-17.6	1.69	0.007		0.85	0.41
4	Heavily weathered, cracked and waterlogged argillite	17.6	2.27	0.1	0.31	1.07	0.59

Monitoring of temperature fields has held for more than 5 years on the site 1. Summarized monitoring results (average monthly temperatures for the period from January 2009 to January 2016) are shown in Fig. 1.

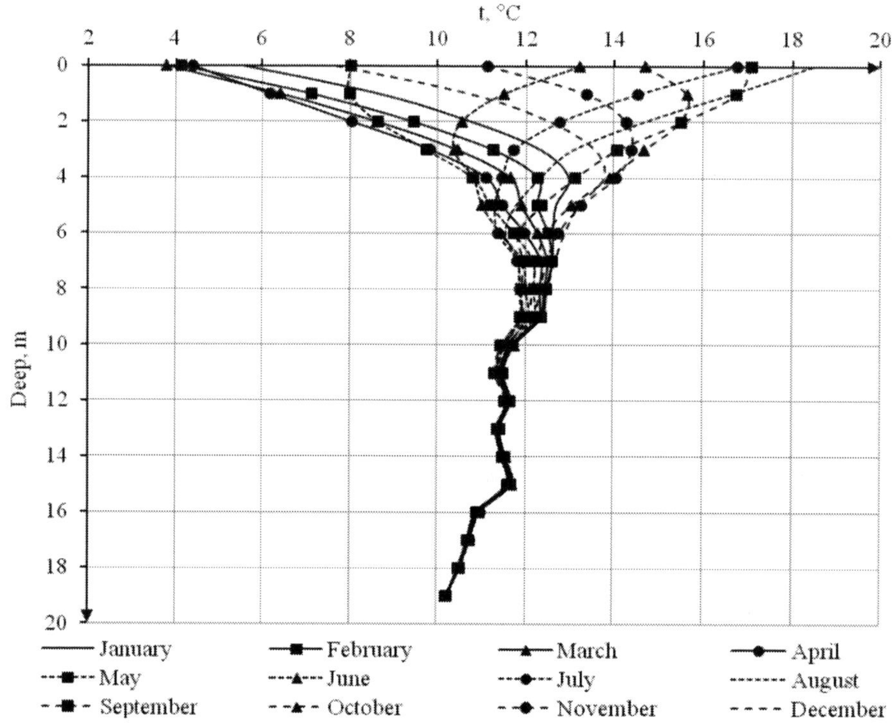

Fig. 1. Site 1. Diagram of depth distribution of temperature in the ground mass.

Under the diagram the temperature of the soil mass from a depth of 8-9 m is practically independent of the seasonal outdoor temperature fluctuations. The temperature is about 12°C, falling to 10°C to the depth of 19 m.

3 Numerical Simulation

To carry out numerical simulation, software-complex GeoStudio was selected. Its basic differential equation is the fundamental heat conduction equation with an internal heat source (1).

Test problems for three main types of underground structures being in contact with ground were preliminary solved:

- a pile with a diameter of 1.2m and 20m long;
- a 24m wide slab foundation, the depth of foundation is 20m;
- a slurry wall of 20m deep.

The temperature at each node in the initial period of time was taken as initial conditions. Boundary conditions were specified for the ground surface and for the lower boundary of the model. The boundary conditions of the lower boundary were taken as time-constant value of the heat flow density. The boundary conditions for the surface were set by applying climatic characteristics in the city of Perm in 2009 [19].

Time parameters of the simulation (number of annual cycles) were taken on the condition of setting a "new" temperature regime of the ground mass taking into account the thermal energy that was extracted.

The boundary conditions for the surface of the construction situated below the soil freezing level as a time-constant temperature +1°C were additionally set. Thus, a maximal extraction of thermal energy through the surface of ground-structure contact was simulated.

To determine the minimal time parameters for the numerical simulations of various types of foundations, calculations were carried out and values of the heat flow density through the contact surface of the ground with the foundations were obtained for several calendar years. Fig. 2 shows the values of the average heat flow density for the heating period.

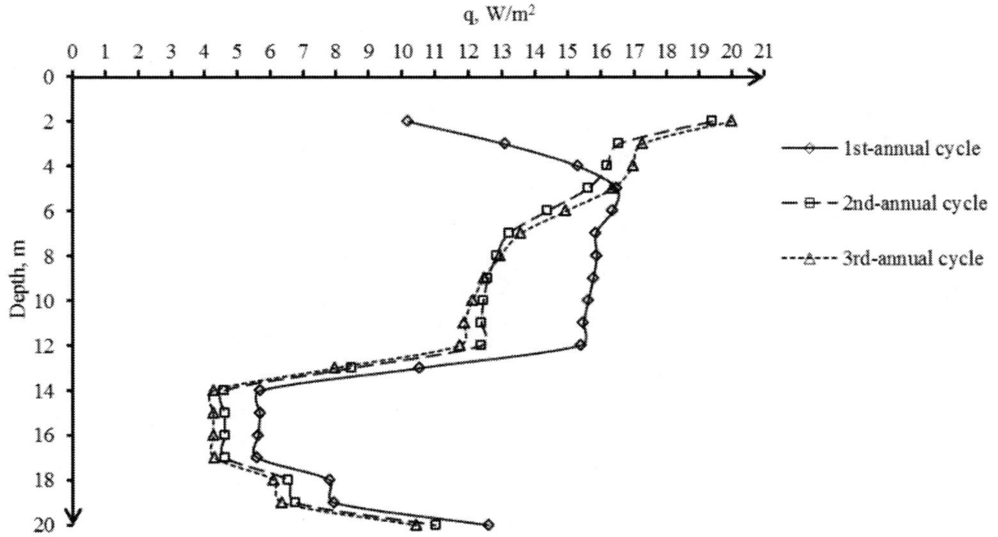

Fig. 2. Diagram of average heat flow density through the contact surface of the soil with a 20 m long pile during the heating period.

The minimal time parameters for the numerical simulation of different energy foundations maintenance were determined on the solution of the test problems. They are the following: 3 years for a single energy pile, 7 years for a slab foundation and 5 years for a slurry wall.

The purpose of the numerical simulation was quantitative evaluation of the thermal energy extracted from different energy foundations under soil conditions in the city of

Perm. Thereto, main types of ground bases typical for the city of Perm were ascertained and numerical experiments were carried out. Based on them regression equations were obtained.

According to the studies done [20], geological conditions in Perm can generally be reduced to two basic types specific to the left-bank (first type) and right-bank (second type) parts of the city. The first type is represented by low plasticity loam and gravel ground with sandy filling aggregate. The second type is medium sand, under which gravel ground with sandy filling aggregate lies. Low-compressible Upper Permian semi-rock occurs below gravel ground.

The experimental site 1 refers to the ground base of the first type. When carrying out the numerical experiment, dependences on various factors of average heat flow density through the contact surface of the deep building parts with the ground were determined.

As two basic types of ground conditions with concrete values of physical and thermal-physical characteristics of the ground were identified, their impact on energy foundations was taken into account through numerical calculations and obtaining the dependences for each type [7].

Therefore, geometrics parameters and underground structure depth were chosen as the main factors, namely:

- for a single pile: a pile radius (r) and pile tip depth (d);
- for a sunk slab foundation: foundation width (b) and foundation depth (d);
- for a slurry wall: foundation depth (d).

The following regression dependences based on numerical simulation results were obtained and nomograms were plotted. An example of a nomogram is shown in Fig. 3.

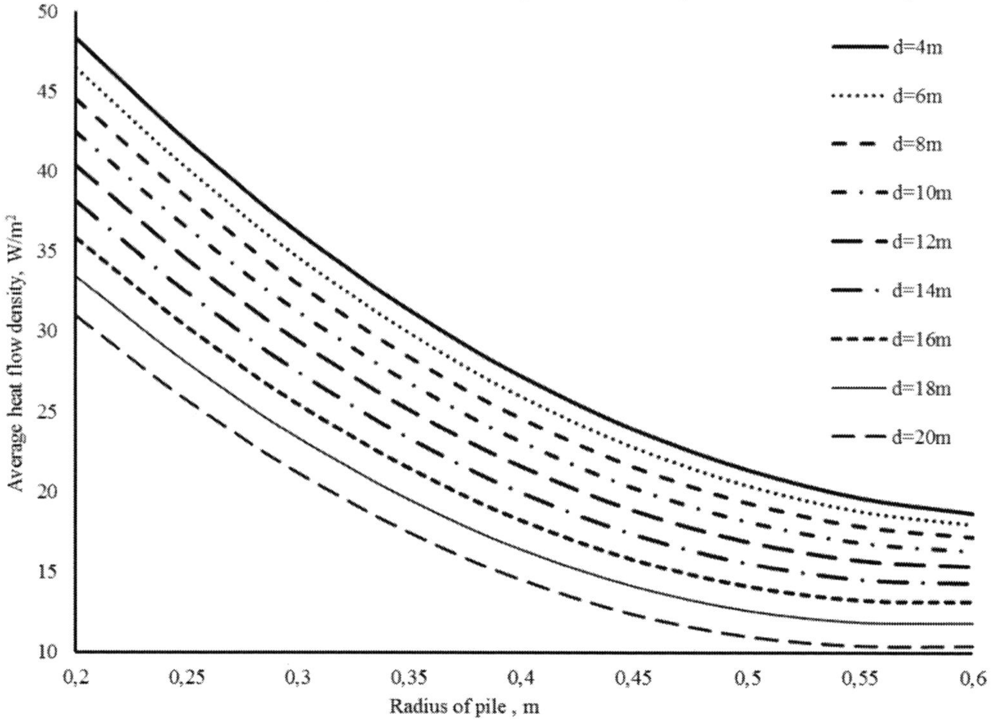

Fig. 3. Nomogram of dependence of the heat flow density ($\bar{q}$) through the pile-ground contact surface on the radius (r) and the pile point depth (d). Engineering-geological conditions of the first type.

On the bases of the analysis of regression equations derived and nomograms plotted we can draw the following conclusions:

1. Heat flow density ($\bar{q}$) through the contact surface of an energy foundation will depend on:

- for a single pile – on its radius (r) and foundation depth (d). In this case, the heat flow density decreases with the increase of the radius and the length of the pile;
- for a sunken slab foundation – on its width (b) and foundation depth (d). In this case, the heat flow density decreases with the increase of the width and the foundation depth;
- for a slurry wall – on the foundation depth (d). In this case, the heat flow density decreases with the increase of the foundation depth.

2. The amount of heat flow density is higher for the engineering-geological conditions of the second type than for those of the first type, namely ≈ 10 % higher for a single pile and ≈ 6 % higher both for a slab foundation and a slurry wall.

3 Monitoring of the temperature fields

For researching of temperature fields of ground bases for second basic type of geotechnical conditions specific to right-bank parts of the Perm city carried out monitoring on the site 2.

The monitoring system is installed on the territory of the university campus in the Leninsky district of Perm. Monitoring has been carried out since August 2015. Currently monitoring data were obtained till October 2016. Site 2 is located on the undeveloped area of the city. The distance to the nearest building is more than 30 m.

According to the results of archival research of surrounding area site 2 is comprised of quaternary alluvial sand and clay of the total thickness of 15 m, with 25% of gravel at the bottom. The bedrock was uncovered at a depth of 17.0 m.

The thickness of the observed soil mass on the site 2 is 37 m. Monitoring results (average monthly temperatures) are shown in Fig. 4.

According to the results of conducted studies the following conclusions were drawn:

1. The temperature of soil mass from a depth of 9 m is constant and equals to 6-7°C. The temperature fall has not been recorded at depths 7-37 m, in contrast to the site 1.
2. The temperature of the soil mass for the site 2 is lower than for the site 1 on average of 4-6°C. The difference of temperatures decreases with increasing depth.
3. The fixed difference of soil mass temperatures of the sites, are probably explained by the presence of additional heat sources at the site 1, in particular site 1 is characterized by dense urban development.
4. When designing power efficient foundations and underground structures the location of the object, the presence of surrounding buildings, engineering services should be taken into account.

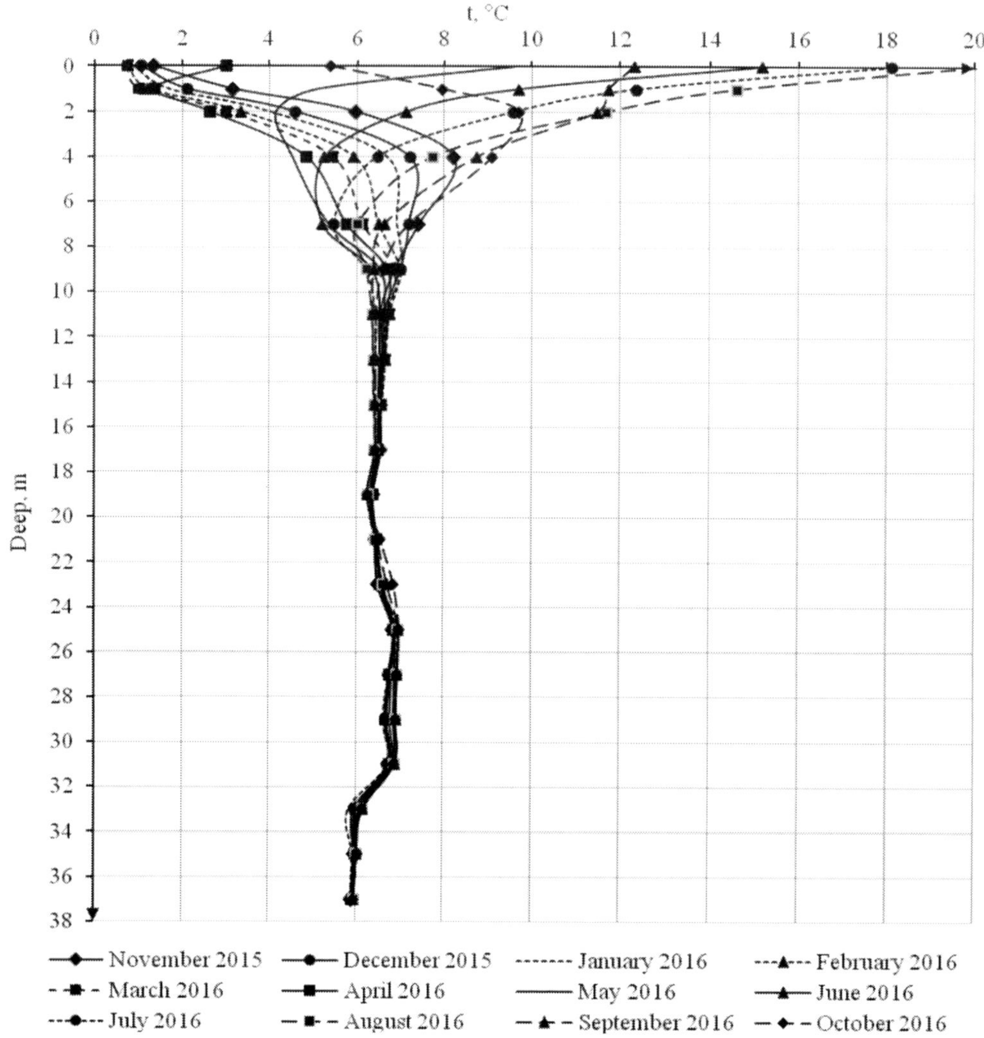

Fig. 4. Site 2. Diagram of average monthly temperatures of soil mass according the monitoring results from November 2015 to October 2016.

References

1. H. Brandl, Geotechnique **56,** (2006)
2. R. Katzenbach, D. Adam, A. Ennigkeit, T. Waberseck, *Innovationen bei der Nutzung geothermischer Energie durch erdberührte Bauwerke, wie z.B. Pfahlgründung mittels Energiepfählen. Geothermie-Symposium Bremerhaven Erdwärme – Energieträger der Zukunft`*(2002).
3. H. Brandl, D. Adam, R. Markiewicz, Acta Geotechnica Slovenica **3,** (2006)
4. H. Lam, H. Wong Proceedings World Geothermal Congress, Antalya, Turkey **1475**, 1-4 (2005)
5. A. Ponomaryov, A. Zakharov, Bulletin of Volgograd State University of Architecture and Civil Engineering. Series: Civil Engineering and Architecture **17(36)**, (2010)
6. A. Zakharov, Bulletin of civil engineers **2(23)**, (2010)
7. A. Zakharov, A. Ponomaryov, A Mashenko, *Energy constructions in underground building, training manual* (Perm National Research University, 2012)

8. V. Shapoval, B. Morklyanyk, *The temperature field in the soil base of heat pumps, monograph* (Porohy, 2011)
9. A. Nevzorov, *Foundations in soil seasonally, training manual* (ASV, 2000)
10. E. Ametistov, V. Grigorev, B. Yemtsov, *Heat and mass transfer. Heat engineering experiment* (Energoizdat, 1982)
11. V. Korolev, *Thermodynamics of soil, training manual* (MGU, 1997)
12. N. Tsytovich, *Mechanics of frozen ground, training manual* (High school, 1973)
13. M. Goldstein, *Questions geotechnics* (Dnepropetrovsk book publisher, 1959)
14. A. Ponomaryov, A. Zakharov, PNRPU Bulletin, Applied ecology. Urban development **4**, (2011)
15. O. Johansen, *Thermal Conductivity of Soils. Ph.D. thesis. Trondheim, Norway.* (CRREL Draft Translation 637, 1977)
16. B. Sanner, GHC Bulletin, (2001)
17. E. Sikora, J. Kossowski, Polish J. Soil Sci. **26(1)**, (1993)
18. D. de Vries, Physics of Plant Environment, North-Holland, Amsterdam (1963)
19. Information on http://thermo.karelia.ru.
20. S. Kaloshina, A. Ponomaryov, Proceed. of the International scientific and technical conference devoted to the 50th anniversary of Bashkir Scientific Research Institute for Construction, Ufa **2**, (2006)

Modified fine-grained concrete for facing and repair of the hydraulic structures

Yuri Bazhenov[1], *Boris* Bulgakov[1,*], and *Olga* Alexandrova[1]

[1] Moscow State University of Civil Engineering, Department Technology of Binders and Concretes, 129337, Russian Federation

Abstract. This article is devoted to improve the performance characteristics of plasticized fine-grained concrete for the hydraulic structures. The results of experimental studies on the selection of the optimal dosage of components. Justified the application of plasticized fine hydraulic concrete for facing and repair of the hydraulic structures. The result of the experimental study is the selection of the optimal composition of the fine hydraulic concrete with improved characteristics.

The Russian Federation has a large number of rivers with a total length of about 10 million km, many of which are navigable. A large part of the river hydraulic structures due to their intensive use is in poor condition. Repair of such structures are generally very time-consuming and expensive.

To solve the problem of increasing the effective performance of hydraulic fine-grained concretes need to optimize their composition and technology of preparation and use the organic and mineral modifying additives [1-5].

The aim of the work was to obtain the fine-grained concrete for facing and repair of river hydraulic structures which has the necessary strength, waterproofness and frost resistance due to seal its structure through the use of the organic and mineral additives in the form of superplasticizer, metakaolin and hydrophobizing silicone fluid.

Table 1. Physical and mechanical properties of fine-grained concrete plasticized "Khimkom F-1".

Composition and properties	Number of composition			
	I	II	III	IV
Cement, kg	550	550	550	550
Sand, kg	1500	1500	1500	1500
"Khimkom F-1", kg	11 (0.6%)	16 (0.9%)	22 (1.2%)	28 (1.5%)
Water, kg	180	156	150	144
Flow of fresh concrete (by the Flow table test), mm	160	115	150	159
Tensile strength at bending, MPa	5.13	7.5	8.8	8.3
Compression strength, MPa	26.4	46.2	55.4	53.2

* Corresponding author: fakultetst@mail.ru

Waterproofness: - concrete resistance penetration of air, s./cm3 - grade of concrete	-	-	355.0 W20	157.9 W20

Note: in parentheses indicate the concentration of "Khimkom F-1" in % by weight of cement in terms of dry matter.

To reduce the water demand of fine-grained sand-cement concrete mixture, while maintaining its desired mobility (the flow of fresh concrete by the Flow table test must be equal 150-160 mm) should be used superplasticizer [6-11]. As the plasticizer used superplasticizer "Khimkom F-1" as a 30% aqueous solution (table 1).

Superplasticizer "Khimkom F-1" represents sulfonated product of the aldol condensation of formaldehyde. Superplasticizer concentration ranged from 0.6 to 1.5% of the cement mass in terms of dry matter additive.

Selecting "Khimkom F-1" as a superplasticizer due to the fact that the use of this plasticizer allow to obtain from highly mobile concrete mix, characterized by high uniformity, necessary persistence of mobility and practically requires not vibration compaction, strong and durable fine-grained concrete with a high frost resistance and waterproofness.

As a result of experimentation, it was found that the maximum compressive strength of the fine-grained concrete (55.4 MPa) and its tensile strength at bending (8.8 MPa) at 28 days of normal hardening, as well as its high waterproofness (W20) obtained by using superplasticizer "Khimkom F-1" in an amount of 1.2% of the cement mass in terms of dry matter (table 1).

One of the main factors softening concrete is its increased porosity [6]. It arises from the concrete degradation during operation, and is expressed in loosening its structure, weakening the connection between the crystal growths in the cement stone and cement stone and the particle aggregates, resulting in reduced strength of the concrete, and also promotes filtration of water and aggressive fluids at concrete volume and, in addition, abrasive and frosty destruction of concrete. The solution to this problem may be a significant seal structure of concrete.

Table 2. Determining the optimal content of metakaolin in strength and waterproofness of plasticized fine-grained concrete.

Composition and properties	**Number of composition**			
	I (control)	**II**	**III**	**IV**
Cement, kg	550	550	550	550
Sand, kg	1500	1472	1445	1417.5
"Khimkom F-1", kg	22(1.2%)	22(1.2%)	22(1.2%)	22(1.2%)
Water, kg	150	150	150	150
Metakaolin, kg	-	27.5(5%)	55(10%)	82.5(15%)
Waterproofness: - concrete resistance penetration of air, s./cm3 - grade of concrete	355.0 W20	325.5 W20	318.4 W20	324.8 W20
Tensile strength at bending, MPa	8.8	8.6	9.0	9.8
Compressive strength, MPa	55.4	55.2	56.2	56.7

Note: in parentheses indicate the concentration of additives in % by weight of cement.

To reduce the consumption of cement and seal of the concrete structure as a finely dispersed mineral additives in concrete mixture was introduced metakaolin (table 2). Moreover, the amount should not exceed 15% by weight of cement, as it contains active silicon dioxide, and aluminum oxide in about equal proportions and therefore stronger than microsilica binds free calcium hydroxide, which reduces the alkalinity of the medium in the concrete and can cause corrosion of steel reinforcement [12]. Furthermore, the introduction of metakaolin as any other from fine fillers in more quantity increases water demand of the concrete mix due to its fineness and a large specific surface area values reaching 15 m^2/g and above [15].

As a result of experiments, it was found that when injected into the plasticized fine-grained concrete mixture of highly active metakaolin with a specific surface area value 15 m^2/g in the amount up to 15% by weight of cement while reducing the proportion of sand was a slight increase in strength of concrete after 28 days of normal hardening when tested tensile strength at bending from 8.8 to 9.8 MPa while a slight increase its compressive strength from 55.4 to 56.7 MPa. According to the experimental data obtained a waterproofness of the fine-grained concrete as a result of the metakaolin introduction changed little and remained at a level corresponding grade of concrete on waterproofness W20 (table 2).

In order to improve a waterproofness and a frost resistance of a hydraulic fine concrete in the concrete mix was introduced a hydrophobizing silicone fluid [9]: “SOFEKSIL-40” which is a 50% aqueous potassium methylsiliconate concentrate.

As a result, it has been experimentally shown that the introduction of plasticized fine-grained concrete mix “SOFEKSIL-40” in the amount of 0.2% of the cement mass effectively increases frost resistance and waterproofness of concrete. When using 0.2% “SOFEKSIL-40” actual concrete frost resistance increased from 492 to 653 test cycles, which corresponds to an increase in its mark on frost resistance from F400 to F600 (table 3).

Table 3. Determination of strength, frost resistance and waterproofness of plasticized fine-grained concrete hydrophobized silicone fluid.

Composition and properties	Number of composition		
	I (control)	II	III
Cement, kg	550	550	550
Sand, kg	1500	1417.5	1417.5
“Khimkom F-1”, kg	22(1.2%)	22(1.2%)	22(1.2%)
Water, kg	150	150	150
Metakaolin, kg	-	82.5(15%)	82.5(15%)
“SOFEKSIL–40”, kg	-	-	1.1(0.2%)
Compressive strength, MPa	55.4	56.7	57.5
Actual resistance to frost, (test cycles)	492	446	653
Grade of concrete on frost resistance	F400	F400	F600

Waterproofness: - concrete resistance penetration of air, s./cm3 - grade of concrete	355.0 W20	324.8 W20	338.5 W20

Note: in parentheses indicate the concentration of additives in % by weight of cement.

Since the modified fine-grained concrete is designed for facing and repair of the hydraulic structures, and it should be laid in thin layers with a thickness of 10 - 30 mm, it is necessary to evaluate the shrinkage deformation. To estimate the shrinkage deformation was used the following compositions:

•control composition - the plasticized coarse-grained concrete (class of compressive strength B 40);

•composition №1 - the plasticized fine-grained concrete (class of compressive strength B 40);

•composition №4 - the plasticized fine-grained concrete with the addition of highly active metakaolin and hydrophobizing silicone fluid "SOFEKSIL-40"(class of compressive strength B 50).

The test results are shown in table 4.

Table 4. The results of determinations of shrinkage deformation of concrete.

Harden ing term	**Control composition**		**Composition №1**		**Composition №4**	
	Absolute shrinkage , mm	**Relative shrinkage, %**	**Absolute shrinkage , mm**	**Relative shrinkage, %**	**Absolute shrinkage, mm**	**Relative shrinkage , %**
1	0.122	0.08	0.106	0.07	0.066	0.04
2	0.170	0.11	0.151	0.09	0.087	0.05
3	0.199	0.12	0.175	0.11	0.096	0.06
5	0.206	0.13	0.186	0.12	0.099	0.06
7	0.206	0.13	0.181	0.11	0.099	0.06
14	0.214	0.13	0.191	0.12	0.102	0.06
28	0.214	0.13	0.202	0.13	0.102	0.06

From the above test results shown that the samples of the developed fine-grained concrete are less prone to shrinkage than the control samples, what makes it possible to use the fine-grained concrete designed as a facing or repair material. Thus, the number and size of shrinkage cracks will be minimal.

Thus, in a result of conducted researches it was established that the introduction into the concrete mixture the organic and mineral additives in the form of superplasticizer "Khimkom F-1" (in the amount of 1.2% by weight of cement in terms of dry matter) with a strong plasticizing and water-reducing action, metakaolin (15% by weight of the cement) as a fine mineral sealing component and hydrophobizing silicone fluid "SOFEKSIL-40" (0.2% by weight of cement) allows to obtain a hydraulic fine-grained concrete having high strength (compressive strength is equal 57.5 MPa), waterproofness (grade of concrete on waterproofness is W20) and frost resistance (grade of concrete on frost resistance is F600). In this case, the optimum ratio of raw materials in 1 m^3 of the fine-grained concrete mixture is:

• cement - 550 kg;
• sand - 1417.5 kg;
• water - 150 kg;
• "Khimkom F-1" (in terms of dry matter) – 6.6 kg;
• metakaolin - 82.5 kg;
• "SOFEKSIL-40" – 1.1 kg.

References

1. S. V. Aleksashin, B. I. Bulgakov, *Collection of Research Papers of the Institute of Construction and Architecture* (KYuG Publ., Moscow, 2012)
2. S. V. Aleksashin, B. I. Bulgakov, Vestnik MGSU **8**, 97-103 (2013)
3. S. V. Aleksashin, B. I. Bulgakov, M. N. Popova, Izv. SFedU. Eng. Sci. **1**, 195-201 (2014)
4. E. S. Ryibintseva, G. A. Tkachenko, Cons. Mat. **7**, 18-20 (2009)
5. M. N. Moroz, V. I. Kalashnikov, V. A. Khudiakov, P. G. Vasil'ev, Cons. Mat. **8**, 55-57 (2009)
6. Yu. M. Bazhenov, D. K.-S. Bataev, S.-A. Yu. Murtazaev, *Energy and resource saving materials and technologies for repair and restoration of buildings and constructions* (Komteh-Print Publ., Moscow, 2006)
7. Yu. M. Bazhenov, N. P. Lukutsova, E. G. Matveeva, Vestnik MGSU **4(2)**, 415-418 (2010)
8. S. P. Shah, S. H. Ahmad, *High Performance Concrete: Properties and Applications* (McGraw-Hill, Inc., 1994)
9. V. R. Falikman, et all, *6th Int. Cong. "GLOBAL CONSTRUCTION", Advances in Admixture Technology* (Dundee, 2005)
10. N. P. Lukutsova, Cons. Mat. **9**, 101—104 (2010)
11. N. P. Lukutsova, A. A. Pyikin, O. A. Chudakova, Vestnik BGTU im. V. G. Shukhova **2**, 67-70 (2010)
12. V. S. Ramachandran, *Concrete Admixtures Handbook Book. Properties, Science and Technology* (Elsevier Science Publ., 1996)
13. RILEM commission 42-CEA, Mater. Struct. **14(6)**, 399-450 (1981)
14. S. A. A. M. Fennis, *Design of ecological concrete by particle packing optimization* (TU Delft, 2011)
15. V. G. Batrakov, *Modified Concretes. Theory and Practice* (Tehnoproekt Publ., Moscow, 1998)

MATEC Web of Conferences **86**, 03010 (2016) DOI: 10.1051/matecconf/20168603010
IPICSE-2016

Thermal stress state of rock massif with a spherical cavity taking into account inhomogeneity of the medium

Irina Frolova[1], *Vladimir* Andreev[1,*]

[1]Moccow State University of Civil Engineering, Yaroslavskoye sh., Moscow 129110, Russia

Abstract. The case of continuous one-dimensional heterogeneity of deformation characteristics of a material of the pedigree array with the spherical cavity, received by means of explosion, are continuous functions of one of co-ordinates – radius is considered. Solution of an initial set of equations concerning displacements is sought in the form of decomposition in Fourier series in Legendre polynomials. Using the method of separation of variables, we can reduce the problem to a system of two ordinary differential equations with variable coefficients, which can be solved numerically. Stress analysis was performed in the homogeneous and inhomogeneous array surrounding the spherical cavity. A comparative analysis of the results is carried out.

1 State of the problem

The paper deals with the problem of elasticity theory in spherical coordinates with the combined effect on the inhomogeneous array with a cavity asymmetrical load (soil resistance) and centrally symmetric temperature field.

Calculation scheme of the problem is shown in Fig. 1.

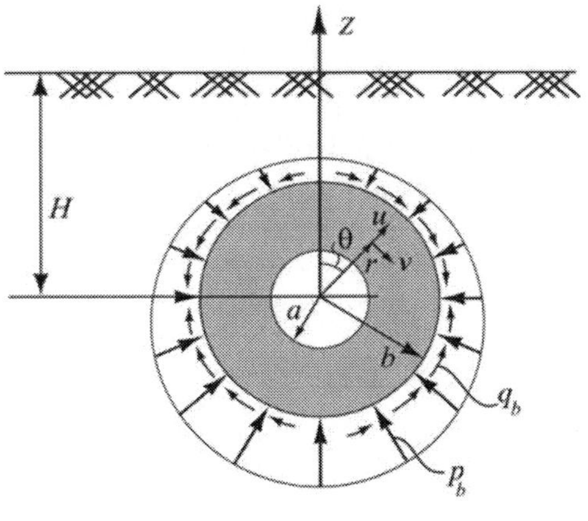

Fig. 1. Calculation scheme.

* Corresponding author: asv@mgsu.ru

Considered rock massif containing the spherical cavity obtained by explosion. When creating underground cavities by explosion releases large amount of heat in the cavity, thus there is a significant warming near a cavity array area with the subsequent distribution of heat in depths of the array and the cooling surface of the cavity [1]. This process is nonstationary, but excluding the initial period and given the relatively slow redistribution thermal field for a sufficiently long time later, the problem of determining the temperature stress can be regarded as quasistationary [2-5]. The calculation was made according the real working conditions of the array (the presence of ground resistance), and depending the mechanical characteristics of the soil on the temperature and on the fracturing of the array.

2 Inhomogeneity of ground

As is known [1, 6] using the method of forming explosive underground cavities, as well as drilling and blasting method for tunneling holes surrounding rock mass undergoing such changes (appearance of microcracks, compaction, sintering) that lead to mechanical inhomogeneity of material. At the same time there is local variation of the modulus of elasticity, which has a central symmetry. To describe the dependence E(r) can be used function obtained on the basis experimental results [7]:

$$E(r) = E^*\left[1+(k_1-1)\left(\frac{a}{r}\right)^m\right], \tag{1}$$

where k1 and m – empirical coefficients, r – the cavity radius, E* – elasticity modulus of undisturbed array.

In solving the problem of thermoelasticity for large temperature gradients should take into account the dependence E* (T), which can be taken in the form:

$$E^* = E_0 e^{-\delta T}$$

Here E0 - the value of the elastic modulus at T =0°C, δ - empirical parameter. Such whole dependence E(r) has the form:

$$E(r) = E_0 e^{-\delta T(r)}\left[1+(k_1-1)\left(\frac{a}{r}\right)^m\right]. \tag{2}$$

3 Solution of the problem

To assess the stress state of the array under the influence of temperature field and external loads, you can use the equilibrium equations in displacements [8]:

$$\begin{gathered}
\mu\nabla^2 u + 3(\lambda+\mu)\frac{\partial \varepsilon_{av}}{\partial r} - \frac{2\mu}{r^2}\left(u + \frac{\partial v}{\partial \theta} + v ctg\theta\right) + 3\varepsilon_{av}\frac{\partial \lambda}{\partial r} + 2\frac{\partial \mu}{\partial r}\frac{\partial u}{\partial r} + \\
+\frac{1}{r}\frac{\partial \mu}{\partial \theta}\left(\frac{1}{r}\frac{\partial u}{\partial \theta} + \frac{\partial v}{\partial r} - \frac{v}{r}\right) - 3\frac{\partial}{\partial r}\left(K\varepsilon_f\right) = 0; \\
\mu\nabla^2 v + 3\frac{(\lambda+\mu)}{r}\frac{\partial \varepsilon_{av}}{\partial \theta} + \frac{\mu}{r^2}\left(2\frac{\partial u}{\partial \theta} - \frac{1}{\sin^2\theta}v\right) + \frac{3}{r}\frac{\partial \lambda}{\partial \theta}\varepsilon_{av} + \frac{\partial \mu}{\partial r}\left(\frac{1}{r}\frac{\partial u}{\partial \theta} + \frac{\partial v}{\partial r} - \frac{v}{r}\right) + \\
+\frac{2}{r^2}\frac{\partial \mu}{\partial \theta}\left(u + \frac{\partial v}{\partial \theta}\right) - \frac{3}{r}\frac{\partial}{\partial \theta}\left(K\varepsilon_f\right) = 0.
\end{gathered} \tag{3}$$

Here λ(r), μ(r) – Lame constants, $K(r)$ – modulus of dilatation, $\varepsilon_f(r) = \alpha T(r)$ – forced temperature deformations; α = const – coefficient of linear thermal expansion; ν – Poisson's ratio; $\varepsilon_{av} = \frac{1}{3}\left(\frac{\partial u}{\partial r} + \frac{1}{r}\frac{\partial v}{\partial \theta} + \frac{2u}{r} + \frac{v}{r}ctg\theta\right)$ – average strain.

The boundary conditions are:

on the inner surface (r=a) $\sigma_r = 0,\ \tau_{r\theta} = 0;$

on the outer surface (r=b) $\sigma_r = -p_b(\theta),\ \tau_{r\theta} = q_b(\theta)$, (4)

where p_b and q_b – surface loads (Fig. 1):

$$p_b(\theta) = \gamma(H - b\cos\theta)\left(\frac{\nu}{1-\nu} + \frac{1-2\nu}{1-\nu}\cos^2\theta\right),\quad q_b(\theta) = \frac{1}{2}\gamma(H - b\cos\theta)\frac{1-2\nu}{1-\nu}\sin 2\theta \tag{5}$$

The solution of system (3) with the boundary conditions (3) is invited to seek in the form displacement of expansions in Fourier series:

$$u(r,\theta) = \sum_{n=0}^{\infty} u_n(r) P_n(\cos\theta),\quad v(r,\theta) = \sum_{n=0}^{\infty} v_n(r)\frac{dP_n(\cos\theta)}{d\theta}. \tag{6}$$

where $P_n(\cos\theta)$ – Legendre polynomial of n-th degree.

Substituting the expressions (5) to the first equation (3) lead to the equation for unknown u_0:

$$u_0''(\lambda+2\mu) + u_0'\left[\frac{2(\lambda+2\mu)}{r} + \lambda' + 2\mu'\right] - u_0\left[\frac{2(\lambda+2\mu)}{r^2} - \frac{2}{r}\lambda'\right] - 3(K\varepsilon_f)' = 0 \tag{7}$$

and for u_n and v_n ($n \geq 1$) to a system of two ordinary differential equations of second order with variable coefficients:

$$u_n''(\lambda+2\mu) + u_n'\left[\frac{2(\lambda+2\mu)}{r} + \lambda' + 2\mu'\right] - u_n\left[\frac{\mu}{r^2}n(n+1) + \frac{2(\lambda+2\mu)}{r^2} - \frac{2}{r}\lambda'\right] +$$
$$+\left[-v_n'\left(\frac{\lambda+\mu}{r}\right) + v_n\left(\frac{\lambda+3\mu}{r^2} - \frac{1}{r}\lambda'\right)\right]n(n+1) = 0; \tag{8}$$

$$v_n''\mu + v_n'\left(\frac{2\mu}{r} + \mu'\right) - v_n\left[n(n+1)\frac{\lambda+2\mu}{r} + \frac{1}{r}\mu'\right] + u_n'\frac{\lambda+\mu}{r} + u_n\left[\frac{2(\lambda+2\mu)}{r^2} + \frac{1}{r}\mu'\right] = 0.$$

Expanding the surface loads $p_в(\theta)$ and $q_в(\theta)$ (5) in series similar to (6), it is sufficient to restrict members to $n = 3$:

$$p_b(\theta) = \sum_{n=0}^{3} c_n P_n(\cos\theta),\quad q_b(\theta) = \sum_{n=1}^{3} d_n\frac{dP_n(\cos\theta)}{d\theta}.$$

The coefficients of these expansions c_n and d_n are determined by the known formulas [9]. Thus, systems (8) are solved for $n = 0 - 3$.

The solution of system (8) for arbitrary dependencies λ(r), μ(r) and K(r) should be carried out numerically. There are different ways of numerical realization of the considered equations. In this paper, method of the orthogonal matrix sweep is used [10].

After finding u_n and v_n, we can calculate stresses σ_r, σ_θ, σ_φ and $\tau_{r\theta}$ according to the formulas:

$$\sigma_r = \sum_{n=0}^{3}\left[\frac{2\lambda}{r}u_n + (\lambda+2\mu)u_n' - \frac{\lambda n(n+1)}{r}v_n - 3Kg_n\right]P_n(\cos\theta);$$

$$\sigma_\theta = \sum_{n=0}^{3}\left\{\left[\frac{2(\lambda+2\mu)}{r}u_n + \lambda u_n' - \frac{\lambda n(n+1)}{r}v_n - 3Kg_n\right]P_n(\cos\theta) + \frac{2\mu}{r}v_n\frac{d^2P_n(\cos\theta)}{d\theta^2}\right\};$$

$$\sigma_\varphi = \sum_{n=0}^{3}\left\{\left[\frac{2(\lambda+2\mu)}{r}u_n + \lambda u_n' - \frac{\lambda n(n+1)}{r}v_n - 3Kg_n\right]P_n(\cos\theta) + \frac{2\mu}{r}v_n ctg\theta\frac{dP_n(\cos\theta)}{d\theta}\right\}; \tag{9}$$

$$\tau_{r\theta} = \sum_{n=1}^{3}\mu\left[\frac{1}{r}u_n - \frac{1}{r}v_n + v_n'\right]\frac{dP_n(\cos\theta)}{d\theta};$$

where $g_0=\alpha T(r)$ at $n=0$, $g_n=0$ at $n\neq 0$.

Calculations were carried out under the following initial data: a=25 m; b=250 m; H=1200 m; γ=2.15 10^3 kg/m^3; E_0=2 10^4 MPa; ν=0.23.

Distribution of temperature in an isotropic medium under the influence of instantaneous heat sources distributed on the sphere of radius a described by the formula [11]:

$$T(r,t) = \frac{Q_0}{8\pi\sqrt{\pi c\gamma a r}\sqrt{a_1 t}}e^{-\frac{(r-a)^2}{4a_1 t}} + T_1 \tag{10}$$

Here T_1 – initial temperature of massif, T – temperature at a point spaced from the center of the cavity by a distance $r \geq a$ at a time t; Q_0 – amount of heat generated; c – specific heat capacity of the environment; γ – density; $a_1 =$ $\lambda/c\gamma$ – temperature conductivity; λ – thermal conductivity of the material; a – cavity radius.

4 Results

In the calculations used the averaged thermal characteristics of the soil: c = 0.96 kJ /(kg·deg.); a_1 = 1.67 10^{-6} m^2/s.

Fig. 2 shows the temperature curves for some moments in time t.

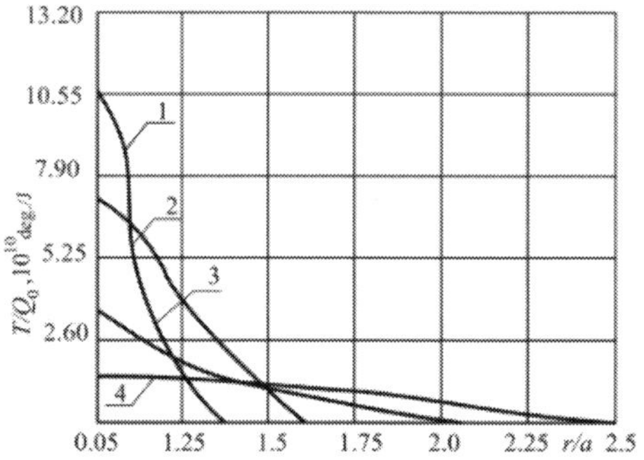

Fig.2. Temperature curves for different time.

1– 3.6·10^6sec.; 2 – 7.2·10^6sec.; 3 – 4.025·10^7sec.; 4 – 1.61·10^8sec.

Of considerable interest is the study of influence temperature inhomogeneity on the value of the components of the temperature stress state of the rock mass. Fig. 3 shows diagrams of thermal stresses σ_θ at time $t = 1.61\cdot10^8$ sec. The parameter characterizing the fractured rock is equal $k_1 = 1$. The graph shows a significant change in the stresses near the contour in an inhomogeneous material as compared with the homogeneous that is associated with a significant drop in the modulus of elasticity due to temperature inhomogeneity.

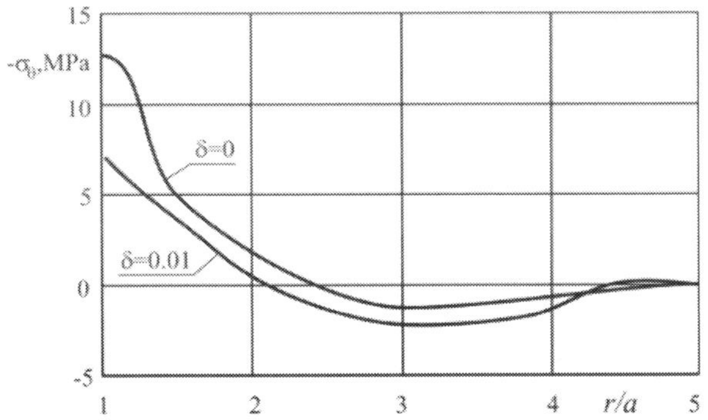

Fig. 3. Diagrams of thermal stresses. $\theta = 90°C$; t=1.61·10^{8}sec.; k_1=1.

Fig. 4 shows diagrams of the total stresses σ_θ (at $\theta = 90°C$), caused by the simultaneous action of temperature and power load on the rock mass taking into account the inhomogeneity both types of material.

Changes stresses σ_θ along the contour of the cavity shown in Fig. 5

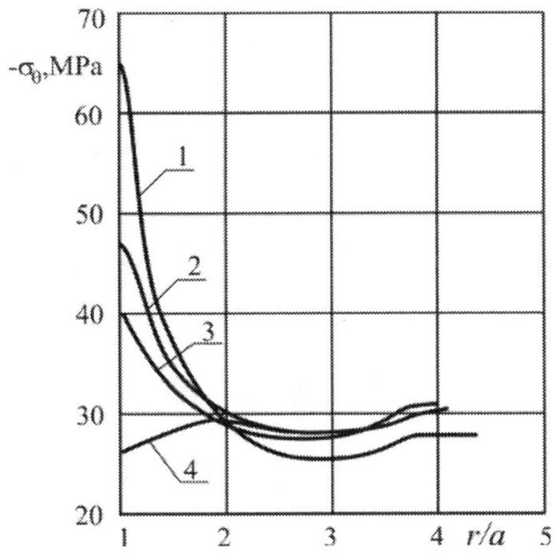

Fig. 4. Diagrams of the total stresses.
σ_θ ($\theta = 90°C$; t = 1.61·10^{8}sec.).

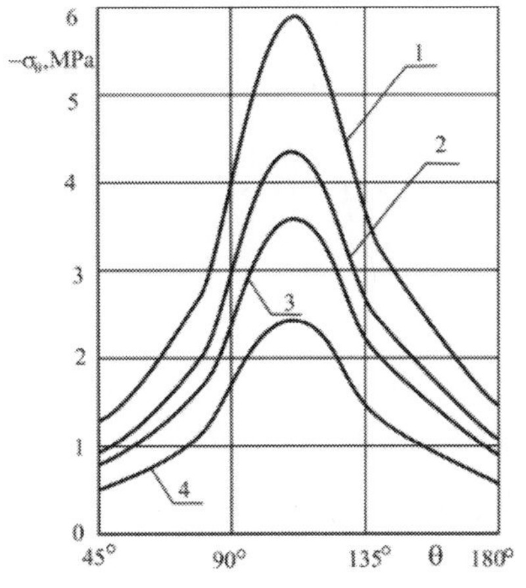

Fig. 5. Changes stresses σ_θ along the contour.
(t = 1.61·10^{8}sec.).

1 – k_1=1, δ=0; 2 – k_1=1, δ=0.01; 3 – k_1=0.5, δ=0; 4 – k_1=0.5, δ=0.01.

5 Conclusions

The Figures 3 and 4 show that allowance for two types of inhomogeneity leads to a more significant reduction the stresses, particularly in immediate vicinity of the cavity contour. Here, this stresses difference is about 50%. Further away from the cavity contour at a distance equal to 2a calculation results in an inhomogeneous and homogeneous cases virtually identical.

Fig. 5 shows that the stresses differ greatly across the range of values for homogeneous and inhomogeneous arrays. Exceptions are zones that are close to the poles of the cavity (and). The biggest difference in the stresses values observed at . But when approaching the poles difference is becoming less marked.

Analyzing the results of the calculations, we can conclude that allowance for the inhomogeneity of the material is necessary in the calculation rock masses as at the power loads as well at temperature influences.

Acknowledgment

This work was financially supported by the Ministry of Russian Education (state task #2014/14).

References

1. *A. A. Bakirov, Application of underground nuclear explosions in the oil industry*, (1981)
2. V. I. Andreev, A. S. Avershyev, Advanced Materials Research **838-841**, 254-258 (2014)
3. V. I. Andreev, A. S. Avershyev, Procedia Engineering **111**, 30-35 (2015)
4. V. I. Andreev, A. S. Avershyev, Structural mechanics of engineering structures and buildings **1**, 30-37 (2014)
5. I. I. Frolova, Scientific Review **12**, 94-99 (2015)
6. D. Rawson, P. Randolf, C. Boardman, V. Wheeler, J. of Geophysical Research **71 14**, 3415-3426 (1966)
7. *I. V. Baklashov, B. A. Kartozia, Rock* mechanics, (1975)
8. *A. I. Lurie, The theory of elasticity*, (1970)
9. *P. I. Romanovsky, Fourier series. Field theory*, (1973)
10. *A. A. Samarskiy, E. S. Nikolaev, Methods for solving grid equations*, (1978)
11. *A. V. Lykov, Heat conduction theory*, (1967)

Innovative geotechnologies is the key to geoenvironmental sustainability of urban areas by reducing the load and control techno-landscapes on the example of innovative water-coal technologies

Tatyana Zommer[1,*] and *Sergey* Chernyshov[1]
[1]Moscow State University of Civil Engineering, 129337, 26, Yaroslavskoye Shosse, Moscow, Russia

Abstract. The relevance of the article lies in the necessity of development of innovative geotechnologies in order to establish the geo-ecological stability of the territories while reducing the negative anthropogenic impact of economic activities on the environment. Therefore, at the present stage innovation policy needs to be fundamentally reoriented in the direction of coordinated development of new technologies aimed at restoring the supply of their biosphere. The authors come to the conclusion that modern innovative geotechnology can not only promote the improvement of ecological state of urban ecosystems, but also to facilitate automatic control techno landscapes on the example of innovative water-coal technology. In this case, technogenesis not the last, apocalyptic, evolutionary phase of human activity characterized by the disturbance of the biotic turnover and balance «biosphere – technosphere» in the direction of the technosphere. Reasonable parity of the biosphere and technosphere marks the transition of innovative thinking to an innovative geo-environmental, geoecosphere thinking, and of the technosphere – to geoecosphere.

1 Introduction

The role of technogenesis in the transformation of the planet as a result of human activities is linked to the development of mining industry, intensive agriculture, urban development, construction of power plants and other types of construction. The concept of technogenesis, in terms of science and technology is man-made changes to natural landscapes as a result of human industrial activity.

At a certain stage the results of the human activities become comparable to geological processes [1].

Technogenesis at comparable geological processes to human activities is becoming a source of anthropogenic hazards. The result of the irrational nature of contemporary technogenesis is characterized by the intensification of human impact on the geosphere of

*Corresponding author: ZommerTV@mgsu.ru

the Earth: the lithosphere, soil, hydrosphere, atmosphere, resulting in randomly created and modified techno-landscapes.

The level of technogenesis is due to the applied technologies and equipment. In our opinion, modern and innovative Geotechnology can and should contribute to the improvement of the ecological state of urban ecosystems. At the same time innovative technologies can be a key to automatic control when you create a techno-landscapes. Example of control action with the projected reduction of the negative anthropogenic impact on geosphere shells of the Earth is the massive introduction of innovative water-coal technologies. According to «Energy strategy of Russia for the period till 2030» the main part of electric and thermal energy to be generated from thermal power plants. At the same time coal mining remains associated with the impact of human activities on the Geosphere of the Earth, comparable to geological processes, and the daily burning of solid coal fuel in thermal power plants and boiler housing is detrimental to the environment and habitats of the ecosystems in urbanized areas.

In the decree of the RF Government also notes that the energy policy is focused on reducing internal gas consumption in the structure of fuel and energy resources. Coal should become the main energy source of thermal and electric power stations. Respectively should be paid great attention to the development of innovative technologies, including innovative coal-water fuel technology.

Innovative public policy is perceived by many only as a means of investment in high-tech manufacturing and knowledge-intensive innovative technologies, and as a consequence – in technogenesis as a product of technology.

In the paper [2] devoted to the study of technogenesis, it is noted that in a global context contemporary technogenesis acquires the features of an unmanaged process is not subordinate to the will and mind of people, is almost chaotic, with unlimited growth of consumption and exploitation of natural resources.

Therefore, at the present stage innovation policy needs to be fundamentally reoriented in the direction of coordinated development of new technologies of production and of the man himself, aimed at restoring the supply of the biosphere.

In the process of conventional coal technology, involving a mass withdrawal from the earth and burning non-renewable coal fuel, is comparable to geological processes of technogenic impact on the Geosphere shells of the Earth. In the structure of the crust changes occur. Coal piles, having the potency to spontaneous combustion, towering like an artificial mountain with a latent flowing volcanic processes. Artificially created voids become a cause of technogenic failures on the undermined territories. In the emptiness in mine workings, rushing underground and surface mine waters.

In the emptiness in mine workings, rushing underground and surface mine water, forming artificial lakes and rivers. Extensive production in open territories provoke dust storms. Emissions of pollutants from coal combustion at thermal power plants and municipal boilers cause acid rain. Emissions of pollutants from coal combustion at thermal power plants and municipal boilers cause acid rain leads to acidification of soils and oceans, promote the self-reproduction of the greenhouse effect and global warming.

In particular for these reasons, according to some pessimistic minded authors [2], technogenesis «is the last stage of the evolutionary human activity», characterized by the disturbance of biotic turnover and balance «biosphere – technosphere» with a clear superiority in the technosphere. In this predictive in its essence, the definition of the word «last» is addressed to humanity as a whole and makes you think about finding a reasonable compromise between the technosphere and biosphere that feeds her. In order to really stage technogenesis was not «the last stage of the evolutionary human activity», it is necessary to develop a new paradigm ecosphere compatible to geo-ecological sustainability of urban areas with decreasing impacts on the Geosphere of the Earth.

A theory, not backed up by concrete technology, may be ephemeral in practice. Therefore, the desired qualitative improvement of the geoecological condition of the urbanized territories as a whole must have an innovative technology with which this improvement can be achieved through scientific and technical progress, based on geosphere compatibility. The word technology we understand the hierarchical structure of some actions, ways and means, which is a production process with the aim of achieving a certain result in the form of products, goods or services that have consumer properties. Innovative technology is inherently carries the notion of critical technology, meaning technological breakthrough in any field of science and technology, requiring the amount of intellectual, material and financial resources.

The system of geo-ecological stability of the urbanized territories as a whole relies on the theory of: philosophical (the laws of development of nature and society) and economic (laws of development of relations of production, theory of automatic control). The purpose of the logistics system principles of geoecological sustainability of urbanized areas is the increase of efficiency of management of innovation technology.

The word innovation implies the introduction of something new. Innovation as a process means investing in the knowledge-based economy in order to change of generations of technologies and technological structures. As the object of the innovation is a new technology arising from scientific and technological achievements.

In scientific development displays a paradigm, or conceptual basis of biosphere compatibility of man and nature, the principles of which is linking «the withdrawal of resources from the natural environment – back to nature waste – the state of health of the population».

Thus, in our opinion, innovative geotechnology as a critical technology should be focused on innovative thinking to ensure optimal use of intellectual, material and financial resources and geoenvironmental built on the principles of biospheric compatibility of man and nature. And in this case, technogenesis becomes not «the last stage of the evolutionary human activity», characterizing the violation of the biotic turnover and balance «biosphere – technosphere» in the direction of the technosphere [2]. Stage a reasonable parity of the biosphere and technosphere marks the transition of innovative thinking to an innovative geoenvironmental, geoecosphere thinking, and the transition of the technosphere – to geoecosphere. Thus, the term geoekosphere we understand as the evolutionary state of the environment and man, which precedes the stage of the noosphere. Accordingly, the term geoekosphere thinking will mean an innovative geo-environmental thinking, which precedes the noospheric thinking.

An integrated approach to geoenvironmental, feasibility, geotechnical and civil engineering issues allows synthesis of exact science with fundamental philosophical and geoecological and ecological approach in studying the effects of the technosphere on the geosphere in connection with engineering structures and the applied technologies.

An integrated approach to geo-environmental, feasibility, geotechnical and civil engineering issues allows synthesis of exact science with fundamental philosophical and geoecological and ecological approach in the study of the impact of the technosphere on the Geosphere in connection with engineering structures and the applied technologies.

At the same time, technology, and engineering structures, perfectly inscribed in the natural environment by minimizing the environmental damage, must be viewed as integrated natural-technical systems in which natural «geo» and «bio» components combined into a single geotechnical and geoenvironmental complex. In this regard, philosophical problems of geoecology, engineering geology and construction as science and engineering can be formulated as the problem of increasing the level of self-organization of natural-technical systems with the steady merging of the natural and technical component in the construction and housing [3].

And in this case before us the example of biosphere compatible innovative coal-water fuel technology is directly associated with nanotechnologies, which are inherently technologies of control structure formation materials at molecular and atomic level.

Especially important is the question of the need for mass introduction of innovative coal-water fuel technology in response to the growing global environmental requirements of the state of the atmosphere, including the reduction of anthropogenic emissions that take into account the global organization.

The solution to the problem of reduction of impact on the Geosphere shells of the Earth in order geoenvironmental sustainability of urban areas through a gradual transition from coal technology to innovative water-coal technologies is important for developed countries.

For mass introduction of innovative coal-water fuel technology in order to ensure long-term energy and environmental modeling geo-ecological resilience in urbanized areas must be agreed upon and subordinate hierarchical geo-ecological and technological problems.

In the modern established practice of the construction of natural-technical systems, the relationship between nature and technology can often be regarded as antagonistic. In order to reconcile nature and technology in urban areas and developed innovative practices for the implementation of innovative technologies.

Conclusions

Therefore, according to the principles of biospheric compatibility [1], under the innovative geotechnologies, we mean only those innovations and technologies that increase rather than reduce the viable capacity of the Geosphere shells.

While in the innovative practice is also necessary to logically link the seizure of natural resources, the throw-in the nature of pollutants, the state of the environment.

The use of innovative coal-water fuel technology allows you to increase the profitability of fossil fuel (coal) by 30-40%. Almost one hundred percent combustion of coal (98 % and above) in the mass implementation of innovative coal-water fuel technology will contribute to a radical improvement in the environment of the urbanized territories.

A corresponding decrease in the number of mines will lead to the predicted reduction of the negative impact on the lithosphere, pedosphere, hydrosphere, including the reduction of disturbed areas, man-made voids, the coal heaps and inflow of mine water.

It is important that the mass introduction of innovative coal-water fuel technology will also contribute to decrease in quotations of anthropogenic carbon dioxide emissions, the elimination of the mechanism of auto-escalation of the greenhouse effect. Including the interruption of this mechanism is provided through the planning of forest plantations - the so-called «carbon trees», depositing carbon from the atmosphere into the pedosphere and lithosphere. That in principle allows us to consider coal as a renewable source of energy.

As a result, we were shown the principle possibility on the example of mass implementation geoekosphere compatible innovative technology of water coal fuel of the transition from the current stage of the uncontrolled phase of technogenesis in the direction of the geo-ecological oriented technogenesis, or geoekosphere of technogenesis.

References

1. V.I. Vernadskij, Noosfera i biosfera, (1989).
2. Lavrusevich A.A., Osnovnye cherty tehnogeneza, Vestnik MGSU, **4-2**, 175-181, (2010)
3. Chernyshev S.N., Jekologija kul'tury – chast' uchenija o noosfere, idejnoe osnovanie vossozdanija zdanij i sooruzhenij, Vestnik MGSU, **12**, 123-130, (2013)

Anti-erosion protection for recreational and sports facilities

Elena Shcherbina[1]*and *Marina* Afonina[1]

[1] Moscow State University of Civil Engineering, 12933, Yaroslavskoe shosse, 26 Moscow, Russia

Abstract. Development of natural lands for recreational purposes generates an increasing human impact. The monitoring results show development of slope erosion processes on the surface of slopes, which can generate gullies and landslides, impeding a sustainable land development and requiring additional operational costs at recreational and sports facilities. The article contains the basic results of a theoretical and experimental research and the monitoring data, which will allow us to justify the design of anti-erosion protection at recreational and sports facilities. The research has shown the efficiency of the use of geocomposite systems represented by a combination of structural elements and a biotic constituent. The efficiency of such approach is proven by the geotechnical monitoring data on specific facilities situated in Moscow Region.

1 Introduction

The growing rates of urbanization require a new dynamic development of recreational areas for satisfying the people's social need for recreation, sports, tourism and entertainment, and for ensuring a sustainable land development. Architecture and construction always respond quite quickly to the current social needs [1,2], which, for example, is shown, today, by the development of recreational areas of the Moscow agglomeration.

Winter sports, such as skiing and snowboarding are gaining popularity. Ski recreation is a combined type of recreation, which includes recreational, amateur-sports, therapeutic tourism and ecotourism. In Moscow region, there are large areas with all natural conditions for the development of recreational, for example, Dmitrovsky district. Rich forest vegetation combined with varied landscape and favorable natural and climatic conditions allow to create a high-performance resort and recreation system, including large-scale facilities of ski and other complexes, in Dmitrovsky district, Moscow Region. The Moscow agglomeration includes the total of more than 20 ski complexes.

Landscape, free space and access to transport [3,4] are very important for selecting a location for recreational and sports complexes. Taking into consideration the high urbanization of the Moscow agglomeration, the natural areas and areas affected by business activity are used for such facilities, for example, the Fox Lodge sports-shooting complex situated on the territory of a depleted pit. Each recreational and sports complex has a set of natural and man-made elements, earth structures, buildings, engineering and transport

* Corresponding author: lev.scherbina@yandex.ru

systems. We can speak of the level of popularity and consumer appeal a complex based on its development andvariety of the services rendered [5,6].

Additional soil embankments and excavations, as well as lake dams are created on the territory of ski complexes, allowing to do water sports andto ensure their use in summertime, for example, Sorochany ski complex. Thus, such recreational facilities essentially modify natural areas or areas affected by business activity, becoming a powerful human impact resulting in the generation of secondary natural landscapes where not only degradation, but destruction of the soil and vegetation cover is observed.On an artificial landscape surface, the process of natural regeneration of the vegetation cover is slow, which is determined by natural and climatic conditions.

The monitoring results show that the slopes surface is affected by the development of slope erosion processes, which may generate gullies and landslides (fig.1a,c; fig.2a); the extent of destruction is proportional to the extent of earth structures. Thus, protecting slopes from the development of slope erosion processes is a prerequisite for sustainable land development andan important goal of the land development aimed at the environmental safety [3,7]. Based on the on-site survey data, it is established that 58% of slopes are reinforced by grass sowing, 27% require soil replacement and grass sowing, 15 % require repair of reinforcement structures.

In order to ensure efficient design of anti-erosion protection,the following main functional tasks have been identified: stabilize the slope surface vegetable soil; create favorable conditions for the root system development; slow down the surface runoff. At present, different technical media are applied for preventing the development of erosion processes, but the experience shows that creating a stable sod layer and planting greenery on the surface, which reduces the flow rate [9,10,11,12,13], remain the most reliable and environmentally friendly one. The implementation of this solution is impeded by a short vegetation period and the lack of fertile layer on the slope surface. In many cases, the slope soil is completely infertile due to insufficient organic components in the composition of soil. In order to plant greenery on such slopes, not only applying a layer of fertile soil with grass seeds, but also stabilizing it is required.

Complex research carried out by MGSU [14, 15, 16] has established that the use of geocomposite systems is an efficient solution in fighting slope erosion processes.The geocomposite systems are represented by a combination of structural elements and abiotic constituent.The structural elements form a system matrix destined for stabilizing the slope surface soil, seeds and plants and creating favorable conditions for the biocommunity development. The matrix may be made of different construction materials, and the composition of the filler includes the following: soil, fertilizers, a variety of seeds, a binder, etc. The specific design solution, the work and stabilization technology depend on the slope steepness, soils, natural and climatic conditions, and are determined based on the data of the predictive calculations of erosive flow rates and system stability. A building structure combined with biological protection media will ensureepy homeostasis of the generated ecosystem throughout the service life of the facility, which is developed within the concept of phytodesign [6,17] consisting in the combination of live plants with artificial material; the vegetation becomes a constituent part of the structure.

Different versions of anti-erosion protection were approved for use at the Sorochany recreational and sports complex (RSC) founded in 2001, which is situated in the floodplain of the Yakhroma river, on Dmitrov highway, 40 km away from Moscow, and is a constituent part of the second-order agglomeration area formed along the Savelovo direction of the Moscow Railway, the Dmitrov highway and the Moscow Canal; its area is of about 324 ha.

The ski slopes are made on an artificial hill. Filled soil mainly composed of sand, less often, sabulous, heterogeneous with multiple inclusions of fragments of brick and concrete, gravel, construction waste, with loam interbeds, slightly wet, with different density and complexity. The total volume of filled soil is of 2 million m^3. The geological structure of the filled hill was determined by 5 engineering and geological sections coinciding with the areas of stabilization work for protecting the artificial filled slope from the development of erosion processes.

Table 1. Types of anti-erosion reinforcement applied at Sorochany RSC.

No.	Structure types	Yearofputting int service	Stabilization area, thousands of m^2
1	Wooden structures	2001 2002	18.8
2	Biogrids	2004	6.03
3	MacMat geogrids	2004	10.553
4	Gabion structures	2004	4.674
5	Geocells Prudon-494	2005	16.57

The region where Sorochany RSC is situated (Kurovo village, Dmitrovsky district) is characterized as typical for Moscow Region, with excessive soil moistening due to a significant amount of precipitation and little evaporability. During the period of construction and commencement of service, development of slope erosion processes was observed on the slopes, so continuous work on slopes stabilization with the application of different methods (table 1) and monitoring were performed during the period from 2001 until 2006. As the result of the regular work, erosion processes were stopped on all areas (fig.1 B,D), which allowed to ensure the stability of the support structures of lifts, stop the destruction of slopes, improve the aesthetics, ensure the stability of the secondary landscape.

Another facility subject to monitoring is Fox Lodge RSC. Today, it is the largest sports-shooting complex in Russia and Europe. The complex is situated in Gurban, a former sand-gravel soil extraction pit which was used during the period from 1982 until 2000; its original area was of 80 ha, the height of the walls was of up to 31 m, the angles of the walls were 28.80 to 32.5^0[9]. Its territory includes over 20 equipped sport shooting areas, Olympic stands and shooting galleries of 10. 25 and 50 meters. In the 300 meters shooting gallery, the highest-level competitions take place. The complex includes a hotel complex, superb sand beaches, a clean lake, boat and bicycle trips, fishing, gulf, badminton and tennis, restaurants. The sportsmen and the guests move within the territory of the complex by electric cars only. Even deers and roes can be seen here.

DOI: 10.1051/matecconf/20168603012

Fig. 1. Sorochany RSC monitoring data. A – one of the slopes before the work on anti-erosion protection (03.2006), B – the first spring after the slope stabilization (05.2007), C –an unexpected landslide (04.2008), D– slope condition in autumn (2011)

The complexity of this facility is due to the fact that it started to form on the territory of a depleted pit with the slopes affected by the development of slope erosion processes (fig. 2 A). According to the calculation data, the slopes were in a stable condition, and the landslides occurred as the result of a water erosion. The anti-erosion protection consisted in a complex of activities, including the following: creating a surface runoff with water discharge into the protected channels of gullies produced, stabilizing the slopes by sowing grass and bushes (fig.2 D). The hillside ditches and the watercourses were protected with the use of geocomposite systems. The system matrix was represented by geosynthetic material, a three-dimensional anti-erosion geogrid with a fiber thickness of 0.5 mm, a porosity of 0.95, a thickness of 10mm and 18mm and a tensile strength of 1.8 kN/m and 1.9 kN/m respectively [9]. The filler was represented by soil with grass seeds stabilized with the use of a bituminous binder in the places of watercourses. As the result, the development of erosion processes was stopped the next year after the work was realized (fig. 2B). The results obtained also allowed to justify experimentally the theoretical model allowing to forecast the erosive flow rates for slopes with unbound soils.

Fig.2 Fox Lodge sports-shooting complex. A – slopes at the commencement of the stabilization work, 2003, B –a fragment of a hillside ditch stabilized with the use of a geocomposite system (2005)C – planting bushes on slopes (2007), D – condition of slopes in spring of 2013 (photographs made by the authors).

Conclusions

The slopes at the sports complexes are stabilized by fulfilling geotechnical activities aimed at preventing the development of slope erosion processes, including the following: the engineering and geological survey; the calculation of the total slope stability; the determination of the critical erosive flow rates; provision of the surface runoff; development of the structure of the geocomposite system; selection of the vegetation type with due regard to the climatic conditions.

Geotechnical monitoring is an indispensable means for providing stable development of recreational and sports complexes; it may also include elements of environmental monitoring, which will allow to ensure environmental safety of the said territory.

References

1. V.P. Strauskas, *Town-planning organization areas and recreation centers*, (1977)
2. M.I. Afonina, M.M. Burova, Proceedings of the international scientific-practical conference *"Modern problems and their solutions in science, transport, manufacturing and education 2012"*. Issue 4 **50**, 18-23, (2012)
3. M.I. Afonina, Proceedings of the 3rd International Scientific and Practical Conference "Energointegratsiya -2013", (2013)
4. M.I. Afonina, Electronic scientific editions "Naukovedenie", **4 (13)**, (2012)
5. V.I. Telichenko, V.M. Rojtma, M.J. Slesarev, E.V. Shherbina, Fundamentals of complex safety of construction, (2011)
6. M.I. Afonina, Ecology of the urbanized territories, 81-86, (2008)
7. E.V. Shherbina, M.I. Afonina, M.V. Otchercov, V.N. Shatohin, Proceedings of the Jubilee Conference, devoted to the 50th anniversary RSSMGFE Moscow. (2007)
8. R. Sullivan, Roadside erosion causes and factors: Minnesota survey analysis. **948**, 47-54, (1983)
9. A.A. Alekseev, Geoecological effectiveness of geokompozitsionnyh screens at restoration of landscapes disturbed by mining activities, (2005).
10. E.V. Shherbina, *Geosynthetics in building,* (2004)
11. P.A. Slepnev, E.V. Shherbina, *The role of long-term strength geosynthetic material in the calculation and design of erosion protection of slopes,* (Moscow, Academy. Architecture and Construction, 2010)
12. P.A. Slepnev, Rationale for effective protection against the erosion of the elements of water recreation facilities in the city, (2009)
13. P.A. Slepnev, Hazard assessment of erosion and fluvial processes developing in the ponds and streams in urban areas, 235-237, (2006)
14. L.A. Volodina, P.A. Slepnev, Integration, Partnership and Innovation in Construction Science and Education, **1**, 397-399, (2011)
15. M.I. Afonina, Ecology of the urbanized territories, 81-86, (2008)
16. M.I. Afonina, Roads of Russia XXI century H4, (2012)
17. M.I. Afonina, V.V. Balaban, Ecology of the urbanized territories, 75-79, (2013)
18. M.I. Afonina, Research fitokonstruktivnyh modules for the rehabilitation of the urban environment and the development of the technology of their creation and implementation, (2000)
19. M.I. Afonina, A.M. Korobko A.M., Statistical analysis of the influence of climatic factors on the development of a sports and recreation complex "Sorochany", Bulletin MGSU, **5**, 24-29, (2011)
20. M.I.Afonina, M.M. Burova, Definition of "Factor attraction" RAC "Sorochany" method kvalimetricheskoj evaluation. *Scientific Readings dedicated to the 100th anniversary of Yu.B.Monfreda,* 293-299, (2013)

Choosing and estimating shear resistance parameters of gravel soils

Anatoly Mirnyy[1,*], *Valeriy* Merkin[1]

[1]Moscow State University of Civil Engineering, 129110, Yaroslavskoe shosse, 26, Moscow, Russia

Abstract. This issue is dealing with new yield criterion for gravel soils development. The most popular criteria for non-cohesive soils are analysed and the parameters they are based on. Proposed yield criterion is based on invariant parameters and concerns friction, cohesion and engagement between particles. It also takes into account second principal stress by using a non-fixed sliding plane. The parameters of this criteria are physically justified and can be determined by standard soil test.

1 Introduction

Under the soil strength its shear resistance is usually recognized. It is a fundamental soil parameter, and it's appropriate esteem is vital for rational design of numerous constructions, especially soil dams. Shear resistance of soil and its strength is determined by: a) inner friction forces, depending on normal stress; b) reversible cohesion of water-colloid nature, c) rigid structural cohesion [1,14-17]. Structural cohesion is normal for rock soils and usually is not considered in disperse soils. Thus, the most rational method for fine-grained soils shear resistance description is the use of two-parametrical yield criterion.

One of the most well-known yield criterion is a Mohr-Coulomb criterion [2]:

$$\sin\varphi = \frac{\sigma_1 - \sigma_3}{\sigma_1 + \sigma_3 + 2c \cdot ctg\varphi} \tag{1}$$

The Mohr-Coulomb theory considers the sliding plane that intersects the plane of the maximum and minimum principal stresses and is parallel to the axis of the intermediate principal stress. Many authors have noted that the Mohr-Coulomb theory is most suitable for solving problems of a flat stress-strain state, slopes stability problems mostly [3, 4].

A significant drawback of this platform is its non-physical hypotesis – intermediate principal stress takes an infinitely large value, its effect is not considered [5]. This leads to an underestimation of the strength, as this plane is the most dangerous, but is no likely to appear in complicated stress-strain state therefore its application to strength and stability calculations of the foundations of most buildings is not recommended.

Experimental studies of different soil in a wide range of stresses σ_3 (from 0 to 4,0 MPa), performed by various researcehrs [6-8] have revealed, that the intermediate principal

* Corresponding author: reg@oconnor.ru

stress has an influence on the shear resistance, hence the yield condition needs to account for this effect.

There was an interesting yield condition (2) performed by A. I. Botkin [5,9], who suggested that the Mohr-Coulomb condition is satisfied on the octahedral plane, equally tilted to the axes of principal stresses:

$$\tau_i^* = \sigma \cdot tg\varphi_{oct} + c_{oct}\,. \tag{2}$$

where $\tau_i^* = \frac{\sqrt{(\sigma_1-\sigma_3)^2+(\sigma_2-\sigma_3)^2+(\sigma_3-\sigma_1)^2}}{3}$ - shearing stress on the octahedral plane; $\sigma = \frac{\sigma_1+\sigma_2+\sigma_3}{3}$ - normal stress on the octahedral plane; φ_{oct} - inner friction angle for this plane; c_{oct}. – cohesion for this plane.

This theory takes into account the intermediate principal stress, however, also has some drawbacks. The magnitude of the normal stress on this plane is equal to the arithmetic mean of the three principal stresses. This decision leads to an overestimation of the normal stress. The shearing stress on this site has a small value, as calculated on the basis of deviations of the values of the principal stresses from the average, and obviously cannot be very large.

Malyshev M. V. [5] noted the "phenomenological" character of this theory, as plane selection is dictated by mathematical simplicity. An important element of this theory is the notion of the angle θ, defining the direction of shearing stress acting on the octahedral plane.

The results of the studies summarized by Rasskazov L. N. [10, 11] confirm a significant influence of intermediate principal stress σ_2 on the shearing resistance (up to 10°); the nonlinearity of the envelope of the Mohr circles in the majority of cases (reduction of up to 15°); the effect on the strength of the loading trajectory and Nadai-Lode parameter (change in the range of 2-4°). In addition, to date there is no possibility of accounting for engagement of particles in coarse-grained soils and the resulting increase in strength.

Especially important these phenomena are in the calculation of foundations and earth structures composed of coarse-grained soils with sand or clay filler. At the suggestion of P. I. Gordienko [12] to assess the strength of coarse-grained soil using not the φ and ***c*** parameters, but the shearing angle ψ, determined with ***c*** = 0. Each value of ψ has appropriate σ_1, σ_n and σ_3 values. Changing of ψ for gravel materials, depending on the stresses at the suggestion of P. I. Gordienko, can be approximately estimated by the formula (3).

$$\psi_\sigma = \psi_0 - 5\lg\frac{\sigma_n}{\sigma_0}\,. \tag{3}$$

Where ψ_0- shearing angle at $\sigma_3 \to 0$, $\sigma_n = \frac{2\sigma_1\sigma_3}{\sigma_1+\sigma_3}$ - normal stress on the sliding plane; $\sigma_0 = 0{,}2MPa$.

Unfortunately, these relationships are markedly empirical and must be adjusted for each site.

At present, a large number of yield and strength conditions where developed for non cohesive soils. Some of them, such as the Matsuoka-Nakai and Lade-Duncan criteria [13], take into account the magnitude of the intermediate principal stress and is applicable at different values of Nadai-Lode parameter, but are single-parameteric. Others, such as the Drucker-Prager criterion, use the same parameters as the Coulomb-Mohr, but have disadvantages, expressed in incorrect results for volume strain and dilatancy.

In this study the authors made an attempt to develop the yield criterion, taking into account all the described above regularities and phenomena.

The yield condition suitable for the description of shear resistance of coarse soils should meet the following requirements:

- it should consider the influence of intermediate principal stress and type of stress state. This is achieved by consideration of the sliding plane, which is not fixed in space of principal stresses and is determined by the current values of stresses;
- take into account the physical nature of the strength of fine-grained soils;
- parameters for this condition should be invariant, i.e., should not depend on the method of their estimation.

The authors deliberately omitted the question of the shear resistance dependence of soil density, since, in practice, coarse-grained soils are often used when the density is close to maximum, and the compaction during loading is negligible.

The task of developing the yield condition can be divided into three fundamental stages. The first step is to define the most dangerous plane, where all the stresses are considered. The second step is to choose invariant parameters, corresponding to the physical meaning of the described phenomena. Finally, in the third stage it is necessary to establish the mathematical relationship between these parameters and the components of stresses at the selected plane, allowing to describe the nature of the envelope of the Mohr circles observed in the experiment. This division greatly simplifies the solution to this problem, since the definition of the stress state and the identification of plane is a mathematical abstraction, but a yield condition, on the contrary, should have reference to the nature of the phenomenon in the medium.

2 Choosing a sliding plane

The account of all three principal stresses is possible when considering the plane, which is determined by the magnitude of the principal stresses at the ultimate state. It is obvious that such a plane will have a different angle relative to each of axes of principal stresses. General view of such a plane is shown in figure 1.

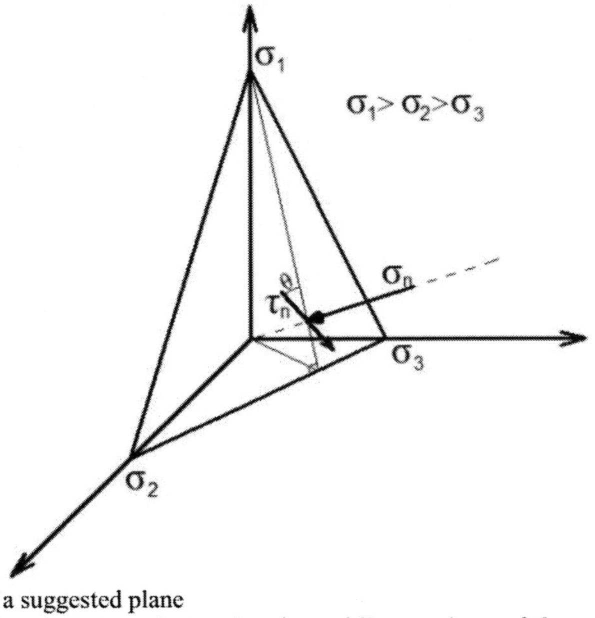

Fig. 1. General view of a suggested plane

The position of the plane may be set by the guiding cosines of the normal vector. The module of this vector is expressed as follows (4):

$$l = \frac{\sigma_1\sigma_2\sigma_3}{\sqrt{\sigma_1^2\sigma_2^2 + \sigma_2^2\sigma_3^2 + \sigma_3^2\sigma_1^2}} \tag{4}$$

From the consideration of triangles values of the guiding cosines can be obtained:

$$\cos\alpha = \frac{l}{\sigma_1} = \frac{\sigma_2\sigma_3}{\sqrt{\sigma_1^2\sigma_2^2 + \sigma_2^2\sigma_3^2 + \sigma_3^2\sigma_1^2}} \tag{5}$$

$$\cos\alpha = \frac{l}{\sigma_2} = \frac{\sigma_1\sigma_3}{\sqrt{\sigma_1^2\sigma_2^2 + \sigma_2^2\sigma_3^2 + \sigma_3^2\sigma_1^2}} \tag{6}$$

$$\cos\alpha = \frac{l}{\sigma_3} = \frac{\sigma_1\sigma_2}{\sqrt{\sigma_1^2\sigma_2^2 + \sigma_2^2\sigma_3^2 + \sigma_3^2\sigma_1^2}} \tag{7}$$

Using the known formulas of solid mechanics to determine the normal and shearing stresses on any plane we can get the expressions for these stresses:

$$\sigma_n = \sigma_1\cos^2\alpha + \sigma_2\cos^2\beta + \sigma_3\cos^2\gamma \tag{8}$$

$$\tau_n = \sqrt{\sigma_1^2\cos^2\alpha + \sigma_2^2\cos^2\beta + \sigma_2^3\cos^2\gamma - \sigma_n^2} \tag{9}$$

After substituting the previously obtained values of the guiding cosines (5-7) these expressions take the form:

$$\sigma_n = \frac{(\sigma_1\sigma_2\sigma_3)(\sigma_1\sigma_2 + \sigma_2\sigma_3 + \sigma_3\sigma_1)}{\sigma_1^2\sigma_2^2 + \sigma_2^2\sigma_3^2 + \sigma_3^2\sigma_1^2} \tag{10}$$

$$\tau_n = \frac{\sigma_1\sigma_2\sigma_3}{\sigma_1^2\sigma_2^2 + \sigma_2^2\sigma_3^2 + \sigma_3^2\sigma_1^2}\sqrt{3(\sigma_1^2\sigma_2^2 + \sigma_2^2\sigma_3^2 + \sigma_3^2\sigma_1^2) - (\sigma_1\sigma_2 + \sigma_2\sigma_3 + \sigma_3\sigma_1)^2} \tag{11}$$

These values of normal and shear stresses are proposed to use in determining invariant sets of the strength parameters.

3 Choosing the invariant strength parameters

Many years of experience in the application of the Mohr-Coulomb criterion in soil mechanics has shown that the two-parameter model is a reasonable representation of the shear strength of fine-grained soils. However, studies on the shear resistance of coarse-grained soils in a wide range of stresses show poor convergence with experimental observations. In the works of many researchers, in particular, Boldyrev, G. G., the discrepancy in the values of the strength parameters obtained by various methods is shown. This fact leads to the conclusion about insufficiency of our ideas about the nature of strength of fine-grained soils.

First and foremost this statement refers to evaluation of internal friction through internal friction angle. Internal friction in dispersed soils is caused not only by sliding friction the Amontons-Coulomb, but the rolling friction when turning the particles, the roughness and roundness of the particles and, consequently, their engagement. Obviously, part of these phenomena does not depend on the normal stress applied and can be related to the cohesion forces.

At the same time, the strength parameters are a function of the loading rate. For example, during very slow loading, internal links partially recover and sliding friction is not transformed into rolling friction.

Thus, as the invariant of the strength parameters of the soil the following should be considered:

- inner friction parameter φ_i, for sliding friction determined by soil mineralogy. If $\varphi_i = f(\dot{\gamma})$ is considered peak strength could be accounted for. As a residual value of φ_i the natural slope angle should be taken;

- cohesion parameter *c*, reflecting the influence of internal relations, regardless of their reversibility and applied stresses. It is constant, due to physical and chemical parameters of the soil;

- engagement parameter, reflectiong particle roughness and its influence on shearing resistance.

The most difficult task is to identify the role of engagement in the shear strength of fine-grained soils. From physical sense of engagement, it follows that its magnitude does not depend on the compression, as even in the absence of compression with only shear forces applied there is a resistance due to engagement. The magnitude of this resistance will be constant for an infinitely strong material, as the destruction of the engagement irregularities will not occur. With decreasing strength of the material the irregularities will begin to split. In the course of the shearing theret will be gradual bridging of the bumps. Finally in the case of extremely fragile material, engagement will not be observed, since the shear plane will pass through the material of the particles. This mechanism is presented in figure 2.

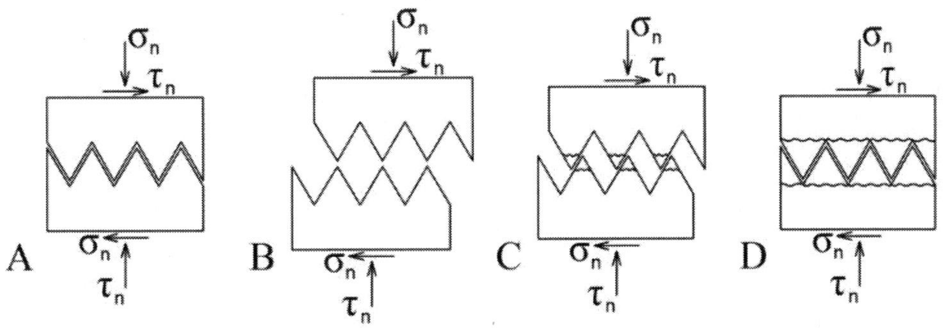

Fig. 2. A) – starting position; B) – infinite strength; C) – finite strength; D) – no shearing resistance.

At the same time, the magnitude of compression will change the amount of friction, the magnitude of the normal force to the surface of the prong is greater than the magnitude of the force normal to the cut surface due to the tilt of prongs. Chipping of prongs will occur when the shear stress exceeds the shear strength of the particles. Thus, the contribution of the engagement in shear resistance is easier to evaluate in the form of increased friction. Engagement is implemented before the start of the shearing prongs, and then begins to decline to zero with smoothing irregularities.

4 The invariant yield criterion development

Yield criterion is taking the form:

$$\tau^* = \sigma \cdot tg\varphi \cdot \left(1 - (1+n)^{\sigma - R}\right) + c, \qquad (12)$$

where R – axial compression resistance of particles material,

n – parameter, depending on shear resistance of particles material.

This yield criterion takes into account the shear resistance due to the engagement and the gradual decrease due to the prongs shearing.

The parameter n can be expressed, provided that we know the shear resistance of the particles, however, it is more convenient to define it on the basis of experimental data. It should be guided by the following values:

When $n = 0$ – there is no friction in shear resistance;

When $n = 0{,}2$ the prongs are nearly indestructable.

On the figure 3 there is a graph of a limit line function $\tau^* = f(\sigma)$ with different n values.

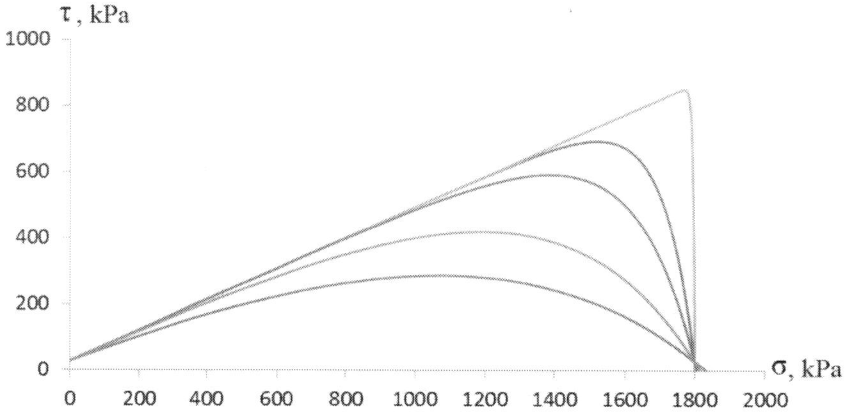

Fig. 3. Limit lines with different values of n.

It is obvious that such a view explains the definition of engagement by many authors as an additional constant component close to cohesion. This is due to the application of the linear form of the limit line. Holding such line through any three points of the curve shows initial shear strength, different from zero, which is interpreted as "engagement".

In its final form, the proposed yield criterion can be formulated in the form of three equations, allowing to determine the stress components for the sliding plane and the interaction between them:

$$\begin{cases} \sigma_n = \dfrac{(\sigma_1\sigma_2\sigma_3)(\sigma_1\sigma_2 + \sigma_2\sigma_3 + \sigma_3\sigma_1)}{\sigma_1^2\sigma_2^2 + \sigma_2^2\sigma_3^2 + \sigma_3^2\sigma_1^2} \\ \tau_n = \dfrac{\sigma_1\sigma_2\sigma_3}{\sigma_1^2\sigma_2^2 + \sigma_2^2\sigma_3^2 + \sigma_3^2\sigma_1^2}\sqrt{3(\sigma_1^2\sigma_2^2 + \sigma_2^2\sigma_3^2 + \sigma_3^2\sigma_1^2) - (\sigma_1\sigma_2 + \sigma_2\sigma_3 + \sigma_3\sigma_1)^2} \\ \tau^* = \sigma \cdot tg\varphi \cdot \left(1 - (1+n)^{\sigma - R}\right) + c \end{cases} \quad (13)$$

It should be noted that the friction angle in this criterion is not numerically equal to the friction angle for other yield conditions, in particular the Mohr-Coulomb and von Mises-Schleicher-Botkin.

The technique of definition of parameters in general is similar to conventional test methods, however, when interpreting data it is necessary to consider the range of stresses in the test.

The angle of internal friction and cohesion should be determined according to the initial part of a limit line, where even in the area of stress concentration the strength of the particles is not exceeded.

5 Conclusions

1. The most common yield criteria have limited application for coarse-grained soils, as engagement and the decrease in shear resistance with increasing normal stress is not fully taken into account.

2. A yield condition suitable for coarse-grained soils must consider the engagement of the particles and enable the strength of the particles as a parameter

3. The proposed yield condition can be used to describe the limit line, taking into account its nonlinearity.
4. The parameters of is yield criterion can be determined on the basis of experimental data.

Acknowledgements

This work was financially supported by the Ministry of Russian Education (state task #7.1576.2014/K)

References

1. N.N. Maslov, *Soil mechanics and engineering geology basics* (Vysshaya Shkola, Moscow, 1968)
2. Z.G. Ter-Martirosyan, *Soil mechanics* (ASV, Moscow, 2009)
3. J. M. Duncan, S.G. Wright, *Soil Strength and Slope Stability* (John Wiley & Sons, Inc., USA, 2005)
4. G. K. Clein, *Structural mechanics of granular medium* (Stroizdat, Moscow, 1977)
5. M.V. Malyshev, *Soil strength and stability of foundations of structures* (Stroizdat, Moscow, 1994)
6. S.M. Ataya, *Gravel soils shear resistance studies in triaxial machine* (VODGEO, Moscow, 1965)
7. T. Mendoza, *The strength of coarse-grained soils with sand filler in terms of spatial stress state* (MISI, 1984)
8. A.L. Kryzhanovskiy, T. Mendoza, E. Ukibaev, Eng. Geol. **2**, (1985)
9. G.G. Boldyrev, *Soils mechanical properties obtaining methods* (PGUAS, Penza, 2008)
10. A.L. Goldin, L.N. Rasskazov, *Soil dams design* (ASV, Moscow, 2001)
11. L.N. Rasskazov, *Gravel soils shear resistance* (VODGEO, Moscow, 1968)
12. P.I. Gordienko, *Some design issues of high stone and soil dams* (MISI, 1960)
13. L. Mengcheng, C. Yufeng, L. Hanlong, Int. Jour. of Rock Mech. and Mining Sc. **50** (2012)
14. A.Z. Ter-Martirosyan, Z.G. Ter-Martirosyan, V.V. Sidorov, Soil Mechanics and Foundation Engineering **1**, 1-8 (2016)
15. Z.G. Ter-Martirosyan, E.S. Sobolev, A.Z. Ter-Martirosyan, International Journal of Applied Engineering Research **11(3)**, 1719-1723 (2016)
16. A.Z. Ter-Martirosyan, Z.G. Ter-Martirosyan, Procedia Engineering **111**, 756-762 (2015)
17. Z.G. Ter-Martirosyan, A.Z. Ter-Martirosyan, Soil Mechanics and Foundation Engineering **51 (6)**, 273-281 (2015)

MATEC Web of Conferences **86**, 03014 (2016) DOI: 10.1051/matecconf/20168603014
IPICSE-2016

The definition of necessary axial force for extension of initial borehole for soft soil compaction process design

Zaven Ter-Martirosyan[1], *Armen* Ter-Martirosyan[1], *Vitalii* Sidorov[1*], and *Nikolay* Aniskin[1]

[1]Moscow State University of Civil Engineering, Yaroslavskoe shosse, 26, Moscow, 129337, Russia

Abstract. The article provides an analytical solution of the soil pile and surrounding soil cylinder interaction problem, with the possibility of extension of the pile shaft in its construction. Presents a closed solution for determination of radial and tangential stresses in the process of expansion of the pile shaft, as well as the minimum vertical force sufficient for the crushing of the pile material and move it in radial direction to the specified value. The problem is most actual for compacted soil bases with use of piles-drains of sand and sand-gravel mixture.

1 Introduction

During construction on weak water-saturated clay soils, as a rule, use bored piles of finite stiffness, as well as jet-grouting piles or made of soil (sand, gravel) [1]. The last is used also as a drain on the preliminary stage of weak water-saturated soils compaction, and as a bearing element for subsequent loading of a compacted soil composed of the foundation [2]. Thus, soil pile – drain plays role of a filter and bearing element in the composition of the compacted weak soil layer. Major supporting part consisting of compacted weak layer is a clay cylinder, containing the pile – drain. Such a "cell" or "system" has the same properties as the surrounding cells of the converted layer. The dimensions of such cells are determined by the distance between their centers in a staggered arrangement of piles – drains. The diameter of piles – drains depends on the technology of their production during deep compaction, i.e. on the expansion of the initial borehole and fill it with sand – gravel mixture.

Thus, as the estimated model is considered a bearing massif consisting of a soil thick-walled cylinder enclosing a pile – drain and serves as the drain and as a supporting element in this column (fig. 1). Under the pressure of the foundation plate in this layer is formed a complex heterogeneous stress-strain state, which is in the process of loading can be transformed as a result of redistribution of load between the pile – drains and the surrounding soil up to the buckling pile – drains. It depends on numerous factors, including:

• diameter and length of piles-drains and the distance between them, i.e. the diameter of the bearing massif;

* Corresponding author: vitsid@mail.ru

• physical and mechanical properties of pile – drains, compacted surrounding and underlying soils;
• boundary conditions under the pile-drains and on the level of contact with the plate, and in contact with the surrounding soil and at the outer edge of the bearing massif.

2 The interaction between soil pile of finite stiffness and surrounding soil with a view to expanding its diameter

Define stress-strain state of the soil around the pile, including the contact between pile and soil. In the first stage is to determine the radial σ_r and tangential σ_θ stresses at the contact of the pile with the ground when you increase the diameter of the initial borehole by a given value u. Thus, by setting the value of the borehole extension u, we determine the stress in the surrounding soil massif.

Formulation and solution of the problems are considered in polar coordinates in terms of plane stress state, i.e. when $\sigma_z=0$. Consider stress-strain state a soil pile and surrounding soil separately, assuming that the contact between pile and soil radial displacement and stress are equal, i.e. when $r=r_c$.

$$u_{rc} = u_{rs}, \; \sigma_{rc} = \sigma_{rz} \tag{1}$$

From the condition of equal deformation the settlement of foundation plate, piles and surrounding soil are equal:

$$S_p = S_c = S \tag{2}$$

On the outer boundary of the soil cylinder there is a condition of the lack of radial movement, i.e. $u_{r2}=0$.

Additionally, there is an equilibrium condition between the load of the slab and stresses in the pile and the surrounding soil , i.e. have:

$$p = \sigma_{z1} \cdot \omega + \sigma_{z2} \cdot (1-\omega), \tag{3}$$

where $\omega = r_1^2 / r_2^2$, r1 and r2 – initial radius of the pile and radius of the cell respectively.

The original equations to solve this problem in cylindrical coordinates are [7]:

geometric equations:

$$\varepsilon_r = \frac{du}{dr}, \varepsilon_\theta = \frac{u}{r}, \varepsilon_z = \frac{dw}{dz}, \tag{4}$$

physical equations :

$$\sigma_r = 2G\left(\frac{du}{dr} + \frac{v}{1-2v} \cdot \varepsilon_V\right), \; \sigma_\theta = 2G\left(\frac{u}{r} + \frac{v}{1-2v} \cdot \varepsilon_V\right),$$
$$\sigma_z = 2G\left(\frac{dw}{dz} + \frac{v}{1-2v} \cdot \varepsilon_V\right), \tag{5}$$

where ε_V - volumetric deformation, is defined as follows , G – shear modulus, v – Poisson's ratio.

$$\varepsilon_V = \varepsilon_r + \varepsilon_\theta + \varepsilon_z = \frac{du}{dr} + \frac{u}{r} + \frac{dw}{dz} . \tag{6}$$

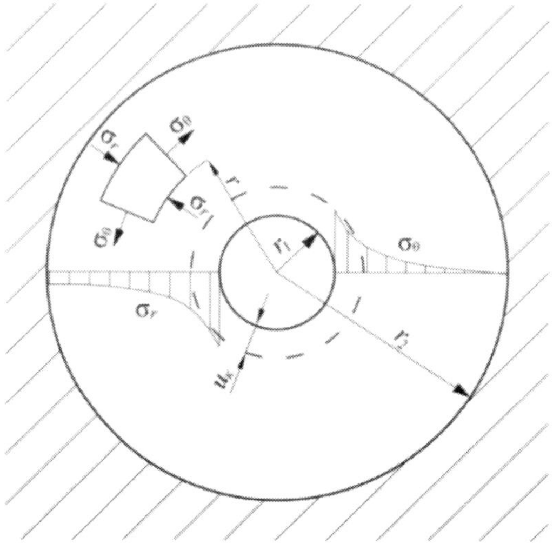

Fig.1. The scheme of increasing the diameter of the initial borehole in the process of pile construction

On the basis of joint consideration of geometric (4) , physical (5) equations and the equilibrium equations (3), the problem of stress-strain state of the soil cylinder can be reduced to the solution of the differential equation in displacements [8]:

$$\frac{d^2u}{dr^2} + \frac{1}{r} \cdot \frac{du}{dr} = \frac{u}{r^2} \tag{7}$$

The solution of this equation is known and has the following form:

$$u = A \cdot r + \frac{B}{r}, \tag{8}$$

where A and B – the constants of integration, that are determined from boundary conditions : when $r=r_1$ $u=u_1$, when $r=r_2$, $u=0$.

Substituting these values in (8) we get:

$$A = \frac{u_1 \cdot r_1}{r_1^2 - r_2^2}, B = -\frac{u_1 \cdot r_1 \cdot r_2^2}{r_1^2 - r_2^2}, \tag{9}$$

and finally :

$$u = \frac{u_1 \cdot r_1}{(r_1^2 - r_2^2)} \cdot (r - \frac{r_2^2}{r}) \tag{10}$$

Given that $\sigma_z = const$, the first two equations (5) can be represented as:

$$\sigma_r = \frac{E}{(1+\nu)(1-2\nu)} \cdot \left[\frac{du}{dr} + \left(\frac{\nu}{1-\nu} \right) \left(\frac{u}{r} + \frac{dw}{dz} \right) \right],$$

$$\sigma_\theta = \frac{E}{(1+\nu)(1-2\nu)} \cdot \left[\frac{u}{r} + \left(\frac{\nu}{1-\nu} \right) \left(\frac{du}{dr} + \frac{dw}{dz} \right) \right] \tag{11}$$

Values $\frac{u}{r}$ and $\frac{du}{dr}$ can be determinated from (8) with (9). Then we have :

$$\varepsilon_r = \frac{du}{dr} = A\left(1 + \frac{r_2^2}{r^2}\right), \ \varepsilon_\theta = \frac{u}{r} = A\left(1 - \frac{r_2^2}{r^2}\right). \tag{12}$$

Then we get expressions for the radial and tangential stresses in final form:

$$\sigma_r(r) = \frac{E}{1-\nu^2} \cdot \frac{u_1 \cdot r_1}{r_1^2 - r_2^2} \left[1 + \nu + \frac{r_2^2}{r^2}(1-\nu) \right] \tag{13}$$

$$\sigma_\theta(r) = \frac{E}{1-\nu^2} \cdot \frac{u_1 \cdot r_1}{r_1^2 - r_2^2} \left[1 + \nu - \frac{r_2^2}{r^2}(1-\nu) \right]. \tag{14}$$

It follows that:

$$\sigma_r + \sigma_\theta = \frac{2E}{1-\nu} \cdot \frac{u_1 \cdot r_1}{r_1^2 - r_2^2}. \tag{15}$$

Then at $r=r_1$ (maximum value) and $r=r_2$ (minimum value):

$$\sigma_{r1} = \frac{E}{1-\nu^2} \cdot \frac{u_1 \cdot r_1}{r_1^2 - r_2^2} \left[1 + \nu + \frac{r_2^2}{r_1^2}(1-\nu) \right], \ \sigma_{r2} = \frac{2E}{1-\nu^2} \cdot \frac{u_1 \cdot r_1}{r_1^2 - r_2^2}. \tag{16}$$

To find the necessary vertical load that needs to be passed through ramming equipment to the soil massif, we have to use the equation of limit equilibrium, where the principal stresses will be present by vertical and radial stresses. The vertical stress must be limit to tamp the layer of material has exhausted its strength, it collapsed and took its proper position in the already enlarged borehole.

The limit equilibrium equation can be written as:

$$\frac{\sigma_z^* - \sigma_r}{\sigma_z^* + \sigma_r + 2c \cdot ctg\varphi} = \sin\varphi. \tag{17}$$

Then we can express the value of the vertical stress:

$$\sigma_z^* = \frac{\sigma_r(1+\sin\varphi) + 2c \cdot \cos\varphi}{(1-\sin\varphi)} \tag{18}$$

Then a necessary load that needs to be applied through the compression equipment to the soil massif, is determined by the formula:

$$N^* = \pi \cdot r_1^2 \cdot \sigma_z^*, \tag{19}$$

where σ_z^* can be found from (18).

Thus, for the design process of material compression into the borehole it is necessary to determine the design diameter of the reinforcing pile-drains and its pitch. Next, we can determine radial stress in the soil in process of extension of the initial borehole to design value and determine necessary compression force for selection the equipment at the construction site.

For example, consider the stress state masiva soil with the following initial data:

r1=0.2 m, *u*=0.1 m, *E*=15 MPa, *ν*=0.3 r2=1 m, *c*=1 kPa, $\varphi = 30°$.

Then, we have values of $\sigma_r = 6456$ kPa and $\sigma_\theta = 19360$ kPa.

And, finally, *N**=2433 kN=243.3 t.

3 Conclusions

1. The stress state of soil around an expanding initial borehole can be determined on the basis of geometrical and physical input equations of the classical theory of elasticity.
2. Using the limit equilibrium equation it is possible to determine the maximum value of vertical stress for compacted column of soil required for crushing on a certain design value.

Acknowledgements

This work was financially supported by the Ministry of Russian Education (state task #2014/14)

References

1. SP 24.13330.2011 (2011)
2. SP 22.13330.2011 (2011)
3. Z.G. Ter-Martirosyan, V.V. Sidorov, P.V. Strunin6 Civil Engineering **9**, 18-22 (2013)
4. V.A. Barvashov, Soil Mechanics and Foundation Engineering **5**, 32-39 (2009)
5. Z.G. Ter-Martirosyan, A.S. Abdulmalek, Soil Mechanics and Foundation Engineering **6**, 8-11 (2007)
6. S.G. Bezvolev, Geotechnics **3**, 30-67 (2011)
7. A.A. Grevcev, Civil Engineering **9**, 2-5 (2012)
8. N.I. Bezukhov, *Bases of theory of elasticity, plasticity and creep* (Visshaya shkola, Moscow, 1968)
9. S.P. Timoshenko, J. Gudier, *Theory of elasticity* (Nauka , Moscow, 1975)

 DOI: 10.1051/matecconf/20168603015

Oscillations and residual displacements of buildings of varying heights and widths, constructed on elastic-plastic base

Zaven Ter-Martirosyan[1], *Evgeniy* Sobolev[1,*], *Armen* Ter-Martirosyan[1] and Aleksandra Bestuzheva[1]

[1]Moscow State University of Civil Engineering, 129337, Moscow, 26 Yaroslavskoe Shosse , Russia

Abstract. Modern regulations in the field of design and construction of buildings in seismically dangerous areas provide for stability calculation of bases, foundations and underground structures, i.e. calculations for the first group of limited states. At the same time, calculations for the second group of limited states, like deformations (yields, relative difference in yields), as a rule, are not fulfilled. This state of affairs is due to insufficient knowledge of the issues of appearance and development of additional stress and strain in ground arrays during seismic influences, that ultimately leads to residual yields and rolls of foundations of industrial and civil buildings. Essentially, residual displacements depend on the type and intensity of dynamic effects, as well as on the model foundation soils. At the same time, the most important factor is the compulsory registration of collaboration between the subgrade and the structure. Therefore, quantification of additional residual displacements of foundations of industrial buildings and structures under dynamic loads, is an actual problem of the modern earthquake-resistant construction.

1 Introduction

The strength and stability of buildings and constructions under static and seismic actions is determined by the nature of the system interaction "foundation - structure". Formulation of the problem is to analyze, under different seismic actions, the forms of vibrations of buildings and structures of different width and height on the two-layer base. At the same time, significant attention is paid to the assessment of movements of residual mass of soil, containing underground part of the structure.

The most significant factor in determining the seismic resistance of a structure is the intensity of seismic effects in the area of its construction. Another major factor is the ratio of primary natural vibration frequencies of "base - construction" system and the seismic effect. In case of coincidence of these frequencies, the resonance phenomenon can lead to loss of stability of structures, even at low intensity of the seismic action [1].

* Corresponding author: e.s.sobolev@mail.ru

2 Method

For analysis of the Stress-Strain State (SSS) of soil mass enclosing the underground part of the building, as well as in order to assess the impact of various factors on the value of the residual deformation of the compressible base, in the Software Package (SP) PLAXIS 2D 2016 a mathematical modeling was performed for the systems consisting of: structures with different geometrical dimensions (width and number of floors), compressed two-layered soil foundation, based on different models (model Mohr-Coulomb and model "hardening soil model with small-strain stiffness").

Quantitative prediction of stress-strain state in heterogeneous base grounds (in the present study defined as a two-layer base) structures under static and seismic action is necessary to ensure the normal conditions of facilities exploitation [2, 3].

In order to do this, one must take into account seismicity of the construction site accommodation area on the basis of engineering-geological conditions as well as physical and mechanical properties of the foundation soil. In this study, seismic effects are defined in accordance with accelerograms recommended by the MGSN (Moscow City Construction Regulations) 4.19-05 Multifunctional High-Rise Buildings and Complexes (Figure 1).

At the same time, in dynamic calculation module the SP PLAXIS 2D 2016 allows to directly provide for the specified accelerogram and to use it in calculation of SSS of subgrade under seismic impacts.

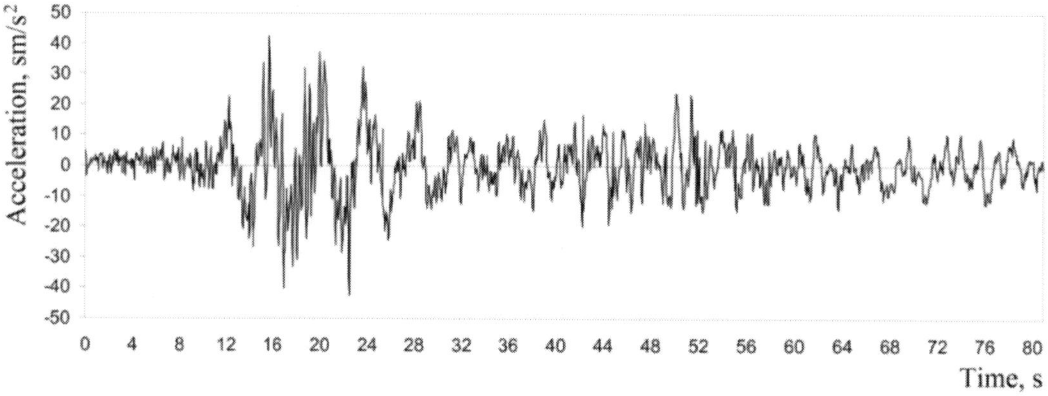

Fig. 1. Earthquake analogue accelerogram, used in FEM (Finite Element Method) calculations (in accordance with MGSN 4.19-05 Multifunctional High-Rise Buildings and Complexes).

In this paper, a two-layer base is analyzed. The top layer with thickness of 15 m is represented by silty-clay soil. The bottom layer consists of sandy soil with thickness of 25 m. The foundation soil properties are presented in Table 1.

The first model used in mathematical modeling is the model of Mohr-Coulomb. It represents a model of the "first level", which includes only a real number of features of the soil behavior. Despite the fact that increase in stiffness can be considered with depth, the Mohr-Coulomb model does not account for dependence of hardness neither on stress nor on the stress trajectory, and also on the deformation anisotropy. The state of stress at break can be described by the Mohr-Coulomb criterion, with effective strength parameters φ' (internal friction angle) and c' (specific cohesion) [4, 5].

The second model used in mathematical modeling - the soil model "hardening soil model with small-strain stiffness" (HSSMALL), while taking into account the soil stiffness at small strains, is based on a fairly well-known elastic-plastic model of hardening soil (hardening soil model) and uses almost the same set of parameters. In fact, description of changes in rigidity due to magnitude of deformation requires only two additional parameters: initial shear modulus and shear modulus of the ultra-low deformations of G_0;

shear strain rate $\gamma_{0,7}$, wherein secant shear modulus G_s decreases to about 70% of the magnitude of G_0.

Table 1. Physical and mechanical properties of soils.

Parameter	Name	Upper clayey layer	Lower sandy layer	Unit
General				
Material model	Model	HS small	HS small	-
Type of material behaviour	Type	Drained	Drained	-
Soil unit weight above phreatic level	γ_{unsat}	16	20	kN/m^3
Soil unit weight above phreatic level	γ_{sat}	20	20	kN/m^3
Parameters				
Secant stiffness in standard drained triaxial test	E_{50}^{ref}	$2.0 \cdot 10^4$	$3.0 \cdot 10^4$	kN/m^2
Tangent stiffness for primary oedometer loading	E_{oed}^{ref}	$2.561 \cdot 10^4$	$3.601 \cdot 10^4$	kN/m^2
Unloading / reloading stiffness	E_{ur}^{ref}	$9.484 \cdot 10^4$	$1.108 \cdot 10^5$	kN/m^2
Power for stress-level dependency of stiffness	m	0.5	0.5	-
Cohesion	c'_{ref}	10	5	kN/m^2
Friction angle	φ'	18	28	°
Dilatancy angle	ψ	0	0	°
Shear strain at which G_s = $0.722G_0$	$\gamma_{0.7}$	$1.2 \cdot 10^{-4}$	$1.5 \cdot 10^{-4}$	-
Shear modulus at very small strains	G_0^{ref}	$2.7 \cdot 10^5$	$1.0 \cdot 10^5$	kN/m^2
Poisson's ratio	ν'_{ur}	0.2	0.2	-

Differences between static and dynamic stiffness of soil, explained by the nature of stressing (e.g., by inertial forces and by strain rate effect), rather than by magnitude of the applied strain, which in dynamic conditions is usually not so large. Since inertia forces and rate of deformation have very little effect on the initial stiffness of soil, the terms "dynamic stiffness" and "stiffness at small strains" may actually be considered as synonyms.

The SSS of an interacting with structures two-layer base is significantly affected by the rigidity of construction, by the upper and lower base layers and their relationship, as well as by the elastic-plastic properties of soils.

SSS calculations completed for double-layer base interacting with buildings of different heights and widths have shown that registration of elastic-plastic properties of soil under static and seismic effects leads to additional residual stresses and displacements. However, frequency of normal vibrations of buildings is significantly different, while depending on the geometrical dimensions of the building [6, 7].

With increase in the number of stores of buildings, the effect of seismic loading on SSS of soil base, enclosing the underground part of the building, is growing significantly. In the contact zone "foundation - base" emerges an additional stress and strain, which leads to additional residual stress and displacement, as well as to formation of local zones of limited state of soil elastic equilibrium.

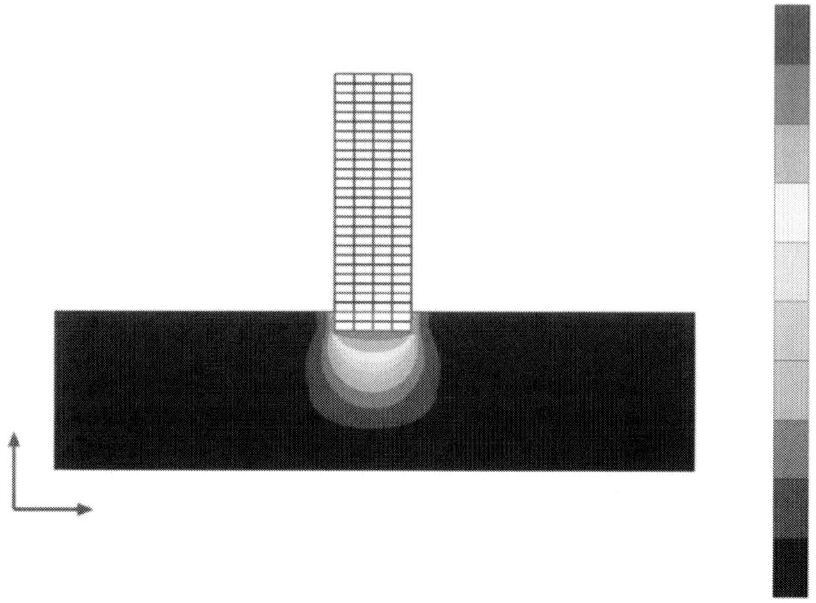

Fig. 2. Isolines of total displacements at the end of the construction phase of construction (dead load). The solution with SP PLAXIS 2D 2016.

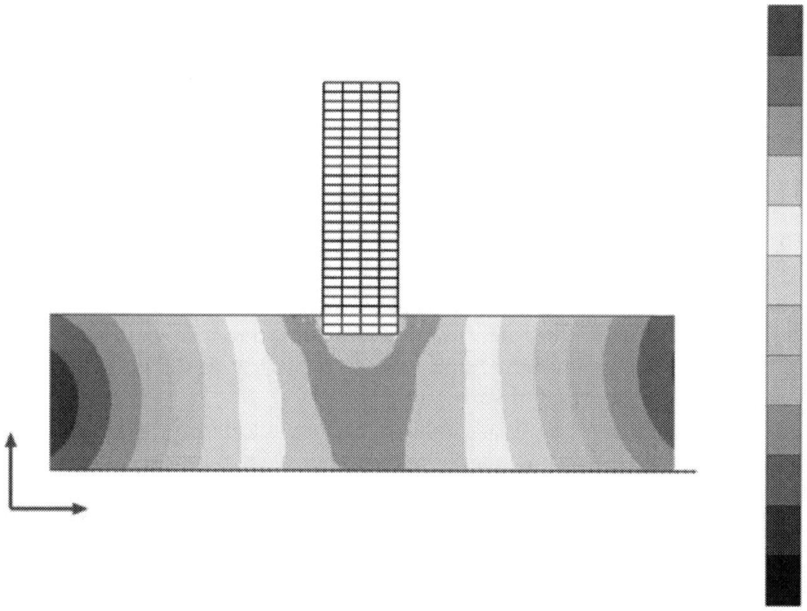

Fig. 3. Isolines of total residual displacements at the stage of completion of the seismic action (dynamic load). The solution with SP PLAXIS 2D 2016.

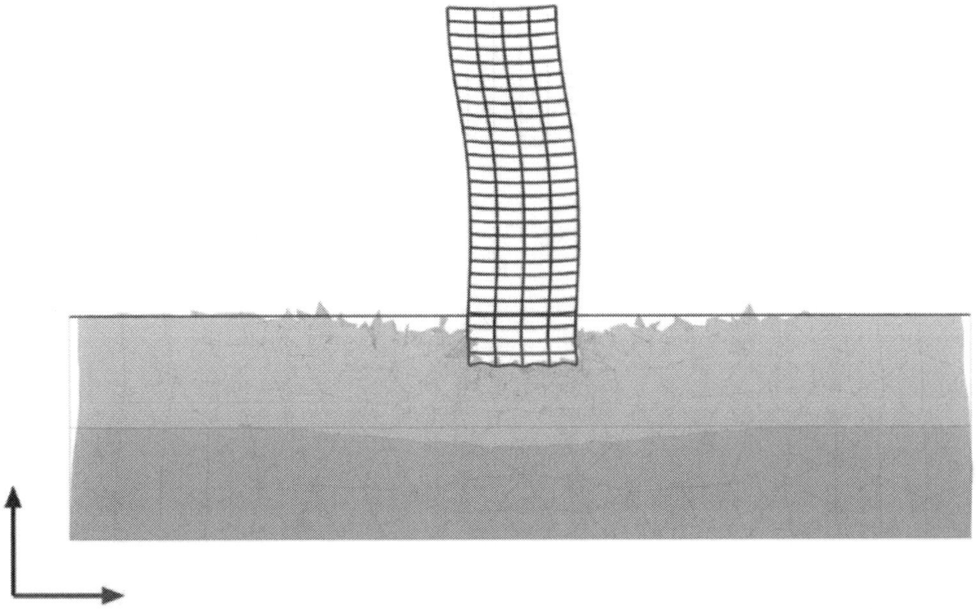

Fig. 4. The deformed grid of finite elements after the stage of the seismic action. The solution with SP PLAXIS 2D 2016.

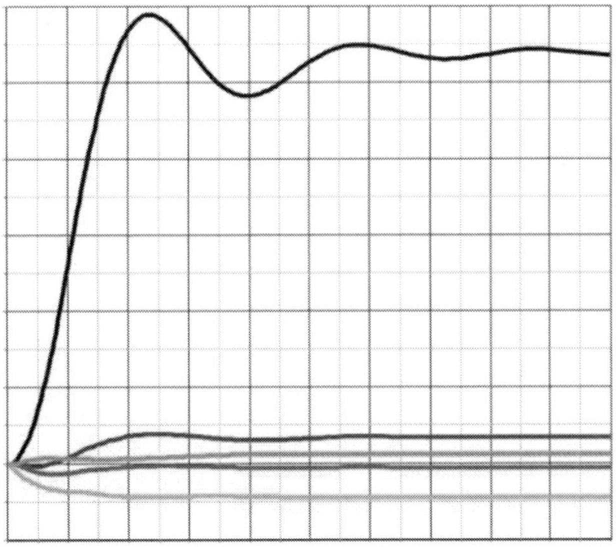

Fig. 5. Graphs of horizontal displacements dependence on duration of the seismic action. The values are given for different points located at different distances from the soil surface: Point A is at a distance of +75 m, Point B is at a distance of 0 m (on the ground-floor floor level), Point C is at 6 meters below the surface (base plate), Point D is 25 m below the surface, Point E is 50 meters below the surface.

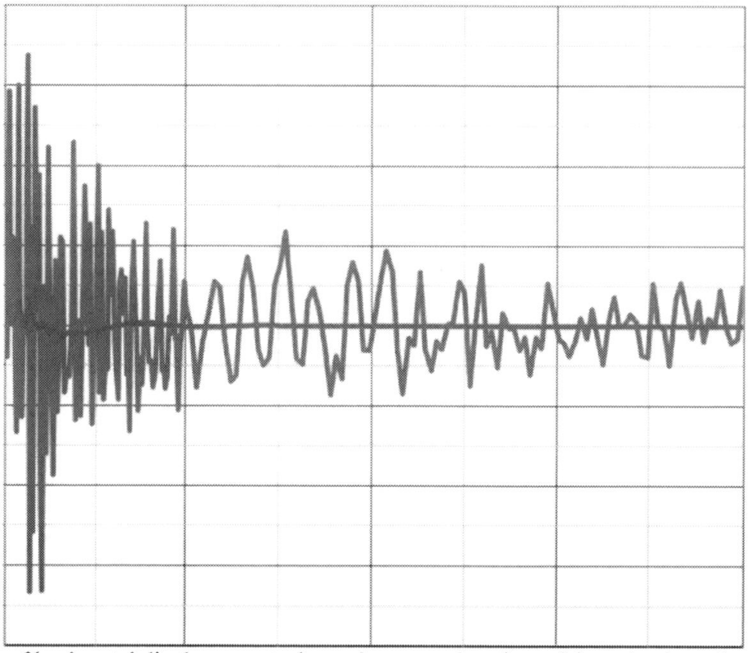

Fig. 6. Graphs of horizontal displacements dependence on duration of the seismic action. Red line marks acceleration at the top of the building (75 m), dark green line - at the level of 25 m below ground surface (-25 m).

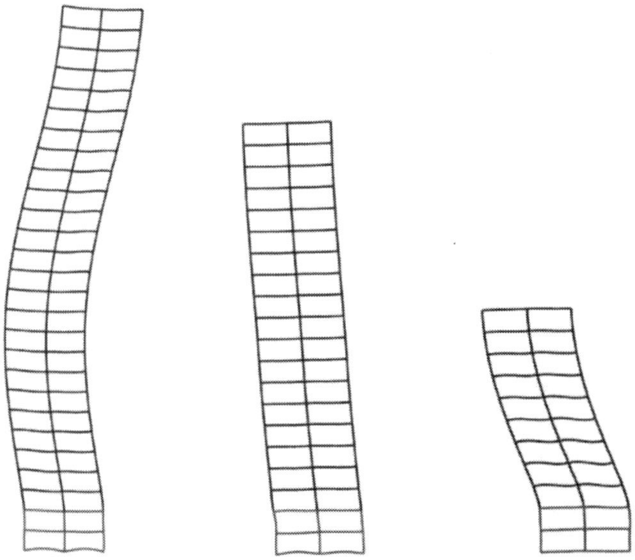

Fig. 7. Waveforms of superstructure parts at different heights. a) height of 75 m; b) height of 54 m; c) height of 27 m.

3 Conclusion

1. The objective of collaboration of the system "foundation - structure" under seismic impacts is solved by taking into account the modeling of bearing structures of aboveground and underground parts of the structure, that has a significant impact on distribution and magnitude of the residual displacement of soil, in contrast to the method of constructions modeling as a solid array with reduced stiffness.

2. In the Stress-Strain State of inhomogeneous soil foundation interacting with the structure under seismic impacts, is significantly affected by registration of stiffness of the structure itself, as well as of upper and lower layers of the base and their ratio (in the case of a two-layer base, considered in this paper) [8, 9, 10].
3. Modeling of soil base under seismic action is performed by using the elastic-plastic hardening soil model for small deformations (hardening soil model with small-strain stiffness (HSSMALL), while taking into account soil stiffness at small deformations and its nonlinear dependence of the strain amplitude.
4. SSS calculations completed for double-layer base interacting with buildings of different heights and widths have shown that registration of elastic-plastic properties of soil under static and seismic effects leads to additional residual stresses and displacements [11].
5. Number of floors affects the waveform of above-ground parts of the building. With the growth of the construction height, the seismic load effect on the stress-strain state of the base soil increases. There are additional residual stresses and strains in the contact zone "foundation soil - foundation" which lead to additional residual stress and displacement, as well as to the formation of local zones of limit equilibrium.

Acknowledgements

This work was financially supported by the Ministry of Russian Education (state task #2014/14)

References

1. H.T. Chan, Thesis on scientific degree competition of the candidate of technical sciences, **174** (2006)
2. K. Ishihara, NGO Georeconstruction-Fundamentproject, **384** (2006)
3. P.L. Ivanov, Gosenergoizdat, **259** (1962)
4. T. Iwasaki, K. Tokida, F. Tatsuoka, S. Watanabe, S. Yasuda, H. Sato, Proc. 3rd Int. Conf. On Microzonation. Seattle **3**, 1319-1330 (1982)
5. M.N. Jaro, Thesis on scientific degree competition of the candidate of technical sciences **173** (2013)
6. I.T. Mirsayapov, I.V. Koroleva, G.Z. Zaripov, Soil Mechanics and Foundation Engineering in Geotechnical Engineering **1**, 31-37 (2015)
7. H.B. Seed, I.M. Idriss, Journal of Soil Mechanics and Foundation Engeneering **97**, 1249-1273 (1971)
8. H.B. Seed, Journal of Soil Mechanics and Foundation Engeneering **105**, 201-255 (1996)
9. L.R. Stavnitser, Building schools Association, **448** (2010)
10. A.Z Ter-Martirosyan, A.Yy. Mirniy, E.S. Sobolev, Geotechnics **1**, 66-72 (2016)
11. E.A. Voznesensky, V.V. Funikova, E.S. Kushnareva, V.G. Kovalenko, Exploration and conservation of mineral resources **12**, 61-65 (2005)

The simulation calculation of acoustics energy transfer through the material structure

Peter Zvolenský[1], *Alžbeta* Pultznerová[2*], and *Juraj* Grenčík[1]

[1]University of Žilina, Faculty of Mechanical Engineering, Univerzitná 8215/1, 010 26 Žilina, Slovak Republic
[2]University of Žilina, Faculty of Civil Engineering, Univerzitná 8215/1, 010 26 Žilina, Slovak Republic

Abstract. The paper deals with the modification of the rail passenger coach floor design aimed at improvement of sound reduction index. Refurbishing was performed by using a new acoustic material with a filamentary microstructure. The materials proposed in research were compared by simulation calculation of acoustic energy transfer trough porous microstructure of filamentary material, and the effect of material porosity on sound reduction index and sound absorption coefficient were observed. This proposed filamentary material can be used in the railway bed structure, too. High degree of noise absorbing, resistance to climate conditions, low specific mass, enable to choose a system of low anti-noise barriers having similar properties as standard high anti-noise walls..

1 Introduction

The paper deals with simulation calculations of acoustics energy transfer through the structure of porous media. The benefit material composition of sandwich character of structure construction offers a wide range of possibilities application of modern acoustic materials. The aim of research is improving the sound reduction index of rail passenger coach body or improving acoustics parameters of railway track structure. Mineral wool provides good acoustic and thermal insulation properties. As a simple and effective solution to improve the acoustic properties of rail passenger coach body it seems to be utilization of the insulation layer from modern acoustics materials. One of the proposed solutions can be utilization of STERED® material in the railway track structure and also in the structure of a rail passenger coach. From this point of view the paper deals with comparison of acoustic properties of porous materials - mineral wool and STERED® material.

2 Simulation of acoustics energy transfer through the porous media

In this case mineral wool and STERED® are porous filamentary materials. The decisive influence on the acoustic properties of porous materials has in particular the porosity (Fig.

* Corresponding author: alzbeta.pultznerova@fstav.uniza.sk

1). By using the simulation computation, the effect of the porosity of a filamentary material on the acoustic properties were studied by changing porosity in the range from 5 to 99.5 % by increment of 0.5% and frequency was changed from 10Hz to 10kHz by increment of 10Hz. The model presents principle of impedance tube and the inlet signal is defined by following function in complex form. [1]

$$1exp^{(-i(kx.x+ky.y))} \qquad [Pa] \tag{1}$$

kx – wavenumber axis x [rad/m],
ky – wavenumber axis y [rad/m],
x – position vector axis x [m],
y – position vector axis y [m].

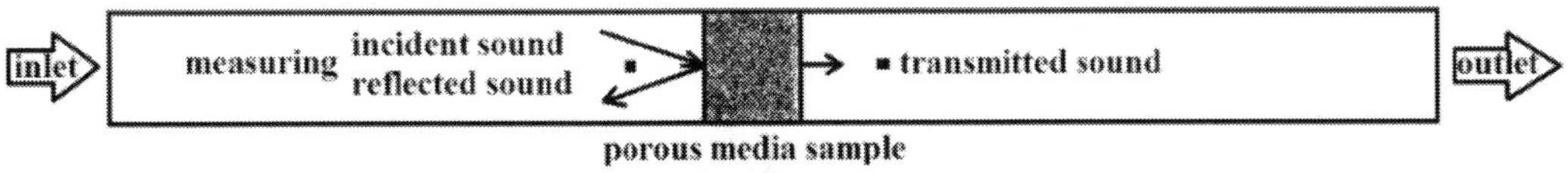

Fig. 1. Model of simulation computation transfer acoustics energy through the porous media [own research].

The results of simulation can be seen in a 3D graph (Fig. 2). The graph was created in MATLAB.

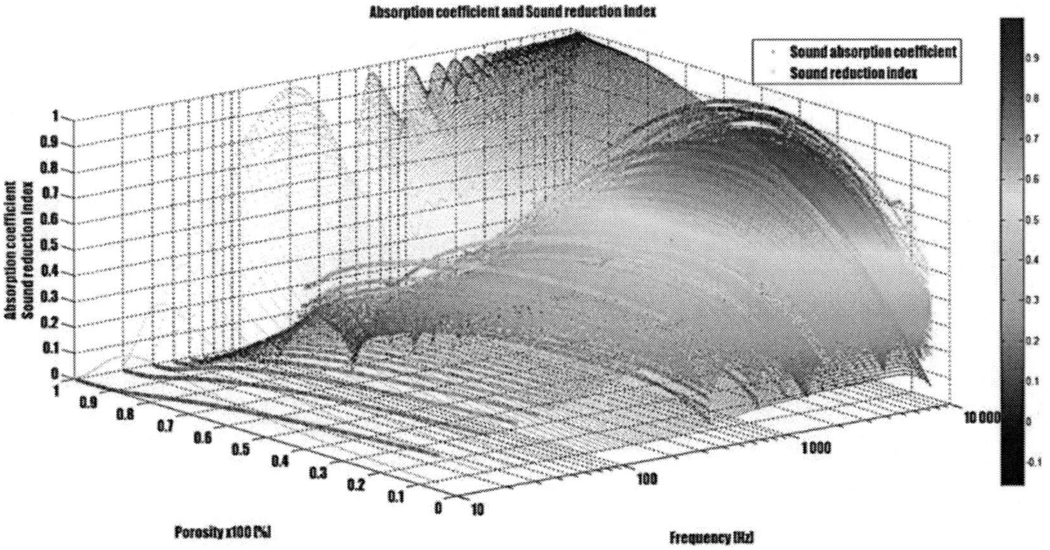

Fig. 2. Frequency dependence of sound absorption coefficient and sound reduction index on the material porosity [own research].

The effect of material porosity on sound reduction index and sound absorption coefficient is better observable on the graph (Fig. 3), which was created by separated specific frequencies. As an example the monitored characteristics of acoustic parameters in frequency 1 kHz are given. The studied relationships are based on energy balance transfer of acoustic energy through the porous microstructure of filamentary material (Fig. 4).

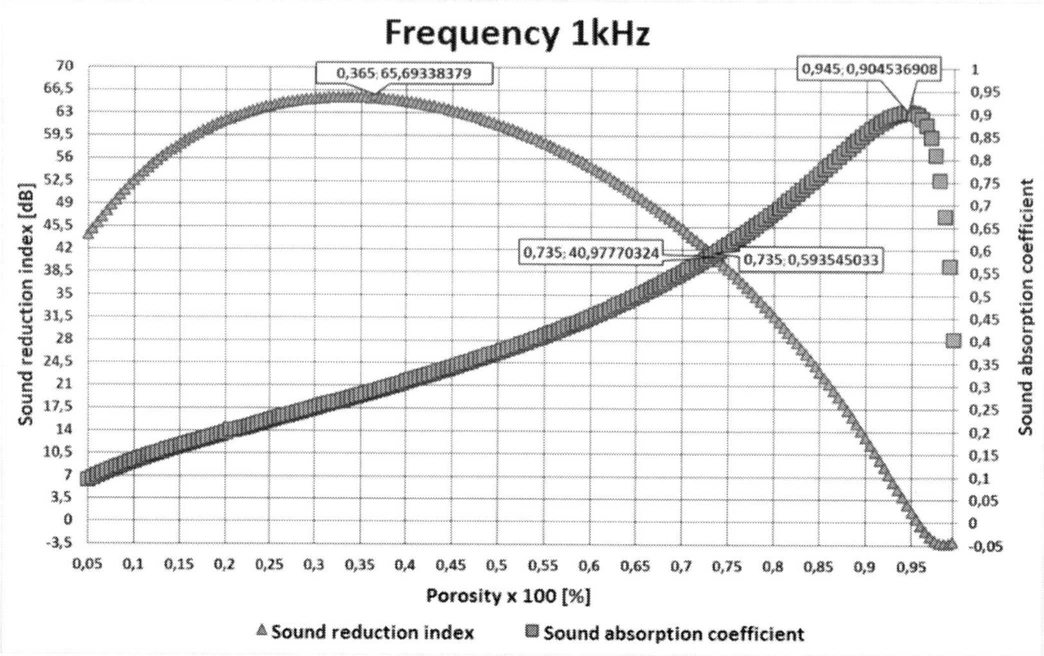

Fig. 3. The characteristics of sound reduction index and sound absorption coefficient depending on the material porosity at 1 kHz frequency [own research].

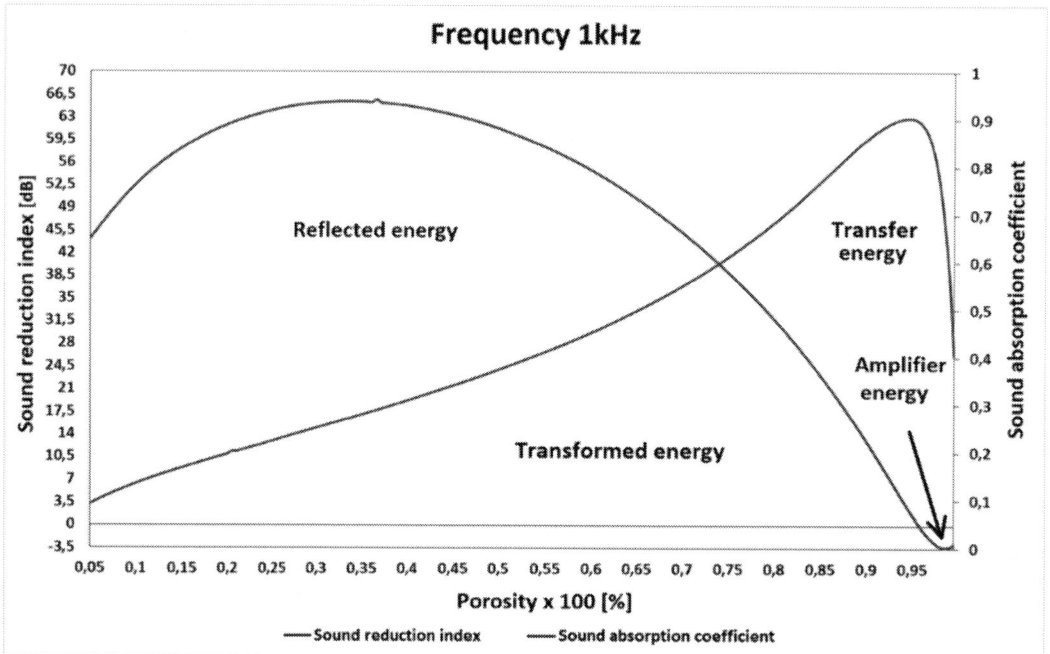

Fig. 4. Energy balance of acoustic energy transfer through porous filamentary microstructure. [own research].

The mathematical model can create a geometrical model of the filamentary material microstructure, which in principle acts as a breaker of acoustic waves. Effect of material porosity is very important, for that a simple porosity measuring of mineral wool, material STERED® 200 and STERED® 250 was done, by using direct method porosity measuring. This method is suitable only for very regular shapes of test samples. The measuring samples must be completely saturated by liquid (water) under vacuum. The volume of pores was directly determined from difference between dry and saturated sample on the base on the know bulk density of water, which was in the pores of porous material. [2, 3]

Table 1. Studied measure samples of porous material. [own research]

Mineral wool	STERED® 200	STERED® 250
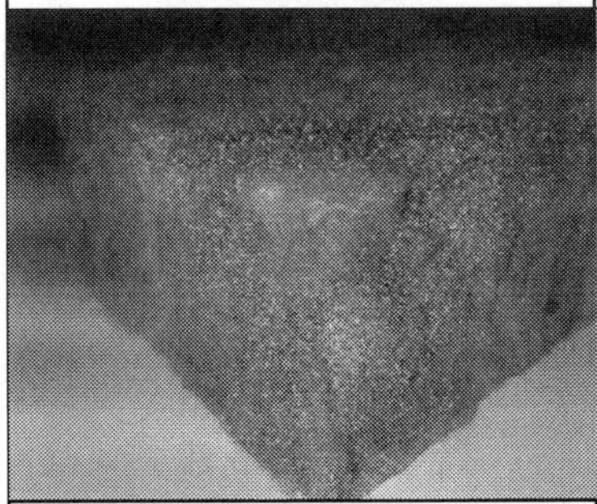		
93.93%	80.28%	77.12%

3 Application of STERED materials on the railway track

3.1 Active noise measures on railway track

Active noise measures on the railway track include mainly: regular maintenance of the railway superstructure, application of modern structures of railway superstructure and types of switches, quality structure of the rail, appropriate structures of bridges, the use of anti-vibration mats and pads into the structure of railway superstructure and in the last years rail dampers fixed to the rail, too.

The rail damper, thanks to its structural composition, has high absorption properties (Fig. 5a, b), which absorbs the vibrations at the wheel/rail interface, and thus prevents generating noise in rail. In the case of train passing, noise is generated in different frequency spectra in the contact between the wheel and rail. Rail damper works on the absorption of vibration for frequencies with the greatest levels of noise. In case of low and high frequencies that correspond to lower levels of noise, the structure of rail damper is designed to have no effects on the amount of noise generated. On the contrary, in the case of frequencies, which are characteristic for most of the noise generated, rail damper is the most effective. Rail damper that prevents the spread of the vibration wave causes its faster absorption. [4]

Studies conducted by German (DB) and French (SNCF) Railways, at the rail track section with rail dampers installed, showed a reduction in noise up to 6 dB (A) and vibration up to 9 dB. [5]. The main advantage is that they can be installed even after the installation of the section at critical locations to improve noise parameters after putting construction into operation. There are no restrictions in the railway maintenance, the rail absorber is maintenance-free and does not affect the overall appearance of railway tracks. [6]

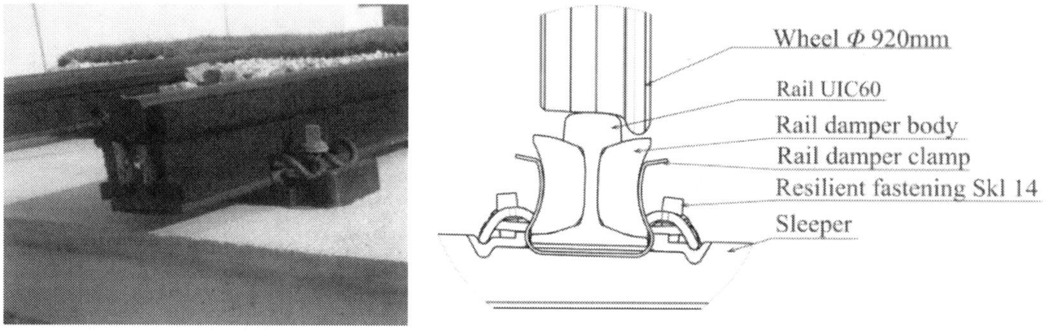

Fig. 5. (a) Rail damper made from STERED [8] ; (b) Rail damper cross-section [own research].

3.2 Passive noise arrangements in the railway track

Passive noise arrangements can be implemented directly on the structure of the railway track or railway vehicle. Noise arrangements, which are part of the railway track include Noise enclosure, Noise barrier (Fig. 6 a)), Near track barrier, Low noise barrier (Fig. 6 b)), Track-bed absorption (Fig. 7 a)), Rail track slab (Fig. 7 b)).

Regarding its excellent and extraordinary acoustic parameters, the STERED product offers extensive application in anti-noise measures. High degree of noise absorbing, resistance to climate conditions, low specific mass, enable to choose a system of low anti-noise barriers along with comparable acoustic results as standard high anti-noise walls exert.

Low anti-noise barriers can contribute distinctly to higher traffic safety – better visibility, better access of rescue services; they also represent distinct landscape forming character. Other distinct contribution is the better visibility for all traffic participants, in particular as for crossing between different kinds of traffic – railway crossings.

Placing a low anti-noise barrier combined with crash barrier of divided road can be a contribution to reducing the solution costs of anti-noise measures behind contra directional traffic lanes. This fact holds for both roadsides of the communication body. Another advantage of low anti-noise solutions for increasing safety (driver's attention) is removing the driving in tunnel effect because of starting occurrence of long road sectors with anti-noise walls up to 4m high. STERED ID, insulating plate, provides high factor of absorption even for low frequencies. Frequency characteristics of noise absorption, applied for traffic noise spectrum according to STN EN 1793-3, prove that the material is suitable for the construction of anti-noise barriers as equipment for traffic noise reduction with high degree of absorption (category A4). Certainly, it is necessary not to exceed the limit values specified in the Decree 549/2007 of the Ministry of Health of the Slovak Republic [7]. Application of STERED material increases noise absorption of traditional anti-noise measures, together with extending of absorbing spectrum thanks to its parameters of high absorption starting from 400 Hz frequency. Examples of STERED solutions are given on following pictures. [8]

a) b)

Fig. 6. a) STERED in high acoustics barrier design **b)** STERED in low acoustics barrier using on the railway [8].

a) b)

Fig. 7. a) STERED in railway bed design **b)** STERED in rail track slab [8].

4 Conclusions

There are several negative environmental factors of railway transport perceived by public, among which air pollution as consequence of energy consumption, especially for diesel traction, and noise produced by railway are the most significant. Numerous efforts have been done in both areas to minimize the effects. This study deals with possibilities of noise reduction on railway wagons and railway track using new materials with enhance acoustic properties.

Among other benefits of the STERED® material, which has been proposed to be used in railway noise reducing systems, is that the material is obtained from recycling of the textile waste used as thermal and acoustic insulators in the automotive industry. The base material consists of synthetic fibres with high resistance against environmental effects. Thus, it can be used even outdoors, and has wide range of applications.

The main aim of using STERED® material is to provide good sound reduction of acoustics barriers in railway track, rail damper and using in others application of noise reduction on the railway.

In the searching suitable technical solutions and methods of noise reduction, it is important to recognize suitability of using a sound absorption material, or sound reduction material, what depends on actual acoustics situation. By using suitable process of simulation calculation it is relatively easy to diagnose vibro-acoustics paths and sources of acoustics energy. Consequently, it is important to find technical solutions for elimination effect of the vibro-acoustic path. For proper choice of the material it is important to know effect of porosity and to find optimum as a balance between sound reduction and sound absorption that are dependent on the porosity. The effect have been studied in the research and results optimum porosity for both parameters have been revealed.

The paper was created within the framework of project VEGA 1/0766/15 " Research of noise emissions sources in the railway transport and methods of their effective reduction ".

References

1. Y. Champoux, J.F. Allard, J. Appl. Physics **70**, (1991)
2. T. Bourbié, O. Coussy, B. Zinszner, *Acoustics of porous media* (Institut francais du pétrole publications, Paris, 1987).
3. M. Matka, J. Gołembiewski, Z. Koza, *Proc. of the 5th Int. Conf. on Porous Media and its App. in Sci. and Eng.* (Engineering Conferences International, Hawaii, 2014).
4. D. Benton, CORUS RAIL, FRANCE SA **5**, (2007)
5. S.Lakušić, M. Ahac, Tehnički vjesnik **19**, 429 (2012)
6. A.Pultznerová, L.Ižvolt, Proc. Eng. **91**, 274-279 (2014)
7. A. Pultznerová, M. Mečár, Logistyka **4(3)**, 3188 (2014)
8. *Stered in traffic infrastructure* Information on http://www.stered.sk/stered-v-dopravnej-infrastrukture?lang=en

Receiving concrete with the superadsorbing component

Aleksander Buryanov [1,*], *Victoria* Petropavlovskaya [2], *Tatiana* Novichenkova [2], and *Kirill* Petropavlovskii [2]

[1]The Department of binders and concretes, Moscow State University of Civil Engineering, 26 Yaroslavskoe sh., 129337 Moscow, Russia
[2]Tver State Technical University, 22 Af. Nikitin, 170026 Tver, Russia

Abstract. Any period of high temperature in which special precautions need to be taken to ensure proper handling, placing, finishing, and curing of concrete. High winds, low relative humidity, and solar radiation all have major effects on concrete. Hot weather conditions can produce a rapid rate of evaporation of moisture from the surface of the concrete and accelerated setting time, among other problems. Generally, high relative humidity tends to reduce the effects of high temperature, because of hot and dry climate in the Iraq has been the addition of the gel to the mixture of concrete for keeping the water into the concrete during solidification and prevent cracks. Through the results, it showed an increase in the moisture content in the samples who containing a certain percentage of the gel in (7-28) days with the increase in the resistance ratio (strength of the concrete).

1 Introduction

While hot weather conditions are commonly encountered in summer, combinations of high temperatures, winds and low humidity could result in conditions leading to problems with concrete placement and finishing at any time [1]. Effect of hot weather conditions on the properties of concrete need the precautions that should be taken, particularly with flatwork, to minimize any potential adverse effects when placing concrete under these conditions [2, 3, 4]. Heat resistant concrete, designed for industrial and construction, with prolonged exposure to high temperatures should be saved in the specified limits their physical and mechanical properties [5]. Heat resistant concrete manufactures cementitious (in which, where appropriate, introduce more active fine and mineral supplement), water (or other a mixing) and refractory aggregates [6, 7]. Results of research and surveillance state test sites show that can reduce the water absorption of concrete up to 0.1-0.3 % (by weight) and at the same time ensure its resistance of the surface layers (impregnation to a depth of 3-5mm) under dry and hot climate according of assessment to some researchers [8]. Hydration is an exothermic reaction, meaning it generates heat, and that reaction goes faster when the concrete is hot. So the main concern with the concrete's strength and set time isn't really the air temperature but the concrete temperature. When cement hydrates it sucks up water and

* Corresponding author: rga-service@mail.ru

grows crystals around the aggregate particles. When it's hot and that reaction is rapid, the crystals grow quickly but don't have time to grow strong. Early strength will be higher but 28-day strength suffers. If the concrete is about 18° hotter than normal (for example, 88° instead of 70°), the ultimate compressive strength will be about 10% lower, in hot weather, as the cement sets up, slump decreases rapidly and more mixing water is needed. This can also contribute to lower strengths (as much as another 10% lower), and in integrally colored concrete, can lead to variations in water content which can result in significant differences in concrete color between adjacent pours [2]. Another potential problem in hot weather is surface drying – although this one comes with a caveat. If the concrete is warm and the sun is shining and there is a hot dry wind, likely to get the concrete more drying and surface shrinkage, and, besides, may be: strength reduction, cracking, shrinkage cracks, and finishing difficulties [3].

2 Main part

There are many types of admixtures available for incorporation in concrete to achieve of a enhances certain properties or to achieve economy or both. Generally, admixtures affect shrinkage of the concrete to a varying degree depending on their formulation, their interaction with the cement, their interaction with other admixtures in the mix, and on the variations or adjustments they bring about in the proportions of the concrete mix. Adding the percentage of the mixture of Super Absorbent Polymers and Water holding inside for a specific period of time and thus compensate during the dry and hot climates, thereby reducing the drought and concrete cracks and similar applications [4].

Fig. 1. Gel: (1) dry gel; (2) 5-8 sec after water added; (3) ready to add more water; (4) 30 sec after adding more water.

Results showed what after 7 days a decrease in density of concrete with adding Gel happens monotonically. But after 28 days found a significant increase in the density ratio of the concrete when the amount of gel 1 %. Gel works in concrete most long as the best super adsorbing polymer for water during process of hardening of the concrete in case of this ratio (refer with: Fig. 2).

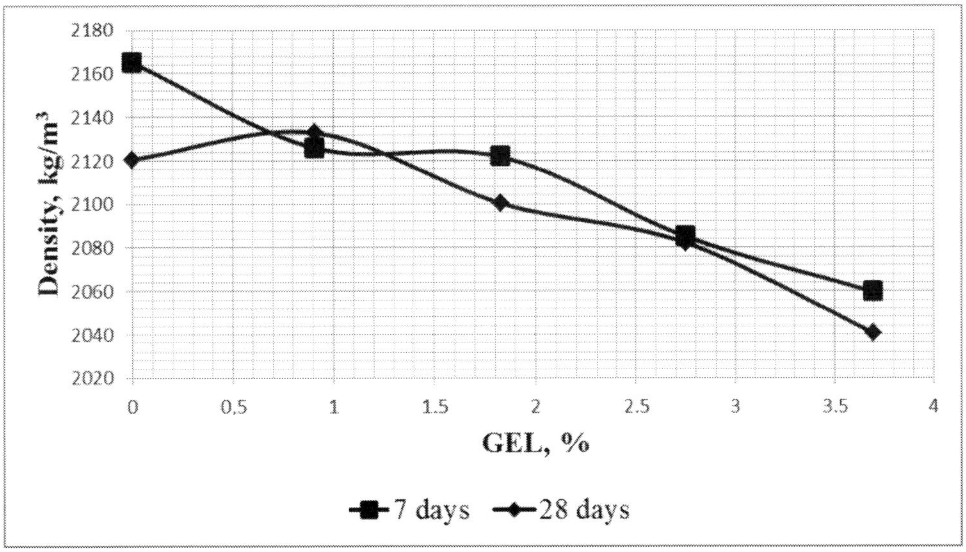

Fig. 2. Relationship between the proportion of Gel added to the concrete mixture with the average density of the concrete in (7, 28) days.

The component of Gel influences concrete strength. The received results of strength at the age of 28 days show that the optimum content of Gel corresponds by Gel ~ 2 % (refer with: Fig. 3). In case of optimum of Gel additions strength of concrete of 23.34 MPa reaches. However concrete density depends on amount of additive a little (refer with: Fig. 2). Dynamics of change of strength from the content of additive is more brightly expressed for concrete at the age of 7 days.

The maximum strength corresponds to the water / solid relation – 0.125 (refer with: Fig. 4).

It is caused by more active curing of a cement stone during this period. Processes of hydration proceed more actively. Therefore maintenance of the modes of humidity during the period up to 7 days is necessary. It will provide the high strength and fuller course of process of structurization.

The general density of system the solvation sheath on cement grains reduces a thickening. It can't but have an adverse effect on properties of cement stone and concrete. When curing such cement stone will be inclined to contraction.

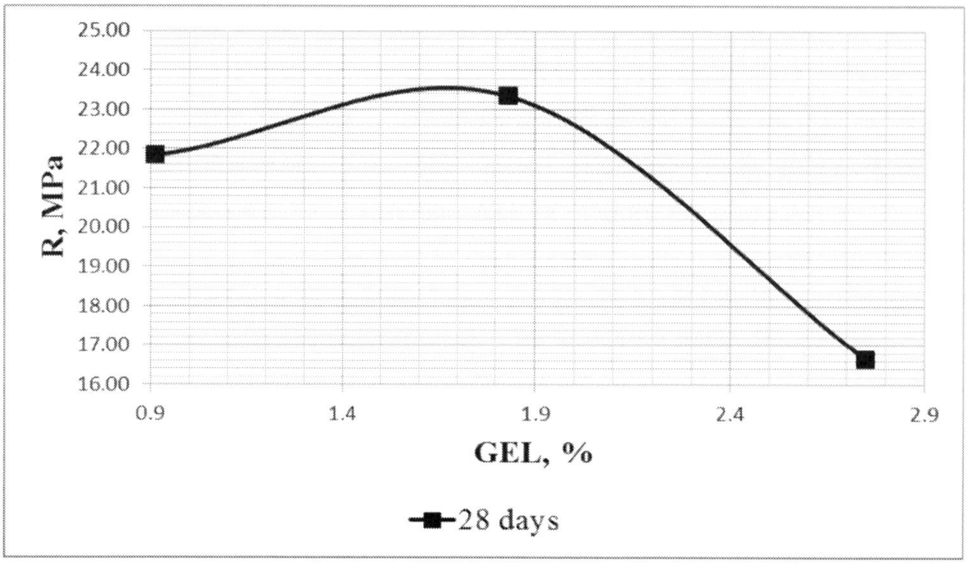

Fig. 3. Relationship between gel additions in the concrete mixture with the strength of the concrete (28 days).

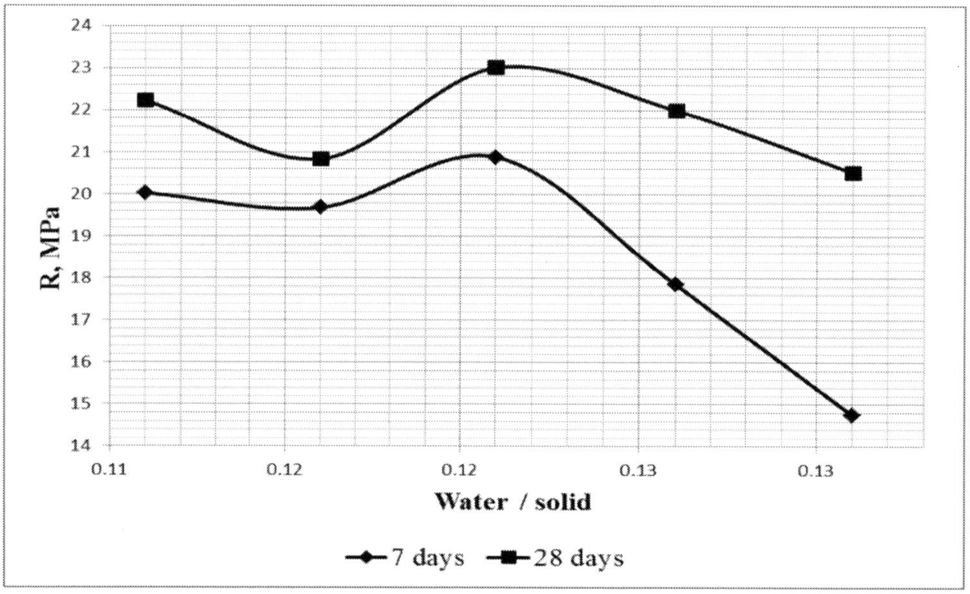

Fig. 4. Relationship between water/ solid relation in the concrete mixture with the strength of the concrete (7, 28 days).

Types of hot concrete structure of concrete after 7 days of hardening (refer with: Fig. 5) and micro-structure after 28 days of hardening was shown (refer with: Fig. 6) through the micro-photo in micro-structure of mixture concrete penetration of Cotton and GEL inside parts of the concrete and connecting parts of the concrete which has contributed to increased strength of concrete and an increase in relative humidity. The additions (GEL, Cotton, Water) in the concrete mixture does not cause dryness concrete, formation cracks but it worked Super Absorbent Polymers and Water holding inside the mixture of concrete for a specific period of time and this led to increase in the efficiency of the work of concrete in conditions of a hot, dry climate.

It is established that use of preliminary saturation leads to education in structure of concrete of the large time reducing its strength (refer with: Fig. 5). Use of the additive of GEL without her preliminary saturation by water more effectively. In concrete additional porosity, which reduces concrete strength, isn't formed (refer with: Fig. 6).

Fig. 5. Structure of concrete with GEL additions after test.

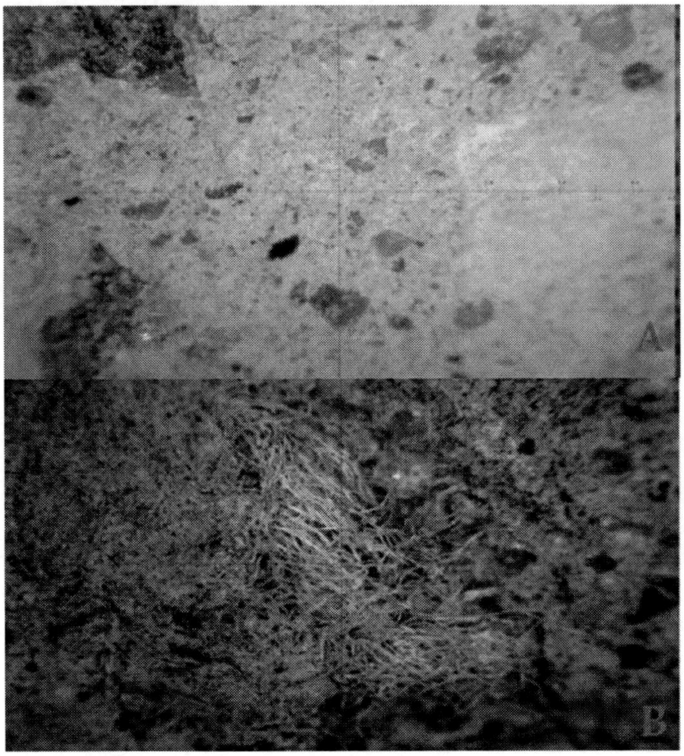

Fig. 6. Microstructure of concrete: (A) without gel; (B) with gel additions.

3 Conclusions

High strength of concrete through the addition of the Gel to the concrete mixture is reached, thereby increasing the work efficiency of the concrete under the hot, dry weather, such as

Iraq's climate. The increase of the humidity of the concrete by addition of Gel leads to increasing water in the concrete for a longer period. So the efficiency of the work of concrete within the hot, dry climate is increasing.

References

1. *AS 1379 Manufacture and supply of concrete Standards Australia* (1997)
2. *ACI Committee 305, Hot Weather Concreting ACI. Manual of Concrete Practice Part 2: Construction practices and inspection, pavements* (American Concrete Institute, Farmington Hills, 1999)
3. V.V. Rusina, Construction materials, 12-13 (2013)
4. *Buchholz and Graham, Modern Superabsorbent Polymer Technology* (John Wiley & Sons, 1998)
5. Information on http://www.amcol.com/fctbook/absor.htm
6. P.V. Smirnova, L.V. Morgun, Construction material, 96-97 (2013)
7. A.M. Scheinin, *Cement concrete for road and airfield pavements* (Transport, Moscow, 1991)
8. *Material journee technical on roads in beton* (Cemsuisse, Fribourg, 2005)

MATEC Web of Conferences **86**, 04003 (2016) DOI: 10.1051/matecconf/20168604003
IPICSE-2016

Optimal design of lattice metal constructions of overhead power transmission lines

Andrey Pustovgar[1], *Anton* Tanasoglo[2,*], *Igor* Garanzha[2], *Liubov* Shilova[1], and *Aleksey* Adamtsevich[1]

[1]Moscow State University of Civil Engineering, Yaroslavskoye sh. 26, 129337 Moscow, Russia
[2]Donbas National Academy of Civil Engineering and Architecture, 2 str. Derzhavin, 86123 Makeyevka, Ukraine

Abstract. The article presents a new concept of design of power lines which allowed us to develop new unified power transmission supports, efficient and reliable ones. For the first time, a method of optimum design of overhead power lines has been developed which considered the power line as a united structural system, the relief, the atmospheric and climatic conditions, the environment pollution, the efficiency requirements, the technological and operational expenditures. The article describes the changes introduced into the requirements for the dimensions and parameters of grid metal works with the minimum mass. The optimum design, the computer-aided assembly process and the use of advanced technologies guarantee the reduction of the construction period and the increase in the maintenance period of power lines.

1 Introduction

The construction of overhead power lines in the Russian Federation, Ukraine, Kazakhstan and Belarus uses traditional unified support structures developed in accordance with the USSR standards [1-4]. Such support structures cannot compete with foreign ones under modern market conditions. So it is necessary to develop new unified structures for overhead power lines. The main reasons for that are following: the international norms for the determination of loads caused by climatic effects became more strict in order to make the overhead power lines more reliable; the consideration of the real work values for space support structures in the power line as a united system as well as that of supporting effects by ropes and cables is necessary [5-7].

The potential solution of the aforesaid problem lies in the improvement of power line support structures with absolute conservation of their maintenance characteristics [8-11]. Therefore, the search of a structural solution with consideration of the requirements to the analysis procedures is a very important problem as it is aimed at the conservation of the competitiveness of power lines at the world market due to the saving of metal in the process of the use of non-typical support structures.

[*] Corresponding author: a.v.tan@mail.ru

The study of the works in the field with respect to the optimum design of structures allowed us to formulate the problem which is dictated by both the modern state in the field of optimization of structures and the needs of practical construction.

The design of structures for high-voltage power lines includes the consideration of many various parameters and requirements: power line length, voltage, relief, region and electrical dimensions, land plot cost, material of structural elements, unification requirements, producer potentialities, assortment [12-16].

In the process of design of overhead power line supports, some design parameters are definitely given: the geometrical dimensions between wires, ropes and structural elements as well as the loads (wind velocity head, ice-deposits on the wires, wire break tensions). The other parameters influencing the metal grid supports vary: the support structure form (gantry support, tower support, guyed support), the span and the support height, the geometric dimensions and forms of certain structural elements (frame, beam, wire supports), the grid types, the nod joint types, the section lengths, the range of rolled metal products, the material of the structures. It is difficult to consider all these parameters within the frames of one new method even theoretically, and it is quite impossible practically. Therefore, it is expedient to solve the support optimization problem using different rational methods at different stages of the optimization process.

2 Optimum design of grid structures

The work [5] describes the optimum one-stage automated system for the design and production of power line structures with consideration of design-and-technology requirements of production plants. This system reduced the production time period and improved the quality of products. The system includes programming package for the strength and optimization analysis of wires, ropes and metal support structures; the full-scale modeling; the computer-aided assembly of structures. The suggested numerical optimization method allows us to optimize the geometrical dimensions and topology of power line supports, their mass, labour-intensiveness and cost simultaneously; the mass and cost parameters are by 20% lower. The well-known "TOWER" program for the analysis of overhead power line supports has been developed in the USA and recommended by the International Electrotechnical Committee. The "TOWER" program has many disadvantages in comparison with the programming package suggested by the authors: the wind loading is considered approximately, the analysis of wires and supports is performed separately; a simple sorting out of geometric parameters is performed instead of optimizing them; there is no overhead power line optimization with respect to the power line track. The authors used the developed programming package [17-19] and developed new efficient support structures with narrow bases and with one stand. The aforesaid support structures showed their good qualities in the Crimea Mountains; the portal support structures with flexible links had the mass by 30% lower than the foreign structures.

The problem of the search of optimum parameters of support structures is based on the numerical methods of minimization of non-linear target functions. The use of non-linear programming methods does not give a well defined mathematical solution of the problem, but these methods consider those necessary factors which make the problem closer to practical needs and to get an appreciable economic effect.

With consideration of the aforesaid facts, the problem of optimum design of overhead power line supports may be formulated like that: to compute the geometrical parameters of the tower with various coordinates and topology parameters for given cross-sections with minimum mass or cost and with maintenance of normative requirements and structural restrictions. The varying parameters xj may be continuous or discrete variables.

The expert estimation of possible optimization methods shows that the most efficient methods are the following ones: the method of improved search, the Hadder-Mid method of deformed polyhedron and the direct search methods by Hooke-Jivce and Davis-Svenn-Campy [20-22]. A successive application of these methods allows us to get closer to the optimum variant.

The authors suggest a search method which at the initial stage allows choosing the descent direction with the determination of optimum geometric parameters. The numerical studies showed that the descent direction was determined in the process of the first iterations. For a successful search direction, a polyhedron with an "improved" vertex is formed, and the search continues until the transition step value becomes less that the given precision value. Otherwise, the search process includes a certain number of iterations. The optimum value of the target function for this method was the initial value at the transition to the next optimization method.

The suggested method of initial optimization is very efficient; it allows us to escape from the "ravines" for essential non-linear discrete problems (Fig.1) in contrast to other methods (deformed polyhedron method, Hooke-Jivce method, etc.).

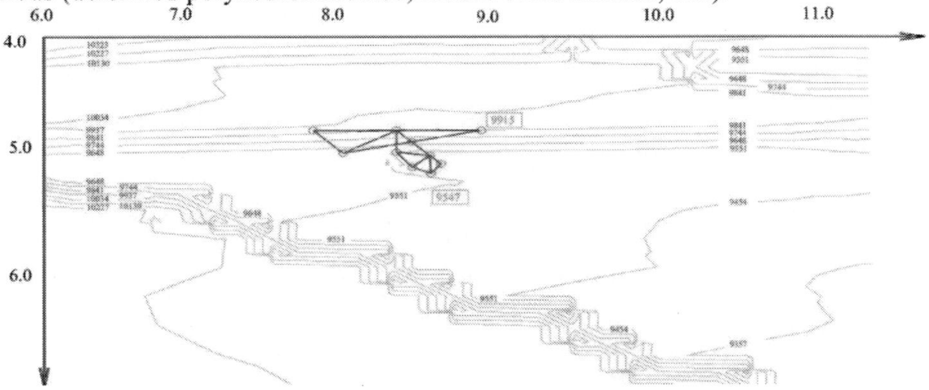

Fig. 1. Efficiency of the search through the coordinate descent method.

As the design optimum criterion, the target function is taken as the mass or the cost of the support structure. The target function is a composite non-linear function of control parameters. The discreteness and essential non-linearity of the target function sets the questions concerning the development of an algorithm providing for the convergence with the minimum mass design.

In the process of solution of the optimization problem for an overhead power line support structure, the initial data are used, the structural and other requirements are considered, the static and optimization analyses are carried out [5,23].

The optimization problem is reduced to the determination of the varying parameter vector (1), which minimizes the target function and satisfies the restrictions (2) – (4) and the conditions (5), (6). The system of restrictions is formed automatically as the test for the target function, the expression of which is a difficult non-linear problem as the number of varying parameters is great [17, 24]

$$Z = f\left(H, L, b, Q, l_t, R_y, K_b, n_p\right) \tag{1}$$

here H, h, b, Q are the given parameters, where H – support height, b –base, L – span, Q –external load vector; the values l_t, R_y, K_b, n_p are the varying parameters of geometric topology, where l_t –panel length, k_b –number of bolts, n_p –grid type.

Write out the main restrictions for the varying parameter vector (1):

- bearing capacity of compressed and tensed structural elements, respectively:

$$N_i \leq A_i \cdot R_{yi} \cdot \varphi_i \qquad \text{and} \qquad N_i \leq A_i \cdot R_{yi}, \tag{2}$$

where N_i –longitudinal force in the i-th element ;

- minimum element area:

$$A_i \geq [A_i], \tag{3}$$

where $[A_i]$ –minimum admissible area of the i – th element with respect to the ultimate flexibility or to the design requirements;

$$J_i \geq \tilde{O}_i \geq j_i, \tag{4}$$

where J_i and j_i-upper and lower limits for the varying parameter range.
The main conditions which the varying parameter vector meets (1):

- equilibrium condition:

$$k(x)\cdot B - P(x) = 0, \tag{5}$$

where k(x) –system rigidity matrix;
B –unit translation vector;
P(x) –external loading vector;

- deformation compatibility condition:

$$\Delta L_{Ni}(N_i, A_i, l_i) = \Delta L_i(\Delta x, \Delta y, \Delta z), \tag{6}$$

where ΔL_{Ni} and ΔL_i–strains of i-th and (i+1) –th elements.

The geometric topology of the power line support is given with consideration of the restrictions (2) –(4) for the variables. These restrictions are directed at the conservation of a certain system form and the correspondence of varying parameters to the design requirements [25-27].

The optimum design algorithm for overhead power line supports is implemented in the "USL" programming package (Fig.2).

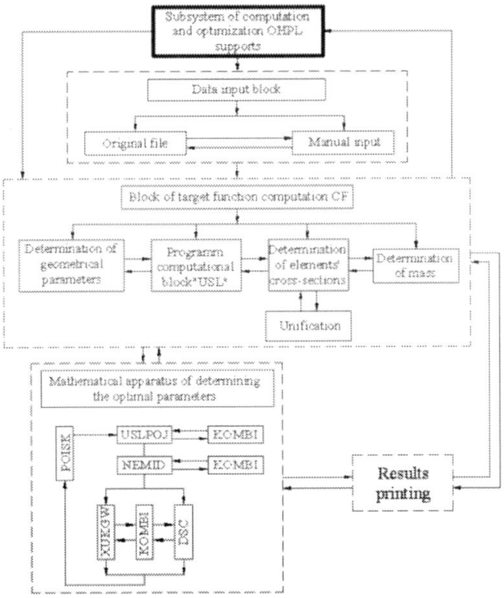

Fig. 2. Principle block diagram of the "USL" program.

The optimization problem has been solved through the numerical methods of non-linear programming which do not use the derivatives. The basic idea of the search is the use of preceding information for the construction of improved solution with the help of iterative procedures. The research work done allowed us to recommend the following methodology of the use of various mathematical methods and their succession. At the first stage, the optimum grid type for given geometric dimensions is determined (for self-supporting supports) with the use of the developed combinatorial algorithm. The value of the target function for the optimum grid type becomes the initial value for the following iterative computations.

Then a regular polyhedron is set which has a form of a regular simplex. The descent direction is determined. For new varying parameter values, the optimum grid type is chosen through the combinatorial algorithm (KOMBI), and the target function value is computed.

Further, a polyhedron is used for the determination of the optimum; one of the vortex of the structure was the optimum one at the precedent stage (USLPOY). The search continues through the deformed polyhedron method (NEMID). Then the optimum grid type is determined through the combinatorial algorithm, and the geometric dimensions of the support are obtained as well.

At the last stage, the efficiency function is optimized through one of the coordinate descent methods. Here the most efficient method is the direct search method by Hooke-Jives (XUKGV) as it gives the "best" (minimum) value of the target function. The iterative search process is completed by the choice of the optimum grid type through the combinatorial algorithm.

The initial design data with the constant values come from the main program (or introduced through the key-board) into the optimization "POISK" program. The target function minimum for the minimum mass structure may be obtained through one of optimization method or by a combination of these methods (Fig.2). the sub-program "CF" for the computation of the target function includes several computation blocks which determine different parameters for each element group: belts, braces, strut frames, diaphragms.

The analysis of optimization results (presented in Table 1) proves the following conclusions: there exists an objective value of the structure mass for the same load values and support dimensions; this mass value may be achieved from different initial design variants with different grid types, diaphragms, nod joints, cross-sections. The grid topology does not influence the end results.

Table 1. Optimization results for the "Y220-2+5" support with different initial states.

Initial support		Optimal support	
Geometric topology	**Diaphragm type**	**Geometric topology**	**Diaphragm type**
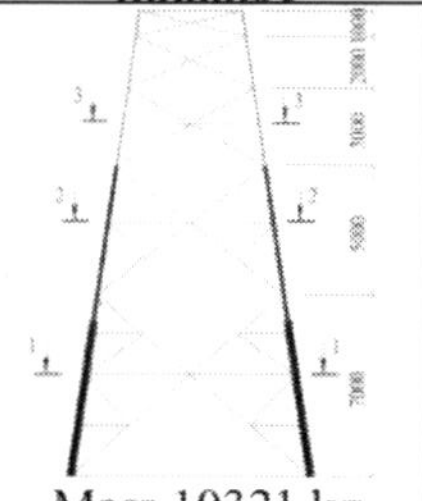 Mass 10321 kg	1-1,2-2,3-3	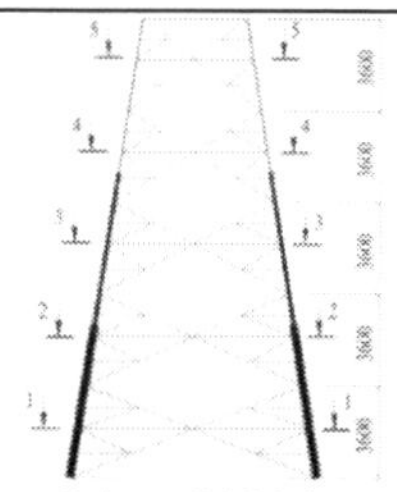 Mass 8429 kg	1-1,2-2,3-3 4-4,5-5
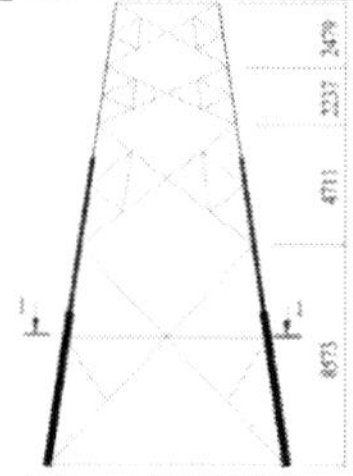 Mass 10197 kg	1-1 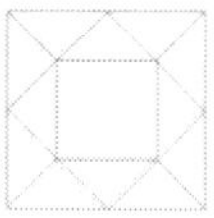	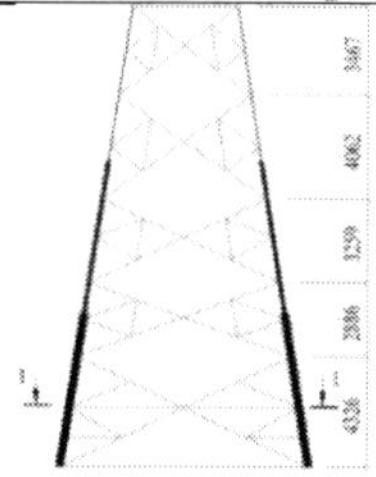 Mass 8233 kg	1-1 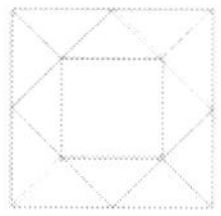

The program may get the information concerning every step of the iterative process of the problem solution for the purpose of its analysis and the correction of some parameters. A graphic representation of the computation process is provided. The optimization results are obtained in the form of tables.

The programming package may be used for the solution of optimum design problems for all types of overhead power line supports; besides, it is possible to add some computation blocks for the solution of other optimization design problems. The optimization computations for various support types for overhead power lines considerably improve the initial design variants. This fact proves the expediency of the aforesaid approach to the support optimization, as this approach gives a rational structure form which meets the optimum criterion. The research works done considered a power line as a united structural system; they took into account the space work of the support structures as well as their technological and economic efficiency.

3 Development of new optimum support structures for overhead power lines with 110 kV voltage

The optimization computation for tower supports (overhead power lines with 110 kV voltage) was carried out through the programming package "USL" [28].

In the process of optimization of anchor support structures, the given parameters were the base and the width of the upper part of the shaft which is determined by the electrical dimensions. The varying parameters were the length and the number of panels, the length of belt sections, the grid type and the shaft diaphragm type, the type of nod mating of struts and belts.

For the development of new supports the following conditions are taken:

-Support material: low-carbon steel C245 (Ry = 240 MP);

-Current-carrying wires (steel-and-aluminum; grade AC 240/32);
-Overhead ground-wire cable C50;
-Wind span lwind=ldim ; weight span [29];
-The climatic conditional are taken in accordance with Chapter 2.5 [30].

The geometric forms of developed optimum anchor supports (overhead power lines with 110 kV voltage) are presented in Fig.3. The comparison of mass values for typical and optimum supports is given in Table 2.

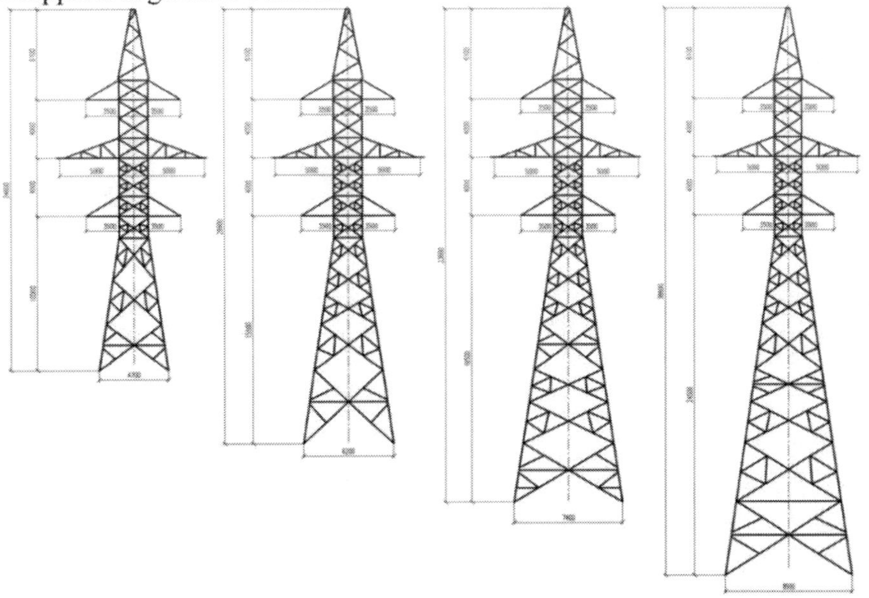

Fig. 3. Geometric forms of optimum anchor supports "Y110-2": a) Y110-2 type support; b) Y110-2+5 type support; c) Y110-2+9 type support; d) Y110-2+14 type support.

The optimization of the "Y110-2+9" support resulted in the mass reduction from 11.391 t to 8.383 t (by 35.9%)/ the belt angle in the structure of the 9m high stand for the optimum support was replaced by another belt angle grade (L 180x11 was replaced by L160x10).

Table 2. Comparison of typical "Y110-2" supports with the optimum ones.

	Support type	**Support mass***		**Saving of material for 1 support**	
		Typical support, KM	**Optimal support , KMD**	**kg**	**%**
1	Y110-2	7704 / 8002	5676 / 5910	2028 / 2092	35.7 / 35.4
2	Y110-2+5	9717 / 10095	7185 / 7481	2532 / 2614	35.2 / 34.9
3	Y110-2+9	11391 / 11834	8383 / 8728	3008 / 3106	35.9 / 35.6
4	Y110-2+14	14643 / 15212	10476 / 10908	4176 / 4304	39.8 / 39.5

*the numerator shows the mass of support without painting; the denominator shows the mass of support with zinc coating.

The results of the research work are introduced into the process of production, erection and maintenance of the "Y110-2+14" optimum support (overhead power line "Kremenching-GPP4"; voltage: 110 kV) at the territory of Poltava mining integrated works, Komsomolsk, Ukraine (Fig.4). The efficiency of the aforesaid support is the mass reduction by 39.5% in comparison with the typical support due to a more precise determination of forces in the space model of the support, the re-computation of the buckling coefficient, the introduction of a composite strut frame grid, the choice of the optimum profile type and the nod mating.

a) b) c) d)

Fig. 4. The optimum two-chain anchor support "Y110-2":The support is being erected; b) The general view of the support; c) The composite strut frame grid of the shaft; d) The nod of fixation of the support on the foundation (shoe).

4 Conclusions

The authors developed a methodology of optimization of parameters of the geometric form and type of the lattice for the structures of overhead power line supports which meets the requirements of construction norms and rules.

The search of minimum target function is performed through a combination of suggested and existing methods of non-linear programming.

The developed method of optimum design of overhead power line supports is universal, and it may be applied to other structures - the supports for wind power units and the overhead supports for radio-relay communication.

The developed system of one-stage automated design of power line structures includes a programming package for optimization computations of wires, ropes, support structures and foundation structures.

The programming package was successfully used for the development of two-chain optimum anchor supports for overhead power lines with 110 kV voltage which are technological with respect to their production and erection and meet the Ukraine climatic conditions. The mass of optimum supports is by 35.2% to 39.8% less as compared with typical "VL 110 kW" supports.

References

1. K.P. Kriukov, B.P. Novgorodtsev, *Constructions and mechanical analysis of power transmission line* (1979)
2. D. Vinogradov, *Construction of overhead power transmission lines 35-500 kV with severe track* (1983)
3. Ye. Gorokhov, S. Shapovalov, Ye. Udod, et al. *Growth of reliability and life of the power supply network construction* (1997)
4. K. Kryukov, *The crossings of transmission lines through of large water areas* (1982)
5. E.V. Shevchenko, *Refinement of steel constructions of power transmission line support* (1999)

6. *IEC 60826. Design criteria of overhead transmission lines* (2003)
7. *EN 50341-1: Overhead electrical lines exceeding AC 45 kV. Part 1: General requirements – Common specifications* (2009)
8. Ya. Nazim, A. Tanasoglo, Met. Const., 49-61 (2015)
9. *Probabilistic design of overhead transmission lines: Companion document to "Improved design criteria of overhead transmission lines based on reliability concepts"* (2000)
10. B.C. Wadell, *Transmission Line Design* (Artech house, Norwood, 2005)
11. H. Li, H. Bai, Prog. in Nat. Sc., 899-911 (2006).
12. Ye. Gorokhov, M. Kazakevich, et al., *Aerodynamics of the power supply structures* (2000)
13. G. Diana, S. Bruni, F. Cheli, F. Fossati, A. Manenti, EACWE, 1759-1766 (1997)
14. M. Gaudry, F. Chore, C. Hardy, E. Ghannoum, *Increasing the ampacity of overhead lines using homogeneous compact conductors* (2011)
15. Y. Chen, M. Farzaneh, E. Lozowski, K. Szilder, IWAIS, 1-8 (2000)
16. A.G. Sokolov, *Support of power transmission lines (calculation and construction)* (1961)
17. Ye. Shevchenko, V. Glukhov, A. Tanasoglo, Metal Structures, 31-39 (2010)
18. Ya. Nazim, Ye. Shevchenko, Mod. Ind. and civ. Const., 83-95 (2014).
19. Ye. Shevchenko, Ya. Nazim, A. Tanasoglo, I. Garanzha, Proc. Eng., 1033-1040 (2015).
20. P. Couneson, J. Lamsoul, D. Delplanque, Proc. CIGRE, 18-76 (2013)
21. G. Arzhakov, *Metal structures. Special buildings and structures* (2002)
22. A. Perelmuter, *Computational models of structures and the possibility of their analysis* (2002)
23. A. Tanasoglo, I. Garanzha, V. Glukhov, Met. Struct., 39-48 (2015)
24. J. Lamsoul, J. Rogier, P. Couneson, A. Van Overmeere, *Belgian experience on initiatives to improve the capability of existing overhead lines* (2000)
25. *DBN B.2.6 – 163:2010 Steel structures. Standards of design, manufacturing and installation* (2011)
26. *Manual for design of steel structures of overhead transmission line supports (OHTL) and power outdoor switchgear (OSG) substations with voltage up to 1kV (to SNiP II-23-81**) (1989)
27. *Design of Latticed Steel Transmission Structures, American Society of Civil Engineers. ANSI/ASCE 10-90, A.N.S.I.* (1991)
28. A. Tanasoglo, Mod. Ind. and civ. Const., 5-14 (2015)
29. M. Gaudry, F. Chore, C. Hardy, Proc. CIGRE, 180-201 (2008)
30. *Rules for electrical installation. Head 2.5 «The overhead power transmission lines voltages above 1 kV to 750 kV»* (2006)

Some prospects for the use of ash and slag wastes of coal power plants for production of building materials

Fedor Bryukhan[1,*]

[1]Moscow State University of Civil Engineering, Department of Engineering Survey and Geoecology, Yaroslavskoe sh. 26, 129337 Moscow, Russia

Abstract. Coal-fired thermal power plants (TPP) generate large amounts of ash and slag wastes (ASW), therefore ecological problems related to ASW storage and disposal for such power plants come to the fore. To obtain new data on ASW properties analysis of a number of ASW samples from ash dump No. 4 of Cherepet' TPP was carried out as part of engineering surveys. The purpose of this study included determiantion of the ASW granulometrical and mineralogical composition, as well as their physical and chemical properties. During field works 4 ASW samples were taken from the surface layer of the ash dump plus one sample of ash froth for further laboratory analysis. Laboratory analysis enabled determination of the granulometrical, mineralogical and chemical composition of ASW. Assessment of the ASW corrosiveness toward various materials determined necessary limitations in utilization of ASW-based bulding materiasls in contact with steel and aluminum. The content of natural radionuclides of K^{40}, Ra^{226}, Th^{232} and man-made Cs^{137} in ASW was determined. The detected increased activity of Ra^{226} and Th^{232} necessitates additional studies of radon and thoron emissions from ASW. Microspheres recovered from ash froth during ASW utilization have special usage value.

1 Introduction

It is known that coal-fired thermal power plants (TPPs) produce large amounts of ash and slag wastes (ASW). Coal-fired TPP in Russia produce about 30 Mt of ASW per year, not more than 10 % of which are recycled [1, 2]. In this regard, ecological problems related to ASW storage and disposal for such power plants have top priority [1, 3, 4].

ASW have many properties typical for different kinds of natural mineral raw materials. Therefore, they have an undeniable resource potential for production of various building materials [1, 5, 6]. In many cases, wastes with similar properties can be interpreted as an equivalent of natural raw materials with consumer properties, and this is also true for ASW. That is why along with the term «ASW» we can also use another definition – «ash and slag materials». Considering large amounts of ASW, there is a range of promising directions of their utilization [1, 4, 7], including production of cement, concrete, cinder-concrete blocks,

* Corresponding author: pniiis-gip@mail.ru

bricks, microspheres, road construction. Implementation of the above methods of ASW usage will enable conservation of land resources, used for slag disposal areas, saving of natural raw materials and improvement of ecological conditions.

To obtain new data on ASW properties analysis of ASW samples from ash dump No. 4 of Cherepet' TPP was carried out as part of engineering surveys carried out for increasing the ash dump capactiy [8]. The purpose of this study included determiantion of the ASW granulometrical and mineralogical composition, as well as their physical and chemical properties.

2 Sample collection and laboratory analysis

The Cherepet' TPP is located in the Tula region, on the Cherepet' River, north-west of Suvorov. Its current installed output power, generated at 8 power units, is 1735 MWt. The power plant works on the Kuznetsk basin coal. The ash and slag handling system is hydraulically operated. The surface area of the ash and slag dump is about 150 ha.

The map of the ash dump and sampling points are shown in Figure 1. During the engineering survey, 4 ASW samples were taken from the surface layer of the ash dump (0-10 cm) plus one sample of ash froth for further laboratory analysis.

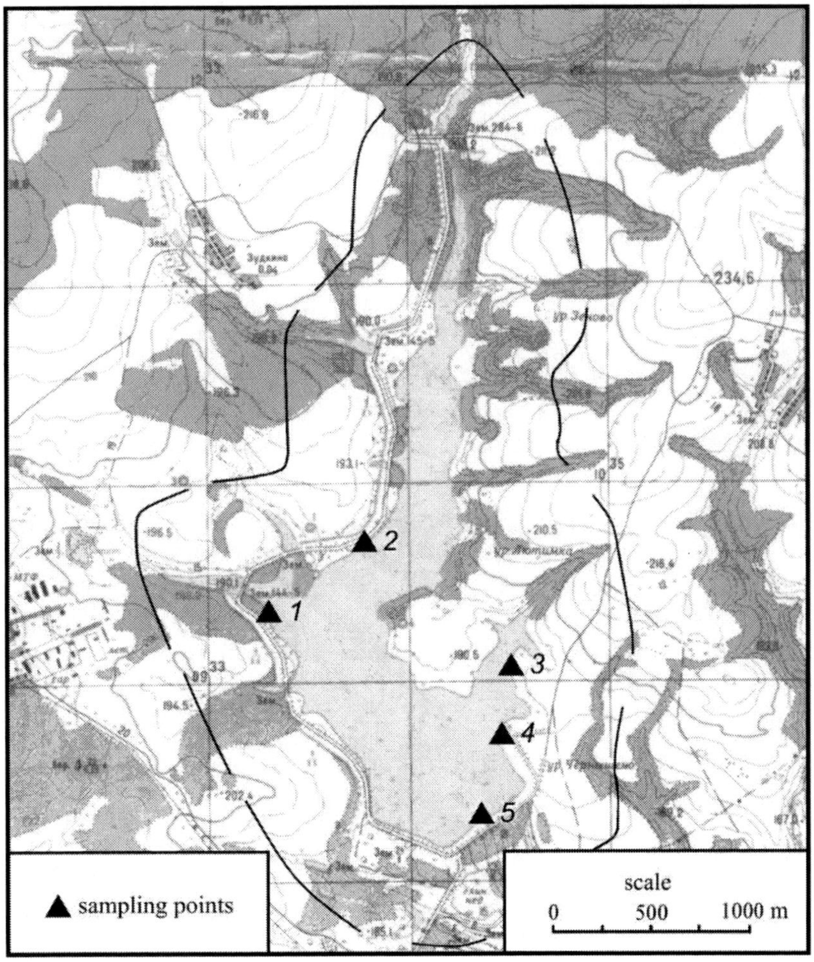

Fig. 1. Locations of ash (1-4) and ash froth (5) sampling points. The dotted line shows the sanitary protection zone of the ash dump.

The following studies were performed during the laboratory analysis:
- granulometric analysis;
- X-ray phase analysis;
- analysis of ASW corrosion activity toward different materials;
- chemical composition analysis;
- radionuclide composition analysis.

Laboratory studies were conducted in the Production and Research Institute for Engineering Survey in Construction, the All-Russian Research Institute of Mineral Resources, the Russian State Geological Prospecting University, the Center of State Sanitary and Epidemiological Surveillance in Moscow.

3 Results and discussion

3.1 Granulometric composition

Average density of ash particles was 2.35 g/cm^3. Their statistical size distribution according to laboratory analysis of 4 samples is shown in Table 1. It is found that ash depositions belong to the dust sand category. According to the Table 1, the share of large particles above 50 μm is 52.8 %, mid-size particles of 5-50 μm – 46.2 %, fine particles (less than 5 μm) – 1.0 %. In ash dump areas not covered by water, medium-size and fine particles (less than 50 μm) are blown off from the ash dump surface.

Table 1. Size distribution of ash particles.

Size, μm	< 5	5-10	10-50	50-100	100-250	250-500
Distribution, %	1.0	8.9	37.3	20.0	7.5	2.5

Size, μm	500-1000	1000-2000	2000-5000	5000-10000	> 10000
Distribution, %	1.1	3.7	9.3	6.2	2.5

3.2 Mineralogical composition

Samples from the ash dump passed the X-ray phase analysis on DRON-3 diffractometer. The following minerals were found in the samples: quartz, hematite, mullite and glass, demonstrating halos on X-ray images. Quartz is present in Kuzbas coals and appears as a detritus mineral. Mullite, hematite and glass occur in the process of coal combustion. A light fraction in the form of specific hollow microspheres of up to 500 μm prevails in ash froth.

3.3 Chemical properties and corrosion activity

Based on ASW chemical analysis, it was found that they contain a large amount of Pb, moderate quantities of Mn and B and fractional amounts of other chemical elements. The content of active forms of Zn, Co, Cu, Ni, Cr is considerably below the maximum permissible values. It is important to emphasize that concentrations of benzo[a]pyrene in the collected samples are very low.

In terms of perspectiveness of ASW application in manufacture of building materials, it is important to determine their corrosion activity toward materials that will be in contact with ASW-based building materials. In this regard, relevant laboratory studies were performed in order to determine the degree of ASW corrosiveness toward different materials. Corrosiveness analysis results are show in Table 2.

Table 2. ASW corrosiveness toward different materials. Water extract mineralization is 240 mg/dm^3, pH 8.6.

Materials in contact with ASW	Corrosiveness level
steel	medium, high
armature of concrete elements	low, medium
lead	low, medium
aluminum	medium, high
concrete	non-corrosive

Thus, the findings indicate the need for limitations in utilization of ASW-based bulding materiasls in contact with steel and aluminum.

3.4 Radionuclide composition

The ash samples were also analyzed for the content of natural radionuclides K^{40}, Ra^{226}, Th^{232} and man-made Cs^{137}. Laboratory analysis have not detected any radioactive contamination in the ASW samples, but found relatively high activity of radionuclides Ra^{226} and Th^{232}. The expert opinion of the Center of State Sanitary and Epidemiological Surveillance in Moscow states that in case of ash utilization in production of building materials it is necessary to assess radon and thoron emissions from ASW.

3.5 Microspheres

As stated above, a light fraction in the form of small hollow microspheres of up to 500 µm prevails in the ash froth. They are relatively rare in the ash – only a few percent. The microspheres are characterized by a range of useful properties contributing to their wide application. Particularly, their dry density is very low – about 130 kg/m^3 at an average. They serve as the basis for low-cost fireproof bricks, heat and sound insulating materials, foam materials. Building materials manufactured using the microspheres also have resistance to vapor penetration, screening capacity against ionizing radiation effects, a high degree of adhesion with many materials (concrete, brick, timber, etc.).

4 Conclusions

• Granulometric analysis of ASW samples showed that the average content of large particles over 50 µm is 52.8 %, mid-size particles of 5-50 µm – 46.2 %, fine particles (below 5 µm) – 1.0 %. Quartz, hematite, mullite and glass particles were detected in the samples.

• According to ASW chemical analysis, they contain a large amount of Pb, moderate quantities of Mn and B and fractional amounts of other chemical elements. The content of active forms of Zn, Co, Cu, Ni, Cr is considerably below the maximum permissible values. The content of benzo[a]pyrene in the ash and ash froth is negligible.

• ASW corrosiveness toward different materials in contact with ASW was analyzed. The assessment showed that the degree of ASW corrosiveness toward the armature of concrete structures and lead is low and medium, steel and aluminum – medium and high. At the same time, ASW are not corrosive to concrete.

• Laboratory tests of ASW samples detected no radioactive contamination of ASW. However, they revealed a relatively increased activity of Ra^{226} и Th^{232}. This necessitates conduction of assessments of radon and thoron emissions from ASW, if they are used in manufacture of building materials.

• Microspheres produced in the process of coal combustion have very useful properties that enable their usage in production of building materials.

• The findings indicate a good outlook and advisability of large-scale utilization of ASW.

References

1. Yu.K. Tselykovsky, New in the Russian Power, **4**, 15 42 (2002)
2. Yu.K. Tselykovsky, Energy: Economy, Technology, Environment **4**, 27 34 (2006)
3. V.Ya. Putilov, Energetik, **4**, 37 40 (1999)
4. V.Ya. Putilov, *Ecology of Energy* (National Research University «MPEI», Moscow, 2003)
5. A.I. Abramov, D.P. Elizarov, A.N. Remezov et. al., *Increasing of TPP environmental safety* (National Research University «MPEI», Moscow, 2001)
6. V. Knat'ko, M. Knat'ko, E. Scherbakova, Ecol. and Sust. Devel. **2**, 30 32 (2004)
7. M. Hanzelova, L. Florekova, B. Stehlikova et al., Proc. of the 6-th Conf. on Envir. and Min.l Proc., Ostrava, 431-436 (2002)
8. A.F. Bryukhan, F.F. Bryukhan, A.D. Potapov, Engineering and environmental survey for the construction of thermal power plants (ASV Publishing House, Moscow, 2010)

Consideration of hazardous and especially hazardous hydrometeorological impacts in design of buildings and structures of nuclear power plants

Fedor Bryukhan[1,*]

[1]Moscow State University of Civil Engineering, Department of Engineering Survey and Geoecology, Yaroslavskoe sh 26, 129337 Moscow, Russia

Abstract. External impacts of the hydrometeorological origin have a significant influence on the safety level of objects of use of atomic energy (OUAE), including nuclear power plants (NPP). Therefore, the existing NPP-related safety regulations demand to consider such impacts at all stages of the NPP life cycle. It is important to make decisions on considering or ignoring certain external impacts while designing NPP buildings and structures. The main criterion for such decisions is the probability of a non-project accident associated with the release of radionuclides into the environment when an extreme phenomena occurs. The aim of this study is to develop a concept for refinement regulatory requirements, considering hydrometeorological factors in organization of NPP engineering protection. Criteria for consideration of hazardous and especially hazardous hydrometeorological impacts for design of NPP buildings and structures were analyzed, and recommendations for refinement of regulatory requirements, considering hydrometeorological factors in organization of NPP engineering protection, were developed.

1 Introduction

External impacts of the hydrometeorological origin have a significant influence on the safety level of nuclear facilities, including NPP [1, 2]. International and Russian OUAE safety codes and regulations state that such impacts shall be considered at all stages of their life cycle [3, 4].

These codes and regulations generally require to consider a wide range of hydrological and meteorological phenomena and processes. However, they do not provide clear criteria determining the need for consideration of relevant hydrological and meteorological loads in calculations. So it is necessary to refine regulatory requirements for consideration of hydrometeorological factors in provision of NPP engineering protection.

* Corresponding author: pniiis-gip@mail.ru

2 Hydrological and meteorological phenomena and processes

The abovementioned normative technical documents [3, 4] require consideration of external hydrometeorological factors listed in Table 1, which also indicates general impacts of such factors on NPP buildings and structures.

Table 1. External hydrometeorological factors and their general impacts.

Factors	General impacts
Flood	Flooding of the NPP site, dynamic impacts on buildings, structures, utilities, foundation subsidence
Tsunami	Flooding of the NPP site, dynamic impacts of tsunami waves on buildings, structures, foundation subsidence, temporary fall in water level within the coastal area
Ice phenomena on water courses (ice jams, gorges)	Flooding of coastal territories due to water rise in water courses, dynamic impacts on hydraulic engineering structures
Coastal wave conditions (run-down, run-up, hurricane wave)	Flooding of the NPP site, dynamic impacts on buildings, structures, utilities, corrosion effects
Seiche	Flooding of coastal territories, temporary fall in water level in water bodies
Tidal surges and ebbs	Intermittent flooding of coastal territories, fall in water level within the coastal area
Changes in water resources (extremely low water yield, abnormal fall in water level)	Shortage of water for NPP cooling systems
Tornado	Dynamic impacts of tornado air flow and pressure drop on buildings and structures, impacts from flying objects
Tropical cyclone (typhoon)	Wind loads on buildings and structures, stormwater loads on roofs of buildings and structures, hail impacts, flooding of the NPP site
Extreme wind	Wind loads on buildings and structures
Extreme precipitation	Stormwater loads on roofs of buildings and structures, hail impacts
Extreme snowfalls and snow cover	Static loads on roofs of buildings and structures
Extreme air temperature	Temperature loads on members of buildings and structures, impacts on communication lines

Table 1 (continuation).

Factors	General impacts
Snow avalanche	Shock action, the NPP site buried under snow
Glaze ice	Ice loads on wires, wire vibration, kinking, swinging
Lightning discharge	Failure of networks, equipment, fire hazard, injurious action on the NPP personnel

According to the table, potential impacts of the listed factors are quite diverse. However, the mentioned extreme hydrometeorological phenomena and processes can cause a non-project NPP accident only in extremely rare occasions. Taking into account the importance of NPP radiation safety assurance, even such cases require detail analysis of possible consequences of a hypothetical accident.

3 Probabilistic safety criteria for NPP

Decision on consideration or ignoring of certain hydrometeorological factors according to [4] shall be based on the threshold probability P_0 (basically, recurrence), which is 10^{-4}/year per reactor. However, taking into account high dangers of tornados and their impacts on NPPs, their threshold probability is assumed equal to 10^{-7}/year per reactor [5, 6]. Lightning impacts on NPP buildings, structures and infrastructure also have specific features. In many instances, parameters of their impacts considered by design shall be determined at the expert level.

Collection and analysis of input data for statistical processing and determination of design hydrometeorological characteristics are carried in the process of engineering hydrometeorological surveys [7]. Statistical data processing is performed using standard methods (e. g., refer to [8, 9]).

In Russia and other IAEA member counties, NPP safety requirements are based on non-exceedance of the non-project accident probability with the accidental release limit of $P_G = 10^{-7}$/year [4]. This value is the main criterion of NPP safety in relation to external impacts of different origin.

The non-project accident probability caused by a specific hydrometeorological factor is

$$P = P_0 \times P_A \tag{1}$$

where P_A – dimensionless probability of an accident resulting in maximum permissible release into the environment upon occurrence of the impact event for this factor. With $P_A = 1$ for tornado impacts and $P_A > 10^{-3}$ for impacts of other hydrometeorological factors, the non-project accident probability will reach or exceed the main safety criteria of $P_G = 10^{-7}$/year. In case of tornado impacts the non-project accident with radionuclides release is not always inevitable. This is due to the fact that according to the tornado zoning scheme of the ex-USSR territory the estimated characteristics of the maximum probable tornado with the NPP impact probability of 10^{-7}/year can be relatively weak [5, 6, 10].

As for impacts of other hydrometeorological factors, their consideration shall be formally based not on the occurrence of the threshold probability of $P_0 = 10^{-4}$/year, as it is provided for by federal regulations [4], but on the following condition:

$$P_0 \times P_A > P_G \tag{2}$$

To ensure appropriate NPP protection against external impacts, including those of the hydrometeorological origin, engineering measures shall be implemented for NPP protection.

If the threshold probability of $P_0 = 10^{-4}$/year is still considered, the NPP engineering protection is required, when the non-project accident probability resulting from the impact of a specific hydrometeorological factor P_A exceeds 10^{-3}. In its turn, the maximum estimated value of a relevant hydrological or meteorological factor, e.g. water level or wind speed, is the probability level quantile of $1 - 10^{-4}$/year and is determined basing on the statistical distribution of this value. In addition, this greatly varying quantile will by no means always result in a non-project accident with a probability above 10^{-3}. Otherwise stated, the potential hazard from the impact of such meteorological factor can be overestimated and cause excessive material costs for NPP engineering protection.

To avoid this, it is necessary to proceed not from the assumption of the threshold level of $P_0 = 10^{-4}$/year, but from the following condition:

$$P_0(x_C) \times P_A(x_C) = P_G \tag{3}$$

In this formula x_C is the estimated extreme value of the variable x (probability level quantile of $1 - P_0 \times$year). Dependence of the non-project accident probability $P_A(x_C)$ on the impact of a given hydrological or meteorological factor x is derived from construction calculations. The cumulative distribution function $1 - P_0(x) \times$year of the variable concerned is determined based on results of statistical processing of observation data. Therefore, it is generally not difficult to determine values of x_C , P_0 and P_A. It may turn out that the quotation (3) will have no solution, if the variation range of the hydrological or meteorological variable x does not result in a potential non-project accident with a probability above 10^{-3}. This means that in some cases the danger of occurrence of a certain hydrological or meteorological event may be overestimated.

The engineering protection of NPPs from hydrometeorological factors provides for significant expenditures for new construction of each object as well as for its expansion or reconstruction. In some cases there will not be necessary to design expensive protection from some factors without reduction of the NPPs safety.

In this regard, it seems advisable to refine regulatory requirements to consideration of hydrometeorological factors, as well as other factors of the natural origin in designing a NPP.

4 Conclusions

- Based on the applicable NPP safety codes and regulations, analysis of consideration criteria for hazardous and especially hazardous hydrometeorological impacts in NPP design.
- Recommendations for refinement of regulatory requirements were developed to consider hydrometeorological factors in organization of NPP engineering protection against such factors.
- It is noted that there may be situations when there is no need to design the engineering protection from some factors without reduction of the NPP safety.

The author thanks Dr A.Yu. Vinogradov, Director General of the Scientific & Industrial Association Gidrotekhproekt for useful consultations.

References

1. I.V. Kaliberda, *Evaluation of the parameters of external actions of natural and technogenic origin: Safety of the objects of use of atomic energy* (Logos Publishing House, Moscow, 2002)
2. A.N. Birbraer, A.Yu. Roleder, *Extreme actions on structures* (St. Petersburg Polytechnic University Publishing House, St. Petersburg, 2009)
3. *Meteorological and hydrological hazards in site evaluation for nuclear installations: Specific safety guide. IAEA safety standards. No SSG-18* (IAEA Publishing House, Vienna, 2011)
4. *NP-064–05. Taking into account of actions of natural and technogenic origin on nuclear power plants* (Rostekhnadzor Publishing House, Moscow, 2005)
5. F.F. Bryukhan, A.D. Potapov, Atomic Energy **115**, 346-350 (2014)
6. RB-022–01, Vestnik Gosatomnadzora Rossii **1**, 59-90 (2002)
7. *SP 11-103-97. Engineering and hydrometeorological surveys for construction.* (Gosstroy of Russia Publishing House, Moscow, 1997)
8. V.N. Malinin. *Statistical methods for the analysis of hydrometeorological information* (Russian State Hydrometeorological University Publishing House, St. Petersburg, 2008)
9. D.S. Wilks, *Statistical Methods in the atmospheric sciences. 2-nd ed* (Academic Press, New York, 2002)
10. F.F. Bryukhan, A.D. Potapov, Bulletin of the Moscow State Regional University, series «Natural sciences» **4**, 48-51 (2014)

Thermal oxidative degradation of wood modified with aminophenylborates

Olga Klyachenkova[1], *Irina* Stepina[1,*]

[1]Moscow State University of Civil Engineering, Department of General Chemistry, Yaroslavskoye sh. 26, 129337 Moscow, Russia.

Abstract. Comparative thermal analysis in the presence of oxygen was carried out for samples of native pine wood and wood samples modified with aminophenylborates. Significant decrease in the amount of heat released during thermal decomposition of the modified samples was established, which is due to the increase of carbonaceous residues on the surface. Reduction of heat release during decomposition of the modified samples may be explained by the lower yield of combustible volatile products as well as by thin film of boron oxide, formed on the surface of the modified wood, that partially reflects heat flow. Produced upon the modifier decomposition water vapor and inert nitrogen oxides dilute gaseous mixture near the wood surface and isolate it from oxygen. This enhances fire-resistance of wood modified with mono- and diethanolamine(N→B)phenylborates. Hydroxyl group at the sixth carbon atom of the glucopyranose ring of cellulose participates in reactions of cellulose modification, which prevents formation of flammable levoglucosan and, consequently, improves the fire-resistance of the modified wood.

1 Introduction

It has been shown in the previous studies that treatment of the wood surface with protective compositions based on aminophenylborates and water [1] ensures wood fire-retardant efficiency of group II (GOST R 53295-2009) [2]. High activity level of such modifiers should be noted – no more than 10% concentration of the modifiers provide the same protection level as 50% concentrations of aminoborate-based modifiers. There is obviously an effect of phenyl radical, present in the modifier composition, on fire-retardant efficiency of the modifier. It has earlier been found [3, 4] that polymers with a high content of phenyl groups are nonflammable.

Wood modification was performed by soaking pine saw dust in aqueous solutions of mono- and diethanolamine(N→B)phenylborates: 10% solution of monoethanolamine-(N→B)phenylborate (composition 1), 5% solution of monoethanolamine-(N→B)phenylborate (composition 2) and 5% solution of diethanolamine-(N→B)phenylborate (composition 3), under constant stirring for three hours, followed by

* Corresponding author: StepinaIV@mgsu.ru

drying out to constant weight in the air. Thermal analysis was carried out after prolonged extraction with water and drying out. To analyze the process of thermal oxidative degradation of the modified wood, differential scanning calorimetry (DSC) and thermogravimetric analysis (TGA) were carried out using SDT Q600 thermal analyzer in dynamic mode of heating in the air (flow rate of 100 ml/min) with a heating rate of 5, 10 and 15°C/min. Thermal analyzer measures heat flow and weight changes accompanying energy transitions and reactions taking place in materials.

2. Results and discussion

Simultaneous calorimetry and thermogravimetric analysis of the same sample on the same device provides enhanced analysis productivity and allows to reduce experimental error. Plotting and processing of thermoanalytical curves were performed using TA Universal Analysis 2000 software. Temperature ranges of wood degradation, weight loss within temperature ranges defined (%), and heat flow maxima (W/g) were found based on the obtained thermoanalytical curves.

Three temperature ranges were defined for thermal degradation of wood in the air (table 1, figures 1-4). Evaporation of surface-adsorbed and capillary condensed water occurs in the first temperature interval (30-150°C) [5]. Weight loss of pine wood samples is insignificant, approximately 4 - 7% (table 1). At this stage, chemical structure of wood remains practically unaffected and volatile products are not produced. Upon wood modification with mono- and diethanolamine(N→B)phenylborates, the wood surface polarity grows, leading to formation of hydrogen bonds between polar amino and hydroxyl groups of the modifiers and water molecules and, consequently, stronger retention of water molecules on the surface of wood. In the first temperature interval at heating rates of 10 and 15°C/min, lower weight loss values are characteristic to the wood samples modified with composition 3 as compared to the unmodified wood. This trend is also observed for the samples modified with composition 1 at 10°C/min heating rate (figures 1-4, table 1). This may be explained by higher energy costs required for breaking of hydrogen bonds [6].

However, weight loss of all modified samples at 5°C/min heating rate and weight loss of wood samples modified with composition 2 at all three heating rates exceed weight loss of native wood samples (figures 1-4, table 1). In this case, presumably, part of energy is expended on wood polysaccharides dehydration, and protolytic -OH-groups on the surface of the modified wood catalyze this process.

In the second temperature range (150-430°C), wood thermal degradation accompanied by formation of low molecular weight products and carbonaceous residues takes place. Reactions proceeding upon heating of wood to 270-280°C are endothermic and require energy. Upon further heating, endothermic process proceeds to exothermic process [7] characterized by reactions of thermal degradation of wood polymers, large amount of heat released, formation of liquid and gaseous decomposition products (carbon oxides, other non-condensable gases, acetic acid, methanol, tars). Charcoal is formed as a remainder of wood thermal degradation [8].

DOI: 10.1051/matecconf/20168604006

Table 1. Characteristics of TGA and DSC curves.

	Heating rate, °C/min											
	native wood			**wood + composition 1**			**wood + composition 2**			**wood + composition 3**		
	5	**10**	**15**	**5**	**10**	**15**	**5**	**10**	**15**	**5**	**10**	**15**
Temperature range of degradation, °C	30-150	30-150	30-150	30-150	30-150	30-150	30-150	30-150	30-150	30-150	30-150	30-150
Weight loss in the range, %	5.98	6.16	5.78	7.82	3.62	6.60	6.52	6.54	5.92	6.30	4.44	4.91
Temperature range of degradation, °C	150-380	150-380	150-400	150-410	150-420	150-430	150-410	150-420	150-430	150-400	150-420	150-430
Weight loss in the range, %	69.91	70.49	72.71	59.34	60.93	59.16	62.45	61.60	61.99	63.17	65.16	65.59
Heat flow maximum, W/g	4.31	8.52	10.15	2.21	3.94	4.94	2.67	4.65	6.07	3.11	5.57	7.53
Temperature range of degradation, °C	380-470	380-470	400-480	410-650	420-670	430-700	410-650	420-670	430-700	400-650	420-650	430-650
Weight loss in the range, %	23.73	23.62	21.48	24.13	24.80	23.78	25.23	25.06	24.94	28.07	26.83	25.11
Heat flow maximum, W/g	10.24	12.23	13.24	0.74	1.70	2.83	0.81	2.92	3.85	1.15	2.99	4.61

DOI: 10.1051/matecconf/20168604006

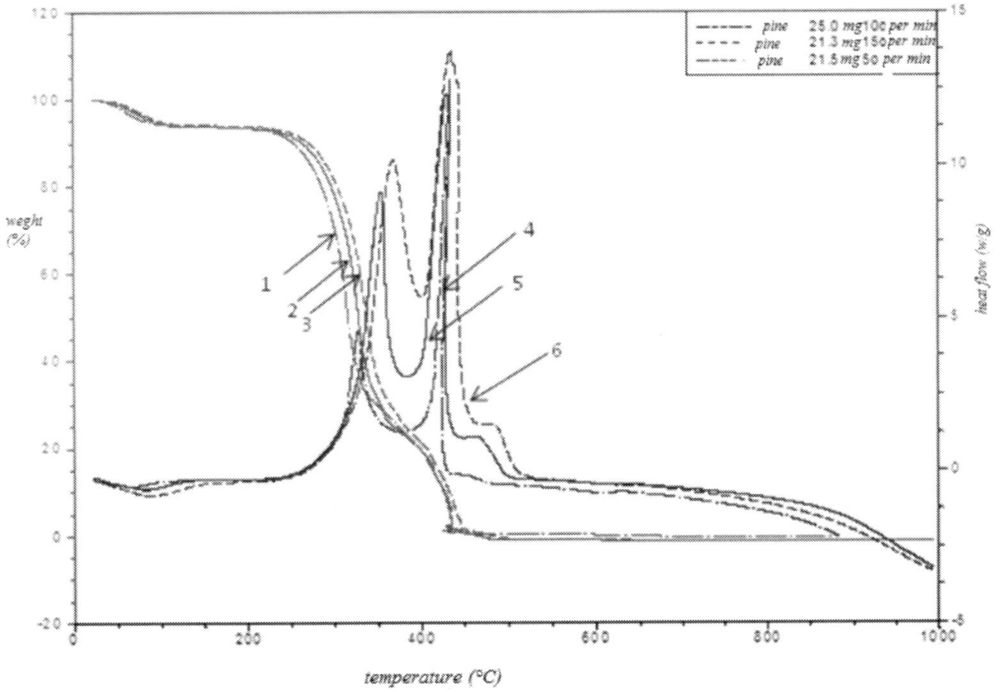

Fig. 1. TGA (1, 2, 3) and DSC (4, 5, 6) curves for the native wood samples (1, 4 – heating rate 5°C/min; 2, 5 – heating rate 10°C/min; 3, 6 – heating rate 15°C/min).

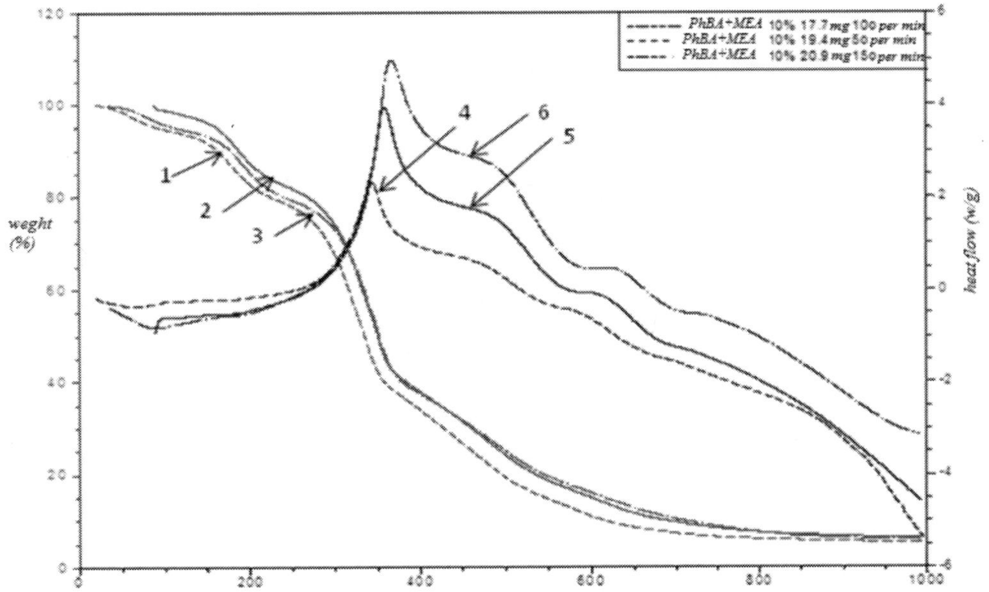

Fig. 2. TGA (1, 2, 3) and DSC (4, 5, 6) curves for the wood samples modified with composition 1 (1, 4 – heating rate 5°C/min; 2, 5 – heating rate 10°C/min; 3, 6 – heating rate 15°C/min).

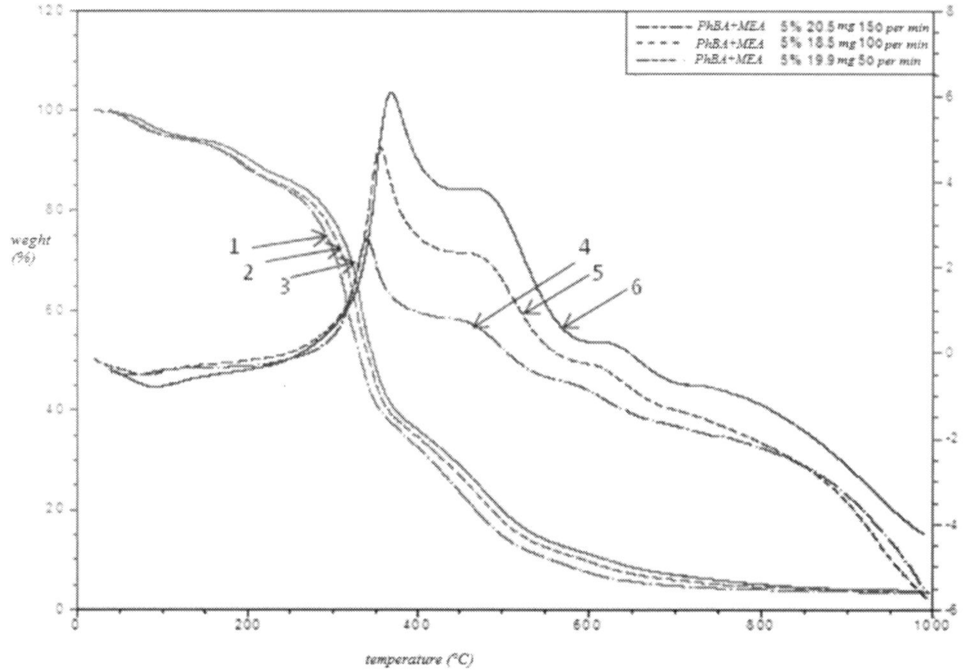

Fig. 3. TGA (1, 2, 3) and DSC (4, 5, 6) curves for the wood samples modified with composition 2 (1, 4 – heating rate 5°C/min; 2, 5 – heating rate 10°C/min; 3, 6 – heating rate 15°C/min).

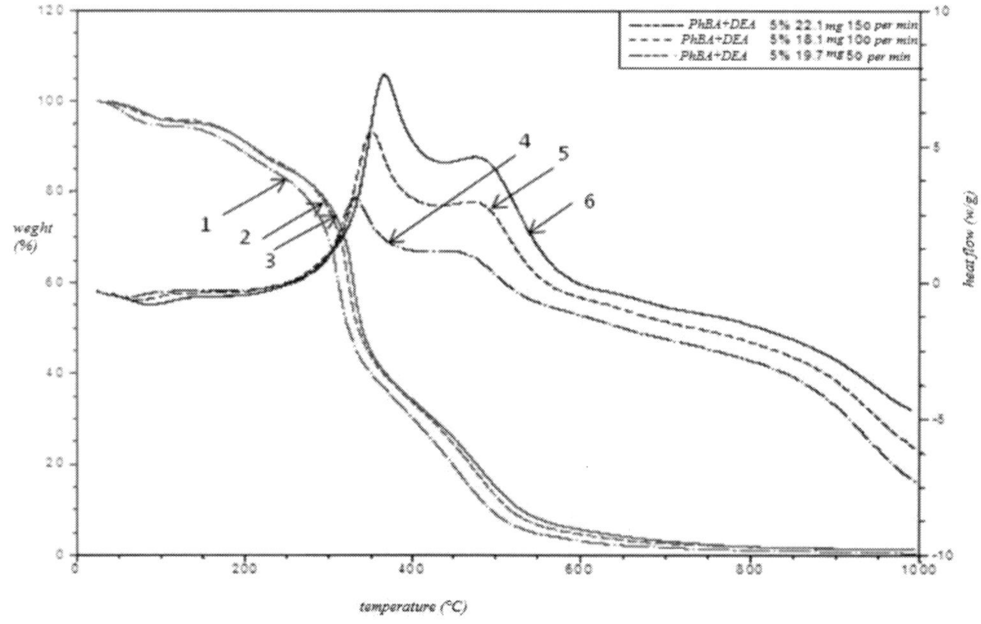

Fig.4. TGA (1, 2, 3) and DSC (4, 5, 6) curves of the wood samples modified with composition 3 (1, 4 – heating rate 5°C/min; 2, 5 – heating rate 10°C/min; 3, 6 – heating rate 15°C/min).

The second temperature range accounts for weight loss of approximately 70-73% for the unmodified wood samples. Heating rate increase causes increase of the weight loss value for the unmodified wood samples and the wood samples modified with composition 3 (table 1). Noteworthy, weight loss in this temperature range is notably reduced for wood samples modified with solutions of mono- and diethanolamine(N→B)phenylborates compared to the unmodified sample: 5-7% for the samples modified with composition 3, 7-10 % for the samples modified with composition 2, and 10-13% for the samples modified with composition 1 (table 1). Modification with solutions of mono- and diethanolamine(N→B)phenylborates changes the scenario of wood thermal decomposition. For the modified samples, heat release temperature maxima of DSC curves corresponding to exothermic reactions of thermal degradation of wood polymers and to production of low molecular weight compounds are shifted to higher temperatures area. This impedes the degradation process at the main stage of wood thermal decomposition and enhances fire-resistance of wood (figures 1-4) [9]. Wood modification with solutions of mono- and diethanolamine(N→B)phenylborates also leads to 1.5-2-fold decrease of heat released at the main stage of thermal oxidative degradation of wood (table 1, figures 1–4). The lowest heat flow maximum is observed upon wood modification with composition 1 at the highest concentration of the modifier (figure 2). Reduced heat release during decomposition of the modified samples may be explained by the lower yield of combustible volatile products [10] as well as by thin film of boron oxide, formed on the surface of the modified wood, that partially reflects heat flow [11]. Formed upon a modifier decomposition water vapor and inert nitrogen oxides dilute gas mixture near the wood surface and isolate it from oxygen [11]. During thermal depolymerization of cellulose, flammable levoglucosan (1,6-anhydro-β-D-glucopyranose) is formed as an intermediate product, at a yield of about 10-13% [8]. Hydroxyl group at the sixth carbon atom of the cellulose glucopyranose ring participates in modification reactions of cellulose, which prevents the formation of levoglucosan and, consequently, enhances fire-resistance of the modified wood [8].

Oxidation of carbonaceous residues takes place in the third temperature range (380-700°C). Modification of wood samples with compositions containing mono- and diethanolamine(N→B)phenylborates leads to significant changes of DSC curve. In contrast to the unmodified samples, heat flow gradually declines with temperature increase during transition from exothermic processes of thermal degradation of wood polymers to oxidation of carbonaceous residues (figures 1-4). Small peak is observed in the DSC curve for the samples modified with composition 3, at that amount of heat is reduced almost 3 folds at a heating rate of 15°C/min and 9 folds at a heating rate of 5°C/min. It should be noted that heat flow increases when heating rate increases both for the control and for the modified samples (table 1). Moreover at this stage, heat release maxima for the samples modified with composition 3 are shifted to higher temperatures as compared to the native wood sample. In this case, shift of heat release maxima is 10, 40 and 30°C at heating rates of 5, 10 and 15°C/min, respectively (figures 1, 4). This points to higher thermal stability of carbonaceous residues resulting from thermal degradation of the modified wood. Greater decrease of heat effect of the process in the third temperature range is characteristic to the samples modified with compositions 2 and 3 (figures 2 and 3), there are no peaks in DSC curves in the range of carbonaceous residues oxidation.

3 Conclusion

Significant heat reduction is due to increase of carbonaceous residues on the surface of the modified samples, which provides thermal insulation and partially reflects heat flow as well as prevents the formation of volatile wood combustion products. It enhances fire-resistance of wood modified with mono- and diethanolamine(N→B)phenylborates.

References

1. I.V. Stepina, O.A. Kolesnikova, *Flame-retardant composition APhB Patent RUS 2552691 22.04.2014* (2014)
2. I.V. Stepina, O.A. Klyachenkova, Nauch. oboz. **12**, 79-82 (2013)
3. V.I. Kodolov, *Flame retardants for polymer materials* (Khimia, Moscow, 1980)
4. Yu.A. Mihailin, *Thermostable polymers and polymeric materials* (Professiya, Saint-Petersburg, 2006)
5. Ye.N. Pokrovskaya, I.I. Pishchik, N.V. Smirnov, Yu.K. Naganovskiy, Stroit. Mat. **9**, 34-35 (2000)
6. V.I. Kondrat'yev, A.N. Zakazov, N.A. Krasikova, E.I. Chupka, Khimiya drevesiny **5**, 90-94 (1987)
7. R.Sh. Yenaleyev, F.M. Gimranov, A.V. Kargin, R.Z. Khayrullin, Yu.S. Chistov, Vestnik tekh. univ. **18(12)**, 74-77 (2015)
8. V.I. Azarov, A.V. Burov, A.V. Obolenskaya, *Chemistry of wood and synthetic polymers: a textbook for high schools* (SPbLTA, Saint-Petersburg, 1999)
9. D.M. Nigmatullina, A.B. Sivenkov, Internet-journal «Tekhnologii tekhnosfernoy bezopasnosti» **3(61)**, 108-115 (2015)
10. N.I. Akinin, N.O. Melnikov, S.A. Maksimenko, Vektor nauki TGU **3**, 28-31 (2013)
11. A.A. Leonovich, *Fire protection of wood and wood materials: tutorial* (LTA, Saint-Petersburg, 1994)

Future trends for the use of nanotechnogenic slurry waste in the production of special building materials

Natalia Chumachenko[1*], and *Alexey* Khlystov[1]

[1]Samara State Technical University, Institute of Architecture and Civil Engineering, Molodogvardeyskaya str. 194, 443001 Samara, Russia

Abstract. The article provides information about the main directions of the use of nanotechnogenic slurry waste, i.e. the industrial waste of Samara enterprises in the production of special building materials, namely, heat-resistant binders and concretes on their basis. It is found that nanotechnogenic wastes exert a polyfunctional impact in heat- resistant compositions (binders, mortars, concretes). High – alumina slurries are a plasticizer in heat-resistant composites on hydraulic cements. They are also a major component in the production of high- alumina binders and in the process of synthesizing phosphate binders which make it possible to obtain high-quality heat-resistant composites (mortars, coatings, concretes, impregnations).

1 Introduction

Currently, the volume of modern construction associated with the production and application of special building materials with the improved performance, expansion of the raw material base by the use of a variety of man-made materials, etc. is rapidly increasing. Organization of the production of highly heat-resistant composites is inseparably connected with the creation of competitive home materials. It is necessary to strive to reduce resource and energy intensity of production, labor costs, to optimize capital investments in the special construction of industrial furnaces and other thermal units. The decrease in the volume of imported refractory products in the Russian Federation will allow creating additional workplaces at enterprises of the country, and the production of domestic equipment will increase the level of no fired material output, i.e. heat-resistant concretes needed as a lining material for many industries. Manufacturing refractory products according to international requirements will increase the interest of foreign firms to Russian producers, which will facilitate the exchange of experience and technical information, strengthening of business ties and positive dynamics in the investment strategy.

Modern technologies for binders and materials on their basis are aimed at using both natural and man-made debris. This fact is due to the acute need to recycle the large-tonnage industrial and slurry- concrete waste that will significantly improve the environment within

* Corresponding author: uvarovang@mail.ru

towns and cities. One of the progressive directions for special concretes is the development of nanotechnologies to apply nanotechnogenic raw materials in manufacturing heat-resistant composites: mortars, concretes, gunning bodies, coatings, impregnations, etc. Technical and technological interest for using nanotechnogenic waste in the production of special concretes is due to the high possibility for fundamental changes in properties with the aim of obtaining materials with desired physical and technical parameters.

2 Research

For over 50 years (since 1956) the department "Production of building materials, products and constructions" of Samara State Technical University and Institute of Architecture and Civil Engineering [1-11] has been actively carrying out research on rational application of multiple, large-tonnage and slurry nanotechnogenic wastes in the production of materials for general construction work and special purposes [6].

Slurry wastes are produced during the chemical interaction of metal products to be processed in alkaline environment or during formation of spent catalysts [11]. Depending on the business type where slurry wastes are formed, they are x-ray amorphous or crystalline. For example, the rapid oxidation of metal powders used as catalysts in the technology of the synthetic rubber production leads to the formation of fine Nano products. Nano disperse sediments resulting from complex chemical processes associated with the non-ferrous metal working (electroplating, machining of aluminum parts in alkaline solutions at metallurgical enterprises, etc.) are of significant interest

The major part of resulting calcium-aluminum and alkaline-aluminum sludge are the hydroxides of aluminum and other metals, as well as hydroxides of calcium and magnesium which are formed after the waste water neutralization with lime [4]. It is found that slurries have a high adhesive strength and in mixtures with different binders they are inorganic nano disperse additives that increase plasticity, the ability to actively interact with various acids, segregation resistance, etc. [11].

In various construction compositions of special purpose nanotechnogenic sludge perform the following functions [9]:

- The formation of a set-up mineralogical composition in composite binders not only for general construction work, but also for special purposes;
- The performing functions of the plasticizer in composites based on hydraulic cements (Portland and aluminous);
- The activation of the cement grain surface energy in compositions with binders that enhance their activity;
- The formation of mortars which are not subjected to sedimentation during their interaction with acids.

One of the most important results of sludge participation in the formation of structure, physic-chemical and mechanical properties of construction materials is a significant increase in durability [9].

High-alumina industrial wastes represent a great interest to be used in heat-resistant concretes instead of the expensive technical and refractory raw materials.

Heat-resistant concretes and mortars is a modern refractory lining material. Unlike a piece ceramic refractory heat-resistant concretes can be applied in construction technologies and repairing linings of industrial furnaces in monolithic and precast options in large volumes.

Various heating units or high-temperature furnaces operate in the variety of sectors of the national economy. Such industries as metallurgy, chemistry, power industry, petro chemistry, oil refining and others use in their technologies thousands of heating units and installations in which heat-resistant lining materials take up one-sided heating at temperature from 350°C to 1800°C. Chamotte, mullite, mullite silica and alumina refractors are known to be used among piece ceramic ones with the increase of their service life. Comparisons of the chemical composition of refractory ceramics' show that with the Al_2O_3 percentage increase their physical and mechanical properties improve.

It is known that the high-temperature stability of thermal unit linings manufactured with the use of refractory ceramics is very low [5]. Depending on the chemical composition of the refractory ceramics used and the type of the aggressive environment in the thermal unit, the lining original life ranges from several months to two years. The use of masonry mortars which differ in the chemical composition from the main lining material promotes the reduction of the brick lining durability. As a result, the mortar does not sinter during the first and subsequent heating of the thermal unit. This fact leads to the formation of thermal stresses due to the rapid temperature change and, consequently, the destruction of ceramic refractory materials in places of their contact with the mortar. Thus, the seams in the lining masonry are a "bottleneck" from which the corrosion process of refractory's starts.

Lining materials made of heat-resistant concretes are free from this effect. Heat-resistant concretes represent the composition of the binder with refractory aggregates. Nowadays, there is a great variety of heat-resistant concretes. Along with hydraulic binders (Portland, calcium aluminate and high alumina cements) chemicals are widely used, namely, alkali silicate and phosphate binders.

Among hydraulic binders only Portland cement in compositions of heat-resistant concretes requires the presence of the finely ground fire additive. It is connected with harmful effects of the hydroxide $Ca\ (OH)_2$. Many of finely ground refractory additives, such as SiO_2 and Al_2O bind a free CaO to be compounds which do not decompose at high temperatures by the following reactions:

$$nCaO + mSiO_2 \rightarrow nCaO \cdot mSiO_2 \quad (1)$$

$$nCaO + mAl_2O_3 \rightarrow nCaO \cdot mAl_2O_3 \quad (2)$$

Physical and mechanical parameters of heat-resistant cement bricks (fire resistance, strength after heating, temperature resistance, etc.) range widely depending on the use of refractory finely ground additives (chamotte, fireclay, chromites, etc.). Studies have shown that the growth of oxides Al_2O_3 and Cr_2O_3 in the additive contributes to the formation of heat-resistant binders based on Portland cement with high operational performance [12].

However, the use of aluminous and high-alumina cements as the refractory binder does not require the use of fine refractory additives. As a result of the aluminous and high-alumina cement mineral hydration highly stable hydro aluminates of $nCaO\ mAl_2O_3\ pH_2O$ type are formed and dehydrate after heating. It is reflected on the decrease in the heat-resistant composite strength after heating them at temperatures of 800 ÷ 1000°C. Therefore, heat-resistant compositions with aluminous and high-alumina cements need only to increase the initial strength by reducing the water-cement factor.

As studies have shown [13,14], hardened compositions based on alkali silicate and phosphate binders have very high physical and mechanical parameters of heat-resistant composites.

Alkali silicate heat-resistant composites are synthesized with soluble sodium silicate in the solid state entitled silicate-block, and the use of the syrupy alkali silicate. Application of sodium water glass in the composition of the heat resistant binder requires the introduction of the fine ground fireproof additive and chemical hardener, e.g. a traditional sodium flu silicate Na_2SiF_6 [13]. Further, sodium flu silicate as a water glass hardener has been replaced by such products with calcium silicates or aluminates as phosphate slag and aluminous cement. This fact has made it possible to considerably increase physical and mechanical parameters of heat-resistant alkali silicate composites [13].

If a silicate block, i.e. a solid vitreous semi liquid glass is used in the synthetic process of heat-resistant binders, a very important technological conversion is needed, that is the mixed grinding of the silicate block with refractory ceramics in a predetermined ratio. To gain the strength of heat-resistant composites with hydrated binders, a dry heat treatment of products at temperatures from 160 to 180°C is required [15]. However, in our opinion, the best way to improve composites of alkali-silicate hardening is their modification by special high-alumina nano additives aimed at the reduction of liquid-solid and water-solid ratio.

Phosphate binders include both liquid water-soluble ones containing P_2O_5 bonds (e.g. alum chrome-phosphate and alumoboronphosphate bonds produced by the chemical industry) and phosphate cements formed by mixing fine ground super refractory oxides of metals or carbonatesc (Al_2O_3; Cr_2O_3; $CaCO_3$; $MgCO_3$) with a phosphoric acid H_3PO_4 of different concentrations [14; 16]. The hardening of heat-resistant composites with the use of phosphate binders is carried out by chemical reactions with the formation of refractory phosphate metals. Therefore, the resulting heat-resistant composites (concretes, mortars, coatings) have higher physical and mechanical parameters (heat stability, compressive strength, deformation of temperature under load, chemical resistance.)

The study of alum chrome-phosphate and alumoboronphosphate formation processes and phosphate cements with the use of H_3PO_4 makes it clear that most of the technical products can be replaced with the industrial waste of similar chemical, mineralogical and granulometric compositions. Such replacement allows not only to reduce the cost of compositions on the whole, but also to improve some technological processes in their production.

The most versatile nanotechnogenic component suitable to be used in the processes of perfection of heat-resistant composites on hydraulic cements and chemical binders is a nana disperse product, namely, a chrome-alumina petrochemical waste, i.e. a dead catalyst IM-2201. In the Samara region it is formed at the Novokyibyshevsk petrochemical plant and in Togliatti at the synthetic rubber plant.

Physical-chemical studies of the chrome-alumina waste showed that it is a nanodispersed powder with a specific surface area of more than 7800 cm^2/g and fire resistance of more than 2000 °C. High fire resistance of the chrome-alumina waste IM-2101 is due to its peculiar chemical composition (Al_2O_3 in the form of corundum is in the range of 72÷75%, and the content of the chromium oxide is 13÷15%).

Mixed high refractory binders containing nanotechnogenic waste, e.g. the alum-alkali sludge, have also been examined to increase the operational performance of many heat-resistant composites. Alum-alkali sludge or sludge of alkali etching of aluminum is a coproduct for manufacturing parts from aluminum-magnesium alloys produced at the Samara metallurgical plant. The as-wasted alum-alkali sludge is of the creamy consistency. Studies have shown that the particle size of the sludge consisting of hydrargillite ($Al(OH)_3$) is 20-80 nm [17].The presence of such nanoparticles in the sludge allows you to use it as a mineral plasticizer in heat-resistant composites based on Portland and alumina cements, water glass and silicate-sodium composite binders. The presence of the nano dispersed sludge in heat-resistant composites with binders mentioned above made it possible to considerably increase their physical and thermal parameters.

Chemical and mineralogical compositions of alum alkali sludge were taken into account in the design of aluminate cement raw mixes, for high-alumina cements in particular. Designing compositions of raw mixtures we were guided by recommendations given in the work [18]. The composition calculated according to the basicity factor representing the following mass relation $M_o = CaO / Al_2O_3$ was taken to be initial.

Chemical composition of the alum-alkali sludge is given in Table 1.

Table 1. Chemical composition of high-alumina alum-alkali sludge.

Content, mass %							
Al_2O_3	Fe_2O_3	CaO	MgO	SO_3	R_2O	loss on ignition	Σ
44.1	0.8	1.4	6.6	0.58	10.73	35.31	10.32

Under factory conditions the sludge of the aluminum alkali dip from the Samara metallurgical plant was used as an alumina-containing component and a lime-containing one was a slaked lime Ca $(OH)_2$ from the local enterprise.The basicity factor was variable from $M_o = 0.275 - 0.55$, because recommendations [18] state that if the value is $M_o = 0.55$, it is possible to obtain clinker of a monoaluminate composition, and when $M_o = 0.275$, the ceramic of a dealuminated composition is formed.

Production tests have resulted in a high alumina cement satisfying GOST 969-91 and having the following physical-mechanical parameters:

- specific surface area, cm^2/g - 3000;
- normal consistency, % - 27.7;
- setting time, hour-minutes
  - beginning - 1-00
  - end - 1-20;
- compression strength in 3 days, MPa - 45.0 – 55.0;
- fire-resistance, °C - 1750 – 1800.

The study of the chemical composition of the sludge of the aluminum alkali dip showed that it is possible to use it for the formation of phosphate binders, i.e. effective additives to heat-resistant composites. In our studies such technical products as $Al(OH)_3$ and Al_2O_3 have been replaced by the nanotechnogenic raw material, namely, the sludge of the aluminum alkali dip. The contact of the sludge with a phosphoric acid results in a chemical reaction with heat release between mineral fillers of nanotechnogenic raw materials containing H_3PO_4. The resulting liquid is an acidic alum phosphate binder consisting of acidic alum phosphates of Al $(H_2PO_4)_3$ and Al_2 $(HPO_4)_3$ type. It is possible to obtain an alum phosphate water-soluble binder of almost any density from 1.13 to 1.66 r/sm^3.

It is also possible to obtain aggregates for heat-resistant concretes from the refractory scrap resulting from the dismantling of old industrial furnace and other thermal units. After crushing the refractory spent lining in the form of bricks and its further classification, it is possible to obtain course and fine aggregates for heat-resistant concretes.

Compositions of heat-resistant binders and composites with high physical-mechanical parameters based on them have been developed on the basis of the synthesized alum phosphate binder, chrome-alumina waste hydraulic cements and alkali silicate binders with the use of refractory chamotte and mullite aggregates. Their compositions and properties are shown in Tables 2-4.

Table 2. Physical-thermal properties of binders with the use of industrial waste.

Binder mixture, %	Compression strength, MPa after heating to temperature, °C				Heat-resistance, water-thermal cycling
	400	800	1200	1400	
Chrome-alumina waste – 45 Portland-cement – 45 Sludge– 10 Water – 30 (above100%)	23.5	14.6	27.2	32.8 (1300°C)	22
Chrome-alumina waste – 45 Aluminate cement – 45 Sludge – 10 Water – 30 (above 100%)	29.6	23.9	33.3	37.1	26

These physical and mechanical parameters of heat-resistant binders will make it possible to obtain concretes with chamotte or mullite aggregates and working temperature respectively of 1300 ÷ 1500°C.

Table 3. Compositions and properties of heat-resistant alkali silicate binders modified by high-alumina sludge.

Binder mixture, kg/m^3	Liquid-solid ratio, l/s	Compression strength, MPa after 3 days of air hardening	Fire-resistance of the modified binder,°C
Chrome-alumina waste – 1350 Sodium flu silicate – 65 Alkali silicate – 650	0.46	28.8	1545
Chrome-alumina waste – 1215 Sludge – 135 Sodium flu silicate – 62 Alkali silicate – 606	0.43	32.9	1565
Chrome-alumina waste – 945 Aluminate cement – 315 Sludge – 140 Alkali silicate – 615	0.44	39.8	1620

Test results of samples of modified binders (Table 3) have shown, that the replacement of the sodium flu silicate on the alumina cement has considerably increased not only the initial mechanical performance, but also operational, i/e/ in the working process. Refractory properties of modified heat-resistant alkali silicate binders have exceeded 1600°C and it allows obtaining composites with the working temperature of 1400 - 1550°C.

Table 4. Physical – thermal properties of heat-resistant concretes on chemical binders.

Concrete mixture, kg/ m^3	Dry specific gravity, kg/m^3	Compression strength, MPa after hardening and heating to temperature, °C			
		20 (7 days)	800	1200	1400
Waste IM-2201 – 440 Crushed stone SHA – 750	2010	6.6	44.2	46.0	45.4

SandSHA – 650 APHC – 260					
Waste IM-2201 – 440 Crushed stone MLS 62-780 Sand MLS – 62-680 APHC – 2600	2100	7.9	47.6	50.4	50.7

3 Conclusion

The use of the synthesized alum phosphate binder to bound carbonate-containing wastes (dolomite screenings from crushing) in compositions with the chrome-alumina waste gave the possibility to obtain phosphate aerated concretes with an average density from 400 to 800 kg/m^3 and a working temperature up to 1500°C. These physical and thermal parameters of heat-resistant aerated concretes may be used as a high-performance monolithic heat insulation of lining in industrial furnaces and other thermal units. Synthesized on high-alumina slurry-like wastes an alum phosphate binder proved to be quite suitable in the compositions of phosphate aerated concretes on the basis of carbonate raw materials instead of expensive chrome-alumina and alumoboronphosphate binders.

References

1. T.B. Arbuzova, S.F. Korenkova, *The use of waste sludges in the production of building materials, ser.2, The use of waste coproducts in the production of building materials and goods, Environmental protection* (VNII ESM, Moscow, 1989)
2. T.B. Arbuzova, *Recycling of aluminiferous sludge of the industrial waste water*, (publishing house of Saratov University, Samara branch, Samara, 1991)
3. T.B. Arbuzova, V.A. Shabanov, S.F. Korenkova, N.G. Chumachenko, *Building materials from industrial waste* (Samara publishing house, Samara, 1993)
4. S.F. Korenkova, T.V. Sheina, *The basics and concept of utilization of chemical industrial waste in the construction industry* (Samara State University of Architecture and Civil Engineering, Samara, 2004)
5. A.I. Khlystov, *Effectiveness increase and improvement of the quality of fire resistant lining maretials, Monograph* (Samara State Universiry of Architecture and Civil Engineering, Samara, 2004)
6. S.F. Korenkova, Nano technologies in construction: scientific Internet - journal **4**, 15 - 18 (2009)
7. S.F. Korenkova, Yu.V. Sidorenko, Nano technologies in construction: scientific Internet – journal **3**, 26-32 (2010)
8. N.G. Chumachenko, Industrial and civil engineering **8**, 3-6 (2010)
9. N.G. Chumachenko, S.F. Korenkova, A.I. Khlystov, Industrial and civil construction **8**, 20-22 (2010)
10. S.F. Korenkova, Yu.V. Sidorenko, Nano technologies in construction: scientific Internet – journal, **5**, 53-60, 71-74 (2011)
11. A.I. Khlystov, Experience of application of high - alumina waste of the Samara region enterprises in heat – resistant concretes, 71st all-Russia scientific and technical conference on results of the scientific research in 2013 " Traditions and innovations in construction and architecture", 554- 558 (2014)

12. A.I. Khlystov, *Theoretical and technological principles of increasing the durability of refractory lining materials, author's abstract of doctor of technical sciences* (Samara, 2004)
13. A.N. Tarasova, *Heat – resistant binders on liquid glass and concretes on their basis* (Stroiizdat, Moscow, 1982)
14. P.P. Budnikov, B. Khoroshavin, *Fire resistant concretes on phosphate binders* (Metallurgy, Moscow, 1971)
15. Yu.P. Gorlov, A.P. Merkin, M.I. Zeifman, B.D. Toturbiev, *Heat-resistant concretes based on compositions of natural and manmade glasses* (Stroiizdat, Moscow, 1986)
16. A.I. Khlystov, V.A. Shirikov, E.A. Chernov, Journal. Scientific review **12**, 196-199 (2013)
17. A.I. Khlystov, S.V. Sokolova, M.V. Konnov, Journal Refractories and technical ceramics **11**, 35-39 (2010)
18. T.V. Kuznetsova, *Aluminate and sulfoaluminate cements* (Stroiizdat, Moscow, 1986)

On research of optimal programming solutions for road construction in modern conditions

Tatiana Dormidontova [1*], *Sergey* Evdokimov [1], and *Natalia* Shekhova [1]

[1]Samara State Technical University, Institute of Architecture and Civil Engineering, Molodogvardeyskaya str. 194, 443001 Samara, Russia

Abstract. The poor condition of roads in most regions of Russia predetermines the need to establish and implement target regional programs of road network development in the country. Depending on the time perspective of the mentioned programs (short, medium or long term), the defining criteria for including roadways in these programs are set. To develop the theory and practice of justification of optimal programming solutions in road construction in the context of multi-criteria assessment, the authors have proposed the method to optimize the programming solutions in road construction and the algorithm of its implementation. The basic principles of comparing alternative variants for the construction of roadways with comparable pavement design as well as recommendations on the selection of the optimal variant have been developed using the technical and economic indexes.

1 Introduction

In many regions of the country the state of roadways could be described as unsatisfactory. In this regard, the implementation of regional target programs of the road network development is currently gathering pace [1-3].

The purpose of the work is to provide the optimal programming solutions for road construction in the context of multi-criteria assessment.

The objectives of the research are:

- to identify the criteria for including roads in the short-, medium- and long-term regional target programs;
- to develop the method for optimizing programming solutions in road construction and the algorithm of its implementation;
- to select the optimal variant of the construction of roadways with comparable pavement designs.

2 Main Part

The in-depth analysis of the roadways condition allowed identifying characteristic criteria for including them in relevant programs of roads development. When creating these

* Corresponding author: adisk63@yandex.ru

programs it is necessary to take into account such factors as completion schedules and the importance of work for road development. Depending on what time perspective these programs are oriented on, they are united into three groups: short-, medium- and long-term ones [4-7].

The main criteria for including roads in the operational program of development are the following:

- the number of road sections with high concentration of accidents that have occurred due to adverse road conditions;
- the number of road sections passing through populated areas with high intensity of transit traffic;
- road sections which are school bus routes that do not meet regulatory requirements.

The fundamental criteria for including roads in the medium-term development program are:

- non-compliance of the actual bridge clearance to regulatory requirements;
- road sections with expired intermaintenance period;
- unfinished objects;
- road sections, which are school bus routes.

The defining criteria for including roads in the long-term development program are the following indexes:

- crossing of a road and a railway at the same level;
- existing transport corridors and prospective strategy of their development;
- municipal streets (in urban areas), which are a continuation of public roads;
- rural settlements that do not have hard surface driveways;
- rural settlements with a high density of transit traffic;
- road sections, where traffic density exceeds the norm for this category.

For the development of the theory and practice of providing optimal programming solutions in road construction in the context of multi-criteria assessment, the following method has been proposed by the authors. The algorithm of its implementation is described below [8-10].

The variety of factors raises difficulties when selecting specific areas of work and prioritizing their implementation. That is why, it is proposed to introduce the point assessment of each factor influence. Based on this assessment system, one can determine the significance of an object in points, which consequently makes it possible to form a ranged list of those objects. This list shows the order of work performance to achieve a multiplier effect.

It is necessary to take into account the factors that require consideration in order to achieve socio-economic effects, bearing in mind the immensity of the tasks.

For example, such criterion as "Connection of a rural settlement with population of 100 people with the existing network of public roads by means of a road" gets one point, similar criterion "a settlement with population of 500 people" - three points. The criterion: "Construction of a roadway up to 1 km" - three points, the criterion for "Construction of a roadway up to 5 km" - one point. The ranged list of roads is formed according to the total of points in order of descending [11].

The proposed principles give grounds for comparison of alternatives for laying roadways with comparable pavement designs, in order to select the best one. The technical and economic indexes are compared.

The main criteria for comparing the roadways are as follows:

- the minimum radius of the curve in the plan;
- the minimum radius of vertical curves;
- the availability and number of artificial structures;
- the specific index of the earth quantities per 1 m^3;

- the length of a road section with the possible longitudinal slope;
- the payback period;
- the construction cost per 1 km of the road.

When introducing the points system for each of these parameters, one can perform the comparison which results in the optimal plan for roadway construction.

One of the urgent problems of the modern road construction is to compare alternative pavement designs. The most important problem that should be solved at the final stages of road construction is the selection of variants of the pavement design from the variety of layers which are optimal for a particular area. When selecting the best option, three possible designs of the pavement are considered: asphalt concrete covering; cement concrete covering; cement concrete base with asphalt covering [12-16].

The efficiency indexes serve as the selection criteria. They are presented in Table 1.

Table 1. Indexes of the selection criteria of the road effectiveness.

Criteria Indexes	**Variant 1**	**Variant 2**	**Variant 3**
Service Life (A)	15years	25years	20 years
Estimated cost (B)	10634.6 thousand rubles.	108468.1 thousand rubles.	27983.0 thousand rubles.
The need for machinery and labor force (C)	103/20	133/140	127/135
Utilization level of local road-building machinery (D)	16 km	91 km	91 km
Intermaintenance period (E)	5 years	12 years	5 years
Maintainability (F)	0.4	0.2	0.4
Operational costs (G)	113.78	74.3	113.78
Pavement wearing capacity (H)	1 ÷ 1.5 mm / year	0.1 ÷ 0.2 mm / year	1 ÷ 1.5 mm / year

In order to define the significance of the particular criteria the relative preference method is used, which is a special case of the expert method. This method is for one expert who is the decision-maker. It includes the following steps.

First. Prioritization of each criterion in the following order:

$$A(8) \geq B(7) \geq C(6) \geq D(5) \geq E(4) \geq F(3) = G(3) \geq 3 \quad (1)$$

Second. Determination of the weight coefficient for each criterion. Each line at its intersection with the column contains the value, corresponding to the order of preference of the line criterion to the column criterion. As a result, each line gives the sum of the importance coefficient, Table 2.

Table 2. Weight coefficients.

	A	*B*	*C*	*D*	*E*	*F*	*G*	*H*	$\sum \beta_{ij}$	A_i
A	-	1	2	3	4	5	5	7	27	0.34
B	0	-	1	2	3	4	4	6	20	0.25
C	0	0	-	1	2	3	3	5	14	0.18
D	0	0	0	-	1	2	2	4	9	0.11
E	0	0	0	0	-	1	1	3	5	0.06
F	0	0	0	0	0	-	-	2	2	0.03

G	0	0	0	0	0	-	-	2	2	0.03
H	0	0	0	0	0	0	0	-	0	0
									$\sum 79$	$\sum 1$

Third. Determination of the standardized values of the partial criteria, Table 3.

Table 3. Standardized values of the partial criteria.

Variants	A_H	B_H	C_H	D_H	E_H	F_H	G_H	H_H
1	1.0	0.83	0.6	0.77	0.18	0.42	1.0	1.0
2	0.15	0.84	1.0	1.0	1.0	1.0	0.5	0.65
3	1.0	1.0	0.75	0.75	1.0	0.42	1.0	1.0

Fourth. The generalized criterion of efficiency is calculated taking into consideration the values of the index and the standardized values of the criteria for each variant. Since all criteria are minimized, the formula (2) is used:

$$Y_1 = \sum_{i=1}^{n} \alpha_i K_i^H \rightarrow \min(\max) \tag{2}$$

Fifth. Determining the best variant, which is the one with the minimum generalized criterion from three given. In the analyzed example, it is the pavement design with asphalt concrete covering.

If you change the importance priority of particular criteria, the best one may be another variant.

3 Conclusion

After considering the results of the research the following conclusions have been made:

1. The criteria for including the roads in the short-, medium- and long-term regional target programs have been identified.

2. The method of optimization of programming solutions in road construction and the algorithm of its implementation have been developed. Studies have shown that the quality of the developed method is determined by the algorithm and depends on a set of the analyzed criteria. Despite all the advantages of this method, it should be remembered that in order to build a high-quality model, it is necessary to understand the nature of the relationship between the dependent and independent variables and prepare a sufficient data set to determine the basic criteria for comparison.

3. Because of the unstable economic situation in Russia companies have to make more thorough and balanced decisions, focusing either on the quality of work performed or on the amount of funds allocated for the construction, prepare development plans, based on pilot tests and practical knowledge, evaluating the reality.

References

1. M.I. Balzannikov, N.V. Ahrameeva, Environmental and hydraulic engineering structures: problems of construction, operation, environment and specialists training: Proceedings of International Scientific and Technical Conference (2014)

2. M.I. Balzannikov, Yu.M. Galitskova, A. A. Bolotova, Bulletin of the Volga regional branch of the Russian Academy of architecture and construction sciences: Proceedings, **16**, 132-135 (2013)
3. V.I. Gvozdovsky, S.V. Evdokimov, Environmental and hydraulic engineering structures: problems of construction, operation, environment and specialists training. Proceedings of International Scientific and Technical Conference, **219-220** (2014)
4. L.G. Goverdovskaya, Environmental and hydraulic engineering structures: problems of construction, operation, environment and specialists training. Proceedings of International Scientific and Technical Conference, (2014)
5. T.V. Dormidontova, N.G. Solkaran, Environmental and hydraulic engineering structures: problems of construction, operation, environment and specialists training. Proceedings of International Scientific and Technical Conference, (2014)
6. T.V. Dormidontova, S.V. Evdokimov, *Complex application of methods of reliability assessment and monitoring of structures and buildings,* (SGASU, Samara, 2012)
7. T.V. Dormidontova, V.V. Kiryakov, Internet-journal "Science of Science" **7(1)**, (2015) information on http://www.naukovedenie.ru/ PDF/ 42TVN115.pdf/
8. T.V. Dormidontova, Tradition and Innovation in Civil Engineering and Architecture: Proceedings of the 70th Scientific and Technical Conference on R & D Results in 2012 (Samara, 2013)
9. T.V. Dormidontova, Procedia Engineering XXIV R-S-P seminar, Theoretical Foundation of Civil Engineering, (2015)
10. E.I. Marinin, T.V. Dormidontova, Environmental safety in construction **4(17)**, 80-84 (2014)
11. S.V. Evdokimov, B.G. Ivanov, V.A. Shabanov, Scientific review **22**, 102-107 (2015)
12. Utility Patent 160415. Protective device for roads, Published 20.03.2016. Bull. № 8
13. N.V. Ovchinnikova, *Institutional support of economic mechanism of externalities regulation* (Samara, 2010)
14. N.V. Shekhova, I International Research and Practice Conference "Innovative development strategies of economy and management", Proceedings, (2014)
15. N.V. Shekhova, Modern economy: problems, solutions, prospects: Proceedings of International Research and Practice Conference dedicated to the 20th anniversary of Economic Department, 156-160 (2014)
16. N.V.Shekhova, P. Surmakh, II International Research and Practice Conference "Economic aspects of management of a building complex in modern conditions": Proceedings, 68-72 (2014)

MATEC Web of Conferences **86**, 04009 (2016) DOI: 10.1051/matecconf/20168604009
IPICSE-2016

Optimal control of ceramsite burning in a rotary kiln

Stanislav Galitskov[1], *Konstantin* Galitskov[1], *Oleg* Samokhvalov[1,*], and *Aleksandr* Fadeev[1]

[1]Samara State Technical University, Institute of Architecture and Civil Engineering, Molodogvardeyskaya str. 194, 443001 Samara, Russia

Abstract. The paper considers various methods and devices of optimal control of ceramsite burning in a rotary kiln. All this ensures ceramsite production with required values of its density or strength at minimum energy consumption under conditions of maximum achievable productivity.

1 Introduction

Ceramsite with different values of density or strength is used in different building spheres [1, 2]. In this regard, ceramsite producing factories should provide a flexible regime of changing technologies and equipmentto provide quick change of type of ceramsite it produces.

Currently known algorithms and kiln control devices are known to have strict There are also difficulties in monitoring parameters of ceramsite temperature field. Papers [3-5] introduce a method of monitoring ceramsite temperature in three so-called kiln bearing cross-sections A, C, F. Papers [3, 6] show that this method makes possible to purposefully change the burning curve within comparatively wide limits by changing the following three control actions: extensive thermal power of the burner – Q_b, the kiln load factor Q_l and the kiln angular velocity ω (see Figure 1). This enables to expand technological capacity of the kiln in regard of density and strength of ceramsite produced. That is why it is important and actual to find optimal conditions of ceramsite burning in regard of its density ρ or strength R in conditions of energy consumption reduction.

2 Burning Ceramsite SAC Structure

Mathematical models of ceramsite burning introduced in Papers [3, 7] made it possible to generate a structure of the model of burning process as a multi-dimensional object with distributed parameters which status is defined by the vector of output coordinates including temperature values in bearing sections $\overline{X}_{вых} = [T_F, T_A, T_C]^T$. Vector $\overline{X}_{вх} = [Q_г, Q_з, \omega]^T$ is taken as input influences. Besides, the following functional relationships are established:

$$\rho = f_2(T_A, T_C, w, \gamma), \tag{1}$$

[*] Corresponding author: indexcitir@gmail.com

$$R = f_1(T_F, T_A, T_C, w, \gamma), \quad (2)$$

where w and γ are raw material moisture and clay generalized characteristics that make possible to determine strength and density of ceramsite produced. In accordance with the object characteristics the authors generated the structure of a three-dimensional automatic control system (see Figure 2) and parametrically optimized egulators of its separate channels.

3 Optimal Control Regimes

While dealing with the problem of burning process optimization the researchers considered two tasks of great practical importance. The first task is connected with light ceramsite of ρ_{req}, required density, and the second deals with high-strength ceramsite production with required strength of R_{req}.

An algorithm for handling the first task involves the following steps. Firstly, exploring the properties of raw material blowup with w humidity (obtained from clay with a known constant γ value), it is possible to build Surface 1 in the space $OT_AT_C\rho$ with the expected density of ceramsite with variations of T_A и T_C [8,9] within the range limited by the top and bottom points of the burning curve (see Figure 3).

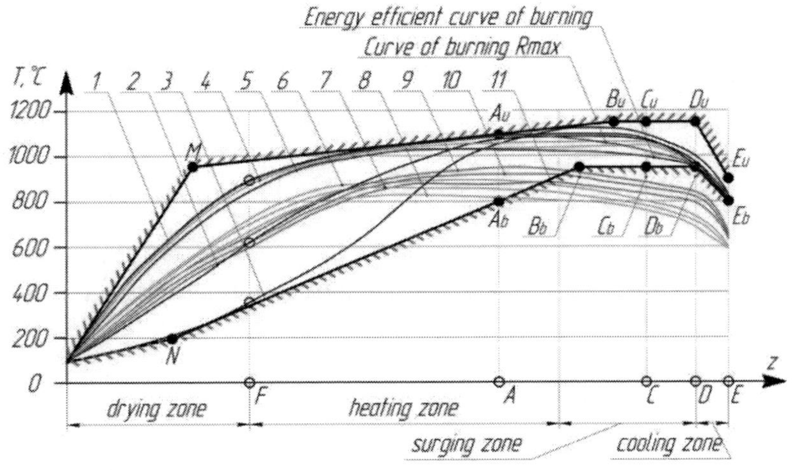

Fig.1. Burning curves area for high-strength ceramsite.

In the second phase, in accordance with the required value ρ_{req} we determine optimal temperature T_A^{opt} and T_C^{opt} along Surface (1). Temperature value here lie on the required burning curve presented as ABC polygonal line. T_A^{opt} and T_C^{opt} values are predetermined for the two corresponding channels of the system (see Figure 2). The third channel operates in a mode of kiln rotation with nominal speed.

To produce ceramsite with required density value R_{req} in conditions of energy consumption reduction, let us consider phase field $OT_CT_AT_F$. In this field we'll use a methodology introduced in Paper [5] on the basis of Dependence (2) and get thermal-stress characteristics of burning ceramsite of requited density in conditions of energy consumption reduction and with maximal possible productivity (see Figure 4):

$$R = f_3(T_F^{opt}, T_A^{opt}, T_C^{opt}, w, \gamma). \quad (3)$$

When we find R_{req} point on this curve, we get the required values $T^{opt}_{C.тр}$, $T^{opt}_{A.тр}$ and $T^{opt}_{F.тр}$, supported by the automatic control system (see Figure 2).

4 Optimal Control Efficiency

To assess the effectiveness of the proposed method of choosing optimum regimes of coordinated control of the kiln work while producing ceramsite with required density of ρ_{req} the authors performed a comparative analysis of gas expense for ceramsite blowup 1 m^3 on conditions of realization of the proposed algorithm (Curve 2, Figure 5). The authors also used a heuristic algorithm, in which the management of blowup is performed by changing power Q_p of the burner with the same loading of the kiln and the same speed of its rotation (Curve 1 , Figure 5). This simulation study [10] is done for the regime, when the constant loading was Q_z = 4.5 t/h and humidity of material w = 15%. This research was carired out for six values Qп = 28000; 30000; 32000; 34000; 36000; 38000 W/m^3. Gas expense for ceramsite blowup 1 м^3 is determined by $\sigma_{0г} = Q_т/P$, where $Q_т$ – is fuel consumption; P – kiln productivity, $P = (q_з\, K_{bl}) / \rho_c$.

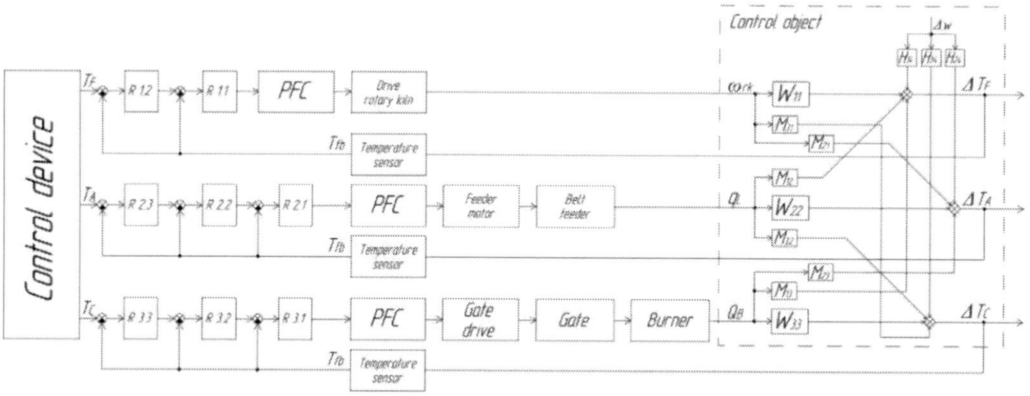

Fig. 2. The structure of a multidimensional system of automatic stabilization of ceramsite strength.

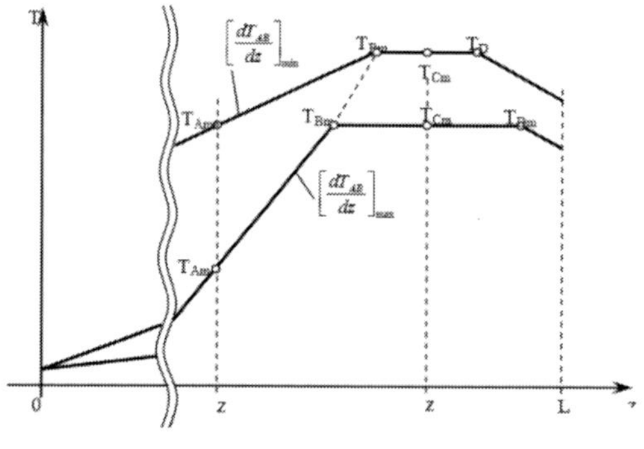

Fig. 3. Boundary curves of ceramsite burning.

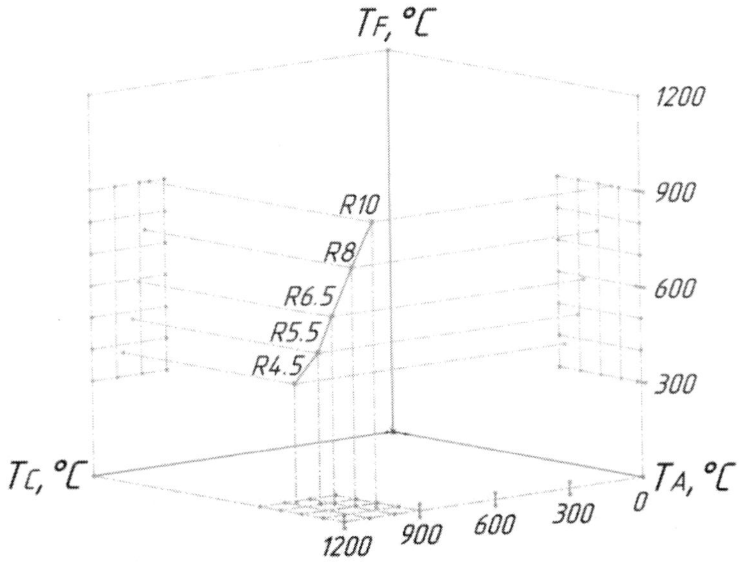

Fig. 4. Temperature and mechanical strength characteristics for ceramsite burning.

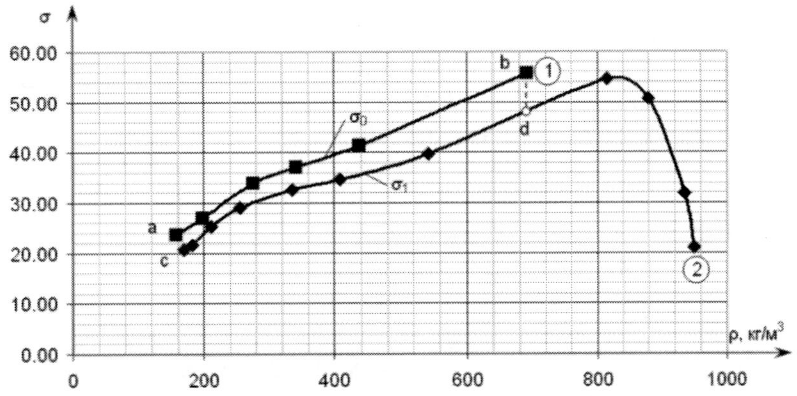

Fig. 5. Energy consumption for ceramsite production in density function.

Figure 5 compares Curves 1 and 2. This comparison shows that for a kiln of 2.5×40 m type while using clay from Smyshlyaevka the system of coordinated control allows to reduce energy consumption for ceramsite burning by 11-15%.

Assessment of this methodology efficiency was carried out on the same basis for selecting optimal regimes for coordinated control of the kiln while producing ceramsite with required strength of $R_{тр}$. Figure 6 presents simulation study results. Here Curve 3 corresponds to the use of optimal control regime of the kiln while producing ceramsite with required strength of R_{req} by a three-channel SAC.

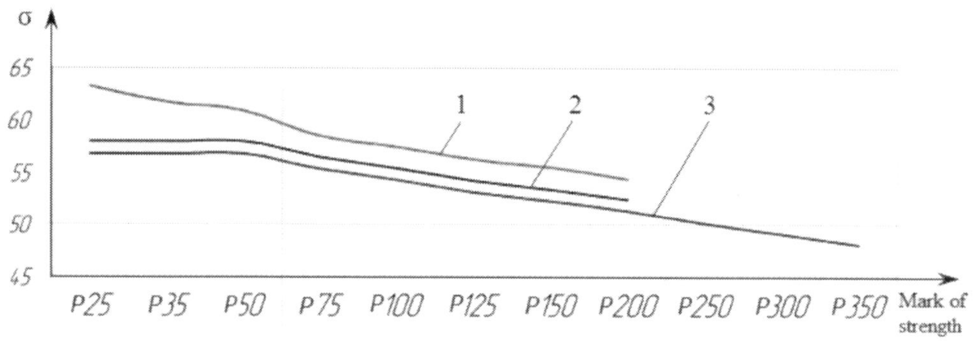

Fig. 6. Energy consumption for ceramsite production in strength function.

5 Conclusions

1. The paper establishes the criterion for optimal control of the burning process: production of ceramsite of required strength R_{req} or required density ρ_{req} on conditions of maximum possible productivity and minimum energy consumption.

2. The paper demonstrates that one of the possible solutions of the task is the coordinated control of three coherent control actions: extensive thermal power of the burner – Q_b, the kiln load factor Q_l and the kiln angular velocity ω. The authors also generated the structure of a multi-dimensional automatic control system of the kiln.

3. The paper introduces a new methodology of calculating optimal temperature in three bearing sections in regard to production of light ceramsite of ρ_{req} required density, and of high-strength ceramsite with required strength of R_{req}.

The paper proves that coordinated control extends technological properties of the kiln and helps reduce energy consumption for ceramsite burning by 11-15% while producing ceramsite with required strength of R_{req} or with required density of ρ_{req}.

References

1. V.M. Gorin, T.A. Tokareva, M.K. Kabanova, Building materials **1**, 9-11 (2010)
2. Yu.S. Vytchikov, M.Ye. Saparev, Construction materials **11**, 12-15 (2013)
3. S.Ja. Galitskov, K.S. Galitskov, O.V. Samohvalov, A.S. FadeevScience Review **7**, 227-237(2015)
4. S.Ja. Galitskov, O.V. Samokhvalov, A.S. Fadeev, Patent 2554964. Method for burning expanded clay in rotary kiln and device for its implementation, BI, 19 (2015)
5. K.S. Galitskov, O.V. Samohvalov, Industrial and Civil Engineering **9**, 91-94 (2015)
6. S.P. Onatskiy, *Ceramsite production, 3-d ed.* (Stroyizdat, Moscow, 1987)
7. A.S. Fadeev, S.Ya. Galickov, A.I. Danilushkin, Vestnik of Samara State Technical University. Technical Sciences Series **2(30)**, 160-168 (2011)
8. S.Ya. Galitskov, A.S. Fadeev, Collected works of 69th all-Russia scientific technical conference, Samara State University of Architecture and Civil Engineering, 514-515 (2012)
9. S.Ya. Galitskov, A.S. Fadeev, O.V. Samohvalov, Procedia Engineering, **111**, 233-235 (2015)
10. S.Ya. Galitskov, A.S. Fadeev, Vestnik SGASU, Town Planning and Architecture **4 (12)**, 95-98 (2013)

Reducing energy consumption for shaping raw ceramic bricks with coordinated control of auger vacuum extruder

Stanislav Galitskov[1], *Maxim* Nazarov[1,*], and *Konstantin* Galitskov[1]

[1]Samara State Technical University, Institute of Architecture and Civil Engineering, Molodogvardeyskaya str. 194, 443001 Samara, Russia

Abstract. This research paper focuses on the issues of enhancing technological reliability and reducing energy consumption while shaping ceramic bricks. It proves that these aims can be effectively achieved by using automatic coordinated control of actuating auger vacuum extruder systems. The paper also assesses the extent of energy consumption reductions when the extruder is equipped with an automatic control system.

1 Problem specification

Reduction of energy consumption while producing ceramic bricks of certain required strength is one of the main mandatory requirements to their manufacture [1]. Paper [2] proves that the quality and durability of bricks depend to a great extent on technologies of shaping raw materials. This technological process is quite energy-consuming. Generating capacity of this process can rise up to *250-250 kW* and above. The extruder is usually controlled with the help of fairly simple algorithms. This significantly limits the ability not only to select optimal power saving modes, but also to ensure stable quality indicators of bricks formation. This paper aims to introduce effective methods and devices focused both on improving this technological process and reducing energy consumption by the installation drivers.

2 Methodology

The authors introduce a mathematical model of the extrusion process considering it to be a nonstationary object with distributed parameters. This model is based on the previous study of factors influencing the technology of bricks shaping as an intermediate stage of production of ceramic bricks and creating the basis for its structure and, consequently, strength [2-4]. As one of the main components of the vector of output coordinates the authors adopted shear rate $\dot{\gamma}$ of clay mixture in pressure head [4, 5]. This is done due to the fact that value $\dot{\gamma}$ determines stability of geometric shapes, homogeneity of the internal

* Corresponding author: nazarovm86@yandex.ru

structure and strength R of fabricated bricks [2, 4]. Papers [2, 6-10] show that the required R value can be achieved on conditions of coordinated control $\dot{\gamma}$ of moisture w of ceramic paste and vacuum pressure of P_v in a vacuum chamber. Let us assume that each of these process values is controlled by a separate automatic system with a minimum value of error Δ_j $j = 1, 2...m$ (where m is the value of controlled technology coordinates of a vacuum extruder). To support the adopted assumption, the authors generated a multidimensional automatic control system of the extruder actuating systems (MACS EAS), such as auger electric drivers, a clay mixer and a vacuum pump. To produce ceramic bricks of an assumed strength grade M_i ($i = 1, 2...n$, where n the number of bricks grades that can be produced with the extruder in question) in conditions of reducing energy consumption generated an algorithm for determining a sustainable control vector $\overline{U}_0$ in the sphere $0w\dot{\gamma}P_v$ of control actions [7, 8]. Disturbances which affect extrusion process can cause displacement of control area boundaries. This might cause either strength error R or increase energy consumption. To ensure that the multi staged control system is invariant, Papers [2, 7, 8] introduced structural synthesis of an intelligent control system for such disturbances. In this system a vector-shaping block of reference-input signal includes a data bank of extrusion process, a technological data bank of an operating unit and a correction algorithm of reference signal vector.

3 Mathematical Model of Energy Processes

Figure 1 presents a mathematical model of energy processes in a processing system. It helps estimate effectiveness of the proposed variant of auger vacuum extruder coordinated control. "A" here designates operators forming links between electromechanical and technological variables, durability and energy consumption: A11, A21, A31 are operators forming motors electromagnetic torques a auger M_{d1}, a vacuum pump M_{d2}, a clay mixer M_{d3}; A12, A22, A32 are operators forming angular velocity ω_a of the auger, ω_{vp} a vacuum pump rotor, ω_s clay mixer shaft; A14, A26, A34 – are operators forming load torque on the drive motor of the auger M_{l1}, a vacuum pump M_{l2}, a clay mixer M_{l3}; A13, A23, A24, A25, A33 – mathematical models of ceramic paste flow in a pressure head of the extruder, a vacuum pump of a suction air duct, a vacuum chamber, and a clay mixer respectively; A4 – forming unit of value R; A51, A52, A53 – evaluation units of specific energy consumption for operations of shaping E_1^y, vacuum processing E_2^y, ceramic paste mixing E_3^y; Σ – forming unit of auger vacuum extruder specific energy consumption E_Σ^y; $\overline{P_0}$ – medium pressure ceramic paste while entering pressure head; S_{vp}, S_{vc} – pumping speed at the vacuum pump output and at the vacuum chamber input; Q_w – water flow for clay watering; w_0 – ceramic paste moisture when entering the mixer loading gate; μ_1, ψ – ceramic paste flow consistency index and flow behavior index at the mixer output; Q_m – clay mixer productivity.

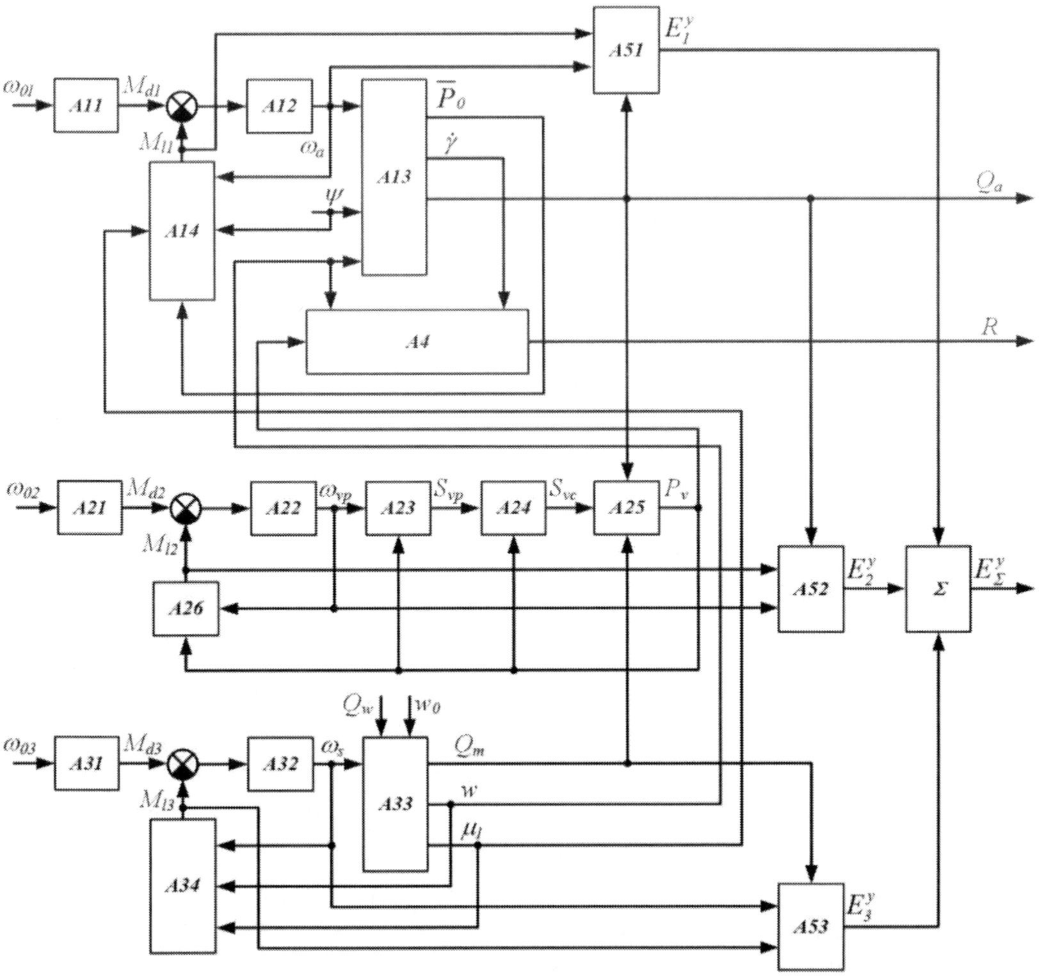

Fig. 1. Mathematical model of energy processes and its structure.

When constructing unit A51 provided a reflection of control area section $0w\dot{\gamma}P_v$ [2] on the plane $0E_1^yQ_a$. Here energy consumption E_1^y and extruder productivity Q_a serve as coordinate axis while producing grade M_i bricks (as Figure 2 shows), where $P_v = const$. The following functions are used in this transition:

$$Q_a = k(\psi)\cdot\dot{\gamma}\,, \tag{1}$$

$$E_1^y = \frac{M_{l1}\cdot\omega_a\cdot i_1}{Q_a}\,, \tag{2}$$

where k – is the coefficient depending on the ceramic paste flow behavior index ψ; i_1 – gear ratio of the auger drivers. ABCD area (as Figure 2 shows) is limited by the clay moisture minimum and maximum values (w_{min}, w_{max}) and by strength (R_j, R_{j+1}) determined by GOST 530-2012 for grade M_i bricks.

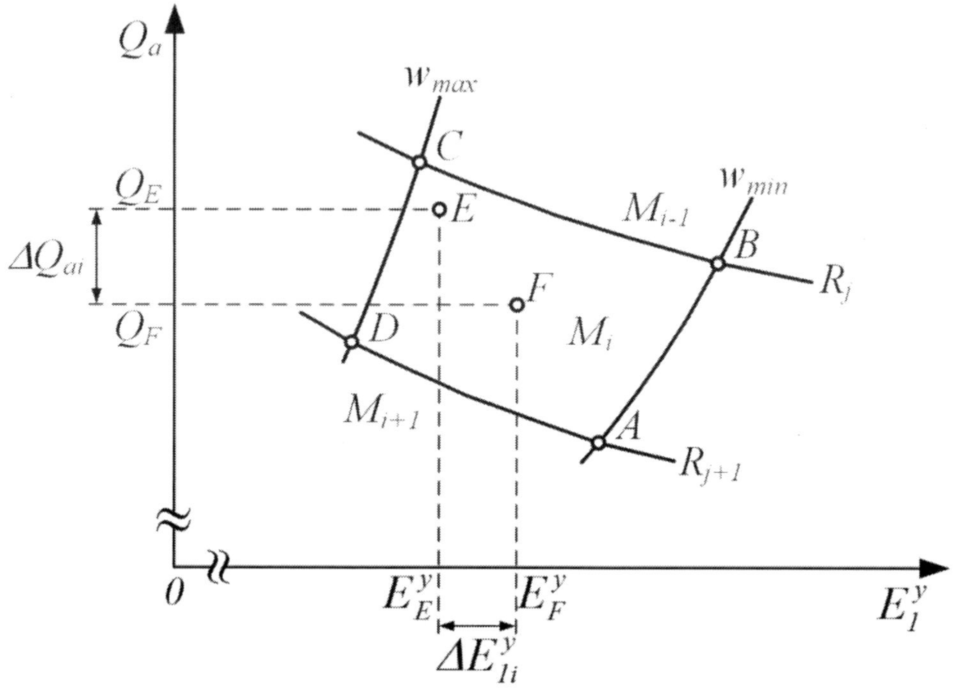

Fig. 2. The auger vacuum extruder control area while producing grade M_i bricks.

Studying energy performance of the designed MACS EAS for two automatic control modes while producing grade M_i bricks. The first mode is an example of traditional control involving automatic pressure leveling $\overline{P_0}$ of ceramic paste in the pressure head. Feed moisture w in the mixer is modified according to $\overline{P_0}$ value, auger velocity is not modified. In the second mode MACS EAS is used. It provides coordinated control $\dot{\gamma}$, w, P_v involving the use of digital observer of strength R.

Control mode 1, in which the working point F is in the center of limited control area ABCD, corresponds to Q_F productivity and E^y_F specific energy consumption. When we switch to mode 2 the working point moves to E-position and is located in ε-vicinity of Point C (ε-vicinity is defined by Δ_j deviations of the extruder actuating systems) which is located on the edge of the area in question. This point corresponds to the possible attainable minimum of energy consumption E^y_E and productivity maximum Q_E while producing grade M_i bricks. To make quantitative assessment of energy performance we introduce a relative coefficient of energy consumption reduction K_{Ei} and a relative coefficient of productivity increase K_{Qi}:

$$K_{Qi} = \frac{|Q_E - Q_F|}{Q_F} \cdot 100\%, \tag{3}$$

$$K_{Ei} = \frac{|E^y_E - E^y_F|}{E^y_F} \cdot 100\%. \tag{4}$$

Assessment of energy performance of the designed MACS EAS as applied to the extruder with drive motors of the auger being *160kW*, vacuum pump of *11 kW*, clay mixture of *90 kW* while producing grades M200, M250, M300 bricks proves advantages of the designed control system (see Figure 3). It has also been proved that reducing energy consumption while increasing productivity of the extruder at the same time depends on grade of bricks.

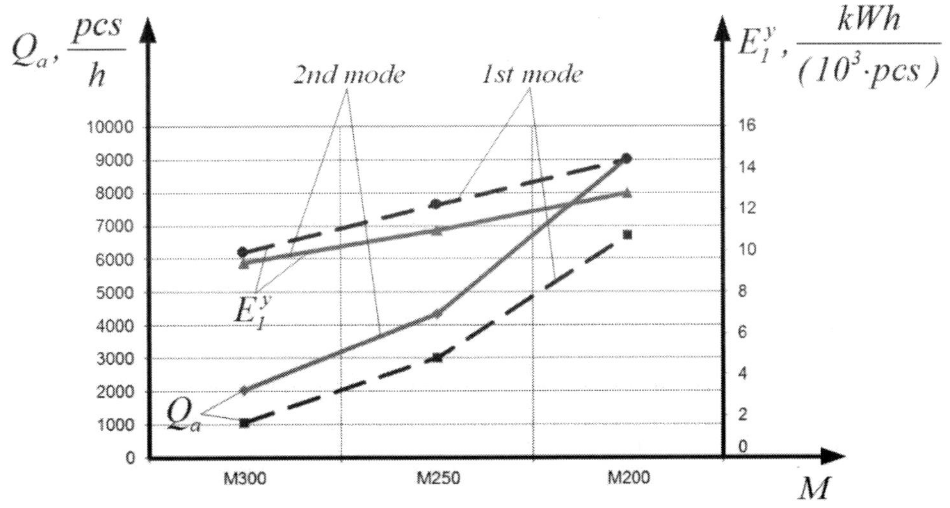

Fig. 3. Comparison of an auger vacuum extruder operation modes.

4 Conclusions

The conducted investigation made possible to design MACS EAS (a multidimensional automatic control system of the extruder actuating systems) for shaping ceramic paste while producing bricks with minimum energy consumption and maximum productivity.

The authors also developed a mathematical model of energy processes observed in an auger vacuum extruder when shaping ceramic paste.

The authors assessed the effectiveness of shaping and proved that K_{Qi} and K_{Ei} coefficients depend on grade M_i of bricks and can rise to *35-93 %* и and *4.6-11.2 %* respectively.

References

1. *GOST R 55646-2013 Resource Saving. Production of Ceramic Bricks and Stones. Guidance on Implementing of Best Available Techniques for Improving Energy Efficiency and Environmental Performance* (Standartinform, Moscow, 2014)
2. S.Ya. Galitskov, M.A. Nazarov, and K.S. Galitskov, International Journal of Applied Engineering, **10(20)**, 40846-40852 (2015)
3. M.A. Nazarov, V.I. Kichigin, and A.S. Fadeev, Procedia Engineering **153**, 509-512 (2016)
4. S.Ya. Galitskov, K.S. Galitskov, and M.A. Nazarov, Industrial and Civil Engineering **3**, 25-29 (2014)
5. S.Ya. Galitskov, and M.A. Nazarov, Fundamental Research **8-1**, 29-32 (2013)
6. S.Ya. Galitskov, M.A. Nazarov, K.S. Galitskov, and A.P. Maslyanitsyn, Science Review **12**, 200-203 (2013)
7. S.Ya. Galitskov, K.S. Galitskov, and M.A. Nazarov, Industrial and Civil Engineering **9**, 27-30 (2015)
8. S.Ya. Galitskov, K.S. Galitskov, and M.A. Nazarov, Science Review **14**, 213-218 (2015)

9. K.S. Galitskov, Industrial and Civil Engineering **6**, 59-63 (2015)
10. K.S. Galitskov, and M.A. Nazarov, Complex systems: control and modeling problems, 351-355 (2016)

Research on the nanolevel influence of surfactants on structure formation of the hydrated Portland cement compositions

Alexander Guryanov[1,*], *Sofya* Korenkova[1], and *Yulia* Sidorenko[1]

[1]Samara State Technical University, Institute of Architecture and Civil Engineering, Molodogvardeyskaya str. 194, 443001 Samara, Russia

Abstract. The research of the structure formation process on a nanolevel of the samples of hydrated Portland cement compositions containing the modifying additives has been conducted with the help of small angle neutron scattering method. Carbonate and aluminum alkaline slimes as well as the complex additives containing surfactants were used as additives. The influence of slimes and surfactants on structural parameters change of Portland cement compositions of the average size of the disseminating objects, fractal dimension samples is considered. These Portland cement compositions are shown to be fractal clusters.

1 Introduction

There is a formation of particles of calcium silicate hydrate (C-S-H) in the course of hydration of cement clinker. Complex gel framework of C-S-H particles is formed as a result of cement stone formation. Structural parameters of particles C-S-H and the cement matrix itself change when shifting from nanolevel to micro level in accordance with the sizes of the formed cement particles. The structural parameters also affect the properties of the formed cement stone. Firmness, durability, water resistance, frost resistance and other properties of cement stone depend on nanoparticles C-S-H form and sizes distribution, a ratio of free water and water filling the space between firm hydro silicates particles as well as pores and their form and size distribution are thus determined by structural parameters of material on micro and a nanolevel.

To create cement compositions with the particular set properties various additives (fillers, modifiers) having both artificial (e.g. carbon materials) and technogenic (slimes) origin [1, 2] are used. The applied additives influence the structure of initial materials and their operational properties [3-8]. It is particularly observed in the presence of nanodisperse component of the additives in cement compositions [9, 10].

The utilization of mineral slimes is a great problem for industrial enterprises. At the same time chemical composition stability and high dispersion allow to use them as additives in cement compositions.

* Corresponding author: gurjanovam@mail.ru

This paper presents the research of the structure formation process on a nanolevel of samples of the hydrated Portland cement compositions modified by additives of technogenic origin. The research has been conducted with the help of small angle neutron scattering method. Structural parameters of Portland cement compositions are determined. Surfactants influence on structural parameters is also considered in this paper.

2 Materials

Slimes are formed as a result of chemical reaction processes, dispersion and adsorption, as well as hashing, upholding and firm particles coagulation of oversaturated sewage solution in purification or water softening in industrial enterprises of metal working, oil processing, thermal power plants. Slimes are the suspension characterized by structure stability, composition homogeneity and composition constituency [11]. Slime formation conditions are as follows: sol-gel technology and content of nanodisperse component allowing to compare them to nanotechnogenic raw materials [9, 10].

In this work slimes which can be referred to as filling ones (carbonate slimes) or structure-forming (aluminum alkaline slimes) by targeting action were used as additives in the Portland cement compositions. Slimes were purified at the final stage of their formation. The chemical composition of slimes depends mainly on the type of formation engineering procedure and conditions [11]. Slimes composition used in this work is given in table 1.

Table 1. Mineral composition of slimes.

Slime sample	Al_2O_3 [%]	CaO [%]	MgO [%]	Fe_2O_3 [%]	SiO_2 [%]	SO_3 [%]	Losses in case of steaming [%]
Carbonate	0-13	36-47	2-14	2-13	2-4	0-6	34-41
Aluminum alkaline	43-59	0-9	0-4	0-7	0-1	0-8	32-57

Cement compositions samples were made on the basis of non additive Portland cement. In case of Portland cement hydration carbonate or aluminium alkaline slimes were added in quantity of 3-5 %. The water cement ratio ranged from 0.27 to 0.30. When adding surfactants the water cement ratio decreased to 0.20.

3 Research Methods

Structural parameters of Portland cement compositions samples were defined with the help of the small angle neutron scattering method (SANS) [12] allowing to examine over atom structure of substance within the distances from 1 nanometer to 100 nanometers. SANS intensity is defined by dispersing ability of non-uniformity and contrast on section phases boundaries allowing to study nano objects of different physical nature and aggregate state.

Dispersion by disperse systems is isotropic and SANS intensity $I(q)$ depends only on the transferred neutron impulse

$$q = (4\pi/\lambda)\sin(\theta/2), \quad (1)$$

where θ – scattering angle. For monodisperse systems SANS intensity is proportional to intensity of one particle dispersion averaged in all directions and bears in itself information about over atom structure of dispersing particle. The experiment results processing concerning SANS is implemented by means of the program complex ATSAS 2.1 [13].

Small angle neutron scattering intensity of by ensemble of particles or non-uniformity can be presented in the form [12]

$$I\,(q) \sim P\,(q)\,S\,(q), \tag{2}$$

where $P(q)$ and $S(q)$ – a shape factor and a structural factor of a particle, respectively. In case of execution of a condition of $qR_c < 1$ (Guinier regime) a structural factor of $S(q) = 1$. In this case intensity of small angle neutron scattering is defined by the characteristic size of particles R_c and shape factor of $P(q)$, that is the form of dispersing particles or non-uniformity:

$$I\,(q) \sim I\,(0)\,exp[\,-(qR_g)^2\,/\,3\,] \tag{3}$$

From data on small angle neutron scattering intensity in Guinier regime it is possible to evaluate the radius of a gyration of R_g of particles and their characteristic size R_c.

There is (Porod regime), degree dependence intensity of dispersion on the transferred impulse providing $qR_c > 1$:

$$I\,(q) \sim q^{-n} \tag{4}$$

From the degree index deviation n from Porod asymptotic ($n = 4$) it is possible to determine fractal dimensionality. In case of volume or mass fractals the index n corresponds to fractal dimensionality of D_V where $1 < D_V < 3$. Providing the dispersion by objects with fractal surfaces the index n is within the interval $3 < n = 6 - D_S < 4$, where D_S fractal surface dimensionality and $2 < D_S < 3$.

Thus, it is possible to determine fractal type by an inclination of straight sections of SANS dependences constructed in double logarithmic scale (volume or surface) and value of fractal dimensionality.

4 Results

Ranges of SANS intensity were measured in Petersburg Nuclear Physics Institute on the diffractometer "Membrana-2" set on the WWR-M reactor [14-16]. Neutron wavelength was $\lambda = 0.3$ nanometers, line width was $\Delta\lambda/\lambda = 0.3$. The range of the registered transferred neutron impulses of q was distributed within 0.03 nm^{-1} and 0.8 nm^{-1}. The neutrons dispersed on samples in the range of angles of $-2° \leq \theta \leq +2°$.

Centers (sources) of dispersion in the considered experimental range of transferred neutron impulses are nanosized non-uniformities of sample material density, that are coherent to wavelength of scattered neutrons.

Such centers in the considered samples of cement compositions are borders of various phases, grains, volume pores and pores on borders of phases or grains.

The composition of the studied cement compositions is brought in table 2. The neutrons dispersion sections of $\sigma(q)$ brought to volume unit of a sample and a single body space angle arose from experimental ranges of SANS. The background and transmission (a passing contribution through a bunch sample without dispersion) were considered. Data were normalized in relation to neutrons dispersion intensity by a standard sample (1 mm of H_2O). Figure 1 represents the experimental ranges of SANS received for cement compositions samples. Sample 2 is the hydrated Portland cement with addition of 5% carbonate slime. Sample 3 differs from previous sample where 0.2% of surfactant is added.

DOI: 10.1051/matecconf/20168604011

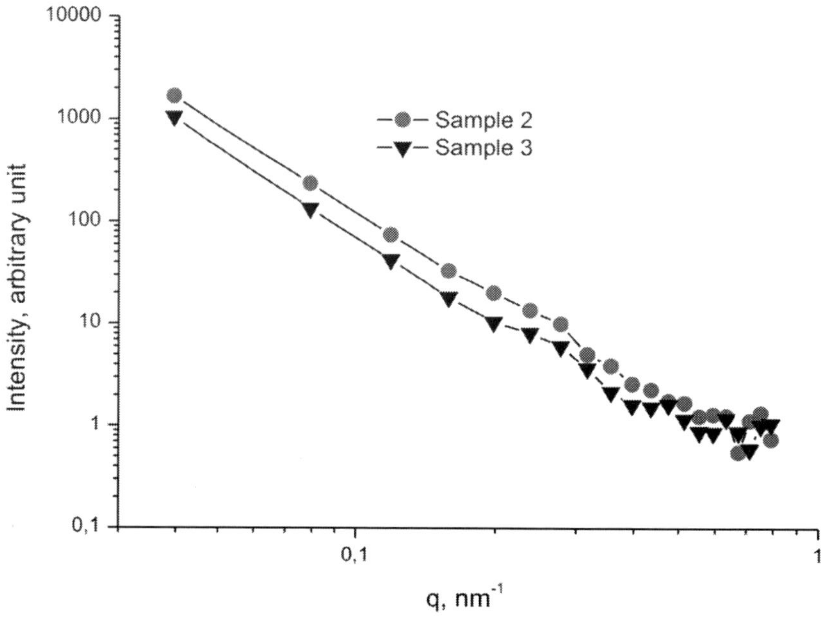

Fig. 1. Experimental ranges of SANS of cement compositions samples.

The characteristics of the studied Portland cement compositions samples, the calculated values of fractal dimension of D, the average size of particles R_c for the hydrated cement compositions samples containing modifying additives are represented in Table 2.

Table 2. Structural parameters of samples in Portland cement compositions.

Sample	**Additive type (slime)/[%]**	**Surfactant [%]**	**Water cement ratio**	***D*, 1 day**	***D*, 28 days**	***R_c*, nm**
Sample 1	-	-	0.30	3.49	3.31	65
Sample 2	Carbonate/5.0	-	0.27	3.40	3.30	60
Sample 3	Carbonate/5.0	0.2	0.20	3.28	3.21	45
Sample 4	Carbonate/3.0	-	0.30	3.42	3.30	58
Sample 5	Carbonate/3.0	0.2	0.22	3.30	3.24	44
Sample 6	Aluminum alkaline/4.5	-	0.27	3.36	3.20	50
Sample 7	Aluminum alkaline/4.5	0.2	0.20	3.25	3.18	40
Sample 8	Aluminum alkaline/3.0	-	0.30	3.38	3.22	52
Sample 9	Aluminum alkaline/3.0	0.2	0.22	3.29	3.19	40

In the table 2 pairs of samples 2 and 3, 4 and 5, 6 and 7, 8 and 9 had the identical chemical composition of added slimes. The average size of particles Rc was calculated for the 28th day after a gauging of cement compositions. The softener of the S-3 brand was applied as a surfactant.

5 Discussion

The use of modifying nanoadditives of technogenic origin (carbonate and the aluminum alkaline slimes) allows to influence directly on the process of hydration and formation of structural parameters of the hydrated Portland cement compositions and as a result on cement stone properties. Nanoadditives influence structural parameters of the hydrated Portland cement compositions concerning nanoparticles distribution by sizes, the average size of nanoparticles and fractal dimension. The studied cement compositions are formed by fractal clusters.

The obtained values of structural parameters of cement compositions show that slimes used as additives promote the increase in durability of contact zone particles as well as the increase water and corrosion resistance of cement stone. Directed change of structural parameters of cement compositions by adding surfactants is caused by the adsorptive modifying mechanism of dispersive systems.

References

1. S.F. Korenkova, Y.V. Sidorenko, A.M. Guryanov, European Journal of Natural History **2**, 34-36 (2012)
2. A.M. Guryanov, *Traditions and innovations in construction and architecture. Construction technologies* (SGASU, Samara, 2015)
3. S.F. Korenkova, Yu.V. Sidorenko, A.M. Guryanov, Scientific review **6**, 61-63 (2014)
4. S.F. Korenkova, Yu.V. Sidorenko, A.M. Guryanov, International journal of experimental education **6**, 102-103 (2012)
5. A.M. Guryanov, Industrial and civil engineering **6**, 72-75 (2015)
6. S.F. Korenkova, A.M. Guryanov, Yu.V. Sidorenko, Dry construction mixtures **3**, 17-19 (2012)
7. S.F. Korenkova, A.M. Guryanov, Yu.V. Sidorenko, *Traditions and innovations in a construction and architecture* (SGASU, Samara, 2013)
8. S.F. Korenkova, Yu. V. Sidorenko, A. M. Guryanov, *Traditions and innovations in a construction and architecture* (SGASU, Samara, 2014)
9. A.M. Guryanov, Procedia Engineering **111**, 283-289 (2015)
10. A.M. Guryanov, Procedia Engineering **153**, 217-222 (2016)
11. S.F. Korenkova, T.V. Sheina, *Foundations and concept of chemical seweage sedimentation utilization in industry* (SGASU, Samara, 2004)
12. D.I. Svergun, L.A. Feygin, *X-ray and neutron small angle scattering* (Science, Moscow, 1986)
13. P.V. Konarev, M.V. Petoukhov, V.V. Volkov, D.I. Svergun, J. Appl. Cryst. **39**, 277–286 (2006)
14. A. M. Guryanov, S.F. Korenkova, V.M. Lebedev, V.T. Lebedev, Yu.V. Sidorenko, *Traditions and innovations in construction and architecture* (SGASU, Samara, 2014)
15. A.M. Guryanov, S.F. Korenkova, V.M. Lebedev, V.T. Lebedev, Yu.V. Sidorenko, *Traditions and innovations in construction and architecture* (SGASU, Samara, 2013)
16. A.M. Guryanov, S.F. Korenkova, Yu.V. Sidorenko, V.T. Lebedev, V.M. Lebedev, *Traditions and innovations in construction and architecture. Natural sciences and technosphere safety* (SGASU, Samara, 2015)

Theoretical and practical aspects of aqueous solution sodium silicate modifying

Sergey Mizuryaev[1,*], *Alexander* Chiknovoryan[1], and *Galina* Solopova[1]

[1]Samara State Technical University, Institute of Architecture and Civil Engineering, Molodogvardeyskaya str. 194, 443001 Samara, Russia

Abstract. This research deals with the use of liquid glass in industry particularly for porous filler production. The aim of this paper is to show the necessity liquid glass modification for the purpose of its rheological characteristics change for raw granules formation and providing given structure after porization. Data on chemical liquid glass modification are provided by adding sodium chloride. Moreover, inert mineral additives influence on porous filler properties are shown in this paper. The basic principles of light concrete composition selection are specified. Test results of light concrete on the developed porous sodium silicate filler are given.

1 Introduction

Water solution of sodium silicate is generally obtained in the process of high pressure steam treatment of silicate lump and water [1-3]. It is like transparent viscous liquid having density 1200-1500 kg/m3 and high adhesive properties.

In Russia water solution of sodium silicate is better known under the trademark "liquid glass" or "soluble glass". Historically there is one more name – "Fuchsian glass" named after the developer of liquid glass industrial technology – J.N. Fuchs [1].

Abroad, this water solution of sodium silicate has also several names - "verre soluble", water-glass, soluble glass, solution of water sodium silicate, soda water glass.

The use of liquid glass is quite various [5-8]. About 30% of its volume is used in mechanical engineering and chemical industry. About 25-27% is used in mining industry and in paper production.

2 Main Part

About 9% of liquid glass is used in construction. In particular it is applied in construction hydro insulation, and can be found in the composition of concrete, solutions, paints, plastering, glues, etc. Despite the wide range of construction materials the use of liquid glass and its importance in construction is obviously underestimated. In particular the valuable property to expand under heat treatment is not practically used. For example, when heated to 250-300 degrees its volume increases almost in 70 times. It makes liquid

* Corresponding author: mizuriaev@gmail.com

glass very attractive, from the point of view of heat-insulating construction materials production, such as porous fillers that are of great demand nowadays [9-14].

High adhesive ability that makes the process of mass mixing more difficult and insufficient plasticity for granules formation as well as considerable unevenness of liquid glass expanding prevent distribution of porous fillers production on the basis of liquid glass from being wide.

One of possible solutions is the directed modification of liquid glass.

The research showed the following results:

- the physical modification of liquid glass which is carried out by addition of granular and (or) powdery inert materials (clay, quartz sand), is inefficient because of high adhesion and decrease in expanding;
- chemical modification due to introduction of acids, different salts is also inefficient since it always leads to instant coagulation and mixture hardening.

During the research of the directed liquid glass modification to develop the technology of porous fillers a number of tests concerning the influence of some chemical compounds on technological properties of liquid glass, in particular, viscosity, molding capacity and character were carried out.

In the conducted tests acids and salts of alkaline metals as modifying additives (MA) to liquid glass were used.

MA influence on liquid glass properties was defined in the following way. MA were added into continuous mixing of liquid sodium glass. After mixing for 30 minutes changes of viscosity and molding capacity were fixed. Molding capacity was defined whenever possible to form a granule with rather steady form.

Tests results showed that the only additive leading to the demanded modification of liquid sodium glass at which viscosity and molding capacity conformed to production requirements was sodium chloride. Sodium chloride interacting with liquid sodium glass reduces its silicate module and, as a result, reduces a number of siloxane bonds. It promotes transition of alkaline metal ions into solution. Thus, chlorine ions acting as a strong oxidizer promote "smooth", i.d. prolonged in time, coagulation of the modified liquid glass.

During thermal tests of the modified liquid glass it was established that heat treatment at 250-300° C of formed granules led to an intensive expanding of granules. Expanding resulted in granules to be close to spherical form. The structure of received expanded granules resembled expanded clay and they had high-porous internal "kernel" and external hard crust.

It should be noted that expanded granules obtained in this way had very low resistibility and were not waterproof.

Further research showed that the durability increase problem was solved due to the adding of various mill ground fillers into the modified liquid glass composition. The most effective additives were those that improve not only the strengthening function but some other characteristics. Thus some types of porous fillers (Figure 1, Table 1) were received. For example, the adding of mill ground chamotte provided heat-resistant filler [11, 12, 13]; clay-containing waste from coal washing provided utilization of similar multitonnage waste and clay gave the expansion of raw materials source for the production of demanded porous fillers.

Fig. 1. Types of porous filler on the basis of modified liquid glass. 1 – high-porous sodium silicate; 2 – heat-resistant; 3 – with the use of coal waste; 4 – with the use of ordinary non expanded clay.

Two main methods to make expanded granules water resistant were developed:

-the adding of clay-containing components into the modified liquid glass composition, for example, ordinary non expanded clay or clay-containing coal waste, and additional clay baking at 800-850 °C;

-processing of expanded granules with hydrophobic solutions [15].

The main physical mechanical characteristics of porous filler with mill ground filler in the form of non expanded clay are given in Table 1.

Table 1. Main properties of porous filler.

Indicators	After expanding at 250 °C	After clay baking at 880 °C	After hydrophobization
Bulk density, g/cm^3	0.34-0.38	0.35-0.4	0.38-0.42
Surface condition	dense crust		
Structure	dense		
Resistance*	1 conventional units	5 conventional units	2-3 conventional units
Water resistance**	less than 0.5 hours	more than 12 hours	more than 12 hours

* - resistance was estimated in conventional units, by comparing the resistance of initial and processed granules;

** - water resistance was estimated by the time spent in water preserving initial durability.

Authors faced the problem of conducting the tests concerning porous fillers in concrete of general construction application.

Filler on the basis of the modified liquid glass with addition of mill ground non expanded clays was taken as a basis. The used porous filler was produced in vitro on the working model of the rotating furnace. The granulated material of fraction of 2,5-5 mm, with a bulk density of 340 kg/m3 was received.

The ordinary Portland cement of resistance brand M400 was used.

Definition of concrete composition (concrete mixture), i.e. the material expense on 1 cubic meter of concrete mixture relied on the following known dependences [16]:

The average density of light concrete is defined as follows:

$$\rho = 1000\,\rho_{з} * \varphi + 1.15C + (1000*(1 - \varphi - (Vc+Va)/100 - C/\rho_{ц} - W) * \rho_{п} \quad (1)$$

where:

- $\rho_{з}$, $\rho_{п}$, $\rho_{з}$ – density according to large filler (more than 5 mm), small filler (sand) and cement;
- φ – volume concentration of large filler;
- Vc, Va – volume of intergrain cells and the involved air, % respectively;
- C, W – cement and water expenses for 1 m^3 of concrete mixture.

Resistance limit at light concrete compression is defined as follows:

$$Rc = 4Rf * (1 - \varphi) \tag{2}$$

where:
- *Rc* – concrete brand;
- *Rf* – branded resistance of large filler;
- φ – volume concentration of large filler.

Taking into consideration an objective to get concrete with the lowest density, porous filler consumption as a part of mixture was accepted as maximum. The testing showed that in this case the average consumption of porous filler was about 1.15 m^3 on 1 m^3 of concrete mixture. The cement consumption for various structures was taken in quantity of 200-250kg on 1 m^3. Water consumption was equal to all the compositions and was W/C = 0.72.

Consolidation of samples was carried out by 20 seconds vibration. The curing of produced concrete samples before test lasted 28 days in the following conditions: curing temperature – 20°C ; humidity – about 95-100% that meets to standard requirements.

Tests of the samples received showed the following results:
- density – 760 … 850 kg/m^3;
- compression resistance – 3.8 … 5.0 MPa;
- frost resistance – not below 25° F

The results show that the tested low density compositions have rather high resistance that is sufficient for the protecting constructions that confirms relevance and importance of results.

3 Conclusion

The technology of liquid glass modification with the purpose of changing its rheological properties, allows to expand considerably its use in production of construction materials, in particular, different porous fillers. Ordinary light concrete on the basis of such porous fillers have sufficient low density resistance for the production of concrete protecting constructions.

References

1. P. M. Grigoriev, M. A. Matveev. *Soluble glass*, (1956)
2. V. I. Korneev, V. V. Danilov. *Soluble and liquid glass*, (1996)
3. Yu.P. Karnaukhov, V.V. Sharova, Construction materials **11**, 14-16 (1994)
4. K.V. Ganechkina. *Receiving liquid glass on the basis of sub-standard raw materials*: Traditions and innovations in construction and architecture, the Electronic resource: materials of the 71st All-Russian scientific and technical conference following the results of NIR, 612-613 (2014)
5. A.M. Guryanov, S.A. Mizyuryaev, V. M. Lebedev, V. T. Lebedev. *The nanostructured porous materials on the basis of soluble sodium glass*, Sat. works V of the Euroasian scientific and practical conference "Durability of Non-uniform Structures", 136 (2010)
6. A.Yu. Zhigulina. *The structured heat-insulating material*: Traditions and innovations in construction and architecture. Materials of the 70th anniversary All-Russian scientific and technical conference following the results of NIR, 110-111 (2013)
7. A.I. Kudyakov, T.N. Radin, M. Yu. Ivanov, Construction materials **11**, 12-13 (2004)
8. S. A. Mizyuryaev [etc.] Industrial and civil engineering **6**, 47-50 2015

9. Yu.G. Ivashchenko, A.A. Surnin, N. V. Zobkova, I.L. Pavlova, Composition for the production of spherical granules for heat-insulating materials, the patent No. 2158716 RU, (2000)
10. S. A. Mizyuryaev, A.Yu. Zhigulina, A.N. Mamonov, N. V. Ivanova, Construction materials **7**, 12-13 (2011)
11. A.Yu. Zhigulina, S. A. Mizyuryaev. Composition for production of porous filler, patent No. 2211196 RU (2003)
12. B. S. Komissarenko, S. A. Mizyuryaev, A.Yu. Zhigulina, Researches in the field of architecture and construction. (1995)
13. S. A. Mizyuryaev, A.Yu. Zhigulin, Construction designs and materials. Express information **16**, (2002)
14. S. A. Mizyuryaev, A.Yu. Zhigulina, Materials VIII international scientific conference, 113-115 (2004)
15. S. A. Mizyuryaev, N. V. Ivanova, Materials of the 63rd All-Russian scientific and technical conference following the results of NIR, 563 (2007)
16. S.G. Vasilkov, S. P. Onatsky, M.P. Elinzon, etc. *Artificial porous fillers and light concrete on their bases*, (1987)

Increased durability concrete for generation of pillars power lines

Grigory Yakovlev[1], *Yuri* Bazhenov[2], *Alexander* Buryanov[2*] and *Igor* Pudov[1]

[1]Izhevsk State Technical University , Studencheskaya 7, 426069 Izhevsk, Russia
[2]Moscow State University of Civil Engineering, Yaroslavskoye sh. 26, 129337 Moscow, Russia

Abstract. In this researches multilayered carbon nanotubes of production of the French corporation "Arkema" were used. It has followed features: diameter of 10-15 nanometers and up to 15 microns long. Multilayered carbon nanotubes were used for increasing of physics and technology properties of cement concrete. It was established that at introduction of multilayered carbon nanotubes in amount of 0.006% of the mass led concrete durability increases by 28%, resistance to frost from F200 to F400, tightness to water from W8 to W14.

Despite keen interest of Russian and foreign researchers in nano-modification technology, many experts are skeptical about the hypotheses describing the mechanism of cement stone structure formation in presence of nanometer-sized carbon particles [1, 2]. One of the main reasons for skepticism about the impact of carbon nanotubes (CNTs) on the structure and properties of cement matrix is their incomplete dispersion in liquid phase.

Due to high surface energy the CNTs form globules during synthesis, the globules size ranges from 400 to 900 microns. The nanotubes are poorly distributed in aqueous medium and require special techniques for their dispersion [3-5]. The problem is not only to disintegrate the initial globules, but also to prevent CNTs coagulation process in water-dispersion system during storage.

Typically, CNTs are dispersed by prolonged mechanical grinding of initial product. However, there is a certain limit of dispersion due to the laws of thermodynamics, since the system tends to reduce free energy, which becomes apparent in reverse aggregation of particles [6-8]. Thus, a crucial technological problem is to disaggregate carbon nanotubes and to distribute them uniformly in concrete mix.

The studies [9-10] conducted in the Kalashnikov ISTU showed a significant improvement in physical and technical properties of concretes modified with carbon nanotubes. It was found [11] that such concretes had better strength, frost resistance and waterproof capacity when modified. All these parameters are crucial for the construction of reinforced concrete poles used for power lines.

Traditional concrete formulations do not provide necessary durability because of deterioration of, first of all, concrete reinforcement protective layer (Fig. 1).

Durability of reinforced concrete poles was increased by modifying cement concrete with multilayer carbon nanotubes Graphistrength Masterbatch CW2-45 (MCNTs) produced

* Corresponding Author: rga-service@mail.ru

by the French corporation "Arkema". To disperse the MCNTs a high-speed bead mill developed Novy dom LLC (Izhevsk city) was used. Currently CNTs dispersion is performed under Fulvec 100 trade name.

Fig. 1. Deterioration of reinforced concrete power line pole protective layer by frost (a fragment of the exposed reinforcement cage is shown in the upper corner).

Usage of the Graphistrength Masterbatch CW2-45 dispersion with 2% content of the MCNTs prepared in the bead mill ("Dispersion 2-BM") with B-30 heavy concrete in the amount of 0.006% of Portland cement weight allowed to increase samples strength by 28% (Table 1).

Table 1. Results of testing B-30 heavy concrete cube samples added with 2% CNTs Graphistrength Masterbatch CW2-45 dispersions on the 28th day under normal curing conditions and on the 1st day after steam curing.

Samples marking	Avg. strength, R_{28}^{c}, MPa	Relative change in strength, %
Reference sample (normal curing conditions)	54.1	-
Added with the "Dispersion 2-BM" (normal curing conditions)	69.3	+28.1
Reference sample (after steam curing)	36.4	-
Experimental sample added with the "Dispersion 2BM" (after steam curing)	41.8	+14.8

Usage of the MCNTs dispersion results in structural changes of cement stone involving formation of dense skin on the surface of solid phases, including cement and filler particles. Contact interactions of structured boundary layers result in formation of spatial frame cells in a modified cement matrix with dense layers of crystalline hydrates. This leads to hardening of the entire modified cement matrix.

High density hydrated calcium silicate-based spatial frame predetermines its greater strength, which leads to higher frost resistance of cement stone in concrete mixture (Fig. 2).

DOI: 10.1051/matecconf/20168604013

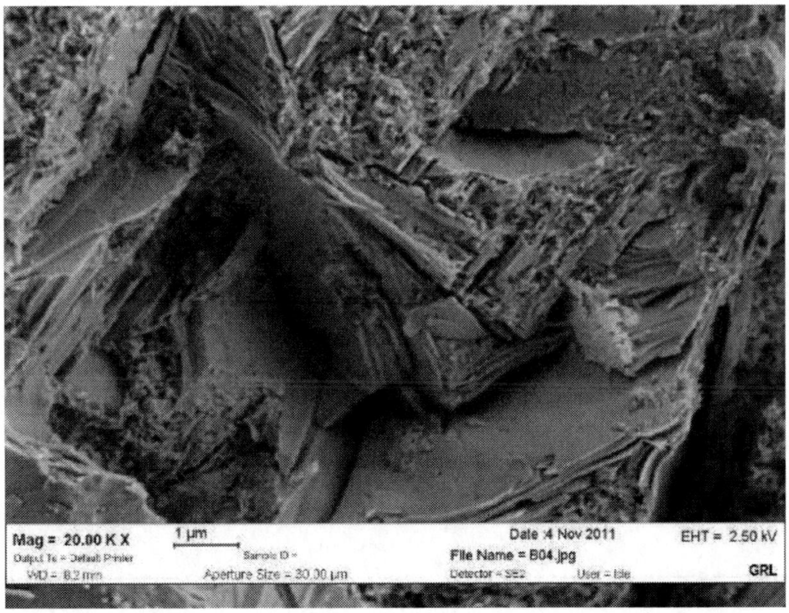

a)

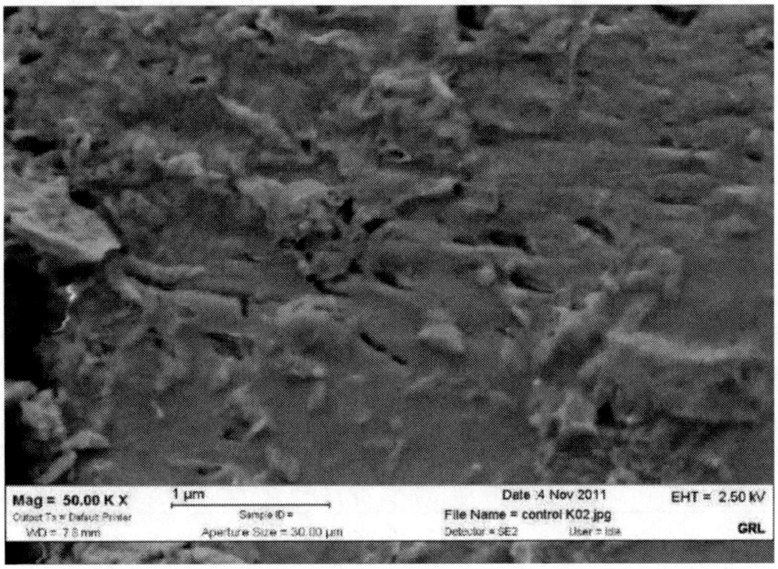

b)

Fig. 2. Fragment of B-30 heavy concrete reference sample surface microstructure after frost resistance test corresponding to F200 grade at 50,000x magnification (a); fragment of sample surface microstructure added with the CNTs after frost resistance test corresponding to F400 grade at 20,000x magnification (b).

CNTs-modified heavy concrete frost resistance test (strength class B-30) showed frost resistance increasing from F200 to F400 (Fig. 2b).

Concrete waterproof capacity is one of the most important parameters affecting its durability.

According to the tests, the reference samples concrete grade by waterproof capacity was W8. Experimental samples made of modified concrete were tested up to 7th stage inclusive at water pressure of 1.4 MPa. No water filtration signs in the form of drops or a wet spot on the upper end surface of the sample were observed. According to the tests, the experimental samples concrete grade by waterproof capacity was W14.

Increasing the grade by waterproofing capacity in experimental samples with the same cement-water ratio, subject to compliance with test procedure, confirms the assumption of a smaller number of interconnected micropores and capillaries making concrete permeable.

Pilot testing of the modifying CNT additive was performed in production of SV-95-3(CB-95-3a) a reinforced concrete pillars for power line VL 0.4–10 kV poles. Bench tests were performed at Reinforced Concrete Products Plant LLC (Izhevsk city). The poles were made of B-30 concrete in accordance with Specification 5863-007-00113557-94 using working drawings of 3.407.1-143 series standard structures. Basic parameters obtained during testing of SV-95-3a (CB-95-3a) reinforced concrete pillars for power line VL 0.4 – 10 kV poles are shown in Table 2.

Table 2. Results of factory tests of the SV-95-3-a reinforced concrete pillars for power lines VL 0.4 – 10 kV.

Stage No.	**Load, kgf**	**Holding time, min.**	**Actual bending, mm**		**Width of cracks opening, mm**		**Requirements of the Specification 53163-007-00113557-94 by strength, hardness, cracks resistance**
			with CNTs	**without CNTs**	**with CNTs**	**without CNTs**	
1	100	10	26	35	not found	not found	-
2	250	10	92	103	not found	cracks less than 0.05	-
3	330	30	142	162	cracks less than 0.05	cracks less than 0.05	cracks are ≤0.05 mm
4	400	30	167	195	cracks of 0.05	cracks of up to 0.1	bending is ≤ f=400 mm
5	500	10	237	277	cracks of up to 0.1	cracks less than 0.15	-
6	560	30	292	348	cracks of up to 0.15	cracks of up to 0.15	no deterioration, bending is ≤ f = 400 mm, cracks are ≤ 0.15 mm

The tests showed that the SV 95-3a (CB 95-3a) reinforced concrete pillars of the VL 0.4...10 kV poles made of heavy concrete modified with the CNTs dispersion comply with the Specification 53163-007-00113557-94 by strength, hardness and cracks resistance. A decrease in bending of the reinforced concrete pillars with the CNTs by 20% on the average in comparison with the reference products made of concrete without modification was also noted.

Economic efficiency calculation showed that with the increase in market value of the SV-95-3a (CB-95-3a) reinforced concrete pillars by 5.0% due to modification with carbon nanotubes, their service life will increase by at least 8 years due to increased durability. The economic effect of concrete modification will be 40% of the cost of a reinforced concrete pillar made without the use of the modifying additive.

References

1. I.A. Pudov, A.V. Pislegina, A.A. Lushnikova, G.N. Pervushin, G.I. Yakovlev, O.L. Khasanov, A.A. Tulaganov, The II International Conference NTC-2010 "Nanotechnology for green and sustainable construction", 34-38 (2010)
2. B.S. Sindu, S. Sasmal, S. Gopinath, Construction and Building Materials **50**, 317-327 (2014)
3. A. Sobolkina, V. Mechtcherine, V. Khavrus, D. Maier, M. Mende, M. Ritschel, A. Leonhardt, Cement & Concrete Composites **34**, 1104-1113 (2012)
4. A. Sobolkina, V. Mechtcherine, C. Bellmann, V. Khavrus, S. Oswald, S. Hampel, A. Leonhardt, Journal of Colloid and Interface Science **413**, 43-53 (2014)
5. O. Mendoza, G. Sierra, J.I. Tobyn, Construction and Building Materials **47**, 771-778 (2013)
6. B. Wang, Y. Han, S. Liu, Construction and Building Material **46**, 8-12 (2013)
7. I.A. Pudov, G.I. Yakovlev, A.A. Lushnikova, O.V. Izryadnova, Intelligent Systems in Production **2**, 285-293 (2011)
8. G.D. Fedorova, A.E. Savvina, G.I. Yakovlev, I.S. Maeva, S.A. Senkov, Construction Materials **2**, 48-54 (2013)
9. G.I. Yakovlev, S. Soliman, G.N. Pervushin, I.A. Pudov, M. Saber, Construction Materials **11**, 3-5 (2011)
10. G. Yakovlev, G. Pervushin, I. Maeva, Ja. Keriene, I. Pudov, A. Shaybadullina, A. Buryanov, A. Korzhenko, S. Senkov, Procedia Engineering **57**, 407-413 (2013)
11. A. Korzhenko, M. Havel, P. Gaillard, G.I. Yakovlev, G.N. Pervuchin, D.V. Oreshkin Procede D'introduction de nanocharges carbonees dans un inorganique durcissable. Patent № 2 969 143. C 04 B 16/12 (2012.01), C 04 B 28/00. Bulletin 12/25 pub. 22.06.12

Determination of gas consumption rate using the thermo technical calculations

Viktor Zhyila[1], *Anna Malyisheva*[1], *Maksim Botnar*[1,*], *Aleksey Ananyev*[2]

[1] Moscow State University of Civil Engineering, Russia
[2] Russian Academy of Architecture and Construction Sciences (NIISF RAASN), Russia

Abstract. At present time the problem of energy efficiency in field of housing and utilities infrastructure is very actual. Gas fuel consumption by population is the main factor influenced on achievement of high values in energy efficiency. One of the stage of optimization and efficiency upgrading of gas consumption is introduction of gas-meter containing smart-cards. This device allows optimization as well as monitoring gas consumption according to real data (not according to data of consumption rate)

1 Introduction

At present time gas-meters generally are absent with consumers. For determination of norm of gas consumption the Building regulations, manuals and recommendations are used. Norms of gas consumption for cooking and water heating when application of gas-stoves and boilers in house buildings should be determined according to SP 42-101-2003 presented in Table 1.

Table 1. nnual gas fuel consumption for house buildings

Gas customers	Gas consumption index	Heat consumption rate, MJ
House buildings		
1. In case, the gas-stove and centralized hot-water supply system present	For a one person per a year	4100
2. In case, the gas-stove presents but centralized hot-water supply system absents	For a one person per a year	6000
3. In case, the gas-stove and gas water heater present but centralized hot-water supply system absents	For a one person per a year	10000

* Corresponding author: botnarmi@gmail.com

Methodology

Calculation of gas consumption in house and public buildings is based on methods of heat losses determination those depend on building type, number of floors and climate conditions.

Gas consumption rates for heating of house buildings are determined taking into account calculated heat losses recovered by local gas heating systems.

Main part. Heat consumption in house building depends on some parameters. The ones of the most important are followings: building type, number of floors as well as realization degree of energy-efficiency procedures when construction process. Calculation of heat consumption it is necessary to accomplished taking into account the followings building types:

1) Buildings constructed before 1985

a) without considering of energy-saving procedures application (1-2-storied, 3-4-5-storied buildings and higher buildings);

b) taking into account of energy-saving procedures application (1-2-storied, 3-4-5-storied buildings and higher buildings);

2) Buildings constructed after 1985 (1-2-storied, 3-4-5-storied buildings and higher buildings).

There is a building type, where local heating systems are set up. In this case the heat loss calculation and gas consumption are determined by following eq.:

$$Q_0^{an} = 24Q_{max}\frac{t_{ext}-t_{avo}}{t_{av-in}-t_0}n_0, \quad (1)$$

where Q_0^{an} is annual heat consumption for house buildings, W;

Q_{max} is maximum heat flow for house buildings with calculated external temperature for heating process design (text), W;

t_{av}-in is average internal temperature in heated buildings, °C;

t_{av0} is average external temperature for average daily temperature up to 8 °C (heating season), °C;

n_0 is duration of heating season, days.

Maximum heat flow for house buildings Q_{max} is determined by the following eq.:

$$Q_{max} = q_0 A \quad (2)$$

where q_0 is integrated value of maximum heat flow for house buildings for 1 m^2 of total area, W;

A is total area of house buildings determined as a sum of all apartment areas in these buildings, m^2.

According to SNiP2.08.01-89 «House buildings» the total apartment area is a sum of their areas taking into account the followings decreasing coefficients: built-in closets – 0.5; recessed balconies and balconies and verandas – 0.3; terraces and refrigerating chambers – 0.1.

Total area is determined by internal measurement of all rooms at floor level without considering a baseboard. At the same time, area occupied by heating stove is excluded of total area.

When heating systems design the value of tav-in is determined according to SNiP 2.04.05-91 «Heating, ventilation and conditioning».

In this case the real average internal temperature can be used.

Values of t_0, t_{av0}, n_0 are determined according to SNiP 2.01.01-82 «Building climatology and geophysics» or according to data of meteorological service.

Annual gas consumption for heating of house buildings is determined by following eq.:

$$G = \frac{3{,}6Q_0^{ron}}{Q_0^n * n}, \qquad (3)$$

where G is annual gas consumption for heating of house buildings, m^3;
Q_0^n is minimum of calorific value of gas fuel, kJ/m^3 (kkal = 4,1868 kJ);
n is coefficient of efficiency for heating system.

For heating stove the parameter n= 0.7–0.75; for heating equipment the parameter n is determined according to machinery certificate.

To simplify the payment procedure for customers the annual gas consumption rate for heating process is equidistributed monthly during a year.

In this case the average monthly of gas consumption rate is determined by follow eq.:

$$P_{im} = \frac{G_i}{12a_i}, \qquad (4)$$

where P_{im} is average monthly of gas consumption for heating of house buildings in i-group (m^3) per 1 m^2 of heated area;
G_i is annual gas consumption for heating of house buildings in i-group, m^3;
a_i is heated area of of house buildings in i-group, m^2.

Gas consumption for local heating systems is determined using average values of heated area by following:

$$P_m = \sum_{i=1}^{re} P_{im} * d_1, \qquad (5)$$

where P_{im} is average monthly gas consumption in oblast (region or republic in Russian Federation), m^3 per 1 m^2 of heated area;
d_1 is part of heated area for heating of house buildings in i-group generally in oblast (region or republic in Russian Federation) vs. area heated with local heating systems;
m is number applied groups of buildings.

Calculation of gas consumption rates realized with selected measures (for exp. the first group of apartment).

According to previously selected 10 proxy objects (n`=10) the followings real gas consumption for a one person are recognized: 9,6; 9,2; 7,8; 8,5; 6,7; 7,0; 6,6; 6,8; 7,5 и 8,3 nm^3 (Table 2).

Using the eq. (2) the parameter σ_E^2.is calculated.

Table 2. Determination of universe variance

Number of proxy objects (i)	Real gas fuel consumption for a one person per a month, nm^3	$(x_1 - x^{\sim})^2$
1	9,6	$(9{,}6 - 7{,}8)^2 = 3{,}24$
2	9,2	$(9{,}2 - 7{,}8)^2 = 1.96$
3	7,8	$(7{,}8 - 7{,}8)^2 = 0$
4	8,5	$(8{,}5 - 7{,}8)^2 = 0{,}49$
5	6,7	$(6{,}7 - 7{,}8)^2 = 1{,}21$
6	7,0	$(7{,}0 - 7{,}8)^2 = 0{,}64$
7	6,6	$(6{,}6 - 7{,}8)^2 = 1{,}44$
8	6,8	$(6{,}8 - 7{,}8)^2 = 1{,}00$
9	7,5	$(7{,}5 - 7{,}8)^2 = 0{,}09$
10	8,3	$(8{,}3 - 7{,}8)^2 = 0{,}25$
n`=10	x=7,8	$\sum_{i=1}^{n`}(x_1 - \sim x)^2 = 10{,}32$

$$O_x^2 = \frac{10{,}32}{10} = 1{,}032 \tag{6}$$

According to eq. (4) the parameter O_x^2 is determined by following:

$$O_x^2 = \frac{1{,}032*10}{1-\frac{10}{280000}} = 10{,}32 \tag{7}$$

where 280000 is total apartment number in the first group.

Then, with eq. (1) the calculation of minimum required apartment sample number in this group is calculated by following:

$$n = \frac{280000*1{,}645^2*10{,}32}{280000*0{,}8^2+1{,}645^2*10{,}32} = 44,$$

where 0,8=10 % of tentative average gas consumption rate that is equal to 8,0 nm^3 for a one person per a month.

Real number of selected proxy objects is 86 (44 at least). Effective gas fuel consumption is based on application of gas-meters with max coefficient of efficiency.

Targeted indicators represented change in measures are calculated relatively to current year vs. previous year. Targeted indicators represented gas-meter availability are calculated according to objects connected with gas-supply system.

Determination of targeted indicators for gas-supply systems as well as efficiency upgrading under equal conditions the variation in structure and volume of requirement in resources those are extra of efficiency upgrading procedures are should be taken account.

Table 3. Calculation of efficiency of gas fuel consumption

№	Parameters	Unit measure	Year		
			2010	2011	2012
2	Gross Regional Product in Moscow oblast	trillion.rub.	2	2,1	2,2
	Total volume of costumed gas fuel	103 m3	450 812,65	443 973,30	324 135,39
4	Volume of costumed gas fuel, taking into account with gas-meter	103 m3	330 808,86	375 849,30	303 822,96
5	Pipeline rates according to gas-meter	rub/103 m3	2,75	3,3	3,795
6	Efficiency of gas fuel consumption	%	73,38	84,66	93,73

The data in Table 3 are actual as at October, 2012.

Conclusions

According to the above data, the volume of gas fuel, consumed by «Odintsovomezhregiongas» (branch of «Mosoblgas» company), taken into account with gas-meters grows annually. At the same time, efficiency of gas fuel consumption also grows by 10 % averagely.

In compliance with it we can conclude the efficiency of gas fuel consumption is significantly increased vs. 2010 due to application of gas-meters with smart-cards.

References

1. V.V. Prokofieva, Z.V. Bagautdinov, *Construction materials based on magnum silicates*

(Stroyizdat SPb, St. Petersburg, 2000)
2. K.I. Lushin, Plumbing equipment, heating, air conditioning **3(102) (**2012)
3. Yu.A. Tabunschikov, D.V. Koptev, V.A. Zhyila, A.K. Klochko, E.B. Solovyova, Bulletin of MGSU **8**, 222–229 (2011)
4. E.S. Gogina, O.A. Ruzhitskaya, O.V. Yantsen, Advanced Materials Research **919-921**, 2145–2148 (2014) DOI: 10.4028/www.scientific.net/AMR.919-921.2145
5. V. Orlov, I. Averkeev, Applied Mechanics and Materials **580–583**, 2384–2388 (2014)
6. A. Volkov, A. Sedov, P. Chelyshkov, E. Kulikova, Applied Mechanics and Materials **584–586**, 761–764 (2014)

Effect of limestone and granite coarse aggregate on drying shrinkage of a concrete

Piotr Woyciechowski[1,*], *Karol* Chilmon[1] , and *Joanna Julia* Sokołowska[1]

[1]Department of Building Materials Engineering, Warsaw University of Technology, 00-637 Warsaw, Poland

Abstract. In this paper the effect of limestone and granite aggregate on drying shrinkage of a concrete is investigated. Mixes containing limestone and granite with different D_{max} and cement paste volume were tested. A two level, full factorial experiment with three independent factors was conducted which allowed not only to determine the influence of every single factor itself but also potential interactions between them. Measurements of shrinkage were performed up to 98 days after demolding on the beams (10x10x50 cm), in accordance with the procedure from polish standard PN-84/B-06714/23 (Amsler's method). The multivariate analysis of variances (MANOVA) was used for the statistical analysis of the results. The obtained results confirm that the cement paste volume has a decisive influence on a drying shrinkage of a concrete. However statistical analysis showed that with the assumed variability, either the D_{max} or type of coarse aggregate has no significant influence on drying shrinkage, still those parameters may potentially be crucial in some cases. A statistically significant interaction between D_{max} and type of aggregate has been found. Furthermore the results of drying shrinkage measurements were compared with estimation results from Bażant-Baweja B3 Model and Eurocode 2. The comparison shows that B3 Model provided more accurate results than Eurocode 2.

1 Introduction

Drying shrinkage plays a significant role in concrete. It causes the vast majority of the total shrinkage strain in concrete [1]. This process, which is caused by migration of water molecules from inside of the concrete, can last up to several decades. The water molecules separate from the surface or from the space between the particles of C-S-H phase, the thickness of the diffusion layer decreases, which is followed by contraction of the cement grains, whereupon the material shrinks. The intensity of the phenomenon depends on many factors, among which the most important are: air relative humidity (RH), water-cement ratio (w/c), cement paste content (by volume), concrete element surface to volume ratio, time and type of curing, concrete pore structure and mechanical properties of concrete.

In general, aggregate enables to reduce the shrinkage strain in concrete. This is due to the reduction of the cement paste content in the concrete mix and by limiting the

* Corresponding author: p.woyciechowski@il.pw.edu.pl

 DOI: 10.1051/matecconf/20168604015

deformation of the cement matrix (stiffening of the concrete structure). The degree of shrinkage deformation limit depends mainly on the geometrical properties of the aggregate and the geomechanical properties of the rock from which aggregate was obtained. The shape of the grains, roughness, size, elasticity modulus of the rock, porosity, chemical composition - all these properties theoretically affect the deformation properties of the concrete which - in a practical way - is determined by the value of elasticity modulus obtained during the compressive strength test. Therefore, it seems reasonable to investigate which of these characteristics in a practical way can be used at the stage of designing the concrete aggregate mix granulation that prevent or limit the shrinkage of concrete. This paper presents the results of research conducted on determining the crucial properties of coarse aggregate that influence the drying shrinkage of concrete.

2 Experimental studies

2.1 Research design

The study was performed according to two-level research design of three independent variables: cement paste content (by volume), maximal size (diameter) of coarse aggregate grains and the type of aggregate. Based on the design eight concrete compositions were designed (Fig. 1). As constants adopted: sand content (%), water-cement ratio (kg/kg), and a method of forming, maintenance and storage of specimens.

Aggregate was characterized using results of tests performed earlier by Pinińska [2,3] – the analyzed property was average static modulus of rock. The value of the modulus was determined during compressive strength test performed on the core specimens of size 5 cm x 5 cm. The test was done assuming the constant increase of circumferential strain compressed specimens. The value of the modulus was calculated on the basis on the straight sections of pre-critical part of stress-strain curves. Because of the different research method, elastic modulus determined using this method cannot be related directly to the secant modulus of elasticity of concrete, and therefore this value is used only to compare the characteristics of various types of aggregate applied into concrete mix in the presented study.

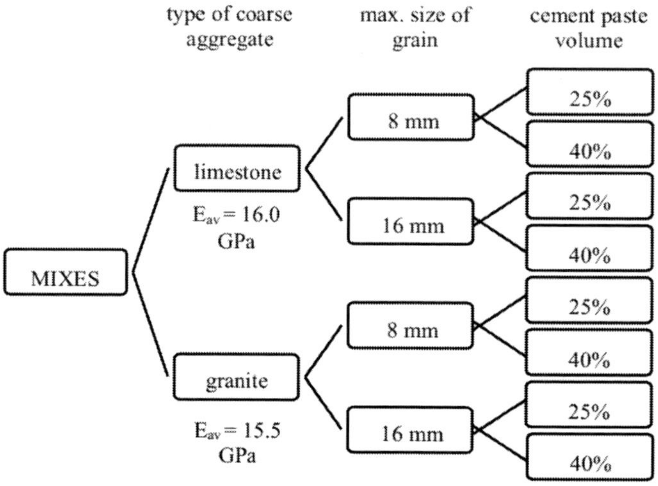

Fig. 1. A two level full-factorial research design used in experimental studies.

2.2 Materials

The cement used to prepare concrete tested in the study was Portland cement CEM I 42,5R, the applied aggregate was river sand and two fractions of gravel: 2/8 mm and 2/16 mm. The coarse aggregate was crushed politic limestone (source: Sitkówka-Nowiny, Świętokrzyskie Mountains) and crushed granite (source: Strzegom-Sobótka, Lower Silesian region). The main mechanical properties of applied aggregates are given in table 1.

Table 1. Mechanical properties of rocks from which coarse aggregate used in this studies was obtained [2,3].

characteristic	granite min – max (average)	politic limestone min – max (average)
elastic modulus (static), GPa	9.6 – 23.8 (15.5)	13.4 – 17.8 (16.0)
compressive strength, MPa	60.7 – 174.4 (120.8)	127.0 – 218.0 (181.0)

2.3 Concrete composition

In the table 2 one can find the quantitative compositions of concretes tested in the presented study. The compositions were designed according to the statistical design described in the paragraph 2.1.

Table 2. Compositions of tested concretes.

		Mix 1-2 0/8 mm 40% cement paste lim. / granite	Mix 3-4 0/8 mm 25% cement paste lim. / granite	Mix 5-6 0/16 mm 40% cement paste lim. / granite	Mix 7-8 0/16 mm 25% cement paste lim. / granite
Cement, kg/m^3		486	304	486	304
Water, kg/m^3		243	152	243	152
Aggregate, kg/m^3	0/2	666	832	666	832
	2/4	297	371	88	109
	4/8	642	802	321	396
	8/16	-	-	530	668
	ALL	1605	2005	1605	2005
superplasticizer, % m.c.		0.91-1.12	1.14-1.28	0.84-1.07	1.07-1.20
slump, mm		90-120	70-100	100-130	80-100

2.4 Drying shrinkage measurements

The shrinkage was tested according to Amsler method; the tested specimens were in the shape of beams of size 100x100x500 mm with special metal grips located on the edges that enabled measurements of the length change in the Amsler apparatus (Fig. 2a) in comparison to metal calibration element (Fig. 2b).

a) b)

Fig. 2. Amsler's test apparatus a) frame with micrometrical screw on its end b) steel calibration element.

The specimens were cured for 24 h in the laboratory conditions in the molds (covered with foil), then after remolding the first shrinkage measurement was performed – the value was considered to be the reference point. Next, the beams are stored for the whole period in a temperature range of 18-22°C and average relative humidity of 80% within the first 28 days, and during the next 70 days – in relative humidity of 50%. Such humidity curing conditions enabled to simulate the variation of relative humidity registered during the year in the central Poland area. Measurements of shrinkage were made in successive time intervals ranging from 1 to 98 days of curing.

2.5 Effect of variables on concrete drying shrinkage

2.5.1 Main effects

For the statistical analysis of the significance of the impact of the volumetric content of the cement paste in the concrete mix, the type of aggregate and the maximum aggregate size the results of shrinkage test obtained after 98 days were used. These values were subjected to multivariate analysis of variance (MANOVA) by which it was possible to determine separately the main variables influence and determine the interactions. The statistical analysis was performed with statistical significance of $\alpha = 0.05$.

Preliminary analysis showed that the distribution of the shrinkage values characterized by a distinct scattering data on both sides of the median (Fig. 3a). Among the data prevailed values located around the three vertices (Fig. 3b), which indicated a strong influence of at least one variable.

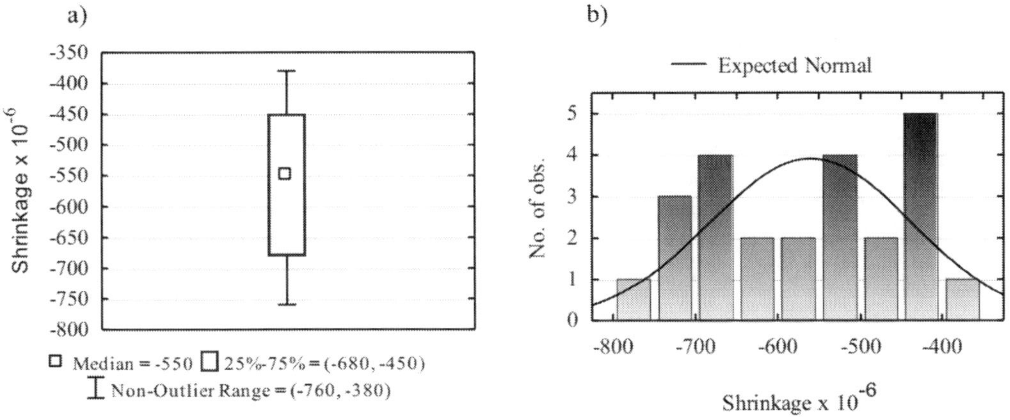

Fig. 3. Basic statistics for obtained data a) Box-Whisker plot, b) histogram.

Before performing the multi-variance analysis, authors examined the conditions for its application. Normality distribution was confirmed using the K-S test corrected for small populations (the null hypothesis accepted with $p = 0.2 > 0.05$) and Shapiro-Wilk (the null hypothesis accepted with $p = 0.6 \gg 0.05$). The homogeneity of variance for all groups was confirmed by the Leven's test.

As a result of multi-variance analysis of the test results it was showed that the strongest influence on the shrinkage value presents the content (by volume) of the cement paste in the concrete mix (see Fig. 4a and Table 3). At the same time it was found that neither the change of the aggregate type – i.e. limestone to granite (Fig. 4b), nor the change of maximal size of aggregate (expressed by maximal diameter of the aggregate grain, D_{max}) from 8 mm to 16 mm (Fig. 4c) do not affect significantly the value of shrinkage (Table 3).

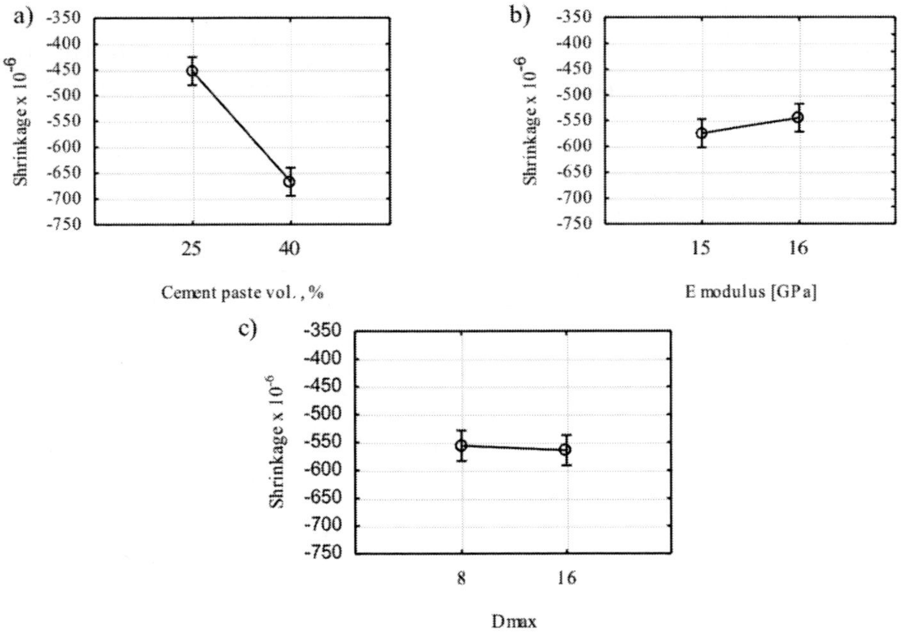

Fig. 4. Relation between variables and shrinkage a) cement paste content (by volume) vs. shrinkage b) E modulus vs. shrinkage (15 GPa limestone, 16 GPa granite c) maximal aggregate size, D_{max} vs. Shrinkage.

Table 3. MANOVA results for main effects (MS – Mean Square , F - F test statistic, p – probability used to determine statistical significance, significance level $\alpha = 0{,}05$).

Effect	MS	F	p-value	observed power
Cement paste vol.	277350	140.73	$< 10^{-6}$	1.00
E modulus	5400	2.74	0.12	0.34
D_{max}	417	0.211	0.65	0.07

2.5.2 Interactions

There was observed one statistically significant interaction – the interaction between the type of aggregate and the maximum diameter of the aggregate (Table 4). The shrinkage deformation value decreased by 5% when the maximum grain size of 8 mm was changed to 16 mm in the case of crushed granite and increased by 11% in case of limestone aggregate.

Table 4. MANOVA results for interactions (MS – Mean Square , F - F test statistic, p – probability used to determine statistical significance, significance level $\alpha = 0.05$).

Effect	MS	F	p	observed power
Cement paste vol. x E modulus	6017	3.05	0.10	0.38
E modulus x D_{max}	16017	8.13	0.01	0.76
Cement paste vol. x D_{max}	8067	4.093	0.06	0.48
Cement paste vol. x D_{max} x E modulus	600	0.304	0.59	0.08

2.5 Drying shrinkage in time

The shrinkage development registered during the period of 98 days in all considered cases were similar. The difference was observed when analyzing the expansion of concrete with various aggregates the noticeable expansion was noted during first days of curing in case of concrete with limestone aggregate of size up to 8 mm and concrete with granite aggregate of size up to 16 mm. This was due to the absorption of water by the aggregate and then water was gradually released back into the concrete. The cement matrix because of lack of water demanded for hydration in the early phase of curing absorbed water from the humid environment causing extension of concrete. The shrinkage increased rapidly during first 10-14 days, and then the rate of process slowed down. Lowering the relative humidity after 28 days (from RH = 80% to 50%) did not significantly affect the development of shrinkage over time.

The obtained results were compared with the results of the estimated shrinkage values with use of two theoretical models: B3 Bażant-Bawej model [4] and the model given in standard EN 1992-1-1 [1] used by concrete structures designers (Fig. 4). The calculations assume an average relative humidity of 65% test conditions and takes into account the compressive strength obtained after 28 days of curing (data for concrete mix of 40% cement paste content - concrete fulfilling the requirements of concrete compressive class C30/37, and of 25% cement paste content - concrete fulfilling the requirements of class C20/25).

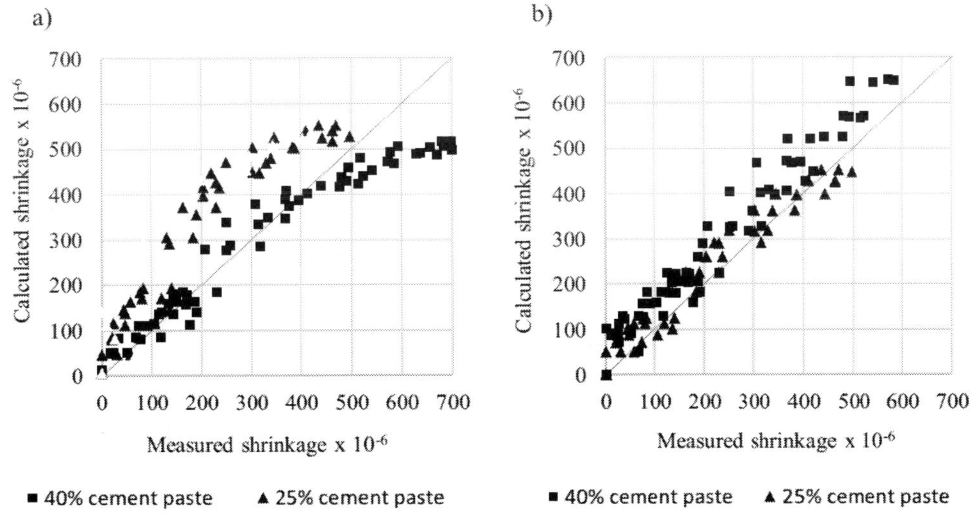

Fig. 4. Theoretical models vs. obtained data a) Eurocode 2 vs data b) Bażant-Baweja B3 Model vs data.

3 Conclusions

The content of cement paste in the concrete plays a decisive role in shaping the drying shrinkage. Although the change of the maximum grain diameter of 8 mm to 16 mm has not influenced the shrinkage deformation value significantly, it has caused different course of shrinkage of concrete with limestone and granite aggregate. This could be related to different adhesion between those two aggregates and cement matrix. Calcium carbonate present in the limestone reacts with the calcium aluminates in Portland cement. As a product of the reaction, the crystals of calcium carbo-aluminate ($C_3ACaCO_3H_{11}$) appear. They clearly improve the mechanical properties of the contact layer between cement matrix and aggregate. Meanwhile the adhesion between cement matrix and granite aggregate is purely mechanical.

However the type of the aggregate has not affected significantly the drying shrinkage value. Concrete with limestone aggregate showed a slightly higher shrinkage than the one with granite aggregate. This could be due to a greater diversity of deformation properties of aggregates than the average values accepted in the study (in case of granite one can clearly find higher range of variation of Young's modulus than in case of limestone). In order to verify this hypothesis there should be determined the secant modulus of elasticity of concrete. Use of aggregates with more diverse characteristics of deformation properties may cause more significant influence of this variable.

It was proved, that lowering the relative humidity from 80% to 50% after 28 days did not result in observable change in the growth rate of shrinkage deformation of tested concretes. Albeit preserving the relatively high relative humidity during the first period, the shrinkage grew rapidly.

Comparing shrinkage measurements results obtained in the experiment with theoretical models shows that B3 Bażant-Bawej model provides a more accurate estimation of shrinkage deformation than Eurocode 2 model. This is due to a greater number of variables taken into account in model B2. In the case of Eurocode 2 model occurred noticeable underestimation of shrinkage deformation of concrete with a higher content of cement paste, especially in the long term, and underestimation for concrete with an average content of cement. In turn, the model B3 overestimated results in case of concrete with higher cement content. Both models deliver better results in case of estimating shorter periods.

References

1. *EN 1992-1-1:2004 - Eurocode 2: Design of concrete structures - Part 1-1: General rules and rules for buildings* (2004)
2. J. Pinińska, *Właściwości wytrzymałościowe i odkształceniowe skał. Część II. Skały magmowe, osadowe i metamorficzne regionu Sudetów. Katalog.* (Wydział Geologii Uniwersytetu Warszawskiego, Warszawa, 1997)
3. J. Pinińska, *Właściwości wytrzymałościowe i odkształceniowe skał. Część I. Skały osadowe regionu świętokrzyskiego. Katalog.* (Wydział Geologii Uniwersytetu Warszawskiego, Warszawa, 1994)
4. Z.P. Bażant, S. Baweja, *ACI Special Publications* (2000)

Cost-effective design of reinforced concrete with use of self-terminated carbonation model

Piotr Woyciechowski[1,*], *Joanna* Sokołowska[1]

[1]Warsaw University of Technology, Department of Building Materials Engineering, al. Armii Ludowej 16, Warsaw 00-637, Poland

Abstract. According to Eurocodes EC0 and EC2 designing of concrete structure durable in terms of carbonation is assured by selection of suitable thickness of reinforcement concrete cover. The selection is done on the basis of structure category and concrete strength class, regardless of the concrete material composition or technological type, thus selected value is an estimation, often exaggerated. The paper presents elaborated "self-terminated carbonation model" that includes abovementioned factors and enables to indicate the maximal possible depth of carbonation. The model, in contrast to parabolic models published in the literature, is a hyperbolic function of carbonation depth in time. The paper explains why such model describes the phenomenon of carbonation better than others. The paper contains an example of calculation of the cover thickness using that model.

1 Introduction

Durability is one of the important determinants of building material sustainability as well as sustainability of the structure. Durability of reinforced concrete structures exposed to any environment depends on the ability of both – concrete and reinforcement – to resist the environmental factors. The most common cause of the reinforced concrete damage is the corrosion of steel due to lack of providing it with efficient protection by concrete cover. The protective abilities of concrete cover deteriorate with time because of the synergistic action of a number of physical and chemical factors. One of the most destructive factors apart from local climatic phenomena (including frost or chemical aggression) is decrease in concrete pH value due to the activity of atmospheric carbon dioxide. Providing the durability of reinforced concrete structure working under certain environment conditions depends on providing proper (1) durability of concrete, (2) proper thickness of the concrete cover, as well as taking into consideration (while designing) serviceability limit states in terms of cracks, namely (3) calculating crack width which would not exceed the Eurocodes limits (EN 1991 *Eurocode: Basis of structural design* – "EC0" and EN 1992 *Eurocode 2: Design of concrete structures* – "EC2"). Concrete elements and structures should meet the design requirements of the expected service life without significantly reducing the serviceability or incurring excessive and unforeseen construction and maintenance costs.

* Corresponding author: p.woyciechowski@il.pw.edu.pl

2 Durability in terms of standard requirements

Principles of material design of concrete durability adopted in Europe are given in European standard EN 206: *Concrete. Specification, performance, production and conformity* and in national complements in relation to the local operating conditions of the structure (eg. in Polish Complement PN-B-06265:2004). From the carbonation threat point of view, principles and requirements are various for four concrete carbonation exposure classes. The criterion of assigning to the particular exposure class XC1÷XC4 is concrete cover humidity and it is corresponding with environment as follows: XC1 – dry or permanently wet environment (eg. building interiors or concrete permanently under water), XC2 – wet, rarely dry environment (eg. foundation), XC3 – medium moist environment (eg. interiors of high RH or exterior surfaces sheltered from the rain), XC4 – cyclically wet and dry environment (eg. zone of water flow in natural water area or fluctuations in water level in reservoirs). For each exposure class there are formulated requirements in terms of water-cement ratio, concrete class and minimal content of cement (Table 1).

Table 1. Requirements for concrete by carbonation exposure class according to EN 206.

Requirement	Exposure class			
	XC1	XC2	XC3	XC4
Maximal value of w/c	0,65	0,60	0,60	0,50
Minimal concrete class	C16/20	C16/20	C20/25	C25/30
Minimal cement content, kg/m^3	260	280	280	300

According to standard EN 206, fulfilling abovementioned requirements ensures the durability of concrete for 50 years. Moreover, in Polish Complement there are given recommendations for the use of cements in the conditions of carbonation exposure class. For instance CEM I is suitable to use in all carbonation exposure classes. However, although all cements are recommended to use in exposure class XC1 and XC2, the cements rich in limestone, fly ash, slag, etc. are not recommended to use in environments described by classes XC3 and XC4 or in case of pre-stressed concrete.

The requirements in terms of the minimal thickness of the concrete cover due to durability formulated in Eurocode EC2 are different in case of reinforced concrete structures and pre-stressed concrete structures; also they are different for each type (category) of structure defined in Eurocode EC0 and exposure class defined in the standard EN 206. Due to Eurocodes recommendations, when determining the structural class the exposure class XC specifics is to be taken into account. The structural class recommended by EC2 for "common" structures designed for service life of 50 years is S4. If the service life of the structure is 100 years, then structural class is to be increased by 2, while in case of concrete strength class higher than C30/37 or in case of the slab elements, or in situation where the concrete special quality control is required - structural class may be reduced by 1.

The analysis of the EC0 and EC2 indicates a certain inconsistency in records concerning the structural class. In EC0 there are defined 5 categories of structure, while in EC2 there are specified 6 classes. The record in EC2 about the need to increase the structural class S4 by 2 in case of assumption of a 100-year period of use leads to structural class S6 that refers to the 100 years of use. However, the same period of use is given in EC0 in relation to the category S5. It seems logical to assume that the record about the need to increase structural class S4 by 2 (i.e. to S6) should apply only to the case of the structure of the required service life of over 100 years (although this is not the case described in EC0).

Knowing the structural class determined according to EC0 and EC2 (Fig. 1) and specific requirements for concrete exposure class XC given in standard EN 206 (Table 1), it is possible to determine the minimal thickness of concrete cover ($c_{min,dur}$, mm) required in case of reinforced concrete threatened by carbonation (Table 2).

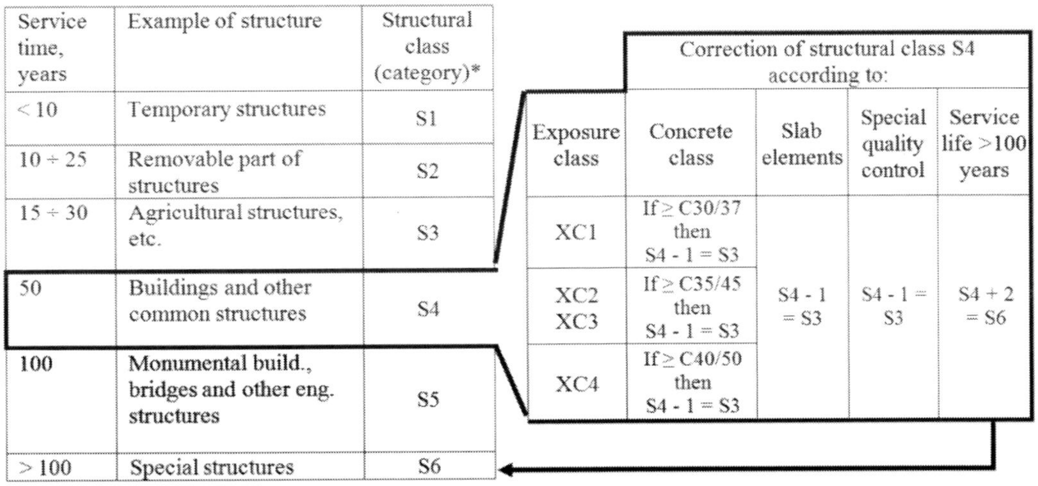

*) basis for determining minimal concrete cover thickness (see Table 2)

Fig. 1. Determining the structural class according to E EC0 and EC2 and specific of standard EN 206.

Table 2. Minimal concrete cover thickness required in reinforced (R) and pre-stressed (P) concrete threatened by carbonation.

Structural class	Minimal concrete cover thickness $c_{min,dur}$ in the exposure class, mm					
	XC1		XC2, XC3		XC4	
	R	P	R	P	R	P
S1	10	15	10	20	15	25
S2	10	15	15	25	20	30
S3	10	20	20	30	25	35
S4	15	25	25	35	30	40
S5	20	30	30	40	35	45
S6	25	35	35	45	40	50

Taking into consideration above it can be concluded that according to Eurocodes EC0 and EC2 the concrete cover thickness in the particular exposure class depends only on the structural class (category) and concrete compressive strength class – it is not differentiated for various cements and additives, nor technological types of concrete. As a consequence, the selected value of concrete cover thickness is in fact an estimation, often exaggerated (too safe or too risky).

3 Research significance

This paper contains authors own [1-5] mathematical model of carbonation. The model defines the carbonation as the process of limited possible range into the concrete and is described by hyperbolic function of time. The presented model of carbonation progress is different from the traditional models described by parabolic functions that were published worldwide so far (for details see paragraph 4), but it was verified in a wide range of material variables, technological variables as well as environmental variables published in the earlier works of the authors. Meanwhile the given algorithm enables optimal choice of the concrete cover thickness, which minimizes the uncertainty occurring during designing the reinforced concrete structures according to the simplified approach discussed in the first part of this paper.

4 Carbonation models

4.1 Traditional approach to mathematical model of carbonation

Research on the development of universal models of carbonation describing its changes in time and taking into account different material and technological variables has been conducted for many years in various research centres [6-17]. In modeling of carbonation a key issue is to determine the intensity of carbon dioxide flow through concrete. The starting point is the first Fick's law that enables to describe the diffusion process under a constant density of the diffusion flux. Final result of carbonation modeling is power function of carbonation depth in time expressed in the form:

$$x = \sqrt{\frac{2D\varphi_{ext}}{a}} \cdot \sqrt{t} \tag{1}$$

where: x - carbonation depth; D - diffusion coefficient; φ_{ext} - CO_2 external concentration; t - time; a - coefficient determining the amount of CO_2 bound in the way of carbonation by unit volume of concrete (in kg/m^3) calculated according to the CEB Bulletin [18]. In practice the model is simplified (it relates to an average constant RH and carbon dioxide concentration in the environment). The most common model used by most of the researchers (e.g. mentioned before [6-17]) as a basic model that determines the depth of carbonation, x after the time of exposure, t can be expressed in the form:

$$x = A \cdot t^{1/2} + B \tag{2}$$

4.2 Model of carbonation as the finite process

The abovementioned models consider the phenomenon of carbonation as process occurring due to exposure concrete to environment containing CO_2 – unlimited in concrete space and unlimited in time. It is assumed that the end of carbonation is related only to the exhaustion of reagents available in the system, including mainly $Ca(OH)_2$ and in the further horizon other hydrates. However, an important issue is the accessibility of CO_2 into the system, especially, in the deeper layers of concrete. Diffusion of CO_2 depends not only on the concentration gradient but also on the concrete microstructure. The abovmentioned models based on the first Fick`s law assume that the medium in which diffusion takes place will not change over time, which allows the reception of a constant diffusion flux in the equation (1). This is a significant simplification of carbonation process description, which does not take into account a number of additional factors, such as changes in diffusivity as a function of humidity, changes in atmospheric concentrations of CO_2 in climatic year, participation in the carbonation of CSH phase and residuals of non-hydrated cement, qualitative and quantitative characteristics of the material composition of concrete (w/c, type of cement, additives, admixtures, aggregates size and content), technological and environmental factors (curing, temperature, state of stress) and, first of all, diffusivity changes resulting from changes in time of the concrete microstructure. The latter effect, resulting from the saturation of the pores with carbonation products, limits the possibility of a direct description of a process based on Fick's law. The result of carbonation is a decrease in porosity, in particular capillarity that takes place in addition to the occurrence of carbonation shrinkage, thus reducing the permeability of the concrete and therefore the possibility of diffusion of gases in concrete. This nature of the phenomenon was mentioned for the first time by Bakker in 1988, [19] and later by i.a. Nilsson (1996) [20], Fagerlund (1997) [21]. Such approach to the carbonation phenomenon was further developed in the DBME (WUT) under Czarnecki guidance – the obtained findings were widely published (e.g. [1], [2-5], [22-24]) concluding that concrete carbonation in urban-industrial conditions can be described with a hyperbolic function of carbonation depth in time (reciprocal square root of time) with an asymptote parallel to time axis. The asymptote value is considered as limit of carbonation depth. Traditional and hyperbolic carbonation models are shown on Fig. 2.

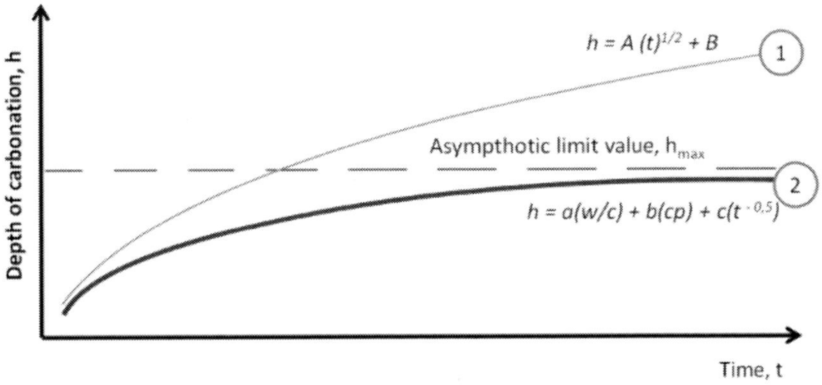

Fig. 2. "Traditional" power (1) and hyperbolic (2) models of carbonation phenomena.

The hyperbolic carbonation model is expressed in the form:

$$h = a(w/c) + b(t_{ec}) + c(t^{-0.5}) \tag{3}$$

where: h - depth of carbonation in mm, w/c - water-cement ratio, t_{ec} - early curing time in days, t - time of exposition in years, a, b, c - coefficients describing relevance of influence of w/c ratio, early curing and exposition time on depth of carbonation. It was stated that parameters a, b, c mainly depend on binder properties, presence of mineral additives and especially on CO_2 concentration. Similar models were elaborated for different types of concrete with use of Portland cement and cement with slag and fly ash. SEM analysis showed different density of concrete structure in carbonated and non-carbonated zones. It was stated that all results are in accordance with following hyperbolic model:

$$h = f(t^{-0.5}) \tag{4}$$

regardless of binder composition, though various function coefficients were obtained for various cements. Determination of carbonation hyperbolic model enabled to specify the maximum carbonation depth and compare it with the thickness of concrete cover of reinforcement in the analyzed element. This enabled to indicate the possible risk of corrosion due to carbonation and to estimate the moment (time) when the carbonation front would reach the reinforcement which can be considered as the time of corrosion initiation.

5 Hyperbolic carbonation model used in cover thickness design

Determination of concrete cover thickness due to durability of construction located in conditions of the carbonation threat includes determination of the XC exposure class (acc. to EN 206) and selection of minimal cover thickness (acc. to EC2). Selecting minimal from the Eurocode, apart from exposure class (XC1÷XC4), one take into account only structural class (S1÷S6) and type of the reinforcement steel (mild steel, prestressed steel). Using the carbonation model obtained in the way of research for a particular type of concrete designed for use in structure, enables to design the thickness of the concrete cover for the individual case on the basis of the actual protective abilities of particular concrete. When designing one should take into account that if the process of carbonation is finite, the adopted thickness of the reinforcement concrete cover is greater than the maximal possible depth of carbonation (in the hyperbolic model the value of asymptote h_{max}) which assures that the initiation of reinforcement corrosion would not arise in the structure. To simplify application of the hyperbolic model into designing practice, authors developed the algorithm [25] describing this experimental-computational step by step. An example of calculation of minimum concrete cover in reinforced element done according to that algorithm is given below.

6 Example of calculation of the cover thickness for reinforced concrete element

The calculation of concrete cover thickness was done for the reinforced concrete column of service life designed for **at least 50 years** in the following exposure environmental atmospheric conditions: relative humidity, RH up to 90%, ambient temperature of +(3÷40)°C, natural CO_2 concentration of c.a. 400 ppm. According to EN 2006, these conditions are adequate to **carbonation exposure class XC4**. The control conditions of production and concrete works on site for the structure were set up as **normal conditions**. The material composition of concrete mix used to produce the analyzed element expressed per 1 m^3 was as following: Portland siliceous fly ash cement of class 32,5 and high early strength: CEM II/A-V 32,5 R – 365 kg, natural aggregate of fraction 0/16 mm (including river sand) – 1927 kg, water – 155 dm^3, superplasticizer admixture – 1,3% of cement mass (i.e. 4,75 kg)._Above gived the water-cement ratio of 0,42 and according to Bolomey equation enabled to obtain concrete of **compressive strength class C25/30**. Laboratory tests performed on concrete specimens (cubes of size 15 cm) cured under suitable moisture conditions for 28 days showed **average concrete compressive strength of 39,1 MPa** and the minimal compressive strength of 35,3 MPa which according to EN 206 confirmed conformity with **class C25/30**. The applied self-terminated carbonation model was elaborated on the basis of results of concrete carbonation tests performed in the laboratory accelerated conditions (1% concentration of CO_2, testing period : 90 days). The final value of minimal concrete cover thickness (after including safety margins and corrections beacuese of technological aspects) was calculated and was **20 mm**, which is 30% less than the value obtained on the basis of Eurocodes EC0 and EC2 (Table 3).

Table 3. Designing of cover thickness according to Eurocodes and with use of hyperbolic carbonation model on the example of for reinforced concrete column.

Designing of cover thickness - steps	according to Eurocodes	calculation with use of hyperbolic carbonation model
Determination of the expected service life of structure/element & adequate structural class	The service life of structure is designed for at least 50 years, which (Fig. 1) indicates the structural class S4	X
Determination of concrete material composition for the particular class	X	Concrete composition per 1 m^3: cement - 365 kg, aggregate - 1927 kg, water - 155 dm^3, super-plasticizer – 1.3% of cm.
Analysis of exploitation environment characteristics & potential structural class correction (Fig. 1)	Element in shape of column (not slab); normal control condition (no special quality control); exposure class XC4, concrete compressive class C25/30 (lower than C40/50) - no correction of struct. class	X
Preparation of specimens, determining concrete compressive class, carbona-tion tests -carbonation depth measurments done after particular times of exposure to CO_2 are data for calculating the carbonation model	X	Average compressive strength of concrete (f_{cm}) was 39.1 MPa and the lowest, minimal registered value ($f_{c,min}$) was 35.3 MPa thus class C25/30. Concrete specimens were exposed to CO_2 of concentration 1% (RH 60%, T = 20°C) for 90 days (accelerated conditions)
Calculation of model - relation «time of exposure to CO_2 - concrete carbonation depth» (Eq. 4.) and its asymptote - limit of carbonation depth*	X	Carbonation hyperbolic model was calculated in the form: $h = 13.6 - 33.8\,(t^{-0.5})$. The asymptote of the function and at the same time the maximal depth of carbonation h_{max} is 13.6 mm*
Determination of reinforcement type and minimal concrete cover thickness ($c_{min,dur}$) required in case of element located in particular environment described by exposure class	According to requirements (Table 2) the minimal concrete cover thickness of reinforced (not prestressed) element of structural class S4 exposed to exposure class XC4 $c_{min,dur} = 30$ mm.	X
Determination of minimal concrete cover thickness ($c_{min,dur}$) – the value must be higher than possible concrete carbonation depth (h_{max}) from the model and should be multiplied by safety coefficient (1.3) and adjusted to the accuracy of stabilization of reinforcement in formwork	X	Maximal carbonation depth (h_{max}) multiplied by safety coefficient of 1.3 gives minimal concrete cover thickness, $c_{min,dur} = 17.7$ mm. Due to accuracy of stabilization of reinforcement in formwork, the minimal concrete cover thickness in analyzed reinforcement concrete element is $1.3 \cdot c_{min,dur} = 20$ mm.
Final result	30 mm	20 mm

*) value 10÷15% higher than value obtained for the same concrete after many years of exposure to natural atmospheric conditions (400 ppm of CO_2) thus h_{max} is a bit excessive (on a safe side)

7 Conclusions

Presented example of designing the concrete cover thickness in reinforced element in two ways confirms that elaborating the precise and accurate mathematical models of concrete carbonation describing the increase of the carbonation depth in concrete in time and application of such models for designing the reinforced structures in terms of ensuring the required durability is useful and reasonable. The example clearly showed that estimating of minimal concrete cover thickness in terms of carbonation threat on the basis of the Eurocodes EC0 and EC2 and European standard 206 is an overestimation and significantly increases the cost of whole structure. However one should be sure about the correctness of the elaborated model. Authors hope that presented analysis of the "traditional" carbonation models in the context of their deficiencies in describing the phenomenon of carbonation, which actually is terminated phenomenon, will encourage others to use more correct models described by hyperbolic functions, as in case of given example.

References

1. P. Woyciechowski, *Scientific works. Buildings* (OWPW, Warszawa, 2013)
2. L. Czarnecki, P. Woyciechowski, ACI Mater. J. **109**, 3 (2012)
3. L. Czarnecki, P. Woyciechowski, B. Pol. Acad. Sci-Tech. **61**, 1 (2013)
4. L. Czarnecki, P. Woyciechowski, B. Pol. Acad. Sci-Tech. **63**, 1 (2015)
5. L. Czarnecki, P. Woyciechowski, *Proc. 4th Int. Conf. on Concrete Repair, Rehabilitation and Retrofitting (ICCRRR-4)* (CRC Press, Leipzig-Capetown, 2016)
6. B. Bary, A. Sellier, Cem. Concrete Res. **34**,12 (2004)
7. O. Burkan Isgor, A. Ghani Razaqpur, Cem. Concrete Comp. **26** (2004)
8. K.M.A. Hossain, M. Lachemi, *Proc. Third Int. Conf. on Constr. Mater.: Performance, Innovations and Structural Implications*, (Vancouver, 2005)
9. T. Ishida, K. Maekawa, M. Soltani, , J. Adv. Concrete Technol. **2**, 2 (2004)
10. K. Maekawa, T. Ishida, Mater. Struct. **35** (2002)
11. Y.H. Loo, M.S. Chin, C.T Tam, K.C.G. Ong, Mag. Concrete Res. **46** (1994)
12. Y. Masuda, H. Tanano, Concrete Res. Technol. **2**, 1 (1991)
13. M.T. Liang, W. J. Qu, C.-H. Liang, J. Mar. Sci. Technol. **10**, 2 (2002)
14. I. Monteiro, F.A. Branco, J. de Brito, R. Neves, Constr. Build. Mater. **29** (2012)
15. A. Steffens, D. Dinkler, A. Ahrens, Cem. Concrete Res. **32** (2002)
16. V.G. Papadakis, C.G Vayenas, M.N. Fardis, ACI Mater. J. **88** (1991)
17. A. Muntean, Meccanica **44**, 1 (2009)
18. CEB Bulletin 238, *New Approach to Durability Design. An example for carbonation induced corrosion* (Comitée Euro-International du Béton CEB 238, 1997)
19. R.F.M. Bakker, *Corrosion of steel in concrete: Report of the Technical Committee 60 – CSC RILEM* (Chapman and Hall, London, 1988)
20. L-O. Nilsson, Constr. Build. Mater. **10**, 5 (1997)
21. G. Fagerlund, *Trwałość konstrukcji betonowych* (Arkady, Warszawa, 1997)
22. L. Czarnecki, R. Więcławski, *Proc. MATBUD`2003* (PK, Kraków, 2003)
23. L. Czarnecki, R. Więcławski, *Int. Congr. on Concrete Durability* (Trondheim, 2012)
24. L. Czarnecki, J.J. Sokołowska, B. Pol. Acad. Sci-Tech. **63**, 1 (2015)
25. P. Woyciechowski, J.J. Sokołowska, *Prof. Of Advances in Civil, Environmental and Materials Research (ACEM16)* (Techno-Press, Jeju, 2016)

Method of calculating heat transfer in furnaces of small power

Pavel Khavanov[1,*]

[1]Moscow state university of civil engineering, Yaroslavskoye sh. 26, 129337 Moscow, Russia

Abstract. This publication presents the experiences and results of generalization criterion equation of importance in the analysis of the processes of heat transfer and thermal calculations of low-power heat generators cooled combustion chambers. With generalizing depending estimated contribution of radiation and convective heat transfer component in the complex for the combustion chambers of small capacity boilers. Determined qualitative and quantitative dependence of the integrated radiative-convective heat transfer from the main factors working combustion chambers of small volume.

Study of heat transfer in cooled combustion chambers and study the effect on the intensity of various factors is difficult and urgent task.

From the analysis of theoretical and experimental data on heat transfer in combustion chambers of various plants it should be that the existing methods do not take into account the peculiarities of heat transfer in furnaces of small dimensions and can't be used for their thermal design.

The process of heat transfer in furnaces from the radiating flow of combustion products is determined by the intensity of convection, molecular and radiative transfer, depending on the conditions of entry, ignition and combustion of the fuel-air mixture, changes in physico-chemical properties of the medium, the hydrodynamic process conditions and other factors. Convective heat transfer, radiation and molecular thermal conductivity as heat transfer components of the complex, depending on the conditions of development of the process and make a different to-quantitative contribution to the resulting heat transfer. Neglecting any of these components without a detailed analysis of the process may lead to a distortion of the actual physical side of heat transfer and, consequently, to incorrect quantitative relations in the process. The same result is obtained and the analysis of the individual contribution of each component without consideration of their mutual influence [1].

Analysis of heat transfer by convection processes, molecular and turbulent heat conduction and radiation in the flow of gas leads to a system of differential equations to be solved simultaneously. The mathematical difficulties in solving this system, make it necessary to search for an approximate solution methods. The basic assumption in this case is connected with the exception of the integral members of the energy transfer equations, which are introduced into it as a result of the presence of radiation. The operation exception

* Corresponding author: roverton@mail.ru

is easiest to implement in extreme cases, the optical thickness of the emitting layer: a very large and very small.

The intensity transfer by radiation heat in a combustion chamber of small volume and the furnaces of steam generators and process furnaces may be different even with the same temperature conditions as the optical properties of the layer emitting the combustion products in the cells of large and small volume differ in their numerical values in relation to those in the flue cells in a small volume of heat exchange formulation objectives of the study becomes important to change the range of the optical properties of the combustion environment, since it is known that the physical model of the process to determine the transfer of energy by radiation and the final results depend on the optical density of the medium to a large extent.

Calculated study the optical properties of the medium [2] in a small size furnace chambers showed that calculation methods regardless of the average absorption coefficient emitting medium is neither optically thin or optically thick.

For intermediate values of the optical thickness of the emitting layer to reveal a natural process requires additional experimental studies, because in some cases it is possible to use absorption coefficients averaged Planck (approximation of an optically thin layer) or the Rosseland (zoom optically thick layer) [3].

Analysis of experimental data has shown that in the investigated range of the optical thickness of a generalization of the experimental data obtained in the installation described below, may be made based on the absorption coefficients averaged Rosseland. Integral heat transfer is possible to determine the number of functional integral heat exchange:

$$K_{\mathrm{T}} = \frac{I_{th} - I_{eg}}{I_{th} - I_{w}}, \tag{1}$$

I_{th} – enthalpy of combustion products at the theoretical combustion temperature, kJ/kg; I_{eg} – enthalpy of combustion products at a temperature of exhaust gases, kJ/kg; I_w – enthalpy of combustion products at the wall, kJ/kg.

In this case, K_{T} determines the performance of the combustion chamber and a heat exchange function between the moving gas stream and radiating heating surface of the combustion chamber [4].

Heat transfer from the volume of the combustion chamber to its walls through the boundary layer due to radiation and turbulent diffusion (molecular transfer in this case can be neglected).

In the boundary layer and the heat transfer wall to the wall is carried out and the resulting thermal conductivity of molecular radiation medium to the heating surface. heat balance equation for the surface corresponding to the unit length of the cylindrical chamber, while using the diffusion approximation for energy transfer by radiation is written as follows:

$$-D_{turb} \cdot \frac{d(\rho c_p T)}{d_b} - \lambda_{rad} \frac{dT}{d_b} = \frac{\sigma_o n^2 (T_c^4 - T_w^4)}{\frac{1}{\delta_w} - \frac{1}{2}} - \lambda_{mol} \frac{dT}{d_{\mathrm{r}}}, \tag{2}$$

D_{turb} – turbulent diffusion coefficient, m^2/s; r – camera radius, m; c_p – isobaric heat capacity of the combustion products, kJ/(kg·K); ρ – the density of the combustion products, kg/m^3; λ_{rad} – radiative heat transfer coefficient of the Rosseland, W/(m·K); σ_o – Stefan-Boltzman constant, W/(m^2·K^4); n – index of refraction; δ_w – the emissivity of the surface; λ_{mol} – molecular thermal conductivity, W/(m·K); d_b – the diameter of the burner, m.

Equation (2) is written for surfaces that obey the law of Kirchgoff. From the resulting heat balance equation on the boundary wall and the boundary layer can be composed of the following criteria:

1. The criterion of similarity of radiation:

$$K_{rad} = \frac{k \cdot d_c}{\dfrac{1}{\delta_w} - \dfrac{1}{2}}[1 + \frac{T_w}{T} + (\frac{T_w}{T})^2 + ... + (\frac{T_w}{T})^n] \tag{3}$$

k – absorption coefficient; d_c – combustion chamber diameter, m

Based on the analysis in (3) the quantities and heat transfer characteristics with regard to small combustors radiation similarity criterion may be converted to the form

$$K_{rad} = f(k \cdot d, \alpha) = f(\tau_o, \alpha), \tag{4}$$

α – excess air ratio; $\tau_o = k \cdot d$ – optical thickness, m.

2. Criterion turbulent and molecular transport:

$$K_{\text{T-M}} = \frac{D_{turb} \cdot \rho(I_{th} - I_w)}{\lambda_{mol}(T_{th} - T_w)}, \tag{5}$$

that under these conditions can be expressed as

$$K_{\text{T-M}} = f_1(\text{Re}_{\text{H}}; \frac{d_b}{d_c}; \frac{l}{d_c}), \tag{6}$$

Re_{H} – Reynolds number; l – length of the chamber, m.

3. The criterion of radiation and molecular transport:

$$K_{\text{R-M}} = \frac{\sigma_o n^2 T^3}{k\lambda_{mol}}. \tag{7}$$

In these conditions, $K_{\text{R-M}}$ has no effect on the heat transfer in the combustion chamber. For homogeneous combustion reactions influence reaction-kinetic factors on the heat transfer can be neglected [5] or to consider them identical for these combustion chambers with similar conditions of entry of reacting masses.

Thus, the similarity of the equation to determine heat transfer in this case can be represented as follows:

$$K_{\text{T}} = f(\text{Re}_{\text{H}}; \tau_o; \alpha; \frac{d_b}{d_c}; \frac{l}{d_c}), \tag{8}$$

The functional dependence of the integral criterion of the transfer of numbers-arguments can most reliably be determined on the basis of an experimental study of complex heat exchange on models and real plants.

Experimental study of complex radiation and convective heat transfer was carried out in the combustion chambers, executed in real (Fig. 1). Installation consists of water-cooled section combustion chambers 1, burners on liquid and gaseous fuels for the different conditions of entry and combustion of the fuel-air mixture 2, devices for feeding and adjusting the air flow, fuel and cooling water, as well as instrumentation to determine pressure, flow rate, temperature and composition of fuel and combustion products.

The combustion chamber is a vertical cylindrical heat exchangers diameter dc – 0.18; 0.24; 0.3; 0.36 m, collected from sections of different lengths in order to obtain middle zone of heat fluxes on surfaces corresponding to the relative length of the section $\frac{l}{d_c}=\frac{1}{3}$ calculated by determining the temperature and coolant flow at points "B".

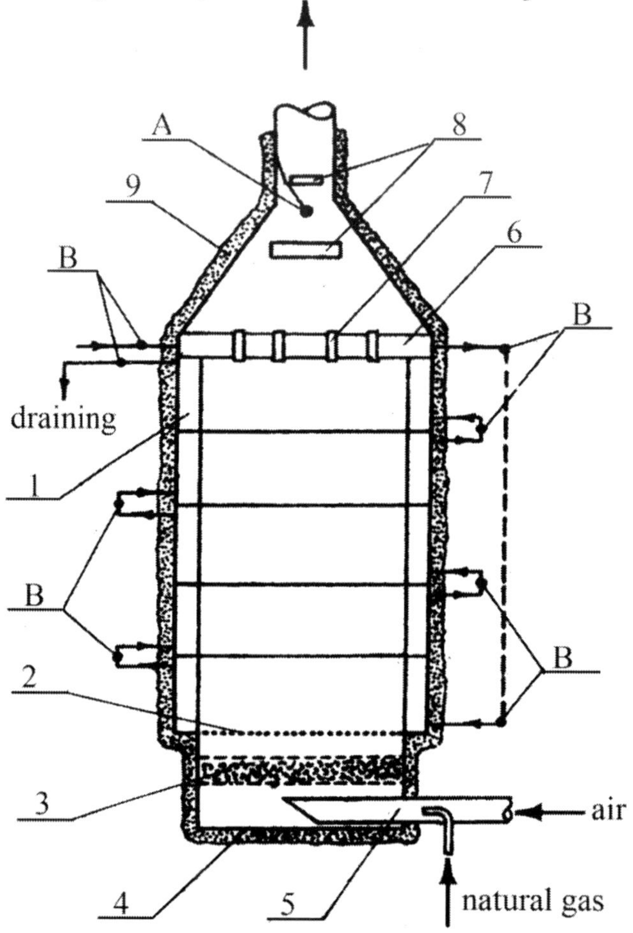

Fig. 1. Experimental stand.

To investigate the effect of heat on the length of the combustion chamber at constant thermal stresses furnace volume q_v experimentally determined integrated heat transfer in cells of $\frac{l}{d_c}=1;\ 1\frac{1}{3};\ 2;\ 2\frac{1}{3}$ in length.

In the upper part of a hollow drive water cooled combustion chamber outlet calorimeter installed fire tubes 7 distributed on the surface so that the ratio of the cross sectional area of the combustion chamber and the outlet section area kept constant flue tubes for cameras of different diameters.

Flue gas temperature at the outlet of the cylindrical part of the definition-fissile as measured gas temperature at the point "A" located between the screens 8 and the largest heat absorption of the calorimeter output.

In one series of experiments the combustion chamber equipped with a burner full air pre-mixing with natural gas. The burner is composed of body 4, a mixer 5, a layer of ceramic chips 3, for leveling the velocity field of air-fuel mixture and heating this mixture, nichrome mesh 2, which was carried out with kinetic combustion gas excess air factor α = 1.05 - 2.0.

This equation (8) allows on the basis of generalization of experimental data to analyze the impact of regime and geometrical factors, as well as the properties of the absorbing medium in the heat in the cold combustion chamber under identical input conditions. A study in this publication and in [6] shows that the absorption coefficient k of the combustion products is calculated on the Rosseland diffusion approximation is inversely proportional to the magnitude of the coefficient of excess air. In this case, radiation similarity criterion may be represented as follows:

$$K_{rad} = \frac{k \cdot d_c}{\alpha} \tag{9}$$

Absorption coefficient of the combustion products at $\alpha = 1.0$ is for natural gas $k_o = 3.34$ m^{-1} and for oil $k_o = 3.92$ m^{-1}.

Excess air ratio has different effects on the individual components of the complex heat flow [7].

With an increase in the excess air ratio is intensification of convective heat transfer due to the increase of Re criterion [8].

At the same time decreases the theoretical combustion temperature of the fuel and the partial pressure of triatomic gases, i.e., excess air ratio affects the magnitude of the thickness of the optical medium.

The introduction of excess air factor as an argument doesn't preclude consideration of Re criterion despite having a dependence of the criterion of α. This is because at constant values α and $\frac{l}{d_c}$ determines the change in Re emitting medium flow mode, i.e. determines the intensity of convective heat transfer. Simultaneously, the criterion is a function of the thermal stress of the combustion chamber volume.

Excess air ratio at constant values of Re and $\frac{l}{d_c}$, but for different specific energy in the volume of the chamber represents the change in the radiation component of heat transfer in a complex.

The experimental data were processed by the method allows to determine the dependence of the radiation and convective components of the air-fuel ratio.

As a result of analysis of experimental data, the following dependence, reflecting the impact of the air excess factor α on heat transfer in the combustion chamber

$$K_{т} = A\left(\frac{k \cdot d_c}{\alpha}\right)^{0,47} \text{ at Re = const; } \frac{l}{d_c} = \text{const}$$

$$K_{т} = B\,\text{Re}_{н}^{-0.33} \text{ at } \alpha = \text{const; } \frac{l}{d_c} = \text{const} \tag{10}$$

As determined by gas velocity takes speed, per unit surface, which determines the size of the diameter of the receiving chamber, for determining the temperature – the theoretical combustion temperature.

Fig. 2 shows the effect of the coefficient α on radiation component at a constant value of the convective heat transfer. Equation (10) describes the dependence of the convective component of heat transfer $\text{Re}_{н} = f(q_v)$ from a small change in the intensity of radiation heat transfer.

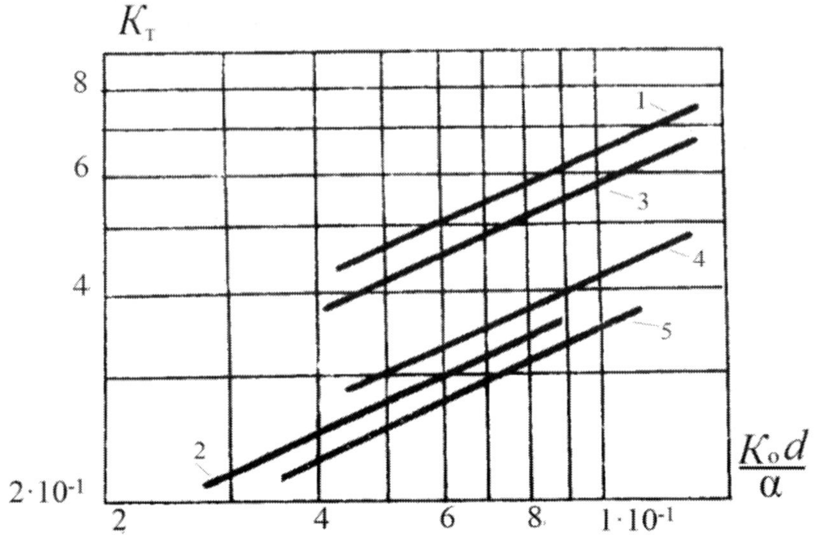

Fig. 2. Dependence of the integral heat transfer from the criterion $\frac{k_o \cdot d}{\alpha}$.

$$1)\ \frac{l}{d_c}=7/3,\ \mathrm{Re}_{н}=67.1-80.1;\ 2)\ \frac{l}{d_c}=7/3,\ \mathrm{Re}_{н}=299.1-315.2;$$

$$3)\ \frac{l}{d_c}=7/3,\ \mathrm{Re}_{н}=121.1-137.1;\ 4)\ \frac{l}{d_c}=7/3,\ \mathrm{Re}_{н}=120.6-144.4;$$

$$5)\ \frac{l}{d_c}=4/3,\ \mathrm{Re}_{н}=189.9-211.3.$$

Generalization of exponential function experimental data results helped brush up on private transport depending on the number of integrated K_T from the criterion of numbers - the arguments included in the similarity of the equation (3).

Processing results of experiments in combustors diameter d = 0.18–0,36 m, equipped with burners and incomplete full premixing of fuel and air, when distributed over the cross section entering fuel-air mixture generalized:

$$K_т=\frac{1}{1+0.1016\cdot \mathrm{Re}_{н}^{0.55}\cdot(\frac{k_o\cdot d}{\alpha})^{-0.86}\cdot(\frac{l}{d_c})^{0.75}}, \qquad (11)$$

In experiments on fuel combustion load is varied in the range of $1.5\cdot10^{-4}$ – $17.5\cdot10^{-4}$ kg/s, the excess air factor α = 1.05 – 2.0, temperature at the exit from a fire chamber T_2 = 1011 – 1752 K, conditional Reynolds number $\mathrm{Re}_{н}$ = 66 – 359, the relative length of the chamber $\frac{l}{d_c}=1-2\frac{1}{3}$, radiation similarity criterion $\frac{k_o\cdot d}{\alpha}$ = 0.301 – 1.145, the number of heat transfer integral $K_т$ = 0.15 – 0.67, specific energy in the chamber q_v = 0.76 – 1.73 MW/m^3.

The resulting dependency generalizes the criterion (11) allows us to estimate the contribution of radiative and convective heat transfer component in the complex for low-power boilers and combustion chambers to identify qualitative and quantitative dependence of the integrated radiative-convective heat transfer from the basic geometrical, physical and regime factors working combustion chambers of small volume.

Evaluation of the reliability of the experimental data showed that the value of the marginal mean square error in the determination of the integral heat transfer K_T will be 3.24%. Correlation generalizing conformity assessment according to (11) was produced in the number of integrated K_T heat transfer and temperature of the gases at the outlet of the furnace T_2. For all experiments, the deviation of the calculated data from the results of the experiment with a probability of 95% is within the confidence interval of ± 9.52%.

References

1. R .Goulard, M. Goulard, Proceding of the Heat Transfer and Fluid Mechanics Institute Stanford University Press. Pelo Alto., Calit. (1959)
2. P.A. Khavanov, *Sources of heat autonomous heating systems: monography* (MGSU, Moscow, 2014)
3. H.M. Abu-Romia, C.L. Tien, Trans, of the ASME, ser. C, Dournel of Heat Transfer **4**, (1967).
4. S.N. Shorin, V.I. Sukhov, *Influence the torches on the characteristic of heat exchange in gas combustion chambers. Theory and practice of combustion of gas. Release 3* (Nedra, 1967)
5. O.N. Bruykhanov, S.N. Shevchenko, *Heat and Mass Transfer. Textbook.* (Infra-M, 2014)
6. N.M. Konovalova, S.N. Shorin, R.M. Shurin. Chemical engineering **1**, (1974).
7. G. N. Delyagin, V. I. Lebedev, B.A. Permyakov, P. A. Havanov, *Heat-generating installation. Studies for higher education institutions. 2nd ed.* (Open company "Bastet's" IDES", Moscow, 2010)
8. V.V. Lemanov, V.I. Terekhov, THT **3**, (2016)

The technology of heat transfer enhancement in channels by means of flow pulsations

Anna Tsynaeva [1,*], and *Maxim* Nikitin [1]

[1]Samara State Technical University, Institute of Architecture and Civil Engineering, Molodogvardeyskaya str. 194, 443001 Samara, Russia

Abstract. The rate and efficiency of curing of concrete can boost when used intense heat. The work is dedicated to the development and research of technologies of intensification of heat transfer in channels by pulsations. The study was conducted by means of numerical methods based on mass and momentum conservation equations (Navier-Stokes) with software Code Saturne. Verification of implemented methods and software was performed. The research of heat transfer enhancement for semicircle-shaped channel exposed to low-frequency pulsations was performed. The pulsation frequency of the flow during the study was in a range of 0...10 Hz. A significant (up to 4 times) increase of turbulent kinetic energy with implementing pulsations was detected. Flow pulsations with frequency of 10 Hz results in 1.21 times increase of heat transfer coefficient.

1 Introduction

Pulsations of the heat carrier flow are common for launch modes of thermal machines in the construction industry. In addition, the rate and efficiency of construction for monolithic building in winter conditions can be increased. This increase is the result of using thermal methods to accelerate concrete hardening. However, the efficiency of heat transfer should be high enough. Herewith the flow might be stabilized by means of controlled pulsations that are mainly used for heat transfer intensification [1, 2]. However sustaining of such pulsations in channels of heat exchangers might be challenging due to damping effects of channels design.

In this regard an effect of channel design on heat transfer efficiency under flow pulsations should be determined. Bended channels should were considered in this study because they are commonly used in design of heat exchangers. The heat exchangers is basic element for heating of constructions of monolithic building. Since experimental research is very expensive authors used numerical simulations (finite volume method with an open source software Code Saturne) as a research tool [3].

2 Numerical study

* Corresponding author: a.tsinaeva@gmail.com

Numerical study of heat transfer in bended channel (fig. 1) with pulsating inlet flow was conducted. Navier Stokes and conservation equations were closed with combined turbulence model (k-omega SST [4]). This model was successfully implemented in the previous numerical study of pulsations in linear channel [5]. An adequacy of implemented model was verified by comparison with adopted experimental data [6] for steady flow in a bended channel. Verification showed reasonable deviations from experimental data in a range of instrumental errors.

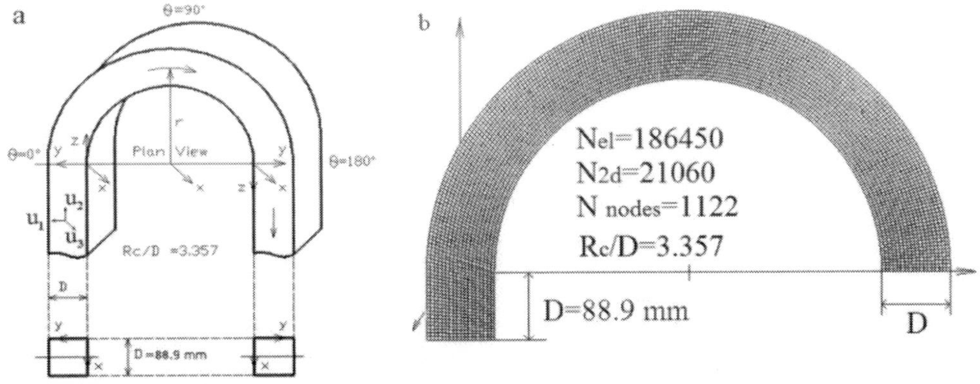

Fig. 1. Bended channel: a – geometry, b – mesh parameters.

Lateral symmetry of computational domain in respect to YZ plane was utilized to reduce computational effort. Computational mesh consisted of 186450 hexahedrons with prismatic layers providing $y^+ < 1$.

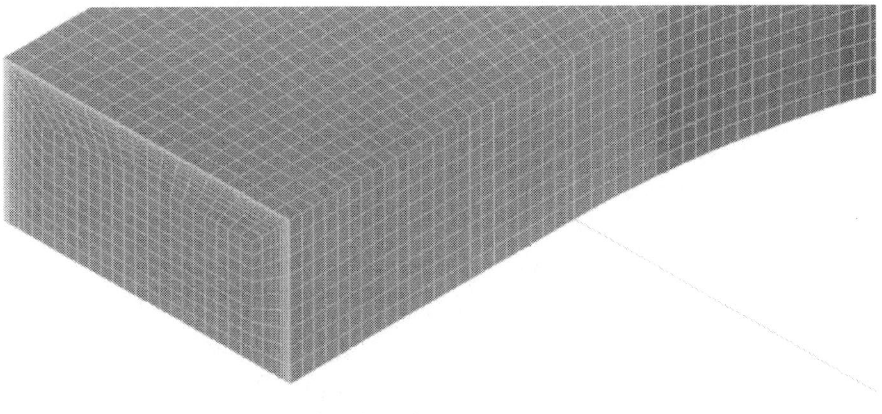

Fig. 2. Prismatic layers of computational mesh.

Gradients were calculated with a general method with handling of non-orthogonality, though mesh was explicitly orthogonal. Numerical algorithm which uses a combination of continuity and momentum equations to derive an equation for pressure wasn't applied for subsonic velocities. However transposed gradient and divergence source terms were included in momentum equation. Pressure relaxation of 0.5 was introduced to stabilize solution. Since there were no significant pressure gradients and head losses near outlet, the standard scheme of pressure interpolation was used for stratified flow. A semi-implicit algorithm (SIMPLEC) was used to speed up convergence.

Combined turbulence model (k-omega SST) was implemented since it appeared to be the most accurate during conducted verification [7]. Linear solvers precision was set at 10^{-8}.

However lowering threshold (down to 10^{-6}) didn't show any significant improvement of adequacy during validation, but notably reduced computational effort.

Second order (Centered) discretization methods were used for solving hyperbolic partial differential equations for velocity and temperature fields. Auto switching to first order scheme (Upwind) was enabled for energy equation to stabilize solution during oscillations. Turbulence parameters (k and omega) were calculated by first order linear solvers (Upwind). Further stabilization was achieved by temperatures range limitation (0...200 ^{0}C).

Simulation was conducted in transient mode since pressure and velocity oscillations observation was the major research objective. Constant time step of 0.0004 s was implemented to ensure Courant number to be less than 5 (max recommended for SIMPLEC algorithms). Moreover this time step had met requirements of minimum resolution for a quarter of period at maximum pulsations frequency. Overall 4 s of pulsating flow were modelled.

Air was used as a working media with constant thermophysical properties at 20 ^{o}C. A constant heat flux (q=1000 W/m^2) was set at channel walls. Inlet velocity was defined with harmonic oscillations:

$$u = \underline{u} \cdot \sin(2\pi\tau) + \underline{u} \tag{1}$$

where $\underline{u}$ – mean velocity, m/s; f – pulsations frequency, Hz; τ – time, s.

Relative amplitude of oscillations was defined by (2) and was equal to 1.0 in this study.

$$\beta = (\underline{u}_{\max} - \underline{u}_{\min}) / (\underline{u}_{\max} + \underline{u}_{\min}) \tag{2}$$

Numerical study was conducted for the following frequencies of flow pulsation: f= 0 Hz, f=5 Hz, f = 10 Hz. While Strouhal number was in range Sh = 0...1.67. This range is consistent with quasi-steady and low-frequency pulsating flows, which were successfully simulated with RANS models before [5]. Harmonic variations of velocity are shown on fig. 3.

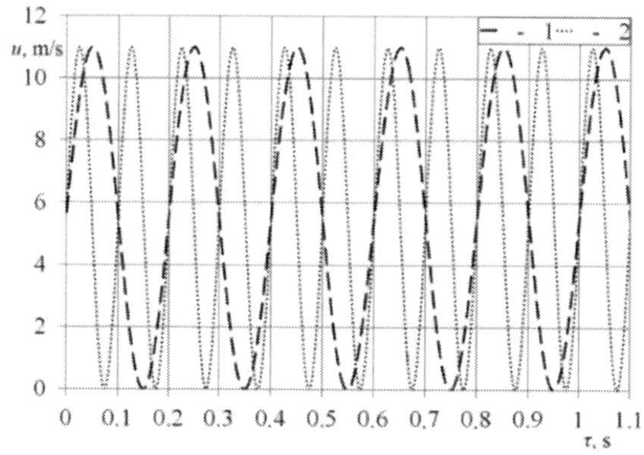

Fig. 3. Velocity variation: 1 – f = 5 Hz; 2 – f = 10 Hz.

Inlet flow velocity without pulsations was calculated as follows:

$$\underline{u}_s = \underline{u}^2 / (1 + \beta^2) \tag{3}$$

Other researches [8] indicate impact of flow pulsations on the flow structure, head losses and heat exchange. Therefore these impacts on flow and turbulence should be considered while conducting research of intensification of heat exchange in channels with pulsating flow.

3 Results and discussion

Fig. 4 shows turbulence kinetic energy profile of pulsating flow at θ = 90o. Depicted simulation data represent the acceleration stage of flow at $\tau \approx 1$ s. Results for steady flow without pulsations are also shown on fig. 4 for comparison.

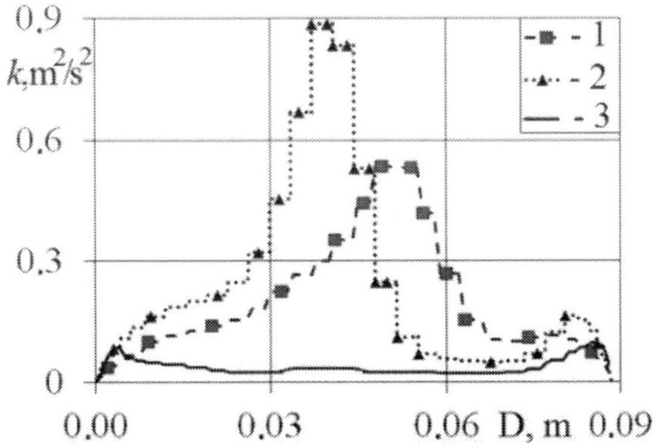

Fig. 4. Kinetic turbulence energy in lateral cross section at $\theta=90^{\circ}$:1 – f = 5 Hz; 2 – f = 10 Hz; 3 – f = 0 Hz.

Numerical simulation results (fig. 4) show increase of turbulent kinetic energy after pulsations application. Herewith at pulsations frequency of 5 Hz kinetic energy increases 1.8 times (at y = 0.09) and 21.2 times (at y = 0.049). For the flow, pulsating with frequency 10 Hz, increase of turbulent kinetic energy can achieve 33.8 times compared to the steady flow.

To determine interdependency of flow pulsations and channel geometry dynamic of turbulence kinetic energy variation at the outlet at $\tau \approx 1$ s should be analysed. This dependency is shown on fig. 5.

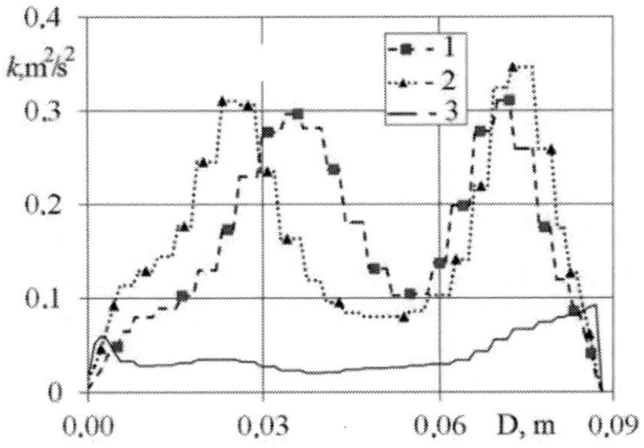

Fig. 5. Kinetic turbulence energy in lateral cross section at $\theta=180^{\circ}$: 1 – f = 5 Hz; 2 – f = 10 Hz; 3 – f = 0 Hz.

Obtained results (fig. 5) are show that presence of pulsations affects turbulent kinetic energy distribution in computational domain. Moreover channel bending decreases turbulence (up to 3 times) and distorts mentioned distribution even more. At the same time introducing pulsations increases peak values of turbulent kinetic energy up to 4 times.

Fig. 6, a) represents velocity profile of channel flow in cross section at θ=180° at time $\tau \approx 1$ s (during the stage of acceleration).

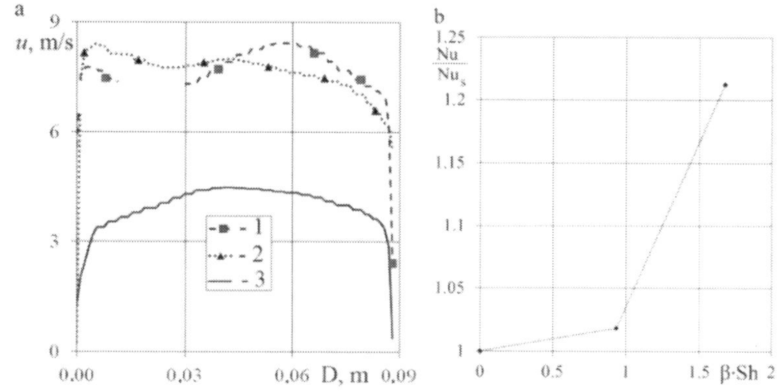

Fig. 6. Simulation results: a - velocity profile in cross section at $\theta = 180^{o}$: 1 – f = 5 Hz; 2 – f = 10 Hz; 3 – f = 0 Hz; b - dependence of heat exchange rate from pulsations frequency.

From fig. 6, a) one can see that velocity profile of steady flow matches the developed turbulent flow. Velocity pulsations distort this profile based on its upstream structure. Fig. 6,b) represents simulation results summary for acceleration stage. Results mainly concern dependence of heat exchange rate from pulsations frequency.

4 Conclusion

Obtained results analysis confirmed strong dependency of heat transfer rate on pulsations frequency. For transient flow in dedicated range of parameters pulsations may increase heat transfer rate up to 21 % compared to steady flow. This provides a sizable potential of gaining efficiency of operating heat exchangers by means of pulsating flow optimisation. This growth of heat transfer coefficient will allow to boost the heating of concrete structures. Consequently, this routine will reduced terms of erection of monolithic concrete buildings.

References

1. R. L. Webb, N. H. Kim, *Principle of enhanced heat transfer* (Taylor Francis, New York, 1994)
2. R. H. Keil, M. H. I. Baird, Ind. Eng. Chem. Process Des. Dev **10(4)**, 473-478 (1971)
3. Page of software Code Saturne, Information on http://code-saturne.org/cms/
4. Page of Langley Research Center: Turbulence Modeling. Information on http://code-saturne.org/cms/http://turbmodels.larc.nasa.gov/sst.html
5. T. Gebreegziabher, E. M. Sparrow, J. P. Abraham, E. Ayorinde, T. Singh, Numer. Heat Transfer **60(10)**, 811-826 (2011)
6. Y. D. Choi , C. Moon, S. H.Yang, *Proc. of the International Symposium on Turbulence Modeling and Experiments* (Dubrovnik, 1990)
7. A.A. Tsynaeva, M. N. Nikitin, Procedia Eng **150**, 2340-2344 (2016)
8. A.E. Goltsman, I. A. Davletshin, N. I. Mikheev, A. A. Paereliy, Thermophys. Aeromech **22(3)**, 319-328 (2015)

MATEC Web of Conferences **86**, 04019 (2016) DOI: 10.1051/matecconf/20168604019
IPICSE-2016

Analyzing screen heat insulation and its effect on energy consumption while heating building envelopes in conditions of intermittent heating

Yuri Vytchikov[1], *Mikhail* Saparev[1*], and *Alexandr* Chulkov[1]

[1]Samara State Technical University, Institute of Architecture and Civil Engineering, Molodogvardeyskaya str. 194, 443001 Samara, Russia

Absract. The paper is devoted to screen heat insulation and its effect on energy consumption while heating building envelopes in conditions of intermittent heating. It also describes the non-stationary process of heat transfer through heat insulated outer walls. The authors introduce calculation results of specific energy consumption for heating insulated and non-insulted outer walls. The paper proves that energy consumption for heating insulated outer walls depends on the thickness of non-aerated air-space insulation The research shows positive effects of using thermal protection systems with screen thermal insulation in outer building envelopes in conditions of intermitten heating.

1 Introduction

Screen thermal insulation, apart from high thermal-shielding properties, has many other advantages. They are low bulk density, good moisture resistance, absence of odour and dust, applicability in the wide ranges of temperatures.

At the moment insulation materials of foamed polyethylene are widely used in civil engineering. These materials are protected by aluminum foil on one or both sides. Papers [1-6] prove that these insulation materials are effective, especially when combined with non-aerated air-space insulations. These papers describe the process of stationary heat transfer through building envelopes which are typical for buildings used in conditions of constant power heating.

Non-steady heat transfer process through building envelopes in conditions of intermittent heating is less known. Papers [7-18] describe these buildings thermal behaviour. To such buildings they refer country cottages, vacation homes and ski resorts and other buildings of full-time residence.

Thus, it is quite actual to introduce a new methodology of calculating building envelopes and assessing these buildings energy performance in conditions of intermittent heating.

* Corresponding author: msx072007@yandex.ru

2 Mathematical Modeling of Heat Transfer Process in Thermally Insulated Outer Wall

In this research the authors study screen heat insulation effect on heating time and building envelopes energy consumption in conditions of intermittent heating. Figure 1 shows a scheme of the heating process through insulated outer walls by use of screen heat insulation.

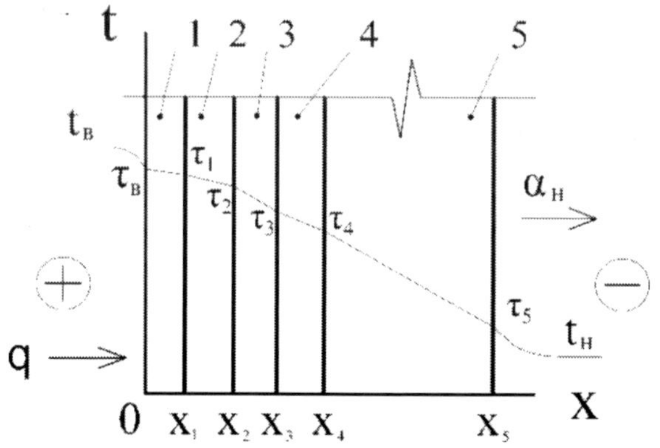

Fig.1. A scheme of the heating process through insulated outer walls: 1 – gypsum plasterboard; 2 – non-aerated air-space insulation; 3 – foamed polyethylene protected by aluminum foil; 4 – sand-cement mortar; 5 – laying of of small building products (bricks, building blocks).

Let us specify the problem of non-steady heat transfer in a insulated outer wall on the following assumptions:

- we neglect heat transfer in the direction if Y and Z axes, i.e. we consider the problem only in one dimension for the outer wall surface;
- we consider that there is a condition of perfect contact between the layers;
- we accept heat-transfer rate coming from heating appliances in the process of the outer wall heating as constant;
- we believe that in the initial time temperature distribution in the outer wall meets the stationary regime of background heating.

With account of these assumptions mathematical setting of the problem is as follows:

$$-\lambda_1 \frac{\partial t_1}{\partial x} = q\,;\ x = 0 \tag{1}$$

$$\frac{\partial t_1}{\partial \tau} = a_1 \frac{\partial^2 t_1}{\partial x^2}\,;\ x_0 \le x \le x_1;\ \tau > 0 \tag{2}$$

$$\frac{\partial t_2}{\partial \tau} = a_2 \frac{\partial^2 t_2}{\partial x^2}\,;\ x_1 \le x \le x_2;\ \tau > 0 \tag{3}$$

$$\lambda_1 \frac{\partial t_1}{\partial x} = \lambda_2 \frac{\partial t_2}{\partial x}\,;\ x = x_1 \tag{4}$$

$$t_1 = t_2 = \tau_1;\ x = x_1 \tag{5}$$

$$\frac{\partial t_3}{\partial \tau} = a_3 \frac{\partial^2 t_3}{\partial x^2}\,;\ x_2 \le x \le x_3;\ \tau > 0 \tag{6}$$

$$\lambda_2 \frac{\partial t_2}{\partial x} = \lambda_3 \frac{\partial t_3}{\partial x}\,;\ x = x_2 \tag{7}$$

$$t_2 = t_3 = \tau_2;\ x = x_2 \tag{8}$$

$$\frac{\partial t_4}{\partial \tau} = a_4 \frac{\partial^2 t_4}{\partial x^2};\ x_3 \le x \le x_4;\ \tau > 0 \tag{9}$$

$$t_3 = t_4 = \tau_3;\ x = x_3 \tag{10}$$

$$\frac{\partial t_5}{\partial \tau} = a_5 \frac{\partial^2 t_5}{\partial x^2};\ x_4 \le x \le x_5;\ \tau > 0 \tag{11}$$

$$t_4 = t_5 = \tau_4;\ x = x_4 \tag{12}$$

$$\lambda_4 \frac{\partial t_4}{\partial x} = \lambda_5 \frac{\partial t_5}{\partial x};\ x = x_4 \tag{13}$$

$$-\lambda_5 \frac{\partial t_5}{\partial x} = \alpha_н(\tau_5 - t_н);\ x = x_5 \tag{14}$$

$$t_i \,\big|_{\tau=0} = f_i(x);\ i = 1 \div 5, \tag{15}$$

where $f_i(x)$ – initial temperature distribution in conditions of background heating regular work; t_{out}, t_{in} – air temperature inside and outside the building respectively, °C; τ_i – temperature in the joint of the layers, °C; t_i – temperatures in the layers of outer walls, °C; a – temperature conductivity coefficient of materials of the layer, m^2/sec; x_i – distance from the inner surface to the joints of the layers and the outer surface of the wall; λ_i – temperature conductivity coefficient of outer walls layers, W/(m•°C); $\alpha_н$ – heat-exchange coefficient on the outer layers of building envelopes W/(m^2·°C).

The equivalent coefficient of air temperature conductivity is determined from the formula

$$\lambda_2 = \lambda_м \varepsilon_к + \alpha_л \delta_2, \tag{16}$$

where $\lambda_м$ is the air thermal conductivity coefficient, W (m·° c); $\varepsilon_к$ - the correction factor, which takes into account the influence of natural convection; $\alpha_л$ - the coefficient of heat transfer radiation, W/(m^2·°c).

Paper [3] gives a detailed methodology for determining coefficient $\varepsilon_к$.

It does not seem possible to get the exact solution for the above differential equation system (1)÷(15).

Paper [19] presents a formula for calculating one-layer building envelope heating time, using an analytical solution for a one-layer wall.

Specific energy consumption for heating a multi-layer construction can be calculated according to [19, 20].

3 Results of Calculation Outer Walls Heating Time and Specific Energy Consumption

Heating time for many-layers building envelopes can be calculated by using an approximate analytical method described in Papers [19, 20].

All the variants which are analyzed in these papers consider gypsum plasterboard with δ_1=0.02 m, air-space insulation (δ_2=0.01 m), foamed polyethylene (δ_3=0.01 m), sand-cement mortar (δ_4=0.015 m). Bearing layer here is a layer of silicate bricks on a sand-cement mortar of 510 and 640 mm thickness, as well as hollow ceramic brick of 510 mm thickness. The internal air temperature (with background heating working) $t_{в1}$ is taken equal to 12°C, design temperature $t_{в2}$ = 22°C; outer air temperature $t_{н}$ = –30 °C.

Table 1 present calculation results of outer walls heating time and specific energy consumption.

Table 1. Calculation results for insulated outer walls.

Variant № for an outer wall	Type of brickwork	R_0, (m^2·°C)/W	τ_{out}, h	Q_{out}, kJ/m^2
1	Silicate brickwork on a sand-cement mortar of 510mm thickness	1.53	37.7	$\frac{2245}{3656}$
2	Silicate brickwork on a sand-cement mortar of 640mm thickness	1.70	56.4	$\frac{3026}{4670}$
3	Ceramic hollow brickwork on a sand-cement mortar of 510mm thickness and 1400 kg/m3 density	1.74	42.9	$\frac{2255}{3318}$
4	Ceramic hollow brickwork on a sand-cement mortar of 510mm thickness and 1300 kg/m3 density	1.84	42.0	$\frac{2088}{2924}$

Note. The denominator here presents values of specific energy consumption for non-insulated outer walls.

Specific energy consumption data given in Table one proves that screen thermal insulation considerably (27-39%) allows to reduce energy consumption for outer walls heating.

The results of Studies [1,4] show that thermal resistance of screened air-space insulation depend on its thickness to great extent. The authors of this paper studied the dependence of air-space insulation thickness on energy consumption for outer walls heating. The study was conducted for one version of an outer wall. The thickness of air-space insulation varied from 0.01 to 0.03 m. Figure 2 shows calculation results as the dependence of energy consumption for outer walls heating and air-space insulation thickness.

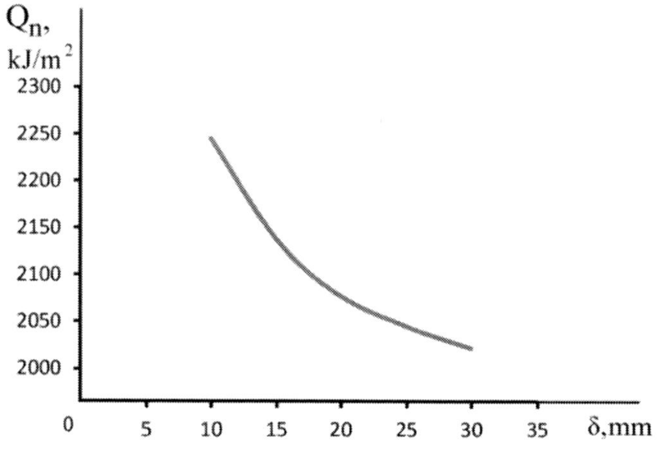

Fig.2. The dependence of energy consumption for outer walls heating and air-space insulation thickness.

4 Conclusions

The research investigates screen heat insulation and its effect on heat-shield performance of outer walls as well as energy consumption for their heating.

The results of results of outer walls heating given in Table 1 prove that when a thermal protection system is installed heat transmission resistance increases approximately 1.5 times, and specific energy consumption is reduced from 27 to 39%.

According to the obtained results, heating time for insulated outer walls (several variants of walls considered) constituted a significant amount of time (from 37.7 to 56.4 h). Therefore, the authors recommend using an automatic remote control system with installed thermal generators for buildings used in conditions of intermittent heating.

When thickness of air-space insulation is increased from 10 to 30 mm, energy consumption for heating of insulated outer walls is reduced by 10%. So, it is recommended to construct outer walls insulated by screen thermal insulation while designing buildings which are planned to be used in conditions of intermittent heating.

References

1. Yu.S. Vytchikov, M.Ye. Saparev, Vestnik SGASU, Town Planning and Architecture **1**, 4-7 (2016)
2. N.P. Umnyakova, House construction **1-2**, 16-20 (2014)
3. V.I. Kovalevsky, G.P. Boykov, *Thermal design of screen insulation and its methodology* (Energy, Moscow, 1974)
4. Yu.S. Vytchikov, M.Ye. Saparev, Construction materials **11**, 12-15 (2013)
5. Yu.S. Vytchikov, M.Ye. Saparev, Scientific Survey **2**, 104-109 (2014)
6. Yu.S. Vytchikov, M.Ye. Saparev, Industrial and civil engineering **3**, 52-55 (2014)
7. V.N. Bogoslovsky, *Thermal behaviour of buildings* (Stroiizdat, Moscow, 1979)
8. B.A. Semenov, *Building envelopes heat insulation: non-steady heat transfer and effects* (Saratov state technical university, Saratov, 1996)
9. E.Yu. Anisimova, Vestnik of South Ural State University, Civil engineering and architecture **38(297)**, 55-59 (2012)
10. E.G. Malyavina, R.R. Asatov, Thermal performance of outer building envelopes and their effect on heating system in conditions of intermittent heating, Academia, Architecture and civil engineering, **3**, 324-327 (2010)
11. E.G. Malyavina, D.Yu. Pertov, House construction **6**, 66-69 (2013)
12. V.I. Panferov, E.Yu. Anisimova, Vestnik of South Ural State University, Civil engineering and architecture **12(112)**, 30-37 (2008)
13. N.A. Datsuk, Yu.P. Ivlev, V.A. Pukhkal, Modern aspects of science and education **5**, 179 (2014)
14. A.S. Gorshkov, P.P. Rymkevich, Magazine of Civil Engineering **8**, 68-82 (2015)
15. A.S. Gorshkov, P.P. Rymkevich, N.N. Vatin, Magazine of Civil Engineering **8**, 38-48 (2014)
16. T.N. Rubashkina, System analysis, Modelling **2**, 188-195 (2014)
17. A.E. Zakharevich, Sanitary engineering, heating, conditioning **1**, 64-67 (2014)
18. V.M. Lapin, AVOK: Ventilation, heating, air-conditioning, heat supply and building thermal physics **8**, 48-51 (2012)
19. Yu.S. Vytchikov, I.G. Belyakov, M.Ye. Saparev, International research journal, Ekaterinburg **6**, 42-48 (2016)
20. Yu.S. Vytchikov, I.G. Belyakov, M. Ye. Saparev, Procedia Engineering **153**, 856-861 (2016)

Pipeline walls hydrophobic properties study and their protective coatings used in trenchless technologies

Vladimir Orlov[1,*]

[1]Moscow State University of Civil Engineering (National Research University), Department of Water Supply and Water Disposal, Yaroslavskoye sh. 26, 129337 Moscow, Russia

Abstract. The results of field and theoretical studies to determine the degree of hydrophobicity of the inner pipe walls in static and dynamic conditions are reported. The water drops spreading nature in contact with the horizontal and concave surface of different materials with the definition of contact angles is determined. The flow pattern formed when the critical mass of liquid miniflow is reached on an inclined surface in the form of an open trough was studied. The geometric parameters of the frontal and coaxial miniflow projections were determined, its dimensions and hydraulic characteristics were studied. The trial stand design for the protective coatings hydrophobic properties study was presented and described and method of evaluation of the pipeline work surfaces hydrophobic properties by applying the relative hydrophobicity ratio. Calculation formulas of the relative hydrophobicity ratio for work surfaces were presented. The relationships of hydrophobicity of various coatings used in the trenchless technologies to relative roughness coefficients were established.

1 Introduction

Nowadays trenchless technologies are widely used for rapid recovery of old water supply and water disposal pipelines [1]. Recovered pipelines must have the appropriate strength parameters determined by static calculations, and ensure a long efficient lifetime [2]. No deterioration of hydraulic parameters compared to the initial for corresponding piping material shall be observed in the recovered pipelines [3]. In addition to the strength and hydraulic parameters the evaluation of such indicators as energy saving, which are derived from hydraulic, can be made. Mechanical parameters – the degree of abrasion of inner pipes surfaces and their coatings, fatigue wear, etc. are subject to accounting [4]. Such parameters as hydrophobicity and hydrophilicity stand apart in the drinking and waste water transportation, and more precisely in the physical and chemical properties study of the inner pipe walls or their protective coatings.

The main objective of ongoing research studies was to examine the liquid contact with the work surfaces and determine the surfaces topology (roughness) impact on the liquid

* Corresponding author: voda@mgsu.ru

flow dynamics. Such works were mainly carried out for thin (mini) channels with ultra hydrophobic surfaces study [5].

Work surfaces hydrophobicity in contact with liquid environments can be encountered in various fields of engineering [6]. Pipes and their protective coatings, where there is a need for the reduction of hydraulic resistance for the energy saving are no exception [7].

The degree of hydrophobicity is evaluated in static conditions by measuring the contact angle of liquid drop applied on the surface of the material to be tested [8]. By contact angle is meant the angle between the tangent to the droplet surface and wetted surface, where the angle vertex lies at the wetting line. With significant quantities of contact angle (90° and above) the surface is considered as superhydrophobic [9].

The researchers have established a classification of surfaces, depending on their degree of hydrophobicity [9]. For many years, the following issues are studied: drops slip from surfaces with different roughness [10], hydrodynamic interaction of flows with superhydrophobic surfaces [11], liquid contact angle measurement with the surface under flow conditions [12], laser research of the superhydrophobic surfaces structure [13], nature of the change in the contact angle during the drop movement on different surfaces [14, 15] measurement of viscous droplets speed moving down on inclined surface [16], use of atomic force microscopy to measure the boundary sliding on hydrophilic, hydrophobic and superhydrophobic surfaces [17] spreading and slipping of liquid droplets on solid surfaces [18] and others. For example, in the article [19] in the observation period of water droplets during its movement the contact angles to the hydrophobic surface normal depending on the magnitude of velocity were investigated. With rapid sliding down the superhydrophobic surface the contact angle was 150°, whereas at the slower sliding and rolling the contact angle to the surface normal ranged from 100 to 79.5°. At the same time it was noted that the drop maintained its shape during movement. This was corroborated by the equality of distances traveled by the advancing and retreating drop edge.

According to the experiments results [20] the drop slides down the superhydrophobic surface without rolling. Some researchers have studied the drops movement and their spreading on the solid surface in terms of three-phase contact [21]. The drops falling and spreading model was proposed to identify conditions when the drop has a long tail.

In this paper some other aspects of fluid-surface interaction were raised:

- water drops behavior (not more than 0.55 ml) in static on different work surfaces in contact with the horizontal and concave surfaces structure;

- behavior of the moving liquid flow layer, i.e. in dynamics, down the inclined surface in the form of a trough, simulating the pipeline with applied repair materials which are used in the water supply and water disposal networks renovation by trenchless technologies [1].

Taking into consideration the well-known basic provisions on the static and dynamic wettability studies [22], the aim of our studies was to determine the most effective protective coating of pipelines identifying the linkages between its degree of hydrophobicity and hydraulic parameters [23].

2 Experiments

2.1 Materials and research methods

The following equipment was used as photo and motion-picture cameras for drops and liquid miniflow configuration: digital reflex camera Sony 550, lens DT 1.8/50 SAM with Kenko Extension Tube Set DG in the continuous multi-frame shooting mode and digital video camera Sony HDR-CX250E. Drinking water was used as liquid for drops formation and its miniflow for simulating the water supply system pipeline operation.

Static studies were made on the most common materials of pipe protective coatings: polypropylene, polymer sleeve by "Per Aarsleff" (Denmark) and "Wawin" (Holland), thin protective film of "3M" (USA) etc. In addition, Smart Surface coating was subject to the study regarding the hydrophobicity, developed by FUJIFILM HUNT (Belgium).

Dynamic studies were carried out on the stand with though length of 1 m, which simplified diagram is shown in Fig. 1. Formation and spontaneous liquid flow movement in the trough was ensured by its slope and liquid critical mass volume.

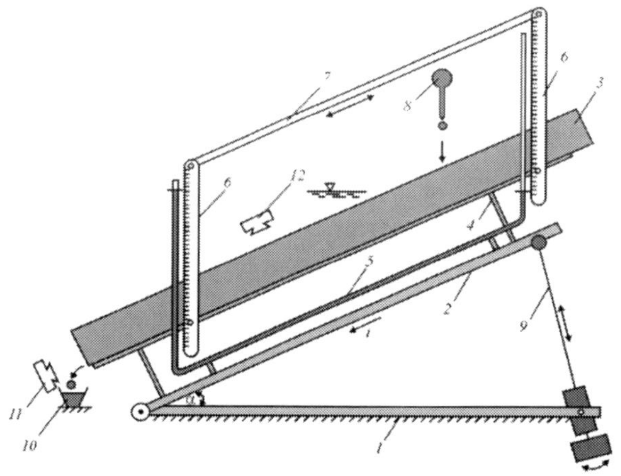

Fig. 1. Layout view of the experimental installation with maximum trough slope of 1 m long 1 – support frame; 2 – rack; 3 – open trough (tray); 4 – stands; 5 – system of communicating vessels; 6 – measuring rules; 7 – control rail; 8 – pipette with water; 9 – mechanical jack rod; 10 – drops collector; 11, 12 – front and coaxial flow cameras, respectively.

2.2 Research methodology

2.1.2 Static studies

During experiments, the contact angles were calculated by determining the derivative of the function describing the front projection of a drop on the work surface, and checked with photographing results. Experiments show that the drops nature for hydrophilic materials is similar in configuration to the inverted parabola, and for hydrophobic to semicircle.

The parabola equation is described by the Lagrange polynomial (1):

$$y = 4hx\,(a - x)/a^2 \text{ or } y = -4h(x^2 - ax)/a^2, \tag{1}$$

where h – parabola height (distance from the horizontal plane to the vertex); a – drop diameter spread out on the horizontal plane.

Values a and h were determined geometrically from the drops front projection photos on the protective coatings. By substituting the values a and h in the formula (1), it takes a particular form, reflecting the front surface nature of spreading drop as a parabola.

Methods of determining the contact angle α for a drop in the form of an inverted parabola is as follows. The coordinate axis with the center in the leftmost its part are imposed on the frontal drop projection. Point coordinates (rightmost) are marked on the abscissa axis through which the tangent passes to the curve described by the parabola equation, for example, $y = -0.050416x^2+0.731x$ (Fig. 2). Then derivative of this function is taken, which for this case is written as $y' = -0.050416 \cdot 2x+0.731$ or $y' = -0.100832x+0.731$. The figure 0.731 characterizes a slope of the tangent to the parabola in the rightmost point C with coordinates (14.5; 0) or slope α of the tangent at this point. Fig. 1 shows that the tangent slope α is approximately 36°. Thus, the contact angle for the analyzed work surface (polypropylene) is significantly less than 90°, indicating a low degree of material hydrophobicity. On other surfaces mentioned above the contact angles were 45-60°.

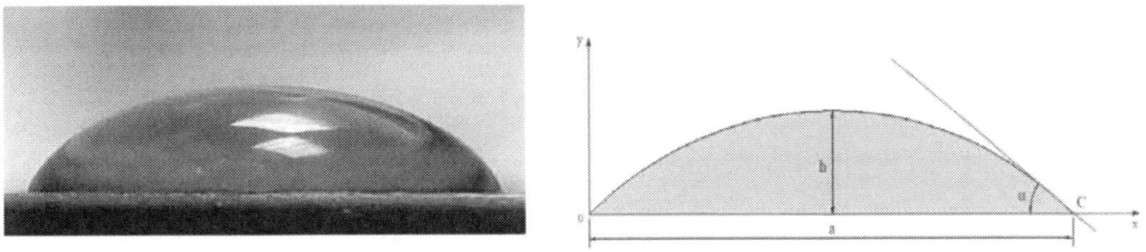

Fig. 2. The drop photo on the horizontal hydrophilic polypropylene surface and the image of its front projection with the contact angle of 36°.

In turn, the circle equation as a complex function can be written in the standard form:

$$y^2 + x^2 = r^2, \tag{2}$$

where r – radius of circle formed by the boundary of drop front projection.

Methods of determining the contact angle at the drop spreading in the form of semicircle is mostly the same as the above-described algorithm for the parabola. Here, the function differentiation through the auxiliary variable is made. In the final form, for example, in the differentiation of composite function $y = \sqrt{r^2 - x^2}$, its derivative is expressed in the form:

$$y' = \frac{-x}{\sqrt{r^2 - x^2}} \tag{3}$$

By substituting in the formula (3) of averaged circle radius r and x value corresponding to the right extreme circle point lying on the horizontal axis the numerical value of the derivative is determined. The absolute value of the obtained derivative is the tangent slope to the circle at the rightmost point or slope $tg\alpha$ to the tangent at this point. Fig. 3 shows the photo and layout view of frontal drop surface in the form close to semicircle on the curved surface of the Smart Surface coating, which falls in the category of hydrophobic.

It must be noted that Fig. 3 shows the ordinate axis passing through the circle center, the abscissa axis coinciding with the horizontal plane, and $OE = a/2$ value is equal to 26.97, i.e. point coordinates where the tangent passing to the semicircle are (26.97; 0). The circle radius value expressed as $r = AO + OB$ is 27 (dimension is not indicated due to the multiple photo and drawing zooming). Visual analysis in Fig. 3 and location of the tangent in point E indicates that it passes at an angle of more than 90°. The derivative is defined as the tangent of adjacent angle β.

$$y' = tg\beta = \left| \frac{-26.97}{\sqrt{27^2 - 26.97^2}} \right| = |-21.119|$$

At the tangent value of 21.119 angle β is about 87°, and the contact angle is determined as the difference: $\alpha = 180° - 87° = 93°$. The obtained values of contact angle indicate a high degree of hydrophobicity of the Smart Surface coating material. This assumes its widespread use as an alternative to other protective pipeline coatings with the lower degree of hydrophobicity.

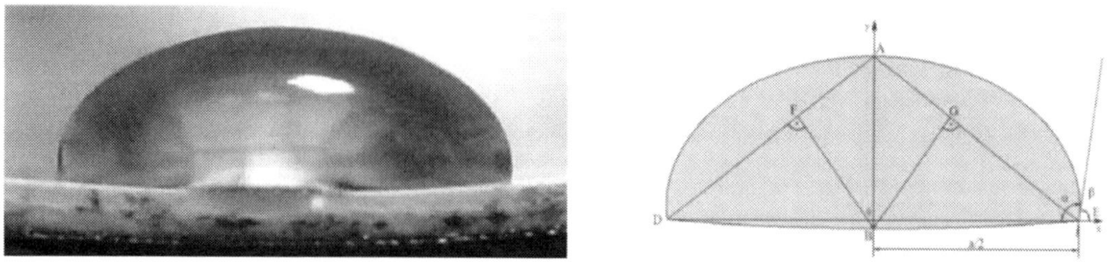

Fig. 3. The drop photo on the concave surface of the Smart Surface hydrophobic material and the image of its front projection with the contact angle of 93°.

2.1.2 Dynamic studies

General view of the miniflow at a certain work surface slope *i* (see Fig. 1) is shown in Fig. 4. The head and tail miniflow regions are presented in the form of volumetric structure (half of the spherical segment), *ABCDA* and *A'B'C'D'A'*, respectively. Their projection on a vertical plane can be displayed in the form of a lens (circumference portion) by analogy with the frontal drop projections. The experiments consisted in the frontal and coaxial photos and filming, and miniflow dimensions determination, mathematical description of shapes and elements of miniflow, by which the degree of material hydrophobicity can be evaluated and identify, where possible, its relation with geometric miniflow dimensions.

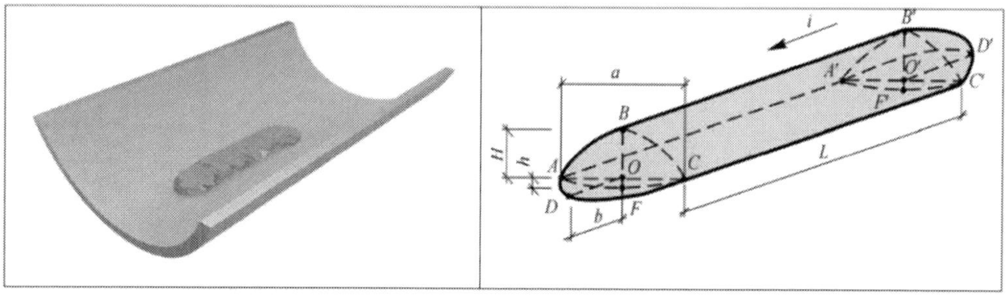

Fig. 4. Axonometric miniflow diagram formed as a result of the critical liquid volume flow on the inclined trough in the form of a tray of 1 m long *a, L* – average values of width and length of the compact miniflow region, respectively; *H, h* – height of the upper and lower lenses, respectively; *b* – height of the half of spherical segment of the miniflow head region; *AFC, AA'C'C* – wetted perimeter and surface of the compact miniflow region (S_{wet}), respectively; *i* – trough slope.

2 Results and discussion

Fragments of the frontal and coaxial miniflow filming on polypropylene troughs and "3M" protective coating are shown in Fig. 5.

Photographs analysis in Fig. 5 shows that both in static and in dynamic the tendency of different degrees of hydrophobicity in the different materials is observed. Based on the obtained data it was suggested to evaluate the degree of hydrophobicity in dynamic both by the contact angle and by relative hydrophobicity ratio determination $K_{rel.}$:

$$K_{rel} = S_{up.}/(S_{wet.} \cdot i),$$

where S_{up} – upper lens area *(ABCOA)*, mm^2; $S_{wet.}$ – trough surface wetting area (*AA'C'C) by* compact miniflow part *L*, mm^2; *i* – trough slope.

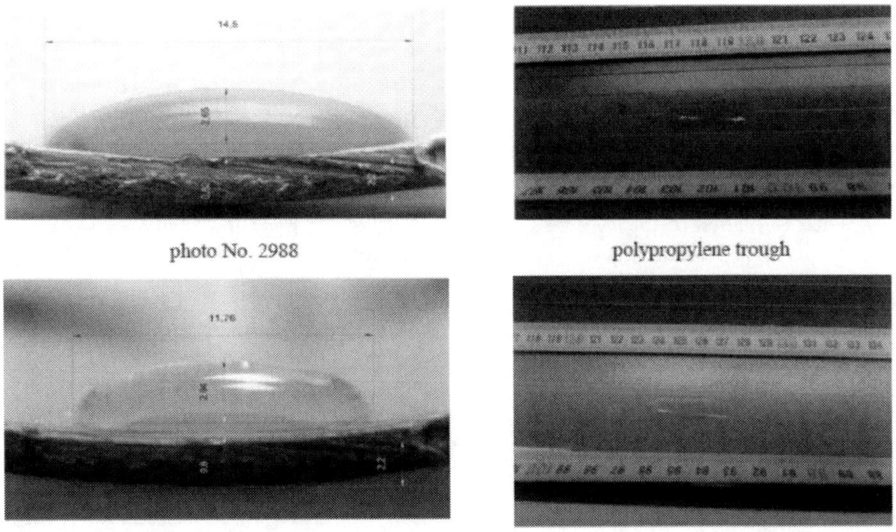

photo No. 2988 | polypropylene trough

photo No. 3185 | trough with "3M" coating

Fig. 5. Selected fragments of the frontal and coaxial photos.

In the course of dynamic field studies of degree of hydrophobicity of the Smart Surface coating (byprocessing of frontal and coaxial photographic materials and filming) the dependence of the flow width and the height of upper lens H and lower lens h on the miniflow N mass in wide ranges of the trough slope i from 0.01 to 0.11 was identified (within the boundaries of optimal pipelines gradient under pressureless mode of operation). Based on the diagrams in Fig. 6, obtained curves are described by linear functions.

With regards to the modification of such geometrical parameters as the average flow width a and height h of the lower lens in the analyzed mass range N it could be stated that these values remain practically unchanged, and compensation for the increase of liquid mass N entering the trough affects the length L of the compact miniflow region towards its increase.

Conclusions from the analysis of dependencies a = f(N), H = f(N) and h = f(N) may be the fact that the experiments for the determination of degree of hydrophobicity of the analyzed materials on the stand can be carried out at any trough slope with the possibility of forming relatively stable length L of compact miniflow region and less time consuming.

Using obtained during the experiments dimensions and average velocity of miniflow, the whole spectrum of hydraulic parameters of analyzed protective coatings with the final roughness coefficient of the corresponding coatings "n" was determined.

As an example, Summary table 1 shows the studies results on the hydrophobicity of coatings obtained as experimental data, designed geometrical parameters of miniflow and hydraulic characteristics of two work surfaces (photo 2988 and 3185 in Fig. 5).

The calculation algorithm of hydraulic exponents is to determine: filling (*h/d*) and wetted perimeter (arc length *AFC*) of miniflow; average hydraulic radius R; velocity of miniflow V, as the quotient of the path P division, passed by the miniflow head in time T, measured by the stopwatch; Chezy's velocity factor C, which depends on the hydraulic radius r, velocity V and slope i $(C=V/(Ri)^{0,5}$; relative roughness coefficient n using Manning's formula:

$$C = \frac{1}{n} \cdot R^{1/6} .$$

According to the presented calculation method the roughness coefficient "n" by Manning for polypropylene based on the results of experiments was $n = 0.0132$ at relative hydrophobicity ratio $K_{отн.}$=1.0655. For other protective coatings the roughness coefficient ranged from 0.009 to 0.0098 and relative hydrophobicity ratios within 1.075-1.098, respectively (for Smart Surface coating 1.098). Thus, the correlation between relative hydrophobicity ratios and relative roughness coefficients was found: the higher the hydrophobicity, the lower the material roughness.

Table 1. The calculation results of geometrical parameters and hydraulic characteristics of miniflow in the tray of d=130 mm in diameter with a slope i=0,11

Photo number	*ABCOA* upper lens area S_{up}, mm^2	Wetted perimeter *AFC, mm*	Wetted surface *AA`C`C* S_{wet}, mm^2	Relative hydrophobi-city indicator K_{rel}	Contact angle α^0	Design value $L_{расч.}$, *mm*
2988	25.62	14.56	218.539	1.065	$36^0 12'$	15.09
3185	25.33	11.89	214.187	1.075	$45^0 18'$	20.07
Photo number	**Average hydraulic radius *R*, *mm***	**Miniflow path *P, mm***	**Travel time *T, s***	**Average velocity of miniflow *V, mm/s***	**Chezy's velocity factor C**	**Relative roughness coefficient *n***
2988	0.38	90	5	18	64.52	0.0132
3185	0.34	83	5	16.6	88.09	0.0096

3 Conclusions

Static studies of water repellent studies (hydrophobic) properties of the protective coatings used in pipelines have shown that most of them have a low degree of hydrophobicity (contact angles are within the range of 36-60^0).

Dynamic studies of the miniflow movement on the Smart Surface coating have shown that it has well-defined hydrophobic properties; this circumstance favors the Smart Surface coating in comparison with alternative work materials of free-flow pipeline walls, because it contributes to the reduction of contact area at the liquid flow and leads to lower friction losses during the flow movement.

Studies of the miniflow liquid movement on the inclined trough have shown that by determining the relative hydrophobic ratio of protective coatings the hydraulic flow exponents can be evaluated (recalculated), in particular, the relative roughness coefficient of the analyzed material; as a result it has been proposed to use the small-scale installations for hydrophobicity study of different materials instead of bulky hydraulic stands for the hydraulic characteristics determination.

References

1. A. Kuliczkowski, *Rury kanalizacyjne Tom II. Projektowanie konstrukcji* (Wyd. Politechniki Świętokrzyski, Kielce, 2004)
2. M. Rameil, *Handbook of pipe bursting practice* (Vulkan-Verlag GmbH, 2007)
3. A. Zwierzchowska, Technologie bezwykopowej budowy sieci gazowych, wodociagowych i kanalizacyjnych (Skrypt Politechniki Świętokrzyskie, Kielce, 2012)
4. S.V. Khramenkov, O.G. Primin, V.A. Orlov, *Reconstruction of pipeline systems* (DIA, 2008)
5. D. Oner, T.J. McCarthy, Langmuir **16(20)**, 7777–7782 (2000)

6. N. Le Grand, A. Daerr, L. Limat, J. Fluid Mech **541**, 293–315 (2005)
7. V.A. Orlov, *Pipelines protecting coatings* (DIA, 2009)
8. A. Kibar, H. Karabay, K. Suleyman Yigit, Ikrime O. Ucar, H. Yıldırım Erbil, Exp. Fluids **49**, 1135–1145 (2010)
9. C. Lee, C.-J. Kim, Langmuir **27(7)**, 4243–4248 (2011)
10. J. J. Bikerman, J. Colloid Sci. **5**, 349–359 (1950)
11. A.V. Belyaev, O.I. Vinogradova, Soft Matter **6**, 4563–4570 (2010)
12. A. Gajewski, Int. J. Heat Mass Transf. **48(23-24)**, 4829–4834 (2005)
13. T. Baldacchini, J.E. Carey, M. Zhou, E. Mazur, Langmuir **22(11)**, 4917–4919 (2006)
14. Reyssat, D. Quere, J. Phys. Chem. B. **113(12)**, 3906–3909 (2009)
15. L. Gao, T.J. McCarthly, Langmuir **22(14)**, 6234–6237 (2006)
16. H.Y. Kim, H.J. Lee, B.-H. Kang, J. Colloid Interface Sci. **247**, 372–380 (2002)
17. Y. Wang, B. Bhushan, A. Maali, J. Vac. Sci. Technol. A. **27**, 754–760 (2009)
18. B. Dussan, Annu. Rev. Fluid Mech. **11**, 371–400 (1979)
19. M. Sakai, J.H. Song, N. Yoshida, S. Suzuki, Y. Kameshima, A. Nakajima, Langmuir **22**, 4906–4909 (2006)
20. M. Sakai, J.-H. Song, N. Yoshida, S. Suzuki, Y. Kameshima, A. Nakajima, Surface Science **600(16)**, 204–208 (2006)
21. F. Brochard, P.G. de Gennes, Langmuir **15**, 3024–3028 (1992)
22. P.G. de Gennes, Rev. Mod. Phys **57(3)**, 827–863 (1985)
23. R. Golestanian, E. Raphael, Phys. Rev. E **67**, 031603 (2003)

MATEC Web of Conferences **86**, 04021 (2016) DOI: 10.1051/matecconf/20168604021
IPICSE-2016

Wooden foundations history and prospects

Lyudmila Cherkasova[1,*]

[1]Moscow State University of Civil Engineering, Yaroslavskoe sh. 26 Moscow, Russia

Abstract. Wooden foundations have been in use since ancient times. With the advent of new materials and technologies they have been almost completely driven out of building practice. The main disadvantage of such foundations is their short service life. However, there are historical examples in which wooden foundations had been used for hundreds of years. In the opinion of the author, if comply with proper rules of operation, wood can be effectively used as foundation and for foundation strengthening in modern foundation engineering.

1 Introduction

Wooden foundations have a long history in European and Asian countries. In Russia such foundations initially were used only for wooden buildings. Builders were aware that the wood in the ground in a fluctuating moisture area is the most prone to rotting, that is why groundsills that were replaced or removed after loosing bearing capacity were used as girder foundations. An example would be a church in the village of Ljavlja [1] that is more than 400 years old. As a result of the gradual removal of groundsills the Church lost 4.5 m of height.

Foundation posts often were made of tree stumps that had the toughest wood. Posts were used for small houses (hut on "chicken legs") and large ones like Ostankino Palace [2], where it was quite difficult to replace girder foundations.

In Russia, foundations were made mainly of coniferous resinous wood as they are prone to rotting less than others, however, longevity was not the primary requirement for foundation design. The main advantage of the wooden foundations was that they were relatively easy to replace without destroying the building.

Wooden piles were often used in marshlands where peaty compressible soils are used for foundation.

For stone constructions foundations where wooden elements were united by rubblework were not intended to be replaced or repaired were built [3]. That is why builders tried to abide by the basic rule for keeping longevity: elements of the foundations must always remain under water. In those cases where this rule was observed buildings have been standing for hundreds of years. Construction on wooden piles was widespread in Holland and Italy (Venice). European experience has been used during the construction of St. Petersburg and spread over Russia. Wooden elements in complex foundations had

* Corresponding author: CherkasovaLI@mgsu.ru

several functions: transmission of loads to denser layers of foundation and sealing with wooden piles - Kazan Cathedral on Red Square, stress equalization in foundation under rubblework boulders (for this horizontally laid logs called sills), compression of the weak layer directly under the raft foundation of massive buildings - Saint Isaac's Cathedral, during the construction of which more than 11 000 piles has been driven in. There are 40 000 piles under the Peter and Paul Fortress. These foundation still retain the bearing capacity.

For comparison, in Venice more than a million oak, alder, hardwood piles are driven in the foundation of the Santa Maria della Salute Church; the Rialto Bridge stands on 12 thousands of stone piles

Alexander Nevsky Cathedral in Nizhny Novgorod is a unique historical example. The massive building, comparable in size to Saint Isaac's Cathedral in Saint Petersburg, is built on the arrow at the confluence of the rivers Oka and Volga. A raft of 3 m thick oak logs is laid on weak water saturated foundation. The raft is immersed in a bath of fat clay to prevent wood from drying out and rotting. The cathedral, built over 150 years ago, is still used, and its foundations successfully perform their function. The Royal Palace in Amsterdam stands on 13659 piles

Pile wood foundations are used in construction of coastal structures or directly on water.

2 Modern Wooden Foundations

Modern wooden foundations are used primarily for traditional structures of country architecture, cottage buildings and small temporary structures. Durability of foundations is extended due to treatment with chemicals and external tarring of logs and bases of buildings (Holland). However, the recommended service life of such buildings should not exceed 10-30 years. Usually pier foundation are constructed, in trenches. Sometimes girder foundation are an available option. The reason for low durability of the proposed foundations lies in their design that typically includes elements in area of soil foundation with a variable level of ground waters.

In construction wood is also used for pile planking of excavation shoring. In modern construction practice wood, as an element reinforcing weak water saturated soils, is almost never used.

Meanwhile, natural durability of logs completely immersed in water and their light weight make wood the most suitable material for reinforcement of weak water saturated soils.

Some wood species not only do not lose, but even gain strength while underwater. Light material almost does not weigh down the weak layer, preventing differential settlement. The specific gravity of pine, even in water-saturated condition, does not exceed the specific gravity of water. Since there is no need to bury pile ends in more durable soils, they can be submerged by the overloading method. Thus eliminates the negative impact of dynamic effects on nearby foundations. Finally, you should take into account the high ecological value of wood. After the use and dismantling of overfoundation structures usable area suitable for further exploitation remains.

3 Comparative ANALYSIS OF VARIOUS METHODS FOR FOUNDATION STRENGTHENING UNDER THE FLOOR OF A WAREHOUSE ON WEAK WATER-SATURATED GROUND

3.1 Description of the Geological Conditions, Structure design and Options for Foundation Strengthening

The warehouse built in the floodplain of the river Yakhroma stands on reinforced concrete plate 30 cm thick on sand-and-gravel cushion 0.5 m thick. The floor plate is loaded with strip load 5.5 t/m^2. Load can be changed during operation. Plate dimensions are 36x114m.

The base is made of fill grounds (EGE 1, 2) with a capacity of 2.5-3.1 m, under which peaty soils are laid (EGE 3) of variable capacity 1.9-2.5 m. Below a layer of plastic sand clay (EGE 4) of average capacity 4.6 m, under which a mass of fine sand of average density EGE 5 and non-determined capacity is located. Minimum groundwater level is set to 134.0. Level of leveled surface is 134.7m (fig.1.).

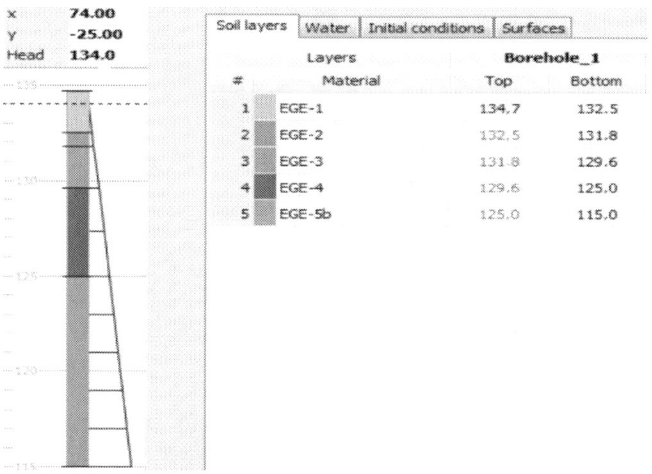

Fig. 1. Geologic column with averaged data on layers's capacity.

After commissioning and loading the project load with following transfer of it wavelike deformation of the floor has been identified, precluding its normal operation and moving of lifting devices.

During development of the method of base reinforcement and reconstruction of the warehouse floor two options were compared.

Option 1. Driving piles of reinforced concrete in sand foundation layer through holes in the floor with the subsequent dismantling and repetition of floor covering works.

Option 2. Pressing-in of wooden piles through holes in the floor with the subsequent dismantling and repetition of floor covering works. Wooden piles are pressed in in such a way so that their heads were at 0.5 m below the minimum water level and cut through the peaty soil layer without immersion into dense sand. Pile heads are covered with sand 0.5 m. A layer of sand is also laid 0.5 m above pile heads (fig. 2).

Fig. 2. Geological section with the mass of the layer compacted by wooden piles (option 2).

3.2 Target Setting. Calculation Results

The goal of calculation was a qualitative comparison of the options, so in the first approximation, influence of the existing pile foundations under the columns of the building was not analyzed. The floor plate is loaded with strip load. Calculated diagram of presented in the fig. 3.

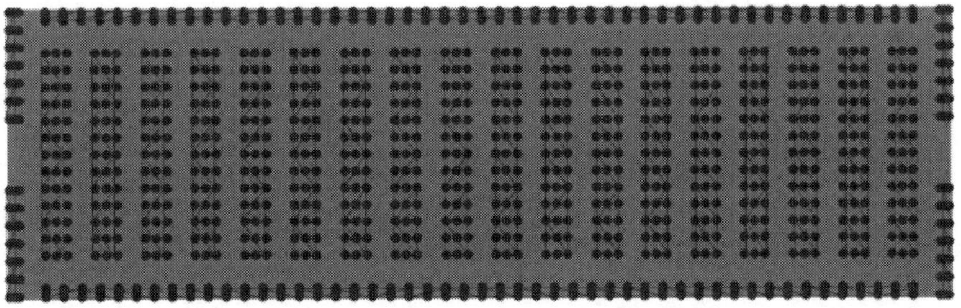

Fig. 3. Diagram of loading the slab of the warehouse floor with strip load.

Option 1. The length of the piles was determined by driving piles into the layer of sand on 0.5 m, embedding heads at the floor level and was 8 m. It should be noted that driving of the 8-meter drive piles indoors was complicated by the need to use composite piles.

Piles space is chosen, basing on project load, on a simple square grid 2.65x2.65

Option 2. Wooden piles 4 m long are pressed in on a triangular grid with the space $2D$ (opt.2a) and $3D$ (opt 2b). Choice of pile space was based on the condition of determination of the provided modulus of deformation according to the formula (1) [4].

$$E_{pr}= (S_{br.}/ S_{ach})\cdot E_d=[\pi/(2\cdot\sqrt{3})]\cdot(D/X)^2\cdot E_d \qquad (1)$$

where $E_{\text{пр}}$ – is a provided modulus of soil foundation with wooden piles;
E_d – wood deformation modulus;
X – distance between the piles;
D – diameter of a wooden pile m;
$S_{br.}$ – pile cross-sectional area;
S_{ach} – low-level cell area, including the pile and surrounding soil.
When positioning the logs with the space $X = 2D$ provided module is E_{pr} = 2724.25 MPa.
When positioning the logs with the space $X = 3D$ provided module is E_{pr}=1212 MPa.

During calculations the deformation module of weak soil is not taken into account.

In the first approximation, we consider: mass compacted by wooden piles as a linear elastic isotropic solid element with provided deformation module E_{pr} = 1212 MPa and E_{pr} = 2724.25 MPa, volume weight 15 kN/m^3, with Poisson's ratio $\nu = 0.4$. Diameter of piles D = is 0.25 m. Compacted mass thickness is 4 m.

The task was solved in three-dimensional setting with maximum simplification of the original data. The purpose of the calculation is to figure out qualitative differences, advantages and disadvantages of 2 options of weak layer fortification.

Determination of maximum, minimum vertical deformation and settlement difference are held for calculation cases.

1. Without strengthening of the foundation
2. After strengthening of the foundation with reinforced concrete piles
3. After reinforcing weak layer with wood piles with the space 3 *D*.
4. After reinforcing weak layer with wood piles with the space 2 *D*.

The results are listed in the Table 1.

Table 1. Results of determination of vertical deformations of floor slab for various options of foundation reinforcement.

Case Nos.	Strengthening the foundation	*min* [cm]	*max* [cm]	Δ=*(max-min)/L*
1	Without foundation strengthening	1.4	8.9	0.0042
2	Piles r.-c.	0.40	1.66	0.0007
3	Wooden piles, space 3*D*	0.56	1.29	0.0004
4	Wooden piles, space 2*D*	0.65	1.23	0.0003

Comparative analysis of 2 main reinforcement options: of reinforced concrete and wooden piles is listed in the Table 2.

Table 2.

Type of reinforcement	Number and (length) of piles	Dynamical loads	Creation of sand gravel layer	Settlement under load, cm	Service life
R.-c.piles (length 8m)	575	Yes (driving in)	0.3m	1.65	Not limited
Wooden piles (ength 4m)	18552 (space 2*D)* 8245 (space 3*D)*	No (pressing-in)	1.0 m	1.29	Limited to maintenance of the groundwater level.

4 Conclusions

In case of reinforcement with wooden piles settlement is less pronounced and more even than in case of reinforcement with r.c. piles. The cause is the more frequent driving and low density of the piles themselves.

Wooden pile space (2*D* or 3*D*) does not affect the degree and uniformity of deformations significantly.

Dynamic loading, which is dangerous at works in the constrained environment, can be excluded at driving wooden piles. Wooden piles are pushed into the soft ground. Driving them into sand is useless.

Construction site area is available in case of dismantling of structure. Wooden piles do not interfere the new constructions.

References

1. M.V. Korolev, L.I. Cherkasova, A.V. Ostiakova, *Specifics of Inspection and Restoration of Orthodox Churches* (MGSU, Moscow, 2016)
2. L.I. Cherkasova, G.V. Alekseev, E.A. Medvedev, Vestnik MGSU **1**, 170-173 (2006)
3. V.A. Il'ichev, N.S. Nikiforova, V.V. Dmitriev, S.V. Devyatov, T.D. Shvets, E.Kostyukov, Soil Mechanics and Foundation Engineering **53(3)**, 180–188 (2016)
4. I.I. Khusainov, *The design of the "structural geomassiv" in difficult geological conditions. The dissertation on competition of a scientific degree candidate of technical Sciences. The specialty 05.23.02 "Bases and foundations, underground structures"* (Perm National Research Polytechnic University, Perm, 2015)

The analysis of the methods of research the influence of nonuniform local stress on a crack resistance of protective coating

Ludmila Frishter[1,*]

[1] Moscow State University of Civil Engineering, Yaroslavskoye shosse, 26, Moscow, 129337, Russian Federation

Abstract. The article deals with a complex analysis methods for studying the influence of local stress on the crack resistance of coatings of buildings and constructions. The consider the computational-experimental methods of research areas of local concentration of stresses structures due to the border shape (angular cuts) under the influence of internally discontinuous deformations entering into a source of geometric stress concentration. Based on data obtained with the analysis of complex naturalistically – fact-based and computational-experimental methods to reliably estimate crack resistance of protective coatings of buildings and constructions.

1 Introduction

The main type of fracture coating surfaces of buildings and constructions is their cracking. Analysis of crack resistance of coatings in the construction and operation of buildings can be reduced to the problem of determining the local stress-strain state caused by the geometrical inhomogeneity surfaces of coatings and their foundations.

Providing the long-term performance properties of coatings of buildings and constructions that perform protective, special, decorative functions, places greater demands on the quality of materials, their performance properties.

During the construction of buildings, the amount of finishing work on labor costs is at least 30% of all labor costs. Maintaining in operating condition of buildings and constructions during the operation determines the application of coating compositions in large quantities, which leads to a rise in the cost of operation. Competition in the market of finishing materials determines the increased demands on the quality of finishing materials and their performance properties.

Create high-quality coatings of buildings and constructions determines the relevance of the analysis of the methods evaluation and predicting of crack resistance of coating, it helps to increase their service life.

The purpose of the work is to analyze the methods for evaluating the crack resistance of the coating buildings and application of theoretical-experimental method for assessing the local stress-strain state caused by the geometric heterogeneity.

[*] Corresponding author: frishter@mail.ru

2 Analysis of methods to solve the problem

The quality of protection-decorative coatings of exterior walls of buildings in accordance with the normative literature is produced by the full-scale fact-based parameters of generalized evaluations of protection-decorative properties [1-10].

In the works Loganina V.I., Makarova L.V., Makridin N.I. and Orentlikher L.P. analyzes the results of many years of inspections condition of painted surfaces of building facades on the example of Penza buildings, which allowed to classify basic types of defects that affect the quality of protection-decorative coatings.

In the works of Makarova L.V. [5] shows the results of inspections of residential buildings according to GOST 9.039-74 "Corrosion aggressiveness of the environment", located in the climatic region IV (moderately cold). Inspection reveals a painted surface buildings [5] the following defects: cracking, spalling, weathering, retention of dirt, wet spots, color variation.

According to the data defect such as cracks of coating along the vertical joint panels, the most common and account for 23-66% of the total number of defects in the coating depending on the type and service life. After five years of operation, a number of major defects include cracks of coating along the vertical joint over the face of the building.

Currently there are no approaches or methods for qualitative and quantitative evaluation of crack resistance of materials [9]. Among the used evaluation methods crack resistance of coatings surfaces, highlighted the following main groups [5]:

- indirect methods of evaluation of the physicomechanical properties;
- methods of direct evaluation crack resistance;
- methods of evaluation, based on the modeling of the structure of the material;
- methods of evaluation of the energy parameters of deformation and destruction of material;

theoretical-experimental methods for solving the problem of stress - strained state (SSS) in the field of local stress concentration due to the form of the border, for example, an angular cut.

Indirect methods estimate of crack resistance by coefficient considering physical-mechanical properties of coatings, or combinations thereof: elastic modulus, the limiting stretchability, the ultimate extensibility, shrinkage deformation and other properties. Such an approach was proposed by Orentlikher L.P., Novikova I.P., Stolnikova V.V., Silaenkovym E.S., Sheykina A.E. and others.

Methods of direct evaluation of crack resistance based on visual inspections of cracks. The most common of these methods - the method rings, standardized in France and the United States. Its essence is to crimp the sensor, cement system the during shrinkage. This takes into account not only the effect of shrinkage, but also creep, the limiting stretchability, elastic modulus, and other factors. The criterion for the crack resistance is a time of occurrence of the first crack at a reduction. The method is simple, but gives only a qualitative assessment of the crack resistance. Developed by authors such as Popov N.A., Orent-Licher L.P., Lhermitte R., Ruse G.M., Skramtaev B.G. and other.

The method based on energy approach, standardized in Russia (GOST 29167-91 "Concretes. Methods for determination of the crack resistance properties (crack toughness) under static loading"). In a computerized facility made automatically recording diagrams of stress-strain and destruction, and calculation of quantitative parameters the crack resistance. Specific power defined per unit cross-sectional area of the sample.

Widespread in the field of fracture mechanics received acoustic emission method based on excitation, recording and subsequent analysis of the acoustic emission signals received in the form of stress waves by an appropriate converter [2,3].

The acoustic emission method makes it possible quantitatively and qualitatively assess the structural changes, plastic deformation, kinetics of defects at various kinds of mechanical action on the material. Given that the main type of the destruction of protection-decorative coating is cracking, the use of acoustic emission method in predicting the crack resistance buildings coatings is one of the ways to solve the actual problem - increase their service life.

With regard to protection-decorative coatings, the problem of crack resistance almost not developed and solved [9].

Currently, evaluation of crack resistance of protection-decorative coatings produced by the method of V. Shneyderovoy [10]. The method consists in modeling stretching reinforced concrete element to cracking under the coating. Coating condition over the crack width shall be measured evaluated for its continuity at 20-30- fold magnification using an optical device. An indicator of the crack resistance are taking the width of the crack opening, when it was observed the formation of the first defect in the coating over crack.

In a phenomenological study covers defects [5] identified the most common and accepted as a characteristic species of cracking of coatings (10 species) and considered the cracks, depending on their density distribution (4 species). Makarova L.V. studied the use of coatings on their properties depending on the composition of the coating [5], the estimation of the adhesive strength of coatings, identified regularities in coating cracking in the aging process.

Based on the analysis of results obtained using the proposed technique of evaluating the crack resistance coatings [5] and the method of acoustic emission observed different character of change stress intensity factor in the coatings, depending on the type of coating, the length and nature of aging, as well as the proposed materials science path increasing the crack resistance protection-decorative coatings.

Makarova L.V. proposed methodology for assessing the crack resistance of protection-decorative coatings make it possible to develop a composition putty intended for leveling concrete and plaster surfaces, which includes: the sludge from the neutralization of a polishing mixture of glass industry, sodium hydroxide, hydrated lime, CMC glue, liquor liquid glass, and determine the basic physical and mechanical properties developed fillings.

Carried out analysis of experimental methods to assess the strength and durability of the coatings, based on natural data, leads to the need to develop theoretical and experimental methods for the determination of the stress-strain state of the coating in a crack zone taking into account the shrinkage and periodic changes in temperature, the strength characteristics.

Research of SSS structures: coating-base in the area of the local concentration of stress caused by the form (angular cuts), the need for further investigation of cracking process.

3 Theoretical foundations of the method

Consider the computational-experimental methods for studying the local concentration zones structures caused by the form boundaries (angular cuts) under the action of the forced rupture strain emerging in the source geometric stress concentration [11,12].

According to theoretical analysis [11,12] solution to the problem of elasticity theory in the neighborhood of a singular point of a plane domain of the boundary, which includes a finite discontinuity (jump) of forced strains represented in the form:

$$\eta = \eta^c + \eta^n, \quad (1)$$

where η^c – "proper" the solution of the homogeneous boundary value problem in a neighborhood of a singular point of the boundary area, characterizing feature of the solution;

$\eta^{(n)}$ – SSS, due to the action of the given loads, depending on the geometrical parameters - "degree of proximity" to the singular point. The representation (1) is valid in the spatial case for points on the special boundary line.

According to the theoretical VAT representation in the form of (1) in the neighborhood of a singular point of a plane domain boundary there are two self- balanced stress state. The first - the radial self- balanced stress state obtained as a solution to a flat homogeneous boundary value problem in a neighborhood of a singular point of the boundary region, passing in a singular state of stress at the approach to the singular point inside the boundary. Another remaining part, self-balanced plane stress in the top of the corner cutout border corresponds to the stresses caused by the action of given loads or general stress field.

An experimental solution by the method of photoelasticity in the neighborhood of a singular point of a plane domain of the boundary is considered a composite on flat model, in one of the areas where (Ω_2) to create a free temperature deformation $\alpha T \delta_{ij}$, while the other region (Ω_1) is free from stress. The line of contact areas out in an singular boundary point - the top corner cutout border plate with different angles of opening.

In the field of non-singular solutions of the homogeneous plane elastic problem possible derive estimates using that, you can extrapolate the data on the cross section, located close to an irregular boundary point, in view of the experimental data and practical measurement accuracy of the method photoelasticity.

According to theoretical-experimental analysis of the state of stress in the neighborhood of a singular point of a plane domain of the boundary, which comes to a jump of forced deformation, proposes the following formula for the extrapolation of the experimental data:

$$m_{i+1} = \left(\frac{r_i}{r_{i+1}} \right)^{1-\lambda_0} m_i, \tag{2}$$

where m_i stripes orders according to an experiment in design section r_i in the neighborhood of a nonsingular solution of the homogeneous boundary value problem, m_{i+1} stripes orders in cross-section of smaller radius $r_{i+1} < r_i$, located in an area with unreadable or "bad" to read a picture izohrom model, $\lambda_0 = \min \operatorname{Re} \lambda$ – the minimum value of the real part of the complex roots of the characteristic equation of the homogeneous boundary value problem for the wedge model, determined settlement [12].

In the above sections are practically identical diagrams orders stripes constructed by the relation (2) and the experimental data directly on the picture stripes of model. Match diagrams stripes orders allows you to apply the formula (2) to extrapolate the experimental data [13].

4 Discussion

The proposed computational-experimental method allows you to obtain and analyze SSS of structures in the area of stress concentration due to the constructive form of boundaries and discontinuous forced deformations. Possible to consider formation of cracks, in particular, angled in the coating and its base during the crack growth as well as crack formation in the region of the end coupling of the coating and base.

5 Conclusions

On the basis of the data obtained with the use of the proposed complex natural-fact-based and computational-experimental methods to reliably estimate the crack resistance protective coatings and bases, as well as the development of recommendations and ways to improve the crack resistance and durability of protection-decorative coating of construction components and structures.

References

1. I.Z. Aronov, Nadezhnost' i kontrol' kachestva **8**, 15-18 (1995)
2. V.A. Greshnikov, Akusticheskaya emissiya, Izdatel'stvo standartov (1976)
3. Y.G. Kudryavtseva, Izvestiya vuzov. Stroitel'stvo **3**, 72-76 (1982)
4. V.I. Loganina, Stoykost' zashchitno-dekorativnykh pokrytiy naruzhnykh sten zdaniy, Izdatel'stvo Assotsiatsii stroitel'nykh vuzov (1999)
5. V.I. Loganina, L.V. Makarova, N.I. Makridin, Prognozirovanie treshchinostoikosti dekorativnykh pokritiy naruzhnykh sten zdaniy. Monografiya, Privolzhskiy dom znaniy (2004)
6. L.P. Orentlikher, Izvestiya vuzov. Stroitel'stvo i arkhitektura **8**, 63-64 (1988)
7. L.P. Orentlikher, Stroitel'nye materialy **10**, 27-29 (1986)
8. L.P. Orentlikher, Izvestiya vuzov. Stroitel'stvo **6**, 128-130 (2001)
9. D.V. Oreshkin, Sed'mye akademicheskie chteniya RAASN, 396-402 (2011)
10. L.Y. Frishter, Academia. Arkhitektura i stroitel'stvo. Rossiyskaya akademiya arkhitektury i stroitel'nykh nauk **4**, 94-97 (2008)
11. L.Y. Frishter, Vestnik MGSU **1**, 165-168 (2008)
12. L.Y. Frishter, Vestnik MGSU **1**, 272-276 (2008)
13. V.V. Shneyderova, Antikorroziynye lakokrasochnye pokrytiya, Stroyizdat (1982)

AUTHOR INDEX

AUTHOR INDEX

AUTHOR INDEX

AUTHOR INDEX

5th International Scientific Conference on Integration, Partnership and Innovation in Construction Science and Education 2016

MATEC Web of Conferences Volume 86 (2016)

Moscow, Russia
16 - 17 October 2016

Part 2 of 2

Editor:

V. Andreev

ISBN: 978-1-5108-3362-3

Printed from e-media with permission by:

Curran Associates, Inc.
57 Morehouse Lane
Red Hook, NY 12571

proceedings.com

Some format issues inherent in the e-media version may also appear in this print version.

Printed by Curran Associates, Inc. (2017)

For additional information, please contact EDP Sciences – Web of Conferences at the address below.

EDP Sciences – Web of Conferences
17, Avenue du Hoggar
Parc d'Activité de Courtabœuf
BP 112
F-91944 Les Ulis Cedex A
France

Phone: +33 (0) 1 69 18 75 75
Fax: +33 (0) 1 69 28 84 91

contact-edps@webofconferences.org

Additional copies of this publication are available from:

Curran Associates, Inc.
57 Morehouse Lane
Red Hook, NY 12571 USA
Phone: 845-758-0400
Fax: 845-758-2633
Email: curran@proceedings.com
Web: www.proceedings.com

TABLE OF CONTENTS

PART 1

PART 2

Green sulfur-extended asphalt concrete: mix design of the complex binder

Vitaliy Gladkikh[1,*], *Evgeniy* Korolev[1], *Valentina* Gladkikh[2]

[1] Moscow State University of Civil Engineering, Nanomaterials and Nanotechnology Research and Educational Centre, Yaroslavskoye hw. 26, 129337, Moscow, Russian Federation
[2] Orenburg State University, Prospekt Pobedy13, 460018, Orenburg, Russian Federation

Abstract. Using the ordinary sulfur extended asphalt leads to emission of toxic gases. The efficient method to prevent the emission is to use other admixtures as suppressors, but this requires the complication of the mix design scheme. In the present work we have extended the well known design method for the case of complex binder with arbitrary number of suppressors. The example of application of the proposed method is given. The results indicate that even for the relatively small amounts of suppressors it is better to use the correct mix design scheme.

1 Introduction and prior work

The researchers from different universities and companies all around the world consider the production of durable, energy-efficient and environmentally friendly asphalt concrete pavements as a significant goal. Enhancement of functional reliability of asphalt concrete pavements is currently associated with application of various modifying admixtures that enhance the operational properties of pavements. Among the mentioned admixtures are thermoplastic elastomers, rubber grains and several others [1-4]. The technical sulfur is also currently accepted as and efficient admixture for asphalt concrete. Asphalt concretes that are prepared with complex binder made of bitumen and technical sulfur are known as *sulfur-extended asphalts* (SEA) concretes. Because of numerous alterations in chemical and phase composition of bitumen, and also due to relatively intensive interaction at the phase interface, such materials are characterized by improved operational properties (especially for increased temperatures of operation), including resistance to rutting, fatigue life, water resistance, etc. [5-12].

However, widespread practical use of ordinary SEA is not possible because of environmental impact. For such SEA the emission of toxic gases (hydrogen sulfide and sulfur dioxide) takes place both during the concrete mix preparation and pavement production stages. The ecological aspects of SEA concretes application were considered quite early [13] and attract even more attention in recent research works [14-22]. Because of this, numerous toxic gas neutralizers that make the application of SEA concretes possible are successfully developed. Unfortunately, in case of many additional emission suppressors,

* Corresponding author: gladkich_87@mail.ru

well known methods that were proposed in, e.g., [23] and [24], can not be directly used to design the binder. Thus, the method of SEA mix design requires modification.

2 Basic mix design scheme

As before, the basis of mix design will be the obvious statement that volume of composition is the sum of components' volumes:

$$\sum_{i=1}^{N} \frac{m_i}{\rho_i} = V, \tag{1}$$

where m_i is the mass of i-th component, ρ_i is the density of i-th component, N is the number of components.

There are numerous local documents that regulate the rules of grain size distribution and volumetric ration of the binder in asphalt concrete; in Russian Federation there are RU GOST 31015-2002 and RU GOST 9128-2013. In case of SEA concretes such documents can still be used as guides during design of grain size distribution.

As it was stated in, e.g., [24], for SEA concretes it is necessary to alter the volumetric ratio between binder and disperse phases since rheological properties of bitumen and sulfur-bitumen mix are different (the viscosity of sulfur at preparation stage is significantly lower). However, below we are going generalize the mix design scheme for *complex binder* – SEA with fine powders of suppressors of toxic gases. In general case, we can assume almost nothing about the viscosity of complex binder. Because of this we state that optimal structure of concrete (minimal porosity) corresponds to constant volumetric ratio between the binder and disperse phases: regardless of the amount of sulfur that is added to the bitumen, the resulting volume of binder should be equal to the volume of bitumen in asphalt concrete without the addition of sulfur:

$$\frac{B_0}{\rho_b} = \frac{SBB}{\rho_{sbb}}, \tag{2}$$

where B_0 is the optimal mass of bitumen in concrete mix (as it can be found during the design according to the local regulations), ρ_b is the density of bitumen, SBB is the amount of sulfur-bitumen binder (ordinary SEA), ρ_{sbb} is density of sulfur-bitumen binder.

Optimal amount of sulfur-bitumen binder in sulfur-extended concrete mix can be expressed as follows:

$$SBB = B_0 \frac{\rho_{sbb}}{\rho_b}. \tag{3}$$

To find out the density of sulfur-bitumen binder we can use Eq. (1):

$$\frac{B}{\rho_b} + \frac{S}{\rho_s} = 1, \tag{4}$$

where B and S are masses of bitumen and sulfur per unit volume, respectively.

Also, for the unit volume the following equation holds:

$$S + B = \rho_{sbb}. \tag{5}$$

Thus, we have:

$$\rho_{sbb} = \left[\frac{1}{\rho_b} - a\left(\frac{1}{\rho_b} - \frac{1}{\rho_s}\right)\right]^{-1}, \tag{6}$$

where a is the ratio of sulfur in sulfur-bitumen binder ($a < 1$).

Substituting Eq. (6) in Eq. (3) we get:

$$SBB = B_0\left[1 - a\left(1 - \frac{\rho_b}{\rho_s}\right)\right]^{-1}. \tag{7}$$

Specific mass of bitumen in sulfur-bitumen binder:

$$B_1 = (1 - a)B_0\left[1 - a\left(1 - \frac{\rho_b}{\rho_s}\right)\right]^{-1}. \tag{8}$$

Specific mass of sulfur:

$$S_1 = aB_0\left[1 - a\left(1 - \frac{\rho_b}{\rho_s}\right)\right]^{-1}. \tag{9}$$

The expressions (8) and (9) differ from the known design scheme [23, 24] only by form (and, obviously, different data are required to use the derived equations).

2 Extended mix design scheme

Now we will take into account the presence of k additional disperse phases – emission suppressors. Again, we can use Eq. (1):

$$\frac{B}{\rho_b} + \frac{S}{\rho_s} + \sum_{i=1}^{k}\frac{N_i}{\rho_i} = 1, \tag{10}$$

where N_i is the mass of i-th suppressor per unit volume, ρ_i is the density of i-th suppressor.

We can also express the mass of i-th suppressor as follows:

$$N_i = c_i S, \tag{11}$$

where c_i is the ratio of i-th suppressor in complex binder.

The specific mass of bitumen in complex binder with k different suppressors is equal to:

$$B_1 = \left[1 - a\left(1 + \sum c_i\right)\right]\rho_{sbb}. \tag{12}$$

Taking into account both Eq. (11) and Eq. (12) we have:

$$\left[1 - a\left(1 + \sum_{i=1}^{k} c_i\right)\right]\frac{\rho_{sbb}}{\rho_b} + \frac{a\rho_{sbb}}{\rho_s} + a\rho_{sbb}\sum_{i=1}^{k}\frac{c_i}{\rho_i} = 1. \tag{13}$$

The last equation can be solved for ρ_{sbb}:

$$\rho_{sbb} = \left\{\left[1 - a\left(1 + \sum_{i=1}^{k} c_i\right)\right]\rho_b^{-1} + \frac{a}{\rho_s} + a\sum_{i=1}^{k}\frac{c_i}{\rho_i}\right\}^{-1}. \tag{14}$$

Optimal amount of complex binder with suppressors of toxic gases is equal to:

$$SBB = \frac{B_0\rho_{sbb}}{\rho_b}, \tag{15}$$

thus, for the specific mass of sulfur in complex binder with suppressors of toxic gases we get:

$$S_1 = \frac{aB_0\rho_{sbb}}{\rho_b}. \tag{16}$$

Taking into account three last equations, we derive the following expression for the specific mass of sulfur:

$$S_1 = aB_0\left\{\left[1 - a\left(1 + \sum_{i=1}^{k} c_i\right)\right] + a\frac{\rho_b}{\rho_s} + a\sum_{i=1}^{k} c_i\frac{\rho_b}{\rho_i}\right\}^{-1}. \tag{17}$$

Specific mass of i-th suppressor:

$$H_i = a\,B_0 c_i \left\{ \left[1 - a\left(1 + \sum_{j=1}^{k} c_j\right) \right] + a\frac{\rho_b}{\rho_s} + a\sum_{j=1}^{k} c_j \frac{\rho_b}{\rho_j} \right\}^{-1} . \tag{18}$$

And, finally, the specific mass of bitumen in complex binder is:

$$B_1 = \left[1 - a\left(1 + \sum_{i=1}^{k} c_i\right) \right] B_0 \left\{ \left[1 - a\left(1 + \sum_{i=1}^{k} c_i\right) \right] + a\frac{\rho_b}{\rho_s} + a\sum_{i=1}^{k} c_i \frac{\rho_b}{\rho_i} \right\}^{-1} . \tag{19}$$

It can easily be proven that for the mix without suppressors ($k = 0$) expressions (17) and (19) becomes expressions (9) and (8), respectively.

3 Examples of mix design

3.1. Example 1: basic design

We need to design the ordinary SEA without any suppressors of toxic gases.

Reference data: density of sulfur is 2070 kg/m^3, density of bitumen is 1000 kg/m^3.

Given data: amount of sulfur in SEA is 30%, specific mass of the bitumen for reference grain size distribution is 52.5 kg per ton.

The density of SEA can be determined by Eq. (6):

$$\rho_{sbb} = \left[\frac{1}{\rho_b} - a\left(\frac{1}{\rho_b} - \frac{1}{\rho_s} \right) \right]^{-1} = \left[\frac{1}{1000} - 0.3\cdot\left(\frac{1}{1000} - \frac{1}{2070} \right) \right]^{-1} = 1183.5 \text{ kg/m}^3.$$

Specific mass of bitumen can be found by Eq. (8):

$$B_1 = (1-a)B_0\left[1 - a\left(1 - \frac{\rho_b}{\rho_s} \right) \right]^{-1} = [(1-0.3)\cdot 52.5]\cdot\left[1 - 0.3\cdot\left(1 - \frac{1000}{2070} \right) \right]^{-1} = 43.5 \text{ kg per ton}.$$

Specific mass of sulfur:

$$S_1 = aB_0\left[1 - a\left(1 - \frac{\rho_b}{\rho_s} \right) \right]^{-1} = 0.3\cdot 52.5\cdot\left[1 - 0.3\cdot\left(1 - \frac{1000}{2070} \right) \right]^{-1} = 18.64 \text{ kg per ton}.$$

The results are summarized in Table 1.

Table 1. Basic mix design of the SEA.

Component	Specific mass, kg per ton
Bitumen	43.5
Sulfur	18.24
Specific mass of the SEA	62.14

As we can see from Table 1, the specific mass of bitumen is decreased (part of the bitumen is replaced by sulfur), while specific mass of the SEA is increased (due to higher density of sulfur). Thus, taking into account prices of sulfur and bitumen, we can expect the SEA concrete of slightly lower price and slightly higher density.

3.1. Example 2: extended design

We need to design the complex binder with single suppressor of toxic gases.

Reference data: density of suppressor is 6310 kg/m^3, other reference data are the same.

Given data: amount of sulfur in complex binder is 30%, specific mass of the bitumen for reference grain size distribution is 52.5 kg per ton, number k of suppressors is 1, sulfur/suppressor mass ratio is 30%.

Specific mass of bitumen (Eq. (19)):

$$B_1 = \left[1 - a\left(1 + \sum_{i=1}^{1} c_i\right)\right] B_0 \left\{\left[1 - a\left(1 + \sum_{i=1}^{1} c_i\right)\right] + a\frac{\rho_b}{\rho_s} + a\sum_{i=1}^{1} c_i \frac{\rho_b}{\rho_i}\right\}^{-1} =$$

$$= [1 - 0.3 \cdot (1 + 0.05)] \cdot 52.5 \cdot \left\{[1 - 0.3 \cdot (1 + 0.05)] + 0.3 \cdot \frac{1000}{2070} + 0.3 \cdot 0.05 \cdot \frac{1000}{6310}\right\}^{-1} = 43.21 .$$

Specific mass of sulfur (Eq. (17)):

$$S_1 = aB_0 \left\{\left[1 - a\left(1 + \sum_{i=1}^{1} c_i\right)\right] + a\frac{\rho_b}{\rho_s} + a\sum_{i=1}^{1} c_i \frac{\rho_b}{\rho_i}\right\}^{-1} =$$

$$= 0.3 \cdot 52.5 \cdot \left\{[1 - 0.3 \cdot (1 + 0.05)] + 0.3 \cdot \frac{1000}{2070} + 0.3 \cdot 0.05 \cdot \frac{1000}{6310}\right\}^{-1} = 18.92 .$$

Specific mass of suppressor (Eq. (18)):

$$H_1 = a\ B_0 c_1 \left\{\left[1 - a\left(1 + \sum_{j=1}^{1} c_j\right)\right] + a\frac{\rho_b}{\rho_s} + a\sum_{j=1}^{1} c_j \frac{\rho_b}{\rho_j}\right\}^{-1} = .$$

$$= 0.3 \cdot 52.5 \cdot 0.05 \cdot \left([1 - 0.3 \cdot (1 + 0.05)] + 0.3 \cdot \frac{1000}{2070} + 0.3 \cdot 0.05 \cdot \frac{1000}{6310}\right)^{-1} = 0.95 .$$

The results are summarized in Table 2.

Table 2. Extended mix design of the complex binder.

Component	Specific mass, kg per ton
Bitumen	43.21
Sulfur	18.92
Suppressor	0.95
Specific mass of the complex binder	62.64

As it can be decided from Table 2, even for the relatively small amounts of suppressors, it is better to use the correct mix design scheme: the difference in densities between sulfur and suppressor leads to noticeable alteration of the specific masses of components.

Summary and conclusion

Despite the fact that the use of sulfur as an extender in asphalt paving mixtures could perform in a satisfactory manner [11], using the ordinary SEA leads to serous problem. The main hazard is the emission of hydrogen sulfide and sulfur dioxide during mixing and dumping operations. The efficient method to prevent the emission of toxic gases is to use numerous suppressors; obviously, this requires the complication of the mix design scheme.

In the present work we have extended the SEA design method for the case of complex binder with arbitrary number of extra phases. The offered method allows precise calculation of specific masses of all components. The example of application of the proposed method is given. The results indicate that even for the relatively small amounts of suppressors, it is better to use the correct mix design scheme: the difference in densities between sulfur and suppressor leads to noticeable alteration of the specific masses of components.

This work is supported by grant of Russian Federation President GP-2627.2016.1.

References

1. V.G. Nikolsky, I.A. Krasotkina, Automotive roads **2**, 37-40, (2010)
2. A.B. Solomentsev, Science and Technology for Road Construction **1**, 14-16, (2008)
3. L.M. Gochman, *Pavements based on Sulfur-Extended Asphalts with Polymers*, Moscow: Infoavtodor (2002)
4. M.A. Vysotskaya, D.A. Kusnetsov, Building Materials **12**, 63-64 (2013)
5. Yu.E. Vasilyev, O.A. Voyeyko, D.S. Tsarkov, Naukovedeniye **5**, (2014)
6. V.A. Gladkikh, E.V. Korolev, D.L. Khusid, Building Materials, Equipment and Technologies of XXI Century, **194**, 30-33, (2015)
7. E.V. Korolev, V.A. Gladkikh, V.A. Smirnov, *Advanced Materials and Structural Engineering: Proc. of the Intl. Conf. on Advanced Materials and Engineering Structural Technology*, 285-288 (2015)
8. V.Z. Gnateyko, *The use of sulfur and sulfur-containing waste in road construction*, Highways: an overview, Moscow: Minavtodor (1990)
9. T.W. Kennedy, R. Haas, P. Smith, G.A. Kennepohl, E.T. Hignell, Transportation Research Record **659**, 12-17 (1977)
10. M. Mazumdar, S.K. Rao, Transportation Research Record **1417**, 144-149 (1993)
11. T.L. Beatty, K. Dunn, E.T. Harrigan, K. Stuart, H. Weber, Transportation Research Record **1115**, 161-170 (1987)
12. R. Hou, Z.-Y. Guo, Journal of Building Materials **16**, 525-528 (2013)
13. D.Y. Lee, Industrial and Engineering Chemistry Product Research and Development, **14**, 171-177 (1975)
14. P. Qian, H. Qiao, Q. Ma, Journal of Southeast University (Natural Science Edition), **39**, 592-597 (2009)
15. J. Colange, D. Strickland, G. McCabe, K. Gilbert, R. May, S. Banbury, Road Materials and Pavement Design **11**, 459-485 (2010)
16. H.A. Kadkhodaei, H. Shabani, M. Shapouri, G. Karimi, *Proc. of the 8th International Chemical Engineering Congress & Exhibition (IChEC 2014)*, (2014)
17. Gladkikh, V.A. & Korolev, E.V, Advanced Materials Research **1040**, 387-392 (2014)
18. V.A. Gladkikh, E.V. Korolev, D.L. Khusid, Building materials, equipment and technologies of XXI century **194**, 30-33 (2015)
19. O.I. Doshlov, I.A. Kalapov, Journal of the Irkutsk State Technical University **11**, 107-111 (2015)
20. G.H. Shafabakhsh, M. Faramarzi, M. Sadeghnejad, Construction and Building Materials **98**, 456-464 (2015)
21. E.R. Souaya, S.A. Elkholy, A.M.M. Abd El-Rahman, M. El-Shafie, I.M. Ibrahim, Z.L. Abo-Shanab, Egyptian Journal of Petroleum **24**, 483-491 (2015)
22. E.V. Korolev, V.A. Gladkikh, V.A. Smirnov, *Advanced Materials and Structural Engineering: Proc. of the Intl. Conf. on Advanced Materials and Engineering Structural Technology*, 285-288 (2015)
23. W.C. McBee, T.A. Sullivan, and J.O. Izatt, *State-of-the-Art Guideline Manual for Design, Quality Control, and Construction of Sulfur-Extended Asphalt (SEA) Pavements, FHWA-IP-80-14, Implementation Package*, US Department of Interior Bureau of Mines with The Sulphur Institute (1980)
24. W.C. McBee, T.A. Sullivan, D. Saylak, ASTM STP **724**, 39-63 (1980)

Properties of sulfur-extended asphalt concrete

Vitaliy Gladkikh[1,*], *Evgeniy* Korolev[1], *Dmitrij* Husid[1], *Ilya* Sukhachev[2]

[1]Moscow State University of Civil Engineering, Nanomaterials and Nanotechnology Research and Educational Centre, Yaroslavskoye hw. 26, 129337, Moscow, Russian Federation
[2]Tuymen Industrial University, Volodarskogo st., 38, 625000, Tyumen, Russian Federation

Abstract. Currently, increased functional reliability of asphalt concrete coatings associated with various modifying additives that improve the durability of pavements. Promising builder is a technical sulfur. Asphalt concrete, made using a complex binder consisting of petroleum bitumen and technical sulfur, were calledsSulfur-Extended Asphalt Concrete. Such asphalt concrete, due to changes in the chemical composition of particulate and bitumen, changes the intensity of the interaction at the interface have increased rates of physical and mechanical properties. There was a lack of essential knowledge concerning mechanical properties of the sulfur-bituminous concrete with such an admixture; therefore, we had carried out the necessary examination. It is revealed that a new material satisfies local regulations in terms of compressive and tensile strength, shear resistance, and internal friction.

1 Introduction and prior work

As practice shows asphalt coating around the world deformed and destroyed before the put and the warranty period and, as a rule, need to be repaired after the first years of operation. The causes of premature destruction and deformities include: high traffic, increasing axial loads, a violation of technology and the production of construction, as well as the poor quality of materials used.

One of the solutions to the problems of extending the service life of road surfaces associated with the task of obtaining the material, operate reliably in a change from positive to negative temperatures on the other hand, under the influence of intense traffic.

Increased functional reliability of asphalt concrete coatings currently in the Russian Federation and abroad associated with various modifying additives that contribute to improve the quality of road surfaces. Among such modifiers, include thermoplastic resin, rubber crumb, and others [1-4]. There exist numerous ways to improve the operational properties of concretes based on thermoplastic binder. By means of application of polymers distributed on the surface of fine filler it is possible to outcome several drawbacks of the traditional material with thermoplastic matrix [5-6]. The disadvantage of such approach is the consequence of extra processing stage during preparation of concrete mix.

The bitumen matrix of the asphalt concrete can be enhanced by polyolefins. In particular, addition of the grinded polyethylene decreases penetration increases the

* Corresponding author: gladkich_87@mail.ru

softening and melting points [7, 8]. As a consequence, the resulting material is characterized by increased rutting resistance and durability. However, admixture of polyolefins makes the production more complex and costly.

One of the most promising admixtures for the asphalt concrete is technical sulfur. Complex of operational properties of sulfur-based and sulfur-extended building materials may lead both to economical efficiency of construction and to reduce load on the environment. The feasibility of the sulfur-extended materials is primarily caused by properties, availability, and low cost of sulfur.

Today, technology of sulfur-bituminous materials develops intensively in different countries. Asphalt modified with sulfur, due to several factors (changes in the chemical composition of bitumen and varying the intensity of the interaction at the interface boundaries) is characterized by increased values of operational properties at elevated temperatures, especially – by high resistance to rutting. The long-term water resistance is also at high level for concretes with binder that includes no more than 45% of sulfur [9].

The sulfur-extended asphalt concretes often have to be characterized by an improved technical and economical efficiency; their usage leads to several benefits: increasing the material's application area, lowering the pollution of the environment (technical sulfur can be considered a waste because in Russia there are numerous sulfur dumps near the oil industry enterprises), decreasing the costs of construction and exploitation of roads (due to long periods between maintenance works).

The main factor limiting the widespread practical use of the mentioned technology was the lack of solutions to neutralize toxic gasses – hydrogen sulfide and sulfur dioxide – formed during the manufacture and installation of sulfur-bituminous concrete. It was shown [10] that oxides of amphoteric metals are effective suppressors for hydrogen sulfide and sulfur dioxide. The obtained experimental data allowed the selection of most prospective suppressors for toxic gasses emitted during the production and laying of sulfur-bituminous concrete.

But till now, there was limited knowledge about physical and mechanical properties of the concrete with sulfur-bituminous matrix and aforementioned suppressors.

2 Experimental setup

The standard samples of asphalt concrete were prepared according to RU GOST 12801 from the concrete mixes with different amount of sulfurs (20%–45%). Quality of the concrete mixes was determined according to RU GOST 31015. The prepared samples were examined for physical-mechanical and operational properties according to RU GOST 12801 (crack resistance), AASHTO TP 63-06 (resistance to rutting), and EN 12697-24 (fatigue life).

Cylindrical specimens of diameter 150 mm and height 75 mm (AASHTO TP 63 test) were compacted with Uniframe device. Six specimens for each composition of asphalt concrete were tested. The value of rut depth for each composition is calculated as an average for the six specimens.

3 Results and discussion

One of the most important mechanical properties of any building material is compressive strength. Experimentally obtained dependencies between the amount of sulfur-based admixture and compressive strength are shown in Fig. 1.

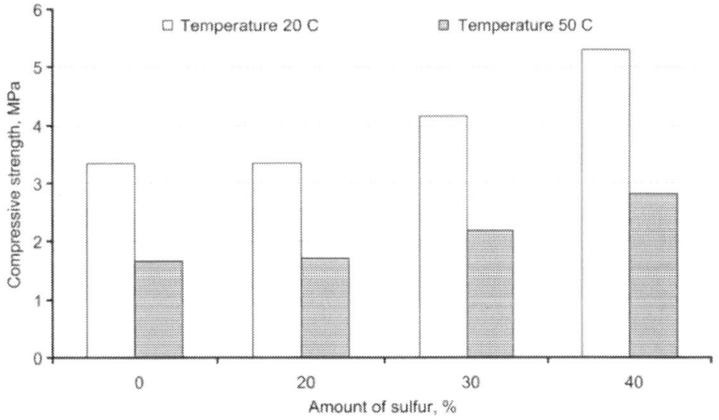

Fig. 1. Compressive strength of sulfur-bituminous pavement.

As it follows from Figure 1, addition of the small amount of sulfur does not change the compressive strength. Starting from concentration of 20%, there is a linear dependence between compressive strength and amount of sulfur. Adding the sulfur notably alters the properties of the asphalt binder (the concentrations above 40% can not be used in practice because several other operational properties of the concrete are out of allowed limits). Such behavior is observed for both temperatures of experiment (20 and 50 ^{0}C).

This can be described as follows. Part of the sulfur is dispersed in the bitumen and leads to only a small improvement of the properties. Another part of the sulfur (when concentration is about 20% and higher) is undissolved and coagulates. Formation of large sulfur droplets takes place; these droplets crystallize upon cooling. The additional links of the crystallization are formed and because of this there is a hardening of asphalt material. At the operating temperatures (both 20 and 50 ^{0}C), sulfur is solid and plays the role of disperse filler. This filler organizes the structure of the bitumen and increases thermal stability, rigidity, and resistance to rutting of asphalt concrete.

At the technological temperatures, sulfur is a highly mobile melt and increases the efficiency of mixing and laying of asphalt concrete mixes; thus, the goals of two different aforementioned groups are achieved simultaneously.

There is one operational property that is tightly correlated with mechanical properties – resistance to rutting. Low values of rutting resistance significantly reduce durability of the pavement under load of heavy vehicles.

The experimental dependencies between number of processing cycles (according to AASHTO TP 63 test procedure) and depth of the rut are presented in Figure 2.

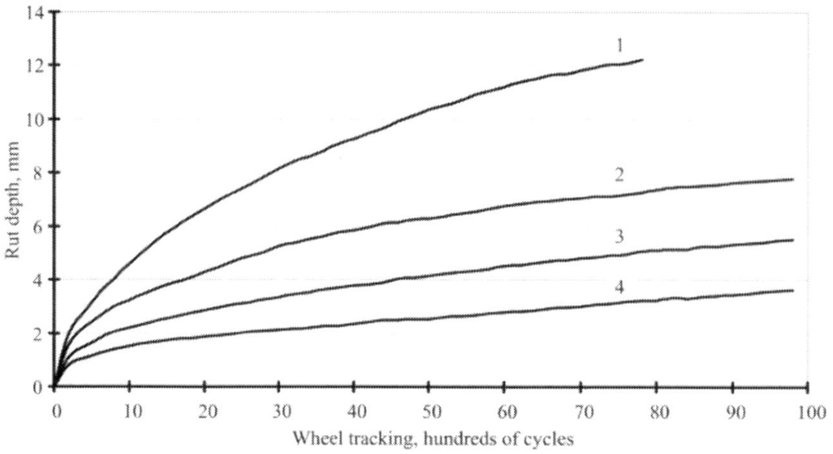

Fig. 2. Dependencies between the rut depth and number of wheel cycles for AASHTO TP 63 test and amount of sulfur: 1 – 0%; 2 – 20%; 3 – 30%, 4 – 40%.

As it follows from Figure 2, even the small amount of sulfur significantly increases rutting resistance. Therefore, the area of primary application of sulfur-bituminous concrete is the construction of pavements with heavy load.

Material's fatigue life, together with rutting resistance, is the primary indicator of suitability for durable and long-life pavement. It is found during the study of fatigue life by means of multiple cyclic loading with indirect tensile scheme that designed sulfur-bituminous materials are of high resistance (Figure. 3).

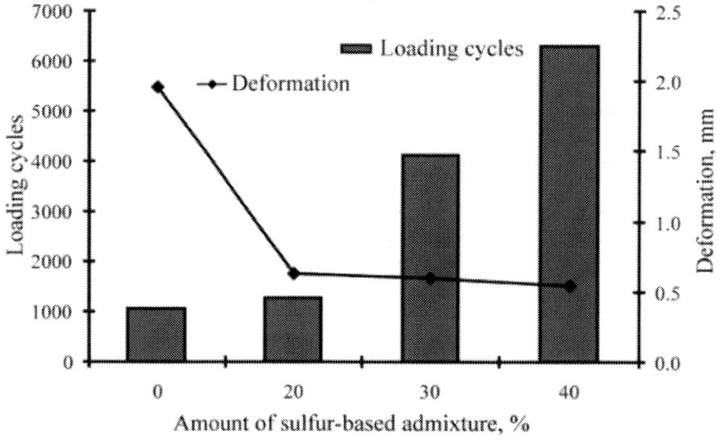

Fig. 3. The dependencies between fatigue limit, horizontal deformation and number of load cycles for different amounts of sulfur-based modifier.

It is clear from the experimental data in Figure 3 that fatigue life increases together with amount of sulfur in sulfur-bituminous mix. As before, starting from 20% concentration, we can observe linear dependence between operational property (in this case – fatigue life) and concentration. The number of cycles preceding the start of irreversible deformation induced by main crack grow, if compared with such number for reference sample, increases significantly for compositions with 30% and 40% of sulfur-based modifier (about 4 and 6 times, respectively). At the same time, enlargement of the content of sulfur modifier leads to decrease of deformability.

So far, all the obtained data clearly demonstrate that there is the ultimate way for production of advanced pavement material – namely, complete elimination of bitumen. Obviously, such a conclusion would be wrong. Some consequences of such decision were already mentioned, and necessity of massive modification in current production facilities is not the only reason for that. Several important operational properties (e.g., water and frost resistance) are negatively affected by excess of the sulfur. Moreover, not every mechanical characteristic are in a positive correlation with amount of sulfur in bitumen. One of such properties is crack resistance. This parameter can be characterized by experimentally obtained values for splitting tensile strength at 0 ^{0}C.

Properties of the sulfur-bituminous materials are summarized in Table 1.

Table 1. Properties of the developed sulfur-bituminous concrete.

Property	RU GOST 31015	Reference sample	Modified concrete, sulfur content:		
			20%	30%	40%
Compressive strength at 20 ^{0}C, MPa	> 2.2	3.3	3.3	4.2	5.3
Compressive strength at 50 ^{0}C, MPa	> 0.65	1.67	1.72	2.19	2.82
Average density, kg/m^3	–	2620	2640	2650	2650
Clutch shear (50 ^{0}C), MPa	> 0.18	0.29	0.29	0.38	0.54
Water resistance	> 0.85	0.90	0.89	0.95	0.90

The experimental results summarized in Table 1 show that compressive strength of sulfur-bituminous concrete at temperatures 20 and 50 ^{0}C exceeds not only the corresponding values for reference concrete without sulfur, but also the values required by RU GOST 31015.

It was also revealed during additional examination that sulfur-bituminous concrete after storage in water shows very low absorption. The absorption is 1.5-1.8 times lower if compared with reference composition. Therefore, freezing and thawing causes only a small negative effect if compared with conventional asphalt concrete. It is established that under identical operating temperatures the modulus of elasticity of sulfur-bituminous concrete is higher than modulus of conventional asphalt concrete. Moreover, with increasing temperature this difference also increases.

Summary and conclusion

In the present work, we have discussed the results of several laboratory tests. The tests were carried out to answer the question concerning dependence between the amount of sulfur-based modifier, mechanical and important operational properties of sulfur-bituminous concrete for pavements.

It is shown that low amounts of sulfur are dissolved in bitumen, the compressive strength of concrete with such matrix is close to the strength of reference sample. Extra amounts of sulfur may lead to the formation of additional spatial cross-linked network. The formation of such a network, in turn, causes the increase of both strength and modulus of the concrete; rutting resistance also increases significantly. Thus, almost all the mechanical properties, including compressive strength and resistance to rutting, are positively correlated with concentration of designed multifunction sulfur-based admixture. Taking into consideration values of such properties as crack resistance (especially at low temperatures), water and frost resistance, and also such factors as necessity for preservation existing production infrastructure, the recommended amount of sulfur-based admixture with gas-suppression additives is about 30–40%.

This work is supported by grant of Russian Federation President GP-2627.2016.1.

References

1. V.G. Nikolsky, I.A. Krasotkina, Automotive roads **2**, 37-40, (2010)
2. A.B. Solomentsev, Science and Technology for Road Construction **1**, 14-16, (2008)
3. L.M. Gochman, *Pavements based on Sulfur-Extended Asphalts with Polymers* (Infoavtodor, Moscow, 2002)
4. M.A. Vysotskaya, D.A. Kusnetsov, Building Materials **12**, 63-64 (2013)
5. E.V. Korolev, V.A. Smirnov, A.I. Albakasov, Nanotechnologies in Construction **4**, 81-87, (2012)
6. E.V. Korolev, S.S. Inozemcev, Vestnik MGSU **10**, 131-139 (2013)
7. T. Ali, N. Iqbal, M. Ali, K. Shahzada, IOSR-JMCE **5** 1-6 (2014)
8. P. Shedame, N. Pitale, IOSR-JMCE **3**, 37-45 (2014)
9. V.A. Gladkikh, E.V. Korolev, Vestnik MGSU **4**, 76-83 (2013)
10. V.A. Gladkikh, E.V. Korolev. Advanced Materials Research **1040**, 387-392 (2014)

The use of sandwich-panels as a safety and easily thrown off designs for internal explosions

Vyacheslav Gorev[1] and *Evgeniya* Salymova[1*]

[1]Moscow State University of Civil Engineering, 129337, Yaroslavskoye Shosse 26, Moscow, Russia

Abstract. New method of explosion-proof ensuring for industrial buildings by usage of sandwich-panels as enclosures and easily thrown off designs by inside combustions is examined in this work.

Possibility of dumping of an overpressure at internal explosion indoors - the obligatory requirement at projection and operation of explosive rooms [7]. The acceptable results are achieved by two options:1) use of the weak protecting designs (glass, membranes); 2) use of loose couplings when fastening of the protecting designs to a building framework. In case of the former the protecting designs collapse, in the second communications collapse and the released protecting designs start moving, opening the aperture blocked by them. Through the opened apertures from the room gases expire that leads to pressure decrease. The best results are achieved at the quasistatic nature of explosion when wave processes are insignificant.

The protecting designs serving for an explosion depressurization are often called the safety designs (SD) [4] or easily thrown off designs (ETD). [9].

Now most the energy industries enterprises are built with use triplex a sandwich panels which at the corresponding fastening can apply, owing to their small weight, to use as ETD.

Full success is achieved, if the area does not allow the rise of the burst pressure to allowable values ΔP_d , which is determined by the carrying capacity of protected structures. This success is possible, if the area of openings, covers easy detachable more than the required minimum, and easily thrown off designs are opened fast enough to provide the required area expires during the development of the explosion.

It is necessary that the time development of the explosion was more than time for the opening of openings. Internal quasi-static explosion devoted many works [1, 2, 3, 4]. Most of the required area is determined by the expiration. Question - how do you ensure this area remains on the sidelines with the exception of destructible structures: glass and membrane [4, 5].

In industrial hazardous locations more convenient to use light panels that have loose connections. The explosion destroyed the first, and the panels fly off under the action of pressure and release they overlap the square. In this paper the process of opening the light panels and the change in pressure directly after the destruction of relationships, retaining panel. The analysis is performed using the equations describing the pressure changes with increasing square of the expiration of the [4] and the equations of motion of the panel.

This assumes that:

* Corresponding author: Kiara_lion@mail.ru

1. The movement of the panel is described by the equation:

$$M\frac{d^2x}{dt^2} = \Delta Pab$$

Where **x**- distance last the panel by the time of time **t**

2. Expire cold gases.
3. The average pressure acting on the panel:

$$\Delta P = \frac{\Delta P_1 + \Delta P_V}{2},$$

where $\Delta \mathbf{P_1}$ – corresponds to the first maximum pressure immediately after opening, $\Delta \mathbf{P_V}$ – pressure opening. From experiments and numerical solutions performed in [6] shows that $\Delta \mathbf{P_1}$ depends on $\Delta \mathbf{P_V}$ and the speed of combustion. And may exceed $\Delta \mathbf{P_d}$.

4. Dealt with the pressure at which the velocity is described by the expression

$$\sqrt{\frac{2\Delta P}{\rho_0}}$$

that overrides the carrying capacity of many buildings. However, this limitation does not matter.

Note the experimental fact that the maximum $\Delta \mathbf{P_1}$occurs at the opening of sandwich panels with incomplete opening of the opening, i.e.

$$x < \frac{ab}{2(a+b)}$$

a and **b** – sides aperture.

Using assumptions 1-4 and condition of maximum pressure

$$\frac{dP}{dt} = 0$$

is the value of this maximum:

$$\frac{\Delta P_1 - \Delta P_V}{\Delta P_V} = \frac{2F^{1/2} - 1}{\left(F^{1/2} - 1\right)^2} \tag{1}$$

$$F = \frac{v_1 (a+b) ab \Delta P_V^{\ 3/2} N}{2^{3/2} \pi M \rho_0^{\ 1/2} U_g (\sigma - 1) \sigma^2}$$

$\mathbf{U_g}$ - the rate of combustion at the time of opening of easily thrown off designs, $\mathbf{v_1}$ – flow coefficient, **N** - is the number of opened panels, **M** – mass of the panel, $\boldsymbol{\sigma}$ - the rate of expansion during combustion, $\boldsymbol{\rho_0}$ - is the initial density of the gaseous medium.

The number of panels is determined from the condition that the area is completely uncovered openings provided for reducing the pressure of the explosion to $\Delta \mathbf{P_d}$, at the time of the maximum speed of energy release, i.e. at the moment when there $\mathbf{U_g}$ and $\mathbf{F_f}$ are maximum (where $\mathbf{F_f}$ - is the area of burning).

Thus, **(N*a*b)** has the necessary space required fully opened openings. This area is defined by the expression [4,6]:

$$S_{treb} = \left(\frac{2\rho_0}{v_2 \Delta P_d}\right)^{1/2} \frac{\sigma^2 (\sigma - 1) U_g K_f V_0^{2/3}}{\left(1 + \sigma^{1/2}\right)} \tag{2}$$

Substituting this expression into the previous one and requiring that would $\Delta \mathbf{P_1} < \Delta \mathbf{P_d}$, after simple transformations, we obtain the condition:

$$\Delta P_d\left[1-\left(\frac{\Delta P_V}{\Delta P_d}\right)^{1/2}\right]^2 > \frac{2\pi M U_g^2\left(1+\sigma^{1/2}\right)}{(a+b)K_f V_0^{2/3}} \tag{3}$$

Here $\mathbf{K_f}$ is the shape factor of the premises, determined from:

$$(F)_{\max} = K_f V_0^{2/3} \tag{4}$$

F – surface area of the flame when deflagration internal explosion.

Using data on the known parameters of an internal explosion and properties of easily thrown off designs, you can pick up $\mathbf{\Delta P_V}$ to $\mathbf{\Delta P_1}$ does not exceed $\mathbf{\Delta P_d}$. Table 1 presents data on $\mathbf{\Delta P_d}$ for different values of $\mathbf{\Delta P_V/\Delta P_d}$.

Table 1. Data on ΔPd for different values of ΔPV/ΔPd.

$\mathbf{\Delta P_V/\Delta P_1}$	0.5	0.6	0.7	0.8	0.9
$\mathbf{\Delta P_V}$**, kPa**	1.17	1.5	2.25	4.4	5.6

The calculations were performed for data: $\mathbf{V_0}$ = 1000 m³, $\mathbf{K_f}$ = 3, σ = 7, $\mathbf{U_g}$ = 2.4 m/sec, **(M/ab)**=25 kg/m.

The table shows that the magnitude of the pressure opening has a significant influence on the resulting blast load. In [6] experimentally studied the autopsy sandwich panels for some cases of attachment. The opening pressure depends on the thickness of the sheet, the thickness of the structure in which twists of «tapping» and the number of attachment points.

Pressure of opening depends on the leaf thickness from design thickness in which «self-tapping screw» and from number of points of fastening twists and in each case is defined experimentally.

Experiments were made on two explosive cameras 1 of m³ and 10 m³. The technique of carrying out an experiment is described in [6,7,8]. The generalized results of experiments are presented in the table 2.

Table 2. Results of experiments.

Series of experiments	**The tested design**	**Thickness of a metal profile, mm**	**Number of self-tapping screws**	**Nature of destruction**	**Pressure of opening, kPa.**	**Load of one self-tapping screw, kN**	**Note.**
1	Sandwich panel, 1200x2000x 50mm	5	4	The «self-tapping screw» in a metal profile	3.2 3.38 3.45	1.92 2.03 2.07	-
2	Sandwich panel, 1200x2000x 50mm	5	4	The «self-tapping screw» in a metal profile	2.4 2.03 2.2	1.44 1.20 1.32	Without washer
3	Sandwich panel, 1200x2000x 50mm	5	4	The «self-tapping screw» in a metal profile	1.2 1.55 1.1	0.72 0.93 0.66	Cut of a metal leaf
4	Part of the sandwich panel, 400x400x50 mm	2	2	The «self-tapping screw» in the panel	11.28 18.8 15.40	0.90 1.50 1.23	-

Experiments were made with use of nominal «self-tapping screws» – for fastening a sandwich panels. The aperture was blocked by a unity element a sandwich panel, results of experiments testify about:

– the nature of opening of the panel depends on thickness of a design to which it to fasten. At a thickness of design of 5 mm - with break of sheets of the panel, at 2 mm – with to pull out «self-tapping screw» from it;

– decrease of the size of washers under a head of «self-tapping screw» reduced opening pressure. In the absence of a washer pressure of opening decreased by 1.5 times;

– the cruciform cut of the panel under a washer, with a cut length to edge of a washer, was reduced by opening pressure by 2.5-3 times;

– at decrease of pressure of opening the maximal pressure immediately after opening according to (3) and table 1 decreased;

– pressure of opening can be estimated on a ratio:

$$\Delta P_{в} = \frac{f * n}{S}$$

– where **f** – effort to a simple point of fastening during the opening, it is defined experimentally at a leaf thickness a sandwich panel, **δ**=0.5mm, **n** – number of points of fastening, **S** - the area a sandwich panels.

Application a sandwich panels of the big size increases load of simple knot of fastening, therefore, more rational is use a sandwich panels of perhaps larger size. In this case there is a possibility of protection not only bearing, but also protecting a sandwich panels of the smaller size.

In the conclusion it is necessary to make two remarks:

1) at installation a sandwich panels they are adjusted to each other by means of the «groove in a groove» lock along the lengthiest party. Carrying out experiments with the block a sandwich panels was impossible. It is possible to consider this effect gathering the block from a sandwich panels so that cross sectional dimension of the block (**n*b**) was common more than length one sandwich panel or to use panels without latch fastening.

2) the second remark arises because opening of an aperture begins not at the time of a start of motion, as in an experiment, and at the moment when ETD passes distance **x =δ** to panel thickness. This circumstance is considered by the amendment to opening pressure to a look:

$$\Delta P_{V1} = \Delta P_{V0}\left[1 - 8.3\frac{\sigma^2(\sigma-1)U_g^3}{V_0}\left(\frac{\delta M}{abP_{V0}^3}\right)^{1/2}\right]^3$$

This amendment is rather small. For typical values $\mathbf{\Delta P_V}=3*10^3$ Pa, $\boldsymbol{\sigma}=7$, $\mathbf{U_g}$ = 1.5 m/sec,$\boldsymbol{\delta}=10^{-1}$m, $\mathbf{V_0}$ = 600 m^3,($\mathbf{\Delta P_V/\Delta P_1}$)=1.004, (**M/ab**)=25 kg/m^2.

Conclusions

1. Triplex a sandwich panel can be efficient easily thrown off designs.
2. Comparison a sandwich panels with a glazing shows advantages of panels before a deaf double glazing so their application as ETD is preferable.
3. Effectiveness of sandwich-panels as easily thrown off designs considerably is defined by opening pressure, that is effort to simple knot of fastening. The opening pressure in relation to the allowed pressure is less, the less robust constructions can be protected.

References

1. D. Bradly and A. Mitcheson. Comb. Flame. **32**, 221-236 (1978)
2. D. Bradly and A. Mitcheson. Comb. Flame. **32**, 237(1978)
3. R. K. Kumar, W. A. Dewit, D. R. Greig, Comb. Science and Techn. **66**, 251-266 (1989)
4. L.P. Pilyugin. *Maintenance Explosion Proof buildings with relief designs* (Fire safety and science, 2000)
5. P. Pritchard. Journal of Occupational accidents. **3**, .69- 85 (1981)
6. Ye. Yu. Salymova. *Dynamics of development of dangerous factors in buildings with enclosing structures made of sandwich panels in fires and explosions* (Candidate dissertation. 05.26.03/Moscow State University of Civil Engineering, 2015)
7. G.G. Orlov, D.A. Korolchenko, A.V. Lyapin, Fire and Explosion Safety **23(11)**, 67-74 (2014)
8. G.G.Orlov , D.A. Korolchenko, Fire and Explosion Safety **24(5)** 62-67 (2015)
9. *SP 56.13330.2011. Industrial buildings (The updated edition of SNIP 31/03/2001 / Approved by order of the Ministry of Regional Development (Ministry of Regional Development of Russia) on December 30, 2010 № 850 and entered into force on May 20* (2011)

Influence of composite fibers on the properties of heavy concrete

Marsel Nurtdinov[1*], *Vadim* Solovyev[1], *Alexander* Panchenko[1]

[1] Moscow State University of Civil Engineering, Department of Technologies of Binding Materials and Concretes, 129337 Moscow, Russia

Abstract. The article presents the results of research an effect of glass-plastic fiber on the properties of heavy concrete. It determined that adding glass-plastic fibers in heavy concrete compositions in an amount of 0.5 to 2.5 % by volume leads to decreasing of slump from 20 to 5 cm, increasing air entrainment on 0.3 ... 0.6 % and reduction in density to about 50 kg/m3. Found dependence of the strength properties of concrete classes B20, B30 and B40 depending on the content of glass-plastic fibers. Introduction of fibers into concretes of different classes in an amount up to 2.0 % by volume leads to an increase in the compressive strength on 7 ... 10 %, tensile strength at bending – up to 14...36%. Defined the perspectives of the use of glass-plastic fiber and ways to improve efficiency in the heavy concrete.

1 Introduction

In present fiber concrete technology as high modulus disperse-reinforcing components, that increase strength and deformation properties of concrete, used steel fiber of different geometric dimensions, as well as glass and basalt fiber. Each of the used reinforcing components has its own drawbacks - steel fiber besides a considerable mass and a propensity to balling also modifies thermal properties of concrete [1], glass and basalt fibers are breaking at mixing and partially dissolved by reacting with the concrete matrix. In this regard, developing in recent years various types of composite fiber - glass and carbon fiber, causing considerable interest and requiring detailed study [2-4].

Glass-plastic composite fiber (Figure 1) is produced by pultrusion - pulling glass thread through a die heated to a certain temperature at which the polymerization process of the matrix occurs. Fiber length is 40 mm, the diameter is 0.7...0.9 mm, the geometric factor 45...60, density 1.54 g/cm3, a modulus of elasticity is about 50 GPa (determined on the samples of armature of 4 mm diameter, made from the same materials and with the same technology).

[1] Corresponding author: nurtdinovmarsel@gmail.com

Fig. 1. Glass-plastic fiber.

2 Experimental part

To estimate the effect of the glass-plastic fiber on properties of concrete mixtures were used three control compositions of concrete classes B20, B30 and B40, with the same slump of 20 cm, shown in Table 1.

Table 1. Test compositions of heavy concretes.

Materials, kg per 1 m^3.	**Concrete class**		
	B20	**B30**	**B40**
Water, GOST 23732-2011	210	210	200
Cement ЦЕМ II/ A-K 42,5 H, GOST 31108-2003	290	370	450
Granite rubble, fraction 5-20 mm GOST 8269.0-97	1060	1060	1060
Sand, fineness modulus 2,1 by GOST 8736-2014	760	710	660
Plasticizer Sika T-34	4.35	5.18	6.32

Influence of glass-plastic fibers on the properties of concrete mixtures was evaluated by slump, air entrainment and the density of concrete mixtures, determined according to GOST 10181. The test results are shown in Figures 2, 3 and 4. The fiber content at mixtures (μ_v) ranged from 0.5 to 2.5 % by volume.

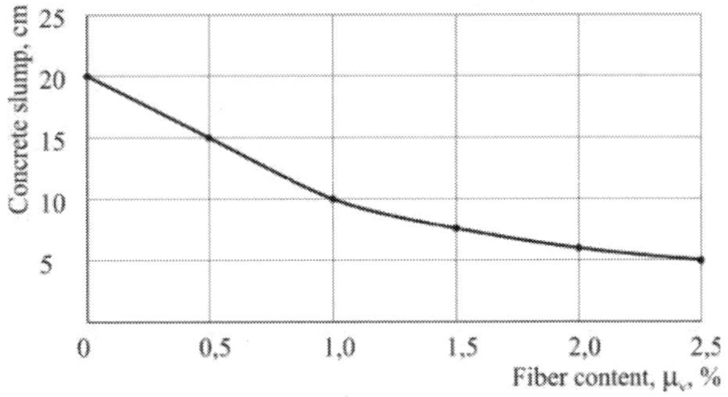

Fig. 2. Concrete slump depending on fiber percentage.

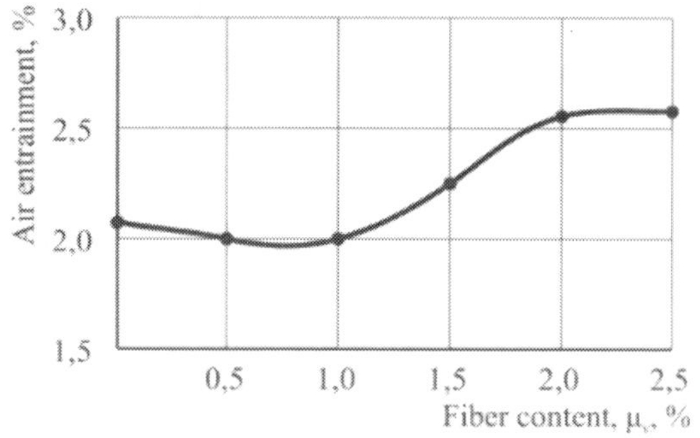

Fig. 3. Air entrainment depending on fiber percentage.

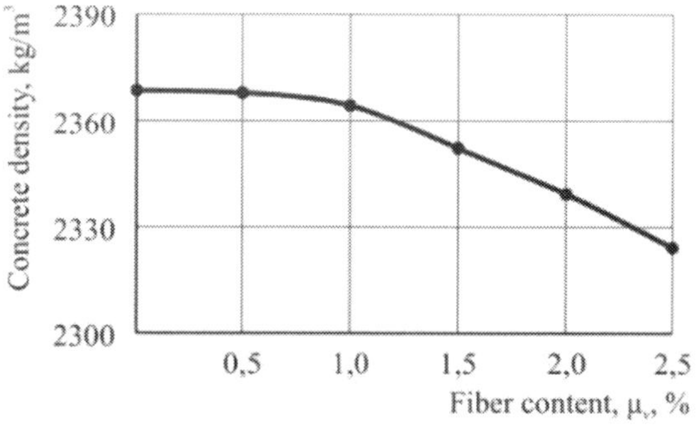

Fig. 4. Density depending on fiber percentage.

When glass-plastic fibers entered to 2.5% by volume (maximum content of the fibers in the mixture, determined experimentally) workability is reduced from 20 to 5 cm, and the obtained dependence almost linear, does not depend on the composition of the concrete mix.

Air entrainment of different mixtures with fiber content up to 1.5% varies insignificantly, due to its smooth surface formed by epoxy resin on which there is no air entrapment during mixing. After content of fiber more than 1.5 % air entrainment increases by 0.3...0.6 % by reducing the fluidity of the concrete mix.

The density of the concrete mix after adding fibers in an amount up to 2.5% is reduced on 50 kg/m3, due to the increase in volume fiber content with lower density than the other components and improve of air entrainment.

It should be noted the identical type of dependencies, slump, air entrainment, and the density of concrete mixtures of various compositions with the content of glass-plastic fibers. That will allow to use this dependencies in correcting the compositions of fiber-reinforced concretes, independently from composition of the concrete matrix.

Effect of glass-plastic fibers on the strength characteristics of concretes performance estimated by the compressive strength test and tensile strength at bending test with specimens sizes 100×100×100 and 100×100×400 mm. Concrete matrix compositions shown in Table 1, the fiber content was taken from 1 to 2.0 % by volume.

Figures 5 and 6 shows the results of tests by determining the compressive strength and tensile strength at bending of samples of different compositions of concrete at the age of 28 days with a glass-plastic fiber content from 0.5 to 2.0 % by volume.

According to the test results found that the tensile strength at bending of concrete classes B20, B30 and B40 with glass-plastic fibers from 0.5 to 2.0 % by volume is increased up to 36, 21 and 14 % respectively. The compressive strength of concrete classes B20, B30 and B40 is also increased up to 10, 9 and 7% respectively. Thus, increasing the concrete class reduces effect from introducing fibers. Tensile strength at bending of concrete class B40 with glass-plastic fiber at different ages almost did not change in percentage to the control composition without fiber (Figure 7), what shows sufficient adhesion of fibers and concrete matrix with the joint efficient action during the curing of fiber concrete.

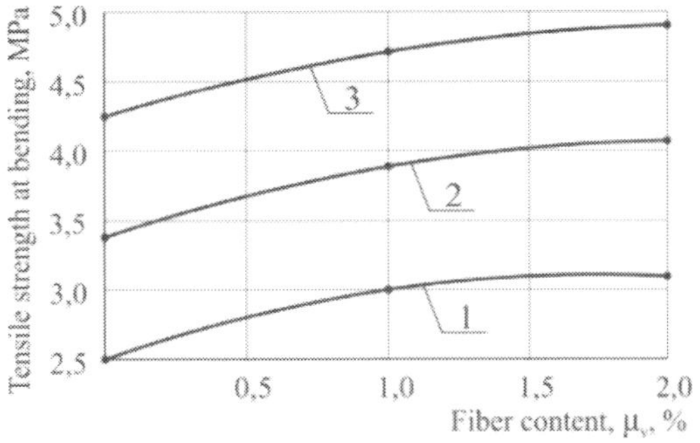

Fig. 5. Tensile strength at bending, 1, 2 and 3- concrete class B20, B30 and B40 respectively.

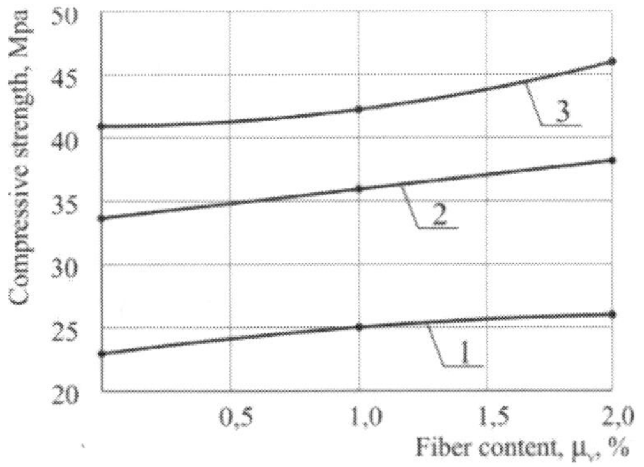

Fig. 6. Compressive strength, 1, 2 and 3- concrete class B20, B30 and B40 respectively.

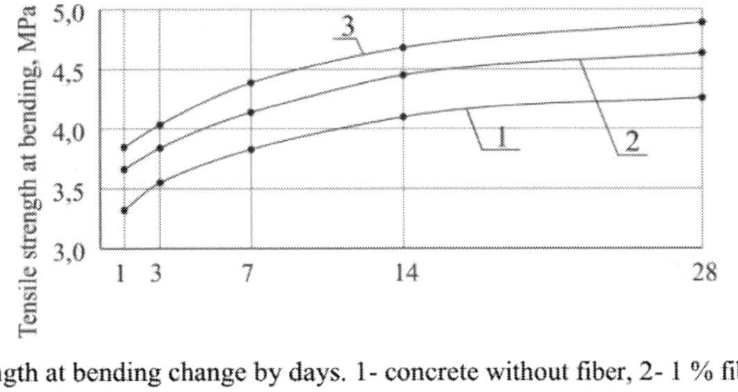

Fig. 7. Tensile strength at bending change by days. 1- concrete without fiber, 2- 1 % fiber content, 3- 2 % fiber content.

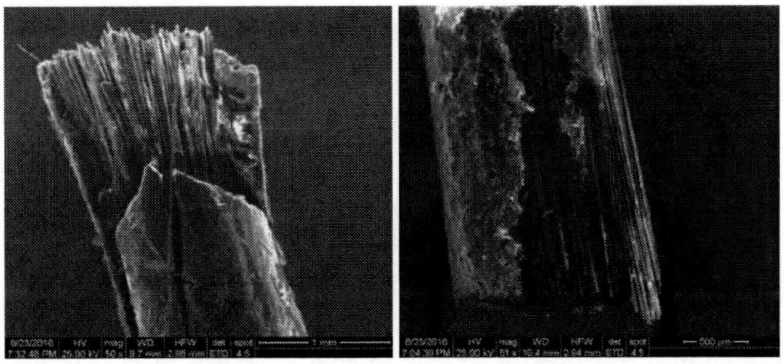

Fig. 8.9 Shell of glass-plastic fiber after the tensile strength test at bending.

The shell of glass-plastic fiber after the tensile strength test at bending in the specimens deformed and partially destroyed - Figure 8, 9. Lower strength of glass-plastic fiber shell does not allow achieving results comparable with steel analogs, because shell destruction leads to slipping out fiber in concrete matrix. Thus, absence of breaks in individual fibers indicates the possibility of improving its efficiency in the concrete matrix. Better results may be obtained by using of glass-plastic fibers with a geometric factor 80...100 [5], either a directed change in the structure and properties of the concrete matrix [6,7], which is due to increasing efficiency of fixing of fiber in the matrix and the properties of the resulting fiber-reinforced concrete in total .

3 Conclusions

These results display that the investigated glass-plastic fibers work in different concrete matrices as a high modulus reinforcing disperse component, and improves the strength properties of fiber reinforced concrete. However, at the present time is necessary to make extra complex studies aimed at improving the efficiency of glass-plastic fibers in concrete, for its wider use as steel fiber analog. Some differences of glass-plastic fiber from the steel fiber - low density and thermal conductivity, absence of balling effect while mixing concrete and corrosion resistance to different environments, all this define huge opportunities for its effective use in the manufacture of fiber reinforced concrete products and structures.

References

1. V.G. Solovyev, A.F. Buryanov, M.S. Yelsufyeva, Stroitel'nye Materialy **3**, 18-21 (2014)
2. M. Rangelov, S. Nassiri, L. Haselbach, K. Englund, Construction and Building Materials **126**, 875-885 (2016)
3. A. El-Nemr, E.A. Ahmed, C. Barris, B. Benmokrane, Construction and Building Materials **113**, 77-89 (2016)
4. E.A. Khoteyev, Transport Construction **1**, 10-13 (2015)
5. F.N. Rabinovich, *The composites on the basis of dispersed-reinforced concretes* (Izdatel'stvo ASV, Moscow, 2011)
6. M.S. Elsuf'eva, V.G. Solovyev, A.F. Bur'yanov, Stroitel'nye materialy **8**, 60-63 (2014)
7. V.G. Solovyev, A.F. Bur'yanov, H.-B. Fisher, Stroitel'nye materialy **9**, 43-46 (2015)

Insulation systems of the building construtions

Boris Rumiantcev [1], *Aleksey* Zhukov[1*], *Dmitry* Zelenshikov[2], *Anatoliy* Chkunin[1], *Kazbek* Ivanov[1], *Yuliya* Sazonova[1]

[1] Moscow State University of Civil Engineering, Moscow, Russia
[2] Modern Roofing Technologies Ltd., Moscow, Russia

Abstract. Constructions of the exterior insulation and decoration combines materials of different functionality and constructive solutions allows to these materials to demonstrate their efficiency to the great extent. Fire safety of buildings is mandatory requirement for building systems. Some insulating material may belong to the group of combustible, but their use in structures so as to minimize the risk of fire. On the other hand, there are special designs, in which non-flammable insulation acts as a flame retardant barrier. In the article carried systematization of construction systems used in the flat and pitched roof during the insulation and wall covering and facades. Taking into account the experience of leading firms were considered the application features of using exterior finish systems: construction solutions, requirements for materials and recommendations about the installation these systems.The article deals with the construction ventilated roofing system of two types: flat roof and pitched roof seam. In the first case, the ventilation system is created using milled insulation boards in the second - by a ventilated gap. In both cases the natural convection of air in the air cavities. Ensuring operational stability insulation is laid on the stages of production of heat-insulating materials. It is important: firstly responsible execution of all process operations associated with providing regulatory properties of materials and secondly, the performance of additional operations associated with the production of materials, working in a specific design. An example of a material whose properties can modify for a particular application, are milled mineral wool (with air channels) for systems of ventilated flat roof.

For Russia, with its harsh climate systems implementation for protect the insulation of the building envelope, aimed at creating a comfortable environment by addressing the following major construction tasks: - minimize heat loss through the insulated perimeter of the building; - exclusion of moisture in building construction and in the building; - creating a

* Corresponding author: lj211@yandex.ru

comfortable environment in the room; - increase the durability of structures; - optimization of logistics and production organization in the implementation of systems [1, 2].

Building systems develop engineering services firms, manufacturing facilities which are located both in Russia and abroad. For example, slabs and mats of glass fiber produces department of insulation «SANT-GOBAIN», located in Russia, Finland, Poland, Sweden. Companies Knauf company released at the Russian enterprises full range of products for complete systems Knauf. Systems companies Rockwool are focused on the use of stone wool products, produced at the company's plant and components from suppliers. Factories Corporation TechnoNICOL produce a whole range of materials — components flat and pitched roofs, ventilated and plaster facades, floors and foundations [3, 4].

Penetration of moisture into the design due to the deterioration of its thermal insulation properties, reducing durability, deterioration humidity of the room and condition of the premises. High humidity of structures increases the risk of biological corrosion. Penetration of moisture in the material prevents a mandatory component of the building system as a vapor and waterproofing.

Comfort of the room, although it is a subjective characteristic, but depends on a number of objective parameters. Firstly, temperature control in the room. Temperatures near the floor and outer walls, as well as the average indoor temperature are indicative of comfort (or discomfort) and cause unregulated air convection [5]. Secondly, the room humidity conditions, which including depends on the water vapor permeability of insulation, humidity enclosing structure. With proper implementation of the insulation of the building envelope, the load on the HVAC system can be reduced to a minimum, which in turn helps to optimize operating costs. Thirdly, the creation of an acoustically comfortable environment due to the fact that the thermal insulation product has good sound insulation capacity and are used to protect the walls, floors and roof structures from external noise [5, 6].

Fire safety of buildings is mandatory requirement for building systems. Some insulating material may belong to the group of combustible, but their use in structures so as to minimize the risk of fire. On the other hand, there are special designs, in which non-flammable insulation acts as a flame retardant barrier. Placing insulation and features of the implementation of design solutions are also linked with the architectural aspects and, in particular, with the design of facades, roofs, interiors.

Accounting for the operating conditions of thermal insulation materials in the construction is necessary because these conditions directly determine the operating insulation resistance and therefore the durability and efficient exploitation of construction.

Technological loads on thermal insulation can divide for 3 groups: the transport and storage loads, the loads during install and exploitation. During the operation of boards exposed to temperature, weather, humid and mechanical stress. Regardless of the type of building systems, its the optimization parameter is the thermal resistance of the structure, optimization parameter of Thermal Insulation Materials (TIM) is the thermal conductivity. At the same time, the specification of TIM exploitation in each from concrete systems imposes additional requirements for thermal insulation - "restrictions", defined as "the normalized indicator".

Thermal Insulation boards which work in *flat roof constructions*, perceive compressive loads from higher layers of the roof pie, temporary seasonal loads and local loads of small area. The normalized indicator is compressive strength at 10% deformation. Moisturization insulation may result in the violation of waterproofing roofing or "failure" vapor barrier, which is located on the ceiling.

If the roof system reliably isolate the material from possible external fire exposure, can be applied in slabs of extruded polystyrene. In roofing systems using slabs of foam glass and stone wool. Successfully proven dual density plates. The topsheet double density boards receives and transmits these loads to the underlying layers.

Thermal insulation in pitched roofs operate with condition "ventilated gap." Mechanical effects on it is minimalMoisturization of insulation possible in the same reasons as in the flat roof. The normalized indicator is breathable insulation. The use of fuel heater is not allowed.

In the case of using mineral-fiber materials (boards, mats) windproof solutions are need, aimed at excluding the removal of fiber erosion products. In designs using glass wool products such a fiberglass (or fiber canvas), the application of products on the basis of basalt fiber - basalt canvas. In the application of stone wool products - dual density plates are recommended.

Insulating the building envelope - the guarantor of energy saving, thus, the efficiency of energy-saving measures is defined and optimized thickness of the thermal insulation layer, and durable construction of the system as such. The durability of the building system with insulation depends on the group of factors: the correct project, quality of installation, performance products. It should be noted that the operational stability, even high-quality insulation depends on the design layout and design conditions in which their operation occurs [7, 8].

Ventilated roof system has a number of design innovations, through which moisture evaporation increases significantly. It circulates air through the channels in the lower layer of insulation. The product might be two and three layer insulation systems with ventilation channels.

Operation of the system ventilated flat roof is based on natural air circulation. The slope of the channels is organized for the correct operation of the system, and can be both constructive and executed with the help of thermal insulation slope-formed plates. Slope-formed plates are stacked under the slab with air ducts. Air under the action of the difference in partial pressures is pulled up via a common transverse channel located along the ridge. The driving force of the process is to create a vacuum in the common channel by the force of the wind [9, 10]. Outer roofing insulation board 40 mm thick evenly distributes the load on the external layer of insulation with ventilated channels.

Milled plates [6], the average density of 115-120 kg / m 3 are made from mineral wool body-oriented structure. Plates are noncombustible, their compressive strength at 10% deformation is 30-35 kPa; thermal conductivity in the dry state at 25 ° C 0,037-0,038 W / (mK); water vapor transmission rate - 0.3 mg / (m·ch·Pa); dimensions 1200 × 600 × (50-200) mm

Owing to the top plate, the air in ducts at 5 ° C above ambient temperature. Exterior mineral wool slab roofing is the basis for the waterproofing layer. Roof slabs milled provide moisture transport along the canals to the ventilation vents (wind vane). Dimensions of the channels are: width 30 mm, depth 20 mm. During installation, connecting channels, as well as technological holes cut out by hand.

Roof installation is performed similar to the technique of flat roofs with polymeric membranes. Thus perform a number of additional operations. To divert water from the layer of insulation in the parapet provide gradients. To provide the necessary air flow, beams, eaves at the device, set with a small gap (air holes). Vane, with the device of the valley, set on both sides of the ridge and gravity layer increments of 6-8 m. Transverse channel 100 mm wide and 20 mm deep cut in the insulation on the roof ridge. To ensure the connection with the channels for deflectors in the upper insulating layer are punched.

Features seam pitched roof that radiation component (on a hot, summer) is presented to a greater extent and the presence of attachment double fold minimizes the possibility of direct leakage, that is, excessive moisture insulation layer. For heat transfer in a pitched roof seam characteristic are the two polar states: in the hot summer and winter conditions.

During the summer, metal roofing accumulates heat from the sun and radiating surface becomes relatively internal layers of roofing system. The surface of the insulation layer accepts: heat flux emitted by the metal roofing and convective flow of air in the ventilated gap. Inside the heat insulating layer is formed convective heat flux due to interconnected porosity insulation and conductive flow - through the matrix material.

The direction of the resulting transfer of heat - inside the room. Mass flow (their effect on the thermal conductivity on the order of less than a heat transfer) due to the difference of the partial pressures of water vapor on the surface of the insulating layer and in a convective flow of air ventilated gap, as well as the partial vapor pressure for steam insulating layer and on the surface of the insulation.

In summer, heat insulation acts as a heat shield the interior from the exterior excess heat. In winter, the insulation acts as a heat shield the interior from the outside cold. Heat transfer in a pitched roof seam causes not only create comfortable conditions in the room (roof space), but also determines the durability of the whole system. For the winter pro-heating of metal roofing due to solar radiation is negligible and radiation component can not be ignored. Heat flow is directed to the outside from the inside premises, and insulation works as a component of heat-insulating building envelope.

There is two systems of air ventilation in the pitched roof systems and roof, it's ventilation of roof and ventilation roof space. Natural ventilation of attic through louvres skylights located on pitched roofs, inefficient due to the irrational arrangement of ventilation holes on the same level in roughly equal aerodynamic coefficients. The organization of ventilation, while ensuring the required air exchange is essential to obtain a complete washing of outside air total roof space. When placing an unproductive ventilation holes in the roof diffuse dormers this position is not satisfied. In the attic are formed zones with stagnant air. Ventilation holes under the eaves of the roof may be arranged in the form of a narrow slit leave between the wall and the roof (slit-like air holes), as well as individual holes placed in the eaves of the wall along the axes of the windows or partitions ("surgical" air holes). The ridge vent holes performed slot-shaped (Fig. 1). With this arrangement, during the calm air vents under the eaves of the roof work on the inflow in the ridge - on the hood.

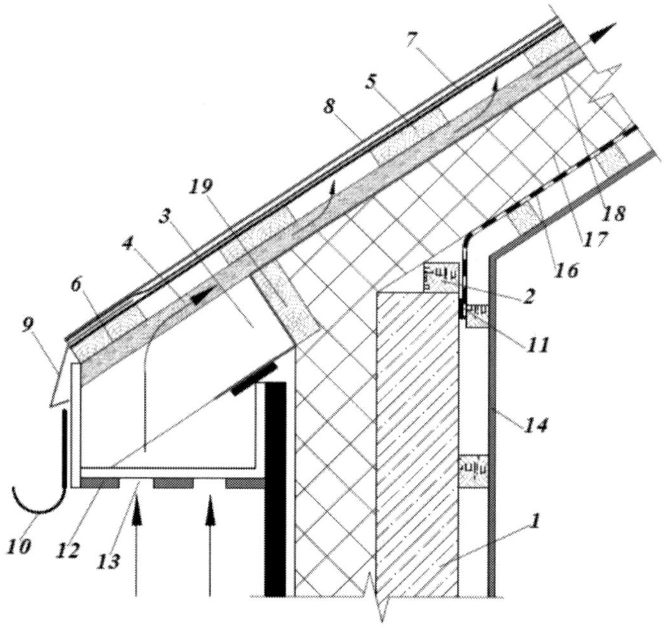

Fig. 1. Schematic diagram of an inflow of outside air through the cornice:
1 — outer wall; 2 — plate; 3 — rafter foot; 4 — counter batten; 5 — sparse batten; 6 — solid base; 7

— carpet backing; 8 — shingles; 9 - Bracket cornice; 10 — water drainage system; 10 — gutter hook is used in case of installation of drainage system; 11 — waterproofing; 12 - Vol siding; 13 - air inlets; 14 - two layers of gypsum board; 15 – counter insulation ; 16 - steam, with sizing joints; 17 - heater; 18 - windscreen; 19 - fixing insulation boards.

The area of ventilation openings a special system of natural ventilation attic set with the calculation, depending on the heat gain in the roof space, its volume and climatic conditions of the area of building construction by a special technique. Cross-sectional area of skylights on the roof and the air holes should be 1 / 300-1 / 500 square attic floor, that is, for every 1000 m2 of attic must be at least 2 m2 of skylights and air holes. With this arrangement, these devices should provide cross-ventilation attic, excluding local stagnation (air sacs). Pressure in the attic should be reduced, so the area of the exhaust vents should be 10-15% more than the supply air.

This is necessary to create a draft of air. To eliminate penetration of precipitation into the attic space above the ridge slot arranged special continuous ridge aerator or made on the spot. For convenience, it is represented in the set with shingles. In order to avoid settling attic birds airways point under the roof overhang closed bars and slit - nets or special straps (spotlights).

Special ventilation of attic is arranged in the roofs of buildings of any purpose and any configuration, with wood, metal or concrete supporting structures and any kinds of roofs (made of roll waterproofing materials, steel roofing, tiles, asbestos cement sheets, etc.). It is particularly important to implement it in the roofs with thick roofing and when the location of the central heating and hot water in the attic.

It should be noted the special role of gap curtain air holes to ensure the safety of the wall portions of the roof (all rafters, plate, battens, roof overhangs), the most disadvantaged conditions. The constant flow of outside air through them is winding design.

Curtain slotted airways facilitate monitoring of the roof in the most damage-prone areas. Their presence contributes to the acceleration of warming over-roof cornice areas during periods of thaw, and with this release and gutters from icing resulting from the melting of snow in the cold days under the influence of solar radiation.

For normal ventilation combined pitched roof it should have three main elements: an opening for fresh air intake channels of thermal insulation for its circulation and exhaust vents at the top of the roof. In the case of under-eaves of firmware siding must use special elements for ventilation - the so-called soffit plate. In the case of lining, a possible option to ensure the ventilated gap is the mounting structure shown in Fig. 1. (subject to de-icing system installation).

A feature of the technology is a special module, located in front of chamber heat treatment. This module includes a drum for unwinding fiberglass mesh, glue application assembly and compression mechanism. Further surfactants carpet laid fiberglass enters the heat treatment chamber. In the chamber resulting in blowing coolant through carpet layer solidifies as a binder in the carpet and the adhesive applied to the grid. In both cases the polycondensation reaction occurs. As a result fiberglass firmly connected to the mineral wool substrate.

The result of research and technological testing was to obtain two types of products based on mineral wool with a reinforced upper layer: mineral wool products MWP VENT Plus and MWP LIGHT Plus. Product non-combustible, organic matter content does not exceed 2.5%. The use of fiberglass mesh (with breaking strength in the operation of at least 350 N / m), which is bonded to the plate material thermosetting adhesive can improve the stability of slabs on vertical and inclined surfaces, to reduce emissions from fiber slabs to a value corresponding to the experimental error. More than 10-fold increase operational stability of products.

Optimization of heat treatment conditions is possible with the use of special software. Computer programs written in C # can perform the following tasks. Selected depending on the characteristics of the mineral wool mat (density, fiber diameter, the content of the binder and the fibrous inclusions, humidity and etc.), optimum differential flow resistance on the sides of the mat, the speed of heat carrier and its temperature, as well as optimize the thermal treatment chamber length and the time required upon fully curing binder.

Mineral wool products VENT Plus installed in a single layer (Fig. 2) in the following sequence. The product is fixed in the upper part of the two dowels and deploy on all length (6 m) to the surface of the wall. Then fix the dowels staggered. Dowels further pressed against the glass wind deflector mesh to the product, thereby increasing the reliability of the system.

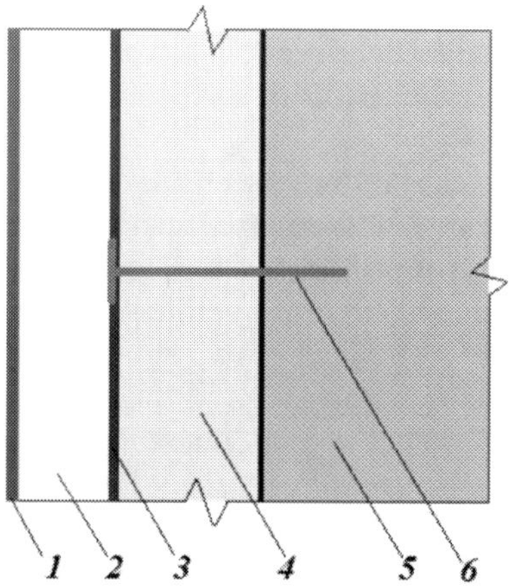

Fig. 2. Scheme of facade system with ventilated gap: 1 - outer facing; 2 - ventilated gap; 3 - fiberglass; 4 - mineral wool slab; 5 - bearing wall; 6 - insulation fastening element.

MWP LIGHT Plus is laid in two layers with a total thickness determined by thermal calculations. For the central region at the required thermal resistance of 5.1 $m^2 \cdot {}^\circ C / W$ thermal insulation layer thickness be 200 mm (Fig. 3).

Length MWP LIGHT Plus up to 6 m, which allows overlap of the roof slope either fully or to a greater part of it. The product is placed between the rafters with a reduction than is achieved and a snug fit to the rafters and fix in position. On the slopes longer than 6 m products are laid with joints offset.

Using the MWP with reinforced outer layer helps to minimize the number of thermal bridges, thus improving the uniformity of heat engineering thermal insulation envelope.

Fiberglass located on the surface MWP facing to ventilated gap also serves as a non-combustible windproof membrane. Thus the possibility of erosion of the fibrous material as a result of the removal of fibers by air circulating in the ventilation gap.

Cost-effectiveness analysis showed certain advantages in the manufacture and use of products with reinforced outer layer. According to technological criteria change in the manufacturing process leads to an increase in capital investment, the size of which does not exceed 2% of the production line. The increase in cost associated with the use of fiberglass and thermosetting glue, as well as depreciation does not exceed 1 %. At the same time in construction using of mineral wool products with reinforced outer layer excludes one production step: packing and fixation windproof layer. The use of products of great length (up to 6 m) allows faster installation of thermal insulation material. These techniques improve manufacturability and reduce the installation costs of construction and installation works (by 10-15 %).

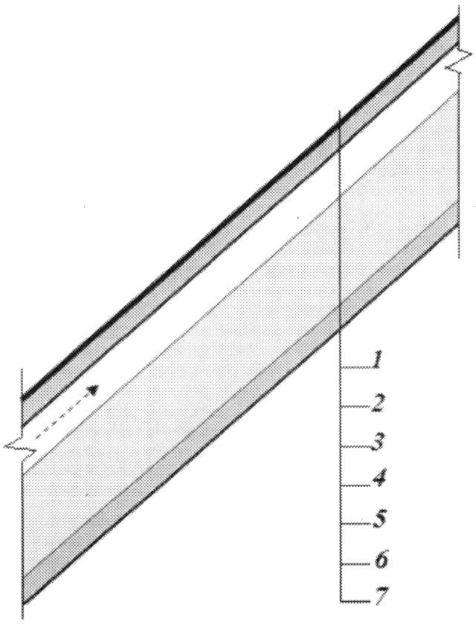

Fig. 3. Scheme of pitched roof with a ventilated gap: 1 - roofing; 2 - waterproofing; 3 - continuous (or rarefied) sheathing; 4 - ventilated gap; 5 - thermal insulation with reinforced outer layer; 6 - vapor barrier; 7 - inner lining.

Conclusion.

During operation the most pronounced are two factors of influence reinforced outer layer. Firstly by fiberglass reinforcement slab significantly increases its tensile strength (tensile). To the greatest extent this factor is manifested for pitched roofs with a slope of more than 35 ° (for example, polygonal or mansard). In the ventilated facades this property is more pronounced when installing insulation. Further, after fixing the thermal insulation products using dowels occurs mechanical fixation of the product as well as its surface layer. Secondly fiberglass is a windscreen that prevents the destruction of the mineral wool layer by exposure to the air flow. Reinforced outer layer can significantly extend the effective exploitation of mineral wool products as in the systems of ventilated facades, roofing and ventilated systems.

References

1. A.D. Zhukov, I.V. Bessonov, A.N. Sapelin, N.V. Naumova, A.S. Chkunin. Composite wall materials. Italian Science Review, 2. 2014, pp.155-157
2. Al. Zhukov, Ye. Bobrova, D. Zelenshchikov, R. Mustafaev, A. Khimich. Insulation systems and green sustainable construction. Advanced Materials, Structures and Mechanical Engineering. Volumes 1025 – 1026. (2014) pp. 1031-1034
3. V.G. Gagarin. Thermal protection and energy efficiency in update version of SNIP “Thermal protection of buildings”. III International Congress. Energy efficiency XXI century. SPb. 2011. pp. 187-191
4. B.M. Rumyancev, A.D. Zhukov, E.Yu. Bobrova, T.V. Smirnova. Technological aspects of the operational stability of the mineral fibers. Industrial and civil building 2015. №1. pp. 32-36.
5. I. Gnip, S. Vaitkus, V. Kersulis, S. Vejelis Long-term prediction of creep strains of mineral wool slabs under constant compressive stress. Mech Time Depend Mater. 2012. № 16, pp. 31–46.

6. I. Ya. Gnip, S.I. Vaitkus, S.A. Vejelis. Prognostic evaluation of creeping deformation of mineral wool slabs with a constant compression on the basis of the initial period of strain. Building materials. 2012. № 12. pp. 40-44
7. A.D. Zhukov, N.V. Naumova, P.M. Mustafaev, N.A. Mayorova. Modeling properties of highly porous materials with combined structure. Industrial and civil construction. 2014. № 7. pp. 48-51
8. A.D. Zhukov, T.V. Smirnova, P.K. Gudkov Dual density thermal insulation in the insulation envelope of buildings. Industrial and civil construction. 2013. № 3. pp. 21-23.
9. A.D. Zhukov, T.V. Smirnova, D.B. Zelenshchikov, A.O. Khimich. Thermal treatment of the mineral wool mat. Advanced Materials Research (Switzerland) Vols. 838-841 (2014), pp. 196-200
10. A.D. Zhukov, E.Yu. Bobrova, T.V. Smirnova. Evaluation of Durability of Mineral Wool Products / // Advanced Materials Research. Vol. 1077 (2015) pp. 109—112

Analysis of heat losses of underground tunnel for engineering utilities with available methods

Oleg Brukhanov[1], *Andrey* Rymarov[1,*], *Anna* Malysheva[1], *Dmitriy* Titkov[1]

[1] Moscow State University of Civil Engineering, 26, Yaroslavskoe Shosse, Moscow, 129337, Russia

Abstract. Methods for calculation of thermal regime and heat loss parameter for underground tunnel into surrounding sub-soil were developed in first half of XX century, when the study of complex heat-mass exchange processes between underground tunnel, sub-soil and outdoor air is started. In tunnel it need to be provided a specific microclimate parameters, that requires a certain load on environmental control system. At present time there are some methods for calculation of heat loss parameter for underground tunnel. In this paper the analysis of real methods of calculation of heat losses of underground tunnel is accomplished.

1 Introduction

At present time the development of engineering communications, the industrialization and urbanization fatally lead to deficit of placement area for engineering utilities in big cities [1].

Underground placement of the communications in special tunnel providing the possibility of maintenance by stuff is good idea.

Network of underground communication tunnel extremely grows and is developed, that requires more detail its investigation in terms of providing with normative microclimate conditions influenced on durability and technical characteristics of communication utilities [2].

One of the key factors affected on tunnel microclimate is heat loss parameter. As underground tunnel is underground building with varied placement depth. Also heat emissions with different intensity take place. So heat losses are unsteady.

2 Method of Kazantcev B.A.

2.1 Features of method

Feature of the method is introduction of instability of heat loss parameter of underground realized due to using of the term «heat loss period» (hours). However the method is limited by depth of tunnel placement up to 10 meters [3].

* Corresponding author: rymarov@list.ru

To apply the method of Kazantcev B.A. the method for calculation of sub-soil temperature around the tunnel depending of required depth of tunnel placement taking into account such parameters as heat loss period, sub-soil heat diffusivity, range of temperature variation [4] at sub-soil surface is necessary. It is realized by the method of Vlasov O.E.

For determination of calculated value of surface area for envelop building in tunnels with rectangular cross-section it is need to take into account the influence of corners in tunnel, increasing its heat loss values. It is achieved by addition to each inner geometrical size of tunnel of some parameter that takes account following parameters: heat loss period, heat conductivity and heat capacity as well as density of material for envelop building based in tunnel.

Heat-transfer coefficient is determined with the equation (1).

$$k = \frac{1}{\frac{1}{\alpha_B} + \frac{2 \cdot H + \sqrt{z \cdot \frac{\lambda}{\rho \cdot c}}}{3 \cdot \lambda_S}}, [W/(m^2 \cdot {}^\circ C)] \quad (1)$$

where: H is distance between ground-level and top of bridging, m; λ is heat conductivity coefficient of envelop structure of tunnel ,W/(m^2·°C); λ_s is heat conductivity coefficient of sub-soil, W/(m^2·°C); ρ is density of material for envelop structure in tunnel, kg/m^3; c is heat capacity of tunnel envelop structure, kJ/ (kg·°C).

2.2 Calculations

On the basis of this method for underground tunnel with rectangular cross-section – 4.83m^2, placement depth up to 10 m and embedment width – 1 m, heat conductivity coefficient for clay loam sub-soil around the tunnel – 1.18 and 1.26 W/(m^2·°C) for dry sub-soil and frozen sub-soil, respectively as well as with constant air temperature in tunnel – 30 °C the calculations of heat loss parameters for Moscow city are accomplished. The results obtained are shown in Fig. 1 and 2.

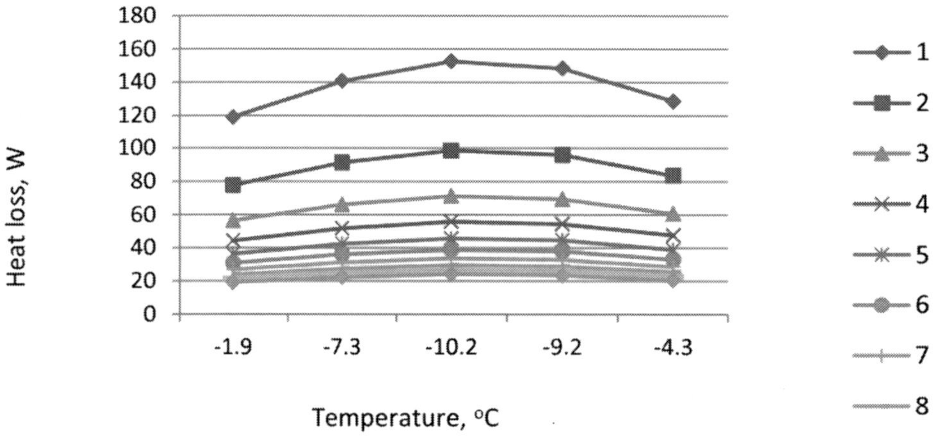

Fig. 1. Variation of heat loss parameter for underground tunnel into sub-soil during cold period depending on outdoor air temperature

Fig. 1 and 2 demonstrate the heat loss values are reduced with increasing of tunnel placement depth as well as season variation of outdoor air temperature. However, calculation method according to Kazantcev B.A. doesn't take account shape and size of building as well as thermal characteristics of surround sub-soil and indoor space heating time period. But it is necessary especially when presence of heating system in tunnel. This method is suitable only for tunnels placed at depth up to 10 m. Also humidity of envelop structures is varied with time and effects on tunnel heat loss parameter. It is difficult to register with this method.

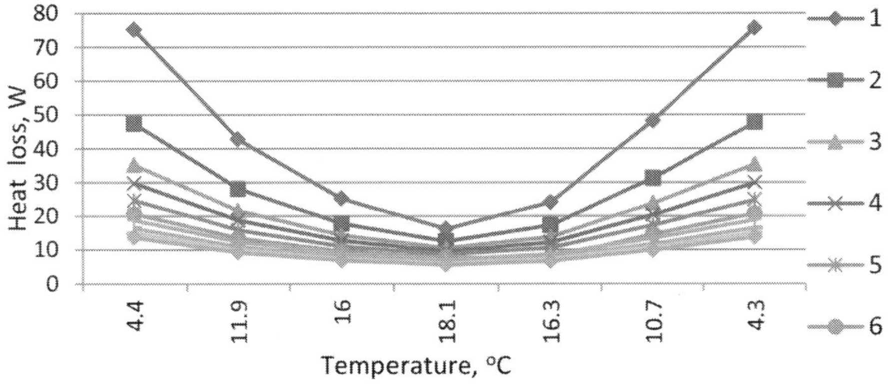

Fig. 2. Variation of heat loss parameter for underground tunnel into sub-soil during warm period depending on outdoor air temperature.

Introduction of air movement in tunnel with varied inner air temperature and heat loss parameter in the above methods was not accomplished [5, 6, 7].

3. Method of Machinsky V.D.

Calculation of heat loss parameter for communication tunnel into surround sub-soil by prof. Machinsky V.D. method is considered. In this case a heat loss parameter is calculated for floor slab, wall and floor separately [3,8].

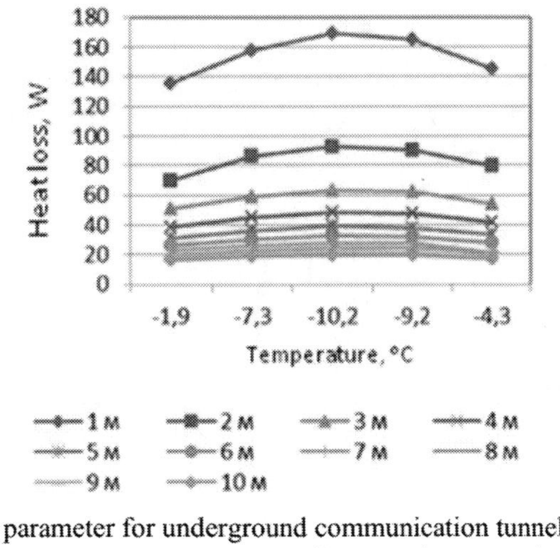

Fig. 3 Variation of heat loss parameter for underground communication tunnel into sub-soil during cold period depending on outdoor air temperature as well as a replacement depth of tunnel.

As an example the dimensions of underground tunnel with fixed cross-section – 4.83 m^2 are used. The tunnel placement depth in range of 1–10 m with 1-meter interval and fixed inner air temperature – +30 °C was studied assuming the presence of heating system in tunnel for town or region.

The calculation of heat loss parameter during a year for Moscow climate taking [9] into account the monthly average air temperature according to SP 131.13330.2012 «Construction climatology» is accomplished. As sub-soil a clay loam with heat conductivity coefficient 1.18 and 1.26 W/(m^2·°C) for dry sub-soil and frozen sub-soil, respectively was used. On the base of results of calculation the diagrams of effect of monthly average air temperature in cold period on heat loss parameter of underground tunnel taking into account different tunnel placement depth are obtained (Fig. 3, 4).

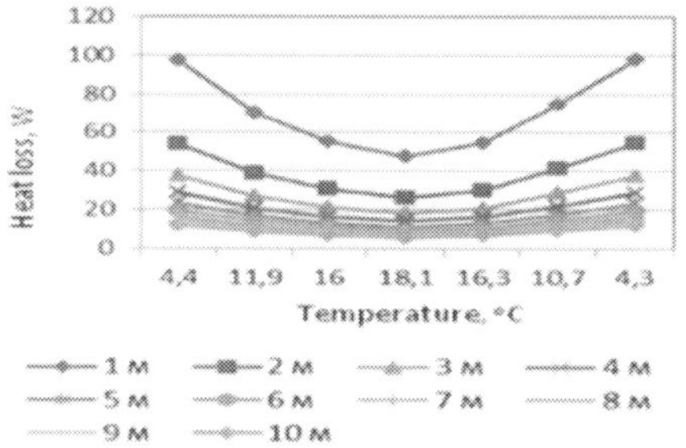

Fig. 4 Variation of heat loss parameter for underground communication tunnel into sub-soil during warm period depending on outdoor air temperature as well as placement depth of tunnel.

Choice of design temperature for heat loss through sub-soil layer depends on sub-soil mass with varied width and thermal properties that is not taken account in this method.

3.1 Analysis of V.D. Machinsky method and method developed in Teploelectroproect Institute

The method developed in Teploelectroproect Institute takes into account climate condition of construction region, ambient sub-soil temperature on depth of tunnel placement as well as shape of tunnel cross-section but depth of tunnel placement relatively the ground level as well as sub-soil heat conductivity coefficient. The V.D. Machinsky method takes into account tunnel shape and sub-soil characteristics depending on tunnel placement depth but climate conditions.

As an example the underground tunnel with cross-section – 4.83 m^2 and fixed inner temperature – +30 °C [10] is used. The calculation of heat loss parameter during a year for Moscow city on the base of monthly average air temperature for each month of a year are curried out. As sub-soil a clay loam with normative depth of soil freezing – 1.3 m for Moscow city was used.

The calculation of heat loss parameter other factors being equal are realized and shown in Fig 5.

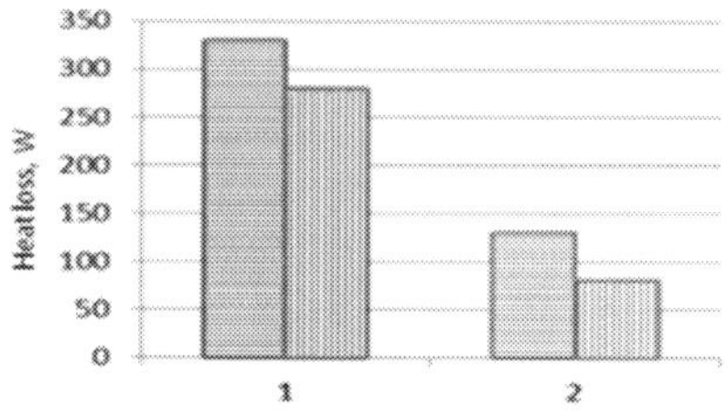

Fig. 5. Heat loss into sub-soil mass in cold period (1) warm period (2), horizontal hatching – according to method of Teploelectroproect Institute, vertical hatching – by V.D. Machinsky method.

Differences between the calculation results for the both methods are within 15 %

4. Summary

Heat balance is dominant factor when required microclimate formation in underground tunnel [11]. Available methods for calculation of unsteady heat losses by underground buildings take into account such factors as depth of underground building placement, and shape as well as sub-soil properties etc. [12, 13, 14] with different limitations. One of the problem of the methods is inability of determination of time period shift of maximal/minimal temperature variations and heat losses by underground tunnel relatively outdoor air temperature varied with time including the building dimensions [15, 16]. It makes difficult of control of equipment operation. In the above methods the air movement in tunnel effecting on inner air temperature and heat losses is not taken into account.

References

1. O. Samarin, K. Lushin, S. Paulauskaite, Technological and Economic Development of Economy **13**, 67-72 (2007)
2. O. N. Brukhanov, D. G. Titkov, Internet-Bulletin VolgGASU **33**, 16 (2014)
3. V. Gagarin, V. Kozlov, K. Lushin, International Journal of Applied Engineering Research **23**, 43438-43441 (2015)
4. A. G. Rymarov, D. G. Titkov, Construction: science and education **4**, 2 (2014)
5. S. A. Isaev, N. I. Vatin, S. V. Guvernyuk, V. G. Gagarin, B. I. Basok, Y. V. Zhukova, High Temperature **53**, 873-876 (2015)
6. V. N. Varapaev, A. V. Doroshenko, Procedia Engineering **91**, 200-203 (2014)
7. A. Volkov, A. Sedov, P. Chelyshkov, E. Kulikova, Applied Mechanics and Materials **584-586**, 761-764 (2014)
8. V. Bodrov, M. Bodrov, K. Lushin International Journal of Applied Engineering Research **22**, 43277-43280 (2015)
9. A. G. Rymarov, Natural and technical sciences **63**, 380-382 (2013)
10. A. G. Rymarov, Academia. Architecture and building **5**, 362–364 (2009)
11. A. G. Rymarov, Natural and technical sciences **62**, 595-599 (2012)
12. A. G. Rymarov, K. I. Lushin, Science, construction, education, (2011) Information on http://www.nso-journal.ru.
13. N. Parfentieva, O. Samarin, K. Lushin, S. Paulauskaite, ICEE 2008 - Conference Proceedings, 854-858 (2008)
14. V. Gagarin, A. Y. Neklyudov, International Journal for Housing Science and Its Applications **39**, 79-87 (2015)
15. Y. Kharkin, E. Korol, A. Davidyuk, Procedia Engineering **117**, 172-178 (2015)
16. A. G. Rymarov, K. I. Lushin, Science, construction, education, (2012) Information on http://www.nso-journal.ru.
17. A. G. Rymarov, D. G. Titkov, Natural and technical sciences **80**, 144-147 (2015)

Expediency of application of explosion-relief constructions to ensure explosion resistance of production buildings

Anton Lyapin [1*], *Anton* Korolchenko[1] and *Evgeniy* Meshalkin[2]

[1]Moscow State University of Civil Engineering, 129337, Moscow, Yaroslavskoye shosse, 26, Russia
[2]NPO PULS, 107014, Moscow, Rusakovskaya street, 28, Russia

Abstract. The article presents a model of economic evaluation and selection of explosion-relief constructions (ERC), as well as determination of explosion protection efficiency of buildings and structures provided on a stage of construction. It has been shown that definition of economic efficiency of ERC is the evaluation of its application for buildings with remote or automatically controlled production. It has been determined that an important role in design of explosive industrial facilities is played by selection of the economically feasible and effective materials for ERC. When selecting materials it is necessary to consider probability and yield of explosions. Necessity to create the methods allow considering such probability has been revealed.

Amount of damages caused to the company as a result of explosion of combustible gases, vapors and dusts inside the building with explosive production depends on the effectiveness of explosion safety measures [1, 2].

The results of inspection of emergency explosion consequences and calculation of explosion loadings for separate buildings, which external protection was designed taking into account the installation of explosion-relief constructions (ERC), have shown that the main protecting and bearing structures of affected buildings and, sometimes, adjacent buildings can be damaged together with destruction of ERC [3].

The measures proposed to ensure explosion protection of industrial facilities differ from existing differentiated approach to choice of types and number of ERC for each specific case [4–9]. Besides, manufacture of ERC, except the specially designed covering elements, does not require additional expenses because the offered constructions are standard and produced on an industrial scale. Application of ERC in the cladding of buildings could limit the damaged area by sizes of destroyed construction. Such destructions minimize the amount of damage due to the absence of additional costs for restoration of the process equipment and the main building structures [10–13]. This can sharply reduce the costs for rescue and recovery operations and decrease the losses from temporary termination of the productive activity.

It should be noted that implementation of explosion safety measures for industrial facilities, which technological processes connected with generation of combustible gases,

[*] Corresponding author: lyapin@mgsu.ru

vapors and dust, could require additional costs. Methods of passive redundancy, such as strengthening of the main bearing structures, increase of their rigidity, change of the building constructive scheme, etc., could also be used to improve explosion safety of buildings [14–20].

Extension of ERC's area as well as increase of bearing capacity of the main structures has improved explosion safety of building. However, initial cost and maintenance costs of ERC have increased as the mounting cost of 1 м^2 of ERC and of main cladding constructions are varied considerably. Allowances for depreciation and construction maintenance have also increased. Moreover, heat losses trough ERV have entailed additional maintenance costs.

The problem of implementation of explosion safety measures should be considered separately for each industrial facility.

It is obvious that industrial facility with production controlled by operated personnel should provide explosion safety measures in order to protect the people, the process equipment and the main bearing structures.

While, for industrial facilities with automatically and remote controlled production, where the probability of emergence explosion is rather small, destruction of the building structures will be economically justified; but only if the restoration costs will not exceed the installation costs of ERC.

To determine the economic feasibility of proposed type of ERV we should solve one of the following tasks:

1) To define the economically justified structural concepts, materials and required area of ERC (in case of presence of operated personnel);
2) To evaluate the feasibility of ERC application taking into account analysis of its mounting costs and the total costs of construction, maintenance and restoration of industrial facility (in case of absence of operated personnel).

Thus, definition of economic efficiency of ERC has been restricted by assessment of its application for buildings with remote or automatically controlled production. In this case it should be determined whether the application of ERC is necessary. If it is necessary we should find the optimal type, which will provide appropriate explosion safety.

When designing explosive production it is important to find the cost-effective materials for ERC manufacturing, and also to minimize total costs of construction and maintenance of building in case of ERC installation. To calculate the sum of costs occurring at different times we have to convert them to equivalent costs at a base time using the compound interest formula (in accordance with «Instructions for definition of economic efficiency of capital investments in construction» (SN 423–71)):

$$B = 1/(1 + E_{dr})^t \tag{1}$$

Where $E_{d.r}$ — discount rate used to convert costs occurring at different times to equivalent costs; $E_{d.r} = 0{,}08$; t — reduction time period, years.

When we choose materials for ERC we should also consider the frequency of emergence explosions. But the current guidance and regulatory literature on economic evaluation of design decisions, using stochastic and probabilistic rates, does not contain certain recommendations. To develop the methods, considering probability of explosions and degree of ERC destruction, we should have statistical data on frequency of explosions at the similar industrial facilities. Such statistical data, have been obtained from industrial facilities (connected with production of butadiene, cumen hydroperoxide, ammonia, etc.), and the probabilistic and statistical methods allow calculating the average time t_{av} between explosions:

$$t_{av} = 1/\lambda_s, \tag{2}$$

Where λ_s — damage rate of structures, measured by amount of damages during one year of operation;

$$\lambda_{\kappa} = \frac{n\lambda}{Nr(t)};\qquad(3)$$

n — total number of a single-type building structures;
λ — frequency of explosions;
N — number of a single-type building structures under observation;

$$N = (r-1)/(Ntr);\qquad(4)$$

$r(t)$ — number of explosions during period of observation.

Materials for ERC can be selected according to criterion of minimal total costs P (thousand of Russian rubles, further "thou. rub.") for fabrication (mounting), restoration and maintenance of structures during the operation period of industrial facility:

$$P = C_{\text{ко}} + \sum_{i=t_{sr}}^{r(t)t_{sr}} C_i - \frac{C_{kd}}{(1+E_{ko})^i} + \frac{E^{year}}{E_{kd}}\qquad(5)$$

Where: $C_{\text{ко}}$ — initial estimated cost of ERC, thou. rub.;

t_{av} — average failure-free operation time, e.g. time between explosions (calculated by formula (2), hours;

C_i — direct costs of ERV restoration, thou. rub.;

C_{kd} — depreciated cost of destroyed structures, thou. rub.;

i — time periods of explosions, years; $i = t_{av}, 2t_{av}, \ldots, at_{av}$;

a — value, which indicates the whole number of building restorations (consider only integer part of ratio);

$$a = (T_{sl} - 1)/t_{av}\qquad(6)$$

T_{sl} — standard service life of building, years;

E^{year} — total annual operating costs, allowances for depreciation and maintenance of structures, thou. rub.

The last item of equation (5) accounts total operating costs for the entire period of operation. As the total annual operating costs E^{year} occurring at different times, so we have converted them to equivalent costs at a base time. The first and the second items of equation (5) account investments of initial application period of ERC; therefore we have accepted initial operation time of ERC to determine the operating costs.

Operating costs can be converted by following formula:

$$E^{year}T_{sl} = \sum_{i=1}^{T_{sl}} E_i^{year} \frac{1}{(1+E_{dr})^i}\qquad(7)$$

If operating costs have not been changed for years, e. g., $E_1^{year} = E_2^{year} = E_3^{year} =$ const, E^{year} value can be taken outside the summation symbol. So the remaining part of expression have been presented as the sum of decreasing geometric progression with the first member a and the denominator g equal to $1/(1 + E_{dr})$, (if $T_{sl} \to \infty$).

$$\sum = a/(1-g) = 1/E_{dr}\qquad(6)$$

In this case total operating costs are equal to

$$\frac{1}{E_{dr}} = E^{year}\qquad(8)$$

Protection from explosions of flammable liquids can be provided by:

- Application of ERC;
- Strengthening of bearing structures;
- Combination of ERC application and strengthening of bearing structures.

Thus we have been faced with the choice of such protective measures, which require minimal investment and provide the specified degree of explosion safety.

The minimum total cost has been determined by following formula:

$$P = C_s + C^{est}\ E_s\ F_{pr}t + \frac{1}{E_{dr}\ \left(1{,}06\ C_i\ \left(a_1 + \frac{a_{2)}}{100}\right) + Z^h\right)} + \sum(Z_i\ + E_s F_{pr}\ t^{susp} / (1 + E_{dr} \quad (9)$$

Where: C_s — estimated cost of works on strengthening the main bearing structures, thou.rub.;

C^{est} — estimated cost of explosion relief constructions, thou. rub.;

E_s — standard index of capital investment efficiency (index of profitability);

$F_{pr.}$ — estimated cost of production funds of the company (-ies), providing mounting of ERC, thou. rub.;

$T_{m.}$ — mounting period, years;

C_i — estimated cost of structures, thou. rub..;

a_1, a_2 — standard renovation deductions and deductions for maintenance of structures (from their estimated cost);

Z^h — annual heating costs, thou. rub.;

Z_i — costs of liquidation of explosion consequences in *i*-year, thou. rub.;

t^{susp} — suspension period of production connected with liquidation of explosion consequences, years.

The cost of strengthening of bearing structures C_s should be calculated for certain building. If $C_s = 0$, then specified degree of explosion safety can be provided only by application of ERC.

Estimated cost of explosion relief constructions have been determined according to the estimated prices on materials, products and constructions taking into account overhead costs. Economic effect E_i of reduction in labor intensity and decrease in time of ERC mounting have also been considered:

$$C^{est} = C_i - E_i\,. \quad (10)$$

Value of economic effect E_i has been determined taking into account the value of labor intensity.

1. The labor intensity m_i (person-days) have been calculated for every type of explosion-relief constructions.
2. The type required highest labor intensity has been accepted as initial.
3. Types required lower labor intensity, under equal production conditions (manpower resources, number of working shifts), have been given the economic benefit in relation to initial type.

For initial type E_i value is equal to 0. For other types E_i value is a sum of tree values:

$$E_i = E_i^1 + E_i^2 + E_i^3 \quad (11)$$

Where $E_i^1 + E_i^2 + E_i^3$— economic effects of reduction of the semi-variable part of overhead costs and the part of overhead costs related to labor intensity and basic wage rate.

Effect of reduction of the semi-variable part of overhead costs has been calculated by formula

$$E_i^1 = 0,5\,C\frac{K_{oc}}{(1+K_{oc})1.06}\left(1-\frac{m_i}{m_{init}}\right) \quad (12)$$

Where: 0.5 — ratio of semi-variable part in the total amount of overhead costs;
C — estimated cost of construction products, thou. rub.;
K_{oc} — ratio of overhead costs; $K_{н.р}$ = 0.12÷0.19;
1.06 — ratio of planned accumulations;
m_{init} — labor intensity by initial type, person-days

Effect of reduction of overhead costs, related to labor intensity has been calculated by following formula

$$E^2 = 0.6\,(m_{init} - m_i) \quad (13)$$

Reduction of overhead costs related to basic wage have been calculated by following formula:

$$E_i^3 = 0.15\,(Z_{init} - Z_i) \quad (14)$$

Where: Z_{init}, Z_i — basic wage (by initial type and i- type accordingly), rub.

The annual heating costs Z^h has been calculated by formula

$$E^h = F_{pk}\,E^h_{m^2} = F_{pk}U_h\,(t^0_{in} - t^0_{av})\,24\,\frac{T_h}{R_{ot}10^6} \quad (15)$$

Where: F_{ERC} — area of ERC, m^2;
$E^h_{m^2}$ — heating costs, 1 m^2/year;
U_h — cost of 1 Mcal of heat, rub.;
t^0_{in} — specified temperature of indoor air, °C;
t^0_{av}— average temperature of outdoor air during heating period, °C;
24 — hours in the day;
T_h — duration of the heating period, days;
R_{ot} — thermal resistance of wall or wall coating, m^2·°C·h·kcal.

Costs related directly to liquidation of explosion consequences *(C)* have been determined as follows:

1. If the area of explosion relief constructions allows preventing complete destruction of the building, then C value can be calculated by formula

$$C = C_{ERC} + C_{ms} + C_{p.e} + C_{adj} + C_{cl.} - D, \quad (16)$$

Where: C_{ERC} — costs of restoration of ERC; determined by estimate standards taking into account design costs;

$C_{m.s}$ — costs of restoration of the main structures (if explosion relief constructions are not able to reduce explosive load);

$C_{p.e}$ — costs of restoration of process equipment, pipelines and networks;

$C_{adj.}$ — costs of restoration of adjacent buildings and facilities;

$C_{cl.}$ — costs related to cleaning of the territory from destroyed structures; actual cost of dismantling and transportation;

D — depreciated cost of damaged structures (can be accepted as 10 % from initial cost of ERC).

Values of $C_{m.s}$, $C_{p.e}$ and C_{adj} have been determined by multiplying the volume of restoration works by the relevant estimate norms. The necessary volume of restoration can

be determined as average value according to statistical data.
2. If design of explosion relief constructions allows complete destruction of a building, then the costs related to liquidation of explosion consequences could be calculated by formula

$$C = C_1 + C_{adj} + C_{cl.} - D, \tag{17}$$

Where C_1 — costs of restoration of destroyed building, thou. rub.

For aggregate calculations C value can be equal to cost of initial mounting of ERC. If the cost of structures have been changed in time then calculation should contain appropriate correction factor K_t.

$$C = 1{,}06\ C_i\, K_t. \tag{18}$$

Conclusions

In accordance with considered model of economic evaluation and feasibility of ERC selection, as well as determination of economic efficiency of explosion protection of buildings and constructions, the following conclusions can be drawn.
1. Protection of buildings against effects of explosions, occurred inside production premises, is connected with additional costs for ERC. Therefore the very important task is to ensure explosion safety with the lowest total construction and maintenance costs of production building.
2. The value of average failure-free operation time is the fundamental parameter for economic evaluation of the selected ERC application expressed by destruction intensity which, as is known, is a random value. For its determination one can use the probabilistic and statistical methods.
3. When selecting materials for ERC fabrication it is necessary to minimize full installation and maintenance costs taking into account probability of emergency explosions. The method presented in this article allows minimizing such costs.
4. The equation (9) used to calculate the economic feasibility of ERC enables to define total costs of construction, operation and restoration of production building considering the probability of explosion. Analysis of this equation as well as further calculations shows that decrease of ERC area is economically justified for remote or automatically controlled industrial facilities, where explosions occur at intervals of 12-15 or more years.
5. The economic benefit of ERC calculated by described method is equal to 3% of total construction and maintenance costs for compressor buildings transporting combustible gases. The economic benefit of this method applying for such facilities will amount to more than 5 million rubles per year. As the list of explosive productions of chemical, petrochemical, oil processing and gas industries includes several dozen items, then economic benefit will amount to several tens of millions of rubles.
6. In order to optimize the building explosion safety coasts it is necessary to use the probabilistic methods which determine the risk of explosion depending on various protective measures. Meanwhile, the both the direct damage caused by explosion and the indirect damage connected with liquidation of a certain type of production and recovery of incapacitated powers should be accounted.

References

1. M. V. Beschastnov, V. M. Sokolov, M. I. Kats,Moscow, Khimiya Publ., **367** (1976)
2. Scientific and technical report, VNIITBKhP Publ., **22** (1978)
3. G. G. Orlov, A. Ye Leybman, A. I.Manasyan,Moscow State University of Civil

Engineering Publ., **1** (1978)
4. I. O. Moen, M. Donato, R. Knystautas, J. H., Combustion and Flame, **39** (1980)
5. C. J. M. Wingerden, J. P. Zeeuwen, Journal of Hazardous Materials, **8** (1983)
6. A. F. Sharovarnikov, D. A. Korolchenko,Applied Mechanics and Materials, **475–476** (2013)
7. V. I. Gorshkov, D. A. Korolchenko, Yu. N. Shebeko, V. Yu. Navtsenya, A. K. Kostyukhin, All-Russian Research Institute for Fire Protection Publ., **800–808** (1997)
8. D.A. Korolchenko, A.F. Sharovarnikov, A.V. Byakov, Advanced Materials Research, **1073-1076** (2015)
9. D.A. Korolchenko, A.F. Sharovarnikov, Advanced Materials Research, **1070-1072** (2015)
10. D. Korolchenko, A. Tusnin, S. Trushina, A. Korolchenko, International Journal of Applied Engineering Research, **10** (2015)
11. L. P. Pilyugin, Pozhnauka Publ., **379** (2010)
12. D. A. Korolchenko, A. F. Sharovarnikov, Fire and Explosion Safety, **23** (2014)
13. D. A. Korolchenko, A. F. Sharovarnikov, Fire and Explosion Safety, **24** (2014)
14. S. M. Frolov, Combustion, Explosion, and Shock Waves, **48** (2012)
15. S. M. Frolov, Torus Press Publ., **6** (2013)
16. G. G. Orlov, D. A. Korolchenko, A. Ya. Korolchenko, Fire and Explosion Safety, **24** (2015)
17. D. A. Cherepanov, A. S. Ermakov, Fire and Explosion Safety, **25** (2016)
18. A. S. Ermakov, D. A. Cherepanov, Fire and Explosion Safety, **26** (2016)
19. O.S. Zosimova, Collection of materials of the XIX International Interuniversity scientific and practical conference of students, undergraduates, graduate students and young scientists, **473-476** (2016)
20. V.V. Holschevnikov, D.A. Samoshin, I.R. Belosohov, R.N. Istratov, I.S. Kudrin, A.P. Parfionenko, Fire-boom-bada-boom-safety, **3** (2011)

Analysis of causes of combustible mixture explosions inside production floor areas

Anton Lyapin [1*], *Anton* Korolchenko[1] and *Evgeniy* Meshalkin[2]

[1]Moscow State University of Civil Engineering, 129337, Moscow, Yaroslavskoye shosse, 26, Russia
[2]NPO PULS, 107014, Moscow, Rusakovskaya street, 28, Russia

Abstract. The work provides a cause analysis for major industrial explosions and a review of the causes of combustive air-gas mixture generation in a production environment. It has been established that during operation of explosive production facilities, it is process equipment that, as a rule, creates explosive environment inside the floor area. A qualitative method for determination of a potential accident has been reviewed. Analysis of the nature of explosion effect on building structures and equipment has shown that exposions characterised by absence of equipment and building structure disintegration normally have a localized character. It has been identified that during explosions inside process equipment, the largest structural damage occurs in spots hit by equipment debris. Complete destruction of building structures and equipment is caused by explosions inside equipment containing large quantities of combustible products. It has been identified that most explosions are accompanied by partial or total destruction of building structures and equipment. Therefore, measures taken to protect equipment and buildings from explosion effects lack efficiency.

In most cases, generation of explosive gas-air mixture within a building is related to leakages and accidental releases of combustible gases to the atmosphere of the floor area. Thus, out of 1,200 industrial explosions registered in the US over a five-year's period, 50% are accounted for by the use of combustible gas/steam-air mixtures in production [1-3]. According to the analysis conducted by the American Insurance Association, the causes of major explosions are distributed as follows: equipment breakdown: 31%, process order violations: 17%, incorrect estimation of explosiveness of products used in the production process: 20% [4]. Moreover, the most frequent cause of combustible substance leakages and releases (50% of all cases) is breakdown or technical failures of process line equipment; 20% can be attributed to errors and negligence of operations personnel during equipment operation.

The examined causes of combustible gas-air mixture generation in a production floor atmosphere have necessitated development of the following explosion-safety measures: prevention of gas leakages or of their movement in an inadvertent direction; implementation of complex production mechanization and automated remote control.

* Corresponding author: lyapin@mgsu.ru

Statistics of intra-building accidental explosions are not systematized; therefore, in order to obtain a cause-based picture of explosion distribution, 134 category accidents have been analyzed, which occurred at different times at domestic chemical and petrochemical facilities.

The analysis of accidents occurring during operation of explosive industrial facilities has been based on accident investigation protocols, in-situ facility examinations and data materials published in a number of open-source research works [5-8].

As a result:

1. The accident causes in terms of explosiveness of an operated production site are: low-quality installation of process equipment; faulty control and instrumentation equipment; errors in design documents; defective equipment parts; faulty elements of main process equipment; errors in development of the process order; violated operation safety rules for process equipment; low-quality preventive maintenance of equipment; errors in development of the process order; violation of equipment repair rules; corrosion in the walls of equipment parts and pipelines; other causes.
2. The main equipment parts causing formation of intra-facility combustible mixtures have been identified: vessels and tanks; process columns; compressors; condensers; cubes; pumps; generators; process kilns/furnaces; scrubbers; mixers; piping; flanged pipeline joints; shut-off and adjustment fittings and other elements, unidentified or omitted from the documentation;
3. The dependence of the explosion number on the type of the gas-air mixture ignition source has been demonstrated: open process fire; open fire used for repair; electric welding sparks; sparks generated during collisions and friction of metal against metal; sparking occurring inside electrical equipment or caused by short circuits in electrical equipment; static discharge sparks; heated surfaces of process equipment; product self-ignition on contact with air, and other causes, unidentified or omitted from the documentation;
4. Distribution of explosions at production facilities is presented based on their destructive effect on building structures and equipment: intra-premise explosions with no destruction of building structures and equipment; same with partial destruction of building structures and equipment; same with complete destruction of building structures and equipment; same with partial destruction of structures of neighboring buildings.

During operation of explosive industrial facilities, process equipment creates explosive environment inside the floor area. In order to develop preventive measures against explosions and to select protection methods, it is essential to know equipment elements, which most often create a combustible environment.

Combustible gases present in process equipment may penetrate the production floor environment in case of equipment destruction. When a device housing is damaged or a cover is blown off, combustible gas may be instantly released into the premises and an explosive concentration of a significant volume may generate within seconds. When a pipe is broken, the combustible medium is emitted in the form of a jet. As a result, an explosive concentration may be generated in the floor environment within minutes.

During a slow gas leakage, e.g. through shut-off fittings, explosive mixture is formed within a time quite sufficient to actuate the emergency exhaust ventilation or to take other fire- and explosion preventive measures.

In each case, the amount of gas penetrating the floor atmosphere may be approximately estimated if the device or the group of devices potentially capable of causing the accidental formation of explosive mixture are known.

The operations personnel must continuously monitor production equipment belonging to potential sources of explosive mixtures in the floor area. Normally operated equipment works without accidents, expensive repairs or outages, which disorganize the production.

Operating machines and devices mainly lose their operability due to destruction of individual parts or their surface layers because of mechanical or corrosion wear whose degree depends on the operating environment. As a result, equipment loses its strength, and its capacity and productivity is decreased. These critical parameters are restored by means of repair.

Chemical and petrochemical facilities deploy a scheduled preventive maintenance system, which has proven itself practically. The repair system is based on the summarized experience of equipment operation and repair at chemical and petrochemical facilities. It uses empirical and statistical repair reference materials of chemical and petrochemical facilities, research and design institutes, and materials related to the design of new equipment and upgrade of operating equipment accumulated by the chemical machine building industry.

The main goal of the scheduled preventive equipment maintenance system representing a combination of organizational and technical measures to monitor and maintain equipment, including utilities, is prevention of equipment accidents. The main parameter for determination of the equipment repair type is operating hours between overhauls (running repair, medium, major overhauls). Knowledge of the group of elements most likely to cause an accident (based on available data on known failure-free operation hours of each element) will bring one closer to tackling the task of determining the potential volume of combustive mixture generated within a floor area during an accident.

All equipment on a production floor able to release combustible gases can be divided into "K" number of groups, each of them in its turn consisting of a range of independent homogenous elements. Since it is impossible to accurately forecast failure of an element due to the incidental nature of this event, it is naturally possible to use the methods of the probability theory in this situation and to construct a probability model [9-15]. We will now consider a model used in the reliability theory.

Given is "K" number of equipment groups, consisting of n_1; n_2..., n_k homogenous independent elements. We will call the sequence of element failures over time a failure stream.

The failure stream of the given i-group shall be characterized by the number λ_i - an average number of failures over a time unit. Due to the independent character of element failures, the average number of faliures in the group of n_i elements shall equal $n_i \lambda_i$. We will call the stream of simultaneous failures in all groups a cumulative failure stream. Due to the independence of the groups, the average number of the cumulative stream failures will equal:

$$\Lambda = \sum_{i=1}^{k} n_i \lambda_i. \qquad (1)$$

The reliability theory has proven that if the failure stream is generated from a larger number of minor intensity streams (λ_i – minor), its qualities are close to the properties of a so-called simple stream. We have a similar picture in the problem under consideration. Therefore, we will assume the resulting failure stream as a simple stream.

The cumulative failure stream is a simple stream with the failure intensity parameter equal to Equation 1. It means that the probability P_m of occurrence of m failures over the time τ is governed by the Poisson law.

$$P_m = \frac{\lambda \tau^m}{m!} e^{-\Lambda \tau} \qquad (2)$$

The probability of no failure over the time τ equals $P_0 = e^{-\Lambda\tau}$. The average failure-free operation hours shall be

$$T_{cp} = \frac{1}{\lambda}. \tag{3}$$

If τ is minor, then

$$P_i \text{ (failure)} = 1 - e^{-n_i\lambda_i\tau} = n_i\lambda_i\tau + 0(\tau), \tag{4}$$

and

$$P \text{ (failure)} = 1 - e^{-\sum_{i=1}^{k} n_i\lambda_i\tau} = \sum_{i=1}^{k} n_i\lambda_i\tau + 0(\tau) \tag{5}$$

The hazard potential of the 1st group of equipment can be estimated as

$$B_i = \frac{P_i}{P} \cong \frac{n_i\lambda_i\tau}{\sum n_i\lambda_i\tau} = \frac{n_i\lambda_i}{n_1\lambda_1 + n_2\lambda_2 + \dots + n_k\lambda_k} \tag{6}$$

However, such qualitative method of determination of a potential accident does not exclude other methods. For this method to be used, the equipment operating modes, operating life and margin of safety need to be clarified in terms of quantity.

During accident investigation, it is often difficult to identify the real source of explosive mixture ignition. Available materials on accidental explosions in most cases indicate a possible ignition source, which significantly complicates development of grounded preventive measures [16-18].

It has been identified that most explosions are accompanied by partial or total destruction of building structures and equipment. This testifies to the importance of the problem under consideration and of the necessity for preventive measures targeted both at prevention of explosions and at mitigating their destructive effect on equipment and building structures.

When analyzing the character of explosion effect on building structures and equipment, the following conclusions can be drawn.

Explosions characterized by absence of equipment and building structures disintegration normally have a localized character. These are explosions of minor intensity occurring inside process equipment or within smaller local volumes of combustible mixtures formed as a result of penetration of small amounts of combustible gases through loose spots in shut-off fittings or pipes.

Explosions involving partial destruction of equipment or building structures occur both inside the equipment and the production floor, and the destruction amount depends on the type of the combustible mixture, its volume and concentration. During explosions inside process equipment, the largest structural damage occurs in spots hit by equipment debris. In this case, destruction is manifested by through-holes or deformation of structure parts at the spots hit by equipment parts.

Destruction in building structures is more severe during explosions of gas-air mixtures inside production premises than during explosions inside process equipment. Most often, lightweight envelope structures are destroyed; while out of main bearing elements reinforced concrete structures are prone to disintegrate more often that metal ones.

It has been identified from past accident protocols that explosions involving complete destruction of building structures are preceded by significant damage of process column vessels, compressors, pumps, piping, shut-off fittings, flange joints and by formation of large volumes of explosive concentrations of combustible mixtures in production floor areas. Here, the most probable ignition sources are open process fire during repair, hot surfaces of process equipment or product self-ignition upon contact with oxygen.

Complete destruction of building structures and equipment is also caused by equipment internal explosions involving large quantities of combustible products.

During explosions of external units and intra-premise explosions, mainly destruction of lightweight envelope structures (i.e. windows, doors or gates) of neighboring buildings is observed. The character and degree of destruction depend of the explosion intensity and on the distance to the explosion site.

Conclusions

Thus, based on the accident cause analysis, the following conclusions can be drawn.

During operation of explosive facilities, most explosions are due to violation of safety operation rules and low-quality preventive equipment repairs, as well as due to imperfect solutions in design documentation and defective equipment parts (15.1%). Moreover, low-quality installation of process equipment and faulty control and instrumentation equipment also cause a significant number of accidents (12.4%).

Most often, explosive mixtures are generated as a result of malfunctioning of shut-off and adjustment fittings and flange joints in pipelines; therefore, the time between preventive inspections and repairs of shut-off fittings and pipe joints should be reduced, and preventive inspections and repairs should be done according to special schedules.

When assessing explosiveness of a production, one should also take into account equipment elements, which cause generation of combustible mixtures.

Statistical data for explosion causes show that most often explosions are caused by open fire (22.1%), product self-ignition upon contact with air oxygen (15.7%), and heated surfaces of process equipment (14.2%). Explosions may also be initiated by faulty electrical equipment (18.5%), spark discharges of static electricity and collision and friction of metal against metal (7.5%).

Most explosions (54.9%) are accompanied by partial or total destruction of building structures and equipment. Therefore, measures taken to protect equipment and buildings from explosion effects lack efficiency. Statistical data related to accidental explosions demonstrate that even the current state of the art does not eliminate explosions at explosive sites as a result of breakdowns or failures of process equipment and some other reasons. Besides, human interference in the production process cannot be eliminated, while people by their nature are not guaranteed against mistakes.

In this connection, in order to ensure explosion safety in production buildings in case of explosions caused by various reasons, special technical solutions should be provisioned.

References

1. O. Pester, Chemische Technik, **11** (1967)
2. E.H. Wilhelm, Lose Preva, Safety Promot. Process., **1** (1974)
3. Lee M. Blast, Bloomberg Businessweek, **10** (2013)
4. A. West, Houston Chronicle, **12** (2013)
5. Topical scientific and technical report, All-Union Research Institute for Safety in Chemical Industry, **118** (1976)
6. N.I. Akinin, Metallurgist, **10** (2004)
7. M.V. Beschastnov, V.M. Sokolov, M.I. Katz, Chemistry, **367** (1976)
8. Texas dotted with residents vulnerable to fertilizer plant accidents, Houston Chronicle, **12** (2013)

9. B.V. Gnedenko, Yu.K. Belyaev, A.D. Solovyev, Nauka, **372** (1965)
10. V.N. Mostochenko, Stroyizdat (1974)
11. G.G. Orlov, Stroyizdat, **200** (1987)
12. G.G. Orlov, D.A. Korolchenko, A.V. Lyapin, Fire and Explosion Safety, **23** (2014)
13. G.G. Orlov, D.A. Korolchenko, A.Ya. Korolchenko, Fire and Explosion Safety, **24** (2015)
14. G.G.Orlov , D.A. Korolchenko, Fire and Explosion Safety, **24** (2015)
15. D.A. Korolchenko, A.F. Sharovarnikov, A.V. Byakov, Advanced Materials Research, **1073-1076** (2015)
16. D.A. Korolchenko, A.F. Sharovarnikov, Advanced Materials Research, **1070-1072** (2015)
17. D. Korolchenko, A. Tusnin, S. Trushina, A. Korolchenko, International Journal of Applied Engineering Research, **10** (2015)
18. I. Urbina, M. Fernandez, J. Schwartz, The New York Times, **9** (2013)

Gas explosion in a room with a window and passage to an adjacent room

Yuri Polandov[2], *Aleksandr* Korolchenko[2*] and *Sergey* Dobrikov[1]

1Orel State University, 302026, st. Komsomolskaya 95 Orel, Russia
[2]Moscow State University of Civil Engineering, 129337, Yaroslavskoye Shosse 26, Moscow, Russia

Abstract. Some publications describe an effect, produced during a physical model experiment, when an adjacent gas-free room influences the gas explosion pressure in a room with a window. The explosion pressure in this case significantly exceeds (2.5 times) the explosion pressure in a room without an adjacent room. This result has been confirmed by our studies. Based on other available information about the influence of the ignition point location on the explosion pressure in one room, it was suggested that this could be true for an explosion in two rooms. In our studies we used a test unit with two connected chambers, each having a volume of 1.125 m^3. It turned out that this influence of the adjacent volume was not so unambiguous as it was described in those publications. It was found out that the maximum effect of explosion pressure amplification by the adjacent room is achieved, when the igniter is located in the chamber filled with a gas-air mixture in the area between the center of the chamber and the window (maximum amplification by more than 3 times). This effect is lower directly by the window (1.8 times) and is practically absent in case of ignition within the area near the passage connecting the chamber with the adjacent room. This suggests that the effect discovered earlier is a special case of the general dependence of the gas explosion pressure in two chambers on the igniter location.

1 Introduction

Much attention has been paid to a gas explosion in a single room with a window. As judged by publications, the idea of the influence of an adjacent gas-free room, connected to the base room, on the gas explosion development was clearly shaped for the first time in works of researchers from the Moscow State University of Civil Engineering [1]. They showed that contrary to expectations the adjacent room serves not an explosion damper, but an explosion amplifier in case of an explosion in the base room. They conducted an experiment using two model chamber (base and adjacent): the base chamber was filled with a gas-air mixture, it had a window to the atmosphere and was connected with the adjacent chamber via a passage. It was found out that in case of ignition of the gas-air mixture in the center of the base chamber the maximum explosion pressure was considerably (2.5 times) higher in the presence of the adjacent chamber than that in a single chamber, not connected with an adjacent one.

[*] Corresponding author: KorolchenkoAYA@mgsu.ru

As for the certainty of the gas explosion pressure amplification effect in a room connected with an adjacent room, where only the air is present, we have recently published materials [2] confirming the experiment results.

2 Working hypothesis

It is quite obvious that in case of a confined gas explosion the maximum explosion pressure does not depend on the ignition location. It is absolutely different in case of an unconfined gas explosion, for example, in a room with a window. It is known that the location of gas-air mixture ignition in a room with a window affects the explosion process and the maximum pressure value [3-7]. According to these data, the farther is ignition location from the window, the higher is the maximum explosion pressure. This effects is more prominent, when the room length-to-width ratio is higher, and it can reach the multiplicity factor of more than 20. But in this case, the window location starts taking effect.

Presumably, in our case the gas ignition location will also affect the maximum pressure value of an explosion in a room with a window and a connected adjacent room. This statement is based on expected variants of the explosion development with different ignition locations. There are three variants:

First variant. The source is located near the passage to the adjacent room. In this case, mainly combustion products will get into the adjacent room rather than the gas-air mixture. There will be no combustion process in the passage or in the adjacent room, i.e. hot gases will just flow into the adjacent room. This room will serve a sort of damper.

Second variant. If the mixture is ignited near the center, it is expected that the unburned mixture will get into the adjacent room, where it will burn down. But in the passage the mixture will become turbulent, which will drastically increase the flame front area when it gets into the adjacent room resulting in the explosion pressure increase.

Third variant. The gas is ignited near the window. In this case, the largest amount of the mixture will get into the adjacent room, but the mentioned effect of the explosion pressure reduction, if the ignition location is near the window, will come into force. The explosion pressure will depend on the relationship between these two factors.

For higher credibility, experiments were conducted on a physical model.

3 Research equipment

In selection of a model unit we considered the influence of such factor as the room length-to-width ratio on the gas explosion process. Keeping in mind the influence of another factor – the adjacent room – on the explosion development, we decided to minimize the influence of the first factor and selected a cubic shape of the both chambers. Photographs of the model unit are shown in Figure 1. It consists of two chambers No.1 and No.2 with the following dimensions: 0.5 x 0.5 x 0.5 m each. There was a disposal partition between the chambers, which separated the chambers in one case and connected them via a passage in the other case. Chamber No.1 had a vent window 0.15 x 0.15 m (facing left in the figure). This window was provided with a door made of 1-mm steel sheet and a damper position sensor. The igniter's positions could change between 5 points (numbered left to right). There is a fan inside the chamber to enable gas and air mixing and mixture preparation for ignition. The passage between the chambers is 0.15 x 0.15 m. While the mixture is being prepared, the passage is blocked from the side of the first chamber with a sheet of paper, which would be destroyed during the explosion. During simulation of explosions in a single chamber, a solid partition was installed between the chambers. Combustion products were removed from the chambers after explosions via the ventilation duct, which supplied the air to chamber No.2.

Propane from a cylinder was supplied to the first chamber. Experiments were conducted at concentrations near stoichiometric.

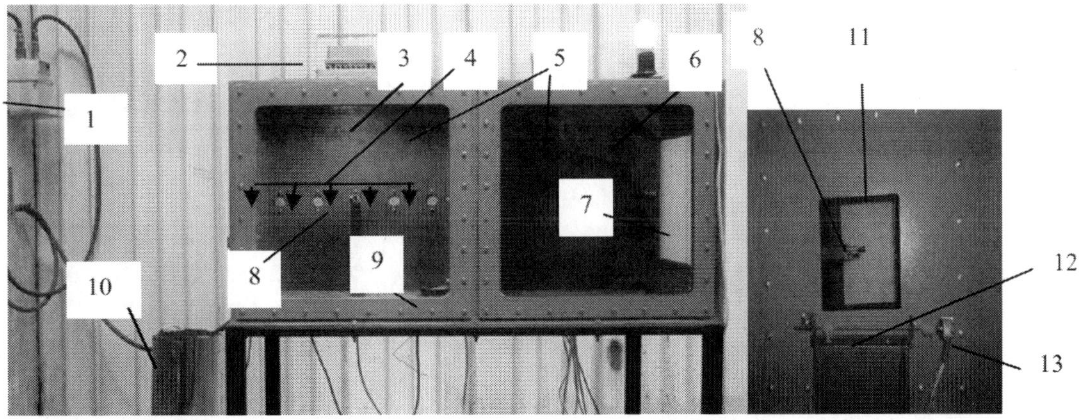

a) Side view of the unit b) Front view of the unit

Fig. 1. General view of the unit: 1 – gas meter; 2 – backlight; 3 – chamber No.1; 4 – igniter locations (5 positions); 5 – pressure sensor locations; 6 - chamber No.2; 7 – back glass; 8 – igniter at position 3; 9 – circulation fan; 10 – ventilation duct; 11 – vent window; 12 – door (open); 13 – damper position sensor.

4 Standard tests

Figures 2 and 3 show video snapshots of the gas explosion development in the single chamber and in two chambers respectively. It can be seen that in both cases, until the door opened, the flame shape was close to spherical, and after the door opened the flame front stretched in its direction. During the door opening the exiting jet is directed upward, deflected by the door. In Figure 2 the igniter is installed in position 3 and in Figure 3 – in the center of the chamber in position 3. The explosion duration in the single chamber is noticeably shorter than in case of two chambers.

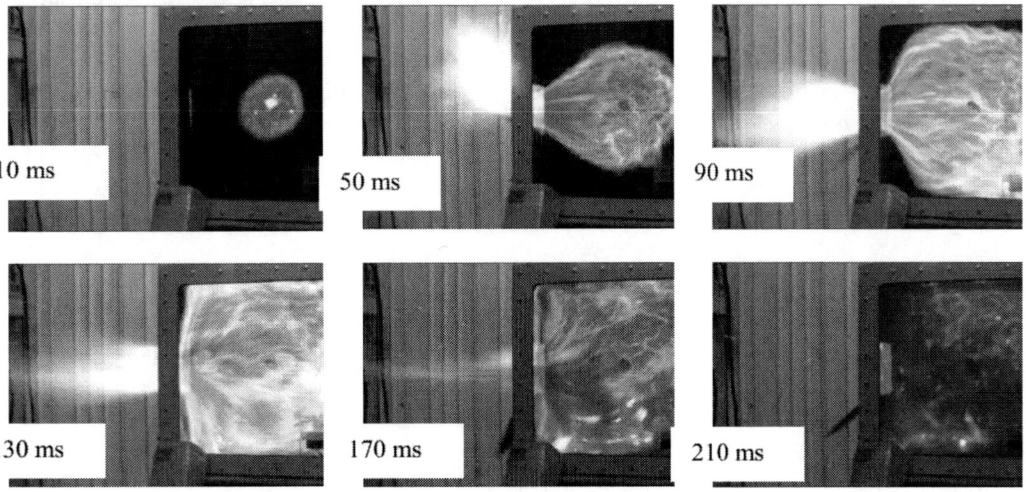

Fig.2. Video snapshots of an explosion in the single chamber.

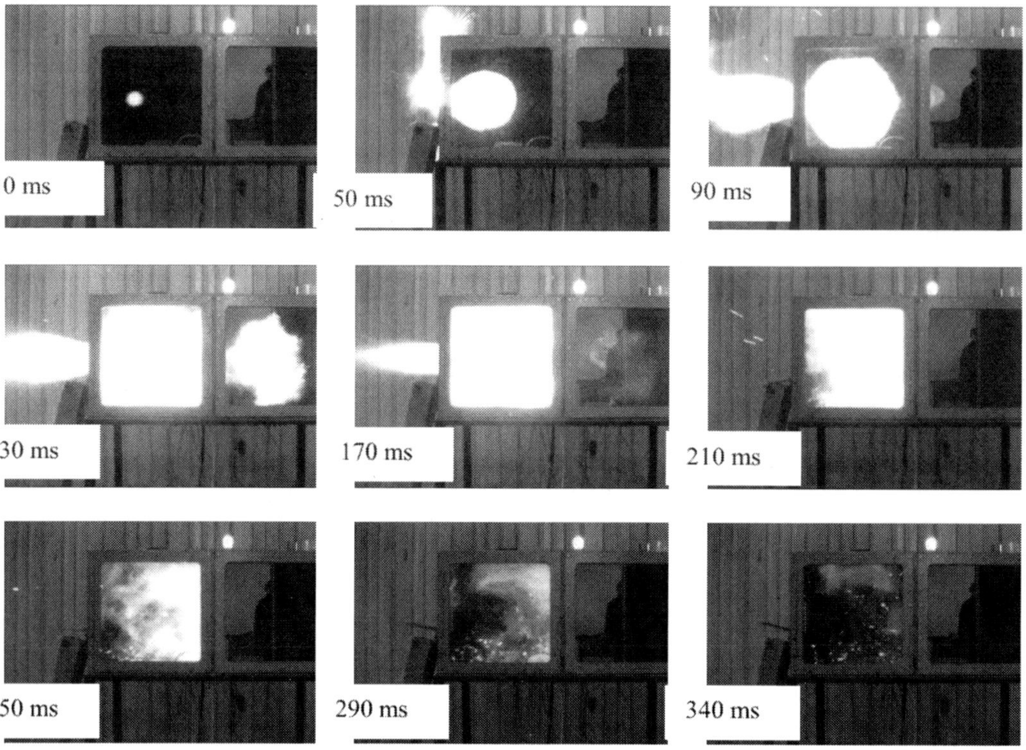

Fig. 3. Video snapshots of an explosion in two chambers.

5 Data analysis

The experiment program included two sets of tests followed by comparison of their results. The first set was conducted to study the dependence of the explosion development on the ignition location in the single chamber, and the second set included the same tests but with the open passage to the second chamber, blocked with a sheet of paper.

Tests with explosions in the single chamber. An explosion in a single chamber and snapshots of this process are not unique – such pictures are well known and we showed them for the sake of comparison with the explosion in two chambers and identification of differences. Test results of the first set with 5 positions of the igniter are summarized in a single diagram in Figure 4. Numbers of positions, at which explosions occurred, correspond to curve numbers. It can be seen that the closer the igniter to the window, the longer the explosion – in our case with reduction of the position number the explosion duration increases from 0.17 to 0.28 second.

Figure 6 shows a plot of maximum explosion pressure values in this set versus the igniter location (curve 2), whence it follows that while approaching the window the maximum pressure tends to reduction: in our case – from 0.1 barg to 0.006 barg.

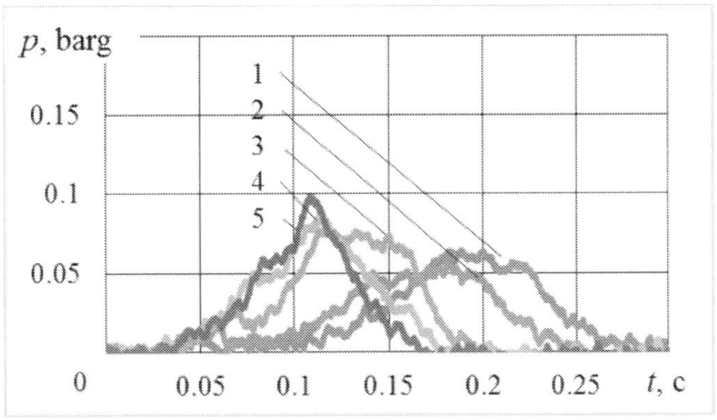

Fig. 4. Pressure trends for the single chamber explosion with different igniter's positions.

Tests with explosions in two chambers. Diagrams of the pressure development for explosions in two chambers at different igniter's positions are shown in Figure 5, and the dependence of the maximum pressures on the igniter's locations – in Figure 6. In this case, the obtained dependence of the explosion pressure against the igniter's location appears somewhat more complicated than in case of the single chamber explosion. When the igniter is placed near the passage to the second chamber (positions 4 and 5), the explosion pressure is even a little less than in case of the single chamber explosion. However, when the igniter is placed in the center of the chamber an near the vent window, the pressure is significantly higher than the single chamber explosion, especially in the second position. The following explanation of this phenomenon is possible. With the igniter installed near the passage, when the sheet is removed, already burnt gases get into the second room, i.e. these combustion products, an inert gas in this case, fill the volume of the second chamber. In this case, the second chamber serves as s damper.

When the igniter is moved deeper into the chamber, during the explosion the unburned mixture first enters the second chamber, becoming turbulent in the passage, and when it is ignited, it is burning across a developed area of the flame front. The further the igniter is moved from the passage, the more unburned mixture enters the second chamber.

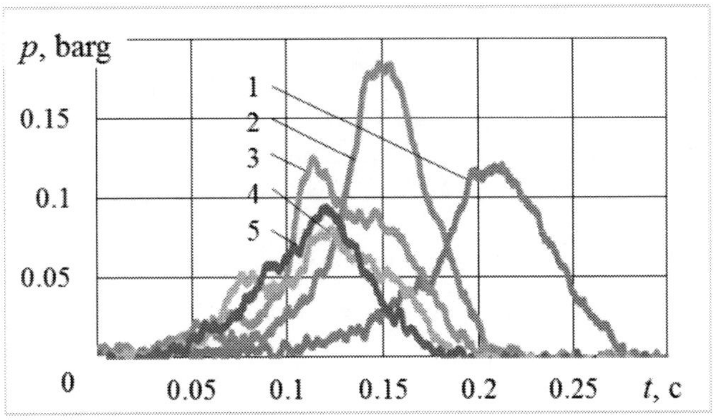

Fig. 5. Pressure trends for explosions in two chambers.

Finally, when the igniter is placed near the vent window, combustion products are discharged through the window, and they move with a high speed like high-temperature gases. Besides that, a part of the flame front is also released through the window into the atmosphere, which also contributes to explosion pressure reduction.

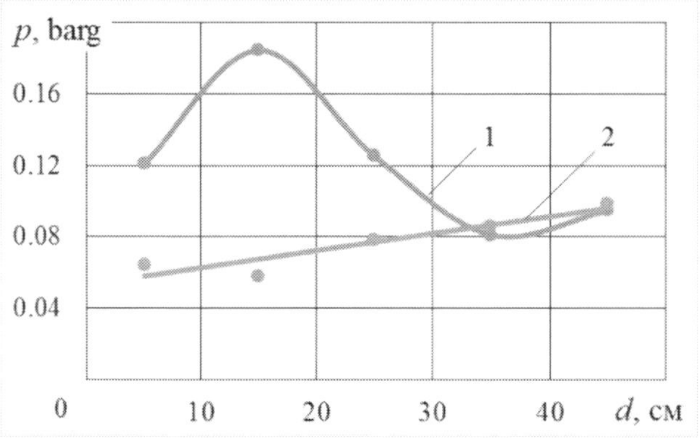

Fig.6. Dependence of maximum pressure values on the igniter's position 1 – with the adjacent chamber; 2 – for the single chamber.

To summarize the research results, we can review the dependence of the maximum explosion pressures on the igniter's location in two cases: first – in the single chamber, second – in the chamber connected with the adjacent chamber. In our case, the relative distance L/L_0 is in x-direction, where L is the distance from the vent window to the ignition location; L_0 – the chamber length. The dependence $p_2/p_1=f(L/L_0)$ is shown in Figure 9. It is understood that the suggested dimensionless values do not fully meet the criteria, nonetheless they enable certain comparison of results obtained under different conditions.

Relation with earlier research results. It is interesting to refer to our earlier results [4], according to which a gas explosion in a single room with the ignition location at the farthest wall is much more dangerous than an explosion near the vent window. This refers to a single room with a window. It is clear that the data provided in this article contradict this statement and make controversial the selection of the gas stove location in a kitchen, which surely has an adjacent room, for example, a hallway.

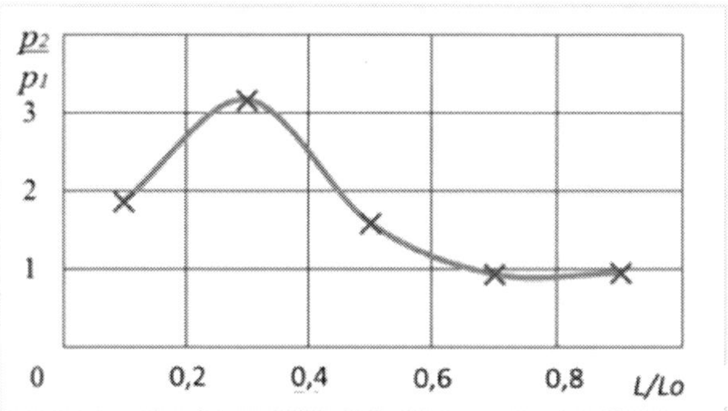

Fig.7. Comparison of maximum pressures of explosions with the adjacent chamber p_2 and without it p_1 at L/L_0.

6 Summary

It was found out that the previously discovered effect of the gas explosion pressure increase in a chamber with a vent window, when it is connected with a gas-free chamber, is highly influenced by the igniter's position in the first chamber. It turns out that this influence is not so unambiguous as stated in certain publications. Thus, with the gas is ignited near the passage to the adjacent chamber, the influence on the maximum explosion pressure is practically absent and even lower than in case of a single-room explosion. In this case, the adjacent room serves as an explosion damper. The influence manifests itself, as the igniter is mover to the center of the chamber and especially within the area between the center and the vent window, where it reaches a certain degree. If ignition occurs in close vicinity to the vent window, the maximum explosion pressure is less than in case of ignition near the center.

References

1. A.V. Mishuyev, V.V. Kazyonnov, A.A. Komarov, N.V. Gromov, A.V. Lukyanov, Fire and Gas Safety, **21** (2012)
2. Yu.Kh. Polandov, V.A. Babankov, S.A. Dobrikov, Fire and Gas Safety, **1** (2016)
3. Yu.Kh. Polandov, S.A. Vlasenko, M.A. Barg, Fire and Gas Safety, **3** (2007)
4. Yu.Kh. Polandov, V.A. Babankov, Zh. Fire and Gas Safety, **3** (2014)
5. C. R. Bauwens, J. Chaffee, S. Dorofeev, Combustion Science and Technology, **182(11)** (2010)
6. G.G. Orlov, D.A. Korolchenko, A.Ya. Korolchenko, Fire and Explosion Safety, **24** (2015)
7. G.G.Orlov , D.A. Korolchenko, Fire and Explosion Safety, **24** (2015)

The role of a technical customer in organization of investment and construction activity at the stage of construction oversight

Sergey Novikov[1*]

[1]Postgraduate, Department of Construction Technology and Organization, 129337, Moscow, Russian Federation

Abstract. This document considers the main tasks promoting development of construction as a branch and defines the correlation of efficiency of using funds with the activity of the technical customer as a guarantor of completing the design and construction works with proper quality, within established periods and planned financing of the facility construction. The functions of the technical customer and developer are defined from the point of view of the law as well as reasonability of their simultaneous work in the form of various legal entities in organization of design and survey works. Goals and objectives of the construction oversight are determined in accordance with regulatory documents, on the basis of which the package of functional procedures of the construction control is identified. Organization and management model of the technical customer's activity at the construction oversight stage is developed in the form of network and linear diagrams aiding in establishing the conditions of the construction oversight function execution with account of assuring the quality of the finished product.

1 Development of the construction sector of the country

The construction sector is assigned one of the leading roles in economies of the developed and developing countries, where the strategic goals of development are determined by increasing the citizens' life quality level [1].

1.1 Development of construction as a branch

Construction as a branch develops within the modern state policy based on formation of a high level of safety and comfort of the living environment supported with high standards of living, efficient financial and economic, legal, technical, and organization and process mechanisms [2].

Creation of conditions for advanced, innovative and effective modernization of the construction sector underlies the efforts on improvement of programs for social and economic development, promotion of national security and spatial development of the

* Corresponding author: novicov.sergey@gmail.com

country. This solves the problems of increasing the level of performance in the sector, increases the volume of construction works along with raising the level of the investment attractiveness, as well as the task of decreasing the building stock depreciation level, saving the budget funds when implementing the projects, and increasing the level of affordability and comfort of residence for the citizens.

Such challenges are the primary criteria for assessment of the state of the art of the sector, while the decisive influence is exerted by the substantiation of the adopted organization and process, economy and management solutions [3].

1.2 Effectiveness of using budget funds in construction

In the light of the past global economic crisis 2008-2009 and other known causes of slowing down of economy growth rates in the developing countries, the Governments of the countries of the Eurasian Economic Union and the European Union consistently take decisions concerning strategic planning, management improvement, enhancement of the efficiency of budget expenditure and state procurement, improvement of conditions of carrying out business activity [4, 5]. In this case, the pursuing of optimal budget policy in implementation of government-owned investment projects is determined by the work on search of the most rational economically justified solutions, including the construction sector.

2 Role of the technical customer at the stage of construction oversight

Economic efficiency of using funds in construction of buildings and structures begins from an effective, proper design, which is based on modern, cutting-edge, advanced equipment, streamlined organization and process solutions of construction operations [6].

Obviously, the efficiency of such multifaceted and multi-task work depends on coordinated work of all participants of the investment and construction activity (investors, developers, designers, contractors, operating organizations, etc.). However, the availability of cost items associated with fulfillment of the technical customer's functions in the design and estimate documentation is a guarantee of manufacture of construction products with appropriate quality and in established periods [7].

2.1 Functions of the technical customer and developer from the point of view of the law

The requirements of regulatory legal documents in the construction activity define the technical customer as the main participant of the investment and construction process [8]. At the same time, a number of regulatory documents does not define the functional responsibilities of the technical customer and the developer in full, and reasonability of their simultaneous work as different legal entities in organization of design and survey and construction works has not been completely uncovered yet [9].

2.1.1 The function of the developer

In the Russian Federation, the developer is a physical or legal entity who performs the following on the land plot owned by it:

• Construction, reconstruction, overhauls of capital construction facilities.
• Engineering survey.

• Preparation of design documentation for their construction, reconstruction, overhauls.

Therefore, in order to hold the status of the developer, one needs to own a land plot on the basis of any right.

All developers are equalized in their rights regardless of the basis for acquiring the right to the land plot, however the responsibilities of developers differ depending on this right, and also on the type of the facility, location of the site, etc. execution of the developer's functions does not require permits of self-regulated organizations, however the expenditures are comprised of the costs for site acquisition and maintenance [8].

Thus, we can draw a conclusion that allocation of a separate legal entity would be reasonable for the function of the developer for implementation of the investment project from the point of view of risk minimization.

2.1.2 The function of the technical customer

In Russia, the functions of the technical customer are defined with the minimal list of authorities and responsibilities for such physical entity acting on the professional basis, or a legal entity, which are authorized by the developer and on behalf of the developer have the right to:

1. Conclude agreements for performance of engineering survey, preparation of design documentation, construction, reconstruction, overhauls of capital construction facilities.
2. Prepare assignments for performance of the specified types of works,
3. Provide entities who perform engineering survey and (or) preparation of design documentation, construction, reconstruction, overhaul of capital construction facilities with materials and documents necessary for performance of the specified types of works,
4. Approve design documentation,
5. Sign documents required for obtaining permission for commissioning of the capital construction facility,
6. Fulfill other intended functions.

The developer is entitled to perform the functions of the technical customer on its own, which follows from definition of the technical customer and is supported by the practice [8]. However, in this case, it will need to obtain a permit corresponding to the minimal list of works on the basis of an order, which incurs substantial costs associated with the need to comply with the minimal requirements, including the technically complex ones, highly dangerous and unique facilities [8, 10, 11].

Aside the requirements regarding the minimal number of employees with the profile education, periodic qualification upgrading, etc., each self-regulating organization (SRO) is entitled to set any additional requirements that do not limit competition, also for the material and technical part [9].

Costs for maintenance of the material and technical base (including costs for licensed software) depend on the company's profile. The cost of verifying the compliance with the ISO depends on the size of the company, on how formally and truly the corresponding certification lasting from several months to half year takes place, and amounts from USD 3,500–5,000 to USD 30,000–50,000, and the amount does not include deductions to the indemnification foundation, which for construction SRO depend on the volume of the contracts, under which a certain organization is going to work, and also on availability or unavailability of insurance, which ranges from USD 5,000 to USD 500,000 per one company [12, 13, 14, 17]. The experience of construction companies shows that optimization of these costs is possible during creation of a technical customer in the form of a separate compact dedicated company.

The responsibility of the technical customer to perform construction oversight is supported by the requirements, in accordance with which the contractor that has improperly performed works may not allege that the technical customer performed oversight and supervision of their execution, except for cases, when obligation to perform such oversight and supervision is placed on the technical customer by the law (8). Therefore, ignoring the obligations on oversight and supervision of the construction process by the technical customer can entail negative consequences for it, which points to the need in dividing the functions of the customer and the technical customer as a means for eliminating the risks of the customer's responsibility for improper supervision and oversight of work execution.

2.2 Goals and objectives of construction oversight

In accordance with the regulatory documents, the goal of construction oversight during construction of capital construction facilities is provision of high quality of construction and compliance with technical specifications of the customer of the building or structure under construction, and performance of the following activities by the customer:
• Checking the fullness and meeting the established terms of incoming inspection and accuracy of documenting its results by the contractor.
• Inspection of control activities performed by the contractor for meeting the rules of warehousing and storage of products used, and accuracy of documenting the results.
• Checking the fullness and meeting of the established terms of control of order and scope of process operations on construction of capital construction facilities and accuracy of documenting its results by the contractor.
• Certification of concealed works and the provisional acceptance of erected building structures influencing the safety of the capital construction facility, sections of utility networks together with the contractor.
• Checking the compliance of the completed construction facilities with the requirements of design and working documentation prepared on its basis with the results of engineering survey, requirements of the land plot development plan, technical regulations together with the contractor.
• Other activities aimed at performance of construction oversight provided by the law and (or) concluded agreement [15].

Therefore, for the technical customer, the stage of construction oversight starts immediately from notification of the construction supervision authorities and local government authorities about the start of the construction and continues till it is finished. Besides, the period of construction is determined by the plan of construction organization (PCO) [16].

3 Organization and management model of technical customer's activity at the stage of construction oversight

3.1 Development of the organization and management model

3.1.1 Package of functional procedures for construction oversight

It is appropriate to pay special attention to functional procedures in the activity of the technical customer when carrying out the stage of construction oversight of construction progress, which can be presented as an organization and management model including a package of functional procedures with indication of their average duration (Figure 1):

• Notification of construction supervision authority of the construction start (formulae 1)

$$t = 0.01t_n \tag{1}$$

Where t — mean time necessary for mastering the type of work under consideration at this stage, $t \in [0,\ t_n]$, where t_n – the mean type required for technical supervision of construction, with account of combination of these functions and the construction period:

- Approval of work performance schedule, $t = 0.02t_n$,
- Control of meeting the work performance schedule, $t \in [0.02t_n,\ t_n]$,
- Approval of the list of persons performing technical supervision, $t = 0.02t_n$,
- Definition of a procedure for maintenance of technical and as-built documentation,
- $t = 0.03t_n$,
- Oversight and technical supervision of construction, correspondence of scope, cost and quality of the works with the designs, cost estimates and contract prices, $t \in [0.03t_n,\ t_n]$,
- Agreement of the list of planned material suppliers, $t = 0.02t_n$,
- Inspection of licenses and certificates of contractors and suppliers of materials, $t = 0.03t_n$,
- Acceptance, accounting, storage, pre-installation revision and transfer of equipment and other material and technical resource for installation and performance of works, $t = 0.03t_n$,
- Agreement with the relative companies about a procedure for installation, testing and registration of process lifting mechanisms and equipment, $t = 0.03t_n$,
- Certification of concealed works and provisional acceptance of critical structures,
- $t \in [0.01t_n,\ t_n]$,
- Conclusion of agreements and control over performance of erection supervision and commissioning operations, $t = 0.02t_n$.

Figure 2 shows linearization of the organization and management model of the technical customer's activity during technical supervision.

It is worth noting that the function of the technical customer at the stage in question includes acceptance of claims for quality from consumers and presentation of claims to contractors (suppliers) in accordance with the law and guarantee obligations (guarantee terms are specified in the agreement).

Pursuant to order the solution on temporary shutdown of the facility (with exception of state-owned facilities) and source of funds for covering the costs associated with temporary shutdown of the facility shall be taken by the developer (technical customer). Consequently, temporary shutdown of the facility is also a part of the functional procedures at the stage of technical supervision performed by the technical customer's department.

When taking a decision about suspension of construction and temporary shutdown of the facility, representatives of the technical customer shall send a notification to the state construction supervision authorities and local government authorities. In this case, the technical customer shall bring the facility and the territory used for its erection to the condition ensuring strength, stability and safe-keeping of the structures, equipment and materials, and also safety of the facility and the construction site for population and the environment [18, 19].

3.1.2 Building of the model in the form of network and linear diagrams

Organization and process model of the activity of the technical customer during construction oversight is design in the form of the network and linear diagrams are represented at the figure 1, 2.

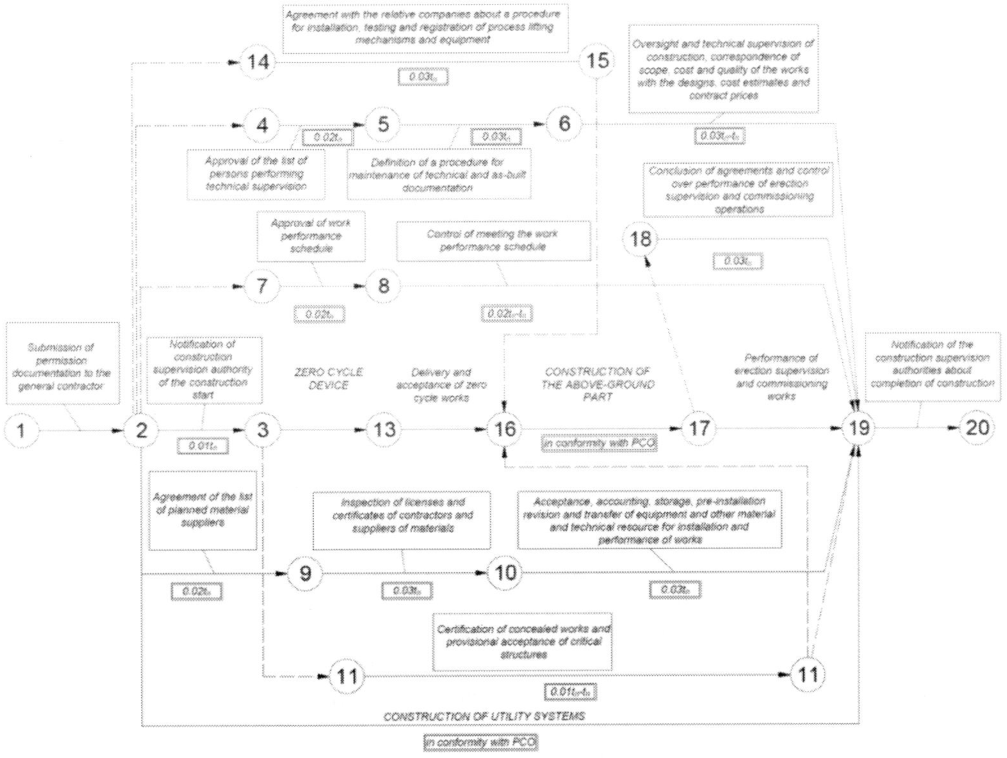

Fig. 1. The organizational managerial model of the technical customer's activity during construction oversight in the form of the network diagram.

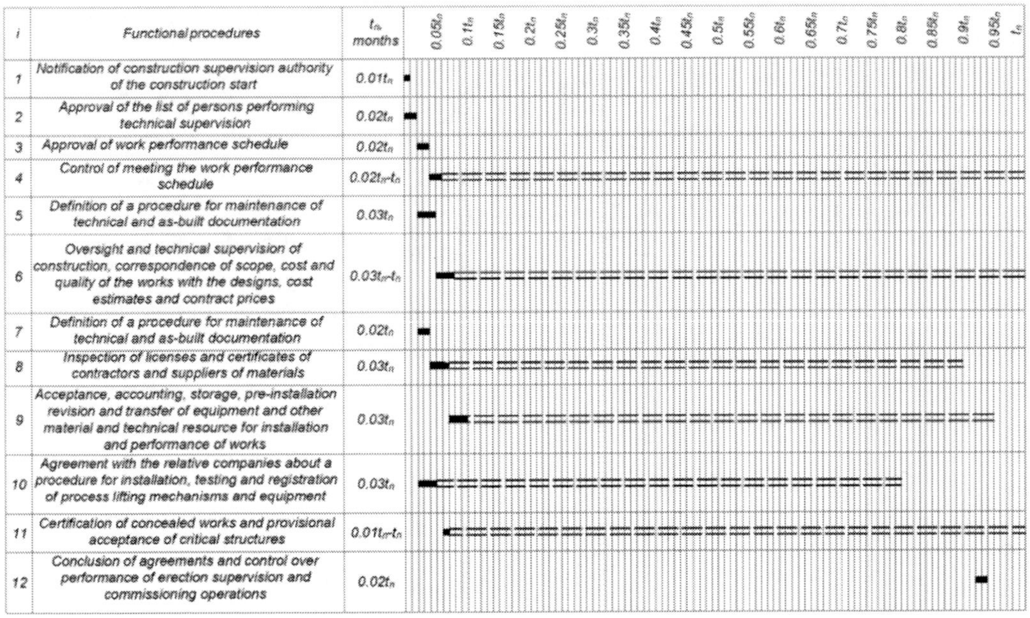

Fig. 2. The organizational managerial model of the technical customer's activity during construction oversight in the form of the linear diagram.

4 Conclusions

The completed study shows that:

1. In selection and justification of rational organization structures and methods for oversight in construction, it seems reasonable that the developer and technical customer, whose functions differ, were different legal entities. The expenditures for support of the activity of two companies separately performing the functions of the developer and the technical customer, as well as possible risks, prove to be lower than those for performance of the activity of the developer and the technical customer by the same entity.
2. Modeling of the technical customer's activity at the stage of construction oversight will allow shortening the erecting time of a capital construction facility and assuring the quality of the finished products in the established periods and within the planned financing.

References

1. M. Legenkova, IJEFI, **6** (2016)
2. S. Bellanca, J. Vandernoot, JMAA, **3** (2014)
3. The Russian Government, Federal law, **72-FL** (2014)
4. K.-D. Borchardt, The ABC of European Union law, **1** (2010)
5. The Russian Government, Order, **2227-o** (2011)
6. S. Novikov, B. Zhadanovsky, Science Review, **21** (2015)
7. B. Zhadanovsky, S. Sinenko, Science Review, **9** (2014)
8. The State Duma, The Town-planning code of the Russian Federation, **190-FL** (2016)
9. The Health and Social Development Ministry, Order, **188** (2008)
10. The Ministry for Regional Development of the Russian Federation, Order, **624** (2009)
11. The Russian Government, Order, **207** (2011)
12. W. Woodhead, M. Rahilly, Computers, Environment and Urban Systems, **10** (1986)
13. The International Organization for Standardization, International standard, ISO 9000 (2005)
14. The International Organization for Standardization, International standard, ISO 14040 (2006)
15. The Russian Government, Order, **468** (2010)
16. A. Volkov, V. Chulkov, R. Kazaryan, M. Fachratov, O. Kyzina, Applied Mechanics and Materials, **580-583** (2014)
17. S. Novikov, A. Sapukhin, Vestnik MGSU, **7** (2013)
18. The Russian Government, Order, **802** (2011)
19. P. Oleynik, T. Kuzmina, Engineering-technical and scientific-educational journal, **11** (2012)

Surface thermodynamic parameters of modified wood

Elena Pokrovskaya[1], *Fedor* Portnov[1*], *Andrey* Dolganov[2]

[1]Moscow State University of Civil Engineering, 129337, 26, Yaroslavskoye Shosse, Moscow, Russian Federation
[2] Zhilekspertiza, 127055, ul. Yablochkov, 21 building 3, Moscow, Russia

Abstract. Energy characteristics of modified wood are studied in the paper. Application of this approach during the study of wooden materials allows forecasting the efficiency of modifiers for surface layer of wood. Phosphites, the efficient fire-retarders, were applied as modifiers. Using the example of a number of ethers with various alkoxy substituents of phosphorus atom, we have made an attempt to associate surface thermodynamic properties of modified wood and formation of properties for fire-, bio- and smoke protection. The dependence of change of energy characteristics and surface structure of wood on the nature of modifiers is determined. To study energy characteristics of wood, modified by various compounds, the following characteristics were used: σ surface tension and ΔG free enthalpy gradient. Easy Drop setting and the corresponding software were used to determine these values. According to the obtained data, the conclusion is made about the influence of modifiers on energy characteristics of wood. The high degree of modification (% P) causes bigger change of Gibbs energy, which determines formation of high-level fire-, bio- and smoke protection. Diethyl phosphite is the most efficient modifier. Formation of fire-protective properties stipulates long-term operation of wood and wood-based materials.

1 Introduction

Wood is a renewable material of high strength and aesthetic qualities. Wood constructions (WC), laminated wood constructions (LWC) and board materials are used as construction materials.

Wood has platelet structure with hydrophilic surface. Hydrolytic processes develop at 18-20 % moisture content of wood, leading to its destruction. Hydrolysis processes accelerate significantly in the presence of wood-destroying fungi, biocorrosion develops [1].

The main wood components are cellulose and lignin - combustible organic polymers. Burning quality and low biostability of wood reduce its operational properties [15, 16].

* Corresponding author: romeo-aleks@yandex.ru

Modification of the materials, including wood, allows enhancing its operational properties, durability of wood and wood constructions. Studying these processes is a task of urgent character.

The papers [2-6] show that wood can obtain fire- and bio-protection and hydrophobic properties due to surface chemical modification by organophosphorous compounds.

Surface modification by organophosphorous compounds proceeds in 1-2 mm surface layer of wood. At the same time, wood acquires enhanced operational and fire-protective properties. Efficient modifiers should react with OH-group supporter of lignin-carbohydrate complex components; the degree of reaction cannot be above 0.8-1.5 % of chemically combined phosphorus [13, 14].

Determining factors that allow increasing modification process is an urgent task.

Surface modification depends on surface energy characteristics; the following thermodynamic parameters are suggested as quantitative characteristics for surface properties: ultimate surface tension and free enthalpy gradient. The factor of surface energy degree is surface tension [7]. The calculation of free enthalpy gradient of modified surface has been carried out on the basis of joint laws of thermodynamics I and II [8-9].

For the calculation of free surface energy for systems with constant chemical composition the joint equation of laws of thermodynamics I and II is the following:

$$dG = -SdT + Vdp + \sigma dS_{yd} + \mu dn \tag{1}$$

Where G is Gibbs energy, S is entropy of the system, T is temperature, V is volume, p is pressure, σ is surface tension (work of surface formation per unit), Sud is specific surface, μ is chemical potential of the matter, composing condensed phase, n is the number of moles.

Under the free enthalpy conditions and at fixed amount of matter the equation is the following:

$$dG = \sigma dS_{yd} \tag{2}$$

The surface tension of a solid body measures the inner energy of the whole system, reduced to the unit of material surface. As the free enthalpy gradient is a function of the system state, its total differential is as follows:

$$\Delta G = \sigma dS_{yd} + S_{yd} d\sigma \tag{3}$$

On the basis of the given equations and published works we set a task to examine the changing energy parameters of the surface of wood after modification as well as examine relationship of the change ΔG with such wood properties like degree of modification, adsorption capacity, smoke generation ability, fire resistance.

The total improvement of these properties relative to the original wood can ensure reliability and durability of wood material and wooden structures.

Study of surface modification of wood provides an opportunity to assess in advance effectiveness of surface layer modifiers using thermodynamic quantities ($\Delta G, \sigma$).

2 Experimental part

Determination of thermodynamic characteristics was conducted on pine sapwood samples with the size of 40x40x10 mm, humidity of samples was 8%. 20% solutions of phosphorous acid esters were used as modifiers (dimethylphosphite (DMF), diethylphosphite (DEP), dipropylphosphite (DPF), dibutylphosphite (DBP)). Increase of wood fire resistance after modification of the listed modifiers is noted in the literature.

The goal of the work was to identify changes in Gibbs thermodynamic potential of modified wood surface depending on the modifiers of the surface layer, as well as in defining link between energy characteristics and physical-chemical properties of the modified samples.

Modifiers were applied to the surface of samples in two layers, at temperatures 20-25 oC, modifier consumption was 300 g/m2. After modification the samples were put into Soxhlet apparatus. The samples were dried in a drying cabinet at temperature 50 oC for 6 hours, after that the samples were put outside and dried at temperature 25 oC for 7 days.

Surface tension σ was determined by neutral drop method [10,11] on EasyDrop station and related software. Values of Sud of wood samples were obtained on NOVA 2200 station by nitrogen sorption.

ΔG calculation was conducted as follows. On the EasyDrop station the dependence cosθ=f(σzh) of contact angle from standard surface tension of a drop of aqueous ethanol solutions of different concentrations in the form of average straight line (Figure 1).

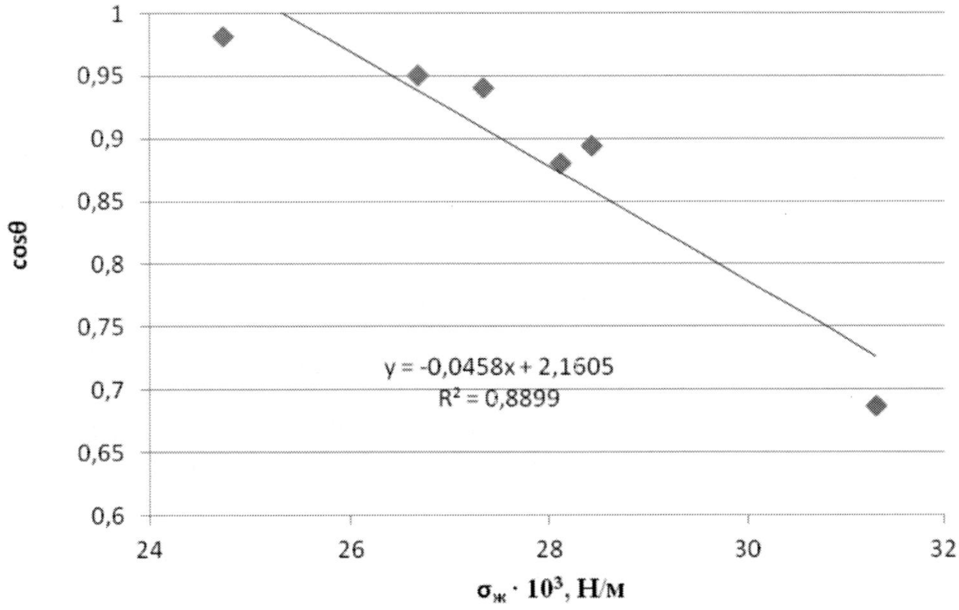

Fig. 1. Diagram of dependence cosθ = f (σzh) for wood surface modified with DMF, where y =-0.0458x+2.1605 - is an equation of a straight line, R2-squared value.

Extrapolating the equation of a straight line up to 1, you can determine the critical value of surface tension σ, which is a characteristic of the surface energy of the surface unit. According to the obtained data Sud and σ the calculation of the free enthalpy gradient is made.

To assess operational characteristics of modified wood data has been obtained on smoke production ability (GOST 12.1.044-89 p. 4.18), fire protection (GOST R 53292-2009, GOST 12.1.044 - 89 p. 4.3, GOST 12.1.044-89 p. 4.19), percentage of phosphorus in a surface level of modified wood (element analysis method), water vapor sorption [12]

The obtained experimental data is collected in Table 1.

Table 1. Summary experimental data.

	Surface modifier					
	DMF	**DEP**	**DFT**	**DBP**	**DFF**	**PFA-1**
%P before thermal decomposition	2.87	3.57	2.6	2.65	2.45	2.11
%P after thermal decomposition	0.85	4.7	0.63	0.71	1.06	0.35
ΔG, J before thermal decomposition	-12.68	-14.5	-4.94	-6.97	-10.16	-2.71
ΔG, J after thermal decomposition	14.31	-14.48	4.05	2.43	-13.69	12.35
$\sigma_k \cdot 10^3$, n/m before thermal decomposition	19.91	25.03	21.97	22.36	23.03	20.05
$\sigma_k \cdot 10^3$, n/m after thermal decomposition	19.07	24.06	116.99	17.36	19.07	22.24
a_m, % before thermal decomposition	155.5	215.8	221.9	193.4	163.52	188.6
Dm, m2/kg	450-470	130-140	640-680	600-640	740-780	400-440
Δm%	6.9	8.5	12.3	11.5	9.5	8.7

As it is shown in Table 1, hydrophilic properties decrease at surface modification by phosphites. The nature of modifier influences hydrophilic properties. The degree of chemical modification of wood also depends on the modifiers nature: the greatest in cases when DMF and DEP are used. Chemical modification is an ester exchange reaction on the hydroxyl groups of cellulose and lignin:

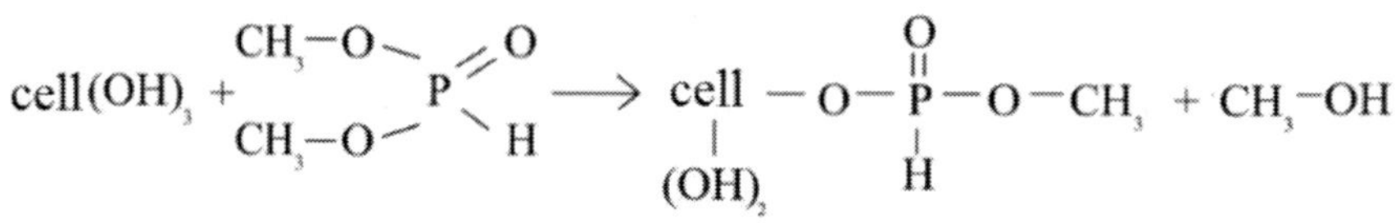

Due to chemical reaction modifier thickens the capillary walls, reducing Sud and water adsorption. Decreased hydrophilic properties result in increase of biostability. Basing on data for loss of mass during thermal decomposition it has been found that DMF and DEF provide the group I of fire-resistance rating.

The change of wood properties at surface modification is determined by surface energy characteristics [7]. The formation of stable chemical bonds, characterized by the most negative values of free enthalpy gradient predetermines the long-term protective effect of modifiers (fire-, bioprotection, ageing resistance, decreased water adsorption, smoke resistance).

References

1. E. Pokrovskaya, Y. Kovalchuk, Biological corrosion, **212** (2013)
2. E. Pokrovskaya, Wooden architecture, **136** (2009)
3. E. Pokrovskaya, Chemico-physical principles,**104** (2003)
4. E. Pokrovskaya, F. Portnov, A. Kobelev, D. Korolchenko, FES, **10** (2013)
5. A. Stenin, CFLE Conference proceedings, **553-559** (2013)
6. I. Koteneva, CFLE Conference proceedings, **191** (2011)
7. V. Rodulgin, Physico-Chemistry of surfaces, **121-127** (2011)
8. A.Tutygin, A.Shinkaruk, A.Aisenstadt, V.Lesovik, Ecological risks, **8** (2014)
9. A. Eisenstadt, IMTCECC, **244** (2014)
10. B.Deryagin, I.Abrikosva, E.Lifshitz, PSSS, **494-526** (1958)
11. B.Deryagin, N.Churaev, Wetting film (1984)
12. N.Keltsev, Fundamentals of adsorption technology, **592** (1984)
13. I.V. Stepina, V.I. Sidorov, Applied Mechanics and Materials, **584-586** (2014)
14. I.V. Stepina, V.I. Sidorov, O.A. Klyachenkova, Applied Mechanics and Materials, **584-586** (2014)
15. I.V. Stepina, O.A. Klyachenkova, Procedia Engineering, **91** (2014)
16. L. Grigorieva, P. Oleinik, Solid State Phenomena, **871** (2016)

Environmental and economic estimation of negative impact of waterproofing works and materials on environment and ability to live of the person

Evgeniy Velichko[1,*], *Edward* Tshovrebov[2] and *Vyacheslav* Semenov[1]

[1] Moscow State University of Civil Engineering, Yaroslavskoye Chaussee, 26, Moscow, 129337, Russia
[2]FBU «The Research Centre on problems of management of savings of resources and a waste», Olympic prospectus, 42, Mytishchi, Moscow region, 142000, Russian Federation

Abstract. In article negative influence of waterproofing materials and works on health of people and environmental pollution is analyzed. Complex measures on maintenance of an ecological security, sanitary-and-hygienic requirements, size standards and rules both during works on a waterproof finish are resulted, and at the subsequent upkeep of buildings and premises, and also action for decrease in negative impact at carrying out of waterproofing works on a natural habitat condition, health of the workers occupied in repair-building manufacture and living. The estimation of components of the environmental and economic damage put to environment by manufacture of waterproofing works is given.

1 Introduction

Questions of protection of a natural habitat and health of citizens both in the course of realization repair and civil work, and at the further exploitation of premises, buildings, structures are functionally connected with level of ecological and sanitary-and-hygienic safety of applied building production at all stages of its reversion.

At carrying out of the waterproofing works providing protection of land and underground building designs (pipelines, the bases, basements and so forth) from impact of water and liquid excited environments, a number of ecologically unsafe materials on the basis of polymers, bitumen imbibition's and the tars, environment causing a significant damage and to health of people is applied. By production of such works the significant amount of the waste demanding processing in the secondary production or salvaging is formed. All these ecologically hazardous processes have negative impact on a condition of environment, natural resources and health of citizens.

* Corresponding author: pct44@yandex.ru

2 Problem statement

Till now in regions of Russian Federation the problem of assembly, processing of a hazardous waste of waterproofing works properly is not solved: bitumen, bitumen mastics, epoxy mastics and other sorts of mastics, the packaging materials, the ground polluted by bitumen, the remnants of the rolled isolation, the fulfilled slag, the asbestos chip, polymeric products, forming an essential ecological problem as a whole for branch [15, 16].

Analyzing an actual problem as a whole, it is necessary to notice, that a complex estimation of influence of various types of a waterproof finish on a condition of natural objects, questions of materials recovery of insulation works, estimations of the damage put by them to environment in modern scientific researches to the full it is not considered [17-32].

Considering absence of such researches in the domestic and foreign literature, and also the urgency and the importance of an existing problem for ecological security maintenance at carrying out repair and civil work, in article are stated results of scientific researches on formation of a complex estimation of negative impact of waterproofing works and materials on environment, health of people with an estimation of a put ecological, economic damage.

3 Results and discussion

Waterproofing products from the point of view of negative impact on health of people and a circumambient are expedient for sizing up in a slit as modular condition: liquid (the impregnating, jet, priming, film-forming); put compacted (plastic insulating concrete and solutions); is plastic-viscous (coating mastics and doughs); glued (on mastics and colloidal solutions); the finishing; elastic-viscous and solid (film, the roll, sheet, bulk), and on sorts of applied materials for various types of isolation: bulk, jet, the impregnating, from sheets, mastics and paints, solutions and hydrophobic powders, metal and polymeric sheets.

With a view of pressure seal of junctures and joints of precast iron concrete designs in modern construction use plastic (bituminous, is bituminous-polymeric, etc. mastics on the basis of bitumen), flexible (the roll strips, tapes, sheets), elastic (profile rubber materials), isolation with single-component (on the basis of polyurethane foam) or the two-component (is bituminous-polymeric) joint sealants, the combined waterproof finish.

With a view of isolation of underground designs, except them, can follow-up be applied is bituminous-latex, it is bituminous-emulsion types and a bitumen felt, and at isolation of surface constructions of a building – all sorts epoxy, bituminous, concrete, cement, glued, the roll waterproof finish [4, 6, 7, 9, 10].

Ecologically hazardous waterproofing coatings from polymer cement and polymeric concrete are introduced. Presence in their composition, in the capacity of furan, polyester and phenolic formaldehyde resins result to allocation in a free air of toxic organic compounds: phenol, methanol, formaldehyde, ammonia, a phthalic anhydride, ethylen glycol, diethylene glycol, styrene and ethyl benzene. In events of application improving properties of concrete of additives (for example, polyvinyl acetate) air medium fouls spikes of acetone, ethyl acetate, benzol, dibutyl phthalate, dioctylphthalate, vinyl acetate, acetic aldehyde. Excess daily average and a work area of maximum concentration limit of these hazardous contaminants in air medium can call negative aftereffects for a human body, in particular: vinyl acetate - irritation mucous or a burn of a cornea of eyes; acetic aldehyde - irritation of mucous membranes, an inflammation and a hypostasis of lungs; acetone - the biological accumulation, leading to chronic poisoning, intoxication [1, 2, 8].

Various sorts of bituminous roll materials in the capacity of applied now izol glass, an izol foil is necessary to evolve a bitumen felt, roofing felt, izol, hydro izol, hydro. As a part of each meaned material there are certain sorts of bitumen's and the bitumastic's which are a source of negative impact on ingredients of a circumambient.

For imbibition of elements of designs of buildings (the bases, slabs, pipes, girders) with a view of maintenance of rust resistance, water tightness, frost resistance of designs in the capacity of the impregnating hydraulic insulating material (Standard of Russian Federation «Mastic the is bituminous-rubber insulating») is used hot, liquid, or heavy asphalt in the form of hot imbibition, and also bitumen emulsions (the maintenance of bitumen more than 50 %) with application of emulsifying agents (1-3 %) and inoculating additives: latex, rubber, polymers. At manufacturing of such hydraulic insulating materials integumentary, impregnating bitumens are applied; for other purposes (mastic, dough) - bitumens of type BN. In the capacity of disperse filling materials for mastics materials of various extent of toxicity are used: a cement kiln dust, ground slag, ash, fibrous fillers - the disintegrated glass-fibre, a mineral wool, asbestos. Roofing and waterproofing mastics (bituminous, is bituminous-polymeric, is bituminous-rubber, etc.) Within the limits of Standart of Russian 30693-2000 «Mastics roofing and waterproofing» make, as a rule, on the site in the open air, mixing in batch agitators bituminous dough, an emulsion with disperse or fibrous filler to the consistencies defined by standards [3, 4].

At the analysis of hazardous impact of oil asphalt with filling materials, it is necessary to take into consideration physical and chemical composition of the oil building bitumen consisting from hazardous to a circumambient and health of people methanoic, naften, aromatic hydrocarbons with an extraneous matter of joints of oxygen, nitrogen, sulphur.

Grading of harmful production factors at works on a warming up of oil asphalts and bitumastics is introduced on Fig. 1. These ecologically hazardous factors simultaneously are a source of negative impact on natural habitat ingredients (a free air, soil, vegetative and fauna, aqueous installations).

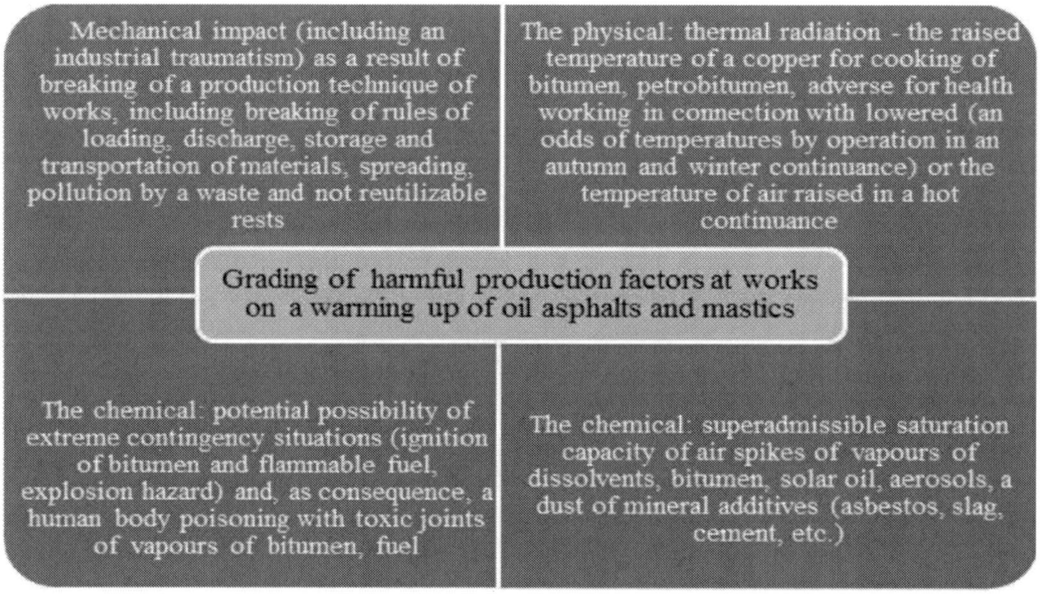

Fig. 1. Grading of harmful production factors at works on a warming up of oil asphalts and mastics.

Bituminous hydraulic insulating materials are capable to evolve a number of toxic joints, thus intensity in an aerosphere and contamination hazard of air sharply increases at their warming up. Bitumen and a road tar can call dermatitises, hyperplasias of a false skin and a leather, a papilloma. Pairs melted bitumen possess irritating act on a leather, mucosas of eyes and the upper respiratory tracts. Pairs mineral hydrocarbon oil introduce health hazard of the person, calling changes in arterial pressure and making negative impact on the central nervous system at breaking of admissible size standards [1, 2].

In places of production of such works the supervisory control of observance of maximum concentration limit daily average and a work area on the toxic joints formed at a warming up of bitumen, fuel burning in the equipments of fusion of bitumen and burners is obligatory: to white spirit, kerosene, alcohol, white damps, nitrogen, to sulphur dioxide; preparation of bituminously-butyl rubber mastic: to benzine fuel (slate), to xylene, toluene; is bituminous-latex mastic: to phenolum, a dust containing asbestos. It is bound to hazardous impact of toxic substances on health of the person. Mineral fillers (finely dispersed cindery, lime-dolomitic), organic solvents have negative impact on breath organs, sight, integuments, a gastroenteric section [1, 2, 8].

With a view of maintenance of an ecological security of work on a bitumen warming up are made on expressly taken away platforms at operating crew minimisation, on a safe passing distance from places of an agglomeration of people on the site or behind its limits (provided that concentration of vapours of contaminants in these working areas does not exceed admissible size standards), in strict observance of sanitary-and-hygienic demands of environment protection and health of the person. Places of a warming up of mastics and the bitumens, placed on expressly taken away platforms, range on distance: from buildings and structures IV-V of extent of flame-resistance - not less than on 30м; III item-20 of m; I and II item-10 of m [11, 12]. Working with bitumens and bitumastics in accordance with Standard of Russian Federation 12.3.040-86 "Works roofing and waterproofing» are provided with individual defence agents: filtrateing breathing masks, suits, shoe, mittens and a goggles.

Among other liquid insulants ecological hazard introduce: epoxy oil, epoxy-furan of mastic, an emulsion on the basis of styrene, phenolum, a formic aldehyde, epoxy, furfurol, urea-formaldehyde resins, oil (a tar asphalt solution in toluene) and perchlorovinyl lacquer from synthetic resins on toxic dissolvents - benzene chloride, butyl acetic aether, xylene, solvent. Except high-toxic styrene, phenolum, a formic aldehyde, such materials contain other ecologically hazardous joints: ethanol, butanol, ammonia, furfurol and benzene chloride. Other joints which are a part of epoxy resins are not less toxic also: diphenyl propane, ethylene diamine, diethanol amide, toxilic anhydride, epichlorohydrin, diamine. Having property in due course to collect in an organism and possessing irritating act, at inhalation, they can lead to a strong boring of mucosas of eyes and the upper respiratory tracts, and at hit in inside organism - to promote origination of a bronchitis, dermatitis's, a derangement of organs of digestion, breaking of nervous system [1, 2, 8].

Health hazard of the person sharply increases in need of drawing of viscous, fast-drying or badly shaded compositions of various waterproofing products by ways of air-operated and hydrodynamic spraying. For prevention of hit of vapours of toxic joints in respiratory system of the person, on integuments and mucosas, by production of waterproofing works presence of agents of an individual defense is obligatory.

Analyzing level of negative impact of hydraulic insulating materials on a circumambient and health of the person it is possible to note, that the basic ecological and hygienic safety materials from a sheet waterproof finish are the zinc-coated and aluminium sheets. In certain extent are harmless applied in the capacity of film materials has flown down a product cloth, the charging waterproof finish which is carried out by placement by beds on the compacted ground or a concrete bed in submersible hollowness's with casing use in case in the capacity of sealers hydrophobic powders from sand or few toxic materials are used: asphalt izol, hydrophobic ash, perlite.

Recently toxic bituminous and oil insulants are in large quantities substituted less hazardous organic polymeric (silicon-organic, methyl acrylate, organic silicone) by emulsions, coating products on the basis of organic compounds: polyurethane, rubber-like, polymers of a type of butyl rubbers (mastic hydrobutyl), with a little toxic filling materials (a chalk, comminuted rubber) [5, 13].

Largely ecologically safety hydraulic insulating material of the class-room of the water-soluble silicates is the water glass in the form of solutions of silicates of sodium and a potassium in water. It is widely applied with a view of a waterproof finish of water resisting, heat-resisting and acid-proof concrete to various designs of walls, floors, floorings, basements, wells, the device of water areas. Acceleration of a dripping of concrete from formation of sodium aluminate and the depressed permeability of a pore space additives for watertight concrete reception have caused wide application of a water glass in the capacity of at carrying out of emergency operations - termination of leakages, junctures and a waterproof finish. The product is a pollution-free antiseptic agent, interfering with a mould growth, decay, funguses [14].

By results of the made analysis of the yielded problem it is possible to note, that, as a whole, negative impact of waterproofing works on a natural habitat, displayed in the form of migration of hazardous making hydraulic insulating materials in a free air, surface and underwaters, soil, depends, first, on a sort of an applied material, presence and the maintenance in it of toxic substances, secondly, from climatic, hydrological and other conditions of a working area of construction (temperature, activity of solar rays, chemical aggressiveness of ingredients of a natural habitat, their background dirtiness the various joints actively entering chemical reactions with formation of toxic substances) and, thirdly, from a site of the hydroisolated design of a construction (land, underground or subsea, depth of a contour interval, presence of contact with soil, subsoil waters, etc.).

Potentially possible environmental and economic injury put to a natural habitat as a result of production of waterproofing works *Unatural habitat* (roubles) will be summed up from following components:

$$Unh = Uwater + Uair + Uland + Usoil + Uvegetation + Ubioresources + Utrouble \qquad (1)$$

U water – the injury put to surface and underground water sources as a result of a translocation and migration of pollutants from a surface of hydraulic insulating materials in an aqueous medium;
U air – the injury put to air mediuim as a result of technological pollutant emissions and (or) burning of a toxic waste;
U land – the injury put to land resources as a result of mechanical impact, and also migration, carryover of pollutants from a surface of hydraulic insulating materials in land and underground beds;
U soil – the injury put to fertile bed of land – to a soil mantle as a result of technological pollution by chemicals, placement of a hazwaste of production and consumption;
U vegetation – injury to flora as a result of impact of pollutant emissions (e.g., a bitumen warming up), and also mechanical impact (a junk pile, a translocation of toxic extraneous matters and so forth);
U bioresources – the injury put to fauna and aqueous biological resources as a result of technological physical, mechanical, biological or chemical action of waterproofing works;
U trouble – the injury put to all ingredients of a natural habitat as a result of the emergency battery waste interceptions, spikes of hazardous substances, explosive and fire-dangerous situations with negative ecological effects.

In the capacity of the basic costs on compensation of deficits to a natural habitat expenses for recultivation of lands, potable water clearing, elimination of junk piles, restoration of trees, liquidation of aftereffects of extreme contingency situations (bitumen straits and so forth) are observed. Economic losses can be attributed to the irreversible loss of insulation products during the ablation of bulk materials in connection with violations of the conditions of storage, transportation, loading and unloading, process marriage, leading to the formation of significant quantities of construction waste.

4 Conclusions

Authors of article investigate scientific approaches on formation of a complex estimation of negative impact of waterproofing works and materials on environment, health of citizens with an estimation of a put ecological-economic damage to natural objects and resources, factors and sources of such impact are defined. Taking into account stated and urgencies of the yielded problem within the limits of scientific researches by authors of article will be developed scientific both methodological approaches and criteria of an ecological security of waterproofing materials on which basis it will be scientifically formed and practically proved optimum complex of nature protection actions at carrying out of civil work.

References

1. T.J. Bindjuk, O.V. Beschetnova, *Professional sickness* (Balashovsky branch,SGU, 2007)
2. *Ecometry* (Standards Publishing House, Moscow, 2004)
3. M.V. Goryachev, Building materials **3**, 8-9 (2005)
4. D.N. Zajkov, Building materials **12**, 20-21 (2003)
5. K. Zaip, O. Krist, Building materials **3**, 25-26 (2009)
6. V.V. Ljapidevsky, Industrial civil building **9**, 35-36 (1998)
7. T.A. Mangusheva, Building materials **3**, 43-44 (2005)
8. Methodical directions 2.1.2.1829-04 (2004)
9. P.I. Meshkov, Building materials **4**, 12-14 (2001)
10. V.A. Olevsky, Building materials **5**, 40-41 (2004)
11. Sanitary rules and size standards 2.2.3.1384-03 (2003)
12. Building size standards and rules 12-03-2001 (2001)
13. A.M. Sergeev, Building materials **3**, 8-9 (2001)
14. E.S. Tshovrebov, E.A. Yaily, Publishing house RGGMU (2013)
15. E.S. Tshovrebov, G.V. Chetvertakov, S.I. Shcanov, Alpha m: INFRA TH (2014)
16. E.S .Tshovrebov, S.V. Sadova, The Bulletin of the Russian Academy of Natural Sciences **2**, 57-59 (2014)
17. A.A. Askadskiy, Gordon and Breach (1996)
18. M. Aslam Bitumen polymer composite products, Journ. Colour. Soc. **21 (3)**, 19-22 (1982)
19. B. Brule, Rapp. Lab. Min. urban **12**, 5-58 (1986)
20. Jo Byung-Wan, Kim Chang-Hyun, Lim Jae-Hoon, ACI Materials Journal **104 (3)** (2007)
21. N. Celik, E. Antmann, X. Shi, B. Hayton, Department of industrial engineering (2012)
22. K.C. Guriya, A.K. Bhattachariya, D.K. Tripathy, Polimer-JAN **39 (1)**, 48 (1998)
23. R. Hoffman, J. Kheol, Chem. Macromol. Symp **36**, 927-965 (1992)
24. H. Korsch, Bitumen **52 (1)**, 10-13 (1990)
25. D. Marković, D. Janošević, M. Jovanović, V. Nikolić, Mechanical Engineering **8**, 65-67 (2010)
26. G.E. Zaikov, M.I. Artsis, A.Y. Polishchuk, Polym. News **21 (9)**, 323-324 (1996)
27. J.D. Nixon, D.G. Wright, P.K. Dey, S.K. Ghosh, P.A. Davies, Waste Management **33**, 2234-2244 (2013)
28. Samuel Huang, Macromol. Sci. A. **32 (4)**, 539-596 (1995)
29. B. Racz, Bitumen Eger **1**, 12-14 (1988)
30. J. Rod-Rigues, G.R. Hamed, Rubber chem. and Thnol **66 (2)**, 286-294 (1993)
31. P. Walstra, Bordeax France, 23-26 (1997)
32. J. Zialenski, A. Bukovski, Rope a uhlie **29 (7)**, 404-414 (1987)

MATEC Web of Conferences **86**, 04035 (2016) DOI: 10.1051/matecconf/20168604035
IPICSE-2016

Reduction of Smoke-Generation Ability of Construction Timber

Elena Pokrovskaya[1] and *Feodor* Portnov[1*]

[1] Moscow State University of Civil Engineering, Department of Integrated Safety in Civil Engineering, 129337, 26, Yaroslavskoye Shosse, Moscow, Russia

Abstract. The problem of reduction of smoke-generation ability for construction timber is of high importance as the smokes and toxic combustion products generating in the course of these materials burning are one of the most dangerous fire factors which threaten life and health of people. According to data presented by Ministry of Emergency Situations of the Russian Federation, rate of mortality due to smoke and toxic combustion gases is more than 70% of total number of people deaths in fires. In this paper, influence of phosphorus-containing modifiers to smoke-generation ability of timber, and also dependence of smoke-generation ability on chemical reactivity of the modifier in interaction with timber has been studied. The problem on fumigation at construction timber burning can be solved by development of chemical compositions efficiently reacting with the surface layer of the timber so that diffusion of solid carbonaceous particles from the timber surface into atmosphere decreases.

The problem of reduction of smoke-generation ability for construction timber is of high importance [1-5] as the smokes and toxic combustion products generating in the course of these materials burning are one of the most dangerous fire factors which threaten life and health of people. Thermal decomposition of timber materials is followed by emission of volatile substances and solid particles. In the atmosphere, solids form smoke aerosol, and the most intense solids emission from the timber surface is observed at glowing [5-8]. At timber materials combustion and at other fire loading, fumigation of evacuation passages causes people disorientation, and also poisoning with fire toxic thermal decomposition products [9-11]. At the moment, solution of problem on fumigation of evacuation passages comes down to the use of automatic smoke removal systems, and the issues of reduction of smoke-generation ability for construction materials is poorly understood. According to data presented by Ministry of Emergency Situations of the Russian Federation, rate of mortality due to smoke and toxic combustion gases is more than 70% of total number of people deaths in fires. In this regard decrease in smoke-generating ability of construction timber is an important factor of people death risk decrease in the fire. Also, it is important to study these issues in the frame of complex reduction of fire-dangerous properties of timber. Efficient fire retardants for construction timber are structures on the basis of organic-phosphorous compositions [12-18].

* Corresponding author: romeo-aleks@yandex.ru

In this paper, influence of phosphorus-containing modifiers to smoke-generation ability of timber, and also dependence of smoke-generation ability on chemical reactivity of the modifier in interaction with timber has been studied.

Pine sapwood timber modified with 20%-solutions of following phosphorus compositions was studied: dimethylphosphite (DMF), diethylphosphite (DEF), dipropylphosphite (DPF), dibutylphosphite (DBF), diphenylphosphite (DFF), ammonia polyphosphate (PFA). Water (for DEF, DMF and PFA-1) and acetone (for other compositions) were used as solvents. Use of obtained composition is 300 g/m^2.

Tests of smoke-generation ability for fire-protected timber were carried out by a reference technique, in accordance with Interstate Standard 12.1.044-89 for glowing state, at heat flow of 5-35 kW/m^2.

As a result (Figure 1), the following is observed: smoke-generation ability of timber in the presence of ethers of phosphorous acid are decreased by 5 times in case of use of diethylphosphite and by 1.5 - 2 times for other ethers of phosphorous acid. These results demonstrate high smoke reduction ability of diethylphosphite.

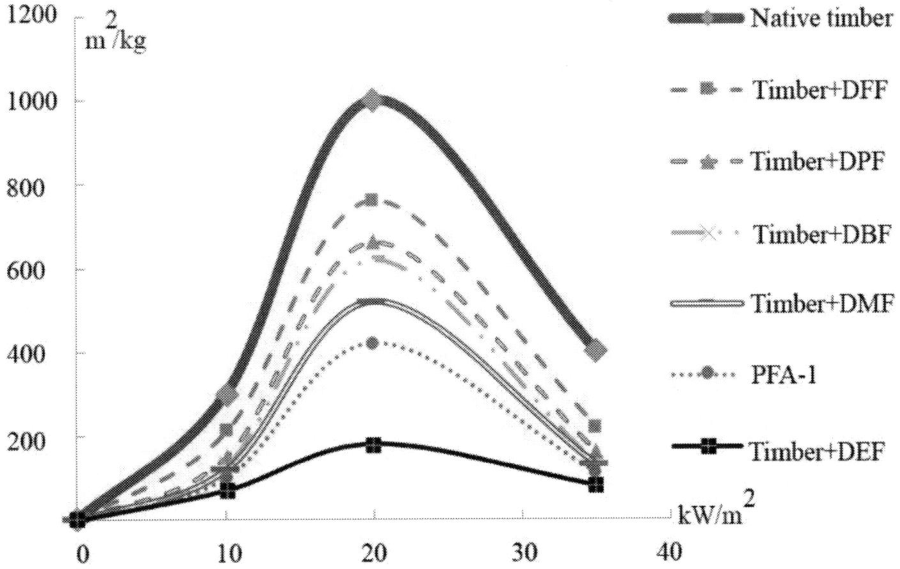

Fig. 1. Smoke-generation ability of timber modified by phosphorus compositions impregnated into timber surface

To clarify mechanism of modifiers action, phosphorus content was measured in the modified timber samples. Chemical composition was defined using scanning Quanta 200 microscope with appliance for elements analysis Apollo 40, using energy dispersive spectroscopy method. For the analysis, a surface layer of impregnated timber of 1 mm thick was cut off with scalpel. On the basis of data obtained as a result of chemical analysis of timber samples impregnated with phosphorus-containing modifiers (Table 1), it can be seen that DEF has the highest reactivity in interaction with timber.

Table 1. Content of phosphorus in the surface layer (1 mm) of the impregnated timber.

Modifier	DMF	DEF	DPF	DBF	DFF	PFA-1
P, %	2.87	3.57	2.6	2.65	2.45	2.11

A process of timber chemical modification with ethers of phosphorous acids can be described in the following way (Fig.2):

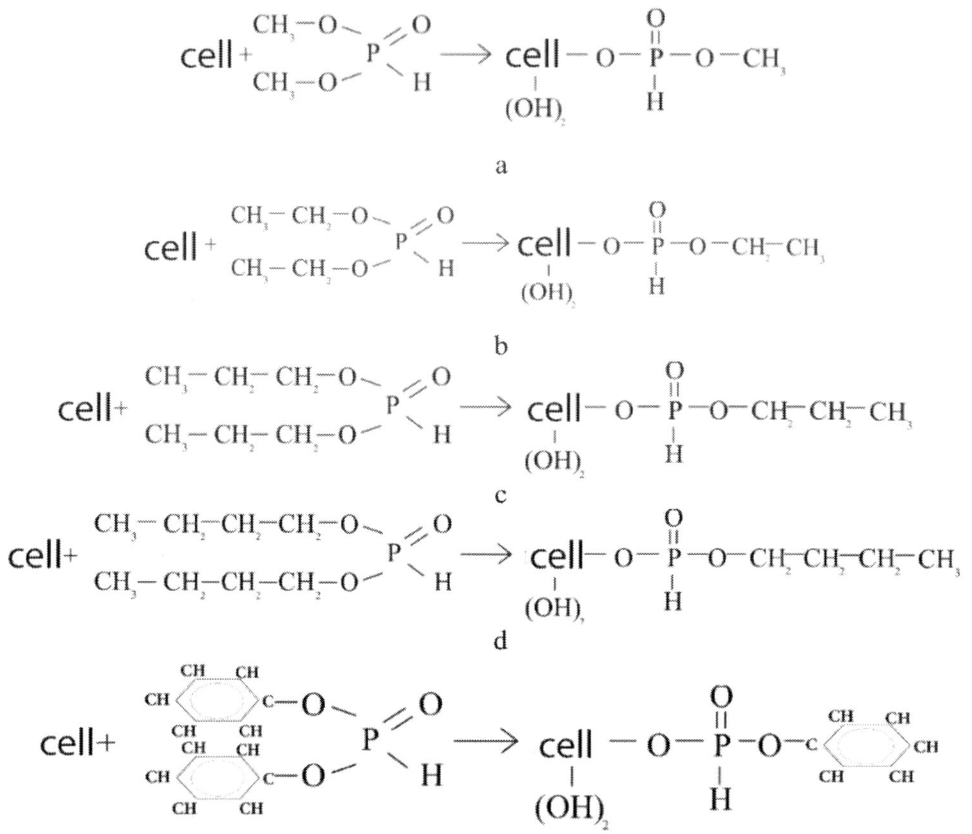

Fig. 2. Schemes of cellulose phosphorylation with use of DMF (a), DEF (b), DPF (c), DBF (d), DFF (e).

To clarify stability of the compositions formed as a result of impregnation, chemical analysis of surface layer of carbonaceous samples was carried out for impregnated timber (Table 2).

Table 2. Phosphorus content in the surface carbonaceous layer of impregnated timber.

Modifier	DMF	DEF	DPF	DBF	DFF	PFA-1
P, %	0.85	4.7	0.63	0.71	1.06	0.35

It can be seen from analysis of the obtained data, that the largest content of phosphorus is found out in timber modified with DEF, and this means that compositions formed as a result of timber modification with this composition has high stability.

It is known that smoke formation is a result of solid carbonaceous particles diffusion from the surface of timber. It is apparent that use of diethylphosphite as a modifier interferes diffusion of solid carbonaceous particles into atmosphere due to formation of stable chemical bonds.

Assessment of surface layer modifiers efficiency for construction timber requires definition of the main fire hazard indicators stipulated by state standards of the Russian Federation. Fire retardants implementation and further application at fire protection facilities becomes expedient if, as a result of tests performed in accordance with reference techniques, required fire protection efficiency is confirmed and reduction of construction timber fire hazard is reached. Main fire specifications are: combustibility, velocity of surface flame propagation, smoke-generation ability, toxicity of combustion gases.

To define combustibility in the presence of the developed impregnation compositions, tests with use of method of slow-burning substances and materials group definition described in Interstate Standard 12.1.044 were carried out. Phosphorus compositions efficiently reacting with surface layer of timber were taken as modifiers.

Table 3. Test results for timber in presence of phosphorus impregnation compositions using of method of slow-burning substances and materials group definition.

Sample	Δt_{max}, °C	τ_{max}, c	Δm, %	Burning index and combustibility index
Timber	500	150	10.2	Combustible, medium combustibility
Timber + DMF	200	280	4.8	Combustible, inflammable
Timber + DEF	220	280	5.2	Combustible, inflammable
Timber + PFA-1	230	270	6.3	Combustible, inflammable
Timber + DFF	220	200	7.1	Combustible, inflammable

During the tests, following parameters were recorded: variation of temperature Δtmax for light gaseous products produced at timber combustion, time τmax to maximal temperature of the timber combustion gases. Results are given in Table 3.

The results obtained show that when using inorganic phosphorus compositions – ammonia polyphosphates and organic phosphorus compositions – protective effect is reached. Parameter Δtmax for timber in the presence of DMF and DEF compositions is 2.5 times lower, than that for native timber. Amount of temperature increase during tests indicates intensity of timber thermodecomposition reactions and, therefore, ability of the fire protection facility to reduce intensity of these reactions. The indicator τmax indicates duration of this protective effect in time. For DMF and DEF this parameter is twice higher than for native timber.

To define group of fireproof effectiveness, tests were carried out in accordance with Interstate Standard 53292-2009. On the basis of results obtained (Table 4) it can be noted that loss of mass for sample of timber modified with DMF is reduced by 11 times, for DEF and PFA-1 - by 9 times.

Table 4. Results of tests on definition of fireproof efficiency group for fireproof composition.

Sample	Mass loss, %	Group of fireproof efficiency
Native timber	79.0	III
Timber, modified with DMF	6.9	I
Timber, modified with DEF	8.5	I
Timber, modified with DFF	9.5	II
Timber, modified with PFA-1	8.7	I

Study of flame propagation process along the timber surface for timber with fireproof coverings was carried out using method described in Interstate Standard 12.1.044-89. During the tests the following parameters were considered: time of passing by the flame front of specified sample areas, maximal temperature of exhaust gases, maximum sample longwise damage caused by flame, time to maximal temperature of exhaust gases. On the basis of the data obtained the flame propagation index I_{RP} – classification parameter - was calculated.

Study was carried out for following phosphorus-containing modifiers: DMF, DEF, DFF, PFA-1. Results of test on flame propagation along the surface of timber construction elements impregnated with fireproof compositions are given in Table 5. These results show that described timber treatment considerably reduces its ability to spread flame along the timber surface. All studied ethers of phosphorus acids considerably reduce parameter I_{RP} (by 12-26 times). The best parameters were found out for timber modified with DMF and DEF.

Table 5. Flame propagation along the timber surface in the presence of phosphorus-containing modifiers.

№	Sample	Time of flame passing initial area (30 mm)		Increase of combustion products temperature, °C	Sample longwise damage, mm	Flame propagation parameter, I_{FP}
		Time (τ), s	Velocity (V_{FP}), mm/s			
1	Native timber	8	3.125	40	300	60
3	Timber+DMF	240	0.12	15	50	2.3
4	Timber+DEF	230	0.13	22	60	3.7
5	Timber+DFF	200	0.15	25	80	4.7
6	Timber+PFA-1	210	0.14	21	65	3.9

The following can be concluded on the basis of the study. The problem on fumigation at construction timber burning can be solved by development of chemical compositions efficiently reacting with the surface layer of the timber so that diffusion of solid carbonaceous particles from the timber surface into atmosphere decreases. For such compositions development, it is recommended to use ethers of phosphorus acids, in particular, diethylphosphite which considerably decreases smoke-generation ability (by 5-6 times), and also considerably reduces fire hazard of timber and this is of great importance for development of fireproof compositions for construction materials.

References

1. D. Bobacz. Behavior of Wood in Case of Fire (2008)
2. E.N. Pokrovskaya, Chemical-physical basis of tread life wood: monograph (2009)
3. D. A. Korolchenko, A. M. Lukyanov, World of Transport **4**, 158 (2012)
4. A.N. Baratov, V.L.Pchelintsev, Fire safety (1997)
5. E.G. Butcher, A.C., Parnell Smoke control in fire safety design (1979)
6. R.M. Aseeva, B.B. Serkov, A.B. Sivenkov, The burning of wood and its flammable properties, monograph (2010)
7. R.M. Aseeva, G.E. Zaikov, The combustion of polymeric materials (1981)
8. V.A. Ushkov, D.I. Nevzorov, A.V. Kopytin, V.M. Lalayan, Fire and Explosion Safety **7**, 27-35 (2014)
9. C.J.Young, J. Moss, J. Clin Anesth **1**, 377-386 (1989)
10. D.J. Shusterman, Occup Med. **8**, 469-503 (1993)
11. J. L. Bryan, J. A. Milke, Project People II. Final Report Health Care Report (1981)
12. Taubkin S.I., RSFSR Publ. **1**, 230 (1944)
13. Polandov Yu.Kh., Dobrikov S.A., Korolchenko A.Ya., Fire and Explosion Safety **25**, 17-26 (2016)
14. G.G.Orlov, D.A. Korolchenko, Fire and Explosion Safety **24**, 62-67 (2015)
15. D.A. Korolchenko, A.F. Sharovarnikov, A.V. Byakov, Advanced Materials Research **1073-1076**, 2353-2357 (2015)
16. D.A. Korolchenko, A.F. Sharovarnikov, Advanced Materials Research **1070-1072**, 1794-1798 (2015)

17. A.A. Kobelev, *Development of complex fire and moisture resistant structure based on connections for surface modification of wood* (Cand. tech. sci. diss, 2012)
18. U.L. Makovskiy, Fire protection of wood materials by phosphorous acid esters (1992)

The systems of insulation and a methodology for assessing the durability

Boris M. Rumiantcev [1,*], *Aleksey D.* Zhukov[1], *Ekaterina Yu.* Bobrova [2], *Irina P.* Romanova[1], *Dmitry B.* Zelenshikov[3] , *Tatiana V.* Smirnova[4]

[1] Moscow State University of civil engineering, 26, Yaroslavskoe, Moscow, 129337, Russia
[2] National Research University. Higher School of Economics (NRU HSE), 20, Myasnickaya. Moscow, 101000, Russia
[3]Modern Roofing Technologies Ltd., 1, Eniseyskaya ul. Moscow, 129344, Russia
[4] Mineralnaya vata" Company ", 9, Zemlyanoy val, of. Company Rockwool, Moscow, 105064, Russia

Abstract. Formation of effective insulation possible only taking into account features of the insulation layer in the construction and use of high quality materials, preserving their characteristics as in the early stages of operation, and for the entire billing period. The operational stability of the heat-insulating material is determined not only by its properties, but also the ability to withstand long-term operating loads without a significant change in properties. Processing load on the heater may be divided into three groups: the load during transportation and storage, loading and mounting load during operation. Methodology for assessing the properties of insulation products includes two main components: the installation for testing and test methods, as well as the methodology for assessing the operational stability. The methodology of the accelerated testing and forecasting durability tested for mineral wool products laminating, corrugating and volume-oriented structure. The test results give good agreement with the methods recommended by the building codes. Accelerated test plates on the compressibility and compressive strength at 10% deformation after exposure to boiling water over the useful spending for operational control properties of materials, as well as when examining the state of the exploited mineral wool insulation in buildings and heat piping

1 Introduction

Increasing gross domestic product energy efficiency in relation to the construction involves the improvement of design solutions (creating systems in which thermal insulation materials demonstrate their effectiveness to the greatest degree), as well as the creation and use of effective thermal insulation materials and methods of evaluation of their properties [1-4]. Energy efficiency in this case is the complex concept. Firstly, the insulating material must provide standard values of thermal resistance design. Second, the energy consumption

* Corresponding author: lj211@yandex.ru

for its production needs to be compensated by his "work" in the design (for the first ten heating seasons).

Thirdly, the material must save its properties (functional, structural, and dimensionally stable) at all times during the construction operation. Fourth, the exploitation is the material should not significantly increase the burden on the environment - the material must be fire safe and environmentally. Fifth, competently executed insulation helps create and maintain comfortable indoor conditions. The operational stability of the heat-insulating material is determined not only by its properties, but also the ability to withstand long-term operating loads without a significant change in properties. Processing load on the heater may be divided into three groups: the loads during transportation and storage, mounting and load during operation [5-6]. In the operation of the plate exposed to temperature, weather, humid and mechanical stress. The character of loads distribution in plates under various conditions of operation is shown in Fig. 1. Regardless of the type of building, system optimization parameter is the thermal resistance of the construction and parameter optimization of TIM is its thermal conductivity. At the same time the specificity of operation of TIM in each of the specific systems imposes certain additional insulation requirements - "restrictions", defined as "normalizing indicators".

2 Problem statement

Insulation boards operating in flat roof constructions, perceive the compressive load on the upper layers of roofing pie, temporary seasonal load and local small area load. The normalize indicator is a compressive strength at 10% deformation. Humidification insulation may in the result of violations of hydro-insulating roofing or "failure" vapor barrier, which is located on the ceiling.

If the roofing system reliably isolates the material from possible external fire exposure, application of EPS boards is real. In roofing systems using plates of foam glass and stone wool. Dual density plates are successfully proved his usage. The top layer takes a double density boards and transmits mechanical loads on the underlying layers, the lower layer - provides effective insulation and uniform adhesion of thermal insulation to the supporting elements. Thermal insulation in pitched roofs running in a "ventilated gap" conditions. Mechanical influences on it are minimal. Humidification heater perhaps for the same reasons as in the flat roof. The normalize indicator is breathable insulation. The use of combustible insulation is not allowed. In the case of mineral fiber materials (tiles, mats) needed windproof decision to exclude the removal of fiber products during erosion. In designs using glass products of such a fiberglass (glass veil), the application of products on the basis of basalt fiber - basalt canvas. In the application of stone wool products - dual density plates are recommended.

The designs of plaster facades are predominant is loads acting in the direction perpendicular to the plane of the insulation. The normalize indicator is the peel strength of the layers of insulation. It is imperative - vapor permeability of the insulation. Do not use EPS or foam glass plates, due to their low vapor permeability.

Allowed to use plates of PPP (or other combustible insulation) to the mandatory fire crosscuts at the height of ceilings and protective inserts on the perimeter of the openings (windows, doors). Optimal are mineral wool boards, including the boards of rock wool double density. Thick layer of dual density plates ensures a secure clutch of thermal insulation and plaster layer and the peel strength of the layers to the product itself.

Slabs of ventilated facades are characterized by three types of mechanical impact: compression under the load point, the gap layer (break down fibers), fiber erosion due to air flow in the material and vibration. Allowed use only non-flammable and vapor-permeable insulation. For wind protection possible to use non-combustible wind protection coating: canvases based on glass or basalt fibers, or use of dual density slabs of stone wool.

Insulation used in the protection of basements systems should be designed for mechanical stress, the penetration of condensed moisture and the possibility of migration of vapor. Normalizing parameters are strength at 10 % deformation, water resistance and water resistance. Insulation design is completely isolated from external fire exposure, so the optimal use of EPS or PIR plates.

3 Results and discussion

Experience has shown that the properties of materials, and, above all thermal insulation as having a low density and strength, change over time. A change more significant than the operating conditions are tougher. The analysis application possibilities insulation shown in various building systems, for most typical application of the thermal insulation on the basis of mineral fibers, so technique express evaluations, first of all, were tested for fibrous materials based on glass and rock wool, as well as basalt fiber.

The traditional criterion adopted humidity. Moisture mineral insulating material (MIM) in accordance with the GOST 17177 assessed by a decrease in their strength (compressibility) after exposure in the environment in the desiccator with a relative humidity of 96-100% and a temperature of 18-20 ° C for 3 days. To intensify this process was first suggested NIISF in the test sample in an environment of water vapor above the boiling water. MIM asked to be considered water-resistant if after 1 hour of testing strength reduction does not exceed 30 %.

The use of this technique not only significantly accelerated the evaluation of the water resistance of various MIM, but also allowed to accumulate practical information for potential improvements in techniques and design of the chamber to determine the moisture resistance MIM with horizontal and vertical orientation of the fibers in their structure. The data showed that between the values of strength reduction of samples-plates at their endurance over boiling water and store in a desiccator in moist air at 18-20 ° C there is a directly proportional relationship. Deeper study of this question showed that effect significantly affected by the type and degree of cure of the binder in the finished plates.

At the same time with a decrease in strength were observed signs of structuring of the uncured binder in the process of a long-term tests, as well as the permanent deformation of the low-density samples (35-70 kg / m^3) when removed from the camera after completing the tests and mounting in clamps of a discontinuous device. Strength reduction occurs at different rates, depending on the density of boards, the type of binders, the degree of cure and product structure [7-8].

With hard factory laboratory, monitoring of the degree of cure of the binder and the test this sample in the cells (Fig. 2 and 4) these disadvantages of the methods can be practically excluded. The camera in Fig. 2 is designed to determine the moisture resistance of samples of hard plates, increased rigidity and hard (by change the compression strength) after exposure of boiling water. The camera in Figure 4 is intended to determine the moisture resistance of soft and semi-rigid boards to change their compressibility after exposure over boiling water.

Camera (Figure 1) consists of unsealed vessel 1 with a lid 13 provided with a heat source 4 and a thermometer 12. Inside the vessel 1 is floating floating mesh tray 2 from the outside - pointers boiling water level 7 and its horizontality 5. During the test sample 10 set in a mesh container 9 and placed on the pallet 2, which to the rack 8 is attached to "floats" 3, providing a constant gap δ = 80-100 mm between the bottom and top of the sample of boiling water. To adjust the horizontal position of the device provided adjusting screws 6, for the convenience of removal cover of hot material 13 in its upper part placed handles 11. The gap between the top of the sample δ2 = 100-150 mm.

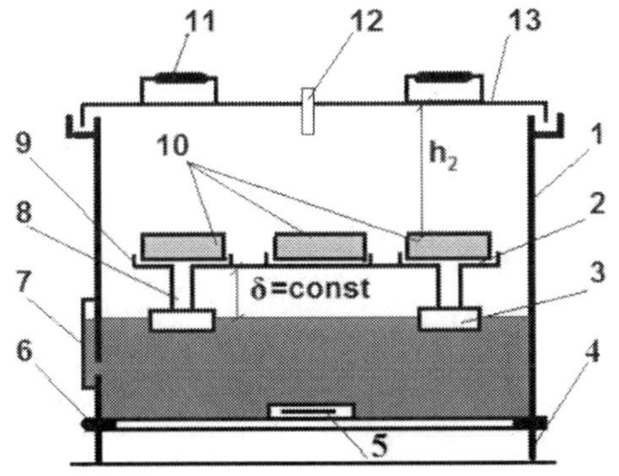

Fig. 1. Schematic diagram of the camera chamber design to determine the change in the compressive strength of mineral wool samples after their exposure over boiling water: 1 - the camera body 2 - floating mesh tray; 3 - floats; 4 - heater-element; 5 - the level of horizontality; 6 - the adjusting screws; 7 - water level indicator; 8 - racks; 9 - mesh container; 10 - prototypes; 11 - the handle; 12 - thermometer; 13 - upper removable cover.

Rapid method of humidity resistance studies is that instead of 3-10-day test sample in a desiccator at a temperature of (22 ± 5) ° C and humidity (98 ± 2) % produce their short term exposure (20 min.) In a water medium vapor at a temperature (98 ± 2) ° C and humidity (98 ± 2) %. Then, the test specimen under the accelerated procedure and build the calibration graph (Figure 3). On the axis of ordinates is strength value (A) or compressibility (A1) obtained by standard tests 3-10 days in a desiccator, and connect it till the intersection with the curve of 1 (2) obtained according to the accelerated tests in the medium of water vapor, and the axis x-value is obtained the required time of accelerated tests (respectively B or B1).

In the future (for example, when the daily trials and quality control boards at the factory), use once build the calibration graph for a particular product, determine the moisture only accelerated method, and compare it with the corresponding value of water resistance in the calibration curve. Change the water resistance may indicate a deterioration of the quality of binder, abuse of technology, etc.

With that, the experience of application of this method showed that the degree of MIM with phenol curing at least 90% over, the duration of exposure of samples upper of the boiling water should not exceed (20 ± 2) minutes. During this period, the temperature at which testing occurs (about 100 ° C) does not significantly affect the validity of the results. Therefore, during the factory laboratory and practical classes in middle and high schools of a building profile to determine the moisture resistance of MTM, you can set the duration of the exposure of boiling water in the range of (20 ± 2) minutes. [9-10].

A feature of the camera shown in Figure 4 is the possibility of it as a definition of the compressibility according to GOST 17177 and its changes after exposure of samples of boiling water without removing samples. To test sample 13 is placed on the lower mesh gasket 7, placed in the middle of the pallet rigid mesh 8. The distance between the bottom surface of the sample and the water surface is (100 ± 10) mm, and is controlled by the water level indicator 12. Then check the horizontality chamber 4 through a pointer 9, if necessary using the adjusting screws 11. On top of the sample 13 is laid mesh gasket 7 and the space frame 14. With the closing the lid 3, on which stationary secured elements of standard device for determining the compressibility by GOST 17177.

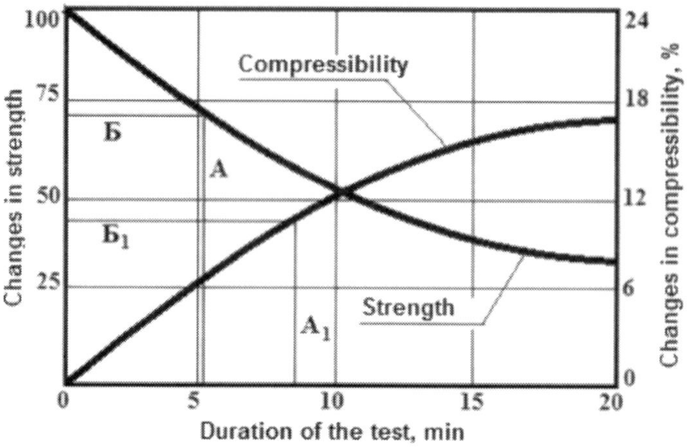

Fig. 2. The calibration curve for the determination of the duration of accelerated tests: 1, 2 - the curves of the changing accordingly compressive strength (σ) and compressibility (Cx) in an accelerated test; A and A1 - values of strength and compressibility after sorption moistening; B and B1 - time value of accelerated tests, the corresponding values of strength and compressibility, obtained after the moisture sorption.

At the start determine the compressibility of a sample 13 (Figure 2) under specific standard load, thus fixing its thickness h1. Then include source and regulator 10 heating the water, the water is brought to boiling by controlling the temperature and humidity by sensor 6 and the sample was kept in an environment of water vapor above the boiling water for a predetermined time: about (20 ± 2) minutes. Fix with the final sample thickness h_2. without removing it from the chamber.

When testing products on synthetic binders should be noted that the degree of hardening of the binder in these products shall be at least 90%. This condition corresponds to the requirements of the standard, according to which the fullness of condensation of synthetic binder should be at least 90 %. At a lower degree of polycondensation not excluded receive inaccurate data due to possible post-curing the binder at a test temperature of about 100 ° C.

The data obtained allow to recommend companies that manufacture MIM, replacing long 3-10 daily water resistance test sample plates by standard methods to express control. Accelerated testing may occasion the procedure (ex-press-technique) IISS (MGRS) in the design of cameras manufactured "MAXMIR".

4 Conclusion

The methodology of the accelerated test of products tested for the layered corrugated and volume-oriented structure. Accelerated test by plates on the compressibility and the compressive strength at 10% deformation after exposure of boiling water is useful to, in cases where the time of shipment party of plates the information about the relative change of their compressibility and strength after sorption humidification in desiccator is not obtained, and at operative factory quality control of manufactured boards (new middleware, the new components of the charge, changes in manufacturing and others. technology), and any disputes in their application, examination of the state of the layer of mineral wool insulation in buildings and heat piping exploited.

Properties of insulation which have arrived at the construction site may be different from embedded in the project. Therefore, to obtain an objective and timely information about the properties of the insulation material is a necessary factor in determining its suitability (or non-suitability, which is a force majeure), as well as the possibility of its using in this system with the permissible deviations from the project. The introduction of rapid methods for assessing the thermal insulation properties allows you to quickly appraise its properties. Information obtained as a result of the express control, of course, is accurate, but cannot be considered final. Full information can be obtained only by the results of laboratory tests based on existing standards.

References

1. S. A. Tkach, V. I. Telichenko, Ecology of urbanized territories in Moscow **2**, 39-44 (2016)
2. V. I. Telichenko, D. V. Oreshkin, Ecology of urbanized areas **2**, 31-33 (2015)
3. A.D. Zhukov, I.V. Bessonov, A.N. Sapelin, N.V. Naumova, A.S. Chkunin, Italian Science Review **2**, 155-157 (2014)
4. I.V. Bessonov, A.V. Starostin, V.M. Oskina, Bulletin MSUCE **3**, 134-139 (2011)
5. V.G. Gagarin, V.V. Kozlov, Architecture and engineering **2**, 60-63 (2006)
6. I. Ya. Gnip, S.I. Vaitkus, S.A. Vejelis, Building materials **12**, 40-44 (2012)
7. A.D. Zhukov, T.V. Smirnova, D.B. Zelenshchikov, A.O. Khimich, Materials Research **838-841**, 196-200 (2014)
8. V.N. Sokov, A.E. Beglyarov, D.Yu. Zemlyanushnov, D.V. Zhabin, Bulletin MSUCE **1-2**, 309-312 (2011)
9. Yu.L. Bobrov, Modern methods of quality control problems and their solutions **1**, 43-44 (1985)
10. B.M. Rumyantcev, A.D. Zhukov, E.Yu. Bobrova, T.V. Smirnova, Industrial and civil building **1**, 32-36 (2015)

Formation of normative-methodical maintenance of quality and safety of campsites

Daniil A. Cherepanov[1,*], *Aleksandr* S. Ermakov[1], *Marina* R. Gozalova[2], *Aleksandr* Ya. Korolchenko[1]

[1] Moscow State University of Civil Engineering, 26, Yaroslavskoye Shosse, 129337 Moscow, Russia
[2]Russian State University of Tourism and Service, Moscow, Russian Federation

Abstract. Organization of campsites construction for auto tourists in Russia is an effective means of ensuring their rest and sleep during travel. Formation of normative and methodological support of campsites fire safety should take into account not only security, but also meet consumers' motivation to visit them. The authors examine conditions and requirements for the infrastructure of campsites and their localization, as well as methods of fire risk assessment in camping. They propose camping localization techniques and expert evaluation of fire risk in them.
Key words: construction of the structure, localization, security, camping, normative and methodological support.

1 Introduction

Constructions easily erected areas for recreation, guest services and cultural, sports, recreational activities, created with minimal anthropogenic influence on the territory and fire risk, are particularly in demand as an effective and economical way of seasonal residence [1, 12].

2 A review of the literature

Also this is the fulfillment of the Decree of the Russian Government dated August 02, 2011 No. 644 (edition date is February 18, 2014) by the Federal target program "Development of domestic and inbound tourism in the Russian Federation (2011-2018)" through the creation of car-clusters with camping. Improving the construction quality of campsites and their safety is the harmonization of systems of technical regulation of Russia and Europe and their regulatory and methodological support. The most important in ensuring quality and safety is to prevent fires in the campsites as the area has the places with open flame (barbecue, bonfire, etc.), different infrastructure options with external and internal threats, as well as with a significant influence on the choice of the location and infrastructure of the its visitors' motivation. Analysis of the causes of fires in the camping shows the cases of death and the victims among the visitors [4]. Choosing the placement of the campsite construction, it's also important to ensure not only safe [1,2] and comfortable accomodation

* Corresponding author: Supfear@yandex.ru

on this site, but the opportunity to visit as many tourist attractions of travel destination in accordance with tourists' motivations and needs. The absence of the normative-methodical maintenance of campsites fire safety with different options for their functionality and composition of infrastructure in the special conditions of natural and engineering communications of their placement requires the normative and methodological support development of their fire safety. Campsites, providing high-quality fire-protected operations and safe people evacuation, have a real reduction of material damage in case of fires.

3 Methodology

They often use systematic analysis, mathematical modeling, and calculations of fire risks, experimental analysis and testing of camping facilities for research.

4 Results

Classification of camping, the level of infrastructure equipment, quantitative scale, location, distance from urban settlements, they are a number of factors influencing fire safety of camping because, in the case of fire in one of the buildings or on one of the camping sites, security of people and property is a really difficult problem [10, 11]. Each characteristic of the camping [8] has its important role in determining aggregate measures of its fire security.

The specific features of the security, and in particular, the camping fire danger in Russia are determined with significant climatic differences on the territory of the country and the fillings of attractions that motivate consumers; features of the structure and composition of the infrastructure [5, 13], which provide various service functions (accommodation, security, rest, sleep, food, entertainment, etc.); area sites with different threats and their diversity and the availability of means for their prevention and use in the system of fire-protection; organization of the territory, functionally performing the services in accordance with the consumer motivation, not always determined with manufacturing sequence and seasonality capacity and temporary accommodation without creating stationary building structures for the individual species [9].

The system security model of campsites constructions requires the creation of normative-methodical documentation, taking into account the accessibility comfort and stay comfort in accordance with their motivation, as well as all possible threats both external and internal threats, their prevention, protection, alarm systems, warning and evacuation for the purpose of preservation of people's life and health, property, infrastructure and environment.

Classification of campsites in the 5-star principle allows taking into account the variety of campsites (their infrastructure, scale, location, purpose and other features) in fire safety: the higher class of camping has the more types of structures and communications. The most rational measure for Russia to create seasonal campsites is the use of mobile homes [5, 6], which are not only able to adapt to changing conditions, preserving the environment [7] of the location and appearance parameters, but also to evacuate easily in emergencies.

The calculations of the structures of the camping, which established their energy load, were carried out to study the level of threats (Fig. 1).

To diagnose security threats in camping, in particular their identification and the assessment of their probability and their consequences, there were compiled diagnostic papers. Diagnostic paper no. 1 depending on the camping location includes information about signs of a threat: external security threats for the platform camping; internal security threats on the camping area; internal security threats in the camping operation. Diagnostic paper no. 2 takes into account the category of risks of threats consequences to people's health and life and security of their property and camping.

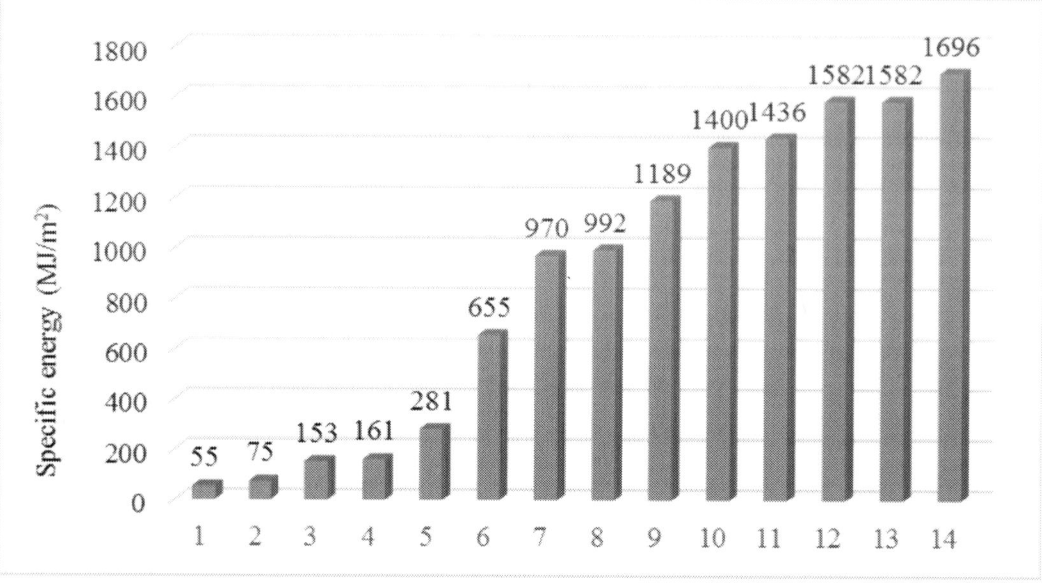

Fig. 1. Comparative analysis of the camping structures on the energy loading: 1 - Two-seat tent (with internal fire load); 2 - Modular shower for 6 people; 3 - Barbecue under a shed with furniture; 4 - Four-seat tent (on the pallet with fire load)); 5 - Modular toilet for 6 people; 6 - Trailer-transformer; 7 - Trailer for 2-4 berthes; 8 - Wooden bungalow; 9 - Middle class car on the Parking; 10 - Security post; 11 – Caravan; 12 - Modular shower; 13 - Modular toilet; 14 - Administrative building of III-IV degree of fire resistance.

The most vulnerable structure in the campsite is the tourist tent that determines the need to study fire safety of the tent materials and methods to ensure its fire safety. According to the results of experimental studies of the tent materials there was installed (Fig.2):
reduced flammability due to the use of flame retardant treatment on cotton tent fabrics (the material is not ignited, but get only the charred edges); the effect of the transmittance of the heat flux T through the tent fabric, applying reflective paint, which in comparison with fabric without reflective paint show a decrease for cotton fabric at a distance of 200 mm from 3.01±0.25 kV/m2 to 2.54±0.23 kV/m2, and at a distance of 300 mm from 1.21 kW/m2±of 0.18 to 1.15±0,17 kW/m2; applying a luminous pigment shows the reduction at a distance of 200 mm from 3.01±0.25 kV/m2 to 2.49±0,2 kW/m2, and at a distance of 300 mm from 1,21±0,18 kW/m2 to 1.09±0,15 kW/m2. The polyester fabric with reflective paint shows the reduction for a distance of 200 mm from 3.2±0.3 kV/m2 to 2.82±0,25 kW/m2 and at a distance of 300 mm from 1.27 mm±0,19 kW/m2 to 1.22±0,2 kW/m2; applying a luminous pigment shows the reduction at a distance of 200 mm from 3.2±0.3 kV/m2 to 2.47±0,22 kW/m2, and at distances from 300 mm to 1.27±0,19 kW/m2 to 1.15±0,12 kW/m2.

Maximum of fire prevention is achieved through the rational camping localization [3, 5] when the target function will be:

$$D = \bigcap_{l=1}^{E} t_1; t_1 \in M^T when(A_l \cdot K_l \cdot B_l) \to \max \quad (1)$$

where D is rational E quantity of camping localizations in the examined destination M^T taking into account the comfort and fire safety, $l=1 \ldots E$, t_l are the coordinates of l the rational point of camping localization in the selected space region, A_l are attractions for l

rational point of camping localization, B_l is evaluation of fire safety for l rational point of camping localization, K_l is the assessment of comfortable localization and functioning for l rational point of camping localization.

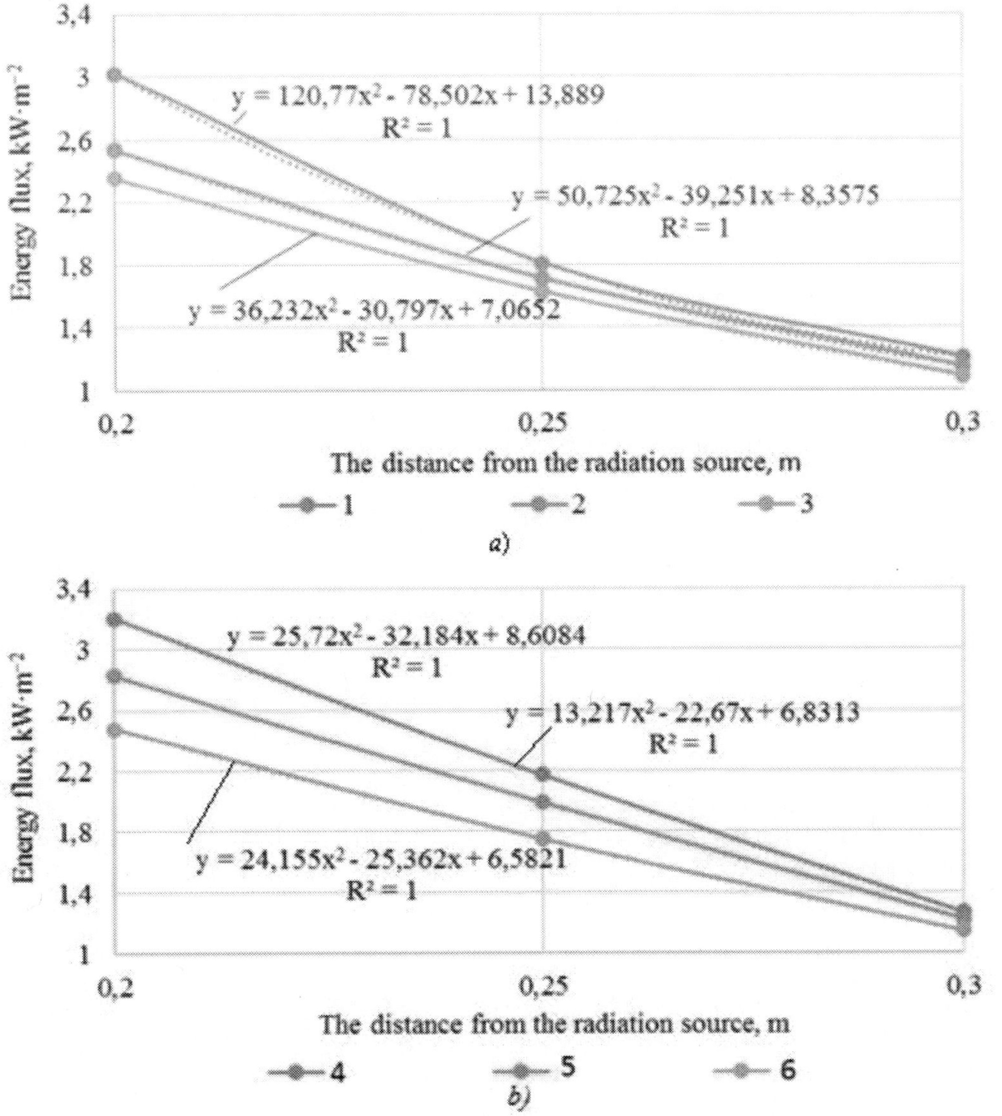

Fig. 2. Dependence of the transmittance of the heat flux from the coating and the distance from the radiation source: for cotton fabric (*a*) and polyester fabric (*b*): 1 - Cotton; 2 - Cotton coated with reflective paint; 3 - Cotton coated with luminous pigments in synthesis paint-and-lacquer coating: 4 - Polyester (polyurethane side coated 2000mm water column); 5 - Polyester coated with reflective paint; 6 - Polyester coated with luminous pigments in synthesis paint-and-lacquer coating.

where D is rational E quantity of camping localizations in the examined destination M^T taking into account the comfort and fire safety, $l=1 \ldots E$, t_l are the coordinates of l the rational point of camping localization in the selected space region, A_l are attractions for l rational point of camping localization, B_l is evaluation of fire safety for l rational point of camping localization, K_l is the assessment of comfortable localization and functioning for l rational point of camping localization.

Assessment of the quality of stay at a particular camping site is formed on the basis of an expert assessment from the implementation of necessary conditions for comfortable stay not only in camping, but transport accessibility of the attractions of the destination, noise pollution of camping sites, the link existence etc.

During the research of fire risk at the campsite for buildings and mobile homes we calculated the time of fire effects under different scenarios and the schemes and time of the evacuation. To ensure evacuation safety it is necessary to provide spare escape routes, highlighting in the night time, to remove all explosive and flammable objects from the living quarters, and recreational areas, as well as to divide the territory into separate functional areas.

We propose the method of expert evaluation to determine the level of fire risk in the camping. This method can be used for pre-project researches and preparation for the camping licensing. A survey of experts is carried out using diagnostic papers 1 and 2 and the subsequent assessment of the reliability of the examination on the value of the concordance coefficient and Pearson's chi-squared test. The fire risk R_i assessment is defined according to the matrix of consequences D_i and probabilities H_i:

$$R_i[H_i,D_i] = \begin{cases} 0 \ \textit{npu} \ k_i = 0; \\ 1 \ \textit{npu} \ H_i \times D_i = \{1\times 1; 1\times 2; 2\times 1; 2\times 2; 3\times 1\}; \\ 2 \ \textit{npu} \ H_i \times D_i = \{1\times 3; 2\times 3; 3\times 2; 4\times 1\}; \\ 3 \ \textit{npu} \ H_i \times D_i = \{1\times 4; 1\times 5; 2\times 4; 3\times 2; 4\times 2; 4\times 3; 5\times 1\}; \\ 4 \ \textit{npu} \ H_i \times D_i = \{2\times 5; 3\times 4; 3\times 5; 4\times 4; 4\times 5; 5\times 3; 5\times 4; 5\times 5\} \end{cases} \quad (2)$$

where k_i =0 - no threat fire.

The decision on the territory zoning of the camping is carried out using the percolation model, when the whole territory is represented by a grid with nodes (camping objects), and the development and spread of fire is assessed with the possibility of fire spreading from one object to another.

The guidance and techniques for assurance of quality and localization security, position of building structures in the form of campsites, which will create conditions for safe and comfortable domestic and foreign tourists' stay, were designed on the basis of the analysis and evaluation of characteristic features of fire protection system for campsites construction and its components of subsystems.

5 Conclusion

To prevent ignition of the tent it is recommended to treat its materials with luminous paint, and not only the tent fabric, but its bracing wires and tent-pegs. Applying reflective and luminous paint on the tent fabric allows escaping of the automobile accident and hooking the tourist for the bracing wires and the pegs in the dark. Also during the evacuation of the campsite in the dark time they apply luminous paint arrows and lines for evacuation routes on the tents, roads and buildings.

Selecting a site for the camping construction it is primarily necessary to install a set of proposals for possible localizations for camping sites, which will allow the tourist to realize their primary motivation, and also to ensure fire safety.

References

1. Almeida M., Barata J., Viegas D. X., *The 7th International Conference on Forest Fire Research*, Portugal, 1-12. (2014) DOI: 10.14195/978-989-26-0884-6_72
2. Cherepanov D.A., Ermakov A.S., *Servis v Rossii i za rubezhom* (Service in Russia and abroad), **1**, pp. 204-210. (2014) (In Russ) DOI: 10.12737/2564

3. Cherepanov D.A., Ermakov A.S., *«Pozharovzryvobezopasnost»* («Fire and Explosion Safety»), **7**, 48-57. (In Russ) DOI: 10.18322/PVB.2016.25.07.48-57. (2016)
4. Cherepanov D.A., Ermakov A.S., *Servis v Rossii i za rubezhom* (Service in Russia and abroad), vol. 10, no. 2 (63), pp. 141-150. DOI:10.12737/19728 (2016)
5. Ermakov A.S., Cherepanov D.A., *Servis v Rossii i za rubezhom* (Service in Russia and abroad), vol. 8, no. 7 (54). (In Russ) DOI: 10.12737/7479, (2014)
6. Ermakov A.S., Kokhreidze M.V., Cherepanov D.A., *Materialy konferentsii Instituta fizicheskoi kul'tury, sporta i turizma Petrozavod-skogo gosudarstvennogo universiteta* (Proceedings of the Conference of the Institute of physical culture, sports and tourism of Petrozavodsk state University), Petrozavodsk, 389-396. (In Russ) (2015)
7. Ermakov A.S., Kokhreidze M.V., Cherepanov D.A., *Sovremennye problemy servisa i turizma* (Service & Tourism: Current Challenges), 9(3), 69-75. (In Russ), (2015)
8. Ermakov A.S., Korneev A.A., Cherepanov D.A., *Servis plus* (Service plus), **8(2)**, 65-70. (In Russ) DOI: 10.12737/3895 (2014)
9. Korolchenko D.A., Sharovarnikov A.F. Advanced Materials Research. **1070-1072**, 1794-1798 (2015)
10. Korolchenko D., Tusnin A., Trushina S., Korolchenko A. International Journal of Applied Engineering Research. **10(21)** , 42541-42548. (2015)
11. Orlov G.G., Korolchenko D.A., Fire and Explosion Safety, **24(5)**, 62-67 (2015)
12. Sakharchuk E.S., Ermakov A.S., Korneev A.A., Cherepanov D.A., *Sostoyanie i perspektivy razvitiya avtoturizma v Rossiiskoi Federatsii* (RSUTS, Moscow, 2014)
13. Zhang Y., Analysis on Comprehensive Risk Assessment for Urban Fire: *Procedia Engineering*, **52**, pp. 618-623. DOI: 10.1016/j.proeng.2013.02.195 (2013)

Influence of spreading structure in an aqueous solution-hydrocarbon system on extinguishing of the flame of oil products

Dmitriy Korolchenko[1*] *and Sergey* Voevoda[2]

[1] Moscow State University of Civil Engineering, 129337, Moscow, Yaroslavskoye sh., 26, Russia
[2]Transneft Research Institute for Oil and Oil Products Transportation (Transneft R&D, LLC),117186, Moscow, Sevastopolskiy prospect, 47A, Russia

Abstract. On the basis of experimental studies it was revealed that adding of electrolyte - ammonium chloride - into solution of film-forming foaming agent leads to decrease of interfacial tension on a border with heptane, and increase of surface tension on a border with air. It is experimentally shown that dependence between specific consumption and flow rate of foaming agent passes through the minimum point that allowed defining influence of electrolyte on a value of minimum specific consumption and optimum flow rate of foam during extinguishing of a hydrocarbon flame. The structure of spreading in a foaming agent solution – hydrocarbon system defines the value of fire extin-guishing efficiency of foam decreasing together with lowering of spreading coefficient and interfacial tension.

1 Introduction

The structure of spreading in a solution-hydrocarbon system mainly defines fire extinguishing efficiency of foam during extinguishing of the flame of oil products [1-4]. Spreading in a solution - hydrocarbon system has a complex structure which is firstly divided by sign and value of spreading coefficient of aqueous solution, flowing over hydrocarbon, and hydrocarbon, flowing over aqueous solution. The structure of spreading coefficient consists of values of surface and interfacial tensions. For calculation of spreading coefficient of aqueous solution – K10 and heptane – K 01 we use the known ratios:

$$K_{10} = \sigma_0 - (\sigma_{10} + \sigma_1) \tag{1}$$

$$K_{01} = \sigma_1 - (\sigma_{10} + \sigma_0) \tag{2}$$

where - surface tension of heptane , mN/m; - interfacial surface tension on a solution-heptane bor-der, mN/m; - surface tension of aqueous solution on a border with air, mN/m.

The positive sign of spreading coefficient of aqueous solution is provided due to application of fluorinated SAS (surface-active substances), whereas spreading coefficient of heptane could have the positive sign in case of application of hydrocarbon SAS. Using of combined compounds, containing hydrocarbon SAS as the main component and fluorinated SAS as the small additives, leads to various situations in mutual spreading and wetting in a solution - hydrocarbon system. Role of interfacial tension in the structure of spreading coefficient can be revealed under conditions when spreading coefficient has positive value

[*]Corresponding author: da-vip@mail.ru

and interfacial tension changes from 2,5 mN/m to 0,2 mN/m. It is known that electrolytes reduce interfacial tension of aqueous solutions of surfactants that sharply increases surface tension of aqueous solution on a border with air. For effective extinguishing of emergency spill of oil product by compact streams of foam given into the bottom of tank or into the fire zone from long distance, it is necessary to provide positive value of spreading coefficient of solution [5-8]. In the course of experiments it has been revealed that decrease of interfacial tension concurrently increases specific consumption of foaming agent.

The aim of this research is to identify the role of interfacial tension value in ensuring of optimum parameters dur-ing extinguishing of the flame of oil products. For this purpose it have been conducted experimental studies during which value of spreading coefficient was supported on a positive level, while value of interfacial tension was gradu-ally reduced due to additives of electrolyte - ammonium chloride. The difficulty of experiment was in searching of stabilizers which could provide constancy of surface tension of aqueous solutions at a high concentration of salt. It became possible to lower the interphase tension from 2,2 mN/m to 0,1 mN/m. As a rule, positive value of spreading coefficient of aqueous solution causes high fire extinguishing efficiency of foam giving onto the burning surface and into the bottom of tank, directly into the layer of fuel [9-16].

Under the terms of research, foaming agent should contain fluorinated stabilizers which provide value of surface tension less than 20 mN/m. As the basis it was used mixture of primary alkylsulfates, with carbon chain of 8 - 12 atoms, and 0,05 m% (mass percent) of perfluorochemical sulfobetaine with radical, containing eight perfluorochemical atoms of carbon. Production of this component is limited in connection with its negative influ-ence on environment. Results of extinguishing of a heptane flame by giving of foam into the bottom of tank are presented in fig. 1. As the concentration of electrolyte - ammonium chloride - increases, the spreading coefficient of solution remains con-stant after slight decrease. The value of spreading coefficient of fuel, flowing over solution, increases. Interfacial tension sharply decreases from 2,0 mN/m to 0,1 mN/m. Curves of dependence of specific consumption on flow rate of foam solution are presented in fig. 2.

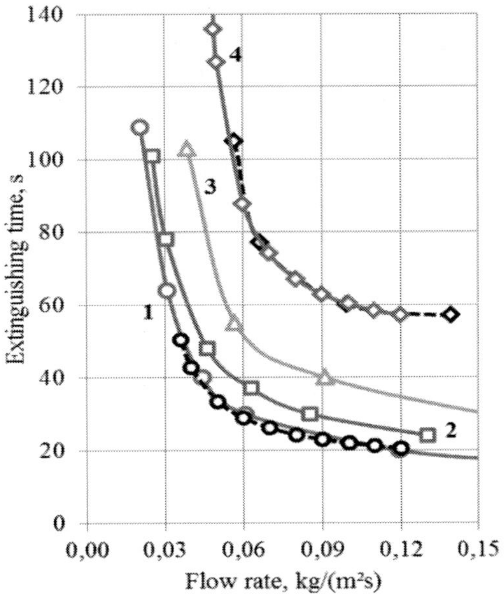

Fig.1. Dependence of extinguishing time on flow rate of foam at a various content of ammonium chloride in aqueous solution, m%: 1- 0,0; 2- 2,0; 3-5,0; 4-10,0; 5- 20,0. (Curves, calculated by

formula (1), are shown by dotted line).

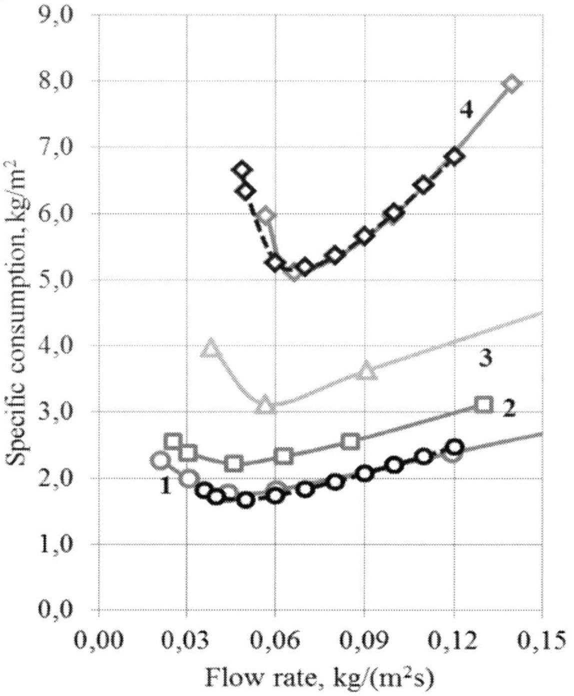

Fig.2. Dependence of specific consumption on flow rate of foam at a various content of ammonium chloride in aqueous solution, m%: 1- 0,0; 2- 2,0; 3-5,0; 4-10,0. (Curves, calculated by formula (2), are shown by dotted line).

Curves of dependence have the obviously expressed minimum point which is displaced to the area of large values in process of increase of electrolyte concentration and decrease of interfacial tension. Decrease of interfacial tension leads to increase of minimum consumption of foaming agent and optimum flow rate of foam. Extinguishing time increases synchronously.

Dependence of surface and interfacial tensions on concentration of foaming agent, at constant content of ammonium chloride, is presented in fig. 3. Increase of surface tension and decrease of interfacial tension can be related to influence of electrolyte additives on dissociation degree of molecules of a surface-active basis of foaming agent. Molecules of amphoteric SAS lose dissociation ability and behave as nonionic SAS at high concentration of electrolyte.

 DOI: 10.1051/matecconf/20168604038

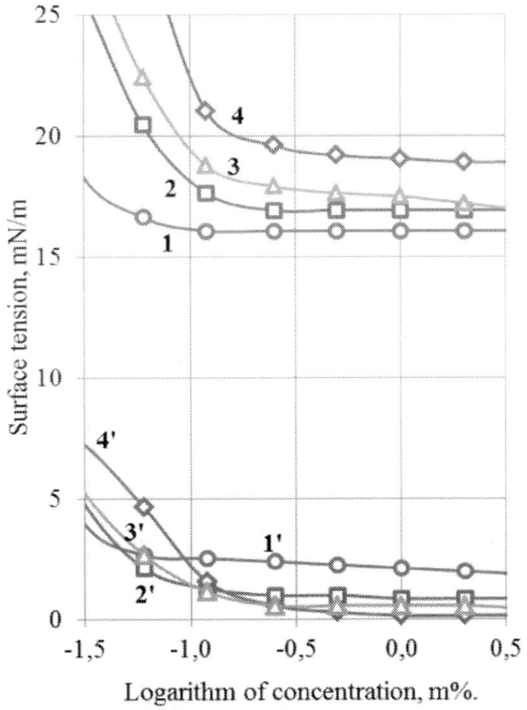

Fig. 3. Isothermal curves of surface (1-4) and interfacial (1′-4′) tensions of aqueous solutions on a border with air and heptane at a content of ammonium chloride, m%: 1, 1′- 0,0; 2, 2′- 2,0; 3, 3′-5,0; 4, 4′-10,0.

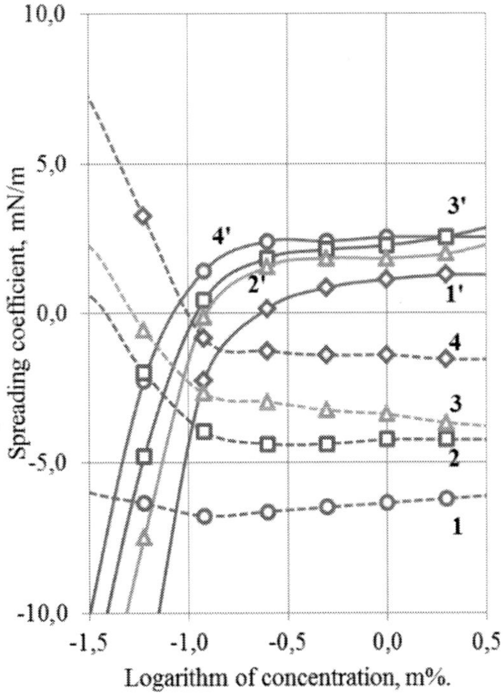

Fig. 4. Influence of ammonium chloride on spreading coefficient of solution flowing over heptane and heptane flowing over foam solution at a content of ammonium chloride, m%: 1, 1*- 0,0; 2, 2*-

 DOI: 10.1051/matecconf/20168604038

2,0; 3, 3*-5,0; 4, 4*-10,0; 5, 5*- 20,0.

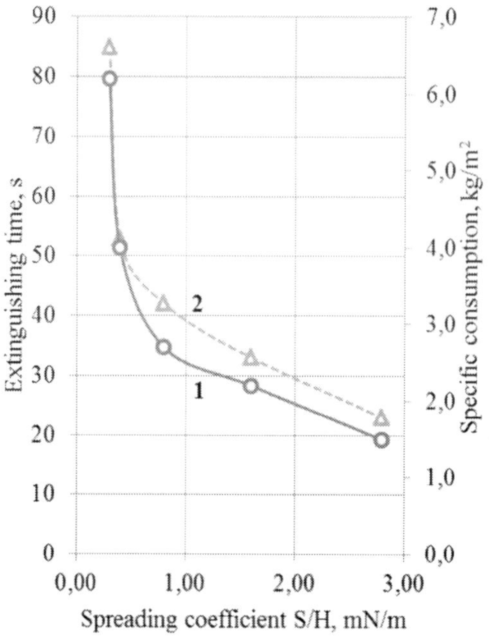

Fig.5. Dependence of heptane extinguishing time (1) and specific consumption of fluorinated SAS solution with electrolyte NH_4Cl (2) on spreading coefficient of solution, flowing over heptane (S/H).

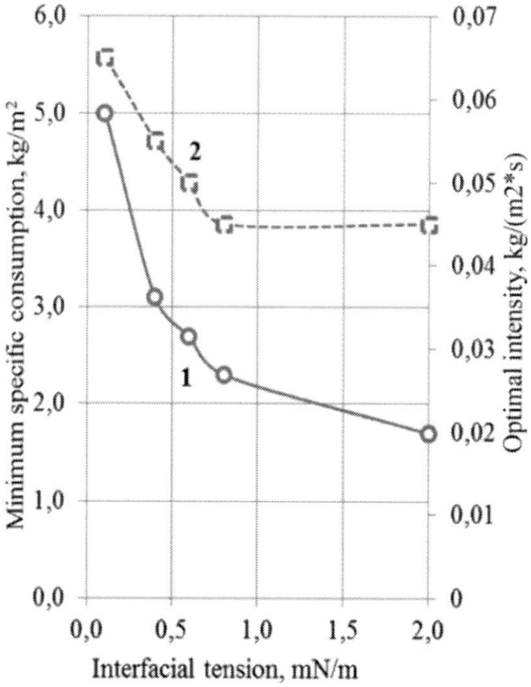

Fig.6. Dependence of minimum specific consumption (1) and optimum flow rate of foam on interfacial tension during extinguishing of a heptane flame by foam solution with ammonium chloride additives.

Dependence, presented in fig. 4, illustrates changes of two spreading parameters concurrently. The spreading coefficient of solution, flowing over heptane, slightly decreases, but keeps positive value. While the spreading coef-ficient of a heptane, flowing over solution, gradually increases from - 6,0 mN/m to - 1,5 mN/m. Increase of concen-tration of a salt leads to decrease of interfacial tension on a border with heptane, and, as a result, fire extinguishing efficiency also decreases. This fact is more important as in this case value of spreading coefficient of solution re-mains positive, i.e. aqueous solution keeps the possibility of spontaneous spreading over hydrocarbon. Dependen-cies of spreading coefficient and value of interfacial tension are presented in fig. 5 and fig. 6. Fire extinguishing ef-ficiency of foam becomes lower together with declining of hydrocarbon spreading over solution and decreasing of interfacial tension.
It is important conclusion that decrease of interfacial tension, if spreading coefficient of aqueous solution is positive, leads to decrease of fire extinguishing efficiency of foam. The value of minimum specific consumption increases from 1,7 kg/m2 to 5,0 kg/m2, and the value of optimum flow rate rises from 0,046 kg/m2s to 0,066 kg/m2s.

Change of the value of parameters, defining fire extinguishing efficiency of foam, is connected with increase of the specific fracture rate of foam and increase of the thickness of minimal isolating foam layer. These values can be calculated on the basis of analysis of dependence of time and specific consumption on flow rate of foam-forming solution using formulas presented in works [10-13].

$$\tau_T = -\frac{\rho h_0}{J_C}(1+(\frac{J}{J_C})^n)\ln\left(1-\frac{J_C}{J}\right) \tag{3}$$

ρ - foam density, kg/m^3; h_0 – minimal thickness of isolating layer, m; J_c- critical flow rate, kg/m^{2}s; J – current value of solution flow rate, kg/m^2, n – exponential quantity, characterizing the rate of rise of an average thickness of extinguishing layer.

For specific consumption Q: $Q = J^*\tau_T$,

$$Q = -\frac{\rho h_0 J}{J_C}(1+(\frac{J}{J_C})^n)\ln\left(1-\frac{J_C}{J}\right) \tag{4}$$

The results of calculation, using formulas (3) and (4), are selectively presented in fig. 1 and fig. 2.

According to results of calculation, values of minimal thickness of isolating layer - h0 and specific fracture rate of foam – Jc increase due to large concentration of electrolyte. Specific fracture rate of foam without additives of electrolyte is 0,028 kg/m²s, and with additives of electrolyte (2,0 m%) is 0,031 kg/m²s. Minimal thickness of foam isolating layer, formed from solutions with large concentration of salt increases from 4,1 mm to 5,1 mm.

On the basis of experimental studies it is shown that addition of electrolyte - ammonium chloride - into aqueous solution of film-forming agent decrease interfacial tension on a border with heptane and increase surface tension on a border with air.

It is experimentally shown that dependence of specific consumption on flow rate of foaming agent passes through a minimum point that allow to define influence of concentration of electrolyte on a value of minimum specific consumption and optimum value of flow rate of foam during extinguishing of a hydrocarbon flame. The spreading structure in a foam solution - hydrocarbon system determines the value of fire extinguishing efficiency of foam which decreases together with lowering of spreading coefficient and interfacial tension values.

The possibility of definition of the minimal thickness of foam isolating layer and specific fracture rate, using calculated ratios which connect the value of time and specific consumption of foaming agent with flow rate of solution, is shown.

References

1. D. White. Industrial Fire World, **21(1)**, 8–10 (2006)
2. K. D. Westwood, Industrial Fire World **21(1)**, 4–7 (2006)
3. A. F. Sharovarnikov, S. S. Voevoda, Proceedings of All-Russian meeting-seminar **1**, 41–47 (1997)
4. A. D. Katayev, All-Russian meeting-seminar **1**, 57–63 (1997)
5. A. F. Sharovarnikov, A. A. Yefimov, Fire Business **12**, 91–96 (1991)
6. A. F. Sharovarnikov, N. K. Grashichev, S. S. Voyevoda, Collection of scientific papers 0000A, 32-35 (1991)
7. A. F. Sharovarnikov, V. P. Molchanov, Fire Business **1**, 83–85 (1995)
8. A. Nakakuki, Haykan gidzyutsu kenkyu kekaysi **21(2)**, 73–77 (1981)
9. N. Nurimoto, Kasay **27(3)**, 11–19 (1977)
10. D. Korolchenko, A. Tusnin, S. Trushina, A. Korolchenko, International Journal of Applied Engineering Research **10(21)**, 42541-42548 (2015)
11. D. A. Korolchenko, A.F. Sharovarnikov, Advanced Materials Research **1070-1072**, 1794-1798 (2015)
12. D. A. Korolchenko, A. F. Sharovarnikov, E. N. Degaev, Science Review **8**, 114–120 (2015)
13. A. F. Sharovarnikov, D. A. Korolchenko, Applied Mechanics and Materials **475-476**, 1344–1350 (2014) DOI: 10.4028/www.scientific.net/AMM.475-476.1344
14. A.D. Korolchenko, A.F. Sharovarnikov, A.V. Byakov, Advanced Materials Research **1073-1076**, 2353-2357 (2015)
15. A.D. Korolchenko, A. F. Sharovarnikov, Fire and Explosion Safety **23(11)**, 84-88 (2014) DOI: 10.18322/PVB.2014.23.11.84-88
16. A.D. Korolchenko, A. F. Sharovarnikov, Fire and Explosion Safety **23(12)**, 59-68 (2014) DOI: 10.18322/PVB.2014.23.12.59-68

Field study method for indoor air of linear underground facilities

Kirill Lushin [1,*], *Natalia* Plyushchenko [1], *Aleksey* Ananyev [2]

[1] Moscow State University of Civil Engineering, 129337, Moscow, Yaroslavskoye sh., 26, Russia
[2] Russian Academy of Architecture and Construction Sciences (NIISF RAASN), Russia

Abstract. The article describes the approach to the organization of field studies of air and thermal regime of underground structures on the basis of experiments carried out in the underground channels with sufficient maintenance.

1 Introduction

Modern cities are complex system with comprehensive relations between all of its components. In everyday practice it is ordinary to meet high requirements to such factors as urban space using efficiency, enhancement of safety level, meeting the ecological requirements and introduction of energy-saving technologies. Enhancement of usage efficiency of urban space is impossible without enhancement of underground facilities network. Among them the transport as well as communication engineering tunnels play a special part.

Underground facilities and indoor air quality is a well-known field for researches in Moscow state University of Civil Engineering [1, 2]. For many years there were field and theoretical researches for different issues in design, construction and maintenance of city tunnels, metro stations, communication channels and so on.

History of global construction and maintenance of such objects has many decades. The number and length of these objects confirms its importance in terms of sustainable development of cities as well as a budget planning for promotion and enhancement of urban functionalities. Improving of energy efficiency of exploitation of urban underground facilities can be a significant reserve of power service at regional and national level. In that context the Moscow engineering stuff from different relevant organizations works towards to optimization and introduction of innovative solutions and scientific justification of real-life experience in construction area as well as service of tunnel for different purpose, different type and varied placement depth [1, 2]. In addition, scientists from the Department of Heating and Ventilation (MGSU, Moscow) carry out theoretical investigations in field of air-thermal conditions of underground facilities. Specific of the last studies is active application of math modeling with using of computer-assisted numerical experiment for building information model (BIM) [3, 4]. For verification of earlier data and correction of further research process some of field studies in Moscow underground buildings is realized.

* Corresponding author: LushinKI@mgsu.ru

The main problem solved by tunnel ventilation system is keeping the air-thermal and gas balance inner space at acceptable level for operating equipment and networks as well as providing of required hygiene and sanitary conditions and safety of staff during revising and work performance. Environment disadvantages influenced on air-thermal and gas balance of the building can be followings: increased and decreased humidity as well as inner temperature and gas accumulation. Herewith the some factors are connected each other's and some factors are incompatible. Thus, solving the problem of gas accumulation in inner space of underground facility by ventilation the negative factor is replaced by others followings factors: increased or decreased air temperature, excess moisture and gas accumulation from exhaust of automobile transport. Efficiency of the tunnel ventilation system could be estimated according to intensity of reduction of influence of negative factors with length of tunnel on distance between plenum ventilation chamber and exhaust shaft. Theoretically it is possible to estimate the efficiency of tunnel ventilation on the base of temperature and humidity data as well as air gas composition. Actually, one of the affordable methods for field study is indirect assessment of ventilation system activity according to temperature factor.

Field study method for indoor air of linear underground facilities

As instrument for investigations was used the hardware-software complex on the base of network of autonomous compact programmed temperature registers Testo 174 –type, high-performance laptop with special software.

Also laboratory machine consists of elements of wire edge to connection with registers as well as package consisting of multi-function apparatus testo 435-4 providing with replaceable probes to measure parameters of temperature, humidity and air movement as well as temperature of solid surface. Working plan supposes subsequent uniform placement of temperature registers along the tunnel length in area from supply grill of ventilation system to input terminal into air flue for a long period of time. Depending of dynamic properties of air-thermal tunnel conditions it is possible the preprogramming of measurement and data recording for different period of time. So, as optimal value of measurement time period for most parts of studies the 3-minute period is used. In this case the register memory size is enough for data recording in non-stop regime during a week. It should be noted the part of data redistricted with equipment in initial and final time period of experiment can be ignored when its analysis due to interferences caused by human factor. All registers when programming was synchronized to each other in time, measurement time period as well as form of data recording.

As study subject the some typical tunnels from different regions of Moscow city are chosen. Among studied objects the facilities with long service history such as tunnels putted into service 3–4 years ago as well as deep placement tunnels are investigated. The special attention is paid to introduction of tunnels with operational heat network as well as only with communication and power cables. Also there are tunnels built by cut-and-cover and shield-driven tunnels [6, 7]. Studies of air-thermal conditions and efficiency of ventilation system are realized for all seasons during a year. All studied objects are in working condition. They are applied regularly and supplied with followings: connection system as well as fire and security signal system; emergency illumination; batch-type plenum mechanical system. Aspiration from tunnels is accomplished naturally as well as by plenum ventilation system.

Activation of ventilation system was realized regularly by engineering staff according to the following regime: an hour in the morning before work day; an hour in the evening after work day. During the study the regime was stable. In case of necessity of realization of large-scale works the opportunity to activate a ventilation system any time and during any time period was organized. During all the experimental period the tunnels normally worked without remedial maintenance, emergencies and failures in operation. It should be noted the technical state of the studied buildings is at high quality level [5] due to the «Moskollektor» company support.

Operational service works efficiently keeping in good order and in a hygienic condition. The strong access mode and safety compliance control onsite is supplied. Technical staff takes required training in area of the studied parameters of air-thermal conditions. The above factors significantly influent on quality and quantity of the data obtained. Partial results are presented in earlier study reports [8, 9].

In Fig. 1 the results of analysis of network register data from the tunnel Grazhdansky obtained for a month (May–June) are presented. Durability of measurement is 3 min. On y-axis the temperature data are presented. L8-register is placed outside in shaded place near the one of the air-intake shafts. L6-register is placed on cable cantilever in height-middle zone in the middle of the tunnel length. L5 and L7-registers are placed in the same cross-section in top and bottom zones of tunnel, respectively. L1, L2, L3, L4-registers are equally spaced in top of tunnel zone from pull-out shaft to plenum shaft. In Fig. 2 the placement of registers along the tunnel is shown schematically. In Fig.3 the example of placement of one of registers of top tunnel zone is presented. In Fig. 1 we can see variation of air temperature in controlled points vs. external air temperature. Also we can observe regular activation of mechanical ventilation system, especially in the morning before work day. Best of all the moment of ventilation system activation is fixed by L4-register (that is next to inflow chamber).

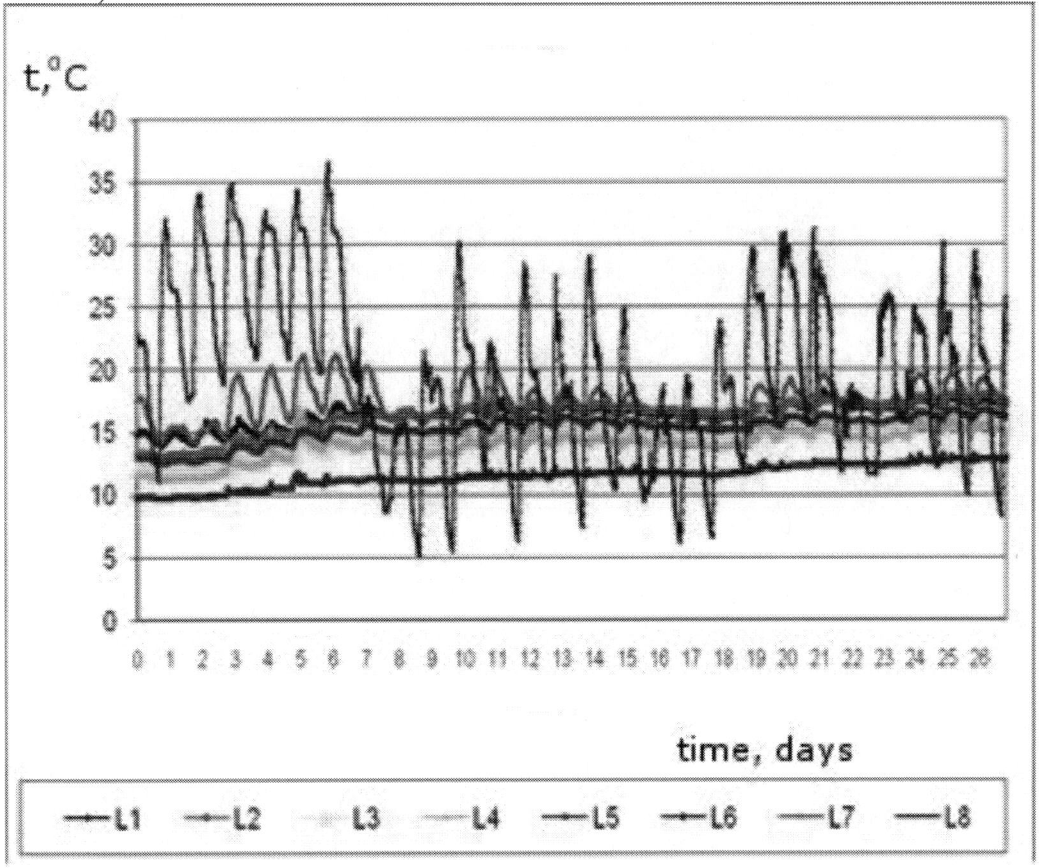

Fig. 1 Variation of air temperature in the tunnel Grazhdansky from 25.05.2007–21.06.2007, 1. Temperature, °C 2. Days

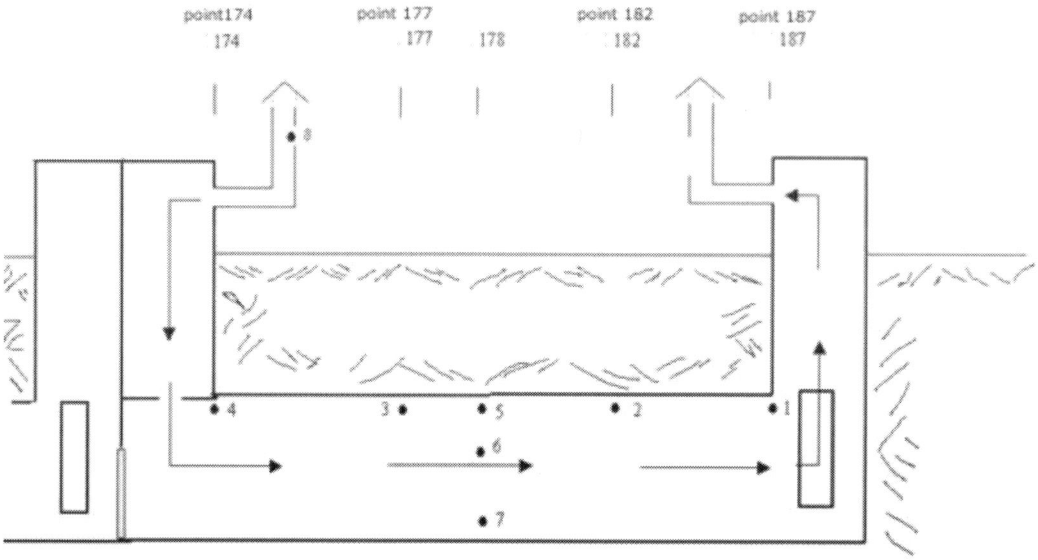

Fig. 2 Scheme of the sensor placement

Fig. 3 Placing the sensor at one of the measuring points.

Registers placed far from an inflow chamber fix weak variations of air temperature at the same moment. It can point out that the air when moving along a tunnel is heated as well as the efficiency of ventilation system is not acceptable to provide air exchange throughout the length of regular section of underground building. L6-register quite expressively fixes the peaks of evening increased loads on electrical cable, placed in tunnel. L2-register is placed in the warmest zone of tunnel. Its values are always 1–1.5 higher than others analogous. According to Fig. 1 irregular temperature variation of external air as well as irregular activation of ventilation system significantly influent on data of underground apparatus.

Conclusion

The presented analytic method allows accumulation of wide base of results of field measurement for linear underground buildings of different type. At present time a correction of earlier applied mathematical models [10] for further improvement of theoretical base for engineering investigations and standardization when design and performance of the work when construction of underground buildings is realized. The developed method allows highly-accurate qualitative monitoring of environment and heat conditions of buildings for different purpose [11-12] and with different properties. Hardware system is easily scaled and adapted for specific engineering tasks. Accumulated experience allows quickly analyzing the measured results as well as estimation and forecasting of air-thermal conditions of buildings and constructions.

References

1. E.V. Kaznin, Preceedings 52. Heating and ventilation **1**, 102-108 (1967)
2. E.V. Kaznin, Water-supply and sanitary engineering **4**, (1966)
3. E.V. Korolev, V.A. Smirnov, Advanced Materials Research **746**, 277-280 (2013)
4. A. Volkov,A. Sedov, P. Chelyshkov, E. Kulikova, Applied Mechanics and Materials **584-586**, 761-764 (2014)
5. N. Makisha, E. Gogina, Applied Mechanics and Materials **587-589**, 644-647 (2014)
6. V. Gagarin, A.Y. Neklyudov, International Journal for Housing Science and Its Applications **39 (2)**, 79-87 (2015)
7. V. Orlov, A. Andrianov, Applied Mechanics and Materials **580-583**, 2398-2402 (2014)
8. Research report, Stage 2, MGSU (2007)
9. Research report, Stage 3, MGSU (2007)
10. Y.Y. Kuvshinov, A.G. Rymarov, Proceedings **1**, 22-25 (2003)
11. Y. Kharkin, E. Korol, A. Davidyuk, Procedia Engineering **117 (1)**, 172-178 (2015)
12. V. Gagarin, V. Kozlov, K. Lushin, International Journal of Applied Engineering Research **10(23)**, 43438-43441 (2015)

MATEC Web of Conferences **86**, 04040 (2016) DOI: 10.1051/matecconf/20168604040
IPICSE-2016

Analysis of the efficiency of the grinding process in closed circuit ball mills

Valery I. Telichenko[1], *Rashid R.* Sharapov[1*], *Svetlana Y.* Lozovaya[2] and *Vladimir I.* Skel[1]

[1]Moscow State University of Civil Engineering, Moscow, Russia
[2]Belgorod State Technological University named after V.G. Shukhov, Belgorod, Russia

Abstract. Ball mills for fine grinding cement clinker and additives are widely used around the world. To improve the efficiency of a ball grinding the ball mills are transferred in closed circuit with air-separators of various designs. In the article the analysis of existing grinding circuits on the basis of closed circuit ball mills made. Introduced a summary measure that allows to evaluate the performance of the grinding system according to various criterias. The conclusion about the prospects of data systems grinding when grinding to ordinary cement, in comparison with the grinding systems based on other principles of grinding.

1 Introduction

Technological system of grinding (TSG) in a closed circuit on the basis of the ball mill (BM), worldwide widespread [1-8]. With the development of dynamic separators (DS) of the third generation dramatically increased the energy efficiency of grinding systems and TSG "BM–DS" began to occupy a dominant position. In Fig. 1 shows the traditional scheme with a return of coarse in the beginning of the process to the I camera. They provide cement production parameters grain structure: $n = 0{,}9...1{,}1$; $d' = 10...25$ μm, which corresponds to the values of specific surface S ≈ 3000...6000 cm^{2}pg [9]. The energy specific consumption in this case is 35...80 kWhpt. TSG "BM–DS" equipped with ball mills with the diameter $D = 3.6...6$ m and a length $L = 12...16$ m with a power level of a single mill 3000...10 000 kW [10].

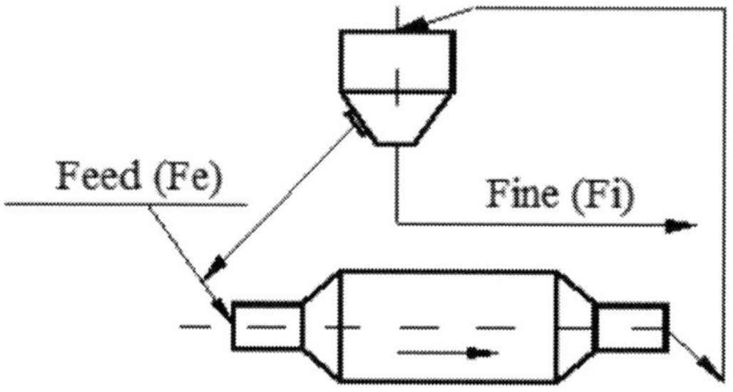

Fig. 1. Technological scheme of grinding "BM–DS"

* Corresponding author: ptdm_zavkaf@mail.ru

One of Russia's largest ball mill in size 4,6×14 m, with an power installation is 4250 kW [11] when the specific surface of the finished product in 3230 cm^2pg reached capacity 126 tph. In this case the specific energy consumption was 35 kWhpt. Metal consumption in the production output amounted to 6.9 tp(tph) at a specific volumetric productivity of 0.62 (tph)pm^3. An important indicator characterizing the efficiency of grinding are the specific energy consumption per unit specific surface area (qpS, kWhpm^2). For this grinding unit this parameter is $1.08 \cdot 10^{-4}$ kWhpm^2.

Experience of operation of ball mills suggests that transfer of mill in closed circuit with a separator of the dynamic type, or modernization of a closed circuits replacement of traditional separators to dynamic is an essential reserve to increase not only the efficiency but also the quality of the end product.

The company Christian Pfeiffer (Germany) carried out a reconstruction of TSG with ball mill open and closed circuits [12] with the installation in all cases the dynamic separator. When transferring to closed circuit a mill size 2.6×13 m at a constant quality cement (3000 cm^2pg) capacity grinding unit was increased from 25 to 38 tph while reducing the specific consumption of electricity from 41.6 to 29.3 kWhpt. The parameter qpS is reduced from 1.39 to $0.98 \cdot 10^{-4}$ kWhpm^2. Due to the more intense the aspiration of the mill and cooling separator air temperature of the finished product decreased from 120..130 °C up to 70...80 °C. The metal content of products made up of 8.95 t/(tph) at a specific volumetric productivity equal to 0,63 (t/h)pm^3 [12].

Replacement of outdated disign of the separator for ball mill 4×13.5 m allowed due to the optimization of the grain composition with constant brand of cement (CEM I 42.5 N) is to reduce the specific surface of the finished product with a 4670 to 3800 cm^2pg, which in turn allowed for a constant power consumption to improve the performance of the grinding unit 57 to 86 tph while reducing the specific consumption of electricity from 51.9 to 37 kWhpt. The parameter qpS also decreased from 1.11 to $0.97 \cdot 10^{-4}$ kWhpm^2. The power density of cement grinding amounted to 21.93 kWpm^3 and the temperature reduction of crushing products made from 105 to 60°C. The metal content of products amounted to 7.7 t/(tph) at a specific volumetric productivity equal to 0.6 (t/h)pm^3.

Analysis of the efficiency of the grinding process in closed circuit ball mills

Improving inner mill accessory (lining mill partition, the output diaphragm and composition of the grinding media) also allows to improve the process of grinding in a closed circuit ball mill. However, the performance of the grinding unit 4×13.5 m with the implementation of these activities is increasing to 98.8 to 117 tph while reducing the specific consumption of electricity from 28.9 to 22.2 kWhpt. The parameter qpS reaches a value of $0.81 \cdot 10^{-4}$ kWhpm^2 in the production of cement CEM I 32.5 N. The metal content of products amounted to 5.65 tp(tph) at a specific volumetric productivity is 0.775 (tph)pm^3 [12].

Requirements to energy efficiency of the grinding process in a closed circuit ball mill cause the necessity of searching the optimal schemes of grinding. As shown by the charts grinding in the first chamber of the ball mill in the area of mill partitions can be up to 40...50 % of the finished size of the material that dampers the energy of grinding media, of course, reduces the efficiency of the grinding process coarse fraction. To solve this problem worldwide [13] is widely distributed TSG "BM–DS" with intermediate separation of the products of grinding, allowing to allocate of the total material flow the fine fraction. The product separation, as a rule, is directed to grind in the first and second chamber, where grind to a standard condition (Fig. 2).

A similar installation has [14] a three-chamber ball mill with size 4×12 m, cyclone separator with a diameter of 5 m, and the power plant is of 3.10 MW. With a capacity of 82 tph specific energy consumption amount to 41.5 kWhpt. The metal content of products amounted to 7.9 tp(tph) at a specific volumetric productivity equal to 0,63 (tph)pm^3. In the production of cement of type CEM I-42,5 R and S is equal to 3100 cm^{2}pg the parameter qpS is $1.34\cdot10^{-4}$ kWhpm^2.

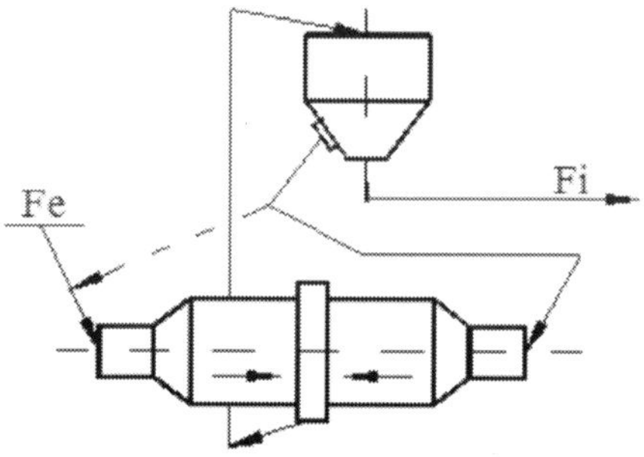

Fig. 2. Technological scheme grinding "BM–DS" with intermediate separation

When the external logic circuits discussed it has a number of weaknesses. First, located in the central part of the mill body window for unloading of grinding products significantly weaken the strength of the mill body, causing this crack, and secondly, the inability of the seal rubbing parts in the central part of the mill drum causes excessive dusting and spillage under the mill.

Also in this scheme the return material of the second camera is the almost finished product, which it is impractical to direct for further separation. To eliminate the last disadvantage developed TSG with intermediate separation with a direction of all coarse to the second chamber, the product on its dispersion is finish. Thus we can see an increase the parameter qpS to $1.16\cdot10^{-4}$ kWhpm^2. This may be explained primarily by the decline of the share over grinded material in the finished product. However, different mechanism of formation of the grain structure of products of both chambers of mills reduces the quality of the final product when it mixing.

The most promising of TSG with intermediate separation is a ball mill with two separators, two bucket elevators and a coarse direction parallel to the first and second chambers with independent circuits "camera–separator" for each camera [15]. This scheme is most economical in terms of specific energy consumption subject to offline configuration of each of the separators, depending on the dispersion of the incoming material. The disadvantage of the scheme is increasing the transport links, the complexity of system configuration and setup automatisation.

Limiting factor, limiting high efficiency ball grinding of clinker is the need to create optimal conditions for each stage of grinding in the drum mill. However, pre-grinding clinker to a particle size of 3...5 mm allows to exclude from the mill drum stage coarse grinding in a ball mill. Some progress got the TSG with pre-grinding cement clinker in a ball mill.

One of the most economical grinding plants of this type is used in Obourg (Belgium) (Fig. 3) [15].

Mill the first stage has a size of 6×3.5 m, power consumption 2020 kW and operates in a closed circuit with a vibrating screen with a mesh size of 2 mm. Mill of the second stage measuring 4.8×13.2 m working with two centrifugal separators with a diameter of 6.7 m for receiving the cyclones, the installed

capacity of the grinding unit 4410 kW. Product of grinding and classification the first stage enters to the separator of second stage.

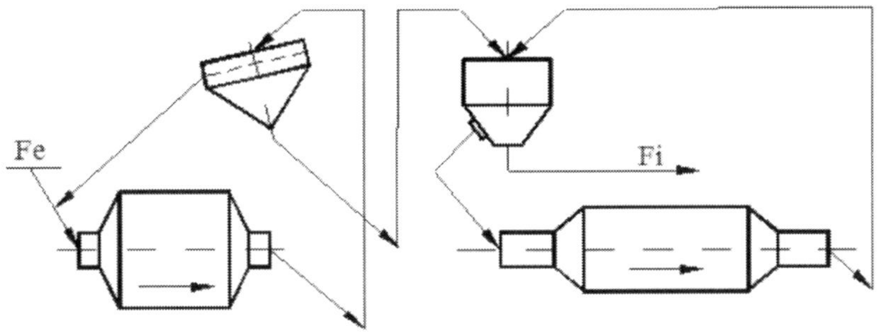

Fig. 3. Dual two-stage technological scheme of the drum mill and ball mills

Further, the product of separation (coarse) grinded till finish product in long ball mill. The plant capacity is 245...265 tph of cement with a specific surface area of 2700 cm^2pg (CEM I 32.5 N) and 155...170 tph cement with a specific surface of 3500 cm^2pg (CEM I 52,5 N). The total specific energy consumption in this case are, respectively, 27.7 and 44.3 kWhpt, including: by the mill the first stage is 7.7...11.9 kWhpt; the mill of the second stage is 16.4… 26.9 kWhpt; for the separation, transport and other auxiliary equipment 3.6…5.5 kWhpt. The parameter qpS respectively was up of $1.02 \cdot 10^{-4}$ and $1.27 \cdot 10^{-4}$ kWhpm^2. The metal content of products in the production of high quality cement was 5.6... 6.33 tp(tph) at a specific volumetric productivity of 0.5...to 0.56 (tph)pm^3.

Higher values of the parameter qpS and low values of specific volumetric productivity, compared with the TSG previously discussed types, with all the external logic of this scheme can be explained, first, low effectiveness of the separators. However, the concept forming the basis of this system is undoubtedly progressive for the following reasons:
– the grinding of clinker size of less than 2 mm increases the grinding efficiency in ball mill;
– in connection with increasing the amount of comminuted material passing through the drum of the ball mill, reduces the wear on the grinding bodies and lining per unit finished product;
– reduced power of ball mill of the second stage.

The drawback of this installation is the increase in items of main and auxiliary equipment requiring a large range of spare parts, therefore there is an increase in operating costs and capital investments.

Application of tube ball mill the first stage of pre-grinding (Fig. 4) allows you to get a fine product at the stage of rough grinding to evade from vibrating screen. Specific energy consumption for grinding of cement, corresponding to type CEM I 52,5 N, is 38.5 kWhpt, and the parameter q/S, respectively, of $1.05 \cdot 10^{-4}$ kWhpm^2, which is much better than a similar indicator of the previous scheme.

Figures and tables, as originals of good quality and well contrasted, are to be in their final form, ready for reproduction, pasted in the appropriate place in the text. Try to ensure that the size of the text in your figures is approximately the same size as the main text (10 point). Try to ensure that lines are no thinner than 0.25 point.

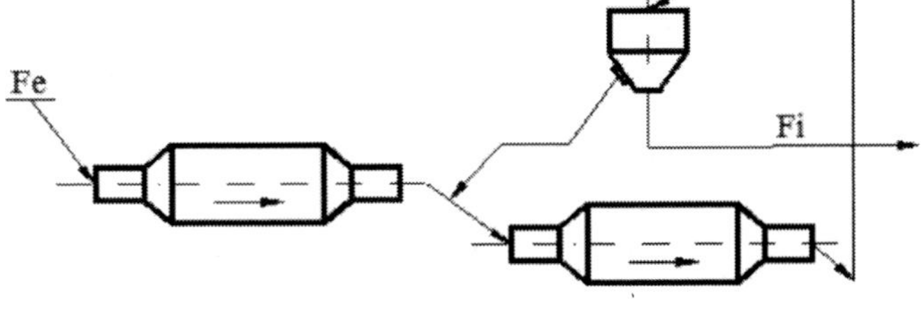

Fig. 4. Two-stage single-circuit of TSG with a tube ball mill under pre-grinding and "BM–DS" in the second stage

However, the intensive grinding effect of grinding media on the comminuted material causes increased wear of the lining tube ball mills of the first stage. Company Polysius created a ball mill "Doppelrotator", which is a drying and grinding installation with a drying chamber and two-way feed of raw material, coarse and hot gases. Material and gases are removed from it in the center across the output lattice. The difference between this mill, from the traditional is the fact that together with the flow of hot gases wet material through the trunnion of the mill hits in the beginning in the chamber of drying, where the lifting blades contribute to a more complete contact of the material with hot gases. From the chamber, drying the material through a slit partition with lifting blades gets into the camera coarse grinding, and then from the central discharge orifice is supplied to the separator. Large grains allocated in the separator, proceed in the chamber of a fine grinding, and some of them mixed with fresh material to improve its yield when entering the drying chamber. To passing a large amount of gas installed bearings with diameter up to 3.4 m under diameter of the drum of the mill 5.8 m Partitions the mills also designed to permit large quantities of gas. Currently operating mills "Doppelrotator" capacity up to 350 tph. With productivity of 270 tph of finished material and fineness of grind, the corresponding 12 % the residue on sieve 009, specific energy consumption accounting is 17.65 kWhpt for raw meal. The speed of gas flow in the coarse grinding chamber is 7.5 mps, and in the chamber of fine grinding – 2.5 mps. the Amount of off-gases from the mill is 400 000 m^3ph [16].

A modern approach to the improvement of TSG "BM–S" requires not only the installation of high-efficiency separators (S), but the separating of air flow and cleaning in separating systems for the dedusting of mill aspiration and technological separation of air. Moreover, increasing use is the scheme of deposition of the separated cement and clean separation of air in a powerful baghouse. Installation of a settling cyclone is not used, which allows to reduce the hydraulic resistance of the system and, accordingly, reduce the capacity of the separator fan.

Washing of crushing products by fresh air (in contrast to the circulating air in the system "separator–fan") allows to reduce the temperature of cement and coarse at 20...25 °C or more, first, improves the construction technological properties of the finished product, and secondly, allows to reduce the process temperature in ball mills.

Company FLSmidth (Denmark) for a similar reconstruction at the existing cement enterprises has developed a separator Sepax modification of IC [17-19]. On ball mill of size 4×12 m with the installation of a separate clean air flow when the finished product with a specific surface of 3400...3500 cm^2pg capacity was increased from 76 to 100 tph, and q decreased from 41.2 to 32.4 kWhpt. The parameter qpS decreased from 1.21 to 0.95 $\cdot 10^{-4}$ kWhpm^2. The separator showed the separation efficiency at S = 3400...3500 cm^2pg about 80 %.

Company Christian Pfeiffer (Germany) in October 2007 in Mykolaiv (Ukraine) [20, 21] reconstructed a closed circuit ball mill size 4,0×13,5 m with with a dynamic separator QDK 29-N. Finished product with separator air is deposited in the bag filter performance 143 000 m^3ph. For aspiration of air is set baghouse performance 55 000 m^3ph. In the same source mentioned that the same project was launched in Russia in October 2007 in Voskresensk on the mill size 4,0×13,5 m with installation of separator QDK 29-N with individual bag filter. The firm mentions the reconstruction of TSG with a ball mill in a number of countries.

The most powerful performance filter installed in 2008 at the cement plant in Malagoj (Poland). Here the existing mill size is 4.4×14.0 m integrated with the newly constructed a closed circuit by means of dynamic separator *QDK* 29-*NZ*. All finished product is deposited in the bag filter, the performance of which is 220 000 m^3ph. The company believes that this is rational reconstruction and notes a decrease in the temperature of the finished cement and coarse to 75...80 °C. Due to the improvement of the separation process in dynamic separators, optimization of grinding process of cement clinker in a ball mill in the production of cement CEM I 42.5 N specific energy consumption amounted to 32...34 kWhpt and the parameter qpS (0.93…0.98) 10^{-4} $kWhpm^2$. The metal content of products is 5.85 to 6.15 tp(tph) at a specific volumetric productivity, component 0.69... 0.74 $(tph)pm^3$, which confirms the economic feasibility of such a reconstruction of modern grinding system with ball mill.

Conclusions

An analysis conducted by TSG "BM–S" with different schemes of their arrangement, showed that the production of ordinary cement types CEM I 42.5 CEM I 32.5 and specific energy consumptions range is 45 to 31 kWhpt, the metal content of products per unit of production ranged from 5.65 to 9.0 tp(tph) at a specific volumetric productivity of 0.6 to 0.775 $(tph)pm^3$.

The parameter qpS ranged from to 0.81 to $1.34 \cdot 10^{-4}$ $kWhpm^2$. This is the most promising of TSG on the basis of ball mills are of TSG, equipped with dynamic separators with the deposition of the finished product in the bag filter and the separator discharge of treated air into the atmosphere. However, opportunities to improve the energy efficiency of grinding process, reducing metal consumption and improving a number of other indicators, existing TSG of the type "BM–DS", are close to the limit, but they are still competitive with the most modern grinding systems.

References

1. S.E. Andreev, V.V Zverevich, V.A. Perov. *Crushing, grinding and screening of minerals*. (Moscow, Nedra, 1980)
2. W.H. Duda. *Cement-data-book. Band 1 – Internationale Verfahrenstechniken der Zementindustrie*, (Wiesbaden, Berlin: Bauverlag. (1985)
3. R.R. Sharapov, V.S. Prokopenko, A.M. Agarkov. Bulletin of BSTU named after V. G. Shukhov. No. 2. p. 84-89. (2015)
4. Kaminsky, A. D., Kaminsky, A. A. Cement. **7**. 8-10. (1980)
5. 5 R.R. Sharapov. *Closed circuit ball mills*. (Belgorod, 270 p. 2008)
6. Grinding Mill – Rod, Ball and Autogenous. Mining magazine. **147**. 9. p. 91. (1982)
7. R.R. Sharapov, A. A. Uvarov, D. M. Annenko. Bulletin of BSTU named after V. G. Shukhov. **3**. 43-45. (2008)
8. M. Poliad, P. Coohet. World Cement. **9**. 395–399. (1990)
9. Aeroplex Fluidised bed opposed YRT Mills. Tiz – Fachberichte Rohstoff – Engeneerlng. **105**(12) 907–909 (1991)
10. A. Nerholm. *Grinding of cement. Symposium on cement manufacture*. (Moscow Niicement Research Institute,1979)
11. A.S. Mikhin, I.Z. Wortman, R. Amert. Cement and its application. **4**. 32–33 (2007)
12. E. Pichlmaier. Cement and its application. **2**. 41–45 (2000)
13. V.Z. Pirotsky, V.S. Bogdanov. Cement and its application. **6**. 12–16 (1998)
14. V.A.Bauman, B.V. Klushantsev, V.D. Martynov. *Mechanical equipment of enterprises of building materials, products and constructions*. (Mechanical Engineering, Moscow, 1981)
15. V.Z. Pirotsky. *Technology of grinding clinker and additives*. (Niicement (Research Institute, Moscow, 1992)

16. P. Tiggeebaumker, G. Blasczyk. Zement – Kalk – Gips. 4. p. 156–161. (1975)
17. R. McDowell, I. Mensz.Pit and Quarry, 12, 60–62 (1987)
18. The Sepax success in cement grinding now continues in raw grinding. Pit and Quarry. No. 12. p. 64. (1987)
19. N.S. Sevryugina. Construction and road machines, **7**, 28-29, (2007)
20. Global Projects. World Cement, 87–90. (2007)
21. N.S. Sevryugina, V.M. Babin. Building and road machines. **9**. 49-53 (2007)

Numerical investigation of some stationary solutions of the Carleman system

Olga Vasil'eva[1,*]

[1] Moscow State University of Civil Engineering, 26 Yaroslavskoe shosse, Moscow, 129337, Russia

Abstract. The negative stationary solutions of the boundary value problem for the Carleman system of equations are investigated. The kinetic Carleman system is a system of two nonlinear partial differential equations. The system describes the interaction of transportation and nonlinear processes. So, it is used for mathematical modelling of problems in various fields: the kinetic theory of gasses, the gas dynamics, the chemistry, ecology, acoustics etc. In particular, the system can be used to describe autokatalys problems for research of building materials. We present and discuss results of numerical investigation of negative problem solution for different values of parameters. There are three problem parameters domains. For the first parameters domain the stationary solution has stable character, for the second parameters domain the stationary solution has stochastic character and for third domain the stationary solution has unstable character.

1 Introduction

We consider the initial boundary problem for the kinetic Carleman system of equations. This system can be used for mathematical modelling problems of the kinetic theory of gasses, chemistry and in particular, for research of building materials. So, the investigation of the Carleman system is very interested now [1-5]. One of the important properties of the initial boundary problem for Carleman system is the stability of the stationary solutions. We consider the negative stationary solutions of the initial boundary problem for the Carleman system equation. The finite-difference methods are used for the investigation of the stationary solution properties.

2 Formulation of the problem

We consider the initial boundary problem for the Carleman system

$$\begin{aligned} &u_t + u_x = \varepsilon^{-1}(w^2 - u^2), \quad t > 0,\ 0 < x < 1, \\ &w_t - w_x = \varepsilon^{-1}(u^2 - w^2), \end{aligned} \tag{1}$$

with initial and boundary conditions

* Corresponding author: vasilijeva.ovas@yandex.ru

$$u|_{t=0} = u^0(x),\ w|_{t=0} = w^0(x)$$
$$u(t,0) = u_0(t),\ w(t,1) = w_1(t),\ u_0(0) = u^0(0),\ w_1(0) = w^0 \qquad (1).$$

Here $\varepsilon > 0$ is the first constant parameter problem.The second parameter of problem we denote by $c < 0$. The function $u^s(t,x) = u^s(x) = c$ and $w^s(t,x) = w^s(x) = c$ is the stationary solution (1) with following initial and boundary conditions $u^0(x) = c$, $w^0(x) = c$, $u_0(t) = c$, $w_1(t) = c$ for any $\varepsilon > 0$ and $c < 0$.

For the investigation properties of the stationary solutions we consider problem (1) with pertrubated initial condition

$$u^0(x) = c,\ w^0(x) = c + \alpha\sin(2\pi x) \qquad (2)$$

and boundary conditions

$$u_0(t) = c,\ w_0(t) = c. \qquad (3)$$

Denote the energy of pertrubation of stationary solution $u^s(x)$, $w^s(x)$ by

$$e_1(t) = \left(\int_0^1 (u(t,x) - u^s(x))^2 dx\right)^{\frac{1}{2}},$$

$$e_2(t) = \left(\int_0^1 (w(t,x) - w^s(x))^2 dx\right)^{\frac{1}{2}}$$

and total enargy by

$$E(t) = (e_1^2(t) + e_2^2(t))^{0.5}.$$

3 Results of numerical investigation

At first we consider the problem (1) – (3) for the $c = -0.1$, $\varepsilon = 0.6$ and $\alpha = 0.1$. Figure 1 shows the dependences of the energy of pertrubation of stationary solution $e_1(t)$, $e_2(t)$. Figure 2 presents the total perturbation energy, which tends to zero. Figure 3 presents phase plane (e_1, e_2). So we can see the stabilization the perturbation solution to stationary solution.

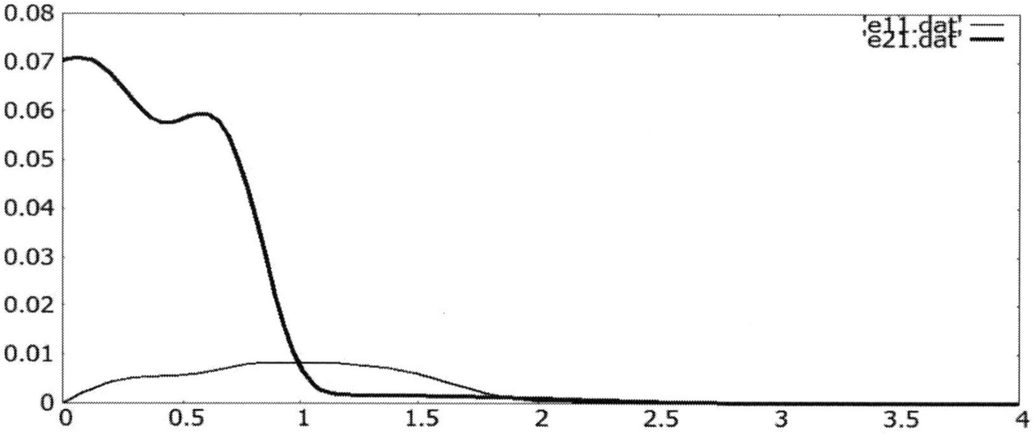

Fig. 1. Energy of perturbations.

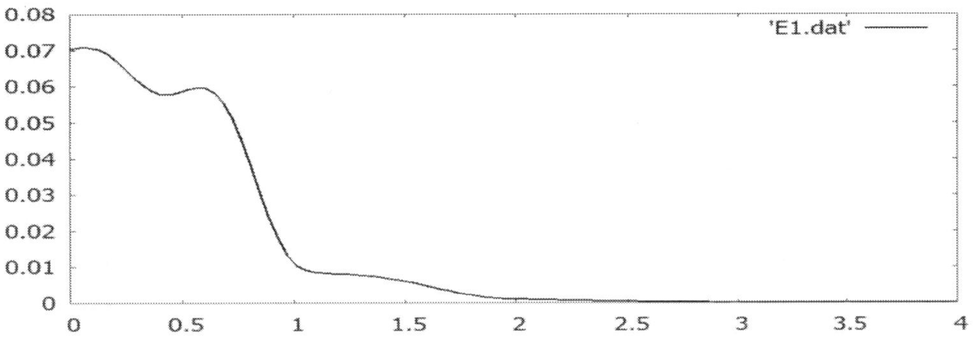

Fig.2. Total energy of perturbation.

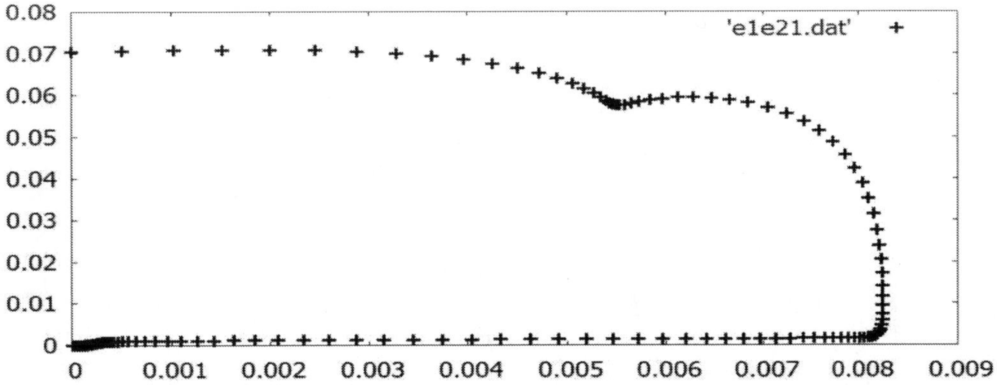

Fig. 3. Phase plane.

Figures 4 - 8 present results of numerical investigation of problem (1) - (3) for the $c = -0.5$, $\varepsilon = 0.6$ and $\alpha = 0.1$. Figure 4, 5 show numerical solution of problem $u(t,x)$ and $w(t,x)$ for the values of the time variable $t_0 = 0$, $t_1 = 0.8$, $t_2 = 1.6$, $t_3 = 2.4$, $t_4 = 3.2$, $t_5 = 4$. Figures 6, 7 present the dependences of the energy of pertrubation of stationary solution $e_1(t)$, $e_2(t)$ and total energy $E(t)$. Dependences have oscillatory nature. Figure 8 presents phase plane (e_1, e_2).

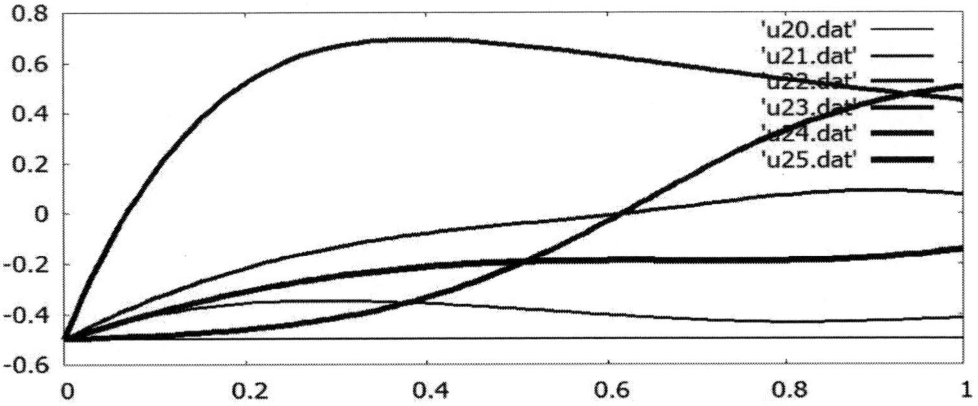

Fig. 4. Numerical solution of problem $u(t,x)$.

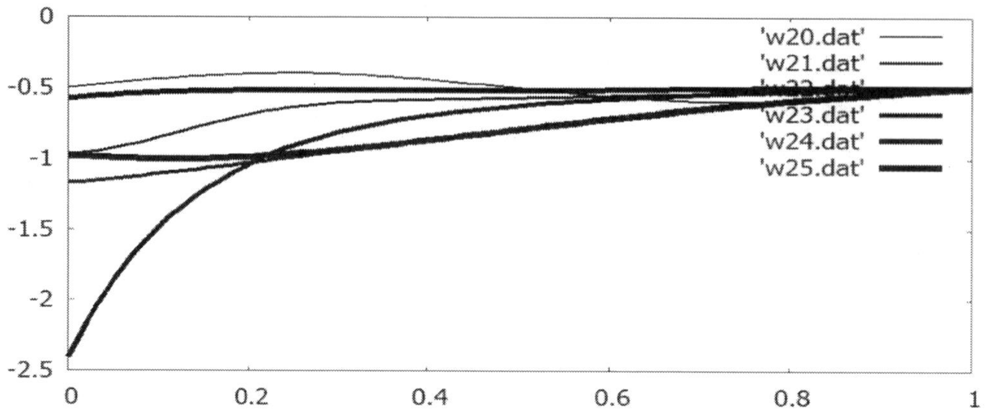

Fig. 5. Numerical solution of problem $w(t,x)$.

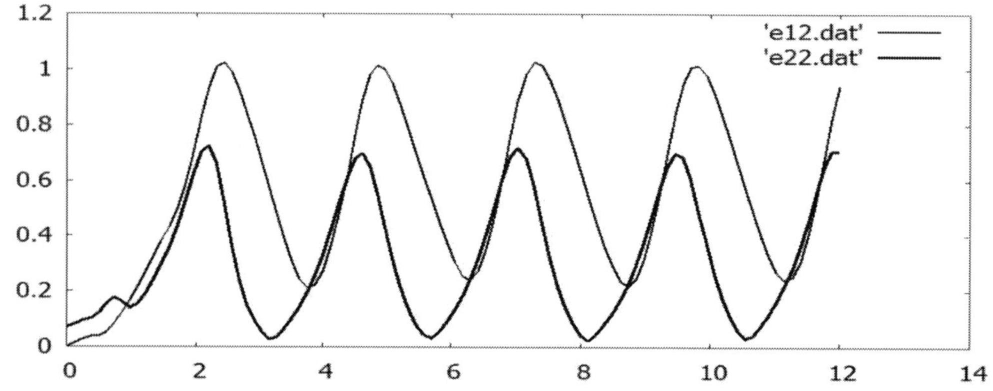

Fig. 6. Energies of perturbations.

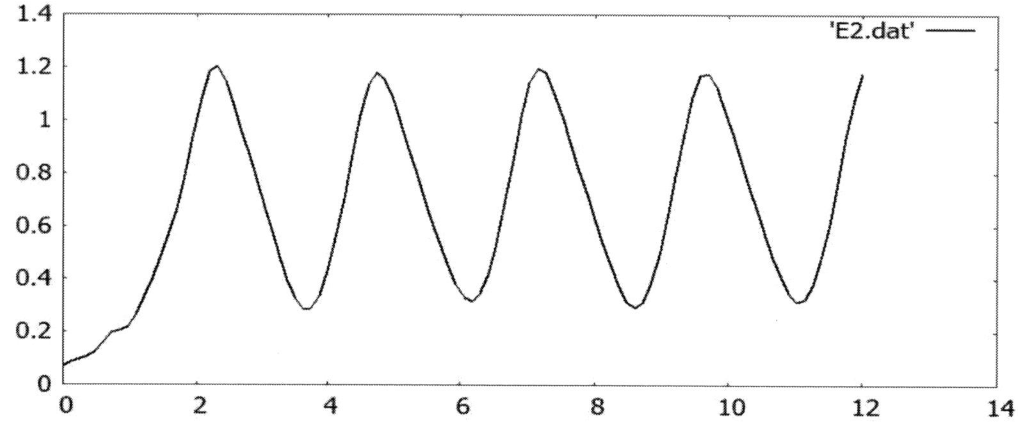

Fig. 7. Total energy

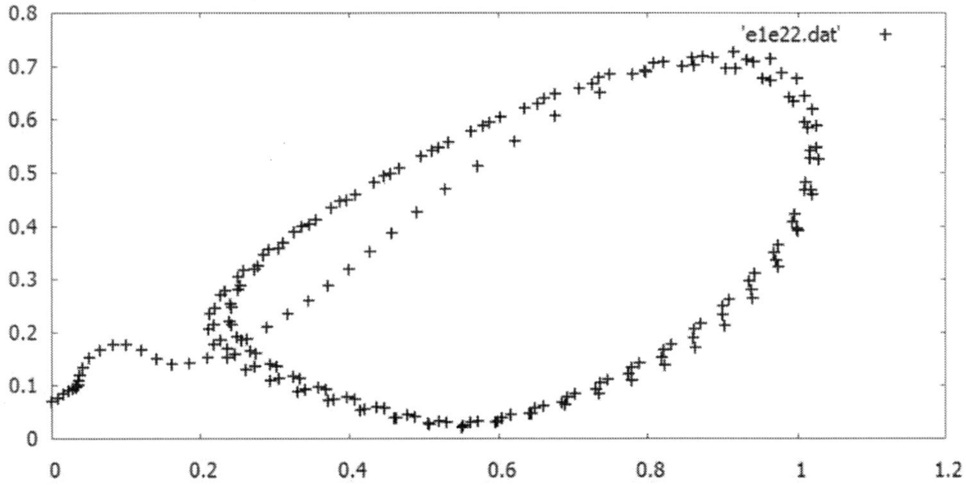

Fig. 8. Phase plane

Figures 9 -11 present results of numerical investigation of the problem (1) - (3) for $c = -1$, $\varepsilon = 0.6$ and $\alpha = 0.00001$. Here we can see energy increasing for small perturbation of the stationary solution ($\alpha = 0.00001$).

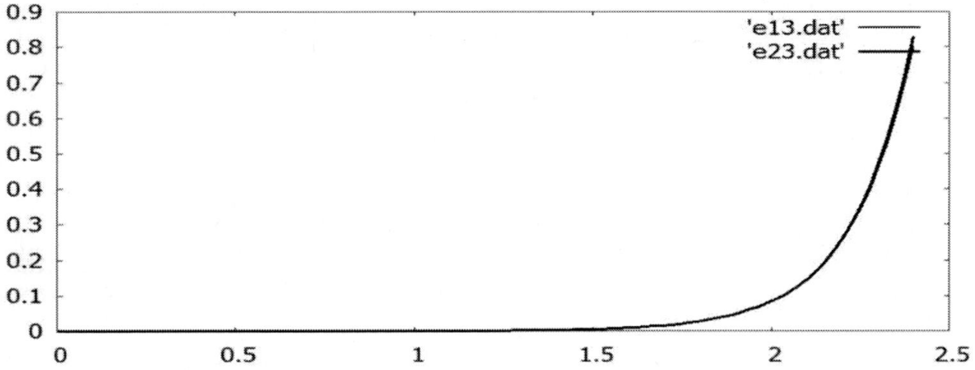

Fig. 9. Energies of perturbations.

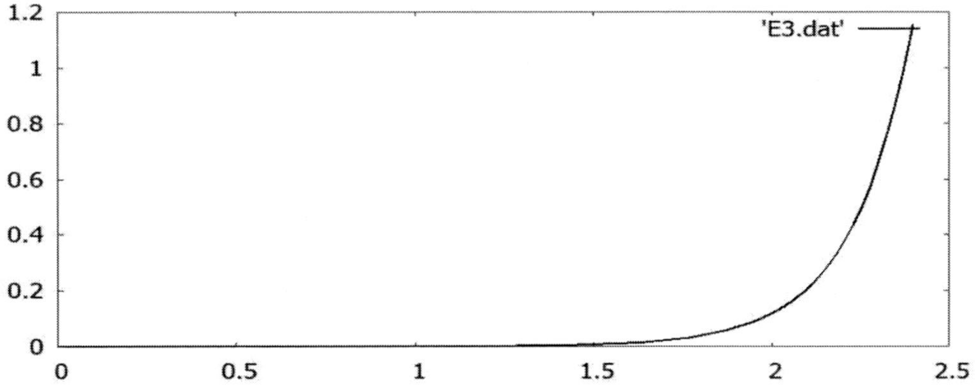

Fig. 10. Total energy.

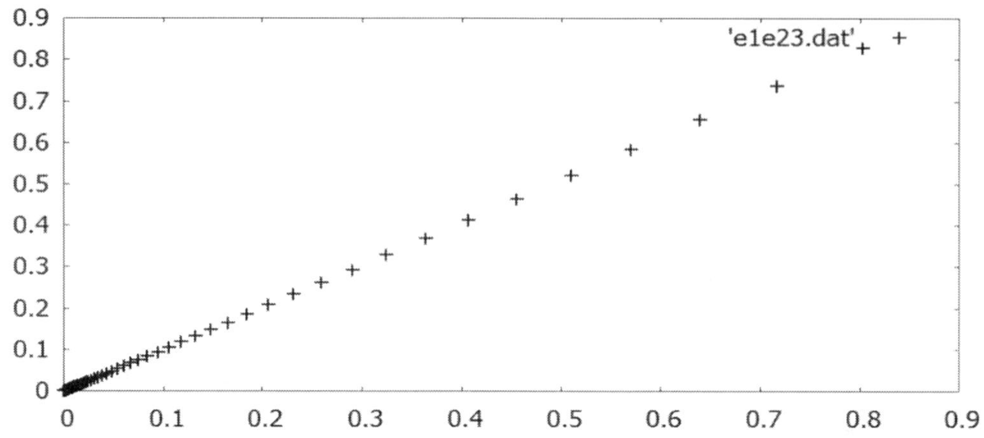

Fig. 11. Phase plane.

4 Conclusion

We present the results of numerical investigation of the perturbed stationary solution of the initial boundary problem for the Carleman system. Based on numerical investigation we can conclude that for some fixed ε there are three different kind of solution behavior. A perturbation of negative stationary solution close to zero leads to stabilization of the perturbed solution to the equilibrium state. With decreasing values of the parameter we have a chaotic nature of perturbed solution. With a further decrease of the parameter c, we can see the instability of the perturbed solution. The author would like to express gratitude to Professor E.V.Radkevich for formulation of the problem and useful discussions and comments.

References

1. S. K. Godunov, U. M. Sultangazin, Math. Surv. **26**, 3 (1971)
2. E. V. Radkevich, J. Math. Sci. **181**, 5 (2012)
3. E. V. Radkevich, J. Math. Sci. **184**, 4 (2012)
4. V. V. Aristov, O.V. Ilyin, Phys. Let. A. **374,** 4381-4384 (2010)
5. E. V. Radkevich, O. A.Vasil'eva, S. A.Duhnovskij, J. Math. Sci. **201,** 32 (2015)

Calcium hydroxide silylation reaction with trimethylchlorosilane

Anatoliy A. Novoselnov[1,*], and *Yuliya V.* Ustinova[1]

[1] Moscow State University of Civil Engineering, 26, Yaroslavskoye Shosse, 129337 Moscow, Russia

Abstract. The silylation reaction of a calcium hydroxide with a trimethylchlorosilane is studied as a silylation model by the gas-liquid chromatography. The silylation process is divided into three stages. A material balance of these stages is calculated. The schemes of the reactions at each stage of the process are proposed. The modified calcium hydroxide obtained at three repetitive stages of the silylation reaction has been investigated by the x-ray phase analysis, IR spectroscopy, thermal analysis, electron microscopy in a combination with the elemental analysis. It has been determined that at the first stage of the interaction the processes of the trimethylchlorosilane hydrolysis and of the hydrolysis products condensation dominate, and at the same time an adsorption process of the trimethylchlorosilane and its derivatives starts. Further, the hydrolysis of the trimethylchlorosilane by the «new» portions of a water formed in the reaction of a calcium hydroxide with a hydrogen chloride takes place, simultaneously the secondary reactions of the Si-O-Ca – ties' formation and cleavage occur including as a silylation-desilylation dynamic equilibrium process.

An interest to the silylation of the mineral components included in the building materials is caused by the fact that it is one of the most of effective chemical methods of the increase of exploitation characteristics of building materials and products and above all – hydrophobicity [1]. It is believed that the effectiveness of this chemical interaction is determined primarily by the chemical nature, both the construction material and the silylating agent.

The researches of the silylation reaction of the various building minerals with the organo-silicon compounds (OSC) have been conducted for several years [2,3,4,5,6,7].

An interaction between the different organosilanes with the functional groups at the silicon atom (Si) of the general formula R_nSiX_{4-n} (wherein: R=CH_3, C_2H_2, C_6H_5; n=1-3; X= Cl, OR; R= CH_3, C_2H_2) and CaO, $(3CaO{\cdot}SiO_2)_n$, $CaCO_3$, $CaSO_4$ and Portland cement, as well as their derivatives of hydrated minerals, has been studied in details. And it will be the subject of our further reports.

Portland cement is a multi-compound system. The main mineral of the Portland cement clinker is alite $(3CaO{\cdot}SiO_2)_n$. The most active part of the hydrated $(3CaO{\cdot}SiO_2)_n$ is $Ca(OH)_2$, which is the main product of its hydrolysis:

* Corresponding author: anovo@inbox.ru

$$(3CaO{\cdot}SiO_2)_n + 2n\ H_2O \rightarrow n\ Ca(OH)_2 + (2CaO{\cdot}SiO_2{\cdot}H_2O)_n \quad (1)$$

Thus the main researches have been aimed to the studies of the silylation reaction of the calcium hydroxide.

An investigation of the silylation reaction of the calcium hydroxide with the $Me_3Si\text{-}Cl$ trimethylchlorosilane (TMCS) has been carried out by the gas-liquid chromatography (GLC) method. The TMCS's concentration change in the reaction mixture has been observed. The silylation reaction has been carried out in a heterogeneous medium, with a molar ratio of mineral : OSC = 10 : 1, and stirring the reaction mixture at 18-20 °C for 300 minutes (hereinafter up to 7 days); inert solvent is nonane; internal standard is toluene [8]. The way of the reaction has been estimated by changes in the TMCS conversion in relation to its internal standard toluene. At the same time, we have conducted a control experiment without minerals. Under carrying out a control experiment any change in a TMCS's concentration in the reaction mixture hasn't been observed.

Trimethylchlorosilane $Me_3Si\text{-}Cl$ is one of the first organo-silicon compounds, which has been applied for the building materials modification [1]. The TMCS conversion change and the extraction of the hexamethyldisiloxane $Me_3Si\text{-}O\text{-}SiMe_3$ (HMDS), the main product of its hydrolysis, make it possible to calculate the material balance of the silylation reaction. The HMDS formation occurs under the hydrolysis of the TMCS with a water, which can be found in the calcium hydroxide, according to the following schemes:

$$Me_3Si\text{-}Cl + HOH \rightarrow Me_3Si\text{-}OH + HCl \quad (2)$$

$$Me_3Si\text{-}OH + HO\text{-}SiMe_3 \rightarrow Me_3Si\text{-}O\text{-}SiMe_3 + HOH \quad (3)$$

$$Me_3Si\text{-}Cl + HO\text{-}SiMe_3 \rightarrow Me_3Si\text{-}O\text{-}SiMe_3 + HCl \quad (4)$$

Hexamethyldisiloxane is a compound with a hydrolytically stable bond, however, it has been impossible to exclude its interaction with a calcium hydroxide under the scheme:

$$Me_3Si\text{-}O\text{-}SiMe_3 + Ca(OH)_2 \rightarrow Me_3SiOH + Me_3Si\text{-}O\text{-}Ca\text{-}OH \quad (5)$$

HMDS's reactivity to the $Ca(OH)_2$ has been examined by GLC method. At carrying out of the reaction (in a nonane, for 14 days) any change in the HMDS's concentration in the reaction mixture hasn't been observed.

The data of the X-ray diffraction, IR spectrum and thermal analysis of the modified sample are virtually indistinguishable from the test results of the initial calcium hydroxide. Even under the calcium hydroxide reflux with a fivefold molar excess of the HMDS in a nonane for 24 hours and at the reflux with a fivefold molar excess of the HMDS in an acetonitrile for 24 hours, any characteristic of the reaction according to the scheme 5 hasn't been found.

A Fig. 1 shows a dependence of the TMCS conversion from the time of the silylation (up to 300 minutes).

The dependence of the TMCS conversion from the time of the interaction with the calcium hydroxide (Fig. 1) shows that its conversion increases to 22% for 300 minutes of silylation.

An analysis of the data represented by the Fig.1 including a material balance of the products of the reaction mixture brings to the following conclusion: there are three periods observed in the silylation process: **I** – up to 60 minutes; **II** – from 60 to 180 minutes and **III** – from 180 to 300 minutes.

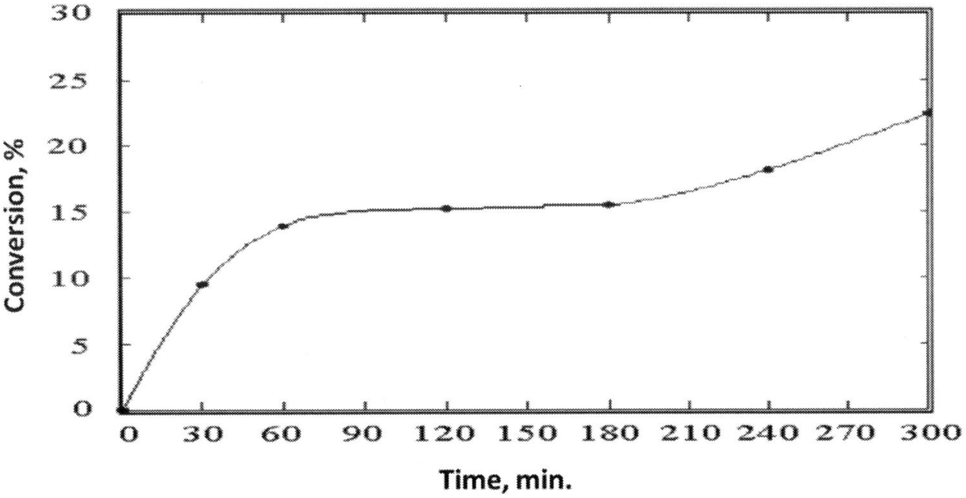

Fig. 1. A dependence of the TMCS conversion on the interaction time with $Ca(OH)_2$

The data shown above approve that an interaction of a TMCS with a calcium hydroxide is very complex and do not much correspond to the data given in [9] and represented for the general estimation of the functional organosilanes with the minerals, where three stages of the process with the different rates can be observed:

1. Stage of the «relatively rapid establishment of the equilibrium sorption».
2. Stage of the «sorption equilibrium», characterized by a slight change in the organosilane's concentration in the reaction mixture.
3. Stage of the «condensation», leading to the formation of the organopolysiloxane coating with its fixing on a mineral's surface.

To evaluate a role of a water adsorbed on a calcium hydroxide's surface, at the first stage of a process a silylation of the mineral with the different preparation degrees has been carried out: a) initial calcium hydroxide; b) calcium hydroxide heated for 3 hours under 140 °C; c) calcium hydroxide heated for 12 hours under 250 °C and vacuum of 0.5 mm. Hg.

It has been discovered that after 30 minutes, the proportion of the reacted calcium hydroxide, transformed to the HMDS is: 95% for (a), 86% for (b), and 14% for (c). Accordingly, the absorption of the TMCS (weight percentage of reacted) account for: (a) - 5%, (b) - 14%, and (c) - 86%. Thus, in a first stage (up to 60 minutes) the TMCS's interaction with a calcium hydroxide can be explained mainly by the hydrolysis of the TMCS by the adsorbed water.

The first period of the calcium hydroxide silylation with a trimethylchlorosilane (up to 60 minutes) is a period, which possibly corresponds to the stage of the «equilibrium sorption establishment». At the end of this period a trimethylchlorosilane's conversion rises to 14%.

Table 1. A material balance of the first period of the silylation

Periods, min.	**0-30**	**30-60**	**0-60**
A part of the TMCS, spent for the HMDS formation, %	70	98	83
A part of the TMCS, spent adsorption, %	30	2	17

The data of the material balance calculation show that in the first 30 minutes of silylation, a hydrolysis consumes and transforms into the HMDS 70% of the reacted TMCS, and 30% of the last one apparently adsorbs on a calcium hydroxide. The calculation of the material balance for the entire period – up to 60 minutes, has determined that a reaction, making the main contribution to the TMCS conversion increase, is its hydrolysis. Thus, at the first period of the reaction two processes – the TMCS hydrolysis and its adsorption on a calcium hydroxide – take place.

A modified calcium hydroxide, obtained from the reaction mixture at this period, has been investigated. The selected samples of the calcium hydroxide suspension in a nonane have been separated from the filtrate, washed with plenty of dry solvent and dried. Then the samples have been investigated by the methods of the physical and chemical analysis (X-ray diffraction, electron microscopy in conjunction with elemental analysis, infrared spectroscopy and thermal analysis).

The X-ray analysis of the sample has showed the presence of a minor amount of calcium hydrooxochloride of a composition of $Ca(OH)_2{\cdot}CaCl_2{\cdot}H_2O$.

The reaction between a TMCS and a calcium hydroxide (due to the TMCS hydrolysis) leads to the formation of the hydrogen chloride which reacts with calcium hydroxide according to the schemes:

$$2\ HCl + HO\text{-}Ca\text{-}OH \rightarrow CaCl_2 + 2\ H_2O \qquad (6)$$

$$CaCl_2 + H_2O + HO\text{-}Ca\text{-}OH \rightarrow Ca(OH)_2{\cdot}CaCl_2{\cdot}H_2O \qquad (7)$$

with a calcium chloride and calcium hydrooxochloride formation

As a modified sample contains of chlorides and crystalline hydrates of calcium chloride an interaction between the HMDS and calcium chlorides can't be eliminated. The possible reaction could correspond to the scheme:

$$Me_3Si\text{-}O\text{-}SiMe_3 + CaCl_2 \rightarrow Me_3SiCl + Me_3Si\text{-}O\text{-}CaCl \qquad (8)$$

A change of the HMDS's concentration in a reaction mixture at a molar ratio of $CaCl_2$: HMDS = 10:1 has been studied by GLC method. The silylation has carried out in a nonane; a calcined calcium chloride containing no water has been applied. In carrying out of the reaction for 14 days, any changes in the content of the HMDS in the reaction mixture are not detected.

The silylation has been carried out also with the watered calcium chloride. The HMDS decline in the reaction mixture at a molar ratio $CaCl_2{\cdot}nH_2O$: HMDS = 10:1 has been investigated by the GLC method, the reaction is carried out in the nonane. The possible reaction could occur according to the scheme:

$$Me_3Si\text{-}O\text{-}SiMe_3 + CaCl_2{\cdot}nH_2O \rightarrow Me_3SiOH + Me_3Si\text{-}O\text{-}CaCl + n\ H_2O + HCl \qquad (9)$$

On carrying out of the reaction within 14 days any changes in the content of the HMDS in the reaction mixture are not revealed also.

There are the absorption bands at: 3550 cm^{-1}, 1650 cm^{-1}, 1280 cm^{-1}, 840 cm^{-1} and 695 cm^{-1} in the IR spectrum of the modified sample.

The presence of the trimethylsilyl derivatives (-$SiMe_3$ group) in the modified sample is determined by the presence of the absorption bands: 1282-1250 cm^{-1}, 841 cm^{-1} and 756-695 cm^{-1} [1]. This spectrum contains absorption bands at 1280 cm^{-1}, 840 cm^{-1} and 695 cm^{-1}, it indicates to the presence of the silylation products in the modified sample.

The electronic microscopy of the modified sample has been carried out in an integrated setting «Comebax» (previously, the sample has been extracted with a toluene for 48 hours under dynamic conditions). The elemental analysis of the particles with a radius greater than 3 microns has been carried out with the «electronic probe» method. Sampling has been conducted on 100 particles. The content of silicon has been. It has been determined that the silicon content is 0.19%, which obviously confirms a chemisorption of the trimethylsilyl groups on a calcium hydroxide.

The endothermic heat effects, which correspond to the removal of adsorbed and structural water from the modified sample, can be found with a maximum degree of precision in the DTA thermogram, prescribed for the range 20-500 0 C.

The exothermic heat effect at 388 °C, which may correspond to the destruction of chemisorbed trimethylsilyl groups, is also observed in the DTA curve of this sample. Pascenco [10] has found that the stability limit of the film (on a substrate of a quartz glass) containing trimethylsilyl groups is in the range of 350 - 550 °C. There is a thermal effect which borders are 310 - 430 °C, in the DTA curve of the modified sample.

Thus, test data indicate that the silylation realization (under these conditions) leads to the chemisorption of trimethylsilyl groups on a calcium hydroxide. In case of a TMCS, we can speak only about one-point chemisorption

Thus, the analysis data indicate that at this stage of the reaction, we have the following processes:

1. The TMCS hydrolysis by the adsorbed (adsorbed water cannot exclude its participation in the TMCS hydrolysis, but its formation in the reaction, obviously exceeds its consumption) and structural forms of the water contained in the initial $Ca(OH)_2$.
2. The reactions between a hydrogen chloride and $Ca(OH)_2$ with $Ca(OH)_2{\cdot}CaCl_2{\cdot}H_2O$ and H_2O formation.
3. Equilibrium reactions can lead to the formation of Si-O-Ca- bonds according to the scheme:

$$Ca(OH)_2 + Me_3SiCl \leftrightarrow Me_3SiOCaOH + HCl \quad (10)$$

$$|\text{-}CaOH + Me_3SiCl \leftrightarrow |\text{-}CaOSiMe_3 + HCl \quad (11)$$

where $|$-CaOH is a schematic representation of the partially reacting $Ca(OH)_2$ surface.

The second period of the silylation (from 60 minutes to 180 minutes) is a period corresponding to the possible stage of the «sorption equilibrium». During this time there is a slight change in the TMCS conversion (from 14% to 17%) observed.

Table 2. A material balance of the second period of the silylation

Periods, min.	**60-120**	**120-180**	**60-180**
A part of the TMCS, spent for the HMDS formation, %	100	excess HMDS	excess HMDS
A part of the TMCS, spent adsorption, %	0	0	0

According to the data of the material balance during this period of the silylation a content of HMDS in the reaction mixture exceeds the number that would be formed from the reacted TMCS. Therefore, it has been impossible to eliminate the reactions taking place under the scheme:

$$(Me)_3SiO\text{-}CaOH + Cl\text{-}Si(Me)_3 \rightarrow (Me)_3Si\text{-}O\text{-}Si(Me)_3 + [HO\text{-}Ca\text{-}Cl] \quad (12)$$

$$(Me)_3SiO\text{-}Ca\text{-}OH + Cl\text{-}Si(Me)_3 \rightarrow (Me)_3SiO\text{-}Ca\text{-}OSi(Me)_3 + HCl \quad (13)$$

$$(Me)_3SiO\text{-}Ca\text{-}OSi(Me)_3 + HOH \rightarrow (Me)_3SiO\text{-}CaOH + (Me)_3SiOH \quad (14)$$

$$(Me)_3SiO\text{-}Ca\text{-}OSsi(Me)_3 + HCl \rightarrow (Me)_3SiO\text{-}Ca\text{-}Cl + (Me)_3SiOH \quad (15)$$

$$(Me)_3SiOH + HOSi(Me)_3 \rightarrow (Me)_3SiOSi(Me)_3 + H_2O \quad (16)$$

$$(Me)_3SiO\text{-}Ca\text{-}OH + HCl \rightarrow (Me)_3SiO\text{-}CaCl + HOH \quad (17)$$

$$(Me)_3SiO\text{-}CaCl + HOH \rightarrow (Me)_3SiOH + [Cl\text{-}Ca\text{-}OH] \quad (18)$$

The data of the XR-analysis of the modified sample does not differ from the analysis of the sample selected in the first period.

Comparative analysis of the IR spectrum of the samples of the modified calcium hydroxide, allocated in the first and second periods of the reaction has showed that in the second period of the silylation the absorption bands at 1280 cm^{-1}, 810 cm^{-1} and 695 cm^{-1}, corresponding to the chemisorption of trimethylsilyl groups on a calcium hydroxide, have disappeared from the IR spectrum.

This indicates the prevalence of desilylation and hydrolysis of SiOCa-bonds by the water formed in the reaction mixture at this stage of the silylation process.

$$|\text{-}CaOSiMe_3 + HOH \rightarrow |\text{-}CaOH + Me_3SiOH \quad (19)$$

The electron microscopy of the particles of a mineral sample has showed the presence of only calcium compounds. The elemental analysis realized by the method of electron probe has showed a significant reduction in the silicon content up to 0.014%.

The survey thermogram DTA for the interval 20 - 500 °C does not differ from the DTA data given above for the sample selected at the first stage of silylation.

The thermogram DTA of the modified sample contains only endothermic effects at 80 °C, 200 °C and 280 °C. The modified sample dehydration starts at 400 °C. The Exothermic heat effect at 388 °C described for the previous stage is absent on this thermogram.

Thus, the analysis data indicate that at this stage of silylation the following processes take place:

1. The hydrolysis of a trimethylchlorosilane by the structural and adsorbed water contained in the initial $Ca(OH)_2$. At the same time the hydrolysis by the water formed in the reaction mixture begins.
2. The reaction of a hydrolytic cleavage of bonds Ca-O-Si-Me_3 including by the water formed in the reaction mixture.
3. The reactions between hydrogen chloride and $Ca(OH)_2$ with a $Ca(OH)_2 \cdot CaCl_2 \cdot H_2O$ and a «secondary water» formation.

The third silylation period (from 180 minutes to 300 minutes) is a period possibly corresponding to the stage of a «condensation». There is a new increase in TMCS conversion (from 18% to 22%) at this period.

Table 3. A material balance of the third period of the silylation

Periods, min.	**180-240**	**240-300**	**180-300**
A part of the TMCS, spent for the HMDS formation, %	72	90	82
A part of the TMCS, spent adsorption, %	28	10	18

The data of the mass balance indicate the resumption of TMCS adsorption on a calcium hydroxide at this period of the reaction. This can be explained by the amount of water sufficient for the hydrolysis predominance. However, in the next period of time (240-300 minutes) the proportion of TMCS gone to HMDS formation rises, which can be explained by starting again increase of the amount of water in the reaction system.

The IR spectrum of the modified sample presents again the absorption bands at 1282 cm^{-1}, 840 cm^{-1} and 695 cm^{-1}, indicating the renewal of trimethylsilyl groups adsorption on a calcium hydroxide (virtually identical to the spectrum obtained for the first period).

The thermogram covers again an exothermic effect at 388 °C, that confirms a resumption of a trimethylchlorosilane chemisorption on a calcium hydroxide in this period.

The elemental analysis also has showed an increase in the silicon content in the light phase to 0.12%, which also confirms the resumption of the trimetilsilinyh groups chemisorption. Consequently, at this stage of the silylation the adsorbed water decreases in the reaction system, due to the continued hydrolysis of TMCS by the water formed in the reaction mixture. But the bound water also increases in the reaction mixture due to the growth of the content of a crystalline calcium hydrooxochloride.

Thus, the analysis data indicate that at this stage of the reaction the following processes take place:

1. A trimethylchlorosilane hydrolysis by the water formed in the reaction mixture and the equilibrium reactions leading to the formation of Si-O-Ca- bonds.
2. Reaction of hydrogen chloride with $Ca(OH)_2$, to the formation of $Ca(OH)_2 \cdot CaCl_2 \cdot H_2O$ and H_2O.

Summarizing the research given above, it can be noted that:

The first silylation period is a stage of a rapid achievement of a sorption equilibrium, the main reaction is the TMCS hydrolysis by the adsorbed water contained in the calcium hydroxide. The reaction system accumulates the water. Simultaneously with the TMCS hydrolysis occurs its adsorption on a calcium hydroxide.

The second period of the reaction is a sorption equilibrium stage; the basic processes are TMCS hydrolysis by the water formed in the reaction mixture. Simultaneously in the reaction system occurring cleavage reaction Si-O-Ca- bonds to form excess HMDS.

The third period of reaction is a condensation stage; a calcium hydroxide silylation by the trimethylchlorosilane recommences. An increase of the content of bound water in the composition of a calcium hydroxochloride continues (a thermal effect on the DTA curve at 200 °C). Apparently, in the third period of the reaction a repetition of the processes occurring in the first period of the reaction takes place again. The third stage of the process seems to be a renewed stage of «hydrolysis-silylation».

Thus, considering the interaction between a trimethylchlorosilane and a calcium hydroxide, we can speak only about two, to some extent repetitive periods of the reaction. In the first period of the reaction trimethylchlorosilane hydrolysis and condensation prevail and hydrolysis products of hexamethyldisiloxane form. At the same time a trimethylchlorosilane chemisorption is in process. In the second period of the reaction desilylation process dominates, mainly due to the secondary reactions of splitting Si-O-Ca-ties. This leads to an increase in hexamethyldisiloxane's content in the reaction mixture (in comparison with theoretical coolers). It is clear that it can only be a dynamic equilibrium processes, where a content of a hydrogen chloride and a water on the surface of the mineral plays an important role.

References

1. A.A. Pashhenko, M.G. Voronkov *Kremneorganicheskie zashhitnye pokrytija,* (Kiev. Tehnika 1969)
2. A.A. Novoselnov, V.I. Sidorov, T.P. Nikiforova, Vestnik MGSU. **1**, 160-162 (2007)
3. A.A. Novoselnov, Vestnik MGSU, **4**, 181 (2010)
4. A.A. Novoselnov, Vestnik MGSU, **4-3**, 133-139 (2011)
5. A. A. Novoselnov Vestnik MGSU, **4**, 174 (2011)
6. A. A. Novoselnov A. A. Vestnik MGSU, **7**, 114-120 (2013)

7. A. A. Novoselnov, Mjasoedov E. M. Estestvennye i tehnicheskie nauki, **11-12,** 56-57 (2014)
8. A. V. Kiseljova, *Jeksperimental'nye metody v adsorbcii i molekuljarnoj hromatografii* (Moscow, MGU 1973)
9. Z. Mahachek Z. Sililirovanie *γ-aminopropiltri- jetoksisilanom kal'cinirovannogo kaolina*. Himicheskaja promyshlennost', Moscow, p. 10 (1981)
10. A. A. Pashhenko. DSc Thesis Kiev (1984)

Nano-modified concretes initial structuring

Lev Alimov[1,*] and *Igor* Engovatov[1]

[1]Moscow State University of Civil Engineering, 129337, Moscow, Yaroslavskoye sh., 26, Russia

Abstract. The article considers the aspects related to initial formation of the structures in concretes containing nano-additives that significantly influence the type of newgrowths and their concentration during the cement hydration. The article depicts the role of the aggregate in formation of the contact zone structure between the cement stone and the aggregate, which mainly consists of ettringite and portlandite. Mathematical models of the cement stone hydrational hardening kinetics and its performance properties have been obtained.

1 Introduction

For cement composites, the initial hardening and the structure formation period (SFP), i. e. transition from the plastic and viscous state to the hard state, may be determined by any method: ultrasound, mixture temperature variation kinetics, electric resistance, contraction, and others. The studies have demonstrated that the values of the variables being measured change abruptly after completion of the initial structure formation period [1, 2, 3].

In "cement with nano-modifier - water" systems, the hydration and heat release kinetics at the same hardening conditions are determined by the water-cementitious ratio. A different situation is, when there is an aggregate. The beginning of the intensive heat release is determined not by the mixture water content, but by W, which occurs in the concrete mixture exposed to the aggregate [4, 5, 6]. Table 1 shows the compositions of concretes with nano-modifiers characterized by different structure formation periods (SFP), and, consequently, different water-cementitious ratio (W) and cement with nano-modifier volumetric concentration in the concrete mixture ($C = 0.001(Ts/\rho_{ts} - W)$).

According to Table 1, when the volumetric concentration is constant (C=0.2) in composites 4-6, the intensive heat release coincides with the end of the structure formation period corresponding to W. The concrete mixtures characterized by constant W=0.2 (composites 1-4), but with different C, had the same structure formation period.

The further heat release process after the transition of the concrete mixture to the concrete state occurs in the compositions that contain nano-modifier and the aggregate in a more intensive manner in comparison with the pure cement stone. This is evident of the fact that the water mechanically bound by the aggregate at an early stage of hardening begins to actively participate in the cement-water paste hydration process [10, 12]. The main structuring component of the concrete mixture is the cement paste with the nano-modifying aggregate, which determines the hydration and cements hardening kinetics [11].

* Corresponding author: tvvib@mail.ru

Table 1. Concrete mixtures compositions

No.	Materials consumption in kg per 1 m^3					V/Vy	SFP, h	C	W
	cement	nano-modifier	water	sand	gravel				
1	1376	344	447	-	-	0.26	7	1	0.26
2	544	136	250	537	880	0.367	7	0.4	0.26
3	413	103	219	625	1025	0.425	7	0.3	0.26
4	274	69	186	716	1170	0.542	7	0.2	0.26
5	290	72	180	716	1170	0.5	6	0.2	0.23
6	222	55	208	716	1170	0.75	9	0.2	0.4

Table 2 illustrates the detailed studies of the cohesives hydration kinetics in the presence of nano-modifiers using the example of TES22 (Moscow) production waste, the fly ash.

Table 2. Composition and properties of the cement composite consisting of portland cement and nano-modifier obtained from the activated fly ash with specific surface area equal to 700 m2/kg.

No.	Specific surface area, m^2/kg	Content of nano-modifier, %	Density, g/cm^3	SFP, h-min.	Strength, MPa
1	265	-	2.10	8-45	49
2	265	-	2.42	9-15	57
3	330	15	2.26	4-15	62
4	395	30	2.12	2-00	47
5	461	45	2.00	2-30	22
6	526	60	1.9	3-00	11

Note: 1 - portland cement, 2-6 - mixtures of portland cement with 0.8 % C-3 and nano-modifier

The analysis according to table 2 has demonstrated that when the nano-modifier is added to the mixture at an early period, the exothermy abruptly growth, and decreases by the end of the period in comparison with the reference composition, which corresponds to the end of the SFP. When increasing the quantity of the additive and reducing the content of cement in the concrete mixture, the exothermy decreases, but does not reach the beginning and the end of the reference composite setting [13].

Adding the nano-modifier to the cohesive composition leads to the reduction of time for the setting start and end in comparison with the cement-water paste with no additives; however, in some extent the increased SFP is observed with regard to the impalpable additive; this may be explained by the fact that the C-3 superplasticizing agent in some extent slows down the setting processes and beginning of the concrete mixture initial structuring.

According to Table 1 and plots (Fig. 1, 2, 3), when the quantity of the nano-modifier increases up to 30 %, the time of setting is reduced; and with further increase of the additive quantity, the structure formation period (SFP) increases. Reduction of the setting time, improvement of the cement stone plastic strength may be explained by the fact that the adding of the nano-modifiers contributes to improvement of the system viscosity and more rapid occurrence of the crystallization centers, which leads to the reduction of the energy demands for the nucleation, and contributes to the formation of the cement stone structure with less stresses and acceleration of the cement systems hardening [14].

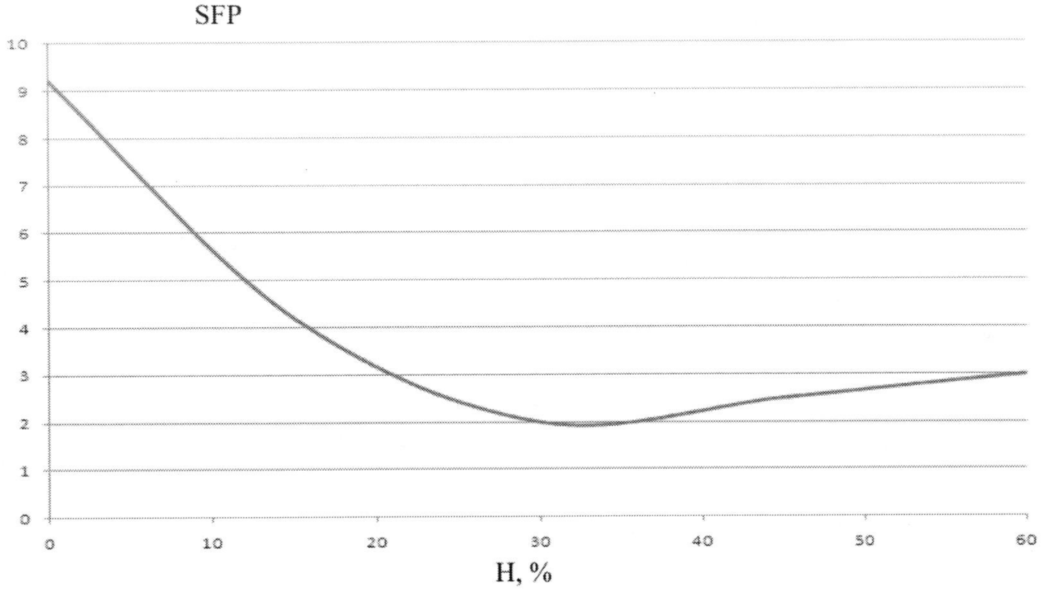

Fig. 1. Cement composite SFP vs. nano-modifier content (H).

The cement-water past structure formation period is consistent with the data related to the gaining of the plastic strength by the cement-water paste with the nano-modifier. When adding the superplasticizing agent, the reference composition structure formation period increases by 0.5 h, which is related to formation of the cement and aggregate grains on the surface owing to the absorption of the thickened enclosure polar groups with significant negative ξ-potential, thus increasing the dispergation effect and particles repulsion [8, 9].

The nano-modifier, from the mixing moment, provides the peptizing and structuring effect, thus accelerating the cement stone hydration and hardening process. Such effect provided by the disperse particles may be explained by that the mineral aggregate particles, when between the separate grains of the cement, move them apart and ensure better access of water to them. The hydration products formed are distributed within the large volume, since they are diverted from their reaction zone to the aggregate particles surface.

The following mathematical models have been obtained:

- Structure formation period vs. the content of nano-modifier based on the activated fly ash from TES 22 (Moscow).

$T = 9.21 - 0.371H + 0.00456\,H^2$ with the correlation coefficient equal to 0.98;

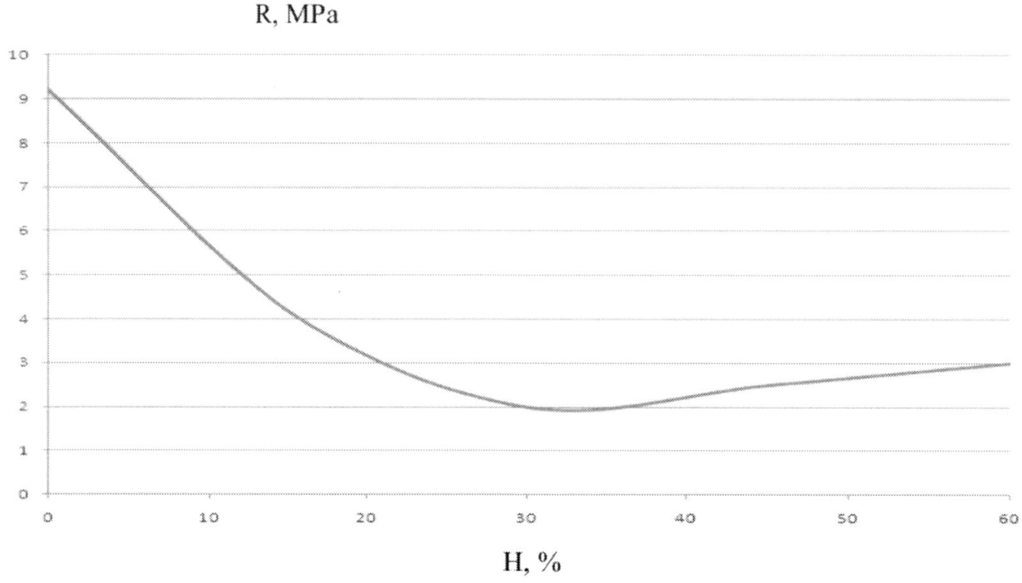

Fig. 2. Cement composite strength (*R*) vs. nano-modifier content (*H*).

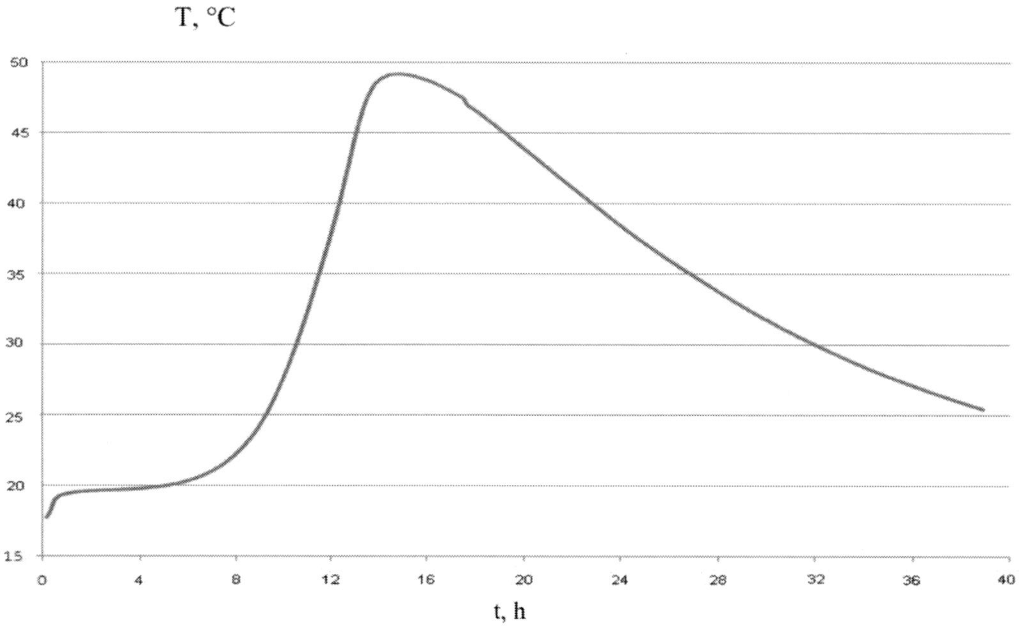

Fig. 3. Temperature (*T*) of the cement composite withnano-modifier (*H*) vs. time from the moment of mixing.

- Cement composite hardness vs. the nano-modifier content:

$R = 60.2857 - 0.1181\, H - 0.0127\, H^{2;}$ with the correlation coefficient equal to 0.97

- Mathematical model of the optimal composition No. 3 of the cement composite with nano-modifier hydration kinetics in Table 2:

$T = 19.218 - 7.819\,\tau + 2.703\,\tau^2 - 0.241\,\tau^3 + 0.008\,\tau^4$; with the correlation coefficient equal to 0.975.

By analyzing the hard cement composite with nano-modifier hydration kinetics, as determined by the temperature measurements provided in Fig. 3, it is possible to track the same regularities as in the ultrasound method; however, some exothermy peculiarities are noticeable in different compositions. As such, when adding 15 % of the additive to the mixture, the temperature drops by 30-50 °C in comparison with the reference mixture temperature; and when increasing the quantity of the additive (up to 60 %), gradual reduction of the exothermy is observed. When using the C-3 superplasticizing agent, the rate of temperature growth decreases, but the heating temperature increases.

A positive quality of the filled cements are their low exothermy, which makes them efficient for use in the massive structures.

Conclusion

Based on the accomplished studies, an active role of the nano-modifiers in the cement stone and contact zone initial structure formation has been established; this is due to the physical and chemical processes that are mainly related to the re-distribution of water and its bonding forms in the concrete mixtures, which are crucial in formation of porosity in the concrete.

References

1. Yu.M. Bazhenov, L.A. Alimov, V.V. Voronin, *Structure and properties of the concretes with production-generated waste nano-modifiers* (EBS ASV, 2013)
2. L.A. Alimov, V.V. Voronin, *Construction materials* (Textbook, Publishing Center, Akademiya, 2012)
3. Yu.M. Bazhenov, *Collection of Round-Table Discussion Reports,* 12-18 (MGSU, 2007)
4. L.A. Alimov, V.V. Voronin, *Concrete works technology* (Textbook, Publishing Center, Akademiya, 2015)
5. Yu.M. Bazhenov, L.A. Alimov, V.V. Voronin, A.A. Buldyzhov, I.V. Romanov, Patent 2474544 Dated Febr., **10** (2013)
6. L.A. Alimov, V.V. Voronin, A.A. Buldyzhov, I.V. Romanov, 'Scien. Rev.' Mag., **9**, 177-181 (2013)
7. L.A. Alimov, S.I. Bazhenova, Vestnik MGSU, **1** (2010)
8. C.J. Benmore, Paulo J.M. Monteiro, Orig. Res. Art. Cement and Conc. Res., **40**, 892-897 (2010)
9. G. Quercia, J.J.G. van der Putten, G. Hüsken, H.J.H. Brouwers, Orig. Res. Art. Cement and Conc. Res., **54**, 161-179 (2013)
10. A.N. Grishina, E.V. Korolev, Key Eng. Mat., **683**, 90-94 (2016)
11. A.N. Grishina, A.V. Eremin, Inorg. Mat., **52**, 973-977 (2016)
12. V. Gladkikh, E. Korolev, D. Husid, Solid St. Phen., **871**, 110-117 (2016)
13. A.O. Adamtsevich, A.P. Pustovgar, S.A. Pashkevich, A.V. Eremin, Inter. Jour. of App. Eng. Res., **11**, 1609-1611 (2016)
14. A. Adamtsevich, A. Pustovgar, S. Pashkevich, A. Eremin, A. Zhuravlev, App. Mech. and Mat., **670-671**, 339-343 (2014)
15. A.S. Inozemtcev, E.V. Korolev, Adv. Mat. Res., **1040**, 176-182 (2014)

Criteria for evaluation of building materials hazard based on their natural radioactivity in Russia and in the European Union countries

Darya Buzina[1], *Igor* Engovatov[1,*], *Lev* Alimov[1] and *Michail* Slesarev[1]

[1]Moscow State University of Civil Engineering, 129337, Yaroslavskoye Shosse 26, Moscow, Russia

Abstract. We hereby have considered the natural radioactivity of building materials as a source of radiation for the human environment. We have considered the Russian and the European evaluation criteria for ensuring radiation safety of the population. We also present an experimental research of the content of natural radionuclides in building materials. We have calculated the effective specific activity and the activity concentration index based on the results presented. We have identified discrepancies between the Russian and the European standards.

1 Introduction

Natural and artificial sources of ionizing radiation determine the radiation effect on humans both directly and through the environment, including, but not limited to the territory of residence, the buildings and the facilities, where they spend most of their life.

The natural radioactivity of the building materials used in buildings is a significant component of the human exposure, and in a series of cases can result in significant exposure doses. All building materials, as well as raw materials for their manufacturing, contain one or another quantity of natural radionuclides (NRN). The most important of them are the ^{238}U radioactive family with its progenies, the ^{232}Th radioactive family with its progenies and potassium -40 (^{40}K).

Virtually, all industrialized countries seek to reduce and limit the radiation exposure of the population by natural radioactivity of building materials, including the application of standards, evaluation criteria and scopes of application.

2 Russian criteria for evaluation

Russia has adopted the fundamental laws and elaborated the standard documents that govern the procedure and the regulations for recording and controlling the use of building materials to ensure radiation safe human environment [1, 2].

In accordance with the current legislation, the classes of building materials and the quantitative limits of the concentration of ^{226}Ra, ^{232}Th and ^{40}K natural radionuclides (NRN)

* Corresponding author: eng46@mail.ru

for them are established using the value of the effective specific activity (A_{eff}). Materials pertaining to classes I and II may be used for building residential and industrial facilities. The quantitative values are: $A_{eff} \leq 370$ Bq/kg for class I and $A_{eff} \leq 740$ Bq/kg for class II.

The value of the effective specific activity (A_{eff}) is calculated using the following formula [1]:

$$A_{eff} = A_{Ra} + 1.31A_{Th} + 0.085A_{K}, \quad (1)$$

where A_{Ra} and A_{Th} are the specific activities of ^{226}Ra and ^{232}Th radionuclides, which are in radioactive equilibrium with the rest of the members of the uranium andthorium series, and A_K is the specific activity of ^{40}K, Bq/kg.

Also, the effective dose limit of 1 mSv a year during any 5 average consecutive years, but no more than 5 mSv a year, is established for the population.

The equivalent dose rate (EDR) of the external gamma radiation and the indoor average annual equivalent equilibrium volumetric activity of radon are the values controlled inside the residential and public buildings within the territory of residence. The indoor EDR may not exceed the outdoor dose rate by more than 0.2 μSv/h. The indoor gamma radiation of radon and thoron present in the indoor air is not considered in this article.

Therefore, the Russian approach to ensuring radiation safety of the population in its environment consists in limiting the NRN concentration in the soil and building materials and the EDR of the external gamma radiation in the residential and industrial buildings. Non-exceedance of the established limits should, in principle, ensure non-exceedance of the annual dose of 1 mSv.

3 European criteria for evaluation

The international organizations, such as IAEA, ICRP and UNSCEAR show strong interests in the regulation and standardization of the radiation safety of the environment [3]. For example, the Commission of the European Communities (European Commission, EU) has issued the regulatory and methodological document "Radiation protection 112. Radiological Protection Principles concerning the Natural Radioactivity of Building Material" [4] which formulates the principles of the radiation protection and proposes to use the following radiological criteria as the basis for the controls of the building materials radioactivity:

a) Dose criterion for controls

Controls should be based on a dose criterion which is established considering overall national circumstances. Within the European Union, doses exceeding 1 mSv a year should be considered from the radiation protection point of view. These doses should be accepted only in specially stipulated exceptional cases where materials are used locally. Controls can be based on a lower dose criterion. In this case, a reasonable practicality study concerning such controls will be required. It is therefore recommended that controls be based on a dose within the range of 0.3 to 1 mSv a year. This is the excess dose to that received outdoors.

b) Level of exemption from all restrictions

Building materials should be exempted from all restrictions concerning their radioactivity if the excess gamma radiation originating from them increases the annual effective dose by more than 0.3 mSv. This is the excess gamma dose to that received outdoors.

We can present the level to be studied as the activity concentration index I. The activity concentration index must consider typical methods and volumes for use of a building material in a building.

The activity concentration index (I) is calculated using the following formula:

$$I = \frac{C_{Ra}}{300Bq/kg} + \frac{C_{Th}}{200Bq/kg} + \frac{C_K}{3000Bq/kg}, \quad (2)$$

where C_{Ra}, C_{Th}, C_K are specific activities of radionuclides ^{226}Ra, ^{232}Th and ^{40}K in a building material respectively, Bq/kg.

The coefficients 300, 200 and 3000 are references for radium, thorium and potassium respectively, which are determined on the basis of the scenarios for the use of materials to secure the dose criterion.

The activity concentration index may not exceed the following values indicated in the Table1.

Table. 1. The values of the activity concentration index for different uses of building materials

Dose criterion	0.3 mSv a year	1mSv a year
Basic structural and finishing materials, e.g. concrete	$I \leq 0.5$	$I \leq 1$
Materials with restricted use: curbs, framing and others	$I \leq 2$	$I \leq 6$

The activity concentration index must be used only as a tool for identifying the materials which has to be focused on. Any final decision on restriction of the use of a material must be based on a separate dose evaluation. Such evaluation must be based on the situation considering direct use of such material.

Normally, changes in the activity concentration are needed only when there are substantial grounds to suspect that the dose criterion for controls could be exceeded.

The manufacturer or the seller is liable for ensuring quality showing that a material complies with the radiological requirements established by the state. However, other approaches can be applied to the controls as well. For example, the builder or the designer of a building can be liable for the compliance of a new building with the radiological requirements.

Materials should be exempt from all elements of controls concerning their radioactivity, if it is shown that the dose criterion for the exemption from restrictions has not been exceeded.

Exempt materials to be used for building without any restrictions concerning their radioactivity must be allowed to enter the consumer market.

4 Analysis and summary

Therefore, the European approach contemplates more varied approaches to the evaluation of the potential hazard and the scopes of application of building materials for their use. And the dose criterion is the main one, which is correct in principle, since it is the annual individual radiation dose that is important for an individual, and not the activity of the building materials in their environment.

For this purpose, it is of undoubted practical interest to carry out comparative evaluations of the radiological hazard of building materials and raw materials for their fabrication, using Russian and European approaches.

The studies of the content of NRN in building materials have been and are carried out both in the RF [5-10] and abroad [11-20]. As the result of the studies carried out, wide qualitative and quantitative material concerning the NRN content in the building materials used and in the raw materials for their fabrication has been obtained.

As the result of the studies, it has been established that in most of the cases, the NRN content is much less than the reference values given in the formula (2). At the same time, it is established that the NRN concentration of some materials and raw materials for their manufacturing can considerably exceed the average typical values.

Table 2 shows the experimental results of determination of the NRN in building materials of the RF and several foreign countries, as well as data concerning the effective specific activity A_{eff} for them, using the Russian approach (formula) and the activity concentration index **I**, using the European approach (formula 2) for the typical and maximum values.

Table 2. Typical and maximum values of the natural radioactivity of building materials

Material	Country	Typical activity, Bq/kg					Maximum activity, Bq/kg				
		^{226}Ra	^{232}Th	^{40}K	A_{eff}	I	^{226}Ra	^{232}Th	^{40}K	A_{eff}	I
Concrete	Russia	27	6	56	40	0.14	53	59	826	204	0.75
	USA	26	18	296	76	0.28	89	118	1147	346	1.27
	Europe	40	30	400	115	0.42	240	190	1600	631	2.28
Light concrete	RF	20	15	62	45	0.16	35	13	162	66	0.24
	Europe	60	40	430	151	0.54	2600	190	1600	2991	10.15
Red brick	RF	11	60	43	93	0.35	55	62	783	206	0.75
	Europe	39	41	560	143	0.52	200	200	2000	640	2.33
	Saudi Arabia	27	21	279	80	0.29	39	27	213	93	0.34
Grey brick	RF	11	60	43	92	0.35	55	62	783	206	0.75
	USA	91	73	596	239	0.86	178	144	1184	472	1.71
	Europe	10	10	330	53	0.19	59	67	670	206	0.76
	India	0	26	152	48	0.18	3	48	412	103	0.39
	Egypt	29	50	167	110	0.40	65	89	376	214	0.79
	Saudi Arabia	15	9	114	38	0.14	32	19	171	72	0.26
Cinder	RF	62	88	367	210	0.77	679	112	577	877	3.02
	USA	706	15	940	809	2.74	3145	7	1850	3321	11.14
	Europe	180	100	650	369	1.32	1100	300	1500	1625	5.67

DOI: 10.1051/matecconf/20168604044

Material	Country	Typical activity, Bq/kg					Maximum activity, Bq/kg				
		^{226}Ra	^{232}Th	^{40}K	A_{eff}	I	^{226}Ra	^{232}Th	^{40}K	A_{eff}	I
Natural stones (granite, marble, etc.)	RF	22	22	472	93	0.34	105	119	103	269	0.98
	Europe	60	60	640	196	0.71	500	310	4000	1263	4.55
	India	74	0	65	80	0.27	82	112	1908	399	1.47
	Egypt	18	24	350	81	0.30	65	60	920	226	0.82
	Saudi Arabia	21	18	276	69	0.25	45	35	821	165	0.60
Plaster	RF	12	2	6	15	0.05	12	9	138	36	0.13
	Europe	10	10	80	30	0.11	1034	43	37	1093	3.67
	India	8	0	27	10	0.04	233	9	22	247	0.83
	Egypt	39	25	226	92	0.33	32	55	116	114	0.42
	Saudi Arabia	29	55	65	106	0.39	58	72	97	160	0.58

Conclusion

The analysis of the table allows us to make the following conclusions:

- Most of the materials comply with both ours and European criteria and can be used without any restrictions.
- A disconformity between the Russian and the European standards can be observed in different classes of materials. Thus, some types of building materials, such as concrete and cinder may be used without restrictions under the Russian legislation, while only restricted use is accepted when applying the dose criterion.
- The dose criterion is more objective, since the annual individual radiation dose is much more important for an individual the human, than the activity of the building materials in their environment.
- Even though complex controls mean considerable financial and time expenses, a great quantitative increase in the nomenclature of building materials and the use of varied natural raw materials, industrial waste and artificial additives do not discard the current task of computational and experimental study of the natural radioactivity of different types of building materials.
- For uninterrupted circulation of building materials on the global market, it is needed to harmonize standard approaches and regulations for the radiation purity of the product (material, raw materials, building).

References

1. SanPiN 2.6.1.2523-09 "Radiation safety standards (RSS-99/2009)"
2. Law of the RF, **3** (1996)
3. *IAEA Safety Standards. Radiation Protection and Safety of Radiation Sources: International Basic Safety Standards. General Safety Requirements Part 3 International Atomic Energy Agency, Vienna*, (2015)
4. *European Commission. Radiation protection 112 Radiological Protection Principles concerning the Natural Radioactivity of Building Materials. - Directorate-General Environment, Nuclear Safety and Civil Protection, Luxemburg*, (1999)
5. E.M. Krisyuk, *Radioactive background* (Energoatomizdat, 1989)
6. O.P. Sidelnikova, *Radiation controls in the building industry* (ASV, 2002)
7. I.A. Engovatov, D.V. Nikolaeva, Nat. and tech. sciences, **11-12**, 394-396 (2014)
8. I.A. Engovatov, D.V. Nikolaeva, Ecol. of urban. Terr. **4**, 60-66 (2015)
9. O.P. Sidelnikova, Yu.D. Kozlov, Int.-Vestnik VolgGASU. Ser.: Multitopic (2013)
10. N.P. Lukutsova, O.Yu. Kozlov, G.I. Krupny, M.A. Maslov, Ya.N. Rastsvetalov, A.A. Yanovich, *Radiation safety of building materials and industrial waste* (Protvino, 2000)
11. *Evaluation of EPA's Guidelines for Technologically Enhanced Naturally Occurring Radioactive Materials (TENORM)* (EPA, 2000)
12. *NUREG-1501. Background as a Residual Radioactivity Criterion for Decommissioning. - U.S. Nuclear Regulatory Commission Office of Nuclear Regulatory Research Washington, DC 20555-0001*, (1994)
13. M.A.M. Uosif, Int. Journal of u- and e- Ser., Science and Tech. **7**, 63-76 (2014)
14. A.H. Korna, S.S. Fares, M. Abd El-Rahman, Nat. Science, **6**, 5-13 (2014)
15. Y. Raghu, N. Harikrishnan, A. Chandrasekaran, R. Ravisankar, Af. Journal of Basic & App. Sciences, **7,** 16-25 (2015)
16. W.R. Alharbi, J.H. AlZahrani, Journal of Am. Science, **8,** 651-656 (2012)
17. E.M. Lee, G. Menezes, E.C. Finch, *Assessment of Natural Radioactivity in Irish Building Materials* // Department of Physics, University of Dublin, Trinity College, Dublin 2, Ireland (2004).
18. S. Pavlidou, A. Koroneos, C. Papastefanou, G. Christofides, S. Stoulos, M. Vavelides, Bull. of the Geol. Soc. of Greece, **XXXVI** (2004)
19. S. Fares, Ali.A.M. Yassene, A. Ashour, M.K. Abu-Assy, M. Abd El-Rahman, Nat. Science, **3**, **10**, 895-905 (2011)
20. A. El-Taher, Rom. Journ. Phys., **57**, **3–4**, 726–735 (2012)
21. A.N. Grishina, E.V. Korolev, A.B. Satyukov, Adv. Mat. Res. **1040**, 351-355 (2014)
22. M.G. Bruyako, V.A. Glukhoedov, D.V. Kravtsova, V.A. Smirnov, V.A. Ushkov, Adv. Mat. Res. **1040**, 730-734 (2014)
23. A.O. Adamtsevich, A.P. Pustovgar, S.A. Pashkevich, A.V. Eremin, Int. Jour. of App. Eng. Res. **11**, 1609-1611 (2016)

Fine-grained concrete with organomineral additive

Vitaly Solovyov[1*] and *Yulia* Sokolova[2]

[1]Moscow State University of Civil Engineering, 129337, Yaroslavskoye Shosse 26, Moscow, Russia
[2]NOU INEP, 115477, Bekhtereva ul, d 13, korp 3, Moscow, Russia

Abstract. The article deals with the issues concerning the formation of the structure and properties of fine-grained concrete with organomineral additive produced through mechanochemical activation of thermal power plant fly ash together with superplasticizer. The additive is produced in a high-speed activator at the collision particles' speed of about 80 m/s. The use of the additive in fine-grained concrete in the amounts of 0.5-1% increased the strength by 30-50% and reduced the size and volume of pores. The cement consumption in such concrete is close to the cement consumption in common concrete of equal resistance.

1 Introduction

The thermal power plant fly ash with the following chemical composition was used for the research: SiO2 - 40...58 %, Al2O3 - 21...27 %, CaO - 4...6 %, Fe2O3 - 4...17 %, NaO-0.4...1.4 %, K2O – 0.4...4.7 %.
Moreover, the composition of the fly ash included, among others, small amounts of SO3, MgO and TiO2.

Highly-homogeneous, chemically active and law water demand nanomodified organomineral additive was produced through mechanochemical activation in a high-speed activator with an annular processing chamber at the speed of impact of material particles of about 80 m/s.

This includes not only grinding of materials, but also change of physical and chemical state and structure.

The crystal lattice is disturbed and transformed up to its transition to amorphous state. Such modification is accompanied by an increase in the specific area and increase in the surface energy of the substance.

The grain size composition of the activated fly ash with the presence of 0.3% S-3 superplasticizer,obtained with the use of a laser granulometer, is shown on Fig.1.

The analysis of the grain size composition of the mechanochemically activated fly ash shows that it contains particles with a size of less than 10-6 m.

The number of particles of less than 500 nm amounts to 2.06% and that of the 500 to 1000 nm particles amounts to 3.12%.

* Corresponding author: sotae@mgsu.ru

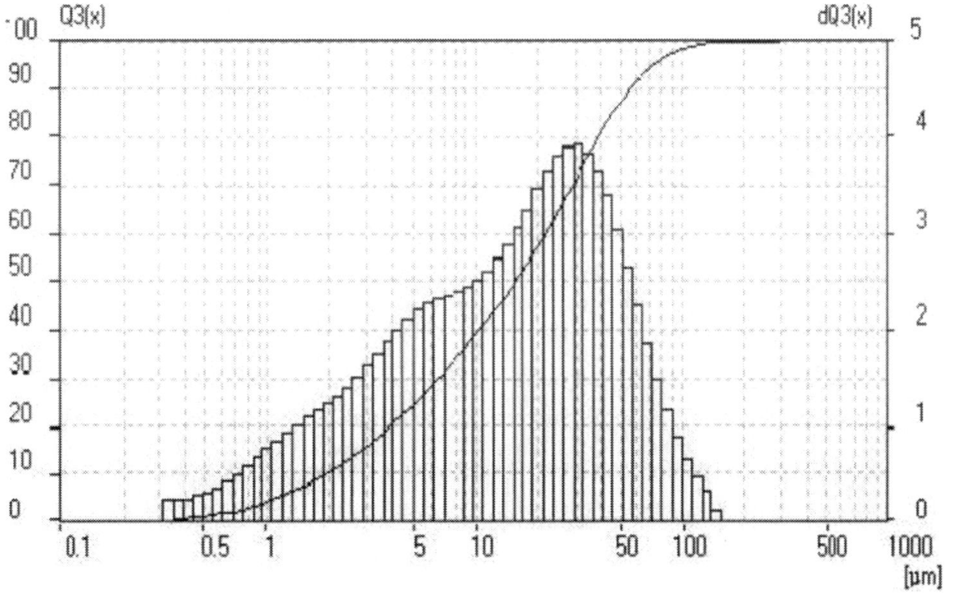

Fig.1. The integral and differential distribution of particles based on the size of mechanoactivated fly ash

The additive was tested in fine-grained concrete (FGC) prepared on the basis of PTsD0 M500 Portland cementand silica sand by weight proportion of 1:2, with the mobility determined on a standard shaker apparatus, and with a spread diameter of 17 cm.

4x4x16 cm samples steam cured under the following regime: 3 + 3 + 5 at 80°C with isothermal warming were made.

The physical and mechanical characteristics of the obtained test resultsare shown in the table 1.

Table 1. Physical and mechanical properties of FGC on the basis of multicomponent binders.

item #	Binder composition	Bending strength		Compression strength,	
		MPa	%	MPa	%
1	PTs	6.0	100.0	32.2	100.0
2	PTs + 25 % AF	5.65	94.2	32.0	99.4
3	PTs + 40 % AFwith SP	5.97	99.5	48.4	150.3
4	PTs + 25 % AFwith SP	5.76	96.0	41.6	129.2
5	PTs + AF(1 : 1)	5.96	99.3	34.4	106.8

Note: AF means activated fly ash

In the context of impact on concrete, fly ash is an active mineral additive; moreover, it fulfills the function of a filler making the concrete denser.

In the process of curing, beside the pozzolanic reaction, amorphous silica and aluminium oxide interreact with calcium hydroxide produced from cement hydration, which results in the formation of law-basic hydrosilicates and calcium hydroaluminates [7,8].

This not only contributes to better strength and density, but to longer life of nanomodified fine-grained concrete.

Conclusion

The research carried out shows that the application of nanomodifiers on the basis of mechanoactivated fly ash in the amount of 0.5-1% has a positive influence on the structure and on physical and mechanical properties of fine-grained concrete, since it contributes to a 35-40% reduction of cement consumption and 30-50% increase in the strength.

References

1. L.A. Alimov, V.V. Voronin, *Construction materials* (Textbook. Publishing center, Academia, 2012)
2. L.A. Alimov, V.V. Voronin, *Concrete work technology* (Textbook, Publishing center, Academiya, 2015)
3. Yu.M. Bazhenov, *Compilation of roundtable presentations*, 12-18 (MGSU, 2007)
4. Yu.M. Bazhenov, L.A. Alimov, V.V. Voronin, *Structure and properties of concrete with nanomodifiers on the basis of technogenic waste* (Nomografiya, EBSASV, 2013)
5. L.A. Alimov, V.V. Voronin, A.A. Buldyzhov, I.V. Romanov, Mag. Nauchnoe obozrenie, **9**, 177-181 (2013)
6. L.A. Alimov, S.I. Bazhenova, Vestnik MGSU, **1** (2010)
7. Yu.M. Bazhenov, L.A. Alimov, V.V. Voronin, A.A. Buldyzhov, I.V. Romanov, Patent 2474544 dated 10.02.2013 *Method for preparation of a nanomodifier from industrial waste for concrete mix.*
8. G. Quercia, J.J.G. van der Putten, G. Hüsken, H.J.H. Brouwers, Orig. Res. Art. Cement and Conc. Res., **54**, 161-179 (2013)
9. A. Krishan, V. Rimshin, V. Erofeev, V. Kurbatov, S. Markov, Pr. Eng., **117**, 211-217 (2015)
10. V. Semenov, A. Pligina, T. Rozovskaya, *IOP Conference Series: Materials Science and Engineering*, **71**, art. no. 012041 (2015)
11. A. Tamrazyan, App. Mech. and Mat., **475-476**, 1563-1566 (2014)
12. A.S. Inozemtcev, E.V. Korolev, Adv. in Cem. Res., **28**, 92-98 (2016)
13. A.S. Inozemtcev, *IOP Conference Series: Materials Science and Engineering*, **71** (2015)
14. A. Adamtsevich, A. Pustovgar, S. Pashkevich, A. Eremin, A. Zhuravlev, App. Mech. and Mat., **670-671**, 339-343 (2014)
15. V.A. Perfilov, D.V. Oreshkin, D.Y. Zemlyanushnov, Pr. Eng., **150**, 1474-1478 (2016)
16. L.N. Rasskazov, A.V. Radzinskii, M.P. Sainov, Power Tech. and Eng., **48**, 358-360 (2015)

Antirivers and their role in the biotechnosphere formation

Antonina Suzdaleva[1*], *Alim* Gur'ev[2] and *Valentin* Tkachev[3]

[1]National Research University "MPEI", 111250, Krasnokazarmennaya 14, Moscow, Russia
[2]JSC «Institute Hydroproject», 125993, Volokolamskoe sh., 2, Moscow, Russia
[3]Moscow State University of Civil Engineering, 129337, Yaroslavskoye Shosse 26, Moscow, Russia

Abstract. In the present-day conditions, we need to stop seeing the creation of antirivers as an environmentally hazardous activity. The threat of a global water consumption crisis forces us to build the hydrotechnical systems that will have a significant environmental impact. The scientists' task consists in prompt development of an interdisciplinary approach to their implementation, combining the resolution of technical, environmental and social issues.

1 Introduction

In the modern world, the ever-growing human impact on water systems takes place on the background of the growing need for water resources. As a result, a deficit of potable water is produced in some regions, which may grow into a global water consumption crisis [1,2]. The main direction in the search for the resolution of these issues is to create river flow management systems [3]. However, this method is not applicable for delivering water to the regions where its deficit cannot be compensated by regulating the flow and reducing the water consumption. It is needed to build hydrotechnical systems redistributing water resources among different regions. Similar projects are already implemented in several countries [4,5,6]. Recently, a system for artificial transfer of river flow is designated in the scientific literature by the term "antiriver"[7,8].

The creation of antirivers is producing great concern on the part of experts in the field of protection and rational use of water resources [9]. It is fully justified, since the implementation of such projects will inevitably lead to significant changes in the environmental condition of not only the river basins and wide territories surrounding them, but also the sea basins transformed river flows fall into.

However, the content of the term "antiriver" still lacks a clear definition, which generates significant difficulties when evaluating the ecological implications of different projects related to redistribution of river flow.

Russian guidelines [10] developed for the evaluation of the sanitary conditions during the river flow redistribution, that was planned, indicate that an antiriver is a reverse flow water body resulting (including the temporary ones) from the functioning of specialized hydrotechnical facilities. A detailed interpretation of the term allows to apply it in a quite

* Corresponding author: SuzdalevaAL@yandex.ru

wide range of meanings; however, it excludes from the antirivers category a significant part of projects the implementation of which is associated with similar ecological implications. From the ecological point of view, the term "antirivers" basically means a river flow transfer resulting in a considerable volume of natural water entering another geographical area (basin, region, etc.), and not the appearance of a reverse flow as such. At the same time, the river flow along the natural channel is reduced, and this is what causes the need to apply the term "antiriver" to all these projects. Based on the definition given above we can consider an "antiriver" a short-term change in the water movement within any limited river section, the effect of which will not go beyond the local scale. Therefore, a clearer and wider definition of this term is needed. Otherwise, if the project contemplates the movement of water in reverse direction along the natural channel, it will be an "antiriver". But if a transfer of the same volume of water is performed by draining water through a channel, it is a "flow redistribution". From the ecological point of view, the implications of these types of activity are equal, but the interpretation and the public resonance may differ fundamentally.

It is appropriate to accept the process aimed at creating similar facilities, namely transfer of the river flow, as the basis of the term "antiriver". However, by no means all types of river flow redistribution may be considered as antirivers. Based on the functions to fulfill, the transfer systems are conditionally divided into three categories [11]:

- intrabasin or local ones: when the transfer system does not go beyond the given river;
- interbasin ones: connecting adjacent river basins;
- interregional and interzonal ones: connecting the river systems of different physical and geographical regions (climatic zones).

Flow transfer systems are also classified based on the scale of the water resources involved [11]. Small transfers may include complexes with anannual volume of flow transfer of up to 1 km^3, the medium ones from 1 to 5 km^3, the large ones over 5 km^3. It is obvious that only the latter should be considered as antirivers.

The fact that the flow transfer system is represented by a controlled natural and technical system (NTS) [12] needs to be considered when evaluating the ecological implications of the creation of antirivers. It is formed by natural and technical facilities the work of which is interrelated and interdependent. The regime of water discharge is determined not only by the needs of the business activity, but by the need to maintain favorable environmental and sanitary conditions, and to provide other aspects of the life-sustaining activity of the population. Based on the above, the following definition can be given: an antiriver is a controlled NTS created for a large-scale interbasin and interregional transfer of water flow.

The controllability of the NTS created on the basis of antiriversis, generally, one-sided in nature and consists in controlling the volumes of water being redistributed. However, the creation of antirivers may be justified environmentally, only if the scope of the NTS control includes timely and efficient influence on all the main environmental components, including soil and vegetation cover and the wildlife. The control of both donor water course regions and recipient water course regions is carried out for the purpose of their maintenance and protection from unfavorable external factors. The latter should include the impact of unfavorable hydrometeorological factors (floods, draughts), the strength and the frequency of which are growing steadily due to global climatic changes. It is obvious that such view implies rethinking of not only the content of the concept of «antirivers», but their roles as well. This, in turn, is impossible without giving up some common stereotypes of thinking which can be overcome only by consistently implanting the innovative approaches to resolving issues [13] in the mind of the scientific community. The main stereotype, which was without any doubt, justified in the past, is the idea of antirivers exclusively as facilities of hydrotechnical construction performed for obtaining commercial benefit while ignoring catastrophic ecological implications.

The second stereotype concerns the foundations of the methodological basis of the environmental safety strategy. At the present stage, there is a traditional prohibitive and isolationist approach to resolving issues in this field. Its essence consists in the prohibition of the activity having a negative environmental impact, as well as in the attempts to isolate the protected natural areas from such impact. Such methodology was efficient in the previous period, when the impact of the industrial activity was local. The processes of self-purification and self-regeneration of natural ecosystems protecting them from unfavorable external impacts could maintain their good condition. At present, on many occasions, the natural homeostasis mechanisms cannot perform their function; as the result, environmental degradation is being registered everywhere. The biosphere is practically transformed into biotechnosphere, the condition of which depends on human factors. Global climatic changes that take place affecting all ecosystems of the planet play a special role in this process. In these conditions, a need of development of an alternative environmental protection strategy [12] arises. It is based on the use of different scale controlled NTS. The end purpose of the development of this strategy is a transition from an uncontrolled biotechnosphere to a controlled biotechnosphere which represents a hierarchy of functionally interrelated controlled NTS. Its prototypes are the existing hydraulic power systems. They regulate not only the river flow, but the pollutant flux as well, provide sanitary and ecological discharges aimed at maintaining good conditions for the existence of people and nature areas, and protect them from emergency situations. Thus, the implementation of ecologically justified antiriver projects will allow to control the processes of the biotechnosphere at a higher, interbasin, and interregional level.

Conclusion

The above materials can be summarized as follows:

1. The separation of the concepts of "antiriver" and "interbasin (interzonal) flow redistribution" is not inappropriate from the environmental point of view. The dissociation of these terms only reflects the specific character of the engineering solutions aimed achieving the same end results; in the present-day world, the creation of an international market of fresh water resources is becoming the main of them.

2. The antirivers are initially created as controlled NTS, but this fact itself is not fully recognized at present. For this reason, the antirivers projects are being considered as purely technical. Their practical implementation is evaluated from the perspectives of economic and geopolitical benefit. The environmental component of such projects is considered as a side effect. Such view of the issue is usual for both "technocrats" and "environmentalists". The difference between these two positions is that the first ones are trying to downgrade the role of the ecological implications (or partially ignore them), and the second ones are trying to justify their significance and achieve the prohibition of the project implementation. Based on the historical experience, we can forecast for certain that the victory of the economical, and especially geopolitical expediency is just a matter of time. This dictates the need to develop a constructive approach to resolving the issue on the interbasin river flow redistribution that would be ahead of the implementation of the projects, including the ecological optimization of the significant aspects of the technogenesis.

3. The development of the antiriver projects must include an evaluation of the ways and possibilities of their ecological optimization [14-16], which means a system of measures aimed at reducing the significant negative environmental aspects of the outlined activity, while increasing the efficiency of the positive ones (for example, protection from extreme hydrometeorological fluctuations). Development of a justified prohibition of the projects

able to result in environmental catastrophes and their replacement by alternative options the environmental risks of which could be considered as admissible should be considered as one of the types of ecological optimization.

References

1. 1. M.A. Duda, M.T. El-Ashry, Water In., **25**, 115-126 (2000)
2. V.I. Danilov-Danilyan, K.S. Losev, *Water consumption: ecological, economic, social and politic aspects* (Moscow, Nauka, 2006)
3. B.D. Richter, R. Mathews, D.L. Harrison, R. Wigington, Ec. App., **13**, 206–224 (2003)
4. G. Golubev, A.K. Biswas, ICID Bulletin, **28**, 18-22 (1979).
5. K.S.S. Murthy, *Interregional Water Transfers: Projects and Problems*, 117-125 (Luxembourg, 1979)
6. F. Quin, *Alberta symposium on Interbasin Water Transfers*, 1-14 (Edmonton, 1980)
7. A. Suzdaleva, S. Goryunova, *Technogenesis and degradation of surface water objects* (Moscow, ID ENERGIYA, 2014)
8. A.L. Suzdaleva, Power Tech. and Eng., **9**, 48-54 (2015)
9. P.P. Micklin, EOS Trans., Amer. Geophys. Union, **62,** 489-493 (1981)
10. *Guidelines for the study of the sanitary conditions of water use under the territorial redistribution of the river flow. Approved* 31.03. **2685-83**, 22, (1983)
11. A.S. Berezner, *Water transfer in the european part of the RSFSR* (Leningrad, Hidrometeoizdat, 1985)
12. M.P. Fedorov, A.L. Suzdaleva, Power Tech. and Eng., **3**, 10-15 (2014)
13. A.L. Suzdaleva, S.V. Goryunova, Biosphere, **7**, 429-449 (2015)
14. A.L. Suzdaleva, V.N. Beznosov, Elshaer Hayam, Hydr. Eng., **4**, (2010) – **1,** 22-25 (2011)
15. Z. Ivanova, O. Yudenkova, J. Naimaviciene, *MATEC Web of Conferences*, **73** (2016)
16. G. Potasheva, *MATEC Web of Conferences*, **73** (2016)

Structuring of nanomodified concrete cured in different temperature and humidity conditions

Viktor Voronin[1*] and *Yuriy* Bobrov

[1]Moscow State University of Civil Engineering, 129337, Yaroslavskoye Shosse 26, Moscow, Russia
[2]National Research University Higher School of Economics, 20 Myasnitskaya Ulitsa Moscow, Russia

Abstract. The article deals with the issues on the influence of the curing conditions on the formation of structure and properties of the concrete with nanomodifiers. Mechanochemically activated waste generated by the production of mineral fertilizers was used as nanomodified additives. It was established that providing favorable conditions for the initial structuring of concrete was the purpose of erecting structures made of concrete mix with said nanomodified additives in different temperature and humidity environments. It is provided that different types of production waste gaining the properties of efficient additives for binders, concrete mixes and compositions made on their basis as the result of the corresponding processing will be used as mineral fillers. Production waste is generally characterized by a heterogeneous composition and structure, by impurity content and by low chemical activity. Thus, one of the methods for increasing the efficiency of the waste is mechanical, chemical or physical activation. As the result of such activation, nanosizeparticles appear, and additional defects of mineral lattices are generated, which leads to accelerating the elementary interaction of the surface layer particles. Data on the research of macro- and microstructure of concrete and composites with nanomodifiers cured in different temperature and humidity conditions is given in the table.

1 Introduction

In accordance with the construction standards and regulations, the following three main climatic regions are identified: cold, moderate and hot [4]. The cold region is characterized by the average monthly temperature of below minus 15°C.

The moderate region is characterized by the average monthly temperature between minus 15°C and minus 25°C. On clear and windless days during the winter period, the atmospheric temperature in the central and southern Europe may go down to minus 35°C, or, during the summer period, go up to plus 40°C.

The hot climatic region has the longstanding average monthly temperature of plus 25°C in the hottest month. The curing of cement composites containing nanomodifieris due to the initial structuring has a significant influence on the concentration of new formations and on the development of porosity [5,6].

* Corresponding author: tvvib@mgsu.ru

Relatively small temperature fluctuations have a significant influence on the rate of curing of concrete as well.

For example, the relative strength at the age of 3 days will be 0.27R28 during concrete curing at 5°C, and at 30°C at the same age, it will be 0.5 R28, which is almost twice more.

The growtn rate of the indicator of concrete strength will gradually slowdown, and its ultimate strength will depend on the curing period when temperature changes take place.

So,if the cement composite containing a nanomodifier is cured at a changing temperature, then by certain age, its estimated strength will be determined as a sum of strengths obtained in the periods of curing at different temperatures:

$$Rc = \Delta R1 + \Delta R2 + \ldots + \Delta Rn, \qquad (1)$$

where $\Delta R1$, $\Delta R2$, ... ΔRn is change in the strength during temperature periods $t1$, $t\ 2$, tn.

Providing favorable curing conditions, especially at an early age, when the future composite structure is formed, will have a positive influence on the concrete curing. It will not be possible to change a structure the formation of which was disturbed during an early period of curing, by providing favorable conditions in a subsequent curing period. Thus, the right curing of concrete during early periods of curing is a condition required for obtaining high-quality concrete containing nanomodifiers. It is known that the normal temperature for concrete curing is15-20°C. At a lower temperature, the concrete strength will increase more slowly than at a normal one.

An interrelation has been established between the curing temperature (t) of the concrete and the period of structuring of the cement composite, which can be expressed as follows:

$$T = 20.1729 - 0.49837\, t + 0.00337\, t2 \qquad (2)$$

The curing of the cement composite with a nanomodifier almost stops at the temperature below 0°C. Frequent rises of temperature above 0°C will have a direct influence on the curing process. A composite frozen after being defrosted will keep hardening, and its strength will be increasing gradually.

Freezing of a cement composite with a nanomodifier at an early age may result in ice build-up accompanied by structure loosening [7]. The water density will be reduced in case of freezing. A porous low-density ice structure will be produced causing an increase in its volume and bond rupture between the filler surface and the cement stone. The strength of the cement composite with a nanomodifier is closer to the normal one, if it was frozen later. Table.5.1. and table 5.2. show the values of the critical strength, which must be reached by the composite before its freezing.

Table 1. The strength of cement composites with a nanomodifier after freezing

Freezing temperature °C	On the basis of Belgorodsky cement and crushed limestone		On the basis of Voskresensky cement and granite crushed stone	
	MPa	%	MPa	%
-20	27.9	77	16.9	82
-10	26.9	75	16.4	80
-5	23.3	62	12.9	62
-2	20.1	56	-	-
Normal curing	35.9	100	20.6	100

Table 2.shows the values of the critical strength which must be reached by the composite before freezing without loss of quality.

Reduced strength of the nanomodified concrete is related to the deformations caused by the occurrence of plastic shrinkage, processes of mass exchange and mass transfer, and contractions. Disturbance of the concrete structure during a short-term curing before freezing will reduce its frost resistance and impermeability. The frost resistance coefficient of the concrete with nanomodifiers turned out to be lower than the reference one [8,9,10].

Table 2. Minimum strength which must be gained by the cement composite with a nanomodifier by the moment of freezing

R_{28}, MPa	Minimum strength, no less than		Time o curing of the composite at 15- 20 °C, day
	% of R_{28}	MPa	
10	50	5	5…7
20	40	7	3…5
30	35	10	2…2.5
40	30	12	1.5…2
50	25	12.5	1…2

Conclusion

The research carried out shows the extreme importance of providing favorable conditions for the curing of concrete with nanomodifiers at an early age, when it is being structured.

It is confirmed by the data obtained when testing concrete curing during 2.5-3 h in unfavorable conditions, and then, 28 days in normal conditions; its strength was reduced by 30-50%, and the frost resistance and the impermeability were considerably reduced as well.

References

1. Yu.M. Bazhenov, L.A. Alimov, V.V. Voronin, *Structure and properties of concrete with nanomodifiers on the basis of technogenic waste* (Nomografiya, EBS ASV, 2013)
2. L.A. Alimov., V.V. Voronin, *Building materials* (Textbook. Publishing center, Academiya, 2012)
3. Yu.M. Bazhenov, *Compilation of roundtable presentations,* 12-18 (MGSU, 2007)
4. SNiP 23.01.99. Construction climatology
5. L.A. Alimov, V.V. Voronin, *Concrete work technology* (Textbook, Publishing center, Academiya, 2015)
6. L.A. Alimov, V.V. Voronin, A.A. Buldyzhov, I.V. Romanov, Mag. Nauchnoe obozrenie, **9**, 177-181, (2013)
7. L.A. Alimov, V.V. Voronin, K.S. Stenechkina, O.V. Aleksandrova, Mag. PGS, **4**, 58-61 (2015)
8. L.A. Alimov, S.I. Bazhenova, Vestnik MGSU, **1** (2010)
9. M.T. Bassuoni, M.M. Rahman, Orig. Res. Ar. Cem. and Conc. Res., **79**, 395-408, (2016)
10. J.J. Gaitero, I. Campillo, A. Guerrero, Orig. Res. Ar. Cem. and Conc. Res., **38**, 1112-1118 (2008)
11. A.S. Inozemtcev, *IOP Conference Series: Materials Science and Engineering*, **71** (2015)
12. M.G. Bruyako, L.S. Grigoryeva, D.V. Kravtsova, Adv. Mat. Res., **860-863**, 1323-1326 (2014)

13. A. Adamtsevich, A. Pustovgar, S. Pashkevich, A. Eremin, A. Zhuravlev, App. Mech. and Mat., **670-671**, 339-343 (2014)

Relationship between gas-dynamic flows and impacts of emergency explosions indoors

Alexander Komarov[1*], *Elena* Bazhina[1] and *Yuriy* Bobrov[2]

[1]Moscow State University of Civil Engineering, 129337, Yaroslavskoye Shosse 26, Moscow, Russia
[2]National Research University Higher School of Economics, 20 Myasnitskaya Ulitsa Moscow, Russia

Abstract. The paper demonstrates that gas-dynamic flows generated by explosive combustion bear a decisive influence on the entire course of explosion emergency. The conclusion is borne out by analysis of calculation results, test data and consequences of real-life emergency explosion in residential housing.

1 Introduction

Emergency explosions indoors occur in the deflagration mode of explosive transformation with specific features [1-6]. Apparent flame speed with deflagration is much below the speed of sound; therefore, the deflagration type of explosive transformation realizes the principle of quasi-static overpressure, understood as independence of the blast load from the spatial coordinate.

During combustion, explosion products expand in ε times (approximately 8 fold). Flame moves at normal combustion speed $U_н$ relative to explosion products. Apparent flame speed (speed at which the flame moves relative to a stationary observer) is the total velocity of expansion of the mix during combustion and its burnout speed (normal combustion speed). Considering that the speed related to expansion of the mix during combustion is much higher than that of normal combustion, one can theorize that gas-dynamic flowsthat accompanya deflagration explosion should mainly determine its further evolution.

Mathematical description of gas-dynamic flows created indoors by explosive combustion should base on two main principles that greatly simplify the task. First, explosive loads indoors must be significantly below the atmospheric pressure due to the bearing capacity (strength) of main structural elements of buildings [7,8].Second: fairly consistent adherence to the principle of quasi-static overpressure; this helps to overlook non-stationary equations of gas dynamics, and to consider the established gas-dynamic flow pattern for each point in time.

Solution to the gas-dynamic problem for each point in time not only describes the dynamics of the flame front,but also the speed rate of fresh mix, as necessary to evaluate the combustion process intensification coefficient.

* Corresponding author: KomarovAA@mgsu.ru

The writers calculated gas-dynamic flows created indoors when the explosive combustion scenario is realized. The results of calculation representing the fields of air flow velocity along with the deflagration explosion are represented in Figure 1.

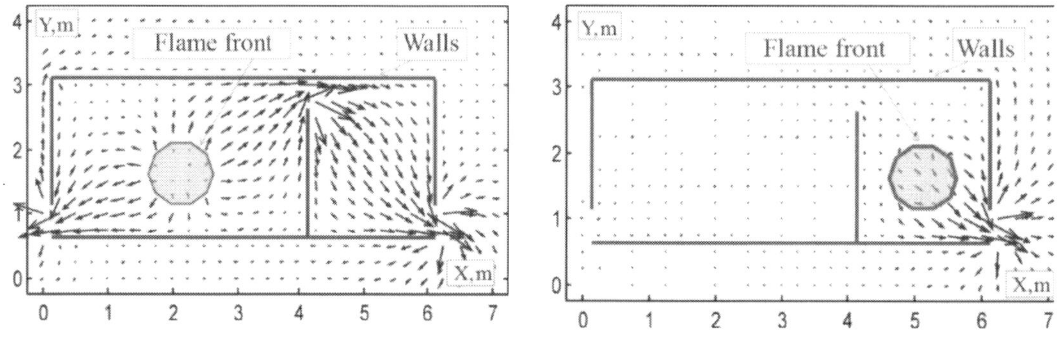

Fig.1. Speed vectors of air flows accompanying explosive combustion indoors

The calculations focused on adjacent rooms and gas-dynamic flows, because significant destruction of buildings during emergency explosions is mainly caused by mix overflowing from room to room, and resulting turbulence. When this mix is ignited, this typically results in a destructive explosion.

The calculations have demonstrated that the shape of flame front bears little effect on gas-dynamics flows indoors. This is particularly true for adjacent rooms. The main factor is the speed of released explosion products (source efficiency).

When explosive combustion occurs in an apartment room (such as the kitchen), air pressure in the adjacent rooms drops significantly. This creates significant inflow of fresh mix to the adjoining rooms and rapid growth of explosive pressure.

Experimental research of the explosive combustion process in adjacent volumes confirmed the main findings of the calculations.

Fig.2 is a photo of a deflagration explosion of propane-air mixin adjacent chambers. We can clearly observea jet stream of the mix from chamber to chamber. Fig. 2 proves that considerable turbulence in the mix occurs, and consequently speeds up its burnoutrate, which in turn causes significant inflow of explosion products and abrupt increase in explosion pressure.

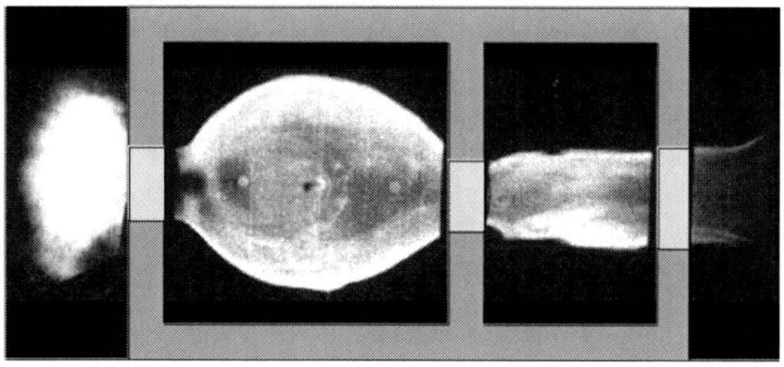

Fig.2. Photo of propane-air mixexplosionin adjacent chambers

Based on the above, let us consider the impact of emergency explosions in residential buildings.

The following observation is necessary at first. The emergency explosions described below had the following shared circumstances: Explosion Safety NTC (of MGSU) participated in expert investigation, and so these writers well know the circumstances of the emergency; in each case, it is known that a gas leak and ignition of the mix occurred in the kitchens of the buildings in question.

Fig.3 is a photo of the emergency explosion impact in the kitchen of a two-room apartment on level 5 of a 9-story building.

Fig.3. Photo of emergency explosion impact from propane-air mixin a two-bedroom apartment on level 5 of a 9-story residential building

A leak of propane gas from the supply hose on the gas stove caused explosion of the propane-air mix. The relay of a refrigerator in the kitchen was the source of ignition. From the picture above we can see that most of the destruction (and, respectively, the greatest impact) occurred in the rooms joined the kitchen.

In fact, total collapse was observed in the room most remote from the kitchen. This indicates that gas-dynamic flows absolutely determine the explosion emergency scenario. In this case, the sequence of events was as follows. The mix gets ignited in the kitchen. Influx of explosion products causes a slight increase of pressure (the apartment's windows and doors were closed); this drives the mix into the corridor and adjacent rooms. Thus causes turbulence and oxygenation of the mix, because near the source of ignition the mix on the kitchen floor is rich: propane is heavier than air and leaks spread over the ground. The triggered process is similar to jet stream in the right chamber as shown in Fig. 2. Explosion pressure rises abruptly due to rapid burnout of the mix and large influx of explosion products. Therefore, the highest pressure and greatest destruction will occur in the furthest room. It should be noted that such effect is only typical for explosion emergencies, completely contrary to fire impacts when the extent of burnout grows towardsthe source of ignition. Therefore, based on their experiences with fire accidents, investigators are often baffled when they consider consequences of explosion accidents.

A similar pattern was observed with emergency explosion of methane in the kitchen of a one-bedroom apartment on level 5 (apartment layout similar to the one above) in a 5-story building. Fig. 4 is a photograph representing the consequences of the explosion. Apparently, explosion pressure came a little short of bringing down the wall panels. The general picture of the explosion was identical with the situation described above.

In conclusion, let us discuss the impact of an explosion accident with a similar scenario, although in a brick building apartment. A photo of the impact is given as Fig. 5. In that case, the owners had merged two apartments into one, but the accident unfurled along a scenario that was generally consistent with above situations.The only difference was that the kitchen had double-glazing windows with enhanced strength characteristics installed that require sufficiently high pressure to open. This example of the explosion impact also illustrates that brick buildings are less resistant to blast compared to bearing-wall structures.

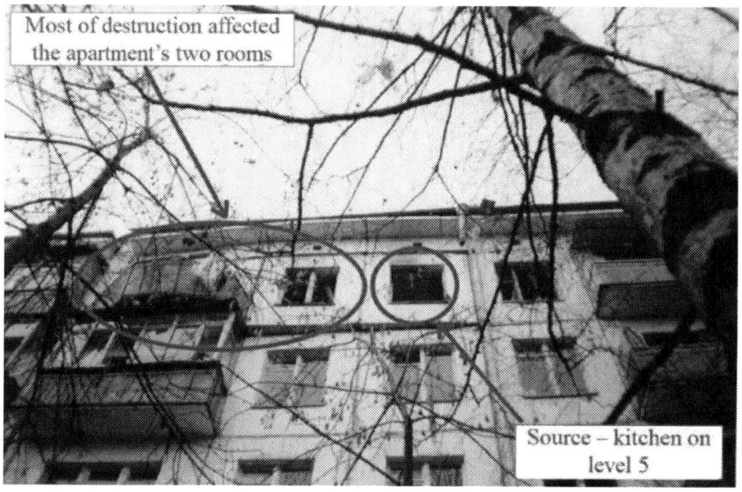

Fig.4. Photo of explosion impact from methane-air mix in a two-bedroom apartment on the 5th level of a 5-story residential building

Fig.5. Photoofexplosion impact from methane-air mix in an apartment on level 3 of a 10 story brick building

The following conclusion results from the above. Gas-dynamic flows created by explosive combustion of gas-air mixburning indoors absolutely determine its evolution, play a crucial role in generation of explosive loads and, therefore, totally determine the nature of the building's destruction.

Conclusions

The paper quotes calculations of the parameters of gas-dynamic flows created by emergency explosions indoors. Analysis of calculated results and test studies demonstrates that gas-dynamic flows generated by explosive combustion are the decisive factor for the entire course of the explosion accident.

Analysis of impacts from real-life emergency explosions inside residential buildings supported the conclusion that gas-dynamic flows created by explosive combustion of air-gas mixindoors absolutely determine the process; they play a crucial role in formation of explosive loads, and so fully determine the nature of destruction in the building.

References

1. A.A. Komarov, *Forecasting Loads and Estimating Consequences of Their Impact on Buildings and Structures* (Moscow State Construction University (National Research University), Moscow, 2001)
2. A.A. Komarov, Fire and Exp. Saf. Pub. (Пожаровзрывобезопасность), **13,** 15-23 (2004)
3. A.A. Komarov, Fire and Exp. Saf. Pub. (Пожаровзрывобезопасность), **8**, 49-53 (1999)
4. A.A. Komarov, G.V. Chilikina, Fire and Exp. Saf. Pub. (Пожаровзрывобезопасность), **11**, 24-28 (2002)
5. A.V. Mishuyev, A.A. Komarov, D.Z. Khusnutdinov, Fire and Exp. Saf. Pub. (Пожаровзрывобезопасность), **10**, 8-19 (2001)
6. A.V. Mishuyev, V.V. Kazennov, A.A. Komarov, N.V. Gromov, I.V. Lukyanov, D.V. Prozorovskiy, Fire and Exp. Saf. Pub. (Пожаровзрывобезопасность), **21**, 49-56 (2012)
7. B.S. Rastorguyev, A.I. Plotnikov, D.Z. Khusnutdinov, *Buildings and Facilities Designed for Emergency Explosion Impacts* (Building School Association, Moscow, 2007)
8. N.N. Popov, B.S. Rastorguyev, *Dynamic Calculation for Reinforced Concrete Structures* (Stroyizdat, Moscow, 1974)

To the Optimal Use of Technological Capabilities of the Excavator – Dragline in Construction

Rustam Khayrullin[1,*]

[1]Moscow State University of Civil Engineering, 129337, 26, Yaroslavskoye Shosse, Moscow, Russian Federation

Abstract. The mathematical model of motion of system "boom on a turning platform – bifilar suspended bucket" was created. As a control function, we choose the moment of force relatively non-movable base. The length of suspension of the bucket is assumed to be constant. The bucket of excavator is suspended bifilar. The bucket and the arrow needs to be moved from initial rest final rest position. The results of solution of the problem of rapid movement of bucket of excavator - dragline are presented. The structure of an optimal control law has been calculated by means of maximum Pontryagin' s principle and the method of control parameterization. A simple technique of calculations of optimal bucket trajectories has been developed. The dynamic of optimal control laws of the bucket and the arrow has been studied. Special software is created.

1 Introduction

At present technological capabilities of the heavy walking excavator - dragline, for a variety of reasons, are used not fully [1-3]. Relevance of a problem of performance improvement of the excavator – dragline doesn't raise doubts.

In [4-8] developed mathematical optimal control theory. In [9] described an effective method for analysis of maneuvering capabilities of controlled mechanical systems and results of its application for solving the problem of optimum maneuvering of aircraft and spacecraft. In [10] described a numerical method for optimization of high dimensionality systems. These methods based on [4-8].

In [1], using methods of [9-10], provides the results of application of the developed technology to the problem of maximum turn angle dragline - excavator boom within a fixed time interval, with finite damping of occurring oscillations of a bucket, which bifilar attached to a boom of excavator.

This article provides the results of application of the method [9-10] to solve the problem about the fastest movements of a dragline excavator bucket and arrow to a given point with

[*]Corresponding author: zrk@nm.ru

finite damping of occurring oscillations of a bucket attached to a boom. The considered problem is reciprocal for the problem of maximum boom swing within a fixed time interval. The results obtained in the article are in complete agreement with the results of solving the problem turn the excavator to the maximum angle within the required time [1].

2 Problem definition

Movement of the model "boom on a turning platform – bifilar suspended bucket" (Fig.1) under certain circumstances may be described by the following system of differential equations [2-3]:

$$\frac{d^2\sigma}{dt^2}+\frac{g}{l}\sigma=\frac{d^2\varphi}{dt^2}\quad,\quad J\cdot\frac{d^2\varphi}{dt^2}=M\,, \qquad (1)$$

where:

φ - a angle of turning of platform around vertical axis,

σ - bucket deflection from the boom plane, normalized to the turntable swing axis,

J - normalized moment of system inertia in relation to the turntable swing axis,

M - moment of force around the axis of the turntable,

l - normalized length of the suspension,

g - gravitational acceleration,

t - time of motion.

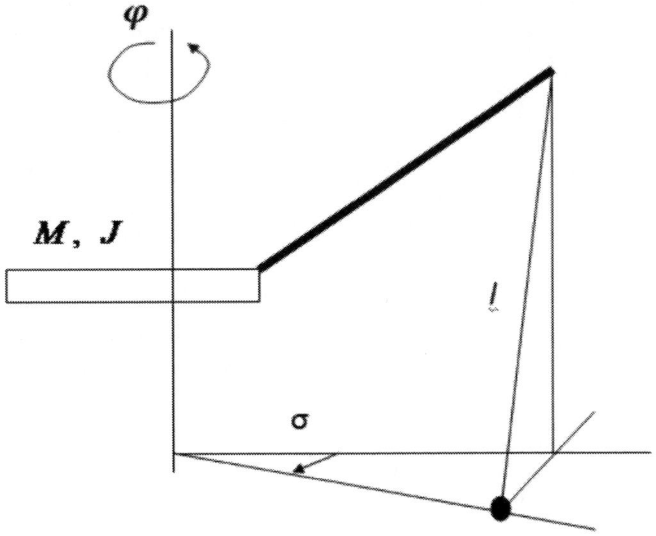

Fig.1. Scheme of the boom on a turning platform.

The system of equations (1) is true under the assumption that the mass of the bucket is negligible compared to the total weight of the boom and driving the platform. The bucket may

not be oscillating in the plane of the boom. The angle characterizing the deviation of the bucket from the plane of the arrows is small enough.
As a control function, we choose $M(t)$ the moment of force relatively non-movable base. The length of suspension of the bucket is assumed be constant.
Let in the initial moment of time $t = 0$ the system was at rest:

$$\sigma(0) = 0, \quad \varphi(0) = 0, \quad \dot{\sigma}(0) = 0, \quad \dot{\varphi}(0) = 0 \ . \tag{2}$$

The control function $M(t)$ should be restricted by the value M_{13}:

$$|M(t)| \le M_{13}, \quad 0 \le t \le T \ .$$

(3)
Its required to minimize the total time of motion:

$$T \to \min \ , \tag{4}$$

from the rest state (2) to the desired rest state:

$$\sigma(T) = 0, \quad \dot{\sigma}(T) = 0, \ \varphi(T) = \varphi_{13}, \ \dot{\varphi}(T) = 0 \quad . \tag{5}$$

Control function $M(t)$ should provide that the maximum angular speed of turn of an arrow shan't exceed the given value C_1 :

$$\max_t |\dot{\varphi}(t)| \le C_1 \quad . \tag{6}$$

The formulated problem of optimum control may be reduced to the standard problem of optimum control with limitations for the control function and phase variable [5,10].
Note that the formulated problem is similar to the problem of minimum time of movement of a mathematical pendulum with a movable suspension point from one equilibrium position to another equilibrium position [8].

3. Algorithms of the method of controls parameterization

Qualitative analysis of the optimum control structure is provided in [1-2]. Optimal controls laws were founded. These functions takes the form as shown at the Fig. 2. In order to get these functions were founded the algorithms of the method of controls parametrization. The provided below algorithms use the essential properties of the symmetry of control function.

The control laws shown at the Fig. 2a will be parametrized as $\xi = t_2$. Let's take $z = \sigma(T)$ as the offset. The corresponding boundary value problem of the maximum principle then reduces to a single non-linear equation $z(\xi) = 0$. This equation was solved using Newton method [5]. The relation $z = z(\xi)$ is calculated basing on the numeric integration (1) with initial conditions (2). Moments of time t_1, t_3, t_4, t_5 will be uniquely determined by the limitations of the movement time T, angular turntable swing velocity and symmetry condition of the required controlling function. Limitations $\dot{\varphi}(T) = 0$ and $\dot{\sigma}(T) = 0$ will be observed automatically due to symmetry.

The control laws shown at the Fig. 2b will be parametrized as $\xi = (t_1, t_2)$. Let's select $z_1 = \sigma(T)$, $z_2 = \dot{\sigma}(T)$ as offset vector $z = (z_1, z_2)$ components. The corresponding boundary value problem of the maximum principle then reduces to a system of second order non-linear equations $z(\xi) = 0$. This equation was solved using Newton method [5]. Moments of time t_3, t_4, t_5, t_6 will be uniquely determined by the limitations of the movement time T, angular turntable swing velocity and symmetry condition of the required controlling function. The limitation $\dot{\varphi}(T) = 0$ will be observed automatically due to symmetry of the controlling function.

The control laws shown at the Fig. 2c will be parametrized by 2 parameters $\xi = (t_2, t_3)$. Let's select $z_1 = \sigma(T)$, $z_2 = \dot{\sigma}(T)$ as offset vector $z = (z_1, z_2)$ components. The corresponding boundary value problem of the maximum principle then reduces to a system of second order non-linear equations $z(\xi) = 0$. This equation was solved using Newton method [5]. Moments of time t_1, t_4, t_5, t_6, t_7 are uniquely determined as described above, condition $\dot{\varphi}(T) = 0$ will be observed automatically.

The control laws shown at the Fig. 2d is limiting law to the laws shown at the Fig. 2a, 2b , 2c.

Calculations show that control laws created by means of method of controls parametrization and corresponding trajectories meet the required extremum conditions specified as Pontryagin's maximum principle.

Note that the described algorithms actually are simple methods of calculation of optimum trajectories and controls functions, since they allow to create optimum regimes basing on solution of algebraic equations of the second order by algebraic numerical methods.

Qualitative analysis of the necessary extremum conditions for the problem (1)-(6) demonstrated that they are equal to the respective extremum conditions for the problem of maximum turn boom swing within a fixed time interval [2]. Therefore, optimal control laws and trajectories for the problems (1)-(6) will be also optimal for the problem described in [1], and vice versa.

4 Results of calculations

Calculations were carried out in case of the following parameter values:
$v^2 = \frac{g}{l} = 0{,}6\frac{1}{c^2}$, $c_1 = 0{,}12\frac{pad}{c}$, $M_{13} = J \cdot 0{,}02$ $н \cdot м$. The main results of solution of the problem of rapid movement of bucket of excavator - dragline are presented in Table 1.

Table 1. Key parameters of optimal motion of bucket and boom.

The time of motion, sec.	16	16,472	18	20
The angle of rotation, deg	67,1	72	74,8	81,2
The form of the optimal control	Fig.2a	Fig.2b	Fig.2c	Fig.2d

5 Conclusion

The main results of the investigation can be summarized as follow:
1. The method of investigation of maneuvering capabilities of controllable mechanical systems was adapted for the problem of increasing performance of excavator - dragline.
2. Efficiency of the developed method was demonstrated through solving the problem of operating speed during swing of boom of dragline excavator to the required angle with finite damping of occurring oscillations of a bucket bifilarly attached to a boom. Obtained results are in complete agreement with the results of solving the problem turn the excavator to the maximum angle within the required time [1].
3. The developed model is the base of special purpose software. The described model is included into the automated management system of enterprise. The model is used for calculation of optimal booms trajectories of dragline - excavator. To solve the optimal control problem at computer it takes less 2 second.
4. The described algorithms of the method of parametrization of control laws actually can be used in practice as simple methods of calculation of optimum trajectories and controls functions

References

1. R.Z. Khayrullin, On the research of maneuverability of excavator - dragline, Vestnik MGSU, **4**, 49-53, (2010)
2. L.D. Pevzner, R.Z. Khayrullin, *XXYIII - International Symposium "Computer applications in the minerals industries"*. October 20-22, 1999, Colorado School of Mines. Golden, Colorado, USA, 863-870, (1999).
3. R. Z. Khayrullin, Automated dynamic load control by Electromechanical systems of excavator - dragline, Vestnik MGSU, **7**, 125-129 (2012)
4. L.S. Pontryagin, V.G. Boltyansky, R.V. Gamkrelidze, E.F. Mishchenko. *Mathematical theory of optimal processes,* (1961)
5. R. P. Fedorenko, *Approximate solution of optimal control problems,*(1978)
6. A. Bryson, Y. Ho, *Applied Optimal Control.* Blaisdell Publishing, Walthman, MA, (1969)
7. R. Vinter. *Optimal Control.* Birkhauser, Boston, (2000)
8. L.D. Akulenko. *Asymptotic methods of optimal control,*(1987)
9. Yu. F. Golubev, R.Z. Khayrullin, On the research of maneuverability of spacecraft during the atmosphere descent, Theory and control systems journal, **4**, 146-151, (1996)
10. R.Z. Khayrullin, Optimization of high dimensionality systems with the use of ILOG software components, Vestnik MGSU, **8**, 157-163, (2013)

Modern efficient methods of steel vertical oil tanks clean-up

Vladimir Nekrasov[1,*]

[1]Tyumen State Oil and Gas University, Volodarskogo st. 38, 625000 Tyumen, Russia

Abstract. The legislative base of the Russian Federation operating in the field of operation of tanks and tank parks is considered, and consecutive stages of technological process of cleaning of vertical steel tanks from oil ground deposits are presented. In work shortcomings of existing most widespread electromechanical mixers are described when using a hydraulic method of removal and prevention of formation of ground deposits in tanks with oil and oil products. For the purpose of increase of efficiency, reliability and decrease in power consumption of washout of oil ground deposits in tanks the new design of system of funneled washout and prevention of formation of deposits is offered.

1 Introduction

The sludge is generated and accumulated in the tanks while storage and transportation of oil or oil products resulting in loss of up to 1/4 of tank storage capacity. Besides, the accumulation of sludge leads to complications while utilization of tank, quality/quantity monitoring of oil volume, and reduction of technical and economical characteristics of oil storage tanks and transportation system in overall.

2 The object of study and methods

The requirement to keep the storage capacity of the tank is strongly followed while creating the design of construction and regulated by authorized documents, such as: GOST 1510-84 «Oil and oil products. Labeling, packing, transportation and storage», guidelines RD 153-39.4-078-01 «Regulations on technical utilization of tanks in oil pipeline», RD 153-39.4-057-00 «Workflow on elimination of generation of tank bottoms and cleaning», RD 153-39TH-012-96 «Regulations on safe cleaning of oil tanks» etc. [1]

Tank bottoms removal is a complex procedure including the following:

- Pumping out of hydrocarbons from the tank till the level required for work of cleaning nozzles and de-sludging by the standard cleaning devices, if available;
- Preliminary degassing by forced ventilation until the concentration of oil vapors is below maximum allowed concentration; and making-up of cleaning equipment;

* Corresponding author: nekrasov-sever@mail.ru

- Oil vapor saturation of gas space in the tank exceeding the maximum allowed concentration;
- Cleaning of the tank walls and bottom by oil stream generated by cleaning device along with simultaneous monitoring of gas space saturation by the hydrocarbons and the level of static electricity;
- Pumping out of dissolved and dispersed sludge together with flushing oil into a special container (or pipeline);
- Cleaning of the tank by stream of water or cleansing agents from cleaning devices;
- Pumping out of emulsified water from the tank;
- Degassing by sorption and (or) forced ventilation and (or) airing;
- Quality monitoring of pumped sludge;
- Monitoring of cleaning quality of internal surface and degassing [2].

At the same time it's required to minimize the generation of sludge in the tank. It will allow decreasing the number of cleaning cycles and increase the storage capacity of tanks and oil transport system in overall.

For the time being the various methods are used to decrease the generation of tank bottoms; the essential of which are classified into mechanical, hydromecanical, thermal and sonic. However, the most power-and-economically effective is the hydraulic method. It's an exposure of the oriented drowned stream of oil at determined velocity to bottom sediments, as a result, separating them into parts and holding them in suspension. [3,4,5,6,7]

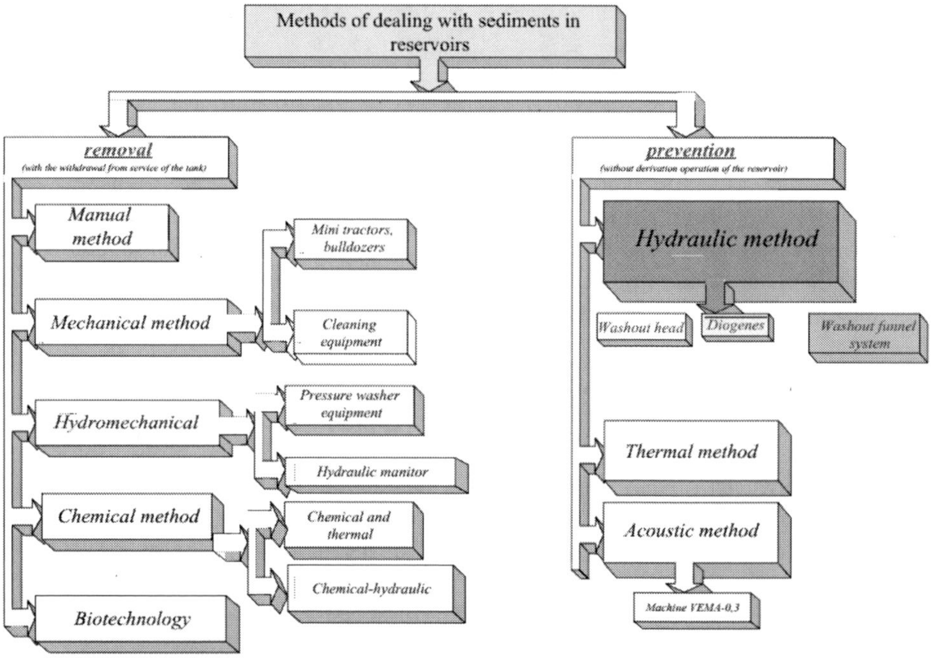

Fig. 1. Methods of tank bottoms generation reduction.

This method is being extensively applied in Transneft Company. In order to create the oriented hydraulic stream for removal of tank bottoms the most widely used are «Diogen»-500/700 and «Taiphun» devices. These are a kind of electromechanical mixers of different design. At the same time they have quite many essential disadvantages.

«Diogen» device is installed on the tanks with 20000 m^3 volume. The operating range of this device at the determined stream velocities is 60° which allows creating so called "dead" zones in the tank where paraffinic-tar-asphaltenic contents are intensively accumulated and compressed and may occupy up to 50% of tank bottom. [8]

Increasing of the revolutions of the device's screw and increasing of the stream velocity leads to increase of pressure on the tank wall which affects the strain-stress state and tanks operation time.

Worth mentioning, the essential power is consumed by the electric motor. The moving parts negatively affect the reliability of the device.

Long term utilization leads to increased gaps in supporting nodes and shaft eccentricity due to irregular collection of multicomponent sediments on propeller being a cause of the additional vibration and local cyclic load on tank walls. [9]

The additional disadvantages of the electro mechanic mixing machine may be the duration of average washing cycle - 98hr, tank leakage and low reliability of system due to the complex design of device.

All of the abovementioned disadvantages of the utilized nowadays devices are giving evidence of the necessity to create a new design preventing from the generation of tank bottoms. Increasing the efficiency, reliability and decreasing the power consumption while removing sludge out of vertical steel tanks can be achieved with utilization of funnel type cleaning system and by prevention of tank bottoms generation. (pic.2). [10]

The construction consists of incomplete curved pipe (5) installed in the tank base (7) which follows the shape of tank walls (8) and equipped with eight ascending pipes (6) of less diameter with 45° angle between them and each of ascending pipes is inclined to tangent circle in horizontal plane at 22°30′ angle and at angle of 45°, 40°, 35°, 30°, 25°, 20°, 15°, 10° in the direction of oil traffic in pipe in vertical plane accordingly. The construction has a shut off valve (3), Γ-shape bend pipe (2) and S-shape pipe (4) which connect the whole pipe chain with the main tank pipeline (1) (pic.3).

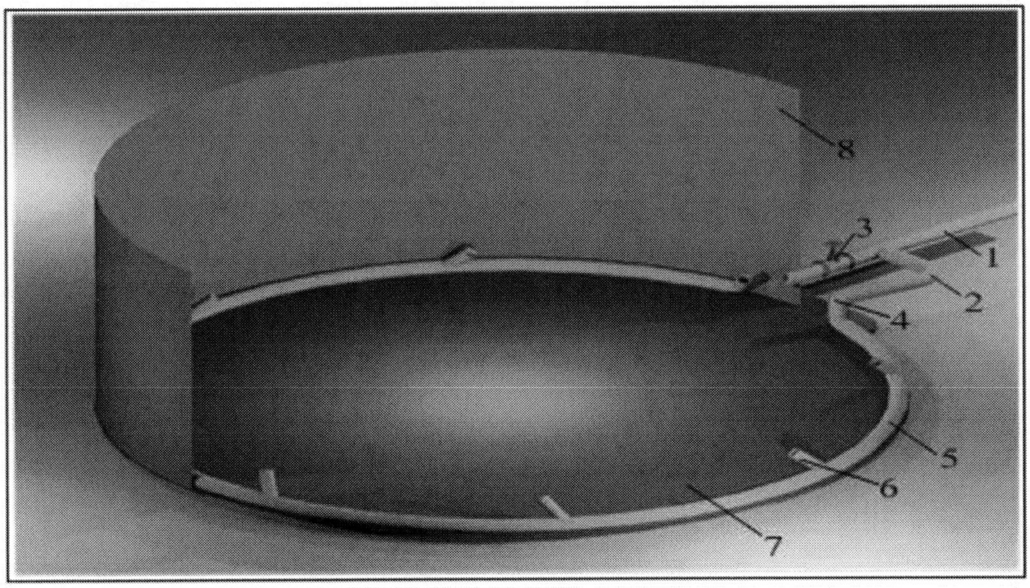

Fig. 2. General view of funnel type cleaning system for sludge removal.

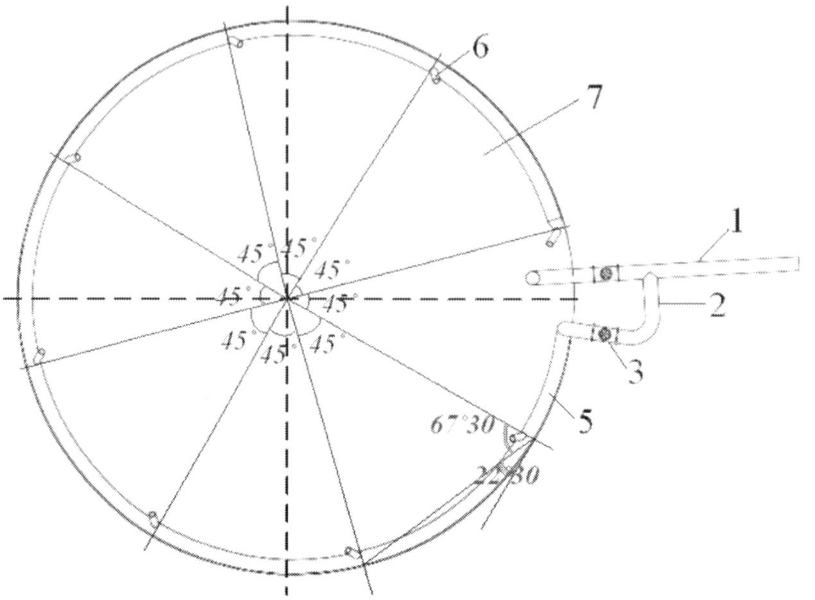

Fig. 3. Design of funnel type cleaning system for sludge removal.

Decreasing of tank bottoms generation and cleaning of the existing ones is done by mixing of oil products. Whereupon the mixing is done by means of energy produced when filling the tank. Some of kinetic energy of the moving oil makes the whole volume of oil in tank circulate as a result of influence of hydraulic drowned streams of oil or oil product.

Location of the pipe at the bottom of tank allows desludging even with low amount of oil or oil products available in the tank. The number of ascending pipes has been chosen on the assumption of the minimum power consumption. As hydraulic pressure is usually lost while desludging due to the resistance of the ascending pipes and during circulation of oil along the length of the curved pipe, the tip angle of ascending pipes in vertical plane is decreased down the direction of oil or oil products in pipe from 45° to 10° accordingly. As a result, the hydraulic pressure losses are decreased and dynamic residual energy, generated from funnel type mixing, is increased and leads to efficient desludging.

As a result, the desludging and prevention of tank bottoms generation is done without moving them out of tank. Whereas, the method provides high efficiency of desludging regardless the tank volume and type of the stored products.

3 Conclusions

Due to wide use in Russia, the perspective of the hydraulic method of tank bottoms removal from vertical steel tanks is proved by efficient de-sludging.

The method is more economically efficient comparing to other hydraulic methods due to the used energy produced when filling the tank.

Application of this technology allows increasing of operating reliability due to absence of additional solid elements and moving parts.

Proposed system doesn't require the maintenance saving the time for maintenance checks.

References

1. Pat. 2445150 RU, MPK B01D53/72. Method of cleaning from vapor-gas hydrocarbons/ E.A. Lyubin, A.A. Korshak (RU). - # 2010133414/05, registered on 09.08.2010 (2010)
2. Y.D. Zemenkov, *Oil and gas transportation and storage in examples. Teaching guide* (Nedra, Saint-Petersburg, 2004)
3. *API Manual of Petroleum Measurement Standards, Chapter 19 - Evaporative-loss Measurement, Section 1 - Evaporative Loss from Fixed-Roof Tanks, Third Edition* (2002)
4. *API Manual of Petroleum Measurement Standards, Chapter 19 - Evaporative-loss Measurement, Section 2 - Evaporative Loss from Floating-roof Tanks, Second Edition* (2003)
5. *API Manual of Petroleum Measurement Standards, Chapter 19.4 – Recommended Practice for Speciation of Evaporative Losses, Second Edition* (2005)
6. *API 650 Welded Tanks for Oil Storage, Eleventh Edition* (2010)
7. *API 2610 Design, Construction, Operation, Maintenance, and Inspection of Terminal & Tank Facilities, Second Edition* (2005)
8. A.V. Salnikov, *Loss of oil and oil products. Teaching guide* (UGTU, Ukhta, 2012)
9. I.A. Ivanov, J.A. Kushnir, S.A. Pulnikov, *Geotechnical problems pipelines. Teaching guide for high school students enrolled in the direction of bachelor 131000 «Oil and Gas Business»* (TSOGU, Tyumen, 2011)
10. Pat. 2500486 RU, MPK B05C7/04, B65D88/74. Device for increasing operational properties vertical steel tanks/ V. O. Nekrasov, R. E. Levitin (RU). - # 2012125478/05, registered on 10.12.13 (2013)

MATEC Web of Conferences 86, 04051 (2016) DOI: 10.1051/matecconf/20168604051
IPICSE-2016

The methodology of the vibrational separating machines bearing structure rigidity optimization and the actuating mechanism parts dynamic forces determination

Valeriy Piven[1,*], *Olga* Umanskaya[2]

[1]Department of machines and equipment of oil and gas industry, Tyumen Industrial University, Tyumen, Russia
[2]Department of applied mechanics, Tyumen Industrial University, Tyumen, Russia

Abstract. The reciprocal motions at the vibrational separating machines of the sieving trunks, transmitted from the actuating mechanism, vibrate frame. The Frame weaving is put over the sieving trunks weaving, disturbing the optimal mode of operation. Therefore, for the bearing structures of separating machines vibration decrease it is necessary to optimize its structures and characteristics of the actuating mechanism, which would provide the minimal mass with the necessarily frame rigidity.

1 Introduction

The Vibrational machines are commonly used in different technical spheres for the granular blends separation on the heterogeneous fractions. Granular blends dosing, its mixing and Compaction.

On the vibrational machines technological process stable behavior is determined by the kinematic regime of the movable operating elements precise adherence. This factor assumes defining significance at the separating processes.

The reciprocal motions at the vibrational separating machines of the sieving trunks, transmitted from the actuating mechanism, vibrate frame. Frame weaving is put over the sieving trunks weaving, disturbing the optimal mode of operation. Therefore, for the bearing structures of separating machines vibration decrease it is necessary to optimize it structures and characteristics of the actuating mechanism, which would provide the minimal mass with the necessarily frame rigidity. Metal consumption of the building structures, on which such machines are installed, would depend on, in its turn, this parameter.

In connection to this, the optimal order of the constructs should be determined within the designing.

*corresponding author: pivenvv@yandex.ru

2 Materials and method

Consider the following case, when elements – horizontal, upstanding or bias Figure 1 are added for the increase of the frame structures rigidity. On their example, it is possible to work out the frame structures' optimization methodology, which is sustaining the vibrational strain. Current indexes of parameters l1 and l2 should be determined in the process of the optimization.

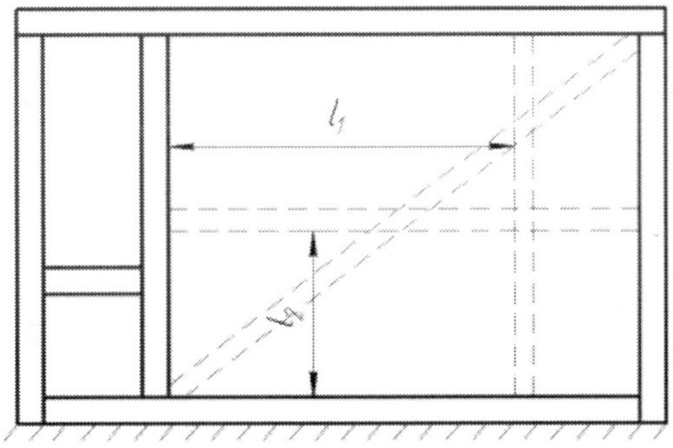

Fig. 1. The scheme of the frame structure with the elements provided its rigidity.

In the sequel, the usage of such approach will allow to optimize the rigidity of any construction of such type within the devised methodology. Thereby, the decompositional scheme of the mathematic model of the vibrational frame motion and the optimization of such a model has a view illustrated on the Fig. 2. It is necessary to explain the objective function during the mathematic modeling, to choose the limitations, imposed on the mathematical model; the power factors of the statically indeterminate frame, of actuating mechanism details must be defined; to specify the risky section's voltage, to state the bearing structure details' the vibration of displacement, the vibration velocity and the vibration acceleration.

The mass of the frame structure should be reasonably accepted as an objective function, this mass would be the function from the moment of the structure's beam inertia of section. The limitations must be imposed on this function: on the structural strength, on the permissible vibration of displacement, on technological process quality. The obtained scheme would allow optimizing the bearing structures of the vibrational separating machines, taking limitations by vibrocharacteristics into account at the stage of design.

By the example of the separating vibrational machine the methodology of the bearing frame structures of the vibrational machines optimization were worked out for the optimization the bearing frame structure rigidity (Figure 2).

The design parameters of the structure itself, kinematic parameters of the actuating mechanism are taken into consideration within this methodology with the help of the theoretical relations (Figure 3), these parameters influence the bearing structure vibrocharacteristics.

The methodology allows taking into account influence on the additional elements of rigidity of the vibrocharacteristics, to define its suitability installation and its seat.

In the first stage it is necessary to state dynamic forces on the actuating mechanism details for the foregoing methodology realization. Let's consider the disturbing influences character on the frame structure from the actuating mechanism on the example of the grain-cleaning machine with the crank-and-rod mechanism. Machine has two sieving trunks, which are moved towards each other. The complication of such task is concluded on that the sieving trunks reciprocal motions are necessary for the separation of the grains blend technological process realization.

DOI: 10.1051/matecconf/20168604051

At the same time the sieving trunks actuating mechanism is the source of the frame structure itself forced oscillations. The frame structure vibrational movements are transmitted to the sieving trunks through the suspension arm mechanism, and they are influenced on the separated components movement process over the sieve. Separation is realized in nonoptimal regime, which is in whole led to the technological separation process quality decrease.

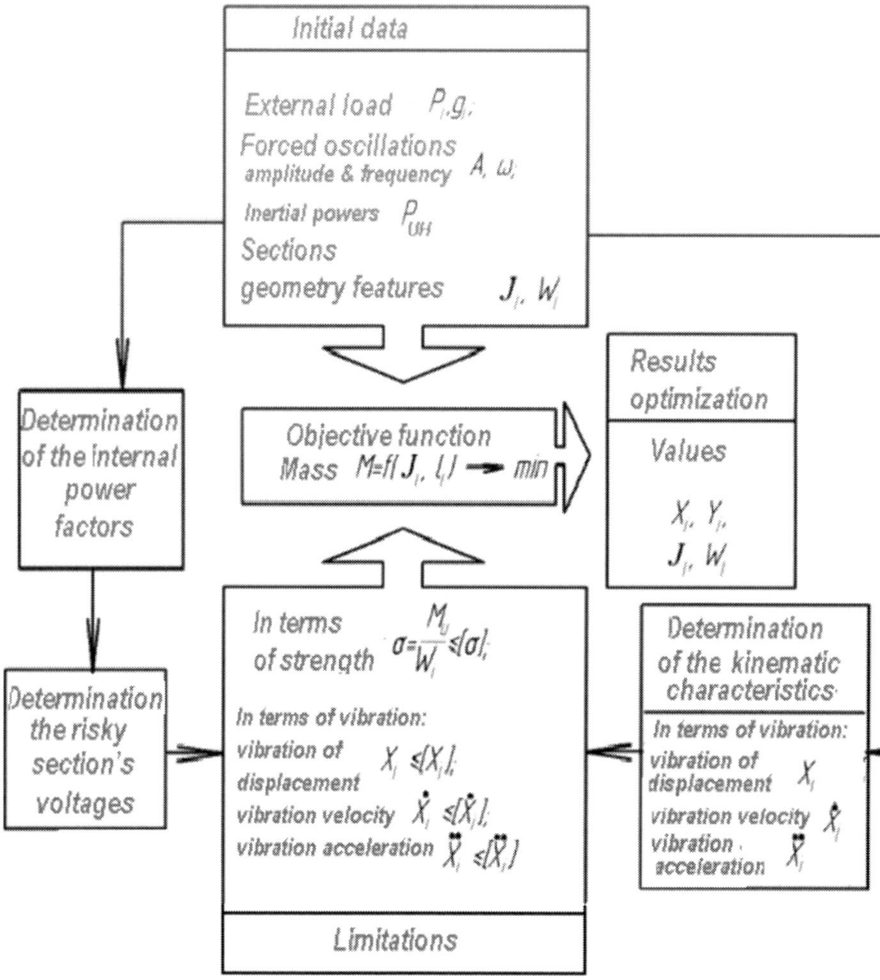

Fig. 2. Decompositional scheme of mathematical model of bearing structure vibrational movement optimization.

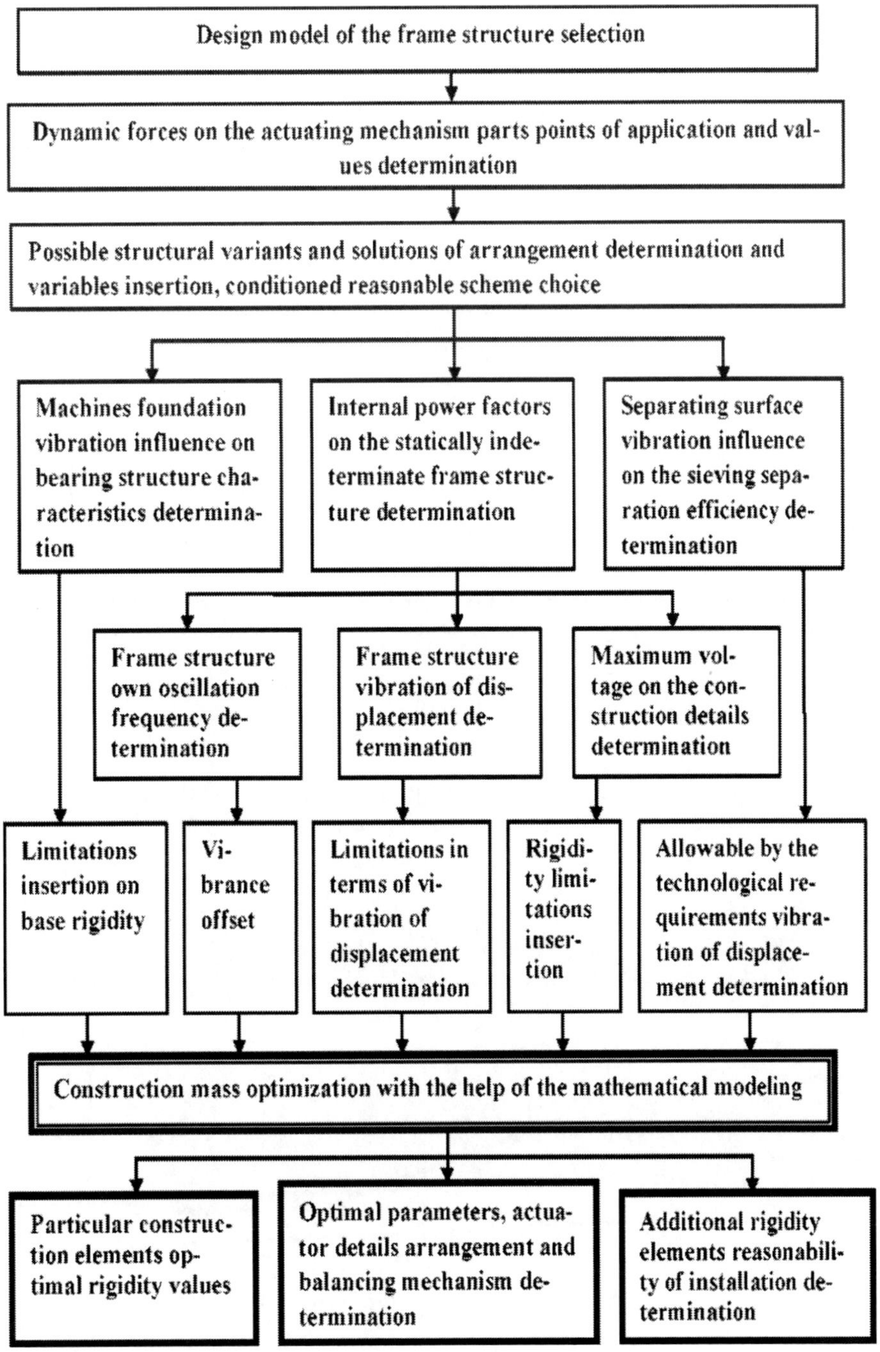

Fig. 3. Structural scheme of the calculating methodology of vibrational machines bearing frame constructions.

Review the bearing frame structure, represented in the Figure 4. The following systems of powers are actuated on the lower sieving trunk on it's the last right position (Figure 5). Examine the reactions, which are occurred at the A and B points in detail. The arm suspensions, made of plywood or steel, are used on the grain-cleaner machines for the rectilinear oscillations transition. In this case let's consider the plywood arm suspensions. Arm suspensions length - 0,53 м. Section – $0{,}06 \times 0{,}01$ м. Separation trunk oscillation amplitude is the maximum bending at the arm suspension end.

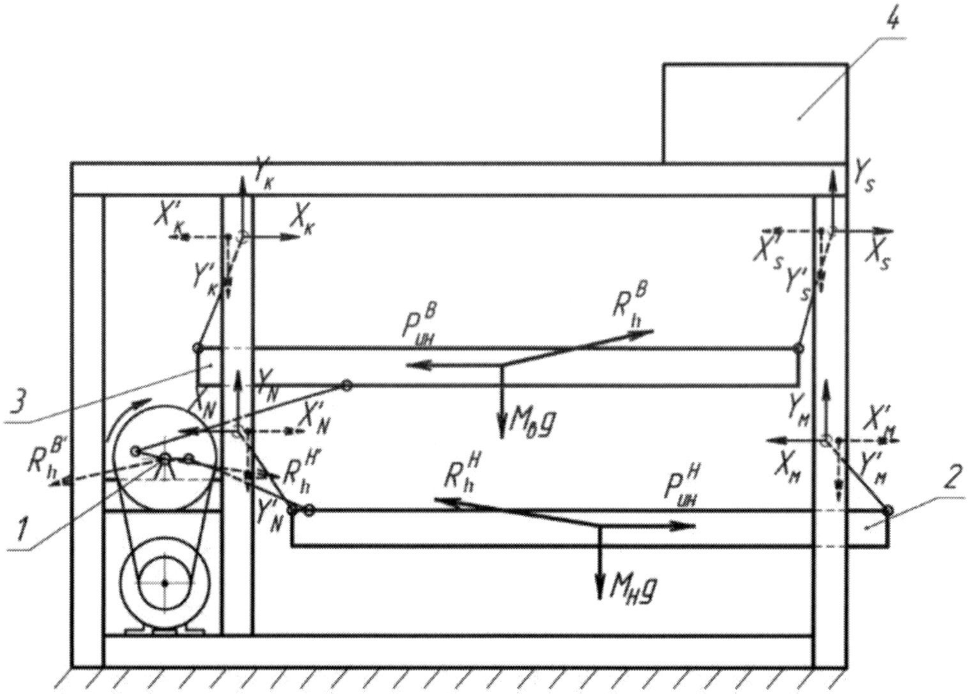

Fig. 4. Separation machine construction scheme (powers, actuating on the bearing structure from the side of the actuating mechanism details are pictured by doted line): 1 – actuating mechanism; 2 – lower sieving trunk; 3- upper sieving trunk; 4 – aspiration system.

The suspension arm reactions are directed along it. Calculations show that the connecting-rod reactive torque considerably lower than the other power factors. It is possible to ignore it.

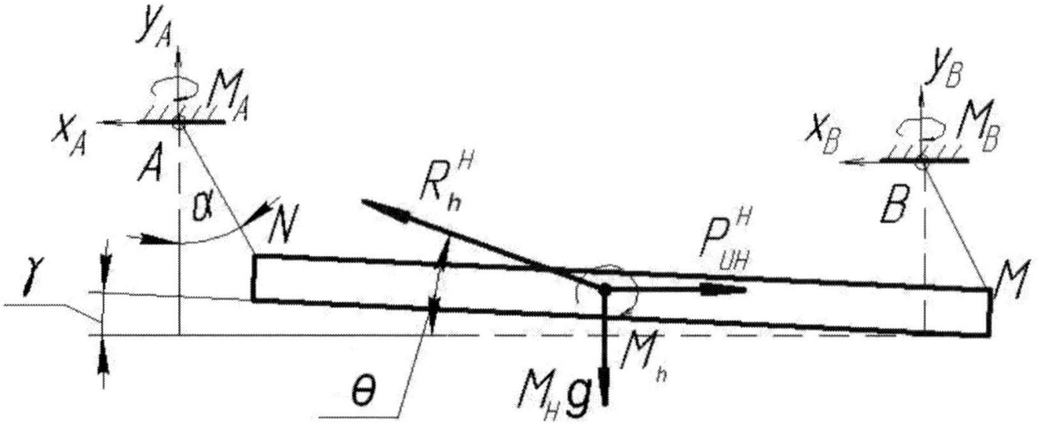

Fig. 5. Scheme of powers, actuating on the lower sieving trunk with suspension arms M_{Hq} - gravity; $P_{UH}{}^{н}$ – inertial force; M , R^{H} – connecting-rod reactions; X_A, Y_A, M_A, X_B, Y_B, M_B – reactions at the suspension restraint point. (points. A and B).

Hereby, taking into account foregoing allowance, the scheme of powers, actuating on the lower sieving trunk, has the following view (Figure 6):

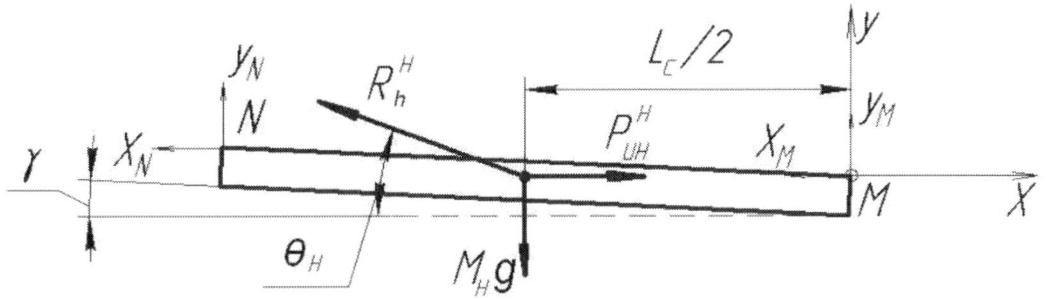

Fig. 6. Scheme of powers, actuating on the lower sieving trunk.

Make up the balance equation for the flat powers system in order to determine the reaction on the points N and M.

$$\Sigma F_{kx} : -x_N - R^{н} \cdot \cos\theta_{н} + P - x \ = 0 \tag{1}$$

$$\Sigma F_{ky} : y_N + R \ \cdot \sin\theta_{н} - M_{н} \cdot g + y_M = 0 \tag{2}$$

$$\Sigma M_M(\overline{F}_k) : x_N \cdot L_c \cdot \sin\gamma - y_N \cdot L_c \cdot \cos\gamma - R^{н} \cdot \sin\theta_{н} \frac{L_c}{2} \cdot \cos\gamma + \\ + R \cdot \cos\theta_{н} \frac{L_c}{2} \cdot \sin\gamma + M_{н} \cdot g \cdot \frac{L_c}{2} \cdot \cos\gamma - P^{н} \cdot \frac{L_c}{2} \cdot \sin\gamma = 0. \tag{3}$$

Inertia $P_{\text{ИН}}^{\text{Н}}$ is determined by the formula:

$$P_{\text{и}}^{\text{Н}} = M_{\text{н}} \cdot a \tag{4}$$

Where: $M_{\text{н}}$ – the lower sieving trunk mass (kg), a – acceleration of masses center of the lower sieving trunk (m/sec^2)
We are used the following relation to determine the acceleration [1]:

$$a = \ddot{x} = r\omega^2 (\cos\omega \cdot t - \frac{e}{L} \sin\omega \cdot t) \tag{5}$$

Where r – crank radius (m), ω - crank angular velocity (rad/s), t – time (s), e – desaxial displacement value (м), L – crank length (м).

Taking into account, that M_A=0, M_B=0 (Figure 5) and the suspension arm reaction is directed along it, get

$$x_N = y_n \cdot tg\alpha \tag{6}$$

$$x_M = y_M \cdot tg\alpha \tag{7}$$

Angle α is varied according to the principle:

$$tg\alpha = \frac{r}{L_n} \cdot cos\omega \cdot t \tag{8}$$

Taking into account the formulas (6) - (8) the powers at the points N and M would be determined by the following formulas:

$$x_N = y_n \cdot \frac{r}{L_n} \cdot cos\omega \cdot t \tag{9}$$

$$x_M = y_M \cdot \frac{r}{L_n} \cdot cos\omega \cdot t \tag{10}$$

Take the following designation: $R^{н} = R_1$; $M_N = M_2$; $L_C = L_1$; $L = L_2$; $L_n = L_3$.

Taking into account the obtained formulas and entering designations the equations system (1) shall have the following view:

$$-x_N - R_1 \cdot \cos\theta_н + M_2 \cdot \ddot{x} - x_M = 0 \tag{11}$$

$$y_N + R_1 \cdot \sin\theta_н - M_2 \cdot g + y_M = 0 \tag{12}$$

$$x_N \cdot L_1 \cdot \sin\gamma - y_N \cdot L_1 \cdot \cos\gamma - R_1 \cdot \sin\theta_н \frac{L_1}{2} \cdot \cos\gamma + R_1 \cdot \cos\theta_н \frac{L_1}{2} \cdot \sin\gamma + \\ + M_2 \cdot g \cdot \frac{L_1}{2} \cdot \cos\gamma - P_1 \cdot \frac{L_1}{2} \cdot \sin\gamma = 0. \tag{13}$$

$$x_N = y_n \cdot \frac{r}{L_n} \cdot cos\omega \cdot t \tag{14}$$

$$x_M = y_M \cdot \frac{r}{L_n} \cdot cos\omega \cdot t \tag{15}$$

Solving this equation system relatively to the x_N, x_M, y_N, y_M and R_1, will get:

$$R_1 = \frac{M_2 \cdot g \frac{r}{L_3} \cdot \cos\omega \cdot t - M_2 r \omega^2 \left(\cos\omega \cdot t - \frac{e}{L_2} \cdot \sin\omega \cdot t \right)}{\sin\theta_H \cdot \frac{r}{L_3} \cos\omega \cdot t - \cos\theta_H} \tag{16}$$

$$y_N = \frac{R_1 (\sin\theta_H \cdot \cos\gamma + \cos\theta_H \cdot \sin\gamma) - M_2 g \cos\gamma + M_2 r \cdot \omega^2 (\cos\omega \cdot t - \frac{e}{L_2} \cdot \sin\omega \cdot t) \sin\gamma}{2 \cdot \left(\frac{r}{L_3} \cdot \cos\omega \cdot t \cdot \sin\gamma - \cos\gamma \right)} \tag{17}$$

$$y_M = y_N \tag{18}$$

$$x_N = y_n \cdot \frac{r}{L_n} \cdot cos\omega \cdot t \tag{19}$$

$$x_M = x_N \tag{20}$$

3 Results and discussion

Solving this equations, relatively to the time and cranks angular velocity, will get the following values x_N, x_M, y_N, y_M and R_1, when e=6,5·10^{-2} м; r=7,5·10^{-3} м; L_1= 1,723 м; L_2=0,5 м; L_3=0,53м:

$$x_N = -1{,}217 + 11{,}513 \cdot \cos\omega t + 1{,}605 \cdot 10^{-4} \cdot \sin\omega t - 1{,}217 \cdot \cos 2\omega t + 0{,}16 \cdot \sin 2\omega t \quad (21)$$

$$y_N = 813{,}5 - 171{,}966 \cdot \cos\omega t + 22{,}562 \cdot \sin\omega t - 0{,}171 \cdot \cos 2\omega t + 0{,}023 \cdot \sin 2\omega t \quad (22)$$

$$R_1 = \frac{2{,}5 \cdot 10\ 3 \cdot \cos\omega t - 329 \sin\omega t}{1{,}013 - 2{,}017 \cdot 10^{-3} \cos\omega t} \quad (23)$$

The last right position of the lower sieving trunk corresponds to the last left position of the upper sieving trunk (Figure 7).

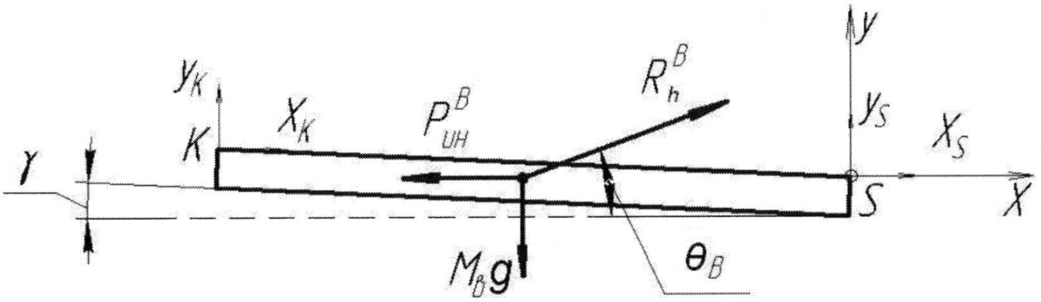

Fig. 7. The scheme of the powers, actuating on the upper sieving trunk.

The balance equation of the upper sieving trunk has the view:

$$\Sigma F_{kx} : x_K + R^{в} \cdot \cos\theta - P + x_S = 0 \quad (24)$$

$$\Sigma F_{ky} : y_K + R^{в} \cdot \sin\theta - M \cdot g + y_S = 0 \quad (25)$$

$$\Sigma M_S(\overline{F}_k) : -y_K \cdot L_c \cdot \cos\gamma - x_K \cdot L_c \cdot \sin\gamma - R^{в} \cdot \cos\theta \frac{L_c}{2} \cdot \sin\gamma - \quad (26)$$

$$- R \cdot \sin\theta \frac{L_c}{2} \cdot cos\gamma + M \cdot g \cdot \frac{L_c}{2} \cdot \cos\gamma + P_{uh}^{в} \cdot \frac{L_c}{2} \cdot \sin\gamma = 0.$$

Taking into account, that

$$tg\alpha = \frac{r}{L_n} \cdot cos\omega t \quad (27)$$

$$x_k = y_k \cdot tg\alpha = y_k \cdot \frac{r}{L_n} \cdot cos\omega t \quad (28)$$

$$x_s = y_s \cdot tg\alpha = y_s \cdot \frac{r}{L_n} \cdot cos\omega t \quad (29)$$

$$P = Mb \cdot r \cdot \omega^2 (\cos\omega \cdot t - \frac{e}{L} \cdot \sin\omega \cdot t) \tag{30}$$

And entering the following designations: R = R_2; M_B= M_1; L_C=L_1 ; L_n=L_3, will get the balance equations system for the upper sieving trunk:

$$x_K + R_2 \cdot \cos\theta + x_S - M_1 \cdot r \cdot \omega^2 (\cos\omega \cdot t - \frac{e}{L} \cdot \sin\omega \cdot t) = 0 \tag{31}$$

$$y_K + R_2 \cdot \sin\theta - M_1 \cdot g + y_S = 0 \tag{32}$$

$$-y_K \cdot L_1 \cdot \cos\gamma - x_K \cdot L_1 \cdot \sin\gamma + M_1 \cdot \frac{L_1}{2} \cdot \sin\gamma \cdot r \cdot \omega^2 (\cos\omega \cdot t - \frac{e}{L} \cdot \sin\omega \cdot t) - \\ - R_2 \cdot \cos\theta \frac{L_1}{2} \cdot \sin\gamma - R_2 \cdot \sin\theta_e \frac{L_1}{2} \cdot cos\gamma + M_1 \cdot q \cdot \frac{L_1}{2} \cdot \cos\gamma = 0. \tag{33}$$

$$x_k = y_k \cdot tg\alpha = y_k \cdot \frac{r}{L_n} \cdot cos\omega t \tag{34}$$

$$x_s = y_s \cdot tg\alpha = y_s \cdot \frac{r}{L_n} \cdot cos\omega t \tag{35}$$

Solving this system relatively to the x_S, x_K, y_S, y_K and R_2, will get the following values unknown power factors:

$$R_2 = \frac{-M_1 \cdot g \frac{r}{L_3} \cdot \cos\omega \cdot t + M_1 r \omega^2 \left(\cos\omega \cdot t - \frac{e}{L_2} \cdot \sin\omega \cdot t \right)}{-\sin\theta_B \cdot \frac{r}{L_3} \cos\omega \cdot t + \cos\theta_B} \tag{36}$$

$$y_K = \frac{-R_2 (\sin\theta_B \cdot \cos\gamma + \cos\theta_B \cdot \sin\gamma) + M_1 g \cos\gamma + M_1 r \cdot \omega^2 (\cos\omega \cdot t - \frac{e}{L_2} \cdot \sin\omega \cdot t) \sin\gamma}{2\left(\frac{r}{L_3} \cdot \cos\omega \cdot t \cdot \sin\gamma - \cos\gamma \right)} \tag{37}$$

$$y_K = y_S \tag{38}$$

$$x_k = y_k \cdot tg\alpha = y_k \cdot \frac{r}{L_n} \cdot cos\omega t \tag{39}$$

$$x_k = x_s \tag{40}$$

Solving this equations relatively to the time and crank angular velocity, will get the following values x_K , x_S , y_K , y_S and R_2:

DOI: 10.1051/matecconf/20168604051

$$x_k = -1{,}103 + 10{,}435 \cdot \cos\omega t + 1{,}436 \cdot 10 - 4 \cdot \sin\omega t - 1{,}103 \cdot \cos 2\omega t + 0{,}145 \cdot \sin 2\omega t \quad (41)$$

$$y_k = 734{,}5 - 155{,}928 \cdot \cos\omega t + 20{,}465 \cdot \sin\omega t - 0{,}155 \cdot \cos 2\omega t + 0{,}02 \cdot \sin 2\omega t \quad (42)$$

$$R_2 = \frac{2{,}76 \cdot 10^3 \cdot \cos\omega \cdot t - 331{,}87 \sin\omega \cdot t}{1{,}237 - 0{,}2 \cdot 10^{-2} \cos\omega \cdot t} \quad (43)$$

Thus, the power factors relations are received x_N, x_M, x_K, x_S, y_N, y_M, y_K, y_S, R_2 and R_1 from the disturbing power frequency and the time. Taking into account, that the suspensions reactions are directed along these suspensions, we shall get the following scheme of disturbing powers application to the frame. (Figure 8).

Reactions components from the connecting rod on the point O we can present as:

$$x_0 = R_1 \cdot \cos\theta_H - R_2 \cdot \cos\theta \quad (44)$$

$$y_0 = -R_1 \cdot \sin\theta_H - R_2 \cdot \sin\theta \quad (45)$$

Will determine the values x_0 and y_0 depending on the time and crank angular velocity.

$$x_0 = 0{,}228 + 229{,}451 \cdot \cos\omega t - 30{,}1 \cdot \sin\omega t + 0{,}228 \cdot \cos 2\omega t - 0{,}03 \cdot \sin 2\omega t \quad (46)$$

$$y_0 = -0{,}652 - 656{,}049 \cdot \cos\omega t + 86{,}088 \cdot \sin\omega t - 0{,}652 \cdot \cos 2\omega t + 0{,}086 \cdot \sin 2\omega t \quad (47)$$

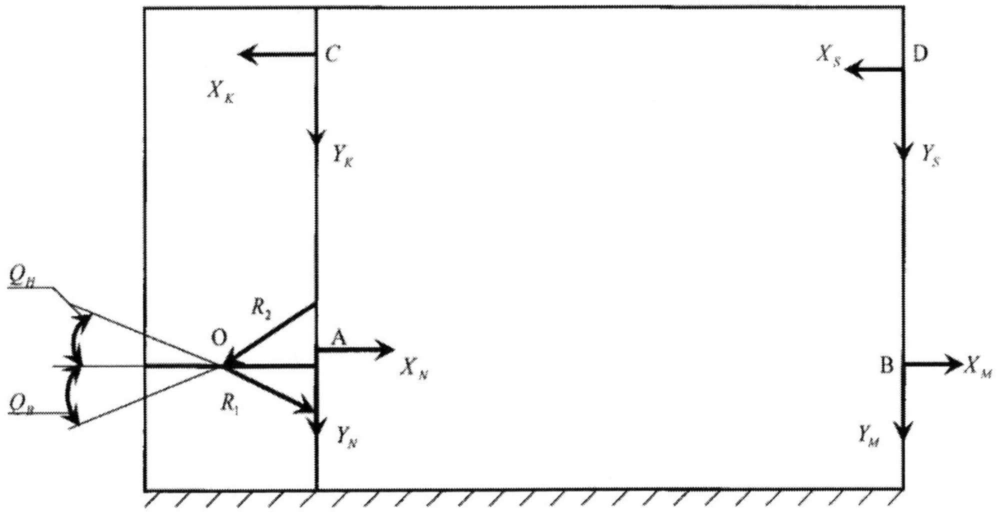

Fig. 8. The scheme of disturbing forces to the bearing structure elements application.

4 Conclusion

The analysis of the equations (41), (42), (46) and (47) shows, that the coefficients of connecting-rods reactions periodic functions considerably exceed the corresponding coefficients at the formulas of the suspension arms reactions determination. Therefore, connecting-rod reactions would effect by means of disturbing influence to the bearing structure.

The obtained simplified formulas (41), (42), (46) and (47) for the dynamic forces on the vibrational separating machine actuating mechanism details determination are the one of the stage of realization of the device methodology of such machines constructions optimization.

References

1. V. G. Chumakov V. G.,Tractors and Agricultural Machinery, **12** (2011)
2. N. M. Ivanov, N. P. Lapshin, International Research Journal, **10-2(41)** (2015)
3. N. P. Lapshin, International Research Journal, **28-30** (2016)
4. V. V. Piven, O. L. Umanskaya, Modern scientific researches and innovations, **3** (2013)
5. V. V. Piven, O. L. Umanskaya, Russian Engineering Research, **27** (2007)
6. V. V. Piven, O. L. Umanskaya, Modern scientific researches and innovations, **3** (2014)

Overview of fire curtains in construction

Oleg Nedryshkin[1*], *Marina* Gravit[2], *Anton* Lyapin[3], *Viktor* Voronin[4]

[1,2]Peter the Great St. Petersburg Polytechnic University, 195251, 29, Politekhnicheskaya st., Russia
[2,3] Moscow State University of Civil Engineering, 129337, 26, Yaroslavskoye Shosse, Moscow, Russia

Abstract. A fire curtain is use where, if there is a fire, it is necessary to create a temporary barrier within an opening, which seals off the area on fire. The curtain descends and prevents any fire and smoke from spreading from one area to another. It also allows people access to protected escape routes without any loss of fire resistance. The paper aims to presents the results of analysis of the scientific literature on the subject of fire curtains.

1 Introduction

In modern scientific, regulatory, and legal and technical literature, the term "fire curtains" refers to the structure, preventing the fire from spreading through the building and its hazards (heat flow, smoke, CO, CO2, HCL et. al.). Fire curtains also known by the terms "fire-retardant curtains," "Smoke curtains", "fire and smoke curtains". Under these terms, in all their diversity we mean the curtains to prevent the spread of fire hazards.

Performing the function of a fire barrier, curtains only when necessary (in case of fire) cut off the individual volumes and escape routes of the total building space. For example, fire curtains with degree of protection EI 60, 120 or 180 do not allow the fire to spread to the neighboring room. Fire curtains are a new type of fire protection. They can be used not only in the design and construction of new buildings, but also in the reconstruction of existing buildings [1].

2 Methods

Fire curtains are certified in accordance with the BS EN 949, BS EN 1634-3, BS 5234-2, BS EN 1363-2, BS EN 1634-1, BS EN 1363-1, BS EN 12605, BS EN 13501-2. In Russia, curtains are certified in accordance with the Technical regulations for fire safety requirements [2], National standards of Russia 30247.0-94, 30247.1-94, 53305-2009, 53307-2009 [3 – 6].

3 Results

The aim of fire protection system's usage is to maintain the temperature of the building component (structural steel element, electrical installation) bellow the critical temperature during fire but also is intended to contain a fire in the origin fire compartment for a limited period of time. In the paper [7] the passive fire protection material solutions were described and their action mode explained. The paper [8] presents findings from comprehensive research on influence of fire barrier on fire performance of External Thermal Insulation Composite System – ETICS with combustible thermal insulation. To confirm findings, tests on identical ETICS systems were conducted in March 2014 and May 2014.

The authors [9] proposed a new way, which consists in the enclosure securable object retardant screen, in which the porous structure materials at creation environment vapor of coolant is shown a continuous physical effect of evaporative cooling.

The paper [10] describes how to use fire-screens and curtains, as part of an automatic fire protection system, in particular in the high-rise warehouse with shelves. Described in detail the action of fire barriers in case of fire in the warehouses. Table 1 provides information of fire curtains.

Table 1. Comparison of fabricators curtains.

№	Fabricator (brand)	Tested according to	Fire resistance	Manufacturer country	Official site
1	Stöbich Brandschutz Gmbh	EN 1634-1 EN 13501-2	EI 90 EI 120 EI 180	Germany	stoebich.com
2	Fogo	EN 1634-1 EN 13501-2	E 60 E 120 EW 120	Lithuania	fogo.lt
3	FireScreen	EN13501-1	E 30	Netherlands	firescreen.com
4	Firetechnics	National standards of Russia [3 – 6]	EI 60 EI 180	Russia	fire-tec.ru
5	Pulsar Doors	National standards of Russia [3 – 6]	EI 30 EI 90	Russia	pulsardoors.ru
6	Door Master	National standards of Russia [3 – 6]	EI 60	Russia	doormaster.ru
7	Bach	EN 1634-1 EN 1634-3	EI 120 EI 180	Spain	bach-sl.com
8	Naffco	NFPA 252, UL 10D, FM Global, BSI, EN.	Optional	United Arab Emirates	naffco.com
9	Bradley Lomas Electrolok	EN 1634-1 EN 13501-2 et.al.	EI 120 EW 120	United Kingdom	ble-smokeandfirecurtains.com
10	Coopersfire	BS EN 1634-1:2008	E 120 E 180 EW 60 EI 120	United Kingdom	coopersfire.com
11	Colt Group	EN 1634-1 EN 13501-2	E 60 E 180	United Kingdom	coltgroup.com
12	FireCurtains	EN 1634-1 EN 13501-2	E 60 E 240	United Kingdom, India, Poland	firecurtainsltd.com
13	Smoke Guard	NFPA 252, UL 10D	Fire resistance up to 2 hr.	United State of America	smokeguard.com
14	Newtex	ASTM D6413, ASTM E-84,	Optional	United State of America	newtex.com

Usually fire curtains are manufactured from a woven glass fibre fabric with stainless steel wire reinforcement. The fabric is coated on each side with silver polyurethane. The complete curtain assembly is rated at 1000°C for a period of up to 240 minutes for Integrity (E), and 60 minutes for Irradiation (W), depending on the size of the curtain. Fire curtains are manufactured in many countries. Fire curtains are popular decision on fire safety in construction.

For Russia this is possible is a relatively new solution for fire safety [1]. For example, the company Stöbich Brandschutz GmbH develops and produces fire curtains for 15 years [11].

Figure 1 – example application of fire curtains.

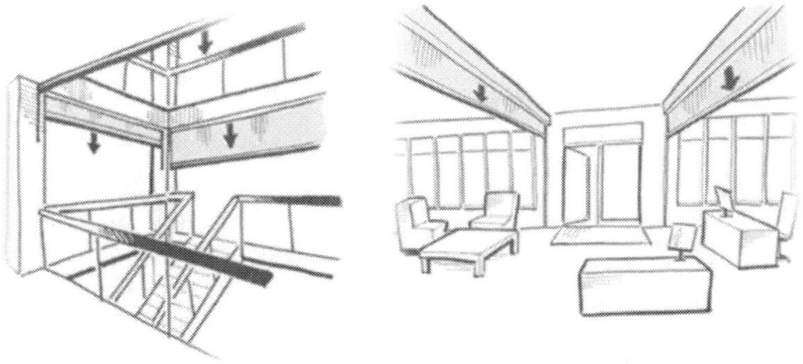

Fig. 1. Example application of fire curtains in construction

Fire curtains are made of silica material and are well suited to prevent the spread of fire hazards in buildings with an atrium. As a demonstration of the spread of fire hazards presented calculation of materials in the GUI FDS PyroSim, Smokeview (figure 2 – 3).

Fig.2. – Distribution of hazardous fire factors elevated temperature

Fig.3. – The spread of fire hazards for the loss of sight

Figure 4 demonstrates the fire resistance ability of the fire curtains it is Fire Dynamics Simulator model.

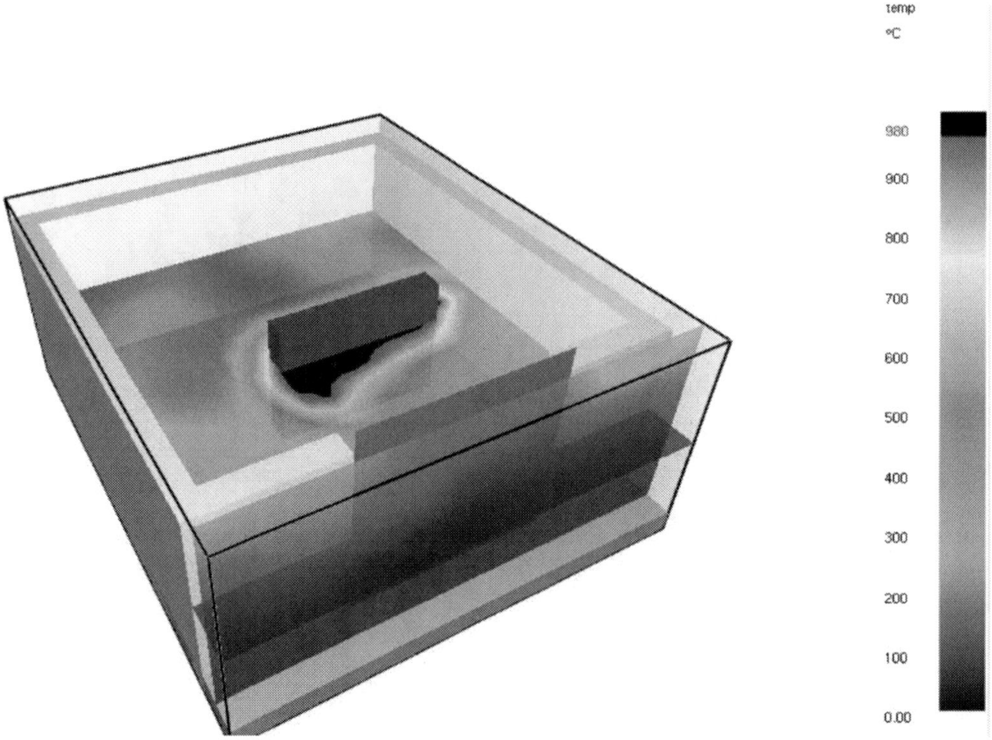

Fig.4. – Fire resistance ability of the fire curtains

CONCLUSION

If the Fire Curtains do not deploy to their operational position, the fire-engineered design solution would be compromised. However, in the event that other fire protection systems or elements do not function, e.g. due to total power failure, the Fire Curtains in the fire-operational position provide fire separation.

Automatic fire-fighting systems to bring the curtains in the operation have probability of failure of 0.99. There are systems, which exclude the electro-mechanical components, which is a more reliable means.

Fire curtains is a good tool for fire protection in civil engineering. The use of fire curtains allows you to keep the aesthetic appearance of structures. Materials used in fire curtains can be used for fire protection of structures. [12 – 13]

References

1 A.YA. Korol'chenko, D.P. Getalo. Fire curtains (review). Fire & Explosion Safety, **4**, 56 – 65 (2015).
2 Technical regulations for fire safety requirements. Federal Law No. 123 (in red. Federal Law of 10.07.2012 No. 117). Rossiyskaya gazeta [Russian Newspaper] (2008), No. 163. (rus).
3 National standard of Russia 30247.0-94 Elements of building constructions. Fire-resistance test methods. General requirements. Information on http://docs.cntd.ru/document/gost-30247-0-94 (on Russian)
4 National standard of Russia 30247.1-94 Elements of building constructions. Fire-resistance test methods. Loadbearing and separating constructions. Information on http://docs.cntd.ru/document/gost-30247-1-94 (on Russian)
5 National standard of Russia 53305-2009 Smoke shields. The test method for the fire resistance. Information on http://docs.cntd.ru/document/1200071868 (on Russian)
6 National standard of Russia 53307-2009 Smoke shields. Elements of building constructions. Fire doors and gates. Fire resistance test method. Information on http://docs.cntd.ru/document/1200071942 (on Russian)
7 K. Mróz, I. Hager, K. Korniejenko. Procedia Engineering **151**, 284–291 (2016) DOI: 10.1016/j.proeng.2016.07.388
8 D. Bjegovic, I. Banjad Pecur, B. Messerschmidt, B. Bojan Milovanovic, M. Alagusic. MATEC Web of Conferences **46**, 1–11 (2016) DOI: 10.1051/matecconf/20164605006
9 S.V. Zaikin, V.L. Strahov. Vestnik MGSU **4**, 107–112 (2009) (on Russian)
10 S. Zharov. Safety algorithm **5**, 6–9 (2006)
11 Information from official site Stöbich Brandschutz Gmbh: http://www.stoebich.com/www/webseite_tpl/ws_tpl1.php?webseite_id=11&sprache_wechseln_get=1&menuepunkt_gewaehlt_get=529
12 J. Ćetković, M. Knežević, M., Žarković, V. Murgul, N. Vatin, AMM, 638-640, 2465-2470 (2014)
13 Yong-Qian Zheng, Lin-Hai Han. Fourth International Conference on Advances in Steel Structures **2**, 1017-1022 (2005)

MATEC Web of Conferences **86**, 04053 (2016) DOI: 10.1051/matecconf/20168604053
IPICSE-2016

Estimation of Emissions During Monitoring of Pipelines in the Dynamic Mode of Operation

Maria Zemenkova[1,*], *Yuri* Zemenkov[1], *Alexey* Gladenko[2] and *Sergey* Podorozhnikov[1]

[1]Tyumen Industrial University, Volodarskogo st. 38,Tyumen, Russia
[2]Omsk State Technical University, st. Mira 11, Omsk, Russia

Abstract. To determine and diagnose the emissions from pipelines operating in the dynamic mode, the model is developed in real time, whereby the determination of the oil leakages volume is possible with sufficient accuracy. The stated methods assume the use of computer equipment and software. The use of sensitive devices and automation of data processing make it possible to achieve high accuracy of calculations. The developed techniques are meant to operate in systems of support and decision-making and expert systems. The rapid assessment results can be used by supervisory agencies to evaluate the damage in case of accidents and failures of pipelines.

1 Introduction

The current emissions control systems on pipelines are based on different physical methods (acoustic, vibroacoustic, estimation methods of estimation of the insulating covering of the pipeline conductivity, using the fluorescence method, etc.) and provide a registration of large leakages inaccuracy up to 5 km. The quality of the models and methods of processing the results of diagnostics is particularly important.

2 Methods

For determining and diagnosing leakages of pipelines operating in a dynamic mode, the models for estimating in real time were developed. With their help, it becomes possible to define the volume of oil leakage with sufficient accuracy [1-4, 7, 10]. The stated methods assume the use of computer equipment and sophisticated software with the use of sensitive devices and automation [5, 6, 8, 9]. Depressurization of the pipeline immediately affects the regime parameters of work preceding and following pumping stations (Figure 1) [12-14]. Thus, if the normal operating mode of the pumping station and pipeline can be characterized by the operating point (F_{r0}, Q_0), in violation of the integrity of the pipeline, Fr0 pressure during the time interval τ decreases to Fr and Q_0 consumption will increase to Q. In this case, the greater the damage is, the more significant changes of Fr0 and Q_0 will take place. Therefore, is possible to judge on the leakage value by the degree change of regime characteristics. Suppose that Q_0, F_{r0} - oil rate and pressure in the initial section of

* Corresponding author:muzemenkova@mail.ru

the pipeline under normal operating conditions of pumping stations [15-18], Z_B and Z_E - benchmark of the beginning and end of the pipeline, L - length of the pipeline (Figure 2.). There is also the pumping mode, i.e. β and m, the kinematical viscosity of the pumped product ν and pipe diameter d.

$$i_o = \beta \frac{Q_0^{2-m} \nu^m}{d^{5-m}}. \quad (1)$$

The hydraulic gradient during normal operation of the pipeline can be determined by well-known Leibenson formula.

Suppose that at a distance x_a from the first pumping station (PSI) depressurization of the pipeline happened [19-20]. Benchmark of point of fault is Z_a. Pumping stations due to the inertia of registering devices and executive mechanisms have been working for some time τ. The length of time τ can be divided into shorter periods of time $\Delta\tau_j$, and for each of these periods of time can determine flow rate $Q_X(\tau)$ in the pipeline section to the point of fault x_a by operating point of the pump station and pipeline (see Figure 1).

Leakage rate can be found as the difference between the flow rate $Q_x(\tau)$ and flow rate $Q_L(\tau)$ in the area $(L - x_a)$, i.e.:

$$Q_y(\tau) = Q_x(\tau) - Q_L(\tau) \quad (2)$$

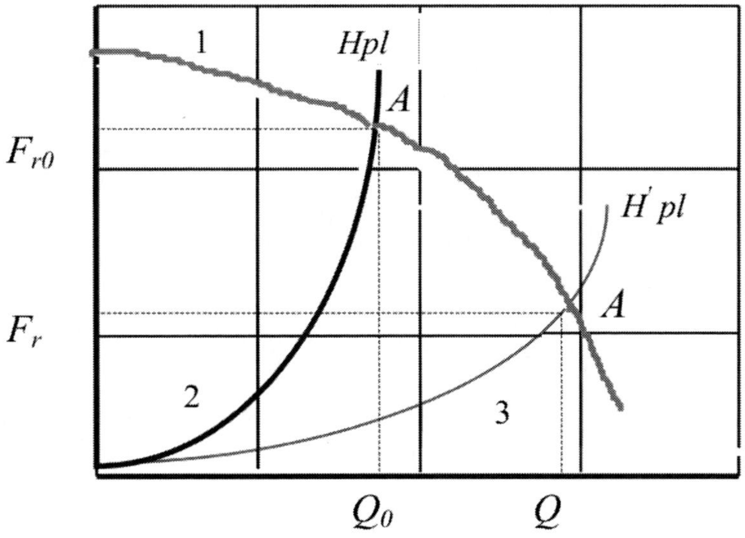

Fig. 1. Change the hydraulic parameters of the pipeline in its depressurization (HPL and H″PL - pipeline characteristics before and after the accident).

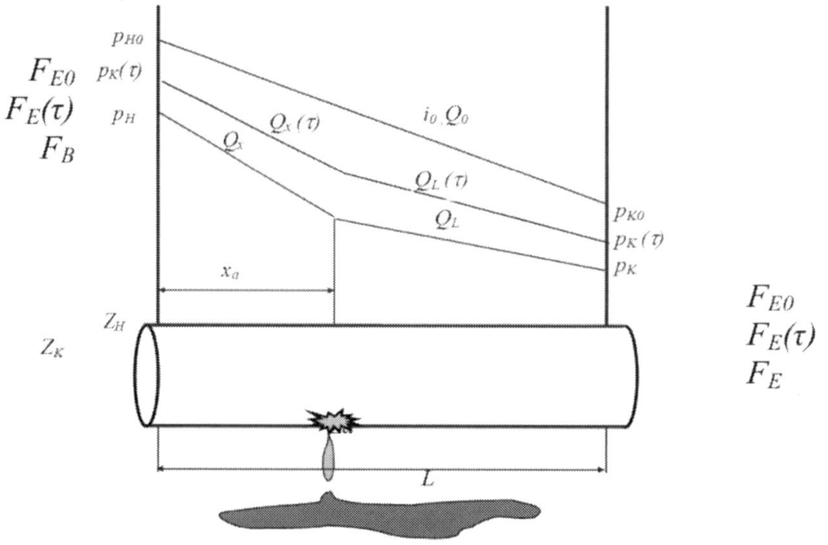

Fig. 2. Measurements of hydraulic parameters of the pipeline in violation of its depressurization (change of hydraulic gradient).

Consumption Q_L (τ) can be determined using the ratio of hydraulic ramps i_o and $i_L(\tau)$, determined by the generalized formula Leibenson. Here, it is assumed that $\beta_o = \beta_L$ and $m_o = m_L$:

$$Q_L(t) = Q_0 \cdot \left(\frac{i_L(\tau)}{i_0}\right)^{\frac{1}{2-m}} \tag{3}$$

where $i_L(\tau)$ – hydraulic gradient at the site *(L - x a)*.

To determine the hydraulic gradient $i_L(\tau)$, draw up Bernoulli equation for the area x_a at working pipe station with a damaged pipeline

$$Z_H + \frac{p_H(\tau)}{\rho g} + \frac{\vartheta^2{}_H(\tau)}{2g} = Z_a + \frac{p_a(\tau)}{\rho g} + \frac{\vartheta^2{}_a(\tau)}{2g} + i_X(\tau) \cdot x_a \tag{4}$$

where $p_a(\tau)$ and $\vartheta_a(\tau)$ - the pressure and flow rate of the product before failure in the j - th time. Since $\vartheta_H(\tau) = \vartheta_a(\tau)$, then we have

$$Z_a + \frac{p_a(\tau)}{\rho g} = Z_H + \frac{p_H(\tau)}{\rho g} - i_X(\tau) \cdot x_a \tag{5}$$

Bernoulli's equation for the pipeline section *(L - xa)* :

$$Z_a + \frac{p_L(\tau)}{\rho g} + \frac{\vartheta^2{}_L(\tau)}{2g} = Z_K + \frac{p_K(\tau)}{\rho g} + \frac{\vartheta^2{}_K(\tau)}{2g} + i_L(\tau) \cdot (L - x_a) \tag{6}$$

where $p_L(\tau)$ *and* $\vartheta_L(\tau)$ - the pressure and flow rate of the transported product after injury in j- th time.

Considering that section *(L - xa) also* $\vartheta_L(\tau) = \vartheta_K(\tau)$ and substituting (5) into (6), we get

$$Z_H + \frac{p_H(\tau)}{\rho g} + i_X(\tau)\cdot x_a = Z_K + \frac{p_K(\tau)}{\rho g} + i_L(\tau)\cdot(L - x_a) \tag{7}$$

Hydraulic gradient $i_x(\tau)$ can be obtained in similar way to (3):

$$i_X(\tau) = i_0 \cdot \left(\frac{Q_X(\tau)}{Q_0}\right)^{2\text{-m}} \tag{8}$$

Then (8) can be represented as following:

$$Z_H + \frac{p_H(\tau)}{\rho\cdot g} - i_0\cdot x_a\cdot\left(\frac{Q_X(\tau)}{Q_0}\right)^{2-m} = Z_K + \frac{p_K(\tau)}{\rho\cdot g} + i_L(\tau)\cdot(L - x_a) \tag{9}$$

Hence

$$i_L(\tau) = \frac{(Z_H - Z_K) + \dfrac{p_H(\tau) - p_K(\tau)}{\rho\cdot g} - i_0\cdot x_a\cdot\left(\dfrac{Q_X(\tau)}{Q_0}\right)^{2-m}}{(L - x_a)} \tag{10}$$

or

$$i_L(\tau) = \frac{(Z_H - Z_K)}{(L - x_a)} + \frac{p_H(\tau) - p_K(\tau)}{\rho\cdot g(L - x_a)} - i_0\frac{x_a}{(L - x_a)}\left(\frac{Q_X(\tau)}{Q_0}\right)^{2-m} \tag{11}$$

Taking into account the dependence (11), $Q_L(\tau)$ flow rate (3) after the transformation can be rewritten as following:

$$Q_L(\tau) = Q_0\cdot\left[\frac{(Z_H - Z_K)}{(L - x_a)\cdot i_0} + \frac{p_H(\tau) - p_K(\tau)}{\rho\cdot g(L - x_a)\cdot i_0} - \frac{x_a}{(L - x_a)}\left(\frac{Q_X(\tau)}{Q_0}\right)^{2-m}\right]^{\frac{1}{2-m}} \tag{12}$$

or after transformation:
Then final amount of leakage in the j - th point of time

$$Q_Y(\tau) = Q_X(\tau) - \left[\left(Z_H - Z_K + \frac{p_H(\tau) - p_K(\tau)}{\rho\cdot g}\right)\cdot\frac{Q_0^{2-m}}{(L - x_a)\cdot i_0} - \frac{x_a}{(L - x_a)}Q_X^{2-m}(\tau)\right]^{\frac{1}{2-m}} \tag{13}$$

The amount of oil leaked from the pipe in a time τ can be determined:

$$\text{V} = \int_0^{\tau} Q_Y(\tau)d\tau \tag{14}$$

or using integral algorithm rectangles approximate formula for calculation:

$$V = \sum_{j=1}^{N} Q_Y^j(\tau)\Delta\tau \quad (15)$$

where $\Delta\tau = \dfrac{\tau}{N}$ – approximate integration step, it is constant in this case; τ - the total duration of the expiration time of the product; N - the number of partitions of τ time on intervals Δτ.

The amount of oil flowing out of the pipe can be found using the approximations method, for which, increasing N by one, the calculation V is carried out as long as it does not meet the following condition:

$$/ VN\text{-}1 \ \text{-} \ VN / \le \delta \cdot VN \quad (16)$$

where δ - is a defined calculation error. For practical solutions it is enough to take $\delta =$ 0.05.

The method was tested on the main oil pipeline with the use of specially prepared programs for PC and route profile [11-14, 21]. The results of experiments carried out on the existing pipeline with a diameter of 1220 mm can be considered the most complete and sufficiently representative (Figure 3) [22-23].

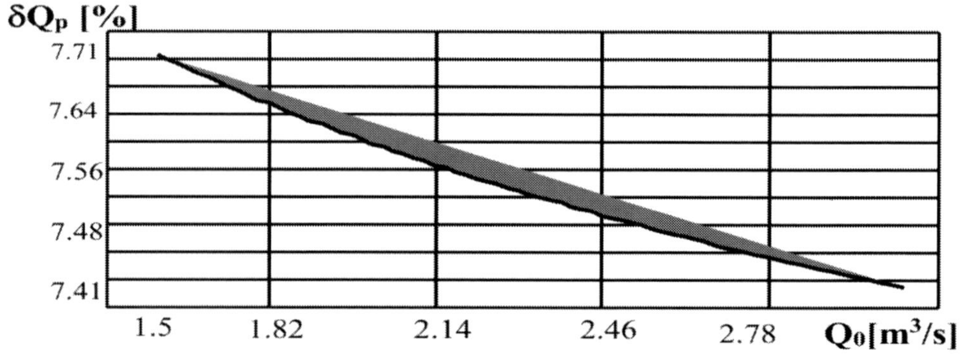

Fig. 3. The dependence of the inaccuracy Q_p from the rate Q_o to the pipeline of diameter D = 1220mm.

3 Conclusions

Development and application of different methods of emission registration, determining the location and magnitude of leaks are caused by different degrees of complexity of the pipeline route, and the various conditions of its bedding in the soil [24]. As far as the detection of imperfections of the method used there is a necessity to find new, more accurate methods of control of the leaks, especially those that are hidden. Thus, the obtained dependence (16) allows determining the amount of leakage in the period when the pumping stations work over the damaged section of the pipeline based on the dispatching data on changes in pump station discharge pressure. In IUT calculation programs are developed.

References

1. *A technique by definition of damage to surrounding environment at accidents on the main oil pipelines. Ministry of Fuel and Energy* (Transpress, Moscow, 1996)
2. V. Antipyev, Yu. Zemenkov, N. Kudryavtseva, *UMN Z and SZS* (Tyumen, 1985)
3. V. Kurushina, Yu. Zemenkov, *WIT Transactions on Ecology and the Environment* **190(2)**, 881-888 (2014)
4. V. Antip'ev, A. Nevolin, Yu. Zemenkov, Neftyanoe Khozyaistvo - Oil Industry **10**, 46-48 (1981)
5. V. Antipyev, V. Arkhipova, Yu. Zemenkov, N. Kudryavtseva, *The instruction on calculation of losses of oil at refusals on the main oil pipelines of the Western and Northwest Siberia* (TSOGU, Tyumen, 1986)
6. S. Dudin, Yu. Zemenkov, A. Shabarov, N. Saranchyn, Higher Educational Institutions News. Oil and Gas **6**, 63-68 (2010)
7. A. Shipovalov, Yu. Zemenkov, S. Toropov, M. Zemenkova, S. Podorozhnikov, I. Tyrylgin, V. Pavlov, *Aspects of technological reliability and economic efficiency of operation of underground storages of natural gas of Western Siberia* (TSOGU, Tyumen, 2012)
8. Yu. Zemenkov, V. Shalay, M. Zemenkova, Procedia Engineering **113**, 254-258 (2015)
9. Yu. Zemenkov, V. Shalay, M. Zemenkova, Procedia Engineering **113**, 312-315 (2015)
10. Yu. Zemenkov, *Dissertation of Doctor of Technical Sciences* (Tyumen: TSOGU, 1999)
11. S. Stepanov, N., Cheremisin, S., Sokolov, A., Altunin, A., Shabarov, *ECMOR 2012 - 13th European Conference on the Mathematics of Oil Recovery* (2012)
12. S. Kudakaev, F. Aminev, V. Galiakbarov, G. Korobkov, V. Kovshov, Gas Industry **5**, 10 (2004)
13. M. Lurie, Neftyanoe khozyaystvo - Oil Industry **9**, 111-113 (2013)
14. M. Lurie, DOI: 10.1002/9783527626199 (2009)
15. E. Kurushina, V. Kurushina, Life Science Journal **11(11)**, 517-521 (2014)
16. V. Kurushina, Yu. Zemenkov, E. Kurushina, Fundamental researches **2-8**, 1632-1636 (2015)
17. E. Kurushina, Canadian Journal of Science, Education and Culture **2(6)**, 378-384 (2014)
18. E. Kurushina, *Problems of formation of a common economic space and social development in the CIS countries: Proceedings of the International scientific and practical conference* (TSOGU, Tyumen. 2011)
19. S. Toropov, Yu. Zemenkov, S. Podorozhnikov, Gas industry **720(720)**, 95-98 (2015)
20. M. Zemenkova, L. Fayzulina, I. Seroshtanov, Gas industry **720(720)**, 59-64 (2015)
21. A. Shabarov, A. Ibragimov, M. Zemenkova, S. Podorozhnikov, Gas industry **12(715)**, 50-54 (2014)
22. A. Pimnev, M. Zemenkova, Pipeline transport: theory and practice **5(51)**, 43-47 (2015)
23. M. Zemenkova, Yu. Zemenkov, A. Pimnev, V. Petryakov, Business magazine Neftegaz.RU **11-12**, 64-70 (2015)
24. R. Mamadaliev, V. Kuskov, Yu. Zemenkov, A. Popova, Applied Mechanics and Materials **770**, 19-22 (2015)

MATEC Web of Conferences **86**, 04054 (2016) DOI: 10.1051/matecconf/20168604054
IPICSE-2016

Monitoring of Phase Distribution when Controlling Safety of Transport Processes and Hydrocarbon Storage

Yuri Zemenkov[1], *Maria* Zemenkova[1,*], *Alexey* Gladenko[2] and *Nikolay* Zakirov[1]

[1]Tyumen Industrial University, Volodarskogo st. 38, Tyumen, Russia
[2]Omsk State Technical University, st. Mira 11, Omsk, Russia

Abstract. Based on statistical processing of an array of information and summarizing the experimental data in IUT, analytical dependences for calculating the coefficient of the phase equilibrium under various pressures and temperatures were obtained. The proposed equations for the first time used the quantitative composition of hydrogen which determines, according to the authors, the volatility of the oil or condensate fraction and greatly simplifies relevant calculations in comparison with known analogues. To find the coefficients a computer program was developed that implements a method of non-linear regression. The developed algorithm was tested and proved to be very convenient and efficient for a rapid assessment of experimental studies.

1 Introduction

Common methods of characterizing volatility of petroleum and petroleum products can be divided into two groups: experimental (direct and indirect) and theoretical. Direct methods include techniques for direct measurement of the volume of liquid before and after evaporation or calculating the volume (mass) of vapor [11-14]. Direct determination of hydrocarbon emissions is extremely difficult and in some cases even impossible, especially for short-term storage encountered in emergency situations. An important task when calculating the storage and transport conditions is correctly defining constants of hydrocarbons phase equilibrium [18-21].

It should be recognized that a priority in the study of hydrocarbon phase equilibrium constants belongs to American Petroleum Institute (API) which charts are widely used in oil refining [6-10]. Also known is a NGPSA technique allowing to determine the value of kei at temperatures reaching 430 °C [1-5]. Published print versions of these techniques are very complicated and inconvenient for practical use, and computerized schemes are the property of the developers: Dowling, Leslie, Stall, Soavva, Todd, Hedden, Sherwood et al. Among Russian publications can be noted developments of A.I. Gritsenko, G.R. Gurevich, V.P. Mamun, T.D. Ostrovsky, G.M. Yaryshev, T.A Omutova, V.I. Shilov, S.F. Barmin, D.P. Vasilev [1-5,10, 11-20]., etc.

* Corresponding author:muzemenkova@mail.ru

Also, the Siberian Research Institute (SibNIINP) developed a technique of determining the coefficient kei of fractions and components depending on the molecular weight of various oils in West Siberia as applied to the oilfield conditions of high absolute pressure [1].

At present, a significantly large amount of experimental material has already been accumulated in Russia [15-17]; however, the value of constant k_{ei} can be used in calculations of vapor-liquid equilibrium only for systems adequate to the laboratory ones. The latter condition greatly reduces the advantages of such information if phase composition constantly changes, for example, when liquid hydrocarbon mixtures evaporate from the exposed surface. It is known [2] that under near-atmospheric pressures coefficients of component phase distribution k_{ei} are functions of the system temperature t only.

For hydrocarbon mixtures of arbitrary nature the equation when modeling dependencies can be $k_{ei} = f_i\ (t), \{x\}$:

$$k_{ei} = f_i\ (t) \tag{1}$$

Based on statistical modeling of an array of information and summarizing the experimental data in IUT analytical dependencies for calculating the coefficient k_{pi} of the phase equilibrium under low pressures (0.1÷1.2 MPa) and temperatures from 273 to 330 °K were obtained. In the proposed equations the quantitative composition of hydrogen is used for the first time determining the volatility of the oil or condensate fraction. This assumption greatly simplifies relevant calculations, compared to foreign counterparts, such as the well-known Stall formula

$$lg\ k_e = A + 1000B/(C+T) \tag{2}$$

where A, B and C - empirical coefficients.

For alkanes dependencies are obtained:

$$A = 6.8 - 2.7\ n_n - p/p_{at} \quad and \quad B = 2 + 0.54\ n_n\ , \tag{3}$$

for alkenes and arenes

$$An = 6.8 - 2.7\ n_{nph} - p/p_{at}, \quad B_n = (2 + 0.54\ n_{nph} - 17(n_n - n_{nph})\ n_{nph} - 1 \tag{4}$$

where n_n – the homolog serial number (alkanes); n_{nph} – the number of hydrogen atoms in naphthenes (cyclanes) or arenes; p and p_{at} – absolute and atmospheric pressure, correspondingly.

A rather high accuracy of calculations should be noted, particularly in relation to hydrocarbons of the methane series. The proposed technique has been tested in determining the values of more than 100 coefficients. Satisfactory results are obtained also when determining the coefficients of various hydrogen-containing compounds such as aniline (C_6H_7N), chloroform ($CHCl_3$), ammonia (H_3N), methanol, ethanol and others, although calculations error is higher and reaches 10%. For practical calculations we can recommended the dependence

$$lg\ k_e = ln\ [(A+1000BT\text{-}1)n_n] \tag{5}$$

The resulting equations (2-5) are adequate according to the Fisher criterion and are characterized by small error (eg, less than 3% for paraffinic hydrocarbons) with the multiple correlation coefficient R=0.98. Deviation of the calculated values of the phase distribution coefficient k_{ei} from those published in reference books is given in Table 1.

Table 1. Values of the phase equilibrium coefficient k_{ei} *(at T=293 K and p=0.098MPa)*

Method of determining k_{ei}	Hydrocarbons					
	CH_4	C_2H_6	C_3H_8	C_4H_{10}	C_5H_{12}	C_6H_{14}
By the nomogram of Zarembo K.S.	165.00	25.50	7.50	3.00	0.53	0.15
By the Cox graph	-	39.50	9.20	2.40	0.60	0.16
By the formula (5)	165.57	24.94	7.83	2.80	0.58	0.17

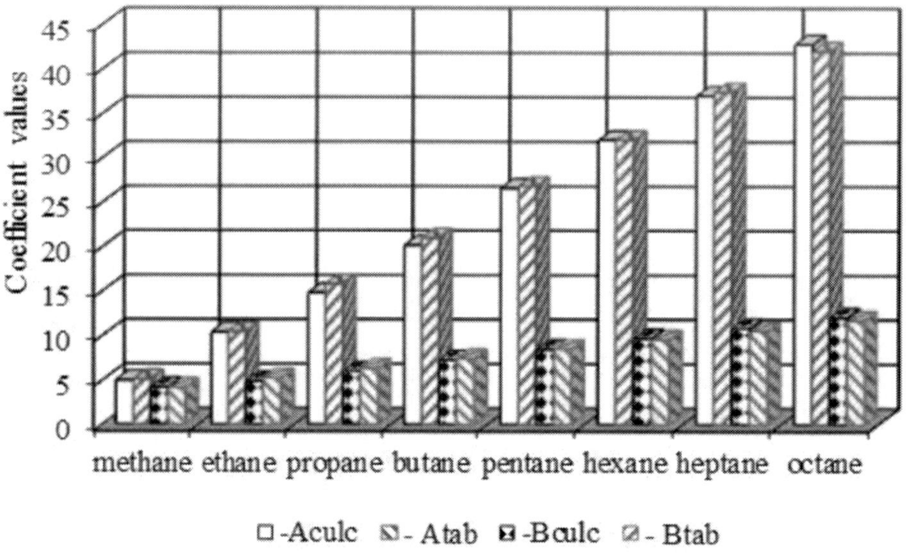

Fig. 1. Coefficient values A and B for alkanes (paraffin hydrocarbons).

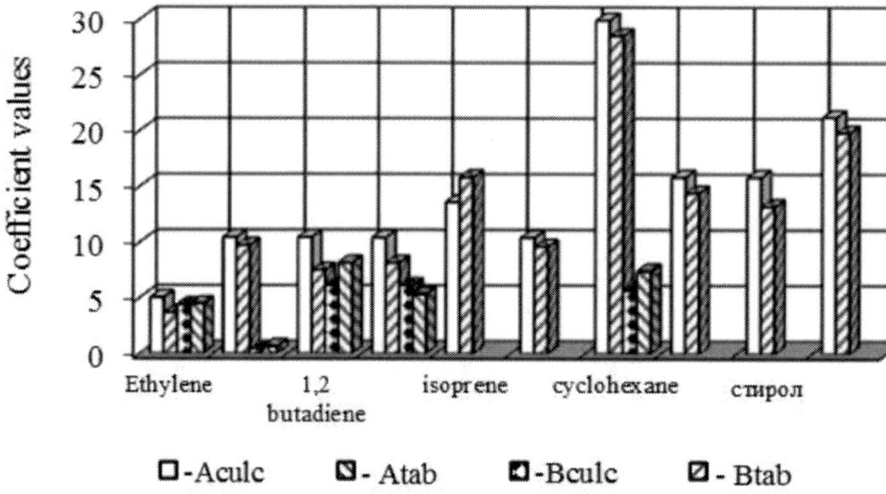

Fig. 2. Coefficient values A and B for alkenes and arenes.

To determine the functional dependence of the coefficients of the phase component distribution k_{ei} on the system temperature t we apply regression analysis (approximation of functions according to the least squares method), substituting k_{ei} for the virtual value Y_i.

Dual regression analysis providing the least mean square error determines parameters of the empirical linear dependence

$$Y(T) = b_1 t + b\,0 \tag{6}$$

characterizing the relationship between a certain number N of value pairs t_i and Y_i.

Graphically, this problem can be represented as follows: – on the field of a set of points t_i, Y_i of the plane t_Y the required function must be built so that the value of all deviations met the condition

$$U = \sum_{i=1}^{N} [Y_i - Y(t_i)]^2 = \min \tag{7}$$

where $Y(t_i)$ is determined by (6).

Thus, non-linear dual regression was reduced to obtaining a predetermined non-linear dependence $Y(t)$ (non-linear in the independent variable t, but linear in the parameters of this dependence).

The pattern of changes in the coefficients of the components phase distribution k_{ei} from the system temperature t defined by experimental data was close to exponential

$$k_{ei} = A_i \cdot e^{0.037 C_i \cdot t} \tag{8}$$

As an example, Table 2 shows the values of the coefficients for pure components of the mixture.

Table 2. Values of coefficients for pure hydrocarbons

Coefficients	**Components**							
	CH_4	**C_2H_6**	**C_3H_8**	**iC_4H_{10}**	**C_4H_{10}**	**iC_5H_{12}**	**C_5H_{12}**	**C_6H_{14}**
A_i	127.14	17.31	4.55	1.57	1.08	0.32	0.24	0.055
C_i	0.324	0.529	0.622	0.757	0.811	0.973	1.027	1.270

The pattern of changes in the coefficients of the phase distribution k_{pi} of narrow fractions from the system temperature t, is also close to exponential and is adequately approximated by the following function:

$$k_{ei} = e^{0.037(C_i t - t_B)} \tag{9}$$

where t_B - the average boiling point of the fraction, ^0C.

Regression models are adequate according to the Fisher criterion and are characterized by small error (not more than 5÷15%) with the multiple correlation coefficient R=0.97. As an example, for a number of hydrocarbons Fig. 3 graphically shows deviation of the calculated values of the phase distribution coefficient from the experimental.

Thus, the developed set of equations allows calculating k_{pi}, which simplifies the tasks of operational estimation of phase distribution, especially if information on the hydrocarbon mixture composition is incomplete.

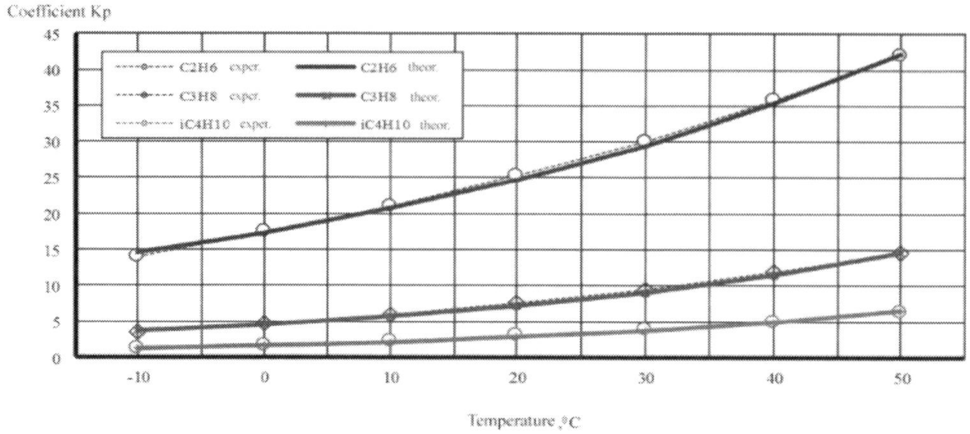

Fig. 3. Deviation of calculated values of phase distribution coefficient from the experimental (on the example of ethane, propane and butane).

2 Conclusions

To find the coefficients Ai and Ci a computer program was developed that implements a method of non-linear regression. The developed algorithm and obtained dependences proved to be very convenient and efficient for a rapid assessment of experimental studies

References

1. V. Shilov, *Abstract of thesis of Candidate of Technical Sciences* (TSOGU, Tyumen, 1984)
2. V. Antipyev, Yu. Zemenkov, N. Kudryavtseva, *UMN Z and SZS* (Tyumen, 1985)
3. V. Kurushina, Yu. Zemenkov, WIT Transactions on Ecology and the Environment **190(2)**, 881-888 (2014)
4. V. Antip'ev, A. Nevolin, Yu. Zemenkov, Neftyanoe Khozyaistvo - Oil Industry **10**, 46-48 (1981)
5. V. Antipyev, V. Arkhipova, Yu. Zemenkov, N. Kudryavtseva, *The instruction on calculation of losses of oil at refusals on the main oil pipelines of the Western and Northwest Siberia* (TSOGU, Tyumen, 1986)
6. S. Dudin, Yu. Zemenkov, A. Shabarov, N. Saranchyn, Higher Educational Institutions News. Oil and gas **6**, 63-68 (2010)
7. I. Alexandrov, *Distillation and rectification in oil processing* (Chemistry, Moscow, 1981)
8. Yu. Zemenkov, V. Shalay, M. Zemenkova, Procedia Engineering **113**, 254-258 (2015)
9. Yu. Zemenkov, V. Shalay, M. Zemenkova, Procedia Engineering **113**, 312-315 (2015)
10. Yu. Zemenkov, *Dissertation of Doctor of Technical Sciences* (TSOGU, Tyumen, 1999)
11. G. Filippov, I. Malkin, L. Vilenkina, Theoretical bases of chemical technology **5**, 772-774 (1989)
12. M. Lurie, Neftyanoe khozyaystvo - Oil Industry **9**, 111-113 (2013)
13. L M. Lurie, DOI: 10.1002/9783527626199 (2009)
14. S. Kudakaev, F. Aminev, V. Galiakbarov, G. Korobkov, V. Kovshov, Gazovaya Promyshlennost **5**, 10 (2004)
15. E. Kurushina, V. Kurushina, Assessment of the smart growth, Life Science Journal **11(11)**, 517-521 (2014)

16. V. Kurushina, Yu. Zemenkov, E. Kurushina, Fundamental researches **2-8**, 1632-1636 (2015)
17. E. Kurushina, Canadian Journal of Science, Education and Culture **2(6)**, 378-384 (2014)
18. V. Shilov, *Abstract of thesis Cand.Tech.Sci.* (TSOGU, Tyumen, 1984)
19. R. Mamadaliev, V. Kuskov, Yu. Zemenkov, A. Popova, Applied Mechanics and Materials **770**, 19-22 (2015)
20. Yu. Zemenkov, Higher Educational Institutions News. Oil and Gas **3**, 78-82 (1998)
21. S. Ueyles, *Phase balance in chemical technology* (Mir, Moscow, 1998)
22. Ya. Gorelik, A. Shabarov,Yu. Sysoyev, Earth's Cryosphere **12(1)**, 59 (2008)

MATEC Web of Conferences **86**, 04055 (2016) DOI: 10.1051/matecconf/20168604055
IPICSE-2016

Testing of bus bar contact under non stationary current impacts

Amankeldy Bekbaev[1,*], *Erlan* Sarsenbayev[1], *Vasiliy* Titkov[2]

[1]Satpaev Kazakh National Research Technical University, Almaty, Kazakhstan
[2]Saint Petersburg State Polytechnic University,Saint-Petersburg, Russia

Abstract. This research paper is devoted to the analysis of non-stationary temperature field near plane bus bar contact. The possibilities of temperature and contact quality estimation by dynamic sensor on the accessible bus bar surface are showed. There is discussed the algorithm of temperature recognising directly on the contact surface. Numerical methods and analytical approaches are used.

1 Introduction

The operating mode of electrical installation determines thermal effect on current-carrying elements [1], and in particular on terminal connections, quality reduce of which can lead to serious consequences (welding, thermal breakdown). Short current pulses that occur in the busbarof power supply equipment as a result of, for example, motor start [2], disconnect of short circuits [3,4], etc. can lead to a dangerous temperature rise in the contact surfaces of busbar.The available experimental findings on plane busbar heating by heavy short-time current describe the average busbar temperature, depending on its cross-section and configuration of the bolt holes [1]. Whereby, instant temperature value of direct-to-contact surface can not be fixed by any means of measurements due to its inaccessibility. At the same time instant temperature values in the area of local increase of the busbar resistivity (contact, holes) can be significantly higher than the average values. This fact is very important for the plane contact connections, changing its state under mechanical, environmental and long-term thermal factors leading to an increase oftransition resistance of the contact. Since the thickness of conduct busbar can be about dozen of millimeters, and the layer thickness with a high level of heat near the contact surface represents a very small fraction of a millimeter, then direct measurements on the busbar surface do not allow fixing directly the true value of the temperature in the contact zone.

However, analysis of the temperature field in the plane contact zone makes possible to establish functional temperature relationship in the contact zone and on the busbar surface, and also the relationship between fixing temperature on the busbar and physical characteristics of contact quality (frequency by transition resistance). Temperature determination of inaccessible surface (heat source plane) is reviewed in this paper. In case

* Corresponding author: bekbaev_a@mail.ru

of stable algorithm for solving of the mentioned inverse problem under conditions of presence of interference and noise, programming of data communications in the controller that receives a signal in the form of timed sequence of temperature values on the busbar surface during the flow of current from the physical sensor, it allows to estimate the temperature and quality of contact in dynamics. It should be noted that using of dynamic temperature evaluation during impulse action as opposed to methods based on the analysis of stationary thermal field [5.6], makes it possible to observe the existence of defects under conditions of a more contrast temperature distribution for a certain length of time [7].

2 Method

As it is well known [8], the temperature field in a solid with heating source is described by the heat equation with respect to the temperature T

$$\rho c \frac{\partial T}{\partial t} = \nabla \cdot (\lambda \cdot \nabla T) + \dot{q} \tag{1}$$

Where ρ, c, λ are the density, specific heat and thermal conductivity, t is time, - volume power density of the distributed heat source. In case of heat analysis in the current-carrying elements of installations the equation for volumetric source is:

$$\dot{q} = \frac{|\bar{j}|^2}{\sigma} \tag{2}$$

Where $\bar{j}$ is the vector of current density, σ is conductivity. The characteristic time for distribution of the thermal field to a depth of l is measured by parameter

$$t_D = \frac{l^2}{D} \tag{3}$$

Where $= \frac{\lambda}{\rho c}$. In particular, if characteristic dimensions in the plane of contact connections comprise several centimeters, then in case of copper busbar (λ = 400 W/m K, ρ = 8960 kg /m^3, c = 380 J/kg K) t_D is equal to few seconds. Therefore, during dynamic temperature assessment with the effects of current pulses with the time interval of 1000 ms the temperature field distribution from the contact zone along the busbar can be neglected.

Thus, the temperature field analysis in the direction perpendicular to the busbar plane at a specified limited time interval is to be performed with the use of one-dimensional heat equation, which is at constant thermal and physical characteristics of busbar material, has the form

$$\frac{\partial T}{\partial t} = D \frac{\partial^2 T}{\partial x^2} + \dot{q} \tag{4}$$

Solution of this equation allows you to set up temperature relation in the contact plane (x=0) and in the temperature measurement point (x=l). Let us consider the distribution of heat release source. In the main cross- section of contacting busbars is taken a uniform distribution of current density j and volumetric heat capacity (2).Despite the diversity of physical and mechanical processes occurring on the contact surface [9,10] increased heat release occurs in the thin transition layer of the contact with the thickness Δ, $\Delta << l$, which has conductivity σ_1, lesser than the busbar material has $\sigma_1 < \sigma$. Volume density of the heat capacity in the transition layer is also determined by the formula (2). However, since the thickness of the layer is smaller than the busbar thickness l, it is more convenient to consider heat generation in the layer in the form of heat flow at the contact boundary ($x = 0$)

$$F = -\lambda \frac{\partial T}{\partial x}\bigg|_{x=0} = \frac{1}{2} j^2 \left(\frac{1}{\sigma_1} - \frac{1}{\sigma}\right) \Delta \tag{5}$$

The multiplier ½ in the last formula takes into account that the heat in the transition layer is distributed by equal parts between the contact busbars.

For further analysis, it is advisable to write equations and ratios (1)–(3) in dimensionless variables: θ is a temperature, ξ is a coordinate, τ is a time.

$$\frac{\partial \theta}{\partial \tau} = \frac{\partial^2 \theta}{\partial \xi^2} \tag{6}$$

Despite the fact that actual forms of current pulses often have a oscillations form [2,3,4] with its relatively short duration (from microseconds during short circuits up to units of seconds during starting currents) heat release can be represented by energy equivalent rectangular pulse of a similar duration.

In case of relatively long current pulses is $t_p \geq t_D$ indication of the active phase termination for the run of dependency dT/dt is easily accomplished by the starting time of this dependence. For the described mode of problem solving for determining of the temperature of inaccessible surface can be made on the basis of analytical solutions for non-stationary thermal field in the plate created by a flat-source on one of its side, that is permissible with substantial degradation in contact quality when volumetric heat can be neglected in comparison with the heat flow arising in the transition layer. In this case, it is advisable to use the dimensionless form of equation of heat transfer that is proposed in [11].

According to [11] the following basic values can be selected: temperature $T_b = F_0 l/\lambda$, where F_0 is constant flow on the surface x=0; time $t_b = l^2/D$. Then dimensionless temperature rise on the surface of the plate over the initial value T_0: $\theta = (T - T_0)/T_b$ are expressed by dependences [11]

$$\theta(0) = \tau + \frac{1}{3} - \frac{2}{\pi^2} \sum_{n=1}^{\infty} \frac{e^{-\pi^2 2 n^2 \tau}}{n^2} \tag{7}$$

$$\theta(l) = \tau - \frac{1}{6} - \frac{2}{\pi^2} \sum_{n=1}^{\infty} \frac{e^{-\pi^2 n^2 \tau}}{n^2} (-1)^n \tag{8}$$

Figure 1 shows the dependences calculated on the basis of formulas (7) depending on the dimensionless temperature at the contact surface $\theta(0, \tau)$ and the busbar surface $\theta(1, \tau)$, and the dependence

$$f_\theta(\tau) = \frac{\theta(1, \tau)}{\theta(0, \tau)} = \frac{T_1(t/t_b)}{T_0(t/t_b)},$$

Which allows us to express the contact surface temperature $T_0(t)$ by the measured temperature on the outer busbarsurface$T_1(t)$

$$T_0(t/t_b) - T_0 = \left(T_1(t/t_b) - T_0\right) f_\theta^{-1}(t/t_b) \tag{9}$$

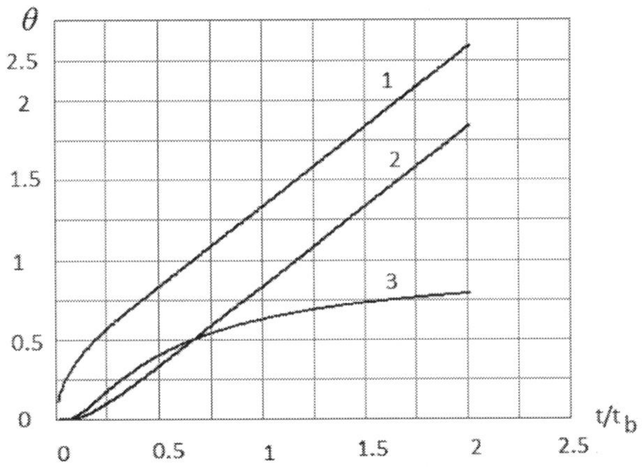

Fig. 1. Dimensionless time dependences of the contact temperature-1 and busbar surface -2, and the calibration curve $f_\theta(\tau)$ - 3.

Universal calibration function $f_\theta(\tau)$can be stored in the memory of sensor controller as a sequence or analytic approximation. For example, in the area 2>τ>0.1 you can use the dependence in the form

$$f_\theta(\tau) \cong 0.1082\tau^3 - 0.5858\tau^2 + 1.159\tau - 0.05514$$

Thus, the contact surface temperature can be calculated from the temperature of busbar surface $T_1(t)$ by the expression

$$T_0(t/t_b) = T_0(0) + \left(T_1(t/t_b) - T_0(0)\right)\cdot\left[0.1082\left(\frac{t}{t_b}\right)^3 - 0.5858\left(\frac{t}{t_b}\right)^2 + 1.159\left(\frac{t}{t_b}\right) - 0.05514\right]^{-1} \quad (10)$$

The value of base time $t_b=l^2/D$, defined by the busbar thickness l, thermal conductivity of its material - D, can be included into the microcontroller program which carries out temperature calculation by the last expression. This circumstance determines the individual adjustment of the measuring device for each controlled contact. However, thicknesses and materials of busbars in power equipment do not vary significantly. In particular, materials are typically copper or aluminum, and the busbar thickness is in the range of 5-20 mm.

Therefore individual adjustment of sensor controller is not an intrinsic problem.The formula (10) used in the algorithm for temperature measure on busbar surface, can be applied only at the stage of temperature growth on contact surface. Therefore, obtained by means of (10) are important up to the biggest values of calculated sequence. Further behavior of analyzed dependencies can not be described by the Beck's formulas (7).Thus, the presented algorithm can estimate the maximum temperature reached by the contact surface.

Figure 3 and 4 show an example of the above described algorithm based on Beck's formulas (7) for the flat contact busbars of 5 mm thick during the break of short-circuit current at conduction of the transition layer contact equal to 0.1% from the conductivity of the base material. (Figure 3, 4).

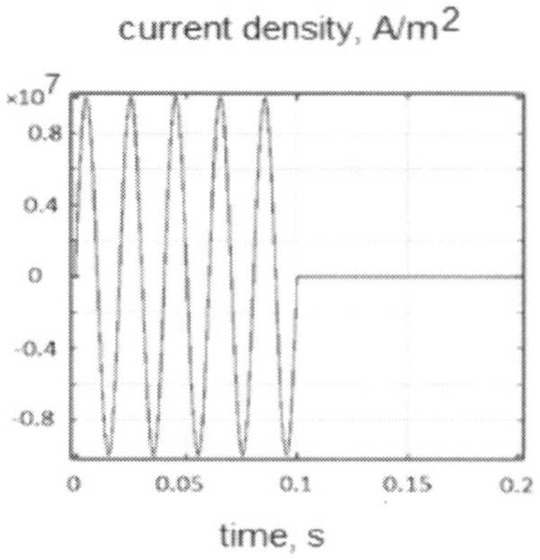

Fig. 3. Dependence of the current density at the contact area during the break of short circuit current.

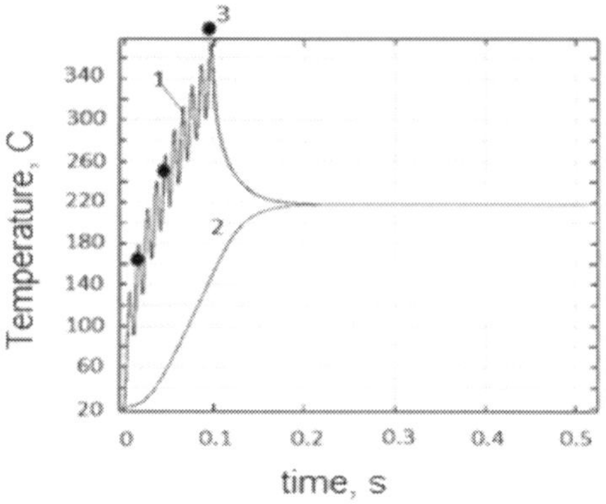

Fig. 4. The calculated temperature dependences from time in the contact surface -1, on the busbar surface - 2, 3 - the contact temperature values, recovered by the formula (7).

It should be noted that, despite the differences of real conditions, characterized by pulsating with doubled industrial frequency heat nature of heat release, from the Beck's model, where it is assumed time - constant heat, the recovered temperature value of the contact surface are completely agreed with the calculated values.

Conclusion

The use of temporary temperature dependences, measured on the busbar surface for a short period of time, in order to neglect the influence of heat distribution along the busbar, make it possible to eliminate the need of consideration of external conditions for contact diagnosis. In case of quasi rectangular heat pulse in the contact the use of the same sensor controller, implementing the proposed algorithm, makes it possible to evaluate the value the maximum temperature direct on the contact surface. It must be noted that in this case there is no need to insert into the data processing algorithm of the temperature curve values on environmental parameters and structural features of the device. It is enough to insert data on the thickness and material of the busbar.

References

1. Y. Vishnevsky, Energoatomisdat, 728 (2002)
2. A. Ivanov, Publishing House **2**, 532 (2006)
3. GOST R 52565-2006, Standartinform, 87 (2007)
4. GOST R 52736-2007, Standartinform, 40 (2007)
5. A. Nazarychev, F. Tadzhibaev, M. Sukhichev, V. Titkov, Energoexspert **2 (37)**, 12 (2013)
6. M. Sukhichev, V. Titkov, Electro **3**, 17 (2010)
7. S.I. Krivosheev, A.I. Tajibav, V.V. Titkov, Scientific proceeddings XXIII **2**, 165 (2015)
8. G.N. Aleksandrov, V.V. Borisov, V.L. Ivanov, Higher School, 312 (1985)
9. E. Kim, V. Omelchenko, S. Kharin, Bibliogr. **235**, 230-232 (1977)
10. N. Myshkin, V. Konchits, M. Braunovich, Intellect, 558 (2008)
11. J. Beck, B. Blekuell, C. St. Clair, Mir, 313 (1989)

Influence of dispersion degree of water drops on efficiency of extinguishing of flammable liquids

Dmitriy Korolchenko[1*] *and Sergey* Voevoda[2]

[1] Moscow State University of Civil Engineering, 129337, Moscow, Yaroslavskoye sh., 26, Russia
[2]Transneft Research Institute for Oil and Oil Products Transportation (Transneft R&D, LLC),117186, Moscow, Sevastopolskiy prospect, 47A, Russia

Abstract. Depending on the size of water drops, process of fire extinguishing is focused either in a zone of combustion or on a burning liquid surface. This article considers two alternate solutions of a heat balance equation. The first solution allows us to trace decrease of temperature of a flammable liquid (FL) surface to a temperature lower than fuel flash point at which combustion is stopped. And the second solution allows us to analyze decrease of burnout rate to a negligible value at which steam-air mixture becomes nonflammable. As a result of solve of a heat balance equation it was made the following conclusion: water drops which size is equal to 100 μm will completely evaporate in a zone of combustion with extent of 1 m if the flying speed of drops is even 16 mps (acc. to Stokes v = 3 mps); whereas drops of larger size will evaporate only partially.

1 Introduction

In automatic fire extinguishing systems as dispersive devices it is usually used drenchers and sprinklers which provide coarse-particle dispersion of water with the average diameter of drops about 2 mm [1-3]. Dispersive devices are installed on ceiling overlaps therefore the finely dispersed fraction of sprayed stream evaporates in heated combustion products.
It is supposed that drops of large sizes practically don't evaporate in a gas phase as because of high flying speed of drops so of its small specific heat-exchange surface. In this case suppression of FL with flash point more than 90 ° C is occurs due to cooling of its surface to a temperature at which the speed of evaporation of a flammable liquid is insufficient for forming of concentration of fuel in gas-air mixture required for combustion. The process of suppression in this case is reduced to following aspects:
- cooling of the surface,
- absorption of the heat flow, moving from flame jet to a surface of flammable liquid, by water drops.

*Corresponding author: da-vip@mail.ru

As the main parameters of fire seat - burning liquid - defining the balance of heat and mass in the process of suppression by water it is considered:

- burnout rate (evaporation) of FL;
- temperature in a zone of combustion (flame jet);
- temperature of a liquid surface;
- water use ratio (during suppression) [4-6].

Depending on the size of water drops extinguishing process is focused either in a zone of combustion or on a burning liquid surface. Drops smaller than 150 μm completely evaporate in combustion zone decreasing its temperature to a critical value called the temperature of extinction [7-11]. Decrease of temperature in combustion zone leads to reduction of the input rate of fuel vapors in it [12].

2 Heat balance equation

Flow of fuel vapors in a gas phase, when the surface temperature is close to the boiling point, is possible to determine by Stephen formula:

$$u_m^0 = \frac{D_0}{RT_0}\left(\frac{T_F}{T_0}\right)^m \frac{P_a}{\delta} \ln\left(\frac{P_a - P_1}{P_a - P_s}\right), \quad (1)$$

where: u_m^0 — specific mass burnup rate of a flammable liquid (FL) in a steady mode of burning, kg·m²/s; D_0 — diffusion coefficient of FL molecules in a gas phase at a normal temperature of T_0, m²/s; R — universal gas constant, J / (mol · K); T_F — flame jet temperature, ^{0}C; P_a — atmospheric pressure, Pa; δ — thickness of diffusion layer, m; P_1 — pressure FL vapors in surroundings at the distance from the surface equal to diffusion layer thickness δ, Pa; P_s — pressure of a FL saturated steam at temperature of a surface T_s, Pa.

The m exponent in a formula (1) for hydrocarbons is approximately equal 2. This ratio considers the molar-convective nature of movement of vapor from an evaporation surface.

Accepting that change of FL evaporation rate is caused by decrease of a flame jet temperature T_F, and considering that for the short period of extinguish of a flame by high dispersive water the surface temperature will not change appreciably, we should find an expression of FL burnout rate in stationary conditions u_m of the extinguishing process:

$$u_m = u_m^0 \, (T_F/T_0)^2 \quad (2)$$

The heat balance equation for a FL extinguishing process by impact of sprayed water on a combustion zone can look as follows:

$$\rho\, C_p V \, dT_F = (n\, u_m^0 \, S_0 Q_1 - Q_w q_f)\, d\tau \quad (3)$$

where: ρ — density of combustion product, kg/m^3 ; C_p — heat capacity of combustion product, kJ/ (kg*K); V — volume of a combustion zone which height is accepted as equal to the luminous part of flame or the distance from sprayer to a combustion surface, m^3; n — coefficient considering heat losses in process of emission; n = 0,6; S_0 — combustion surface area, m^2; Q_1 — low heat value of FL, kJ/kg; Q_w — specific quantity of heat taken by water from surface layer of FL during suppression, kJ/kg; q_f — mass rate of flow, kg/s; τ — time of extinguish, sec.

3 Two variants of solution of the heat balance equation

Let us consider two alternate solutions of a thermal balance equation. The first solution allows us to trace the decrease of a flammable liquid surface temperature to a temperature lower than fuel flash point at which combustion is stopped. And the second solution allows us to analyze the decrease of burnout rate to a negligible value at which steam-air mixture becomes nonflammable.

3.1 The first embodiment of the heat balance equation solutions

The first alternate solution can look as follows. We express u_m through the temperature in combustion zone. For this purpose we use a formula for heat-mass exchange in a stationary mode:

$$u^0_m = \frac{\alpha(T_F - T_S)}{Q_{fl}} \tag{4}$$

α — heat transfer coefficient, kg/(м² · K) ; Q_{fl} — specific heat of FL evaporation, kJ/kg.

If we substitute an expression for specific mass burnup rate (4) into the thermal balance equation (3), we will get:

$$\frac{\rho C_p\ V dT_F}{d\tau} = \frac{n\alpha Q_{fl}\ S_0(T_F - T_s)}{Q_l} - q_f\, Q_w \tag{5}$$

$$q_f / S_0 \equiv J, \tag{6}$$

we should rewrite the formula (5):

$$\frac{\rho C_p\ h dT_F}{d\tau} = \frac{n\alpha Q_{fl}\ S_0(T_F - T_s)}{Q_l} - JQ_w \tag{7}$$

J — intensity of water supply, kg/m^2·s;*h* — height of the luminous part of flame over FL surface, m.

We reduce the equation (7) into the form suitable for integration:

$$d\tau = \frac{\rho C_p\ hQ_{fl}}{n\alpha Q_l} \frac{dT - JQ_w Q_{fl_g}(T_F - T_s)n\alpha Q_l}{JQ_w\ Q_{fl}\ (T_F - T_s)n\alpha Q_l} \tag{8}$$

We integrate formula (8) ranging from τ = 0 to τ = τ_T and from T_F^0 to T_F:

$$\tau_T = \frac{\rho C_p\ h}{n\alpha Q_l} \ln \frac{J - J_{cr}\alpha(T_F - T_s)}{J - J_{cr}} \tag{9}$$

where T_F^0 — temperature in combustion zone before extinguishing, ^{0}C.

The critical condition of extinguishing when the time of suppression become endless, i. e. $\tau_T \to \infty$, is expressed so:

$$J = J_{cr} = g\frac{n\alpha Q_l\ (T_F - T_s)}{Q_w\ Q_{fl}} \tag{10}$$

At the same time $\alpha(T_F - T_s)/Q_{fl} = u^0_m$, therefore

$$J_{cr} = g\frac{nQ_{fl}\ u_m^0}{Q_w} \tag{11}$$

If we enter expression for J_{cr} (12) into the formula (11) we will get:

$$\tau_T = \frac{\rho C_p\ h}{n\alpha Q_l}\ln\frac{J-(T_F-T_s)n\alpha Q_l\ /Q_w\ Q_{fl}}{J-J_{cr}} \tag{12}$$

3.2. The second embodiment of the heat balance equation solutions

Now we consider the second alternate solution.

Using the formula (4) we get:

$$dT_F = \frac{Q_{fl}}{\alpha}du_m \tag{13}$$

Now we substitute the expression (13) into equation of thermal balance (3):

$$\frac{\rho C_p\ hQ_{fl}}{\alpha}du_m = \left(nu_mQ_{fl}\ -JQ_w\right)d\tau \tag{14}$$

We solve the equation within such limits: from $\tau = 0$ to $\tau = \tau_T$ and from $u_m = u_m^0$ to $u_m =$ 0.

Condition when $\tau = \tau_T$ at $u_m = 0$ is real only for suppression of flammable liquids with high flash point. As practically all water evaporates in a gas phase, the surface temperature of FL during suppression does not change appreciably; therefore the executed earlier transformation in a formula (14) is reasonable for a case of FL suppression.

The solution of an equation (14) will be the formula

$$\tau_T = \frac{\rho C_p\ h}{n\alpha Q_l}\ln\frac{J}{J-(\frac{n\alpha Q_l}{Q_w\ Q_{fl}})} \tag{15}$$

We should find the value of critical intensity, when $\tau_T \to \infty$:

$$J = J_{cr} = g\frac{nQ_{fl}\ u_m^0}{Q_w} \tag{16}$$

By substituting the value of critical intensity (16) into (15) formula we can get:

$$\tau_T = \frac{\rho C_p\ h}{n\alpha Q_l}\ln\frac{J}{J-J_{cr}} \tag{17}$$

The formula (14) considers parameters of highly flammable liquid better and and the ratio (17) is suitable for description of the extinguishing process of high-boiling FL and solid combustible materials (SCM) which surface temperature is close to T_F.

It is necessary to define how the specific burnup rate is changing during extinguishing.

On the basis of interrelation of u_m and T_F parameters through the ratio (4) we constitute the equation of thermal balance for combustion zone which width and height are defined by fire seat diameter:

$$\frac{\rho C_p \ hQ_{fl} \ du_m}{\alpha \ d\tau} = nu_m Q_{fl} \ - JQ_w \tag{18}$$

And now we execute an integration of differential equation (18) ranging from $\tau = 0$ to $\tau = \tau_T$ and from $u_m = u_m^0$ to $u_m = 0$:

$$\mathrm{d}\tau = \frac{\rho C_p}{JQ_w \ - nQ_{fl} \ u_m} \tag{19}$$

$$\tau_T = \tau_0 \ \ln \frac{J}{J - J_{cr}} \tag{20}$$

We should take into account that the quantity of water evaporating in process of suppression is reduced as the temperature in combustion zone is decreased.

We make an equation of thermal balance:

$$h\rho C_p \ dT_F \ = [\ nu_m Q_{fl} \ - \frac{\alpha f_\Sigma}{S_0}(T_F - T_s)] \ \mathrm{d}\tau \tag{21}$$

where f_Σ — total surface area of water drops participating in heat exchange, m^2.

We can take the thermal balance equation for process of suppression by impact of sprayed water on combustion zone:

$$\rho C_p V \ \mathrm{d}T_F = (nu_m \ S_0 \ Q_{fl} - Q_w q_f) \ \mathrm{d}\tau. \tag{22}$$

Now we can find an expression for critical rate of supply of sprayed water:

$$J_{cr} = 2/3 \frac{\rho r_k^2 n Q_{fl} \ u_m^0}{T_F^0 \ \tau_0} \tag{23}$$

This formula takes into account dispersion of an initial water stream.

It is necessary to define part of water evaporated in a flame jet.

To define the stay time of water drop in a zone with increased gas temperature it is necessary to calculate the speed of its falling. If we regard a drop as a solid ball than with small Reynolds numbers (Re) the steady falling speed of a drop is determined by Stokes formula

$$v = 2/9 \, g \, r_k^2 \frac{\rho_v \ - \rho_g}{\eta} \tag{24}$$

where: $\rho_г$ — gas density, kg/m^3; η — gas dynamic viscosity, Pa·s ; g — gravitational acceleration, m/s^2.

If the water drop with an initial radius r_0 heats up while passing through the gas layer, heated up to the temperature of flame jet T_F in this case, changes of drop temperature in time can be determined by this ratio

$$\frac{4\pi r_0^3 \rho_v C_w \; dT_0}{3d\tau} = 4\pi r_0^3 \alpha \left(T_F - T_0\right) \tag{25}$$

where C_w — heat capacity of water, kJ/kg-K; T_0 — initial temperature of drop, ^{0}C.

Now we find relation of an evaporation time to a time of drop heating:

$$\frac{\tau_T}{\tau_l} = \frac{3Q_w}{2C_w(T_F - T_k)} / \ln\frac{(T_F - T_0)}{T_F - T_k} \tag{26}$$

Specific heat of water evaporation $Q_w = 2{,}35 \cdot 10^6$ J/kg. Specific heat capacity of water $C_в = 4{,}2 \cdot 10^3$ J/ (kg· K); $T_F - T_к = 1100$ K; $\ln[(T_F - T_0)/(T_F - T_к)] = 0{,}068$;

$$\frac{\tau_н}{\tau_н} = \frac{2{,}35 \cdot 10^6}{2 \cdot 4{,}2 \cdot 10^3 \cdot 1100 \cdot 0{,}068} = 3{,}8.$$

Therefore, time of drop evaporation is in 3,8 times more than time of its heating while passing through a flame jet. If we suppose that the drop motion speed can exceed in 150 times the speed calculated for conditions of falling without initial impulse ($v = 3{,}0$ m/s — steady speed), i.e. it will be 15 m/s, than heat transfer coefficient should be determined by a formula (52) if $T_F = 1500$ K, $v = 225 \cdot 10^{-6}$ m/s, $\rho = 0{,}27$ кg/m^3, $d_к = 1 \cdot 10^{-4}$ m, Re = 9,4, from where $\alpha_2 = 1{,}35\lambda/r_0$. Therefore, increase of drop motion speed in 150 times changes the α value in 1,35 times and increase the time of heating and evaporation of water drop in a flame jet equally.

If the distance from sprayer to a burning surface is 1 m than the drop flying time at a speed of 16 m/s makes $\tau = 0{,}065$ s, and at a steady speed — $\tau = 0{,}33$ s.

Conclusion

Thus, drops with size of 100 μm will evaporate completely in a burning zone with a length of 1 m if the flying speed is even 16 m/s (by Stokes v = 3 m / s).

Water drops have an influence on different area of a burning object depending on their size. Finely dispersed water with an average drop size smaller than 100 μm has predominant influence on combustion zone. Water drops which size is larger than 0.5 mm are only partially evaporate in a gas phase and have influence mainly on a burning surface of solid combustible materials or flammable liquids.

The first solving method of the thermal balance equation of process of suppression of a flammable liquid by supplying of dispersed water into combustion zone allows us to determine the limiting temperature at which combustion is stopped. The second solving method allows us to determine the blowout rate which decreased to a negligible value when steam-air mixture becomes nonflammable.

It is possible to conclude that the formula (14), in case of solving of the thermal balance equation by first method, has better accounting of FL parameters, whereas the ratio (17) is fully describes the process of suppression of a high-boiling FL and solid combustible materials, which surface temperature is close to a flame jet temperature while burning.

References

1. P. Nash Fire Prev. No. **118**. 21–45 (1977)

2. D. A. Korolchenko, V. Yu. Gromovoy, O. O. Vorogushin Vestnik MGSU — Proceedings of Moscow State University of Civil Engineering, **1-2**, 331–335 (2011)
3. R.S. Volkov, Kuznetsov G.V., Strizhak P.A. Journal of Applied Mechanics and Technical Physics. **56(2)** 248-256 (2015)
4. V. I. Blinov, Khudyakov G. N. *Diffuzionnoye goreniye zhidkostey [Diffusion burning of liquids]*. (Moscow, Russian Academy of Sciences Publ., 1961)
5. Ya. B. Zel'dovich, et al. *Matematicheskaya teoriya goreniya i vzryva [Mathematical theory of burning and explosion]*. Moscow, Nauka Publ., 480 p. (1980)
6. Rasbash D. J. The extinction of fire with plain water: a review // Fire Safety Science. —**Vol. 1.** — P. 1145–1163. DOI: 10.3801/iafss.fss.1-1145 (1986)
7. Frolov Yu. G. *Kurs kolloidnoy khimii. Poverkhnostnyye yavleniya i dispersnyye sistemy [The course of dispersoidology. Surface phenomena and disperse systems]*. Moscow, Khimiya Publ. 400 p. (1982)
8. Semenov P. The Russian Journal of Applied Physics, **14, 7–8**, 427–437 (1994)
9. Friedrichsberg D. A. *Kurs kolloidnoy khimii [The course of dispersoidology]*. Leningrad, Khimiya Publ., 352 p. (1974)
10. A:F: Sharovarnikov, et al. Pozharovzryvobezopasnost — Fire and Explosion Safety, **22, no. 12**, 69-74 (2013)
11. D. A. Korolchenko Pozharovzryvobezopasnost — Fire and Explosion Safety, **vol. 21, no. 5**, pp. 79-80 (2012)
12. N.M. Kovalchuk, et al. Advances in Colloid and Inter-face Science, **210**, 65-71, doi:10.1016/j.cis.2014.04.003 (2014)

MATEC Web of Conferences **86**, 04057 (2016) DOI: 10.1051/matecconf/20168604057
IPICSE-2016

Reliability of buildings with internal network

Andrey Dolganov[1*], *Denis* Kaleev[1], *Galina* Malyha[2], *Ilkhomzhon* Shukurov[2]

[1] ZHILEKSPERTIZA, 127055, ul. Yablochkov, 21 building 3, Moscow, Russia
[2] Moscow State University of Civil Engineering, 129337, Yaroslavskoye Shosse 26, Moscow, Russia

Annotation. During the course of performance, reliability of a building with internal network located in Western Administrative District of Moscow was examined and estimated. Reliability means probability of no failure in the initial time.

1 Introduction

Reliability value makes not as many as one [1, 3, 6] and is recommended as 0.99865 in building systems with normal level of reliability. Earlier [4], calculation of reliability of monolithic multi-storey buildings using logical-and-probabilistic methods was provided. Thus, formulas are derived for the frame-type, flexible rods.

Slightly different approach is required to assess reliability of buildings with internal network, but the main provisions of reliability calculation remain unchanged. According to [3], the main property that determines reliability of a building is faultless of its performance. Reliability means probability of non-occurrence of the first limit state of load-bearing structures. Let us select strength of normal sections of all possible states of the first group. Other possible states of the first group are not considered as far as they are improbable and constructively impossible. For example: at maximum flexibility of columns in this building being equal to seven their loss of stability is possible only after normal cross-sectional destruction [9].

Thus, according to the accepted concept, reliability means probability of no failure in the initial time. We consider the first group of limiting states, for which required value of reliability makes 0.99865 (3σ). We consider that the main power factors of flexible systems are bending moments, and bending moments and longitudinal forces are the main power factors of beam columns.

Within the concept adopted, failure does not occur in the context of deflections, crack opening width exceeding permissible standard values, which are mainly determined by psychological, performance or other factors. For example, when designing columns for buildings only the first group of limiting states is mainly considered.

In this paper, during reliability calculation, variability of mechanical properties is only taken into account, because it is the most studied and categorized by the relevant reference documentation. We shall also take into account that average values of geometric parameters (their mathematical expectations) are inserted into SNiP and Code Design formulas, so reliability is evenly affected by them.

[*]Corresponding author: zlx777@yandex.ru

2 Subject and methods of research

Three-section building is considered. We assume that reliability of the building will be determined by the minimum reliability of one of three sections. Load bearing capacity of each section is provided by six-storey (including basement) frame. Extreme poles of these frames are masonry separation walls. Middle poles are columns. Basement columns are made of reinforced concrete. Columns of the building superstructure are masonry. Basement girders are made of steel, superstructure girders are made of reinforced concrete.

In addition to its main task (flooring), floor slab serves as horizontal brace. Building reliability calculations take into account sequential operation of frames and floor slabs. Therefore, failure (event being opposite to reliability) occurs upon floor slab or frame failure.

Linearization method was chosen to calculate reliability of floor slabs and girders. This method involves formula relationship of random arguments function, and this dependence should be differentiable.

Monte Carlo method was used to evaluate columns reliability. Versatility and high accuracy are the main advantages of Monte Carlo method. The disadvantages are sometimes encountered procedural difficulties in calculation of small values, as well as relatively high labor intensity when debugging programs requiring special care and rigor of their developers.

Let us calculate reliability of girder section by means of linearization method. Bending moment M_x is selected as girder reliability criterion. Transverse force in this case is structurally taken into account. The girder span makes 5.405 *m*.

Concrete resistance and resistance of stretched and compressed reinforcement are accept as random variables: $\tilde{R}_b$, $\tilde{R}_s$ and $\tilde{R}_{sc}$. As a result of surveys, we know average resistance and reinforcement variation coefficients: $R_{sm} = 384$ *MPa*, $V_s = 0.07$.

3 Research data

Statistic data on concrete are indicated in Table 1.

Table 1. Results of testing by means of "SCHMIDT" hammer, GOST 22690-88.

1	2	3	4	5	Mean strength, MPa		Coef. of variation	Concr. grade, MPa	Test location
					instrumen	actual			
41.56	36.52	44.35			40.8	37.4	0.097	31.4	Above 2-nd floor
42.14	45.23				43.7	42.2	0.050	38.7	Above 3-rd floor
43.03	40.66	27.37	39.7	47.0	39.6	35.4	0.186	24.6	Above 4-th floor
3802	51.3	50.69	38.1		44.5	43.7	0.168	31.1	Above 5-th floor
40.26	50.42	38.02			42.9	40.9	0.154		Above 5-th floor

Average acquired concrete strength variation coefficient (V_b) on girders makes 0.164. Average girder concrete grade makes 26.9 *MPa*. Average cube strength makes R_m – 40 *MPa*.

Average prism strength is defined according to the following formula (1):

$$R_{bm} = R_m \cdot (0.77–0.001R_m) = 40\cdot(0.77 – 0.001\cdot 40) = 29.20\ MPa \tag{1}$$

Design strength of reinforcement and concrete: $R_s = 340\ MPa$; $R_b = 29.8 \cdot (1–3\cdot 0.164) = 15.1\ MPa$.

It is deduced from experiments, that the girder is reinforced by 2∅25*A*-III (positive reinforcement) and 2∅10*A*-III (negative reinforcement). The instrument has shown reinforcement diameter and 12 *mm*, but, as a reserve, we have taken diameter of 10 *mm*. Reinforcement coefficients of positive and negative reinforcement are correspondingly equal to: $\mu = 0.010957$ and $\mu/ = 0.0017531$, i.e. double reinforcement case is considered. Coefficient Cμ: Cμ = $\mu/ / \mu = 0.16$.

Resisting moment M_x:

$$M_x = R_b\, bh_0^2\alpha_m + R_{sc}\, A_s{}'(h_0–a') \tag{2}$$

$$\alpha_m = \xi(1–0,5\xi) \tag{3}$$

$$\xi = (R_s\, A_s – R_{sc}\, A_s{}')/R_b \tag{4}$$

$$\xi_R = 0.8 / (1+ \varepsilon_{s,el} / \varepsilon_b) = 0.8/[1+(340/200000)/0.0035] = 0.538 \tag{5}$$

Let us transform expressions (2) – (4) into one, as well as divide left and right parts of equation (2) by bh_0^2:

$$m_Y = (R_s\, \mu – R_{sc}\, \mu')[1–0.5\times (R_s\, \mu\ – R_{sc}\, \mu') / R_b] \tag{6}$$

Further, let us take partial derivatives on R_s, R_{sc} and R_b. Taking into account that $(h_0 – a') / h_0 = 0.9$, we will get the following result

$$\partial m/\partial R_s = \mu[1–(R_{sm}\mu – R_{scm}\mu')/R_{bm}] = 0.01096\times [1–384\times(0.01096–0.001753)/29.8] = 0.00966 \tag{7}$$

$$\partial m/\partial R_{sc} = –\mu'\ [1–(R_{sm}\mu – R_{scm}\mu')/R_{bm} – 0.9] = –0.001753\times [1–384\times(0.01096–0.001753)/29.8+0.9] = -0.003123 \tag{8}$$

$$\partial m/\partial R_b = 0.5\times(R_{sm}\mu – R_{scm}\mu')^2 / R_{bm}^2 = 0.5\times 384\times (0.01096–0.001753)^2 / 29.8^2 = 0.000002 \tag{9}$$

Let us define mean square deviations of reinforcement and concrete resistance, *MPa*:

$$\sigma_{Rs} = M_s \times K_{sm} = 0.07 \times 384 = 26.88\ \sigma_{Rb} = M_b\times K_{bm} = 0.164 \times 29.8 = 4.89 \tag{10}$$

Section load bearing capacity under average values of material resistance m_Y and mean square deviation of load bearing capacity m_Y thus will be defined according to formulae (6) and (11). In formulae (6) and (11), R_{sm} shall be inserted instead of R_s and R_{sc}, and R_{bm} shall be inserted instead of R_b.

$$\sigma_Y^2 = \sum_{i=1}^{n}\left(\frac{\partial\varphi}{\partial x_i}\right)_m^2 \sigma_{x_i}^2 + 2\sum_{i<j}^{n}\left(\frac{\partial\varphi}{\partial x_i}\right)_m\left(\frac{\partial\varphi}{\partial x_j}\right)_m r_{ij}\sigma_{x_i}\sigma_{xj} \tag{11}$$

Where r_{ij}, σ_{x_i}, σ_{x_j} – are correspondingly correlation coefficient and mean square deviations of random variables. In our case, correlation coefficient r_{ij} is equal to zero for random independent variables (concrete and reinforcement resistance).

$$\sigma_Y^2 = 0.00966^2 \times 26.88^2 + (-0.003123)^2 \times 26.88^2 + 0.000002^2 \times 4.887^2 = 0.745,\ \sigma_Y = 0.86$$

$$m_Y = 384 \times (0.01096 - 0.001753) \times [1 - 0.5 \times 384 \times (0.01096 - 0.001753)/29.8] = 3.33$$

Thus, we will receive m_Y = 3.33 MPa, σY = 0.86 MPa. Let us define load bearing capacity m_0 with probability of three standards (0.99865):

$$m_0 = m_Y - 3\sigma_Y = 3.33 - 3 \times 0.86 = 0.75\ MPa$$

With load of 10.0 *kPa* on a floor slab we acquire relative moment equal to 1.61 *MPa*. Reliability of a girder under load of 10.0 *kPa* makes: *P*[(3.33–1.61)/0.86 = 2.29] = 0.97725. Acquired value of reliability has probability of less than 3σ, thus load shall be limited down to 8.70 *kPa*. At floor load of 8.70 *kPa*, the girder moment makes 0.75 *MPa* and the girder reliability under this load makes: *P*[(3.33–0.75)/0.86 = 3.0] = 0.99865. Acquired reliability value makes probability of 3σ. In this case, the girder reliability is comparable with requirements of GOST 27751-2014.

Reliability of concrete columns is defined by means of Monte Carlo method.

The columns are solid-cast. Section – 0.80×0.60 *m*. Effective depth, h_0: 0.80 – 0.05 = 0.75 *m*. Concrete strength of columns corresponds to *B*25 grade. The columns are reinforced symmetrically along the contour of 8∅25: $A_{s,tot}$ = 39.27 cm^2. Here, $A_s^{/} = A_s$ = 14.726 cm^2.

Reinforced concrete column slenderness ratio makes:

$$\lambda = l_0 / h = 3.3 / 0.8 = 4.13 \tag{12}$$

The first group of limiting states is the most important within calculations of beam columns. We assume that failure of an element does not occur if the following three conditions are met:

$$\begin{cases} N \le \tilde{N}_c & (13) \\ \tilde{N} \le \tilde{N}_{cr} & (14) \\ Ne \le \tilde{M} & (15) \end{cases}$$

In formulas (13) – (15) N – axial force, deduced from static analysis; $\tilde{N}_c$ – axial force, perceived by the element during direct compression and deduced from probability calculation; $\tilde{N}$ – accident force, determined as a result of probability calculation; $\tilde{N}_{cr}$ – axial buckling load (Euler's force), also defined by means of probability calculation; $\tilde{M}$ – resisting moment, deduced from probability calculation.

Symbol "~" (twiddle) over variable value shows its random nature.

Let us mark the events, indicated by conditions (13) – (15), by the letters *A*, *B*, *C*. Then, no failure operation of an element shall be expressed through conditional probability formula:

$$P(ABC) = P(A)\, P(B/A)\, P(C/AB) \tag{16}$$

Or

$$P(C/AB) = P(ABC) / [P(A)\, P(B/A)] \tag{17}$$

Probability calculations were carried out according to (17), Concrete grade and slenderness were used as initial data; $\alpha_s = R_s A_s / (R_b bh_0) = 3400 \times 14.726 / (148 \times 60 \times 75) = 0.075$;

$C_\mu = \mu' / \mu = 1$; ξ / ξ_R; γ_{bi} $(i = 1, 2, 3, 4)$; φ_l; Henceforward, SNiP 52-01-2003 symbols are applied.

Using general case formulae (13) – (15), we shall carry out deterministic calculation of load bearing capacity under calculated material parameters.

According to random values distribution laws and correlation dependencies, we shall specify sets of realizations $\tilde{R}_s, \tilde{R}_b, \tilde{E}_b$.

Using each set of random value realizations, $\tilde{R}_s, \tilde{R}_b, \tilde{E}_b$ we shall perform computations under (10)–(12), adopting that b=h_0=1. As a result of one computation: $\tilde{v} = N/\tilde{N}, \tilde{u} = N/\tilde{N}_{cr}, \tilde{w} = \tilde{M}/M$.

Now, we shall fit acquired results $\tilde{v}, \tilde{u}, \tilde{w}$ with Pearson curves. Probability density is acquired $f(\tilde{v}), f(\tilde{u}), f(\tilde{w})$.

According to (17) we shall calculate $P(C/AB)$.

Diagrams shown on Fig. 1 were plotted for flexibility elements 4, 13 according to results of calculations. Maximum load duration coefficient was specified as a reserve for durability and reliability: $\varphi_l = 2$.. Results of deterministic calculations according to SNiP 52-01-2003 under calculated resistance values are shown with full lines. Results of probability calculations with probability of 0.99865 or 3σ (three standards) are shown with dotted lines. Reliability diagrams are plotted for *B*25 grade concrete and reinforcement – *A*-III.

4 Evaluation

Based on diagram 2 on Figure 1, reliability of reinforced concrete columns within the range 1600 *kN* to 2100 *kN* is just over 3σ and maximum reliability value makes 0.9998. With forces indicated, two materials - concrete and reinforcement - are operating simultaneously. "Bad" concrete (design resistance R_b) is compensated by "good" reinforcement (for e.g., average resistance R_{sm}) and vice versa.

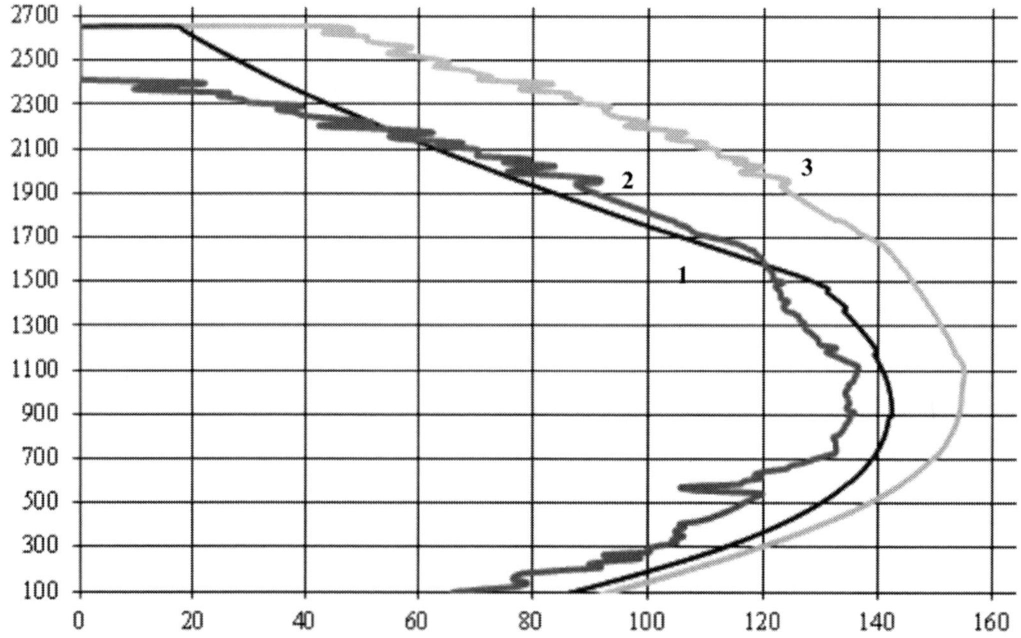

Fig. 1. Load bearing capacity of reinforced concrete columns, B25 grade concrete, A400 (A-III) grade reinforcement: 1 – deterministic calculations; 2 – bearing capacity diagram with probability of

0.99865 (integrating according to Pearson curve); 3 – bearing capacity diagram with probability of 3σ (3 is deducted from average value 3σ).

The column reliability quickly decreases with increase of axial force *N*. It is affected by the influence of initial concrete modulus of elasticity taking into account the column stability. This is the case of concrete failure, i.e. when structural reliability is mainly determined by one material - concrete. Concrete is known to have higher strength variation than reinforcement, so these elements have the lowest reliability while remaining factors remain the same.

Reinforcement is the main operating element in case of large eccentricities, so reliability is rather lower.

Thus, the highest reliability for the columns considered is seen in cross sections, which provide uniform operation of reinforcement and concrete.

At floor load of 8.70 *kPa*, the column section reliability at the top edge of foundation ($N = 1099.3$ *кН*, $M = 24.8$ *кН×м*) makes $0.9^6713 \approx 1$

Let us estimate reliability of separation partition of the first floor of the masonry. Separation partitions are the vertical load bearing frame elements. Calculated separation partition dimensions: depth of cross section, $h = 0.51$ *m*; width of cross-section, $b = 1.70$ *m*; insertion depth of the girder into the masonry makes 200 *mm*. Masonry resistance with *M*100 bricks and M4 mortar makes 0.9 MPa. Masonry elastic response is determined via Table 15* [8]: $\alpha = 500$.

Load transfer from the crossbars to the masonry is done through the spreading beam. 0.20 *m* of the girder is supported by the masonry. Stress block under the girder is quadrangular.

Load bearing capacity of the separation partition is carried out using the formula (18):

$$N \le N_0 = m_g \varphi_1 R A_c \omega \quad (18)$$

Where m_g – coefficient taking into account prolonged stress effect, $m_g = 1$; $\varphi_1 = (\varphi + \varphi_c)/2$; φ – buckling coefficient; φ_c – buckling coefficient for compressed part of section, determined by actual height of element *H* acc. to Table 18 [8] within bending moment plane; *A* – element sectional area; $A_c = A(1-2e_0/h)$ – area of compressed part of section at rectangular stress diagram, determined taking into consideration that its gravity center coincides with point of application of calculated axial force *N*; area border position A_c is determined if first moment of this moment relatively to its gravity center equals to zero.

Masonry flexibility: $\lambda_h = 3.3/0.51 = 6.471$and $\lambda_{hc} = 3.3/0.12 = 27.5$. Calculated parameters: $m_g = 1$; $e_0 = (51/2-20\times1/2) = 15.5$; $\lambda_h = 3.3/0.51 = 6.471$ and $\lambda_{hc} = 3.3/0.2 = 16.5$; $\varphi = 0.89$, $\varphi_c = 0.35$; $\varphi_1 = (0.89+0.35)/2 = 0.62$; $\omega = 1 + e_0/h = 1+15.5/51 = 1.3 \le 1.45$; $A_c = 51\times170\times(1-2\times15.5/51) = 3400$.

Axial force being under the influence of external load and own weight of the masonry will make:

$$N = [18.00\times0.51\times(3.3\times4)] \times 1.0 + 8.70\times3.1\times6/2\times4 + 6.00\times3.1\times6/2 = 500.62\ kN$$

Load bearing capacity of the masonry: $N_0 = 1\times0.62\times0.09\times3400\times1.3 = 256.12$ *kN*. Load bearing capacity of the masonry under this supporting method is insufficient to accommodate floor slab load, which makes 8.70 *kPa*.

Reliability of separation partition of the first floor of the masonry makes 0.5. Upper floors were not considered, because maximum axial forces are acting on lower floors. Thus, frame reliability with load of 8.70 *kPa*, will make: $0.5 \times 0.999999713 \times 0.99865 = 0.499324857$.

Post footing reliability is determined by reliability of the slab and the column footing. The slab in this case is the flexural member with single reinforcement. Post footing slab failure is determined as event Q_1. As for the column footing, we will consider possibility of destruction caused by eccentric compression (event Q_2, selection of vertical reinforcement area) and oblique section bending (event Q_3, selection of horizontal reinforcement area).

Reliability of column footing shall be determined as follows:

$$R_{2\text{-}3} = R_2 \times R_3 = 0.99903 \times 0.99918 = 0.99821 \tag{19}$$

Foundation reliability will be provided, if slab and column footing will not fail at the same time: $R_f = R_{2\text{-}3} \times R_1 = 0.99821 \times 0.99886 = 0.99707$.

Correspondingly, foundation failure can be calculated as follows: 1 – 0.99707 = 0.00293.

Thus, the building (system) reliability, R_c, under 1-5 floor slab load of 8.70 *kPa* will make

$$R_c = 1 - Q_f \times Q_P \times Qp = 1 - (0.00293 + 0.5 + 0) = 0.49707 \tag{20}$$

Thus we have abnormally low reliability value. Total load acting on the floor slabs shall be lowered down to 5.70 *kPa* or separation walls of floors 1–3 shall be reinforced.

Low reliability of the building is confirmed by presence of vertical and horizontal cracks in external walls of brick masonry. The crack width reaches 0.005 *m*. Cracks can be ssen across full height and width of exterior walls. Actual building reliability is slightly higher, since actual load in separate premises is lower, than 8.70 *kPa*. With load on floors making 5.70 *kPa*, the building reliability rises up to 0.9987.

References

1. G. Augusti, A. Baratta, F. Kashiati, *Probabilistic approaches in construction design* (Strojizdat, Moscow, 1988)
2. *Concrete and reinforced concrete constructions without reinforcement prestressing* (DEAN Publishing House, St.Petersburg, 2005)
3. *Reliability of building structures and foundations* (Standartinform, Moscow, 2014)
4. A. Dolganov, Industrial and civil construction **8**, 50–51 (2010)
5. A. Dolganov, *Reliability of braced reinforced concrete structures* (MAOBTI, Magadan, 2001)
6. V.D Rajzer, *Methods of reliability theory in normalization of design parameters of building structures* (Strojizdat, Moscow, 1986)
7. Federal Law No. 384-FZ (2009)
8. CD 15.13330.2012 (2012)
9. O.V Mkrtychev, G. A. Dzhinchvelashvili, M.S. Busalova, Procedia Engineering **111**, 550-555 (2015)

Increase of road tunnels safety using Tunnel Traffic & Operation Simulator

Eva Jančaříková[1,*], *Peter* Danišovič[1], and *Juraj* Šrámek[1]

[1]University of Žilina, Faculty of Civil Engineering, Department of Construction Management, Univerzitná 8215/1, 010 26 Žilina, Slovakia

Abstract. Tunnels, constructions used to shorten transport routes and to improve road safety, are specific engineering structures where the safe operation is very important. Tunnel Traffic & Operation Simulator (Simulator) was created as a solution of research task focused on increasing safety in road tunnels within the framework of the project "Centre of Transport Research (CVD)" funded by European Regional Development Fund (ERDF). The Simulator is located on Faculty of Civil Engineering of the University of Žilina (Slovakia) and is adapted to control the tunnel in two operator's mode which are functionally convertible. They can during the operation of the simulator control the tunnel's technology and operation in normal conditions and in exceptional situations. Thanks to this, they could work under the pressure with non-performance threat to the tunnel technology and safety of their users. The coordinator checks and assesses the correctness of their conduct.

1 Centre of Transport Research

The objective of this project, Centre of Transport Research (CVD), is to increase economic efficiency at the macrolevel in transport through direct cooperation of VÚD Žilina (Transport Research Institute, Žilina) and University of Žilina. Responsible workplace is a Department of Construction Management of Faculty of Civil Engineering at the University of Žilina [1].

Transport Research Institute is an organization of science and research with a long tradition and the main area of activity is the organization of research in transport.

The history of the University of Žilina (UNIZA) began on 1 September 1953 when the University of Railway Transport was founded by separating from the Czech Technical University in Prague. It gradually became an important pillar of education in the fields of transport, which resulted in the change in the name to the University of Transport. It was moved to Žilina in 1960 where it underwent many transformations, and in 1980 it was renamed again to the University of Transport and Communications. Later in 1996, it was renamed the University of Žilina. Currently, University consists of 7 faculties, scientific and educational institutes and centres and has approximately 1,500 employees. Faculty of

* Corresponding author: eva.jancarikova@fstav.uniza.sk

Civil Engineering has 12 departments and each department is focused on scientific research activities [2].
Specific objectives and activities of the project:
- Applied research of new technologies to improve operational safety of road tunnels.
- Expansion planned research of advanced technologies in the field of law enforcement in road transport.
- Completion of the research activities on innovative approaches to measuring physical and chemical properties of road signs.
- Increasing the rate of research activities in the field of land fragmentation by transport infrastructure on progressive technologies.

The first specific objective includes instrumental and technological workplace building and its commissioning, applied research and integration in the field of road tunnels operations management, system´s calibration and then extraction and processing of data for the increase of road tunnels safety and prediction models as a tool for safety improving in road tunnels operation.

2 Tunnel traffic & operation simulator

This research site for road tunnels operation is realized as a real operator´s workplace. It shows virtual 1 km long highway two-tube tunnel [1]. The main objectives of the simulator are:
- Increase of transport´s safety in road tunnels.
- The testing possibility of different scenarios in tunnel
- Collection and processing of data from tunnels operation, risk analysis.
- The creation of prediction models
- New knowledge attainment for new tunnels project.

Basic composition of the Simulator consists from central control system (CCS – part of automatic tunnel equipment control), manual control module (MCM – part of manual control as separated module, it serves to a manual tunnel operation in the case of the central control system´s accident), software for prediction of phenomena (EMUT) - evidence of tunnel incidents.

Fig. 1. Tunnel Traffic and Operation Simulator – operators' workplace monitors

Thanks to Tunnel Traffic & Operation Simulator it is possible to verify operating conditions by real operation at the visualization on each operating condition. Verification of operators' response in operating conditions in the tunnel tube, also throughout a formation of fire, is possible only on the Tunnel Traffic & Operation Simulator in the Transport Research Centre. In the case of operating condition with dangerous goods transportations vehicles it is very necessary and required by the TP 082 a preventive verifying of operators professional competence.

Fig. 2. Some possible simulations of various incidents

2.1 Manual control module (MMR)

The role of this operators and tunnel maintenance specialists department is to inform operators and reporters of data collection with any new technical and technological tunnel equipment following changes in tunnel operation legislation [3].

Fig. 3. Manual control module

2.2 Software for prediction of phenomena - EMUT

EMUT is a special program for evidence and evaluation of incidents in the tunnel. It is registered every change compared to the tunnel´s operating conditions under normal operating conditions, when an intervention of technology´s operator and traffic´s operator is not required. It is possible to create a graphic representation of incidents in tunnels at a different frequency assessment of a monitored parameter.

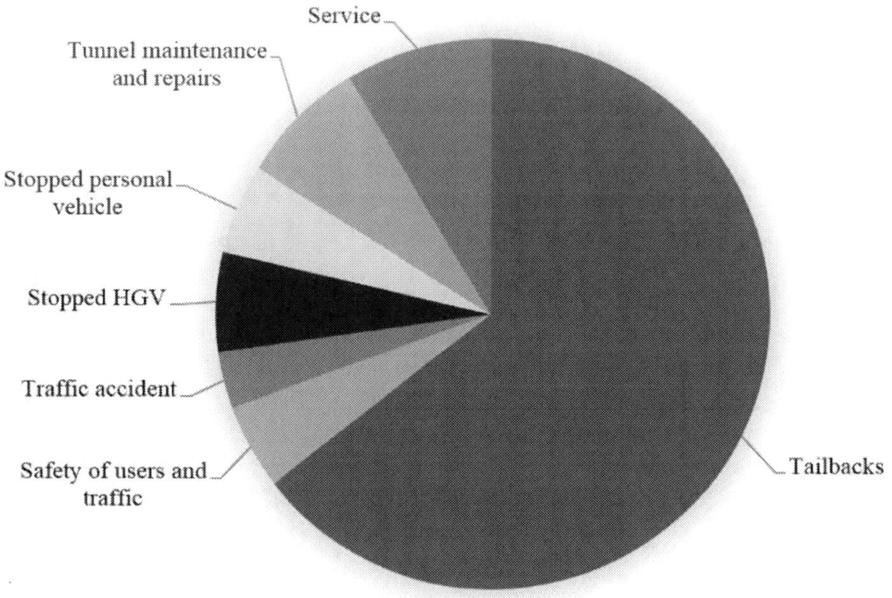

Fig. 4. Sectoral graph of events in a one tube tunnel (per year)

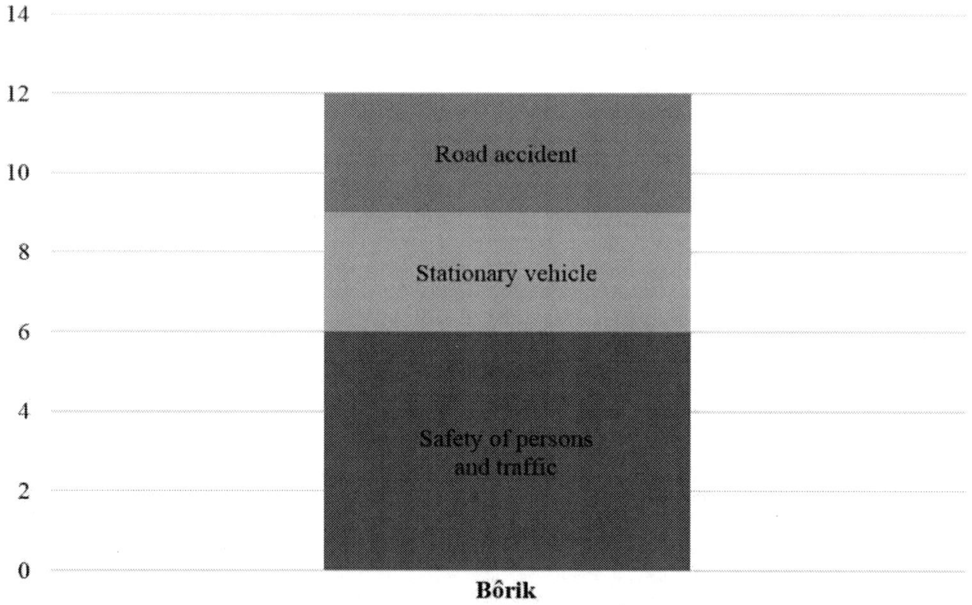

Fig. 5. The number of incidents in both tunnel tubes for a period of 1 month

3 Risk analysis

The risk analysis of road tunnels in Slovakia is performed according to specifications TP 041 (02/2011) Risk Analysis for Slovak road tunnels. This regulation was created primarily to the definition of the exact methodology for risk analysis of the safety of road users in tunnels. For the calculation of risks in road tunnels is in TP precisely defined and explained the risk model, which was developed based on the Austrian model for risk analysis of tunnel TuRisMo [4].

Table 1. Overview of the Slovak tunnels in use [5]

	Šibenik	**Bôrik**	**Sitina**	**Horelica**	**Branisko**
Tunnel´s length	588 m	995 m	1 440 m	605 m	4 975 m
Number of tunnel tubes	2	2	2	1	1
Category	2T 8,0	T 7,5	T 7,5	T 9,0	T 7,5
Route	Jánovce - Jablonov	Mengusovce – Poprad, západ	BA, Polianky – BA, Mlynská dolina	Oščadnica – Čadca, Bukov	Beharovce – Široké
Start of construction	06/2012	06/2006	10/2003	04/1998	05/1997
Used since	11/2015	12/2009	06/2007	10/2004	06/2003
Traffic load	13 573 veh/24h	12 342 veh/24h	50 351 veh/24h	13 011 veh/24h	9 433 veh/24h
The proportion of passenger vehicles	75,00 %	80,13 %	87,90 %	68,96 %	75,24 %
The proportion of trucks	23,08 %	18,21 %	11,41 %	30,52 %	23,14 %
The proportion of bus	1,95 %	1,66 %	0,69 %	0,52 %	1,62 %

Table 2. Accidents in highway tunnels.

Tunnel	**Incident inside the tunnel**	**Traffic intensity**	**Number of accidents/1million of vehicle x km**
Horelica	6	13 011	0,19
Bôrik	6	9 433	0,30
Branisko	30	12 342	0,11
Sitina	56	50 350	0,27

Statistics for tunnel Šibenik are not available yet for because a short time of its use.

Table 3. Risk analysis values of Slovak tunnels in use [6].

Tunnel	Statistically expected value of the number of death toll per year	Danger Category
Branisko	0,2156	III
Horelica	0,0705	II
Bôrik	0,0529	II
Sitina	0,2327	III
Šibenik	0,0361	II

The basic number of accidents in tunnels in Slovakia with two-way traffic is 0,025 and with one-way traffic is 0,036 accidents/1million of vehicle x km. Just for comparison with some EU countries, in Czech Republic this number is 0,312, in Germany 0,228 and in Austria 0,0112 accidents/1million of vehicle x km (updated in 09/2015).

Conclusion

One of the primary objectives of each tunnel is to advance the high level of traffic and users safety. Therefore, it is crucial to do the necessary engineering adjustment and install special technological equipment designed to improve a safety. Compared to the accidents on the roads outside the tunnel, even a minor incident in the tunnel can escalate to catastrophic proportions. So it is important to optimize and simplify the security system of the tunnel to limit the possible human or technical error. The objective of the system is to communicate early and reliably all the necessary information about the current status of traffic conditions in the tunnel and to allow a safe evacuation of users from the tunnel in the case of an accident. The Tunnel Traffic & Operation Simulator in our University of Žilina is used for simulating critical events in road tunnels. We can simulate the various incidents and they can serve for prediction of these events in real tunnels.

Acknowledgement

"This work was supported by the Slovak Research and Development Agency under the contract No. APVV-15-0340".

This contribution is the result of the project implementation: "Centre of Transport Research" (ITMS 26220220135) supported by the Research & development Operational Programme funded by the ERDF.

"We support research activities in Slovakia/project co-founded from the resources of the EU"

References

1. http://svf.uniza.sk/cvd/index.php
2. http://www.uniza.sk/menu/inc.php?id=9
3. P. Danišovič, Š. Šedivý, M. Rázga, E. Jančaříková, *Tunnel Traffic & Operation Simulator* (13th International Conference Underground Construction Prague 2016, Czech Republic)
4. TP 041 (02/2011) *Risk Analysis for Slovak road tunnels* (in origin: *Analýza rizík pre slovenské cestné tunely)* (technical specifications, 2011)
5. http://www.ndsas.sk/tunely-nds/44745s
6. M. Rázga, *Technical and technological aspects of road tunnels operation safety* (in origin: *Technické a technologické aspekty bezpečnosti prevádzky diaľničných tunelov)* (PhD. thesis, University of Žilina, 2016)

MATEC Web of Conferences **86**, 04059 (2016) DOI: 10.1051/matecconf/20168604059
IPICSE-2016

Construction of a complex object

Pavel Oleynik [1] , *Sergey* Sinenko[1], *Boris* Zhadanovsky[1], *Victor* Brodsky[1], and *Marat* Kuzhin[1,*]

[1]Moscow State University of Civil Engineering, 129337, Moscow, Yaroslavskoe shosse, 26, Russia

Abstract: The peculiarities of application of nodal method in the construction of large and complex industrial systems, the principles of the spatial division of the object on the technological, building and public spaces, the basic documents on the use of nodal method. As an example, we consider the use of nodal method in the construction machinery manufacturing enterprises. To this end, provides a general description of space planning and designs of single-floor and multi-floors buildings of this industry, the composition of industrial complexes and sub complexes engineering, their basic units and plots, schemes interlinkages units and energy supply. It includes steps for the development of complex aggregated network diagrams network diagrams and working with the justification of continuous work flow.

1 Introduction

An effective method of construction of large industrial complexes is the nodal method.
The essence of the nodal method is that the composition of the starting complex stands out in a constructive and technological separate parts - spare parts for organizations focused and technologically sound production work and achievements in the shortest possible time frame for technical readiness of autonomous testing and commissioning of individual production lines, offices and plants.
The use of nodal method allows you to:
- clearly coordinate within each node and the whole complex;
- to create a reliable basis for planning the construction and installation works, acquisition of material, technical and human resources, operational management and supervisory control over the course of construction;
- to provide the necessary details of the organizational and technological documentation for all building production management levels;
- maximize the combination of work and organize protection production of construction and installation works on the basis of long-term specialized streams;
- Concentrate and most efficient use of material and technical manpower;

*Corresponding author: kuzhinmf@mail.ru

determine the intensity of loading and rhythmic work of the organizations and participants in the construction period.

Node - structurally and technologically separate part to be erected industrial complex (object), the technical readiness which, after completion of construction and installation work allows for commissioning and testing machines, tools and devices.

When forming the nodes must be considered:

- structural completeness of the allocated industrial complex (object);
- ensure spatial stability of the building or structure is part of the assembly;
- completeness single technological cycle in total industrial production technology;
- possibility of production start-up and stand-alone unit for delivery of the working act of the customer;
- the possibility of mounting on the nodes lead singer given the predominance of its work profile;

creating conditions for the mass production of works;

- the efficient operation of construction machines and mechanisms;
- the possibility of opening in the shortest possible time to the front of works related organizations.

When forming the nodes must be considered the characteristic features of the complexes under construction in various industries:

- completeness of technological conversion of feed components, prompro-product or process;
- consistent capacity building.

2 Application procedure of Node method

On a functional purpose sites are divided into technological, construction and Organization of the work sites.

Technology unit is a structurally separate part of the process line or installation within the boundaries of which are made to RMC technical readiness required for the commissioning and testing of assemblies, mechanisms and devices. The structure of the processing unit mainly includes - foundations for process equipment, technological equipment, related to the site underground utilities (flume, Electro cable, transport tunnels) and underground structures (pumping of all kinds, oil), technological metal structures and pipelines built premises the main production facilities (control panels, electrical devices), flooring and fine finish.

The construction site includes building (structure) of the main production facilities or a structurally separate part, within which are made of construction and installation work to the technical readiness necessary for the transmission assembly by mechanical assembly work. The boundaries of the construction unit can be placed one or more processing units. The most characteristic nomenclature construction site is under the frame of the building foundations, supporting and protecting the building structure, roofing and finishing works, built facilities and ancillary service facilities, overhead cranes, electric building.

Organization of the work sites nodes are - preparing the construction site, the objects of administrative-residential and utility-of subsidiary purpose, power, energy, water recycling, transport, agriculture and the improvement of the industrial site.

The composition of the industrial complex of nodes includes:

technological components: boiler, gas station, hardware management framework, wood and sawmill, Wood Factory, refrigerator, laundry factory, factory cleaning, fire station, billet plant, training center, medical center, a waste treatment plant, mechanization base The foundations of the oil sites, asphalt plant, playground equipment base, base production and

processing equipment, concrete testing ground to impregnate the wood shop, shop insulating pipes and fittings, etc.;
Construction units: industrial buildings and units, office buildings, dining room, garage, a dispensary, a cattle-breeding complex, pension, fire station;
Organization of the work sites components: transport rack, sewerage system, interior and exterior (industrial, household, rain, etc.), Mobile (inventory) buildings and structures, work on site preparation, construction and electricity networks, roads and driveways, buildings and network communications (internal and external), water supply system (industrial, household, drinking and fire), heating networks (internal and external trunk), the communication tunnel, gas supply network, air and fuel, work on landscaping, traffic management facilities, complex of warehouses, security measures, the base of sanitary cleaning, universal supply base, the base operation of utilities services, nursery with a greenhouse, jetty and basic storage, a playground for the reception and storage of materials, water intake from underground sources, the station iron removal, water intake from the reservoir, the station purification of river water, ash dump TPP, wastewater treatment plants, the main pumping station, storage pond, active storage of chlorine, the connecting path, industrial plant, CHP station, factory station, railway track.
At the same time, the application method hub has the following drawbacks: the formation of knots is often one construction site area includes both of the process, and the construction Organization of the work sites nodes; within the boundaries of the construction site can be located several process units; when fixing at the sites of the leading performers of the ability of their uncoordinated actions.
Many years of experience in the application node of the method in the construction of such complex objects like plate mill "3000" metallurgical plant. (Mariupol), Pellet Plant mining complex, oxygen-converter shop of the Dnieper Metallurgical Plant them. Dzerzhinsky, and others. Showed that the elimination of these drawbacks is achieved by the formation of technological units, combining technological lines, buildings and structures, as well as the adjoining utility networks. At the same time the construction and installation work on the construction of the object considered as a stage of the corresponding processing unit, and the objects of administrative-residential and utility-auxiliary facilities, engineering and transport communications - both Organization of the work sites. Thus, the existing nodes of the method remains the same, but provides a significant simplification of the design and estimate documentation and reducing the complexity of its design, creates the necessary conditions for the formation of long-term flows and increases the efficiency of the operational management of the construction industry.

Construction Management Project consists of four parts - I - the text part; II - graphical part; III - part design; IV - application.

The main documents required for the use of nodal method, developed at the stage of the project of works (PPR); they contain passport units and operating a rolling roadmap.

Passport complex components - a set of documents consisting of a breakdown of the circuit nodes on the list and the composition of components, circuits, technological coherence and energy supply units.

The initial data for the development of the passport sites are:

- title list;
- design and estimate documentation;
- breakdown of the scheme on the list of nodes and structure nodes;
- technological interlinkages units and energy supply circuit;

Typical solutions for nodes:

In drawing up the passport scheme specified breakdown on the list of nodes and the composition of components and circuit technology interlinkages nodes developed at the design stage of the project and working as part of PIC.

Work node roadmap is an integral part of the design work and production is developed in order to:
- identifying and linking technological sequence of performance of works on the sites;
- linking the work of all artists (sites, teams) participating in the construction site;
- clarifying and specifying the timing of the work fronts related organizations;
- ensure at all stages of the construction site of a single continuous planning and management in all production units with the use of computer technology;
- rational distribution of logistical and human resources;
- organization of operational control over the progress of construction and installation works;
- focusing on the work of the leaders, lying in the critical zone.

The source document for the development of the working node network schedules are as follows:
- nodes passport;
- organizational and technical measures;
- working project documentation;
- projects or production work flow diagrams;
- Categories of standard routings and standard flow sheets linked to a specific node.

Working nodal network diagrams develop general contracting trusts together with specialized organizations with the involvement of design institutes.

In the development of the working node and the network schedule is necessary to strive for a level of detail that the work had one unit and one artist.

During the construction of complex objects and large industrial complexes there is the need to establish operational service management start-up complex construction. The main objective of launching complex is to ensure the effective operation of general contractors and subcontractors, providing enterprises and farms on the basis of systematic monitoring and control at all stages of construction - from the development of design estimates, technological, organizational and technical documentation to the delivery of objects into operation.

3 Results of Nodes method

Practical experience in the construction of industrial complexes using the nodal method has gained in the construction of the enterprises of ferrous metallurgy, machine building, gas, chemical and other industries. By branches of engineering built of heavy and transport engineering, electrical industry, chemical and oil engineering, machine tool and tool industry and others.

Production engineering buildings made, usually in the form of large dimensional blocks by combining various plants based processes, unification and unity of design solutions. The most typical are the four types of buildings - with bearing metal structures with application of effective profiles (type 1), with the bearing and enclosing structures made of reinforced concrete (type 2), made of light metal structures complete delivery (type 3), with the bearing and enclosing structures AI Seriesmbe-04 (type 4). In the development of space-planning decisions envisaged scheme of carcasses of buildings with a minimum number of height differences.

Buildings can be a multi-floors and single-floor. For single-floor buildings spread a net 24x12m columns and 30x12m. Building frames span up to 24m are executed, as a rule, in precast concrete structures, and a span of 30m and more with bridge cranes with lifting capacity of 50 tons and a height from the floor to truss bottom exceeding 14 4m, - metal structures.

Administrative buildings are designed largely free-standing as close to the production. For the device of foundations and structures used piles 12m (for collapsible soils), bored piles with a diameter of 600-1500mm and more than 12m in length (for heavy loads), a natural base (for unsettled sandy soils). As the foundations for increasing application framework are prefabricated foundations weighing up to 25 tons, and for foundations for technological equipment - prefabricated monolithic foundations with enlarged blocks.

More and more are beginning to be used for type 1 metallic columns of buildings in the form of Wide-I-Beams at full height (mounted bezvyverochnym way), and for buildings of type 2 and 3 - reinforced concrete rectangular construction (instead of the traditional two-branch).

Covers the main production buildings are carried out in large blocks, assembled to complete construction readiness on conveyor lines.

The exterior walls of buildings consist of light ("sandwich") for light (for example, "KZHS" 3x24m) panels, and for administrative and residential buildings the panels have a textured finish. Roofs are made roll with booking the bed bindings window openings - from roll-formed closed section profiles at 1.436-6 series or series of joint ventures with wooden windows.

Roof lights have sizes 3x6 and 3x12m.

Flooring according to the technological requirements can be concrete, because of high oil-resistant plates with disperse reinforcement and wear-resistant coating of silica sand and grit; of large-sized slabs of refractory concrete, pressed mosaic tiles for administrative and household premises.

As partitions are used large-industrial, with disperse reinforcement of structural glass, lightweight aggregate panels and blocks.

Engineering support for the main production building at the expense of the location of the main networks in the area and cross-farm areas of crane tracks.

The initial data for the development of technological schemes interlinkages units and energy supply are: a list of nodes and composition; description of technological processes; general plan; schemes supply facilities with electricity, water, gas and other essential energy resources.

Launch Complex are based on technical - economies-who study for construction. Each launch complex should provide for the implementation of the scope of work, ensuring the normal operation of the facilities put into operation. For example, in the starting housing complex #1 includes objects and structures listed in Table 1 below.

Table 1. The composition of the starting complex of housing #1

Nodes	plots
1	2
Node 1.1 Case #1	Thermal circuit - 1.1.01 Internal engineering network - 1.1.02 workshop #1 – 1.1.03 workshop #2 – 1.1.04 workshop #3 – 1.1.05
Node 2.1	ABC-1, ABC-2 - 2.1.01 Factory ATS 600 rooms - 2.1.02
Node 4.12 - off-site water supply	HVL 200 kV p / o 500 kV to kV RP Prom site - 04/12/01 Substation GPP-220/110 kV - 12/04/02 110 kV overhead line between 220/110/10 kV GLP - 12/04/03

Node 1.8 - intra-electricity 1.9 Node - inside networks and communication underground installation	110/10 kV modular package #1; GPP- 1 220/10 kV RP-5 conductors - 01/08/01 Rainwater and industrial sewage around the body and from the tunnel to the exit from the industrial site - 01/09/01 Industrial and domestic sewage around the body; potable water main collector sewer - 01/09/02 The communication tunnel with the networks of drinkable water pipe water recycling - 09/01/03 Onsite networks and facilities of water supply, a network of drinkable running water tanks - 09/01/04
Node 3.18 - inside networks of heat and steam	Network Elevated gaskets; piping, condensers - 3.18.01 underground installation network - 3.18.02
1	2
Node 1.11 - onsite road	Onsite road around the body - 1.11.01
Node 1.12 - intrasite railway tracks	Access track to the body - 12/01/01
Node 4.21 - off-site water supply system and water intake facilities	Water intake pump 1 lift - 4.21.01 The pump station of lifting II - 4.21.02 Pumping station III lift - 4.21.03 Water pipeline from the pumping station lifting III - 4.21.04
Node 4.22 - offsite utility fecal sewage network and wastewater treatment plant	Pumping pumping station - 4.22.01 Off-site sewer - 4.22.02 Sludge ponds with sludge pumping water - 4.22.03 Storm sewage - 4.22.04
Node 5.24 - off-site roads	The expansion of the road - 5.24.01 Viaduct interchange - 5.24.02 North Road and the road to the industrial base - 5.24.03
Node 5.25 - off-site railway tracks	Development of the connecting station - 5.25.01 Access track - 5.25.02 The bypass route - 5.25.03
Node 4.26	External communications network - 4.26.01
Node 1.20	Landscaping - 1.20.01

Once the start-up complex objects compiled statements: the estimated cost of the starting complex objects; the volume of work on the launch complex; the need for building structures, components, semi-finished products and the basic materials on the launch complex, on the launch complex of the equipment.

Comprehensive enlarged roadmap is developed with a degree of detail that allows to link the West the process of creating complex in time from design to putting it into operation.

Operating a network schedule for construction of the facility is the organizational and technological model that reflects the interconnection of all types of work. Scheduling is carried out in stages.

In the first phase set volume installation and terms of their performance. At the second stage - the entire volume of construction and installation work is divided into separate parts, and assigned to the responsible executives. In the third phase the responsible executors specify initial data for the completion of the network schedule. In the fourth stage are calculated the main parameters of the chart - foundations under the framework of buildings, installation of frame buildings, installation of wall fencing, installation of built-in premises, roofing, etc. (Table 2).

Table 2. The parameters for calculating the network schedule

Code phase	works	event #		work code	Code resources
		I	II		
1	2	3	4	5	6
200	Vertical planning	1	2	24	190
200		2	11	24	190
201	Driving of reinforced concrete piles	2	3	42	150
201		3	4	42	150
202		4	7	42	237
202		7	12	42	237
201	Arrangement of bored piles	2	5	46	150
202		5	14	46	150
202		14	21	46	150
201	Concrete preparation device	3	6	70	11
201		6	15	70	11
202		15	22	70	11
202		22	24	70	11
201	Foundations for the frame	6	8	54	11
201		8	9	54	11
201		9	16	54	11
202		16	25	54	11
202		25	30	54	11
201	Backfilling	8	10	20	150
201		10	13	20	150
201		13	18	20	150
202		18	28	20	150
202		28	32	20	150
201	Columns	10	17	146	234
201	Communication			147	234
201	Crane beams			167	234
201	Columns	17	26	146	234
201	Communication			147	234
201	Crane beams			167	234

In addition, due to the low duration of work on the site flows provided by the foundation for the equipment, concrete preparation, installation of process equipment, Ventilation equipment installation, installation of secondary and technological metal structures. In turn, the rhythmic flow is largely determined unmounted enlargement of building structures and technological equipment.

Thus, for the large-assembly coatings are widely used conveyor lines for the assembly of reinforced concrete structures, construction of complete delivery with structural coatings, metal structures.

Construction of the Krasnoyarsk plant of heavy excavators with capacity of 140.000 tons per year mehanoizdely all nodes were built parallel to the five streams - streams of 1, 2, 3 on the main body, flow 4 administrative buildings, stream 5 on buildings and facilities auxiliary facilities. The first three streams served conveyor assembly line coatings, equipping them with electrical, plumbing and industrial installations and communications: prefabricated buildings (housing optional equipment), from designs complete supply structural coatings (building service shops), of metal structures (body smith workshop #1). Conveyor line for the housing cover assembly smith workshop #1 (MSC-1), a total area of

331.6 thousand m² had a U-shaped arrangement in the plan to reduce the extent and ease of installation of plumbing and electrical communications and included 18 sites:

- Assembling the main load-bearing metal block designs #1, 2
- Installation runs and roof lights #3
- Lapping units, conducting sanitary works #4
- Painting. #6, 7
- Installation and mounting corrugated #8, 9
- Installation of roof fans. #10
- Electrical work #12
- Glazing of the roof lights and device #13-17
- Delivery unit for mounting #18
- Reserved parking #5-11

Moving blocks on the conveyor carried by individual groups (up to four blocks) special winches and K-701 tractor. Supply and installation of tower blocks were carried out using the installer load capacity of 120 tons, moves along the span on the crane tracks. For the unit to lift installer used railway crane MRS-2600 load capacity of 130 tons. Total weight of the complete unit reached 120 tons.

Widespread use in the construction of facilities for the Krasnoyarsk water-heavy excavator received conveyor for manufacturing and installation of complete supply of buildings. One such object is the CEC, which is a building in terms of size 521x253m. The grid columns in the main spans accepted 24x12m. The load-bearing structure are the spatial coverage of cross-core building blocks (of tubular elements) with the size in terms of 24x12m. Enclosing cover designs are structurally composed of a steel corrugated galvanized sheet and a layer of insulation. The coating applied skylights series 3230.

Coating to be assembled on the line, there has been a total area of 123.000 m^2. The total number of blocks is equal to 423 pieces. Clearing the pipeline capacity was 3 units a day in two shifts at his job. Driving conveyor line consists of 8 work sites, the drive pre-engineered, post unit overload of conveyor trolleys for transportation. Conveyor line provides access to the complete installation of the unit construction readiness, fully equipped with technological equipment.

Stacking carried on stand representing arranged in a particular order established stationary pedestals for supporting the lower tier node elements during assembly and the entire unit prior to installation of the assembled conveyor trolleys. As you move the coating unit on the working conveyor line parking areas it is saturated with sanitary and electrical equipment, and arranged roof unit is brought to a complete construction readiness.

After installation, the blocks in the design position remains sealing of joints between the blocks and the connection of communications, being in a previously assembled unit.

Served two conveyor line tower crane - one directly in the production and installation works and other (gantry) at a construction warehouse. Lifting and installation of blocks in a design position carried out by crane crawler ICG-100 in the tower-boom performance. Rigging block made with the help of four-branch balancer sling load capacity of 25 tons.

At present, the development of large-block construction method the installation is carried out primarily in the direction of increasing the size and weight of the blocks. So, if at the plant were applied coating blocks span of 36m and weighing up to 200 tons and an area of 300m^2, on the construction of the industrial complex of the Ulyanovsk coating units already had 24x72m size with an area of about 1.800m^2 and a weight of up to 500 tons.

With the construction of engineering systems widely used and large-block installation of technological equipment. For example, in the construction of the first stage of the plant "ATOMMASH" 51 bridge crane was installed in the spans of the main building, located on two levels of 30 and 42-meter spans. All possible operations on the installation of cranes have been moved down to the assembly area. As a result of mounting blocks half-bridges have a weight of up to 150 tons, and trucks up to 70 tons. The unit includes the main beam half-bridges, service platforms, end carriages with balancing trolleys, drives, and in some cases, and electrical systems. A number of overhead cranes were installed fully assembled. In particular, the bridge crane 160/32 tons and a span of 28m was set at design elevation jib crane rail MRS-2600 all over the two lifting. Most of the crane weighing 97 tons was collected on a free site and then delivered on trailers under jib crane hook and raised on the runways. trolley was installed the second rise.

4 Conclusion

Hub is the recommended method to use in the design, organization and management of construction and reconstruction of complex objects and large industrial complexes. The degree of complexity of objects is determined based on the number of constructed buildings and structures that make up the complex, and construction conditions, the level of unification, typing and standardization of design solutions, the need for special support structures, devices and systems, a variety of construction processes, the number of contractors and subcontractors, involved in the construction industry. An indicative list of complex objects and large industrial complexes are listed in Table 3.
Industrial objects and facilities, design and construction (reconstruction) which is advantageously carried out using a nodal method, determined jointly by the building organizations and the contracting authority with the participation of general designer and design organizations.

Table 3. An indicative list of complex objects and large industrial complexes

Name of industrial construction industry	Name of industries, buildings and structures
1	2
Ferrous metallurgy	Iron and steel Sinter plants. Product coke plant. Complexes of blast furnaces. The oxygen-converter shop. Continuous steel casting installation. Electric smelting plant. Rolling mills. Tube welding mills.
	Seamless pipe mills. Oxygen station. Workshops for the production of ferroalloys. Factories pelletizing concentrate mining and processing. Workshops for roll forming.
Wood and paper industry	Sawing and woodworking enterprises. Pulp Mills. Pulp and paper and cardboard factories. Forest chemical plants.
Chemical and petrochemical industry	Various large production of basic chemistry products. Factories rubber products. Plants asbestos products. Major production of chemical fibers. Nitrogen fertilizer plant. Plants of varnishes and paints. Production of plastics. Tyre factories.
Mining industry	The mines with open pit development. Land mines mining complexes. Crushing-sorting-paid-factory. Concentrating factory ferrous and nonferrous metallurgy.
Non-ferrous metallurgy	Aluminum plants. Plants of secondary non-ferrous metals.
Electrical power	Thermal power plants (TPP, TPP). Nuclear power plants. Hydroelectric power plants.
Oil and gas industry	Refineries. Plants for the processing of oil for various purposes at existing refineries. Gas processing plants.
Coal and oil shale industry	Land complexes coal and shale mines. Coal and shale sections. coal preparation plants.
Mining and Chemical industry	potash plant
Building materials industry	Cement plants. Plants asbestos. Plants roofing materials. Factories and production of glass and glass products. Plants for the production of precast concrete. House-building plant.
Engineering	Plants for the production of cars, machines and equipment for various purposes. Complexes of large plants and production at the existing plants.

References

1. V.P. Abarykov, Optimal design of the system in construction, *Grail.* Moscow, (2000)
2. V.A. Afanasyev, Flow Construction Organization, *Stroyizdat.* Moscow, (1990)
3. V.I. Brodsky, The main conditions of mobile building systems organization, *Modern science: current problems and solutions* #3 (16). **30-32**, (2015)
4. A.V. Ginsburg, Automation of design of organizational and technological reliability of construction. *RIA SIP*. Moscow, (1999)
5. R.A.Grebennik, V.R. Grebennik, Organization and technology of erection of buildings and constructions. *High school.* Moscow, (2008)
6. K.H.Grey, E.W. Larson, Project Management, a practical guide. Translation from English. *Business and Services.* Moscow, (2003)
7. L.G. Dickman, The organization of building manufacture. *DIA.* Moscow, (2006)
8. B.V Zhadanovsky., S.A.Sinenko, M.F. Kuzhin, Rational organizational and technological schemes of construction and installation works in a reconstruction of the enterprise. *Technology and organization of construction industry.* #1, **38-40**, (2014)
9. B.V. Zhadanovsky, Organizational-technological preparation of reconstruction of civil and industrial buildings and structures. *Industrial and civil construction* #10, **59-60**, (2009).
10. V.A.Zarenkov, A.Y. Panibratov, Modern design solutions, technology and management practices in construction. *Stroyizdat.* St. Petersburg, (2000)
11. T.K. Kuzmina, The investment of the customer-builder activity. *Industrial and civil construction* #10, **31-32**, (2000)
12. T.K.Kuzmina, A.M. Slavin , Modeling activities. Technical customer at the stage of technical supervision. *Industrial and civil construction* #4, **62-66**, (2015)
13. Y. Kulikov, Assessment of quality solutions in construction management. *Stroyizdat.* Moscow, (1990)
14. A.A. Lapidus, Organizational design and management of large-scale investment projects. Moscow, (1997).
15. A.A.Lapidus, L.P. Demidov, A study of factors affecting the indicator-building construction site. *MSUCE Herald* #4, **160-166**,(2014)
16. P.P. Oleynik, Analysis and development of standards the duration of construction of residential buildings typical series. *Mechanization construction* #2, **18**, (2008)
17. P.P.Oleynik, S.P. Oleynik, Organization and technology of building production. The preparatory period. Moscow, (2006)
18. V.I. Telichenko, O.M., Terentyev, A.A. Lapidus The technology of erection of buildings and constructions. *High school.* Moscow, (2004)
19. V.A. Kharitonov, Underground buildings and constructions of industrial and civil objects. *DIA.* Moscow, (2008).

Problems and prospects of cottage construction in Russia

Vitaly Kasyanov[1*], *Olga* Landysheva[1*]

[1] Moscow State University of Civil Engineering, 129337, 26, Yaroslavskoye Shosse, Moscow, Russia

Abstract. Sustainable development of building complex in suburban territories and reorientation of building construction to low-rise construction is the actual direction of the construction industry. Countryside real estate in the cottage settlements is a new popular housing format in Russia. It is already an established market in Moscow and St. Petersburg: it has clear rules, criteria, participants, fixed offer and demand. The article deals with the sustainable development of suburban territories. The article is carried out qualitative and quantitative assessment of cottage construction industry in the Moscow region, problems of development of cottage construction as part of sustainable development for cities and suburban territories. The article focuses on the legal status of the cottage and the cottage settlements, it proposes some elements of cottages classification. The article describes the possible prospects of cottage construction development as one of the main directions of low-rise construction development. The presented results in the article can serve for further development and improvement of methods and tools, and also for researches materials that conducted in this area.

1 Introduction

Sustainable development of cities and suburban territories is aimed at the preservation and modernization of urban environment to improve people's standard and quality of living. Architectural and planning concept is relevant for suburban territories development. It includes:

- preservation and reconstruction of the architectural and landscape heritage;
- implementation of a clear and rational territorial zoning;
- achievement more rational and environmentally friendly planning and functional organization of suburban territories.

World trends in residential development focused on the formation of a healthy, clean and comfortable life environment of population. Modern regional planning is destined for the priority of suburban low-rise construction in ecologically clean zones.

Modernization of mass residential development is one of the main directions in the sustainable development of cities and suburban territories. The quality of residential development is a complex concept, which includes the concept of rationality, functionality, comfort, economy, security, solidity.

According to numerous sociological survey most Russians want to live "closer to the ground" - in cottages or in areas of low-rise construction [1]. And this is understandable: unlike a city apartment

* corresponding e-mail LandyshevaOE@mgsu.ru

own house is conceptually new quality of life. A priority for the population is the comfort of living, but not volume of commissioned lodging space.

Data of Federal State Statistics Service of Russian Federation (Rosstat) shows the general trend of the Russians preferences: the share of commissioning of individual houses in the total commissioned lodging space has increased annually over the past 10 years (table. 1) [2].

Table 1. Dynamics of commissioning of individual houses

	2005	2006	2007	2008	2009	2010	2011	2012	2013	2014
Commissioning of individual houses, thousand	123,9	140,5	184,6	196,4	206,8	188,6	197,7	208,5	225,0	266,5
% to previous year	108,9	113,9	131,7	104,0	104,3	89,4	104,9	106,0	108,1	116,7
Share in the total area of commissioned lodging space,%	40,2	39,5	43,0	42,7	47,7	43,7	43,0	43,2	43,5	43,0

The last decade of cottage construction is successfully developing in the suburban territories of large cities in Russia [3].

The number of private houses in the informal settlements is much more than in organized cottage settlements. This trend is caused by the fact of organized cottage settlements mainly belongs to the premium and business class residential property segments (table. 2).

An analysis of the market of cottage settlements in Moscow region was carried out. More than 20 cottage settlements located in the central region of Russia were considered. For the analysis it was used the sources of information and data of Rosstat, print advertising, media, analytical and scientific articles, Internet.

The criterions of estimation for cottage settlements were average cost, location, distance from a city center, site plot , environmental setting, infrastructure development.

The assumed scheme of classification is presented in table. 2.

Table 2. Classification of cottage settlement

Criterion of estimation	Premium class segment	Business class segment	Economy-class segment
Average cost	17-30 million rubles	8-20 million rubles	5-7 million rubles.
Locaton	Eclogically clean and picturesque area, located close to the forest and water bodies	Ecologically clean and good natural area	Propitious area
Distance from a city center	Up to 30 km. Easy transport accessibility to a center city	Up to 30 km.	Up to 50 km.

 DOI: 10.1051/matecconf/20168604060

Architectural style	Original architecture, landscape design	Modern architecture	Absent or in the separate elements
Infrastructure	Secured area, access mode, facilities of the social infrastructure and services sector	Facilities of the social infrastructure and services sector	Children's and sports playgrounds, shop. Infrastructure is primarily outside the settelments

The analysis identified major criterion of estimation for cottage settlements.

Chaotic uncontrolled construction in suburban territories is a serious problem for the city. Chaotic suburban territories development damages not the city image, but also the environmental situation (Fig. 1).

Fig. 1. Chaotic uncontrolled construction in suburban territories/ private sector of suburban territories

The private sector of suburban territories is a special area that requires centralized planning and building development.

As part of the sustainable development of construction complex houses in organized cottage settlements have to be made in the same architectural style, harmoniously fit into the surrounding landscape, avoiding confrontation with nature.

It is necessary to create and implement new specific conditions for private construction of suburban territories.

High cost of private houses in suburban territories is the most important factor hinder the wide development of cottage construction, especially in economy-class segment. Also the significant obstacles of cottage construction are:

— high competitiveness of high-rise construction companies [4];
— difficulties with allocation of land plots for the low-rise construction development;
— high cost of engineering infrastructure.

It should be noted that the legal status of "cottage" is not registered anywhere in the Russian legislation, as well as the concept of "cottage settlements", although there are numerous cottage settlements of different classes around the major cities.

However, all of people understand what is "cottage" intuitively. It is one, two or three-storey building designed for year-round living and supplied with necessary engineering services and sanitized house territory. The two important characteristics of cottage are the modern architecture of the façade and needs to be built in the organized cottage settlements (Fig. 2).

Fig. 2. Cottage settlements «Zarechny», Istra district.

Due to the development of cottage construction it may to solve the following problems:

1. Increase the proportion of square meters per capita.
2. Costs reduction for cottage in suburban territories because of the state programs for development of low-rise construction.
3. Improving the quality of life for people. It will have a positive impact on the demographic processes in the country.

The mentality of Russian citizens is gradually approaching to European standards of comfort living [4]. Living comfort level determined by such factors as space-planning decisions, sanitary norms, reliability, safety and efficiency of power supply, the level of automation of engineering systems and others [5].

Conclusion

As part of the sustainable development in the suburban territories, the strategic goal for building complex modernization should be the creation of social environmental residential development, material and spatial conditions to ensure environmental safety and quality of living [6- 8]. For planning the building complex modernization in suburban territories consideration must be given to the priority of the interests of people and the long-term interests of society.

From the results it is concluded that the dominance of the construction industry is reorientation of building construction to low-rise construction.

It may be stated that the analysis data prove the necessity to develop and build new private housing that meets the modern requirements for living comfort.

References

1. I.V. Malyshev, Dissertation abstract, 15-17 (2011)
2. Rosstat, *Statistical book on the housing construction in the Russian Federation* (2014)
3. A.S. Ermakov, Science and education in modern society, 33-34 (2015)
4. A.S. Ermakov, O.E. Landysheva, Vestnik of Donbas National Academy of Civil Engineering and Architecture **3 (113)**, 22-24 (2015)
5. V.F. Kasyanov, Publisher of Building universities Association, 224 (2015)
6. V.S. Kazeikin, Baronin S.A, A.G. Chernykh, A.N Androsov, Problematic aspects of the development of low-rise housing construction in Russia, 17-20 (2011)
7. V.F. Kasyanov, B.S. Strigin, Natural and Technical Sciences **9-10 (77)**, 468-470 (2014)
8. V.F. Kasyanov, N.A. Tabakov, Scientific Review **2**, 159-165 (2012)

Rheological properties of the polymeric blends

Tatyana Matseevich[1*], *Andrey* Askadskii[1]

[1]Moscow State University of Civil Engineering, Yaroslavskoe sh. 26, Moscow, 129337 Russia

Abstract. The calculation scheme for storage modulus and loss modulus for compatible polymers as a function of the temperature, stress state, and composition has been suggested. The relations for calculation of the storage modulus at high frequency for polymer of any chemical structure containing carbon, hydrogen, and oxygen atoms are suggested. Calculation scheme allows evaluation both of the storage modulus and loss modulus for compatible blends of polymers. The calculations are performed for mixtures of polystyrene, poly(methyl methacrylate), polycarbonate, and polyethylene oxide used for the manufacture of building materials.

1 Introduction

Within the frame of the papers [1] it was shown that the storage modulus at high frequencies is described by the following equation

$$G = \frac{\sum_i \Delta V_i}{\sum_i g_i \Delta V_i + \sum_j f_j}, \quad (1)$$

where $g_i = \frac{S_i}{k_i l_i}$; the values of g_i characterize an average contribution of each atom into the value of $\frac{S_i}{k_i l_i}$; f_i is the selection of constants which characterize the influence of strong intermolecular interaction (dipole-dipole interaction, hydrogen bonds, etc.); S_i is the Van-der-Waals surface of the i-th atom, through which the intermolecular interaction occurs; κ_i is the elasticity coefficient of the i-th atom bond; l_i is the characteristic size of the bond; $\sum_i \Delta V_i$ is the Van-der-Waals volume of i-th atom entering into repeating unit of polymer.

For compatible blend of polymers the equation (1) is transformed to the form:

$$G = \frac{\alpha_1\left(\sum_i \Delta V_i\right)_1 + \alpha_2\left(\sum_i \Delta V_i\right)_2}{\alpha_1\left(\sum_i g_i \Delta V_i + \sum_j f_j\right)_1 + \alpha_2\left(\sum_i g_i \Delta V_i + \sum_j f_j\right)_2}, \quad (2)$$

where α_1 and α_2 are the molar parts of polymer 1 and 2, respectively;

*Corresponding author: tmats@rambler.ru

$\left(\sum_i g_i \Delta V_i + \sum_j f_j\right)_1$ and $\left(\sum_i g_i \Delta V_i + \sum_j f_j\right)_2$ are the selections of constants for the polymer 1 and 2, respectively.

Taking into account that in accordance with the equation (1) the values of $\left(\sum_i g_i \Delta V_i + \sum_j f_j\right)$ is equal to $\left(\sum_i g_i \Delta V_i + \sum_j f_j\right) = \frac{\sum_i \Delta V_i}{G}$ we can rewrite the equation (2) as:

$$G = \frac{\alpha_1\left(\sum_i \Delta V_i\right)_1 + \alpha_2\left(\sum_i \Delta V_i\right)_2}{\alpha_1\frac{\left(\sum_i \Delta V_i\right)_1}{G_1} + \alpha_2\frac{\left(\sum_i \Delta V_i\right)_2}{G_2}}, \quad (3)$$

where G_1 and G_2 are the moduli of the component 1 and 2, respectively. Because $\alpha_1+\alpha_2=1$, obtain:

$$G = \frac{\left(\sum_i \Delta V_i\right)_1 + \alpha_2\left[\left(\sum_i \Delta V_i\right)_2 - \left(\sum_i \Delta V_i\right)_1\right]}{\frac{\left(\sum_i \Delta V_i\right)_1}{G_1} + \alpha_2\left[\frac{\left(\sum_i \Delta V_i\right)_2}{G_2} - \frac{\left(\sum_i \Delta V_i\right)_1}{G_2}\right]}. \quad (4)$$

The equation (4) allows description of the shear modulus G depending on molar concentration of Polymer 2.

Let us examine the example of calculation using the polymers used in the building industry. It is well known that fully compatible blend of polymers is as following: "polystyrene + polycarbonate". The Table 1 of the values g_i and f_j is borrowed from the paper [1]:

Table 1. Values of constants g_i and f_j

Atom or type of intermolecular interaction	Values of g_i or f_j	
Carbon	g_C, cm^2/kg	-1.277
Hydrogen	g_H, cm^2/kg	4.376
Oxygen	g_O, cm^2/kg	116.95
Double bond	$f_{\neq}$, Å^{3}cm^2/kg	30.925
Dipole-dipole interaction*	f_d, Å^{3}cm^2/kg	84.472

* *Parameter f_d is introduced for each branching in the back bone or side chain. This parameter is introduced also in the presence of a polar group of any type.*

Using these parameters as well as the Van-der-Waals volumes of the atoms [2-5] we obtain for polystyrene

$\left(\sum_i g_i \Delta V_i + \sum_j f_j\right)_{PS}$ = −1.277(13.1+8.7+8.4+5·12.7)+4.376(8·2.0)+84.472=34.8. Van-der-

Waals volume of Polystyrene is $\left(\sum_i \Delta V_i\right)_{PS}$ = 110 Å^3 (calculations are produced by the software Cascade). So, the value of storage modulus is G_{PS} = 3.15 kg/cm^2 (it should be noted that the experimental value of G = 3.16 kg/cm^2).

For polycarbonate $\left(\sum_i g_i \Delta V_i + \sum_j f_j\right)_{PC}$ = −1.277(4·16.9+6·8.4+2·11.6+4·12.7+4.5+ +2·12.7+19.0)+4.376(22·2.0)+116.95(2·2.7+5.95)+84.472 = 1285.3. Van-der-Waals volume of polycarbonate is $\left(\sum_i \Delta V_i\right)_{PC}$ = 305.3 Å^3. So, the value of storage modulus is G_{PC} = 0.24 kg/cm^2. The dependence of storage modulus of the blend under consideration on the molar part of polycarbonate is shown on Figure 1.

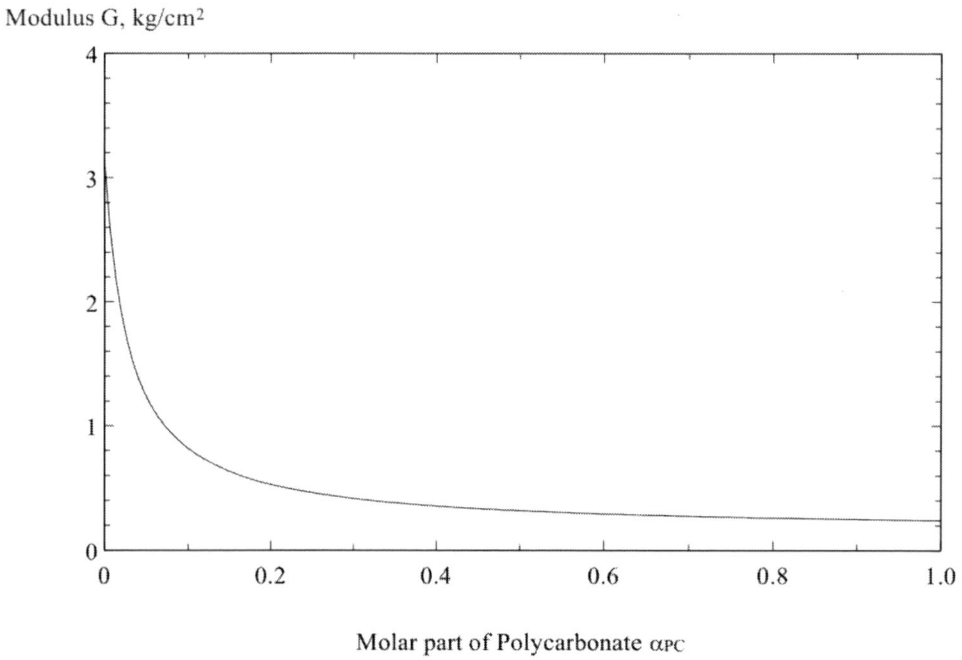

Fig.1. Dependence of storage modulus on the molar fraction of polycarbonate.

The major curvature of the dependence is due to the great difference between the Van-der-Waals volumes of the components.

The other compatible blend is "poly(ethylene oxide) + poly(methyl methacrylate)". For poly(ethylene oxide) $\left(\sum_i g_i \Delta V_i + \sum_j f_j\right)_{PEO}$ = −1.277(2·16.2)+4.376(4·2.0)+116.95·3.4= =391.26 cm^2/kg·Å^3. Van-der-Waals volume of poly(ethylene oxide) $\left(\sum_i \Delta V_i\right)_{PEO}$ = 43.7 Å^3. So, the value of storage modulus is G_{PEO} = 0.112 kg/cm^2. For poly(methyl methacrylate) $\left(\sum_i g_i \Delta V_i + \sum_j f_j\right)_{PMMA}$ = −1.277(13.1+5.0+17.2+15.9+20.3)+4.376(8·2.0)+116.95(5.85+ +3.4)+84.472·2=1229 cm^2/kg·Å^3. Van-der-Waals volume of poly(methyl methacrylate)

$\left(\sum_i \Delta V_i\right)_{PMMA} = 96.4$ Å^3. So, the value of storage modulus is $G_{PMMA} = 0.08$ kg/cm^2.

The dependence of storage modulus of this blend on the molar part of poly(methyl methacrylate) is shown on Figure 2.

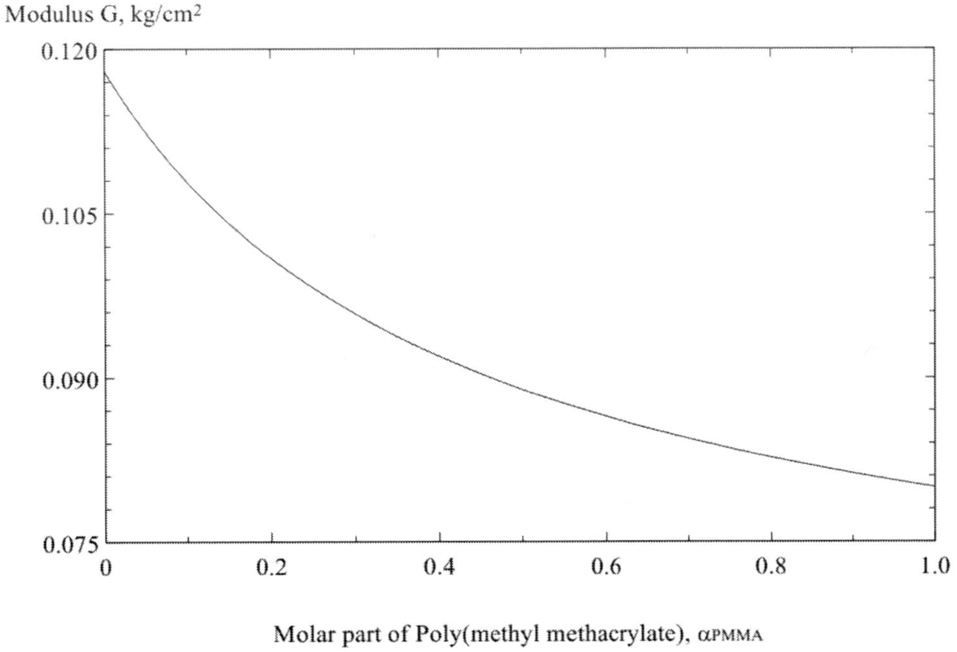

Fig.2. Dependence of storage modulus on the molar part of poly(methyl methacrylate)

It is well known that molar α_m and weight α_w fractions are related by

$$\alpha_m = \frac{1}{1 + \frac{M_2}{M_1}\left(\frac{1}{\alpha_w} - 1\right)}, \qquad (5)$$

where M_1 and M_2 are the molecular weights of the repeating units of polymer 1 and polymer 2, respectively.

The dependence of the storage modulus on the weight part $\alpha_{w,2}$ of the polymer 2 is described by the following formula

$$G = \frac{\left(\sum_i \Delta V_i\right)_1 + \frac{\left[\left(\sum_i \Delta V_i\right)_2 - \left(\sum_i \Delta V_i\right)_1\right]}{1 + \frac{M_2}{M_1}\left(\frac{1}{\alpha_{w,2}} - 1\right)}}{\frac{\left(\sum_i \Delta V_i\right)_1}{G_1} + \frac{\left[\frac{\left(\sum_i \Delta V_i\right)_2}{G_2} - \frac{\left(\sum_i \Delta V_i\right)_1}{G_1}\right]}{1 + \frac{M_2}{M_1}\left(\frac{1}{\alpha_{w,2}} - 1\right)}}. \qquad (6)$$

For poly(ethylene oxide) $M_{PEO} = 44.1$; for poly(methyl methacrylate) $M_{PMMA} = 100$; for polystyrene $M_{PS} = 104$; for polycarbonate $M_{PC} = 310$. Substituting all the physical parameters mentioned above obtain the dependence of storage modulus on the weight part of poly(methyl methacrylate) (Figure 3):

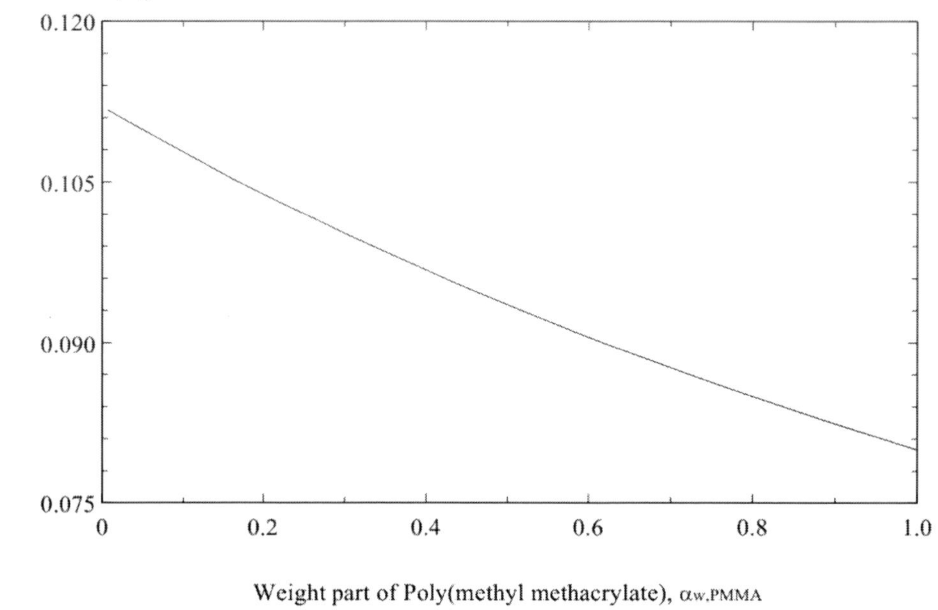

Fig.3. Dependence of storage modulus on the weight part of poly(methyl methacrylate).

The dependence of storage modulus on the weight part of polycarbonate is shown on Fig. 4.

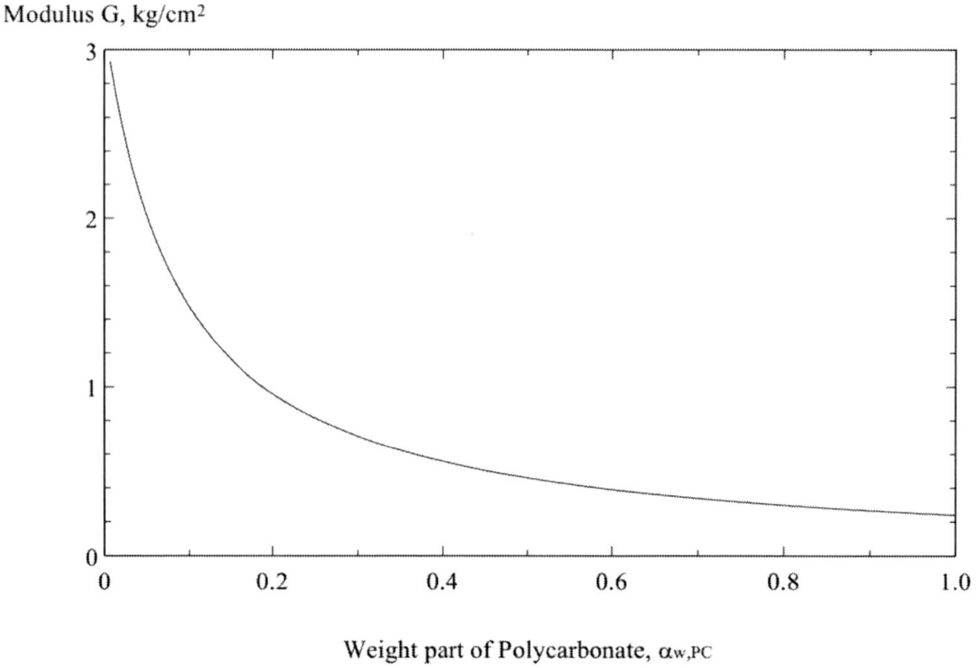

Fig.4. Dependence of storage modulus on the weight part of polycarbonate $\alpha_{w,PC}$.

The storage modulus and loss modulus of the polymer blends

Now we consider the calculation of storage and loss moduli $G'(\omega)$, $G''(\omega)$ of the polymer blends. It is well-known that in the Maxwell regime (i.e., when there is a single typical relaxation time τ) the storage an loss moduli are given by the following formulae

$$G'(\omega) = G\frac{(\omega\tau)^2}{1+(\omega\tau)^2}, \tag{7}$$

$$G''(\omega) = G\frac{(\omega\tau)}{1+(\omega\tau)^2}, \tag{8}$$

where G is the high-frequency limit of the storage modulus and the relaxation time τ is given by

$$\tau = \frac{\eta}{G\Gamma\left(1+\frac{1}{k}\right)}, \tag{9}$$

where $\Gamma(x)$ is the Euler gamma function, and the parameter k is due to the non-exponential relaxation in polymer systems.

As an example of calculations we consider a mixture of poly(ethylene oxide) and poly(methyl methacrylate). Using values $\left(\sum_i \Delta V_i\right)_{PMMA}$, $\left(\sum_i \Delta V_i\right)_{PEO}$, G_{PMMA}, and G_{PEO} for these polymers, the resulting dependencies on the concentration of PEO are given in Table 2.

Table 2. The values of modulus G for the compatible blends of PEO and PMMA.

α_{PEO},%	α_{PMMA},%	G, N/m^2
0	100	8000
20	80	8240
40	60	8570
60	40	9050
80	20	9810
100	0	11200

Viscosity of the mixtures calculated by Bicerano method [6] as a function of temperature is given in Table 3.

Consider first the temperature dependence of the storage and loss moduli. Assume, for definiteness, that $\alpha_{PEO} = 40\%$, $\alpha_{PMMA} = 60\%$, and the weight average molecular weight to be $M_w = 40000 g/mol$. Then the high frequency storage modulus is $G = 8570\,\mathrm{N/m^2}$ and the relaxation time as a function of temperature is given in the Table 3.

Table 3. The values of η_0 and τ at various temperatures.

T, K	400	415	430	445	460	475
η_0, N·sec/m^2	1387974	105633	283	147	79	45
τ, sec	81.0	6.16	0.0165	0.00855	0.00463	0.00262

The corresponding frequency dependences both of the storage and loss moduli are presented in Figure 5. Storage moduli are plotted in full lines, and loss moduli are in dashed lines.

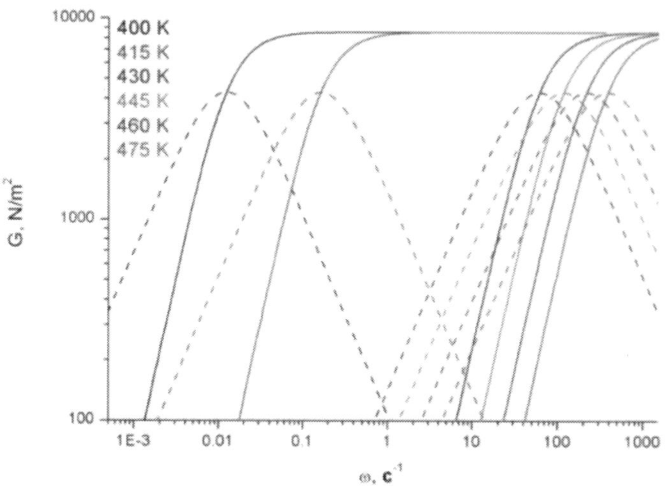

Fig.5. The dependencies of storage (G') and loss moduli (G'') on frequency at various temperatures.

To study the concentration dependence of the moduli, assume T = 445 K, and, once again $M_w = 40000 g/mol$. Then, depending on the composition of the blend one has the following results for the high frequency storage modulus, the viscosity, and the relaxation time are given in Table 4.

Table 4. The values of G, η_0 and τ at various concentrations of the components.

α_{PEO},%	0	20	40	60	80	100
G,N/m^2	8000	8240	8570	9050	9810	11200
η_0, N·sec/m^2	130.4	125.0	146.5	245.8	447.7	419.3
T, sec	0.00815	0.00758	0.00855	0.0136	0.0228	0.0187

The corresponding frequency dependences of the storage and loss moduli are presented in Figures 6 and 7. Storage moduli are plotted in full lines, and loss moduli are in dashed lines.

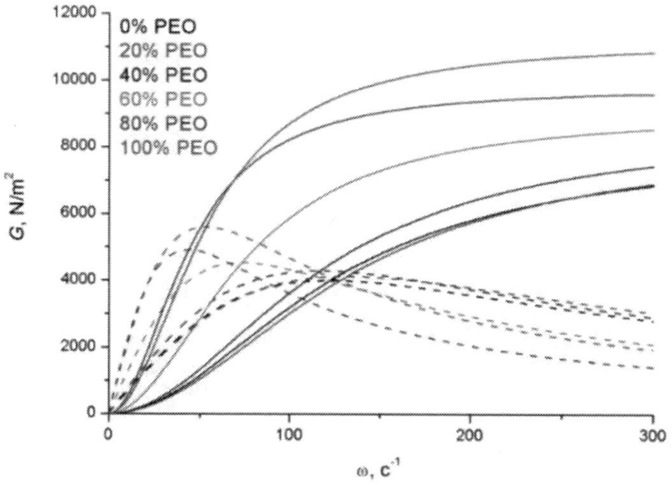

Fig.6.The dependencies of storage (G') and loss moduli (G'') on frequency at various contents of PEO.

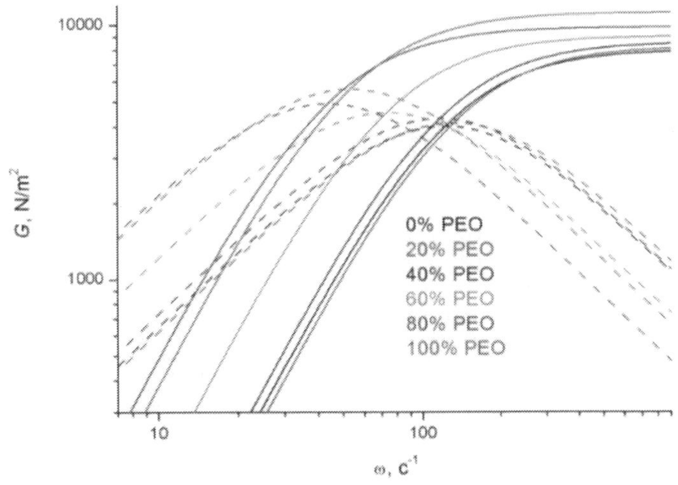

Fig.7. The dependencies of storage (G') and loss moduli (G'') on frequency at various contents of PEO.

Conclusions

The possibility of calculation of the modulus of elasticity at high frequency has been demonstrated. All the physical parameters we need for calculations can be predicted by the software Cascade (INEOS RAS). The dependencies of modulus of elasticity on both the molar and weight part possess various forms depending on the Van-der-Waals volume of the components. The proposed calculation scheme may be useful for predicting the rheological properties of polymeric materials and the search for optimal process conditions of production of building materials.

References

1. T. Matseevich, A. Askadskii, M. Popova, O. Serenko, S. Qi, Appl. Mech. Mater. (to be published).
2. `A.A. Askadskiy, V.I. Kondrashenko, *Computer Polymer Materials* (Atomic and molecular level, Moscow, 1999)
3. A.A. Askadskii, *Computational materials science of polymers* (Cambridge International Silence Publishing, Cambridge, 2003)
4. A.A. Askadskii, A.P. Hohlov, *Introduction to Physics and chemistry of polymers* (Moscow, 2009)
5. A.A. Askadskii, M.N. Popova, V.I. Kondrashenko, *Physical chemistry of polymeric materials and methods of their study* (Moscow, 2015)
6. J. Bicerano, *Prediction of Polymer Properties* (Marcel Dekker Inc., New-York, 1996).

Automated management of engineering infrastructure of pools of different function

Lev Shirokov[1], *Pavel* Chelyshkov[1], and *Evgeny* Romanenko[1,*]
[1]Moscow State University of Civil Engineering, 129337, Moscow, Yaroslavskoe shosse, 26, Russia

Abstract. The main problems of operation of locations with the difficult air-and-water environment are considered in the article. Questions of maintenance of parameters of a microclimate in the pool are considered. Need of usage the control system of a microclimate, its efficiency, profitability and social effect of its implementation is described. The mathematical model of the thermal mode of location is constructed. Process of regulation of indoor temperature of the pool is considered.

1 Introduction

In the modern world, an important role in processes of people's activity is played by pools. They satisfy their requirements during rest, in everyday life, positively influencing their health. At the same time, pools are important also for an industrial application and are included in the construction industry. Studying of parameters influence of a microclimate on construction materials and designs is very important for various tasks of a construction. Pools as specialized grounds for testing of construction materials and designs are for this purpose demanded in waters are various environment conditions, including impact conditions directly. Effective use of pools provides a need for fixed maintenance for them of certain hydroclimatic parameters. Treat such parameters: humidity and air temperature, water temperature, chemical composition of water and air [1-2]. In specialized grounds or cameras for testing construction materials and designs, it is necessary to create and support still certain tasks in parameters of aggression of water and air environments. For ensuring the required parameters of environmental condition of pools, these objects need to be equipped with the corresponding engineering systems. Systems of water treatment, heating, ventilation, etc. concern to them. An important problem of these systems is not only stabilization of parameters, but also the solution of questions of optimization and adaptation in process of management of engineering systems in the conditions of numerous external and internal indignations [3-4]. The pool is a hydroclimatic object with difficult air-and-water indoor environment. Different types of pools differ to destination, on structure, on structures and pickings of engineering systems. In general, pools can be divided, as shown in fig. 1, into set of.

* Corresponding author: roni1968@yandex.ru

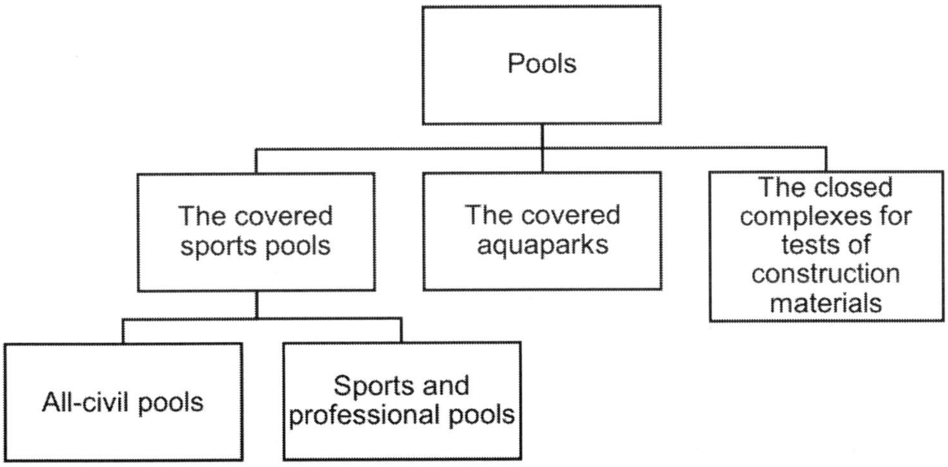

Fig. 1. The block diagram of pools on tasks

Constructions, climatic parameters and concentration of aggressive substances it is supported according to the requests made by technologists individually on each test. Besides, unlike swimming pools or aqua parks as from the point of view of temperature, so from the point of view of humidity and aggression of the environment the range of the supported parameters in a complex of tests is wider.

Systems of water treatment and heat supply are responsible for maintenance of water temperature in bowls of pools. The chemical composition of water in swimming pools and aqua parks is regulated by means of filters of rough and thin cleaning, and automatic systems of disinfecting [5-8]. In a complex for test of construction materials and designs, change of aggression of the environment is provided by the special installations dosing aggressive substances, spraying them in air or adding to water.

The wide range of the engineering systems applied on hydroclimatic objects makes increased requirements to a control system [4]. In complexes for tests, it is very important to support precisely the set parameters and to change them in time according to the set programs [7]. In aquaparks and sports pools, the problem of stabilization of climatic parameters is important. Along with the most important problem of providing the technological modes in all views of pools, it is necessary to resolve optimization issues. It is explained by high power consumption of the considered objects, large-scale consumption of resources and therefore it is important to economic expenditure of these resources to pay special attention.

Effective approach to creation of a management system capable to solve the problems considered above, synthesis of the integrated management system is. Implementation of the integrated adaptive systems provides need of development of the corresponding software and hardware complexes [8].

2 Creation of mathematical model of the thermal mode of the room

For synthesis of the integrated adaptive control system, it is necessary to conduct previously a complex research of an object, including making mathematical models of the operated processes.

Tasks of automation of maintenance of temperature conditions of air and water environments, which are determined by the current thermal modes in them, are most urgent for all types of pools.

The mathematical model of the thermal mode of rooms of [9-13] various pools can be generally provided by system of the equations:

$$\begin{cases} \sum_i G_i + \sum_j Q_j = 0 \\ \sum_i M_i + \sum_j M_j = 0 \end{cases} \quad (1)$$

where

Q_i – The convective heat transferred to internal air from internal surfaces of the protections and the surfaces of the equipment washed by this air;

Q_j – The convective heat which is directly transferred to room air, for example, from heaters;

M_i – Air streams through the protecting designs (an ex-filtration and an infiltration)

M_j – The air streams, which are directly transferred to the room or deleted from him.

In system of the equations (1) the first equation - the equation of thermal balance of internal air and the second is the equation of air balance of the room.

For providing in the pool of the set thermal mode we will define amount of heat of Q_{HEAT}, which needs to be brought to the room a ratio from the equation of thermal balance:

$$Q_{HEAT} = Q_{WALL} + Q_{INF} - Q_{IN.H.} = Q_{ALL} - Q_{IN.H.} = \frac{\mu+1}{\mu} Q_{INF} - Q_{IN.H.} \quad (2)$$

where

Q_{WALL} – Losses of heat through the protecting designs due to heat conductivity, W;

Q_{INF} – Losses of heat at the expense of an infiltration, W;

$Q_{IN.H.}$ – Internal thermal emissions, W;

Q_{ALL} – Total losses of heat, W;

$\mu = \frac{Q_{INF}}{Q_{WALL}}$ – infiltration coefficient.

Internal thermal emissions include:

$$Q_{IN.H.} = Q_{PEP} + Q_{COM} + Q_{EQUIP} + Q_{MAT} + Q_{RAD} + Q_{TP} \quad (3)$$

where

Q_{PEP}– The thermal emissions the people having a rest in the pool considering intensity of rest in water, W;

Q_{COM}– Thermal emissions communications and from the surfaces of the equipment, W;

Q_{EQUIP}– Thermal emissions the electric equipment, W which is defined with the power of the equipment, coefficient of her use and simultaneity of his use, and a share of transition of electric energy to thermal.

Q_{MAT}– Thermal emissions heated materials and products, W;

Q_{RAD}– Warmly from the solar radiation getting into the room, W;

Q_{TP}– Thermal emissions at technological processes (moisture condensation, exothermic chemical reactions, etc.), W.

At some technological processes, additional heatlosses can be created. It, in particular, occurs at liquid evaporation, which are considered in balance with a minus sign [12-17].

We will determine amount of heat spent for ventilation by a formula:

$$Q_{VENT} = G_{SP} * C_{SP} * (T_{SP} - T_0) \quad (4)$$

where

G_{SP}– Mass of affluent air, kg/h;

C_{SP}– Specific heat of air, (W*h/kg*°C);

T_{SP}, T_0– Respectively temperature of affluent and external air, °C.

For temperature of affluent air, we will write down:

$$T_{SP} = T_{EX} + \frac{Q_{SP}}{G_{SP} * C_{SP}} \tag{5}$$

where

Q_{SP}– Amount of heat compensated by system of air heating, W;

T_{EX}– Air temperature, leaving the room, °C.

For determination of mass of affluent air, we will accept a formula:

$$G_{SP} = V * \rho * \lambda_V + G_{SP}^{+} \tag{6}$$

where

V– Volume of the building, cbm.;

ρ – Density of air, kg/cbm.;

λ_V – Ventilation rate, h^{-1};

G_{SP}^{+} – The mass of the external air coming to the room at the expense of an infiltration, kg/h:

$$G_{SP}^{+} = \sum_{i=1}^{n} G_{SP.i}^{+} \tag{7}$$

Within, the objects considered in this article we would note that in grounds for test of building constructions, the room for carrying out test is configured by the principle of a thermos. Therefore, the thermal balance depends entirely on heat inflows of systems of ventilation and heating which cope automatic system of regulation, proceeding from the requirements specified by technologists in the application for test.

In swimming pools and aqua parks the main sources of heat receipt are systems of heating and ventilation, heat receipt from people and water in a bowl of the pool, an electric lighting. The main heat losses are connected with evaporation of water from the surface of the pool and roundabout paths, loss through the protecting designs and a glazing, and also from exhaust system of ventilation [16-18]. For minimization of losses of heat through exhaust system, it is equipped with heat utilizers, which transfer a part of heat (to 60-80%) of exhaust air to affluent, reducing thereby load of system of heat supply and increasing profitability of system.

The corresponding equations of dynamics defining communications between input and output parameters are necessary for design of system of automatic control of temperature of the air environment of rooms of pools. In the case under consideration values of adjustable temperatures of rooms, and entrance – the brought heat streams are output. For rooms of pools we will consider them as uniform a tel. In this case is accepted that distributions of temperatures in these rooms is provided identical in all points [18-20].

The thermal energy of Q coming to the room is spent as was considered above, on heating of the room Q_{ROOM} and on a covering of heatlosses through the protecting designsQ_{WALL}, i.e.:

$$Q = Q_{ROOM} + Q_{WALL} \tag{8}$$

For an infinitesimal interval of time $d\tau$, to the room energy will be brought $Qd\tau$, spent 1) on heating of $Q_1 d\tau$ of the placement on dt of degrees:

$$Q_1 d\tau = G_{ROOM} * c_{ROOM} dt \tag{9}$$

and on a power consumption $Q_2 d\tau$ covering through the protecting designs

$$Q_2 d\tau = k_{WALL} * F_{WALL} * \Delta t_A d\tau \tag{10}$$

Then the differential equation of thermal balance of the warmed room of the pool can be written down in a look

$$Qd\tau = Q_1 d\tau + Q_2 d\tau$$

or

$$Qd\tau = G_{ПОМ} * c_{ПОМ} dt + k_{ОГР} * F_{ОГР} * \Delta t_B d\tau \tag{11}$$

where

F_{WALL}– The surface area of the protecting structures of the pool;
τ – Time;
G_{ROOM} – Mass of the room of the pool;
c_{ROOM} – Specific heat of material of the room;
k_{WALL} – Coefficient of a heat transfer of the protecting structures of the pool;
Δt_A – Thermal head, $\Delta t_A = t_A - t_O$, t_A – air temperature in the pool; t_O – outside temperature.

Based on the received equation (11) it is possible to write down:

$$Q = G_{ROOM} * c_{ROOM} \frac{dt}{d\tau} + k_{WALL} * F_{WALL} * \Delta t_A$$

or

$$\frac{Q}{k_{WALL} * F_{WALL}} = \frac{G_{ROOM} * c_{ROOM}}{k_{WALL} * F_{WALL}} * \frac{dt}{d\tau} + \Delta t_A.$$

Entering designation

$$T_{ROOM} = \frac{G_{ROOM} * c_{ROOM}}{k_{WALL} * F_{WALL}},$$

for the last equation we will write down

$$\frac{Q}{k_{WALL} * F_{WALL}} = T_{ROOM} * \frac{dt}{d\tau} + t_A - t_O,$$

or

$$\frac{Q}{k_{ОГР} * F_{ОГР}} + t_H = T_{ПОМ} * \frac{dt}{d\tau} + t_B, \tag{12}$$

where

Q – The thermal energy coming to the room;
T_{ROOM} – Constant of time of the warmed room,

In the equation (12) we will divide both parts into value t_A. As a result, we will receive

$$\frac{Q}{k_{WALL} * F_{WALL} * t_A} + \frac{t_O}{t_A} = \frac{T_{ROOM}}{t_A} T_{ROOM} * \frac{dt}{d\tau} + 1. \tag{13}$$

From the equation (13), passing into the area of complex variable p, and designating

$$K = \frac{Q}{k_{WALL} * F_{WALL}} + t_O,$$

Let us receive transfer function of the warmed room of the pool on the channel amount of the brought heat – internal temperature indoors.

$$W(p)_{ROOM} = \frac{K}{T_{ROOM} * p + 1}. \tag{15}$$

Similarly, considering multiple layers of wall designs, generally it is possible to write down

$$W(p)_{\Sigma} = \prod_{i=1}^{i=n} \frac{K_i}{T_{ROOM,i} * p + 1}, \tag{16}$$

where n – quantity of layers of wall structures of the room of the pool.

3 Conclusion

Classification of pools of different function as objects with the difficult air-and-water environment is given in the article. Hydroclimatic parameters, which need to be controlled in rooms, including sanitary and hygienic requirements to parameters of a microclimate of the main rooms of the closed swimming pools, are considered. Problems of operation of the engineering systems providing maintenance of these parameters are considered. Systems of ventilation, heating, water treatment, heat supply, etc. concern to them. The stated mathematical model generally presented by expression (16) can be used for design of system of automatic control of temperature condition in the pool. In mathematical model, the main technological and design features of the pool as object of automation, which have to be considered by development of systems of automatic control, are considered. At the same time, it is necessary to take into account that the thermal state is influenced by losses of heat through ceiling overlapping that is considered in model (16), but also heat losses through the opened doors, influence of mass of water of a bowl of the pool and visitors of the pool. These questions have to be considered independently and add cumulative model of the considered object.

References

1. V.F. Drozdov, *Otoplenie i ventilyaciya* (Vysshaya shkola, Moscow, 1984)
2. B.V. Barkalov, H.H. Pavlov, S.S. Amirdzhanov, *Vnutrennie sanitarno-tekhnicheskie ustrojstva* (Strojizdat, Moscow, 1992)
3. Yu. Ya. Kuvshinov, Stroitel'stvo i arhitektura **4**, 13-16 (1993)
4. A. Volkov, P. Chelyshkov, A. Sedov, Applied Mechanics and Materials **409-410,** 630-633 (2013)
5. A. Volkov, A. Sedov, P. Chelyshkov, Applied Mechanics and Materials **409-410,** 630-633 (2013)
6. A. Volkov, A. Sedov, P. Chelyshkov, Engineering **1**, 27-29 (2014)
7. A. Volkov, A. Sedov, P. Chelyshkov, Procedia Engineering **2**, 17-21 (2014)
8. A. Volkov, A.S edov, P. Chelyshkov, A. Doroshenko, Applied Mechanics and Materials **580-583**, 3231-3233 (2014)
9. A. Volkov, A. Sedov, P. Chelyshkov, E. Kulikova, Mechanics and Materials **584-586**, 753-756 (2014)
10. A. Volkov, A. Sedov, P. Chelyshkov, E. Kulikova, Mechanics and Materials **584-586**, 757-760 (2014)
11. A. Volkov, A. Sedov, P. Chelyshkov, E. Kulikova, Mechanics and Materials **584-586**, 760-763 (2014)
12. A. Volkov, A. Sedov, P. Chelyshkov, International Journal of Applied Engineering Research **10 (22)**, 43106-43107 (2015)
13. A. Volkov, P. Chelyshkov, A. Sedov, Robotics and Automation **1**, 1193-1196 (2015)
14. A. Volkov, P. Chelyshkov, A. Sedov, Robotics and Automation **1**, 1196-1199 (2015)
15. A. Volkov, A. Sedov, P. Chelyshkov, D. Lysenko, A. Doroshenko, International Journal of Applied Engineering Research **10 (22)**, 43269-43272 (2015)
16. A. Volkov, A. Sedov, P. Chelyshkov, E. Radzievskiy, International Journal of Applied Engineering Research **10 (22)**, 43273-43276 (2015)
17. A. Volkov, A. Sedov, P. Chelyshkov, B. Titarenko, G. Malyha, E. Krylov, Research Journal of Pharmaceutical, Biological and Chemical Sciences **7(3),** 2416 (2016)
18. A. Volkov, A. Sedov, P. Chelyshkov, A. Pavlov, L. Kievskiy, International Journal of Applied Engineering Research **11 (3)**, 1645-1648 (2016)
19. A. Sedov, P. Chelyshkov, A. Afanasev, M. Vainstein, Y. Vilman, Y. Grossman, Research Journal of Pharmaceutical, Biological and Chemical Sciences **7(2),** 1046 (2016)
20. A. Volkov, P. Chelyshkov, D. Lysenko, Theoretical Foundation of Civil Engineering, (2011)

MATEC Web of Conferences 86, 04063 (2016) DOI: 10.1051/matecconf/20168604063
IPICSE-2016

The analysis of the functionality of modern systems, methods and scheduling tools

Ivan Abramov[1], *Tatiana* Poznakhirko[1*], *Alexandr* Sergeev[1]

[1] Moscow State University of Civil Engineering, 129337, Moscow, Yaroslavskoye sh., 26, Russia

Abstract. Calendar planning is a key tool for efficient management applied in many industries: power, oil & gas, metallurgy, and construction. As a result of the growing complexity of projects and arising need for improvement of their efficiency, a large number of software tools for high-quality calendar planning appear. Construction companies are facing the challenge of optimum selection of such tools (programs) for distribution of limited resources in time.The article provides analysis of the main software packages and their capabilities enabling improvement of project implementation efficiency.

1 Introduction

The method of area development by building complexes is used in Moscow nowadays. Also, the building of the central part of the city is actual too. Often it includes reconstruction and restoration adapted for housing or commerce of existing buildings. The increasing competition is observed in modern conditions, so the technical customer must organize the building production effectively. It will enable to realize the maximum benefits at minimum costs of time and resources. Besides that, the failure of the construction period is a serious problem in construction today. It leads to increasing costs of the project implementation and consequently to the profit's loss of construction participants. The construction of complex of objects requires long-term planning and so appears the problems of formation of long-term construction flows, which includes the issue about the optimization of priority development of objects of the construction flow. Therefore there is a need for tools, that allows to select the optimal sequence of construction, which is resistant to distributing factors at the same time. Currently the solution of this issue is only at the level of methodological foundations, which are quite general in nature. Therefore, there is a need to solve the problem, which is focused on modern instrumental-based scheduling. [1]

Scheduling refers to the fundamental principles of management and it is used in many industries, such as: energy, oil and gas sector, metallurgy, construction. The reason for the development of scheduling is the increasing complexity and the need in upgrading of

* Corresponding author: top1977@yandex.ru

effectiveness of their implementation. The task of the optimal allocation of scare resources in time is faced before the construction companies.[6]

Investment-construction project is a complex system, which moves from one condition to another through the life cycle. These conditions are called as phases of project. Each phase has it's own characteristics and risks. That's why the scheduling problems differs for each phase. [2]

The initial phase (pre-investment phase) – is characterized by the fact, that there is no accurate information about the project. Only aggregated indicators can be known, and only the general conception of the project can be presented. In this phase the scheduling solves the following tasks: the assessment of the project's effectiveness and feasibility, the timing of the project and it's cost, the drafting of the investment process.[4]

The investment stage (the main stage of project) includes the negotiation and conclusion of contracts, design, construction, marketing. Scheduling allows to obtain the optimal tender offer, when the trades are carried. It allows to solve the following tasks: development of the condensed schedule of works, the schedule of funding, the statements in need of human and material resources [6]

2 The analysis of the functionality of modern systems

In terms of planning it can be divided into two stages: project development and control ever it's execution. The tasks of the first stage includes defining the scope of work, development of work and cost calendars, development of timetables, resource assignment, calculation and optimization of the target dates, making schedules of workforce and schedules of machines and mechanisms, the formation of the schedule of material procurement and material's delivery, defining the costs on different stages and risk assessment. [2]

In accordance with the tasks, which are lying ahead scheduling nowadays, there are many modern systems, which should have the following capabilities:

- The creation of organizational-technological scheme;
- Mean of calculating, according to the method of the critical path;
- Means of resource scheduling;
- Means of monitoring of the project's implementation process;
- Tools for creating reports and graphical presentation of the project structure;
- Means of cost analysis.

Means of creating of the organizational-technological scheme includes:

- The description of the global planning parameters of the project;
- The description of the structure of complex of works;
- The support of the project calendar.

Means of resources and costs planning for individual works and about the whole project includes:

- The support of resource calendars;
- Resource assignment;
- Scheduling with limited resources;
- Maintain a list of resources and cost items.

Means of verification of the execution process includes:

- Fixation of planned parameters of the project schedule;
- Input real indicators of assigned tasks;
- Comparison of planned and actual indicators;
- Forecasting the performance of future works.

Tools for creating reports and graphical representation of the project structure includes:

- The Gantt chart;
- The network diagram;

- PERT chart (Program Evaluation and Review Technique).

In ninety fifty eight was created (1958) a technique of PERT-method analysis of time for each individual task and for minimum time of the project at whole for countering the uncertainty.

The most common part of PERT is the critical path method, which was appeared at the end of twentieth century. It's novelty lies in the fact, that the timing of the project's phases are cut by fifty per cent, which means that fifty per cent probability of their achieving are established. Also the possibility of the parallel using of resources excludes, and with help of the buffers of resources the critical chain protects from non-performance of terms. However, if you turn to major projects, you will see, that method failed, because the terms were exceeded and the budget increased. In result, the shortcomings were revealed. For example, the project manager, who knows, that the timing of the project is artificially increased (resource buffers), allows you to waste time, and as a result, the failure of the timing happens when the problem occurs.[4]

Methods of resource scheduling allow us estimate the project's completion time, project's costs and build the optimal resource schedule and control the state of affairs.

In the basis of the first programs for project management, which have appeared about forty years ago, the algorithms of the network planning and calculating of time parameters of the project on the critical path method were lying. These programs allow to calculate early and late stages of the project and display them on the Gantt chart. In the future, with help of the development of information technologies, the systems were supplemented with resource and cost planning and with means of monitoring the process of works.[5]

Currently, there are many software packages on the market, that are divided into professional and "desktop" (non-professional) The difference between them consists in the complexity in use, because the professional programs requires time-consuming, but they are able to provide more flexible results while the non-professional can be used single. You can obtain enough informative result without spending a lot of time and effort. [2]

In addition to universal programs, the additional software packages became common in the last time. They allow to approach to the project in more individually way. They solve such tasks as:

- Risk analysis, planning the best risk management strategy (PalisadeCorp@RISK);
- Scheduling with accordance with the objectives, calculation of wages (HMSSoftwareTimeControl);
- Scheduling with limited resources (ParsifalSystemInc.BestSchedule for Project);
- The integration of management system into the corporate systems (MarinReseachInc.Project Gateway, Time Line Solutions Corp. Project Management Integrator etc.).

The most common universal professional software packages, which costs more than one thousand dollars, are: Artemis Project View, Open Plan Professional, Primavera Project Planner.

Among non-professional universal software packages the most frequently are: OpenPlanDesctop, Project 98, ProjectScheduler, SuperProject, Time Line.

The result of the choice of a program depends on the outcome, which you want to achieve. Let's consider the main features of individual software packages.

Open Plan Professional belongs to a category of professional software packages and accordingly has powerful tools for resource and cost planning. In this regard it is recommended to use for large projects, which requires detailed planning and integration with other automated system in the organization. The program allows to create the hierarchical structure of resources and allows to perform the automatically assign of resources for the tasks in accordance with their qualification. Also, here is an opportunity to analyze risks, access their likelihood and consequences according to possible delays in delivery of the materials. Also you can analyze the budget overruns. [3]

TimeLine belongs to the category of "desktop", but it has a fairly powerful tools of working with resources, including means of alignment of overload resources, the possibility of describing a specific schedule of resource's work. The program provides the opportunity to add any indicators and thus to set the model of the project in accordance with the required needs. However, TimeLine doesn't have enough flexibility, because it doesn't allow to generate the hierarchical structure of resources, in contrast to the professional program, which was described before. [3]

Microsoft Office Project represents the best set of tools to manage the projects. This program is one of the most common on the market today. Microsoft project allows to managers the possibility to control the scheduling and resources in dynamic, monitor the project status and analyze the data. Besides that, with help of integration of this program with Microsoft Exel or Outlook, it is possible to convert tasks into schedules.

Primavera Project Planner is a professional software product. It used for large projects. Besides the ability of the monitoring resources and performance control, the program allows to manage a group of projects, distribute the information in a multiuser environment and manage the complex projects. Also, it is possible to conduct the analysis "what-if", enter and update data with help of the PERT view and with calendar and network schedules. The program is easy to use, because it involves the integration into any corporate structure. [3]

DefSmetaLight is used in small organizations and prefers the following tasks: drafting or reviewing of estimates on the base of estimate-normative base, which was adopted in two thousand one; also with help of this software product you can perform the operational planning of financial costs and the consumption/supply of resources during the project execution. Besides this program, there are similar software packages.

Microsoft Enterprise Project Management was designed for using in large companies. Due to the fact that the information is stored in a single database, you can link the projects between themselves; plan the supply of resources with giving the fact that they can be used in other projects. Moreover, you can control the implementation of the project with help of this program. Also, it allows to users to enter extra information and allows to manage the document circulation on different levels. Many companies besides the independent software packages produce the scheduling programs, which are implemented only in certain software applications. For example, there is a set of components like TMSPlannerComponents, which let to implement scheduling functions in the applications, which are developed with help of means likeDelphyVCL+VCL.NET. [2]

3 Conclusion

Many companies use tools of scheduling, which are oriented on the work in the internal LAN. One scheduling programs operate with resources, like personnel equipment, the other makes charts. One of these products is a program WelcomHome. It provides interaction of all project participants and helps to plan and control projects in real time.

So, as a result of conducted analysis, the following conclusions can be done:

1. There are many modern ways, methods and means of scheduling, which allow to solve various tasks and which give to the methodology of scheduling the constant growth and improvement (this is an attribute of the dynamic).

2. There are many universal software systems, single programs and applications which have been already designed. They allow to solve general and specific scheduling tasks.

3. Each project has it's own specific and there is a need to develop algorithms, which can solve specialized tasks with possibility of including them into integrated systems.

References

1. V. I. Voropayev, J. D. Gelrud, Project and program management **1**, 62-71 (2014)
2. T. F. Morozova, N. N. Bokovaya, Xia Tszyamin, Construction of unique buildings and structures **1(6)**, 36-46 (2013)
3. V. A. Pavlova, M. V. Timofeeva, International research journal **7-3(14)**, 49 (2013)
4. A. A. Lapidus, M. N. Yershov, P. P. Oleynik, O. P. Lyang, A. V. Ishin, V. I. Telichenko, D. K. Tumanov, O. A. Feldman, Construction Technology and Organization **4**, 6-11 (2014)
5. A. A. Lapidus, T. K. Bidov, International Journal of Applied Engineering Research **11(8)**, 5972-5974 (2016)
6. P. P. Oleynik, V. A. Grigoryev, Construction Technology and Organization **2(7)**, 42-44 (2014)

Temperature conditions of hothouses during a warm season

Valery Bodrov [1,*], *Mikhail* Bodrov[1], *Kirill* Lushin[2]

[1] Federal State Budgetary Educational Institution of Higher Professional Education « Nizhny Novgorod State University of Architecture and Civil Engineering (NNSUACE)», Russia
[2] Moscow State University of Civil Engineering, Russia

Abstract. On the basis of justification of the heat and mass transfer processes inside the hothouses during a warm season authors developed methods and means that control dynamics of temperature and humidity parameters and air conditions with the help of complex systems of removal of overheat in the hothouses during the all year round and diurnal operations at minimum power inputs.

1 Introduction

Under the climatic conditions of sharply continental climate in most regions of Russia growing vegetables in the cold season is only in winter hothouses. Exploitation of winter hothouses in the warm season is difficult due to overheating of the air inside of them because of the increased intensity of solar radiation. Crop losses in this period can reach 50... 80 %, and sometimes there is a destruction of plants.

The authors have developed a complex system of removal of overheat in the hothouses in the warm season. Given technique consists in use for the longest period of natural (passive) systems providing microclimate parameters (transoms, technological openings, aeration shaft) and for short-term inclusion of active systems, their key elements are the systems of water aerosol cooling (SWAC). While the system of water aerosol cooling the water is spraying in the whole volume of the hothouse through the nozzles, and during the adiabatic evaporation water lowers the temperature of the internal air.

We studied six typical modes of a complex system of heat removal to ensure acceptable temperature parameters of air in hothouses: mode I – an organized natural ventilation; mode II – co-operation of the systems of organized natural ventilation and active aeration; mode III – co-operation of the mechanical ventilation and natural ventilation; mode IV – co-operation of the SWAC and natural ventilation; mode V – co-operation of the SWAC, natural ventilation and active aeration; mode VI – co-operation of the SWAC, natural ventilation and mechanical ventilation.

Dynamics of changes in air temperature in an unventilated, with closed openings and transoms, hothouse with plants in the warm season is shown on the *i–d*-diagram of humid air (fig. 1).

* Corresponding author: unirs@nngasu.ru

In the early morning hours the air temperature in the hothouse with plants in the phase of cultivated biomass t^{m}_{int} is close to or slightly higher than the outside air temperature t^{m}_{ext} (point H^{m}), relative air humidity $\varphi^{m}_{int} \rightarrow 100$ % (point B^{m}). On the inner surface of translucent enclosures of hothouses we can observe condensate. In this period of the day the temperature t^{m}_{int} can be below, equal to or higher than the design temperature, shown in figure 1 by the hatched area (1) with temperature gradient Δt_{calc}.

The point B^{d} with the parameters t^{d}_{int}, φ^{d}_{int}, the moisture content d^{d}_{int}, g / kg of dry air, enthalpy of i^{d}_{int}, kJ / kg, shows the air condition in the hothouse during daylight hours in the absence of ventilation (area 3). The air temperature t^{d}_{int} is increased by solar radiation and significantly exceeds the allowable values. Such temperature conditions in hothouses arise in the climate conditions of central part of Russia, starting from mid-April.

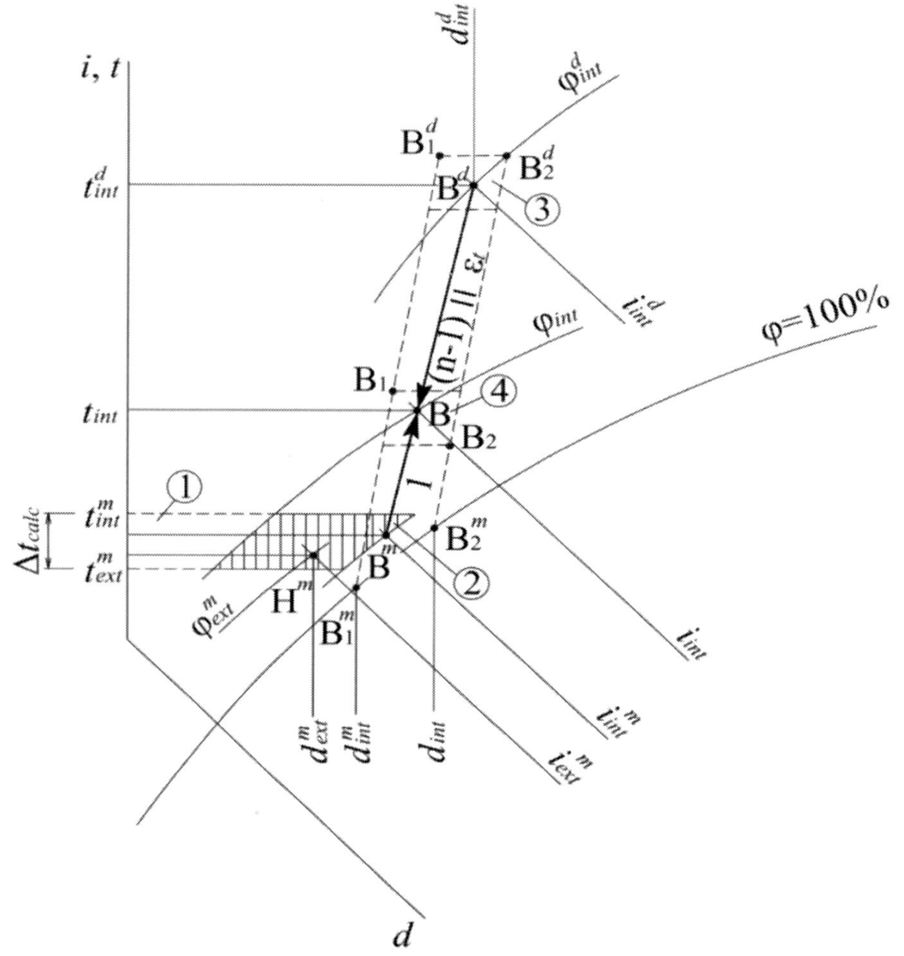

Fig. 1. The area of self-regulation of the temperature in the hothouse when the untreated air supply: *1* – the area of technological temperatures in the hothouse; *2* – the area of the temperatures during the morning hours; *3* – the area of maximum temperatures in the absence of ventilation; *4* –the area of possible temperatures when the outdoor air ventilation

In the first approximation consider that the parameters of internal air (point *B*) lie on the line $B^m - B^d$, that is the geometric position of the points of the air state in the unventilated hothouses.

The first stage of reducing the air temperature is organized natural ventilation of the hothouse by opening transoms (mode *I*). A further reduction of the air temperature in the hothouse is achieved by opening the technological openings and exhaust aeration shafts (mode *II*). Then additionally turn on the exhaust fan (if any), located in the exhaust aeration shaft (mode *III*). The position of point *B* on the line $B^m - B^d$ during the light period of the day depends on the ventilation rate n, h^{-1}, the ability of biomass to self-regulate the temperature around the plant. A ventilation rate changes from $n = 5 ... 10$ h^{-1} under the natural air-change to $n = 25...30$ h^{-1} under the mechanical ventilation [1]. Therefore, the location of the point *B* (on the *i-d*-diagram of the moist air) will strive along the line $B^m - B^d$ to the point B^m in a ratio of 1: (n – 1).

So the area, bounded by the points $B_1^m - B_1 - B_1^d - B_2^d - B_2 - B_2^m$, characterizes the borders of temperature and humidity condition of the air in the daytime hours in the hothouse during the warm period of the year at the expense of natural and mechanical ventilation by the untreated atmospheric air and self-regulation of plants.

After exhaustion of the cooling effect of outdoor air to maintain the technological parameters of microclimate arises the need for artificial cooling of the air in the hothouses. Such a period of time usually occurs during the daylight hours, when the total solar radiation reaches 450 W/m^2 or more (from May till October).

Figure 2 shows the processes of changing the air state in the hothouse with the biomass during the operation of system of water aerosol cooling. After switching SWAC there is adiabatic cooling of the air in the volume of hothouse according i_{int} = const to $\varphi = 97...98$ % (fig. 2 a process schematically shows to $\varphi = 100$ %) from the temperature t_{int} to the temperature of the wet the roomette $t_{w.t}^{int}$, which corresponds to the parameters of the point *B*. Concurrently moisture content in the hothouse increases from d_B to d_A according to the relative humidity in the point *A* (97 ... 98 %). If the amount of the heat of solar radiation in the hothouse exceeds the amount of heat absorbed during evaporation of water aerosol droplets, so process of the air cooling is terminated at the point *C* during the temperature t_C, the relative humidity φ_C and the moisture content of air d_C.

The air temperature after its processing by SWAC to the point *A* or to the point *C* (fig. 2) may be reduced to a range of technologically necessary temperatures in the hothouse (beam *B-A*) or be above them (the beam *B–C*). Specific parameters of the air in the hothouse with volume V_h, m^3, during the ventilation rate n, h^{-1}, are determined by increment of its water content Δd, which depends on the flow rate of sprayed water $G_{s.w}$, kg/h, and air density ρ_{int}, kg/m^3:

$$\Delta d = d_C(d_A) - d_B = \frac{1000 \cdot G_{s.w}}{V_h \cdot n \cdot \rho_{int}}. \qquad (1)$$

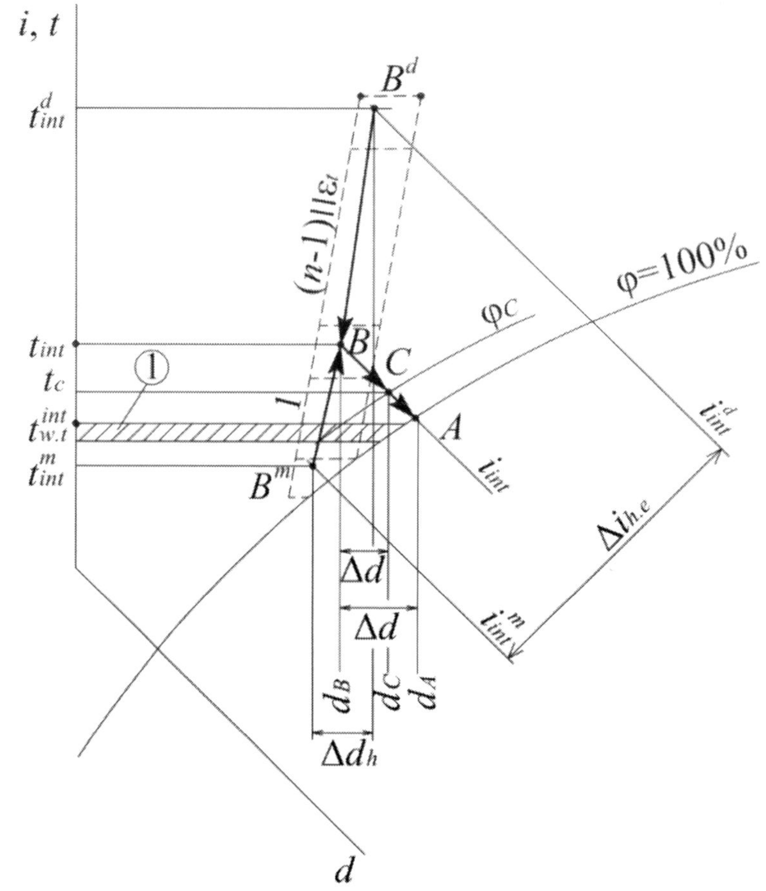

Fig. 2. The processes of changing the air state in the hothouse with the biomass:
1 – the area of technological temperatures in the hothouse

The reducing of intensity of solar radiation in the afternoon is accompanied by a decrease of internal air temperature. SWAC switched off, but the exhaust fan built into the shaft continues to operate, transoms and technological openings are opened. The parameters of internal air are transformed into the state of the point *B*, corresponding to the parameters before the activation the adiabatic cooling system. The relative humidity and moisture content of the air decrease and the temperature increases. It is this thermodynamic process is inevitable, because the heat capacity of the equipment in hothouses is negligible and plants can self-regulate the air parameters around themselves in the process of adaptive reactions during transpiration of moisture.

The process of removing the intensity of solar radiation in the day period after switching off SWAC and further regulation of the air parameters in the hothouses within a standardized technological values will be carried out by changing the position of the point *B* on the line $B^m - B^d$. These processes are possible when changing the ventilation rate n by switching - off the exhaust fan, opening - closing the technological openings and transoms. In the evening when the internal air temperature decreases to the design temperature in the hothouse and below all the technological openings and transoms are closed.

The direction of the beam of the process on *i–d*-diagram $\varepsilon_t = \Delta i_{h.e} / \Delta G_{m.r}$. The total heat emissions in the hothouse $i_{h.e}$ are determined according to the intensity of insulation heat gains. Moisture release in the hothouse $\Delta G_{m.r} = \Delta d_h V_h \rho_{int}$ are composed of secretions of plants and evaporation from soil surface. Analysis of literature data and our field studies recommend during the peak of heat load for central part of Russia take ε_t = 3700…4800 kJ/kg.

As an indicator of the efficiency of removal of overheat for a day (N = 24 h) during the period of the field studies was adopted a coefficient of probability of operation of cooling systems K^d_{pr}. Its value shows the share of the total number of hours for a day, which does not allow to exceed the internal air temperature in a hothouse relative to the design temperature:

$$K^d_{pr} = (N - m)/N, \quad (2)$$

where m – the number of hours of temperature rises for a day.

Figures 3 and 4 show generalized experimentally obtained temperatures in the hottest period of sunny day. The shaded part indicates the area of optimal temperatures of internal air for growing crops. Mode IV of operation of complex system of removal of overheat is not typical in practical terms and, therefore, has not been studied.

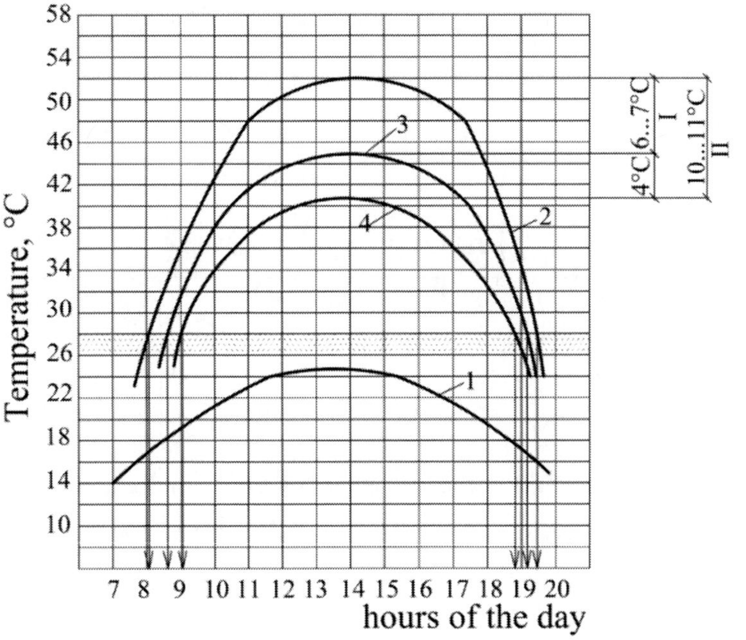

Fig. 3. Changing of air temperature in a hothouse during the period of fruiting during daylight hours: *1* – outdoor air; *2* – internal air without operation of systems of removal of overheat; *3* – operation of the systems of organized ventilation (mode *I*); *4* – co-operation of the systems of organized ventilation and active aeration (mode *II*)

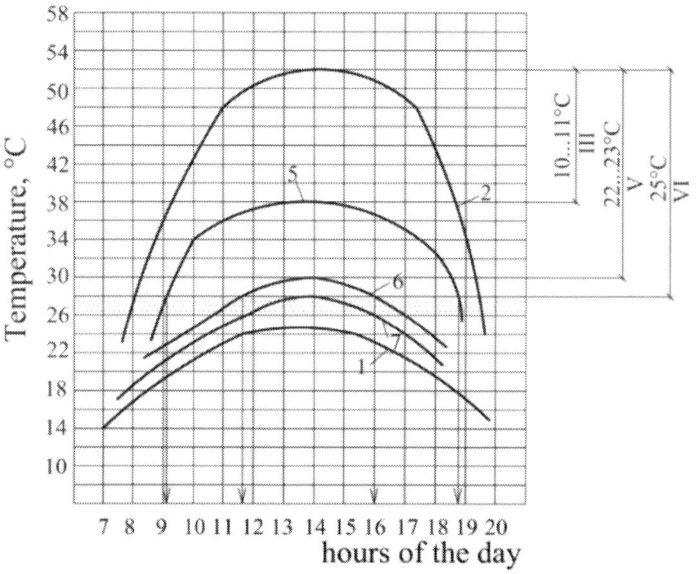

Fig. 4. Changing of air temperature in a hothouse during the period of fruiting during daylight hours: *1* – outdoor air; *2* – internal air without operation of systems of removal of overheat; *5* – co-operation of the systems of organized ventilation and mechanical ventilation (mode *III*); *6* – co-operation of the systems of organized ventilation and SWAC (mode *V*); *7* – co-operation of the systems of organized ventilation, active shaft or mechanical ventilation and SWAC (mode *VI*)

The results of determining the values of a coefficient of probability of operation of cooling systems during the day K_{pr}^{d} and quantitative indicators of reduce the temperature in each of the operating modes of the complex system of removal of overheat Δt_{int} are shown in Table 1.

Table 1. Values K_{pr}^{d} and Δt_{int}

Operating modes	I	II	III	V	VI
Δt_{int}, °C	6…7	10…11	14…15	22…23	up 28
K_{pr}^{d}	0,567	0,60	0,63	0,833	1,0

A coefficient of probability of temperature conditions in the hothouse during the warm period in the annual cycle K_{pr}^{an} is necessary to determine for the period of fruiting of tomatoes and cucumbers, which lasts from April 1st to August 1st. The values of the coefficient are calculated by the formula:

$$K_{pr}^{an} = 1 - \left(N^{an} - m^{an}\right) / N^{an}. \tag{3}$$

The number of days of fruiting for the central part of Russia is $K_{pr}^{an} \approx 120$ days. The duration of the period of maintenance of maximum permissible internal air temperature ($t_{int} \leq 28$ °C) m^{an}, days, in the hothouses from April to July, depending on the operation modes of systems of maintenance of parameters of technological microclimate calculated by figure 5 that are obtained as a result of field studies. Changing the temperature t_{int} in the hothouse under the co-operation of SWAC with active shaft aeration and mechanical ventilation (mode *VI*) can be traced by the curve *6*. The maximum decrease of the air temperature in the hothouse during operation of complex system is 23…25 °C. It is almost always provides the necessary temperature regime for all the phenological phases of vegetables, including the fruiting stage (April - July).

Therefore, the value $K_{pr}^{an} \approx 1 - (120 - 120)/120 \approx 1{,}0$.

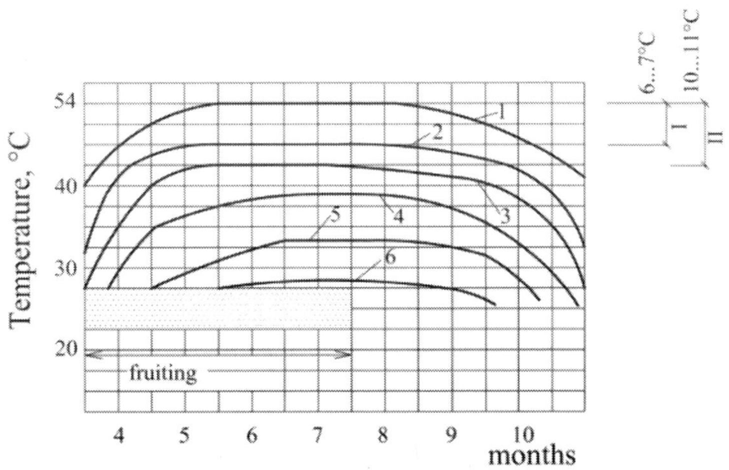

Fig. 5. Changing of air temperature in a hothouse during the period of fruiting: *1* – without operation of systems of removal of overheat; *2* – operation of systems of organized ventilation (mode *I*); *3* – co-operation of the systems of organized ventilation and active aeration (mode *II*); *4* – co-operation of the systems of organized ventilation and mechanical ventilation (mode *III*); *5* – co-operation of the systems of organized ventilation, active aeration and SWAC (mode *V*); *6* – co-operation of the systems of organized ventilation, active shaft or mechanical ventilation and SWAC (mode *VI*)

The results of determining a coefficient of probability of the operating mode for the annual cycle of fruiting period are summarized in Table 2.

Table 2. The values of K_{pr}^{an} for the period of fruiting

Operating modes of the complex system	*I*	*II*	*III*	*V*	*VI*
N^{an}, сут.	120	120	120	120	120
m^{an}, сут.	0	0	6	3,0	120
K_{pr}^{an}	0	0	0,05	0,25	1,0

Conclusion

Graphic-analytical and field studies of the thermodynamic processes in the hothouses showed that the internal air temperature during the period of maximum heat gain from solar radiation can be reduced by 23 ... 25 °C. The authors obtained quantitative values of coefficients of air temperature parameters in the hothouses during the warm period of the year for the climate of central part of Russia in the daily and annual cycles of operation of hothouses.

Reference

1. V.I. Bodrov, Nizhny Novgorod State University of Architecture and Civil Engineering, 623 (2008)

Some aspects on improvement of water resistant perfornace of gypsum binders

Ekaterina Fomina [1], *Elena* Voitovich[1,2,*], *Oleg* Brukhanov [2]

1 Belgorod State Technological University, Russia
[2] Moscow State University of Civil Engineering, Russia

Abstract. The study reports that control of water resistant and strength performance of composite binder can be accomplished by using nanostructured gypsum silicate binder and portland cement and reduction of calcium sulfate dihydrate. Formation of an arranged more dense packing in crystal matrix of binder structure can be achieved through polydispersity of nanostructured silicon component. Highly reactive silica particles participate in synthesis of microcrystalline compounds such as hydroxylellestadite and serve as nucleation sites providing an increase of surface contact area between crystal structures. Cement in the system provides conditions to synthesis of stable flat-like hydroxysulfoaluminates. Upon hydration of a binder crystals of calcium sulfate dihydrate impound to a mass of water resistant crystalhydrate new formations, type and morphology of which promotes decreasing of intercrystal porosity followed by density, strength and water resistance increase of the composite.

Introduction

Gypsum materials can be attributed to progressive construction materials thought their simplicity, feasibility and low energy demand during production.

Gypsum-based products are usually with a low mass, relatively low thermal and sound conductivity, they can be easily shaped into any architectural configuration of any color [1].

The orientation of modern technologies of construction materials production to nanosystem concept allowed the design principles of highly efficient binder composites to be developed due to application of nanodispersed modifiers [2-5].

The studies related to modification of anhydride composites with ultra- and nanodispersed admixtures were done [6]. As a main binding component and an effective nanostructured modifier can be used nanostructured silica component [7].

Nanostructure silica component (NSC) is a mineral polydispersed system of dominantly silicate composition with high concentration of active solid phase and up to 5-10% of nanodispersed particles.

NSC is produces using eco-friendly technology based on mechano-chemical synthesis of silica component in wet conditions at high concentration of solid phase, high temperature

[*] Corresponding author: e.voitovich@mail.ru

and a margin dissolution. Polydispersed particle size distribution of NSC have an impact on gypsum binder system at different scale (macro-, micro- and nanoscale).

Microparticles of NSC (1-100μm) act as a microfiller in the composite binder and promote a maximally compact defect-free structure [8].

It was found that nanodispersed particles (0,02-0,1 μm) involved to chemical interaction processes with a synthesis of buffer nanosized formations such as calcium sulfosilicates and new formations such as hydroellestadite [9, 10].

This helped to reduce portion of gypsum binder in the composite up to 30% and increase a quality of gypsum composites as well as materials for different application on their base.

Regardless of high strength the obtained binder has a low water resistance, which restricts its outdoor application as well as indoor constructions with a high humidity.

Low water resistance of gypsum binder driven by high solubility of calcium sulfate dihydrate, high permeability and wedge effect of water molecules when penetration into the intercrystalline cavities (Rehbinder effect). This leads to strength reduction and failure of gypsum materials under the water impact [11, 12].

Nevertheless, water resistant performance of nanostructured gypsum-silicate binder can be improved by incorporation of an optimal quantity of cement, which due to its hydration will contribute to a synthesis of slow-soluble phases, and cover calcium sulfate dehydrate in the binder structure, protecting it this way of water impact.

Materials and methods

The development of water resistant gypsum composite was focused on study of the effect of NSC and Portland cement concentration on physical and mechanical performance of gypsum binder according to Russian standard GOST 23789-79 "Gypsum binders. Testing procedure" [13].

Standard beams of 16×4×4 cm was prepared. Water in the 1 m^3 of gypsum mix was adjusted in order to achieve a standard flow.

X-ray diffraction of the specimens was done using diffractometer DRON-4. WorkStation with $\lambda Cu_{K\alpha 1,2}$ with Ni-filter attenuation of β-radiation components.

The full-width at half-maximum (FWHM) of these patterns was 0,05°.The time of intensity measurement of the at the scan point – 1s.

IR-spectroscopy of aluminosilicate materials and obtained cellular concrete was performed with IR-Fourier spectrometer VERTEX 70/70v.

Porosity was defined using concrete cubes with the size of 7×7×7 cm, after 28 days of storage at natural conditions and drying till a constant mass, according to the Russian standard GOST 12730.4-78 [14].

After the test, all specimens were weighted.

Experimental part

Nanostructured silica component (NSC) is a mineral silicate polydisperse binder system with high concentration of active solid phase and up to 5-10% of nanodispersed particles. The binder was produced by wet milling of quartz sand in ball mill.

The main characteristics of the nanostructured silicate binder presented in Table 1.

Table 1. Main characteristics of nanostructured silicate binder

№	Parameters	Units	Spec. limit	Test result
1	Compressive strength	MPa	≥ 3	4,5
2	Flexural strength	MPa	≥ 1,5	2,5
3	Density	kg/m^3	2000–2250	2140
4	Dispersity	%	≤ 1	0,89
5	Viscosity	Pa·s	≤ 25	16,3
6	pH	pH	7–10	8
7	Water content	%	14–20	14,5
8	Specific efficiency of natural radionuclides	Bq/kg	370	210

Gypsum binder (GB) without NSC and Portland cement was considered incorporation as a reference specimen. Percentage of NSC and Portland cement varied of 10 to 3% with pace variability of 10% and of 1 to 5% with pace of variability of 2% respectively.

Table 2.Relation of physical-and-chemical characteristics of composite binder vs. concentration of NSC and Portland cement

№	Composition, %			Setting time, sec.		Compressive strength, after 28 days, MPa	Density, kg/m^3	Malaxation coefficient
	GB	NSC	PC	Initial	Final			
1	100	-	-	535	985	10,19	1201,1	0,46
2	90	10	-	566	1082	11,88	1263,2	0,50
3	80	20	-	805	1330	13,97	1307,5	0,54
4	70	30	-	1025	1634	9,46	1274,3	0,56
5	89	10	1	571	1089	11,62	1324,5	0,55
6	19	20	1	825	1344	13,76	1330,2	0,59
7	69	30	1	1036	1628	9,89	1334,9	0,61
8	87	10	3	554	1102	12,45	1278,2	0,59
9	77	20	3	835	1372	15,16	1332,5	0,64
10	67	30	3	1039	1701	12,26	1337,3	0,68
11	85	10	5	536	1054	13,54	1298,1	0,65
12	75	20	5	834	1354	16,48	1342,4	0,70
13	65	30	5	1103	1704	15,73	1351,1	0,72

Replacement of 10% gypsum binder by NSC (mix 2, Table 2) impacted the rate of hydration followed by retardation of setting time of the composite. Further change in hydration was observed with additional incorporation of 1% of PC (mix 5, Table 2). In this case, initial setting time was retarded to 7 %, and final setting time to 11 %.

The results of physical and mechanical tests showed, reduction in gypsum content by incorporation of highly dispersed hydraulic admixture NSC and PC provides a better density of the composite from 1201 kg/m^3 to 1351 kg/m^3.

At the same time an improvement of water resistance of the composite was observed, where the maximal malaxation coefficient was 0,72 for mix 13 (Table 2), which is 1,5 times higher than that for reference specimens. The best specimens were obtained by replacing of 20-30% NSC and 5% PC (mixes 12 and 13, Table 2). The strength was increase up to 55-60% and malaxation coefficient was 0,70-0,72.

Malaxation coefficient describe water resistance of composite, that can be defined though water absorption, whereas, water absorption related to porosity and structure of material. In this regard, an analysis of an open porosity based on kinetics of water saturation (discrete method) was performed. Results of the calculations presented in Table 3.

Table 3.Porosity of gypsum systems vs. composition

Mix ID (according to Table 2)	Total porosity cm^3/g	Pore distribution by radius, %				Volume of pores with R less than 94,6 nm, cm^3/g
		<0,01 μm	0,01–0,1 μm	0,1–1 μm	>1 μm	
1	0,3065	0,6	16,6	27,5	55,3	0,02
4	0,2071	17,6	36,4	30,6	15,4	0,04
13	0,1798	20,2	41,3	26,8	11,7	0,04

Based on the data, structure of gypsum stone for reference specimen is macroporous with dominate pore radius > 1 μm. The amount of pores with radius < 0,01 μm is extremely low (mix 1, Table 3). Shift from macroporosity towards mezoporosity in the binder system 'GB-NSC-PC' was observed.

Pore space structure of nanostructured gypsum silicate binder with cement can be characterized as microcrystalline with prevalence of pores with radius 0,01-0,1 μm (about 40%) and with radius < 0,01 μm (about 20%). This caused by formation of more dense crystal packing as well as by shifting microporosity towards nanoporosity, which results in improvement of water resistance of composite.

Study of phase composition of the binder with incorporation of cement onto nanostructured gypsum silicate binder was performed with analysis of X-ray diffraction of the specimens.

Considering the fact that main formulation of the composite was presented by gypsum and NSC, the most intensive X-day diffraction peaks belong to those components. For identification of new formations a quantitative full-prof XRD using Rietveld refinement method was carried out.

As structural models for mineral components were used: gypsum $CaSO_4{\cdot}2H_2O$, bassanite $CaSO_4{\cdot}0,5H_2O$, α-quartz SiO_2, anhydride $CaSO_4$, dolomite $CaMg(CO_3)_2$, hydroxylellestadite $Ca_5(SiO_4)_3(SO_4)_3(OH)_2$, calcium hydroxysulfoaluminate $3CaO{\cdot}Al_2O_3{\cdot}CaSO_4{\cdot}12H_2O$.

Based on the obtained data, the entire hydrated system comprises interaction products of gypsum and NSC in the form of hydroxylellestadite and the interaction produts of portland cement and gypsum in the form of calcium hydroxysulfoaluminate (Fig. 1 b, c). Due to low content of portland cement (up to 5%), a small amount of calcium hydroxide is formed in the binder system, which is just enough for binding of calcium hydroxysulfoaluminate into steady low sulfate phase.

Based on IR-spectroscopy analysis, the absorption band at 900 – 1000 cm^{-1} confirms a presence of calcium hydroxysulfoaluminate in the system (Fig. 2). Variation in intensity fluctuation and some frequency shift at the range of 1100 – 1400 cm^{-1} also validate a crystal structure rearrangement of hydrate phases with the participation of ion SO. Taking into account, that deformation fluctuations of adsorbed water molecules take place in the interval of 1600 – 1700 cm^{-1}, it is possible, that the cement portion takes part in building-up of crystalhydrate new formations.

DOI: 10.1051/matecconf/20168604065

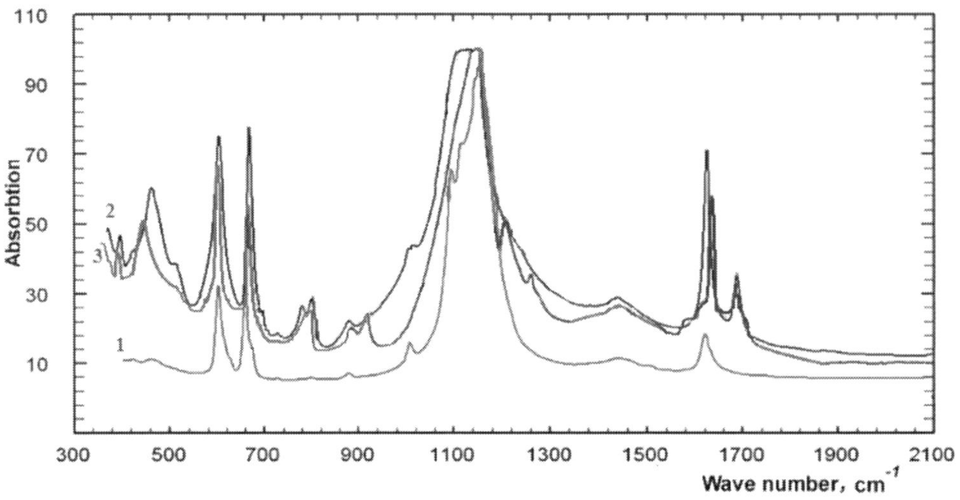

Fig. 2. IR-spectrum for: 1 – mix № 1; 2 – mix № 4; 3 – mix № 14 (see Table 2).

Comparing the IR-spectroscopy and XRD results, it can be concluded that reduction in gypsum content as well as the presence of NSC and optimal portion of portland cement in combination make it possible to synthesis of slow-soluble phases of hydroxylellestadite $Ca_5(SiO_4)_3(SO_4)_3(OH)_2$, as well as the phases of hydroxysulfoaluminates such as $3CaO \cdot Al_2O_3 \cdot CaSO_4 \cdot 12H_2O$, which boost water resistance of the binder. SEM analysis of the reference composition demonstrated, that its structure comprises bigger defected gypsum crystals with a few intergrowth contact points (Fig. 3, a). Formation of bigger crystals leads to porosity increase, which is 0,31 cm/g^3 for reference composition (Table 3).

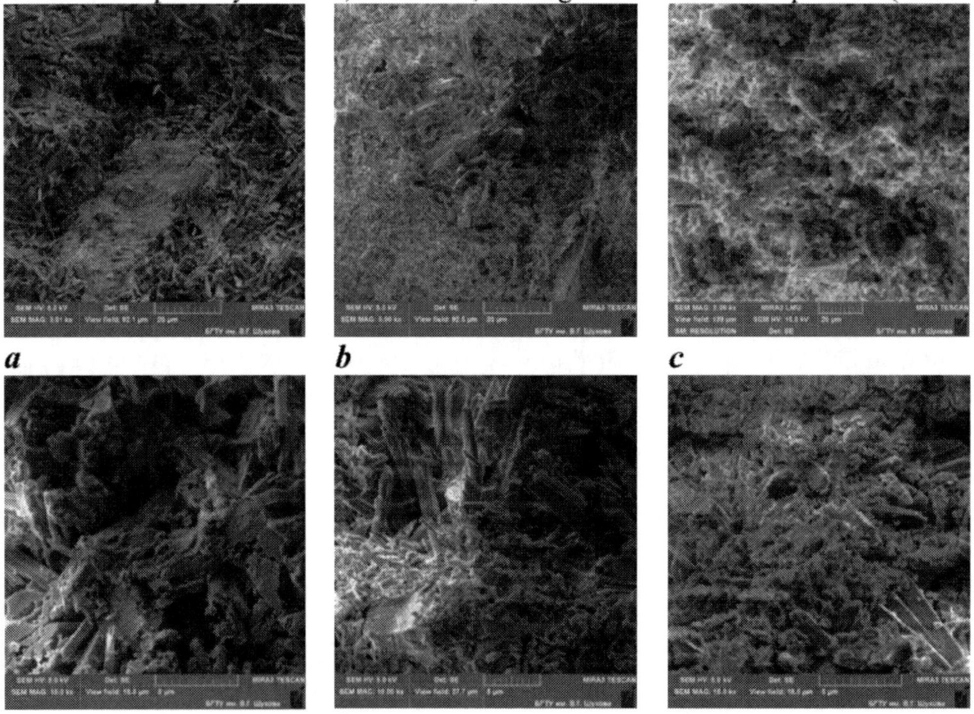

a *b* *c*

Figure 3. SEM images of gypsum composites:
a – mix № 1, b – mix № 4, c – mix № 14 (see Table 2)

Incorporation of NSC with finely-crystalline structure helps to densify packing of grains in crystal matrix. Moreover, NSC particles serve as nucleation sites, providing an increase of contact area between crystal structures and finely-crystalline new formations such as hydroxylellestadite (Fig. 1, b). In contrast to reference specimen, porosity of the composite dropped to 0,21 cm/g^3, at the same time, the density was increased up to 1274 kg/m^3, which boosts strength and water resistance of developed compositions.

Incorporation of portland cement creates a favorable conditions for additional formation of hydrates in form of flake-like aggregates – hydroxysulfoaluminates (Fig. 1, c). It is known, that low water resistance of gypsum binder driven by high solubility, high permeability of calcium sulfate dihydrate and by wedge effect of water molecules when penetration into the intercrystalline cavities (Rehbinder effect) [15]. This leads to strength reduction and consequent failure of gypsum materials under the water impact.

The developed binder compositions with reduced amount of gypsum, showed, that the structure framework of calcium sulfate dihydrate becomes overgrown by new formations synthesized due to interaction of the system 'GB-NSC-PC', which provide an improvement in density of crystal matrix from 1201 to 1351 cm/g^3, increase strength up to 61%, reach a malaxation coefficient up to 0,70 – 0,72 with consequent improvement in water resistance of the composite.

Key insight

It was found out, that its possible to control of gypsum binder performance by engineering of multicomponent binder and modification of conditions for phase- and structure formation. It was determined in this study, that water resistance and strength performance can be adjusted by incorporating nanostructured gypsum silicate binder and portland cement with reducing amount of calcium sulfate dihydrate.

Use of finely-crystalline NSC provide a better packing density of grains in crystal matrix of binder. For the system 'GB-NSC-PC' macroporosity range was shifted towards mesoporosity with formation of high dense impermeable structure.

Hydration of gypsum binder in presence NSC and PC was found to slow down. With reduction of gypsum portion in the hydrated system less soluble new formations such as hydroxylellestadite $Ca_5(SiO_4)_3(SO_4)_3(OH)_2$ and calcium hydroxysulfoaluminate $3CaO \cdot Al_2O_3 \cdot CaSO_4 \cdot 12H_2O$, in comparison with calcium sulfate dihydrate, were observed. The presence of such phases in the system boost water resistance of the binder.

Composition optimization of gypsum binder using NSC and portland cement allows changing the phase- and structure formation kinetics followed by increase in matrix dense of the harden binder and by boost of water resistance and strength of the composite.

Conclusion

Improvement in water resistance of gypsum binder can be achieved by varying of phase formation with at simultaneous reduction of gypsum binder portion up to 35% by replacing it with NSC and portland cement. Then, the hardening process goes without significant fluctuations in hydration kinetics and destructive deformations. High disperse NSC promote more arranged and dense structure and participate in a synthesis of finely-crystalline phases such as hydroxylellestadite.

The presence of cement in the system contributes to additional synthesis of steady flake-like hydroxysulfoaluminates. As a result, the crystals of calcium sulfate dihydrate inclosed into the mass of water resistant crystalhydrates, the morphology of which determine intercrystal porosity, providing an improvement in dense to 12,5% , strength to 61%, malaxation coefficient of 0,70-0,72 and as a consequence increases water resistance of the composite. The developed binder meets the requirement for its application in construction materials for external walling.

Work is performed under the strategic development program BSTU. V.G. Shoukhov.

References

1. A.V. Volzhensky, V.I. Stambulko, A.V., Stroyizdat, 316 (1971)
2. Yu.M. Bazhenov, V.S. Demyanova, V.I. Kalashnikov, Assosiations of construction High Establishments, 368 (2006)
3. A.V. Cherevatova, E.O. Gashchenko, Bulletin of Belgorod St. Technological University named after V. G. Shoukhov **1**, 25–30 (2007)
4. V.S. Lesovik, V.V. Potapov, N.I. Alfimova, O.V. Ivashova, Construction materials **12**, 60–62 (2011)
5. N.I. Kozhukhova, E.V. Voitovich, A.V. Cherevatova, I.V. Zhernovsky, D.A. Alehin, Construction materials **6**, 65–69 (2015)
6. I.S. Maeva, Abstract of a PhD thesis, 154 (2010)
7. I.V. Zhernovsky, V.V. Strokova, A.V. Cherevatova, LAM LAMBERT Academic Publishing GmbH & Co. KG, 170 (2011)
8. E.V. Voitovich, A.V. Cherevatova, I.V. Zhernovsky, H.B. Fisher, K. Sobolev, Journal of the Society for American Music **2**, 1611 (2014)
9. I.V. Zhernovsky, A.V. Cherevatova, E.V. Voitovich, N.I. Kozhukhova, E.I. Evtushenko, International Journal of Applied Engineering Research **11(12)**, 7732–7735 (2016)
10. V.V. Strokova, A.V. Cherevatova, I.V. Zhernovsky, E.V. Voytovich, Construction materials **7**, 9–12 (2012)
11. V.V. Babkov, V.M. Latyipov, L.N. Lomakina, V.S. Asyanova, R.I. Shigapov, Construction materials **7**, 4 (2012)
12. V.S. Lesovik, N.V. Chernyisheva, V.G. Klimenko, News of Higher Educational Institutions **4**, 3–11 (2012)
13. Russian standard 23789–79 «Gypsum binders. Test methods»
14. Russian standard 12730.4–78 «Concretes. Methods of determination of porosity parameters»
15. A.V. Ferronskaya, ACB, 488 (2004)

MATEC Web of Conferences **86**, 04066 (2016) DOI: 10.1051/matecconf/20168604066
IPICSE-2016

Risk monitoring for the city in geological hazards conditions

Alexander Ginzburg[1,*], *Sergey* Kachanov [2,3] and *Gennagij* Nigmetov[2]

[1] Moscow State University of Civil Engineering, *129337, 26, Yaroslavskoye Shosse, Moscow, Russian Federation*
[2] All-Russia Scientific Research Institute on Problems of Civil Defense and Emergency Situations (federal scientific and high-tech solutions center), *121352,7, Davidkovskaya str., Moscow, Russian Federation*
[3]Moscow Institute of Physics and Technology, *141700, 9, Institutskiy per., Dolgoprudny, Moscow Region, Russian Federation*

Abstract. Article looks on the main risks of geological hazards for potentially dangerous, unique and technically complex facilities and given proposals for their monitoring

1 Introduction

Nowadays, the development of modern socio - economic system and growth of urban areas leads to the fact that more and more potentially dangerous, unique and technically complex facilities are located in geologically hazardous areas. In addition, soil mass that is in state of relative equilibrium before construction, can exit this state of equilibrium due to effects dynamic and static loads from construction and transport objects, located nearby. Loss of equilibrium state and stability of the soil mass leads to subsidence, landslides, disruption of hydrological regimes, suffusion, karstic phenomena, excessive vibration, and, as a consequence, to an increased risk of collapse of buildings and structures .

2 Main part

Specialists of VNII GOChS designed technology for dynamic-geophysical monitoring of the "soil-structure" system that lets you see the possible changes in the system according to the impact of external factors.

Examples of implementing of this technology by VNII GOChS on the objects with geological hazards are:

-monitoring of residential buildings in Kaliningrad city with tilts and deformation cracks caused by exposure to vibration during the bank protection works on the embankment of the river;

-monitoring of the facilities during subsidence of foundation slabs against each other during construction of Olympic facilities in Imereti Valley in Adler city;

* Corresponding author: ginAV@mgsu.ru

-monitoring of the school in Moscow city, affected by the influence of excessive vibration from a nearby subway tunnel;

-monitoring of the building with deformation cracks caused by construction of a nearby facility with deep foundations;

-monitoring of the slope in the zone of sliding roadway in Ulyanovsk, caused by the influence of inefficient drainage, excessive vibration from passing railway transport and heavy vehicles (Fig.1).

We have accumulated many of such examples in our practice. In this regard, besides automated monitoring of engineering structures of buildings in geologically hazardous areas, we advise to monitor geological hazards, which may cause damage to the objects located in risk areas of such hazards. Such monitoring should be carried out at the stage of research and construction, and after the completion of construction in potentially dangerous areas.

The technology of complex dynamics and geophysical monitoring consists of the measurement and analysis system (Fig. 2 and Fig. 3) that is installed in the zone of possible geological hazards and consisting of:

- a computer with specialized software that allows to receive and, according to specific criteria, to process and analyze digital data from sensors installed in a controlled area;
- multi-channel analog-to-digital converter (ADC);
- three-way acceleration sensors, tilt sensors, water level sensors and soil pore pressure sensor;
- radio channel or cable transmission system from the sensors to the ADC.

Fig.1. Sliding roadway in Ulyanovsk, caused by the influence of inefficient drainage, excessive vibration from passing railway transport and heavy vehicles

Fig.2. Installation of a mobile dynamic-geophysical monitoring system in landslide hazardous road section in Ulyanovsk.

Criteria, developed with participation of the authors, for assessing stability of soil mass with dynamic-geophysical monitoring results, make it possible to determine the onset of unstable soil equilibrium at an early stage from a day to a few hours. Oscillations are sensitive parameters for stiffness of structural systems. Oscillations in soil mass, as well as in structures depend on its mass and stiffness.

Solution of the differential equation that describes the oscillations of an object with length l (for example a beam) appears as follows [1-8]:

$$T_1 = \frac{l^2}{\pi^2}\sqrt{\frac{m}{EJ}}, \qquad (1)$$

where T_1 is a period of oscillation of the object (s); l is a length of an object (m); m is a mass of an object (kg); E is a module of elasticity (N/m^2)and J is a moment of inertia of an object (m^4).

Equation for soil mass, relating its oscillations with geometrical, physical, and mechanical properties, can be represented as follows:

$$T_1 = 2.63 * H\sqrt{\frac{\rho}{G}}, \qquad (2)$$

where ρ is a density of the considered soil mass unit; G is a shear modulus of the soil mass and H is a height of the soil mass unit.

Therefore, by controlling the oscillation period of an object or soil mass, we can control rigidity including the degree of water content of the soil mass.

Fig.3. Data acquisition position for dynamic-geophysical tests of the impact of dynamic and baseline loads on soil mass in the area of the construction site of residential complex and the road at the site of landslide.

Table 1.Results of dynamic-geophysical tests of the soil mass with different kinds of dynamic effects.

Object and types of measurements	**A_x , m/s²**	**A_y , m/s²**	**A_z , m/s²**
Soil mass, baseline effects	**0,003** 0,015 0,03 0,004	**0,003** 0,01 0,01 0,004	**0,004** 0,005 0,005 0,003
Soil mass, trams passage	**0,004** 0,01 0,03 0,008	**0,004** 0,01 0,01 0,003	**0,004** 0,02 0,005 0,004
Soil mass, passage of 35 t truck	**0,01** 0,042 0,03 0,006	**0,017** 0,05 0,02 0,006	**0,01** 0,09 0,005 0,002
Soil mass, pile driving with 2.5 t diesel hammer	**0,005** 0,022 0,022 0,15	**0,004** 0,02 0,015 0,3	**0,005** 0,01 0,01 0,2
Soil mass, pile driving with 5 t hydraulic device	**0,006** 0,045 0,03 0,52	**0,0041** 0,03 0,03 1,25	**0,0052** 0,015 0,01 0,6
Soil mass, passing train	**0,02** 0,01 0,02	**0,005** 0,01 0,015	**0,005** 0,0075 0,0075

Table 2. Period of natural oscillations of the soil mass in wet state and after extensive drainage works:

№	**Condition of the soil mass during dynamic-geophysical measurements.**	**T_x, s.**	**T_y, s.**	**T_z, s.**
1	Heavily watered	0,078	0,078	0,078
2	After drainage instalment, soil moderately moist	0,05 Shifted to the high frequency area	0,05 Shifted to the high frequency area	0.05 Shifted to the high frequency area

Analysis of the data presented in Table №1 shows that the amplitude of accelerations of landslide slope caused by the passage of railway freight train, exceeds baseline fluctuations of the slope by **6.6 times**, the passage of a heavy trucks increases amplitude of accelerations of the slope exceeds by **3.3** times. Slope oscillation periods during the passage of trains and trucks shifts to the resonance frequencies (0.1-0.2) s, which increases the risk of loss of stability of slope soil mass.

Results of comparison of the dynamic-geophysical parameters of moistened soil mass (Table 2) with the soils with installed drainage system, and with reducing the groundwater level at landslide-prone area show that in all axes of measurement periods of natural oscillations shifted to higher frequencies.

These comparisons show how sensitive dynamic-geophysical parameters are to the change of the soil state. Effectively performed work on lowering the groundwater level increased stiffness of the soil mass, reduced deformability processes and increased stability of landslide-prone slopes.

Measurement results show that the greatest accelerations and therefore greatest displacements of the soil mass-produce dynamic loads generated by the movement of freight trains and heavy trucks. Maximum acceleration occurs in the X-axis (along the river Volga, maximum soil displacements occurred in this direction).

Accelerations occurred due to driving of foundation piles during construction of the building on the top of the slope are registered in the area of the landslide at the baseline level and could not be the cause of this landslide.

The main cause of the landslide was a strong hydration of the soil mass, which led to its oscillations with a period of natural oscillations equal to 0,078 in all axes, and caused by resonant dynamic effects of passing trains and heavy trucks.

3 Conclusions

Example, presented in the article, shows that the decrease of the landslide hazard of soil mass on slope can be provided by:

1. Installment of drainage ans storm-water removal systems to ensure effective withdrawal of groundwater;
2. Control of the soil mass state for the timely definition of dynamic-geophysical parameters and for elimination of resonance phenomena from dynamic effects;
3. Comprehensive monitoring of "soil-structure" systems.

Our experience shows that application of such complex monitoring systems can provide timely warning of the urban population about the risk of geological hazard. Monitoring measurements will allow to ensure the correct choice of rational engineering measures to improve the stability of the "soil-structure" system on the design stage, and will allow to ensure monitoring of the effectiveness of the work on the operational phase.

References

1. A.Ginzburg, S. Kachanov, IJAER **10(20)**, 40869-40872 (2015)
2. A.Ginzburg, S. Kachanov , IJAER **11(3)**, 1660-1665 (2016)
3. A.Ginzburg, A. Ryzhkova, IJAER **10(21)**, 41828-41830 (2015)
4. A. Volkov, AMM **838-841,** 2973 (2014)
5. A. Volkov, V. Chulkov, R. Kazaryan, R. Gazaryan, AMM **584-586,** 2685 (2014)
6. G. Nigmetov, M. Chubakov, Seysmostoykoe stroitel'stvo **4**, 51-55 (2011)
7. G.M. Nigmetov, Cataloque of seismoforecasting research, 147-154 (2009)
8. G.M. Nigmetov, Cataloque of seismoforecasting research, 99-104 (2010)

Assessing Transaction Environment Conditions for Investment and Construction

Marina Mishlanova[1,*] *and Vadim* Kharitonov[2]
[1]Moscow State University of Civil Engineering, 129337, Yaroslavskoye Shosse 26, Moscow, Russia
[2]ANPO NGS-Orgproyektekonomika, 115172, ul. Narodnaya, Building 4, Moscow, Russya

Abstract. This paper contains characteristic description of the transaction-based approach and identifies problems of studying the transaction environment for investments and construction. It also discusses the results of transaction cost analysis in the real estate sector, and assessment of the business environment in Russia. The authors present indirect assessment of transaction environment through the economic freedom index, detailed particularly for Russia; they demonstrate the transaction environment structure and offer methodological recommendations on evaluation of transaction environment for investment and construction.

Introduction

Evolution and the modeling of any socio-economic system as a phenomenon can be regarded through the following basic elements: sanctions, operations and transactions [1]. The phenomenon of transactions holds a significant position in the modern theory and practice of management, in the theory of databases and software engineering. The definitional basis of the "transaction" category requires at least a logically meaningful operation that makes sense and could be effected only in entirety. In these terms, the economic definition of a transaction assumes a deal executed to meet the parties' requirements and encumbered with reciprocal concessions. The information definition of a transaction understands a whole group of logically united sequential steps used to process data. An integrated approach considers a transaction in broad sense as a series of activities to exchange commodities, information being one of such.

Each transacting player has their own status agreed with the other players. Acting jointly, the players maintain a certain condition of their shared field of relations, as this corresponds to some distributed mutually accepted capabilities and limitations. The phenomenon of transaction represents coordinated transition of economic relations to a new status, which generally moves the entire system to a new state [2]. A set of various arguments helps to consider some individual, complex and systemic transactions as a subsystem in the investment and construction sector, as an involved composition of functional and non-functional requirements, restrictions and controls. In terms of the systemic approach, a significant role in implementation of transactions is played by the environment that creates conditions for long-term stable interactions, enables implementation of compensatoryrestorable transac-

* Corresponding author: mishlanova_m@mail.ru

tions, allows use of alternative transactions, shapes conditions for monitoring and auditing, and guarantees security of transactions [3].

Principal Section

The main reason for focusing on the environment in which economic activities are pursued is that transaction costs tend to grow against the total gross domestic product [4].This problem is relevant for investments and construction because total transaction costs can reach up to 70-200% of construction projects costs [5etc.]. According to research by Center for Strategic Research Fund, evolution of the housing market is constrained by high transaction costs that exist in the housing market and mortgage loans [6]. Vertically integrated construction companies register transaction costs of 50% of other costs and 30% of total sales [7]. These factors obstruct growth of investment and construction activities.

According to Global Property Guidenational consultants [8], Russia is the European leader in transaction costs of real property sales. To quote Global Property Guide, such costs account for 20% to 25.5% of the ultimate value of the property in the long term (see Fig. 1). This includes: agency fees – about 2-5%, legal fees - 0.7-1%, registration costs - 0.01-1%, and taxes - 18%.

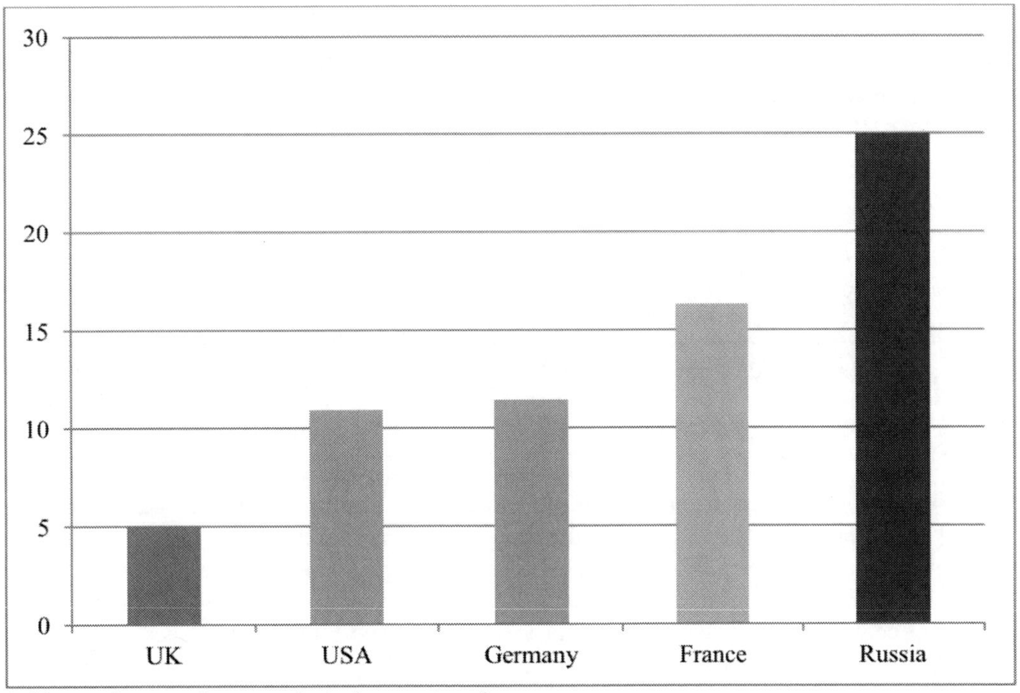

Fig.1. Share of transaction costs in real property sales

The transactional environment for investment and construction includes components of commercial and institutional nature. Business environment is the economic player's immediate surrounding, it reflects direct influences and interactions. Therefore, one can use the business environment index as one of the evaluation components to evaluate the condition of the transactional environment [9]. To generate the index, we use seven indexing sections: product market, logistics and infrastructure, B2B, B2G, financial markets, individual assessment of business climate, investment and social activity [10].

The economic crisis in Russia has affected assessment of business environment [9]. Late in 2014, all index indicators dropped and the decline continued in January 2015, when the composite index reached its historic low of 34.1. The annual average business climate index coasted down to 41.7 in 2015 from 44.5 in 2014, thus entering the negative value area.

The index of business climate individual assessment represents subjective attitude of business players to the domestic economic situation and business environment; this is in clear correlation with the transactional environment. The indicator is highly sensitive to both economic and non-economic factors. Throughout the year of 2015, the index of business climate individual assessment stayed in the negative assessment area. The annual mean of the index was 31.9, or 1.8 points below the 2014 figure. It is the first time the index drops so low since 2012 ever since the Russian Union of Industrialists and Entrepreneurs (RUIE) started its survey polls on business climate in Russia.

In 2016, the Russian Union of Industrialists and Entrepreneurs polled businesses again, and the results are presented in Table 1. The composite index of business climate was 46.9 in September 2016.

Table 1. 2016 Business Environment Index by RUIE

RUIE Business Environment Index Components	July	August	September
Product market index	44.6	46.7	44.6
Logistics and infrastructure index	44.3	48	46.4
B2B index	47.8	46.4	46.9
B2G index	45.6	47.9	48.6
Material markets index	44.4	43.5	46.2
Business climate individual assessment index	45.2	41.1	45.3
Investment and social activity index	55.6	48.3	50.2
Business environment index	46.8	46	46.9

The transaction environment can be assessed indirectly with the economic freedom index on the scale of 100 [11] covering four segments: supremacy of law (categories: rights of ownership and corruption level); constraints on government (categories: fiscal freedom and government spending); regulatory efficiency (categories: freedom of enterprise or business, freedom of employment, monetary freedom); and market openness (categories: freedom of trade, freedom of investment, and financial freedom). According to the research, during late 2014 to early 2015, based on retrospective data in certain cases (see Fig. 2), Europe had five "essentially unfree" nations (Croatia, Bosnia and Herzegovina, Moldova, Greece and Russia). Russia ranked 153 (on the list of 178) on the global rating, bordering on "repressed economy". The reason for concern was the growing trend in the categories "rule of law", "freedom of employment", and "freedom of investment". The researchers registered irregular nature of market reforms, growing inflationary pressure, capital outflow, corrupt government institutions, red tape handicaps for business, and some other negative factors. For details about the economic freedom index in Russia, see Fig. 3.

 DOI: 10.1051/matecconf/20168604067

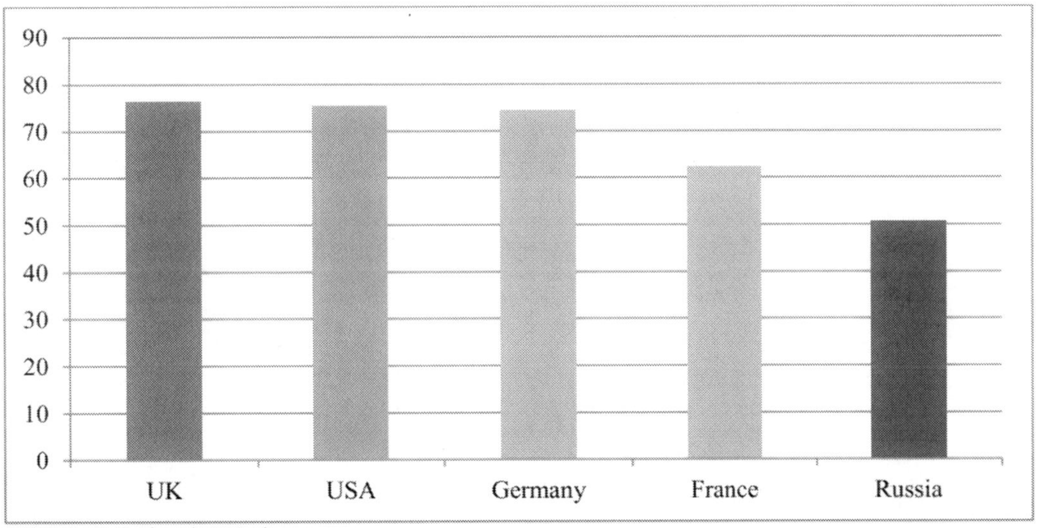

Fig. 2. Economic freedom index

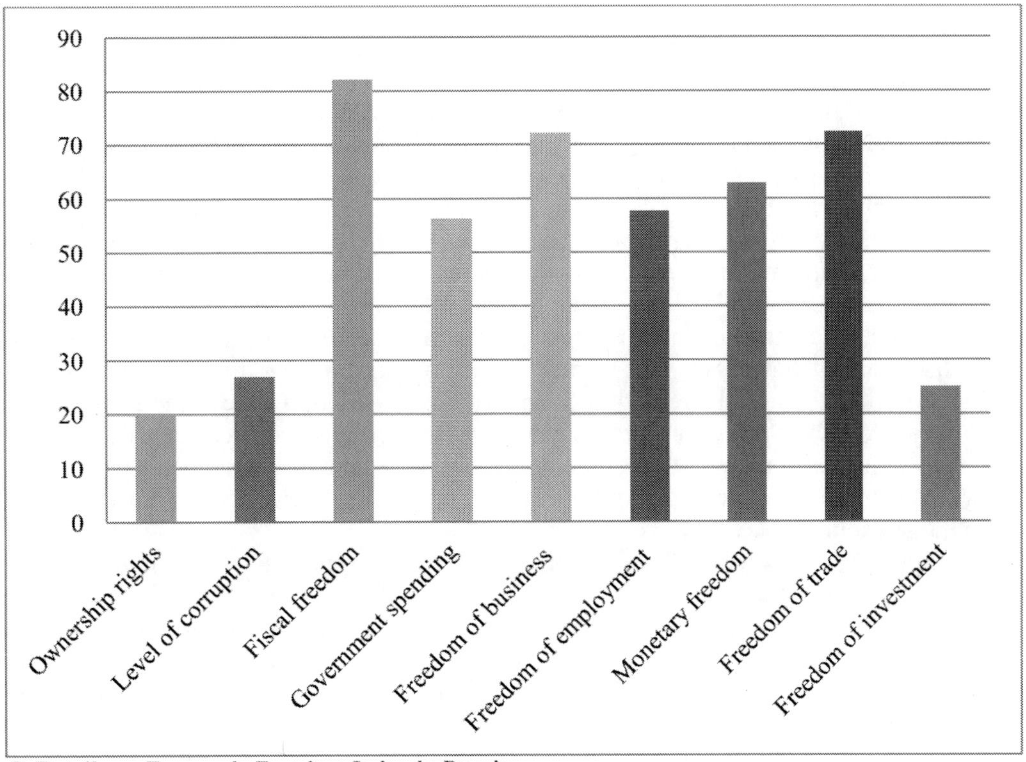

Fig. 3. Details on Economic Freedom Index in Russia

A sub-component of the Economic Freedom Index beyond all doubt is the ownership rights index (Fig. 4) that assumes assessment of the legislative and political environment (categories: independence and impartiality of courts, rule of law, political stability, level of corruption); rights to physical property (categories: protection of physical property rights, property registration, accessibility of loans); intellectual property rights (categories: protection of intellectual property rights, protection of patent rights, level of piracy) [12].

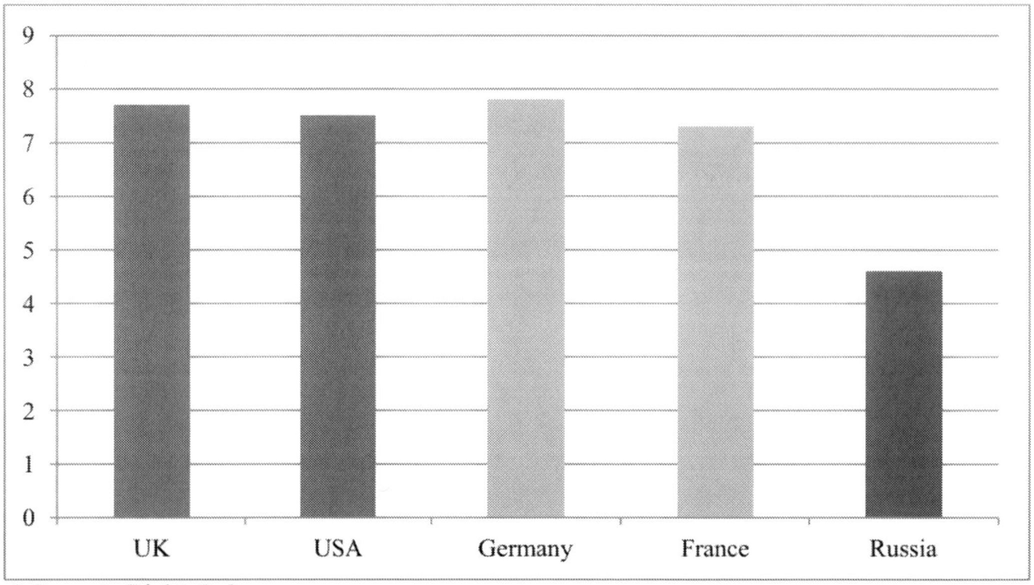

Fig. 4. Property Rights Index

Indirect assessment of the transaction environment can be shifted towards assessment of the economic entities' interests; to do so, we suggest that the indexes of business confidence and business expectations be taken into account. The business confidence index (RF Statistics Board, Center for Economic Analysis under RF Government) for construction is found as the arithmetic mean of balances of the actual order portfolio and the expected change in the number of employees in the companies. The index can be amended with results of other methods. The Finance market index CIF is computed based on indicator dynamics: investments in fixed assets, the share of businesses with "satisfactory" financial condition, order portfolio index (supply) etc. The business expectations index (Russian Economic Barometer) includes the results of polling among businesses, and finds the ratio of the share of businesses who expect higher resource prices, to businesses who expect higher product prices; the ratio of the share of businesses who expect higher product prices, to those who expect higher employment pay rates; the share of businesses who expect increase in production output; the share of businesses who expect new jobs; the share of businesses who expect increase in equipment costs; the share of businesses who expect increase in the portfolio [13].

Issues of assessment and control of transaction environment are examined respective to specific sectors and complexes; this is caused by different structure of transactions, and consequently by different approaches to research them. To validate the substance of transactional environment for investment and construction in the economic space, let us perform actual structuring. As our premises, let us assume that transaction means social interaction that results in transfer or appropriation of ownership rights and privileges in respect of certain resources, intermediate and ultimate construction products. We assume also that in the course of this interaction, the actor of investment and construction will adapt their activities to changing system conditions and quantitatively maximize their well-being.

Transactions in investment and construction based on their subject can be represented in this way: information support, negotiations, contracting, prevention of opportunistic behavior, warranty support, support to production processes, assurance of construction product quality, specification and protection of property rights. Having examined the works of John Commons [14 et al.] we can distinguish three main types of transactions by areas and levels of interaction, which we have to slightly alter in order to achieve the objectives of our research:

- commercial transaction (contract) – horizontal type of symmetrical relations, occurring in the sphere of market relations;
- control transaction (intra-system) – vertical type of asymmetrical relations, occuring in the sphere of system hierarchy;
- rationing transaction (external to system) - diagonal type of asymmetric relations, occurring the sphere of rights specifications authority.

The transaction types listed above position various relationships as vectors arranged in a certain way in some part of the economic space - in the transaction environment has its unique coordinates. Such approach lays down the principles of economic modeling in respect of control and optimization of transaction environment, as it helps to visualize the interaction processes. Analysis of structured factors of investment and construction activities [13] suggests the following axes for the part of economic space in question: political determinants – business determinants - investment determinants. To work out a methodology for evaluation of the transaction environment for investment and construction, in the first approximation one can propose the following factor structure of putative coordinate determinants.

Political factors:

- level of political stability (overall political situation, political events, public order, institutional environment, target programs);
- a set of legislated rules to govern economic relations (elaboration of the regulatory framework, relevance of amendments, legal compliance);
- taxation system (acceptability, extent of tax burden, relevance of amendments, system of benefits, activities of tax authority);
- relation of the authorities and society with business (administrative climate, degree of business support, level of corruption, government-business interaction).

Investment climate:

- ability to attract investors (extent of financial systemdevelopment, investorappeal);
- protection of investors and lenders to minimize risk of lost income/investment (investment stability, investment risk);
- efficient investment processes (effective demand for investments or capital intensity, investment activity, investment efficiency);
- available credit for businesses (requirements to borrowers, procedures, loan terms);
- available internal sources of investment (actual depreciation and profits);
- advanced mortgage loans (requirements to borrowers, procedures, loan terms).

Business environment:

- incorporation procedure (costs, number of steps, time limits);
- construction permits (costs, number of steps, time limits);
- connection to power grid (costs, number of steps, time limits);
- registration of title (costs, number of steps, time limits);
- business activity (working capital turnover, payback of core business activities);
- development of business community (business confidence, business ethics, access to information, transparent business processes, in particular, tendering processes);
- enforcement of contracts and payments (intervals and timing of payment delays, insolvency, settlements).

Conclusion

The results of assessment of the business environment, economic freedom and property rights presented herein can be regarded as fragmented components of the transactional environment assessment for investment and construction. In addition, the authors suggest that indexes of business confidence and business expectations should be used to achieve the objectives of the study. The inductive logic underlying this paper leads to methodologically general provisions that are at the same time specific for investment and construction activities. The proposed coordinate determinantsshould become the basis of assessment of transactional environment for investment and construction. The writers represent their vision of the transaction environment that brings close together the information and economic contents of the category of "transaction" and enables methodological transition to economic modeling through assessment of transaction environment for investment and construction.

References

1. A.A. Piskunov, A.V. Suvorov, Software and Systems **1 (113)**, 96-100 (2016)
2. I.V. Artamonov, Business Informatics **2 (20)**, 29-34 (2012)
3. A.A. Afanasev, V.F. Kasyanov, I.G. Lukmanova, D.N. Silka, International Journal of Economics and Financial Issues **5 (3S)**, 121-124 (2015)
4. J.J. Wallis, C.N. Douglass, University of Chicago Press, 1870–1970 (1986)
5. A.N. Asaul, S.N. Ivanov, IPEV (2008)
6. S.N. Ivanov,Bulletin of Civil Engineers **3 (12)**, 82-88 (2007)
7. M.Yu. Lukichev, *Transaction costs management for building contractors* (Moscow, 2011)
8. http://www.globalpropertyguide.com
9. http://рспп.рф
10. A.Mottaeva, MATEC Web of Conferences **1**, 73 (2016) DOI: 1051/matecconf/20167307020
11. http://www.heritage.org
12. http://www.internationalpropertyrightsindex.org
13. M.Yu. Mishlanova, Economics and Business **1 (54)**, 745-749 (2015)
14. J.R. Commons, Terra economicus **10 (3)**, 69-76 (2012)
15. A.D. Ishkov, M.Yu. Mishlanova, K.P. Grabovyi, International Journal of Applied Engineering Research **11 (3)**, 1676-1679 (2016)

Method of systematic determination of specific thermal characteristics of building

Vitaliy Prokhorov [1,a]

[1] Moscow State University of Civil Engineering, Russia

Abstract. In the paper the classical term «specific thermal characteristics of building» proposed by professor Chaplin V.M. for heating systems developing also for others heat-consuming system for building such as ventilation, air conditioning, hot-water supply system is observed. The followings attributing values are established: external building volume, temperature potential in annual crevice cycle.

Introduction

The term «specific thermal characteristics of building» was proposed by prof. Chaplin M.V. more than 100 years ago. Primarily this characteristics was applied for pre-project evaluation of followings parameters:

a) capacity of heating systems and equipment;
b) tentative fuel requirement of building;
c) sufficiency and optimality of thermal protection of buildings.

This characteristic are widely used at present time including educational programs because the specific capacity of heating systems per 1 m3 of building measured by external dimension. The new application fields of this parameter are proposed for normalization of building energy consumption reflected its energy efficiency.

Taking into account the modern building has not only heating system but also three types of thermal-consuming systems (mechanical ventilation, air conditioning, hot-water supply system) that are not completely described in regulatory documents the technical reasonability of extending of the term «specific thermal characteristics of building» for each heating system type takes place. It allows correct estimation of annual total energy-consuming in the building.

This idea and concept of the calculation procedures are proposed in 2002 [1] in form of equations.

Further development of this idea is reflected in research works [2, 3 etc.], regulatory documents [4, 5], works of developers of the document «SP50.13330-2012» [6] as well as the principle paper about «boundary of analysis» [7], required addition and refinement of earlier data.

The paper is devoted to realization of calculation method and the above concepts when design process.

---

[a] Corresponding author: ProkhorovVI@mgsu.ru

The necessary requirements are followings:

1) application for every type of heat consumption systems using external building volume Vext for calculation of all rooms [7];

2) justification of temperature differential for each engineering system (potential heat exchange method by V.N. Bogoslovsky) for every design condition.

Heating systems

In this work the classical form of parameter «specific thermal characteristic of building» is following:

$$q^{c}_{heat} = \frac{Q^{CB}_{heat}}{V_{ext} \cdot (t_{int} - t_{ext})}, \text{ W/m}^3\,^{o}\text{C} \quad (1)$$

where: q^{c}_{heat} is calculated specific thermal characteristic of building, applying for heating systems;

Q^{cb}_{heat} calculated balanced building heat losses, W.

V_{ext} is external building volume including all envelopes having a heat exchange with the environment (ventilated facades, heated underground floors, glazed balconies), m^3.

t_{int} is computation average internal air temperature of building, ^{o}C.

t_{ext} is computation external temperature for climate zone of construction, ^{o}C.

Determination of parameters q^{c}_{heat} and Q^{cb}_{heat} taking into account transmission heat losses and heat load on natural ventilation: controlled and uncontrolled (infiltration and exfiltration) is the basic procedure when design of energy balance of building in case of heating system.

The main idea of equation (1) is ratio of capacity of heating system Q^{cb}_{heat} to calculated value (tint –text) in working medium (generally, it is air) for internal and external side of building envelope per unit of heat exchange potential. Ratio of heat consumption value (per 1 oC) or «specific» capacity to V_{ext} is specific thermal characteristics of building in case of heating system.

Next stage is to determine analogue values for others heat consuming systems of building by calculation method. Realization of this stage demonstrates comparable (or higher) values of analogues vs. value of q^{c}_{heat}. Thus, they can be used according to common recommendations of energy performance certificate, particularly, for determination of annual heat consumption.

So, extending of term «specific thermal characteristic of building» consists in individual application not to heating system only but for others mechanocaloric systems of building that follows in the Chaplin V.M. theory.

It should be noted, in the bellow methods the common denominator in «external task» (heat-mass exchange of building with environment medium according to prof. V.P. Titov) must be used for constant value of external building volume V_{ext} according to eq. (1), but not internal building volume served by certain mechanocaloric system only. These methods are suitable for «internal tasks» (according to prof. V.P. Titov) only, for exp. when design of air-heat microclimate system as well as when comparison of performance indicator (for exp., energy consumption) for certain systems and served buildings.

Calculated value of (tint –text) should be in compliance with heat conductor and operating conditions of certain system.

More detail description of methods is presented bellow.

Systems of mechanical ventilation

$$q^{c}_{vent} = \frac{Q^{c}_{vent}}{V_{ext}(t_{in} - t_{ext})}, \text{ W/m}^3\,{}^{o}\text{C} \tag{2}$$

Q^{c}_{vent} thermal capacity of systems with mechanical ventilation in building, (W);

V_{ext} – the building volume, m3; t_{in} – calculate temperature of inlet air, °C; t_{ext} – calculate temperature of external air, oC for ventilation systems during winter season.

During a summer season the ventilation systems work without air heating, i.e. only electric energy consumption takes place.

Air conditioning system

Heat consumption of conditioner depends of applied air handling scheme. In this work the simplified system (straight-through arrangement) is studied.

As air heating in central air-conditioner is accomplished annually («reheat») the calculation of «specific thermal characteristic» value for building is accomplished for winter and summer seasons separately.

Winter season

$$q^{wc}_{ac} = \frac{Q^{wc}_{ac}}{V_{ext}(t_{in} - t_{ext})}, \text{ W/m}^3\,{}^{o}\text{C} \tag{3}$$

where Q^{wc}_{ac} thermal capacity of air heating during winter season («preheat» and «reheat» if it takes place) taking into account calculated parameters, W;

t_{in} – temperature of inlet air, °C.

During winter season the air heating process with aspirator in many cases (exp. in air conditioners for textile industry) is not taken account. In others cases the difference between t_{in} and t_{ext} $(t_{in} - t_{ext})$ is reduced. Required capacity for «reheat» effect is described with the below scheme for summer season.

Summer season

To provide the temperature of inlet air equal to heat-to-humidity ratio in the building for achievement a required humidity after cooling and drying process the air should be heated even during a summer season (Fig. 1).

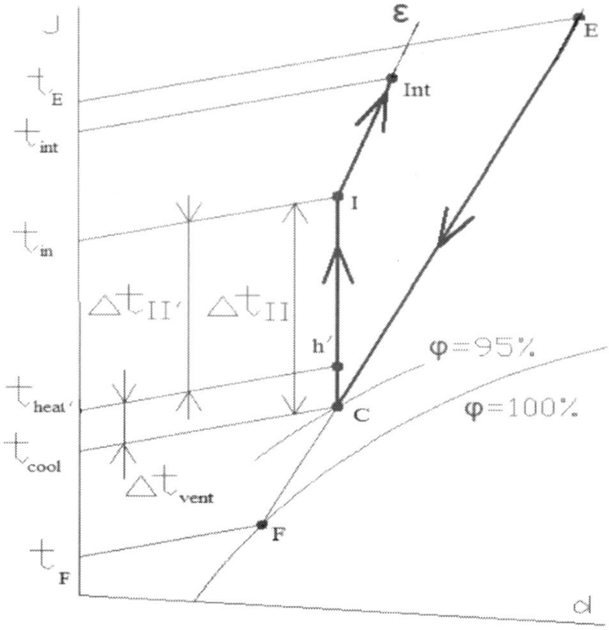

Fig. 1. Explanatory scheme for heat consuming processes during a summer season.
Δtvent = (theat –tc) – potential air heating with ventilator;
ε is heat-to-humidity ratio in room;
$\Delta t_{II'} = (t_{in} - t_{heat'})$ is air heating with heating element («reheat») taking into account its partial heating with ventilator;
I–C – internal process;
E–C–F –cooling and drying of external air

$$q_{ac}^{sc} = \frac{Q_{ac}^{sc}}{V_{ext}(t_{in} - t_{cool})}, \text{ W/m}^3\,{}^{\circ}\text{C} \quad (4)$$

where: Q_{ac}^{sc} calculated thermal capacity of air heating («reheat») for summer season, W;

t_{in} – temperature of inlet air, oC;

t_{cool} – temperature of air after cooling, oC.

If in this scheme a ventilator is placed after air-cooling system the value of Δt_{vent} (difference between tc and theat) should be taken account (see Fig. 1).

To simplify the equations, hereinafter the parameter tc is used only (i.e. Δt_{vent} =0).

Also, the ratio of mass flows for both interacting mediums should be considered as air heating is realized in heat exchanger with water. So, the value Q_{ac}^{sc} can be determined for air and water.

Balance equation of heat-exchange is following:

$$Q_{ac}^{sc} = G_{in} \cdot c_{air} \cdot (t_{in} - t_c) = G_{in} \cdot c_{air} \cdot \Delta t_{II} = G_w \cdot c_w \cdot \Delta t_w \quad (5)$$

where: Gin and Gw are mass flow of heated inlet air and heating water in air heater, kg/s;

c_{air} and c_w are heat capacities for air and water;

Δt_w is temperature difference between input and output of heated water in heat exchanger.

Return temperature of water in heat exchanger depends of initial temperature of interacting flows, air and water flow rates in obstruction-free flow areas and obtained by traditional calculation method. Input water temperature during a summer season is equal to temperature in hot water supply system.

In scheme where water in «reheat» conditioner after water heat exchanger is used for hot water supply system, its temperature can be equal to 55 oC according to SP30-13330-2012 for water supply system in building (i.e. water temperature in water tap).

Advantage of this scheme is highly-accurate monitoring in case of low heat load and small temperature differences in air flow that typical for «retreat» effect. In case of «retreat» effect, air-heater in conditioner is directly connected with flow line on heat system, the max temperature during summer season is 70 °C (during winter season this value is higher that requires reconnection of heat exchanger to return line of heating system).

Temperature differences in these cases presented a heat-change potential for water are followings: $\Delta t_w = 55 - t^s_{wh}$ and $\Delta t_w = 70 - t^л_{wh}$ where

$\Delta t^л_{wh}$ –average temperature of return temperature of water calculated for «reheat» air heater.

Temperature differences of air and water flows in «reheat» heat exchanger are related in followings equations:

$$\Delta t_{II} = \Delta t_w \frac{G_w c_w}{G_{in} c_{air}} \quad (6)$$

$$\Delta t_w = \Delta t_{II} \frac{G_{in} c_{air}}{G_w c_w} \quad (6')$$

According to eq. (6) and (6') as well as at $t_w = 55$ the value of specific thermal capacity can be shown by following:

$$\frac{Q^{sc}_{ac}}{t_{in} - t_{heat}} = \frac{Q^{sc}_{ac}}{55 - t^s_{wh}} \cdot \frac{G_{in}}{G_w} \cdot \frac{1}{4.19} \quad (7)$$

$$\frac{Q^{sc}_{ac}}{55 - t^s_{wh}} = \frac{Q^{sc}_{ac}}{t_{in} - t_{heat}} \cdot \frac{G_w}{G_{in}} \cdot \frac{4.19}{1} \quad (7')$$

Hot-water supply system

For these systems the determination of specific thermal characteristics of building is differed for summer and winter seasons due to variety of cold water temperature.

Winter season

$$q^{wc}_{hw} = \frac{Q^{wc}_{hw}}{V_{ext}(55 - t^{wc}_{cw})}, \ W/m^3\,^{\circ}C \quad (8)$$

where Q^{wc}_{hw} – calculated thermal capacity of hot-water supply system for winter season, W; t^{wc}_{hw} – computation temperature of cold (tap) water during a winter season, °C (for exp. 5–8 °C).

Summer season

$$q^{sc}_{hw} = \frac{Q^{sc}_{hw}}{V_{ext}(55 - t^{sc}_{cw})}, \ W/m^3\,^{\circ}C \quad (9)$$

where: Q_{hw}^{sc} – calculated thermal capacity of hot-water supply system for summer season, W;

t_{cw}^{sc} – computation temperature of cold (tap) water during a summer season, °C (for exp. 12–15 °C).

Note, the values Q_{hw}^{wc} and Q_{hw}^{sc} include heat losses through pipes due and to daily and seasonal irregularity of hot water consumption.

As the denominator in the eq. (8) and (9) for specific thermal characteristics of building is the temperature difference of water vs. others heat consuming systems where temperature difference of air is used, the reasonability of introduction of universal analog parameter for air heating process with the same thermal capacity with using eq. (6) and (7) takes place.

Data obtained with calculation methods demonstrate the application of mathematical analysis is not effective because the values of Q_{hw}^{wc} and Q_{hw}^{sc} in the eq. (8) and (9) are equal as well as value of temperature differences for air and water (heat exchange potential) in real heat exchanger are varied not significantly (in case of subregeneration) or are equal (in case of ideal heat exchange process).

Conclusion

Mass air and water flows correlate with their heat capacities and are used when Q_{hw}^{wc} and Q_{hw}^{sc} determination.

Mathematical form for winter season is following:

$$q_{hw}^{sc} = \frac{Q_{hw}^{sc}}{V_{ext}(55 - t_{cw}^{sc})} = \frac{Q_{hw}^{sc} G_{hw}}{V_{ext}(t_{final} - t_{initial}) G_{cond}} \cdot \frac{4.19}{1} \quad (10)$$

where: G_{hw} – mass hot water flow, kg/s;

G_{cond} – mass conditional air flow, kg/s when air heating from calculated external air temperature for winter season t_{initial} to final temperature t_{final} for heating system.

Value of this temperature can be determined by using statistical value of subregeneration for real air-water heat exchanger and following determination of nominal air flow Gcond using the following parameters: standard hot water flow G_{hw}, t_{cw}^{wc}, t_{ext} and heat capacities of medium:

$$Q_{hw}^{sc} = G_{hw} \cdot 4.19(55 - t_{cw}^{wc}) = G_{cond} \cdot 1(t_{f-cond} - t_{\text{ext}}) \quad (11)$$

Analogous equations with parameters of t_{cw}^{sc}, $t_{initial}^{s}$, t_{f-cond} are can be applied for determination of G_{cond} for summer season.

$$G_{cond} q_{hw}^{sc} = \frac{G_{hw} \cdot 4{,}19(55 - t_{cw}^{sc})}{(t_{f-cond}^{s} - t_{initial}^{s})} \quad (12)$$

References

1. V.I. Prokhorov, Stroitelni Expert **13(16)**, 127–131 (2002)

2. V.I. Prokhorov, Advanced systems of heat and gas supply and ventilation (2003)
3. V.I. Prokhorov, House construction **1** (2012)
4. Organizations standard STO17532043-001-2005 (2006)
5. SP 50.13330.2012 (2012)
6. V.G. Gagarin, V.V. Kozlov, Association of HVAC Engineers. Northwest **1** (2012)
7. V.I. Prokhorov, Web–Bulletin VolgGASU. Multidisciplinary series **2** (2014)
8. Yu.P. Grigoriev, Stroitelni Expert **1** (2003)

MATEC Web of Conferences 86, 04069 (2016) DOI: 10.1051/matecconf/20168604069
IPICSE-2016

Development of mechanisms for innovative projects management based on their identification and results forecast in the risk environment

Ella Okolelova[1], *Natalya* Trukhina[2*]

[1]Voronezh State Technical University, Russia 394026, Moscow Avenue, Voronezh 14, Russia
[2]Moscow State University of Civil Engineering, 129337, Yaroslavskoye Shosse 26, Moscow, Russia

Abstract. The article presents the essential characteristics of innovative projects and classifies them by grade of novelty and types of innovative products to be created. The classification features allow us to identify the categories of tangible innovative products and innovative technologies. The article contains the main challenges in evaluating, forecasting and managing innovative projects in the risk environment. An innovative project is considered as a complex of possible implementations, i.e. as a random function. The investors' goal is to determine the most likely outcome of project implementation for making further decisions.

Introduction

The innovative projects vary in terms of functional purpose, tasks and characteristics. The functional purpose of an innovative project also varies on the basis of its "novelty carrier" extent, i.e. whether the product underlying the economic feasibility study of an innovative project is fundamentally new for the market, or the consumer sees improved characteristics of an existing product that allow to extend the range of the functions performed and tasks that were fulfilled previously.

Before considering the mechanisms for management of an innovative project, it is needed to identify the subject of management as an innovative project, determine the main evaluation criteria, systematize the project characteristics, identify possible risks, evaluate the development strategies, etc.

To determine the characteristics of an innovative project, first, it is needed to evaluate the grade of novelty of the product, the implementation of which it is aimed at. Therefore, it is required to evaluate the novelty of the original product and determine to which extent a new product differs from those previously existing in the market, which new or additional tasks it allows to fulfill from the user's point of view, to which extent its characteristics meet the consumer preferences of the market, etc.

* Corresponding author ntruhina@list.ru

Classification of innovative products

The following three innovative levels which determine the grade of novelty and the functional and qualitative level of an innovative project can be identified:

- absolutely new products (grade of novelty I);
- substitute products (grade of novelty II);
- analog products (grade of novelty III).

Absolutely new products pertaining to the first category are the products which do not have any analogs on the market and do not substitute any existing products, are designed to perform new functions and fulfill new tasks, have significant qualitative differences from the products positioned on the market, and are designed to perform fundamentally new functions.

Substitute products are characterized by a significant improvement in technical, qualitative and other characteristics and are designed to provide a substitute on the market of obsolete goods not meeting the up-to-date consumer requirements.

Analog products enter the market in parallel with the related products and perform the same functions, but have a significant difference in their qualitative characteristics and extend significantly the range of tasks to fulfill.

Table 1 Classification characteristic of innovative products based on grade of novelty

Grade of novelty	Functions	Tasks
Absolutely new products	Fundamentally new functions, new purpose of good.	New tasks which were either impossible or quite hard to fulfill before
Substitute products	The same functions, extension of the range of functions and improvement of the basic functions are possible	Fulfillment of the basic tasks, extension of the range of tasks, including fulfillment of new ones.
Analog products	The functions remain the same, addition of new functions is possible	Fulfillment of the basic tasks with possible addition of new ones

An innovative product can be represented by new goods or by a new material used for creating the final market product, having qualitative indicators and characteristics changed significantly. The innovative products also include intangible objects which represent development of new technologies, building of a production management system, formation of a product quality management system, etc.

The category of intangible innovative products can include fundamentally new intercorporate systems, connections and relations aimed at the improvement of the mechanisms of economic interaction, development of efficient communications and mediation connections.

Therefore, the innovative products can be differentiated by their essential content as follows:

- tangible objects;
- technologies;
- structural and management systems and relations;
- organizational and corporate systems and relations.

Formation of innovative solutions

The common purpose of creating an innovative product of any of the presented categories is profit-making, i.e. presence of business effect. Other types of effects are also possible, for example, social, environmental, etc. determined based on the purposes of implementation of the innovative project. The essential content of the classification features of tangible innovative products is shown on Fig. 1.

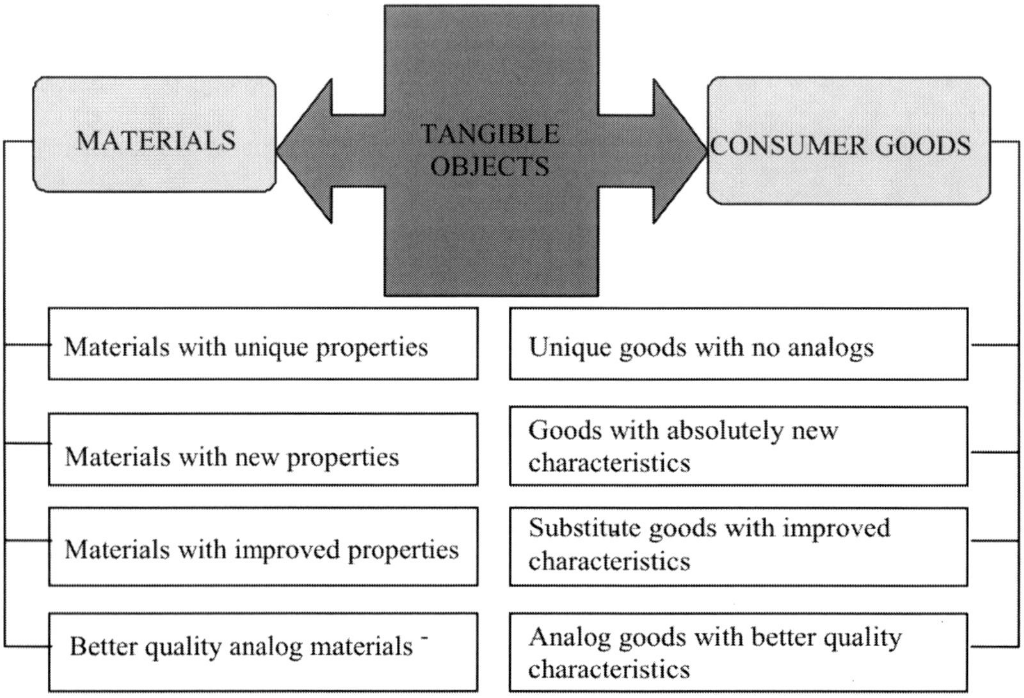

Fig. 1. The essential content of the classification features of tangible innovative products.

Another type of innovative product to which implementation can be aimed at are technologies.

The following classification of innovative technologies is proposed based on the purposes of creation of innovative technologies:

- manufacturing, aimed at creating a tangible product;
- information technologies;
- information and communication technologies.
- organization and management technologies;

The grade of innovativeness and creativity of technologies is determined by the novelties or improvements they carry. An innovative technology can have features of a fundamentally new, newly applied technology, if it contains innovativeness as to the application of materials and resources, technical performance, use of improved equipment, application of new manufacturing flowcharts, methods, processes, etc., as shown on Fig. 2.

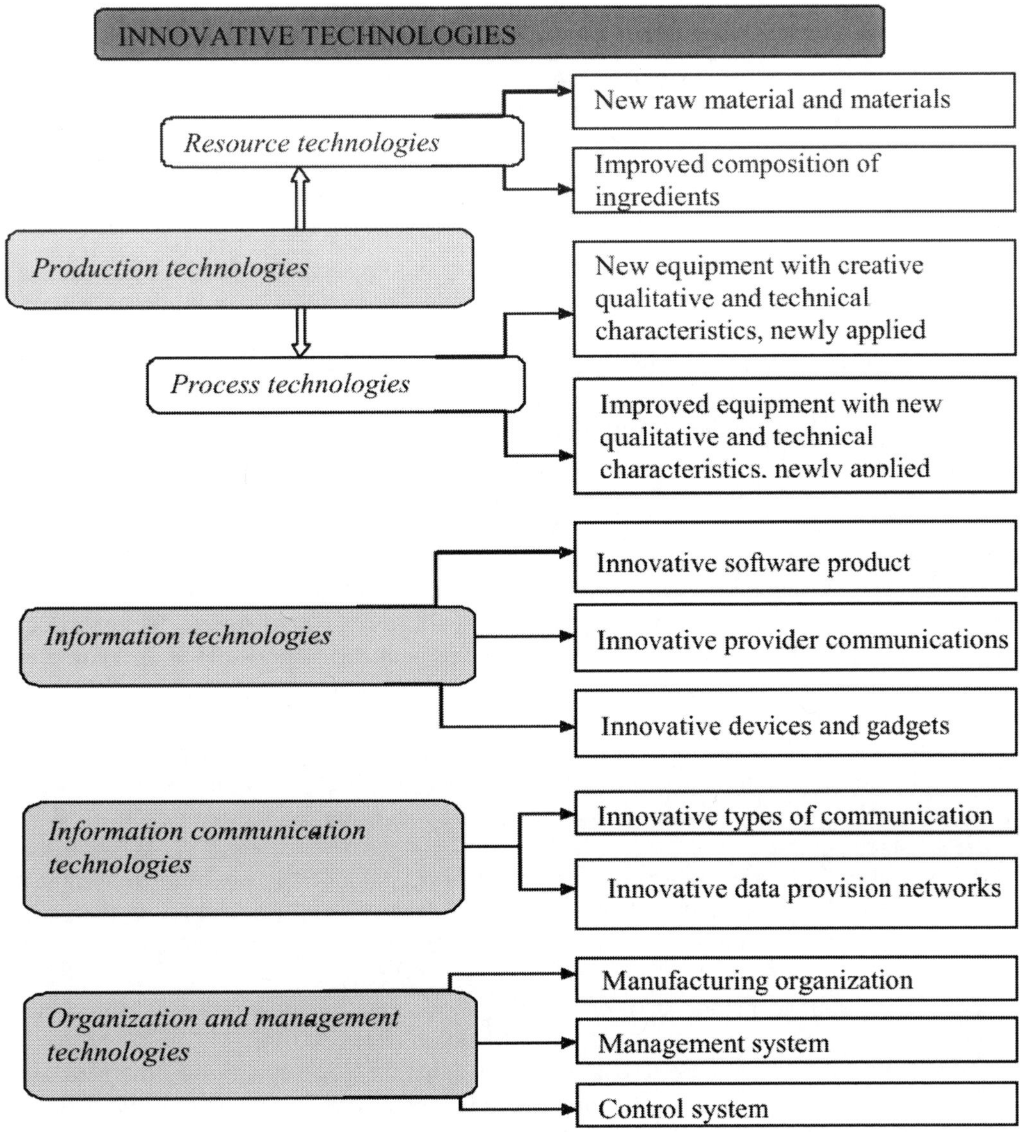

Fig. 2. Content of the classification features of innovative technologies.

It is especially shown in manufacturing technologies. Thus, a technology can have fundamentally new differences in terms of using raw material components, ingredients proportionand include new components, etc. This type of improvement of manufacturing technologies is defined as *resource technologies*.

The resource technologies imply significant change in the raw materials for the manufacturing of the final product.

The machines and equipment for manufacturing the final product are also subject to improvement. This type is defined as *process technologies*.

Contentwise, the information technologies can be oriented for development of a new software product, can be based on innovative provider communications, as well as on improved and fundamentally new devices and gadgets.

In terms of innovation, the information and communication technologies represent the creation of new possibilities for provision and receipt of data, development of new types of communication, etc.

The innovative organization and management technologies can include creative structural changes in the manufacturing process management, changes in the control functions, management systems of both the manufacturing and separate processes.

This classification category represents innovative enterprise management mechanisms distinguished by the building of fundamentally new structural communicative networks proving not only the presence a new approach to the manufacturing process management and control functions, but its efficiency as well.

The methodology for project identification allows to determine possible strategies of its development, assess the risks of the project and forecast the economic effect of its implementation.

The innovative projects, generally, present high risks. It is quite difficult to forecast their development, their implementation in the manufacturing process and the assessment of efficiency of their implementation. Risk is inevitable when implementing innovative projects under uncertainty. Therefore, it is preferable to make decisions while having certain safety reserve, the essence of which is to formulate arguments allowing to affirm with reasonable certainty that the investor will not incur any critical losses, even if the key parameters were set too high (too law) during the forecast.

Regardless of the forecasted project profitability value (for example, the volume of implementation and the value of income), there will always be different factors affecting the accuracy of the final forecasts. An innovative project can be considered as a complex of possible implementations, i.e. as a random function.

Therefore, speaking of an innovative project, we can affirm that there is a functional dependency of the economic effect on a series of governing parameters over time. Based on the actual practical conditions of the environment, we will define an innovative project as a random function of a nonrandom argument t, which is a random value at each fixed value of the argument.

If we present an innovative project as a dynamic model, it is possible to assume that such model will be stable under certain conditions depending, for example, on the value of financial investments, forecasted level of income from the implementation of the project, inflation, etc.

In case of change in one or more of the above-mentioned parameters, the dynamic system can move to a whole new level, i.e. bifurcation changes arise [4].

We define an innovative project as a random function, i.e. a function of the nonrandom argument t, which is a random value at each fixed value of the argument. If the value P of the expected income from the implementation of the project is a random value, then the function of profitability of the investment project is also random:

$$F(t) = S(P;t). \quad (1)$$

Let us consider the model of a dynamic system with random perturbations. Such models describe a wide class of phenomena of the real world, including economic processes, and represent systems of ordinary stochastic differential equations of the nth order complemented by the corresponding initial conditions which can always be transformed into an equivalent system of n differential equations of the 1st order as follows:

$$y_\nu(t) = Y_\nu(y_1, y_2, \ldots, y_n), \qquad \nu = 1,2,\ldots n. \quad (2)$$

Here, $y_v(t)$ are any dependent variables related with "motion", i.e. with temporary (dynamic) course of the process.

The system motion which we will call an unperturbed motion, as opposed to another motion which we will define as perturbed motion $y_v(t)$ [1], will correspond to a partial solution $f_v(t)$ of one of the system of equations (2)

$$f_v(t)=Y_v(t_1,f_1,f_2 ...,f_n), \quad v=1,2,...n . \tag{3}$$

The difference between the values of the perturbated $y_v(t)$ and nonperturbed $f_v(t)$motions at each moment of time t will be called perturbation $x_v(t)$:

$$y_v(t)-f_v(t)=x_v(t). \tag{4}$$

The nonperturbed motion $f_v(t)$ must satisfy the system of equations under the following conditions:

$$|x_v(t=t_0)| \leq \delta; \tag{5}$$

$$|x_v(t>t_0)| \leq \varepsilon(\delta); \tag{6}$$

$$\lim_{t\to\infty}|x_v(t)| = 0. \tag{7}$$

An unperturbed motion is called stable, if for any small positive number δ >0 another such number $\varepsilon(\delta)$ can be found, so that the inequality (5) could be carried out for all perturbed motions $y_v(t)$ for the initial moment of time $t=t_0$, and the inequality (6) would be true at all subsequent moments of time $t>t_0$. Otherwise, an unperturbed motion will be unstable [4].

The definition of the asymptotical stability is more rigorous than the previous definition. In case of an asymptotically stable motion perturbed at the initial moment of time $t=t_0$, the perturbations not only remain within the neighborhood of the initial condition $\varepsilon(\delta)$, as in case of normal stability, but, in addition, eventually fade to zero [4]. To formalize further consideration of the issue, let us introduce a series of concepts which will allow us to quantitatively characterize the perturbations of the mathematic models of the following type

$$y=P_n(s)u, \tag{8}$$

where $P_n(s)$ is a nominal transfer function [2]. The object with a mathematic model (8) will be stabilized by the regulator

$$u=-K(s)y, \tag{9}$$

where $K(s)$ is a transfer function of the system (8).

Suppose that the transfer function $K(s)$ of the regulator does not change in the process of functioning, and the transfer function $P_n(s)$ of the object is affected by nonparametric type unstructured perturbations. As the result of such impact, the regulator (9) will actually complete a different object and not the object with the model (8)

$$y=P(s)u, \tag{10}$$

the transfer function $P(s)$ of which will differ from the nominal one.

Let us determine the indicator $\Delta_A(s)$ of the absolute perturbation of the model (or the perturbation of the nominal transfer function $P_n(s)$).

$$\Delta_A=P(s)-P_n(s). \tag{11}$$

Accordingly, we will call the following rational fraction a relative perturbation of the model or of the nominal transfer function:

$$\Delta_0 = [P(s) - P_n(s)]P_n^{-1}(s). \tag{12}$$

The following formula will be considered a weighted relative perturbation of the model or of the nominal transfer matrix:

$$\Delta_p(s) = [P(s) - P_n(s)][P_n(s)W_d(s)]^{-1}, \tag{13}$$

where $W_d(s)$ is the prescribed weight fractional rational function.

The introduction of the weight function $W_d(s)$ in the determination of the perturbation (13) is conditioned by the amplitude frequency characteristics (AFCh) $A_n(\omega) = |P_n(\omega)|$ and $A(\omega) = |P(\omega)|$ for the nominal and the perturbed objects respectively. Let us introduce an admissible boundary of the nominal mathematic model perturbation, determining it by an upper bound (for each frequency) on the value of the relative change module of the AFCh by a positive number $|W_d(\omega)|$. In other words, the introduction of the function $W_d(s)$ determines the condition

$$\left|\frac{|P(\omega)| - |P_n(\omega)|}{P_n(\omega)}\right| \le |W_d(\omega)|, \quad \forall \omega \in R^1, \tag{14}$$

prescribing the admissible "corridor" for AFCh variations of the actual (perturbed) object (8), which is shown on Fig. 2.

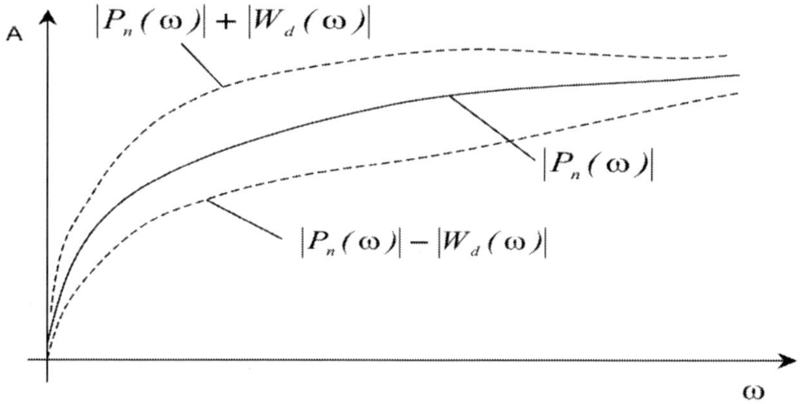

Fig. 3. An admissible "corridor" for AFCh variations of the actual perturbed abject

Therefore, the frequency function $|W_d(\omega)|$ is the relative width of the admissible corridor for the AFCh of the perturbed object. Note that it is not quite convenient to work analytically with the condition (14). Instead of it, we can use the relation of the following type

$$\left|\frac{P(\omega) - P_n(\omega)}{P_n(\omega)}\right| \le |W_d(\omega)|, \quad \forall \omega \in R^1, \tag{15}$$

which is stronger than (14), which results from the following inequality

$$\big||P(\omega)| - |P_n(\omega)|\big| \le |P(\omega) - P_n(\omega)|, \quad \forall \omega \in R^1, \tag{16}$$

true for any two complex numbers P and P_n. Therefore, carrying out (15) will cause carrying out (14).

Conclusion

Therefore, it is possible to determine the "corridor" of the values of the forecasted parameters. Economically, these will be the limit values of the impact of both external and internal factors on the result of the investment project implementation. The instability conditions for an economic system can be determined quantitatively with the conditions of the tolerances, providing an opportunity to forecast the results of the process subject to nonparametric impact.

References

1. E.Yu. Okolelova, O.V. Kornitskaya, N.I. Trukhina, Management of Economic Systems scientific electronic journal **60** (2013)
2. N.I. Trukhina, I.I. Chernyshikhina, Proceedings of Moscow State University of Civil Engineering **9**, 227-233 (2012)
3. E.Yu. Okolelova, Hannover, 96-101 (2016)
4. E.Yu. Okolelova, A.V. Shabanov, News of the Universities of the Black Earth Region **T.1**, 96-99 (2006)
5. I. Karakozova, Y. Prokhorova, MATEC Web of Conferences **07015**, 73 (2016)
6. V.S. Morozov, N.N. Taskaeva, Journal of Internet Banking and Commerce **21 (4)**, 20 (2016)
7. G. Alexander, R. Anastacia, International Journal of Applied Engineering Research **10 (21)**, 42410-42411 (2015)
8. G. Alexander, R. Anastacia, International Journal of Applied Engineering Research **10 (21)**, 41828-41830 (2015)
9. G. Bubnov, B.Titarenko, S. Titov, R. Titarenko, Information Processing and Control Engineering **1** (2015)

The quality evaluation of the ballastless track construction in the area of Bratislava tunnel no. 1

Libor Ižvolt[1], *Michal* Šmalo[1,*] , and *Jana* Malchová[2]

[1]Department of Railway Engineering, Faculty of Civil Engineering, University of Zilina, Univerzitna 8215/1, SK–010 26 Žilina, Slovak Republic
[2]Institute of Continuing Education, University of Zilina, Univerzitna 8215/1, SK–010 26 Zilina, Slovak Republic

Abstract. The ballastless track design is the perspective for the construction of tunnels and bridge structures on the modernized track sections in Slovakia. Thanks to its indisputable advantages this construction is also used within the tunnel reconstructions. The Department of Railway Engineering and Track Management of the Faculty of Civil Engineering of the University of Zilina carries out the diagnostics of an experimental section with ballastless track construction in Bratislava tunnel No. 1 twice a year.

1 Introduction

The ballasted track is composed of rails, rail supports, rail fastening and ballast. The track superstructure design where the ballast function is substituted by the reinforced materials placed on the concrete supporting slab or the bituminous supporting slab is called the ballasted track [1], [2]. In general, the ballastless track structure is implemented especially on the high speed lines, the tracks with high traffic load and the modernized sections of the standard lines (operating speeds up to 160 km.h-1), respectively the tracks for higher speeds (operating speeds 160 km.h-1 $< V \leq$ 200 km.h-1) nowadays. The ballastless track structure is built particularly on the track sections conducted in tunnels because of properties of the foundation soil which does not observe any settlement. This ballastless track structure is also used on the bridges. Moreover, the usage of the ballastless track structure has a positive influence also on the amount of investment costs for the building of tunnel tube regarding the smaller tunnel section in case of its new construction resp. by elimination of the expensive tunnel widening (e.g. in case of its electrification on the existing lines) [3].

* Corresponding author: michal.smalo@fstav.uniza.sk

Nomenclature	
MSO	measurements before putting section into operation
PK	track superelevation [mm]
PO	operational measurement
QM	quality mark
QN	quality number
RK	track gauge [mm]
RP	velocity zone
SDV	standard deviation of variable
SL	alignment of left rail [mm]
SR	alignment of right rail [mm]
VL	rail top level of left rail [mm]
VR	rail top level of right rail [mm]
ZK	quasi-twist on a short base [$mm.m^{-1}$]
ZR	change of gauge [$mm.m^{-1}$]
ZSR	Railways of Slovak Republic

2 The characteristics of the experimental section

The experimental section in the area of Bratislava tunnel No. 1 is located from km 53.063 to km 53.710 the double track Devinska Nova Ves – Sturovo. The ballastless track structure of ÖBB-PORR type was built within the tunnel modernization realized over the period from October to December 2014 (reconstruction of the adjacent tunnel tube – of Bratislava tunnel No. 2 – was realized in the year 2009 and there was built the ballastless track structure with the Y-type sleepers in the reinforced concrete slab of SATO type) [1]. The aim of the modernization of Bratislava tunnel track No. 1 was the removal of the insufficient technical condition of the railway superstructure resulting from the complicated maintenance of the fastenings on the wooden sleepers (up to a half of their height built into the reinforced concrete slab) occurring in bad technical state (cracks and rotten material of the sleepers) and the removal of bad technical condition of the contact system caused by the water flowing through the vault into the tunnel [4].

The design of the track superstructure of Bratislava tunnel No. 1 is composed of the precast reinforced concrete slabs of ÖBB-PORR type that are monolithically joined to the invert vault of the tunnel tube. The reinforced concrete slabs have dimensions 4 760 x 2 400 x 160 mm and there are eight pairs of fastening points in the cross-distance 600 mm on each slab. The fastening points are dimensionally produced for the fastening system of UIC 60 E2 type in the gradient of the rail surface 1:40 and with the fastening Vossloh W 300-1. There are altogether 124 slabs of ÖBB-PORR type imbedded in the tunnel on the sanitized bottom part of the tunnel tube. There are imbedded steel clips with the diameter 20 mm and the length 300 mm into the bottom tunnel tube construction to fasten the level regulating concrete course and the invert of the tunnel.

In comparison to the monolithic ballastless track structure is the advantage of the construction system ÖBB-PORR in the possibility of damage elimination in the case of the occurrence of an accident (derailment). In this case the slabs can be individually replaced. The flexible layers, which are a part of every slab´s equipment (the walls of technological holes and the bearing surface), enable with the use of special appliance and after applying sufficient force to raise up the slab and to separate it from the concrete sealing coat. After the realization of the repairs of the concrete sealing coat a new slab is built in and embedded into the concrete and the track can be put into operation within several hours.

The ending of the ballastless track structure is in regard to the present state (track speed 60 km.h^{-1}) composed of the combination of glued ballast and track skeleton with concrete sleepers of ZPSV type. The transition area in the vicinity of the Lamac portal in the length of 18.020 m is composed of 30 sleepers of BV 08 type with the fastening Vossloh W 14T. The track ballast of different thickness (from 100 to 350 mm) is embedded on the top surface of the sub-ballast composed of 200 mm thick concrete slab. In the vicinity of Bratislava main station tunnel portal is the transition area placed directly behind the station head of the railway station Bratislava main station. In regard to the limited area conditions and a high track stiffness is the transition area realized in the length 5.592 m by means of seven concrete sleeper pieces with the fastening Vossloh W 14T and two pieces of wooden sleepers. The track ballast of size fraction 31.5/63 mm was in both transition areas glued in volume by the double compound epoxy resin Kryolit – the first stage was realized within the building-up (under the bottom surface of the sleepers) and the second stage was realized after one year of operation (in the space between sleepers and behind the sleeper) [4].

3 Methods and means of the diagnostics of the track geometry

The diagnostics of track geometry is carried out by measuring trolley KRAB™–Light. The measurement is referred to as continuous, but in fact, the data is recorded with the measuring step of 250 mm [2], [5], [6]:

- gauge deviation *RK* (after calculating is also recorded change of gauge *ZR*),
- alignment of right rail *SR* (after calculating is also recorded alignment of left rail *SL*),
- rail top level of right rail *VR* (after calculating is also recorded rail top level of left rail *VL*),
- track superelevation *PK*,
- quasi-twist on a short base (calculating to a quasi-twist on a base of 1.8 m long – *ZK 1.8*, on a base of 6.0 m long – *ZK 6.0* and on a base of 12.0 m long – *ZK 12.0*),
- track distance.
- After the transmission of the measured data into the operating computer and after the defining of the tolerance there is following generated for the purposes of evaluation [6]:
- graph of the measured parameters of the track geometry,
- table of the local errors, i.e. the summary of the tolerance overrun,
- table of the track section evaluation which is based on the standard deviations and the quality numbers.

4 The results evaluation of the diagnostics of the track geometry

The results of the diagnostics are evaluated in terms of the rules in operation and the regulations of the railroads in Slovakia:

- STN 73 6360 Track Alignment Design and Track Geometry of Normal-gauge Tracks (1999) and Amendment 1 (2002), this is valid for the line tracks with the operating speed up to 60km/h including (i.e. for *RP* 1) [5],
- Regulation ŽSR SR 103-7 (S) Measurement and Evaluation of Track Geometry by Measuring Trolley KRAB (in Slovak) (2008) [6].

The diagnostics of the track geometry was realized as [7]:

- the initial measurement before putting the construction into operation (signed *MSO*) executed on 29 April 2015,
- the first operational measurement (*PO*1) executed on 22 October 2015,
- the second operational measurement (*PO*2) executed on 11 April 2016.

The measured parameters were evaluated according to the installation requirements for track parameters with the use of new material (*MSO*), resp. according to the maintenance limits and the safety limits for velocity zone No. 1 /*RP*1/ (*table 1*).

The individual parameters evaluation of the track geometry is insured in the sectional track evaluation by elaboration of the standard deviations. These are calculated of the deviations from the central line which is computed of the dynamic component of the parameters of the track geometry (*SK*, *RK*, *PK*, *VK*) in the zone of wave lengths of 1 m to 25 m measured after 250 mm. The standard deviation (*SDV*) expresses the course irregularity of the parameter of the track geometry in the evaluated section.

The state of the constructional track layout and the track geometry in the evaluated section is expressed by [6]:

- the sectional quality number (*QM*) of the individual geometric parameter (*SK*, *RK*, *PK*, *VK*),
- the general quality mark (*GQM*) which serves as the evaluation of the track geometry in the evaluated track section as a whole,
- the tamping mark (*TM*) which is used to decide whether to use the tamping machine and it differs from *GQM* in that, that it does not include the track gauge influence (*RK*) which is not repaired by the tamping machines,
- the quality number (*QN*).

The sectional evaluation is realized according to [6] and represented by:

- the quality number (*QN*) of the evaluated section which serves as the quality examination of the evaluation of the track geometry in the evaluated track section as a whole, and which is calculated of the standard deviations of the individual parameters of the evaluated section (*SK*, *RK*, *PK*, *VK*). For the quality evaluation of the track geometry are specified the recommended values of the standard deviations of the individual parameters and quality numbers for the velocity zones. The limiting values of the standard deviations and the quality numbers for *RP*1 are introduced in the *table 2*,
- quality marks (*QM*) of the individual geometric parameters (*SK*, *RK*, *PK*, *VK*). The evaluation results according to the quality marks are the orientation and the additional evaluation and they are not obligatory for the state evaluation of the track geometry. The actions which are specified by the particular quality mark intervals have a recommending character (*table 3*).

Table 1. The parameters tolerance of the track geometry for velocity zone No. 1 [5].

Measured parameter	**Installation requirements**		**Maintenance limits**		**Safety limits**		**Note**
RK (mm)	-2	3	-4	10	-5	20	straight track
	-2	3	-4	26	-5	35	transitioned simple curve
ZR (mm/m)	3		4		5		
PK (mm)	-3	3	-15	15	-20	20	
Measured parameter	**Limiting values**		**Maintenance limits**		**Safety limits**		**Note**
ZK (1:n) (mm/ base)	1:250 (7.2; 4.0)		1:250 (7.2; 4.0)		1:167 (10.8; 5.99)		on the measuring base 1.8 m
	1:320 (18.8; 3.13)		is not evaluated		1:250 (24.0; 4.0)		on the measuring base 6 m
	is not evaluated		is not evaluated		1:333 (36.0; 3.0)		on the measuring base 12 m

Measured paramete r	Relative installation requirements		Relative maintenance limits		Relative safety limits		Note
VL, VP (mm)	-4	4	-15	15	-20	20	–
SL, SP (mm)	-5	5	-15	15	-20	20	–

Table 2. The parameters tolerance of the track geometry for RP1 [6].

Installation requirements				Safety limits			
SDV_{SK}	SDV_{RK}	SDV_{PK}	SDV_{VK}	SDV_{SK}	SDV_{RK}	SDV_{PK}	SDV_{VK}
1.5	1.5	1.8	2.0	2.7	2.9	2.9	3.4
QN				QN			
4.1				7.2			

Table 3. The scale of quality marks (QM) according to quality section evaluation [6].

Interval of quality marks	Verbal assessment section according to the quality mark	Colour of the quality mark in printed output
$0 < QM \leq 2$	state of track geometry is satisfactory in section evaluated	no colour marking
$2 < QM \leq 3$	it is recommended to design the repair of track geometry in the section evaluated into maintenance work plan	green colour
$3 < QM < 4$	it is recommended to perform the repair of track geometry in the section evaluated to the nearest inspection	violet colour
$4 \leq QM \leq 6$	it is recommended to perform immediate measures in the section evaluated to ensure the safety of operation	red colour

5 The quality of the diagnosed sections

Nowadays the results of the initial measurement, the first operational measurement and the second operational measurement are available before putting the section into operation. There were executed repair works by the builder with no provided records after the execution of the initial measurement. There was carried out an exchange of two reinforced concrete slabs on the Bratislava main station tunnel portal before the second operational measurement. Both slabs were damaged by the derailment at the switch No. 59.

In regard to the curve parameters the whole section is qualified according to the rules for the straight sections (regarding the maintenance tolerance and the safety limits of the parameter *RK*).

There were identified 10 local errors within the initial measurement before putting the construction into operation (4 errors out of them are classified as irrelevant errors). There were no local errors diagnosed within the operational measurements after the execution of the repair works [7].

The quality mark values of the individual evaluated parameters are presented in the *Fig. 1.* The limiting values indicating the impaired section state are according to the *table 3* the values $2 < QM \leq 3$. In this case, in the evaluated track section is recommended to design the repair of track geometry in the section evaluated into maintenance work plan. In the intervals $3 < QM < 4$, where is in the evaluated section recommended to perform the repair of track geometry in the section evaluated to the nearest inspection and $4 \leq QM \leq 6$, which indicates the infrastructure provider the recommendation to execute the immediate measures to ensure the operation safety. In the last two intervals were no quality mark values detected for any of the parameters of the track geometry.

The quality mark of the geometric parameter alignment of right rail QM_{SK} (after calculation also of left rail) was in the interval $2 < QM_{SK} \leq 3$ within the initial measurement (the recommendation to design the repair of track geometry in the section evaluated into maintenance work plan). After execution of the repair works the quality mark values decreased to the interval $0 < QM_{SK} \leq 2$ (the state of track geometry is satisfactory) in both realized operational measurements. The quality marks of the other geometric parameters - gauge deviations, superelevations and the rail top level of right rail (after calculation also the left) line were by all measurements in the interval $0 < QM \leq 2$ (the state of the track geometry is satisfactory in the evaluated track section) [7].

The quality numbers are presented in the *Fig. 2.* The quality number from the initial measurement (2.08) did not overrun the limited value specified for the installation requirements and the quality numbers from the first operational measurement (1.97) and the second operational measurement (1.97) did not overrun the values of the safety limits. The repair works executed before the first operational measurement improved the quality of the ballastless track and after the execution of the second operational measurement the quality number value steadied [4].

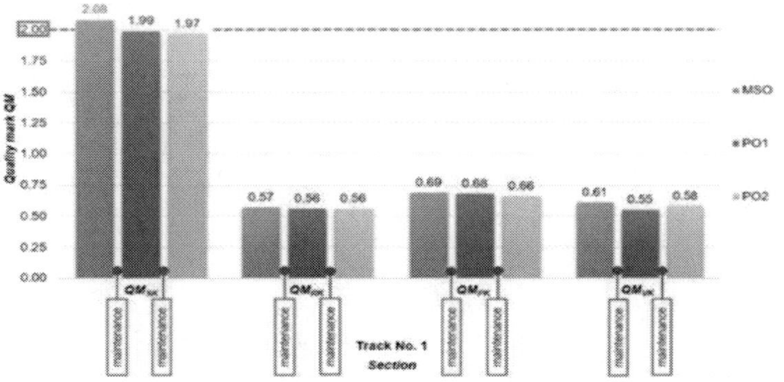

Fig. 1. Quality marks QM.

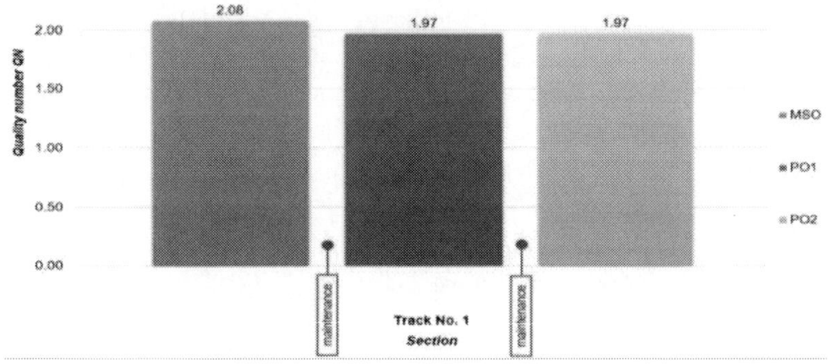

Fig. 2. Quality numbers QN.

The paper contains partial results of the grant VEGA 1/0597/14"Analysis of methods used to measure the unconventional railway track construction from the point of view of accuracy and reliability".

References

1. L. Izvolt, M. Smalo, Civ. and Env. Eng. **10(2)**, 78-93 (2014)
2. *ZSR SR 103-8 (S) General Requirements for the Design, Construction, Repair, Maintenance and Acceptance of Construction Repair and Maintenance Works on the Slab Track Construction* (GR ZSR, 2012)

3. J. Sestakova, Civ. and Env. Eng. **11(1)**, (2015)
4. L. Izvolt, J. Sestakova, M. Smalo, Proc. of Sem. Res. for Railway Infr.Const., 9-10 March Decin, Czech Republic., 98-110 (2016)
5. *STN 73 6360 Track Alignment Design and Track Geometry of Normal-gauge Tracks, and Amendment 1* (SÚTN, Bratislava, 2002)
6. *ZSR SR 103-7 (S) Measurement and Evaluation of Track Geometry by Measuring Trolley KRAB* (GR ZSR, 2008)
7. L. Izvolt et al., *Monitoring of Transition Areas in Bratislava Tunnel, Track No. 1 of Line Devinska Nova Ves - Sturovo – 1st and 2nd Stage* (KZSTH SvF: University of Zilina, Zilina, 2015)

KASS v.2.2. scheduling software for construction

Michal Krzemiński[1,*]

[1]Warsaw University of Technology, Faculty of Civil Engineering, Al. Armii Ludowej 16, 00-637 Warsaw, Poland

Abstract. The paper presents fourth version of specialist useful software in scheduling KASS v.2.2 (Algorithm Scheduling Krzeminski System). KASS software is designed for construction scheduling, specially form flow shop models. The program is being dedicated closely for the purposes of the construction. In distinguishing to other used programs in tasks of this type operational research criteria were designed closely with the thought about construction works and about the specificity of the building production. The minimal time, the minimal slack of brigades, the minimal slacks of the chosen working brigade and costs of the transfer operation of working fronts are included in operational research criteria between work centers. It is possible to enter data into the program both by hand as well as to load the Excel from files, similarly is with results, they are presented on-screen as well as a possibility of enrolling them in the file exists Excel. An element is very valid for it since allows for further simple processing of received results. In providing software for performing operational research calculations a technique of the complete review and simulation technology are being exploited. Described algorithms a program is using which will stay in the article as well as shown computational examples will remain.

1 Introduction

Construction scheduling problem is particularly important when the work accomplishment process is being prepared. Scheduling is a fast-growing area of scientific research in construction production engineering. It has applied methods of artificial intelligence, such as genetic algorithms, ant colony optimization algorithms, simulated annealing, neural networks. But also fuzzy logic techniques/elements of fuzzy logic are being widely used in order to determine time [1], [4].

This paper investigated a flow-shop scheduling problem, which consists in determining the optimum working sequence for brigades at workspaces. The model is based on assumptions that it is necessary to perform a sequence of technological operations, that the sequence is determined in advance and cannot be changed, that every brigade works at chosen workspace only once and at the exact time of its transition to another workspace, the next brigade comes in [7], [8].

* Corresponding author: m.krzeminski@il.pw.edu.pl

There is a lot of computer software which can help with scheduling process. Most popular are MS Project and Primavera. If we looking for software dedicated for flow shop models we can find programs like AIMMS - Flow Shop [9] and LEKIN -- Scheduling System [6]. These are very good and efficient programs but for mass production. For unique construction production we need something different. The construction has special conditions during production process so software need to have number of specified assessment criteria to give the best result. Answer for this problem may be KASS software.

2 KASS v.2.2 program description

The KASS v.2.2 software was written in the Java programming environment. It has a bilingual(Polish and English) interface, it does not require installation, it can be downloaded for free from the website www.ipb.edu.pl (bookmark - Programy).

The program uses the complete overhaul, when no more than 14 workspaces are considered. If there are more workspaces, the simulation was applied.

There are four optimization criterion: time, continuity, continuity of chosen brigade and brigade transition costs.

To show how the program works, we will use the example of optimization for chosen task. The picture below depicts the welcome window that appears after we start the program.

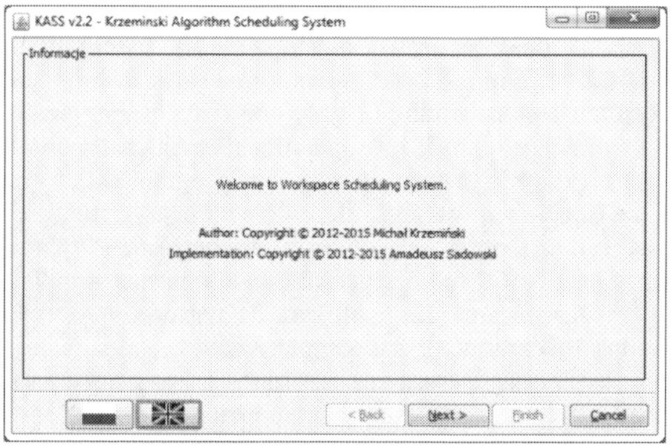

Fig. 1. The welcome window of the KASS v.2.2 program.

After we press the button Next, we move to the next window and we must choose workspace count (seven in this example), as well as working brigades count (four in this example).

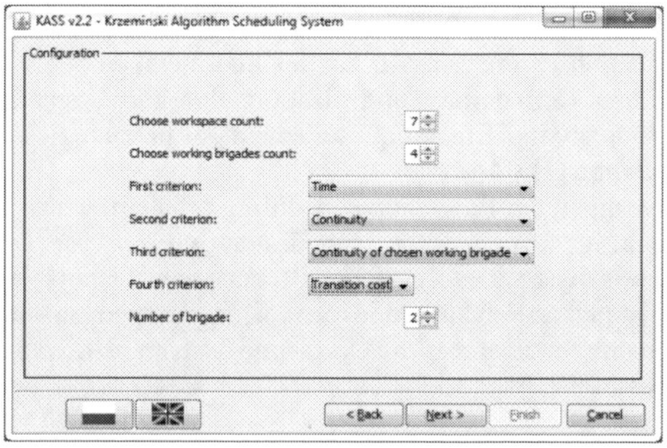

Fig. 2. The configuration window of the KASS v.2.2 program.

It is also necessary to choose the assessment criteria which will be taken into account during optimization calculations. It is possible to choose from one to four criteria In this example shortest work accomplishment time was chosen as the main criterion, continuity is second criterion, continuity of the brigade number 2 is the third criterion and the lowest transition cots is the last criterion. The criteria sequence is important, as it sets its precedence. The first one will influence the most the results of optimization.

On the following pictures you can see two matrix . The first matrix should be completed with duration of activity on individual working plots. In the second matrix user need to enter the cost of teams transition between working plots. It is possible to load data from excel file.

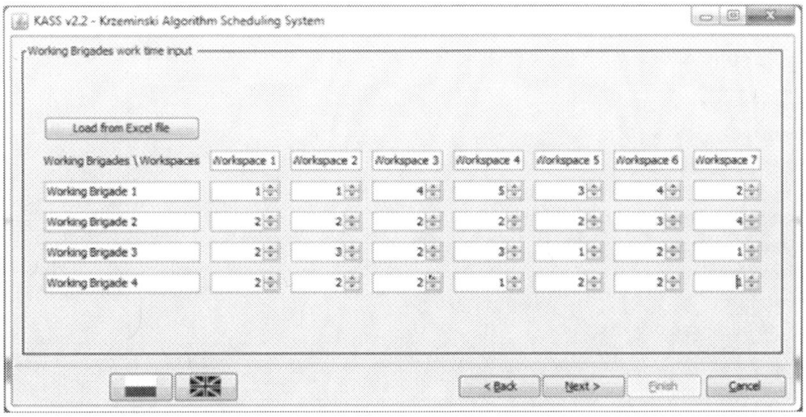

Fig. 3. Working time matrix.

The matrix of the transition costs don't need to be symmetrical. For example, equipment transport in up direction may be more expensive than transport in down direction.

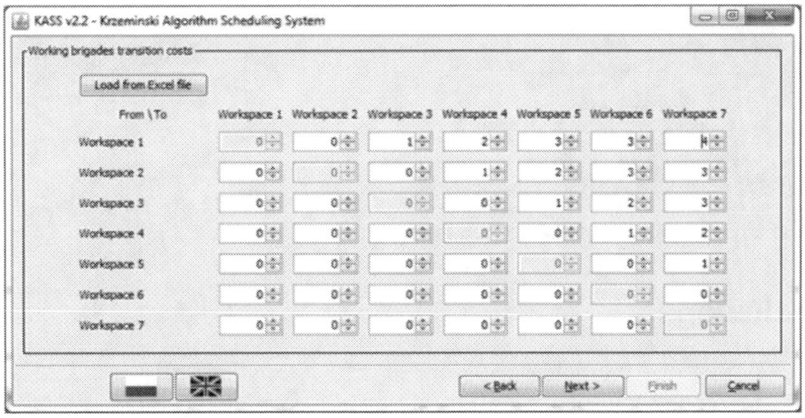

Fig. 4. Transitions costs matrix.

The last two picture number 5 and 6 depicts the results. The window presents the right scheduling and it is also possible to read off data concerning total time, total brigades work stoppages, chosen brigade work stoppage and brigades transition cost.

It is also possible to save the results as *.xlsx files, which are readable in the Excel program among others. It is useful, because it enables the KASS v.2.2. users to work further in an easy and comfortable manner on data that were obtained.

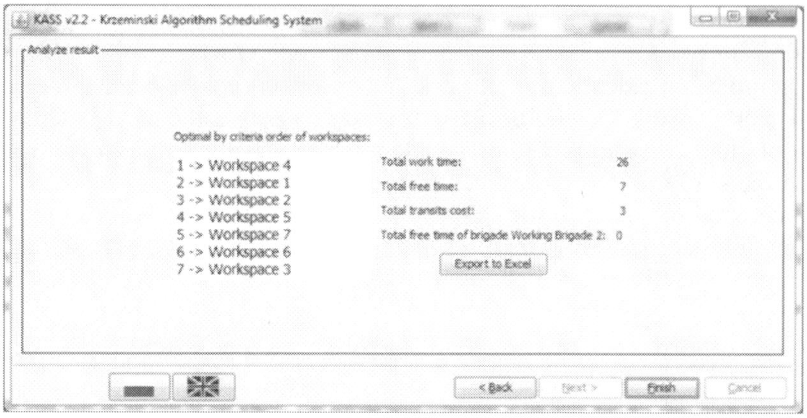

Fig. 5. The results window of the KASS v.2.2 program.

KASS results xlsx	
KASS version:	v2.2
Computation time [ms]:	31
Total operation length:	26
Total transition costs:	3
Total free time:	7
Chosen brigade:	Working Brigade 2
Free time of chosen brigade:	0

Day	**1**	**2**	**3**	**4**	**5**	**6**	**7**	**8**	**9**	**10**	**11**	**12**	**13**	**14**	**15**	**16**	**17**	**18**	**19**	**20**	**21**	**22**	**23**	**24**	**25**	**26**
Working Brigade 1	4	4	4	4	4	1	2	5	5	5	7	7	6	6	6	6	3	3	3	3						
Working Brigade 2						4	4	1	1	2	2	5	5	7	7	7	7	6	6	6	3	3				
Working Brigade 3								4	4	4	1	1	2	2	2	5		7			6	6	3	3		
Working Brigade 4											4		1	1		2	2	5	5	7			6	6	3	3

Workspaces' legend:	
Workspace 4	4
Workspace 1	1
Workspace 2	2
Workspace 5	5
Workspace 7	7
Workspace 6	6
Workspace 3	3

Fig. 6. The Excel results of the KASS v.2.2 program.

3 Conclusions

The article presents the latest version of the specialist software KASS. Software is dedicated for construction scheduling. Used assessment criteria and the possibility of cooperation with the MS Excel environment make program appropriate to support the real construction scheduling process. Previous versions of the software ware wildly tested. Tests were conducted for finishing works in the mall [3], for roadworks [5] and for monolithic construction works [2].

The author plans to work further on the KASS software and enlarge criteria, introduce heuristics, as well as elaborate the probabilistic version.

References

1. N. Ibadov, J. Kulejewski, M. Krzemiński, Fuzzy ordering of the factors affecting the implementation of construction projects in Poland, AIP Conf. Proc. **1558**, (2002)
2. P. Kowalik, M. Krzemiński, Logistyka NR **4,** (2015)
3. M. Krzemiński M, E. Krasnodębska, Theor. Found. Of Civ. Eng. Polish-Ukrainian Trans. **22**, (2014)
4. M. Krzemiński, AIP Conference Proceedings **1648**, edited by T. Simos et al. (2015)
5. M. Krzemiński, A. Wypysiak, Scheduling complete review application for road works, Procedia Engineering **91** (2014)
6. A. Madureira, I. Pereira, N Sousa, P. Avila, J. Bastos, Comp. Intel. for Eng. Syst. **46**, (2010)
7. M. L. Pinedo, *Scheduling: Theory, Algorithms, and Systems* (Springer, 2012)
8. M. L. Pinedo, *Planning and Scheduling in Manufacturing and Services* (Springer, 2009)
9. E. Zondervan, T. P. J. van Boekel, J. Fransoo, A. B. de Haan, Comp. Aid. Chem., Eng. (2014)

Fuzzy sets on step of planning of experiment for organization and management of construction processes

Azariy Lapidus [1], and *Aleksandr* Makarov [1,*]

[1]Moscow State University of Civil Engineering, Yaroslavskoye Shosse 26, 129337 Moscow, Russia

Abstract. In this article, problems of mathematical modeling and experiment planning of the organization and management of construction. The authors designated the basic restrictions and the difficulties in this field. Concluded that the planning of research experiment is possible in the information sphere with using of heuristic, graphical, mathematical models, as well as neural networks and genetic algorithms. The authors note the need for use of expert information in the case of the formalization of quality parameters. The article presented an overview of the translation methods of qualitative information into mathematical language. Comparison of methods the qualimetry of USSR scientists, the analytic hierarchy process and fuzzy set theory were performed. The benefits of the latter for interpretation of qualitative parameters were identified. The authors have given many examples of application fuzzy sets for formalization of organizational factors of construction processes. Finally, there conclusion was made about progressiveness and effectiveness of fuzzy set theory to describe the qualitative parameters of organization and management of construction.

1 Introduction

The development of construction which involves improving the basic criteria of its evaluation such as efficiency, quality, safety, reliability, durability, maintainability, is directly related to the perfection of scientific provision of its composing processes. The formalization of the real processes of the actuality with the assistance of variety of modeling means allows you to move them out of the area of the subjective perception of the human mind into a public area of the objective knowledge. Currently, there is a wide variety of modeling methods [1, 2, 3, 4, 5], a common feature of the most of them is usage of a mathematical approach that allows to implement input, processing and output of information. The degree of coincidence of the predicted results with the actual and the effectiveness of management based on modeled data depends on the measure of adequacy of the model. In this regard, the most important issue for the scientific presentation of construction processes modeling is search of the method, which more comprehensively reflects them with the help of mathematical formalization.

* Corresponding author: anmakarof@yandex.ru

To describe the construction processes through the formation of empirical relationships, refinement of mathematical models, and searching of the optimal conditions, usually scientists use statistical data analysis of the experiments. The main mathematical tools are the methods of the factorial, regression and variance analysis. These approaches are widely used in the scientific study of building materials [6], structures [7] and of various stages of building design [8]. The presence of a large database of theoretical and practical use of these methods and a high degree of reliability contributed to their widespread application. The processes to which factor and regression analysis are applicable, allow:

- carrying out multiple experiments under specified conditions;
- availability of the statistical database about the object of research;
- registration of the numerical results of parameter measurements constituting a statistical sample, the analysis of which leads to the determination of the sought regularities.

2 Determination of experiment conditions

Construction operation includes a large number of processes and phenomena that could be radically different regarding a set of control parameters and the complexity of their functioning. For example, researching of the compressive strength of concrete [9] and modeling of organizational and technological processes of erection of building structures [10] describe the elements of having different formalized models.

For the study of the first object, researcher himself can create and control external and internal factors and fix them definite numerical values by various instruments. In the end, he gets a statistical sample of values for system parameters, between which the relationship must be installed basing on the chosen criterion.

In the second case the studied system is much more complex, it has more factors and parameters with the implicit nature of the bonds. The life cycle of such process is different by large expenditure of time, material and human resources. As part of the experiment, it is hard to simulate it in artificial controlled conditions, and given the fact that we are using the exponential dependence between the number of required experiments and the number of study parameters for achievement of the statistical significance, it sometimes becomes absurd. Reproduction of this process in the laboratory conditions will not allow to create an adequate model, according to the system theory. The investigated construction process is one element in the system complex of construction of the building, and therefore, by law of the integrity of the system, it will change part of properties when coming out of it. To receive the correct model, the target process should be studied in conditions of the real construction site. At the present time, in a market economy, the access of researchers to the construction site is restricted. The possibility to conduct field experiments for determination of the organizational factors of construction processes, guided only by the purposes and problems of carrying out a scientific experiment depends on the desire of the investor or the project manager, and it is usually absent. Therefore it is virtually impossible to recruit the necessary number of actual construction objects to generate the model of research.

On the basis of the above, in case the scientific interpretation of the processes of organization and management of construction, the planning of research experiment is possible only in the information sphere with using of heuristic, [2] graphical [3] and mathematical models [11], as well as neural networks [4] and genetic algorithms [5]. The development and formation of the models under consideration is based on the application of expert estimate methods. Its center inevitably is a professionally trained specialist, an expert in the investigated area. His experience forms the basis of statistical data, the equivalent of the number of experiments. Using this method the researcher may invite the required number of experts to participate in the research whose wide construction experience will allow to satisfy the requirement of statistical reliability of the received results.

3 Methods of measurement of qualitative parameters

The next step after determining the conditions of the experiment is the determination of the input and output parameters of the investigated system and methods of their measurement. These parameters can be divided into two groups, viz. quantitative (number of workers - Nw, time - T, coast - C etc.) and qualitative (quality - Q, safety – S, resource conservation – R, construction supervision - Cs, information technology - It, weather conditions - Wc, qualification of workers – Qw, etc.) Approaches to measuring and processing of quantitative indicators are well studied, working on their formalization a researcher works with mathematical figures, i.e. on the accessible language for model.

Qualitative parameters traditionally represent a class of factors that are difficult to integrate into mathematical models, as they are the prerogative of the human - expert perception of the world. The application of the stochastic noise or multifactor analysis of variance proposed by R. Fisher in the first half of the 20th century give a partial unilateral solution of the problem. For a more extensive analysis of the qualitative properties of the object, the researchers somehow turn to the help of an expert from investigated area. His consciousness is a mechanism of output information on qualitative parameters of the system, the equivalent of a measuring instrument. But scale of measurement of technical devices is not characteristic of the human mind. A man evaluates any event more complex, due to the its perception of the world through the senses, and the further formation of the output in the form of concepts, images, conclusions. In this regard, the main problem of integration of qualitative parameters in the mathematical models is a translation of the information by human perception into the strict language of mathematical symbols.

In 1968, USSR scientists introduced the term "qualimetry" [12], implying a methodology for quality assessment in various fields of human activity. Approach of formalizing these systems consisted in creating a "tree of properties" of the research object (Fig. 1) with further comparison of its components with each other or with a sample. At that, the rank classification of parameters or conditional measure of their compliance with the basic pattern was used. With the help of qualimetry it is possible to successfully determine the best object out of a certain series, its place relative to the etalon, as well as a comprehensive indicator of the quality (P) through the weighted function of properties.

Table 1. Matrix of comparison of the parameters on the criterion "Safety".

	Wc	**Cs**	**Nw**	**Qw**	**It**
Wc	1	1/8	6	1/7	1/3
Cs	8	1	9	3	6
Nw	1/6	1/9	1	1/8	1/7
Qw	7	1/3	8	1	4
It	3	1/6	7	1/4	1

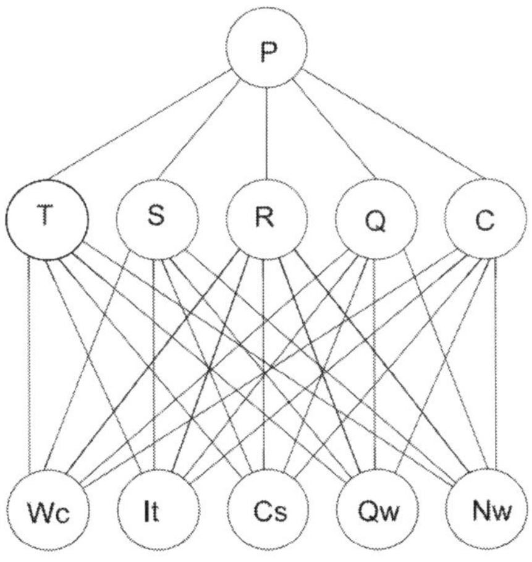

Fig. 1. "Tree of properties" (up-down).

In the 70s of the last century, American scientist T. Saaty explored the decision-making methods and suggested searching of the most faithful alternative through their pair-wise comparison by several criteria. The analytic hierarchy process is more simple and accurate tool in the field of expert information because a comparison of the two objects only simplifies the decision and increases the reliability of the assessment. When comparing pairs of objects that have a variety of quantitative and qualitative characteristics, the expert associates in his mind the images, enabling him to make more practical and actual conclusions. The comparison results are integrated into the matrix (Table 1.), the main eigenvector of the matrix after normalization becomes a vector of the desired priorities on the selected criteria [13]. The complex of simple mathematical apparatus and reliable transfer of intuitive expert information into it made this approach widely used in the practice of management decision-making at present, as well as in determining the weight coefficients of parameters of investigated system.

In 1965, L. Zadeh began the development of fuzzy set theory [14] the most efficient solution for the analysis of qualitative parameters from the point of view of the authors. The term linguistic variable [15] contains a fundamental property of qualitative factors to determine their value is primarily a function of the human mind, which it transforms into information through language. Levels of varying the parameter are determined by "granules" of human language or by the terms: "low", "medium", "high" or "poor", "satisfactorily", "good", "excellent" etc. Names of the terms reflects images of the qualitative parameter arising in the mind of the expert, the number of terms depends on detalization of the evaluation.

Qualimetry offers translation of these terms into mathematical language by using rank scale: maximal term is assigned to "1", and the rest of terms are assigned to values from the interval [0, 1], for example, "low" is "0", "medium" is "0.5", "above average" is "0.75". This transition is single-valued and strict which leads to the inevitable loss of information when assessing the qualitative factors. The fuzzy set theory represents a term not by one number, but by set with the grade of membership of all values to term through membership function.

On figure 1 the linguistic variable "qualification of workers" of the organization parameter of the construction process was graphically presented. This linguistic variable is divided into three terms: "low", "medium" and "high", which accept values from 0 to 100. The range of values for the linguistic variable is the quantity of terms, the range of values for a specific term is a fuzzy set, a region of space, which is formed in accordance with the membership function μ (x). The membership function defines the boundaries of a fuzzy set,

"0" means the point not belongs to the linguistic term, "1" is the maximum grade of membership, values in the interval (0,1) represent intermediate states of the point.

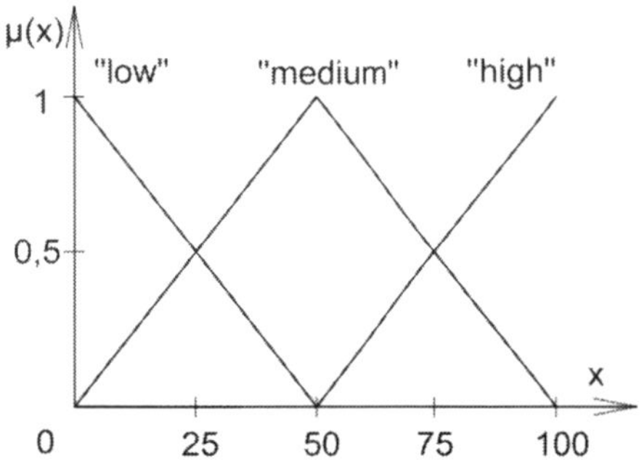

Fig. 2. Parameter "qualification of workers".

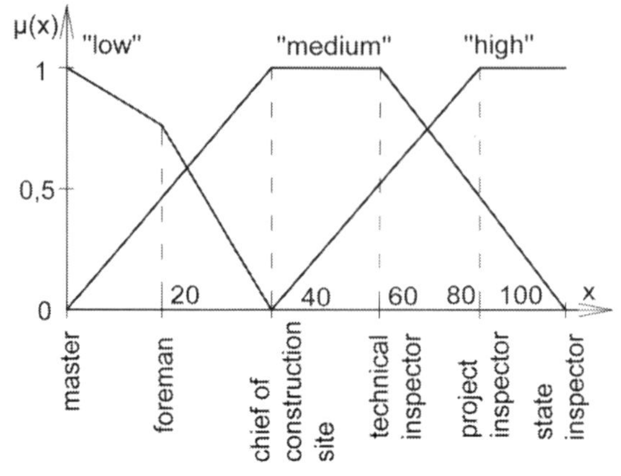

Fig. 3. Parameter "construction supervision".

In general, the term of a linguistic variable is denoted by a fuzzy set

$$A = (\mu^*(x)|x), \forall x \in X \quad (1)$$

where X is continuous (x_0, x_n) or discrete area $\{x_1, x_2, \dots x_n\}$, x is a value of specific point of term, $\mu^*(x)$ is the grade of membership x to the set A, defined with the help of the membership function $\mu(x) \in [0,1]$. Thus, based on Fig. 1, the term "low qualification" represents by fuzzy set (1|0; 0,5|25; 0|50), "medium" by (0|0; 0,5|25; 1|50; 0,5|75; 0|100), "high" by (0|50; 0,5|75; 1|100).

4 Fuzzy sets of organizational variables

The fuzzy set theory can overcome the strict limits of the traditional approach and gives freedom in the interpretation of qualitative parameters through linguistic variable and the membership function, the distribution of which is determined by the creator of the model of the construction supervision (Fig. 3). Also, this approach allows the expert to give a more reliable assessment of the measured variable of the construction supervision. The expert, in contrast to the machine, can detect not only the number of links in the construction supervision chain on construction site, but also to analyze the capacity of the fulfillment by each of them of their duties. We have to keep in mind the problem of indifferent, passive, performing their duties "in so far as" subjects, as well as regulatory authorities assigned for the construction project formally, not having real powers or executing their functions not in full volume. The expert has methods for the detection and analysis of this problem. Fuzzy sets allows him to assign to the parameter of construction supervision "low" or "medium", "medium" or "high" level with a definitely predetermined membership function under the same conditions.

Changing parameters may be in accordance with the Gaussian distribution curve as shown in Figure 4 for analyzing the quality of construction products.

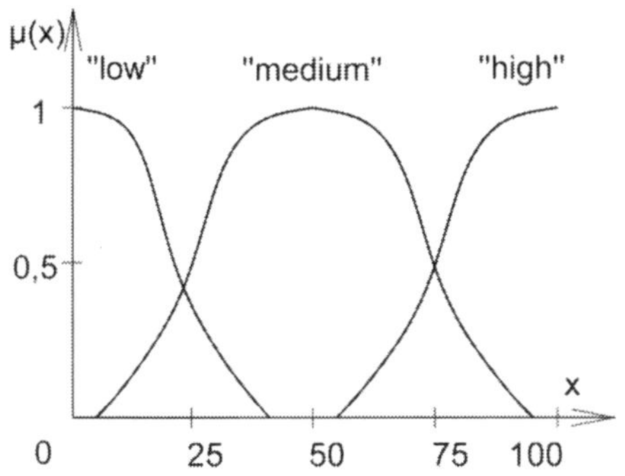

Fig. 4. Parameter "quality".

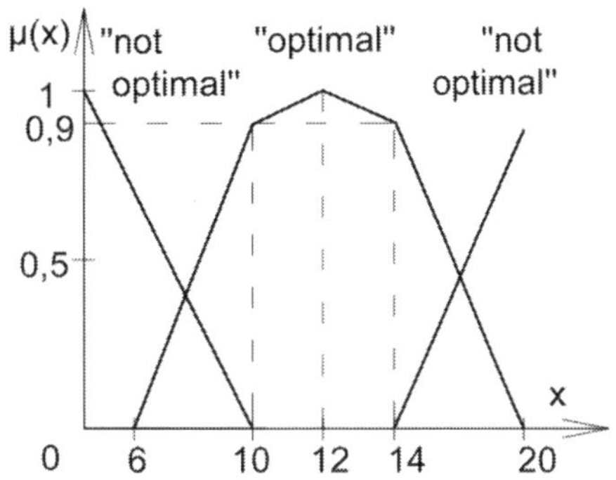

Fig. 5. Parameter "construction supervision".

Fuzzy sets along with qualitative works well with quantitative parameters under uncertainty [16]. As noted above, construction processes are characterized by a large parametric complexity, inability for accurately measure all components of the system, by significant influence of the human factor, long life cycle. These properties oblige researchers investigate construction processes with help of chaos theory and Heisenbergs uncertainty principle which states the impossibility of accurately predicting and measuring the components in complex systems. For example, the optimal distribution of organizational parameter "number of workers" for certain construction process is shown in Fig. 5. The number of workers is 12 people counted on the basis of performance taken from the normative documents and project time of works. But the expert knows that first of all, the performance of the same brigade on different objects is always different, secondly, the brigade often may have changes during the construction process, and thirdly, the decision-maker can change the number of workers for regulating criteria of resource conservation or efficiency. Thus, a fuzzy set shown in Fig. 5 reflects the real situation more fully, when the number of workers in the brigade is "10" or "14" can also be attributed to the linguistic variable "optimal" with the grade of membership 0.9.

Consider another example, winter conditions for concrete works were established by normative documents with an average daily temperature below 5 °C and conditions with high temperatures were limited to above 25 °C. But for the decision-maker, this border has always "fuzzy", he turns it into an interval (a fuzzy set), guided by the grade of the design responsibility, by his experience and intuition and in conclusion by the impossibility to determine the exact average daily temperature at the time of the decision. For this reason fuzzy representation of the conditions of application of special measures of organizational process is a better way: winter conditions are (1 | 0; 0.9 | 2; 0.5 | 3; 0.2 | 5; 0 | 8), summer conditions are (1 | 30; 0.9 | 25; 0.5 | 23, 0 | 20).

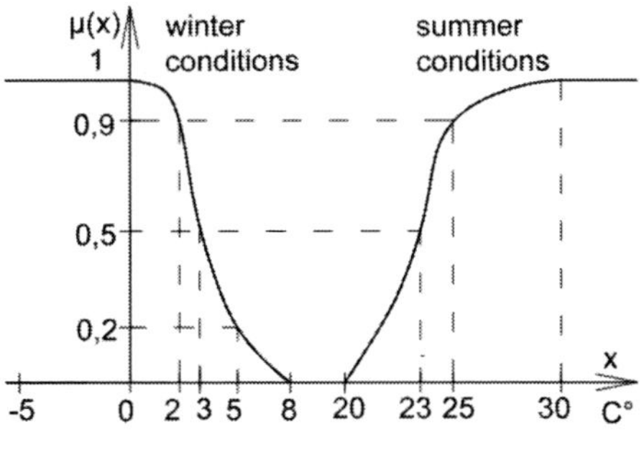

Fig. 6. Parameter "weather conditions".

5 Conclusion

The article presented two stages of experiment planning in the field of organization and management of the construction processes. In the first stage, the definition of the experimental conditions shows that obtaining statistically reliable sample of the experiences on construction site is almost impossible. It is proposed to plan an experiment based on heuristic, graphical, mathematical models, as well as neural networks and genetic algorithms with help of expert assessments. Within the second stage, the order of measurement of parameters, several methods of translation of quality characteristics of the investigated object into the mathematical language of were considered. Of these, preference is given to fuzzy set theory, which is a more reliable tool for the evaluation of the parameters of organizational models. Also it is not contrary to chaos theory at the description of complex dynamic systems due to the “fuzziness” of input and output information on the system parameters. Based on the above, authors concluded that there is a need for further application of fuzzy sets in research of the organization and management of construction processes.

References

1. A.A. Lapidus, AMM **584-586**, 2230-2232 (2014)
2. M. Buitrago, J.M. Adam, Y.A. Alvarado, J.J. Moragues, I. Gasch, P.A. Calderon, ES **111**, 1-10 (2016)
3. C.-F. Chi, S.-Z. Lin, R.S. Dewi, AA&P **72**, 359-369 (2014)
4. D.A. Patel, K.N. Jha, SS **89**, 240-248 (2016)
5. Y. Shin, H. Cho, K.-I. Kang, AiC **20-5**, 550-558 (2011)
6. J.-S. Choua, C.-F. Tsai, AiC **24**, 52–60 (2012)
7. A. Silva, J.L. Dias, P.L. Gaspar, J. de Brito, AiC **30**, 151–160 (2013)
8. H. Lulic, A. Civic, M. Pasic, A. Omerspahic, E. Dzaferovic, PE **69**, 902-910 (2014)
9. S. Chithra, S.R.R. Senthil Kumar, K. Chinnaraju, F.A. Ashmita, CaBM **114**, 528–535 (2016)
10. A.A. Lapidus, A.N. Makarov, PE **153**, 378-383 (2016)
11. S.S. Amiri, M. Mottahedi, S. Asadi, EaB **109**, 209–216 (2015)
12. G.G. Azgal'dov, A.V. Glichev, Z.N. Krapivenskij, J.P. Kurachenko, V.P. Panov, M.V. Fjodorov, D.M. Shpektorov, ZSiK **1**, 34-35 (1968)
13. T.L. Saaty, L.G. Vargas, MM **5-5**, 309-324 (1984)
14. L.A. Zadeh, IaC8, 338-353 (1965)
15. L.A. Zadeh, IS8, 199-249 (1975)
16. A. Salah, O. Moselhi, ECaAM **22-2**,214-241 (2015)

Ranking of actions to reduce risks in QMS of construction enterprises

Nikolay Ivanov[1,*]

[1]Moscow State University of Civil Engineering; 26, Yaroslavskoe sh. 26, 129337 Moscow, Russia

Abstract. In conditions of economic crisis, businesses of construction industry lack the resources to eliminate real risks of non-conformities due differences between construction products and consumers' expectations. In this situation when classic models based on the CAPA become ineffective, the management of non-conformities requires some new acceptable, efficient, and low-cost approaches. The paper describes the original ranking algorithm for measures aimed at reducing the risks in business environment with financing constraints.

1 Introduction

In the history of any company comes a time when in front of its leadership raises the question about what to do to improve company's performance? This question is relevant as to a successful enterprise, which has quite a large market share, as well as for a company on the verge of survival. ISO 9001:2015 answer the question this way: the company is more successful when its management system and its products meet the requirements of all stakeholders. The process approach, which applied in the Standard and includes the cycle PDCA ("Plan - Do - Check - Act") and risk-based thinking, allows the organization to plan its processes and their interactions. Implementation of a PDCA cycle allows the organization to ensure its processes with the necessary resources and manage them as well as identify and implement opportunities for improvement. Risk-oriented thinking enables an organization to identify the factors that may lead to a deviation from the planned results and to apply preventive controls to minimize negative impacts and maximize the use of emerging opportunities.

As a rule, the aforementioned deviations with negative consequences are called nonconformities. An organization is required to fix detected inconsistencies, identify their causes, plan and implement corrective actions to eliminate the causes of non-conformities.

Any non-conformity appeared during construction of buildings or structures leads to a non-conforming product. Additional investment of time and resources of the organization are the result of processing or disposal of such products. Therefore, activities related to the elimination of nonconforming products (corrections) and with the elimination of the causes of non-conformities (corrective actions) should be considered as sources of additional costs. In some cases, the additional costs of the elimination of nonconforming products may be 28-33% of the cost of construction. Any construction organization will seek to minimize

* Corresponding author: IvanovNA@mgsu.ru

these costs. A design of effective corrective action plan can be considered as one of the ways to reduce the additional costs. Consider a plan as an effective plan when the cost of its implementation is significantly lower than potential cost of removal of nonconforming products due non-conformities, which were not removed.

Ideally, any construction enterprise must address the causes of all identified discrepancies, which lead to the emergence of nonconforming products. However, it is not a case in everyday practice. As previously mentioned, the problem lies in the lack of funds for implementation of all necessary corrective actions. Analysis of practice used by building organizations of Russia for the management of non-conformities revealed that most often implemented two options. The first variant is the choice of the least expensive measures from the list of corrective actions. Second variant is the choice of measures aimed at eliminating the causes of the most resonant non-conformities that cannot be ignored. In both cases, the efficiency of the funds used on the corrective actions is not defined and is not considered.

Questions of data collection and data processing related to non-conformities and the reasons for their occurrence are widely enough considered in the scientific literature [1, 2]. These issues are reflected in a number of previous articles of the author [3, 4]. However, in the scientific literature, problems of formation of corrective action plan are not considered. Separate studies provide only general information about the role of this plan as part of CAPA. Thus, the study aimed at the development of practical recommendations on the formation of the corrective action plan seems very important. In papers published during 2014-2016 [5, 6] the author took the steps to formalize the problem. In particular, in [5] describes the algorithm, which selects the primary non-conformity based on priority rank of corrective actions associated with that the non-conformity. Additionally, [5] discusses the algorithm for creating a set of corrective actions to address the root causes of non-conformity. In [6] the author carried out the original ranking algorithm for corrective actions included in the action plan, which aimed at addressing the identified non-conformities and their causes.

This article is another attempt to formulate a methodical approach, which is based on an effective governance mechanism of non-conformities, detected in the process of QMS.

2 Description of the problem

2.1 Verbal description

Let there be a set of non-conformities, each of which contains a list of reasons that led to the emergence of non-conformity. Let as well, for each of the identified non-conformities there are fixed costs, which would be used on elimination of nonconforming products.

Assume that frequencies of occurrence of different non-conformities were identified. Furthermore, so called chains of non-conformities, which helped to define root causes for all non-conformities, were recorded. Assume also that an expert defined a set of corrective actions to address each of the causes of each of non-conformity. Additionally, the expert identified potential costs associated with the implementation of each of the corrective actions.

It is necessary to determine a procedure to eliminate the causes of non-conformities. The procedure should reduce the risk of occurrence of non-conformities and minimize the costs of the removal of potential nonconforming products.

For simplicity, consider the task in terms that are not directly related to the QMS. Imagine that new completely clean carpet was laid out in receptionist head-manager of the organization. Eventually, the carpet will get all sorts of stains, different sizes and for

different reasons. It would be nice to take the carpet to the cleaners every night; but, the organization does not have the time or resources. Instead the organization hired a maid, which has the task of keeping the carpet clean. When a stain on the carpet is found, the maid has to document the type of stain and the cause of its occurrence. Then she/he has to clean the carpet with special cleaning products and try to remove the spot entirely to the best of her/his effort. The worst result is that she/he reduces its area and/or the degree of damage. Acquisition costs of cleaning products are included in the enterprise's product cost, thus lowering its competitiveness. The organization made a decision to create a plan of measures on elimination of the known causes of spots in order to prevent re-staining and thus to reduce non-productive expenditures. Note, that the amount of money that can be spent on the elimination of known causes of spots is limited.
We assume that each detected non-conformity *j*, which is caused by the reason *i*, leaves the "spot" of a certain type. The cost of such "spot" removal will be considered equal to the value of the cost of removal of non-conforming products associated with the identified non-conformity - z_{np}^{ji}. The graphical representation of the term "spot" is illustrated in the form of a rectangle with sides, equal to 1 and z_{np}^{ji} (Fig. 1a). If any nonconformity *j*, due to the reason *i*, is detected repeatedly, the total area of "spots" of one type ("area of contamination") will be equal to the sum of the areas of elementary rectangles (Fig. 1b) or the area of a rectangle with sides N_{ij} and $\bar{z}_{np}^{ji}$ (Fig. 1c). Value N_{ij} is the number of times when non-compliance *j* due to reason *i* happened. Value $\bar{z}_{np}^{ji}$ is the average cost of the nonconforming products elimination, which is calculated according to the formula (1):

$$\bar{z}_{\text{нп}}^{ji} = \sum_{k=1}^{Nij} z_{np}^{jik} \;/\; Nij \tag{1}$$

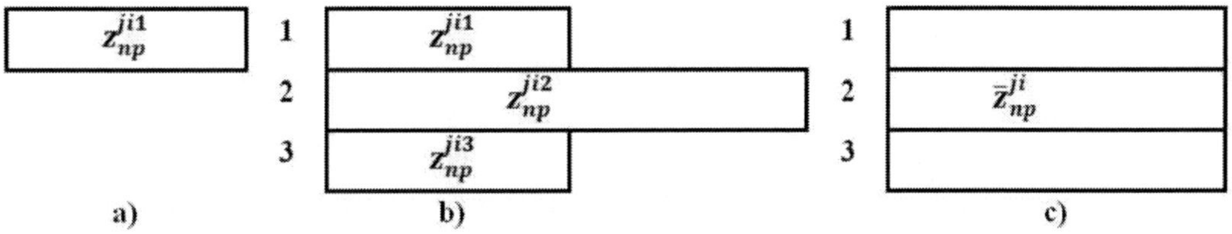

Fig. 1. The cost of removal of nonconforming products ("area of contamination") which is a result of non-conformity j due to reason i; a) for a single detection of non-conformity; b) with repeated detection of non-conformity; c) model "average cost".

If there are a few causes of non-conformity *j*, the general "area of contamination" can be defined by the formula for this discrepancy (2):

$$z_{np}^{j} = \sum_{i=1}^{M} \sum_{k=1}^{Nij} z_{np}^{jik} \tag{2}$$

M is the number of identified causes of non-conformity *j* (the number of "spots" types).

Then the overall cost of the removal of "spots" (full carpet cleaning) **Z** can be determined by formula (3):

$$\mathbf{Z} = \sum_{j=1}^{L} \sum_{i=1}^{M} \sum_{k=1}^{Nij} z_{np}^{jik} \tag{3}$$

L is the number of identified inconsistencies.

The cost of removal of any *i*-th cause of the *j*-th non-conformity is determined by the cost of corrective actions to eliminate this *i*-th cause. Let us call them as f^{ij}.

The cost of removal of any non-conformity is determined by the cost of corrective actions to eliminate all causes of the non-conformity. Let us call these costs as a F_{ca}^{j}. The volume of these costs can be determined by formula (4):

$$F_{ca}^{j} = \sum_{i=1}^{M} f^{ij} \tag{4}$$

2.2 Formal description

We need to create a set of corrective measures, for which the conditions (5) - (8) can be satisfied.

$$\sum_{j=1}^{L} Z_{np}^{*j} \rightarrow min \tag{5}$$

with constraints

$$L^{*} < L \tag{6}$$

$$\sum_{j=1}^{L} \sum_{i=1}^{M} f^{ij} \leq \text{Splan} \tag{7}$$

$$F_{ca}^{j} < Z_{np}^{j} \tag{8}$$

L – the number of non-compliances identified in the previous period;
L^{*}– the number of possible non-conformities of the planning period (the number of non-conformities of the past period, the cause of which is not completely eliminated);
M – the number of different reasons that resulted in non-conformity j;
f^{ij} – the costs of elimination of the reason i, which led to the emergence of non-conformity j;
Splan – the amount of funds allocated for corrective measures following the results of the past period;
Z_{np}^{j} – the costs of removal of nonconforming products related to the identified non-conformity j;
Z_{np}^{*j} – the potential costs of the removal of non-conforming products associated with the re-emergence of non-conformity j;

3 Algorithm of designing an effective corrective action plan

The original algorithm of designing an effective corrective action plan, "Carpet Cleaning" algorithm, was a result of our research aimed at studying the efficiency of construction companies before and after the introduction of the QMS and the effectiveness of decisions taken to reduce the risk of non-conformities.

The algorithm consists of the following sequence of actions:

1. Identify all types of stains that appeared on the carpet for a certain period of time, and establish the cause.

2. For each of the causes of the spots define an event aimed at their elimination, the costs for the implementation of this action, as well as the total cost of cleaning the carpet from stains that appeared as a result of the considered reason.

3. If the cost of eliminating the causes of any type of stain is greater than the cost of removing all the stains of this type, then limit the actions by only the removal of stains

("carpet cleaning"); otherwise, in addition to removing the stains, plan to carry out activities to address the causes of spots.

4. Determine the effectiveness for each scheduled event. The efficiency of measures is the ratio of the cost of stains removal, where the particular cause is the reason of stains, to the sum of the all the costs for the elimination of that particular cause. The higher the value of this index, the better the event.

5. Determine the amount of funds needed for the least costly of all the activities included in the plan. Assign the value found for the indicator "Minimum Cost".

6. Sort the planned activities in order of decreasing effectiveness.

7. Set the value of the index "Action Priority" to 1.

8. Choose the most effective measure out of the possible measures that don't have implementation priority.

9. If the costs of the measure are more than the remaining amount of funds allocated for the elimination of the known causes of stains, the measure is removed from the plan and the process proceeds to step 8.

10. Keep the event in the plan. Set its "Action Priority" index equal to the current value of "Action Priority" index.

11. Reduce the remaining amount of funds allocated for the elimination of the known causes of stains for the cost amount of conducted activities.

12. If the remaining amount of funds allocated for the elimination of the known causes of stains is less than the indicator of "Minimum Cost", go to step 14.

13. Increase the value of the "Action Priority" indicator by 1 and go to step 8.

14. All activities with no sequence of their execution are excluded from the plan.

4 Summary and conclusions

Each company has own way of treating risk, and it's the task of risk-team of company to determine the best way to interpret risk levels [7]. Once this is done, risk-team needs to take action on the risks management. They can to make use known quality management processes such as corrective and preventive actions (CAPA) to address the risks. Reviewed above algorithm can be considered as one of the possible ways of solving the problem of minimizing the risks. The advantages of the algorithm are simplicity and scalability. The algorithm can be modified by changing the level of details of the taken decisions.

References

1. M. Donauer, P. Peças, A. Azevedo, Robot. and Comp.-Integr. Manuf. **36**, 84-92 (2015)
2. D. Okes, *Root Cause Analysis: The Core of Problem Solving and Corrective Action* (ASQ Quality Press, 2009).
3. N. Ivanov, Mod. App. Sc. **10(1)**, 47-51 (2016)
4. N.A Ivanov, Int. J. Appl. En. Res. **10(21)**, 42415-42418 (2015)
5. N. Ivanov, J. of Ec. and Entrepreneurship **11-2**, 621-626 (2014)
6. N. Ivanov, Proc. Eng. **153**, 228–231 (2016)
7. T. Lozier, Quality Digest Magazine, Information on http://www.qualitydigest.com/inside/risk-management-column/030216-what-risk-based-thinking.html

Innovations in Construction Personnel Education

Aleksander Nicał[1,*], *Paweł* Nowak[1], and *Jerzy* Rosłon[1]

[1]Civil Engineering Faculty Warsaw University of Technology, Armii Ludowej 16, 00-637 Warsaw, Poland

Abstract. The Erasmus+ programme, financed by the European Union, encourages international projects focused on education and recognition of professional qualifications. Projects presented in this paper deepen the ideas of Directive 89/48/EWG, which will lead to the creation of a proper European system of comparison, certification and mutual recognition of qualifications in construction. Results of the projects will improve didactic process and qualifications recognition of European Union managers in construction. Paper presents 3 Erasmus+ projects: "The Augmented Reality Formwork Assembly Training" (Project number 2016-1-PL01-KA202-026102), "Common Learning Outcomes for Construction Managers in EU, part IV" (Project number 2015-1-PL01-KA202-016454), and "Continuing V.E.T. Training Programme in Lean Construction to support the transition of the Construction Sector towards a more competitive and productive model" (Project number 2016-1-ES01-KA202-025694).

1 Introduction

Projects, within "Erasmus +" European Union's Programme, are expected to enhance modernisation and reinforce education and training systems in response to the main challenges of today's world (employment, economic stability and growth, as well as active participation in democratic life and free labour market) [1]. According to European Commission: "The construction industry is very important to the EU economy. The sector provides 20 million direct jobs and contributes to about 10% of the EU's GDP. It also creates new jobs, drives economic growth, and provides solutions for societal, climate and energy challenges. The goal of the European Commission is to help the sector become more competitive, resource efficient and sustainable" [2]. Commission also lists the main challenges facing construction. Two of these challenges are:

• training: improving specialised training and making the sector more attractive, in particular for blue-collar workers, technical colleges and universities.

• innovation: more active uptake of new technologies.

Paper presents three Erasmus+ projects: "The Augmented Reality Formwork Assembly Training" (Project number 2016-1-PL01-KA202-026102), "Common Learning Outcomes for Construction Managers in EU, part IV" (Project number 2015-1-PL01-KA202-016454),

* Corresponding author: a.nical@il.pw.edu.pl

and "Continuing V.E.T. Training Programme in Lean Construction to support the transition of the Construction Sector towards a more competitive and productive model" (Project number 2016-1-ES01-KA202-025694)

2 ARFAT - Augmented Reality Formwork Assembly Training

Developed training systems, based on modern IT technology - Augmented Reality (AR), are being constantly improved by most influential companies all over the world. Being used, among others, in trainings for pilots, F1 drivers and military personnel, it is safe to say, that AR is the future of high-quality work-based VET [3]. "The Augmented Reality Formwork Assembly Training" (Project number 2016-1-PL01-KA202-026102) deals with AR (figure 1) and another important issue - Health & Safety.

Fig. 1. AR system for formworks (source: own & PERI).

Falls from height are the most common causes of serious accidents, often fatal, not only in Poland but throughout European Union. As human life is the most important value, there is a great need of elevating H&S level. This problem has been raised in European Directive 2001/45/EC, which obliges to take appropriate measures to improve safety and health at work [4].

The project will address H&S issues as it stems from the need of prevention of accidents on construction sites. These accidents are mainly caused by falling from height, especially from scaffoldings. Works at heights (i.e. formworks, scaffolding works) are one of the most dangerous among construction works. One of the main reasons is: because current training methods are insufficient. As the research shows, workers are not interested in classic trainings. Use of mobile devices (smartphones, tablets), modern AR technology and supporting media files will make ARFAT training much more interesting and unforgettable. The need for the projects was also confirmed by the questionnaires and previous project (ARCW – Health and safety procedures for curtain walls with the use of Augmented Reality Technology). Both construction employees and employers stated that there is a great need for construction trainings with the use of AR [5].

As a result of the project, training system will be created together with paper manual, application for mobile devices and supporting media files. Training will be open for all interested parties, it will use innovative methods that suit in the best way modern, digital era. Never before was AR used for formwork training. Use of mobile devices and AR technology will make the training innovative and interesting to the participants. Foreseen manual, training system and augmented reality software will contain best practices from different European countries in the field of formworks and scaffoldings, and series of exercises for users. ARFAT can also help to improve language competences (Polish, Spanish, English and German versions of manuals, software and training will allow users to improve their language skills).

After the completion of project, ARFAT outcomes can help to develop appropriate assessment and certification methods for formwork and scaffolding workers in EU. International contents of the training prepared on the basis of experience of international Partnership will strengthen possibility of construction staff mobility across EU. It will also help them to act in the multicultural environment - relevant to the construction site and multicultural global construction companies. Project will strengthen cooperation between industry and education sector. ARFAT will promote access to and learning through Open Educational Resources (OER), as the training will be available for all interested parties on the internet and in libraries.

The project is planned to last for 24 months, starting on the 1st of December 2016 and finishing on the 30th of November 2018. The following organisations will be involved in the Project: Warsaw University of Technology (Poland) (Project's Promoter), Polish Association of Building Managers (Poland), Technische Universitat Darmstadt (Germany), Fundacion Laboral de la Construccion (Spain), Universitat de Valencia (Spain), PERI Polska Sp. z o.o. (Poland).

The outputs containing objectives (O) of the Project are following:

• O1: Evidence based learning outcomes. This output comprises learning outcomes on formworks and scaffolding works namely statements of what learners should know, understand and be able to do upon completion of the ARFAT training, in the form of definitions of specific knowledge, skills and competences.

• O2: ARFAT training system.

• O3: ARFAT manual.

• O4: ARFAT application (software). It is foreseen to prepare two versions of application for two most common systems for mobile devices in EU: android OS and iOS.

• O5: ARFAT AR markers.

• O6: ARFAT instructional movies.

The objective of these outputs is to develop up-to-date, tailor-suited to sectoral needs, modern formworks and scaffolding works training, appropriate to be integrated into existing VET offerings or to serve European community as a stand-alone training. These outputs will address modern skills needs of construction engineers, construction workers, stakeholders and associations in the construction sector, SMEs and companies (construction sector), VET providers, and technical universities, delivering a European solid, reliable and comprehensive pedagogical tool.

3 CLOEMC - Common Learning Outcomes for Construction Managers in EU, part IV

Project addresses the vocational education and training sectoral priority by improvement of vocational education and training didactic materials of the construction project management and horizontal priority by facilitation of qualification recognition system for managers in construction [6].

There is a great migration of engineers, construction managers in the construction sector due to the different state of economic development of the EU countries. For this reason, it is extremely important that construction managers' qualification and skills are being recognized and certified in the same way all over EU [7].

The Project is complementary to previous three CLOEMC projects, where 19 manuals (the set is also known as Construction Managers' Library - shown on figure 2) in several languages were successfully created, namely:

• (CLOEMC I) - Project Management in Construction, Human Resource Management in Construction, Partnering in Construction, Business Management in Construction Enterprise, Real Estate Management, Economy and Financial Management in Construction, Construction Management;

• (CLOEMC II) - Risk Management, Process Management – Lean Construction, Computer Methods In Construction, PPP Projects in Construction, Value Management IN Construction, Construction Projects – Good Practice;

• (CLOEMC III) - Due-Diligence in Construction, Motivation and Psychology Aspects in Construction Industry, Professionalism and Ethics in Construction, Sustainability in Construction, Health and Safety in Construction, Managing Building Pathology and Maintenance.

In CLOEMC IV project the following manuals are under preparation: Revitalisation and Refurbishment in Construction, Building Information Modelling – BIM, Optimisation of Construction Processes, Diversity Management in Construction, Mechanics of Materials and Structures for Construction Managers, Corporate Social Responsibility in Construction [8], [9].

The CLOEMC IV is promoted by Warsaw University of Technology (Division of Production Engineering and Construction Management) (Poland), partners in the project are: Technische Universität Darmstadt (Germany), Reykjavik University (iceland), Polish Association of Building Managers (Poland), Chartered Institute of Building (UK), Association of Building Surveyors and Construction Experts (UK), AWBUD S.A. (Poland).

Fig. 2. Construction Managers' Library - 19 manuals printed so far (on the shelf).

In order to suit the content of the manuals to the need of construction industry, and academic and scientific needs, series of questionnaires were conducted among construction managers, civil engineering professors, and government representatives. Partial results of the questionnaires are shown in tables 1-3 and figures 3-5.

Table 1. Is the presented scope of knowledge important for Building Managers for qualifications recognition? (1 - not important; 5 - very important).

1	2	3	4	5	Number of observations	Average
1	0	7	29	21	58	4.19

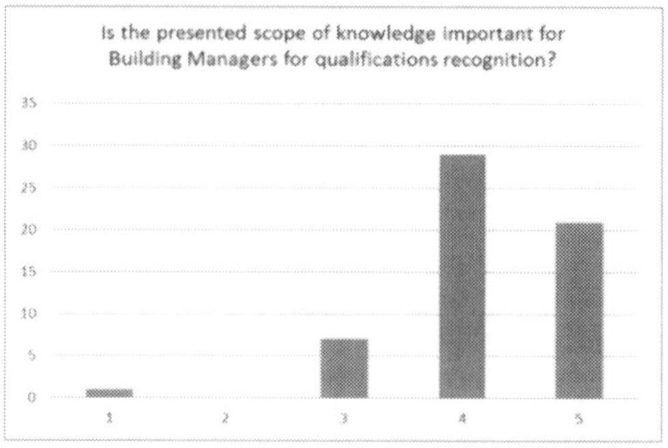

Fig. 3. Results of the survey part a.

Table 2. Are the manuals properly responding to the industry needs in respect of the managerial knowledge? (1 - not important; 5 - very important).

1	2	3	4	5	Number of observations	Average
0	3	9	29	17	58	4.03

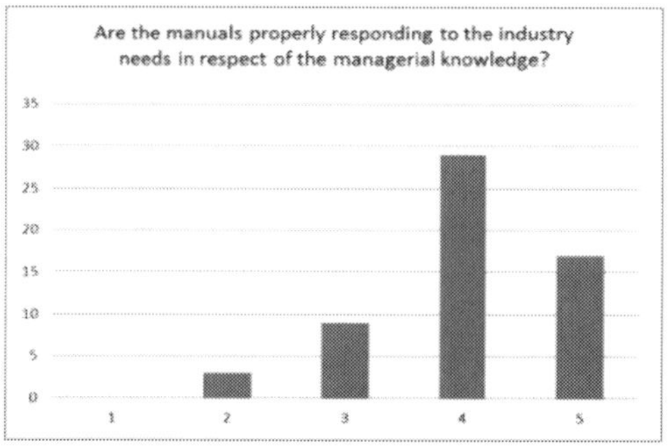

Fig. 4. Results of the survey part b.

Table 3. Will you participate (or delegate the personnel) in the Postgraduate Studies based on the created manuals? (1 - not important; 5 - very important).

1	2	3	4	5	Number of observations	Average
6	4	18	12	14	54	3.44

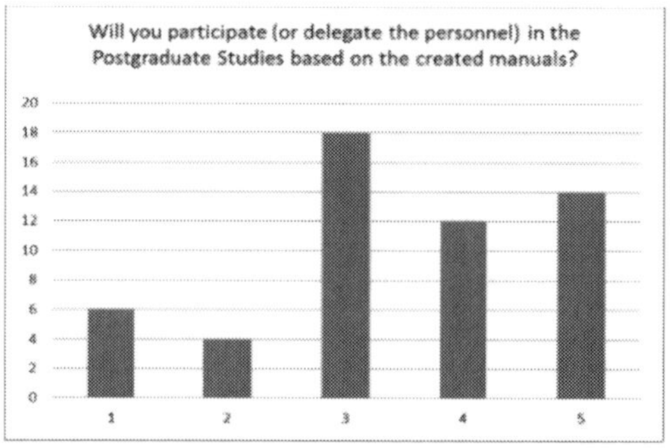

Fig. 5. Results of the survey part c.

4 LEANCO - Continuing V.E.T. Training Programme in Lean Construction to support the transition of the Construction Sector towards a more competitive and productive model

LEANCO Training Programme is about training professionals from the Construction sector in Lean Construction, and bringing the necessary knowledge and competences to Professionals from the Construction sector to let them ensure the implementation of Lean Construction approach and management models within the companies of the construction sector. Figure 6 presents collectively the principles and factors decisive for success of the lean construction methodology.

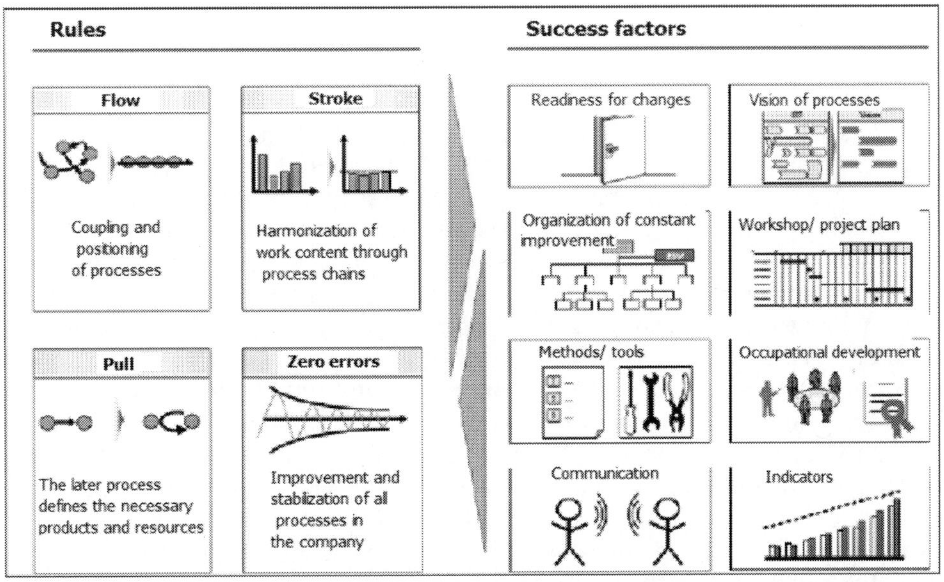

Fig. 6. Application of rules and factors that are decisive for successful implementation of lean construction [10].

The promoter of the LEANCO project is Fundacion Laboral de la Construccion (Spain), Partners are as follows: Asociación de Constructores Promotores de Navarra (Spain), ESKAL Eureka (France), Associacao Plataforma para a Construcao Sustentavel (Portugal), Civil Engineering Faculty of Warsaw University of Technology (Poland).

As a result, to reach this objective, the LEANCO Training Programme will mainly focus on mastering the principles of Lean Construction as well as its specific and related tools/methods/techniques; and on how transferring and materializing it within the companies. To do so, the LEANCO Training Programme will be made of 240 hours and will be composed of the following parts and modules:

Part 1. Conceptual Level (30 hours) aiming at providing theoretical and general knowledge about the Lean Construction in order to allow Professionals to know and understand the Lean Construction philosophy and its principles. This part will be focused on the concept of Lean Construction and its main characteristics:

- Module 1.1. Lean Principles (6 hours);
- Module 1.2. Systems Thinking (6 hours);
- Module 1.3. Lean Enterprise (6 hours);
- Module 1.4. Lean Management (6 hours);
- Module 1.5. Integration to the Construction Industry (6 hours).

Part 2. Operative Level (160 hours) aiming at providing the specific skills to master the different tools/methods/techniques related to Lean Construction in order to allow Professionals to apply and implement the principles and practices of the Lean Construction approach within companies of the construction sector. This part will be concentrated on the main tools/methods/techniques related to Lean Construction and will be composed of the following modules:

• Module 2.1. Integrated Project Delivery – IPD (30 hours);
• Module 2.2. Value Stream Mapping – VSM (20 hours);
• Module 2.3. Building Information Modeling – BIM (60 hours);
• Module 2.4. Last Planner Simulation – LPS (50 hours).

Part 3. Transversal Level (30 hours) aiming at providing complementary training to broad and enhance competences and skills of Professionals:

• Module 3.1 Transversal issues (e.g. Kaizen Lean, 5S, etc.) (30 hours).

Part 4. Implementation LEVEL (20 hours) aiming at providing Professionals with methodology to let them transferring and implementing Lean Construction in the context of their companies:

• Module 4.1 Methodology development for Implementation (20 hours).

Regarding the methodology, the LEANCO Training Programme – a theoretical-practical training – will be implemented through a mixed format (classroom-based and e-learning sessions), with at least one classroom-based session per module. The classroom-based sessions will be based on a practical methodology – workshops (simulations with models, legos and group dynamics), in a way that key concepts will be learnt through instruction, discussion, simulation and collaborative learning [11]. The e-learning sessions will also be based on a practical methodology with projects and case studies that will allow Professionals to learn training contents in an integrated way [12]. Moreover, the LEANCO Training Programme will be sustained by a didactical support all along the learning phase: the LEANCO Platform. The LEANCO Platform will be a specific tool that will act as a methodological instrument to support the LEANCO Training Programme, and as an informative and networking area in the field of Lean Construction. Thanks to the LEANCO Platform, all Professionals participating to the LEANCO Training Programme will have access to the following elements:

• consultation and exchange forums;
• e-mailing and tutorials with trainers;
• chat;
• documentation: programs, contents, learning materials and additional documents;
• directs and records of the classroom-based sessions;
• news in Lean Construction (e.g. conferences, technical seminars, press releases, etc.).

The LEANCO Platform will be developed using Blackboard tool which offers the possibility to build an intuitive, user-friendly and customizable platform, to create dynamic online trainings and to give private access to users.

5 Summary

The main effect of presented projects is the recognition of needs for qualifications of construction personnel, recognition of applied systems of education, certification of the personnel and accreditation of studies and courses (formal and informal) in European Union. Projects aim to develop European system, which would assure standardization of levels and transparency of construction personnel qualifications. ERASMUS+ projects are proper tools for creation of the vocational courses for all professions, including construction. The innovative aspects of the projects is that: ARFAT application will be available for mobile devices such as tablets and smartphones (Android and IOS systems);

CLOEMC IV manuals will be used for the development of postgraduate studies in construction; LEANCO training will be available on Blackboard platform. All the described projects will significantly facilitate education process in construction sector.

References

1. A. Nicał, J. Sobieraj, Oficyna Wydawnicza Politechniki Warszawskiej, 134-141 (2013)
2. M. Książek, A. Nicał, P. Nowak, J. Rosłon, Materiały budowlane **6**, 176-177 (2016)
3. A.G. Martín, A. Minasowicz, F.H. García, J.G. López, J. Rosłon, M.F. Marín, M. Książek, P. Nowak, INTED2015 Proceedings, 2124-2134 (2015)
4. M. Książek, P. Nowak, J. Rosłon, Logistyka, **6** (2014)
5. M. V. Książek, P. Nowak, J. Rosłon, Materiały Budowlane, **12** (2015)
6. P. Nowak, J. Rosłon, Technika Transportu Szynowego – koleje, tramwaje, metro, TTS **10**, 471-477, (2013)
7. A. Minasowicz, P. Nowak, J. Rosłon, Proceedings 27th IPMA World Congress, Croatian Association for Project Management, 116-131 (2013)
8. A. P. Minasowicz, P. O. Nowak, J. J. Zawistowski, I-ESA 2012, Interoperability for Enterprise Systems and Applications "Shaping Enterprise Interoperability in Future Internet", conference proceedings published: M. Zelm, R. Sanchis, R. Poler, G. Doumeingts: "Enterprise Interoperability", Wiley, UK, pages. 331-338, (2012)
9. A. Minasowicz, J. Kulejewski, A. Foremny, P. Kluczuk, A. Nicał, International Conference Creative Construction, 432-441 (2012)
10. J. Kaiser: *Lean Construction.* In: 18. Kasseler-Darmstädter Baubetriebsseminar, Kassel (2008)
11. M. Książek, P. Nowak, J. Rosłon, Logistyka (2015)
12. A. Minasowicz, P. Nowak, E. Pellicer, J.C. Teixeira, INTED2014 Proceedings, 2639-2646 (2014)

Method of statistical data processing safety ecological monitoring combined heat and power station in the megalopolis territory

Valeriy Telichenko[1,*], *Michail* Slesarev[1], *Tatiana* Kuzovkina[1]

[1] Moscow State University of Civil Engineering, Institute of Hydraulic Engineering and Power Plant Construction, Department of Construction of Thermal and Atomic Power Stations, EDP Sciences, Yaroslavskoye sh. 26, 129337 Moscow, Russia

Abstract. The first developed method allowed to determine the combined effects of emissions of pollutants into the atmospheric air in the city by combined heat and power stations group, as well as to assess the interference impact of combined heat and power station ion the child population in different districts of Moscow. It has been proven emissions of pollutants by combined heat and power station interference are distributed on the territory of the city and does not equally affect the disease incidence of children in different administrative districts with emissions of each of the combined heat and power stations individually and do not exceed maximum allowable concentrations.

1 Introduction

Analyzed the possibility of known quality assessment methods presented in the list of references under the numbers [1 - 3] and [4], which, as it turns out, are inappropriate for environmental impact assessment of stationary objects energy. The individual characteristics of the emissions of pollutants by objects energy in these methods cannot be regarded. Were also reviewed the methodology of ecological assessment of the quality of the territories represented in the bibliography numbered [5 – 8] and [9], which were also inappropriate for environmental impact assessment of stationary objects on quality of the environment.

Therefore, the authors developed a new method of environmental assessment and evaluation of the quality of the territory based on correlation dependencies between emissions of pollutants from objects energy, air condition, and disease incidence of the population.

The methodological basis of the investigational study was mathematical statistics methods, correlation data analysis, information modeling, methods of assessment ecological safety in the construction industry and municipal services.

* Corresponding author: President@mgsu.ru

The accuracy of the study results justified by applying adequate scientific practice research and analytical apparatus, using data obtained from official records of urban medical services and statistical data published by city services environmental monitoring, as well as approbation of results. When making an assessment of ecological safety of municipal services approved applied scientific and theoretical positions. Upon receipt of new data and the results used the standard number of measurements required to get results with a margin of error of no more than 5% when the probability belief of 0.95.

Environmental assessment of impact by energy objects in the work presented for example combined heat and power stations in Moscow.

2 Main part

Correlation analysis of data in environmental assessment of combined heat and power stations of the city of Moscow has allowed to determine the contribution of emissions from combined heat and power stations in air pollution, and also revealed the relationship with disease incidence of children living near the location of stationary objects energy. Source data taken from official sources in the form of «2-TP (air)» for the period from 2006 to 2013 years. Environmental data for analysis taken from atmospheric air pollution monitoring stations in the vicinity of each of combined heat and power station, located on the territory of the city of Moscow for a similar period of time. Correlation analysis of data was carried out using paired correlation between emissions of pollutants from stationary objects of power and disease incidence of children of Moscow aged 0 to 14 years. Study period from year 2005 to 2011 biennium. The results of correlation analysis are divided into 14 stationary objects energy, 4 types of disease incidence and 6 types of pollutants.

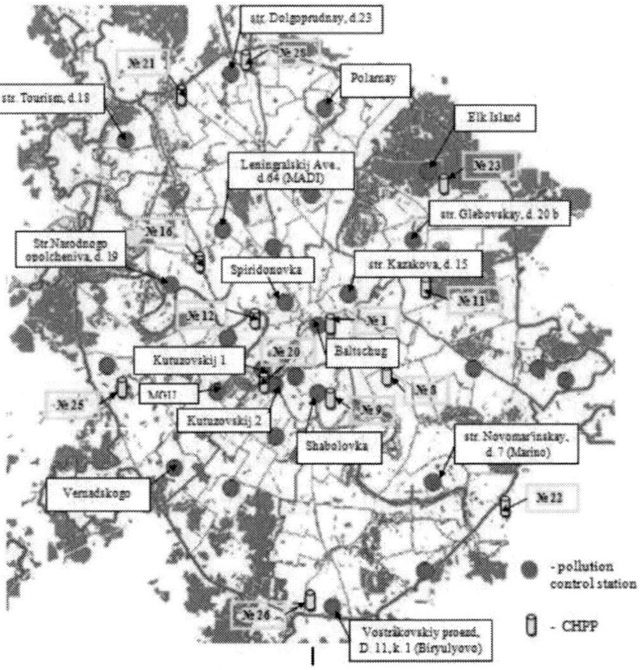

Fig. 1. Location of the stations monitors pollution and emissions of pollutants and locations of combined heat and power stations on the territory of the city of Moscow.

Assessment method of combined heat and power stations effects on environmental quality using correlation analysis data of ecological monitoring is to contained the

 DOI: 10.1051/matecconf/20168605006

correlation analysis of pollutant emissions each combined heat and power stations with disease incidence of the population in each of the districts: f (p_i; z_1); f (p_i; z_2) ... f (p_i; z_n).

Graphic display method for assessing the impact of combined heat and power stations on the environmental quality using correlation analysis presented in Figure 1, where P_1... P_i - sources of air pollution; Z_1...Z_n - disease incidence of the child population in administrative districts of the city.

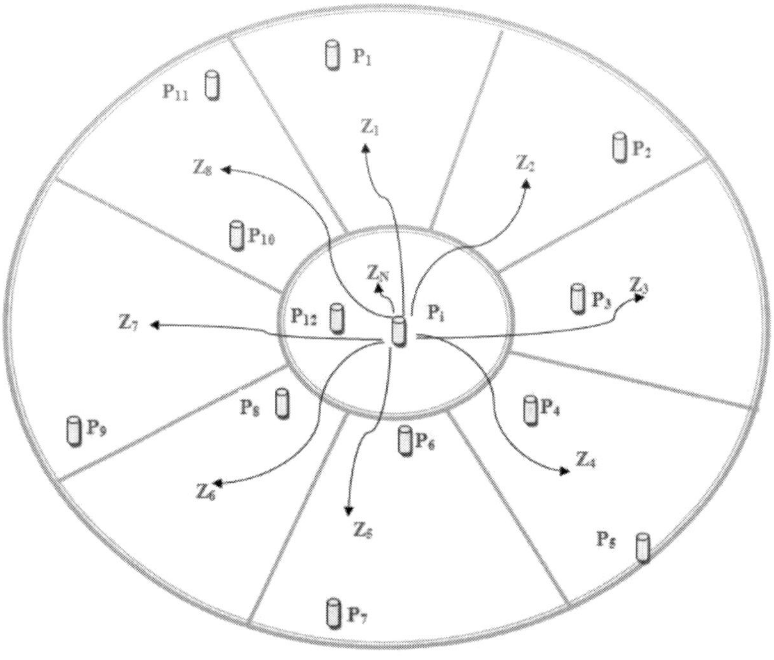

Fig. 2. Scheme of correlation analysis of pollutant emissions combined heat and power stations and disease incidence of child population.

Analysis of correlation of dependencies of pollutant emissions from combined heat and power stations and air condition was conducted according to the scheme shown in Figure 3.

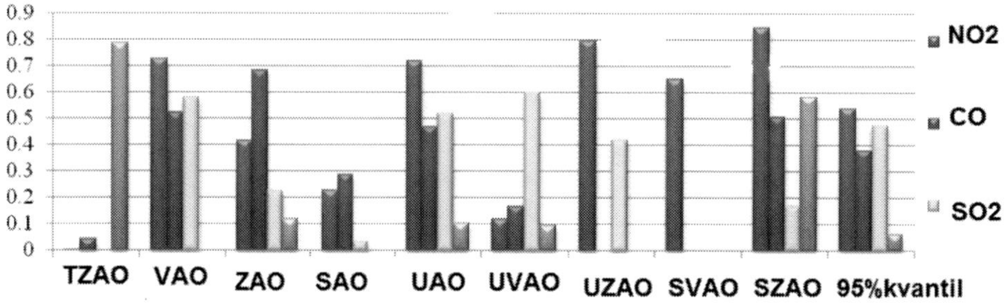

Fig. 3. The magnitude of the correlation between emissions of pollutants the combined heat and power stations and the air condition of the city of Moscow.

Analysis of data on the diseases incidence of upper respiratory tract organs (AIDP) was conducted on emissions from combined heat and power stations to nitrogen dioxide and solids. Density (power) correlation between values of pollutant emissions from combined heat and power stations and diseases incidence of the organs of the upper respiratory tract of children in Moscow are shown in Table 1.

Table 1. Interference emissions of pollutants combined heat and power stations (CHPP) and the correlation with the disease incidence of upper respiratory organs of the child population in administrative districts of Moscow.

Name title	Pollu-tion	TZAO	VAO	ZAO	SAO	UZAO	SVAO	SZAO	UAO	UVAO	Arithmetic middling	95 % quantile
CHPP-1	NO2	0.993	0.335	0.953	0.831	0.991	0.831	0.902	0.713	0.376	0.769	0.415
	Solids	0.984	0.040	0.819	0.627	0.906	0.627	0.733	0.472	0.083	0.588	0.525
CHPP-8	NO2	0.755	0.806	0.960	1.000	0.896	1.000	0.989	0.983	0.831	0.913	0.834
	Solids	1.000	0.213	0.906	0.753	0.966	0.754	0.841	0.619	0.256	0.701	0.292
CHPP-9	NO2	0.700	0.851	0.934	0.996	0.857	0.996	0.974	0.995	0.873	0.908	0.831
	Solids	0.987	0.375	0.964	0.854	0.996	0.854	0.920	0.742	0.415	0.790	0.455
CHPP-11	NO2	0.981	0.025	0.810	0.616	0.900	0.616	0.723	0.459	0.068	0.578	0.513
	Solids	0.403	0.804	0.015	0.292	0.159	0.292	0.151	0.464	0.777	0.373	0.196
CHPP-12	NO2	0.918	0.589	1.000	0.955	0.988	0.955	0.988	0.883	0.624	0.878	0.620
	Solids	0.347	0.991	0.707	0.875	0.573	0.875	0.796	0.950	0.996	0.790	0.448
CHPP-16	NO2	0.978	0.013	0.803	0.606	0.894	0.606	0.715	0.448	0.056	0.569	0.502
	Solids	0.984	0.392	0.969	0.863	0.997	0.863	0.927	0.755	0.432	0.798	0.472
CHPP-20	NO2	0.970	0.024	0.780	0.576	0.877	0.576	0.688	0.415	0.020	0.547	0.470
	Solids	0.124	0.995	0.527	0.742	0.371	0.742	0.637	0.853	0.990	0.665	0.359
CHPP-21	NO2	0.804	0.403	0.483	0.222	0.628	0.222	0.360	0.037	0.363	0.391	0.227
	Solids	0.311	0.996	0.679	0.856	0.540	0.856	0.772	0.937	0.999	0.772	0.553
CHPP-22	NO2	0.983	0.396	0.970	0.865	0.998	0.865	0.929	0.757	0.436	0.800	0.475
	Solids	0.959	0.065	0.754	0.543	0.857	0.543	0.658	0.377	0.021	0.531	0.434
CHPP-23	NO2	0.953	0.086	0.740	0.525	0.846	0.525	0.642	0.358	0.042	0.524	0.415
	Solids	1.000	0.207	0.904	0.749	0.964	0.750	0.837	0.614	0.250	0.697	0.285
CHPP-25	NO2	0.796	0.416	0.455	0.208	0.617	0.208	0.347	0.023	0.376	0.383	0.217
	Solids	0.719	0.836	0.943	0.998	0.871	0.998	0.980	0.992	0.859	0.911	0.835
CHPP-26	NO2	0.976	0.002	0.796	0.598	0.890	0.598	0.707	0.439	0.046	0.561	0.493
	Solids	0.719	0.836	0.943	0.998	0.871	0.998	0.980	0.992	0.859	0.911	0.835
CHPP-28	NO2	0.224	1.000	0.610	0.806	0.463	0.806	0.712	0.902	0.999	0.725	0.472
	Solids	0.000	0.000	0.000	0.000	0.000	0.000	0.000	0.000	0.000	-	-

Thus was developed and tested method that includes approach «each object energy with each subject territory». This method allowed to determine interference picture of inter influence pollutant emissions from combined heat and power stations, based on the air condition impact on disease incidence of child population in administrative districts of Moscow.

3 Conclusions

The developed method allowed to determine the combined effects of combined heat power station group emissions on air condition in the city, as well as to assess the interference effect of combined heat and power plants on the child population in different districts of Moscow.

The dependence of the emission of pollutants from combined heat and power stations and air condition is a linear function for emissions of nitrogen dioxide and carbon monoxide, the emissions of sulphur dioxide, solid particles is a nonlinear function.

Proven that data correlation coefficients of pollutant emissions from combined heat and power stations and air condition depending on the type of substance: nitrogen dioxide - 0.537; sulphur dioxide - 0.039; carbon monoxide - 0.378; solid particles - 0.063.

It has been proven that emissions from combined heat and power stations interference are distributed on the territory of the city and have an impact on the disease incidence of child population irregularly. Emissions of each of the combined heat and power stations alone not exceeding maximum permissible concentrations (MPC), but the overall impact from several combined heat and power stations interference is obtained and significantly increases the disease incidence of children locally of the city.

References

1. *Gost R ISO/TS 21667-2009, Computerization of health. Conceptual model of health indicators* (Standartinform, Moscow, 2010)
2. *2.1.10.1920-04. Guidelines for the assessment of risk to human health when exposed to chemicals polluting in the environment* (Moscow, 2004)
3. *Methodology of evaluation criteria of ecological situation of the territories to identify environmental emergency zones and ecological disaster zones* (Moscow, 1999)
4. *OND-86 technique for calculating concentrations in ambient air of harmful substances contained in the emissions* (Moscow, 1986).
5. G.A Kachaev, *Methods of evaluating the ecological territory State on topographic maps: extended abstract of Cand. Sci*. (Moscow, 2011)
6. G.E. Asylbekova, *Estimation of ecological state of Pavlodar urban ecosystems using plant sites: extended abstract of Cand. Sci.* (Novosibirsk, 2010)
7. S.N. Kirillov, J.S. Polovinkina, Vestnik of Volgograd State. University, part 3, Econ. Ecol. **1(18)**, 239-245 (2011)
8. M.V. Lewinsky, *Ecological-genetic environmental assessment and the health of the population in some industrial centers of the Republic of Moldova: extended abstract of Cand. Sci.* (Voronezh, 2008)
9. Z.M. Mustafayeva, *Use of GOSTS, environmental pathology and stability of plant development as biorecognition parameters of quality environmental: extended abstract of Cand. Sci.* (Kaluga, 2005)

Construction audit as a modern instrument of investment and construction process management

Olga Didkovskaya[1], *Olga* Mamayeva[1] , and *Alexey* Bocharov [1,*]

[1]Samara State Technical University, Institute of Architecture and Civil Engineering, Molodogvardeyskaya str. 194, 443001 Samara, Russia

Abstract. The problem of methodical approaches formation for implementation and use of the construction audit as a new control instrument in investment and construction projects is discussed in this paper. The authors analyzed and generalized the theory and practical experience of construction and technical as well as economic expert assessments concerning capital construction projects. A complex methodological basis of construction audit adapted for Russian legislation conditions and applied to investment and construction projects of various readiness level is also given in this article. The main reasons defining the relevance of the subject are the increasing requirement and lack of methodical support of this expert activity type in construction providing insufficiency conditions of its standard and legal regulation.

1 Introduction

Today, a range of target investment construction programs confined to important events is realized on a nationwide scale. The audit of realization processes is becoming a perspective activity. In this regard the state undertakes different measures to introduce some control instruments over investment and construction processes. At the same time, the scope of available legitimate tools is limited. Construction audit as expert study is very complex and universal and it includes the examination and control of investment construction projects (capital construction projects) from the legal, construction technical, financial and economic points of view.

Construction audit feature provides control over project execution of each process (including incomplete one) and each division participating in its realization, but techniques (rules, standards, regulations) of the organization of such control can be found neither in modern domestic scientific literature, nor in open resources of the profile companies and professional communities.

* Corresponding author: kafedra_cen@mail.ru

2 Methodical Basis and Principles of Construction Audit System Formation

In practice, in the course of construction complex functioning are (investors, customers, builders) often face the necessity to solve problems connected with the violation of standard investment construction procedure and expert study methods.

A set of such methods can be titled as "construction audit" depending on the purpose, stage, nature of the study and its results.

Construction audit is the synthesis of construction technical, legal and financial economic expert studies carried out for the assessment and optimization process of investment construction implementation projects at any stages. The listed types of research can be carried out separately or all together.

The development of this type of professional activity is relevant for several reasons:

- growing need for internal and external auditor services concerning the realized investment construction projects (first of all, state ones);
- drawbacks in standard and legal regulation of professional area;
- absence of regulations, rules, standards, implementation techniques of construction audit;
- existence of "critical bulk" of projects (capital construction projects) which are potentially an audit subject.

The analysis of standard and legal base showed that the construction audit as a type of professional activity is subjected to legal co - regulation of urban-planning, investment and expert activity. The construction audit provision in the present legal boundaries is closest to provisions of the following documents:

Clause 53 of the Urban-Planning Code of the Russian Federation and relevant by laws regarding the organization and implementation of construction control procedure;

federal law of 31.05.2001 No. 73-FL "About state judicial and expert activity in the Russian Federation" regarding procedures of judicial construction technical expert studies;

resolutions of the Government of the Russian Federation of 30.04.2013 No. 382 "About carrying out of public technological and price audit of large investment projects with the state participation and about the modification of some acts of the Government of the Russian Federation" regarding large investment projects audit at predesign and design stages.

At the same time, special legislative regulation of construction audit as universal system is absent.

Methodical support development of construction audit needs to be carried out depending on investment construction process stages (Table 1).

Table 1. Types of expert studies at various stages of investment construction process.

Pre -design stage	**Design stage**	**Construction**	**Setting into operation**
Technical economic reasoning. Technological and price audit	Project documentation expert study. Cost expert study	Construction technical expert studies. Construction control. Incomplete construction assessment.	Actual expenditure control. Budget funds expenditure control.
Construction audit			

Use of construction audit instruments allows to reduce or to avoid technological and economic hazards of investment construction decisions concerning incomplete construction objects. Condition creation task of construction audit implementation is extremely important.

Thus, audit should be divided into some directions depending on subjects types of investment construction activity: for an individual builder - technical and financial control (today this niche is occupied by construction and technical experts); for the state and municipal builders the formation of normative legal acts of construction audit is necessary taking into account the level and the size of the budget.

To give this institute power is necessary for:

- investment potential assessments;
- credit investments involvement for construction objects;
- controversial issues solutions between subjects of investment and construction activity;
- control over possible unreliable accounting documents;
- prevention of inappropriate use of monetary funds;
- exclusion of unplanned delay time;
- protection against unplanned actions;
- necessity of wear research.

3 Construction Audit Perspectives

Researches for the current investment construction projects and objects of incomplete construction are the most demanded and at the same time least methodically developed field of construction audit. Approaches to construction audit conducting in such cases have to be provided in the complex methodical documents on construction audit.

The generalized scheme of construction audit of the investment and construction project can be submitted according to the following expert actions:

– analysis of initial and permissive documentation;
– analysis of project and detailed documentation;
– analysis of executive documentation;
– adherence to specifications;
– organization and implementation of construction supervision;
– analysis of current construction state;
– analysis of construction deadline;
– assessment of structure, bulk and quality of the performed works;
– analysis of the contract relations;
– analysis of project budget implementation;
– monitoring of target expenditure of capital investments;
– deviations identification of actual indicators of design and their reasons;
– project risks assessment;
– project optimization recommendations, etc.

Methods for the incomplete construction objects include some positions listed above, as well as the delay reasons analysis of construction, an assessment of object completion degree, technical expertise of its state, cost assessment. And the most important thing is recommendations of technical capability, economic suitability and optimum ways of construction completion.

The largest project in Samara region is the stadium construction for the Football World Championship in 2018. It may be considered to be the example of the construction audit. It is known that there is a change of a contractor at the object there. In this connection it is necessary for a customer to receive full and detailed expert information about current construction state with the purpose of taking further management decisions.

The audit study roadmap of the project is shown in Table 2.

Table 2. Roadmap of construction audit performance of "Samara Arena" project.

Samara Arena	Project analysis	Amount and types of work data analysis
		Preliminary cost analysis with deflator index
	Contract analysis	Collaboration scheme defining and recommendations
	Technical /finance analysis. Project supervision analysis	Cost analysis and materials expense norms
		Finance structure analysis
		Condition analysis of work implementation
		Materials and items purchase activity analysis
		Machine exploitation analysis and machine work at construction objects
	Final conclusion concerning performed work, developed optimization recommendations	

4 Conclusion

Mechanisms realization of construction audit in the Russian Federation becomes difficult because of the absence of standard legal base, lack of precise understanding of a subject and absence of experts in this branch. In foreign countries the construction audit realization is quite accurate, and unites professional communities (institutes of construction audit and cost control).

The development of legal base and methodological base is the first stage of construction audit effective system as a required element in construction management.

References

1. *Town-planning code of the Russian Federation (federal law of 29.12.2004 No. 190-FZ)* (2004)
2. *The federal law of 31.05.2001 No. 73-FZ "About the state judicial and expert activity in the Russian Federation"* (2004)
3. *The resolution of the Government of the Russian Federation of 30.04.2013 No. 382 "About carrying out public technological and price audit of large investment projects with the state participation and about modification of some acts of the Government of the Russian Federation"* (2013)
4. O.V. Didkovskaya, O. A. Mamaeva, M. V. Ilyina, Procedia Engineering **153**, 131-135 (2016)
5. Yu.P. Sologubov, Traditions and innovations in construction and architecture. Construction: collection of articlesSamara, 308-312 (2016)
6. M.V. Ilyina, Problems and prospects of the development of construction cost engineering system: collection of materials, 107-110 (2015)
7. O.V. Didkovskaya, Cost engineering and examination of real estate: problems and prospects development: Collection of articles, 5-10 (2016)
8. M.A. Konovalova, Cost engineering and examination of real estate: problems and prospects development: Collection of articles, 11-17 (2016)
9. M.V. Ilyina, Cost engineering and examination of real estate: problems and prospects development: Collection of articles, 18-25 (2016)
10. N.S. Belomyttseva, Economic aspects of a construction complex management under modern conditions, 162-165 (2014)
11. O.V. Didkovskaya, Budget and contractual work in construction **11**, 23-33 (2014)
12. A.Yu. Bocharov, XXII of the International scientific and practical conference, 165-168 (2013)

Strategic directions of personnel potential forming of a building complex

Marina Simonova [1,*] *Natalia* Shekhova[1], and *Sergei* Kolesnikov[1]

[1]Samara State Technical University, Institute of Architecture and Civil Engineering, Molodogvardeyskaya str. 194, 443001 Samara, Russia

Abstract. The analysis of directions of strategic approach forming of labor potential management of a building complex is carried out in this paper. On the basis of this analysis the system of actions for strategy forming divided into consecutive stages is offered. The development of the personnel forecast is a strategic planning basis. One of personnel forecast variants is the correlation of needs estimates in personnel of a building complex with available allowances. On the basis of the personnel forecast strategic analysis it is possible to compose working programs for the stated goals of implementation. Operational assessment of personnel requirements of a building complex is proved to be combined with strategic objectives. Some assessment approaches to qualitative and quantitative need for specialists of a building complex are offered. The fact that high-quality labor power supply system of a building complex with should be based on industry development forecast and increase in construction products competitiveness is revealed in the article. Strategic management priority will allow to react immediately to the current situation changes, to introduce amendments both into tactical, and operational management.

1 Introduction

Construction industry crisis solution is confirmed by increase in amounts of housing constructions taken into use in 2014-2015 [1] and increase in demand for a wide range of specialists and professions and, first of all, highly qualified workers in case of significantly smaller labor power offer, especially of a required qualification. The tasks set by the Government of the Russian Federation for construction industry assume the growth of housing construction volumes to 100 000 000 m^2 annually that requires the corresponding supply of the all branches of industry including staffing. However, professional education has been suffering for several years some consequences of demographic recession of the 1990s and in several years it is necessary to be ready to essential offers reducing in the labor market. Such aspects as low level of labor migration, some peculiarities of emigration and immigration processes, training duration of qualified specialist and many others should be taken into account. Short-term programs of the labor market regulation do not always give necessary effect that leads to demand contraction and unemployment growth [2].

* Corresponding author: m.simonova@mail.ru

Perspective programs allow to perform alternative demand and supply scenarios of labor power, employment structure according to gender and age [3]. It is necessary to estimate current personnel demand as well as a long-term perspective to accomplish stated plans, determine forthcoming tasks and their timely solutions increases and that assumes determination of the main objective of such planning and its decomposition.

2 Main Part

The main objective is to provide necessary amounts of construction and quality of the construction objects being built. Such a broad objective definition requires formulation of sub goals and is true for a number of tasks having various nature, resources and solution sequence.

According to the objective set it is possible to formulate top sub goals that help to determine the directions of subsequent solutions and actions:

- Regulation of professional education structure and content: order forming for specialists training of all the levels of professional education according to quantity and quality. The order initiators can be: the state; self-regulatory organizations; target orders from organizations; population (for-profit education).
- Regulation of professional labor market: reconciling demand and supply taking into account innovative production development, automation, work mechanization that can result in occupational skill structure change in terms of professional groups and specialties.
- Reasoning opportunities for implementation of investment and construction projects on certain territories: assessment of the existing personnel and human resources, cost and time assessment to hire labor power attraction of required quality and quantity.
- Reasons for planification, site and territory development: formulation of top solutions development directions of certain resettlement and industrial project zones based on the planned construction amount.
- Providing interrelation of economic, social planning with implementation of investment and construction objects: calculation of needs for social, medical, educational services depending on the changing resettlement density.
- Regulation of labor migration, emigration and immigration processes: forecasting the need for labor migrants of various territorial entities of the Russian Federation, involvement of qualified personnel, target training and national processes management.
- Professional orientation for the population of all age groups.
- Ensuring the continuity of generations of builders, designers, researchers.

Taking into account strategic personnel goals, it is possible to develop operational and tactical plans to provide investment construction programs with human resources. Both operational and strategic assessments of personnel requirements of a building complex are necessary since operational forecasting is not sufficient for the majority of stated sub goals and all the objectives can't be achieved without strategic planning [4].

There are essential distinctions in staffing needs planning on the near and longer time lag thus only confirming the urgent need in both types of planning.

Features of human resources planning for a period of 1 up to 3 years:

- Relative calculations accuracy owing to the available data on the objects being built and planned to construction, financing sources, applied technologies, approved wage rate in the estimated project cost, a personnel situation in the region.
- Discrepancy of planning terms and investment building cycle volume.

Features of advanced and strategic human resources planning for a period of 3 up to 10 years and more:

- Compliance of planning terms and investment building cycle volume.

- Difficulty in forecasting the structure of qualification groups, specialties owing to technological mode change, acceleration of innovative process, scientific technical progress.
- Difficulty of accounting demographic and migration factors on personnel resources of regions.
- Educational system lag effect, long period of training and adoption of continuous education system during life.
- Approximation of calculations owing to a large number of unknown factors and expert evaluations prevalence.
- Dependence of predicting assessment accuracy on political and economic situation in the country and in the world.

It is possible to offer several approaches to the assessment of qualitative and quantitative need for specialists of building complex:

- The forecast "from achieved" can be conducted on the basis of data correlation of the construction amount being carried out currently and the number of personnel. These data can be obtained from self-regulatory organizations and construction organizations building low-rise buildings and constructions. These data should be correlated to sociological polls data, statistical service, employment service, recruitment agencies for reliability assessment of personnel situation in construction organizations. If human resources available today cope with implementation of the declared amount of construction works, so it is possible to estimate how many specialists in qualification groups and categories are necessary to accomplishing the set construction amounts. However, services, trade organizations, operational structures in the system of housing and public utilities and other accompanying systems are not considered here. The forecast based on work payment share of the square meter cost of housing, industrial and civil engineering planned to construction. In case of known and predicted ratio of qualification groups, positions and professions, work payment "range" it is possible to estimate quantitative limits of personnel which can be attracted to create of finished construction product that has consumer cost. It is necessary to consider not only quantitative aspect, but also a ratio of the required labor power quality and the amount of work payment for which in case of the limited range influences a ratio of quantitative and qualitative personnel structure. This approach allows to take into account staff costs of all building complex, including project cost, engineering survey and construction materials production, except the government employees and professional education workers having budget financing.
- The forecast for labor input of building objects planned to be constructed. In case of known technological schemes and set labor productivity level it is possible to calculate the required number of various qualification workers and hence, technical, functional and administrative personnel. Such method is more suitable for operational planning.

The offered variants are conceptual and show the possible directions of personnel requirements assessment which don't exclude, but supplement each other. These data can be used to form an order to professional education of major training programs, the number of state funded places created both from public funds, and by means of self-regulatory organizations. The personnel forecasts need is usually based on the assumption that the discrepancy between demand and supply doesn't lead to reorganization in the labor market [5]. It is necessary to take into account that the number of students obtaining professional education should prevail the number of required specialists of a building complex received according to the calculations based on the following essential factors:

- Approximation of personnel requirement calculations on strategic prospect which should consider training period.
- Students excluding from training process because of family, personal circumstances.

- The absence of graduates distribution system doesn't oblige the young specialist to work according to the specialization even if training was state funded.
- Citizen rights to get desired profession which should not be limited by the availability of sate funded places.
- Implementation of the competitive environment in the professional labor market allowing to redistribute specialists with the identical education level according to capabilities and business qualities.
- Availability in structure of economy of workplaces which are directly not connected with construction production but which assume professional education. These are the government, municipal institutions, elected structures, higher-education teaching personnel, self-regulatory organizations, the intermediary, trade organizations and other structures connected with a construction complex not participating directly in construction production.

Correlation of calculated need rates for construction complex personnel with the available labor reserve shows personnel potential development of a construction complex on basis of which it is possible to elaborate working programs for implementation of the specified purposes [6]. Economic science operates with various methods of calculation of required quantity of labor power which need to be adapted for the construction specifics connected with seasonality of the performed works, a wide range of the built construction objects, and mobility of production structures depending on construction nature.

Taking into account development tendencies of construction complex management in our country it is necessary to mind who should take care of the need for strategic providing of a construction complex with personnel potential. We should not deny significant influence of the industry on different social processes being within the sphere of state interests. It is necessary to remember, however, that there is relatively small number of the construction organizations of the state-owned legal entity which basically provide strategic objects construction. The rest commercial organizations are self-government now.

Innovative construction complex development process should be ahead the appropriate level labor power training. The construction complex supply system with a high-quality labor power should base on industry development forecast as well as on competitiveness increase of construction products. Use of a forecasting model with use of the IBS models which is based on the analysis of balances of key sectors of economy [7] is of great interest. It is also necessary to mention the model which includes the demand equations, training period and labor market variables; the equations for forecasting labor market variables which influence demand; and procedures for calculation of statistically confidential intervals in the forecast registration [8].

Special attention should be paid to the qualified professionals capable of carrying out a research and experiments, the industry managers capable to analyze the market, to choose the priority investment directions, the technologies providing level increase, to organize the specified work, be able to get profits from materials and constructions.

This professional category also includes research and educational personnel of the industry capable to train specialists and managers able to adopt to the varying conditions of equipment and technologies evolvolution. Industry progress is impossible without information technology and database specialists capable to provide reliable information development.

Basic elements of the organizational and financial mechanism of system functioning of training of high-quality specialists are oriented to a long-term outlook, system management process being included into the category of the complex economic challenges requiring serious scientific basis. A number of state and public organizations as well as business structures are included into management system. Public organizations and commercial enterprises of the construction materials industry act as the expert bodies of the system responsible for plans and programs corrections. These organizations are labor power consumers interested in its quality increase.

In case of elaboration of long-term development programs of a construction complex it is reasonable to allocate several stages allowing to choose forms and methods of quality regulation to timely estimate effectiveness of each stage and to introduce necessary amendments (figure 1). Tendency assessment and development directions of a regional construction complex become an important factor in case of organizational and economic allowances implementation of labor power quality improvement.

The analysis will allow to determine necessary labor power quality regulation methods involved in production and planned for perspective use. The most significant methods are the use of organizational managerial and standard and legislative regulation in case of economic incentives impact. At the following stage it is necessary to exercise operational, tactical and strategic control of labor power quality development by means of the chosen forms and methods. The priority of strategic management will allow to react quickly to the current changes, to introduce amendments both in tactical and in operational management. Managerial actions should be followed by the periodic assessment of economic result allowing to control efficiency of the made actions. In case of efficiency decrease it is necessary to provide a possibility of adjustment of process from the first stage, depending on complexity of the arisen situation. Ensuring high level efficiency it is possible to increase planning horizons and move to longer strategic plans development. Effective implementation of all stages of organizational and economic process of labor power quality improvement will allow to increase skilled labor force inflow to the construction complex, raising thereby the general labor power quality level of the industry, promoting products quality improvement and production profitability.

The current trends in the labor market are determined by the increasing influence on the market of such negative factors as "...structure subjects deformation of the labor market; marginalization of a labor power; low wage rate; cheap labor power demand; "brain drain"; inflow low skilled labor force; work capital replacement; secondary employment development and working hours duration increase; increase of "shadow" sector scales; "revival" of discrimination, forced labor, human trafficking; transformation from labor-excessive to labor-deficit" [9].

3 Conclusion

Standard and legislative providing completely depends on government institutions and national associations of self-regulatory organizations. There should a rational system of legal acts stimulating the enterprises to labor power quality improvement in the country since now capital investment in advanced training and education is economically unprofitable and there is no system of professional standards and the principles of determination of professional competences uniform throughout the country.

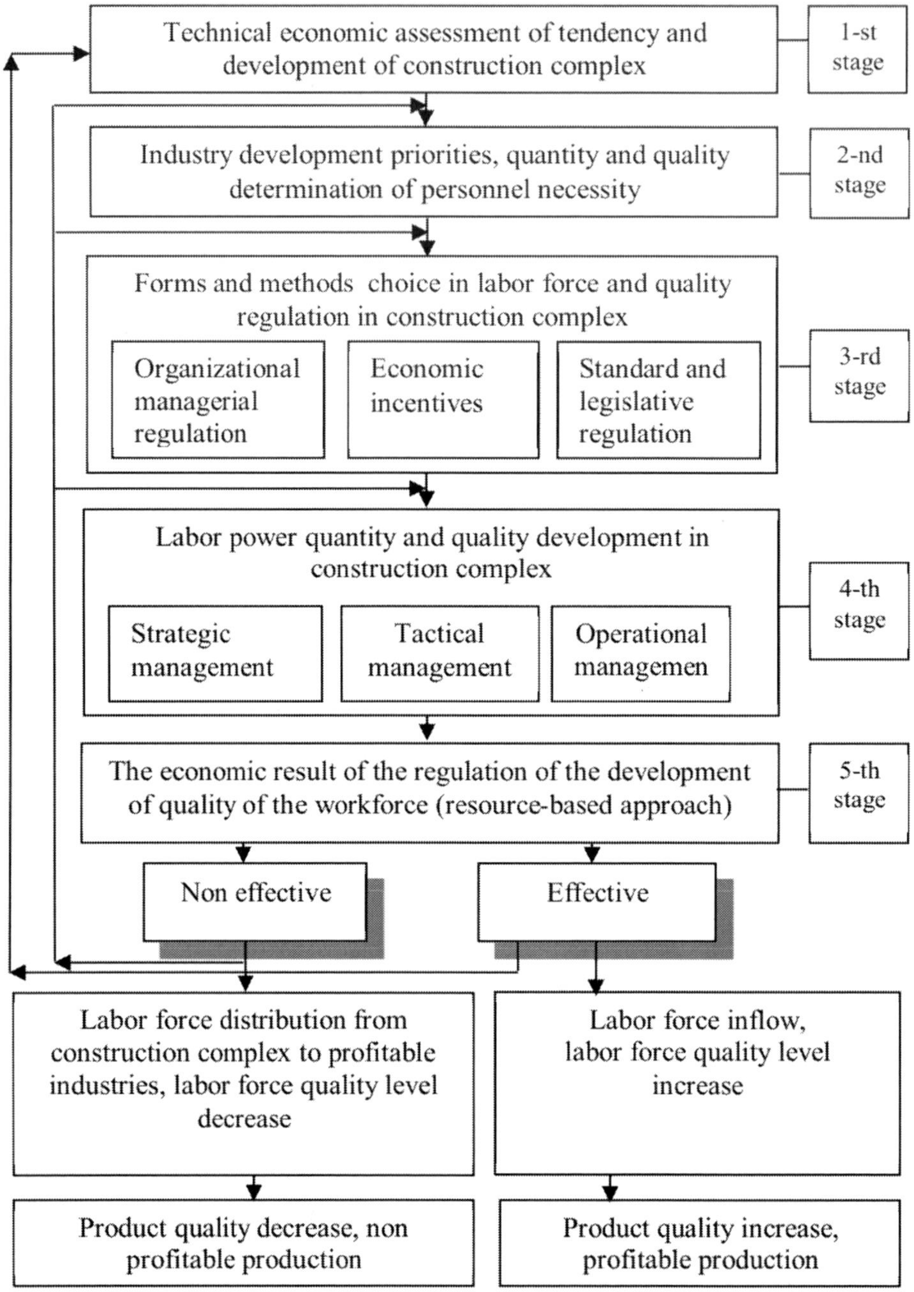

Fig. 1. Stages development and implementations of the strategic plan forming of construction industry personnel potential.

References

1. *The report of the head of Ministry of Construction of Russia Mikhail Men at a meeting of the State Council of the Russian Federation concerning "Development of a Construction Complex and Enhancement of Town-planning Activities in the Russian Federation". Accessed October 24* (2016) information on

http://www.minstroyrf.ru/press/doklad-glavy-minstroya-rossii-mikhaila-menya-na-zasedanii-gosudarstvennogo-soveta-rossiyskoy-federat/
2. M. Nakajima, International Economic Review **53**, 399-432 (2012)
3. M. Bruni, Labour **2**, 55-116 (1988).
4. M. V. Simonova, L. A. Ilyukhin, I. V. Bogatyreva, *Bases of personnel forecasting at the regional level* (ITS, INFRA-M, Moscow, 2015)
5. L. Borghans, H. Heijke, Journal of Applied Econometrics **25(4)**, 720-754 (2007)
6. M.V. Simonova, materials of the International scientific and practical conference, 198-203 (2010)
7. J. Wayne,. Labour **2(1)**, 55-116 (2015)
8. S. Hoenack, W. Weiler, American Journal of Economics and Sociology **74(2)**, 419-456 (2007)
9. V.Yu. Lapshin, *Functioning and market development of work in modern conditions of Russia: deviation problem* (TGU of G. R. Derzhavin, Tambov, 2006)

Some criteria of critical infrastructures stability

Andrey Volkov[1], and *Liubov* Shilova[1,*]

[1]Moscow State University of Civil Engineering, 129337, 26, Yaroslavskoye Shosse, Moscow, Russian Federation

Abstract. Critical Infrastructures (CIs) are complex systems with a set of individual elements, such as buildings and facilities, technological equipment and energy supply networks, engineering, technology and other communications. The system criteria of Critical Infrastructures stability and their threshold are shown in the paper. The main criterion of CIs protection is engineering and functional stability. And the special criteria are the next: the criterion of work safety; the criterion of environmental safety; the criterion of energy safety; the criterion of resource safety (resource using in CIs); the criterion of CIs' automation level; the criterion of technological equipment safety; the criterion of reliability of project design solution; the criterion of engineering systems reliability; the criterion of engineering safety; the criterion of explosion and fire safety. The algorithm for evaluation of Critical Infrastructures stability is formulated in this paper.

1 Introduction

For many years the basis for the development of standards and regulations of critical infrastructures stability was the concept of absolute security, which began on the dawn of the technological revolution [2]. The idea of absolute security concept is based on the strict regulatory approach and it is a logical analysis of possible initiating events of emergency situations at the facility. It helps to model the factors influencing at the facility and to develop a system of compulsory measures to reduce possible losses. The main advantage of the absolute security concept is the fact that over the time it has evolved and nowadays it includes statistics of accidents and their consequences. In addition, when we use this approach, the security measures are installed as required, and their implementation is supported legally, so their performance are easy to control. For example, if a set of measures for particular class facilities is made, it is believed that this facility is absolutely secure, but the reliability of the technical means is implemented and staff actions aren't always correct. So the adequacy of selected measures can be evaluated only by experience, because otherwise the concept of absolute security contradicts to probabilistic laws of nature.

Currently one more concept is used widely in the industrialized countries. It is the concept of acceptable risk.

* Corresponding author: ShilovaLA@mgsu.ru

In foreign and Russian literature [2], as well as to the authors' research work, preference is given to the second probabilistic approach of acceptable risk. This concept helps to define a quantitative evaluation of risk level; to simulate ways of transition emergencies situations to accidents accompanied by affecting factors; to establish acceptable levels of risk and etc.

According to [3,14] the authors consider that risk is a hazardous events measure which characterize the probability of arise accident or disaster and scale of their consequences. The probabilistic concept of acceptable risk bases on the evaluation of the security index. And the adequacy of the chosen measures can be estimated only by using this security index.

So it is necessary to develop some index, which can help us to make a clear conclusion about CIs stability level. In practice this index is called a criterion.

CIs are complex systems with a set of individual elements, such as buildings and facilities, technological equipment and energy supply networks, engineering, technology and other communications, so we have to make a system of criteria, which helps us to determine the composition of protected and secured working space for Cis [6,7,16].

2 Methodologies

In our research work a criterion of CIs stability is a limit value of quantitative and qualitative indexes which characterizes state of CIs and the conditions of their operation.

The analysis of actual research works in the field of our study [1,4,5,8-11,12,15,18] helps us to identify the main requirements for the criteria of CIs stability. They are the following:

- the criterion should be complete and reflect main types of security threats;
- the criterion should be consistent and integrated;
- the criterion should have a minimum dimension.

Than we use method of expert evaluation for determining main criteria of critical infrastructures stability. This method is a part of decision science, and there are two groups of expert evaluation. The first group is based on individual opinions of different experts, and the second group consists of the collective experts' opinion.

In our study the group of experts from different organizations was formed. They are:

- the specialists of Russian and foreign universities;
- the heads of fuel and energy complex enterprise;
- the specialists of EMERCOM of Russia and etc.

The stability of CIs is a complex notion, that's why we decide to divide stability of CIs for two types: engineering and functional. We formed list of 70 criteria which could characterize engineering and functional stability of CIs. Later we used a ranking method to determine the most important of them. The experts arranged all criteria in the descending order of properties characterized stability of CIs. At the end of the experiment we got 10 criteria. But these criteria characterize stability of CIs different, and later we determine what criterion is more important and calculate their specific weight. Specific weight of criterion of CIs stability is a number from 0 to 1, which is proportional to its importance.

There is an algorithm of expert evaluation method in the fig.1.

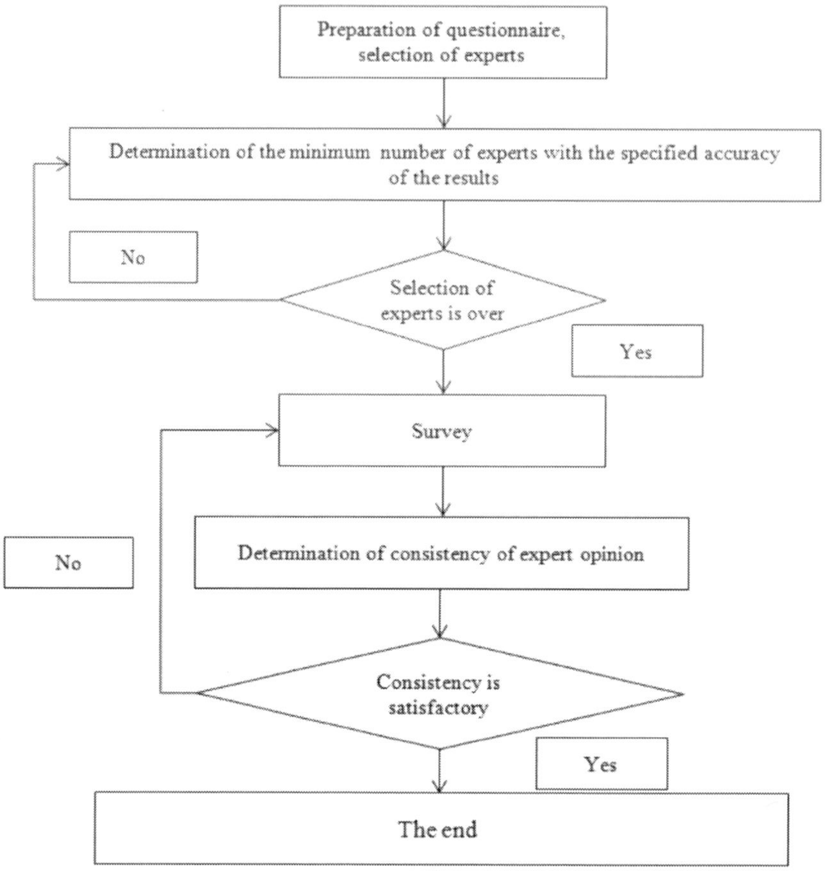

Fig. 1. An algorithm of expert evaluation method.

The number of experts can be determined through coefficient of mistake of expert analysis result, which lays in the range from 0 to 1:

$$n \geq 0.5*(0.33/b+5) \quad (1)$$

where *n* is the number of experts; *b* is the coefficient of mistake of expert analysis result.

When b=10%, n≥0.5*(0.33/0.1+5) ≥4.15, so there must be about 4 people in the expert group. The level of decision importance is very high, so we adopt the specified accuracy of the results about 95%, *B*=0.95.

The minimum number of experts we can determine by next formula (2):

$$N = \frac{B^2 \cdot r_a \cdot r_0}{\Delta^2} = \frac{0.95^2 * 0.95 * 0.05}{0.05^2} = 17.1475 \quad (2)$$

where r_a is the proportion of sample elements with the presence of the specified attribute; r_0 is the proportion of sample elements without the presence of the specified attribute; B is the accuracy of the results and Δ is a coverage error.

So the minimum number of experts is about 17 people.

In practice the minimum number of experts leads to unreliability opinion of the expert group, but maximum number of experts is not good too. In our research work we use next conventional formula for determination experts' number [17]:

$$n \approx \frac{mM_0(M_0 - 1)}{mM_0 - \sum_{i=1}^{M_0} \mu(i)} + 1 \tag{3}$$

where M_0 is the original number of experts; m is the number of experts are called by each respondent.

In case when expert from the original number of experts calls the person who is not from this group, than $\mu = 1$, otherwise $\mu = 0$.

For determination of specific weight of criterion of CIs stability 15 people were invited and $M_0 = 15$. Each expert called some new experts, so we determined 12 more people. And the final number of experts became:

$$M_0 + \sum_{i=1}^{M_0} \mu(i) = 15 + 12 = 27 \tag{4}$$

3 Criteria of critical infrastructures stability

According to the first stage of statistical research (statistical observation) and using statements of the official statistical government [13] we make three threshold values for each criterion.

The first threshold value is for normal state of CIs stability and it takes the value 1.

The second threshold value is for potentially dangerous state of GIs stability. This value describes boundary state of CIs stability between normal and crisis and it takes value 0.

The last one threshold value is for crisis state of CIs stability. It takes value "-1".

The main criterion of CIs protection is engineering and functional stability. And the special criteria are shown in the table 1.

Table 1. The criteria of critical infrastructures stability and their threshold value.

№	Threshold value		
	«Crisis state of CIs stability »	**«Potentially dangerous state of GIs stability »**	**«Normal state of CIs stability»**
	-1	**0**	**1**
	Engineering stability		
1.	The criterion of explosion and fire safety		
	Constructions of CIs have the 4th or the 5th degree of fire resistance.	Constructions of CIs have the 3th degree of fire resistance.	Constructions of CIs have the first or second degree of fire resistance.
2.	The criterion of engineering safety		
	State of bearing structural elements is emergency, state of not bearing structural elements is dilapidated	All structural elements must be repaired	All structural elements have some removable defects
3.	The criterion of engineering systems reliability		
	Engineering systems have fatal physical deterioration of	Engineering systems have disposable physical deterioration	Engineering systems are new or renovated.
4.	The criterion of reliability of project design solution		
	Longevity of project design solution is about 20 years	Longevity of reliability of project design solution is about 50 years	Longevity of reliability of project design solution is more than 20 years

5.	The criterion of technological equipment safety		
	Technological facilities have fatal physical deterioration	Technological facilities have curable physical deterioration	Technological facilities are new or renovated
Functional stability			
6.	The criterion of CIs' automation level		
	CIs are not automated	CIs are partly automated	CIs are automated
7.	The criterion of resource safety (resource using in CIs)		
	There are dangerous resources (substances or materials) in the CIs. Regulatory requirements are not carried out.	There are dangerous resources in the CIs, but all regulatory requirements are carried out.	There are not dangerous resources in the CIs
8.	The criterion of energy safety		
	There are any power supply interruptions and no backup power supply in the CIs.	There are no power supply interruptions and no backup power supply in the CIs.	There are backup power supply and no power supply interruptions and no in the CIs.
9.	The criterion of environmental safety		
	The CIs generate waste of I, IV hazard class during its functionality.	The CIs generate waste of III hazard class during its functionality.	The CIs generate waste of IV, V hazard class during its functionality.
10.	The criterion of work safety		
	The requirements of GOST R 54934-2012/OHSAS 18001:2007 are not carried out	The requirements of GOST R 54934-2012/OHSAS 18001:2007 are carried out partly	The requirements of GOST R 54934-2012/OHSAS 18001:2007 are carried out

There are values of specific weight for all criteria of CIs stability according to chosen methodology in the table 2.

Table 2. The values of specific weight of CIs stability criteria.

The criterion of Cis stability according to table 1.	**Specific weight of criterion**	**The criterion of Cis stability according to table 1.**	**Specific weight of criterion**
1	0.0816	6	0.0783
2	0.1987	7	0.0870
3	0.1493	8	0.0844
4	0.0778	9	0.0734
5	0.1111	10	0.0584

4 Discussions

In our research work we try to determine some criteria, which can describe engineering and functional stability of CIs. But we have to understand that CIs are complex system. And after getting all information about our criteria we can't make an evaluation of engineering and functional stability of CIs without special formulas. So according to foreign experience we make next algorithm (5) for evaluation stability of CIs:

$$R_{stab.i} = \begin{cases} N, \sum_{i=1}^{n} V_i^N \geq \delta_{Ni} \\ P.D, \sum_{i=1}^{n} V_i^{P.D.} \prec \delta_{Ci} \quad and \quad \sum_{i=1}^{n} V_i^N \prec \delta_{Ni} \\ C, \sum_{i=1}^{n} V_i^C \geq \delta_{Ci} \end{cases} \quad (5)$$

where $R_{stab.i.}$ is the evaluation of engineering or functional stability according to the criteria of CIs stability; V_i^N, V_i^C are the specific weights of criteria, which numerical value is in the tolerance range of normal and crisis values; $\delta_{Ni,}\delta_{Ci}$ are the coefficients that characterize the level of the normal and the crisis states of CIs.

If total weight of all criteria in the range of normal state is not less $\delta_{Ni,}$ the state of CIs stability is normal and $R_{stab.i.}$ =*N*. In case when total weight of all criteria in the range of crisis state is δ_{Ci} and more, then the state of CIs stability is critical and $R_{stab.i.}$ =*C*. In other cases $R_{stab.i.}$ =*P.D.*

The value of the normal state of CIs coefficient depends of the criteria's specific weights. And according to our study the coefficient for engineering stability (δ_{Neng}) is in the range from 0.348 to 0.6185. The coefficient for functional stability (δ_{Nfunc}) is in the range from 0.1627 to 0.3815. The value of the crisis state of GIs coefficient depends of the criteria's specific weights too and the coefficient for engineering stability (δ_{Ceng}) is in the range from 0 to 0.1594. The coefficient for functional stability (δ_{Cfunc}) is in the range from 0 to 0.1318.

If we get other values for engineering and functional stability of CIs, than this CIs have potentially dangerous state.

5 Conclusions

In this paper we have studied the available tools for evaluation of CIs stability. To this end we have reviewed advantages and disadvantages of the concept of absolute security and the concept of acceptable risk. Moreover through an analysis of recent literature and according to expert evaluation method we have identified the main criteria of CIs stability.

But it is obviously that for different types of CIs we have to correct these criteria using calculated values and information which we can get during exploitation CIs. Future work should be devoted to developing optimal system of criteria of CIs stability according to types of CIs.

References

1. Z. Banschikova, V. Kamzolkin, V. Mitrofanov, TCS **2**, 55-58 (2006)
2. *Security of Russia. Risk analysis and security problems* (2006)
3. G. Belyaev, I. Teterin , V. Yatsunenko, VTPU **5**, 147-150 (2008)
4. A Volkov, HMSSU **3**, 30-35 (2007)
5. A. Volkov, ICC **6**, 34-35, (2000)
6. A. Volkov, IMFS, 133-160 (2004)

7. A. Hussars, Proceedings of the II International Scientific and Practical Conference "Technical means of impending disaster", 27-36 (2006)
8. V. Durdenko, A. Rogozhin, HVIMIR **1**, (2012)
9. S. Kondratiev, SS, 103-106 (2006)
10. M. Lyubimov, BMETXXI **8**, 57-58 (2006)
11. B. Rodionov, Nanotechnology and integrated security BMETXXI **5**, 60-63 (2009)
12. Y. Sdobnov, B. **4**, 4 (2007)
13. *Statistics: Textbook* (2005)
14. F. Kadri, B. Birregah, E.Chatelet, JHSEM **11-2**, 217-241 (2014)
15. A. Abbasi, N. JHSEM, 9-1(2012)
16. E. Casalicchio, M. Caselli, A. Coletta, I. Fovino, IJCI **9-1/2**, 111-129 (2013)
17. Yu. Geraskin, *Improving the efficiency of construction industry in the monolithic housing construction based on an evaluation of organizational and technological solutions* (2004)
18. A. Ginzburg, S. Kachanov, IJAER **11(3)**, 1660-1665 (2016)

Numerical simulation of extended structures aerodynamic with high level of responsibility

Anastasia Fedosova[1,*], *Yulia* Gribach[1], and *Dmitry* Gribach[1]

[1] Moscow state university of civil engineering, Yaroslavskoye sh. 26, 129337 Moscow, Russia

Abstract. Using numerical simulation the paper investigates the aerodynamics of extended airport complex; according to the classification of existing regulations the complex is related to buildings and structures of high level of responsibility. We present the technique used for numerical simulation and the results obtained by the application of numerical simulation, the results were verified with the results obtained during physical modelling in wind tunnel of architectural type. According to the results of the research the conclusions about the pattern of flow around the building wind flow were made.

1 Introduction

Airports buildings are classified as buildings and structures of high level responsibility. The majority of these facilities are unique space-planning solutions. Construction of these facilities is a complex engineering task.

Fig. 1. Extended airport complex.

The considered airport (Fig.1) is a large-span structure in the form of a rectangular with volume terms (axes dimensions are 50 * 70 m) and varying elevations. The roof has two ramp from two opposite angles; minimum height is 13.45 m and a maximum reach is 19 m. The given design features allow to categorize the complex as unique extended structure of buildings and structures of increased level of responsibility. For this class of

* Corresponding author: mgsu@broll.ru

building there is no data on the distribution of wind load and acting domestic and foreign building codes and regulations prescribe wind parameter assignment by the results physical modelling. In this paper a study of wind loads will be produced using the numerical simulation and the verification of the obtained results with the results of physical modelling.

2 Numerical simulation

The creation of three-dimensional geometric model is preceded the process of calculating the aerodynamic characteristics of the building. 3D geometric model must correspond to the original configuration of the object. 3D-body was projected on the basis of the analysis of space-planning and constructive solutions to the passenger terminal on the principle of similarity, which, in consequence of different scale transforms unstructured grids (Fig. 2). The grid is defined by the necessary number of refinements (in this case: the number of volumes of more than 5 million).

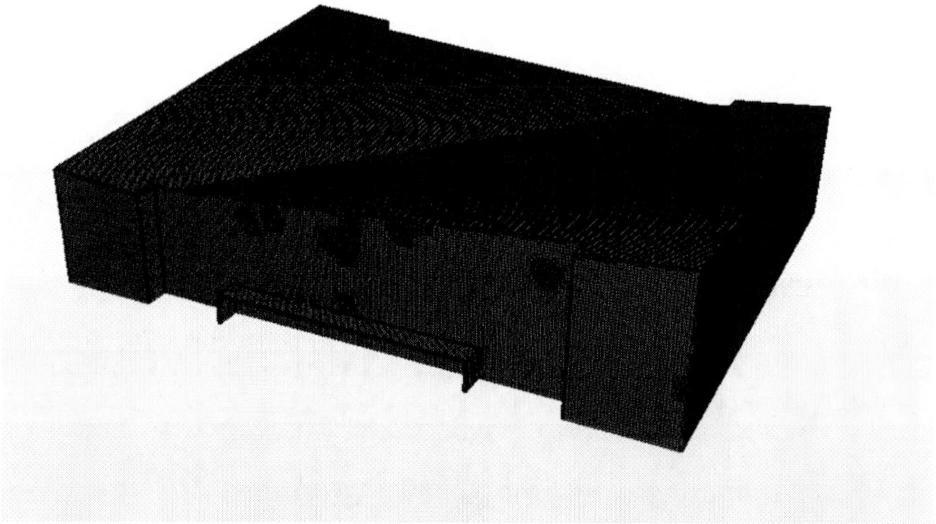

Fig. 2. Detail of the grid near the building.

The next stage of the study is setting the boundary conditions: the entrance of the computational domain (inlet) is defined by the velocity flow U (z), the intensity of turbulent pulsations I and the scale of turbulence; the orientation of the wind direction is defined as well; the condition of constant pressure is set on the output of the calculated area.
– the entrance of the computational domain (inlet) is defined by the velocity flow U (z), the intensity of turbulent pulsations I and the scale of turbulence
– the orientation of the wind direction is defined as its shows at Fig. 3;
– the condition of constant pressure is set on the output of the calculated area;
– on the bottom, on top of the remote and on the internal borders (on the facades of buildings and surrounding objects) - no-slip wall - zero speed (parietal cells height was $2.5 \cdot 10^{-4} H$ that provided y + about 1 and made it possible not to use the model wall functions).

In this study, for determination the aerodynamic characteristics with a numerical modelling software package Ansys Fluent 14.5 is used. The structure of the performing numerical simulation of wind influence has the following order: the three-dimensional configuration of the body flows about the turbulent flow of an incompressible air, having an average distribution rate adjustment calculation domain in accordance with the experimental velocity profile and intensity of turbulent pulsations. As the calculated value a dimensionless pressure coefficient Cp is used, which is distributed on the facades around the perimeter of the horizontal building section. The turbulent motion of the air environment near the volume described by the system of Reynolds equations.

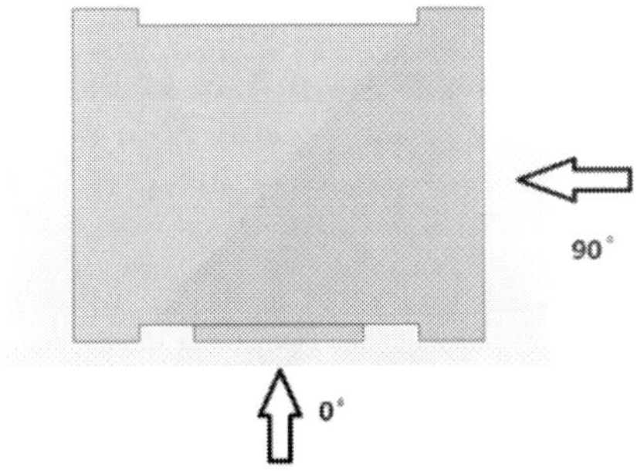

Fig. 3. Directions wind model in numerical simulation (top view).

The essence of this method consists in that all the instantaneous flow values of studied parameters are presented as the sum of the averaged value of the pulse and its component, i.e. the resulting value is random, which in time averaging gives her expectation, and the fluctuating component of which is the variance of the random variable. However, using only the solution of the basic equations makes the system appears to open, and finding the unknown quantity becomes impossible. For solving the system the system closure is made by bringing additional equations that govern the laws of turbulence development.

3 Verification

Obtained by computer simulation the distribution of average pressure on the facades of opportunity should be verified with the data that was determined experimentally. For this a number of drainage points on the model and layout was chosen. A series of preliminary calculation was performed for a number of areas, and then a comparative analysis of the results was performed (Fig. 4).

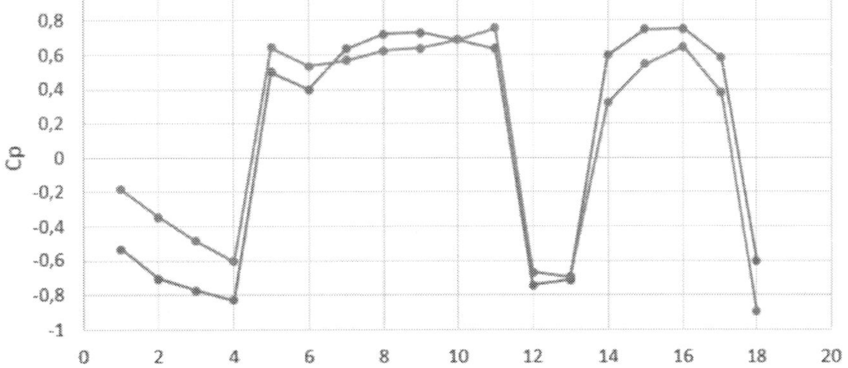

Fig. 4. The distribution of the average coefficient of aerodynamic pressure at verification points along the perimeter of the airport terminal complex with flow angle equal to 0° (read – experimental, blue – numerical).

Carrying out data verification allows choosing the most rational parameters of the computational grid, turbulence model and its characteristics, providing relatively little difference between the calculated and experimental results. In this paper, it does not exceed 20%.

4 The results of numerical simulation

The complex configuration of the airport leads to the formation of numerous vouchers and jet vortex zones, which form a non-monotonic pattern of distribution of wind pressure on the facades of buildings. Fig. 5 gives a picture of the distribution of the streamlines at a height of 15 m for the flow angles of 0 ° and 90 °, and in Fig. 6 shows a picture of the distribution average coefficient of aerodynamic pressure on the facades of the airport complex.

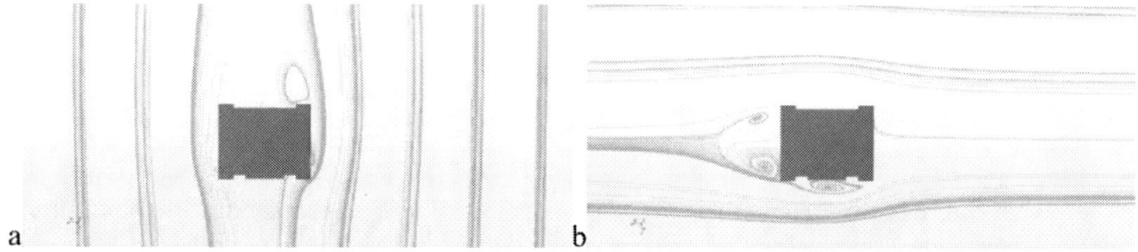

Fig. 5. Streamlines of 15 m at flow angle 0° (a) and 90° (b).

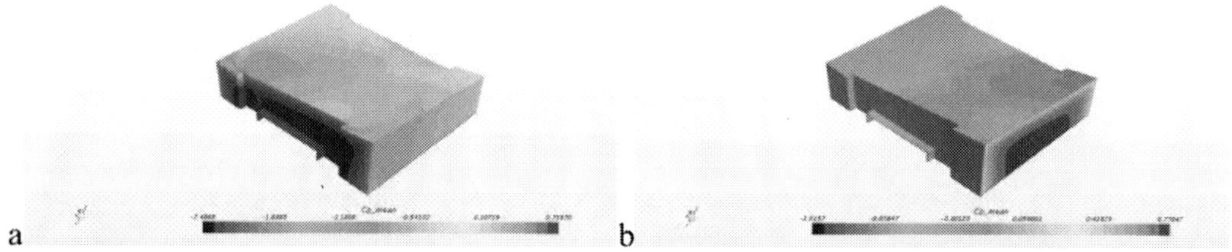

Fig. 6. The distribution of the average aerodynamic pressure on the facades rate of flow at an angle of 0° (a) and 90° (b).

Simulation of flow around the building using the computer technology provides information on distributed wind loads on the facade of the object just as it is done by physical experiment in conducting drainage tests of object models in a wind tunnel. To present the results of a "virtual experiment" "drainage point" were selected virtual, which was defined wind pressure coefficient (average component) according to the aerodynamic formula

$$C_{\text{p mean}} = \frac{w_m}{0.5\rho U^2(H)}, \tag{1}$$

where w_m is calculated average wind flow pressure in the "virtual drainage point " at the facade of the building in averaging several temporary quasi-period fluctuations; ρ is an air density; *U (H)* is depending on the height H the average flow velocity in the wind lay.

Based on the pattern of distribution of the average coefficient of aerodynamic pressure on the facades it is impracticable to determine the dimensionless coefficient *Cp,* that will be applied in the calculation of wind loads on the building of the airport complex.

Fig. 7 is a picture of the distribution of minima and maxima peak on the facades of the airport terminal complex.

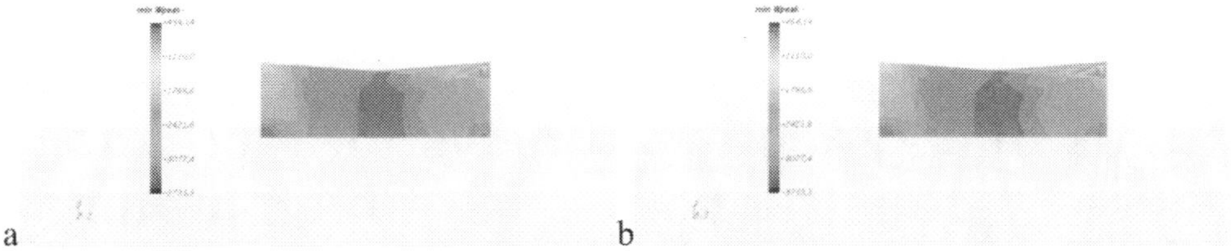

Fig. 7. Distribution minima (a) and highs (b) peak (Wpeak- and Wpeak +) on the facades of the airport terminal complex.

5. Conclusions

Simulation of flow of the airport terminal using provided in this article computer technology shows information on the distribution of the wind effects on the building facades. The data medium aerodynamic coefficient in computational research provide a good consistency with the experimental results and allow us to determine the design wind loads on the facade of the airport complex in accordance with the SP 20.13330.2011 "Loads and effects". The accepted method of determining the coefficient Cp corresponds to the requirements of normative documents on the implementation of such works.

This work was financially supported by the Council for Grants of the President of the Russian Federation for state support of young Russian scientists in the framework of the research MK-6194.2015.8.

References

1. *Building codes SP 20.13330.2011 Loads and effects. The updated edition of SNIP 2.01.07-85* (2011)
2. A. Kubenin, A. Fedosova, O. Poddaeva, Scientific Review **21**, 368-372 (2015)
3. A. Kubenin, A. Fedosova, The National Association of Scholars **X(15)** 99-102 (2015)
4. Al Zoubi Feras, Li Zheng-liang, Wei Qi-ke, Sun Yi, Journal of Chongqing University **1(9)** 47-53 (2010)
5. *The use of Wind tunnel experiments for wind loads on structures. C.P.W.* (Geurts TNO Built Environment and Geosciences, Delft)
6. *Wind Tunnel Testing of High-Rise Buildings* (An output of the CTBUH Wind Engineering Working Group, 2013).
7. O.O. Egorychev, S.I. Dubinsky, A.N. Fedosova, Applied Mechanics and Materials **713-715**, 1729 (2015)
8. O.I. Poddaeva, S.I. Dubinsky, A. Fedosova, Industrial and civil engineering **9**, 23-27 (2014)
9. V. Gagarin, S.V. Guvernyuk, A.S. Kubenin, Housing construction **7**, 3-8 (2014)

Features of engineering surveys for recreational facilities construction in high-mountain environment

Olga Krinochkina[1,*] and *Andrey* Lavrusevich[1]

[1]Moscow State University of Civil Engineering, Yaroslavskoye sh. 29, 129337 Moscow, Russia

Abstract. This article discusses the features of engineering surveys for recreational facilities construction in the high-mountain environment. Basic survey types are considered. Special attention is paid to danger geological processes and phenomena, and the condition of basic components of mountainous landscape for the planned recreational facilities. The study is intended for construction designers, ecologists, geologists and surveyors. Key words: engineering survey, geodesy, dangerous geological processes, recreational facilities, high-mountain environment.

The features of an engineering survey will be considered by the example of surveys conducted under the supervision of the authors in the valley of rivers Mamisondon and Adaikom, North Ossetia – Republic of North Ossetia-Alania, for construction of the Mamison mountain and sports recreational facilities. The engineering survey included all the known basic types: engineering geodetic, engineering geological surveys (EGS) and engineering ecological surveys (EES). This study discusses the features of the latter.

The engineering surveys for construction of the Mamison mountain and sports recreational facilities were carried out in compliance with the Russian regulatory documents in force.

Development of high-mountain areas for recreational purposes is quite a new trend resulting from the rising interest of modern recreation lovers to extreme sports and ecotourism.

Engineering surveys (ES) carried out for construction of recreational facilities (RF) in the high-mountain environment should consider the features involved. First, this is explained by the special status of recreation areas, and, second, by the very high-mountain environment. These conditions predetermine the high dynamics of all processes, both geotechnical and geochemical, which, in their turn, condition the features of geotechnical and ecological surveys. See below for more details.

The status of a recreation area implies a presence of corresponding potential, i.e. such an area should contain recreation resources. The latter firstly indicate the components of natural environment of proper quality. And it's the components of the environment constituting part of recreation resources that we will focus on.

Construction of recreational facilities requires the presence of a few natural factors, namely: aero-, hydro-, phyto- and certainly lithocomponents. These form peculiar

* Corresponding author: vdovinaok@mail.ru

mountanious landscapes being most valuable targets of recreational nature management [1]. So, the challenge of an ecological survey is to determine if the quality of these natural components meets sanitary standards. Besides, their geochemical situation is also important for recreation areas [2, 3]. High-mountain areas are considered a priori clean since all long-livers of the planet usually live there. Yet, local communities for the lives of many generations tend to adapt to the local geochemical situation, so even considerably exceeded concentrations of harmful substances in the components of the environment over the maximum permissible thresholds are not detrimental to them. However, it was A.P. Vinogradov who found that even relatively small concentrations of some elements (2-3 times the Clarke values) could cause serious diseases [4]. This is why a geoecological assessment of a recreational area should consider its potential for causing ecological and geochemical danger to the environment conditioned by the area geochemical situation. The latter is usually determined by the peculiarity of the chemical composition of bedrock. Mountanious areas may contain low thickness or no loose sediments often resulting in bedrock outcropping. The high content of toxic elements could make them accumulate in depositing components of the landscape. A potential natural environmental and geochemical danger (NED) is estimated by the total accumulation of chemical elements of Danger Classes 1-3, the level of which exceeds the Clarke values considerably [5].

Apparently, the feature of an ecological survey for RF construction planning is assessment of the NED of the lithological component of the recreational resources within the area.

The other afore-mentioned feature of engineering surveys for recreational facilities construction in the high-mountain environment deals with the highland itself and refers to a geotechnical survey. The latter should consist in a more detailed study of dangerous geological processes (DGPs) and phenomena because these are highly intensive in the highlands environment [6].

Since elevations at the RF under study vary from 2,000 to 3,800 m, there is every ground to relate it to highlands. High intensity of DGPs is observed here emphasizing the importance of ensuring safety of the facility units.

The technique of investigating DGPs in the course of EGS, at the stage of the investment feasibility study for the Mamison mountain and sports recreational facilities, also had some peculiarities.

First of all, a 1:25,000 preliminary sketch map of Quaternary sediments was generated after data gathering and summarizing, and based on satellite image interpretation and remote sensing. During the map compilation, special attention was paid to identification of basic genetic types of DGPs based on their final result – occurrence of corresponding accumulations and analysis of the degree of their activity from the point of morphological features, terrain fragmentation degree and other attributes.

Upon verification of the interpretation results with a reconnaissance survey, the final Quaternary sediments map was generated, the analysis of which revealed some patterns of DGPs occurring at the planned construction site.

For example, a pattern of activation of exogenic processes was identified related to erosional network development and a decrease of erosional bases in Holocene period, to neotectonic factors and high seismicity within the Nar zone.

Note that the river valleys forming the basin of river Mamisondon are deeply cut into the slopes of ridges Bokovoy and Glavny preserving their U-shapes at the tops and having a complex structure of glacial, sloping and alluvial-proluvial accumulations. All these disperse sediments are the material of collapses, mudflows and landslides under certain conditions. The latter are most common at the area under study.

Landslides are contributed by the widely spread Jurassic clayey rocks and periodic overwetting of this material. The neotectonic factor is also important being expressed in the intense fragmentation of rocks in thin zones of most recent tectonic movements.

The field study revealed that there were 2 types of landslides based on the composition of displaced masses: deluvial and structural gravitational.

The deluvial landslides comprise deluvial loam consisting of abundant fine and coarse crushed bedrock. These are individual landslide bodies not widely spread within the area.

Besides the loose deluvial sediments, the structural gravitational landslides are composed of lumps and piles of bedrock crushed to blocks displaced downslope, along with the deluvial rocks. Based on the results of natural observations, largest landslides of type ES were mapped.

Some of the landslides were related to the Upper Quaternary quiet stabilized landslides, the movements of which are not known.

In order to consider the geodynamic and technical condition of mudflow catchment areas, we made a qualitative assessment of the area by the degree of modern erosional fragmentation of the basins of tributaries of river Mamisondon. It was found that, from the point of mudflow danger, the territory of the basins of rivers Adaikom and Mamisondon could be assigned a medium danger level. Mudflows with a simultaneous efflux of hard material under 50,000 m^3 are observed here.

It was also found that the mudflow origination centers developed at >2,300-2,500 m elevations, so all basins of the main tributaries may be considered potentially mudflow danger. The study results were used in the identification of zones of DGPs requiring safety measures to be taken during the construction of the facility units.

In consideration of the ecological survey at the high-mountain areas to be developed for recreational purposes, it is worth restating that their specificity consists in assessment of the natural environmental and geochemical danger (NED) of the landscape components. In order to assess it quantitatively, I.A. Morozova suggested using a natural environmental danger index [7,8]. It was calculated by analogy with the cumulative pollution index (CPI) of Ye.Yu. Sayet [9] and characterizes the effect of a group of toxic elements Classes 1-3 of Danger at the sampling point.

It is calculated by the following formula:

$$NED = \sum Cc - (n-1), \tag{1}$$

whereKk – the Clarke value of concentrations, n – number of chemical elements with Cc>1 ($Cc = C_f/C$).

C_f – situation content of chemical element (Danger Classes 1-3), C – the Clarke values for global soils.

Note that, approbation of this index should involve rationing (normalization) of bedrock to the Clarke values for the crust (C) per Bowen [10], and the Clarke values for global soils should be used for rationing of bottom sediments. The Cc value we used exceeded 2 since, in abruptly changing high-mountain conditions, Cc≤2 reflects fluctuations of natural situation. Besides, sampling average (C_{av}) should be used instead of C_f.

By analogy with CPI [9] and depending on the NED value, the areas under study were related to those of medium danger (8-16), danger (16-32) or highly danger (>32) pollution levels; we have also introduced a permissible pollution level (<8).

Then, the territory was zoned based on the nature management restriction for recreation arrangement. The basin-landscape concept developed by S.V. Zotov was used as the basis [11]. Unlike in the concept, the zoning criteria laid in the level of NED and geodynamic conditions within the basins (see Table 1). It was determined using a two-point system:

strained (if at least one of the DGPs occurs within the basin) and unstrained (if only stabilized DGPs occur or are absent within the basin).

Thus, the main features of engineering surveys at the stage of the investment feasibility study for the recreational facilities construction in the high-mountain environment are associated with arrangement of geotechnical and ecological surveys. For the first group, they consist in more detailed investigation of DGPs both during the preparation (applying satellite image interpretation and remote sensing), and the fieldwork period; for the second group – in assessment of NED components of the survey area landscape.

Table 1. Zoning of the territory by level of NED and geodynamic conditions.

Seq. No.	Basin	Zoning criteria				Regulation type and recommendations
		Geodynamic conditions*	NED			
			NED<8	8<NED<16	NED>16	
1	Gibitandon	-	+	-	-	No limitations
2	Kozydon	+	+	-	-	No limitations
3	Zemegondon	-	+	+	-	Limited regulation for using materials within 8<NED<16 areas
4	Ruchatdon-Bubudon	+	+	-	-	No limitations
5	Upper Mamison	+	+	+	-	Limited regulation for using materials within 8<NED<16 areas
6	Lagatkom	+	-	+	+	Completely limited use of materials within NED>16 areas. Limited regulation for using materials within 8<NED<16 areas. Surface water microcomponent composition quality control
7	Kaitakomdon	+	+	-	-	No limitations
8	Kasaikomdon	+	+	+	-	Limited regulation for using materials within 8<NED<16 areas
9	Khalatsa	+	+	-	-	Surface water microcomponent composition quality control (NH^{+}_{4})
10	Kuatedon-Retsidon	+	-	+	+	Completely limited use of materials within NED>16 areas. Limited regulation for using materials within 8<NED<16 areas. Surface water microcomponent composition quality control
11	Dnuidon	-	-	+	+	same
12	Adaikom	+	-	+	+	same
13	Tib-Tsargas	+	-	+	+	same
14	Pus	+	-	+	+	Completely limited use of materials within NED>16 areas. Limited regulation for using materials within 8<NED<16 areas. Surface water microcomponent composition quality control

*Geodynamic conditions: + strained, - unstrained.

References

1. *Federal Law On Natural Healing Resources, Health-Improving Areas and Resorts, No.26-FZ, dated February 23* (1995)
2. O.K. Vdovina, A.A. Lavrusevich, R.V. Vysokinskaya, I.M. Evgrafova, K.S. Polyakova, MGSU Bulletin **8**, 98-106 (2014)
3. O.K. Vdovina, Ye.N. Malinina, A.N. Popova, Subsoil Exploration and Preservation **7**, (2012)
4. A.P. Vinogradov, Proceedings of the USSR Academy of Sciences **18(4/5)**, 283-286 (1938)
5. O.K. Vdovina, Bulletin of the Peoples' Friendship University of Russia. The Engineering Studies series **3**, (2009)
6. V.L. Poznanin, O.K. Vdovina, Subsoil Exploration and Preservation **9-10**, 139-142 (2006)
7. I.A. Morozova, *Geochemical Landscapes and Environmental Threat, The Applied Geochemistry collection, 1 edition. Geochemical Mapping* (IMGRE, Moscow, 2000)
8. I.A. Morozova, Subsoil Exploration and Preservation, 29-30 (1994)
9. Yu.Ye. Sayet, B.A. Revich, Ye.P. Yanin et al., *The Environmental Geochemistry* (Nedra Publishers, Moscow, 1990)
10. H.J.M. Bowen, *The Environmental Chemistry of the Elements* (Academic Press., London-New-York-Toronto-Sydney-San Francisco, 1979)
11. Proceedings of the USSR Academy of Sciences, the Geography series **6**, 55-65 (1992)

Systematization of the major stages of the client in certain branches of construction production

Tatyana Kuzmina[1,*], and *Nadezhda* Cherednichenko[1]

[1] Moscow State University of Civil Engineering, Department of technology and organization of construction production, Yaroslavskoye sh. 26, Moscow 129337, Russia

Abstract. The article contains substantiation of the need for investment and construction projects in various sectors of the construction production. On the basis of domestic and foreign experience it describes the concept of the client (customer) and his functions and tasks in the implementation of such projects. It also systematizes the main stages of the customer's activities, which are presented as a preparation for the construction, development of construction sites, monitoring the progress of the construction, commis-sioning and commissioning of facilities. There is also a determination of time intervals at each stage, the final formula for calculating the duration of the investment project due to the given the shared stages. The article also gives optimization variants in some sectors of construction production.

1 Introduction

Since the beginning of the 21 century the development of the construction industry in Russia allowed to implement major investment and construction projects in various spheres of social and housing construction.

Compared with Europe and North America, housing in the Russian Federation after the 2008-2009 crisis recovered faster. Resulting in 2014 year a record in terms of the volume of housing construction has been achieved (1.08 million completed construction of new residential buildings, that is 20.3% more than in the year 2013) [1].

In recent years, the dynamics of the Russian economy is influenced by two factors. On the one hand, under the influence of increasing sanction pressure drastically worsened the conditions of Russia's economic cooperation with key partners in trade and investment and technological cooperation. On the other hand, a steady decline in world commodity prices, first of all energy prices, led to a significant reduction in export earnings and revenue [1].

The negative impact of economic sanctions affected the ability to attract external resources for the financing of the Russian economy and international investment-construction projects. Along with this drop in investment attractiveness of the country which took place against the backdrop of sustained improvement in its position in the international ranking of Doing Business [1, 2, 3].

*Corresponding author: KuzminaTK@mgsu.ru

At present, measures to improve the investment climate in the country should be comprehensive, taking into account both the economic and the administrative and foreign policy determinants of investor's decisions [3, 4, 5].

In these circumstances, the tip goes to import substitution, but there is a high level of capacity utilization and labor, as well as the hard deficit of investment resources in conditions of financial sanctions and the restriction of access to Western technology and the suspension of international cooperation projects.

In such circumstances, there is a need for the development of agricultural complex and all branches of construction production, modernization of existing and construction of new industrial complexes. Such projects could attract investors, but they need to understand timelines and outputs of their objects of construction onto the starting power.

2 Procedure

To determine the terms of implementation of investment projects by branches of the construction production, it is necessary to highlight the basic steps for creating an object of construction, identify changes in modern conditions and determine factors affecting the total duration.

Handling arrays of expert assessments developed by specialists within the survey study "Adaptation of client activity in market conditions", made it possible to highlight the main stages of the realization of investment projects and determine their relevance in the present (fig. 1.) [3, 6, 7].

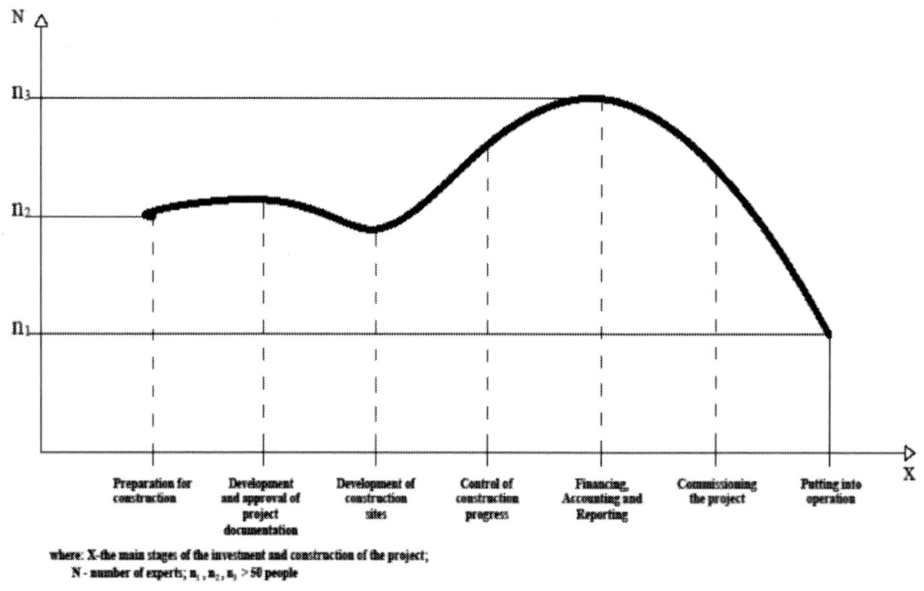

Fig. 1. Enhanced features on the major stages of creation of an object of construction.

It should be noted that an important consideration in the construction is the role of the client (customer) and who performs his functions [2]. In England, Sweden and Finland the representative of such a service is mainly an architect. He picks up the necessary expertise that define the provisional construction cost, perform work under the sub-contract, collect other data required for construction. In the United States, the client creates a group, which includes economists, estimators, contract managers, engineers and manufacturers of constructional works. Then they develop all the necessary documentation and diagrams of construction production. In Germany the general contractor receives the work order, he chooses the specialized organizations and bears full responsibility for the timing of construction and the quality of the finished product. In Japan, when construction companies organize research laboratories and institutions with scientists as the main staff. Thus instead of the concept of "client" there is used the notion of "developer", whose responsibilities include the acquisition of venues, financing, construction and sale of the finished product [8, 9, 10].

In Russia the functions and tasks of the clients when implementing construction projects define the forms of investment (state, municipal, private, etc.).

Therefore, depending on the scope of investment and scale of the project, there are several forms of organizations that implement client functions [11, 12, 13].

Compare changes in the function of the client on the main stages of construction, in terms of state and commercial orders (fig. 2).

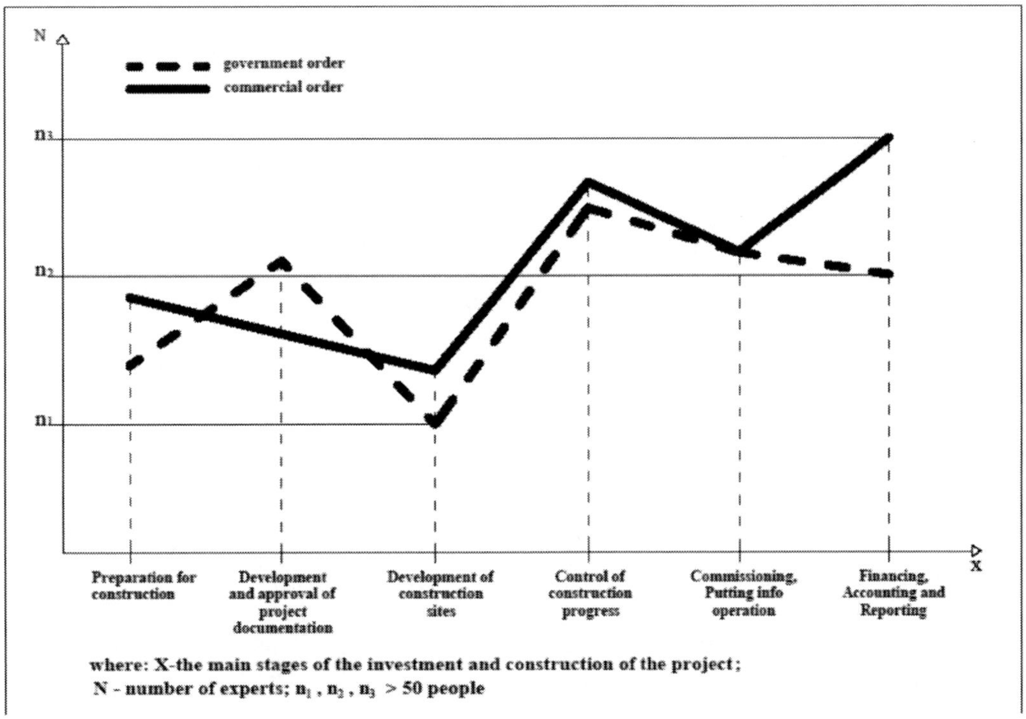

Fig.2. Comparison of the functions of the client on the main stages of construction, depending on the forms of investment.

Let's define the coherence of expert opinions [3].

The coefficient of the concordance can be found according to the formula:

$$W = \frac{12 * \sum_{i=1}^{m} D_i^2}{m^2 \cdot (n^3 - n)}, \tag{1}$$

$$\text{Where: } D_i = \sum_{j=1}^{m} r_{ij} - \frac{m(n+1)}{2} \tag{2}$$

Got: W=0.848214

to find the value: $m(n-1)W^3 = 17.26722$

Tabulated value: $\chi^2{}_{1-\alpha,n-1} = 12.6$ when the significance level is: ά=0.95

Because $m(n-1)W^3 = 17.26722 > \chi^2{}_{1-\alpha,n-1} = 12.6$, so the W factor is significant at the level of:ά=0.95

Therefore, the opinions of experts are coherent.

3 Main part

Let's organize the main stages of client's activities, influencing the duration of erection of the objects of construction:

-preparation of construction (including formation and approval of project documents);

-development of construction sites;

-monitoring of construction process;

- commissioning and putting into operation.

Time intervals of considered steps are summarized in table 1.

Table 1. Intervals of mastering the basic steps of creating an object of construction.

Name of the stages	Time intervals corresponding to the development phase	Average value of expert performance
Preparation for construction	$0.15T \le t_p \le 0042T$	0,37T
Exploration of construction sites	$0.05T \le t_o \le 0.35T$	0.25T
Monitoring of construction	$0.75T_c \le t_n \le 0.95T_c$	$0.85T_c$
Commissioning and putting into operation	$0.02T \le t_c \le 0.15T$	0.08T

Thus, the resulting formula for calculating the total duration of the erection of an object of construction can be expressed as follows [14]:

$$T = t'_p + t'_o + t_n + t_c, \quad (3)$$

where T is the duration of the investment construction project of the erection of an object of construction;

t'_p - part of the duration of the preparatory phase of the construction, not shared with the stage of development of building sites;

t'_o - part of the duration of the development phase of construction sites, not shared with the stage of preparations for the construction;

t_n - duration of construction monitoring;

t_c – duration of the phase of putting into operation and commissioning.

While the average duration of each stage, bearing in mind the linkage is as follows:

$t_p=0.26T$; $t_o=0.39T$; $t_n=0.62T$; $t_c=0.08T$.

The options for optimizing client's activities were designed for a number of objects of the following sectors: electric power engineering, gas refining and petrochemical industry, gas industry, ferrous metallurgy, heavy and transport machine building, chemical and petroleum engineering, the automotive industry, building materials industry.

The options for optimization of the client's activity on the main stages of the realization of investment construction projects of some sectors of construction production are presented in table 2.

Table 2. Options for optimization of the client's activity in certain branches of construction production.

Name of object	Characteristic	Time of realization of investment construction projects according to the stages of construction, month				
		t_p	t_o	t_n	t_c	T
Electric power engineering						
Expansion of nuclear power plant RBMK-1000	Closed type, single circuit, on nuclear fuel, with revolving water with pond-cooler or cooling tower	30	19	78	6	123÷125
Gas refining						
Installation of complex gas preparation	At gas condensate deposits. Power, bln. m3/year: 5	11	4	28	4	55÷57
Chemical and petroleum engineering						
Plant for the production of oilfield equipment	Production capacity of 50-70 million. rub. per year. The total area of all industrial buildings 70-100 thousand m2. The main building with an area of 45 thousand m2, height of 20 m, equipped with overhead cranes with lifting capacity of 50 t	17	12	38	5	73÷75
Automotive industry						
Plant for the production of trucks with capacity up to 4 tons	Production capacity of 100 thousand cars per year. The total area of all industrial buildings is 620 thousand m2, height of 12.8 m, equipped with overhead cranes with lifting capacity of 20 tons Without the production of engines and transmissions	21	16	75	8	109÷112
Building materials industry						
Cement plant	Production power of 3450 thousand t per year comprising: three technological lines with ovens of 6.4/7m in diameter, 95 m in length and auxiliary buildings, constructions and communications	22	14	50	8	94÷96

4 Conclusion

It is important to note that in the construction of industrial buildings, special attention is paid to environmental monitoring and industrial safety management of construction waste. However, the implementation of industrial environmental control and ecological monitoring requires leveraging additional financial resources. Industrial construction is one of the most complex areas of construction activity in the whole [14, 15].

Studies show that at present, there has been an increase in all the major stages of construction investment projects. At the same time, in the context of work with commercial orders, special attention is paid to ensuring the quality of construction products and the equitable distribution of financial flows. That is, in particular, because in most cases, large holdings investing is provided from their own funds for further use of constructed objects to fit their needs. In addition, the payment is usually made in fact of fulfilled works and financing of such services is usually continuous.

Ordering of the main stages of the client's activity helps to determine the duration of the implementation of investment projects in some sectors of construction, that provides investors with the opportunity to plan the output of enterprises in the start-up power, the payback period of capital investments

References

1. A.A. Dynkin, V.G., Baranovskii, *Russia and the World: 2016. Annual Forecast: Economy and Foreign Policy* (IMEMO RAN, Moscow, 2015)
2. T.K. Kuzmina, P.P. Oleinik, S.A. Sinenko, *Technical client activity in market conditions* (ASV publishing, Moscow, 2015)
3. T.K. Kuzmina, Technology and organization of construction production **1**, 50-53 (2012)
4. V.A. Lukinov, D.K. Vishnyakov, Real estate: economy, management **1**, 71-76 (2012)
5. A. Lapidus, N. Cherednichenko, V. Smotrov, Procedia Engineering **153**, 371-377, (2016)
6. P.P. Oleinik, L.S. Grigorieva, V.I. Brodsky, Applied Mechanics and Materials **580-583**, 2294-2298 (2014)
7. P.P Oleinik, V.I. Brodsky, Technology and organization of construction production **1(2)**, 27-31 (2013)
8. E.A. Korol, P.B. Kagan, T.A. Barabanova, I.A. Bunkina, International Journal of Applied Engineering Research **11(3)**, 1691-1693 (2016)
9. S.A. Sinenko, B.V. Zhadanovsky, International Journal of Applied Engineering Research **11(3)**, 1724-1727 (2016)
10. P.P Oleinik, V.I. Brodsky, Technology and organization of construction production **4(5)**, 40-45 (2013)
11. T.K. Kuzmina, S.A. Sinenko, Scientific review No7 **222-227** (2016)
12. A.A. Lapidus, N.D. Cherednichenko, Scientific review No 21, **338-341** (2015)
13. S.A. Sinenko, T.K. Kuzmina, Scientific review, No18, **156-159** (2015)
14. P.P Oleinik, T.K. Kuzmina, Civil and Industrial Engineering, **11**, 42-43 (2012)
15. L.S. Grigoreva, P.P. Oleinik, Procedia Engineering **153**, 208-216 (2016)

Specifics of system of external influences on the life cycle of a construction object

Aleksander Aleksanin [1,*], and *Sergey* Sborshikov [1]

[1]Moscow State University of Civil Engineering, Moscow, Russian Federation

Abstract. There is a very important issue today which includes the harmonious and effective development of the system 'man –environment'. Construction is a branch of material production, which has a significant negative impact on the world around us. It is necessary to plan and operate processes of construction at all stages of the life cycle of a building without exception, to prevent of ecological threats. The article describes the concept of 'life cycle' as applied to various fields of knowledge, analyzes existing in the scientific literature division of the life cycle of buildings in the periods, proposes own approach to the division of periods of the life cycle on the basis of resource-saving. The article proposes the creation of a unified organizational system for the effective management of all periods with the constituent phases and formulates the main external influences on the building life cycle.

The present stage of development of the construction industry is characterised by close inter-relationship with the natural environment. The harmonious development of the construction and the natural environment is the key to sustainable development of the system 'man - environment'. To date, there is the task of organization all the processes accompanying the creation of the construction object in a single control system, starting with the stages of initiation and pre-investment and ending with decommissioning or its liquidation. One of the main criteria for the operation of this system, in the first place, should be the focus on the balanced development of the environment. The decision of this task is based on the need to develop a set of measures to streamline and improve the existing mechanisms of formation and functioning of the processes in the context of the life cycle of a construction object. The process of creating a construction object can be divided into sub-processes, the interaction of regulation which will contribute to the harmonious coexistence of the construction object and the surrounding world.

Let's consider stages of life cycle of the object, most often meeting in modern scientific works. The concept of 'life cycle' is widely used in many fields of knowledge, such as in biology, this term means the consequent shift of all generations, characteristic for this type of living organisms. In economics, there is the concept of 'life cycle of the organization' - a set of developmental stages which runs the firm during the period of its existence.

For our research the most interesting definition is 'the life cycle of a building '. The life cycle of a building is a period during which carried out engineering survey, design,

[*] Corresponding author: aleks08007@mail.ru

construction (including conservation), operation (including current repairs), reconstruction, major repairs, demolition of a building or structure.

There is a lot of works of foreign authors, which devoted to scientific researches of the life cycles [1-5], many of them are directly related to energy efficiency and resource saving [6-9]. In the scientific and technical literature distinguish the following periods of a building's life cycle [10]:

the 1-st period is the process of developing a feasibility study for construction of the building;

the 2-d period is the process of developing of construction and design decisions;

the 3-d period is process of building, together with the development of technology, organization and technological regulations of work;

the 4-th period is the process of pre-service development;

the 5-th period is directly operation of buildings and work aimed at ensuring payback of funds invested in their creation;

the 6-th period is processes of manufacture scheduled preventive and major repairs to maintain in a good technical condition of constructive elements and engineering systems of the building;

the 7-th period is the stage, both physical and moral deterioration of the building, when necessary to carry out works on modernization, reconstruction or demolition;

the 8-th period is reconstruction, restoring physical, mechanical and operational characteristics of the buildings. This period include also 1-st and 2-d periods of feasibility study and development of technical documentation.

There are scientific papers in which the authors identify some types of life cycle of object of the real estate: economic, legal and physical. From an economic point of view there are three periods: construction (step up investments and investment stage); exploitation to the point where the object is fully pay for themselves (the entrepreneurial stage of the project); operation in the subsequent growth of the results on investment (innovation, facility closing). From a legal point of view, allocate periods of registration of ownership; possession, use, disposal; change of ownership; registration of liquidation [11].

It is necessary to distinguish the concept of 'life cycle of a construction object' and 'the life cycle of a construction project'. The authors offer the following life cycle stages of construction project: pre-investment stage, stage design, stage management, stage of implementation [12]. The life cycle of a universal project can be represented by the following phases:

- definition of goals and objectives;
- development concept;
- design;
- examination;
- decision on the implementation of the project;
- implementation of the project;
- monitor the implementation;
- evaluation and analysis of results;
- completion of the project.

Within the framework of scientific research suggests to integrate described above periods of the life cycle of the construction object and to allocate only three major:

- creation of an object;
- exploitation of an object;
- elimination of an object.

The period 'creation of an object' combines phase feasibility study, design and construction. This stage also includes the phase of production of building materials, as it accumulates a huge potential resource in the implementation of the next phase.

So, each of the periods of the life cycle of the construction object includes certain phases.

The period of 'Creation of an object' contains a phases:

- production of construction materials;
- predesign development and researches;
- design works;
- construction and installation works.

The period of 'Exploitation of an object' contains a phases:

- works to improve the energy efficiency of the facility;
- maintenance works on the facility;
- maintenance repairs;
- capital repairs.

In this period a special place is occupied by improving energy efficiency, because during the operation of the building development of scientific and technological progress does not stop, and develops new energy-saving systems.

The period of 'Elimination of an object' contains a phases:

- the dismantling and demolition of buildings;
- works on the disposal of construction and demolition waste;
- works on the processing of construction and demolition waste.

For the effective management of all periods with the constituent phases, it is proposed to create a unified organizational system. The system has its own life cycle, comprising the steps of the process, covering a variety of state of the system, starting from the moment of the need for such a system, and ending with its complete decommissioning [13].

This organizational system is closely related to the impacts of external environment in the form of federal, regional or municipal authorities, the population, the level of development of scientific and technical progress, etc. The main external influence is the administrative and legal impact, economic impact, social impact, and the impact of science and technology. Scheme of impacts in terms of construction object lifecycle is shown in Figure 1.

Administrative and legal influence is due to the publication of new administrative regulations, standards, regulations, etc. The economic influence is a powerful regulatory tool, the essence of which is the availability of funds for works on a particular period. The social impact is the attitude of the society to the life cycle of the building, for example, the protests at a hearing on new construction projects in areas with unique natural landscapes, the lack of demand for products, realized within the functioning of buildings (space rental, attendance at trade and entertainment buildings, etc.). The impact of scientific and technological progress is the creation of new building materials, the development of modern technical, technological and organizational methods of the construction of buildings and structures.

The indispensable condition to demolish buildings and structures it is almost total recycling of construction waste and their subsequent use in the construction of new facilities, which reduces the impact on the environment in the extraction of raw materials [14; 15]. Thus, the life cycle of a single object of construction is obtained related to the life cycle of other objects. The high level of development of the market of construction and demolition waste contributes to the effective provision of recyclable material of new objects [16; 17]. The basis of sustainable development of natural and man-made environment is a balanced functioning of the system life cycle of construction projects based on resource conservation policy.

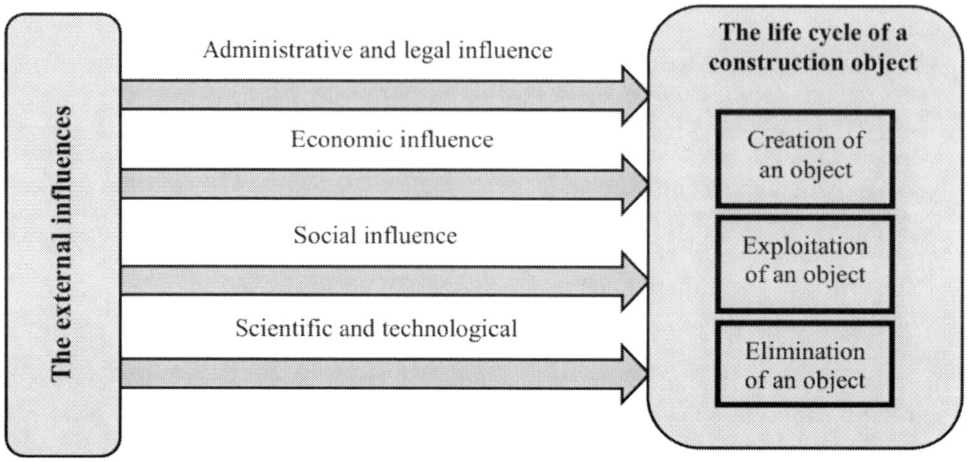

Fig.1. Scheme of the system impacts in the context of construction project life cycle.

It is necessary to take into account the fact that continuous improvement of production of construction materials and the emergence of new innovative technologies in the creation of the organizational life cycle management system of buildings. If we analyze the building, built in the XX century, it is possible to get an idea about the changes of the basic materials of building structures. For example, the buildings of the first half of the 20th century were wooden, clay and stone; in the second half of the 20th century building built of brick, block, panel; in the late twentieth and early twenty-first century is monolithic. The architects proposed many variants of the buildings of the future today. It is believed that the main trends of the construction will be ecology, mobility, self-reliance and full automation of all systems running artificial intelligence. Already practiced production of bricks from waste rice husk, the use of smart glass, which are capable of generating electricity, the introduction of a shell building facades, regulating indoor climate [18-20]. Perhaps, in the not too distant future it will be impractical to construct buildings from recycled designs objects built hundreds of years ago.

References

1. A.F.A. Rashida, S. Yusoffb. Renew. Sustainable Energy Rev. **45**, 244 (2015)
2. C.K. Chau, T.M. Leung, W.Y. Ng. Appl. Energy. **143**, 395 (2015)
3. I. Kovacic, V. Zoller. Energy. **92**, 409 (2015)
4. S. Butera, T.H. Christensen, T.F. Astrup. Waste Management. **44**, 196 (2015)
5. J. Honga, G. Q. Shena, C. Maob, Z. Lia, K. Lib. Journal of Cleaner Production. **112**, 2198 (2016)
6. V. Motuzienėa, A. Rogožaa, V. Lapinskienėa, T. Vilutienėb. Journal of Cleaner Production. **112**, 532 (2016)
7. M. S. Tokarik, R. C. Richman, Energy Build. **118**, 160 (2016)
8. S. Proiettia, P. Sdringolaa, U. Desideria, F. Zepparellia, F. Masciarellib, F. Castellania. Energy Build. **64**, 463 (2013)
9. L. F. Cabezaa, L. Rincóna, V. Vilariñob, G. Péreza, A. Castella. Renew. Sustainable Energy Rev. **29**, 394 (2014)
10. A.A. Afanasev, E.P. Matveev. *Reconstruction of residential buildings* (Moscow, 2008)
11. P.V. Fedyaeva *Organizational and technological support for improving energy*

efficiency in the housing sector of the Russian Federation [Text]: the thesis of Cand.Tech.Sci. 05.23.08. (2015).
12. V.V. Kuschenko, *Development. The modern concept for the development of Real property* (Moscow, 2005).
13. B.S. Blanchard, W.J. Fabrycky. *Systems engineering and analysis* (Prentice Hall, 2006)
14. A.V. Aleksanin, S.B. Sborshikov. Industrial and Civil Engineering **11**, 84 (2014)
15. A.V. Aleksanin. Scientific and Technical Volga region Bulletin **5**, 94 (2014)
16. A.V. Aleksanin. Industrial and Civil Engineering **10**, 79 (2014)
17. A.V. Aleksanin, S.B. Sborshikov. Vestnik MGSU **6**, 229 (2013)
18. C. Ionescu, T. Baracu, G-E. Vlad, H. Necula, A. Badea. Renew. Sustainable Energy Rev **49**, 243 (2015)
19. R. Volk, J. Stengel, F. Schultmann. Automation in Construction **38**, 109 (2014)
20. M. L. Rosenzweig. Israel Journal of Ecology & Evolution **62**, 7 (2016)

Standardization of green building technologies for environment design

Andrey Benuzh[1, *], *Ekaterina* Orenburova[2]

[1]Moscow State University of Civil Engineering, Department KBS, 26, Yaroslavskoe Sh., Moscow, 129337, Russian Federation
[2]Moscow State University of Civil Engineering, Department OSUN, 26, Yaroslavskoe Sh., Moscow, 129337, Russian Federation

Abstract. The article describes the structure and field of standardization ISO / TC 205 "Building environment design", provides examples of green building technologies. The main purpose of the article is to show the interaction between international ISO / TC 205 "Building environment design" and created in Russia in 2016 the Technical Committee of Standardization № 366 "Green technology of the build environment and green innovative products". Both of these technical committees promote green building technologies for environment design, thereby deal with the negative impact on the environment and the reasons of global warming. Instead of buildings that attempt to suppress and overcome nature, why not design buildings that integrate with the environment, on every possible level? The international standardization work which ISO/TC 205 "Building environment design" performs seeks, in addition to lowering trade barriers for engineering design, to promote and facilitate the design of high performance buildings: higher performing as economic assets for their owners, higher performing as buildings that provide amenable indoor environment for their occupants, and higher performing with respect to resource utilization and environmental impact.

1 International standardization of innovative green technologies

The standardisation related to the energy efficiency of heating and cooling systems as part of the overall energy use of buildings and has to take into account innovative green technologies. In Europe, The Energy Performance of Building Directive (EPBD) accelerated in Europe the standardisation activities in this field. Products are now no longer evaluated as products but as part of system. On the European Committee for Standardization (CEN) level a coherent structure has been put in place between product Technical Comities (TC), providing the basis for individual testing, system TC's dedicated to specific uses as heating and cooling, and a "horizontal" TC responsible for the overall energy use in buildings for the aims of sustainable development at all.

In 2008 the decision was made to bring the CEN EPBD standards, as produced under the mandate 343 EPBD, to the ISO level, the working group 9 "Heating and cooling

[*] Corresponding author: ABenuzh@gmail.com

systems" was established within ISO TC 205 "Building environment design" under the chair of Johann Zirngibl / France [1].

The scope of ISO technical committee ISO/TC 205 is standardization in the design of new buildings and retrofit of existing buildings for acceptable thermal and visual comfort, indoor air quality, and energy conservation by the green technologies (Fig. 1) [2]. Because of this scope, the committee treats the employment of materials and techniques in their optimal way in the design of buildings, which is intricately related to overall sustainability in building design. The unique opportunity afforded ISO/TC 205 is the holistic treatment of the many aspects of building environment design, which ultimately helps to determine the indoor environment.

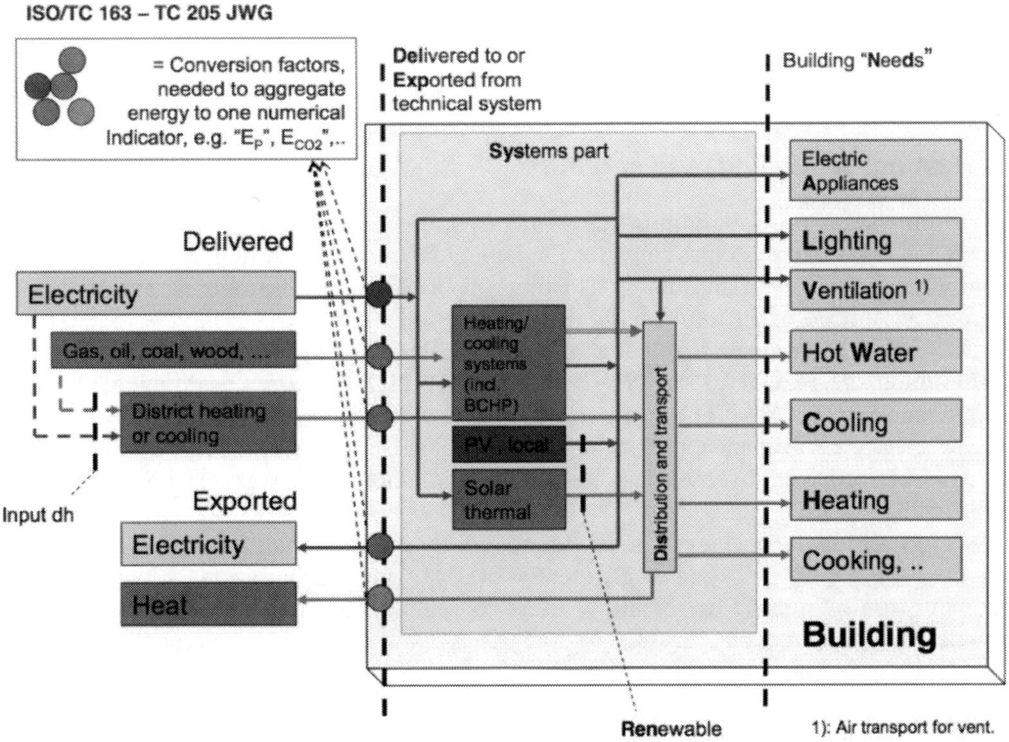

Fig. 1. Scope of standardization the technical committee ISO / TC 205.

The main aspects of standardization ISO/TC 205 include the followings:

- sustainability related to indoor environmental quality and energy that can be addressed in the design of buildings and the design of retrofits of existing buildings;
- general principles of building environment design;
- design of energy-efficient buildings;
- building automation and control systems in building and retrofit design;
- indoor air quality in building and retrofit design;
- indoor thermal environment in building and retrofit design;
- indoor acoustical environment in building and retrofit design;
- indoor visual environment in building and retrofit design;
- design of heating and cooling systems including radiant; and
- application of methods of testing and rating the performance of building environmental equipment in the design of new buildings and retrofits.

Furthermore, the work of ISO/ TC 205 recognizes that architectural engineering – the design of building systems – can no longer be undertaken separately from the overall design of buildings. The system of standards under development is intended to integrate both across engineering sub disciplines, and to vertically integrate architectural engineering with the entire building design process. With the standards developed by ISO/TC 205 providing the proper framework, the results of this integration can be buildings that respect the greater environment and provide safe and comfortable indoor environments.

ISO/TC 205 is organized into six active working groups, with two addition- all task groups currently evaluating future work items. Truly an international effort, the convenorships are distributed amongst Australia, Egypt, Republic of Korea, the United Kingdom, and the USA.

In ISO TC 205 several working groups were already dedicated to technical building systems, like WG3- "Building control systems design" (with a very successful cooperation with CEN TC247) and WG8- "Radiant heating and cooling systems". ISO TC 205 is also dealing with indoor environment which is related to the performance of heating and cooling systems [3]. Because buildings must respect the environment by minimizing their impact on the external environment whilst providing safe and comfortable indoor environments [4].

1.1 Heating, ventilating, and cooling

Heating systems in buildings represent an important market sector in Europe and worldwide. Heating systems constitute a key element in providing a comfortable and healthy environment in buildings. Heating of buildings in Europe is accounting for approximately 35% of the total energy use [5].

On the international level, calculation of heating systems for annual energy consumption is related to the Work program of ISO/TC 205 "Building environment design" and ISO/TC 163 "Thermal performance and energy use in the built environment". For the energy performance of buildings using holistic approach, a Joint Working Group "Energy performance of buildings using holistic approach" was established between ISO/TC 205 and ISO/TC 163.

HVAC system efficiency can be improved by adding equipment that can convert delivered gas or electric power efficiently or by using economizers, which allow the automatic use of outside air or allow users to regulate space conditions. Energy codes provide minimum criteria for the size of HVAC systems and equipment, taking into consideration the energy demands of the building space [6, 7].

1.2 Building envelope

Local climate plays a role in the energy code requirements for the material selection and techniques used to construct the building envelope. Code requirements specify the insulation levels in the floor, ceiling, and walls and are intended to seal the building against air leakage and moisture migration. The defined energy-efficiency levels of doors and windows take into consideration heat loss and gain, depending on whether heating or cooling of the building is the predominant concern, and daylighting [8]. Designers and contractors must make sure that the building materials and installation are completed as specified for the building to comply with the standards from ISO/TC 205.

2 Russian Technical Committee on standardization green technologies

In 19.01.2016 Moscow State University of Civil Engineering (MGSU) initiated the establishment at the Federal Agency for Technical Regulation and Metrology (Rosstandart) new Technical Committee on Standardization "Green technology of the build environment".

As a result of serious researches on harmonization the relevance of the standardization green technologies, conducted by experts MGSU and partners, then Rosstandart by order №1315 from 15.09.2016 decided to establish a Technical Committee for Standardization "Green technology of the build environment and green innovative products" (TC 366).

Created in Russia new TC 366 on standardization opens new areas of research in the development of science and technology, provides an improvement of existing results and normative-technical documentation, because "green technologies" - this is one of the basic systems of production activity, providing at all stages of life cycle the product (an object), its maximum compatibility to natural environmental, technogenic conditions and parameters in order to improve the quality of the human living environment [9].

Moreover on the rules of the Rosstandart to the Russian TC 366 was appointed international appropriate committee ISO / TC 205. So that Russian green technologies, engineers and scientists would be involved to the responsible decision on the build environment. Because regarding to the global climate change the "green" technologies are priority way of standardization to all over the world [10].

4 Conclusion

Equations The mistakes of the past in building design cannot be sustained into the future. The mistakes of the 1970's can inform us today. Standardization of energy efficiency alone is not a worthy goal, if it results in dissatisfied and unproductive occupants. To ensure the future, buildings must respect the environment by minimizing their impact on the external environment whilst providing safe and comfortable indoor environments by using green technologies.

To assure that technical building systems and services of building energy engineering are considered in appropriate way in the standardization work, it is important that the heating industry and service companies cooperates actively in the reorganization of the international standardization committees ISO / TC 205.

As the recast EPBD requires more specified energy performance declarations for heating, cooling and ventilation systems, it is expected that the EN or EN-ISO standards to be developed to support this, will satisfy the legislation need on national level. This expectation will, with support of a second mandate from the EU commission to CEN, enable the EU experts to work on these issues on the European level and at the same time contribute to the ISO level.

The position of the heating and cooling systems within the system standardization has now well been established, especially on ISO level. The standardization of heating and cooling systems has an important position in the overall energy efficiency evaluation of the building and standardization work on ISO level is progressing.

The Brundtland Commission of the United Nations in 1987 stated that development is sustainable "...if it meets the needs of the present without compromising the ability of future generations to meet their own needs." Therefore Russian Technical Committee for Standardization № 366 "Green technology of the build environment and green innovative products" together with ISO/TC 205 "Building environment design" would be seeking how to standardize building environment design, including all aspects of sustainability and environmental responsibility.

References

1. J. Zirngibl, *CEN and ISO activities for heating and cooling systems standards* (2011)
2. S. Turner, *Building environment design: energy conservation and efficiency* (ISO Focus, 2005)
3. D. Dijk, E.E. Khalil, *Practical illustration of the concept of a systemic approach: Energy efficient buildings*
4. V.I. Telichenko, V.M. Roitman, A.A. Benuzh, *Integrated safety in construction* (MISI-MGSU, 2015)
5. Y.A. Tabunschikov, "Green" building in Russia **5**, 14-15, (2010)
6. R. Thakore, J. Goulding, A. Benuzh, Housing and society: need for progressive sustainability indicators for societal sustainability, 877-890 (2013)
7. A.A. Benuzh, E.N. Orenburova, Housing construction **2**, 14-16 (2015)
8. V.I. Telichenko, A.A. Benuzh, *Environmental glossary for a builders* (MISI-MGSU, 2016)
9. V. Telichenko, A. Benuzh, Advanced Materials Research **1065-1069**, 2169-2172 (2015)
10. V.I. Telichenko, A.A. Benuzh, Industrial and Civil Engineering **10**, 40-43(2014)

Computer module for scheduling of transportation of composite beam bridge structures

Wojciech Bożejko[1], *Zdzisław* Hejducki[1,*], *Mariusz* Uchroński[1] , *Mieczysław* Wodecki[2]

[1]Wrocław University of Science and Technology, Wyb. Wyspiańskiego 27, 50-370 Wrocław, Poland
[2]University of Wrocław, Pl. Uniwersytecki 1, 50-137 Wrocław, Poland

Abstract. The paper presents the theoretical basis, an algorithm and a computer module system supporting scheduling of transportation and assembly of structures of the composite beam bridge implemented in a Just In Time system (JIT). Tabu search method has been used in the optimization procedure.

1 Introduction

During an organization of work in the JIT system it is necessary to synchronize the supply (prefabrication transport) with the installation, avoiding the need of storage. It requires solving of many difficult (strongly NP-hard) optimization problems at the same time. There are currently no optimal algorithms for solving the problems in polynomial time. Therefore, we propose to use a fast approximate algorithm of polynomial time computational complexity – tabu search method.

The paper concerns an issue of transportation of prefabricated bridge components (by rail and road) to ensure collision-free paths. In recent years, there have been developed new technological solutions, using polymer composites (Fibre Reinforced Polymers called for short FRP).They consist of the composites of plastics reinforced with various fibers. Most of them are carbon, glass or aramid fibers. Due to a number of the beneficial properties, such materials find more and more practical applications, including bridge construction.

The establishment of the EN 13706 European standard for the production of polymer composites with the use of pultrusion method enabled the development of new technology, methods of design and installation of bridge load-bearing components. The proposed technology provides:

- Lower costs of construction associated with prefabrication and low weight of elements,
- Lower costs of maintenance during the exploitation of the structures (elements surfaces do not corrode and do not require painting),
- Less disruption to traffic flow by reducing the installation time.

It is assumed that the assembly of composite structures takes place on concrete support with bearings in the Trans-IND developed technology and construction system. In this system, work division structure includes:

[*] Corresponding author: zdzislaw.hejducki@pwr.edu.pl

1. Transportation of beams from the factory to the installation site.
2. Installation of beams.
3. Transport of mechanical fasteners and resin for fastening designed to connect elements at the site.
4. Transport of modular composite bridges to be installed at the site.
5. Leveling and connecting bridge decks.
6. Connecting of bridges and beams.
7. Execution of concrete slab on the surface of bridges.
8. Building of hard asphalt pavement.
9. Installation of the bridge equipment (barriers).

Figures 1-3 depict some phases of bridge construction using composite structures.

Fig. 1. Concept of a composite bridge (by BSiR Mostostal SA).

Fig. 2. Scheme of beams and slab (by BSiR Mostostal SA).

In the paper [1] there was, among other issues, the problem of scheduling of bridge structure construction described, including girders installation. This paper presents a description of the problem along with the organizational and technical considerations, the theoretical basis for model building, the construction algorithm based on the tabu search method, and the results of computational experiments. On the basis of the schedule set by the algorithm for girders installation there are time windows determined, i.e. the earliest and the latest required delivery of individual beams. At their base there is determined optimal (or nearly optimal) schedule of transportation from the place of prefabrication to the construction site. A model of this problem and the method of its solution are presented below.

2 Girder transportation problem

One of the important aspects of the installation of bridge structure, using a load-bearing components made of composites, is the transportation and fitting of bridge girders. On the basis of developed and agreed with all subcontractors the schedule of construction: dates and order of delivery of beams (in the order of technology) is determined. The basic assumption of the developed methodology relies on a specific logistic condition, that is, assumption concerning assembly and transport in the Just In Time system. As a result, there is no storage capacity for girders on site - possibly only on vehicles. Therefore, the girders should be directly placed on the supports. The problem of optimal transportation of beams (in short the problem TB) is to determine such delivery dates of girders that they minimize the assumed criterion (e.g., the cost of non-compliance with delivery dates, the number of necessary vehicles, etc.). The following restrictions must be met:

1. Beams should be delivered in accordance with the order of construction (technology policy),
2. The vehicle can carry only a single beam (or part of beams) at once,
3. After loading, the beam can be removed immediately prior to installation only.

In the following part, we present some variants of beam transportation problem.

2.1 Transportation with the earliest and the latest delivery dates

We assume that on the basis of work implementation dates and technology requirements there has been a detailed schedule of construction and assembly works determined. Among many other issues, it contains the starting and ending dates of girders installation. On this basis, there are the earliest and the latest possible delivery dates of individual girders determined (taking into account also the technological order). In determining the latest delivery date the unloading time should be taken into consideration, so that launching of girder installation take place according to the schedule. It is recommended, moreover, to introduce a small buffer in time, taking into account the fact that in the course of proceedings some slight complications may occur. The earliest delivery date result from the fact that the beams cannot be stored on site. Therefore, the vehicle which arrives too early cannot be discharged before that date. The earliest and latest date of delivery form so called time window. This is the time interval in which the delivery generates no extra cost. A mathematical model of this problem is presented below.

Let:

$$B = \{B_1, B_2, ..., B_m\}, \quad (1)$$

Be a set of girders provided for installation which have to be delivered from the prefabrication site to the construction site. For any beam $B_i \in B$ we introduce the following symbols:

z_i – loading time,
t_i – time of transportation to building site,
r_i – time of unloading at construction site,
p_i – returning time (time taken from the construction site to the prefabrication site),
e_i – earliest required delivery to construction site,
d_i – latest required delivery to construction site,
v_i – penalty function coefficient of too early delivery to the construction site,
w_i – penalty function coefficient of too late delivery to the construction site.

On the basis of the schedule of construction and assembly works there was the order of girders installations determined (delivery to the construction site). To prove the statement, we assume that this is the sequence (B_1, B_2,..., B_m). We also assume that the loading of the first beam starts at zero moment. By S_i we denote the delivery date of B_i girder to the site. The dates S_1,S_2,...,S_m must satisfy constraints (a)-(c).

If the transport and assembly took place on a continuous basis, the date of delivery of the girder B_i is equal to

$$S_i = \sum_{i=1}^{n}(z_j + t_j + r_j + p_j) + z_i + t_i. \qquad (2)$$

Girders transportation problem boils down to determining of delivery dates S_1,S_2,...,S_m satisfying constraints (a) - (c) that optimize certain adopted optimization criterion. In JIT system it is usually minimization of a number of exceeded delivery dates, the sum of penalties for late delivery or for maximum delays, etc.

3 Support system for scheduling of transportation and assembly of composite elements

Support system for scheduling transportation and installation of prefabricated composite bridge structure elements was written in C ++Microsoft Windows. It consists of three main modules, of scheduling:

1. Construction works.
2. Transportation.
3. Works and transport in uncertain conditions.

After booting the system there is the main window displayed containing various system options: Scheduling, Transportation, Uncertainty and About. Figure 4 shows the window, after clicking the scheduling of construction works option (Scheduling).

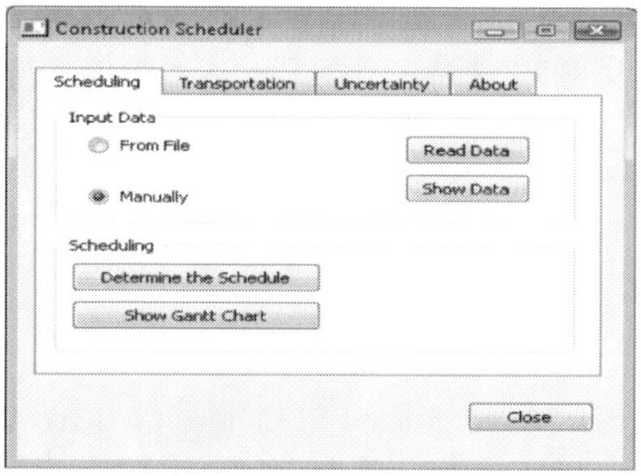

Fig. 3. Main window, scheduling of construction works.

3.1 Transportation scheduling

Transportation scheduling module allows us to determine the delivery dates of girders to the construction site. These terms are based on the basis of the construction schedule, set in the Scheduling module. For each girder there is a time window determined (the earliest and latest date) in which the girder must be delivered to the construction site. The theoretical basis of the transportation problem was presented in Chapter 2. There was an algorithm based on the tabu search method [2] used to solve this problem. Installation of girders was done in the JIT system, therefore, as optimization criterion there was minimization of the sum of the penalties for late delivery (1) adopted.

After selecting the Transportation option from the main window (Figure 4) there are shown the menu options for this module which allow us, among other things, to introduce and correct data (they were thoroughly described in the previous chapter). The parameters necessary for determining the schedule of girders delivery to the construction site can be entered manually or read from a text file. The window displaying transportation scheduling module was shown in Figure 4.

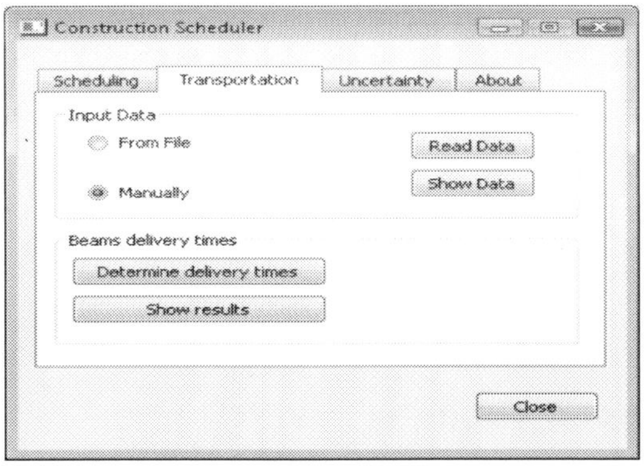

Fig. 4. Beam transportation scheduling window.

The individual modules, in particular the optimization algorithms were tested on the examples taken from the construction practice and from the literature. The experimental results were entirely satisfactory. The optimum solution, or little different from the optimal (average relative error for a large set of examples in the literature did not exceed 3%) was obtained in a very short period of time (maximum in a few seconds).

4 Conclusions

Modules of the support system for scheduling of construction projects, presented in the paper, provides a simple-to-use graphical user interface. This allows for a rapid determination of solutions for problems of different sizes, parameters and characteristics. The obtained results are presented in graphic form (Gantt chart) or in the form of text (tables of starting and ending dates of individual work performance). Application of a fast approximate algorithm (tabu search) for scheduling of construction projects allows us to obtain high quality solutions (in terms of the value of the objective function) in a short time.

References

1. W. Bożejko, Z. Hejducki, P. Rajba, M. Wodecki, Management and Production Engineering Review, **2** (2011)
2. W. Bożejko, Z. Hejducki, M. Uchroński, M. Wodecki, Journal of Civil Engineering and Management, **20(5)** (2014)

MATEC Web of Conferences **86**, 05016 (2016) DOI: 10.1051/matecconf/20168605016
IPICSE-2016

On the System of Management Process Indices in Building Sector Enterprises

Roman Chernov[1], *Svetlana* Shilkina[1,*]

[1] Moscow State University of Civil Engineering, 26 Yaroslavskoye Highway, Moscow, Russian Federation, 129337

Abstract. Analyzing the structure of the management process indices in building sector enterprises. Quality level of the system of enterprises managing in the building sector.

1 Introduction to the subject

In our opinion, the set of indices for the management process in building sector enterprises shall depend upon the system's three main factors: its performance results, resources used in the course of such performance and management system's functioning quality.

First of all, let us dwell upon the results. It is not our intention to deem the number of adopted managerial decisions as such. Moreover, we are sure that the indices characterizing the performance result of the management system for an economic entity in the building sector shall be such object's performance indicators, i.e. the scope, structure and quality of the products and services manufactured and performed by it and the efficiency of resources using.

Speaking of resources, their structure and scope shall depend upon the components and needs of the management system and, as a rule shall include the elements as follows: methods, information, personnel, right, technique, technology, living environment. Each of the aforementioned shall be in kind (real number) and in value terms. Sometimes they are referred to as supporting subsystems.

In its turn each and every supporting subsystem shall be characterized by its performance results, required resources scope and components, and the quality and efficiency of the services being rendered. For instance, the subsystem of information support shall collect, store, process and provide to the management system's employees any and all necessary information in due time and required form. The goal of legal coverage subsystem is observing the managerial decisions legal purity. For this purpose, they shall have their own personnel, methods, and finance at last. And all such elements with their specific indices shall be the elements of the management system. This is also true about any other supporting subsystem. Thus, the complex of indices characterizing the management system's resources shall include the aggregate of all performance indices of all supporting subsystems.

* Corresponding author shilkina@bk.ru

2 Methods of making management decisions

Here it is pertinent to draw your attention to one of supporting subsystems, namely the technology of managerial decisions making. Such subsystem had some bad luck. It is rarely if ever and rather passingly is mentioned in the publications dedicated to management theory and practice. For some reason everyone supposes that if one has the information and personnel, is aware of the methods to transform it into managerial decisions, nothing shall prevent the management system's appropriate functioning. However, let us imagine the situation when instead of such system's resigned employee a new one comes. In order to get on the inside he shall need the manager's and colleagues' advice and hints, let alone the time. Being aware of the human nature, we may be sure that such hints more than seldom show right way to go. All these may lead and as often as not actually lead to both current failures in the whole system's functioning and long-term conflicts among the staff by no means improving the efficiency of functioning thereof.

This is where the role of technological support steps forward, providing for each manager's workplace a developed instruction for routine problems solving. It specifies where, when and what information may be received to settle a task, when, who to and what information shall be delivered following the results of settling thereof. And the matter of methods choosing to settle such tasks shall be the employee's responsibility.

Let us pass on to the management system's quality level. Here it is appropriate to formulate our interpretation of this definition. We suppose that the management system's quality shall be characterized by timeliness, scientific and formal validity of managerial decisions being adopted. The formal validity is the compliance of decisions being adopted with the current legal regulations.

The specific values of the management system's quality level shall depend upon the requirements made by contractors, i.e. products and services consumers and resources suppliers. Apart from such contractors, the external environment representatives for the management process are the Russian Ministries and Departments, and superior management authorities, and each of them in its own way influences the management process's indices set and values.

From Russian Ministries and Departments the management process shall receive (fig1):

- technical and sanitary standards and criteria regarding both the products and the processes of manufacturing thereof;
- environmental standards also referring to both the products, and the processes of manufacturing thereof;
- statutory acts determining the special aspects of various kinds of citizens' and legal persons' business activity.

The main goal of the superior management authorities for the process under consideration is to form its milestones.

Finishing to consider the elements of the system of the management process indices we shall analyze its last element being the feedback from the object under management and external environment to the management system.

We remind that the management process consists of the following consequence of its functions: analyzing the state of the object under management, forecasting any possible options of its development, designing the plan of the object's development, informing the executives of the scheduled figures, administering and analyzing the plan implementation. It is quite natural that in the course of performing each of the aforementioned functions we use a certain set of management indices. However, in order to provide continuity when passing from a function to function and to lock in the whole chain from analyzing the plan implementation to analyzing the object's state we shall form a certain indices subset specific all at once for all functions.

 DOI: 10.1051/matecconf/20168605016

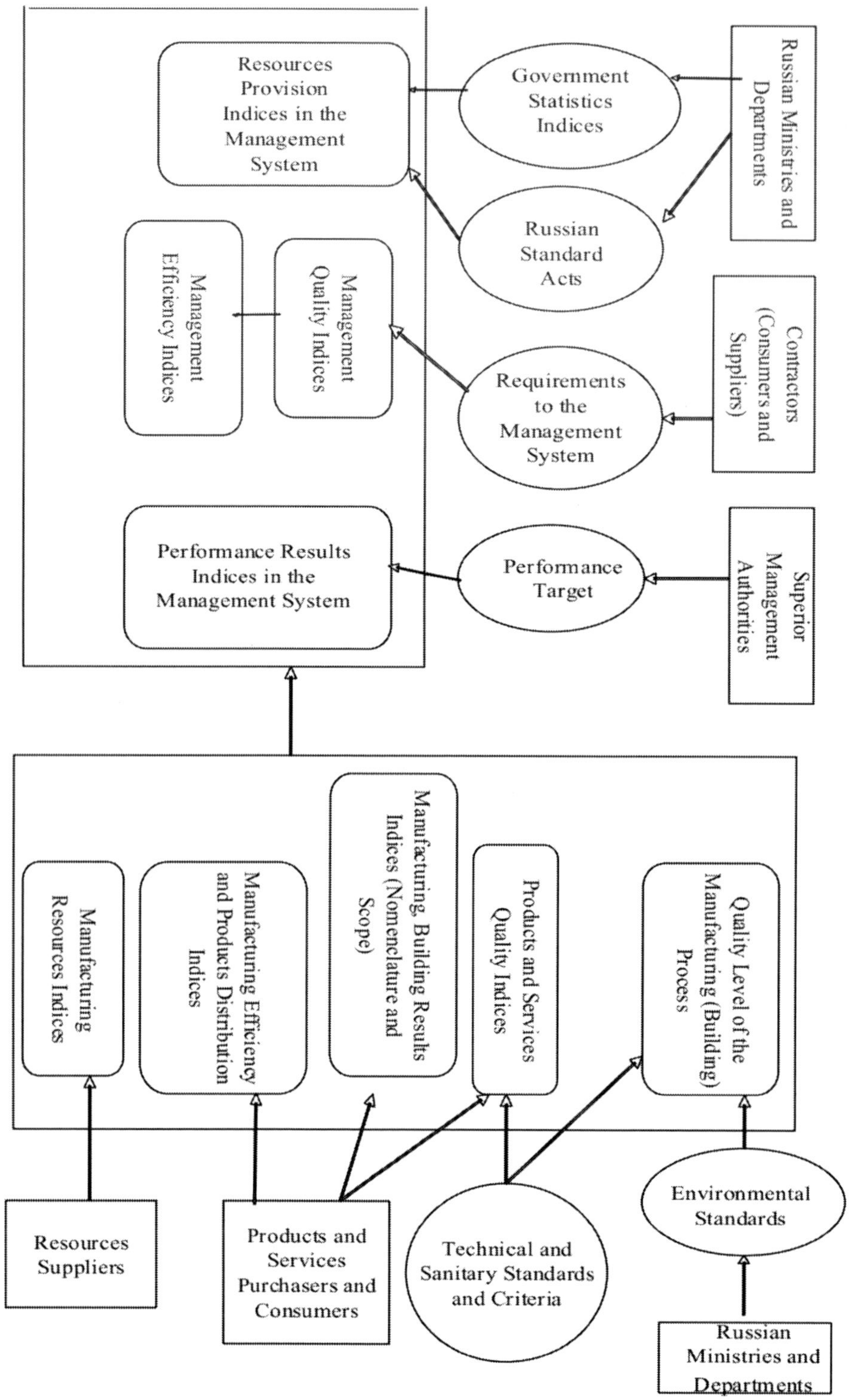

Fig. 1. Structure of the System of the Management Process Indices in Building Sector's Economic Entity.

DOI: 10.1051/matecconf/20168605016

The simplified diagram is given in fig. 2. To this particular subset we refer the indices of feedback from the object under management. Their principal goal is to provide to the managers (management subject) in brief, structured from as the complex of indices the most important information enabling to estimate the management process state and make any required decisions. In this case depending on the enterprise's kind and changing external environment the set and elements of the feedback indices may change.

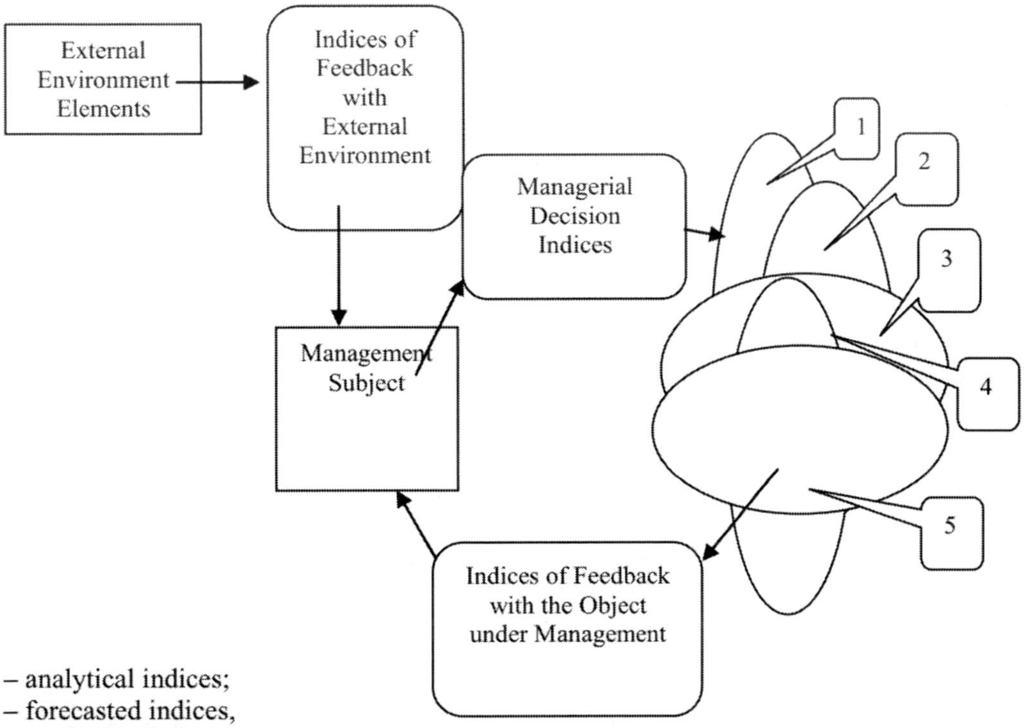

1 – analytical indices;
2 – forecasted indices,
3 – scheduled figures,
4 – indices to be notified to executives,
5 – indices to be administered and analyzed.

Fig. 2. Scheme of Forming Feedback Indices.

3 Figures and tables

At the same time, it is important to emphasize that to make managerial decisions the management subject besides the information regarding the indices of the object under management received via feedback channels shall need to have data on the degree of contractors contentment with the management system's quality. The concept of the indices balanced system (IBS) being so popular over recent years is dedicated exactly to the matters of forming the system of feedback indices in the process of an enterprise managing.

The IBS was developed in the early 1990s by American economists David Norton and Robert Kaplan. According to the developers, such system shall have the information reflecting any and all principal edges of an enterprise's performance. Such information is arranged into four groups reflecting the four performance sphere for a building sector's enterprise:

- financial sphere dealing with the efficiency of the company's activity in the context of return of capital employed and attractiveness for its own shareholders;

- company's inner potential, inner operating activity development;
- consumers' contentment with the company's goods and services practicability, its corporate image;
- knowledge, skills, capacities and personnel training influencing an enterprise's ability to perceive new ideas, its flexibility and constant improving.

The authors of this article did not set a goal to closely look at the essence of such concept. However, we shall make certain notes directly due to marked euphoria of Russian IBS admirers.

First of all, how can they talk about the balanced system of the management process indices when the IBS deals only with one of its elements namely, feedback indices? Secondly, among the four enterprise performance spheres which according to the IBS developers shall be of the manager's interest include both the object's indices and the management system's indices (that is quite natural). But all at once at the same level with them appear not indices but the factors determining them.

Falling back upon our article's content, we shall accentuate the fact that the elements from the feedback complex in the course of strategic goals settling rather substantially differ from those used in the course of current operating activity.

References

1. V. P. Silin, R. O. Chernov. Herald of the Moscow University of MIA of Russia **6**, (2011)
2. R. S. Kaplan, D. P. Norton, B. Rugelsjoen, Harvard Business Review **88**, (2010)
3. A. Volkov, P. Chelyshkov, A. Sedov, Applied Mechanics and Materials **409-410**, 1620-1623 (2013)
4. A. Volkov, A. Sedov, P. Chelyshkov, A. Doroshenko, (2014) Applied Mechanics and Materials **580-583**, 3231-3233 (2014)

Aspects of transport transit hubs construction management in coordination with object lifecycle projecting

Nina Danilina[1*], *Denis* Vlasov[2]

[1] Moscow State University of Civil Engineering, 129337, 26, Yaroslavskoye Shosse,
[2] Genplan Moscow, Russia Institute of Moscow, 125047, 2/14, 2-nd Brestskaya str., Moscow, Russia

Abstract. The article involves the results of the research in the field of the initial phase in the project of the object lifecycle for urban public transportation transit hubs (TTH). The main purpose is to provide definition and justification of area parameters for spatial solution of a hub with consideration of its construction viability and sustainability of urban development. In the article there are presented the results of analytical studies of the data available in the approved Planning documentation for TTH objects listed in the short-term program of TTH construction in Moscow. The point of interest was to obtain correlations between the total area of TTH area, areas of commercial and technological use and passenger traffic value in order to define territory balance providing request from city authorities and private investors for construction project management. Also, there is presented the theoretical model as an evaluation guide that allows evaluating not only the demand of a city and its population in the development of TTH territory but also and especially the requirements for placing commercial areas inside TTH.

1 Introduction

The evolution of modern town planning relies on improving quality of urban environment according to people's diverse needs and abilities. Organization and management of construction processes that take place on the territories of the cities become the topical issues and require the development of long-term prospects. Currently, construction management system integrates a new object-oriented approach for the developing cities in the Russian Federation. It consists in the implementation of life cycle contracts for Construction project management [1]. Experts underline the effectiveness of such form of public-private partnership for design, construction, finance and operation of the key city facilities. The first stage of this contract consists in the development of project planning and design solutions and involves the development and approval of the Territory Planning Project and initial architectural and construction solutions of the object [2,3]. Information received with the completion of this stage serves the starting data for further stages of

* Corresponding author: nina_danilina@mail.ru

project development necessary to provide the highest quality of design decisions.

This article presents the results of the scientific research in the field of spatial and planning design solutions for the most complex urban integrated facility – public transport transit hub. The complexity of such object is determined by large variability of its components: different types of transport and objects of social, public, commercial, recreational infrastructures. That's why, the development of the regulatory support allowing to applicate the most rational and effective design solutions becomes one of the most important tasks that define the economic performance of the project needed both for private investors and city authorities [4,5].

2 Project objectives

Spatial planning project of a transport transit hub (TTH) is intended to define the most efficient proportion of the spaces used by different types of facilities to the total area. Internal role of TTH within public transportation system and its placement in the functional urban areas composition determine the value of passenger traffic and its construction scale [6,7]. On this basis, it seems necessary to distinguish three basic groups of TTH:

1. Local public transport transit hubs. These are unimodal public transit hubs with a minimum rate of passenger traffic. They occupy small land area and their formation requires just local arrangements on the territories of road network and transport infrastructure objects.

2. Municipal public transport transit hubs. These are multimodal hubs with medium or high rate of transit passengers. The need for passenger transfer between different modes of transport requires a significant territorial resource in walking radius from the central passenger transport node – the station of the main type of passenger transport, especially when you need "Park and Ride" lots.

3. City, regional and other high-level public transport transit hubs with high and very high rates of passenger traffic. The central constructive element of these types of hubs is a transit terminal, which provides all internal transfer processes. The location of such terminal requires big spatial resource, and also causes the necessity for epy development of adjacent territories with a view to creating a high-quality urban space.

The third group of TTH is the most problematic one from the point of view of a number of transport , social, environmental and urban development problems caused by high rates of passenger traffic. So, these hubs are urban transport facilities of a priority interest for reconstruction or construction as for municipal authorities so for private investors. From investor's point of view such transport transit terminals are most attractive for development, offering prospects of the construction or reconstruction of a large scale object, group of objects, or zones of «transit oriented development». Alignment of interests makes such hubs most attractive for evaluation of their investment potential on the first level of construction project management by means of space planning and project design.

The availability of technological space defines the performance of the main transport service. Commerce areas are optional services to population, but their presence determines the investment interest in construction as thus its viability. The correct area ratio determines the sustainability of the entire facility in general. So, the main issue of the research becomes to determine the ratio between commercial and technological areas that attracts both city authorities and private sector investment. Composition of the areas is presented on figure 1.

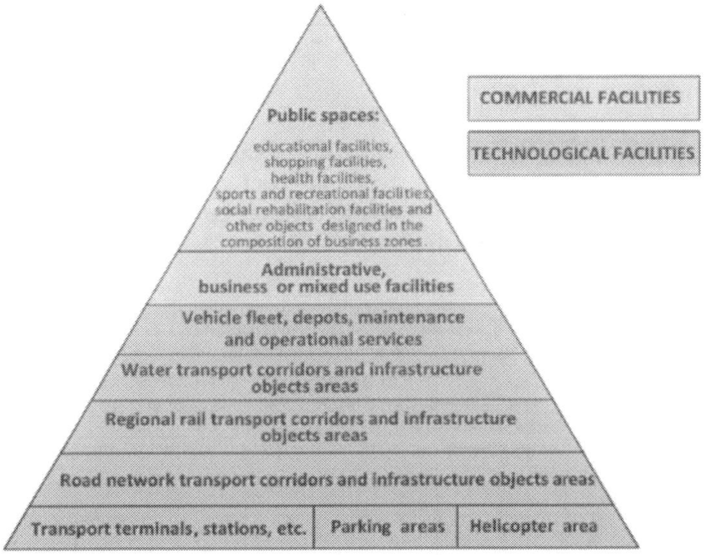

Fig. 1 Composition of transport hub areas.

The main data for the research was collected from official and approved planning projects of public transport hubs, designed in recent years in Moscow. Comparison of theoretical data and experimental survey results has provided interesting results. In the overall list (fig.1) the proportion of transport technological facilities amounts to 85%, so the percent of commercial facilities doesn't exceed 15 %. However, the survey analysis shows that in 22 approved projects with a total planning construction area over 1.8 million sq.m., only 16% of the area is allocated for technological purpose (fig. 2).

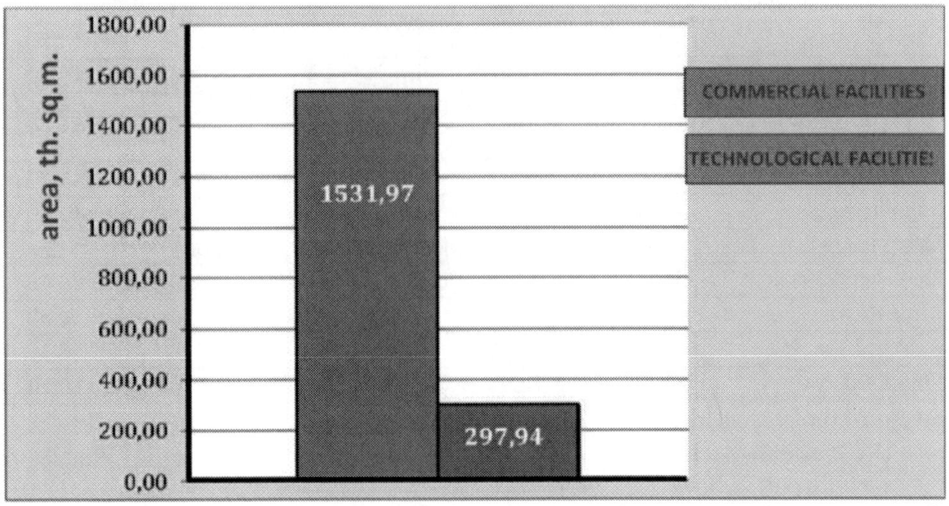

Fig. 2 Distribution graph of commercial and technological areas (according to approved Planning projects of transport hubs in Moscow).

Such data are an indicator of the interest of city authorities in attracting private investors for object life cycle management on the earliest stage of projecting process. In that case, it becomes necessary to find such correlation for the areas of the overall lot that allows both to satisfy public demand for transport services and to provide the investment attractiveness of the construction project. The practical significance of results obtained from such research studies is that they can be used for improvement of the regulatory framework for the development of hubs and for the elaboration of further planning projects.

3 Data study

In the experimental part of the study there was conducted the analysis of existing planning projects of transport hubs. To date, there've been approved 22 planning projects for transport hubs that constitute almost 10% of the total number of hubs in Moscow (table 1). So the aim of the study was to identify patterns in the spatial distribution of TTH resources.

Table 1.

	Public transport transit hub	Total hub area, ha	Facility area, ha			Passenger traffic, 1000 persons /morning peak hours
			Technological area, th. sq. m	Commercial area, th. sq. m	Total area, th. sq. m	
1.	Alma - atinskaya	57.24	0.64	84.37	85.01	26.20
2.	Botanicheskiy sad	15.52	36.43	141.30	177.73	44.00
3.	Varshavskoe shosse	23.73	13.47	25.50	38.97	16.20
4.	Volgogradskaya	44.71	9.76	0.00	9.76	18.80
5.	Dubrovka	44.71	11.62	0.00	11.62	25.60
6.	ZIL (Park legend)	13.87	17.63	93.09	110.72	17.00
7.	Koptevo	21.93	6.72	0.00	6.72	22.80
8.	Kosino	43.30	9.88	221.55	231.43	50.20
9.	Lujniki	6.90	6.05	0.00	6.05	24.90
10.	Lubereckaya	46.60	8.88	192.94	201.82	36.40
11.	Novokosino	47.40	8.14	94.22	102.36	58.00
12.	Novochochlovskaya	33.21	7.18	17.03	24.21	18.30
13.	Paveleckaya	13.43	0.56	131.96	132.52	67.80
14.	Pyatnickoe shosse	77.88	0.40	51.24	51.64	19.40
15.	Sevastopolskiy prospekt	14.32	14.47	0.00	14.47	26.60
16.	Seligerskaya	28.25	66.11	128.50	194.61	41.70
17.	Sokolinaya gora	19.70	8.05	40.00	48.05	12.00
18.	Horoshevo	7.56	7.66	0.00	7.66	9.20
19.	Horoshevskaya	53.00	44.60	141.60	186.20	46.60
20.	Mitino	21.49	7.98	26.86	34.84	14.80
21.	Ryazanskaya	30.10	29,60	121.00	150.60	78.30
22.	Shipilovskaya	13.80	12,04	20.76	32.80	10.50

The analysis of the presented data reveals the following patterns:

- there is a certain correlation between technological and commercial areas and total area of a transport hub;
- there is a certain correlation between technological and commercial areas of a transport hub;
- there is a certain dependence of the commercial component volume on the value of passenger transit traffic through a hub during morning rush hour.

Each of the studied regularities is the supporting material for the initial phase of development of lifecycle management strategy for a particular public transport hub construction. Commercial component, its architecture and construction design as well as users projected passenger traffic are classified as techno-economic parameters of the project and define the investment attractiveness of the project for the private investor.

At the same time, the execution of transportation and urban formation functions by TTH is provided by sufficient amount of technological areas, ensuring passengers arrival, departure, transfer and waiting as well as the operational functioning of TPU and different types of transport infrastructure.

Value of facility area is received in the result of the Planning project data processing for every transport hub. Passenger traffic data refers to the total volume of passengers in arrival and departure for all modes of transport calculated and used by Moscow Master Plan Institute for the development of planning

Figure 3 provides the graph of correlation between technological and commercial areas and total area of transport hub, understood as the area within the boundaries of the site planning project. The problem is that these correlations are hardly to be used as a performance indicator, because today there are no clear guidelines on the definition of the boundaries of the area development plan and they are determined in negotiation with relevant city authorities. However, it is obvious that big proportion of commercial areas within TTH provides the interest of private investors and, in fact, the feasibility of project construction. That is why commercial areas are planned in 16 of 22 transport hubs, provided with documentation on territory planning.

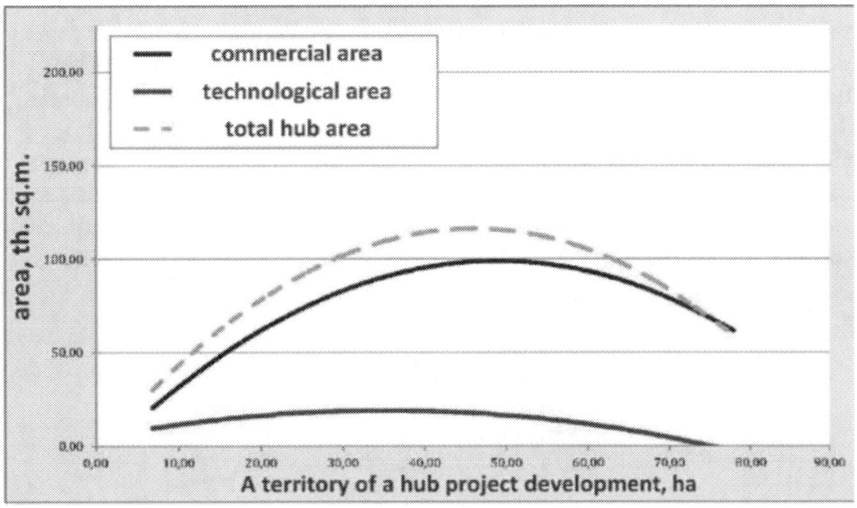

Fig. 3 Correlation between technological and commercial areas and total area of transport hub.

Therefore, the more rational is to use graph 3, reflecting the ratio of TTH area components to the passenger traffic passing through a hub during morning rush hour. As the previous one, it also looks as a parabolic curve. It should be emphasized that the curve referred to commercial facilities area and passenger traffic (shown in blue) has more pronounced curvature than the corresponding curve defined by technological area indicators (shown in red). This could be explained by the relatively small difference between minimum and maximum rates of technological facilities area. The nature of this ratio distribution is more uniform, and therefore, its value can be predicted more precisely than a ratio of commercial facilities area. Also worth noting is the overall lower values of technological facilities areas in comparison with the values in the areas of commercial facilities inside TTH.

Analytical method used above, like all has advantages and disadvantages. The main advantage of this method seems to be that it clearly reflects a balance of two main groups of TTH areas, stimulating investment interest for the project both of city authorities and private investors. Such information obtained at object construction level is of significant value to guarantee for the viability of the project and transition to the next level of the life cycle contract development - further design stages of construction project.

The disadvantage of this method is that it does not offer a scientific approach to the problem of how to determine spatial decisions and focuses only on the importance of economically attractive factors, thus have limited application. To overcome this limitation, the development of theoretical and scientific method becomes useful based on systematic analysis of the complex urban planning factors that can ensure sustainable development of city objects and territories. The authors propose a model that reflects the dependence of the relationships of commercial and technological area spatial decisions on three integrated factors, such as:

- transport factor, that is estimated by the integrated index with consideration of territorial limits of the road network and passenger transport infrastructure objects;
- social and urban planning factors, that include main demand-oriented indicators concerning the demand of the city population for various functional facilities on particular urban areas;
- economic factor, that indicates the balance of expenditure and revenue parts of the project.

It should be emphasized that the sustainable development of urban transport system and creation of comfortable and safe environment on the territory of TTH depend directly on the functionality and convenience of technological-oriented area of a hub [8]. Regarded in that light, the task to determine the volume of commercial areas without harming the implementation of major transport, social and technological functions becomes one of the major issues on the initial development level of the life cycle of the construction project. The proposed model (fig.5) allows to divide the whole system into clusters, each of which may determine the availability of location, limits and functionality of commercial facilities.

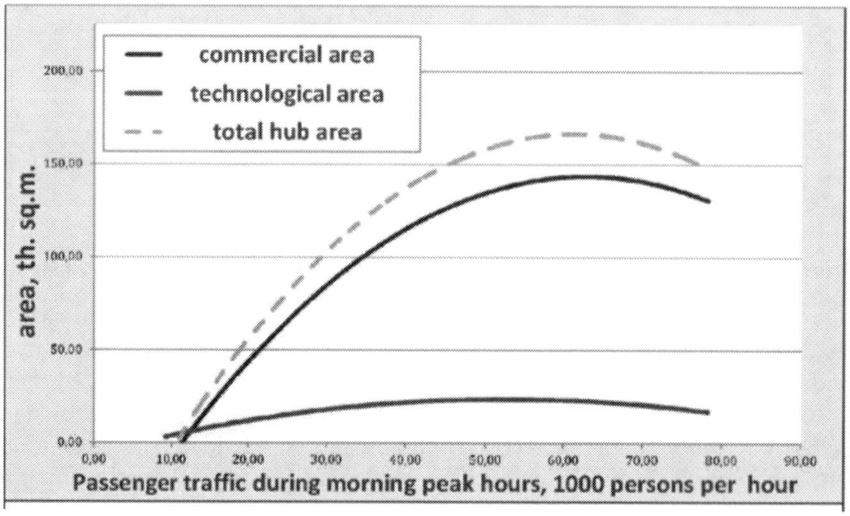

Fig.4 Dependence of the area components on the value of passenger transit traffic.

The rationales for placing the commercial component are the maximum values of all factors specific for the 3rd group of TTH with a need for transit terminal placement. This model is justified as it is based on the methods of scientific research as well as multifactorial analysis of the functioning of the construction sites in urban environment. A clear advantage of this model is that it offers the possibility to determine balance of interests between the city authorities and private investors – on the one hand, and population on the other, which is extremely important for development of modern high-quality urban environment [9,10].

Conclusions

The results of a research study determine the dependencies that can be used at the initial stage of the development life cycle of a city public transport transit hub aiming to make reasonable decisions for spatial-planning project design. Such stage is absolutely necessary to calculate the main technical and economic parameters of the project and its transition to further stages of construction development and project documentation design. The main results obtained in the study can be summarized in the following:

1. Analytical dependencies of ratios of commercial and technological areas within TTH, serving urban authorities and private investors a source data to show investment attractiveness of the construction project and mutual interest and understanding. It should be underlined that these data surely need to be updated in the next phase, addressing the urban planning as well as social needs of the population.
2. Theoretical model of commercial and technological areas distribution ensuring sustainability of urban construction and quality, and safety of the environment.

The results of these studies may provide adequate scientific justification at the development of regulatory support documents having regional and Federal value and allow for the methodical insurance of urban and transport planning process and transport transit hubs design. These data also forms the basis of further study, or experimental research.

References

1. E. Dorohina, Intern. journal of appl. and fung. **4-3**, 652-652 (2016)
2. D. A. Kudelina, Conf. paper, Econom. and manag.: actual asp. of theory and practice, 92-95 (2015)
3. E. G. Irwin, Journal of Regional Science **50(1)**, 65–91 (2010)
4. J. Bröcker, A. Korzhenevych, C. Schürmann, Transp. Res. Part B **44(7)**, (2010)
5. N. Daniina, Conf. paper, Social – economic probl. and tend. of territ. develop., 39-43 (2016)
6. S. Gelareh, R. Neamatian Monemi, Nickel, Transp. Res. Part E **75**, 67–94 (2015)
7. M. E. O'Kelly, H. J. Miller, Journal of Transport Geography **2(1)**, 31–40 (1994)
8. E. Sherbina, N. Danilina, D. Vlasov, Int. Jour. of Appl. Engin.Res. **10(22)**, 973–4562 (2015)
9. D. Vlasov, Modern probl. of science and educ. **4**, 65 (2013)
10. D. Vlasov, Real estate: econom., manag. **1**, 57-61 (2011)

The Management of Park Equipment and Machinery Used in the Construction Industry and Housing Sector

Rustam Khayrullin[1,*], *Pavel* Marichev[2]

[1]Doctor of Physics and Mathematics, Professor, Moscow State University of Civil Engineering, Moscow, Russia
[2]Ph.D, Head of Department, Main Scientific Metrological Center, Mytishchi, Moscow region, Russia

Abstract. The mathematical model was created, applicable to elaboration of options of development of a park of equipment and machinery used in construction industry, housing and utility management. The model based on solving the linear programming problem by means of simplex method. The mathematical model allows develop the rational plans and quasi optimal plans. The model is based on the created in the article the financial management algorithms of procurement and of repairs of equipment and machinery. The algorithms allow to take into account joint consideration of indicators of modernity and operability of equipment and machinery. The basic model is developed. The model is included into the automated planning system. It is used for planning and preparation of proposals of draft policy documents for development of park devices, equipment and machinery.

1 Introduction

The problem of management of park of industrial equipment, computer facilities and control – measuring devices is an urgent practical task in mechanical engineering, industry and information technologies [1-6]. In the construction and housing sectors used a wide variety of equipment and machinery (EM): lifting mechanisms, water meters, heat meters, air-conditioners and so on. So the problem of management of park of equipment and machinery (PEM), undoubtedly, is an urgent practical task in the sphere of a construction and housing sector.

To assess the level of technical excellence and technical condition of the PEM and some types of EM it is proposed to use an indicator of modernity and an indicator of operability. Maintenance of indicators at the required level is possible at the expense of the balanced combination of purchases and repairs of some types and samples of EM.

2 Model of the modernity and operability

We introduce the notation:
N - total number of samples of EM,
N_{op} - number of operating samples of EM at the initial time of planning,

N_{mo} - number of modern samples of EM at the initial time of planning,
N_z - number of the purchases,
N_p - number of repairs,
S_z, S_p - the average price of purchase and repair,
S - amount of the funds allocated for purchase and repair together,
k_1, k_2 - operability coefficients for modern and outdated samples of EM, $0<k_1\leq k_2<1$.
Modernity indicator:

$$P_{mod} = \frac{(N_{mo} + N_z)}{N} \quad (1)$$

Operability indicator:

$$P_{opb} = \frac{N_{op}}{N} = \frac{k_1 N_{mo} + k_2(N - N_{mo}) + N_z + N_p}{N} \quad (2)$$

Amounts of financing for purchases and repairs:

$$S = N_z S_z + N_p S_p \quad (3)$$

It is supposed that the purchased sample of EM is modern and operable and remains to this state during the period under review.

If to express from (1) and (3) variables of N_z, N_p and to substitute them in (2), then we will receive linear relationship between indicators:

$$P_{opb} = aP_{mod} + b \quad (4)$$

Where

$$\begin{cases} a = 1 - \frac{S_z}{S_p} \\ b = \frac{S}{S_p N} + \frac{N_{mo}(k_1 - k_2 - 1 + \frac{S_z}{S_p}}{N} + k_2 \\ a < 0, b > 0 \end{cases} \quad (5)$$

Therefore:

1. In case of the fixed amount of financing *S* and increase the indicator of the modernity P_{mod}, the indicator of operability P_{opb} will decrease and vice versa,
2. In case of increase *S* and the fixed value of P_{mod}, the indicator of operability P_{opb} will increase,
3. In case of increase *S* and the fixed value of P_{opb}, the indicator of modernity P_{mod} will increase.

3 A problem of maximum the indicator of the modernity of PEM in case of restrictions for indicators of operability and the amount of financing

It is required to maximize the indicator of modernity of PEM:

$$\frac{1}{I}\sum_{i=1}^{I} P_{mod}^i = \frac{1}{I}\sum_{i=1}^{I} \frac{N_{mo}^i + N_z^i}{N^i} \to max \quad (6)$$

Under restriction from above for the total amount of financing directed to purchase of modern samples and repair of defective modern and defective outdated samples of EM:

$$\sum_{i=1}^{I} N_z^i S_z^i + \sum_{i=1}^{I} N_{pmn}^i S_p^i + \sum_{i=1}^{I} N_{pyn}^i S_p^i \leq S \qquad (7)$$

And condition that operability of each type of EM is limited from below and from above:

$$k_{min}^i N^i \leq N_{opo}^i + N_z^i + N_{pmn}^i + N_{pyn}^i \leq k_{max}^i N^i, i = 1,2,\dots,I \qquad (8)$$

We assumed that number of purchases and repairs cannot be negative:

$$N_z^i \geq 0, N_{mno}^i \geq N_{pmn}^i \geq 0, N_{pyn}^i \geq 0, i = 1,2,\dots,I \qquad (9)$$

Here the upper index means parameter belonging to the corresponding type of EM;
I - number of types of EM;
N_{opo}^i- number of operable samples;
N_{mno}^i- number of modern defective samples;
k_{min}^i, k_{max}^i - restrictions from below and from above on a operability indicator;
N_{pmn}^i, N_{pyn}^i - the number of repairs of modern defective samples and outdated defective samples of EM correspondingly.

The problem formulated in this chapter's is a standard integer linear programming problem (LPP) in unknowns: $N_z^i, N_{pmn}^i, N_{pyn}^i$[7] , where $N_p^i = N_{pmn}^i + N_{pyn}^i$. It solve by means of simplex method [7].

4 The results of the calculation

Calculations were performed with the following data values:
$I = 3$; $N^1 = 41503$; $N^2 = 24000$; $N^3 = 52000$; $N_{mo}^1 = 8469$; $N_{mo}^2 = 2300$; $N_{mo}^3 = 21000$; $N_{opo}^1 = 23365$; $N_{opo}^2 = 12300$; $N_{opo}^3 = 24000$; $N_{mno}^1 = 847$; $N_{mno}^2 = 200$; $N_{mno}^3 = 2000$; $k_{min}^1 = k_{min}^2 = k_{min}^3 = 0.8$; $k_{max}^1 = k_{max}^2 = k_{max}^3 = 0.95 - 1.00$; $S_z^1 = 0.55$; $S_z^2 = 1.0$; $S_z^3 = 0.14$; $S_p^1 = 0.055$; $S_p^2 = 0.20$; $S_p^3 = 0.028$
Calculations were performed for different values of $S, k_{max}^i, i = 1,2,3$.
In Table 1 shows the indicators of modernity and operability. The table shows that in case of the small amounts of financing: $S < S_1$ all funds should be allocated for repair of EM (on achievement by indicators of operability of each of types of EM of the admissible lower bound). The corresponding LPP has no solution.

At $S = S_1$ indicators of operability of all types of EM reaches to the lower admissible bound of 80% due to repairs.

At $S_1 < S < S_2$ a condition of operability are carried out strictly on the lower bound, and funds redistribution with increase in purchases and reduction of repairs of EM of the third type (EM -3) is made.

At $S = S_2$ amount of the funds allocated for repair decreases up to the size necessary for repair of the available 2000 modern defective samples and there is the modernity indicator exit on EM -3 to the level of 95%.

At $S_2 < S < S_3$ means go for purchase of EM-1, and the funds allocated for repairs of EM-1 decrease.

At $S = S_3$ the funds allocated for repairs of EM-1 decrease up to the size necessary for repair of the available 847 modern defective sample, and there is a modernity indicator exit on EM-1 to the level of 95%.

At $S_3 < S < S_4$ means the funds allocated for repair of EM-2 decrease go for purchase of EM -2.

At $S = S_4$ the indicator of operability of EM-2 reaches to the level of 95%. The funds allocated for repair decreased up to the size necessary for repair of the outdated 200 available modern samples of EM-2.

Since the number of samples of different types of PEM and the corresponding values of purchases and repairs are represented by numbers of a different order of magnitude, then for ease of perception of information in Table 2 shows the amount of funds for purchase and repairs.

Table 1. Results of calculations indicators for operability range from 80% to 95%.

Notation	Funds	Indicators	Performance indicators			Criterion
			EM-1	EM-2	EM-3	
	1000	P_{mod}	No solution	No solution	No solution	0
		P_{opb}	No solution	No solution	No solution	
S_1	1856	P_{mod}	0.22	0.1	0.44	0.253
		P_{opb}	0.8	0.8	0.8	
	4000	P_{mod}	0.22	0.1	0.8	0.373
		P_{opb}	0.8	0.8	0.85	
S_2	4695	P_{mod}	0.22	0.1	0.89	0.403
		P_{opb}	0.8	0.8	0.95	
	8000	P_{mod}	0.39	0.1	0.89	0.460
		P_{opb}	0.8	0.8	0.95	
	12000	P_{mod}	0.57	0.1	0.89	0.520
		P_{opb}	0.93	0.8	0.95	
S_3	12569	P_{mod}	0.59	0.1	0.89	0.527
		P_{opb}	0.95	0.8	0.95	
	16000	P_{mod}	0.59	0.28	0.89	0.587
		P_{opb}	0.95	0.81	0.95	
S_4	19297	P_{mod}	0.59	0.42	0.89	0.633
		P_{opb}	0.95	0.95	0.95	

The general indicator of the modernity of PEM constitutes 0.63.

Note that further increase in level of the modernity for each type of EM and for PEM at all in general is possible only by means of the purchases some number of modern operable samples with replacement of outdated operable sample in the sequence of EM - 3, EM -1, EM-2.

In the Fig. 1 shows changes of an indicator of the modernity of PEM depending on amount of financing in case of restrictions for an indicator of operability of 80%-95% and 80%-100% are provided. Let's note that in case $S = 22003$ it is possible to increase an indicator of the modernity of PEM to level 0,68.

5 Stages of management of funds allocated for PEM development

On the basis of the analysis of calculations results it is possible to allocate the following consecutive stages of financing management for development of midterm and long- term plans:

Table 2. Results of calculations of founds for indicator operability range from 80% to 95%.

Notation	Funds	Purchase / Repairs	Amount of financing		
			EM-1	EM-2	EM-3
	1000	Purchase	No solution	No solution	No solution
		Repairs	No solution	No solution	No solution
S_1	1856	Purchase	0	0	0
		Repairs	541.035	822	492.8
	4000	Purchase	0	0	2580.2
		Repairs	541.035	822	56
S_2	4695	Purchase	0	0	3276
		Repairs	541.035	822	56
	8000	Purchase	3671.8	0	3276
		Repairs	173.855	822	56
	12000	Purchase	7799	0	3276
		Repairs	46.585	822	56
S_3	12569	Purchase	8368.25	0	3276
		Repairs	46.585	822	56
	16000	Purchase	8368.25	4213	3276
		Repairs	46.585	40	56
S_4	19297	Purchase	8368.25	7510	3276
		Repairs	46.585	40	56

1. Funds shall be allocated only for repairs both modern defective and outdated defective sample of EM in sufficient quantity to achieve the lower bound on a operability indicator on all types of EM in the sequence: EM-3, EM-1, EM-2.

2. Funds shall be re-allocated from repairs on purchases of EM, at the same time funds for repairs decrease to the amount sufficient for repair of modern defective samples. At first restriction on a operability indicator on the lower bound is maintained. Then the indicator of operability increases to the upper bound. The clause is implemented for all types of EM in the sequence: EM -3, EM-1, EM-2. The indicator of the modernity increases up to the target objective.

Note that if the amount of financing is rather big, and financial risks are small, then in point 1 it is reasonable to direct funds only for repairs of modern defective samples and purchase of modern operable samples. Then proceed to clause 2.

6 Conclusion

The provided model shall be applied for development of mid-term and long- term plans of purchases and repairs (and financial plans too) [1]. At each planning interval the following shall be performed:

1. Clarification of initial data as of the commencement planning time,
2. Solution of basic LPP,
3. Adjustment of initial data using the model of modernity and operability,
4. Clarification of initial data using the model of non-availability of funding and impossibility of implementation of the required development option [1].

The described model is included into the automated PEM management system. It used for creation of development plans of PEM, development of refurbishment strategy, procurement write-off, repair strategy and so on. It used for middle - term planning and long-term planning. It can be used for simulation of PEM development in case of limited and unlimited funding.

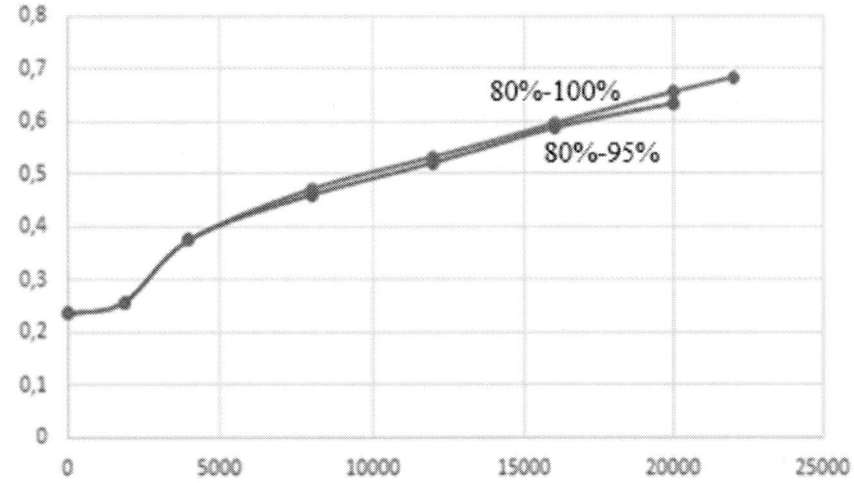

Figure 1. Indicator of the modernity depends on the amount of found.

To obtain results suitable for practical application the of generally LPP solution is sufficient [7]. Solving the LPP using a computer takes less 1 second. Solving an integer values linear programming problem takes 3-4 seconds.

References

1. V.M. Burenok, R.N. Pogrebnyak, A.R. Skotnikov, Mashinostroenie, **1** (2010)
2. V.N. Burkov, G.S. Javakhadze, Russian Academy of Science, **69** (1996)

3. I.S. Lifanov, N.G. Sherstyukov, Moscow, **225** (1979)
4. R.Z. Akberdin, Mashinostroenie, **184** (1987)
5. R.Z. Khayrullin, Scientific review, **10** (2016)
6. A. Mas-Collel, M.D. Whinston, J.R. Green, Univ. Press, **981** (1995)
7. R.Z. Khayrullin, Vestnik MGSU, **8** (2013)

Intensification of the investment process of construction

Pavel Oleinik[1], *Tatyana* Kuzmina [1,*], and *Zenov* Viktor[1]

[1] Moscow State University of Civil Engineering, Department of technology and organization of construction production, 129337, 26, Yaroslavskoye Shosse, Moscow, Russia

Abstract. The article provides negative trends of braking development of the investment process of creating buildings and structures and describes the dynamic system of the investment process and its components given the basis of domestic and foreign experience. It also categorizes the factors influencing the duration of the investment process, which are presented as procedural, engineering design and manufacturing. The article discloses the accounting procedures of these factors by measuring their information content with access to the formation of a rational structure of the investment process.

1 Introduction

Improving the efficiency of the investment process and reducing of its duration is one of the main tasks at the present stage. In recent years, Russia adopted indigenous measures for the restructuring of the whole process of design and construction at the expense of wide introduction of scientific and technical progress, resource conservation, maximizing the use of intensive factors of growth rates of construction [1, 6].

But there are negative trends of braking the emerging positive phenomena – the absence of target orientation to shorten the duration of the investment process, the separation of the functions of activities of the participants of creating the object, weak convergence of the stages and elements of the investment process, the irrational decision making. The ineffectiveness of the applied in certain cases schemes of erection of enterprises, buildings and constructions manifests itself in the face of the principles of full cost accounting and self-financing [2, 3, 4]. For example, system analysis of construction of enterprises buildings and constructions held in eleven branches of industrial construction (electric power engineering, metallurgy, oil refining and petrochemical industry, chemical industry, mechanical engineering, etc.) indicates that about 65.6% from 100% of all failures in the construction were because of the imperfection of the planning, design, logistics and financing of construction works, and 29.8 % were because of poor organizational and technological solutions.

Russian and foreign experience, progressive approaches and solutions of reduction of the length of the investment process proposed by a number of construction, design and research organizations have shown the reality of shortening of the speed of creating objects in 1.5-1.6 times due to intensive factors, which do not require any additional resources [5, 6, 7, 9].

* Corresponding author: KuzminaTK@mgsu.ru

2 Methods

The investment process of objects creation should be seen as a complex dynamic system, consisting of four main functional subsystems - feasibility study, design, preparation of an object for construction process and construction itself, which are in the certain relations and connections with each other.

Each of the subsystems in turn consists of a set of elements (as indivisible parts of sub-systems), with both the internal relationships between elements, subsystems and external links with elements from other subsystems.

Existing in the system of the investment process the flexible relationships between elements and subsystems are subdivided into recursive synergies, and cyclic.

Recursive relationships are irreversible in the system, they install the specific reason of a lengthening of the timing of the investment process (according to the functions of the participants creating the objects of construction) and the resulting consequences.

The synergistic relationships characterize the strengthening of the data flow in this element and, therefore, such an element is treated as a multifunctional.

Cyclic relationships describe periods of enabling the participants object creation into the investment process.

The ideology form of the structure of the investment process of an object creation is outlined in Figure 1.

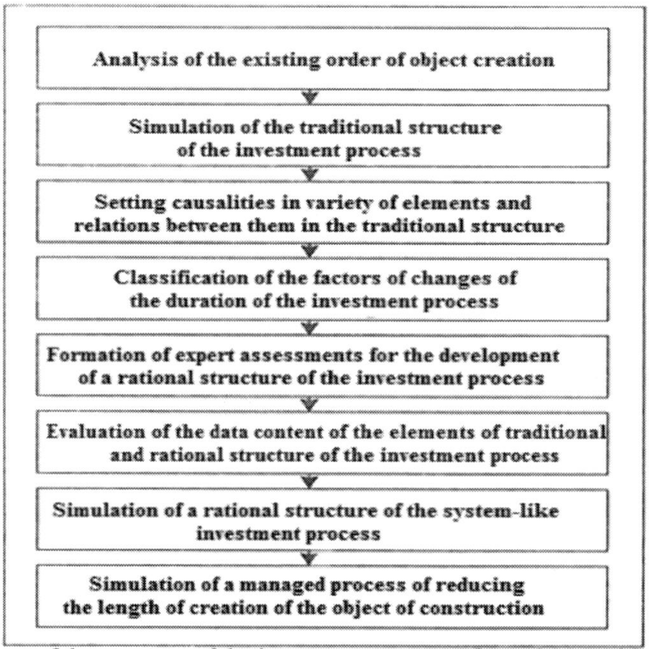

Fig. 1. Ideology form of the structure of the investment process of creating an object of construction.

3 Main part

Development of the system of the object creating is determined by the combination of sustained actions made in order to achieve the quantified results – solution development, coordination and approval of documents, construction of buildings and structures, etc. Such actions on the basis of their specific products are divided into three types of processes:

1. Procedural processes characterizing features of the object creation order and forming on this basis the relationship between the parties;

2. Engineering and computational processes, including the development of a space-planning, structural, organizational, technological, technical and other solutions;
3. Production processes, providing the preparatory and main construction works.

The driving force of a quantitative manifestation of the processes is the set of factors that determine the nature of the changes in the system or its separate subsystems. Some factors directly or indirectly reduce the length of the investment process, while others extend it. A number of factors is acting throughout the whole stage, and a number of others run only in a very short period of time. At the same time the properties of the factors are also various – deterministic and stochastic [8, 10, 11].

Therefore, in order to effectively manage the entire coherent system of creation of the objects of construction it is needed to identify almost all the major influencing factors. Moreover, it should also be borne in mind the fact that in some circumstances, there may have been a negative impact, and in other conditions – a positive influence. In order to take account of the totality of the factors influencing on the duration of the investment process, there was developed their classification (Table 1), including five classifications. In the process of grouping factors there were identified more than 50 of them, extending both the length of separate phases and the investment process as a whole (Table 2).

Table 1. Classification of the factor changes for the duration of the investment process.

Level indication	Classification indicator	Factors and their groups
I	Belonging to the parts of the process	Procedural (organizational management, procedural-performing, economical). Engineering and design (design and engineering, organizational and technical). Production (common site, object, nature and climate).
II	Degree of influence on the process duration	Shortening the duration. Prolonging the duration
III	Impact duration	Short-term. Long-term.
IV	Nature of manifestations	Deterministic. Stochastic.
V	Accounting functions	Customer-developer. General Designer. General contractor. Subcontractor. Bank. Project organizations. Bodies of material maintenance. Transport organizations.

Table 2. Groups of factors, prolonging the duration of the investment process, in accordance with "the belonging to a constituent process".

Components of the investment process	Groups of factors	Factor	Code of factor
1	2	3	4
Procedural	Organizational and managerial $(h^{п1})$	Inconsistency of participants creating an object of construction	$h_1^{п1}$
		Disruptions of supply of material and technical resources	$h_2^{п1}$

		Failures in supply of technological equipment	h_3^{n1}
		Inefficient use of manpower (downtimes, non-production costs)	h_4^{n1}
		Inappropriate use of technical resources	h_5^{n1}
	Procedural-performing (h^{n2})	Irrational use of material resources	h_6^{n1}
		Failure to comply with the order of development, agreement and approval of documents	h_7^{n1}
		Irregularities in the selection of constructional sites	h_8^{n1}
		Violations of approval of constructional site	h_9^{n1}
		Irregularities in the transfer of the object of construction (part of the object) to subcontracting organizations	h_{10}^{n1}
		Improper design specifications, specifications for developing design and preparation works, etc.	h_{11}^{n2}
		Delayed correction of documents	h_{12}^{n2}
		Transfered documentation is not full	h_{13}^{n2}
		Non-arrival of the representatives of oversight services	h_{14}^{n2}
	Economical (h^{n3})	Delayed opening and uneven funding	h_{15}^{n3}
		Improper allocation of funds and limits	h_{16}^{n3}
		Incorrect definition of contract prices	h_{17}^{n3}
		Failure to fulfil contractual obligations	h_{18}^{n3}
		Delays in signing of contracts	h_{19}^{n3}
		Irregularities in the registration of orders, requests, etc.	h_{20}^{n3}
Engineering calculations	Design and engineering (h^{p1})	Poor quality of design solutions:	h_{21-23}^{p1}
		technological	h_{21}^{p1}
		space-planning	h_{22}^{p1}
		constructive	h_{23}^{p1}
	Organizational and technical (h^{p2})	Violations of the duration of the construction	h_{25}^{p2}
		Lack of continuity solutions	h_{25}^{p2}
		Irrational timing of labor, material and technical resources	h_{26}^{p2}
		Irrationality of the deadlines of the nomenclature and volumes of preparatory works	h_{27}^{p2}
		Disunity of the projected production processes	h_{28}^{p2}

		Inefficient distribution of volumes of works on the projects and years of construction	h_{29}^{p2}
		Approximate calculation of parameters of work implementation	h_{30}^{p2}
		Lack of consistency in the development of building sites	h_{31}^{c1}
Productional	Common ground (h^{c1})	Separate construction of underground parts of buildings and constructions and engineering communications	h_{32}^{c1}
		Separate laying of engineering communications	h_{33}^{c1}
		Large amount of temporal engineering and transport communications, buildings and structures	h_{34}^{c1}
		Violations of priority works	h_{35}^{c1}
		Performing primarily material-works	h_{36}^{c1}
		Errors in binding mounting and storage sites	h_{37}^{c1}
		High labour intensity of works	h_{38}^{c1}
		Low level of sanitary and domestic service for workers	h_{39}^{c1}
	Object h^{c2}	Unpreparedness of the front work performed by general contractor	h_{40}^{c2}
		Delayed execution of works performed by subcontracting organizations	h_{41}^{c2}
		Lack of combining of the primary and preparatory periods of construction process	h_{42}^{c2}
		Transfer of the start of the main construction to a later date	h_{43}^{c2}
		Significant proportion of manual labor	h_{44}^{c2}
		Mismatch in number and qualification of the working staff to the structure of works	h_{45}^{c2}
		Inconsistency of construction machinery to structure works	h_{46}^{c2}
		Failure of construction machinery	h_{47}^{c2}
		Weak interaction of labor resources	h_{48}^{c2}
		A large volume of work in the year of commissioning	h_{49}^{c2}
		The discrepancy of the actual rhythm of building to a settlement	h_{50}^{c2}
	Natural and climatic (h^{c3})	The loss of working time due to low temperature, snow, strong wind, fog, heat, etc.	h_{51}^{c3}

The process of forming a rational structure, providing in outlet a sharp reduction in the duration of creation of the objects of construction represents a connection of united into some coherent whole optimized elements, including their integration, summation, and re-combining, and is held in the three procedures.

The first procedure involves selecting elements of the new structure, based on the experience of design and construction of facilities and expert evaluations.

Generalization of experience in the design and construction using high-speed methods shows that the decisive factor to reduce their terms of creation were - parallel implementation of the design and construction, type-ahead start of the launch complexes, the introduction of innovative solutions of construction organization, continuous financing of the construction on the basis of contractual prices, concentration human and material resources with the use of a rotational form of work organization and collective contract [12, 13, 14].

The second procedure is to establish linkages between the elements of the new structure, based on assessments of information messages.

At the beginning of the procedure a topology of relationships is generated (recursive, synergistic, circular) between elements of the structure throughout the investment process.

As a measure of the information content there is used an informational message (IM)-table, chart, list, clearly formulated position etc. The main provisions of the pragmatic approach include:

1. Mandatory exclusion of duplicated and derived information;
2. Consideration of im as a complete unit (without any initial data, techniques of receipt);
3. Ability to change the im in the elements, regardless of the existing regulations;
4. Evaluation of textual materials on a qualitative level only.

The third procedure directly generates efficient structure of the investment process, based on the results of the analysis of the information content of elements.

The evaluation of information flows for more than 50 objects resulted in a rational structure of the investment process through validation and development of the rational priority in construction and its duration, timing of working documentation development, composition and timing of preparatory and basic works, requirements in labour and technical resources, etc.

The resulting calculation formulas for determining the timing of the initial and final stages of the investment process for the average facility are summarized in Table 3.

Table 3. Calculation formulas for determining the timing of the initial and final stages of the investment process.

Stages	Calculation formulas	
	initial, t	**final, t^0**
Feasibility study	$t_1 = 0$	$t_1^0 = 0.1\,T$
Designing	$t_2 = 0.06\,T$	$t_2^0 = 0.6\,T$
Preparation of a building for construction	$t_{no}^{н} = 0.1\,T$	$t_3^0 = 0.81\,T$
Construction	$t_3 = 0.22\,T$	$t_4^0 = 0.97\,T$
Putting into operation	$t_4 = 0.97\,T$	$t_5^0 = T$

The final formula to calculate the overall length of the investment process for the average object of construction is expressed as follows:

$$T = t_1^0 + (t_3^0 - t_3) + (1 - u_4) + (t_4^0 - t_4) + (t_5^0 - t_5) \tag{1}$$

Where T – the duration of the investment process of creation of an object of construction;

u_4 – coefficient of the combination of a construction stage with other stages ($u_4 \approx$ 0.79).

This model is fairly compact and allows to fully or partly exclude the influence of the most part of factors which can negatively influence on the duration of the investment process (Table 4).

Table 4. Composition factors, the influence of which is completely or partially excluded because of the implementation of the concept of express creation of the objects of construction.

Components of the investment process	Groups of factors	Codes of factors
Procedural	h^{n1}	$h_1^{n1}, h_2^{n1}, h_3^{n1}, h_4^{n1}, h_5^{n1}, h_6^{n1}, h_7^{n1}$
	h^{n2}	$h_{12}^{n2}, h_{13}^{n2}, h_{14}^{n2}$
	h^{n3}	$h_{15}^{n3}, h_{16}^{n3}, h_{17}^{n3}, h_{18}^{n3}, h_{19}^{n3}$
Engineering calculations	h^{p2}	$h_{24}^{p2}, h_{25}^{p2}, h_{26}^{p2}, h_{27}^{p2}, h_{28}^{p2}, h_{29}^{p2}, h_{30}^{p2}$
Productional	h^{c1}	$h_{31}^{c1}, h_{32}^{c1}, h_{33}^{c1}, h_{34}^{c1}, h_{35}^{c1}, h_{36}^{c1}, h_{37}^{c1}, h_{38}^{c1}, h_{39}^{c1}$
	h^{c2}	$h_{40}^{c2}, h_{41}^{c2}, h_{42}^{c2}, h_{43}^{c2}, h_{44}^{c2}, h_{45}^{c2}, h_{46}^{c2}, h_{47}^{c2}, h_{48}^{c2}, h_{49}^{c2}, h_{50}^{c2}$

4 Conclusion

For the effective management of the system of creation of the object of construction, it is required to take into account all the influencing factors both extending and reducing the duration of the investment process.

The analysis shows that such a combination of factors represents more than 50 factors, classified by various characteristics of their influence.

The process of accounting of the influencing factors involves three procedures, the end result of which is the rational structure of the investment process.

As an example, there is shown a model for determining the length of the investment process for an average object of construction specifying a number of decisions mentioned in the text. This model demonstrates the possibility of statistical accounting of various factors and building the investment process for the uniform objects both of productive and non-productive function.

References

1. V.V. Buzyrev, V.D. Vasiliev, A.A. Zubarev, Publishing house of Saint Petersburg State University of Economics, (2001)
2. Clifford f. Gray, Erik W Larson, Business and Service Publishing house, (2003)
3. F. Kotler, G. Armstrong, J. Saunders, B. Wong, Williams Publishing house, (2000)
4. P.P Oleinik, T.K. Kuzmina, Civil and Industrial Engineering **17**, 42-43 (2012)
5. P.P Oleinik, V.I. Brodsky, Technology and organization of construction production **7**, 40-45 (2013)
6. A.I. Romanova, Publishing house of Kazan State Academy of Architecture and Civil Engineering **9**, 180 (2001)
7. V.D. Sapiro, DvaTrI Publishing **13**, 256 (1996)
8. H.N. Ahuja, John Wiley & Sons **23**, 505 (1994)
9. P.P Oleinik, L.S. Grigorieva, V.I. Brodsky, Applied Mechanics and Materials **6**, 580-583 (2014)
10. Frame J. Davidson, A Wiley Company **8**, 360 (2002)

11. A.A. Lapidus, N.D. Cherednichenko, Scientific review **36**, 338-341 (2015)
12. S.A. Sinenko, T.K. Kuzmina, Scientific review **5**, 156-159 (2015)
13. T.K. Kuzmina, S.A. Sinenko, Natural and technical sciences **7**, 637-639 (2015)
14. P.P Oleinik, V.I. Brodsky, Technology and organization of construction production **1**, 27-31 (2013)

Specifics of system of external influences on the life cycle of a construction object

Aleksander Aleksanin [1,*], and *Sergey* Sborshikov [1]

[1]Moscow State University of Civil Engineering, Moscow, Russian Federation

Abstract. There is a very important issue today which includes the harmonious and effective development of the system 'man –environment'. Construction is a branch of material production, which has a significant negative impact on the world around us. It is necessary to plan and operate processes of construction at all stages of the life cycle of a building without exception, to prevent of ecological threats. The article describes the concept of 'life cycle' as applied to various fields of knowledge, analyzes existing in the scientific literature division of the life cycle of buildings in the periods, proposes own approach to the division of periods of the life cycle on the basis of resource-saving. The article proposes the creation of a unified organizational system for the effective management of all periods with the constituent phases and formulates the main external influences on the building life cycle.

The present stage of development of the construction industry is characterised by close inter-relationship with the natural environment. The harmonious development of the construction and the natural environment is the key to sustainable development of the system 'man - environment'. To date, there is the task of organization all the processes accompanying the creation of the construction object in a single control system, starting with the stages of initiation and pre-investment and ending with decommissioning or its liquidation. One of the main criteria for the operation of this system, in the first place, should be the focus on the balanced development of the environment. The decision of this task is based on the need to develop a set of measures to streamline and improve the existing mechanisms of formation and functioning of the processes in the context of the life cycle of a construction object. The process of creating a construction object can be divided into sub-processes, the interaction of regulation which will contribute to the harmonious coexistence of the construction object and the surrounding world.

Let's consider stages of life cycle of the object, most often meeting in modern scientific works. The concept of 'life cycle' is widely used in many fields of knowledge, such as in biology, this term means the consequent shift of all generations, characteristic for this type of living organisms. In economics, there is the concept of 'life cycle of the organization' - a set of developmental stages which runs the firm during the period of its existence.

For our research the most interesting definition is 'the life cycle of a building '. The life cycle of a building is a period during which carried out engineering survey, design,

* Corresponding author: aleks08007@mail.ru

construction (including conservation), operation (including current repairs), reconstruction, major repairs, demolition of a building or structure.

There is a lot of works of foreign authors, which devoted to scientific researches of the life cycles [1-5], many of them are directly related to energy efficiency and resource saving [6-9]. In the scientific and technical literature distinguish the following periods of a building's life cycle [10]:

1. The 1-st period is the process of developing a feasibility study for construction of the building;
2. The 2-d period is the process of developing of construction and design decisions;
3. The 3-d period is process of building, together with the development of technology, organization and technological regulations of work;
4. The 4-th period is the process of pre-service development;
5. The 5-th period is directly operation of buildings and work aimed at ensuring payback of funds invested in their creation;
6. The 6-th period is processes of manufacture scheduled preventive and major repairs to maintain in a good technical condition of constructive elements and engineering systems of the building;
7. The 7-th period is the stage, both physical and moral deterioration of the building, when necessary to carry out works on modernization, reconstruction or demolition;
8. The 8-th period is reconstruction, restoring physical, mechanical and operational characteristics of the buildings. This period include also 1-st and 2-d periods of feasibility study and development of technical documentation.

There are scientific papers in which the authors identify some types of life cycle of object of the real estate: economic, legal and physical. From an economic point of view there are three periods: construction (step up investments and investment stage); exploitation to the point where the object is fully pay for themselves (the entrepreneurial stage of the project); operation in the subsequent growth of the results on investment (innovation, facility closing). From a legal point of view, allocate periods of registration of ownership; possession, use, disposal; change of ownership; registration of liquidation [11].

It is necessary to distinguish the concept of 'life cycle of a construction object' and 'the life cycle of a construction project'. The authors offer the following life cycle stages of construction project: pre-investment stage, stage design, stage management, stage of implementation [12]. The life cycle of a universal project can be represented by the following phases:

- Definition of goals and objectives;
- Development concept;
- Design;
- Examination;
- Decision on the implementation of the project;
- Implementation of the project;
- Monitor the implementation;
- Evaluation and analysis of results;
- Completion of the project.

Within the framework of scientific research suggests to integrate described above periods of the life cycle of the construction object and to allocate only three major:

- Creation of an object;
- Exploitation of an object;
- Limination of an object.

The period 'creation of an object' combines phase feasibility study, design and construction. This stage also includes the phase of production of building materials, as it accumulates a huge potential resource in the implementation of the next phase.

So, each of the periods of the life cycle of the construction object includes certain phases.

The period of 'Creation of an object' contains a phases:

- Production of construction materials;
- Predesign development and researches;
- Design works;
- Construction and installation works.

The period of 'Exploitation of an object' contains a phases:

- Works to improve the energy efficiency of the facility;
- Maintenance works on the facility;
- Maintenance repairs;
- Capital repairs.

In this period a special place is occupied by improving energy efficiency, because during the operation of the building development of scientific and technological progress does not stop, and develops new energy-saving systems.

The period of 'Elimination of an object' contains a phases:

- The dismantling and demolition of buildings;
- Works on the disposal of construction and demolition waste;
- Works on the processing of construction and demolition waste.

For the effective management of all periods with the constituent phases, it is proposed to create a unified organizational system. The system has its own life cycle, comprising the steps of the process, covering a variety of state of the system, starting from the moment of the need for such a system, and ending with its complete decommissioning [13].

This organizational system is closely related to the impacts of external environment in the form of federal, regional or municipal authorities, the population, the level of development of scientific and technical progress, etc. The main external influence is the administrative and legal impact, economic impact, social impact, and the impact of science and technology. Scheme of impacts in terms of construction object lifecycle is shown in Figure 1.

Administrative and legal influence is due to the publication of new administrative regulations, standards, regulations, etc. The economic influence is a powerful regulatory tool, the essence of which is the availability of funds for works on a particular period. The social impact is the attitude of the society to the life cycle of the building, for example, the protests at a hearing on new construction projects in areas with unique natural landscapes, the lack of demand for products, realized within the functioning of buildings (space rental, attendance at trade and entertainment buildings, etc.). The impact of scientific and technological progress is the creation of new building materials, the development of modern technical, technological and organizational methods of the construction of buildings and structures.

The indispensable condition to demolish buildings and structures it is almost total recycling of construction waste and their subsequent use in the construction of new facilities, which reduces the impact on the environment in the extraction of raw materials [14; 15]. Thus, the life cycle of a single object of construction is obtained related to the life cycle of other objects. The high level of development of the market of construction and demolition waste contributes to the effective provision of recyclable material of new objects [16; 17]. The basis of sustainable development of natural and man-made environment is a balanced functioning of the system life cycle of construction projects based on resource conservation policy.

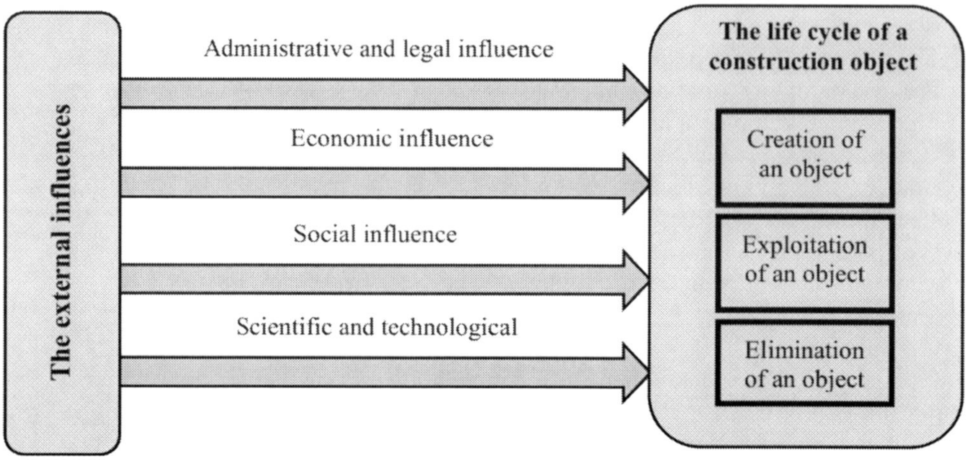

Fig.1. Scheme of the system impacts in the context of construction project life cycle.

It is necessary to take into account the fact that continuous improvement of production of construction materials and the emergence of new innovative technologies in the creation of the organizational life cycle management system of buildings. If we analyze the building, built in the XX century, it is possible to get an idea about the changes of the basic materials of building structures. For example, the buildings of the first half of the 20th century were wooden, clay and stone; in the second half of the 20th century building built of brick, block, panel; in the late twentieth and early twenty-first century is monolithic. The architects proposed many variants of the buildings of the future today. It is believed that the main trends of the construction will be ecology, mobility, self-reliance and full automation of all systems running artificial intelligence. Already practiced production of bricks from waste rice husk, the use of smart glass, which are capable of generating electricity, the introduction of a shell building facades, regulating indoor climate [18-20]. Perhaps, in the not too distant future it will be impractical to construct buildings from recycled designs objects built hundreds of years ago.

References

1. A.F.A. Rashida, S. Yusoffb. Renew, Sustainable Energy Rev. **45**, 244 (2015)
2. C.K. Chau, T.M. Leung, W.Y., Ng. Appl. Energy **143**, 395 (2015)
3. I. Kovacic, V. Zoller, Energy **92**, 409 (2015)
4. S. Butera, T.H. Christensen, T.F. Astrup, Waste Management **44**, 196 (2015)
5. J. Honga, G. Q. Shena, C. Maob, Z. Lia, K. Lib, Journal of Cleaner Production **112**, 2198 (2016)
6. V. Motuzienėa, A. Rogožaa, V. Lapinskienėa, T. Vilutienėb, Journal of Cleaner Production **112**, 532 (2016)
7. M. S. Tokarik, R. C. Richman, Energy Build **118**, 160 (2016)
8. S. Proiettia, P. Sdringolaa, U. Desideria, F. Zepparellia, F. Masciarellib, F. Castellania, Energy Build **64**, 463 (2013)
9. L. F. Cabezaa, L. Rincóna, V. Vilariñob, G. Péreza, A. Castella. Renew. Sustainable Energy Rev **29**, 394 (2014)
10. A.A. Afanasev, E.P. Matveev. *Reconstruction of residential buildings* (Moscow, 2008)
11. P.V. Fedyaeva *Organizational and technological support for improving energy efficiency in the housing sector of the Russian Federation* (Cand.Tech.Sci., 2015)

12. V.V. Kuschenko, *Development. The modern concept for the development of Real property* (Moscow, 2005)
13. B.S. Blanchard, W.J. Fabrycky, *Systems engineering and analysis* (Prentice Hall, 2006)
14. A.V. Aleksanin, S.B. Sborshikov. Industrial and Civil Engineering **11**, 84 (2014)
15. A.V. Aleksanin, Scientific and Technical Volga region Bulletin **5**, 94 (2014)
16. A.V. Aleksanin, Industrial and Civil Engineering **10**, 79 (2014)
17. A.V. Aleksanin, S.B. Sborshikov,Vestnik MGSU **6**, 229 (2013)
18. C. Ionescu, T. Baracu, G-E. Vlad, H. Necula, A. Badea. Renew, Sustainable Energy Rev. **49**, 243 (2015)
19. R. Volk, J. Stengel, F. Schultmann, Automation in Construction **38**, 109 (2014)
20. M. L. Rosenzweig, Israel Journal of Ecology & Evolution **62**, 7 (2016)

Prediction of parameters affecting the safety of tank farms in case of emergency

Vyacheslav Gorev[1,*], *Vladimir* Roytman[1]

[1] Moscow State University of Civil Engineering, 129337, Moscow, Yaroslavskoye shosse, 26, Russia

Abstract. The current article demonstrates that the physical nature of blow-out of oil products burning in the tank is connected with overheating and fast vaporization of bottom water accumulated in the tank or on a surface of internal floating roof. It is shown that the period of time, during which the homothermal layer of oil, heated up to the boiling point while burning in the tank, reaches the level of bottom water or the water added in the tank in the process of extinguishing depends on the heat losses into environment through the side surface of the tank. It is determined that blow-out time depends on the water cooled surface area of the tank. It is shown that intensive water cooling extremely decreases the rate of formation of homothermal layer with increasing of its thickness.

1 Introduction

Fires in tanks usually begin with explosion of steam-air mix and disrupt of a roof or damage of integrity of its separate parts [1]. Depending on force of explosion in a vertical metal tank events can develop by several scenarios:

• the roof of tank is damaged, completely disrupted and throwed away on a distance of 20-30 m; liquid burns on all area of the tank;

• the roof of tank is a little raised, completely or partially disrupted, but it is still stay in a burning liquid in semisubmerged condition;

• the roof is deformed in attaching points with the tank wall, small cracks are formed in weld seams of the roof.

In conditions of any of these scenarios fire extinguishing systems of tanks mounted on the top of constructions are unable to work.

In case of combustion in cylindrical horizontal and spherical tanks it is usually disrupt the bottom therefore liquid spreading on a large area creates danger for neighboring tanks and constructions.

Protection of crude oil in drain tank is reached by presence at its bottom of a gas-salt layer of water according to oil production technology.

Displacement of oil from under the ground is made using the water with various additives which flows into the tank together with crude oil. The watery oil supplied for cleaning into the tank where the mixture is stratified with time and oil floats above the water. Water with big concentration of inorganic salts dissolving in it during passing

* Corresponding author: ikbs@mgsu.ru

through oil layers, is periodically drained. As a result it is formed two-layer system: oil layer is above, gas-salt layer of water is below. Height of the last can reach 6–8 m.

The physical nature of blow-out of oil products burning in the tank is connected with overheating and fast vaporization of bottom water accumulated in the tank or on a surface of internal floating roof [2].

The period of time, during which the homothermal layer (HTL) of oil, heated up to the boiling point while burning in the tank, reaches the level of bottom water or the water added in the tank in the process of extinguishing, depends on the heat losses into environment through the side surface of the tank [3-5].

2 Formation homothermal layer

In order to detect the time of blow-out it is necessary to define the water cooled surface area of the tank [6-8].

Heat flow from a flame jet is spent generally on oil vaporization and formation of a flow of combustible vapors in combustion zone [9-15]. Dispersion of heat through the side area of the metal tank is proportional to a thickness of homothermal layer – h.

Heat spent on vaporization of oil is defined by product of burnup mass rate and combustion value:

$$Q_T = m\,Q_s + U_n F_o \rho C(T_n - T_o) + \alpha_1 F_b (T_n - T_o) + \alpha_2 F_o (T_n - T_o), \quad (1)$$

where: Q_T – flow of heat from a flame jet to a surface of the tank liquid, J/s;
m – mass burnup rate of a flammable liquid (FL), кg/s;
Q_s – heat value, J/кg;
U_n – FL warm-up rate, m/s;
F_o – surface area of FL in the tank, m^2;
ρ – FL density, кg/ m^3;
C – FL heat capacity, J(кg· K);
T_n – FL surface temperature in process of burning, K;
T_o – initial temperature of FL equal to the ambient temperature, K;
α_1, α_2 – heat transfer coefficients: through the tank wall and from homothermal layer deep into FL,; J / m^2(s·K)
F_b – lateral area of HTL,m^2;

The mass burnup rate m we express through the specific u_m (кg·(m^2 ·s)), and lateral area of HTL – through its thickness h (m):

$$m = u_m F_o; \quad (2)$$
$$F_b = 2\pi R h, \quad (3)$$

where R – radius of the tank, m .
We express U_n variable value in an explicit form:

$$U_n = dh/d\tau \quad (4)$$

The given below figure presents the diagram of the oil tank where the homothermal layer is formed (Figure 1).

In an equation (1) we transfer time-constant members from the right to the left part and denote so:

$$\Delta T = T_n - T_o \tag{5}$$

Substituting the expression (2)-(5) into equation (1), we obtain:

$$Q_T - u_m F_o Q_c - \alpha_2 F_o (T_n - T_o) = F_o \rho\, C \Delta T \frac{dh}{d\tau} + 2\pi R\, \alpha_1 h\, \Delta T \tag{6}$$

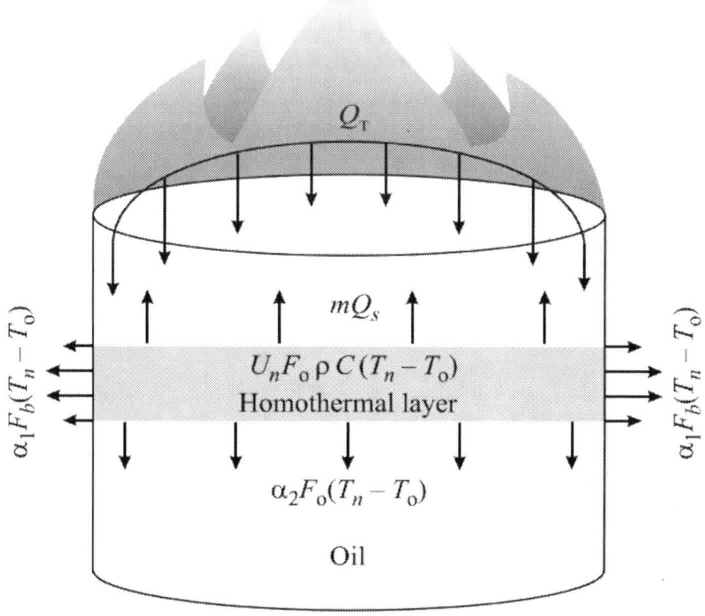

Fig 1. The schema of main heat flows in process of homothermal layer formation in the tank with burning oil product: → — main directions of heat flow.

Designating the left part through q, we receive an initial differential equation depicting change of thickness of a layer in time:

$$F_o \rho C \Delta T \frac{dh}{d\tau} = q - 2\pi R\ \alpha_1\ \Delta T h; \tag{7}$$

$$a \frac{dh}{d\tau} = q - bh, \tag{8}$$

where: $a = F_o \rho\ C \Delta T$; $b = 2\pi\ R\ \alpha_1\ \Delta T$.

The solution of equation (8) with the initial conditions of $\tau = 0$ and $h = 0$ is the correlation:

$$\tau = \frac{a}{b} \ln \frac{q}{q - bh} \tag{9}$$

Let's enter the designation:

$$\tau_0 = \frac{a}{b} = \frac{\rho c R}{2\alpha_1}. \tag{10}$$

Let's remove the *h* value from the equation (9):

$$h = \frac{q}{b}\left(1 - e^{-\tau(b/a)}\right) \tag{11}$$

In conditions when $\tau \to \infty$ the *h* value becomes maximal

$$h\ h_m = a/b, \tag{12}$$

where: h_m - maximum possible thickness of the homothermal layer, m.
Let's find $dh/d\tau$ derivative:

$$-\left(\frac{\mathrm{d}h}{\mathrm{d}\tau}\right) \equiv U_n \equiv \frac{q}{a} e^{-\tau(b/a)} \tag{13}$$

If $\tau \to 0$, than:

$$U_n = U_0 \equiv \frac{q}{a} \tag{14}$$

where *Uo* - initial warm-up rate of liquid, m/s.
From the correlation (14) we shall receive the formula for calculation of *q* value:

$$q = aU_0. \tag{15}$$

If the fraction under the logarithm in the formula (9) will be divided by *b* after replacing *q* by aU_0 from formula (15), we shall receive:

$$\tau = \tau_0 \ln \frac{\tau_0 U_0}{\tau_0 U_0 - h} = \tau_0 \ln \frac{h_m}{h_m - h} \tag{16}$$

where:

$$h_m = \tau_0 U_0 \equiv U_0 \frac{a}{b} \equiv \frac{q}{b} \tag{17}$$

τ_0 - starting time of liquid warm-up, h.
We shall similarly simplify the formula (11)

$$h = h_m(1 - e^{-\tau(b/a)}) \tag{18}$$

Formulas (16) and (18) describe dynamics of change of homothermal layer thickness and can be used for prediction of oil product blow-out time during long-continued fire. The applicability of the (16) correlation is limited by condition of $h_m >> h$, that is the maximal possible thickness of homothermal layer should be must be appreciably greater than the thickness of a layer of fuel in the tank.

Numerical values of the α_1 heat-transfer coefficient on the "tank wall - air" boundary can be defined from criteria correlation for free turbulized convection near the heated wall [3]:

$$Nu = 0.1(Gr \cdot Pr)^{1/3}, \tag{19}$$

where: Nu — Nusselt number; Gr — Grashof number; Pr — Prandtl number.

In our case Pr = 0.75 [3], consequently:

$$1 = 0.09\lambda \left(\frac{\beta}{v}\right)^{\frac{1}{3}} \tag{20}$$

where: λ. — heat conductivity coefficient of heat-transfer agent, W/(m·K);
β — coefficient of thermal expansion, K^{-1};
ν — kinematic coefficient of viscosity, m^2/s.

If surface of the tank is cooling by water heat-transfer coefficient can be defined by this correlation:

$$Nu_2 = 0.25\ Ar^{1/3}, \tag{21}$$

where Ar — Archimedean number.

This correlation allow for a mode of film evaporation of water in case of contact with heated surface of the tank:

$$\alpha_1 = 0.25 \frac{\lambda_f}{\nu_f} g^{1/3} \tag{22}$$

where:
λ_f—heat conductivity coefficient of a film, W/(m·K);
ν_f—kinematic coefficient of viscosity of a film, m^2/s;
g — free fall acceleration, m/s^2.

If the cooling is conducted with a small consumption of water average value of α can be reduced in proportion to watered surface.

In case of intensive water cooling the speed of homothermal layer formation is extremely reduced with the growth of its thickness. The maximal thickness of homothermal layer is determined by correlation of heat quantity going out through the tank walls which increases in process of warming-up the liquid and value of heat flow from the flame jet to the combustion surface.

If the maximal thickness of homothermal layer is less than distance from its surface to the aqueous layer at the bottom of the tank than the time through which the homothermal layer reaches the water will be determined by warming-up rate as well as by burn-out rate of oil product.

Let's output the calculated correlation for the case, when the maximal thickness of homothermal layer is less than flammable liquid h_g, i.e. $h_m < h_g$ (the h_m value is calculated by (17) formula).

In the general case the lower bound of homothermal layer will be determined as by warming-up of the flammable liquid so by lowering of its level in process of burnout. If as the reference point will be taken the level of liquid before combustion than coordinate of the lower boundary of homothermal layer can be defined by such correlation:

$$z = U_1\tau + U_n\tau, \tag{23}$$

where U_1 – linear rate of burn-out, m/s.
We express U_n value in an explicit form:

$$z = U_1\tau + \frac{q}{b}(1 - e^{-\tau(a/b)}) \tag{24}$$

If $z >> h_m$ then value $e^{-\tau(a/b)}$ is negligible. So we can get such example:

$$z = U_1\tau + \frac{q}{b} = h\,. \tag{25}$$

Taking into account that $q = aU_0$ and having the α and b values we shall receive an equal determining the time from the beginning of warm-up to the blow-up of FL, if $h_m <<$ h_g:

$$\tau = \frac{h_g}{U_1} - \frac{U_0}{U_1}\frac{\rho R}{2\alpha} = \frac{h_g - \tau_0 U_0}{U_1} \tag{26}$$

To define the time of blow-out the oil product we use two correlations:
if $h_m >> h_g$ or $\tau_0 U_0 >> h_g$

$$\tau = \tau_0 \ln\frac{\tau_0 U_0}{\tau_0 U_0 - h_g} \tag{27}$$

where

$$\tau_0 = \rho\, CR/(2\alpha); \tag{28}$$

if $h_m << h_g$

$$\tau = \frac{h_g - h_m}{U_1}. \tag{29}$$

From the obtained correlations we can see that warming-up value of liquid, its linear burn-out rate, initial level of flammable liquid in tank and the thickness of homothermal layer have an impact on the blowing-up time of the oil product burning in tank.

3 Determinig the thickness of homothermal layer

For determination of the thickness of homothermal layer at the set time point (h) it is possible to use two variants of formulas:

From the ratio (27) we omit intermediate transformations:

If $h_m >> h_g$, than we receive

$$h = \tau_0 U_0 (1 - e^{-\tau/\tau_0}); \quad (30)$$

From the ratio (23)

if $h_m << h_g$,than we receive

$$z = U_1\tau + U_0\,\tau_0. \quad (31)$$

Accordingly, distance to water layer δ at the bottom of the tank at the set timepoint (if $h_m << h_g$) is determined by formula:

$$\delta = h - z = h - (U_1\tau + U_0\,\tau_0). \quad (32)$$

Let's define the time before blow-up of oil product (gasoline) in case of fire in partially water-cooled tank by substituting in a formula (27) the following parameters [4]: $R = 26$ m, $\rho = 700$ kg/m^3, $C = 2\cdot10^3$ J/(кg·K), initial level of FL in tank $h_g = 9.5$ m, the thickness of water layer under the FL- 0,5 м, burn-out velocity of gasoline $U_1 = 0.4$ м/ч, initial time warming-up rate $U_0 = 1.8\cdot10^{-4}$ m/s, heat transfer coefficient in conditions of free combustion $\alpha_1 = 10$ W/(m·K), $\alpha_2 = 730$ W/(m·K).

To select the calculated correlations let's define the h_m value:

$\tau_{02} = 700\cdot2\cdot10^3\cdot13/(2\cdot730) = 12.5\cdot10^3$ s;

$h_{m2} = 12.5\cdot10^3\cdot1.8\cdot10^{-4} = 2.2$ m.

Consequently, $h_m << h_g$, therefore for the calculation we use the formula (29):

$$\tau_2 = \frac{h_g - h_m}{U_1} = \frac{9.5 - 0.5}{0.4} = \frac{9}{0.4} = 22.5h$$

4 Conclusion

Thus, in considered fire conditions the blow-out of oil product will occur after 22.5 hours, so the fire should be localized and liquidated during the day.

So, as a result of the conducted analysis and calculations it was formulated the prognosis of a time period during which the homothermal layer (HTL) reaches the level of the tank bottom water or water supplied into the tank during fire suppression.

Based on analysis of heat flows coming from flame to a burning oil surface and dispersing into environment it was deduced the calculated ratio of dependence of thickness of the homothermal layer on the burning time of oil in the tank.

The intensive water cooling is sharply decreasing the speed of formation of homothermal layer together with increase of its thickness. The maximal thickness of homothermal layer is determined by ratio of heat losses which increase in process of warming-up of a liquid and heat flow from the flame jet to a burning surface.

If the maximal thickness of homothermal layer is less than the distance from the surface to water layer located at the bottom of the tank, than the time during which the homothermal layer reaches the water lavel will be determined not only by speed of warming-up but also by burn-out rate.

References

1. *P. Nash, D. Hird, R. J. French, Base Injection of Foam for Fuel Storage Tanks,* (1960)
2. *V. I. Blinov, G. N. Khudyakov, Diffusive burning of liquids. Academy of Sciences of the USSR,* (1961)
3. *V. I. Blinov, G. N. Khudyakov, I. I. Petrov, V. Ch. Reutt, On movement of the tank liquid while mixturing by the air stream. Mechanism of flame suppression in oil tanks. Prod. Ministry of Housing and Communal Services of RSFSR,* (1958)
4. *W. H. MacAdams, Heat transfer. Translation from engl. vers. Leningrad,* (1936)
5. *H. S. Mahley, Mobil Oil Corporation,* (1967)
6. S. A. Sharovarnikov, D. A. Korolchenko, A. V. Lyapin, Fire and Explosion Safety **23(6)**, 76–80 (2014)
7. *B. Persson, A. Lönnermark, H. Persson, D. Mulligan, A. Lancia, M. Demichela, FOAMSPEX - Large Scale Foam Application - Modelling of Foam Spread and Extinguishment, SP Swedish National Testing and Research Institute, SP Report 2001:13,* (2001)
8. D. A. Korolchenko, A. F. Sharovarnikov, Fire and Explosion Safety **23(11)**, 84–88 (2014)
9. D. A. Korolchenko, S. A. Sharovarnikov, A.V. Byakov, Advanced Materials Research **1073-1076**, 2353-2357 (2015)
10. D. Korolchenko, A. Tusnin, S. Trushina, A. Korolchenko, International Journal of Applied Engineering Research **10(21)**, 42541-42548 (2015)
11. A. F. Sharovarnikov, D. A. Korolchenko, Applied Mechanics and Materials **475-476**, 1344-1350 (2014)
12. I. F. Bezrodnyi, Fire and explosion safety **12**, 49-55 (2011)
13. S. Korzeniowski, T. Cortina, Industrial Fire Journal, 18-20 (2008)
14. Q. Zhang, L. Wang, Y. Bi, H. Zhi, P. Qiu, Journal of Hazardous Materials **287**, 87-92 (2015)
15. H. Ranjbar, B. H. Shahraki, Chemical Engineering and Technology **36**, 295-299 (2013)

Geo-ecological approach to evaluation of ecological condition of water bodies

Vladimir Smetanin[1*], *Ilya* Tsvetkov[1] and *Andrey* Nasonov[1]

[1]Russian State University of Agriculture, 127550 Timiryazevskaya st., 49, Moscow, Russia

Abrstract. The paper proposes a new approach to evaluation of ecological condition of water bodies. The method is based on the use of fractal dimension of the temporal dynamics of water environment pollution components as a quality indicator of state dynamics for a specific type of pollution or an ecosystem in its entirety. The method enables both common integrated and component-based differential evaluation. The writers have identified fractal threshold values that describe the extent of environmental stability of a water system, its resistance to industrial pollution, and probability of further anthropogenic loads. The paper offers a methodology to forecast the ecological condition of water bodies on the basis of quantitative estimation of dynamic fractal parameters representing pollution dynamics of water system components.

By definition, an ecosystem is a functional amalgamation of living beings complete with their habitats that persists as long as its self-repairing properties are not compromised. The habitat is directly exposed to anthropogenic impacts that can impair such properties causing reversible or irreversible changes. Therefore, generally an ecosystem should be regarded as a non-equilibrium system of exchange interactions with the environment,where its condition is determined by the degree of its openness (intensity of exchange interactions).

If an ecosystem enters an irreversible state, an environmental disaster ensues; a reversible state causes a crisis. Balance equilibrium exists if the system has not lost its self-repairing ability. This is the basis of classification of geo-environmental changes in ecosystems triggered by anthropogenic impacts.

Evaluation of ecological condition of water bodies rests on identification of inclusions of negative anthropogenic factors that cause loss of self-repairing properties: the researchers must build a comprehensive description of the water body's functionality and structure, determining the change dynamics of its adaptation resources. The language of fractals is used to describe the ecosystem's change dynamics.

As of now, there is no alternative to the method of evaluating the quality of water bodies other than the "Comprehensive Method of Surface Water Pollution Evaluation by Hydrochemical Parameters" RD 52.24.643-2002, which calculates the unified combinatorial index of water pollution (UCIWP) [1].

The main objective of the UCIWP method is to identify a set of manmade pollutants that persistently impair health and hygiene standards of water quality and consumer

* Corresponding author: smetanin2000@yandex.ru

standards, and to establish a pollution scale: thus, the method generates a comprehensive functional description of the water body to meet established expert requirements and official quality standards.

The main purpose of the method of Integrated Evaluation of Water Body Ecological Condition (IEWBEC) is to evaluate the ecological condition of the water body by identifying inclusions of adverse anthropogenic factors that cause loss of its self-repairing ability. So, the method not only assesses the quality of aquatic environment, but also the changes dynamics of its adaptive ability through comprehensive description of its functionality and structure, as represented in Table 1.

To describe a water body on the function-logical and logical-physical levels, researchers use a fractal concept according to which complex systems with different scales (hierarchies) exhibit scaling (scale-invariant) properties, and thus the hierarchy, just as geometry of system divisibility, acts as a physical factor that determines the said properties.

Items of technogenesis embedded in the environment represent non-equilibrium flow-type open systems in terms of physics. Flows of energy and matter passing through the systems give rise to self-consistent structures of fractal/scaling type within a broad range of parameters and scales inside the systems, [2].

Table 1. Comprehensive functionality-structure description of a water body that determines the change dynamics of its adapting ability.

Water body description level			Water body description character
IEWBEC	UCIWP	Functionality level	Represents the nature of exchange interactions between the water body and its environment through time series of control parameters
	Functionality logic level		Represents the nature of adaptation processes through manifestation of scaling (scale-invariant) system properties. Gives an idea of complementary behavior of constituent subsystems exposed to anthropogenic activities.
	Functionality physics level		Represents the comprehensive structure of adaptation properties of the water body. Gives an idea of the nature of the water body's adaptive resource of redistribution responding to changes of external anthropogenic activities.

Virtually every natural system, including biological ones, can be described as an open non-equilibrium structure created by a passing flow of matter and energy. Optimal distribution of such factors is ensured by scaling (scale-invariant) structures that act as sort of distributor channels, [3].

Described in terms of mathematics, fractal dimension as a quantitative measure of a structure's scaling disorder connects two types of system characteristics: integral ones that describe the properties of the object in question complete with its local parameters, which emerge after its breakdown. This feature allows representing same-level system descriptions in terms of higher-level models. For example, a water body's differential description on the functional logic level can be easily described to represent its integral properties on the physical logic level.

Thus, fractal description of a water body by its scalability/hierarchy stands alone as a separate coordinate pivotal to register changes in the system's properties at the level of its structure, [4].

Changes of fractal dimension values in time illustrate how the system uses its internal resources (natural potential) to adapt when influenced by random changes of external loads.

From this perspective, we can argue that the fractal is a flexible natural mechanism that helps highly involved natural systems with inherent structural heterogeneity achieve most effective distribution of matter and energy under non-equilibrium conditions through changes in their structure.

With every change of the environment's parameters, the system needs to choose the most efficient configuration for its interaction with the environment.

Among transitional form structures, state can be isolated with characteristic values of fractal dimensions D_o, D_d, and D_k, which were identified as markers that indicate changes of environmental conditions, [5, 6].

On this basis, the classification of environmental conditions considers the degree of disorder in the water body structure within the limits of its self-repairing ability. Generalized directed graph (digraph) of conditions shows physical orientation of structural phase transitions respective to identified characteristic fractal dimensions, as seen in Fig. 2.

The above classification suggests that ecological reclamation of a water body (as a set of environmental protection measures) is perceived as the system's forced structural transition from conditions M2 and M3 to conditions M1. The transition is forced by nature because in conditions M3 and M2 the ecosystem loses its self-repairing properties.

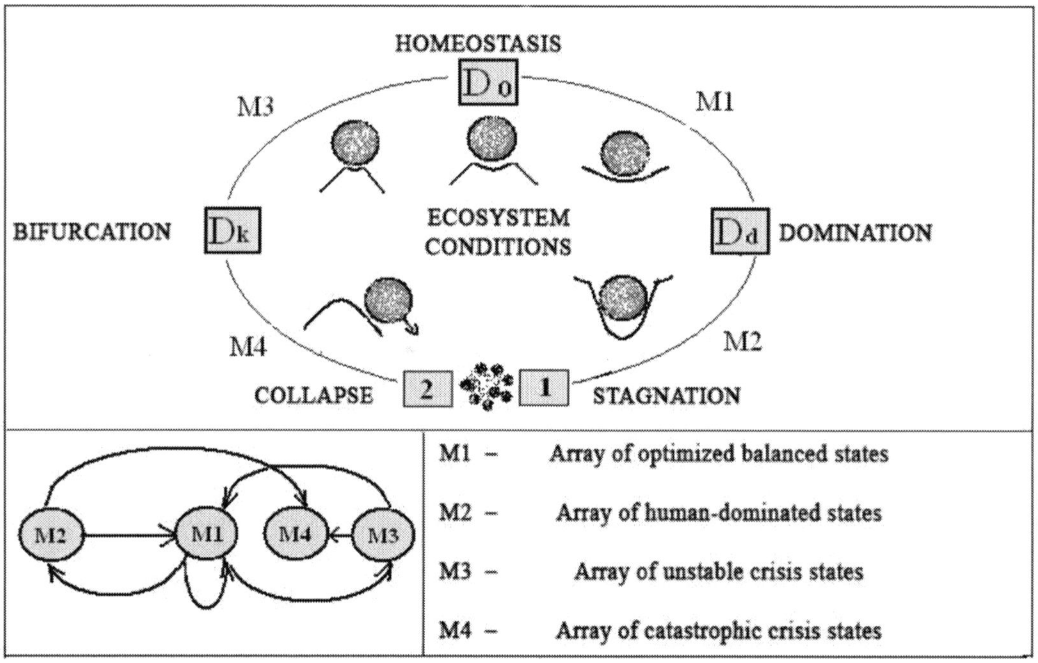

Fig. 2. Classification of water body conditions, and generalized digraph of structural phase transitions

The forced transition $M_3 \Rightarrow \mathrm{M}_1$ is generally achieved through reduction of anthropogenic activities in the list of engineering measures that enhance the adaptive capacity of the water body and thus localize the existing excessive anthropogenic load. If the system's resource capabilities to maintain M_1 condition are insufficient, the system can spontaneously (inadvertently) revert to its previous stable condition: $M_1 \Rightarrow \mathrm{M}_3$. Forced transitions $M_2 \Rightarrow \mathrm{M}_1$ are achieved by increasing system diversity (more intense interaction with the environment); this also contributes to better adaptation ability. Just as in the previous case, spontaneous phase transitions in the other direction $M_1 \Rightarrow \mathrm{M}_2$ also occur.

The spontaneous nature of phase transitions $M_1 \Rightarrow \mathrm{M}_3$ and $M_1 \Rightarrow \mathrm{M}_2$ and the forcible nature of reverse transitions are signes of comparatively low anthropogenic potential of conditions M_3 and M_2 compared to conditions M_1 that are more complex in terms the degrees of freedom and ensure optimized load distribution thanks to structure scaling.

Therefore, the use of fractal dimension in the classification registers both stable environmental conditions and their intermediate forms, which is impossible for conventional topological dimensioning (UCIWP). When the latter is used as a quantitative evaluation unit for conditions, this invariably results in loss of significant solutions.

In the particular case, characteristic fractal dimensions are found individually for each dynamic system with multifractal dynamics; however, studies have demonstrated that their size is rather common for complex dynamic systems of vastly different types (natural, biological, socioeconomic), [5].

See Table 2 for characteristic values of fractal dimensions of a water body ecosystem, and specifics of its dynamics.

Table 2. Characteristic values of fractal dimensions of a water body ecosystem, and specifics of its dynamics

Fractal dimension	Character of ecological condition	Specifics of system dynamics
$D_d = 1 - 1.2$	Human-dominated state, causing loss of self-repairing properties, stagnation and lack of inherent growth	System behavior is strictly regulated by unidirectional dominant factors of anthropogenic influences
$D_0 = 1.4 \pm 0.15$	Optimally balanced condition with the environment, no loss of self-repairing properties.	Steady and stable dynamics with gradual variations of system characteristics over a fairly narrow range
$D_k = 1.7 \pm 0.15$	Unstable state of crisis, loss of self-repairing properties, reduced diversity in the original ecosystem.	Unstable ecosystem dynamics, with abrupt leaps of system characteristics

In terms of mathematics, each subset of classified conditions $M_s = \langle \mathrm{M}_1 \cup \mathrm{M}_2 \cup \mathrm{M}_3 \cup \mathrm{M}_4 \rangle$ possesses isomorphism – a similar structure within a limited range of fractal dimension values that indicate persistence of specific properties/attributes of a water body interacting with the environment:

1. Optimality of environment: fractal dimensions of the water body focus in the interval (D_d – D_o) – a sign of self-regulating conditions where the environment structure corresponds to intensity of anthropogenic impact;
2. Criticality of environment: fractal dimensions of the water body focus in the interval (D_k – D_o) – a sign of severe manifest instability of conditions, while diversity in the original water body (partial "burning" of the environment) is reduced to compensate for preserved stability against excessive activity of human impact;
3. Stagnation of environment: fractal dimension of the water body focus in the interval (*1* – D_d) – a sign of stagnation and lack of evolution in the ecosystem, as degeneration with manifest anthropogenic domination;

4. Ecological disaster: fractal dimension of the water body focus in the interval ($D_k - 2$) – a sign of uncontrollable destruction/collapse of the water body ecosystem and transition to other qualitatively different ecological forms of existence.

As an analytical form is a complex description of the water body, we propose a ratio in which the entire array of ecological conditions $M_s = \langle M_1 \cup M_2 \cup M_3 \cup M_4 \rangle$ is described with a digraph that determines physical orientation of structural phase transitions:

$$M_s = \langle A, F, S, I^*, \Pi\ (F, S, I^*) \rangle \quad (1)$$

Where A is the object structure of a complex-structure system; F- submodel that presents the system's material properties or functions, showing the nature of exchange interaction with the environment over a certain period of observation; S- submodel to describe the structure of adaptation processes through manifestations of the system's scaling (scale-invariant) properties;

I^* - integrated structure of the water body in an active environment; $\Pi_f(SI^*)$ – limits to functional integrity of the water body.

Formally, the ratios are determined by the terms of F imaged against S to ensure integral distribution of adaptation potential $\Pi_f(I^*)$ relative to critical signs of fractal system violations.

In digraph (1), submodel (F) that represents essential properties or functions of the system that represents the nature of exchange interactions with the environment for a specific period of observation, is represented with time series of the anthropogenic load parameters in question, which are complex curves of irregular interaction between the water body and the environment. The structure of parameters is regulated by GOST 17.1.307-82 "Rules of Quality Control for Water Reservoirs and Ducts".

The submodel (S) that describes the structure of adaptation processes through manifestations of the system's scaling (scale-invariant) properties by its scalability coordinate, is expressed analytically by measuring the lengths of time series curves invariant to rescaling (fractal curves). A fractal curve in time interval $t \in [a,b]$ is defined as a continuous and differentiable curve, the length of which depends on the averaging scale, [7].

Fractal feature of the water body generates structural trends (gradients) of its adaptability, expressed numerically through fractal dimension.

Integrated structure of the water body in an active environment (I*) is graphically expressed through the multifractal phase portrait: a geometric image of the water body in the point of monitoring.

Limits to functional integrity of the water body [$\Pi_\Phi(SI^*)$] are set by the limits of self-recovery of its structure, and expressed through system coefficients of differential and integral stability.

To calculate coefficients of differential stability, the following ratios are used.

Differential rates of inertia

$$\Delta K_d = K_d(D_i^{t+1}) - K_d(D_i^t) \quad (2)$$

$$K_d(D_i^t) = D_i^t - D_d$$
$$K_d(D_i^{t+1}) = D_i^{t+1} - D_d$$

Where ΔK_d - differential coefficient of inertia hydrochemical indices; $K_d(D_i^t)$; $K_d(D_i^{t+1})$ - differential coefficients of stability of hydrochemical indicators for successive time intervals t and (t+1) respectively; D_d - fractal indicator of inertia (anthropogenic domination)

Differential rates of optimality:

$$\Delta K_0 = K_0(D_i^{t+1}) - K_0(D_i^t); \tag{3}$$

$$K_0(D_i^t) = D_i^t - D_0$$
$$K_0(D_i^{t+1}) = D_i^{t+1} - D_0$$

Where ΔK_0 - differential coefficient of optimal hydrochemical indicators; D_0 - fractal indicator of balance stability

Differential rates of criticality:

$$\Delta K_k = K_k(D_i^{t+1}) - K_k(D_i^t); \tag{4}$$

$$K_k(D_i^t) = D_i^t - D_k$$
$$K_k(D_i^{t+1}) = D_i^{t+1} - D_k$$

Where ΔK_k - differential coefficient of critical hydro-chemical indicators; D_k - fractal indicator of anthropogenic criticality.

To calculate integral coefficient of stability, we use the following ratios.

Integral coefficients of inertia:

$$\Delta K_d^n = K_d^n(D_i^{t+1}) - K_d^n(D_i^t) \tag{5}$$

$$K_d^n = \frac{\sum_1^n (D_i^t - D_d)}{n}$$
$$K_d^n(D_i^{t+1}) = \frac{\sum_1^n (D_i^{t+1} - D_d)}{n}$$

Where ΔK_d^n - integral coefficients of inertia by hydrochemical indicators; $K_d^n(D_i^t)$, $K_d^n(D_i^{t+1})$ - integral differential coefficients of inertia for successive points in time t and (t+1) respectively; n – number of factors (control components).

Integral coefficients of optimality:

$$\Delta K_0^n = K_0^n(D_i^{t+1}) - K_0^n(D_i^t) \tag{6}$$

$$K_0^n(D_i^t) = \frac{\sum_1^n \left(D_i^t - D_0\right)}{n}$$

$$K_0^n(D_i^{t+1}) = \frac{\sum_1^n \left(D_i^{c+1} - D_0\right)}{n},$$

Where ΔK_0^n - integral coefficients of optimality by hydrochemical indicators; $K_0^n(D_i^t)$, $K_0^n(D_i^{t+1})$ - integral differential coefficients of stability for successive points in time t and (t+1) respectively.

Integral coefficients of criticality:

$$\Delta K_k^n = K_k^n(D_i^{t+1}) - K_k^n(D_i^t) \tag{7}$$

$$K_k^n(D_i^t) = \frac{\sum_1^n \left(D_i^t - D_k\right)}{n}$$

$$K_k^n(D_i^{t+1}) = \frac{\sum_1^n \left(D_i^{c+1} - D_k\right)}{n},$$

Where ΔK_k^n - integral coefficients of criticality by hydrochemical indicators; $K_k^n(D_i^t)$, $K_k^n(D_i^{t+1})$ -integral differential coefficients of criticality for successive points in time t and (t+1) respectively.

Limits to structural stability $\Pi_f(F,S,I^*)$ are used to evaluate the resulting condition of the water body and the underlying anthropogenic load parameters.

In particular, the resulting ecological condition of a water body is evaluated by integral coefficients of stable condition through the following formulas:

$$M_{\text{рез}} = \min\langle \Pi_f(I)\rangle ; \langle \Pi_f(I)\rangle = \begin{matrix} K_d^n(D_i^t) & K_d^n(D_i^{t+1}) & \Delta K_d^n \\ K_0^n(D_i^t) & K_0^n(D_i^{t+1}) & \Delta K_0^n \\ K_k^n(D_k^t) & K_k^n(D_i^{t+1}) & \Delta K_k^n \end{matrix} \tag{8}$$

The physical sense of mathematical condition (8) is that the system remains in the subset of conditions $M_s = \langle M_1; M_2; M_3; M_4 \rangle$ [9], in which the system's adaptation potential is used in the most efficient way; this determined by proximity of the integral coefficient to one of such subsets that formally determine similarity of structural parameters (and therefore of the system's manifest properties). The sign of the differential coefficient points to either positive or negative dynamic that determines stability of the system's condition/properties [10].

The anthropogenic risk factors are measured by differential coefficients of stability to meet the condition:

$$F_r = \min\langle \Pi_f(S)\rangle ; \langle \Pi_f(S)\rangle = \begin{matrix} K_d(D_i^t); K_d(D_i^{t+1}); \Delta K_d \\ K_k(D_i^t); K_k(D_i^{t+1}); \Delta K_k \end{matrix} \tag{9}$$

The physical sense of mathematical condition (9) is that stability of the selected condition can be compromised by specific anthropogenic risk factors, each of which is determined by proximity of their differential coefficients to their critical values.

Existence of anthropogenic risk in the factor system can result in energy-inefficient integral redistribution of its adaptation potential, thus causing undesirable simplification of the system: the environment "burns out" or loses its ecological diversity.

The extreme of such simplification is when the ecosystem loses its fractal properties – this means transition from the fractal to the uniform mode, [8].

An integrated model formally describes the water body as an ecosystem with a non-uniform and non-equilibrium structure. The model's fractal component helps to register differences in the water body's adaptive responses over the range of hydrochemical parameters, thus helping to measure the development level of its structure. The more developed is the water body structure (hierarchical micro-levels of the system), the more manifest is its self-consistent behavior with respect to critical fluctuations of the environment: the stronger are the system's scaling/fractal properties S.

This is submodel that describes system by its scaling coordinate. It is analytically expressed by measured lengths of time series curves that are rescaling-invariant (fractal) curves.

In the time interval $t \in [a,b]$, a fractal curve is defined as a continuous undifferentiated curve of which the length depends on the averaging scale, [7]. Fractal property of the water body generates structural trends (gradients) of its adaptability, expressed numerically with fractal dimension.

In this way, one generates a set of differential descriptions of adaptive response systems and their limitations $<S;\Pi_f(S^*)>$ on the impact of respective pollution parameters.

Integral properties of the water body and their limits $<I^*;\Pi_f(I^*)>$ are described on the logical-physical level through a multifractal phase image that clearly represents the dynamics of redistribution of adaptation potential responding to changes of interaction with the environment.

Structural limits $\Pi_f(F,S,I^*)$ [11] help to quantify redistribution of the water body's adaptation resource (natural potential) responding to changes of anthropogenic activities, and to see how this redistribution brings the system closer to adverse environmental conditions or on the contrary maintains optimized balanced mode of interaction with the environment.

In this way, the method comprehensively evaluates the water body's ecological condition and highlights the risk factors that jeopardize it with their positive trend. Quantification of such factors helps to assess their critical impact on the system's structural balance.

Fractal description of the water body generates uniform descriptions of various processes of interaction with the environment that are different by their physical nature. The method can be used successfully to create descriptions of both biotic and hydrochemical parameters of the ecosystem; the descriptions can be later correlated.

The model clearly visualizes the dynamics of inclusion of new or exclusion of old factors of anthropogenic impact on the water body, and then can assess criticality of these inclusions. This feature of the integrated model is useful in natural environment management, planning and examination of efficiency of rehabilitation activities.

To match the integrated model (1), we propose the following generalized algorithm for Integrated Evaluation of Water Body Ecological Condition (IEWBEC) – see Figure 3.

DOI: 10.1051/matecconf/20168605022

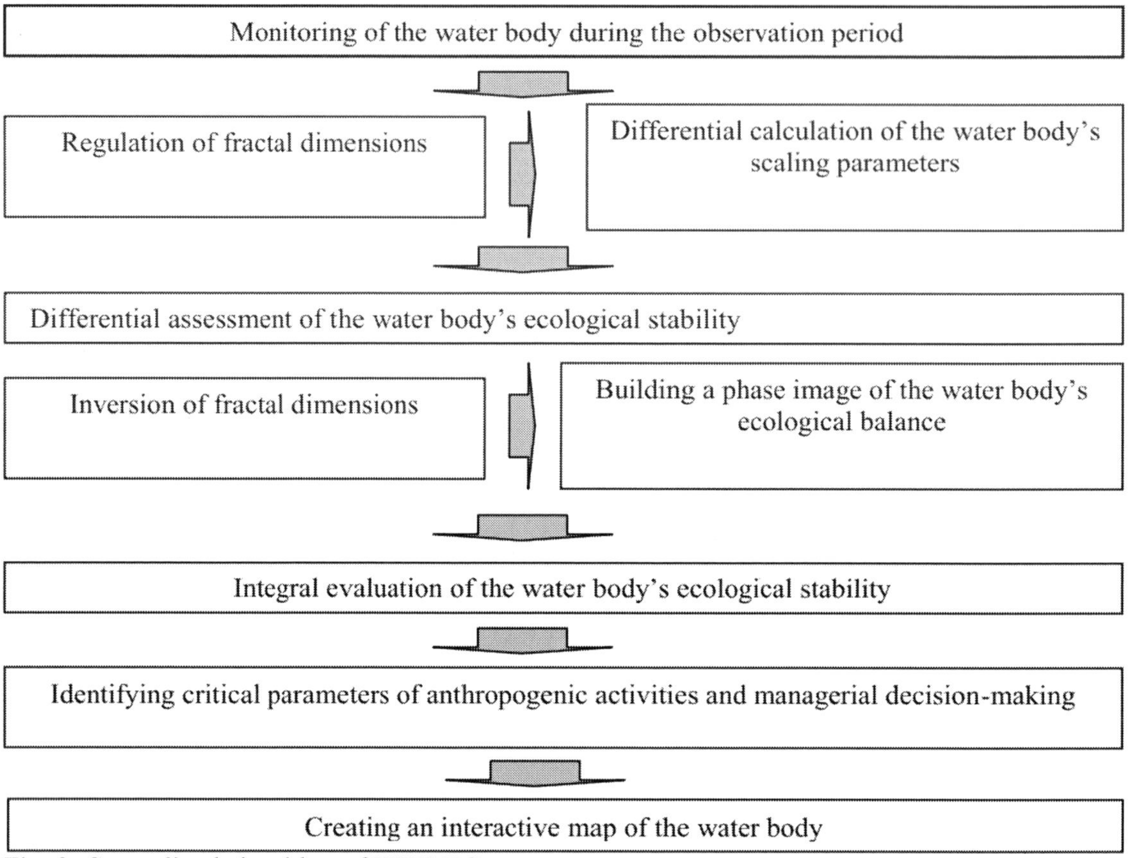

Fig. 3. Generalized algorithm of IEWBEC

References

1. RD 52.24.643-2002, Rosgidromet (2003)
2. D.I. Iudin, Bulletin of Samara Scientific Center under Russian Academy of Sciences **7**, 248-251 (2005)
3. V.N. Yakimov, L.A. Solncev, G.S. Rosenberg, D.I. Iudin, D.B. Gelashvili, Ontogenes **45**, 207-216 (2014)
4. A.D. Titov, F.I. Mavrikidi, Korolev SGAU **1**,127 (2011)
5. A.N. Kudinov, V.P. Tsvetkov, I.V. Tsvetkov, Russian Journal of Mathematical Physics **18**, 149-155 (2011)
6. A.N. Nasonov, I.V. Tsvetkov, A.V. Nikiforov, Prirodoobustroystvo Publishers **1**, 69-78 (2016)
7. A.N. Kudinov, O.I. Krylova, V.P. Tsvetkov, I.V. Tsvetkov, Mathematical Journal **5**, 52-60 (2014)
8. D.I. Iudin, A.V. Chechin, O.V. Kaschenko, Information and Cosmos Publishers **1**, 84-87 (2014)
9. V.I. Smetanin, A.N. Nasonov, I.V. Tsvetkov, Nelineyniy Mir **5**, 67-69 (2015)
10. A.N. Kudinov, D.Y. Lebedev, V.P. Tsvetkov, I.V. Tsvetkov, Mathematical Models and Computer Simulations **3** (2015)
11. I.V. Tsvetkov, O.I. Krilova, Software and Systems Publishers **4** (2015)

12. A.P. Ivanov, A.N. Kudinov, D.Y. Lebedev, Mathematical Models and Computer Simulations **1** (2016)
13. A.N. Kudinov, S.A. Mikheyev, I.V. Tsvetkov, Inseries: Applied Mathematics **1** (2014)
14. V.I. Smetanin, A.N. Nasonov, I.M. Zhogin, I.V. Tsvetkov, Prirodoobustroystvo Publishers **5** (2013)

MATEC Web of Conferences **86**, 05023 (2016) DOI: 10.1051/matecconf/20168605023
IPICSE-2016

Modelling of organizational and technological parameters of the informational model of municipal infrastructure facilities subject to reorganization

Olga Kuzina[1,*], *Evgeniy* Pankratov[2] and *Valentin* Tkachev[1]

[1]Moscow State University of Civil Engineering, 129337, Yaroslavskoye shosse, 26, Moscow, Russia
[2]Plekhanov Russian University of Economics, 117997, Stremyanny lane, 36, Moscow, Russia

Abstract. This article describes constructing methods of an information reorganization facility model based on parametric mathematical methods. The principles of reorganization parameter selection at the stage of municipal infrastructure facility pre-operation are described. The algorithm of information model building of facilities reorganization, as well as information model of organization processes of reorganization are being shown.

Introduction

The main goal of using the information modelling technology at the stage of facility reorganization is to collect data and documentation including all the resources and project solutions from the previous stages of the building lifecycle. According to this technology, information model in the process of building reconstruction is available for the analysis, and then for introduction of reasonable modifications [1].

A significant number of facilities in the municipal environment system are project-oriented, with the possibility to use the advantages of their parameterization [2-4]. Consequently, having undergone slight changes, they can be reused in similar projects involving historical sites reconstruction. Currently, most of the simulation information models of historical sites are created for the purposes of one project. It is necessary to extend the existing modelling technology using modern approaches and technologies (including software) to create more realistic and informative models [5].

Organizational and technological parameters of the information model of municipal infrastructure reorganization facilities will be the models of functional necessity in the facility reorganization, constructional wear, organizational and technological processes. The goal of an ideal reconstruction informational model is a workflow automation for reconstruction (or other types of remodelling) of buildings and constructions. This approach will allow to estimate explicit costs, as well as time and labour costs for buildings reconstruction as early as at the stage of facility pre-operation (figure 1) [6,7].

[*] Corresponding author: KuzinaON@mgsu.ru

Functional modelling of municipal infrastructure reorganization facilities allows to determine facility parameters and represent them in a form of the informational model, which reflects the processes of selecting the required organizational and process parameter. Mathematical modelling includes three interconnected stages: preparation of mathematical description of a facility under study (physical, regime, physicochemical and structural parameters); selection of a method of solving the mathematical descriptions equation system, and its implementation in a form of simulation program; compliance establishing (model adequacy to the facility) [8].

Creation of generalized building information model based on BIM-technology at the planning stage of a building lifecycle is the main tool to increase the efficiency of the organisational process of reconstruction (municipal environment reorganization). While creating the functional model of municipal infrastructure reorganization facility, it is possible to prepare a work description of the created system and analyse actuality of the upcoming improvements, modifications, amount of the necessary costs of all types of resources, as well as to most accurately coordinate work of subsystems. One of the common methods of solving this problem is a parametric design (or just parameterization) based on a model creation using parameters of model elements and correlation between these parameters [1, 6, 9].

All software systems involved in particular projects have their own model variants. To store the parameters of model elements, a database of geometric parameters of facilities; physical parameters of facilities; assigned (allocated) parameters of facilities should be formed. The parametric model of a building integrates three-dimensional model and elements behaviour model (edit history). All the construction documents are formed based on such informational model [10]. The model documentation is updated automatically when the smallest changes are being introduced. The agreed model modification reminds of the table cells change, the values in which are assigned by formulas. The formulas allow automating the calculations, and parametric modelling systems of the buildings automate the building documents acquisition.

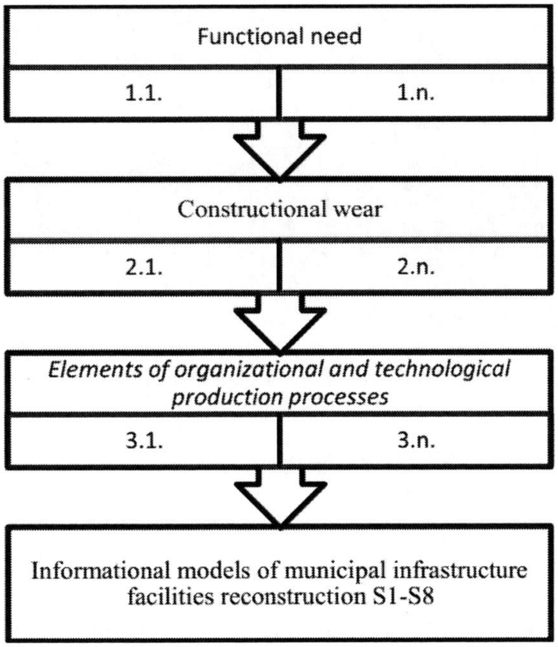

Fig.1. Algorithm of building the informational model of facilities reorganization

 DOI: 10.1051/matecconf/20168605023

In the process of hierarchic parameterization (parameterization based on scheme history), "scheme tree" is formed for all the model elements with their decomposition into elements (a group of elements – element). Variational, or dimensional, parameterization is based on sketching (with application to the facilities of a sketch of different parametric correlations) and application by the user of restrictions in the form of equation systems determining connections between the parameters [11].

Mathematical model of parametric design of generalized building information model is determined by: *Ni* elements set and $\{D1, \ldots, Dn\}$ set of design decisions.

$$Ni = (E, P, Vr, C, R, Pr, cf), \text{ where} \tag{1}$$

E – is a set of structure elements $\{e1, \ldots, en\}$;

P – is a set of parameters $\{p1, \ldots, pn\}$;

Vr – is a parameter values set $\{V1, \ldots, Vn\}$, where $Vi=\{vi1, \ldots, vin\}$ is each parameter value set;

C – is a value constraints set $\{c1, \ldots, cn\}$;

R – is a set of requirements for $\{r1, \ldots, rn\}$ model;

Pr – is a set of preferences $\{Pri, \ldots, Prj\}$;

cf – is a global cost function.

Ei structure element determines a primitive element of the design model. The *pi* parameter is a characteristic of the *i* structure element. Each parameter is connected with the range of *vi* values predetermined by a set of values assigned in *pi*. There are *m* possible values for the parameter and *n* parameters. The total amount of the design space is $N = m$*n. The values of functionally dependent parameters are uniquely determined by functional dependencies or requirements. Functionally independent parameters are called the key ones, and their values are determined by degrees of freedom in the design process, i.e. the actual size of the design space [12].

For each specific type of reorganization, informational model (Figure 2) of organizational processes, functional requirements and the level of constructional wear is formed [13].

S2 (Reconstruction)

1	1.1	1.2	1.3	1.4	1.5	1.6	1.7	1.8	1.9	1.10	1.11	1.12	1.13												
	2.1	2.2	2.3	2.4	2.5	2.6	2.7	2.8	2.9	2.10	2.11	2.12	2.13	2.14	2.15	2.16	2.17	2.18	2.19	2.20	2.21	2.22	2.23		
3	3.1	3.2	3.3	3.4	3.5	3.6	3.7	3.8	3.9	3.10	3.11	3.12	3.13	3.14	3.15	3.16	3.17	3.18	3.19	3.20	3.21	3.22	3.23	3.24	3.25
	3.26	3.27	3.28	3.29	3.30	3.31	3.32	3.33	3.34	3.35	3.36	3.37	3.38	3.39	3.40	3.41	3.42	3.43	3.44	3.45	3.46	3.47	3.48	3.49	3.50
	3.51	3.52	3.53	3.54	3.55	3.56	3.57	3.58	3.59	3.60	3.61	3.62	3.63	3.64	3.65	3.66	3.67	3.68	3.69	3.70	3.71	3.72	3.73	3.74	3.75

1	To choose functional needs
level	To choose structure damage
3	To choose organizational and technological production processes

Fig.2. Informational model of reconstruction organizational processes

Conclusion

BIM-Manager enters the required modification parameters to the building information model, determines the hit ratio of the selected parameters with the results received on levels. He gets chains – modules for each of the reorganization types, which allows preparing documents for achieving the objectives and needs of the modifications made.

Informational model of municipal infrastructure reorganization facilities based on BIM-technology includes all project lifecycle phases (taking into account operation process and possible building reconstruction) [14]. That is why the main task of BIM-Manager is to develop a working automated system with proven efficiency, which is required in the process of operation of a building and its systems. Such automated system should be based on the information received at the previous stages of the building lifecycle [15].

References

1. J. Tulke, M. Nour, K. Beucke, Proceedings of the 7th European Conference on Product and Process Modelling in the Building and related Industries (Sophia Antipolice, France, 2008)
2. Anatol'Evich, V.A., Vladimirovich, S.A., Dmitrievich, C.P., Andreevich, D.L., Valer'Evna, D.A. Int. Jour. of App. Eng. Res. **10**, 43269-43272 (2015)
3. A. Volkov, A. Sedov, P. Chelyshkov, E. Kulikova, App. Mech. and Mat. **584-586**, 761-764 (2014)
4. A. Volkov, A. Sedov, P. Chelyshkov, E. Kulikova, App. Mech. and Mat. **584-586**, 753-756 (2014).
5. Russian innovation development strategy for the period up to 2020 (approved by the Federal Government of 8 December 2011 № 2227-r "On approval of the Russian Federation Innovative Development Strategy for the period until 2020").
6. O.N. Kuzina. Forming the conceptual functional model of housing and communal infrastructure. Integration, Partnership and innovation in building science and education collection of materials of the International Scientific Conference.
7. A.A. Volkov, E.I. Batov, Vestnik MGSU, **10** (2015)
8. A.A. Volkov, Stroit. Mat., oboryd., tehnol. XXI veka (Constr. Mat., equip., tech. of the XXI cent.), **7** (2001)
9. A. Volkov, V. Chulkov, D. Korotkov, Adv. Res. Mat. **1065-1069 (**2014)
10. J. Underwood, U. Isikdag, *Handbook of Research on Building Information Modelling and Construction Informatics: Concepts and Technologies* (New York, Inf. Science Publ. 2009)
11. A.A. Volkov, S.N. Shulghenko, A.I. Shirokov, E.N Kulikova, Vestnik MGSU, **6** (2011)
12. K.A. Shreyber, *Variantnoe proektirovanie pri reconstrukcii jilih zdanii [Alternative design at reconstruction of residential buildings]. Seria Zhilishche - 2000* (Moscow, Stroyizdat, 1991)
13. O.N. Kuzina, V.O. Chulkov, PGS, **11** (2009)
14. V.O. Chulkov, Pereystroistvo. Organizacionno-antropotehnicheskaia nadejnost stroitelstva [Reorganization. Organizational and anthropotechnical reliability of construction]. Seria «Infograficheskie osnovi funkcionalnih system (IOFS)» [Series «Infografic bases of functional systems» (IOFS)]. Pod redakciei V.O. Chulkova [Under the editorship of V.O.Chulkov]. (SVR-ARGUS, 2005)
15. A.V. Guinsburg, D.A. Semernin, I.A. Lobireva, Nauch. Obozr. **16** (2015)

MATEC Web of Conferences **86**, 05024 (2016) DOI: 10.1051/matecconf/20168605024
IPICSE-2016

Strategy development management of Multimodal Transport Network

Natalia S. Nesterova[1,*], *Sergei M.* Goncharuk[1], *Vladimir A.* Anisimov[1], *Aleksandr V.* Anisimov[2]

[1]Department of Survey and Design Railways and automobile roads, Institute of Transport and Civil Engineering, Far Eastern State Transport University, 680021 Khabarovsk, Serysheva 47 , Russia
[2]Skolkovo Institute of Science and Technology, 143026 Skolkovo Innovation Center, Building 3, Moscow, Russia

Abstract. The article gives a brief overview of works on the development of transport infrastructure for multimodal transportation and integration of Russian transport system into the international transport corridors. The technology for control of the strategy, that changes shape and capacity of Multi-modal Transport Network (MTN), is considered as part of the methodology for designing and development of MTN. This technology allows to carry out strategic and operational management of the strategy implementation based on the use of the balanced scorecard.

1 Introduction

In the Transport Strategy of the Russian Federation [1] formulated the main objectives of the Russian transport system development for the period till 2030.

One of the most important tasks is to create globally competitive transport corridors for the successful integration of Russia into the global transport system that will contribute to the implementation of the transit and export potential of Russia, repaying investments.

To solve this problem in [2] proposes to allocate from the unified transport network (UTN) multimodal transportation network (MTN), as a set of multi-modal transport corridors (MTC), consisting of a multi-modal transport nodes and transport links of the various modes of transport, to address the strategic objectives of UTN development on the main transit and export routes.

2 Methods

* Corresponding author: mer-maid2@yandex.ru

Analysis of existing studies has shown that over the past decades numerous works about an effective development of transport infrastructure for multimodal system appeared [2-10 etc.].

[3] deals with the issues of multi-modal transport corridors formation, based on a systematic approach. In [4] one can find an example of investigating of formation questions using Novosibirsk multimodal transport node. [5] analyzes the history and current state of the Euro-Asian corridors, placed through the territory of Russia. A lot of attention is paid to the strategy of the creation of a core terminal complexes and logistics centers in major nodes and seaports. Moreover, the issues of ITC formation in Russia and ensuring the national security were investigated.

[2, 6, 7] deals with the integration of Russian transport system into international transport corridors. The favorable geographical position of Russia between the growing economies of Europe and Asia-Pacific Region gives a chance to the Russian transport to become a major trade link.

In [8] proposes a method of estimating the competitiveness of transport corridors using the method of the ideal point.

[9, 10] describes the methodology of addressing the formation problem, and then development problem of Trans-European Transport Network (TEN-T). Unfortunately, this approach does not take into account the problems and specifics of Russian transport and economy, mainly focusing on the objectives of the EU transport infrastructure.

The influence of external and internal factors on the formation of MTN and choosing its development strategy is considered in [11, 12]. The conceptual proposals for the development of "Asia-Pacific Region - Europe" ITC are also formulated in this work. Based on these, it was suggested the method of justifying and economic assessment of ITC projects.

In studies [13-19] proposes models, methods and techniques developed within the framework of a unified methodology for the MTN design. It allows you to determine the area of effective strategies that change MTN shape and capacity, and, finally, using this basis choose the optimal solution.

This article deals with the technology for management of accepted MTN developing strategy, ensuring the successful implementation of it. This technology is designed based on positions of strategic management [20-24], using the main universal management functions – planning, accounting, control and analysis – and includes the following [25]:

1) Planning the implementation of adopted MTN shape and capacity change strategy;
2) Organization of resource provision of finance, time, materials and contractors;
3) Monitoring performance of action plan and achieving the target indicators of the strategy [15];
4) Deviation analysis of actual quality indicators, cost and work time by monitoring of data (see item 3.);
5) Correction of the action plan for the development of MTN;
6) Definition of the regulatory technical, operational and economic performance of MTN taking into account corrected plan.

Implementation of the adopted strategy should be regularly analyzed in order to timely detect deviations from the scheduled plan and evaluate their impact on target indicators. Regular measurement of project parameters and identification of emerging deviations relates to performance monitoring.

According to the data collected during implementation monitoring, makes analysis for assessing of condition and prognosis of strategy execution success in accordance with the criteria and limits. These constraints are defined at the design and planning stage based on the Balanced Scorecard (BSC) [14,15]. The basic limitations include targets indicators, deadlines, quality and cost of the work. In the case of negative prognosis, we need to make corrective action, the choice of which is carried out in the change management process.

The main processes of analysis are directly related to the above restrictions

1) Timing analysis - determination of conformity of actual and projected deadlines;
2) Cost analysis - determination of conformity of actual and forecasted cost;
3) Quality analysis - monitoring results to check on compliance with relevant quality standards and identifying ways to eliminate causes of the declining quality indicators values;
4) Confirmation of the purposes - acceptance of the contribution of participants (investors, consumers, etc...) in the implementation results.

According to the analysis of the results, makes the decision to continue previously scheduled plan or determine corrective actions.

Determination and application of the necessary control actions with a view to the successful implementation involve the following steps:

1) If the execution of the strategy is in accordance with the agreed plan, the management actually need to deliver to the participants their targets and monitor their implementation;
2) If in the process of implementation any crucial deviations appear, it is required to find the optimal corrective action, adjust the plan of remaining work and to agree on the pending changes with all stakeholders.

At this stage the main management processes are:

1) General change management – definition, coordination, approval and acceptance to perform corrective actions and coordination of changes in strategy;
2) Resource management - changes in the composition and the assignment of the resources;
3) Management of objectives - adjustment of target indicators using analysis results;
4) Quality management - development of measures to address the causes of unsatisfactory execution.

Thus, the technology of strategy development management based on BSC [14,15] allows to efficiently manage the trajectory of changing of MTN shape and capacity. Along with that, it is made in accordance with the overall objectives and taking into account changes in the implementation conditions throughout the estimated period. The practical use of the proposed technology allows solving following problems:

1) Support of the action plan implementation;
2) Effects analysis of the plan implementation for the mtn development and its subsequent adjustment for the further period.

At the same time a prerequisite for the successful implementation of the strategy is its support of the expert group, which should include:

- MTN strategy developer;
- Specialists in modes of transport, which are part of the ITC;
- Specialists in the organization of multimodal transportation and logistics services;
- Representatives from customs control;
- Experts in foreign economic activity and international communication;
- Investment managers;
- Representatives of administrative structures and legislative power from areas through which ITC goes;
- Ecologists;
- Specialists in regulatory support and standardization.

One of the coordinators of the expert group should be the strategy developer, who has access to all sources of information provided to MTN designing.

The expert group is involved in the development of the investment project on the chosen strategy and coordinates the actions of all participants of the project.

3 Conclusion

Proposed technology of MTN strategy development management is the final element of the designing methodology. It allows to identify possible deviations from specified purposes [15] in the implementation of the strategy caused by the influence of external and internal factors [11,18,22,26], evaluate the variation data and make the necessary adjustments.

The external factors include:

1. The political environment in the countries of the region under study (change of power, war, sanctions, etc.).;
2. Economic conditions (market conditions, financial and credit policies, the economic interests of the participating member-states and private investors);
3. Infrastructure links between the participants of the strategy, that determines the conditions of rational distribution of productive forces, environmental management, environmental protection and others.
4. Social conditions (demographics, the level of employment, etc.).;
5. Natural and climatic disasters;
6. Technological conditions (the development and spread of new technologies).
7. The internal factors include:
8. Technical (changing the technical condition of the mtn due to failure of its elements);
9. Economic (financial instability, capital intensity growth in the construction process due to unaccounted factors associated with the uncertainty of the source data);
10. Availability of human resources and their level of qualification.

References

1. Transport Strategy of the Russian Federation for the Period Through to 2030 **1734**, 22 (2008)
2. S.M. Goncharuk, V.A. Anisimov, N.S. Nesterova, N.A. Lebedeva, Izdatelstvo DVGUPS, **227** (2012)
3. V.I. Galakhov, B.A. Lyovin, V.N. Morozov, V.V. Shashkin, Izdatelstvo Transport, **71** (2001)
4. K.L. Komarov, S.A. Maximov, Izdatelstvo SGUPS **5**, 43 (2003)
5. S.M. Rezer, T.A. Prokofieva, S.S. Goncharenko, Russian academician of sciences, **312** (2010)
6. S.G. Kovalev, Supplement to the World of transport Journal **1(9)**, 62-67 (2015)
7. D.V. Shapoval, Technique and technology of construction **1(5)**, 8 (2016)
8. Y.A. Bykov, A.B. Vasilyev, V.A. Fadeeva, Transport of the Urals **1(44)**, 53-56 (2015)
9. Infrastructure – TEN-T – Connecting Europe, http://ec.europa.eu/transport/themes/ infrastructure/ studies/ten_t_en.htm (2016)
10. New EU transport infrastructure policy – background, http:// europa.eu/rapid/press-release_MEMO-14-525_en.htm (2016)
11. N.S. Nesterova, V.A. Anisimov, The International Forum on New Industrialization Development in Big-data Era, **460-463** (2015)
12. Vl. A. Anisimov, N.S. Nesterova, N.V. Bondarenko, S.M. Goncharuk, Modern technologies **1(49)**, 166-170 (2016)
13. V.A. Anisimov, M.V. Nechiporuk, Izvestiya PGUPS **3(40)**, 9-15 (2014)
14. M.V. Nechiporuk, V.A. Anisimov, Transport of the Urals **3(42)**, 13-17 (2014)

15. N.S. Nesterova, V.A. Anisimov, Izvestiya PGUPS **13**, 197-205 (2016)
16. N.S. Nesterova, Volga Region Transport Herald **1**, 66-74 (2016)
17. N.S. Nesterova, Modern technologies **2(50)**, 173-179 (2016)
18. N.S. Nesterova, Modern technologies **3(51)**, 146-153 (2016)
19. N.S. Nesterova, V.A. Anisimov, Transport of the Urals **2(49)**, 26-29 (2016)
20. V.D. Markova, *Strategic Management* (Izdatelstvo INFRA-M, Novosibirsk, 1999)
21. A.N. Petrov, L.G. Demidova, G.A. Burenina, S.M. Klimov, G.Z. Scherbakovsky, G.M. Gvichiya, M.V. Vlasova, *Strategic Management* (Izdatelstvo Piter, Saint Petersburg, 2005)
22. A.T. Zub, *Strategic Management. Theory and Practice* (Izdatelstvo Aspect Press, Moscow, 2002)
23. J.L. Thompson, *Strategic Management: Awareness and Change* (Chapman and Hall, London, 1990.
24. G. Jonson, K. Scholes, *Exploring Corporate Strategy. An Aproach to Strategic Management* (Pitman, London, 1992)
25. V.A. Anisimov, *Theory and practice of design of the development of a regional network of Railways taking into account change of shape and capacity of stations and hubs* (DVGUPS, Khabarovsk, 2005)
26. N.S. Nesterova, *Decision making to change railway direction capacity in terms of the uncertainty of its technical condition* (DVGUPS, Khabarovsk, 2003)

Cost Reduction Ability by Electricity Tariff Selection for Construction Facilities Located in Non-price Areas

Svetlana Makasheva[1,*] and *Pavel* Pinchukov[1]

[1]Far Eastern State Transport University, Seryshev st. 47, 680021, Khabarovsk, Russia

Abstract. State of the art in the tariff type selection in the retail and the wholesale electricity market for construction or reconstruction facilities located at Far East are considered. As a management solution the electricity tariff's selection for enterprises at stage of construction or reconstruction is accented.

1 Introduction

The Far Eastern Region is one of two non-price areas with regulated wholesale electricity and power market, which has intensive construction growth by strategy of priority development areas [1-3]. Where industrial, agricultural, tourism and sports facilities of construction and reconstruction are situated [2].

Construction and start enterprises growth is accompanied with increase in electricity consumption. As a result, the electricity consumption in the Far Eastern Region is approximately 5% of total electricity consumption in Russia [3].

According to the amendments, which adopted by Resolution of the Government of the Russian Federation on May 17, 2016 № 433, the ability of selection one of the 6 types of electricity tariff category on the retail electricity market is provided for consumers. At the same time special tariff conditions wholesale electricity market are approved for consumers with capacity from 670 kVA to 10 MVA, which are located in non-price areas [4]. Therefore, the electricity tariff's selection for enterprise should be considered as a management solution in the facilities construction or reconstruction.

2 State of the art inn the tariff type selection in the retail electricity market

Since 2006, a part of electricity sold to the consumers (except population) at non-regulated, free market prices, was determined on the basis of the market's demand and supply in the majority of regions in Russia [3-8]. Nevertheless, the Far Eastern Region including the Primorsky Territory, the Khabarovsk Territory, the Amur region and the south of Yakutia is one of two special non-price areas categories of the wholesale electricity market [9]. In

* Corresponding author: dee@festu.khv.ru

contrast to the central part of Russia there are no free market electricity prices in non-price areas. Competition in the wholesale electricity and power market among generating stations is not possible because there are only two generating company in the Far East. It is the RusHydro [5] and the JSC «Far East Energetic Company» (JSC «FEEC») [6].

Resolution of the Russian Government defined the unified buyer and the unified guaranteeing supplier of electricity for the Far Eastern Region [1,5,6]. This status is assigned to JSC «FEEC». Purchase of electricity on the wholesale market has been allowed only for JSC «FEEC». That is why it is called «the unified buyer». JSC «FEEC» sells electricity to consumers not only in wholesale market, but in the retail market too. However. because of absence of free market electricity prices in the Far Eastern wholesale electricity market, there are none of them in the retail markets too.

All the electrical energy in the Far Eastern Region is bought and sold on the set rates. Rates are set by the state. Absence of competition rules destroys the possibility of free choice of electricity supplier. This situation doesn't permit to reduce the cost of electricity to consumers by using free market prices, but there is a possibility of selecting one of 6 tariff categories in the retail electricity market. Possibility of selecting electricity price category in the retail market for consumers depends on the following factors: (Figure 1):

1. Consumer's power agreed upon in the contract with the power supplier.
2. Consumer's absence or presence of electricity integral hourly metering system.
3. Possibility of electricity consumption planning.

The way to be more effective on the retail power market for consumers includes creation of an automatic metering system and electricity consumption planning. But the electricity consumer has a choise even there is no integral metering system or consumption planning system. In other words, the consumer can to select one-part tariff or two-part (differentiated or zone) tariff. For example, if the electricity consumption is evenly without large fluctuations throughout the day, the consumer has more benefits through one-part tariff because of simple and quick calculation and no need of additional equipment.
In contrast, if the electricity consumption is evenly without large fluctuations throughout the day the consumer has more benefits in one-part tariff.

A considerable difference the consumption of electricity during the day-time and night-time (more then 30-40%) determine the desirability of application two-part (differentiated) tariff, which is lower than the one-part tariff. For example, two-part tariff more beneficial if the electrical energy consumption of the enterprise in the evening and night-time increases as compared to the day-time by 30-40%. In this case, cost of electricity for the night hours is below [from 2.7 RUR/1 kW*h to 1.55 RUR/1 kW*h] [6].

This approach also gives consumers the opportunity to change their hour high consumption at the stage of designing the operating mode of enterprises, sliding them to time with lower cost of electrical energy. It should also be noted that 1 and 2th tariff category are integral. These rates apply to the total volume of electricity consumption per month or per day zone. This is the price for the total electricity (or power) volume during this whoul time period.

At the same time 3,4,5 and 6th tariff categories are interval. It means that the fixed set price to be applied to the different hourly volumes of consumption and the price is different for each hour [6-8, 10]. 3,4,5 and 6th tariff categories provide interval different fixed price for electricity for each hour of the day by the Law of the Russian Federation [1]. Meaning that cost of electricity to set for each hour of the day each month for each tariff category. It considerably complicates the calculation of the final efficiency of use different tariff categories for a particular company, because it requires time and expertise in this sphere. In addition, modern measurement systems are rather expensive and require considerable installation costs and installation [11-18].

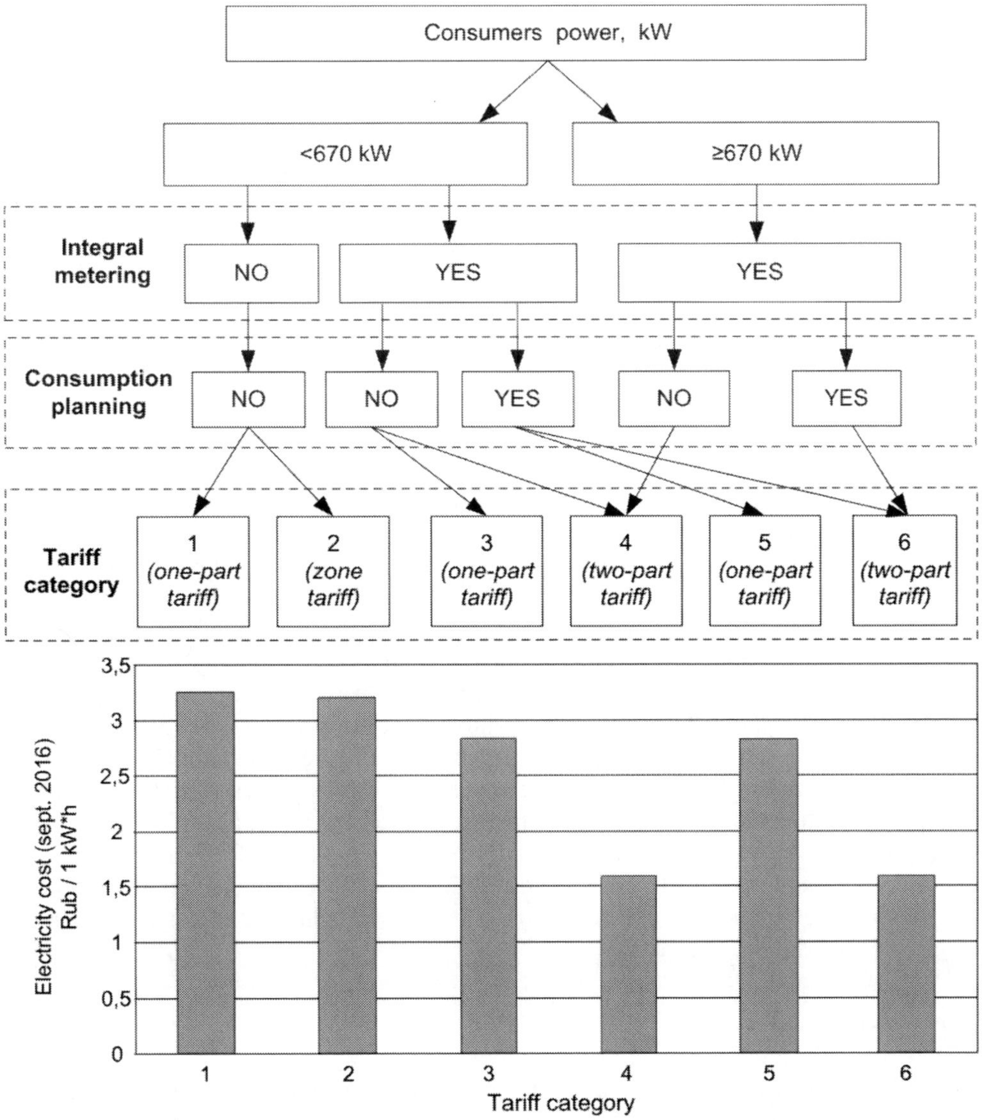

Fig. 1. Principles of selection price category in the electricity retail market in non-price areas.

Nevertheless, there is a positive experience of transition companies of Far Eastern region to new tariffs categories of retail market. Transition has the payback period from 3 to 5 years. It depends on specific conditions of the enterprise – maximum power and electricity (or power) consumption. The accuracy corresponds to the consumption of electric energy to the volume, declared at the stage of its forecast, is also important. According to available statistics, these factors allow large retail electricity consumers to reduce electricity cost for 5-10 percent. This significantly reduces the cost of their products [2,6,19-22].

3 State of the art inn the tariff type selection in the wholesale electricity market

However, there is another way to reduce electricity cost by electricity tariff selection. Starting 07.01.2016 the company with a power capacity more than 670 kVA can buy electricity from the wholesale market in wholesale price. But in fact this way is not as simple as it looks. Entering and working on the wholesale electricity market can only residents of this market. In the wholesale market of electric energy purchase is carried at wholesale prices set by the state. But the cost of electric energy on wholesale market is naturally lower than the retail market [2,4,6,7,].

The development strategy of the JSC «Energy Systems of East» involves the future creation the technological conditions and the same market relations like in price areas of the wholesale electric power market of Russia for introduction them in the non-price areas of the electricity wholesale market [2,6].

However, company wishing to buy electricity on the wholesale market can't do it by itself, if it is not a resident of that market. Become a resident of wholesale market is expensive on time and need administrative and financial investments. This option is often not necessary to company whose main activity is not the generation or sale of electricity. Therefore, the company can buy electricity from the wholesale market for the wholesale price itself does not directly, but through an intermediary who is a resident of the market.

There are three types of entities in the wholesale market of electric energy and power, which interact with each other: sellers, buyers, infrastructure organizations. In the wholesale market of electric energy trade is conducted in three sectors, which differ in terms of buying and sales transactions and terms of delivery of electric energy. These sectors include: day-ahead market, the balancing market, regulated by contract [2,3,7,8,10].

Such purchase from the wholesale market, even with the fixed allowance intermediary-resident market for the sale of electric energy, allows the company to reduce production costs by 30%. The non-price zone of Far East has experience of large companies in the wholesale market entry. This experience shows that the process of coordination and entry on the wholesale market of electric energy takes more than six months, often up to two years. Such complexity and duration of the entrance procedures to the wholesale market stopped more than half of companies willing to implement it [2,6].

As a whole, knowledge of ways to reduce the payment for electricity is very important at the design stage and construction of the facility. The necessity change in metering system in operation will require considerable investments, up to changes in design and engineering systems of buildings and structures.

4 Conclusion

Despite the availability of several cost reduction ability by electricity tariff selection the choice of the optimal variant is unique in each case and is accompanied by a lengthy review and approval procedures, which reduces its attractiveness for consumers.

Therefore, think about the electricity tariff selection for new or existing enterprise is desirable as soon as possible - for example, at the construction stage or even at the design stage.

References

1. Official Internet Resources of the President of Russia, http://en.kremlin.ru/acts/news/47349
2. Ministry of Far East Development, http://minvr.ru/press-center/news_minvostok/?ELEMENT_ID=4690
3. M. Bystrov. *Electricity and Capacity Market: Today and Tomorrow*
4. Association «NP Market Council», http://www.en.np-sr.ru/
5. RusHydro, http://www.eng.rushydro.ru/

6. JSC «FEEC», http://www.dvec.ru/electric_market/
7. RAO «Energy Systems of the Eas», http://www.rao-esv.ru/en/
8. A. Wasilewski, *The Russian energy sector: The way for future development* (AAP, Moscow, 2015)
9. L. Chaika, Ec. and Soc.Ch: Facts, Trends, Forecast **2,** 104-115 (2015)
10. Inter RAO UES, http://www.interrao.ru/en/activity/traiding/tradesb.php
11. J. Torriti, Energy **44**, 576 (2012)
12. P. Lombardi, T. Sokolnikova, K. Suslov, N. Voropai, Z.A. Styczynski, Renewable Energy **90**, 532–541 (2016)
13. A. Kirillova, S. Beliakov, A. Krylova, MATEC Web of Conferences **73**, 07009 (2016)
14. I. Daminov, E. Tarasova, E. Gladkikh, T. Andreeva, MATEC Web of Conferences **37**, 01017 (2015)
15. M. Doostizadeh, H. Ghasemi, Energy **46** (1), 221 (2012)
16. P. Faria , Z. Vale, Energy **36**, 5374 (2011)
17. H. Allcott, Resource and Energy Economics **33**, 820 (2011)
18. Ipsos MORI, http://en-mart.com/energo-consultant-en/
19. Energomart. National Power Sales Company, http://en-mart.com/energo-consultant-en/
20. Kyrov Energy Supply company, http://www.eskzgroup.ru/en/energy-supply/
21. JSC « Komi Energy Retail Company», http://www.komiesc.ru/legal_entities/tariffs/
22. http://ufo-stroy.ru/vyibor-varianta-tarifa-na-elektroenergiyu-tsenovoy-kategorii.html

MATEC Web of Conferences **86**, 05026 (2016) DOI: 10.1051/matecconf/20168605026
IPICSE-2016

Methodology for analysing capacity and level of service for signalized intersections (HCM 2000)

Ivana Nedevska[1], *Slobodan* Ognjenović[2,*] and *Elena* Gusakova[3]

[1]GEING Krebs und Kiefer-Skopje, Macedonia
[2]University "Ss. Cyril and Methodius", Faculty of Civil Engineering – Skopje, Macedonia
[3]Moscow State University of Civil Engineering

Abstract. The augmentation of motorization level leads us to the need for mobility and demands better infrastructure, in urban and suburban areas. The complexity of this problem is esspecialy notable in urban areas where the need for mobility must be considered, as well as the integration of infrastructure network in the urban plan of the city. Besides the space delimitations, functional characteristics and space demands connected with different transportation in cities.
This methodology analyses the needs of the urban traffic network in highly populated areas, where space limitation and movements in peak hours must be considered. That is why the analysis of capacity and level of service is key part of the urban traffic planning.
Highway Capacity Manual gives the opportunity to analyse the capacity and level of service of the roads in urban or rural areas, by defining the delay of the analysed facilities. To obtain these information, we must have data for geometric, traffic and signalization parameters, when the intersection is signalized.

1 Introduction

In urban areas, the flow is analysed as interrupted, because of the signalisation, pedestrians or bicycles. With these interruptions, the moving time for traffic participants increases. Bigger moving time leads to bigger delay, which is the main indicator for capacity and level of service (LOS).

If the delay has lower value, the LOS is better, and if the delay has higher value the LOS decreases.

In order to obtain these information, there must be information for geometric, traffic and signalization parameters.

First step in determining the capacity and LOS is Lane Grouping. This means grouping the lanes by the movements (left, right turns and through movements).

* Corresponding author: ognjenovic@gf.ukim.edu.mk

2 Methodology

Each lane group is analysed separately. There are two approaches for this capacity analysis. The first one allows quick estimation for one cycle length and green times, in order to constitute a signal timing plan. The second estimation procedure is more complicated and provides the basis for the design of signal plan that equalize the degree of saturation on the critical approach for each phase.

The first approach allows quick analyses with minimal input field data, which makes it very simple to use, considering the default values for the required traffic and control parameters.

This methodology includes few steps:

- Lane grouping and Demand flow rate
- Saturation flow rate
- Capacity and v/c
- Performance measures

Capacity is defined by the saturation flow and saturation flow rate, and the LOS is defined by control delay per vehicle. Signalized intersection methodology is presented in figure 1.

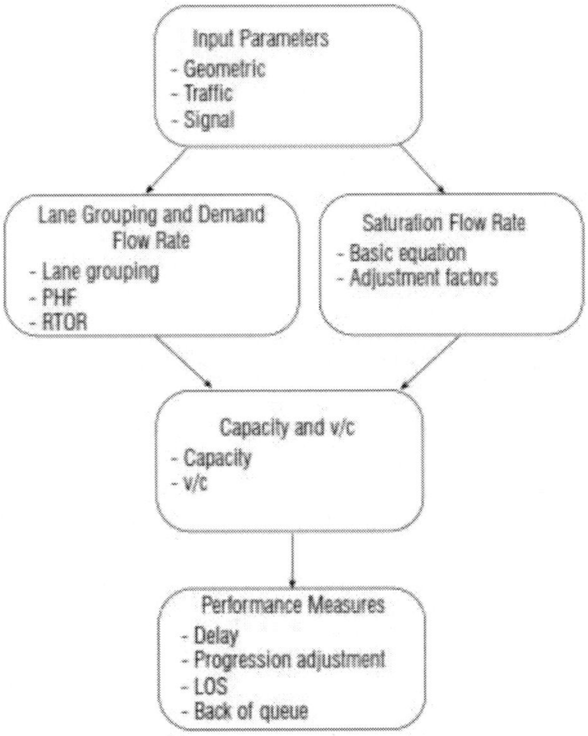

Fig. 1. Signalized intersection methodology.

2.1 Input parameters

In order to conduct an analysis for signalized intersection we must provide some input parameters, such as:

a) Geometric conditions:

- Area type

- Number of lanes N,
- Average lane width W (m),
- Grade G (%),
- Existence of exclusive LT or RT lanes,
- Length of storage bay LT or RT lane, L_s (m),
- Parking

b) Traffic conditions

- Demand volume by movement V (veh/h),
- Base saturation flow rate s_0 (pc/h/ln),
- Peak-hour factor PHF,
- Percent heavy vehicles, HV (%),
- Approach pedestrian flow rate v_{ped} (p/h),
- Local buses stopping at intersection N_a (buses/h),
- Parking activity N_m (manoeuvres/h),
- Arrival type AT,
- Portion of vehicles arriving on green P,
- Approach speed S_A (km/h)

c) Signalization conditions

- Cycle length C (s),
- Green time G (s),
- Yellow – plus – all – red change – and – clearance interval (intergreen) Y (s),
- Actuated or pretimed operation,
- Pedestrian push button,
- Minimum pedestrian green G_p (s),
- Phase plan,
- Analysis period T (h)

2.2 Lane grouping

First step in this analysis, after providing input data, is lane grouping, for each approach separately. Lane grouping is very simple procedure, depending from geometric conditions and traffic movements distribution. The lane group describes the manoeuvres in the intersection area:

- Lanes for left turns should be considered separately, unless there is lane for left – through shared lane. In this case, portion of left turns should be computed.
- When there are exclusive left and right – turn lanes, the other lanes go in one lane group.
- If the lane is used for both left – turning and through vehicles, the portion of left turns will determine whether the lane will be considered as exclusive left lane.

2.3 Determining flow rate

Traffic demand is determined by average flow in analysed period (usually 15 minutes). Flow rate during peak 15 minutes period is computed with following equation, using the hourly volume (veh/h) and peak – hour factor.

$$v_p = \frac{V}{PHF} \tag{1}$$

if right turn is allowed on red light, this volume should be corrected by the traffic volume of right turns on red light.

2.4 Determining saturation flow rate

Saturation flow rate should be determined (computed) for each lane group, and it is considered to be the flow in veh/h in a lane group, assuming the green phase is displayed 100% of the time.

$$s = s_0 \, N \, f_w \, f_{tv} \, f_i \, f_p \, f_{bb} \, f_a \, f_{LU} \, f_{LS} f_{RS} f_{Lpb} \, f_{Rpb} \tag{2}$$

Besides the base saturation flow rate per lane (s_0) and the number of lanes (N), there are included some adjustment factors for:

- Lane width f_w,
- Heavy vehicles in traffic stream f_{HV},
- Approach grade f_g,
- Existence of a parking lane and parking activity f_p,
- Blocking effect of local buses in intersection area f_{bb},
- Area type f_a,
- Lane utilization f_{LU},
- Left turns in lane group f_{LT},
- Right turns in lane group f_{RT},
- Pedestrian and bicycle factor for left and right – turn movements f_{Lbp} and f_{Rbp}.

2.5 Capacity and v/c ratio

a) Capacity

Capacity of signalized intersection depends on saturation flow and saturation flow rate. The capacity, for each lane group, can be calculated by the following equation:

$$c_i = s_i \frac{g_i}{C} \tag{3}$$

Where:
s_i – saturation flow rate for lane group i,
g_i/C – effective green ratio for lane group i.

b) v/c Ratio

The volume to capacity ratio is presented by the symbol X and is referred to as a degree of saturation.

$$X_i = \left(\frac{V}{C}\right)_i = \frac{Vi}{Si\left(\frac{gi}{C}\right)} = \frac{ViC}{Sigi} \tag{4}$$

In the previous two equations, few inputs are inevitable:
v_i – demand flow rate for lane group,
s_i – saturation flow rate for lane group,
g_i – effective green time for lane group,
C – cycle length.

If the value of X = 1, the flow rate equals capacity. Demand excesses capacity if the value of X is above 1.

c) Critical lane groups

When analysing an intersection, it is very important to determine the lane group with highest v/s ratio. This is the lane group that demands longer green light. This means, each phase will have one critical lane group. The critical Xc is calculated with this equation:

$$X_c = \sum\left(\frac{v}{C}\right)_{ci}\left(\frac{C}{C-L}\right) \quad (5)$$

In this equation L represents the total lost time per cycle.

2.6 Determining delay

Values given by the calculated delay represent the average control delay of all vehicles that arrive in the analysed period. The average control delay per vehicle is calculated as:

$$d = d_1(PF) + d_2 + d_3 \quad (6)$$

Where:
d_1 – uniform control delay assuming uniform arrivals (s/veh),
PF – uniform delay progression adjustment factor,
d_2 – incremental delay for random arrivals and oversaturation queues, adjusted for duration of analysis period and type of signal control. This delay assumes that there is no initial queue for lane group at start of analyzed period
d_3 – initial queue delay for all vehicles in analyzed period. This delay appears due to initial queue at start of analyzed period.

2.7 Determining level of service (LOS)

Level of service of an intersection is connected to the average control delay per vehicle. By the value of the delay, we can determine the LOS, according to the following table:

Table 1. LOS for signalized intersection.

Level of service	**Control of delays of the vehicles (s/veh)**
A	≤ 10
B	> 10-20
C	> 20-35
D	> 35-55
E	>55-80
F	>80

With this method we provide two key outputs:

- v/c ratio for each lane group and for every critical lane group at the intersection

- Average control delay for each lane group, and each approach as well as the intersection as a whole, with appropriate LOS.

Values of v/c ratio higher than 1.0 indicate to actual or potential breakdown. In this case multiple period analyses are recommended. The v/c ratio is bigger then 1.0 usualy as result of inadequate capacity because of signal and geometric characteristics.

In some cases, v/c could be less than 1.0 for the intersection, but v/c for some lane groups can be higher than 1.0. This means that the green light is not appropriately apportioned and the existing phasing should be attempted.

There might be cases where v/c is less than 1.0, but the delay is high because of inappropriately long cycle length, poor progression or both.

When v/c is around 1.0 or exceeds this value, the delay may stay in acceptable limits if the time frame where v/c has high value is short.

3 Conclusions

The planning process is inevitable in city development, and it is connected with the designing process. Traffic volume and geometry are very important in order to achieve the needed capacity and level of service of every street or intersection as a whole. In future, with industry and technology development, the planning process will become more complexed. Therefore, the problem with capacity and level of service should be solved, first in the area of intersection, and then at the approaches. If the offered solution for the intersection does not accomplish the needs, another intersection design should be analyzed.

If none of the solutions resolves the problem with capacity and level of service, some changes in the traffic organization should be proposed, such as:

- To transfer some of the vehicles to the existing traffic net,
- Vehicles of public city transportation should have exclusive lane, which would decrease their influence to the other vehicles,
- The city center may be closed for vehicles, only pedestrians and bicycles should use these streets.

Every important intersection in the city should be analyzed, in order to determine the capacity and level of service.

References

1. R. Bankovich, *Planning of public urban passenger transport, Construction book* (Belgrade, 1984)
2. M. Bublin, *Planning movement and its proximity to the main roads* (Univerzitetsko graduation from the Construction faculty in Sarajevo, Sarajevo, 2007)
3. *Board of international research, Standards on geometric features and traffic intersections in urban road* (Rome)
4. C.R.O.W., Recommendations for traffic provisions in built – up areas, C.R.O.W., (The Netherlands, 1998)
5. R. Doncheva, *Urban sobral* (Faculty of civil engineering, Skopje, 2013)
6. M. Maletin, *City roads* (Faculty of civil engineering, Belgrade, 1996)
7. M. Maletin, *Planning and design due to the proximity to the main roads in cities* (Faculty of civil engineering, Belgrade, 2005)
8. M. Maletin, *Planning, movement and space* (Faculty of civil engineering, Belgrade, 2004)
9. NSRA, *Guidelines for the geometric design of urban roads – rigu73* (National Swedish Road Administration, Stockholm, 1973)

10. *Transportation research board, Highway capacity manual 2000* (National Research Council 2000)
11. *Transportation Research Board, Highway capacity manual 2010* (National Research Council, 2010)

Construction of detours of rail nodes, as an effective way to increase the throughput

Denis Kuklev[1], *Natalya* Kukleva[1,*]

[1]Far Eastern State Transport University (FESTU), 680021 Khabarovsk, Russia

Abstract. In the article the question of the justification of expediency of the construction of bypasses junctions, intended for the admission of transit freight trains. According to the authors, on the basis of the most important factors, were considered the possible combinations of the length of the bypass and the intra-site the main ways. On the example of the conventional rail node, the results of the calculations, which allowed us to make inferences that require expansion and refinement in future studies.

1 Introduction

Increase of the capacity of railway nodes is an important task, the solution of which operate the railroad around the world. One of the frequently used methods is to increase the capacity of the node device through the construction of additional tracks (main, receiving and departure track) [1], which leads to fairly high costs both in investment and technology, because it requires to conduct construction works in the period of operation of the facility. To achieve the results and in other ways, for example, in Italy in railway stations in Rome and Milan by 2018 it is planned to introduce a system to reduce significantly the intervals between trains [2, 3]. One of the most effective ways to increase the capacity of railway nodes, and the elimination of difficulties in their work is the construction of detours for the passage of transit trains. Construction of bypasses is one of the only activities that increases the capacity of the railway nodes. It applies equally to passenger and cargo movement. For example, to pass high-speed passenger trains without entering the Paris node be the 1994 was built a bypass that allowed to accelerate the promotion of high-speed trains of the southern and Northern high-speed railway lines, reducing travel time for passengers between Lille and Lyon-3 hours, and between Lille and Marseille – up to 6 hours. The commissioning of the bypass Nimes – Montpellier it is planned in 2017 [4]. In Germany, the bypass railway junction Hamburg helped to make the routes of freight trains from the Central areas of the city. Bypassing of nodes was based in Hannover (via Elze to Braunschweig), the stowage of the new bypass in Bebra has eliminated the need to turn trains the message from Frankfurt to Berlin, the travel time from Frankfurt to Erfurt was reduced for 40 minutes, in Leipzig – 55 minutes. These examples and many others (Brasil [5], France [6], Slovenia [7], China [8], USA [9]) indicate that bypassing of nodes, among other activities, successfully applied to decongest the existing railway junctions, pushing deadlines necessary to strengthen them.

* Corresponding author: Kukleva.natalya@bk.ru

However, hasn't resolved yet the question of scientific substantiation of expediency of the construction of the railway crawls sites that are designed at the first stage, for facilitating the transit of freight trains without entering as into the separate areas and station node and the railway junction as a whole. This article focuses on the feasibility of constructing a bypass given the breakdown of the development of the conventional railway junction (Figure 1) at each stage.

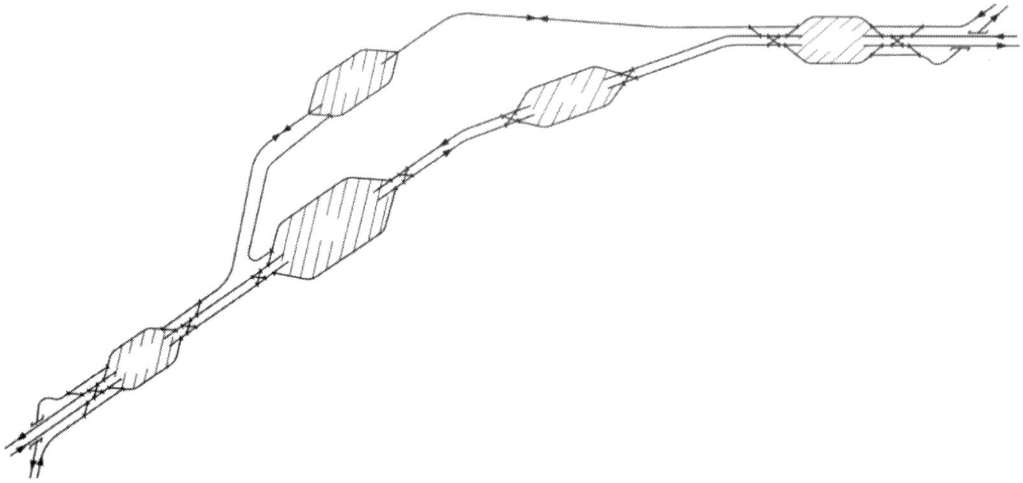

Fig. 1. Scheme of conventional railway junction with the bypass: PS - prenodal station, S - switchyard; P - passenger stations; PS- the precinct station on the bypass.

2 Statement of the problem

The issue of construction of bypass railway junction occurs if the bandwidth of the main ways intrasite or receiving and departure track designed for transit freight trains is not sufficient to permit flow of these trains without delay. The construction of the bypass railway junction at a certain stage of its development is considered as an alternative solution to strengthen motion progress with the aim of increasing the bandwidth of its stages and receiving and departure tracks to handle transit freight trains in the station, most of the sorting.

The boundary values of the sizes of movement of transit cargo trains, in which the construction of the bypass node becomes cost-effective, depends on a number of factors, the main ones are [10]:

- The size, nature and timetable of passenger trains in the node and on approaches to it in the period of maximum transport (summer). This factor determines the time spent in the host cargo transit trains, their delays due to crossing of passenger trains;
- Overall dimensions of cargo movements, of which less transit trains, the number of turns of the trains that cannot be rejected on the bypass (for trains arriving to and dissolution of its formation);
- Bandwidth of the main ways of intrasite course;
- The number of ways in receiving and departure tracks, dedicated for the transit of trains;
- The ratio of the length of the bypass and intrasite turn, determines the run of the rolling stock through the following options;
- Scheme of traction service freight trains, to determine whether construction on the bypass devices for maintenance and equipment train locomotives, or the absence of such devices;

- Topographic, urban planning and other local conditions affecting the cost of construction of the bypass path and the node device during the reconstruction of the intrasite course (stations, spans);
- The volume and conditions of production of construction works in the construction of a bypass and installation of additional mainline track on the progress and intrasite receiving and departure track's ways for transit trains in the switchyard.

As one of the original provisions in determining the terms and boundaries of the feasibility of the construction of the bypass is assumed that the gain of the device node to the processed traffic volume, and development of other stations will be the same as in the construction of the bypass, and the strengthening of intrasite course.

3 The characteristics of possible variants in the construction of the bypass

One of the factors that significantly affect the ratio of capital expenditures and operating expenses in the construction of the bypass and strengthening intrasite speed is the ratio of their lengths.

To design crawls of the railway junctions the two options are possible:
option I – the length of the bypass is less than the length of intrasite course;
option II – bypass length is greater than the length of intrasite course.

If the length of the bypass is less than the length of intrasite course, as a rule:
- Investments in the construction of the bypass is less than for option gain of intrasite course;
- Operating costs associated with mileage transit freight trains and the time spent in the node is less than the bypass.

If the length of the bypass is greater than the length intrasite course, then the two options are possible:

1. Construction costs of the bypass, there is less gain intrasite courses;
2. The cost of construction of a bypass are more than the cost for the strengthening of intrasite course.

However, current expenses associated with mileage of transit freight trains will tend to be smaller in the embodiment of the gain node, and the expenses associated with the residence time of these trains in the railway border node can be either higher or lower. This will depend on the fact of how the length of the bypass is greater than the length of the intrasite course.

Among the options considered in the ratio of investment in the second option will be beneficial to strengthening intrasite course.

In the first construction of the bypass will be beneficial in the case that additional investments to strengthen intrasite progress will not be repaid in a reasonable period due to savings on operating costs associated with mileage transit freight trains and the time spent in the node.

Possible ratios cash (existing) and required (necessary) capacity of receiving and departure track's for transit trains $(N_{ex}^{r-d}, N_{n}^{r-d})$, hauls intrasite course (N_{ex}^{ic}, N_{n}^{ic}), as well as lengths of bypass (L_{bp}) and intrasite course (L_{ic}) are given in table 1. In table. 2. the characteristics of each of the options and the possible benefits are shown that point to the benefits of bypass railway junction for the passage of transit trains.

Table 1. Possible ratios.

№ var.	Ratios N_{ex}^{ic} & N_{n}^{ic}	Ratios N_{ex}^{r-d} & N_{ex}^{r-d}	Ratios L_{bp} & L_{ic}
1	$N_{ex}^{ic} > N_{n}^{ic}$	$N_{ex}^{r-d} > N_{ex}^{r-d}$	$L_{bp} > L_{ic}$
2	$N_{ex}^{ic} > N_{n}^{ic}$	$N_{ex}^{r-d} > N_{ex}^{r-d}$	$L_{bp} < L_{ic}$
3	$N_{ex}^{ic} > N_{n}^{ic}$	$N_{ex}^{r-d} < N_{ex}^{r-d}$	$L_{bp} > L_{ic}$
4	$N_{ex}^{ic} > N_{n}^{ic}$	$N_{ex}^{r-d} < N_{ex}^{r-d}$	$L_{bp} < L_{ic}$
5	$N_{ex}^{ic} < N_{n}^{ic}$	$N_{ex}^{r-d} > N_{ex}^{r-d}$	$L_{bp} > L_{ic}$
6	$N_{ex}^{ic} < N_{n}^{ic}$	$N_{ex}^{r-d} > N_{ex}^{r-d}$	$L_{bp} < L_{ic}$
7	$N_{ex}^{ic} < N_{n}^{ic}$	$N_{ex}^{r-d} < N_{ex}^{r-d}$	$L_{bp} > L_{ic}$
8	$N_{ex}^{ic} < N_{n}^{ic}$	$N_{ex}^{r-d} < N_{ex}^{r-d}$	$L_{bp} < L_{ic}$

Table 2. The characteristics of each of the options.

№ var.	**Due to what can be justified the construction of the bypass railway junction**
1	To reduce the time spent by transit trains in the node
2	To reduce of the run and time of transit trains in the bounds of the node
3	The Reduction in time of transit trains in the node; saving investments, investments for laying of additional tracks in receiving and departure track's parks for transit trains in the rail yard
4	To reduce of the run and time of transit trains within the boundaries of the node; - saving investment for laying of additional tracks in receiving and departure track's parks for transit trains in the rail yard
5	Reduction in time of transit trains in the node; saving investments in laying more tracks intrasite course
6	Reduction in time of transit trains in the node; saving investments in laying more tracks intrasite course; reduced run transit freight trains
7	Savings in laying more tracks of intrasite course and additional ways in receiving and departure track's parks for transit trains in the rail yard
8	Saving in laying more tracks intrasite course and additional ways in receiving and departure track's parks for transit trains in the rail yard; reduced run of transit freight trains

The question is: how to develop a railway junction in terms of growth of transit and other traffic volumes to achieve acceptable economic costs. To justify the period of construction of the bypass railway junction, it is expedient to use the method of dynamic programming [11].

4 Example of calculations to determine the feasibility of building of a bypass in the node

Initial data for calculations to determine the feasibility of the construction of the railway bypass of the conditional node are:

- Dimensions of freight trains;
- Investments and operating costs for construction of the bypass;
- Investments and operating costs for the option of strengthening the railway junction.

For the study, it is assumed that in the node there is annual increasing of the size of the movement of all categories of freight trains with pace in 4 pairs of trains per year. Length of intrasite speed is assumed to be constant and equal to 35 km, and the length of the bypass varies from 25 to 45 km.

Below for the calculations were taken the following dimensions of the movement at the beginning of the reporting period: 48 pairs of freight trains per day with a ratio of transit and other freight trains 0,5/0,5 increments of 2 pairs of trains every year.

In each of the basic options of the scheme of node were chosen five sub-options of a technical condition corresponding to the size of the traffic of transit freight trains from 24 to 64 pairs of trains. After continuous numbering of sub-options, the first five of them describe the original version of the technical state of the node (with double-track intrasite course) and a sequentially increasing number of receiving and departure track's ways for transit trains from 3 to 7 in each direction. Modifications from the sixth to the tenth characterize in the second variant the technical condition of the node of four-tracked intrasite speed and the same number of in receiving and departure track's ways for transit trains from 3 to 7. When comparing the first with the third option sub-options from the sixth to the tenth characterize the node with the bypass and the number of in receiving and departure track's ways for transit trains at the precinct station of bypass from 3 to 7 in each Park.

The results of the calculations showed that under other equal conditions, the greatest impact on the economic feasibility of the construction of the bypass has a ratio of length of the bypass and intrasite course, which can be seen from the example in figures 2 and 3, where there are diagrams of transitions when the length of the intrasite speed of 35 km and the length of the bypass. From figure 2 it is seen that, for example, when the length of the bypass is 25 km, the feasibility of its construction comes at the 6th year of operation, when the size of the movement of transit trains made 36 pairs of trains a day. If the length of the bypass will be 45 km long, term of profitability of its input is extended to 13 years of operation and occurs with the transit of 50 pairs of trains per day.

From figure 3 it can be seen that at the same lengths the transition from the existing technical state of the node to the option of strengthening intrasite progress and construction of the bypass occurs in almost one year, with 48 pairs of transit trains.

It was found that the increase in the extent of bypass by an amount equal to 40% of the length of the intrasite turn will lead to excess costs in not more than 5 %, indicating the feasibility of the construction of the bypass.

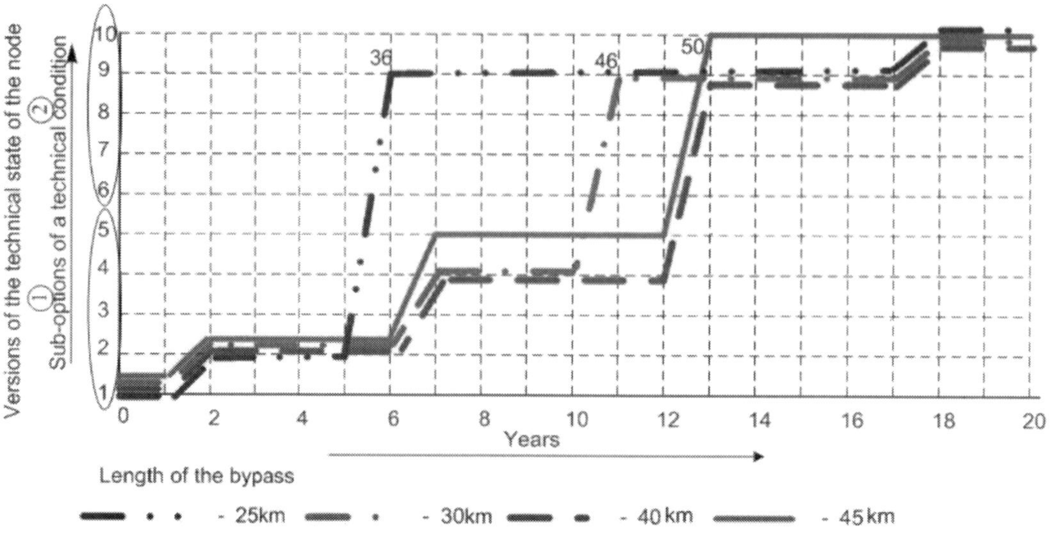

Figure 2. The dependence of the technical state of the node from the dimensions of motion at different length of the bypass and intrasite course.

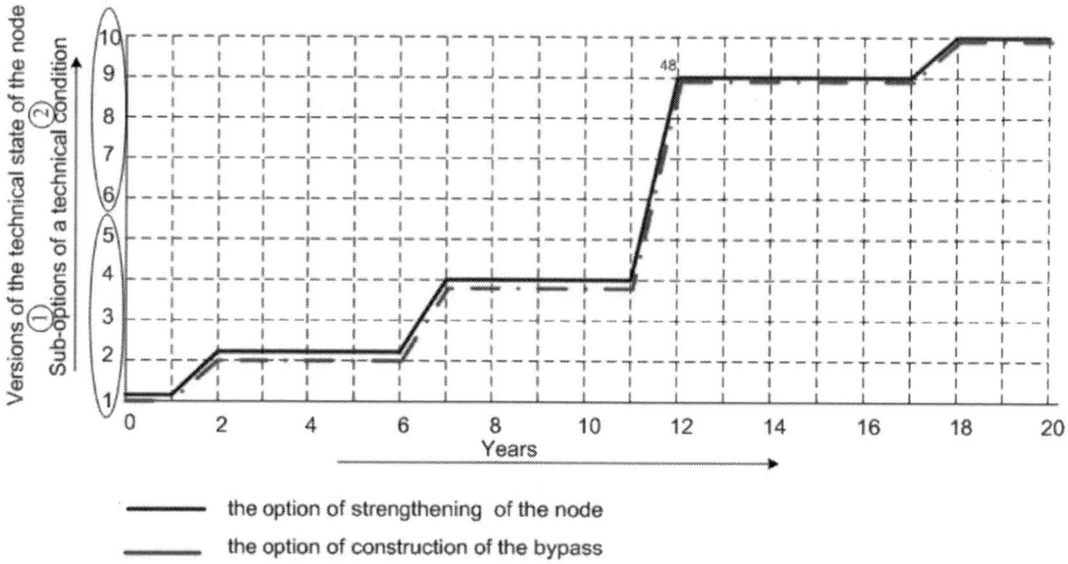

Figure 3. The dependence of the technical state of the node from the size of the movement with the same length of the bypass and intrasite course.

Conclusion

Received on the example of the conventional rail node, the results of justification of the construction of a bypass to skip the transit of freight trains, show that if the length of bypass node is less than the length of intrasite course, the construction of the bypass is undoubtedly effective. When the length of the bypass is more than the length of the intrasite turn the final decision should take into account the additional factors, and in excess of the cost of the bypass to 10 % in comparison with the strengthening of intrasite course, should recommend the construction of a bypass, as this cost difference allows us to consider options for the development of the node, in terms of performance is virtually equal. However, the obtained results can be considered only as intermediate, a special case, because you have not defined the boundaries of the feasibility of building crawls with a significant number and range of influencing factors, the study of which can have great value and will be the subject of future research of the authors.

References

1. D.N. Kuklev, International Railway Journal **9**, 20–23 (2015)
2. D.N. Kuklev, Railway Gazette International **4**, 41–43 (2016)
3. D.N. Kuklev, International Railway Journal **11**, 42–44 (2015)
4. J. Kolodziejski, Railway Gazette International **7**, 32–34 (2016)
5. J. Kolodziejski, Railway Gazette International **8**, 498–501 (2007)
6. J.P. Charlanne, Revue Générale des Chemins de Fer **149**, 44–51 (2006)
7. Yu.I. Efimenko, Railway Gazette International **10**, 785–790 (2008)
8. A.V. Sugorovskiy, Chinese Railways **1**, 63–71 (2007)
9. J. Stagl, Progressive Railroading **4** (2009)
10. D.N. Kuklev, RGOTUPS **4**, 37 – 41 (2006)
11. Yu.I. Efimenko, A.V. Sugorovskiy, Nauka i tekhnika transporta **4**, 6 – 12 (2010)

Approach to Organizational Structure Modelling in Construction Companies

Igor V. Ilin[1], *Olga* V. Kalinina[1,*], *Anastasia* I. Levina[1], *Oksana* Yu. Iliashenko[1]

[1] Peter the Great Saint Petersburg Polytechnic University, 195251 Polytechnicheskaya str. 29, Saint Petersburg, Russia

Abstract. Effective management system is one of the key factors of business success nowadays. Construction companies usually have a portfolio of independent projects running at the same time. Thus it is reasonable to take into account project orientation of such kind of business while designing the construction companies' management system, which main components are business process system and organizational structure. The paper describes the management structure designing approach, based on the project-oriented nature of the construction projects, and propose a model of the organizational structure for the construction company. Application of the proposed approach will enable to assign responsibilities within the organizational structure in construction projects effectively and thus to shorten the time for projects allocation and to provide its smoother running. The practical case of using the approach also provided in the paper.

1 Introduction

The change in approach to running a business in today's information society results in the fact that the advantage is possessed by those participant of the market who are better adapted to the changing environment and provide innovative solutions to meet the market challenges. Many companies show a growing interest in project management as a method by which they can work with unique challenges both for implementation internal organizational changes and for running core activities.

Many companies are project-oriented by the nature of their business, and their activity can be considered as a portfolio of projects in various stages of execution. Project-oriented companies can be found in such industries as engineering, construction, IT-sector, machinery manufacturing, consulting, banking, and many others. Such companies need to implement a unified corporate standard for project management in order to provide the quality of each project and effective management of a portfolio of projects as a whole. [1] Implementation of a unified project management standard in such companies is a prerequisite for competitiveness, and project management principles are to form a foundation of the organizational management system of the company.

*Corresponding author: olgakalinina@bk.ru

A construction company is a classic example of a project-oriented company. During last two decades, Russia has experienced a significant growth of industrial and civil construction, as well as the need for the reconstruction and modernization of outdated facilities. Many construction companies are aware of the need to revise their internal management system in order to bring into compliance strategic business objectives, business processes and organizational structure to form a balanced enterprise architecture, which in turn can enable a high level of business performance. As a result, many companies of the industry have started to pay serious attention to the need of organizational structure reforming, which means management architecture restructuring in the mean times.

As the personnel is one of the key resources of any business, the way a company organizes and uses its human capital defines its performance indicators. In order to perform construction projects effectively, construction companies need to form an appropriate organizational structure that from the one hand would allow to run core activities (engineering design, constructing, servicing) and from the other hand will be flexible enough in order to form effective temporary project teams. The paper examine the project-oriented nature of construction business and declares that it is reasonable to form organizational structure of construction companies basing on the typical project management team structure. The approach of adopting the existing organizational structure of the construction company to those of the project management team is presented in the paper. Such a project-oriented structure will effectively support both business process system of a company and its project portfolio execution. An effective organizational structure, ready to meet challenges of the market, can form one of the main prerequisites of sustainable development of the company.

2 Method

Nowadays construction business in big cities of Russia is actively developing in spite of the several global financial crisis of the last decade. The competition on this market is rather high and the competitive advantage is possessed mainly by big companies that have a convincing story of survival during difficult financial periods of 21st century. [2] The main reasons which force construction companies to revise their enterprise architecture, and specifically to pay attention to organizational structure, are the following: [3]

1. Absence of the precise strategy of management architecture development
2. Absence of an integrated architecture adaptability to market conditions
3. Discrepancy between the organizational structure and increased business demands
4. Discrepancy between the organizational structure of companies and organizational structures of projects/
5. Absence of a common corporate standards of project management
6. Absence of precisely prescribed roles and responsibilities in the current organizational structure
7. Absence of detailed and transparent business processes

To face business challenges effectively, companies try to build their enterprise-wide management system in such a manner that would allow a company to be in the mean time stable from the inside and flexible in reactions to the external challenges. Development of such a management system requires an integrated approach in order to create a solid foundation for the whole management system. The concept of the enterprise architecture as a complex management tool has become a sort of mainstream management concept in recent decades. One of the definitions states that "enterprise architecture is a coherent whole of principles, methods, and models that are used in the design and realization of an enterprise's organizational structure, business processes, information systems, and infrastructure" [4]. Traditionally, the enterprise architecture can be represented as a set of components which are grouped into layers. Different authors state for different names, numbers and composition of layers [4-8], but from the point of view of the whole set of the components and their interdependencies the main idea in all the sources is more or less the same (the comprehensive review can be found in [9]). For example, [5] focuses on the following three layers:

- Corporate mission and vision, strategic goals and objectives;
- Business architecture: business processes, organizational structure, workflow system;
- System Architecture (IT architecture): applications, data, hardware.

Some researches confess that the business architecture is the primary element that shapes the effectiveness of the company [4, 10-13].

Enterprise architecture is a system view of the key structural sections (certain key components and their relationships), applied for various practical problem solving of the organization [14]. The feature of the enterprise architecture is its heterogeneous composition – business processes, functions, organizational structure, document flow, information technology, etc. While forming the enterprise architecture it's necessary to solve the following tasks in coherent and interconnected manner [1]:

1. Mission and strategy of the company, strategic goals and objectives;
2. Business architecture «as is» and «to be», including business process system and organizational structure as key elements;
3. IT-architecture «as is» and «to be», including information system, data bases, technical tools and solutions;
4. Structure of project portfolio for enterprise architecture migration from its current state («as is») to the planned state («to be»), including business processes reengineering projects, organizational structure reengineering projects, IT-systems implementation or reengineering projects, technological equipment projects.

A large number of companies in various fields of activity (including construction business) must address the unique business tasks that can not be resolved by the operation of standard business processes – the project approach is required to solve such business challenges. [15] It means that such kind of companies in order to be effective have to introduce project approach as a part of their enterprise architecture. Project orientation influences the business process system of the company, its organizational structure, document flow. Thus the objective of modeling the enterprise business architecture based on project activities becomes very relevant for such companies.

The basis of the business of the construction company is the portfolio of contracts for execution of construction projects. Construction companies decide separate business tasks by means of temporary organizations – projects and programs. Project approach for business management has its features:

- The project is considered as a unique combination of project delivery processes;
- Rights and responsibilities for project results delivery belong to project manager and project management team;
- Certain budget of the project;
- Implementation of specific project organizational structure and motivation of project management team members;
- Development and implementation of specific standards of project processes performing [14].

The mission of the company as the main objective of its development defines the strategy, including the strategy of architecture development management. [16] The first step in the formation of the enterprise architecture is analysis and reengineering (if needed) of business processes with the subsequent formation of the organizational structure. The organizational structure of the construction company, on the one hand, should match the system of its business processes, providing its effectiveness, and the other hand – it should match the corporate standard of project management. For effective implementation of project management activity, it seems reasonable to introduce a single enterprise-wide project management standard. Its presence is intended to ensure a common understanding of objectives and procedures of project management by all participants of the project, to provide all project participants with a common methodology and uniform terminology, to make more effective communication within and outside the project team. [17] As a basis for the corporate project management standard in a company one of the standard methodologies, well-known and highly referenced by practitioners, can be adapted.

Currently, there are a number of world-wide accepted project management methodologies, developed by leading professional associations and organizations. These methodologies are the result of analysis, synthesis and formalization of best practices in project management. The most famous in the world of professional community are the approaches developed by organizations such as the AXELOS (successor of Cabinet Office, UK), PMI (USA), IPMA (Switzerland), Microsoft (USA), etc. The methodology of each organization is documented in the form of a guidelines – Managing Successful Projects Using PRINCE2 (AXELOS), PMBoK (PMI), ICB (IPMA), MSF (Microsoft) – and is associated with a certain certification system. [18]

The authors propose the following approach of developing the project-oriented organizational structure of construction companies (and related sectors):

1. Examine process model of the company;
2. Make correlation between business process system of the company and typical project stages and project management processes;
3. Develop or adopt a corporate project management standard;
4. Examine organization structure of the company;
5. Make correlation between organization structure of the company and project team structure from the chosen standard;
6. Reengineer existing organization structure (in terms of organizational units, roles and responsibilities) according to the requirements of project team structure.

The step-wise algorithm of organizational structure development, described above, can be considered as a main contribution of the paper.

3 Results

The top management of a construction company (hereinafter referred as a Case Company) initiated a project of organizational structure reengineering. Roles and responsibilities in the existing organizational structure were designed to provide the quality of particular functions within the construction project, but did not take into account the overall project success. Thu, it was a lack of attention to project portfolio management level, where company's business interests are defined and controlled. The top management of the Case Company has identified the goal of the reorganizational project: to revise the existing business processes and organizational structure in order to optimize the latter for the sake of sustainable business development. This objective implies a re-engineering of business processes and reforming the organizational structure, which on one hand would provide compliance with the principles of the organizational structure of project management, on the other hand – would create conditions for the further optimization of business architecture based on several criteria: the exclusion of redundant business processes and their segmentation within the various organizational units, minimizing organizational interfaces, improved internal and external communications, increasing flexibility on the market.

To realize this goal the following consequence of objectives was set, according to the organizational structure development approach, described in Chapter 2 "Method" of the paper:

1. To adopt the corporate project management standard;
2. To model and analyze the business architecture (including business processes and organizational structure) «as is»;
3. To analyze roles and responsibilities in the existing organizational structure in order to make correspondence between project team structure of the adopted standard and the existing practice of roles and responsibilities distribution in projects;
4. To model the business architecture (including business processes and organizational structure) «to be»;
5. To implement the updated system of business processes and organizational structure.

The application of the algorithm, presented above, should result in optimized organizational structure from the project management effectiveness point of view.

The Case Company has adopted PRINCE2 as a corporate project management standard as the most procedural one, that includes the precise process-oriented instructions on how to manage projects step by step [19]. PRINCE2 (Projects in a Controlled Environment) is a structured method of project management based on thousands of best practices of successfully realized projects. This method has the following advantages:

1. Includes the best practices that has proved its effectiveness;
2. Can be implemented for any kind of the project;
3. Is widely known and provide the common language for all the members of the project;
4. Is oriented on strategic goals achieving;
5. Sets certain roles and responsibilities for project management;
6. Is oriented on delivery of project results.

Basic principles of project management according to PRINCE2 are [19]:

- Continuous business justification;
- Learning from experience;
- Defined roles and responsibilities;
- Management by stages;
- Management by exceptions;
- Focus on products;
- Tailoring to the environment.

For the purpose of architecture forming of engineering company based on the revision of management processes and organizational structure, the principle of defined roles and responsibilities is particularly important. PRINCE2 allows to create a system of hierarchy and interaction of the participants of the project which form a well-functioning structure that takes into account the interests of the three interested parties in the project – business, future users and suppliers. Such a structure provides certain subordination levels of project management, each of which controls the interests of different levels, ultimately subordinate to the strategic goals of the business. Responsibility delegation to the higher level is performed according to the «management by exception» principle. This provides the lower levels with more management freedom and the higher ones are not involved into routine processes of lower level processes.

PRINCE2 prescribes to have three levels of project management authority within a project management team – Directing, Management, Delivering (Fig.1). The first one is responsible for overall success of the project, the second one provides day-to-day project management and coordination of activities, while the third one manages directly project products delivery. In addition to the distinction between levels of project management, the organization of the project team according to PRINCE2 implies specific roles and responsibilities, which allow to avoid function duplication, to provide a clear procedure of project control at all stages, as well as ongoing expert and administrative support to the project management team.

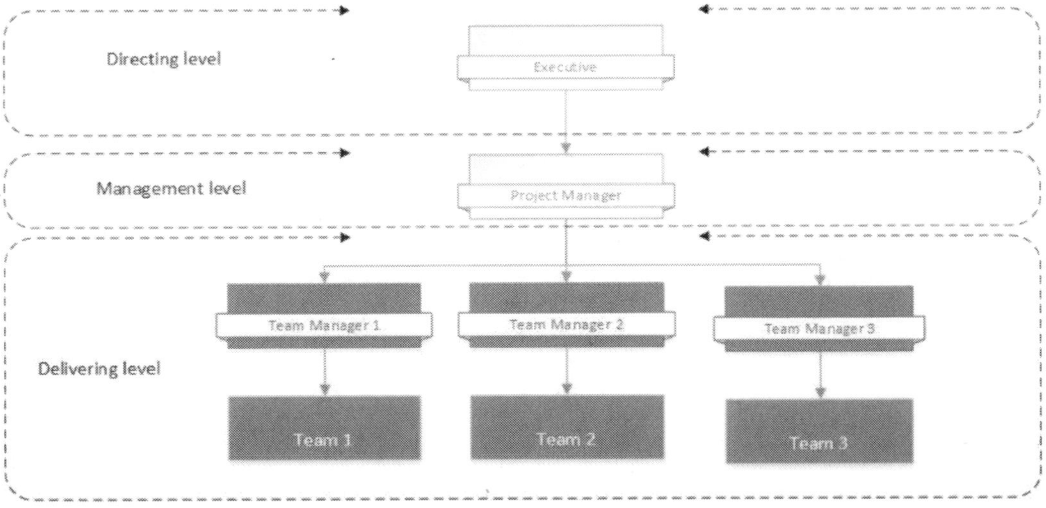

Fig. 1. PRINCE2 project management team structure.

After staff interviewing and document analysis, the diagram of added value chain, containing the business-processes of the upper level and the correspondence organizational structure were developed. Analysis of the detailed process breakdown structure allowed to classify processes for core and supporting business processes, to identify «bottlenecks» in terms of duplication and inconsistency of prescribed responsibilities for different roles. The analysis of the upper-level processes showed the correspondence between sequence of core processes and stages of the typical construction project: pre-project and initiating stages (incoming order processing), project delivering stages (engineering design, construction and installation), post-project activities.

Analysis of the departments and key positions of the existing organizational structure revealed the following distribution of functions and responsibilities while managing contracting projects:

1. Sales Director, the head of Contract Department – responsible for overall success of each construction contract, concluded by the Case Company;
2. Head of Designing Department and Chief Project Engineers (hereinafter referred as CPE) from Designing Department – responsible for managing everyday activities of particular projects from the contract conclusion till the end of construction works and project transfer to the customer; CPE can participate in managing designing stage of the project;
3. Employees of Contract Department, Designing Department and Construction Department – perform the appropriate part of project work.

Analyzing the existing organizational structure and current roles and responsibilities, it was identified the inconsistency of responsibilities distribution between the management levels and the discrepancy between the management organizational structure and project organizational structure. As a result, on the one hand there is a lack of proper control of the division from the strategic business objectives point of view, on the other hand – duplication of functions and control at lower levels. Thus, managers, responsible for the overall management and coordination of the work of the design («Directing» level) in reality is involved not only in the management of individual projects («Managing» level), but also in the control of the projects' execution («Delivery» level). At the same time, the CPE («Managing» level) factually does not have sufficient authority to carry out project management and is substantially involved in the control of direct execution of projects («Delivery» level).

The analysis of the «as is» organizational structure revealed the "bottlenecks" of the existing project management approach. The following changes to the organizational structure and system of roles and responsibilities were proposed:

1. To provide a clear demarcation between the «Directing» and «Managing» levels of project, prescribing the responsibility for the business interests control to the Sales Director and giving more management authority for individual projects to the CPEs;
2. To increase the number of managers of the «Managing» level (CPE) to enable the execution of greater number of projects;
3. To form project delivering teams from the employees of the appropriate departments.

Taking into account all the proposed changes for organizational structure after revising the existing roles and responsibilities, a new «to be» model was performed (Fig.2).

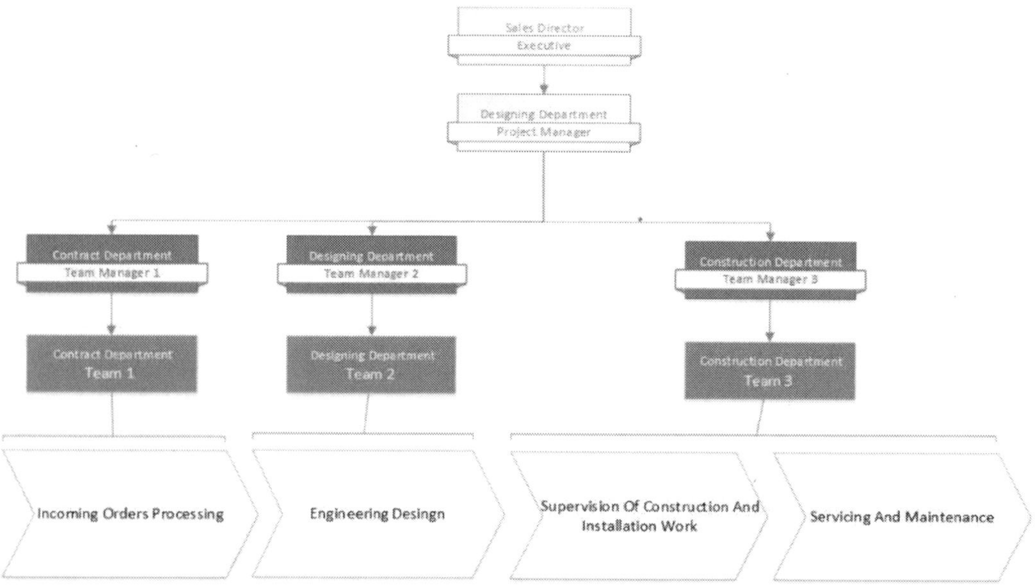

Fig. 2. Project tem roles and responsibilities.

The proposed model of the «to be» organizational structure has a number of advantages:

- It enables the use of common and proven approach to project management (such as PRINCE2 for the Case Company);
- It provides a basis for distinguishing between the roles and responsibilities in the project management team and the organizational structure of the whole unit;
- It forms the organizational basis for further growth of the company in accordance with the strategic goals of the company.

4 Conclusion

The company as a system requires a systematic approach while implementing any organizational changes. The approach for the organizational structure development of the construction company, proposed in this paper, means using of project management methods for the sake of improvement of the team performance. Development of organizational structure according to the developed approach allows filling of different management levels of project management with real responsibilities, to provide correspondence between system of management business processes and organizational structure of the engineering company from one hand and its strategic goals and growing business interests from the other hand. The approach can be adopted as for full-cycle construction projects (those which include construction from Greenfield to operations) and for projects for only certain parts of engineering or constructing services.

The quantitive estimation of the effectiveness of organization structure modelling approach is beyond the scope of this paper as it requires a specific research and a wide experimental database. It could be the subject of further development of the topic arisen in the paper.

References

1. I.V. Ilyin, J.L. Levchenko, A.I. Levina, St.Petersburg State Polytechnical University Journal. Economics **163**, 48-54 (2013)
2. A.Khrustalev, RIA Realty, URL: http://riarealty.ru/analysis_interview/20150120/404182124.html
3. I.V. Ilyin, A.I. Levina., Journal of Business and Economics **5**, 698-708 (2014)
4. M. Lankhorst, Springer-Verlag **1** (2013)
5. G. Kalyanov, Enterprise architecture and instruments of its modeling **13** (2011)
6. The Open Group, TOGAF **9** (2009)
7. Gartner, IT Glossary **174** (2016)
8. MIT Center for Information Systems Research, Enterprise Architecture **16** (2016)
9. D.V. Kudryavcev, M.Yu. Arzumanyan, L.Yu. Grigorev, Polytechnic University Publishing **27**, 427 (2014)
10. M. Op't Land, E. Proper, M. Waage, J. Cloo, C. Steghuis, Springer Verlag **1** (2009)
11. Y. Bekker, L. Vilkov, V. Taratukhin, M. Kugeler, M., Exmo **84**, 384 (2010)
12. M. Weske, Springer Verlag. **1** (2007)
13. V.V.Glukhov, I.V. Ilin, O.Yu. Iliashenko, Springer **9870**, 512-524 (2016)
14. V.V. Kondratiev, Exmo **504** (2007)
15. R. Gareis, Happy Projects **657** (2005)
16. A. Bril, O. Kalinina, O. Valebnikova, Springer **766**, 766-775 (2016)
17. R. Gareis, M. Stummer, Processes and Projects **75** (2008)
18. I.V. Ilin, A.I. Lyovina, A.R. Antipin, Results and Experiences **51**, 51-63 (2016)
19. The Office of Government Commerce, Managing Successful Projects with PRINCE2, 329 (2009)

MATEC Web of Conferences **86**, 05029 (2016) DOI: 10.1051/matecconf/20168605029
IPICSE-2016

The role of the concept of LEAN management in modern business

Boban Melović[1,*] *Slavica* Mitrović[2], *Andrey* Zhuravlev[3] and *Natalia* Braila[4]

[1]University of Montenegro, Faculty of Economics Podgorica, Jovana Tomasevica, 37, Podgorica, Montenegro
[2] University of Novi Sad, Faculty of Tehnical Sciences, Trg Dositeja Obradovica 6, Novi Sad, Serbia
[3] Moscow State University of Civil Engineering, Yaroslavskoe shosse, 26, Moscow, 129337, Russia
[4] Peter the Great St. Petersburg Polytechnic University, 195251, 29, Politekhnicheskaya st., Russia

Abstract. LEAN management is the concept of production management, product development and services that aims to create a true step change specifically targeted to the buyer or the goods or service. LEAN management is a dynamic process that uses knowledge and is focused on the customer. This is a process by which companies are continuously working to eliminate losses and value creation.

1 Introduction

"Edit your processes and increase the value of your company."
James Womack and Daniel Jones,
experts in LEAN manufacturing

Although the concept of LEAN (LEAN philosophy) appeared in Japan at the end of XIX century, wider application is entered only after the Second World War. During this period, the country Japan, or relatively poor in natural resources, located on earthquake prone area and destroyed by war activities, with the aim of economic recovery and competitiveness in the market, should be distinguished from competitors with high quality products and shorter delivery time, with the possession of very small amounts of necessary resources (natural, financial, human). For this reason, it was necessary to engage in scientific and technological research at a high level, whose results were later to be implemented in industrial processes.

LEAN concept represents Toyota's Production System (TPS - Toyota Product System) or the mode of production and its creator Taiichi Ohno of is considered, the head of Toyota's manufacturing plant in the postwar period. Today the name of the LEAN concept, is defined by Professor James Womack (James P. Womack), Daniel Jones (Daniel T. Jones) and Daniel Rosa (Daniel Roos), and is presented in the book "The machine that changed the world". Generally speaking, LEAN represents a way of thinking and a set of principles, defined for the purpose of effective and efficient organization of the production process (or services) and supply, with the elimination of all activities that do not add value to the product, or represent losses as well as striving for continual improvement of the system. In

* Corresponding author: bobanm@ac.me

addition to this, one of the definitions of this concept as follows: the LEAN production system, which allows the supply only what is needed, when needed and in the required quantity [1]. Storage excessive amount of inventory, or the amount of inventories that are not required to support basic processes, is considered the main loss and primary aim of elimination [2].

The leader Toyota, Taiichi Ohno and Shigeo Shingo (hired consultant) have defined the seven largest group losses incurred in the business processes of the organization and the effect of raising the cost of products, and they do not add value.

Womack and Jones believe that less than 5 percent of activities that add value to the product, and that about 35% of activities that do not add value, but they are necessary. In the end, about 60 percent of the activity is in a completely unnecessary activities and those that are the complete loss.

These seven groups of loss are [3]:

- Overproduction,
- Waiting time in the processes and inactivity of workers and equipment, waiting time for a product,
- Transport,
- Excessive processing,
- Stocks of raw materials, semi-finished products,
- Unnecessary movement of labor, equipment and materials and
- Frequent errors, the occurrence of rejects and finishing.

These losses added some authors and inadequate processes of product design, which does not meet the requirements of users, untapped intellectual potential of workers and missed opportunities.

2 Key kind of wastage in the work process

LEAN management experts have defined seven key types of waste. These types of scattering are [4]:

1. Excessive production
- Create a product that can not be placed on the market.
- Perform the operations that are not necessary
- Creating documentation that no claim or which shall not be later used
- Poor estimates sales ie. market demand
- Sending instructions by too many people (or vice versa)
- Production of "just in case"
2. Transpot
- Unnecessary movement of materials between operations or between storage areas
- Poor communication and information flow
- Communication failures, data loss, incompatibility, the unreliability of information
3. Waiting / congestion
- Waiting time between operations of materials, machines that workers hold are available or hold material (poor planning of production)
- Waiting for data, test results, information, decisions, signatures, permits and the like.
- Awaiting delivery (e.g., late raw materials and the like.)
4. Excessive processing
- Oversized facilities, poor or no technological equipment, preparatory-final time, cleaning between processing
- Previsto processing
- In deep processing

- Poor design i.e. design products that require too much processing steps (too complex product)

5. Inventories

- Large inventories associated with excessive production ("frozen capital" in warehouses)

6. Unnecessary movements

- Improper machine, unnecessary displacement of workers
- People who have to leave their positions to get to the necessary information
- Handmade in order to compensate for a lack of production

7. Broke

- Interruption flow due to an error, unnecessary time, cost and space for analysis and elimination
- Incomplete, inaccurate or incomplete information
- Taking account of that waste, we conclude that the conditions for the implementation of this concept:
- Establishing a system of direct supply by a small number of trusted and reliable supplier,
- Smaller-sized production batches (lots)
- Establishing and aligning production capacities,
- Multifunctional (universal) trained workers,
- Specialized production lines,
- Preventive maintenance system, and
- Uninterrupted flow of cases through the system work.

3 The benefits of the application of LEAN management

There are many benefits of using this concept, and the authors are distinguished:

1. Reduction in inventory levels and therefore bound capital,
2. Less scrap and finishing,
3. Reducing the number of suppliers, the ability to standardize procurement and harmonize the quality of procurement,
4. Reduced paperwork and administration and
5. To increase the level of productivity and overall efficiency of workers and equipment.

In order to eliminate losses, i.e. those activities that do not add value to the product and more efficient development of the process, these authors Womack and Jones, defined five key principles in the implementation of LEAN concepts [5]:

Value – determining the value of products, defined by the user; focus on meeting customer demands for products or services, at the right time and in an appropriate manner;

Value stream – establishing a series of activities over the work material and information from 'the supplier's supplier' to 'member users, "the purpose of efficient implementation of the process of determining user requirements, determine the possible product defects and implementation of improvements, ie removing all steps that does not add value to the product or service;

Flow – synchronized and a steady flow of products through the value chain to the end user through all the activities that add value to the product (service) in a manner which eliminates barriers, poor quality, spoilage and delays in the way of the product;

Pull – start production when users expressed the need for the product; signaling the need for the product next factor in the supply chain (in the direction of users - manufacturer - supplier);

Perfection – continuous pursuit of eliminating all losses; cyclic trajectory seeking perfection through the elimination of the losses, which the user does not want to pay the cost of the product and put added value of the same product.

Lean management, therefore, represents a major step forward. Observing the concept of project management, then the transition from management on lean management, was as great as the step transition from project management to portfolio management. The biggest advantage of LEAN methods the connection between reducing costs and improving productivity and quality. In LEAN-in, all the energy and creativity have been organized so as to strive to better optimize and add value to the user side. This includes other things such as price, delivery time, maintenance costs, environmental taxes and more. Also, it happens to improve and internal factors such as safety, hygiene and ergonomics.

4 Conclusion

By applying LEAN management improves the flexibility of the organization and the quality of operations, reduces the goods in stock, improves the functioning of the process, eliminating activities that do not add value, it is better utilization of human resources, improving the organization of work and employee satisfaction, and ultimately, the most important, increases customer satisfaction.

The organization, therefore, should be oriented towards customers and coordinate all activities that do not bring value to customers [6]. Taking care of the natural resources, LEAN management avoids all superfluous in the chain of production. This approach is the administration of which, based on the fact that for a 'which more with less ". The purpose of the LEAN management is to lead the organization of the world's leading organizations in terms of quality, through a constant effort that the number of defective products is reduced to the lowest possible level. However, about LEAN management has different opinions, for some this is the perfect model of governance for any organization, while others believe that this is a bad approach to management, because after many years of application do not achieve satisfactory results.

Based on the foregoing, it can be concluded that LEAN management is the business philosophy of the organization, aimed at reducing operating costs, i.e. all losses, to thereby increase the success of the organization.

References

1. Lean Lexicon. Lean Enterprise Institute Inc (2008)
2. T. Goldsby, R. Martichenko, J. Ross Publishing **1**, 25-28 (2005)
3. D. T. Jones, P. Hines, N. Rich, International Journal of Psychology **27**, 13-19 (1997)
4. http://project-management-srbija.com/upravljanje-kvalitetom/osnove-lean-proizvodnje (2016)
5. J. P. Womack, D. T. Jones, A division of Simon & Schuster **57**, 85-88 (2003)
6. B. Melović, Faculty of Economics Podgorica **13**, 32-37 (2007)

Water-resistant gypsum binder

Alexander I. Panchenko[1,*], *Nikita* Kozlov[2]

[1]Prof. Chair of binding agents and concrete, Moscow State University of Civil Engineering, 26, Yaroslavskoye Chausse, Moscow, Russian Federation
[2]Director, "Promressurs" Co. Ltd. Moscow, Russian Federation

Abstract. The authors developed a multi-component gypsum binder (MGB) with an improved (by 1.8 to 2.2 times) water resistance value in comparison with the initial gypsum and a softening factor value within the range of 0.85 to 0.91 through the introduction of a complex additive containing carbide mud and bio-silica. The use of a complex additive provides for the formation of a denser structure due to the formation of low-base calcium hydrated silicates at early stages of the hardening process under wet or dry air conditions. The developed binding agent has a lower (by 2.5 to 3 times) open porosity, a greater compressive strength (by 1.4 to 1.6 times for a dry state and by 2.6 times in a water-saturated state) and does not require any special curing conditions in contrast to the other multi-component gypsum binders. The authors show the influence of mud-and-silica additive on the properties of hardened MGB and suggest a method of computation of the optimum MGB composition for given components.

1 Introduction

The expansion of the field of the use of gypsum binder (GB) as the materials for fencing structures is rather promising due to the abundance of gypsum stone deposits, waste products containing gypsum as well as a simple production technology and a low energy consumption. One more advantage of gypsum binder is a higher hardening rate under dry air conditions in comparison with the Portland cement. The disadvantages of gypsum binder are: a sharp decrease in strength when moistening, a low resistance to environmental impacts and a high creep value, i.e. a low operation reliability [1].

One of the ways to improve the operation reliability is the reduction of gypsum solubility in combination with the modification of the porosity. The use of carbide mud (a waste by-product from the acetylene production process) and bio-silica as the components guarantees considerable engineering, economic and environmental effects as the acetylene production in the Russia (400,000 tons in 2013) gives more than 1,000,000 tons of carbide mud yearly (1,140,000 tons in 2013) [2].

2 Experimental studies

2.1 Raw materials

*Corresponding author: alex250354@gmail.com

The research works were carried out with the use of the following materials:
1. Semi-hydrated gypsum;
2. Silica containing components (micro-silica and bio-silica);
3. Carbide mud;
4. Portland cement;
5. Retarder agent;
6. Plasticizer.

2.2 Process of the experiment

In the process of experimental studies, the authors prepared the samples from gypsum binder, carbide mud, silica containing components (micro-silica and bio-silica) and plasticizer. The samples were subject in various environments according to experimental conditions within the curing period until 28 days. The produced samples were mainly beams (4x4x16 cm), the curing process lasted 28 days. All the gypsum binder compositions were produced in accordance with requirements of the standard GOST 23789-79* "Gypsum binding agents. Test methods". The softening factors for gypsum binder were determined in accordance with the TU 21-31-62-89 normative document.

3 Results and discussion

The experiments showed that the optimum ratio values (with respect to the improvement of water resistance and strength of the material) turned out to be the following: the micro-silica to carbide mud ratio in the mud-and-silica additive equaled 1.0 to 1.2 and the bio-silica to carbide mud ratio equaled 0.5 to 0.7 (Fig 1,). Thus, the consumption of bio-silica was by 1.8 times less than that of micro-silica [2].

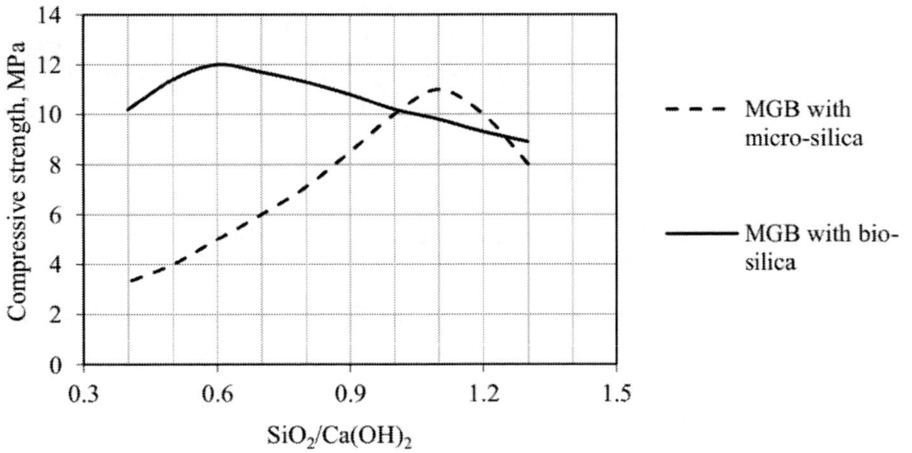

Fig. 1. Compressive strength in a water-saturated state for the MGB (multi-component gypsum binder) samples depending on the $SiO_2/Ca(OH)_2$ ratio.

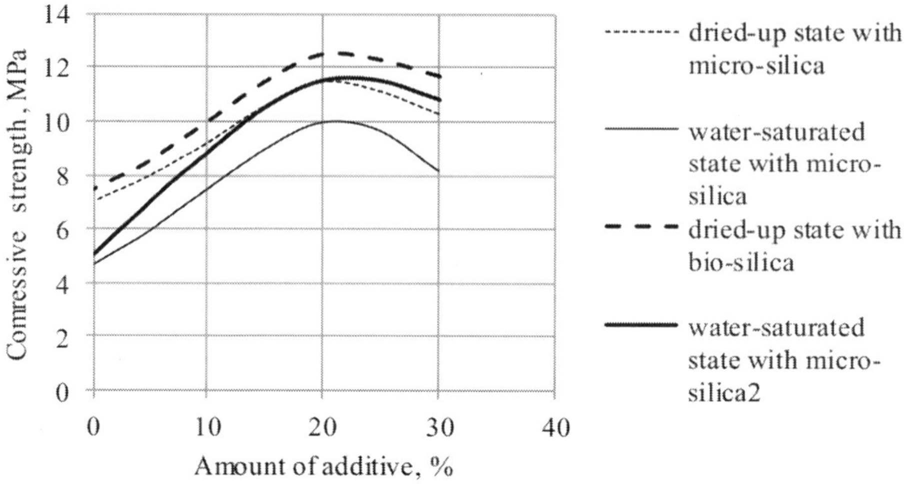

Fig. 2. Compressive strength depending on the amount of additive.

The optimum amount of additive in gypsum equals 18 to 22% (mass) (Fig. 2). The use of multi-component gypsum binder as a part of the micro-silica additive may improve the strength of the material by 1.6 times (water-saturated state) and by 1.1 times (dry state), respectively. The use of the bio-silica additive improves the material strength by 2.6 and 1.4 times, respectively (Fig. 3).

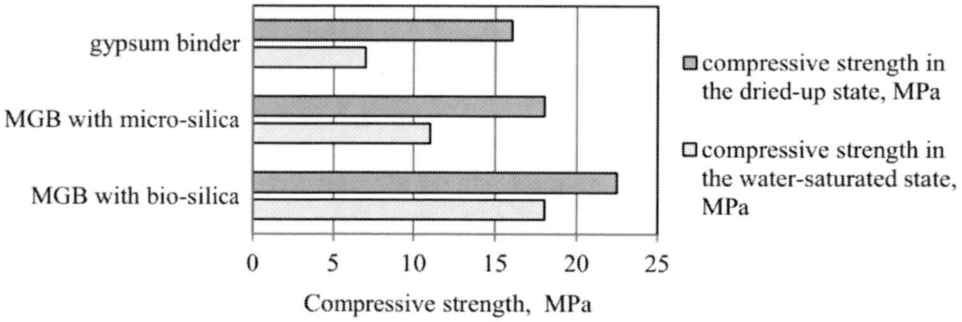

Fig. 3. Comparison of efficiency of the use of micro-silica and bio-silica as the components in a mud-and-silica additive.

The authors studied the influence of the hardening process on the MGB properties. The following curing procedures were used:

1. Natural dry atmosphere (NDA) during 28 days;
2. Thermal-and-moisture treatment (TMT) at the temperature of 80^0C during 8 hours;
3. Normal conditions (NC) (normal hardening chamber) during 28 days.

The production of a material with an improved water resistance requires a thermal-and-moisture treatment of the material or its cure under normal conditions. The use of mud-and-silica additives based on the bio-silica does not require any special hardening conditions. In the process of hardening of these additives under natural dry conditions, the softening factor only slightly decreases, and its value is not less than 0.85 (Fig.4).

Fig. 4 shows that the use of bio-silica as an additive component reduces the change in the softening factor by 3.6 times with respect to the micro-silica. The reduction of the softening factor value for a bio-silica additive equals 5%, while that for a micro-silica additive is 18%. Thus, the use of MGB with bio-silica does not require any special hardening conditions.

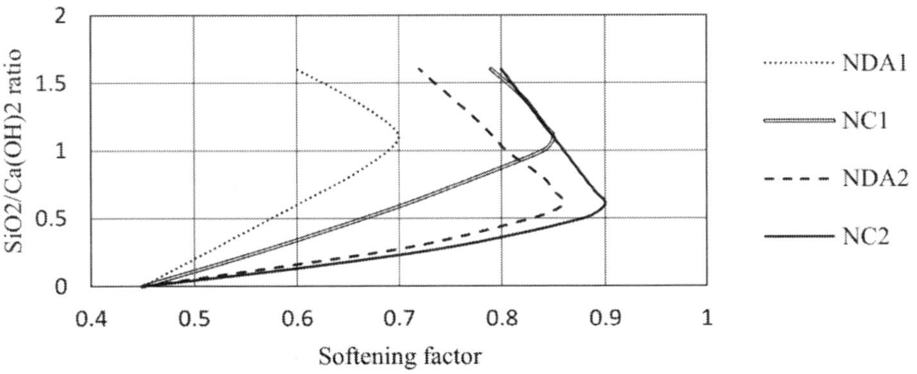

Fig. 4. Change in the strength values for water-saturated samples: 1 - MGB with micro-silica; 2 – MGB with bio-silica for various hardening conditions depending on the $SiO_2/Ca(OH)_2$ ratio in the additive.

In addition to that, the use of bio-silica reduces the material cost and so improves its economic efficiency. The cost of the mud-and-silica additive based on the bio-silica is by 2.4 times lower than that of an additive based on the micro-silica. The studies proved the efficiency of bio-silica, therefore the further research works consider the bio-silica to be the main silica component [3,4].

A slow drying rate for MGB samples may prove the fact that the material structure has less pores and capillaries, i.e. the overall porosity is smaller. So the moisture loss from the material is less and the reactions with the formation of calcium hydrated silicates proceed earlier. It may be explained by a high rate of absorption of the carbide mud by the bio-silica; the absorption value after 30 days is by 4 times higher than that for natural active additives, and it is by 60% higher than the micro-silica activity. The studies showed that in addition to a high activity value for the age of 30 days, an intensive absorption of carbide mud during the first 3 days for the bio-silica takes place with the formation of calcium hydrated silicates. It causes the formation of a finer porous structure and the reduction of the rate of water removal from the hardened stone during the initial curing period, which, in turn, provides for a more complete proceeding of reactions between bio-silica and carbide mud without wet hardening conditions in contrast to the hardening gypsum stone with a micro-silica additive.

The studies of porous structure of the MGB samples at the age of 3 days with the use of a specific surface analyzer as well as the studies of the pore size values and their volume distribution performed through the standard porosimetry method proved the hypothesis of a fine porous material structure.

Table 1. Porosity of produced samples

Samples	Porous structure			
	Total pores (P), %	Open pores (P_O), %	P_O/P	Predominant pore size, μm
GB	36.2	25.2	0.7	0.003 - 0.1
MGB	37.1	15.3	0.42	0.001 - 0.005

The porous structure of MGB considerably differs from the initial gypsum structure: the number of open pores in a hardened MGB is by 1.5-3 times less than that in hydrated gypsum. The hardened MGB has a fine porous structure: the predominant pore size is 0.001-0.005 μm in comparison with 0.003-0.1 μm for hardened hydrated gypsum. So, the micro-pore concentration for hardened GB is about 50% (volume), and for hardened MGB it is about 80% (volume); this fact may be a precondition for the improvement of water resistance and frost resistance of MGB as well as of the concrete based on them.

The X-ray phase analysis (Figs 6, 7), the scanning electron microscopy (Fig. 8) and the thermo-analysis showed new formations including low-base slightly soluble calcium hydrated silicates consolidating the structure of the MGB material, which improves its water resistance and reduces the creep value in comparison with GB materials.

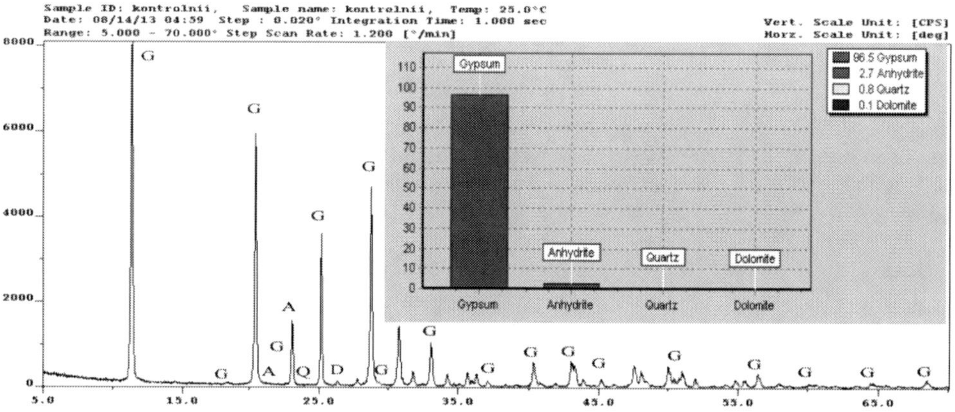

Fig. 6. X-ray diffraction pattern for a GB sample combined with its phase structure:
G - hydrated gypsum; A - anhydrite II; Q - quartz; D – dolomite.

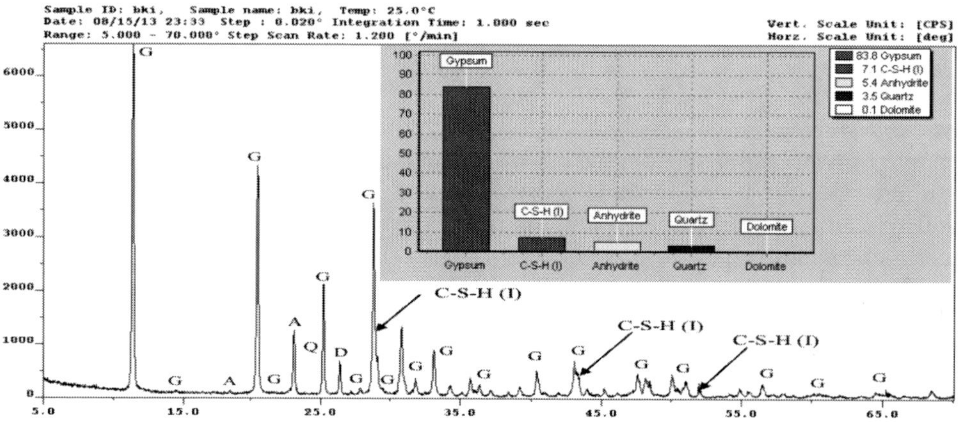

Fig. 7. X-ray diffraction pattern for a MGB sample combined with its phase structure:
G - hydrated gypsum; A - anhydrite II; Q - quartz; D - dolomite; C-S-H (I) - two-base calcium hydro silicate.

A) B)

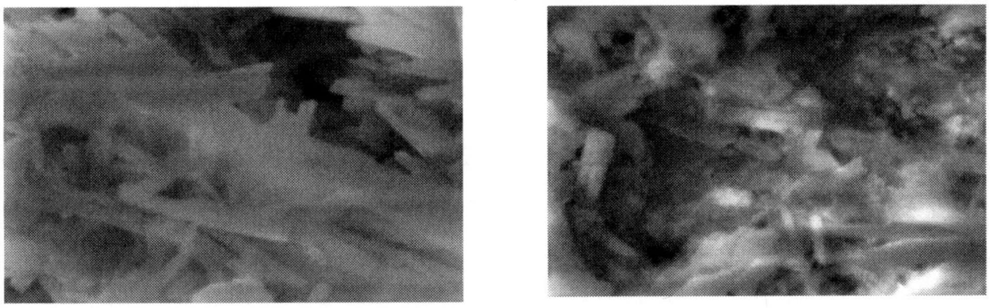

Fig. 8. Scanning electron microscopy of the sample structure (magnification: 5000x):
A - GB; B – MGB.

The study of the influence of curing conditions on the properties of hardened MGB showed that there are several variants of the cure of materials: natural dry conditions, normal hardening conditions and thermal-and-moisture treatment. The steam curing is the best way of improvement of water resistance of products, but this method is rather difficult and not always available. Almost the same effect may be achieved by the process of curing of MGB samples under normal conditions during the period of 14 days. In the process of hardening of MGB samples under dry air conditions (14 days), the softening factor only slightly decreases, and its value is not less than 0.85. Thus, any method is good for the production of water resistant materials.

4 Conclusions

The introduction of a complex additive with carbide mud and bio-silica into the two-water gypsum provides for the production of a gypsum binder with improved water-resistance which allows producing structural elements and structures for the operation under wet conditions including the elements for fencing (wall) structures. Here the hardening process for MGB elements does not require any special cure conditions in contrast to the other known gypsum binders.

The use of a complex additive provides for the formation of a more dense structure due to new formations (low-base calcium hydrated silicates), the reduction of open porosity by 2.5 to 3 times and the improvement of water resistance by 1.8 to 2.2 times with the softening factor of up to 0.91.

Here the compressive strength of the material in water-saturated state becomes by 2.6 times greater.

The intensive interaction of carbide mud with bio-silica during the first 3 days with the formation of low-base calcium hydrated silicates provides for the formation of a fine pores structure and considerably reduces the rate of water removal from the hardening stone during the initial hardening stage which causes a more complete proceeding of reactions between the bio-silica and the carbide mud without wet hardening conditions.

References

1. Korovyakov V.F., Construction Materials, **3** (2008)
2. Kozlov N.V., Panchenko A.I., Bur'yanov A.F., Solov'ev V.G., Buldyzhova E.N., Gal'tseva N.A, Nauchnoe obozrenie, **9** (2013)
3. Panchenko A.I., Bur'yanov A.F., Solov'ev V.G., Kozlov N.V., Pashkevich S.A., Construction Materials, **12** (2014)
4. Patent 2562621 Russian Federation, Binder, (2014)

Integrated approach to industrial sewage water treatment - a way of water resources rational use and protection

Evgeny Alekseyev[1,*] and *Valery* Volshanik[1]

[1]Moscow State University of Civil Engineering, 129337, Yaroslavskoye Shosse 26, Moscow, Russia

Abstract. Ion-exchange softening plants wash water treatment studies are provided in order to bring them back to the process. Poorly soluble Ca and Mg compounds formation conditions have been studied during the course of such water caustic treatment using the potentiometric titration method. Information has been provided about the Ca and Mg hydroxide sludge sorption properties with regard to the characteristic contaminants of the sewage water generated by the textile industry enterprise. Suitability of the hydroxide sludge has been established for use in the sewage water treatment technology in order to remove the persisting organic compounds, such as the synthetic dyes.

Introduction

The waterworks system in multiple enterprises is a consumer of large quantities of softened water. Depending on the enterprise profile, the demand in the softened water may reach as much as 40 % of total water consumption.

At the ion-exchange softening plants, the cation exchanger regeneration after depletion of its exchange capacity is accomplished using 5-10 % NaCl solution, as follows:

$$[K]2\ Ca + 2NaCl \leftrightarrow 2[K]\ Na + CaCl2, \quad (1)$$

$$[K]2\ Mg + 2NaCl \leftrightarrow 2[K]\ Na + MgCl2 \quad (2)$$

These reactions result in that the Cl anions and Ca & Mg cations that form the hardness salts are transferred to the wash water. Their content in the wash water may reach as much as 600 mg·equiv/l and more. The income of the large quantity of the hardness salts into the enterprise sewage water limits its repeated use in the production processes and stipulates the need to extract them during the course of the common sewage water flow treatment. It should be appreciated that the aqueous system formed by this water is related to the stable type containing the conservative mineral and persistent organic substances. Treatment of such water is accompanied by large-scale consumption of reagents and energy [1-2]. Owing to that, it is reasonably practicable to consider a possibility to recover and bring back or return the ion-exchange plants wash water into the process by means of hardness salts

*Corresponding author: AlexeevE@mgsu.ru

sedimentation. This excludes their mixing with the common sewage water flow and negative impact on its chemical composition. On the other hand, the possibility of using this resulting sludge as sorbents for extraction of organic contaminants from the other local industrial sewage water flows shall be evaluated as well [6-8].

Extraction of Ca and Mg from the wash water is principally possible in the form of sediments using various reagents that form poorly soluble compounds [3-4].

For these studies, the sedimentizing reagents have been chosen that are rather cheap, available, and applicable in the textile industry enterprises processes [9,12]. Meanwhile, it is undesirable to use the substances that enrich the water with chemical interaction products that deteriorate its quality in significant manner. Therefore, the following substances have been adopted: Na_2CO_3, NaOH, and $Ca(OH)_2$.

The Ca and Mg compounds contained in the wash solutions form the poorly soluble compounds with these adopted reagents, as follows:

$$CaCl2 + Na2CO3 \rightarrow CaCO3\downarrow + 2NaCl \quad (3)$$

$$MgCl2 + Na2CO3 \rightarrow MgCO3\downarrow + 2NaCl \quad (4)$$

$$CaCl_2 + 2NaOH \rightarrow Ca(OH)_2\downarrow + 2NaCl \quad (5)$$

$$MgCl_2 + 2NaOH \rightarrow Mg(OH)_2\downarrow + 2NaCl \quad (6)$$

In general case, these reactions provide simultaneous extraction of Ca and Mg sludge within a single stage. Meanwhile, weight ratio between $Ca(OH)_2$ and $Mg(OH)_2$ in the sludge or sediment will depend on their content in the wash water, hydroxide form solubility, and sufficiency of the sedimentizing reagents quantity [5].

The summarized data concerning the solubility properties of the sludge formed through the reactions (1) are provided in table 1.

Table 1. Solubility of Ca and Mg hydroxides and carbonates

Poorly soluble compound	Solubility product (SP)	Water hardness after sedimentizing, mg-equiv/l
$CaCO_3$	$4.8 \cdot 10^{-9}$	0.035
$Ca(OH)_2$	$3.1 \cdot 10^{-5}$	5.55
$MgCO_3$	$1.0 \cdot 10^{-5}$	2.54
$Mg(OH)_2$	$5.0 \cdot 10^{-12}$	0.065

Since the sewage water of the water softening plant in the most general case contains both Ca salts, and Mg salts, in the experiments accomplished NaOH has been used as the reagent efficient for both ions sedimentizing.

In addition to the main objective - determination of optimal conditions for sludge formation - another point of interest was the use of neutralization potentiometric titration method for identification of the content values and conversion of the water permanent hardness components [10,11,13].

Potentiometric studies of the complex sediments formation process have been accomplished using the titration of the Ca and Mg salts solutions blends with NaOH. These measurements have been accomplished for different ratios of Ca and Mg in the blends, as well as for the samples of water delivered to the water softening plant of the worsted and wool products enterprise. The results of measurements are illustrated in Fig. 1A.

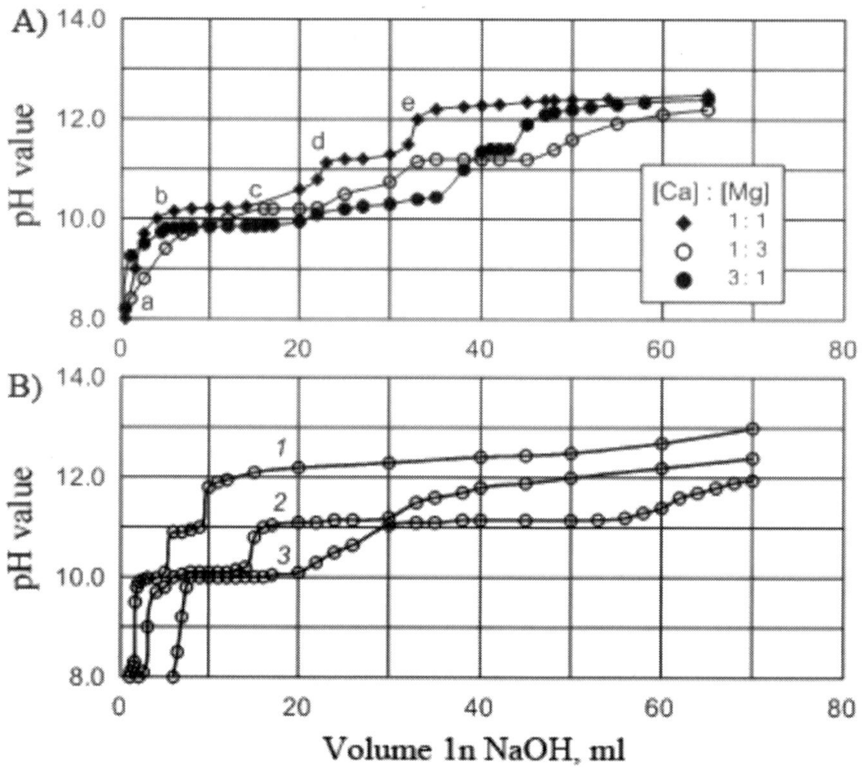

Fig. 1. Titration potentiometric curves: A - Ca and Mg salts blends with various [Ca]:[Mg] ratios; B - wash water of the Na cation exchanger with the following hardness values: 1 - 50 mg-equiv/l; 2 - 260 mg-equiv/l; 3 - 480 mg-equiv/l.

The analysis of the given dependencies allows revealing main regularities of the sedimentizing from the solutions that contain both Ca salts, and Mg salts. The resulting curves are characterized by sufficient selectivity for quantitative estimate of the formed compounds with the acceptable degree of accuracy. As such, the ratio between the standard solution (titrant) volumes at *b-c* and *d-e* sections rather well accords with the ratio between the Ca and Mg concentrations established by chemical analysis.

Ion-exchange filters wash water for the study has been taken from the enterprise softening plant draw-off conduit. The results of the hydroxides sedimentizing from the wash water having the permanent hardness equal to 50, 260, and 480 mg-equiv/l are illustrated in Fig. 1B.

During the course of the wash water titration, the point when the flakes begin to form occurs with the pH change within 10.1-10.3 range. This value remains unchanged when the volume of the added caustic increases, while the significant increase in the formed dispersion quantity is noted. Further titration leads to the pH increase up to as much as 11.0-11.2; after that, the pH value is stabilized and forms the second semi-wave. In this range, the structure of the formed sludge changes. By comparing the resulting dependencies of the wash water titration with the Ca and Mg standard solutions titration curves (see Fig. 1A), one can see rather close position of the semi-wave in terms of pH value that correspond to the Ca and Mg hydroxides formation. Meanwhile, good harmonization of the caustic volumes spent for titration within the first and second semi-waves with the corresponding water permanent hardness value should be noted. The analysis of the resulting curves allows determining the equivalent ratio between the Ca and Mg ions. For the studied wash water, this ratio approximately equals to 1:2.

The potentiometric titration curves for the wash water from the ion-exchange filters regeneration operations enable to determine the optimal pH area and the required quantity of caustic for the Ca and Mg hydroxides sedimentizing.

The evaluation of the suitability of the hydroxide sludge obtained from the wash water for use in the industrial sewage water reagent-based treatment technology has been accomplished on real-life dyeing and finishing sewage water. The sorption parameters of extraction from the sewage water of the industry-wide used dyes, such as active, sulfix, direct, and disperse, have been studied.

The processing of the dyes sorption results with the Ca and Mg hydroxides using the Langmuir isotherm has enabled to determine the marginal adsorption average values, as illustrated in table 2.

Table 2. Marginal adsorption average values for various types of dyes on Ca and Mg hydroxides

Dye class	Marginal adsorption, mg/mg-equiv	
	$Mg(OH)_2$	$Ca(OH)_2$
Active	2.04	3.86
Sulfix	-	0.615
Direct	10.63	1.73
Disperse	23.84	13.68

The most of efficiency in extraction of dyes from the sewage water is noted when using the $Mg(OH)_2$ sludge. The exception is the sulfix class dyes, where rate of extraction by the sludge disperse phase is rather poor.

By comparing the marginal adsorption values, one can see that the deepest cleansing of water from active, direct, and disperse dyes may be provided when using the Mg and Ca hydroxides simultaneously.

In order to evaluate the efficiency of the sludge extracted from the ion-exchange filters exhausted regenerative solutions, a number of experiments using the settling tank as the separating stage have been accomplished in the textile industry enterprise real-life sewage water treatment system. The quality of the treated water has been evaluated using the standard methods in the samples taken after 60 min of settling (Fig. 2).

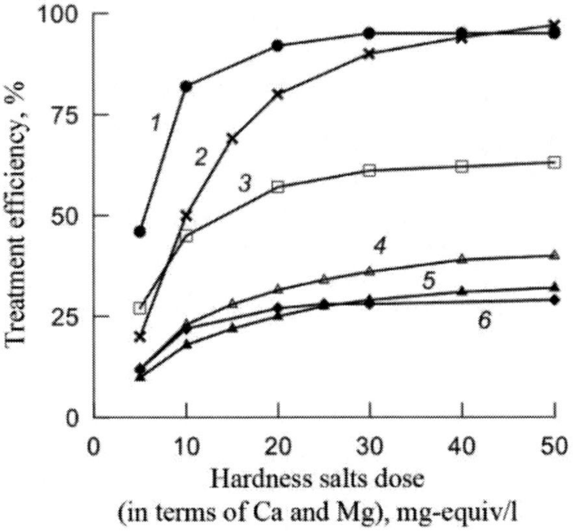

Fig. 2. Influence of the hardness salts dose onto the industrial sewage water treatment efficiency: 1 - color upon dilution; 2 - suspended solids; 3 - chemical oxygen demand (COD);4 - anionic surfactants; 5 - non-ionogenic surfactants; 6 - petroleum products.

The analysis of the resulting dependencies leads to a conclusion that the interaction of the real-life sewage water contaminants with the fresh Ca and Mg hydroxides sludge occurs via the adsorption mechanism. The most of the treatment efficiency (over 90 %) has been achieved for hard disperse impurities adsorbed by the hydroxides at the structure formation moment, and for the dyes. Extraction of hydrocarbons, such as the petroleum products and surfactants, does not exceed 40 %, which may be explained by the hydrophilic nature of Ca and Mg hydroxides. The change of the COD value is adequate to the extraction of the organic part of the contamination.

The results obtained have confirmed the possibility to use the ion-exchange softening plants wash water for treatment of the sewage water flow generated by the textile industry enterprises dying shops before the mixing of such flow with the common flow.

Conclusion

The reasonability of the separation of the softening plants sewage water flow from the common flow and Ca & Mg salts sedimentizing has been justified using the experimental approach. It has been shown that the potentiometric titration curves for the wash water from the ion-exchange filters regeneration operations enable to determine the optimal pH area and the required quantity of caustic for the Ca and Mg hydroxides sedimentizing. The separation of the sewage water flow from the water ion-exchange softening plants in the industrial enterprises water disposal system enables to use such flow as the sedimentizing solution in the sewage water treatment processes, when such sewage water contains dyes. This integrated technological solution provides for the reduction of water consumption owing to the return or bringing back the wash water into the water treatment process, reduction of costs spent on the reagents for the industrial waste water common flow treatment, and reduction of the contaminants or impurities ingress into the environment.

References

1. E.V. Alekseyev, Vestnik MGSU, **2**, 140-144 (2012)
2. E.V. Alekseyev, Water and Ecol.: Probl. and Solut., **4**, 68-79 (2015)
3. L.S. Grigoryeva, *Physical and chemical evaluation of the natural water quality and natural water treatment* (Publ. House of the Const. Higher Educ. Inst. Ass., 2011)
4. V.I. Sidorov, E.P. Aganesyan, T.P. Nikiforova, *Chemistry in construction* (Publ. House of the Const. Higher Educ. Inst. Ass., 2007)
5. E.V. Alekseyev, *Physical and chemical treatment of sewage water* (Publ. House of the Const. Higher Educ. Inst. Ass., 2007)
6. E. Gogina, A. Pelipenko, MATEC Web of Confer. **73** (2016)
7. A.G. Pervov, A.P. Andrianov, T.P. Gorbunova, A.S. Bagdasaryan, Petrol. Chem. **55,** 879-886 (2015)
8. N. Makisha, Y. Voronov, E. Poupyrev, V. Volshanik, Inter. Journal of App. Eng. Res. **10**, 41919-41922 (2015)
9. I. Kantardgi, V. Maderich, E. Terletskaya, I. Brovchenko, L. Prokhoda-Shumskikh, *MEDCOAST 2013 Joint Conference*, **2**, 989-1000 (2013)
10. N.V. Suikova, Y.V. Bryanskaya, V.S. Borovkov, Water Res. **39**, 200-207 (2012)
11. A.G. Pervov, A.P. Andrianov, E.B. Yurchevskiy, Petr. Chem. **55**, 871-878 (2015)
12. N. Zaletova, S. Zaletov, Inter. Journal of App. Eng. Res. **10**, 43873-43874 (2015)
13. N. Makisha, V. Scherbakov, A. Smirnov, E. Scherbina, Inter. Journal of App. Eng. Res. **10**, 44347-44349 (2015)

Improvement of the system of inorganic waste recycling to man-made soils

Konstantin Chertes[1], *Olga* Tupitsyna[1], *Vitaliy* Pystin[1,*], *Aleksey* Savelyev[1], and *Elena* Martynenko[1]

[1]Samara State Technical University (SamSTU), Russia, 443100, 244 Molodogvardeyskaya St.

Abstract. Elements of the complex evaluation of inorganic waste condition to use it as a source of raw material for production of man-made soils with application of staged criterial selection are introduced. The proposed elements allow identification of waste which can be recycled to man-made soils within the waste-forming area of a region. Complex technology of sludge treatment prior to disposal is developed. The proposed system allows for reduction of expenditures on purchasing natural soils for recovery of disturbed territories, as well as waste placement costs.

1 Introduction

The growth of urban agglomerations leads to the extension of waste-forming area, which in this paper refers to a combination of waste sources and logistics of waste transportation. The most common large-capacity types of waste in waste-forming areas of the Central Russia are as follows: water industry sludge (WIS), sewage sludge (SS), oil sludge and oil-contaminated soils (OS & OCS), overburden rocks (OR), demolition waste (DW), municipal solid waste (MSW) and drilling slurry (DS). Most of these types of waste contain mainly inorganic compounds. At the present time the vast majority of inorganic waste is disposed to storage pits.

Waste storage pits withdraw from use dozens of thousands of square kilometers of lands. Strengthening of the land and environmental legislation, as well as the need to develop disturbed lands, require creation of the methods for waste processing and disposal which ensure minimum impact on the natural environment components (NEC).

A promising direction of inorganic waste disposal is recycling into man-made soils (MMS) used for recovery and improvement of disturbed territories. This reduces the required amount of natural inorganic soils.

In order to involve inorganic waste into utilization to the maximum extent a complex system of waste processing and disposal with simultaneous land recovery is proposed. The system includes the following:

1. Stepwise evaluation of waste suitable for recycling taking into account resource, environmental and technological criteria of given waste-forming area;

* Corresponding author: vitaliy.pystin@yandex.ru

2. Development and adaptation of the technology for waste treating in order to make it as similar as possible to the geoenvironment being recovered in terms of physical, chemical and geomechanics characteristics.

3. Use of multivariate data processing techniques for evaluation of suitability of waste deposits for utilization as a source of raw material for MMS production with division into groups by labor intensity at the stage of preparation to recycling.

4. Design of facilities for recycling of inorganic waste into MMS and feasibility study (FS)

Application of the complex system contributes to the efficient determination of waste disposal directions, including recovery of geoenvironment using MMS.

Utilization of MMS produced from neutralized inorganic waste covers the following directions of disturbed geoenvironment development:

- recultivation of landfills of municipal and industrial waste;
- remediation of lands degraded as a result of accidental spills of toxic substances;
- preparation of industrial sites for new construction;
- development of temporary transportation system in remote deposits of raw materials.

Borrow pits can often be located at a considerable distance from the areas being reclaimed, and their utilization is associated with additional impact on NECs. An alternative to natural borrow pits is man-made deposits of inorganic waste accumulated in storage pits, such as sludge dumps, drilling slurry pits, tailing dumps and mineralized MSW deposits. In this case systemic approach to evaluation of the condition of these objects can give grounds for possible use of waste for production of MMS, which will further be used for recovery of geoenvironment.

The first stage of the feasibility study is a criterial selection of the objects applying resource, environmental and technological criteria.

Resource criterion (1) shows practicability of the use of inorganic waste as a raw material for MMS production.

$$R = \frac{V_{was.} \times k}{V_{req.}}, \qquad (1)$$

where $V_{req.}$ - required volume of soil to perform the works on recovery of the land being developed;

$V_{was.}$ - actual total volume of inorganic waste within the boundaries of given waste-forming area;

k (0...1) - coefficient corresponding to the portion of inorganic component in waste, which constitutes a basis for shaping structure and individual soil-forming properties of MMS.

The optimal value of resource criterion is 1, when the demand for natural soils is compensated by the MMS produced. If R <1, then the amount of MMS is insufficient for recovery of the geoenvironment, and additional natural soils are required. If R >1, there is risk of surplus production of the waste-based MMS, which will not be demanded further on.

Evaluation of waste-forming areas of the territories under study using resource criterion in most cases reveals excess of demand for soils over resources availability within the range of the geoenvironment being recovered. In its turn, the deficiency of natural soils can be covered by using low-toxic inorganic waste, and the second stage will provide environmental assessment of this waste.

At this stage geoenvironmental peculiarities of the territory under study are also evaluated, and the selection is made taking into account encumbrances restricting the works. This refers to the areas under special natural resource management, recreation areas, water conservation areas, close proximity of residential areas and a number of other restrictions.

Environmental criterion (2) gives grounds for the use of inorganic waste for MMS production in terms of its toxicity and the condition of the geoenvironment being recovered:

$$E = \frac{Ic_{was.}}{Ic_{geo.}}, \qquad (2)$$

where $Ic_{was.}$- waste contamination index;
$Ic_{geo.}$ - geoenvironment contamination index.

Contamination index of the disturbed geoenvironment $Ic_{r.c.}$, being recovered using MMS was assumed according to the existing regulatory documents [1], and it equals to 16 and 32, which, respectively, corresponds to acceptable and moderate contamination categories of grounds and soils. If E < 1, the waste is non-hazardous for the geoenvironment and can be used without restrictions. If E > 1, the waste requires neutralization prior to recycling. Complexity of processing and complex of methods for waste treating are associated with its evaluation by the technological criterion.

Table 1. Basic types, characteristics and processing technologies of large-capacity inorganic waste (Samara Region case study).

P. No.	Type of waste	Portion in waste-forming area, %	Basic waste characteristics*					Key component determining the method of treating	Proprietary technologies
			Ash content, %	Moisture content, %	Density, t/m³	Porosity coefficient,	Deformation modulus, MPa		
1	Landfill soils	30-50	40-60	35-65	0.25-1.25	9.0-12.0	up to 1	Biodegradable organics content; Deformation modulus	1. Isolation [3]; 2. Acceleration of waste deposit settlement [4]; 3. Surface cementation [5]
2	WIS	15-20	70-98	25-85	1.1-1.3	1.4-3.0	1-3	Moisture content; Ash content	1. Geo-containerized dewatering; 2. Mineralization; 3. Strengthening [6];
3	OS & OCS	5-25	60-95	30-60	0.7-1.4	4.0-7.0	3-7	Oil products concentration	1. Biodestruction [7]
4	DS	up to 10	85-99	40-80	1.1-1.8	1.2-4.0	2-5	Oil products concentration; Salts content	1. Geo-containerized dewatering; 2. Reagent treatment [8]
5	SS	up to 5	20-60	65-95	0.8-1.2	-	up to 0.5	Moisture content; Ash content; Pathogenic microbial flora content	1. Mechanical and geo-containerized dewatering; 2. Biothermal treatment with thermophilic phase extraction [9]

A key principle for selection of treatment technology is to achieve maximum similarity between composition, properties and structure of inorganic waste and the geoenvironment being recovered. For this purpose the use of proprietary technologies of waste preparation is proposed with taking into account separate and combined application of field and mechanized methods of treatment. The complex of these methods shall provide minimum impact on the natural environment components. Selection of technologies and technological sequence depends on natural and mane-made characteristics of waste, its origin, as well as similarity between the content of key components in the waste and the geoenvironment [2] (Table 1).

According to research data of Samara State Technical University in 2000-2015.
Analysis of the table shows that the main obstructions for the direct use of waste as MMS are high moisture content and presence of organics. These characteristics along with fine texture and low density of inorganic component of waste prevent direct use of the waste as MMS without preliminary treatment. When using MMS, their properties shall comply with the requirements for natural soils. Level and complexity of the required treatment of waste in order to bring its properties into line with the parameters of the geoenvironment being recovered are evaluated using technological criterion (3).

$$I_{(E,\tau)} = \left\{ I_W = \frac{W_{was}}{W_{MMS}}; I_\rho = \frac{\rho_{was.}}{\rho_{MMS}}; I_M = \frac{M_{was.}}{M_{MMS}}; \ldots, \right. \quad (3)$$

where $I_{(E,\tau)}$ – integral criterion of mechanical properties of MMS;
$I_{w,\rho,M}, \ldots$ – criteria of dewatering, density and fineness modulus, respectively;
w,ρ,M – moisture content, density and fineness modulus of inorganic component of waste and man-made soil, respectively.

Combined evaluation of waste condition using technological criterion allows determining methods, their sequence and level of controllable treating for achieving the required characteristics of the MMS produced. As a result of this stage of selection types of waste are divided into groups differing by the methods of treating. Multivariate evaluation in this case allows determining the required complex of methods for treating waste in order to transform it into MMS with desired characteristics (Fig. 1).

The Figure shows a flowchart for production of MMS with specified level of responsibility from the waste with initial characteristics (governing parameters) which involves application of one or another method of treatment (controllable treating).

The flowchart also represents the suggested directions of use of MMS obtained by sequential application of controllable waste treating. In this case the treating (technologies) combination is determined by both initial characteristics of the waste being treated and the intended use of MMS. Generally, large-capacity inorganic waste, in most cases, is not acutely toxic (WIS) and can be used as replacement for natural soils after geo-containerized dewatering. Some types of waste contain organics and can be neutralized using the methods of mineralization (SS, OCS), some of them can be processed using binding and adsorbing additives (DS). MSW and dewatered WIS can be involved into construction operations through strengthening of waste deposits.

For designing of waste recycling facilities, as well as in order to create and adept treatment technologies, an integrated survey of the condition of waste deposits and associated disturbed lands is required using the methods of engineering surveying. Evaluation of the condition refers to the need of obtaining reliable information about structure and properties of the waste deposits. Generally, these data are collected for qualitative and quantitative evaluation of adverse effects of waste for the geoenvironment and do not allow considering the waste as a source of raw material for MMS production.

Condition of the disturbed territories, in general, and waste deposits, in particular, are evaluated through a large number of parameters which determine selection of treatment methods and directions of further development of lands. For easier evaluation of the condition of multiple waste storage pits and associated disturbed territories, as well as selection of methods of recovery, a classification is proposed to divide them into groups by the combination of characteristics using multivariate data analysis [10].

Multivariate data analysis was successfully used by the specialists from Samara State Technical University for evaluation of MSW, OCS, WIS and SS waste deposits [11,12,13].

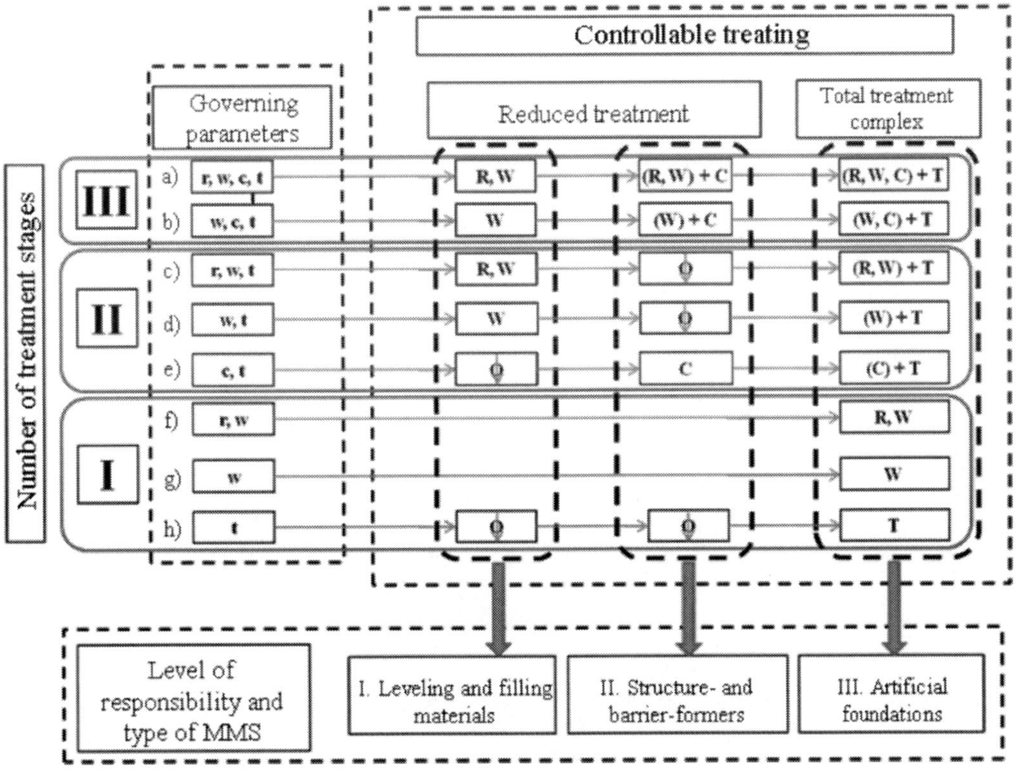

Explanation:
r, w, c, t (high specific filtration resistance, high moisture content, presence of degradable organics and low strength, respectively) – parameters, which determine the need for the various methods of controllable waste treating.
R, W, C, T (conditioning, dewatering, mineralization and strengthening, respectively) – methods of controllable initial waste treating.
O – possibility of using waste as MMS without additional treatment.

Fig. 1 Flowchart for classifying initial sludge by types of man-made soils after selected treatment stages.

This approach allows selecting qualified directions of disposal for the accumulated waste and classifying waste deposits by possible prospects of using them for MMS production.

Apart from that, special aspects of this evaluation method allow identifying waste deposits with excess moisture, organics and toxic components contents, which require treatment and neutralization.

Known methods of treating such waste involving mechanized techniques usually require construction of capital facilities, purchasing expensive equipment and considerable energy costs. In many cases negative economic effect from utilization of such technologies is accompanied by minimum environmental effect, and the result is generation of waste which requires burying. In addition, the limiting factor for the implementation of these methods is that the waste deposits are often locate at a large distance from the developed engineering infrastructure. Therefore, sludges preparation shall be carried out using technically available and environmentally efficient methods. These methods include geo-containerized dewatering, mineralization, reagent neutralization and strengthening [6, 7, 8].

Sequential selection of waste by resource, environmental and technological criteria within the implementation of the proposed system makes it possible to develop efficient and technically available technologies of MMS production using geo-containerized dewatering, stacked layers mineralization and strengthening, as well as realize this technological sequence in the form of waste recycling facilities (Fig. 2).

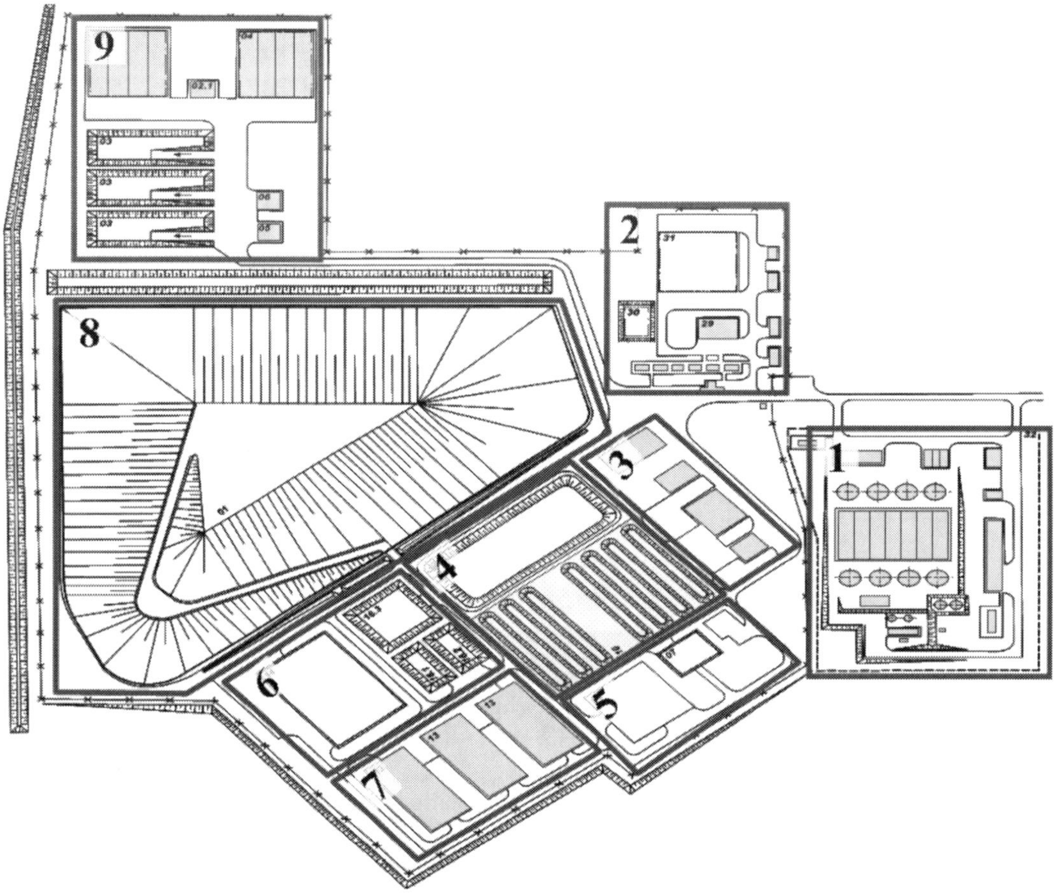

Explanation:
1 – Area of delivery, incoming inspection and preparation of waste; 2 – Area of auxiliary facilities; 3 – Area of temporary accumulation of reagents and materials; 4 – Area of OS & OCS treatment; 5 – Area of MMS strengthening; 6 – Area of DS reagent neutralization; 7 – Area of WIS geo-containerized neutralization; 8 – Area for placement of neutralized fractions of waste; 9 – Area of temporary accumulation of toxic waste components.

Fig. 2 Facilities for inorganic waste recycling with MMS production

Conclusion

The provisions represented in this paper were used in designing the facilities for sludge waste recycling and recultivation of disturbed geoenvironmental compartments in Novokuybyshevsk urban agglomerations in Samara Region. Treatment of inorganic waste and its recycling to MMS used for recovery of the geoenvironment will allow reducing required amount of natural soils down to 20,000 ths. m^3/year.

References

1. SanPiN 2.1.7.1287-03. Sanitary and Epidemiological Requirements on Quality of Soil (2003).
2. D.E. Bykov, Complex Multi-Level System of Examination and Recycling of Industrial Waste, 102 (2003)
3. Patent 2318619 RF, Method for Coating Formation on Waste Storage Pits No.2006124064/03 (2013).
4. Patent No. 2487767 RF, B09B1/00. Method for Disposal of Municipal Solid Waste and Device for Its Implementation, No. 2011117255/13 (2013)
5. K.L. Chertes, A.A. Savelyev, O.V. Tupitsyna, E.G. Martynenko, Proceedings of the 5th International Ecological Congress (the 7th International Scientific and Technical Conference), 283-290 (2015)
6. V.N. Pystin, Bulletin of Moscow State University of Civil Engineering **2**, 110-129 (2015)
7. D.V. Zelentsov, Ecology and Industry of Russia, 33-34 (2011)
8. V.N. Pystin, Ecology and Industry of Russia **6**, 13-17 (2014)
9. Patent No. 1816742 RF Method of Biothermal Treatment of Sewage Sludge (1991)
10. K.H. Esbensen, *Multivariate Data Analysis In Practice 4-th Ed.* (2000)
11. V.V. Ermakov, *Classification of Oil Sludge Dumps and Prediction of Waste Biodestruction Process during their Liquidation* (2010)
12. V.N. Pystin, Bulletin of Moscow State University of Civil Engineering **6**, 88-101 (2015)
13. O.V. Tupitsyna, *Development of Natural and Man-made Systems of Urban Agglomerations* (2014)

Development of ecologically safe technology of recycling of industrial waste in the production of modified non-autoclave aerated concrete

Evgeniya Tkach[1,*]

[1]Moscow State University of Civil Engineering, Yaroslavskoe sh. 26, 129337 Moscow, Russia

Abstract. The results of environmental monitoring for environmental security within the framework of territorial administration of the Central Federal district of the Russian Federation showed the necessity of developing a new ecological mechanism for rational control of the process of recycling of technogenic wastes. It is established that at increase of industrial production and the growth and accumulation of industrial waste and increases the negative impact on the environment. Determined that the production of phosphorus mineral fertilizers in Russia as a whole is formed 25 million tons per year of phosphogypsum, and utilized only a tenth. In Russia on distilleries waste DDGS is 9-14 million tons per year, not utilized – about 1 million tons. In Russia milk manufacture gives rise to waste of whey in an amount of about 6 million tons per year. Warehousing, industrial waste dumps occupy thousands of hectares. They are washed into sewers, groundwater, reservoirs, result in the earth, causing serious ecological damage to the environment. This requires the disposal of such waste. The most promising method of disposal is considered as the placement of industrial waste in the production of construction materials, particularly aerated concrete products.

1 Introduction

When the analysis and synthesis of contemporary research and experience in the production of cellular concrete is determined that it is necessary to continue the search for new environmental and technical ways to expand the raw material base of their production. Moreover, increased production of cellular concrete by 2020, and environmental degradation due to the growth of volumes of industrial waste requires the adoption of new solutions. Best way disposal is a disposal in the production of effective and environmentally-friendly porous concrete. This is confirmed by the analysis of existing studies. However, if environmental and technical analysis found that existing technology of production of cellular concrete have a number of disadvantages that affect environmental safety. These include: energy and explosive pressure production with a high pressure and temperature, the possibility of explosion by release of hydrogen from the concrete mass after swelling [1-2]. A set of problems associated with the construction technical properties

* Corresponding author: ev_tkach@mail.ru

of concrete mixtures arises: the excess of mixing water during the injection molding technology is that the increase of the set of structural strength and manufacturing process, the presence of waste after cutting and cutting crusts of the problem of interrelation between the viscosity of the system and the limiting shear stress of concrete mixtures [3-4].

2 Problem statement

The goal of the presented research is to develop two-stage environmental mechanism for large-scale disposal of industrial wastes in the production of concrete products with improved physico-technical properties. It is established that two-stage ecological mechanism of large-scale disposal of industrial wastes in the production of concrete products includes: environmental assessment of industrial waste; the definition of the full life cycle of products based on industrial waste; the two-stage method of waste disposal while obtaining complex organic-mineral modifier and modified non-autoclave gas-concrete products; environmental assessment of production technologies of concrete products based on industrial waste; the total ecological rating of technogenic wastes, technology of production and operation of concrete products based on industrial waste throughout the entire life cycle of products. Based on a proposed block diagram and the theoretical basis has been developed ecologically safe technology of production of structural and heat-insulating modified aerated concrete. It includes several main stages of technological process: preparation of raw materials, preparation of concrete mixture, molding; heat treatment.

The method of disposal of anthropogenic wastes comprises of the production technology of modified non-autoclave aerated concrete products, operating in a closed cycle and includes the main stages: pre-autoclaved aerated concrete mixer in the mixing of the batched by weight components: sand , cement with water and phosphogypsum for 3-5 min.; cooking in the mixer slurry of aluminum in an aqueous solution of a surfactant (organic modifier) to remove paraffin from the surface of the aluminum powder; mixing the autoclaved aerated concrete mixer is obtained a slurry of aluminum in an aqueous solution of a modifier with the basic components of aerated concrete mixture for 1-3 minutes; pouring the thoroughly mixed concrete of a mixture forms when the air temperature in the shop +18-20 ° C; the swelling of the concrete mass in a fixed form (9) within 30-50 min.; cut layer and cutting the array in blocks of special cutting machine with metal strings, reciprocating and rotational motion, grinding, cut "salmon", i.e., extending beyond the dimensions of the array in its upper part, and the return of this waste in the autoclaved aerated concrete mixer for mixing with the main components of the mixture; heat treatment of the cut blocks in the chamber with insulating cover. Cap allows to reduce energy consumption compared to traditional thermal processing in autoclaves. Curing takes place within 5-6 hours. The use of thermal insulation cap allows to protect the aerated concrete products from drafts and heat loss, which enables to obtain a stable quality of the blocks; stacking, packing of ready-made blocks and sent to the warehouse of finished products; defective product sent to the grinding with the subsequent return to mixing in the autoclaved aerated concrete mixer with the components of the mixture.

On the ecological safety of the environment will influence the quality of the structure and the quantitative values of the indicators of construction and technical properties [5-7]. The definition of such properties of the modified non-autoclaved aerated concrete based on industrial wastes has allowed to establish the suitability standard, i.e. the possibility of their use in construction. The quality of concrete structures depends not only its strength, but also such important characteristics as environmental safety, high density, thermal insulation, water absorption, capillary leak, and others [8-10].

3 Results and discussion

The results of x-ray diffraction and microstructural analyses found that the use of modified aerated concrete based on industrial wastes will provide ecological security by improving the uniformity, structure hardening, the cement consumption and energy production and home heating while reducing the release of heat into the atmosphere. To conduct microstructural analysis of the samples obtained modified concrete was used to install an electronic raster microscope-microprobe FEI Quanta 200 SEM(fig.1-2).

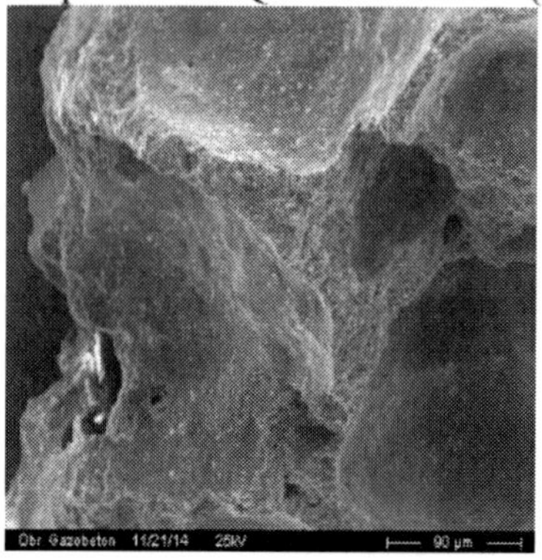

Fig. 1. Structure of modified non-autoclave aerated concrete: macrostructure.

Fig. 2. Structure of modified non-autoclave aerated concrete: microstructure.

The types and distribution of the pore diameters in the control composition and a modified concrete made of the proposed technology are shown in table 1.

Table 1. Porosity of modified aerated concrete.

The types of pores	The pore size, cm	Share in total porosity, %
The control (traditional concrete)		
Gel	$1\cdot10^{-6}-5\cdot10^{-6}$	10 – 15
Capillary	$8\cdot10^{-5}-2\cdot10^{-4}$	15 – 25
Wire mesh	$10^{-4}-(0.12-0.15)$	60 – 75
Modified aerated concrete		
Gel	$1\cdot10^{-6}-5\cdot10^{-6}$	7 – 9
Capillary	$8\cdot10^{-5}-2\cdot10^{-4}$	10 – 15

Wire mesh	10^{-4} – (0.05 – 0.08)	76 – 83

To confirm the conformity of the modified non-autoclaved aerated concrete based on industrial wastes it is necessary to conduct not only the microstructural, but also x-ray studies. The results of x-ray diffraction studies are presented in figure 3 and 4.

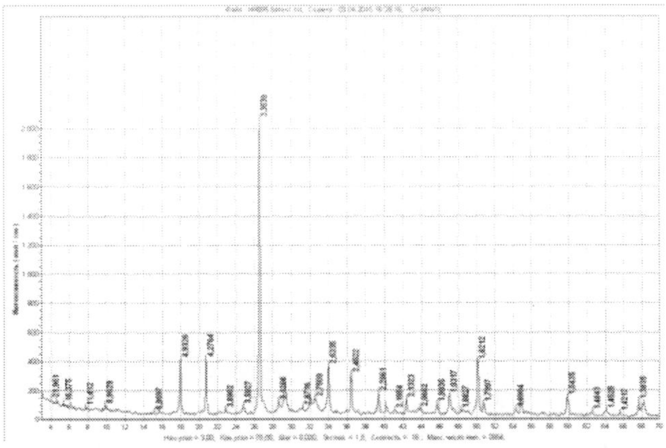

Fig. 3. Diffraction patterns of aerated concrete of the control composition (traditional).

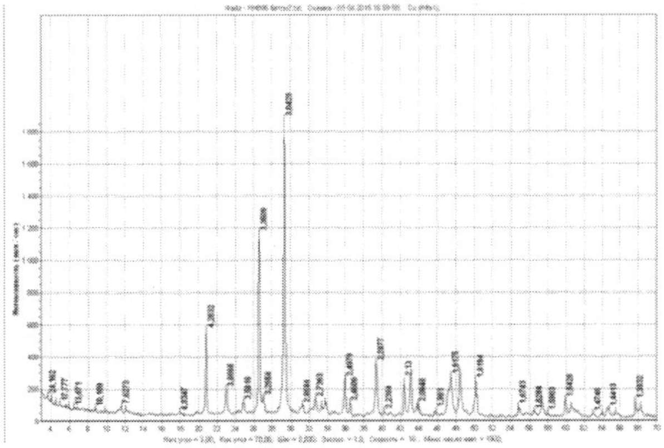

Fig. 4. Diffraction patterns of aerated concrete of the modified composition.

X-ray analysis in figure 1, allowed the identification of the following compounds: alit C_3S with the interplanar distance with the peak d=(3.03; 2.78; 2.75; 2.61; 2.18; 1.76; 1.49 etc.)· 10^{-10} m. At the peak d=1.76 assesses the degree of hydration of cement. Modified aerated concrete it is equal to 91 %,and it assesses the aerated concrete control or traditional composition-67%.

Therefore, the first higher strength and other properties better than the second: C_3A with the peak d= (4.23; 4.08; 2.05; 1.91)· 10^{-10} m;, Belit β_L-C_2S with the peak d= (2.748; 2.609; 2.189; 2.047; 1.98; 1.696)· 10^{-10}m; C_4AF with the peak d = (7.32; 2.67; 2.04; 1.92;1.81;1.39) 10^{-10} m; quartz: β-SiO_2 with the peak d=(4.26; 3.35; 1.82; 1.54)· 10^{-10} m. This suggests that quartz sand in a modified aerated concrete became part of the matrix of interporous partitions, reinforcing them and reduced shrinkage, natural gypsum $CaSO_4{\cdot}2H_2O$ with the peak d=(7.56; 4.26; 1.99; 1.66)· 10^{-10}m, calcite $CaCO_3$ with the peak d= (3.828; 3.04...3.06; 2.511; 2.301; 1.88)· 10^{-10} m. CSH(I) the partially crystallized calcium silicate hydrate with the peak d = (12.5;3.07; 2.8; 1.83)· 10^{-10} m, CSH(II) with the peak d = (9.8; 3.07; 2.8; 2.0; 1.83; 1.56)· 10^{-10} m., $Ca(OH)_2$ with the peak d = (4.93; 3.11; 2.63; 1.78...1.79; 1.315)· 10^{-10} m. The portlandite is a weak link cement material to the action of corrosion.

According to the peak at d=4.93 we can say that the composition of the modified concrete contains almost 7 times less portlandite than conventional aerated concrete. It is known that soluble substance can leach from the material. Ettringite is a high strength joints in cement systems formed with the participation of gypsum and aluminates. Modified aerated concrete it is formed with the participation of phosphogypsum. With the peak at d=9,73 we can say that the composition of the modified concrete contains 2 times more ettringite than in traditional concrete: -calcium silicate hydrate – $3CaO{\cdot}2SiO_2{\cdot}3H_2O$ with the peak d = (6.46; 5.74; 3.19; 2.84; 2.74; 1.704; 1.604)$\cdot 10^{-10}$ m ;- calcium hydrated ferrite –$3CaO{\cdot}Fe_2O_3{\cdot}6H_2O$ with the peak d = (5.18; 4.50; 3.402; 2.07; 1.71)$\cdot 10^{-10}$ m; - calcium hydrated aluminate – $3CaO{\cdot}Al_2O_3{\cdot}6H_2O$ with the peak d = (5.14; 4.46; 2.3; 2.23; 2.04; 1.68; 1.60)$\cdot 10^{-10}$ m.

Optimal spatial structure of aerated concrete consists of cement-sand matrix, filled with small pores (less than 0.8 mm in diameter). The maximum porosity of aerated concrete with a thickness of interporous partitions 80-100 microns and a pore diameter of 0.8 mm up to 76 to 83 %. The use of complex organic-mineral modifiers of the type of GM in the modified non-autoclave aerated concrete products with man-made waste significantly increases the probability of receiving quality products by controlling the processes of gas evolution and structure formation due to the organic modifier.

From table 1 and figure 1 it is evident that the modified concrete is different from the traditional more developed and uniform porous structure with increased cellular porosity. In this case, the diameter of the cellular pores is less than 0.8 mm, or 800 microns, ie, decreases in 2 times. A reduction in the gel and capillary porosity in 1.5-2 times, which is explained by the positive influence of hydrophobe effect of ion modifier. On the one hand, capillary porosity and porosity at the air entrainment, reduces the frost resistance of saturated concrete. On the other hand, with air entrainment, porosity and multi-porosity with pore sizes larger than 1 mm increases the conductivity. The heat loss through the walls caused by convection method of heat transfer during the heating season due to the increase of thermal conductivity. Therefore, reduction of the gel and capillary porosity reduces the risks of destruction of concrete walls with alternating freezing and thawing. The diameter reduction of cellular porosity reduces the heat loss through the walls.

4 Conclusion

Analysis of the results shows that the application of the modified non-autoclaved aerated concrete based on industrial wastes will provide environmental security by significantly reducing heat loss through the walls of buildings in the formation of cracks under the action of frost and inhomogeneous pore structure of traditional concrete. Environmental safety will be improved due to saving of energy for heating of residential premises and due to reduce release of heat into the atmosphere. The removal of heat in the atmosphere, as is known, contributes to the greenhouse effect. Moreover, the environmental safety of buildings and the environment is impossible without reliable operation walling - concrete wall blocks. X-ray diffraction peak at the C3S evaluated the degree of hydration of cement: modified aerated concrete as 91 %, while the traditional concrete composition – 67%. This is also confirmed by the peak of quartz, which is 40% lower than that of conventional composition. The use of technogenic wastes in a modified aerated concrete and organic-mineral modifier allowed 7 times to reduce the amount of soluble and susceptible to corrosion portlandite compared to traditional concrete. 2 times increases the amount of ettringite.

Therefore, recycling of technogenic wastes in a modified aerated concrete and organic-mineral modifier significantly increases the environmental safety of the environment due to the hardening of the structure, increasing its uniformity and thermal properties.

References

1. S.A. Tkach, V.I. Telichenko, Ecology of urbanized territories in Moscow **2**, 39-44 (2016)
2. V.I. Telichenko, D.V. Oreshkin, Ecology of urbanized areas **2**, 31-33 (2015)
3. D.V. Oreshkin, Building materials **11**, 6-8 (2010)
4. D.V. Oreshkin, O.B. Lyapidevskaya, K.I. Kirillov, Construction of oil and gas wells on land and at sea **10**, 41-46 (2006)
5. E.V. Tkach, S.A. Tkach, R.F. Serova, E.A. Stasilovich, Modern problems of science and education **1-2**, 83-88 (2015)
6. D.V. Oreshkin, Building materials **11**, 6-8 (2010)
7. E. Tkach, Procedia Engineering **153**, 761-765 (2016)
8. A.V. Sumin, V.V. Strokova, V.V. Nelyubova, S.A. Eremenko, Building materials **1-2**, 70-75 (2016)
9. B.M. Rumyantsev, A.D. Zhukov, D.I. Aristov, Scientific review **13**, 128-131 (2015)
10. Yu.M. Bazhenov, L.A. Alimov, V.V. Voronin, Vestnik MGSU, 204-210 (2013)

Design and application of environmentally effective concrete with usage of chrysotile-cement waste

Lada Egorova[1,*], *Vyacheslav* Semenov[1], *Anna* Pligina[1], *Aizat* Askhadullin[1]

[1]Moscow State University of Civil Engineering, Yaroslavskoye Sh., 26, Moscow, 129337, Russian Federation

Abstract. Construction is resource-demanding industry, characterized by a large volume of waste. Particularly chrysotile cement waste obtained both in production and in dismantling over age chrysotile-cement products: corrugated asbestos boards and flat sheets, chrysotile-cement tubes. We propose to use dry chrysotile-cement waste as recycled aggregate for concrete. Based on developed compositions and identified properties of heavy concrete with chrysotile-cement waste introduce this technology to the production of foundation wall blocks. The studies confirmed the possibility of using chrysotile-cement aggregate and fine screening of crushing as a secondary coarse and fine aggregates for concrete with proper quality without increasing the cost of the product. Environmental safety of the obtained products was ensured. The direction for implementation of the research project was proposed.

1 Introduction

Nowadays experts of various spheres of business, including construction, design solutions to environmental pollution problems. Construction is resource-demanding industry, characterized by a large volume of waste [1-3]. Particularly during the existence of the asbestos cement industry in the Soviet Union and then in Russian Federation, there is appeared a problem of recycling chrysotile cement waste obtained both in production and in dismantling over age chrysotile-cement products: corrugated asbestos boards and flat sheets, chrysotile-cement tubes. According to Rosstat, in 2014 it produced 793,000 tons chrysotile-cement products, and waste volumes at chrysotile-cement products factories amounted to 80 000 tons.

2 Problem statement

Among the modern methods of recycling, we could highlight the use of wet waste in the production of building materials with their mechanochemical activation mainly offered in Russia. As well as developed abroad methods for chrysotile-cement waste utilization by of

*Corresponding author: EgorovaLV93@gmail.com

 DOI: 10.1051/matecconf/20168606005

heat treatment to fully crystal-chemical transformation of asbestos fibers, with subsequent disposal or use in the production of cement, ceramic, etc. [4-8]. The existing methods of their utilization are energy-intensive and expensive. Therefore, it calls into question the rationality of such technologies. We propose to use dry chrysotile-cement waste as recycled aggregate for concrete mixes. It will free up landfill area of and broaden raw material base of building materials, thus contributing to solve important environmental and economic problems [3-8].

The aim of research is to develop compositions and to identify properties of heavy concrete with chrysotile-cement waste (scrap and fine screening of crushing) for introduction this technology to the production of foundation wall blocks.

3 Materials and methods

From scrap chrysotile-cement flat sheets we obtained secondary coarse aggregate and identified its strength and performance properties. Mean density – 2028 kg/m^3, bulk density – 843 kg/m^3, the emptiness of material – 58.4%, the water absorption by weight – 19.2%, and the water absorption by volume – 38.9%.

A by-product during the fragmentation of the scrap became fine screening which contain sand and dust-like fractions. An analysis of grain composition graph of fine screening (Figure 2) showed the possibility of its reuse as the secondary fine aggregate.

The value of true specific density – 2400 kg/m^3; bulk density – 940 kg/m^3, water demand – 10.5%, fineness modulus – 2.93.

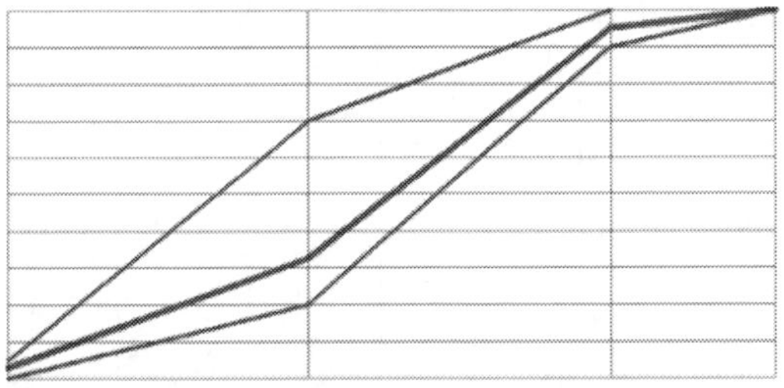

Fig. 1. Grain composition graph of coarse chrysotile-cement aggregate.

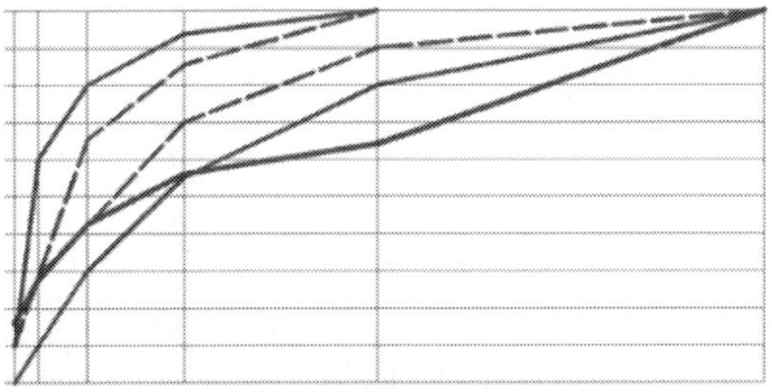

Fig. 2. Grain composition graph of fine chrysotile-cement aggregate.

X-ray diffraction analysis (Figure 3) revealed the existence of cement minerals. Mineral composition of fine screening consists of chrysotile asbestos, product of Portland cement hydration, calcspar, cement minerals. Strength test of samples made from mixing screening with water showed that dust approves cementing properties. Ultimate bending strength of samples– 0.85 MPa, ultimate compressive strength – 3.9 MPa.

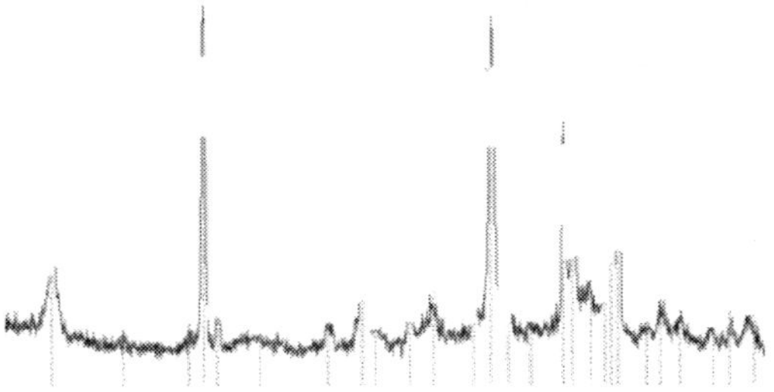

Fig. 3. X-ray diagram of cc fine aggregate.

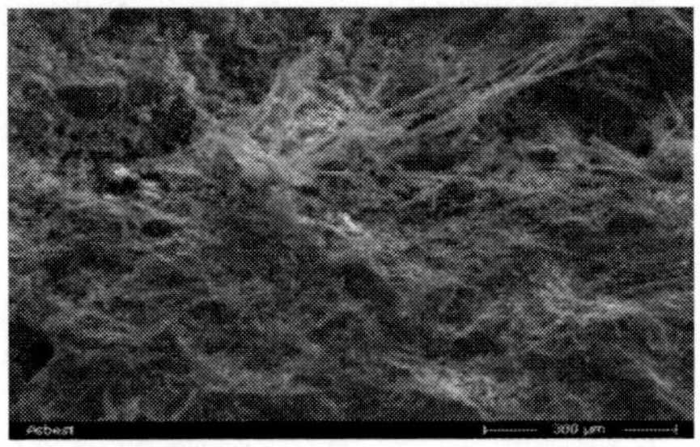

Fig. 4. Microstructure of coarse chrysotile-cement aggregate.

4 Calculation of concrete mixture

The composition of concrete was calculated with usage of law of concrete strength formula and absolute volume equation taking into account the increased water demand of secondary aggregates and their proportion in the whole mass of aggregate. The basic concrete composition with coarse chrysotile-cement aggregate calculated for concrete class B15, with fine aggregate – B20. As binding substance was taken Holcim Portland cement with slag CEM II / B-SL 32.5 N, fine aggregate – silica sand with average grain size, coarse aggregate – granite rubble fractions 5-20 mm. To reduce actual water consumption we used superplasticizer C-3.

5 Results and discussion

Properties of concrete with different contents of recycled coarse and fine aggregate were investigated. Concrete samples with replacement from 9% to 100% of crushed granite to coarse secondary aggregate were prepared. Next, we made samples with replacement from 10% to 50% of sand to fine aggregate.

Table 1. Compositions and properties of concrete with usage secondary coarse aggregate.

Features		Control comp.	The share of CC aggregate in all coarse aggregate				
			9	23	33	50	100
The composition of concrete per m^3, kg	Portland cement	213	227	244	256	285	323
	Water	160	170	183	192	214	242
	Sand	754	786	827	842	842	873
	Coarse granite aggregate	1285	1075	809	652	432	0
	CC coarse aggregate	0	107	243	326	432	658
	Superplasticizing admix C-3	1.6	1.7	1.8	1.9	2.2	2.4
Medium density, kg/m^3		2425	2395	2280	2310	2235	2120
Compressive strength of concrete, MPa		19.5	22.6	21.5	22.2	26.8	27.3
Strength-to-weight ratio, MPa		8.0	9.4	9.4	9.6	12.0	12.9
Water absorption, %		2.9	3.0	3.2	3.3	3.6	3.9
Frost resistance		300	300	300	300	300	300
The cost of materials per 1 m^3,		4053	4045	4034	4025	4027	4008

With increase of content of secondary porous aggregate in concrete, the average density of the concrete reduces, but the strength increases up to 40%. Accordingly, the specific strength of concrete with chrysotile-cement coarse aggregate increases up to 60% compared to the control composition.

This effect is due to increased consumption of cement, because it is necessary to ensure constant water-cement ratio.

Note that coarse porous chrysotile-cement aggregate plays a positive role in the concrete structure formation. By absorbing water at the initial stage of hardening, it consolidates contact area "hardened cement paste – concrete aggregate". Then giving the absorbed water during the concrete consolidation it leads to increased degree of hydration of cement.

Table 2. Compositions and properties of concrete with usage secondary fine aggregate.

Features		Control comp.	The share of CC fine aggregate		
			10	30	50
The composition of concrete per m^3, kg	Portland cement	231	260	290	328
	Water	146	164	183	207
	Sand	682	582	411	265
	Gravel	1351	1318	1258	1193
	CC fine aggregate	0	65	176	266
	Superplasticizing admix C-3	1.7	1.9	2.2	2.5
Medium density, kg/m^3		2410	2390	2320	2260
Compressive strength of concrete, MPa		27.6	26.9	28.3	33.1
Strength-to-weight ratio, MPa		11.4	11,3	12,2	14.6
Water absorption , %		2.9	3.0	3.5	3.8
Frost resistance		300	300	300	300
The cost of materials per 1 m^3,		4084	4117	4068	4061

As in the case of coarse secondary aggregate, higher water demand of fine aggregate also leads for increase of cement consumption. In case of replacement up to 50% of sand to chrysotile-cement fine aggregate the actual increase of cement consumption will amount up to 42%. However, the actual consumption of coarse aggregate is reduced by 11%, sand - 61%. The average density of the concrete is reduced by 6% and the compressive strength increases by 20%. CC fine aggregate exercises cementing properties and contributes to increase strength of concrete.

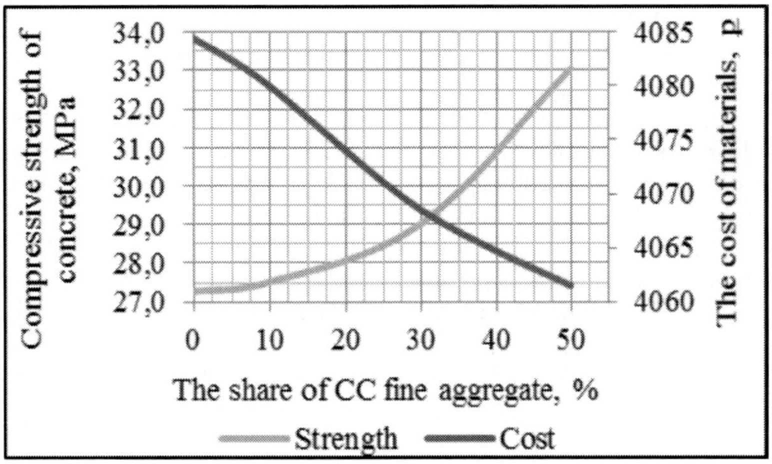

Fig. 5. The graph of the compressive strength of concrete and the cost, depending on the share of fine chrysotile-cement aggregate.

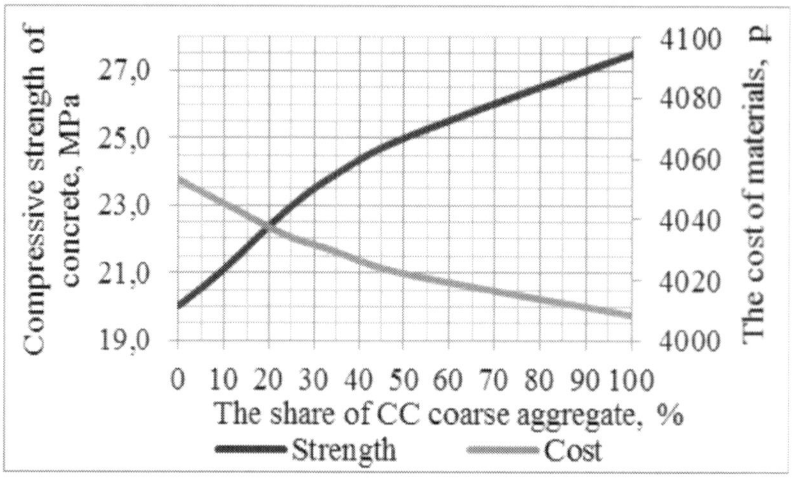

Fig. 6. The graph of the compressive strength of concrete and the cost, depending on the share of coarse chrysotile-cement aggregate.

6 Assessment of environmental safety

There are several scientific publications, which shows the possibility of the safe usage of chrysotile-cement [10, 11]. Opening chrysotile fibers due to mechanical damage are covered with products of hydration and subsequent carbonization of cement and have a reduced biological activity. The usage of dry chrysotile-cement waste as aggregate excludes fiber emission due to its dual "conservation" by the concrete. Foundation blocks during operation are located in the ground, the negative impact of fiber by respiratory route on human is impossible.

We additionally carried out an environmental assessment of the resulting material and technology in general. The radiological examination of chrysotile-cement aggregate and concrete with secondary aggregate was made. The effective specific activity of natural radionuclides suits for Russian regulation norms. Fiber emission from the concrete and from coarse chrysotile-cement aggregate tests in repeated freeze-thaw in the water, in solution of magnesium chloride and sulfuric acid, which conforms to sulphate, magnesia and chloride high corrosive ground for concrete waterproofing capacity W4. During microscope analysis, emitted chrysotile asbestos fibers were not observed in filtration residues (Fig. 7.). Furthermore, we decided to determine the frost resistance of the obtained samples. According to Russian norms of methods for determination of frost resistance on the basis of 37 cycles performed 3 accelerated method obtained mark on frost resistance 300. Reducing weight of concrete samples was less than 2%.

DOI: 10.1051/matecconf/20168606005

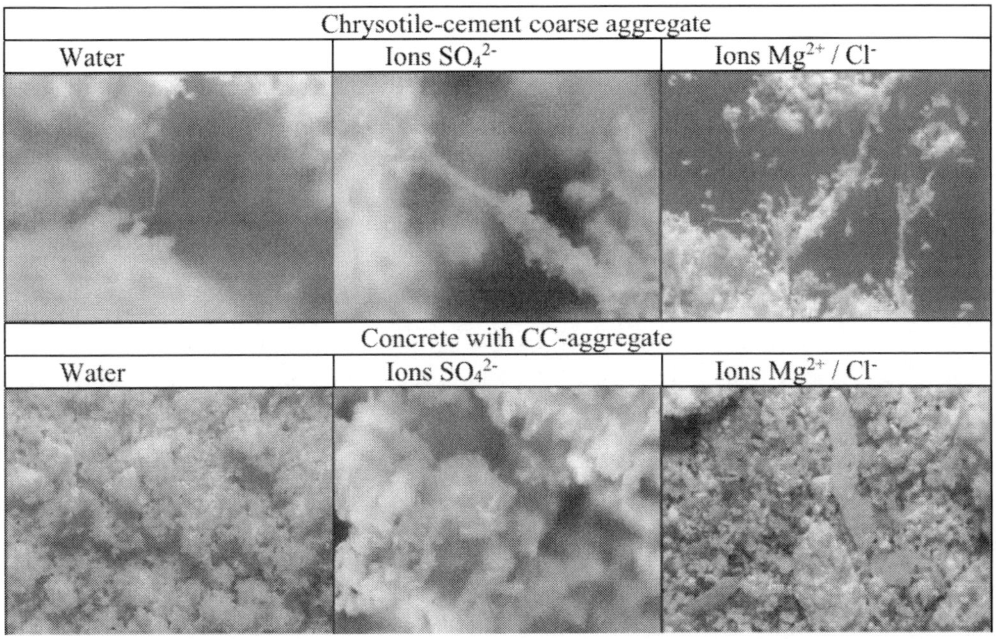

Fig. 7. The degree of influence of high corrosive ground at fiber emission.

7 Conclusions

Increased water demand of chrysotile-cement aggregate leads to increase water and cement consumption. In the same time, the consumption of other concrete components reduced. Comparing the market prices for concrete components, we can see that the usage of such chrysotile-cement waste allows us to obtain concrete without increasing the cost of the product. The usage of chrysotile-cement waste allows obtaining concrete in the range of the required strength of 15...30 MPa depending on the proportion of secondary aggregates and cement consumption. Factory production of some types of products contributes the complete replacement of natural aggregates by chrysotile-cement aggregates without reducing products' physical, mechanical and exploitation properties.

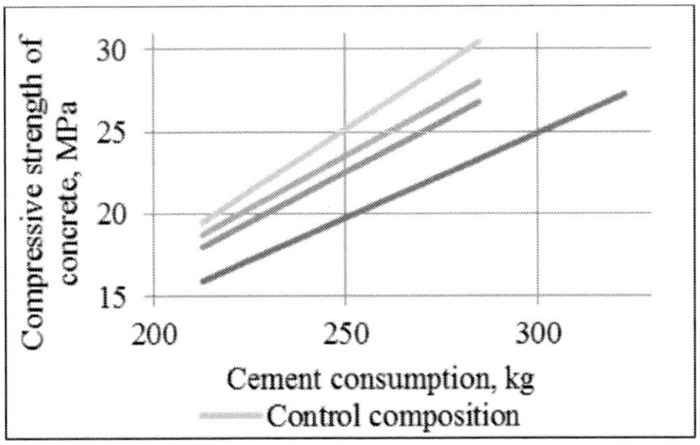

Fig. 8. The graph of the strength of concrete, depending on the flow of cement, grade 32.5.

As result of research, the effective concrete with chrysotile-cement waste was developed and environmental safety was confirmed. Environmental effectiveness of technology consist in comprehensive utilization of chrysotile-cement waste: coarse aggregate – gravel and fine screening of crushing. The research results are presented for "Chrysotile association" and received positive feedback from the chrysotile industry.

The inactivity of construction industry in relation to new technologies and materials, normative limitations of the use of waste, ensuring uninterrupted receipt of secondary resources and uniformity of product properties, and justification of economic efficiency of developments complicate the implementation of waste management technologies in the production process.

The development of starting local manufacturing heavy concrete with chrysotile-cement waste in chrysotile-cement products factories is more expedient because large concrete manufacturers have their own long-standing supply lines of traditional aggregates [12]. Eventually based on formed experience introduce this technology to the production of foundation wall blocks [3, 9].

Concluding all, we propose to implement the following plan for technology realization in the construction industry:

1. Evaluation of chrysotile-cement waste volumes at a specific landfill or a chrysotile-cement -products factory dump.
2. Making a treaty about delivery chrysotile-cement waste with companies which dismantle buildings.
3. Obtaining scrap and fine screening by crushing and sorting chrysotile-cement waste.
4. Creating a production line and certification of products.
5. The introduction of products to the market.

References

1. D.V. Oreshkin, Construction materials **11**, 6 (2010)
2. D.V. Oreshkin, K.N. Popov, A.I. Liljak, A.G. Mezhov, Vestnik MGSU–MSUCE Herald **2**, 296–298 (2011)
3. V.S. Semenov, A.I. Pligina, T.A. Rozovskaya, Materials Science and Engineering **012041**, 71 (2015)
4. V.V. Kozlov, K.N. Popov, A.G. Mezhov, A.I. Liljak, Vestnik MGSU - MSUCE Herald **2**, 284–287 (2011)
5. P.P. Gujumdzhjan, M.L. Kashnikova, T.N., Construction materials **9**, 20–21 (2006)
6. M. Radvanec, L. Tuček, J. Derco, K. Čechovská, Z. Németh, Journal of Hazardous Materials **252–253,** 390–400 (2013)
7. T. Prasauskas, D. Martuzevicius, E. Krugly, L. Kliucininkas, M. Kireitseu, A. Zerrath., Building and Environment **46**, 2295–2302 (2011)
8. A.F. Gualtieri, C. Cavenati, I. Zanatto, M. Meloni, G. Elmi, M. Lassinantti Gualtieri, Journal of Hazardous Materials **152**, 563–570 (2008)
9. A.I. Pligina, V.S. Semenov, L.V. Egorova, A.A. Ashadullin, Science Review **2**, 84-88 (2015)
10. N.S. Makarova, S.V. Kashansky, E.G. Plotko, Construction materials **9**, 19–20 (2001)
11. S.M. Nejman, I.G. Luginina, A.I. Vezencev, V.V. Turskij, L.N. Naumova, L.L. Nesterova, A.I. Gorshkov, A.V. Sivcov, Construction materials **4**, 30-31 (2002)
12. G.G. Lunev, Internet Journal Naukovedenie **3**, 22 (2014)

MATEC Web of Conferences **86**, 07001 (2016) DOI: 10.1051/matecconf/20168607001
IPICSE-2016

Traffic seismicity loaded historical building frequency parameters identification due to most commonly used truck in Slovakia

Daniel Papán[1,*], and *Veronika* Valašková [1]

[1] Department of Structural Mechanics and Applied Mathematic, Faculty of Civil Engineering, University of Žilina, Univerzitná 8215/1, 010 26 Žilina, Slovak Republic

Abstract. Experimental investigation and combination with numerical modelling is one of the progressive method in many scientific areas. The structural dynamics including traffic seismicity effects are also becoming an increasing topic. The aim of this paper was to realize the numerical analysis of heritage Upper Gate in Modra - Slovakia and FEM simulation of the lorry T-815 natural vibration. These subsystems are dominant processes in traffic seismicity vibration effects in buildings. For this purpose the variants of FE model by computing program Scia engineering has been created for building and software ADINA for lorry. These models are important for the assessment of the dynamic vibration transmissibility due to mechanical impact load properties. The results of this simulation was evaluated in frequency area. Next part of the investigation was the realisation of the experimental measurement. The results obtained from the experiment were compared with FE analysis. Using of the theoretical analysis, experimental procedures results and FEM simulation of the natural vibration it seem to be the practical application for engineering practice in prediction and assessment buildings vibration due to seismicity induced by traffic.

1 Introduction

The roads and the transport structures are subjected to direct dynamic effects of moving load. Unevenness in the pavement is the main source of kinematic excitation of the vehicle. It significantly effects the size of the contact forces between the pavement and vehicle. This is the reason to deal with it. The dimension of road unevenness is the higher order then pavement deflection or tire unevenness. So it is fully reasonable to take into account the unevenness located on the road only as a source of kinematical excitation of vehicle. The real load acting on the roads is variable in time and in frequency composition. This should be known for the solution of many engineering tasks as lifetime, design, reliability, structure development, micro tremor, etc.

The reliable building structures design in seismically active regions needs a detailed analysis of SI (Soils-structure interaction) under technical seismicity load. Nowadays

* Corresponding author: daniel.papapn@fstav.uniza.sk

technical seismicity is part of the environmental issues. The increase in vibration induced by different traffic has an unfavourable impact on building constructions and the people living there. The objects placed near the road and railway communications are highly influenced by vibrations due to technical seismicity. The comprehensive theoretical and experimental analysis of the dynamic response of the structure must be solved as a part of the task.

2 Representative vehicle fem frequency analysis

One of the most important part of the process of numerical simulation is to create a proper computing model. For this case the quarter model of TATRA 815 lorry was chosen as the most representative vehicle (Fig.1). The theoretical solution is based on the FEM. FEM has been mostly used since computers calculation started as one of the possible way to find solution on systems of differential equations. In this case the vehicle moved on the pavement and overcome obstacle belongs among non-stationary dynamic actions. These actions are described by the following differential equation:

$$[\boldsymbol{M}]\cdot\{\ddot{\boldsymbol{u}}(t)\}+[\boldsymbol{C}]\cdot\{\dot{\boldsymbol{u}}(t)\}+[\boldsymbol{K}]\cdot\{\boldsymbol{u}(t)\}=\{\boldsymbol{F}(t)\} \tag{1}$$

Where ***[M]***, ***[C]*** and ***[K]*** are mass, damping and stiffness matrices.

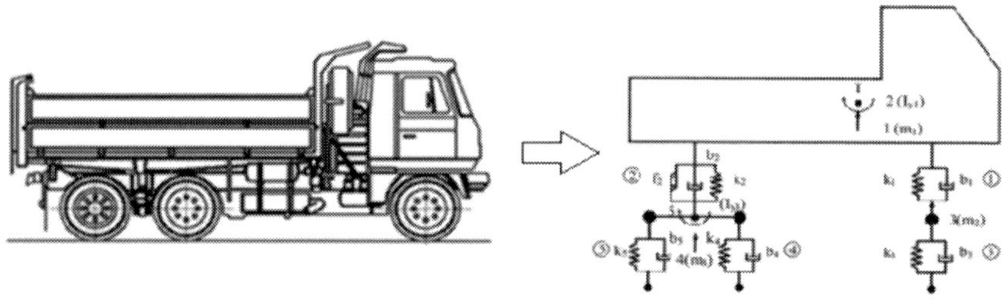

Fig. 1. Half-part model of lorry T-815.

Important part of simulation is to choose the appropriate model of vehicle. Model has to be represented with appropriate mathematical formulation. Half-part model are usually preferred and used for 2D analysis as a combination of mass, spring and beam elements. This discrete model of the vehicle with finite degrees of freedom makes solution easier from the mathematical point of view. This assumption changes partial differential equations to the ordinary differential equations [2,3].

The main characteristic of the half part model is defined by three diagonal matrices. They are the mass **{m}**, stiffness **{k$_i$}** and damping **{b$_i$}** matrices which contain experimentally measured values [3].

$\{\mathbf{m}_i\}_D = \{m_1, I_{y1}, m_2, m_3, I_{y3}\}_D = \{11475; 31149; 455; 1070; 466\}_D$ [kg, kg.m^2]
$\{\mathbf{k}_i\}_D = \{k_1, k_2, k_3, k_4, k_5\}_D = \{143716.5; 761256; 1275300; 2511360; 2511360\}_D$ [N/m]
$\{\mathbf{b}_i\}_D = \{b_1, b_2, b_3, b_4, b_5\}_D = \{19228, 260197, 2746, 5494, 5494\}_D$ [kg/s]

Determined as a result of the calculating natural frequencies.
$\{\mathbf{f}\} = \{f_{(1)}; f_{(2)}; f_{(3)}; f_{(4)}; f_{(5)}\} = \{1.13; 1.45; 8.89; 10.91; 11.71\}$ [Hz]

3 The affecting monumental building fem frequency analysis

These case study presents the frequency domain analysis of one monumental building in the area Modra – Slovakia. For the purpose of the case study was chosen Upper Gate (Fig.2). built in 1610 – 1646. The Upper Gate is located near the roadway line that produced dynamic load due to technical seismicity. The house of this type was selected as the most frequent building located near the roadway line of 2nd class (no. 502, Bratislava - Smolenice) in the town center in Modra, Slovakia [1].

Fig. 2. The view on the historical Upper Gate building.

The analysis of SI applied in Modra locality types of FEM model were created in Scia engineering software. This computing models was used with dynamic parameters evaluation in frequency domain. For the vibration dominant frequency bands and the vibration transfer 3D FEM model was used. Natural frequency values and modes of vibration were calculated for this computing model type [4].

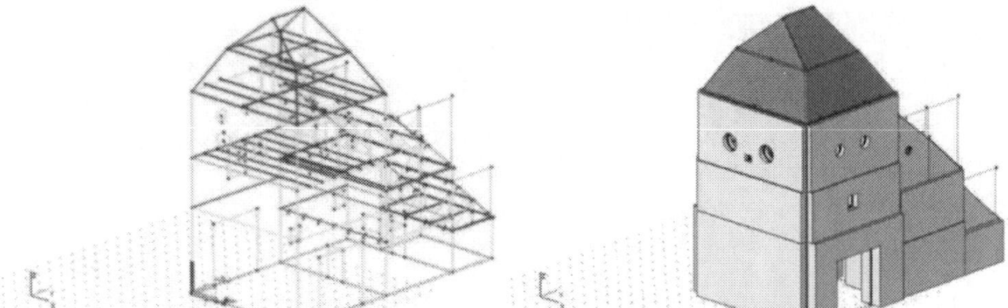

Fig. 3. The computational model of historical building – Upper Gate, framed and general.

In the calculation of own shapes we considered 4613 of 2D elements, 976 of 1D elements, and 5169 of network nodes. The calculation required assembling of 31,014 equations. For the mass group combination CM1 we used 10 natural frequencies. Amount of materials for x-direction is 411,678.05 kg, the y-direction is 411,678.05 kg and the z-direction is 411,682.24 kg. The calculation was carried out on the basis of Lanczos and Mindlin bending theory.

Table 1. FE model – Example of the natural frequencies results.

Natural mode no.	Natural frequency [Hz]	Vibration type
3	6.46	asymmetrical
4	7.11	asymmetrical
5	8.48	asymmetrical
6	10.76	symmetrical
7	12.09	asymmetrical

4 Experimental analysis of si in frequency domain

Dynamic response of real tangible environment (half-space structure) by random excitation was measured by 2 lines of accelerometers with a frequency range of 1 ÷ 4000 Hz. (Brüel-Kjaer), placed in terrain and roof of the object Fig. 4 Accelerometers measure the response in the vertical direction. Dynamic response in the observed points were measured in the form of vibration velocities (m/s), in three orthogonal directions x, y, z. The measurement was performed using the "off - line". The, recorded signals are simultaneously stored on storage media PC/DX4 (program DISYS) and PCs AMILO - PC FS (NI CompactDAQ program). Evaluation of the measured data was carried out in laboratory conditions KSM - FCE - ZU based on the evaluation of software lines KSM (VL - KSM).

The measuring unit consisted of:

- piezoelectric accelerometers BK 8306 (Bruel - Kjaer) - 5pcs
- integration amplifier BK-2693-014 (Brüel-Kjaer), measuring PC/DX4 Notebook PC and FS File Format: MS Word
- NI cDAQ™ -9191NI CompactDAQ One-Slot Wireless Chassis, Moxa AWK-3121 With NI CompactDAQ

Fig. 4. The accelerometers positions in-situ experiment – B1.

Mechanical movement in the respective measuring point has been transformed by means of accelerometers and an electrical signal after amplification and integration from the acceleration signal - *a(t)* to vibration velocity - *v(t)*, and it was conducted by means of Wi-Fi devices to the measuring central of the Department of Structural Mechanics (KSM) ZU University (ZU). These signals were recorded on laptops and PCs FS. Laptops are working in conjunction with the software NI Compact DAQ. The analog signal was recorded by parallel system DISYS and NI Compact DAQ, while the sampled frequency was $f_s = 500$ Hz, with respect to the required criteria for sampling signals ($\Delta t < 1/2 f_{max}$).

Table 2. Experimental and theoretical natural frequencies comparison.

FEM Natural Frequency (Hz)	Calculated Natural Frequency T815 (Hz)	Experimental identified frequency PSD (Hz)	Exper. CPSD (Hz)	Exper. GF (Hz)
		B1		
	1.13			
	1.45			
4.34				
5.3				
6.46		6.3; 6.5; 6.7;	6.1; 6.2; 6.3; 6.5;	
7.11		7.7; 7.8;	7.6; 7.8;	
8.48	8.89	8.6; 8.7; 8.8;	8; 8.1; 8.7;	8.6; 8.7;
10.76	10.91	10.2; 10.7; 11.0; 11.4;	9.9; 10.1; 10.6; 10.7; 11.3;	10.0; 10.1; 10.4; 10.8; 11.1; 11.4;
12.09	11.71	11.7; 11.8; 12.2;	11.5; 11.7; 11.8;	11.7; 11.9;
13.92		14.0; 14.1; 14.2;	14.2;	13.9; 14.0; 14.1; 14.2; 14.5;
15.52		15.5;	15.5;	15.4; 15.6;

As an example Figure 5. shows the results from signal spectral and amplitude analysis. Frequency peak finding are showed too. Identification of calculated FEM natural frequencies in experimental spectral analysis results are summarised in Table 2. In this table was summarised peak detection for the PSD (power spectral density), CPSD (cross power spectral density) and GF (gain factor) functions. The records of TATRA 815 lorry passing were used for this spectral analysis.

DOI: 10.1051/matecconf/20168607001

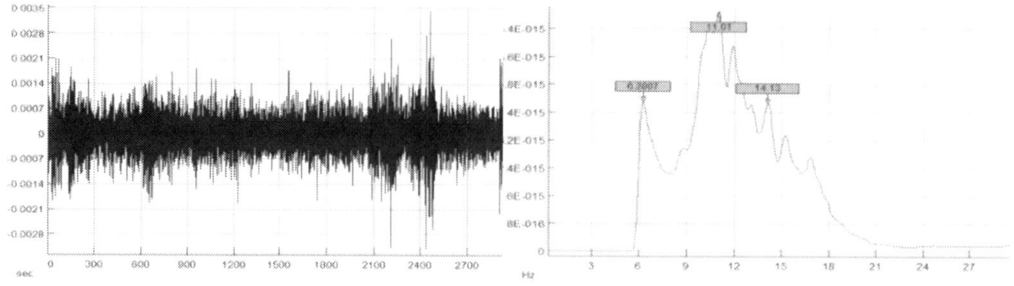

$v_1(t)$; $B1_z$ PSD G11 (f); $B1_z$

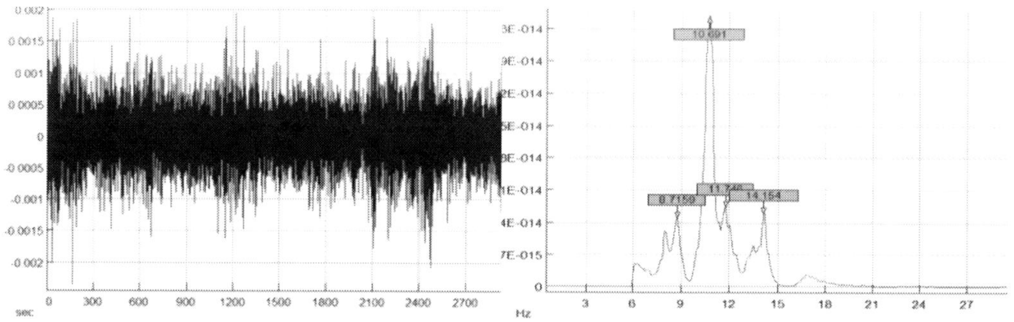

$v_1(t)$; $B1_z$ PSD G11 (f); $B1_z$

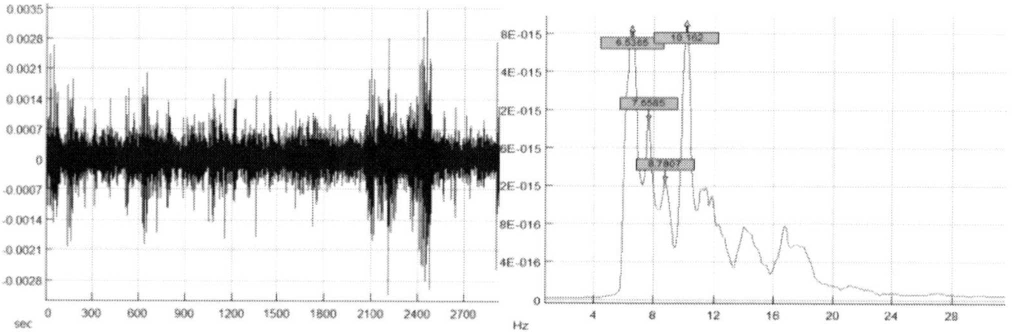

$v_1(t)$; $B1_z$ PSD G11 (f); $B1_z$

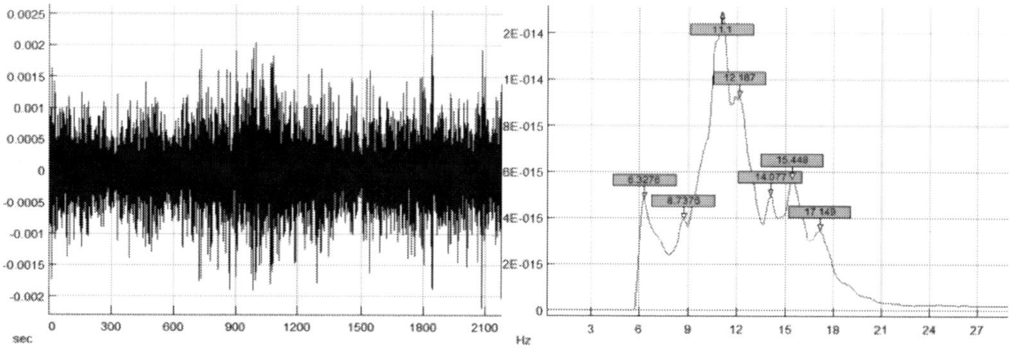

$v_1(t)$; $B1_z$ PSD G11 (f); $B1_z$

Fig. 5. The spectral analysis example.

5 Conclusions

These case study contain the SI dynamic response frequency domain identification in three area. The first area of interest was numerical model of representative vehicle TATRA 815 natural frequencies identification. The second area of interest was the natural frequencies identification of the one important heritage building in south Slovakia. This building is traffic seismicity attacked intensively. The third and the most important area of this analysis was the experimental measurement evaluation. The experimental frequency detection was realised via spectral analysis. The main spectral functions was used for the peak detection. In SI area are expected the natural frequencies of vehicle and building. The important frequency band is between 4-30Hz. The identified natural frequencies of both numerical models are in highlighted frequency band 8-13Hz. Lower and upper frequencies are dumped. These experimental results verified numerical models results. The combination of numerical and experimental approach can be used for the traffic seismicity effect prediction. This prediction is significant for road reconstruction planning and monument restoration [5,6,7,8].

References

1. K. Kantek, *Modra mesto plné histórie - Sprievodca dejinami a pamiatkami* (Modra, 1997)
2. Ľ. Daniel, V. Valašková, J. Kortiš, Stav. a env. Inž. **10(2)**, 108-112 (2014)
3. I. Martinická, In: Poz. Komun. a dráhy. **6(1-2)**, 41-50 (2010)
4. V. Valašková, D. Papán, J. Fričová, Proc. Earth and Plan. Sc. **15**, 193-198 (2015)
5. M. Marschalko, I. Yilmaz, K. Kubačka, T. Bouchal, M. Bednárik, M. Drusa, M. Bendová, *Utilization of ground subsidence caused by underground mining to produce a map of possible land-use areas for urban planning purposes* (Arab J Geosc, 2009)
6. M. Decký, M. Drusa, Ľ. Pepucha, K. Zgútová, *Earth Structures of Transport Constructions: Essex CM20 2JE, Pearson Education Limited* (Edinburg Gate, Harlow, 2013)
7. J. Melcer, App. Mech. and Mat. **617**, 29-33 (2014)
8. J. Melcer, D. Kuchárová, *Frekvenčné charakteristiky nákladného vozidla. New trends in statics and dynamics of buildings : conference proceedings* (Slovak University of Technology, Bratislava, 2013)

MATEC Web of Conferences **86**, 07002 (2016) DOI: 10.1051/matecconf/20168607002
IPICSE-2016

Transformation of Belarus and Russian agricultural settlement system in the new economic conditions (post-Soviet period)

Elena Shcherbina [1], *Elena* Gorbenkova[2,*]

[1] Moscow State University of Civil Engineering, Architecture and Urban Planning Department, 129337, 26, Yaroslavskoye Shosse, Moscow, Russia
[2] Belarusian-Russian University, Highways Department, 212000, 43, Prospect Mira, Mogilev, Belarus

Abstract. The transition to new economic relations in 1990, the collapse of the Soviet Union, the formation of new states caused the processes of transformation of the settlement system. That determined the need to analyse the current situation at this stage of development. The greatest changes in the structure has undergone a rural settlement that has identified the need for research in Belarus and Russia, including the justification for the new approaches to assessing the level of development of rural settlements in the structure of the current scheme of administrative divisions. Today both in Russia and in Belarus adopted a system of local government areas of rural settlements, which are included in the municipal districts that are component territorial units of federal subjects. However, the system of rural settlement of these republics have differences, which occurred because of the transformation, due to historical processes, economic and political conditions the existing States. The demographic situation in these countries is characterizing by population decline. In Russia, among federal districts the greatest population losses typical for the Far East - 6%, Siberia - 4%, and for the Volga Federal District - 4%. That is, in absolute terms, respectively, 0.4; 0.8 and 1.2 million people.

1 Socialist theory of the USSR settlement system

The collapse of the Soviet Union was the beginning of building a new settlement in the former republics, which was caused by the transition to new economic relations and a paradigm shift (a theoretical concept) of the theory of "socialist" settlement covering all territorial levels (Federal, regional, local) and all functional elements (work, recreation, everyday life) of the resettlement process and emerging as a result of its implementation of the rural network.

In the USSR the regulations system was developed (construction norms and rules SNIP, sanitary rules SanPiN), which have been used in urban design and regional planning. The methodological basis of this paradigm is the "method of employment balance", according to which were defined perspectives of development of individual settlements, and in the

* Corresponding author: gorbenkowa@yandex.ru

aggregate, and the whole structure of long-term settlement, depending on the development scheme and a placement of industry objects and productions.

The main forming elements of the country's support framework in the territorial and settlement structure of the USSR were large (100-250 and even 500 thousand people) and the largest cities. Companies of a military-industrial complex performed a city-forming function in smaller settlements. The voluntaristic approach was present in the development of regional planning schemes of rural areas in the system of rural settlement. It was basing on a private paradigm theory of "promising" and "unpromising" residential areas. According to this theory, the existing settlements were divided into promising to be coarsening and improvement, and the unpromising to be gradual elimination as housing fund depreciation. The existing standards forbidden for capital construction in the unpromising settlements. This paradigm was basing on the collective farming concept, the development of collective and state farms (1960). The result of this policy was reduction the number of villages and rural population [1, 2].

In the Soviet period, the settlement system formation based on the principle of creating a reference settlement frame, which is based on interconnection of two coupled trends - the centripetal (the development of large cities and urban agglomerations) and linear-swift (the development of a country's transport system). This identified the various forms of a spatial territories organization; the three major ones are wireframe, linear, agglomeration (Figure 1). In fact, these structural forms are typical for many countries and determine a current spatial organization trends.

a b c

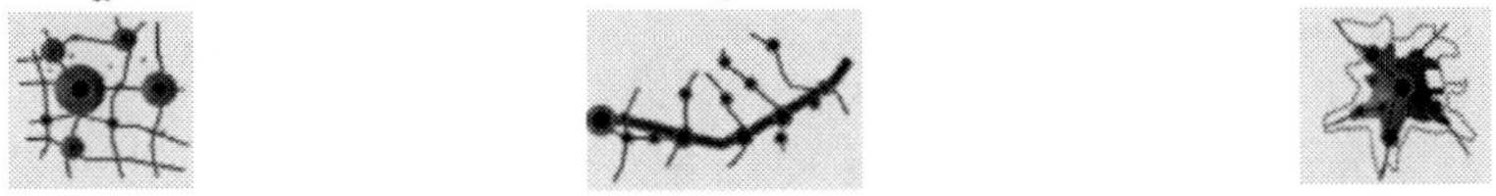

Fig. 1. The forms of a spatial organization of a settlement system: a – wireframe, b – linear, c – agglomeration.

The development of one or another form caused by a multitude of factors. Natural environment as well as spatial and historical processes of settlement Russian territory had the greatest impact on the character of rural settlement system. This is typical for the development of a rural settlement system formed under the influence of soil and climatic conditions, as well as historical labour population skills. That led to the non-uniformity of a rural settlement in the territory.

The simple principle was determined to form another settlement paradigm. «People go where a favorable environment not only provides a comfortable accommodation, but also allows to realize their potential as individuals as well as citizens responsible for their own destiny and a worthy public status of towns and villages» [3]. It is this approach best meets the requirements of an innovative model of economic development, where the main resource - human capital with high intellectual potential.

2 Features of the current Belarusian settlement system

The highest intensity of urbanization processes (compared to other republics of the former USSR) was one of the main features of of the BSSR (Byelorussian Soviet Socialist Republic) (Table 1 [3]). That had a significant influence on the settlement structure.

Table 1. Belarusian population dynamics.

The area's name	Population size (thous. persons) * urban population are above the line,	The proportion of urban/rural population, %

	rural population are below the line							
	1979	1989	2009	2015	1979	1989	2009	2015
Brestskaya	177	258	1404.5* / 491.9	1 388.9 / 425.4	45	57	65 / 35	69.4 / 30.6
Vitebskkaya	297	350	1237.5 / 341.1	1 198.5 / 283.1	56	64	72.4 / 27.6	76.4 / 23.6
Gomelskaya	383	500	1443.2 / 397.8	1 424.0 / 336.9	52	64	72.4 / 27.6	76.4 / 23.6
Grodnenskaya	195	270	1076.7 / 341.8	1 052.6 / 287.0	43	56	68.3 / 31.7	72.8 / 27.2
Minskaya	1262	1589	1431.1 / 642.1	1 407.9 / 608.1	53	64	55 / 45	56.6 / 43.4
Minsk			1814.3 / -	1 938.2 / -				
Mogilevskaya	290	356	1106.3 / 271.5	1 070.8 / 224.4	34	43	75.5 / 24.5	79.1 / 20.9
Total			9 513.6 / 2486.5	9 480.9 / 2 155.9			73.9 / 26.1	77.3 / 22.7

Currently, Belarus settlement system represented by three levels, which based on the existing administrative and territorial structure and conditions of transport accessibility of cities - centres of various hierarchical rank: Republican (national); Regional; Local (Figure 1) [4]. Each level corresponds to its own settlement system.

The basis of Belarus territorial at the **national level** was the functional zoning principle, which allows to allocate the following areas: urbanized (23%); agricultural (31%); environmental and recreation (34%); special (12%).

Agricultural areas are characterizing by a predominance of agricultural and forestry purposes lands. They based on medium and small urban settlements and including agro-towns and other rural settlements, territories of various functional use. The main resource of rural areas economic development are the earth and its resources; the leading functions are in production and processing of commercial agricultural and forestry products.

Local settlement system based on existing administrative regions, and including urban and rural settlements at the head of the town-centre, which function performed by the centre of an administrative district [4].

The basis of the rural settlement formed a group of rural settlements united to each other and with the urban settlements - the district centre in a single system, using the administrative, organizational, economic, socio-cultural and communication connections. Rural settlements are classifying according to their role in the system of settlement and population (Table 2).

Table 2. The classification of rural settlements in Belarus.

The rural settlements types	
role in the settlement system	**the number of population (people)**
Reliance local centers (agro-towns) 1449 rural settlements	large – over 1000
	big – from 500 to 1000
	medium – from 100 to 500
Ordinary rural settlements	large – over 1000
	big – from 500 to 1000
	medium – from 100 to 500
	small – to 100

The existing network of rural settlements of the republic preserves finely dispersed structure due to historical and natural features. Thus, the greatest number of small villages (with a population up to 100 people) are in the north (Vitebsk region), where they account for 83 per cent of rural settlements. The largest rural settlements are in Brest and Gomel regions. The settlements proportion with a population between 200 and 1000 is 36% in Brest region and 31% in Gomel region. That caused by natural conditions influencing the agricultural activity.

The rural settlement centres of Belarus are agro-towns - comfortable rural settlements, administrative, organizational and economic, social and cultural centres that have established production and social infrastructure to ensure social standards to agro-town's population and surrounding areas residents within a radius of 15 kilometres [5]. Such a scheme is optimal for a rural settlement. That reduces the services radius to the public social infrastructure facilities, as well as providing higher living standards to rural population. It seems that this will create a basis for improving the demographic situation.

Nowadays agro-towns are supporting settlement points, the number is 1449 settlements, and they are evenly distributing across the regions of the country.

It seems that the agro-towns sustainable development has to promoting the rural settlement optimization, formatting the profitable agricultural production, improving the infrastructure of rural settlements, increasing the cultural and o rural population living standards, creating the preconditions for demographic situation improvement, and enhancing the prestige of living in rural areas. Therefore, it is necessary to evaluate agro-towns/villages development level for the sustainable development of Belarus rural areas.

However, despite the measures taken at the national level (Program of sustainable development of rural areas) there is a rural population decrease, in addition to which the network polarization processes of settlements is occur. On the one hand, the share of medium-sized settlements, passing into the small decreases, on the other - there is a concentration of the remaining population in a few large settlements (usually in agro-towns).

Thus, we can identify the main elements in the core network of Belarus rural settlement - agricultural towns - as local centers of rural councils (governing bodies) and agricultural organizations. The basis for their further development should be a build-up of the socioeconomic potential.

2 Features of the current Russian settlement system

In the post-Soviet period, there have been significant changes in the structure of a rural settlement, which is associated with the authority transfer to the formation of spatial development plans to local authorities, according to the Law «On local self-government». Today, Russia has adopted a two-tier system of local self-government. The first level includes the urban and municipal districts; the second level – the urban and rural settlements, included in the municipal districts.

This led to the transformation and disorganization throughout the settlement system; in this case, there was an asymmetry in the living standard of urban and rural areas [6]. In addition, the outflow residents from the sparsely populated rural settlements into a larger brought to the widening of the largest settlements and crushing the rest, especially the smallest ones.

Implementation of the principles of local self-government and provisions of the Town Planning Code of the Russian Federation faced a number of difficulties preventing sustainable development. The uneven economic development of the regions, metropolitan regions, urban and rural settlements identified modern direction of a control system, which based on the administrative allocation of "promising" and "unpromising" villages, as well as in Soviet times. Today there is a trend of disappearance of rural settlements and municipal areas that converting to urban districts.

The crisis of rural areas adversely affected the small towns - undermined the economic basis from the. The dynamics of Russia population in the period from 1979 to 2015 in relative terms shows that the rural population share in relation to the urban remains unchanged. But the absolute values show the rural population decrease of 3.5 million in 1979 compared to 1989 (the collapse of the USSR), and 1.0 million people to 2015, i.e. less rapid decline in the rural population, both in Belarus (Table 3). This is due to initially higher urbanization level and migrants influx.

Table 3. Dynamics of a Russian population in the period from 1979 – 2015.

Year	Total population, million people	Including those		In the total population, %	
		urban	rural	urban	rural
1979	137.4	94.9	42.5	69	31
1989	147.0	108.0	39.0	73	27
2009	142.7	104.9	37.8	74	26
2015	146.3	108.3	38.0	74	26

The existing system of Russian settlement has its own specifics. On the one hand, about 74% of the population is concentrated in urban areas; on the other hand, the area occupied by cities is only 5% of the country. Rural settlements divided into groups according to the population (Table 4).

Table 4. The rural areas classification by population.

Groups	Population, thousand people
	Rural settlements
Large	over 5 from 3 to 5
Big	from 1 to 3
Medium	from 0.2 to 1
Small	from 0.05 to 0.2 before 0.05

A distinctive feature of the Russian settlement system is the presence of different natural and geographic zones. Therefore, Russia characterized by unevenness of rural settlement in the territory: the largest concentration falls on a favorable for living and farming regions.

The population concentration in a few large settlements led to the settlement landscape polarization: increasing the share of the smallest and large settlements, while reducing the number and the share of medium. Thus, the rural settlement system characterized by a high depopulation level of the rural population; polarized areas, formed under the influence of processes of center-periphery development model; unemployment; low incomes, as well as other problems [7].

Reforms related to the liquidation of "collective-state farm system", contributed to the reduction process of small rural settlements and the increase in the rural population concentration in large settlements. These reforms were initially focusing to the cost optimization in the social sphere - the elimination of inefficient social facilities. All this has led to a violation of the ratio of the density and rural settlements size and the further polarization progress of the rural settlement structure. The output of this situation is representing in the development of a specific development concept of a rural settlement system - enhancing in agricultural production, creating the co-operative farms and their specialization based on local conditions [8].

However, the rural settlements network reduction is not always associated with a rural population decrease. Increasing the number of large rural settlements influenced by a variety of factors: transformation of the urban settlements into the rural ones and migration from medium and small villages. This has led to the rural population concentration in large settlements with the development potential, and to the formation of a new settlement framework of rural settlement [9].

Thus, today the Russian rural settlement system is located at the bifurcation point. In this regard, various development ways are possible for this system based on a variety of factors, both external and internal. Therefore, the study of forecast development options and optimization ways of rural settlement system seems particularly relevant today.

3 Le Corbusier settlement system

Le Corbusier settlement system includes three forms of settlement (Figure 1) [10]: Residential unit of agricultural production (food); Linear industrial city (production); Radial-ring city of social life (government, science, trade).

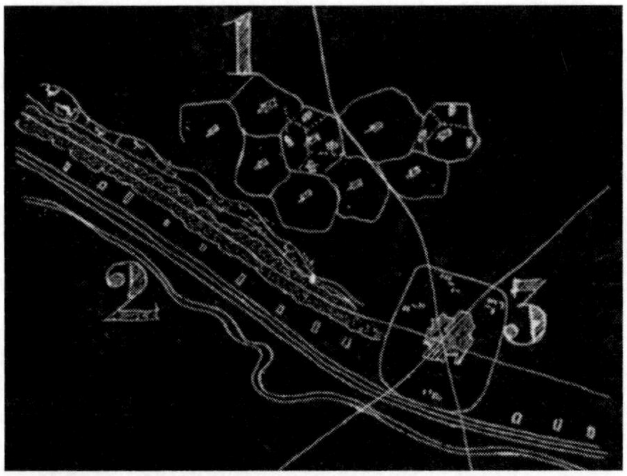

Fig. 2. The three forms of settlement of Le Corbusier: 1. Residential unit of agricultural production. 2. Linear industrial city. 3. Radial-ring city of social life.

We will consider the agricultural production unit.

Countryside unit. A rural unit management is carrying out by the mayor or by another administrator. A rural unit size exceeds the ordinary village size. Transport infrastructure must solve the needs of the agricultural production and cooperative center (Figure 3).

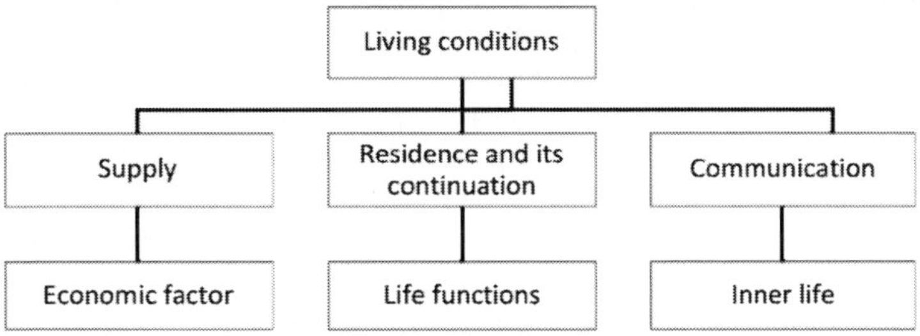

Fig. 3. The needs of agricultural production and cooperative centre.

The constituent parts of agriculture include cattle and the pastures, cattle yards and their equipment, feed storage, silage storage and feed kitchen, shepherds houses.

The cooperative center objects are dairy farm, agricultural products storage, mechanical workshops, a garage for agricultural machinery and motor vehicles, local enterprise for agricultural products processing; in addition, houses, cooperative store, school, youth workshops and social club with a sports complex.

Le Corbusier offers three options for cooperative centers. ***A.*** Rural Cooperative Center for a sufficiently large village; ***B.*** Cooperative United Center for a number of nearby villages; ***C.*** Cooperative center for a new agricultural production unit. The first two are composed of similar elements grouped into a single building "Cooperative Center" - a unified complex, headed by the leader, well positioned, well organized and maintained. Thus, there is a villages regrouping around a cooperative center.

Le Corbusier proposed two hypotheses. According to the first hypothesis, the rural unit with optimally short distances is in close proximity to the cultivated land, and the initiative of agricultural labor moves from the individual farmer to the group. This hypothesis is suitable for co-operative center serving the several nearby villages (B). The second hypothesis is basing on the organization of the cooperative land cultivation. The co-operative center may be located outside the settlements, while serving several villages. The transport infrastructure has to be well develop in this case.

To reduce the rural population outflow requires the following. 1) Creation the opportunity for satisfying the increased needs, while remaining to live and work in their village; 2) Creation the favorable living conditions; 3) Establishment a sufficient number of labor application places in the villages; 4) Provision food possibility by qualitative farm produce. The most important of the above seems in ensuring the of population employment, i.e. enhancing productive capacity. The presence of a permanent industry, where the rural population is engaging in throughout the year, can be barrier on a way of peasants escape from countryside.

Thus, the Co-operative Centre should be considering as a mechanism for bringing provision and welfare to the rural population. Thus, the farmer continuing to live and work in his usual rural surroundings can enjoy the advantages and achievements of a modern civilization.

The cooperative village. The elements of the cooperative village ("Radiant village") include the following. *1. Club. Youth House.* The club is the real administrative and educational complex, community center of the commune. *2. Communal syndicate. Cooperative warehouse.* The cooperative communal warehouse owned by an agricultural cooperative. His presence is a guarantee of security of the village and the country. *3. Cooperative Production Company.* Companies of this kind should be placing directly in the village to meet the ever-increasing rural population requirements. *4. Cooperative allocator.* It serves to organizing supply the essentials, food, dry goods, clothing, manufactured goods, and so on through the establishment of commercial enterprises. *5. Residence.* The cooperative village is located away from large highways, near local roads. Therefore, all houses are located at the distance from the transport arteries. *6. Farm.* It is proposing a uniform land distribution in farms.

Comparing reviewed settlement system, it is possible point out the following.

1. In Belarus, the support system of rural settlement network consists of agro-towns, evenly distributed across the country. The main objects of the socio-economic and industrial infrastructure, available for residents of nearby settlements, are concentrated in agro-towns. The current settlement system formed in the Soviet period and was further developing in the post-Soviet years. The majority of rural settlements belongs to agricultural activity type. This is due to the agrarian orientation of the state's economy.

2. The distribution unevenness and settlement types' variety, formed under the influence of agro-climatic conditions and economic demands, is a characteristic feature of the Russian agricultural settlement system. The system of rural settlement in Russia's regions has its own features, which complicates carrying out of a uniform policy.

Agricultural Reforms of the 1990s contributed to the settlement system transformation:
• Transformation of the state and collective farms into joint stock companies, partnerships,

cooperatives, etc.; • State lands monopoly cancellation and transfer it to small private landowners by dividing a large part of state-owned land used by agricultural enterprises on the land shares and permits for buying and selling of lands; • Creation the conditions for the farms and private households development;• Large agribusiness corporations' development.

3. Le Corbusier's settlement system based on optimal conditions' creation for agriculture. The ideal is a "Cooperative Village". It performs the functions of agro-industrial and socio-cultural center of the rural settlement system, which contributes to the inter-settlement connections' development.

Thus, the greatest similarity seems in Belarus rural settlement systems, and the "ideal" Le Corbusier's system. That suggests the correctly chosen development direction of Belarus rural settlement system. Considering the similarity of Belarus territorial division's scheme with a territorial structure in Russia, is possible to talk about the feasibility of using the experience of Belarus under Russian conditions.

References

1. O.V. Usol'ceva, Vestnik Tomskogo gosudarstvennogo universiteta, **397**, 178-186 (2015)
2. V.V. Vladimirov, *Problemy razvitija teorii rasselenija Rossii.* (2003)
3. V.Ja. Ljubovnyj, Ju.A. Sdobnov, Sbornik materialov po problemam razvitija gorodskih aglomeracij v stranah SNG k nauchno-prakticheskoj konferencii «Nauchnye i prakticheskie aspekty formirovanija gorodskih aglomeracij», 13-24, (2011)
4. Nacional'nyj statisticheskij komitet respubliki Belarus' http://www.belstat.gov.by/ofitsialnaya-statistika/ssrd-mvf_2/natsionalnaya-stranitsa-svodnyh-dannyh/chislennost-naseleniya-na-1-yanvarya-po-oblastyam-respubliki-belarus/
5. *Gosudarstvennaja shema kompleksnoj territorial'noj organizacii Respubliki Belarus* (2007)
6. *Gosudarstvennaja programma ustojchivogo razvitija sela na 2011-2015 gody*, (2012)
7. P. N. Davidenko, E. V. Gorbenkova, Gradostroitel'stvo **5 (39)**, 37-70, (2015)
8. N.V. Chugunova, Ju.V. Litovchenko, S.A. Ignatenko, Geografija i region. Materialy mezhdunarodnoj nauchno-prakticheskoj konferencii, 283-288, (2015)
9. R.A. Galin, Jekonomika i upravlenie **5(127)**, 28-33 (2015)
10. Le Corbusier, Les éditions de minuit, Paris, (1959)

UV-decontamination of potable and sewage water in the city with population over one million

Aleksandr Smirnov[1*], *Evgeniy* Pupyrev[2], *Ilkhomzhon* Shukurov[2] and *Konstantin* Chertes[3]

[1]AO "NIIVODGEO", 19435, B. Savvinsky per., 9, p. 1., Moscow, Russia
[2]Moscow State University of Civil Engineering, 129337, Yaroslavskoye Shosse 26, Moscow, Russia
[3]Samara State Technical University, 443100, 244 Molodogvardeyskaya st., Samara, Russia

Abstract. The waterworks system in a modern city is a complex challenge. From the one hand, it is necessary to provide high-quality potable water to the residents with observance of all sanitary and hygienic requirements; from the other hand, the sewage water discharged from the city should not affect the environment. Meanwhile, the microbiological safety is the top-priority and crucial parameter for evaluation of any work and any project. In Novosibirsk, solutions have been found for both of them by using the cutting-edge approaches in the decontamination technologies. The UV-decontamination enabled to create a multi-barrier efficient protection when dealing with the potable water treatment and ensure environmentally-friendly decontamination of the sewage water.

The location of Novosibirsk near one of the biggest rivers in Russia, the Ob river, imposes certain duties concerning the use of water in the city. Ob is the source of centralized potable water supply in Novosibirsk, and the sewage water discharged into this river may significantly affect the ecological status of the river downstream.

In order to solve these problems, several modern technological solutions have been consistently applied; these solutions ensured the observance of all microbiological requirements for the potable water and sewage water.

Novosibirsk is the largest city in the Russian Federation which has implemented environment-friendly and safe method of UV-decontamination in the municipal water treatment plants and sewage treatment plants.

1 Potable water supply

During the course of the 'Water supply and water disposal systems development in 2007-2013' program implementation, in 2011, a municipal water UV-decontamination plant has been launched at the Pumping & Filtration Station No. 1 (NFS-1) with the rated capacity amounting to 350 ths. m^3/day. According to the expert opinion issued by 'Novosibirsk Oblast Hygiene and Epidemiology Center' Federal State Healthcare Institution in 2008, the

* Corresponding author 5171434@mail.ru

river water is affiliated to the 2-nd class in terms of microbiological contamination. Therefore, in order to create an efficient multi-barrier technology, the UV-radiation decontamination and chlorination (chlorine-ammonia) has been used in order to provide long-lasting disinfecting effect.

In two years, the next stage of the city water treatment system modernization has been implemented: UV-decontamination plant has been launched at the Pump & Filtration Station No. 5 (NFS-5). The rated capacity of this plant amounts to 500 ths. m^3/day. At this station, the multi-barrier decontamination system has also been implemented.

The multi-barrier decontamination technology that combines the UV-radiation and chlorination enables to use the advantages of each of the methods. UV-radiation is characterized by maximum decontaminating efficiency even when dealing with the microorganisms resistant to conventional chlorination (such as viruses, and protozoa cysts). Chlorination allows to ensure the long-lasting decontamination effect, which preserves the quality of water inside the distribution grids. Reduction of the actually used chlorine doses is beneficial for the reduction of the organochlorine compounds concentrations in the potable water [7].

Fig. 1. UV-equipment at NFS-5

2 Sewage water treatment

The observance of the state requirements for the sewage water treatment (SanPiN 2.1.5.980-00 'Hygienic requirements for the surface water protection') implies the mandatory decontamination of the city treated sewage water in order to prevent the ingress of the potentially pathogenic microorganisms into the environment and water bodies that may be later used as the potable water sources [4].

The desired degree of the potable water decontamination may be ensured by multiple methods; the most widely used one is the reagent-based treatment (chlorination), and the use of UV-radiation[1-3].

The use of chlorine or other chlorine-containing reagent (hypochlorite) observing the recommended concentrations for ensuring protection against the bacterial contamination does not guarantee the protection against the viruses and other chlorine-resistant microorganisms (such as parasitic protozoa). The ingress of the residual chlorine into the water body - sewage water receiver - may significantly deteriorate the ecological status of such water body up to complete seizure of the self-purification processes and death of all life forms. The sewage water chlorination is also accompanied by the active formation of by-products (various organochlorine compounds) that may be characterized by high toxicity, mutagenic action, and cancerogenic action. Even one-time contamination with such products may lead to long-lasting harm to the water body biocoenosis.

Modern applicable Russian requirements for the sewage treatment facilities (SP 32.13330.2012 'Updated version of SNiP 2.04.03-85 'Sewage. External grids and facilities") require the mandatory stage of dechlorination after the decontamination accomplished with chlorine or other chlorine-containing reagent. Dechlorination allows to remove the residual chlorine, but cannot prevent the formation and ingress of the organochlorine compounds into the water body, which have been formed during the chlorination process [6].

Decontamination accomplished using the UV-radiation is characterized by a number of rock-solid advantages:

- High efficiency with regard to all types of microorganisms (bacteria, viruses, and protozoa), even in their resistant cyst forms;
- The water temperature and pH-factor do not affect the UV-decontamination efficiency;
- Absence of consequences excludes the adverse effect for the water body and environment;
- No decontamination by-products;
- Easy control and management of the decontamination process; automation is possible;
- The method is safe for the public, there is no need to arrange storage facilities with toxic chlorine-containing reagents that require observance of the dedicated process safety and environmental safety measures, which improves the reliability of the sewage systems in general;
- High energy efficiency;
- UV-radiation decontamination leads to lower operational costs in comparison with the chlorination-dechlorination, especially when using safer, but more expensive sodium hypochlorite.

For efficient and complete fulfillment of the requirements concerning the sewage water microbiological quality parameters of the sewage water discharged into the water body, the UV-decontamination technology has been adopted, which is based on the amalgam ozone-free low-pressure lamps.

In February 2016, the Sewage Water UV-Decontamination Station was put into operation at the Municipal Sewage Water Treatment 1-st Train, Novosibirsk, with the nominal flow rate amounting to 26,000 m^3/h (480,000 m^3/day).

The domestic UV-equipment (NPO 'LIT') allows to efficiently decontaminate the sewage water, the UV-radiation dose in accordance with MU 2.1.5.732-99 and MUK 4.3.2030-05 amounts to at least 30 mJ/cm^2.

The UV-station comprises three working channels. Each channel includes three working sections with two 88MLV UV-modules in each of them. Further, channels comprise one redundant section in each of them; these redundant sections are to be used during the routine maintenance. The power consumption of the entire UV-station amounts to as much as 495 kW.

In order to bring down the sewage water UV-decontamination station footprint, this station is located within the running contact tank. The well-established two-level configuration has been used for the UV-station: the modules with the UV-lamps are located on the lower level in the concrete channels of the former contact tank, and all cabinets containing the UV-lamps control and power systems are located above the modules in a new easily-erected structure building. The entire UV-station is located inside the building 14 × 13.5 m.

Vertical orientation of the UV-lamps enabled to reduce the area occupied by the UV-station and minimize the scope of service work, since there is no need to remove the module from the channel for the scheduled maintenance procedures (UV-lamps replacement). Modern high-performance amalgam low-pressure lamps are characterized by high efficiency and good service life (over 12,000 hours).

An important feature of such station is the UV-lamps radiation power regulation integrated system. Such systems enabling the significant savings of the consumed electric power depending on the quality and quantity of the sewage water being treated are rather new within the domestic practical applications. One of the features of the system used is that the power regulation or adjustment occurs not only depending on the sewage water flow rate, but also based on the integral value of the sewage water quality - sewage water transmission coefficient with regard to the UV-rays. In order to measure this coefficient, the station is equipped with an additional device - taumeter; it measures the transmission coefficient in real-time and automatic manner with no lag, and with no need to engage service personnel. Such a system used for the UV-lamps power adjustment enables to reduce the electric power consumption by as much as 30-40 % of the UV-station power rating.

Fig.2. UV-station building, Novosibirsk

The UV-decontamination allowed to ensure efficient observance of the sanitary & epidemiological and environmental safety of one of the most important water supply source in the region - the Ob river.

It should be noted that the UV-decontamination station complies with the most up-to-date requirements concerning the use of the best available technologies (BAT) as stated in the Informational and Technical Manual ITS 10-2015 'Sewage water treatment using the centralized water disposal systems in settlements and urban districts'. This Manual specified that the use of chlorination, even with the follow-up dechlorination, is not the best available technology for the decontamination applied in the facilities related to such scale ('major' - according to the classification provided in the Manual) [5].

Fig.3. Redundant UV-module

Conclusion

The biggest Siberian water service (Novosibirsk water service is in the 3-rd place in production capacity after Moscow and Saint Petersburg) has completely been transferred to the UV-decontamination, used both for potable water, and sewage water.

The potable water undergoes the treatment using the multi-barrier decontamination technologies (combination of UV andchlorine-ammonization), which enables to provide safe and tasty water to every citizen living in the city.

The UV-decontamination station built on the municipal sewage water treatment facilities is a reliable barrier that prevents the proliferation of pathogenic microorganisms and viruses, while causing no harm to the water bodies and environment.

The complex use of the modern UV-decontamination technology both in potable water treatment, and in sewage water treatment enabled to fulfill the most strict requirements of both the sanitary, and environmental legislation.

References

1. O.A. Guseva, M.M. Gerasimov, A.D. Smirnov, Water Supply and San. Eq., **2**, 59-62 (2010)
2. L.M. Vasilyak, A.D. Smirnov, Water Supply and San. Eq., **9**, 49-56 (2014)
3. S.Yu. Shishov, E.A. Troshkova, V.Ya. Zhukova, A.D. Smirnov, R.A. Davlyaterova, V.A. Smagin, Water Supply and San. Eq., **6**, 32-38 (2014)
4. N. Zaletova, S. Zaletov, Int. Jour. of App. Eng. Res., **10**, 43873-43874 (2015)
5. N. Zaletova, Y. Voronov, N. Makisha, Int. Jour. of App. Eng. Res., **10**, 42454-42455 (2015)
6. E. Gogina, N. Makisha, App. Mech. and Mat., **587-589**, 636-639 (2014)
7. V. Chukhin, A. Andrianov, V. Orlov, The steel pipe corrosion in drinking water distribution systems and its rehabilitation techniques (2014) International No-Dig Madrid 2014.

New approaches to the electromagnetic monitoring for improving the regulatory framework "green" building

Marina Grafkina[1], and *Evgeniya* Sviridova[1*]

[1] Ecological safety of technical systems, Moscow Polytechnic University, Moscow Russia

Abstract. In the regulatory legal base of a "green" construction existing now in the Russian Federation for assessment of an ecological safety of rooms of residential and public buildings criteria of electromagnetic comfort aren't used. Its completion with obligatory assessment of indicators of electromagnetic safety is reasonable. Suggestions for improvement of the existing documents regulating assessment of conformity of real estate objects to ecological requirements are given in article. For the purpose of further development of system of regulation of negative factors of the habitat and electromagnetic safety, authors offer new approach to environmental control of low-frequency electromagnetic fields – transition from measurement of amplitude characteristics to determination of energy parameters. Sanitary and hygienic assessment and regulation of low-frequency electromagnetic fields it is performed separately on intensity of electric and magnetic field that doesn't give an idea of an overall picture of spatial distribution of energy. The energy approach offered by authors will allow to estimate total energy impact of electromagnetic fields per capita, and also to develop new system of regulation of low-frequency electromagnetic fields in energy parameters.

1. Introduction

Currently, in the Russian Federation with the aim of improving the quality of life of the population, including through the formation of human comfort for the external and internal environment, as well as the safety of life is under development and implementation of "green" technologies in all sectors of the economy, including construction.

The main objective of "green" building is not only a reduction in the consumption of energy and material resources throughout the life cycle of the building, but also to improve the quality of buildings and the comfort of their indoor environment.

For the development of "green" technologies in construction is necessary to improve the regulatory framework and the legislation right. In Russia, developed and used certification

* corresponding e-mail: evg_sviridova@mail.ru

systems that take into account the requirements of national standards, construction and sanitary norms, rules and guidance documents, as well as the basic provisions of the international LEED rating systems, BREEAM, DGNB and the HQE, including:

- GOST R 54964-2012 "Conformity assessment. Environmental requirements to the objects of real estate ";
- SRT NOSTROY 2.35.4-2011 "Green Building". Residential and public buildings. The rating system for assessing the sustainability of habitat;
- SRT NOSTROY 2.35.68-2012 "Green Building". Residential and public buildings. Taking into account regional peculiarities in the rating system of evaluation of the stability of the environment.

Environmental requirements to properties defined combination of the following basic categories:

- environmental management;
- infrastructure and the quality of the environment;
- the quality of the architecture and layout of the facility;
- comfort and environment of the internal environment;
- the quality of health protection and waste management;
- the rational use of water and stormwater management;
- energy conservation and efficiency;
- protection of the environment during the construction, operation and disposal facility;
- life safety.

In this category of "comfort and internal environment of the environment" includes the following criteria: air-thermal comfort; light comfort; acoustic comfort; protection of the premises from the accumulation of radon; monitoring and control systems engineering support buildings; monitoring and control of air environment.

None of these documents to assess the environmental safety of the habitat is not being used criteria for electromagnetic pollution premises of residential and public buildings, despite the fact that the well-known adverse effects of electromagnetic fields (EMF) on the human body. The list of normative references is present SanPiN 2.2.4.1191-03, which regulates exposure to EMF in a production environment. The application of these sanitary regulations for premises properly, as it does not guarantee favorable conditions habitat for a 24-hour stay of people in residential areas, especially in the light of the main tasks of the "green" construction.

Thus, the need for further improving the regulatory framework for the right "green" construction.

2. Methodology

First, we propose to amend the GOST R 54964-2012 on the basis of the existing regulations for electromagnetic safety. In particular, in paragraph 4 "Internal environment Comfort and Ecology" Table 1 "Basic categories and criteria" must add paragraph 4.7 (see. Table 1).

Table 1. Improving Table 1 GOST R 54964-2012 in terms of the criteria of the electromagnetic comfort

Basic categories	Criterion
4. Comfort and Ecology of the internal environment	4.7. Electromagnetic comfort

In addition, n. 4 "Comfort and Ecology of the internal environment" Table 2 "Basic category criteria, the minimum environmental requirements and methods of their determination." Table 2 below shows some additional indicators for premi

Table 2. Improving Table 2 GOST R 54964-2012 in the part of the electromagnetic comfort indicator

Criterion	Indicator	Minimum environmental requirements	Recommended indicators	Methods for determining the minimum environmental requirements
4.7.Electro-magnetic comfort	The electric field inside the premises	SanPiN 2971-84	0.5 kV / m	Evaluation of the measurements. The measurements can be carried out by specialized instrumentation. Analysis of project documentation, calculations to reduce them as part of the project; measurements
	The magnetic field inside the premises	GN 2.1.8 / 2.2.4. 2262-07	4 A / m	
	EMF RF remote control for the population in different frequency bands	SanPiN 2.2.4 / 2.1.8.055-96	25 V / m (30 kHz - 300 kHz) 15 V / m (0.3 MHz - 3 MHz) 10 V / m (3 MHz - 30 MHz) 3 V / m (30 MHz - 300 MHz) 10 mW / cm^2 (300 MHz - 300 GHz)	

One of stages of process of the organization of a construction directed to increase in an ecological safety of construction objects is carrying out engineering-ecological researches. Environmental monitoring is a basis for objective assessment of state of environment and an integral part of implementation of basic provisions of engineering-ecological researches.

Well-known negative impact of low-frequency electromagnetic fields on the person, environment, the built-up territories. The main sources of low-frequency electromagnetic fields in the urban environment are: power lines, power transformer substations, distribution points of system of power supply, air electric networks, transport stream, networks of power supply of the land electric transport, etc.

According to the SP 11-102-97 engineering-ecological researches for a construction which part the research of electromagnetic fields is are carried out for assessment of the current state and the forecast of possible changes of the surrounding environment under the influence of anthropogenous loading for the purpose of prevention, minimization or liquidation of harmful and undesirable ecological and related social, economic and other consequences and preserving optimum living conditions of the population. In the period of a construction, to operation and liquidation of construction objects environmental monitoring provides control of dynamics of an ecological situation (a condition of natural and technical systems, efficiency of protective and nature protection measures, etc.).

The research of energy pollution shall be performed first of all in case of development of town-planning documentation and designing of housing construction in the mastered territories. At the same time the main sources of harmful effects, its intensity shall be fixed and discomfort zones with excess of admissible level of harmful physical impact are revealed.

Development of new approaches to carrying out and assessment of results of environmental monitoring acquires special relevance at detection of real and potential health hazard of the population of pollution of the habitat. The complex indicator of anthropogenous loading which consists of the most significant factors of negative impact has to be the criterion of quality of the environment defining the state of health of the population.

In previously published papers [2-3], the authors demonstrated the possibility of environmental monitoring of energy parameters of the low frequency of anthropogenic impact, confirming the advisability of further improving the existing legal framework for Electromagnetic Safety and environmental requirements to the objects of "green" construction.

In accordance with the existing methods of evaluation and environmental monitoring of low frequency electromagnetic fields of their level of determination includes the study of amplitude spectrum of the electric and magnetic fields. The research practice is also not customary to define the energy parameters of low-frequency electromagnetic fields, because it is believed that the electromagnetic wave radiation source in the near zone has not yet been formed, so in a study of low-frequency electromagnetic fields EMF energy parameters are not evaluated. Rationing Energy indicators currently holds only for EMF radio frequency spectrum (in this case, the normalized energy exposure and the energy flux density).

However, there are certain techniques [3] to determine the noise energy source parameters in the near zone. Given the similarity of wave processes occurring in the sound and electromagnetic field, the authors propose a new approach to assessing the level of low-frequency electromagnetic fields - the transition from the measurement of the amplitude characteristics to the definition of energy parameters.

In the near-field zone of EMF, two processes that are qualitatively different in energetic terms take place. The first process is the process of periodic energy exchange between the energy source and the near-field zone (the reactive intensity I_i). Energy is either taken from the source and accumulated in the near-field zone electromagnetic field or is directed back to the source. [4]

The second process is the process of energy radiation (the active intensity I_a). It describes the wave process in the near-field zone. The radiated energy has a rather small value in comparison with the energy periodically accumulated in the electromagnetic field of the near-field zone and then directed to the power supply.

As well as in the general theory of wave processes, there is a concept of the integrated intensity in the electromagnetic field theory, which includes the active and reactive components:

$$\vec{I}_k = \vec{I}_a + i\vec{I}_i, \qquad (1)$$

where $\vec{I}_a$ is the vector of the active intensity of EMF, $\frac{W}{m^2}$;

$\vec{I}_i$ is the vector of the reactive intensity of EMF, $\frac{W}{m^2}$.

Determination of the EMF energy performance (integrated intensity) will allow gaining a new strong effect of monitoring – electromagnetic energy spatial directional distribution.

3. Results

Integrated intensity can be determined by means of the cross spectrum function, which spectrum is computed by multiplication of one spectrum by the complex conjugation of the second spectrum.

The cross spectrum is a complex value. The cross spectrum amplitude characterizes the total energy of EMF in the given point, and the phase is a difference of phases between the intensity of the electrical field and the intensity of the magnetic field.

Thus, the integrated intensity of EMF is equal to:

$$I_k = E \cdot H^* = (|E| \cdot cos\varphi_1 + i|E| \cdot \sin\varphi_1) \cdot (|H| \cdot \cos\varphi_2 - i|H| \cdot \sin\varphi_2 =$$
$$= |E| \cdot |H| \cdot (\cos(\varphi_1 - \varphi_2) + i \cdot |E| \cdot |H| \cdot \sin(\varphi_1 - \varphi_2), \quad (2)$$

where E is the electric field intensity, V/m;

H is the magnetic field intensity, A/m;

H* is the complex conjugate value of the magnetic field intensity, A/m.

At time averaging (for a period), the reactive intensity is converted to zero, and there remains only the active intensity, which can be measured.

For determination of the reactive intensity, 90° is added to the initial phase of the electric field intensity and its value is gained at the spectrum analyzer's output:

[illegible] (3)

where I_a is the active intensity of EMF, $\frac{W}{m^2}$;

I_i is the reactive intensity of EMF, $\frac{W}{m^2}$;

φ_E is the electric field intensity initial phase, grad.

Thus, in any point of the low frequency EMF, it is possible to find the integrated intensity – the energy parameter of EMF.

For the experimental determination of the energy of low-frequency EMF authors parameters measuring system was developed, which consisted of antenna measurement, oktafon, analog-to-digital converter (ADC) and dual-channel spectrum analyzer.

When measured simultaneously measured the electric and magnetic fields, then converted by the ADC signal is transmitted to the spectrum analyzer and the intensity of the active was determined (the angle between the antennas was 90 degrees).

The experiment was obtained spectrum intensity active in the low-frequency EMI radiation source near zone in absolute units. Active intensity is positive and it has a direction from the radiation source. The maximum value at a frequency of 50 Hz up to 8.7 kW/m^2.

4. Conclusion

Most have to deal with multiple sources of low-frequency EMF, while not all sources are obvious. Active intensity in this case points to a radiation source, introducing a maximum contribution to electromagnetic pollution at a given point and thus, will develop the most effective methods of protection.

The proposed new authors new approach will provide a comprehensive assessment of comfort objects "green construction" for several indicators of energy (noise, EMF, thermal radiation, vibration) and allow to solve essentially new problems of improve in ecological safety of the built-up territories

References

1. M.V. Grafkina, B.N. Nyunin, E.Y. Sviridova, Bulletin MGSU, **4**, 116 (2014)
2. M.V. Grafkina, B.N. Nyunin, E.Y. Sviridova, Ecology of the urbanized territories, **3**, 66 (2014)
3. Bryul and Kjer, *Intensity of a sound* (Denmark, 2000)
4. L.A. Bessonov, *Theoretical bases of electrical equipment: electromagnetic field (The higher school*, Moscow, 1998)

Analysis of selected factors that generate the costs of accidents at work using the Polish construction industry as an example

Anna Hoła[1], *Bożena* Hoła[1], *Marek* Sawicki[1], *Mariusz* Szóstak[1,*]

[1]Wroclaw University of Science and Technology, Faculty of Civil Engineering, Poland

Abstract. The paper presents analysis of selected factors that generate the costs of accidents at work using the Polish construction industry as an example. The individual components of the cost of accidents have been identified. Using the statistical data published by the Central Statistical Office, the impact on the size of the cost of accidents at work of such factors as the lost time of an injured person, the lost time of other people involved in the removal of accident effects and also material losses caused by an accident, was analysed. On the basis of the conducted analysis, conclusions regarding economic losses due to accidents were formulated.

1 Introduction

The implementation of construction works is an activity that is burdened with a high level of risk of adverse events such as construction catastrophes, accidents at work or events that can potentially cause accidents. The construction industry is one of the most dangerous sections of the economy in most countries of the world. This is evidenced by published statistical data and reports of organizations and agencies dealing with security issues in the world such as, among others, the National Safety Council (NSC), the Bureau of Labor Statistics (BLS), the National Institute of Occupational Safety and Health Administration (OSHA), the Central Statistical Office of the European Union (EUROSTAT) and the Central Statistical Office (GUS) in Poland.

It can be concluded from statistical data published by EUROSTAT that in 2012 the construction industry was classified in second place in terms of the severity rate of accidents at work which cause more than 3 days of absence from work, and also in second place in terms of fatal accidents at work [1, 2]. According to statistical data published by the Central Statistical Office (GUS), 6264 accidents happened in the construction industry in Poland in 2014, including 55 fatal accidents and 74 serious accidents [3]. In the ranking of individual sections of the economy in terms of the incidence rate of accidents, the construction industry was in a very high eighth place. The results of research published in many publications indicate that the scale of the problem related to accidents at work in the construction industry is significant.

[*] Corresponding author: mariusz.szostak@pwr.edu.pl

An accident at work, in addition to human losses, also carries tangible economic losses [1,3,4] related to, among others: the treatment and rehabilitation of an injured person, the suspension of production, the need to replace an injured employee with another employee, the obligation of paying compensation to an injured person, the repair or replacement of damaged equipment and sometimes the rebuilding of an entire destroyed job post.

The article attempts to analyse the accident phenomenon with regards to selected factors that generate costs. Based on the subject literature, the identification and classification of individual cost components have been made. Using statistical data published by the Central Statistical Office (GUS), analysis of selected factors that generate costs was carried out. The conducted analysis is of an estimated nature because a significant part of the costs related to an accident is difficult to calculate and not recorded.

2 Identification of costs associated with an accident at work

In the subject literature, the costs that generate accidents at work are often presented as a "drifting iceberg". Its visible part, located above the surface of the water, symbolizes the costs that are possible to be estimated. The much larger and invisible part of the iceberg is under the surface of the water and symbolizes the costs that are difficult to estimate and often go unnoticed. [5] The estimation of the costs of accidents at work is the subject of scientific research conducted around the world. The basis for this estimation is the identification of cost components.

Thus, the authors of the study [6] divide the costs of accidents into three groups: direct costs, indirect costs and quality of life costs. Direct costs include payments for hospitals, physicians, allied health services, rehabilitation, nursing home care, home health care, mental health payments for treatment, police, fire and emergency transport. Indirect costs include victim productivity losses, which include wage losses and household production losses, and also administrative costs, which include the cost of administering worker compensation wage replacement programs and sick leave. Quality of life costs refer to the value attributed to the pain and suffering that victims and their families experience as a result of an injury.

In turn, the authors of studies [7,8] divide costs of accidents at work into two categories, namely: primer costs and secondary costs. The primer costs include the loss of productive employment as a result of premature death or injury measured in terms of time loss or wage loss. Secondary costs include employer financed fringe benefits, household production losses and payments for medical services.

In Poland, according to [3.9], there are three categories of costs of accidents at work: the costs incurred by an enterprise (CE), the costs incurred by a victim and family (CVF) and the costs incurred by the whole society (CIS). It can be expressed by the following formula:

$$CAW=CE+CVF+CIS \quad (1)$$

The costs incurred by an enterprise are the difference between the sum of the individual components of costs and compensation (C) received from insurance companies [3,9] and are expressed in formula (2). The designations used in the formulas (1 and 2) are described in Figure 1.

$$CE=(CLT+CMCaT+CO+CS+CDP+CML+CR+COR+OC)-C \quad (2)$$

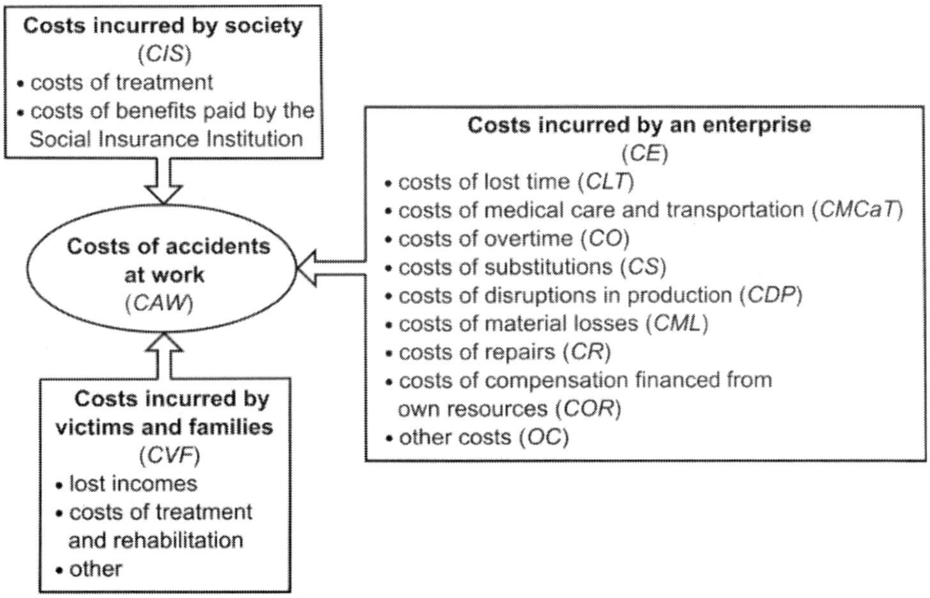

Fig. 1. Classification of costs of accidents at work

It should be noted that an accident at work causes a break in work, not only for the victim but also for other people. This break generates the so called costs of lost time (CLT), which include: the time lost by an injured person on the day of an accident and during the days of absence; working time of people organizing first aid, conducting the first aid and those re-organizing the work; time required to carry out an accident investigation and also the time necessary for planning preventive actions.

Ensuring production continuity requires the involvement of additional people on both the day of an accident and also during the days of the injured employee's absence (CO, CS). Production interruptions cause malfunctions of machines and whole production lines and the fact that they are not in use generates costs due to production losses (CDP).

An accident at work generates costs associated with providing medical help (CMCaT). This can be: transport of an injured person to a doctor or other place of medical care, the cost of bandages and medicines taken from the enterprise's infirmary or the point of medical care on a building site and also the cost of medical staff employed in the enterprise's infirmary.

Losses of property (CML) are often related to accidents at work. They include the value of damaged raw materials; semi-finished products; finished products; the lost value of machinery, equipment and vehicles; as well as the cost of buying new machines, equipment and vehicles, or the cost of their repairs (CR).

An employer often also bears the costs of financial compensations (COR), which are vested on the basis of the Labour Law by a court on the basis of a civil action. Concerning legal cases, the costs of court expert opinions and court expenses must also be taken into account.

Other costs (OC) relate to, e.g. rescue actions and penalties incurred by an enterprise due to a failure to comply with a signed agreement or contract as a result of an occurred accident. Immeasurable costs caused by the loss of image of a company in which accidents happen frequently cannot be omitted.

3 Factors that generate costs of accidents

On the basis of available statistical data published by the Central Statistical Office (GUS) the following factors were analysed:

• lost time resulting from the treatment and rehabilitation of people injured in heavy and light accidents at work in the construction industry in 2014,

• the course of the accident severity rate in the time period covering the years from 1992 to 2014 and time resulting from substitutions at work,

• the cost of property losses caused by an accident at work.

3.1 Lost time

The consequence of accidents at work, beyond the effects causing fatal or serious injuries of the body, is the absence of employees due to their treatment and rehabilitation. Statistical data published by the Central Statistical Office indicates that the absence of employees caused by accidents at work in the construction industry is very large. Figure 2 shows the distribution of the number of people injured in accidents at work in the construction industry in 2014 in relation to the number of sick leave days [10]. This figure shows that in the construction industry accidents with a negligible degree of severity that are not eligible for sick leave also occur. However, a significant amount of accidents, represented by 54.6% of all accidents, are eligible for sick leave that lasts for more than one month.

The total number of sick leave days per total number of accidents that occurred in 2014 in the construction industry amounted to 325,364 days [5]. Assuming that the average number of working days in a month is equal to 22, it can be calculated that 1232 employees from the construction industry were on sick leaves that lasted for a year.

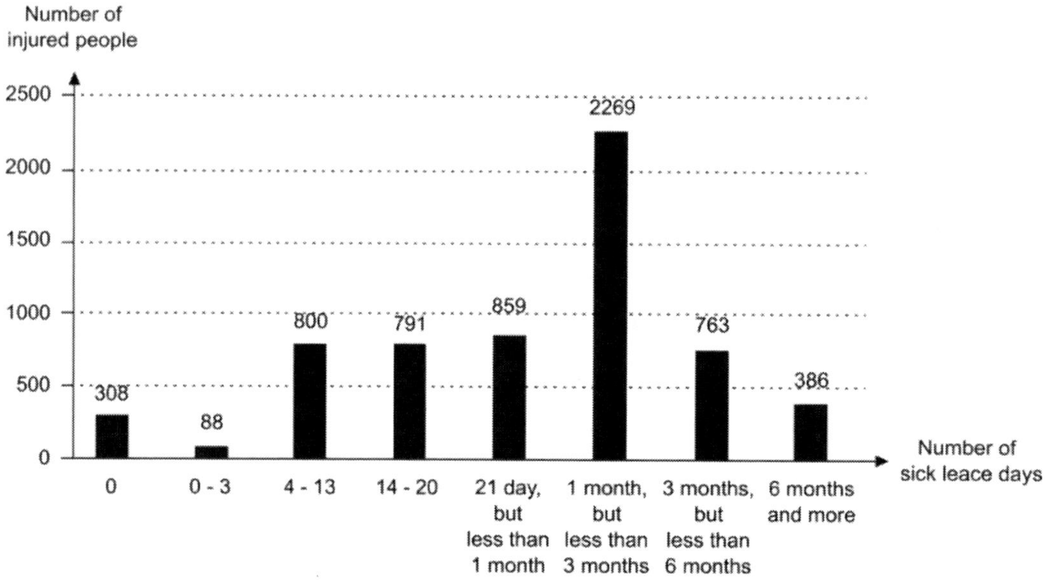

Fig. 2. Distribution of the number of people injured in accidents at work in the construction industry in 2014 in relation to the number of sick leave days.

3.2 The severity of accidents at work

The phenomenon of absence from work caused by accidents is described by the severity rate of accidents, which indicates the average number of workdays lost due to sickness per one person injured in heavy and light accidents [10]. The course of changes in the value of this rate during the period between 1992 and 2014 is shown in Figure 3. After a period of dynamic growth the rate in 2006 reached a value of 56.9 days per one accident. From 2007 the severity rate had lower values and in 2014 it was equal to 52.4 workdays per injured person. The value of this rate confirms that injuries caused by accidents in the construction industry are serious and require long treatment that generates high costs. In terms of the value of this rate, the construction industry placed second in the ranking of all sections of the economy, just after mining and extraction for which the severity rate of accidents was equal to 76.8.

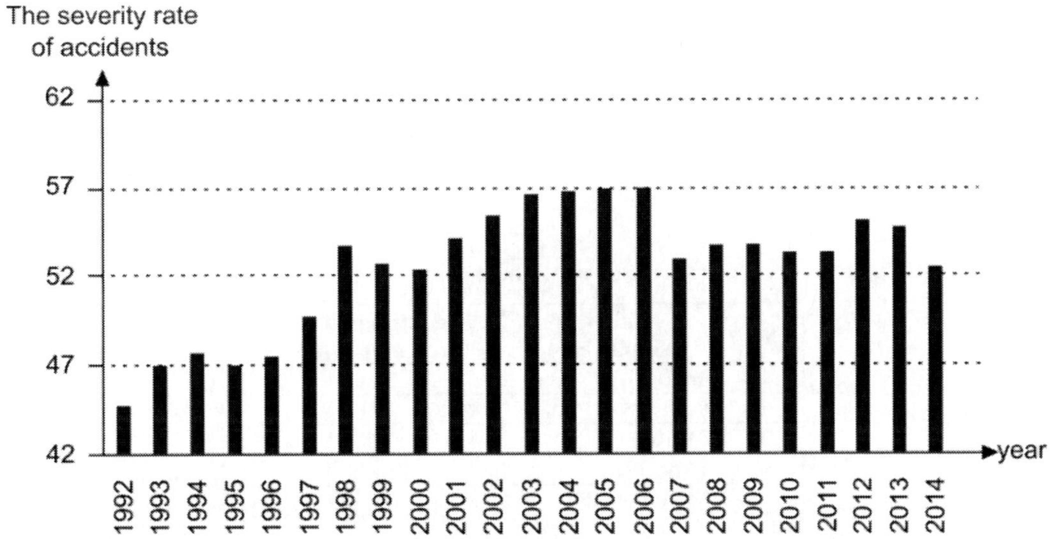

Fig. 3. The course of changes of the severity rate of accidents at work in the construction industry.

3.3 Time losses of other people caused by accidents at work

Removing the effects of accidents often requires the involvement of additional people, which generates additional costs. When calculating the costs of an accident it is necessary to consider the working time of employees providing first aid, the working time of workers defining the causes and circumstances of an accident at work, and in the case of the suspension of the working process, also the working time of people covered by downtime. Employee absence due to treatment and rehabilitation is often associated with the need of replacing such a person, and therefore with the employment of another person or by paying overtime to people already employed.

In 2014 in the construction industry, additional working time of such people amounted to 8437 working hours. [10]. When converting this into an 8-hour working day, it results in about 48 working months. This means that four employees from the construction industry throughout the whole year of 2014 were involved in removing the effects of various accidents at work.

3.4 Material losses

As a result of an accident at work, construction equipment, raw materials and finished goods are often destroyed. According to [10], the estimated material losses caused by accidents at work in 2014 in the construction industry, which include the total value of damaged machinery, equipment, tools, raw materials, finished products and buildings with their contents, amounted to PLN 1813.2 thousands. This represented 4.2% of the costs of material losses incurred in all sectors of the national economy, which amounted to PLN 42 780 thousands. In the ranking of all sections, the construction industry was in 5th place. The greatest material losses due to accidents at work were reported in the following sections: transportation and warehousing and also manufacturing, which amounted to PLN 15 179.9 thousands and PLN 14 859 thousands respectively.

4 The estimated total costs of accidents at work

Cost analysis of 330 accidents at work that happened in different sections of the Polish economy [11] showed their dependence on the number of people employed in an enterprise. The data included in Table 1 indicates that the costs of an accident increase with the size of an enterprise. This is caused by the fact that in large enterprises, due to the application of more sophisticated technologies and expensive technical equipment, the costs of production disruptions and repairs are much higher.

Table 1. The average estimated cost of an accident in relation to the number of employees [11]

No.	Number of employees	The estimated cost of an accident [PLN]
1	up to 5	12 000
2	6-20	19 000
3	21-100	29 000
4	101-250	49 000
5	More than 250	63 500

In turn, Table 2 summarizes the costs incurred due to fatal, serious and collective accidents. It was calculated on the basis of 368 accidents that occurred in the years 2004-2009 in construction companies, healthcare, industrial processing plants and waste management plants [1].

Table 2. The costs of accidents at work [1]

Type of accident	The average total cost (CAW) [PLN]	The costs incurred by an employer		
		The average total cost of an employer (CE) [PLN]	The cost of an accident "cheapest-most expensive" (CAWmin/CAWmax) [PLN]	The average cost of prevention [PLN]
fatal	109 395	58 897	11 685 / 344 176	1 149
serious	236 904	213 470	2 564 / 1 394 456	1 712
collective	77 207	32 435	143 970	796

The average cost of a fatal accident incurred by a company amounted to PLN 58,897. The average costs of serious and collective accidents amounted to PLN 213,470 and PLN 32,435 respectively. The last column gives the estimated costs of the preventive measures that should be taken to prevent an accident [1]. These costs were disproportionately low in relation to the total cost of an accident and ranged from 0.8% to 2.45%. The included data only refers to the costs incurred in the first year after an accident. However, it should be noted that costs associated with an accident are also incurred in subsequent years and these costs are: a pension paid to an employee or a person authorized by the Department of Social Insurance and also rehabilitation costs

5 Conclusions

The conducted research and analysis are the basis to draw the following conclusions regarding economic losses due to accidents that occurred in Poland in 2014:
• Measurable economic losses are related to accidents at work, as well as human losses in the form of injuries of varying severity. The costs of accidents at work are incurred by a company, the victim and his family and also the whole of society.
• The consequence of accidents at work, beyond the effects causing fatal or serious body injuries, is an absence of employees. As many as 54.6% of the victims of accidents in the construction industry were qualified for sick leave lasting for more than one month. The phenomenon of absence from work caused by accidents is described by the severity rate of accidents. In 2014 the rate was equal to 52.4 workdays per one injured person. Regarding the value of this rate, the construction industry placed second in the ranking of all sections of the economy, just after mining and extraction.
• Removing the effects of accidents often requires the involvement of additional people. Based on statistical data, it can be calculated that in 2014 an average of 4 construction employees were involved in removing the effects of accidents at work throughout the whole year.
• In terms of material losses caused by accidents at work, the construction industry placed fifth in the ranking of all sections of the economy. Percentage participation of material losses in the construction industry in relation to losses incurred in all sectors of the economy amounted to 4.24%.
• Analysis of accident costs has shown that the costs of an accident increase with the size of a company. This is due to the fact that large workplaces use expensive equipment and technologies, and for this reason the costs of production disruptions are bigger. The estimated hypothetical costs that should be incurred for prevention, in order to ensure occupational safety, were disproportionately low in relation to the total cost of an accident and ranged from 0.8 to 2.45%.

The study is the result of the implementation by the authors of the research project No. 244388 "Model of the assessment of risk of the occurence of building catastrophes, accidents and dangerous events at workplaces with the use of scaffolding", financed by NCBiR within the framework of the Programme for Applied Research on the basis of contract No. PBS3/A2 /19/2015.

References

1. H. Purwin, „About the costs of occupational security - the costs of accidents at work", *Przyjaciel przy pracy*, **07**, (2011)
2. B. Hoła, M. Szóstak, ACE, **61**, 19-34 (2015)
3. J. Rzepecki, The social costs of accidents at work in Poland". *Bezpieczeństwo Pracy* , **5**, (2012)
4. T.C. Haupt, K. Pillay, Journal of Engineering, Design and Technology, **14**, 373-419 (2016)

5. R. Studenski., „The organization of occupational safety in an enterprise", Publishing House of Silesian University of Technology (1996)
6. G. M. Waehrer, Accident Analysis and Prevention, **39**, 1258-1266 (2007)
7. Y. Feng, Journal of Construction Engineering and Management, **141** (2015)
8. T. E. Cressler, J. R. Moore, Safety Science, **88**, 44-53 (2016)
9. www.ciop.pl (dostęp październik 2016)
10. „Accidents at work in 2014". Central Statistical Office, Warsaw (2015)
11. E. Kowal. I. Gabryelewicz, A. Kowal, „Estimation of the costs of accidents at work in mines", 548-561 (2014)

Height–zoning change of residential buildings in contemporary urbanism with respect to daylight and insolation

Agnes Iringová[1,*], *Ján* Rybárik [1]

[1]University of Žilina, Faculty of Civil Engineering, Slovakia

Abstract. The paper focuses on legislatively allowable amount of external shade by proposed height-zoning change of a structure and its impact on distribution of daylight and direct sunlight, light comfort, and insolation time in affected areas in dependence on their orientation. It also deals with a volume solution in terms of hygienic conditions in such residential buildings.

1 Introduction

Daylight is one of the primary aspects of a healthy environment. Any restriction under the allowable limit worsens hygienic condition of the affected area. In addition, of course, to sell such flats is not easy. External shading has a major impact on the availability of diffused light in shaded areas, and on light comfort in residential or work areas. Extensions, additions, and new housing development are barriers to already existing buildings and its collector areas, keeping them from transfer of global solar radiation. It causes reduction in daylight intensity, insolation, and passive solar gain in affected rooms.

The paper is aimed at examining to what extent the actual legislative limits are "hygienically correct" in relation to the residential environment and occupants of affected areas, i.e. the extent of required hygienic limits that are to be met after shading and taking apartments to use. It will try to find a compromise of a volume solution in case studies in order to optimize the relationship between the right of occupants to daylight and sustainable economic interest of developers entering into the residential area. The issue is very topical, since densification of urban structures by extensions or additions in the cities as well as in the country is still ongoing and affects all ages.

2 Legislatively allowable amount of shade by designed structure in terms of daylight

The effect of designed buildings on the existing environment in terms of availability of diffused light on the facade is specified in Regulation 259/2008 determining criteria and

* Corresponding author: agnes.iringova@fstav.uniza.sk

methodology of assessment given in STN 73 0580 Daylight in Buildings 1,2, Change 2 from 2000. The criterion for allowable daylight amount is the equivalent shade angle depending on the position and angle of the assessed area. Affected windows in the living and working areas are subjects of assessing.
Regulation 259/2008 requires the implementation of designed buildings not to make interior daylight conditions in nearby buildings worse than it is set in STN 73 0580 Daylight in Buildings 1,2, Change 2 from 2000. The allowable change or maximal amount of shade, determined by maximal equivalent shade angle, or, in cases where allowable amount of shade goes beyond the limit - e.g. gap sites, the criterion is equivalent sector shade angle in the boundary area of the building site. Values of allowable equivalent angles are based on existing conditions of urban planning. The highest allowable amount of external shade in historic city zones is 42 °; camber ratio to the distance of the opposite building 1: 1 was applied in historic buildings, and subsequently clearances and window heights in the lowest-floor rooms were adjusted. Window heights from ground floor to higher floors were typically changed. [1]

The allowable amount of external shade in the second zone is 36 °. It comprises facades of buildings situated in highly attractive locations, which brings owners´ requirement to increase the value their estate. It primarily refers to zones near urban historic centres and busy streets attractive to investors, whilst the amount of shade allowed there is only towards the street; as for yards, it is up to 30 °. Generally, applicable amount of external shade for all other areas is maximally 30°. Exceptions are affected faces of nursery and primary schools, where the maximal allowable equivalent shade angle is 25 °. Such amount of external shade, if the window and room geometry is optimally designed, produces a presumption that the value of daylight factor after shading will be met.

3 Volume solution of an apartment building with a superstructure in an urban area within legislative limits and its impact on light comfort in affected flats

Allowable amount of external shade is one of the criteria that have significant impact on light comfort in affected or designed areas in terms of daylight intensity. It is possible to define it as the primary assumption affecting optimization of window area, its positioning and slope in the wall. Lower shade by built horizon also affects the structure and positioning of façade elements as well as room geometry itself. As for new buildings, in dependence on the site with allowable amount of external shade, the parameters can be optimized so that the required criteria of light comfort would be met. Buildings raised by 2000, when designs went out of the boundary conditions of urban density of that time, contain window design, room geometry, and face structure optimized for their specific criteria. The change in these criteria - currently allowed higher amount of external shade - often worsens light comfort in affected flats bellow the hygienic limits.

3.1 Optimization of a superstructure volume solution within legislative limits

The developer´s requirement was to design legislatively maximal volume of a superstructure, built on existing apartment building, in respect to affected apartment buildings in the suburb. Boundary conditions were as follows: spacing between the affected opposite apartment building and the designed superstructure should be about 28 m; it should affect northern face of the building.

The one side of an extended apartment building is parallel with a similar existing apartment building; its opposite side faces the enclosed yard with a playground. It is one of four residential buildings enclosing and bordering the yard area.
The building with superstructure is located in the residential zone in Ružinov, where allowable amount of external shade is currently max 30°. The previous allowable amount of external shade in original urban disposing of residential buildings was about 20°. The superstructure volume solution, as requested by the investor, was optimized to the maximal allowable amount, which, according to legislation, enabled to build the 3-storey superstructure with the receding top storey. Side shading by other buildings, in terms of their distance, is insignificant for assessed flat areas.

3.2 Impact of allowable amount of external shade by superstructure on light comfort in affected flats

The example to illustrate the impact of allowable amount of external shade on light comfort in affected flats is the first floor in the apartment building raised in 1970s in housing area. Windows of one room in each 3-room flat in this building face the planned superstructure; the flats are east/west-oriented.

Room and window geometry was designed to comply required hygienic limits; current conditions in all affected rooms met required daylight factor - see Fig. 3a. After the implementation of a 3-storey superstructure having allowable equivalent shade angle 30 °, the value of daylight factor (DF) in the middle of the affected rooms will reduce by approximately 50% when compared to original conditions. The required hygienic limit will be met in the shallowest room only, its depth is 3.5 m. Light conditions in all 10 flats on the first floor will be worse.

Fig.1 . Floor plan section of affected flats on the first floor

Affected flats on the 2nd floor, wherein equivalent shade angle under allowable amount of external shade is 28.2 °, will have the DF value in critical rooms also below the limit - approximately 0.8 %, which is 11% below required limit.
Affected flats on the 3rd floor, wherein equivalent shade angle under allowable amount of external shade is 23.8 °, will have the DF value in critical rooms within the limit with an average DF higher than 0.9%. Daylight reduction in affected flats below the current legislative limits defined by equivalent shade angle is significantly higher than its specification in the original restriction before 2000. It weighs against their occupants.

Table 1 Daylight reduction in critical rooms on the 1st floor

Amount of shading equivalent shade angle (ESA)	Floor	Average daylight factor (DF) in %		Daylight reduction - from minimal standard DF= 0,9%	
		Room 1	Room 2	Room 1	Room 2
ESA=20° original situation	1st	1.00%	1.9%	-	-
	2nd	1.35%	2.30%	-	-
ESA=30° proposed situation	1st	0.50%	1.0%	50%	-
	2nd	0.785%	1.54%	13%	-
ESA=25° optimized situation	1st	0.80%	1.5%	11%	-
	2nd	0.95%	1.8%	-	-

4 Legislatively allowable amount of external shade by designed structure and its impact on insolation time in affected rooms

The second criterion restricting setting and volume solution of the designed home building - amount of external shade in the area in relation to the housing environment - is the minimal required insolation time in affected flats. This is specified by Regulation 259/2008 and criteria and methodology to assess is defined in STN 73 43 01 Residential Buildings. Affected windows in the residential environment and in pre-school buildings are subjects of assessment. According to Regulation 259/2008, the implementation of new buildings is required not to worsen light conditions in living rooms, rooms in children´s and retirement homes, as well as rooms in similar facilities used for long-term stay of persons by more than it is admitted in STN 73 43 01 Residential Buildings.

Shaded living rooms are required to have minimal insolation time 1.5 hours over a period from 1st March to 21st June when the sun is higher than18 °.

The exception can be only applied on residential buildings located in historic parts of cities, where insolation can be reduced to 1 hour in particular cases (construction work, building in gap sites etc.) as defined in Section 4.2.1.2 of STN 73 4301.

For pre-school facilities, the required minimal insolation time must not drop under 3 hours within 10:00 am - 15:00 pm due to external shading, whilst the implementation of newly built structures must not reduce the size of sunlit surface area of pre-school facilities by less than 50% of its total area. If the current sunlit window surface of playrooms is smaller than 50%, it not allowed being more shaded.

5 Allowable amount of external shade and its impact on insolation of affected rooms

Depending on the orientation of the affected facade, its insolation time changes according to allowable amount of shade. In the particular case, the equivalent sector shade angle in checkpoint 1.2 m above the floor is 32.7 °. Here is the example of the situation where west-facing windows in shaded rooms of affected flats would have equal amount of external shade. Insolation time in a critical position - with the largest sector of shading depending on window orientation is in Table. 2. Effect of an obstacle on shading the affected facade is in Fig. 4, wherein the dotted line means the sun´s path according to solar time. It follows that on the critical day 1st March, east- and west-facing windows are strongly affected by shading collector areas and insolation time is about 1 hour, i.e. hygienic limits would not be met.

When the room is ideally oriented and contains south-faced windows, insolation time is only 2.25 hours. When facades are east-or west-oriented as well as when they are northeast- or northwest-oriented, the insolation time is under hygienic limits.

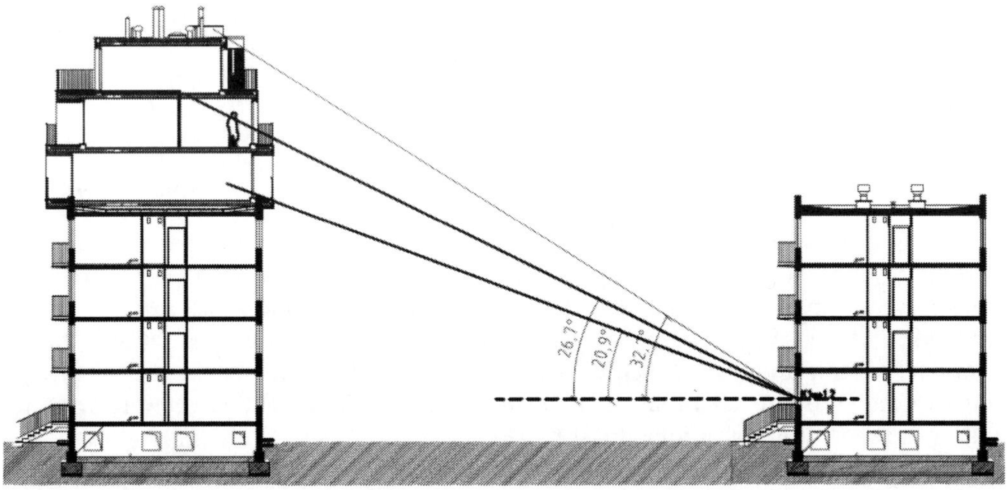

Fig. 2 Insulation - cross-section of superstructure and affected building - shade angles under the current, designed, and optimized conditions

Table 2 Insolation time at allowable amount of external shading in terms of daylight distribution

Amount of shading equivalent shade angle (ESA)	Critical checkpoint position	Window orientation				
		Insolation time (hours), 1.3. 49 north latitude				
		south	southeast	southwest	east	west
ESA=20.9° original situation	K1	8	4	4	1.75	1.75
ESA=32.7° proposed situation	K1	2.5	2	2	0.5	0.5
ESA=26.7° optimal situation	K1	6	2.75	2.75	1.5	1.5

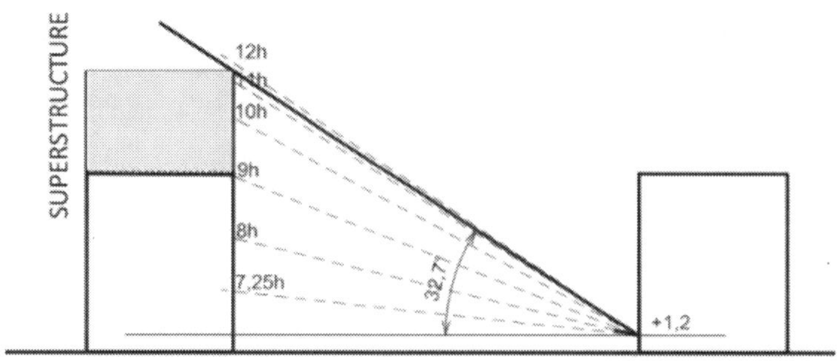

Fig. 3 Cross-section of sun height depending on solar time

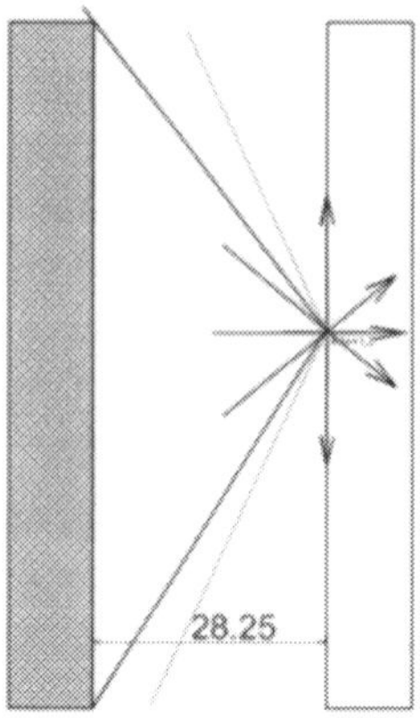

Fig. 4 Situation

In the specific project solution, the affected facade was east-faced and affected flats were those with more than 1/3 west-oriented floorage. In this case, since the west-facing windows were free of the horizon, and required insolation time in rooms with windows oriented to this cardinal point met hygienic limits; shaded side did not have to be sunlit according to legislation. To design the superstructure volume, only amount of shade on the window wall was crucial in terms of diffuse light, allowing building the superstructure within allowable limits according to STN 730,580. [2]

6 Optimization of the superstructure volume solution in terms of sustainable hygienic quality in affected flats

Considering aforementioned statements, it is of all-society interests to re-evaluate current limits for external shading and to specify legislatively acceptable hygienic level, at least in relation to the residential environment.

The optimal solution for assessed condition and for similar boundary conditions of housing development in suburbs and individual built-up zones, allowable amount of external shade would be up to 25 °.

Based on this assumption, rooms on the first floor in the assessed building would have daylight intensity reduced by approximately 25% when compared to standard limits - see Fig. 5. The impact on final insolation time would be also positive when east/west-oriented.

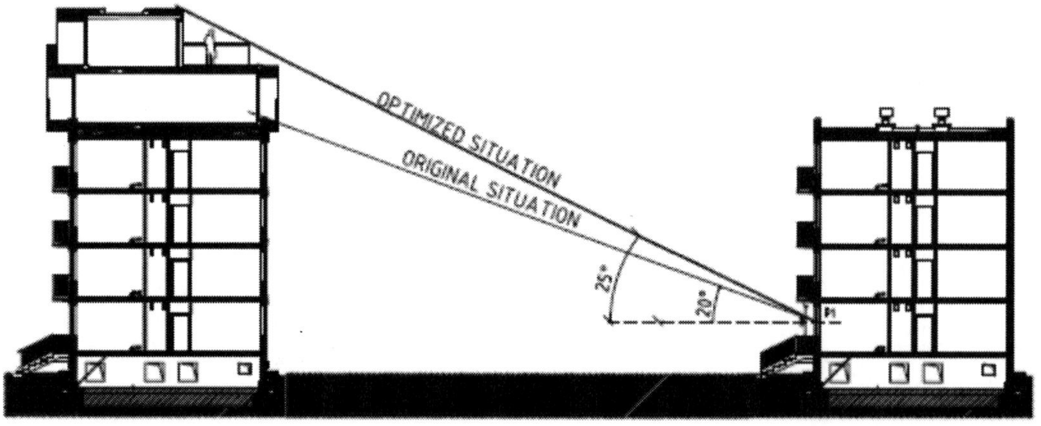

Fig. 5 Cross-section with optimized superstructure volume in relation to affected flats

7 Conclusion

In terms of daylight, allowable amount of reduced daylight is significantly higher on a comparative level than admitted in legislation by 2000. Up to that time, maximal allowable reduction of daylight factor was 15% on a comparative level in contrast to hygienic limits. At present-day methodology, the current amount of daylight reduction is really depending on boundary conditions compared to a standard limit of about 50%. In zones with higher allowable amount of shade - wider city centre and central urban areas - the impact on light comfort in affected flats is even more considerable. Flats on the 1st to 3rd floors often become the flats without daylight.

According to contemporary methodology, the only criterion to assess through equivalent shade angle is external illuminance on the outside in the middle of a shaded window. The legislation admits high-loyal limits to investors, but less-loyal ones to occupants and owners of such estates. It is of all-society interests to optimize these conditions in such a way that hygienically acceptable level would be reached to benefit of all concerned taxpayers.

The optimal solution for assessed condition, as well as for similar boundary conditions in housing development and zones with individual housing construction, would be strictly set requirements for maximal equivalent shade angle up to 25 °.

In that situation daylight intensity in shaded rooms on the first floor in affected building, investigated in the case study, would be reduced approximately by 11% from limit value - see Table 1. On the upper floors, the value of daylight factor would meet hygienic limits.

Insolation in rooms with east- or west-oriented windows at this amount of external shade would be about 1.5 hours, which is the lower line of hygienic limits. The impact on current hygienic conditions in existing flats would not be so critical. This shading restriction would largely influence intensity of direct irradiance in terms of passive or active making use of collector areas in reducing energy demands of these buildings.

Acknowledgement
Presented results were obtained with the support of the grant project VEGA no. 1/0945/16 (KEGA no. 041ŽU-4/2014).

References

1. P. Ďurica, *Failure buildings* (EDIS, Žilina, 2012)
2. M. Halahyja, *Building heat engineering, acoustics and lighting* (ALFA/SNTL,1985)

MATEC Web of Conferences **86**, 07007 (2016) DOI: 10.1051/matecconf/20168607007
IPICSE-2016

Scaffoldings for nonstandard facades

Marat Kuzhin[1], and *Bulat* Kuzhin[1,*]

[1]Moscow State University of Civil Engineering, 129337, 26, Yaroslavskoye Shosse, Moscow, Russian Federation

Abstract. This article dedicated to structure of scaffoldings on facades with difficult shapes, which sometimes makes impossible for using usual facade equipment. Exactly, in this article, will be learnt angled facades. Considered experience of solutions in such works and suggested new variants of scaffoldings for sloping facades. Reviewed safety questions on the facades. Were given definitions of efficiency and laboriousness for that facade system, and enumerated factors influencing on them. The article is actual for today, because there is not any normative documents in construction sphere for buildings with irregular shapes.

1 Introduction

Nowadays facades can have almost any shape and restricted only by architecture's mind. Often it challenges builders. Because it is not an easy job to transfer facade workers in a high levels of modern buildings. For example, it is impossible to use ordinary falseworks on sloping facades. Therefore, it need to be resolved.
Below was given few pictures of buildings, situated in Moscow, with sloping facades and variants of using scaffoldings.

Fig. 1. Sandglass form building **Fig. 2**. Sloping in the down part of building

* Corresponding author: kuzhinbf@gmail.com

Fig. 3. Sloping with negative angle in the top part of building

Fig. 4. Transfer workers on sport complex

2 Models of scaffoldings for sloping facades

Basing on that experience, suggesting new facade system, shown in following drawings (Fig. 5, 6 and 7).

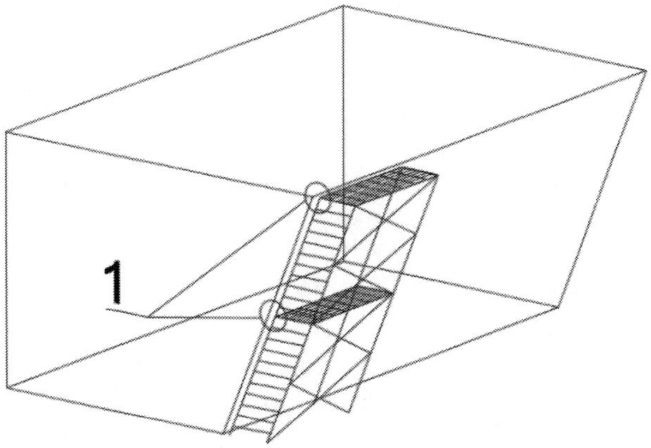

Fig. 5. Scaffolding for façade with negative angle (not existing facade system)

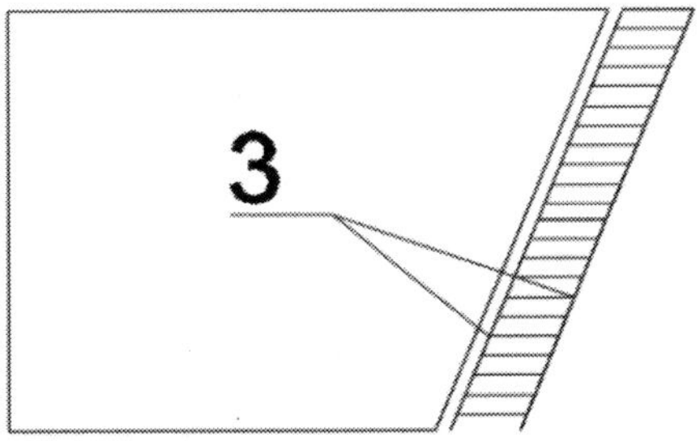

Fig. 6. Side view of the system

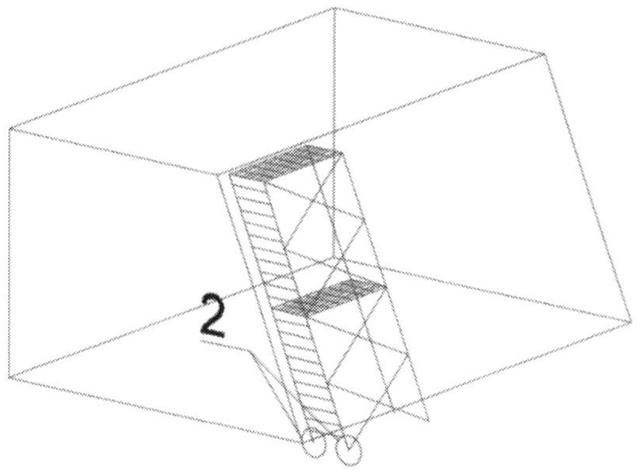

Fig.7. Scaffolding for facade with positive angle

where 1 – nodes for fixing with the wall, 2 – basements of scaffoldings, which can be rotated and fixed in different angles, adopting to sloping facades, 3 – scaffoldings columns, which can change length, also adopting to different angles.

3 Safety on the scaffoldings

Speaking about safety, constructors will be in more dangerous situation than in the scaffoldings for strict vertical facades. Therefore, safety belts and other fixing equipment should be improved and adopt to those systems. In addition, it is important to install safety nets, for example, on each three levels, in case of falling people and subjects from working horizons.

4 Factors influencing on feasibility

Efficiency application of such scaffoldings is restricted by the height. Laboriousness is changing due to angle and height of facades (1):

$$q = q_0 + \Delta q \ (1),$$

where q_0 – basic laboriousness, $\Delta q = f(\frac{h}{tg\alpha})$ – additional laboriousness, which depends on h [meters] - height of building and α - angle of facade, shows change of laboriousness along the height.

With intensive development of monolithic concrete constructions, unification of architectural planning solutions are depart to the background. Buildings are getting different shapes in planes and in facades. Particularly, there are buildings with different curvatures and slopes, broken facades and with different protruded parts. While making overhaul, the difficulty of such works depends on presents of balconies, their amount and character of surface of the walls. When working on such constructions the laboriousness is increasing, In that case there should be introduced coefficient of facade complexity (table 1).

Table 1. Groups of facade complexity of buildings according to geometrical specificity.

Facade complexity	Group of facade complexity	Coefficient of complexity
Flat facades	1	1
Facades with few protruded elements of the building	2	$1 + \frac{2S_1}{S_2}$
Facades with average broken lines equal to “n”	3	1+n

Note for table 1: "n" is determine as a number of broken facade's planes to unit of length of building on perimeter. S_1 – square of protruded elements and windows, S_2 – square of exterior walls.

The ledge, such as balcony also influence on laboriousness. And present as a coefficients on formula. One of the most important things is feasibility of such systems. Below is given diagram, which shows range of factors, influencing on feasibility (fig. 8). [1]

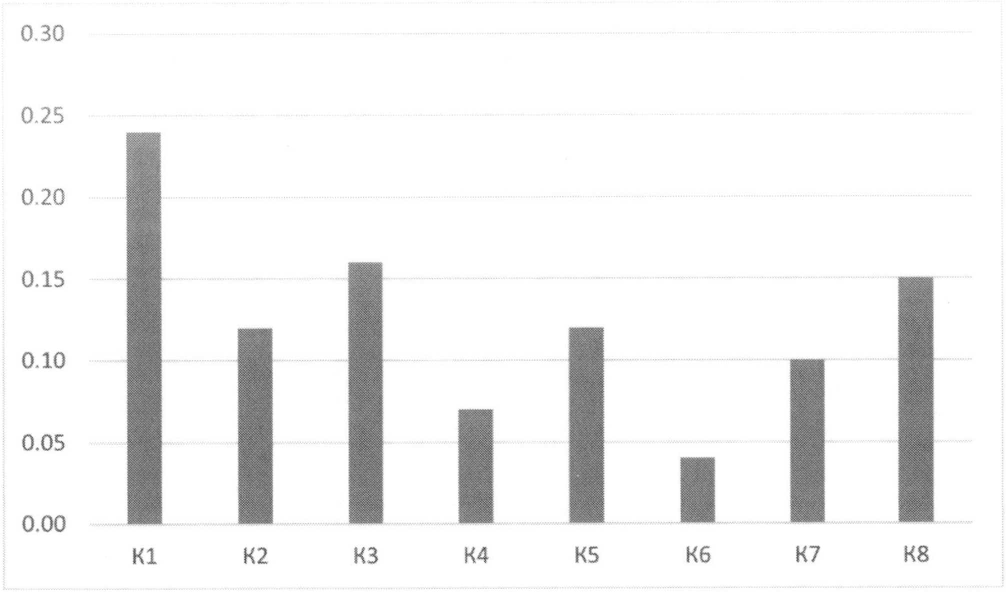

Fig. 8. Range of factors, influencing on feasibility: K1 – geometry of building; K2 – logistic of contract organization; K3 – organizational and technological parameters of work; K4 – spatial and time limits; K5 – climate conditions; K6 – supply conditions; K7 – type of facade system; K8 – finance supply.

In depending on power of influence of certain factors, can change organizational and technological parameters of works. Detailed analysis is necessary for making reasonable decisions.

Because unreasonable choice of organizational and technological parameters of works can lead to significant increase of cost, increase of duration and decrease of quality of works.

According to diagram on fig. 8 the highest influencing coefficients has following factors: geometrical characteristics of building, financial support of construction, organizational parameters of works.

To geometrical characteristics of building, influencing on technical and economical parameters related:

1. The height of building;
2. Complexity of facade, including presence of protruded elements, nonlinearity of facade.

Existing normative documents suggest work on facades from scaffoldings on height until 10 meters. Over 10 meters to labor input should be used next coefficients in depending on height: until 30 m – 1, 2; until 50 – 1,35; until 60 – 1,4; over 60 – 1,5.

According to graph on fig. 9 determined efficiency of using facade systems:

- for buildings with a height until 20 m – scaffoldings;
- with a height from 20 to 58 m – lifting platforms;
- with a height over 58 m – facade lifts.

5 Conclusions

1. Analysis of Russian and international experience with working on facades with irregular shapes showed that existing normative and technical documents does not fully disclose organizational and technological requirements for such works. Meanwhile, unreasonable choice of organizational and technological solutions can lead to significant decrease of feasibility index in works.
2. Defined, evaluated and ranked groups of factors, influencing on feasibility of works. Identification of rational area of organizational parameters concludes in reasonable choice of scaffolding, choice a number of workers and rational splitting to divisions. And also is essential to count indirect factors, which also can impact on organizational and technological parameters of works. They are cramped on construction side and climate conditions of region of construction.
3. In this article were given drawings of new achievable and helpful scaffoldings, which allows working on sloping facades.

References

1. http://dlib.rsl.ru/01005061649
2. S.M. Glikin, PGS, **18**, 24 (2009)
3. O.G. Ivanov, PGS, **3**, 35 (2008)

The effectiveness of the implementation and automation of heat pump on public buildings

Artur Trusov[1,*], and *Svetlana* Shilkina[1]

[1]Moscow State University of Civil Engineering, 129337, 26, Yaroslavskoye Shosse, Moscow, Russian Federation

Abstract. This article is devoted to consideration of effectiveness of the implementation and automation of heat pump on public buildings, in particular, at public catering. This article shows the advantages of this equipment compared to other heating systems, as well as the future development of these systems in Russia. The article presents the analysis of prices for the cost of the heating period, analyze the operating parameters of the gas, electric boilers and heat pumps. Also, there is a heat pump automation scheme, which shows the functional relationship of the devices of automation, regulators and actuators, which are controlled heat pump. According to the calculations the authors come to the conclusion that the best is the heat pump.

1 Introduction

Today, more and more urgently the question of increasing the energy efficiency of residential buildings in Russia arises. There are various ways to increase their energy efficiency, which is usually realized by increasing the efficiency of engineering systems of buildings and structures. Heat supply systems are some of the possible directions of modernization in public buildings. Heat pump system has aroused great interest among specialists, because of it is economy and energy efficiency, reliability and simplicity of technical solutions. Let us consider some of the issues associated with the system work, fields of application and automation of thermal processes. This article describes the use of the heat pump at a public catering.

2 Heating system analysis

A heat pump (HP) is a device that provides heat energy from a source of heat to a destination called a "heat sink". Heat pumps are designed to move thermal energy opposite to the direction of spontaneous heat flow by absorbing heat from a cold space and releasing

* Corresponding author: trusov.artur@gmail.com

it to a warmer one. A heat pump uses some amount of external power to accomplish the work of transferring energy from the heat source to the heat sink.

Energy efficiency assessment methodology is actively being developed nowadays. Compare the effectiveness of the introduction of the heat pump in relation to other heating systems. Below we present the most popular equipment used for heating: gas boilers, electric heaters and heat pumps. The table 1 provides a comparative analysis of the heating costs for the object, which is located in Leningrad region. The heat consumption of this object is 30kW.

Initial data: heat demand facility of 300 m2 (depending on insulation) is 30 kW; water temperature in the heating system must be 35 ° C; Minimum flow temperature is 0 ° C. For heat the building the heat pump capacity of 36.6 kW is selected, a power consumption of 6.6 kW, therefore the heat capacity is 30 (Q) kW.

Tabel 1. Heating systems analysis

Type of equipment	Capacity/ smount	Tarrif of the city	Tarrif of the village	Equioment price
Electric heater	31,5 kW	3,14 rub kW/h	2,3 rub kW/h	80 000 rub.
Gas boiler	3,6 m^3/h	5,1 rub m^3/h	5,1 rub m^3/h	150000 rub.
Heat pump	6,6 kW	3,14 rub kW/h	2,3 rub kW/h	350000 rub.

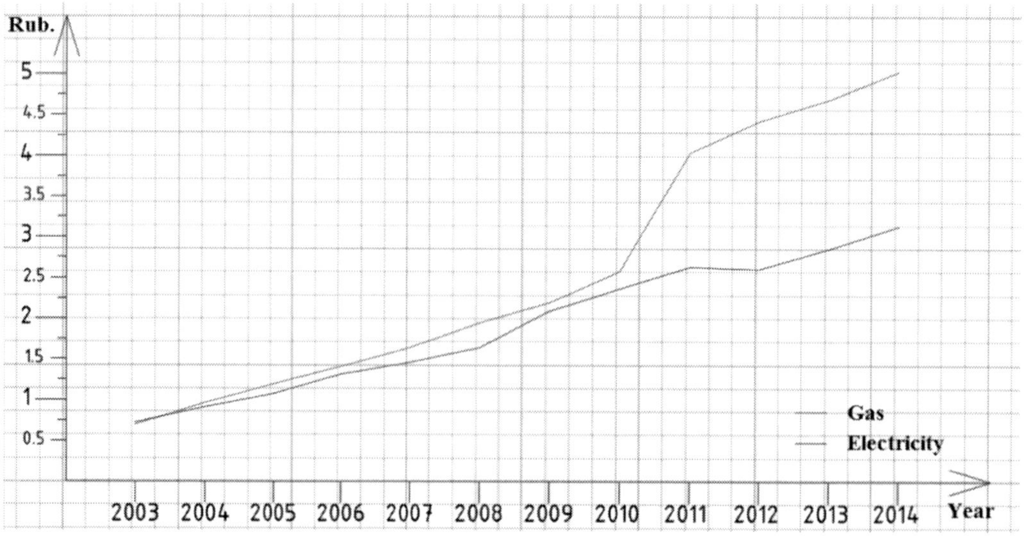

Fig. 1. Electricity and gas cost behaviour analysis

Define the cost of heating.

The heating period in Russia lasts 7 months, take into account that all the equipment work 7 months, 24 hours per day, at rates which are indicated in Table 1. Thus, we can conclude that now, a relatively cheap form of heating may be carried out using the heat pump, and over the years while maintaining the current trends it will become cheaper and cheaper relative to gas heating.

The costs of heating with an electric boiler is 498 thousand rubles per year at the rate of the city, and 365 thousand rubles at the rate of the village. Expenses with gas boilers is 92 thousand a year. Finally, the cost of heating with a HP will be at 4 thousand for the city and 76 000 for the village. Thus, we can conclude that the cheapest equipment for heating is the heat pump, but with the rates of the village. According to forecasts in the next two or three years, heating with heat pump, while the city'c tarrif, will be cheaper. Below is an analysis of the electricity and gas prices for the period 2003-2014. (Fig. 1).

After analyzing the prices of these fuels in the period from 2003 to 2014, we can see that gas prices are rising faster than electricity and this trend will increase year by year. According to forecasts, in 2018 the use of heat pumps will be cheaper than gas boilers.

Consider the heat pump automation scheme (Fig.2)

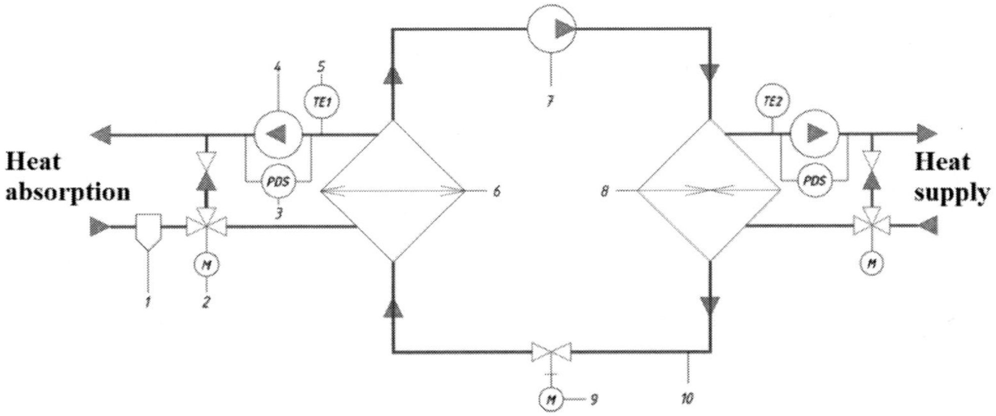

Figure 2. Heat pump automation scheme

Operating principles: the working fluid, in its gaseous state, is pressurized and circulated through the system by a compressor. On the discharge side of the compressor, the now hot and highly pressurized vapor is cooled in a heat exchanger, called a condenser, until it condenses into a high pressure, moderate temperature liquid. The condensed refrigerant then passes through a pressure-lowering device also called a metering device. This may be an expansion valve, capillary tube, or possibly a work-extracting device such as a turbine. The low pressure liquid refrigerant then enters another heat exchanger, the evaporator, in which the fluid absorbs heat and boils. The refrigerant then returns to the compressor and the cycle is repeated.

Since the kitchen and hot shops have great heat radiation, it makes sense to apply heat pumps in these areas (if using air source HP). But it would be better to use HP with heat recovery wastewater. Wastewater temperature of the dishwasher, and other areas of the restaurant reaches 20-25 degrees, which in turn brings greater energy efficiency. Applying such heat recovery scheme, it is possible to save on the expensive installation of geothermal analogues. The use of HP in the ventilation system preferably by the fact that it saves on installation of air conditioners, since during the summer HP will work in cooling mode. In addition to these advantages simple automation system should be noted. Below are items included in this system.

Temperature sensors are installed in return coolants to monitor and control coolant in the low grade heat circuit and in the consumer circuit. Differential pressure switches are installed on the pump to monitor their health. Three-way valves are used to control the coolant. These valves are controlled by a PI controller.

3 Conclusions

The advantages of heating by using HP are: environmental friendliness (CO2 emissions is significantly less than with gas boiler), durability and safety in operation. The service life is 20-25 years, while the life of the gas boiler is 10-15 years. The question arises - why installation of HP in our country is not very popular? Of course, because of the high cost of equipment. In this regard, it is less preferable to gas boilers. But in Europe, this problem was solved by subsidies in favor of the introduction of heat pumps, so that part of the cost of installing HP borne by the State. This reduces the dependency on the use of non-renewable natural sources of energy, which in turn reduces the amount of CO2 emissions, and to improve environmental quality.

References

1. E.S. Bondar, A.S. Gordenco, V.A.Mikcalov, G.V.Nimich, *Ventilation and air condition systems' automation*, p. 560, (2009)

2. E.M. Belova, *Central air condition systems in the building*, (2006)

3. S.V. Shilkina, Methodical instructive to make economic part of diploma paper, (2007)

4. A.V. Livchak, M.A. Malachov, S.N.Khorev, *Hot shop's ventilation in the public catering*, (2007)

5. A. Trusov, S. Shilkina, The effectiveness of implementing automated ventilation system in premises intended for public use, Applied Mechanics and Materials, *Advances in Civil and Industrial Engineering IV*

The methodology of technical due diligence report preparation for an office, residential and industrial buildings

Beata Kutera[1], and *Hubert* Anysz[1] *

[1]Warsaw University of Technology, Civil Engineering Department, Armii Ludowej 16, 00-633 Warsaw, Poland

Abstract. The methodology of a technical due diligence preparation is presented in this paper. It comprises actions that have to be undertaken prior to formal agreement with party ordering due diligence preparation, building a team of consultants, data collecting, preparing analysis and handing over the report to the client. All important issues were described and supported by examples. As there are many types of building objects this paper is limited to office, residential and industrial buildings.

1 Introduction

The term *due diligence* has been primary applied, with its literally meaning, to describe the set of rules, requirements and good practices applied by every party engaged in signing insurance contracts in United States [1]. By the time it was extended for all cases where a detailed analysis (prior to signing a contract) is required by law or by a party engaged in a transaction. The aim of due diligence analysis is to limit the risk i.e. protect a buyer from any unexpected costs arisen from facts that can be disclosed after finalizing the transaction, when the ownership rights could not be returned.

2 The term of technical due diligence

The other reason is to recognise business opportunities related to the property being bought. The higher value of a transaction, the higher risk and willingness for preparing a due diligence. Hence due diligence is a natural and desired process, the immanent part of transactions on a property market. The change of landlord can arise from the change of the owner of a company possessing the property too. Usually, due diligence prepared for aforementioned purposes comprise three detailed analysis focused on following aspects of property: law, technical, financial [2]. This technical part is called technical due diligence (TDD) and it refers to a building together with the lot it is located in. The Roman rule of law *caveat emptor* (let the buyer beware) is widely applied nowadays. It means that the seller is not responsible for undisclosed defects of a good. In practice, a seller is not responsible for defects at all, they don't have to show all of defects known for , but is

* Corresponding author: h.anysz@il.pw.edu.pl

 DOI: 10.1051/matecconf/20168607009

obliged to make the building accessible for potential buyer in order to check its technical condition. Due to experience and educational background of authors this paper is limited to the methodology of preparing technical due diligence report for office, residential and industrial buildings.

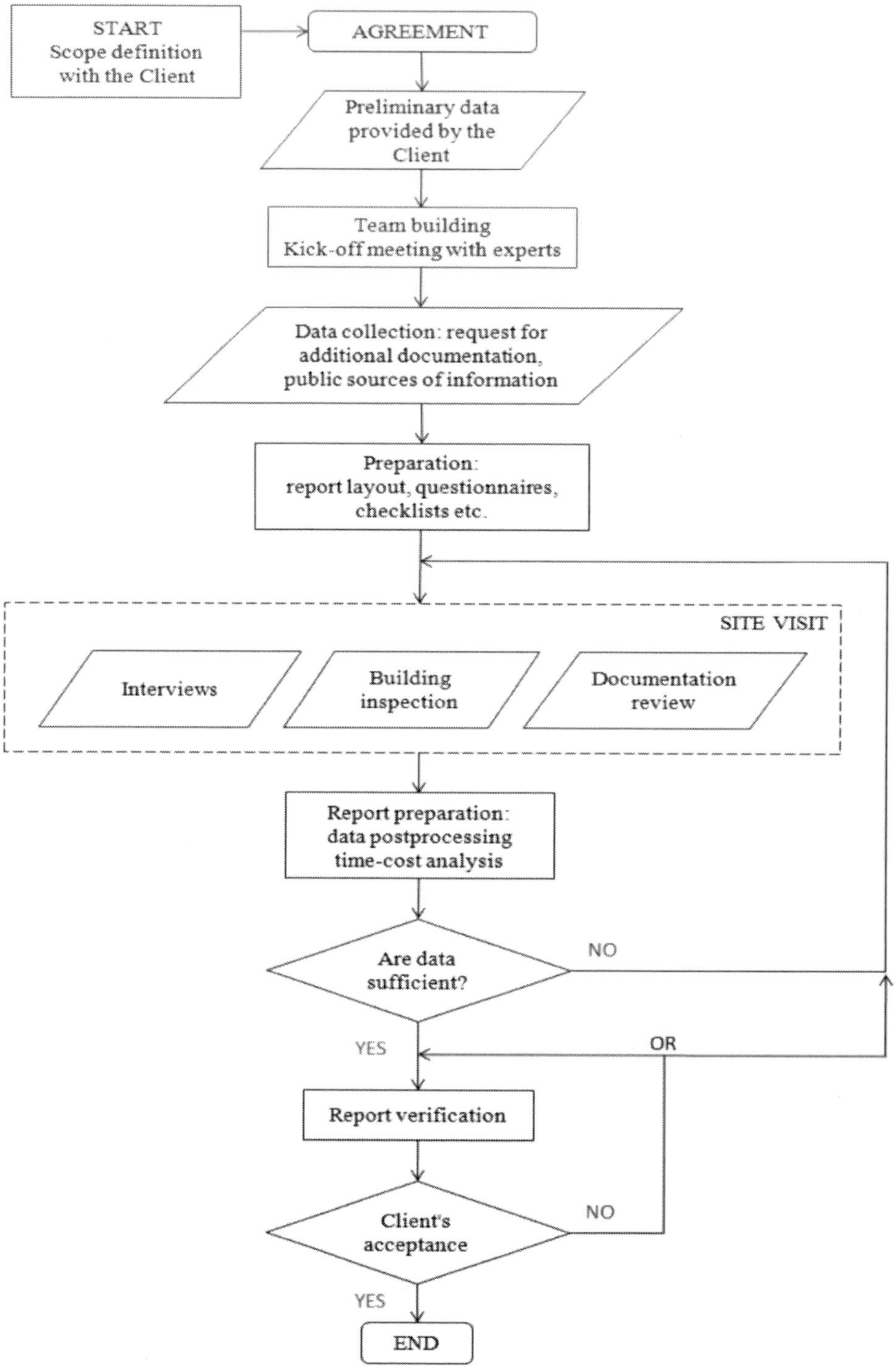

Fig. 1. The complete process of technical due diligence.

3 Methodology of TDD preparation

The complete process of technical due diligence analysis is presented in fig. 1. and its most important parts are described below.

3.1 The scope of TDD

There are no legal requirements for TDD in Poland. If they exist in other countries, they must be fulfilled. The basis for defining the scope of technical due diligence are the client (the buyer of a building) needs, in any case. Prior to signing the contract between client and consultant following technical matters should be clarified [3]:
- detailed scope of TDD,
- address of property to check,
- present and planned function of a building together with planned schedule of the change,
- sources of information known for the client and their reliability (if known),
- if the client hires branch experts for TDD or is it a matter of consultant,
- other important issues arisen from planned usage of the building (health and safety, environmental, fire safety etc.) that can influence analysis for TDD,
- form of TDD report,
- limitations (e.g. arisen from availability of sources of information, limited time for checks and analysis etc.),
- the total time for preparing TDD report.

It is strongly recommended to prepare a questionnaire (shortened example is shown in tab. 1) when clarifying the scope of due diligence (and TDD too, as a part of it). It helps to protect consultant from problems during handing over TDD report, as well as it is very useful during TDD analysis to systemize the process.

Table 1. Shortened example of a questionnaire [3].

service / scope	Yes	No	by third party	remarks / details
General description of a building				
Description of the construction				
Building materials				
Description of ventilation system				
Description of heating system				
...	...	...	...	...
Legal status				
- owners' rights				
- rental agreements				
- service agreements				
Tax analysis				
Analysis of construction process				
...	...	...	...	...

3.2 The consultant team

Irrespectively from the organizational scheme chosen by the client, (three exemplary solutions are shown in fig.2) there are two recommendations for the team preparing technical part of due diligence i.e. TDD. An expert, apart from their core competence, should be familiar with subjects covered by other team members. It helps a lot at every stage of TDD analysis, but especially during inspection of a building so as not to omit any defects and their influence on elements of other building systems (e.g. services defects

DOI: 10.1051/matecconf/20168607009

influencing state of construction elements). Secondly, strong analytical skills should be required from team members as well as cost estimation skills in every technical branch being analysed (or at least ability to estimate scope of works to be carried out to restore particular element of a building to proper condition).

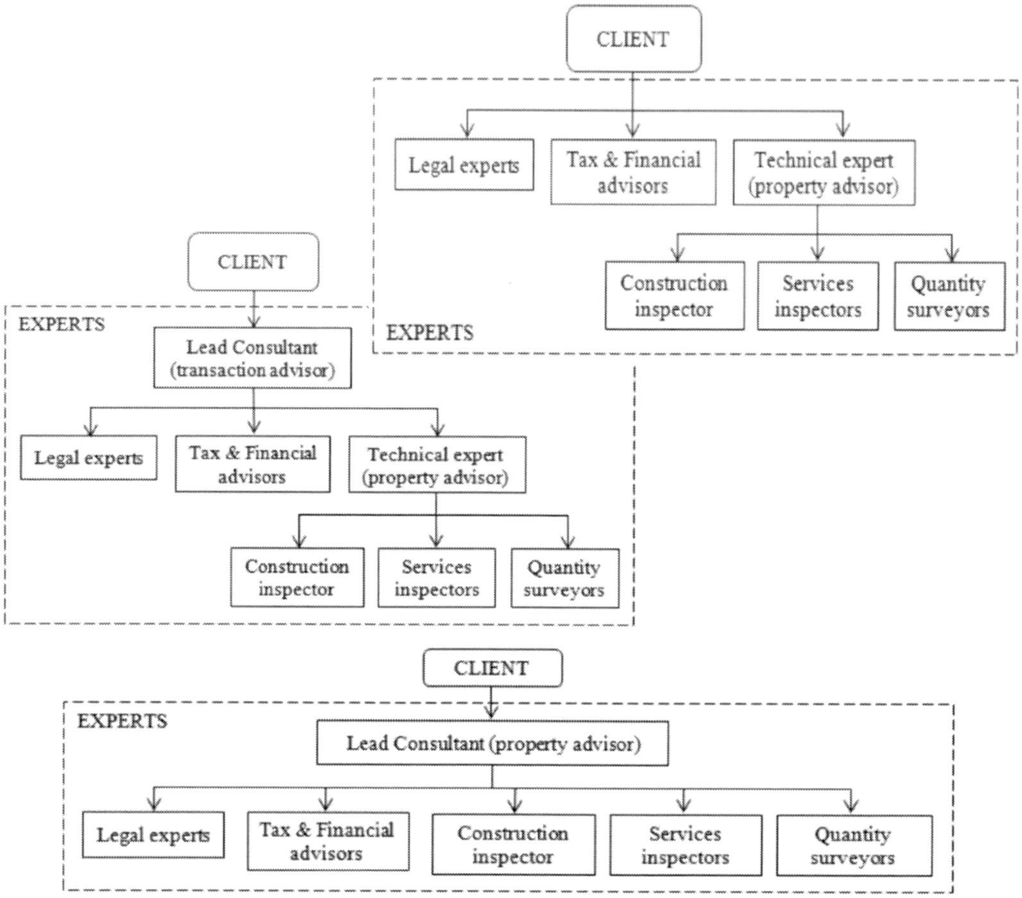

Fig. 2. Possible organisational schemes of consultant team.

Coming back to the name of report i.e. due diligence, experts engaged should present scrupulousness, reliability, impartibility. The feeling of team spirit is required from experts too.

3.3 Information collecting

There are three main groups of data sources concerning any building:
- interviews with people who know the building, were involved in its design, construction, maintenance, and with its neighbours,
- inspection of the building
- documentation concerning the property.

3.3.1 Interviews

The deepest knowledge about the building object (especially about problems arisen during operation) is possessed by facility manager and the staff/service providers. hired by them.

Comparing the interview with documentation checks, it is very time-saving process of data collection. Acquired information should be verified using the other data sources. If it is possible and if consultant would recognize the need, interviews with designer, construction site manager, construction supervisor (Engineer), neighbours, users of the building can be helpful too. As for preparation process, the set of questions (matters to be clarified) should be prepared. Questionnaire form is useful.

3.3.2 Inspection of the building

There are some key areas that consultants should pay attention to during inspection.

Overall impression s is important to evaluate the standard, aesthetics and prestige of the building, as well as its comfort (thermal comfort, freshness of air, time of waiting for warm water from a tap etc.).

The fire safety and evacuation ways (required by law) are important part of the checks, as lack of them is a subject of penalization [4]. Any inconsistency with the fire protection plan should be noticed.

Utilities for handicapped people are necessary part of public buildings (required by law in Poland). The list of required conditions to fulfil and utilities to install is enumerated in the relevant ordinance.

Soil subbase (its type, water content, underground water level and flow), weather conditions, internal environment (humidity, steam condensation, aggressive gas, fluid and dust), biological environment (plants, algae, insects, birds, microorganisms), mechanical loads, extreme situations (e.g. fire, serious damages) are agents influencing the building much [5], so naturally these influences are a subject of inspection.

The most often examples of improper operating the building are [6]:
- lack of maintenance and repairs of internal systems,
- cursory periodic checks of the building,
- overloaded floors,
- changes made to structural elements of the building without prior analysis,
- changes of a function of the building without checking the influence of the change to the structure or services,
- installing utilities – a source of vibrations – without protecting the structure from vibrations.

Visual evaluations of recognized defects. These tasks of consultant require taking into account the age of the building, so the evaluation should differ depending on age of the building. The diagram of visual evaluation of defects shown on fig. 3 can be helpful [7].

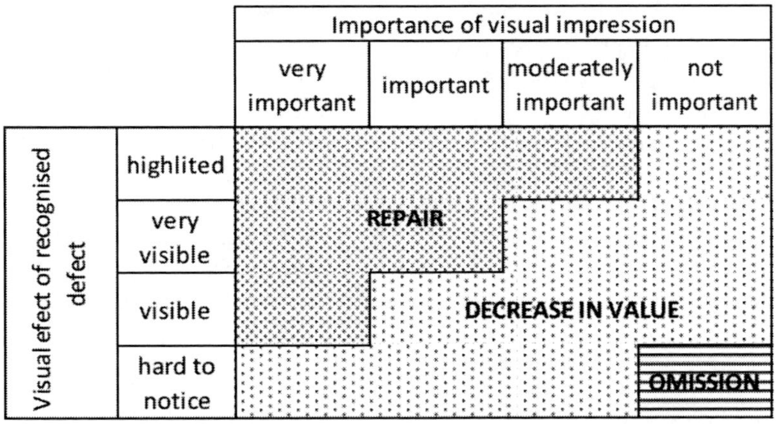

Fig. 3. The diagram of visual evaluation of defects

3.3.3 Documentation

Legal side of investment process in construction sector should be studied prior to searching documents for technical analysis. There are several kinds of documents required by Polish law at different stages of investment process.
Following documents are being prepared during pre-construction stage:
- conceptual design,
- construction design,
- detailed branch designs (plus workshop drawings),
- local plan,
- report on influence of the building object on environment,
- building permit,
- detailed specification of executing and handing over building works.

When it comes to construction phase, another set of documents is required to create or to collect and keep:
- construction log,
- assembling works log (if performed),
- time schedule (not required by law, but commonly prepared),
- hand-over protocols,
- as-built documentation,
- certificate of energy efficiency of the building,
- building operation permit.

A contractor is obliged to hand over all these document to a client (investor) and a client should keep them for the time a building physically exists. All aforementioned are followed by the required record of operating a building consisting of:
- the log book of operating a building with enclosures as
    - reports from periodical checks of a building (required by Polish law too),
    - reports of experts (if periodical check recommended preparing them),
    - documentation of modernisation, maintenance and refurbishment works,
- operation and maintenance manuals of utilities installed,
- protocols from periodic official controls of utilities installed (e.g. of elevators);
- contracts with water supplier, electricity supplier etc.

The number of documents and their volume make it almost impossible to analyse them in a short time. It is one reason why the other sources of information are equally important. The second one is the fact that even for newly built objects lacks in documentation happen.

3.4 Necessary analysis to include

Collected data should be included in a final TDD report. Without experts' comments it is still set of information not very usable for the client who is usually also an expert, but in business area. It is necessary to make analysis that converts collected data to usable information and recommendations.
Risk analysis of problems detected. Probability of incident can be matched with the level of negative impact on client needs [8]. Showing the result of analysis in a risk matrix (the example shown in fig. 4) makes the risk easy to read for the client. Here is the next important matter when the client needs should be well recognized by a consultant. Otherwise TDD will not be accepted by a client. Following risk assessment the cost estimation should be prepared. Every recognised defect should be analysed taking into account client's needs as well as real (technical) necessity of the repair.

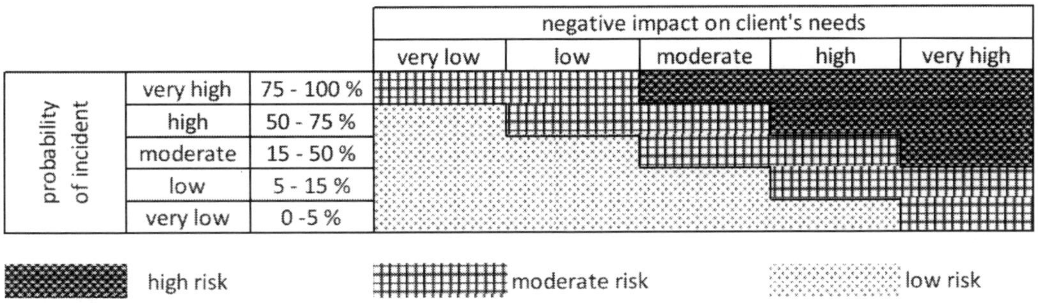

Fig. 4. The example of risk matrix]

Again the table form is suitable. Every defect or problem that cannot be omitted (see fig. 3) should be photographed, labelled and described. Labels of photos should be given as well as proposed solution, evaluation of risk associated with the problem and the cost of fixing it. The example is shown in tab. 2.

Table 2. Shortened example of a table for risk-cost analysis [3].

#	Defect / Problem	Photo #	Recommendation	Risk	Cost [PLN]
...	...	...	...	...	...
23	Too low pressure in a sprinkler system	17	Sprinkler pumps to be changed	high	150 000
24	Anti-smoke roof flap doors are not remote operated	21, 25	Change of the flap doors to remote operated	moderate	40 000
25	Lack of operating and maintenance manuals for utilities of sewage system	n/a	Collecting manuals	low	2 000
26	Windows in the front elevation do not have sound insulation properties and they are not aesthetic	from 30 to 41	Change of windows on the elevation	moderate	650 000
...	...	...	...	...	...

Having tab. 2 prepared consultant can concentrate on a time schedule for fixing problems and repairs of defects together with a plan of expenditures necessary for above mentioned purposes. The client may need some scenarios of repairs, depending on different functions of the building planned by prospective owner of a building.

3.5 General remarks

Irrespectively from placing a given defect in the table (see tab. 2) every defect or problem should be presented to the client, even if it was evaluated as to be omitted (see fig. 3) – it is due diligence. Every information gathered from interviews, inspection of a building or from documentation should be written down and labelled. The same is with the data that was not collected. If any room or any place, or any archive stayed closed for consultant, the statement about that should be written down in TDD and presented to the client. This is what really scrupulousness, reliability, impartiality of a consultant mean.

4 Summary

Presented methodology of technical due diligence preparation may vary between countries, as their legal systems differs. However the differences will concern the details (e.g. set of documents required by law in a building investment process, legal requirements concerning TDD itself, legally required utilities for handicapped people, scope of the design of a building etc.). General rules, necessary sequence of activities, possible organisation scheme of consultant team will remain unchanged and are applied worldwide. The reason is simple. The set of good practises, scrupulousness, reliability, impartiality of a consultant combined with deep and interdisciplinary knowledge of consultant team members allow to create their expert opinion about building object called technical due diligence.

References

1. J.P. Duffy, *Some Thoughts on Due Diligence*, The OFC Reports (1995/1996) http://www.bergduffy.com/Personnel/Articles/95ddatl.htm
2. A. Kiziniewicz, *Due Diligence majątku jako mądrość przed szkodą*, C.H.Beck Nieruchomości (02/2015)
3. *Technical Due Diligence of Commercial & Industrial Property*, Royal Institute of Chartered Surveyors, Australia (2009) http://sgaproperty.com/docs/RICS-Due-Diligence-Guidance-Note.pdf
4. B. Latos, *Utrzymywanie budynków administracji publicznej w należytym stanie techniczno-użytkowym*, Presscom Sp. z o.o., Wrocław (2013)
5. L. Lichołaja (red.), *Budownictwo ogólne. Tom 3. Elementy budynków. Podstawy projektowania*, Wydawnictwo Arkady, Warszawa (2008)
6. W. Buczkowski (red.), *Budownictwo ogólne. Tom 4. Konstrukcje budynków.* Wydawnictwo Arkady, Warszawa (2009)
7. O. Rainer, A. Ruth, *Wady i usterki w budynkach: rozpoznanie, zapobieganie, usuwanie*, Instalator Polski, Warszawa (2000)
8. J.C. Teixeira, J. Kulejewski, M. Krzemiński, J. Zawistowski, Zarządzanie ryzykiem w budownictwie, Guimares, Warszawa (2000)

AUTHOR INDEX

AUTHOR INDEX

AUTHOR INDEX

AUTHOR INDEX